D1044223

MECHANICAL DRAWING

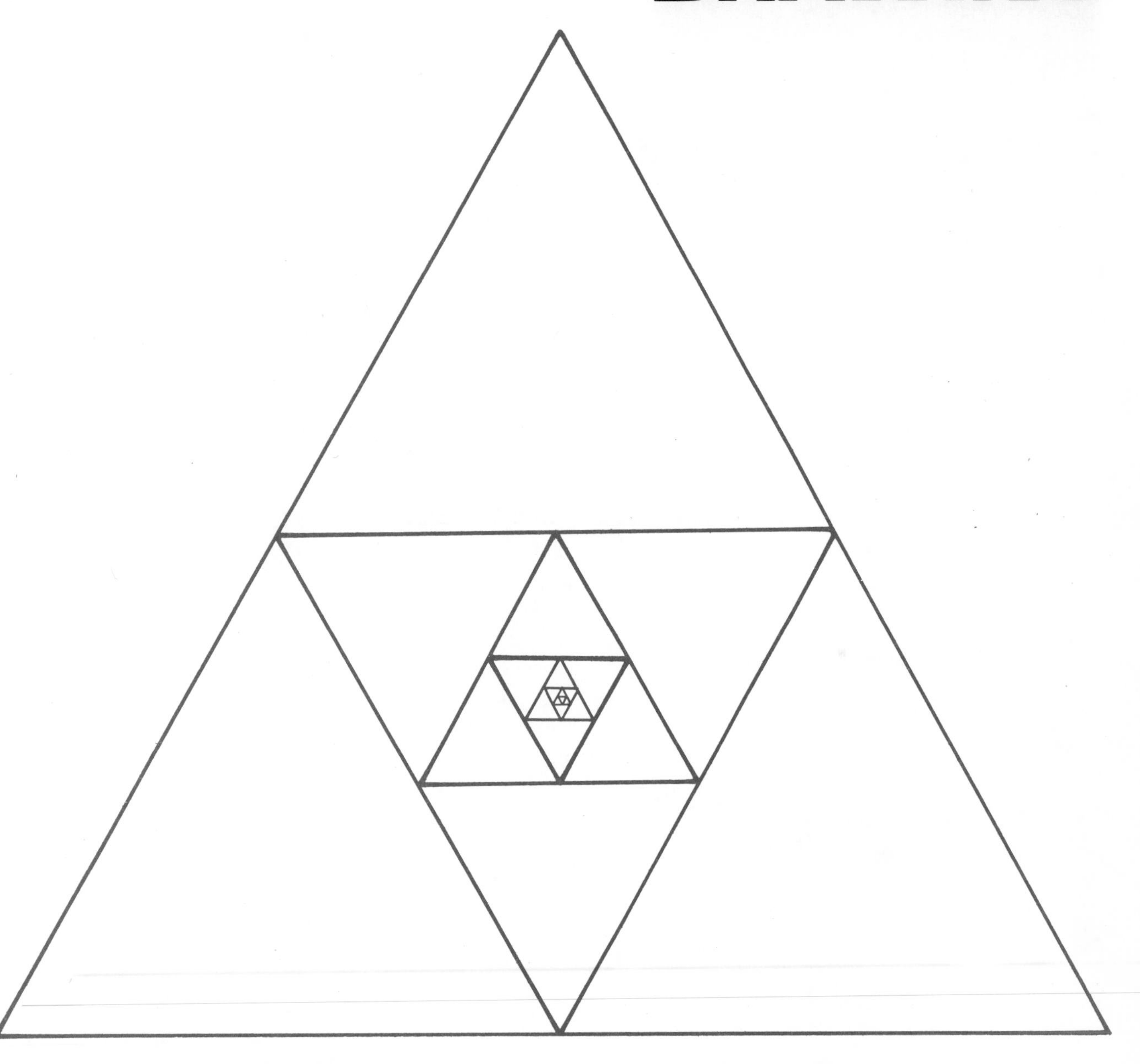

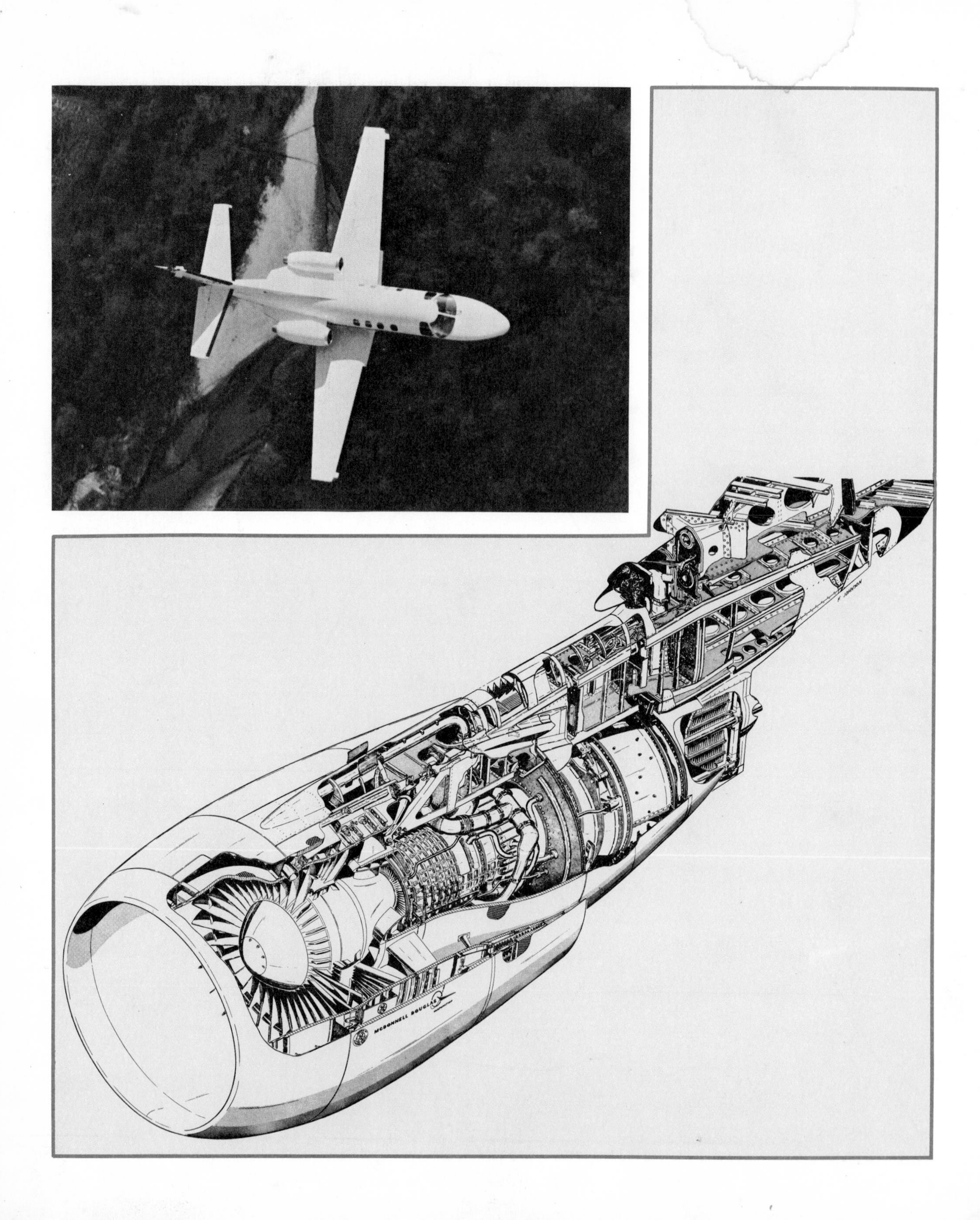
McDONNELL DOUGLAS

MECHANICAL DRAWING

Ninth Edition

Thomas E. French
Carl L. Svensen
Jay D. Helsel
Byron Urbanick

McGraw-Hill Book Company

New York St. Louis San Francisco Dallas Auckland Bogotá Düsseldorf Johannesburg London
Madrid Mexico Montreal New Delhi Panama Paris São Paulo Singapore Sydney Tokyo Toronto

Editor: Hal Lindquist
Associate Editor: Ronald Grigsby Kirchem
Managing Editor: Alma Graham
Coordinating Editor: Mary Ann Jones
Design Supervisor: Margaret Amirassefi
Production Manager: Karen Romano

Outside Editing: Charles Wall
Cover Design: Jenny Morrill
Cover Photograph: George Senty
Layout Designer: Nancy Mattimore
Additional Photo Research: Alan Forman

Figures 10–21 and 10–22 are adapted from *Mechanical Drawing*, Canadian Edition, copyright © 1977, with permission of McGraw-Hill Ryerson Limited.

Library of Congress Cataloging in Publication Data
Main entry under title:

Mechanical drawing.

Includes index.
SUMMARY: A textbook introducing the basic theory, techniques, and uses of drafting for industrial arts and vocational high school students.
1. Mechanical drawing. [1. Mechanical drawing]
I. French, Thomas Ewing, 1871-1944.
T358.M48 1980 604'.2 79-17003
ISBN 0-07-022313-0

Copyright © 1980, 1974, 1966, 1957 by McGraw-Hill, Inc. All Rights Reserved

Copyright 1948, 1940 by McGraw-Hill, Inc. All Rights Reserved. Printed in the United States of America. No part of this publication may be reproduced, stored in a retrieval system, or transmitted, in any form or by any means, electronic, mechanical, photocopying, recording, or otherwise, without the prior written permission of the publisher.

Copyright 1934, 1927, 1919 by McGraw-Hill, Inc. All Rights Reserved.
Copyright renewed 1968, 1962 by J. F. Houston and C. L. Svensen.
ISBN 0-07-022313-0

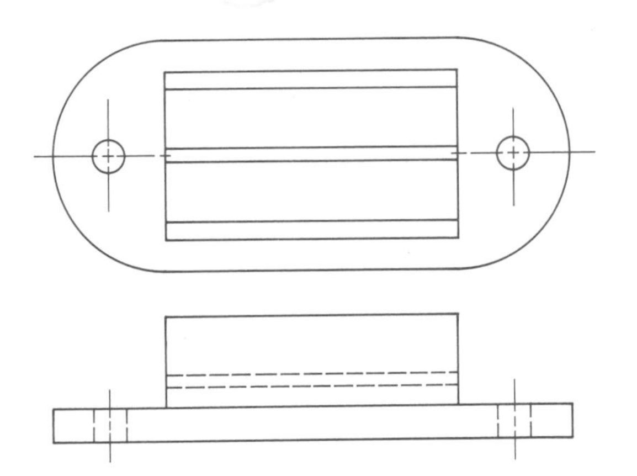
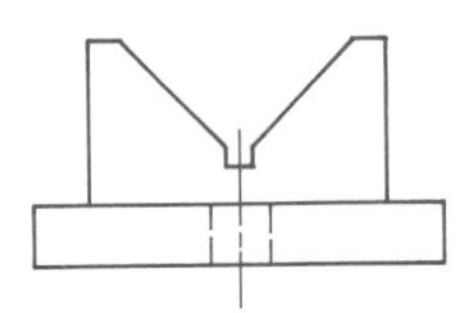
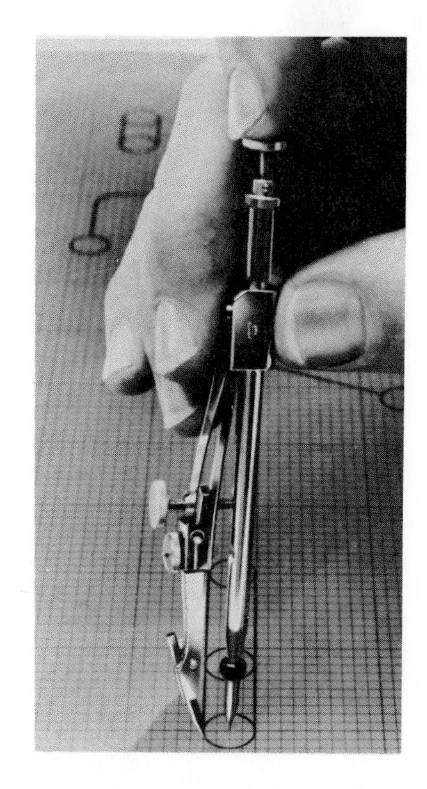

ABOUT THE AUTHORS

JAY HELSEL is a professor of industrial arts and vice president for administrative affairs at California State College in Pennsylvania. He completed his undergraduate work in industrial arts at California State College and was awarded a master's degree from Pennsylvania State University. He completed a doctoral degree in educational communications and technology at the University of Pittsburgh.

Dr. Helsel has worked in industry and has taught drafting, metalworking, woodworking, and a variety of laboratory and professional courses at both the high school and college levels. During the past fifteen years, he has worked as a freelance artist and illustrator. His work appears in a great variety of technical publications.

Dr. Helsel is also coauthor of *Engineering Drawing and Design*, *Programmed Blueprint Reading*, *Reading Engineering Drawings through Conceptual Sketching*, and *Automotive Transparencies*. In addition, he is the author of the series, *Mechanical Drawing Film Loops*.

BYRON URBANICK is chairman of the Industrial Technology Department at Oak Park–River Forest High School in Oak Park, Illinois, where he has taught technical, architectural, and engineering graphics for twenty-two years. He has also taught drafting and graphics at the Illinois Institute of Technology, Northern Illinois University, College of DuPage, and for the U.S. Naval Ordnance staff.

Mr. Urbanick's professional experience includes work for the architectural firms of Friedman, Alschuler and Sincere; and Fairbanks, Morse and Company; and for Starme Industrial Engineers and Malco Tool and Manufacturing Company. He also has his own design and drafting practice.

Mr. Urbanick earned his B.S. and M.S. degrees from Northern Illinois University. He is a member of the American Society for Engineering Education, the American Industrial Arts Association, the American Vocational Association, and the Illinois Technical Drawing Teacher's Association, where he has served as secretary and president.

CONTENTS

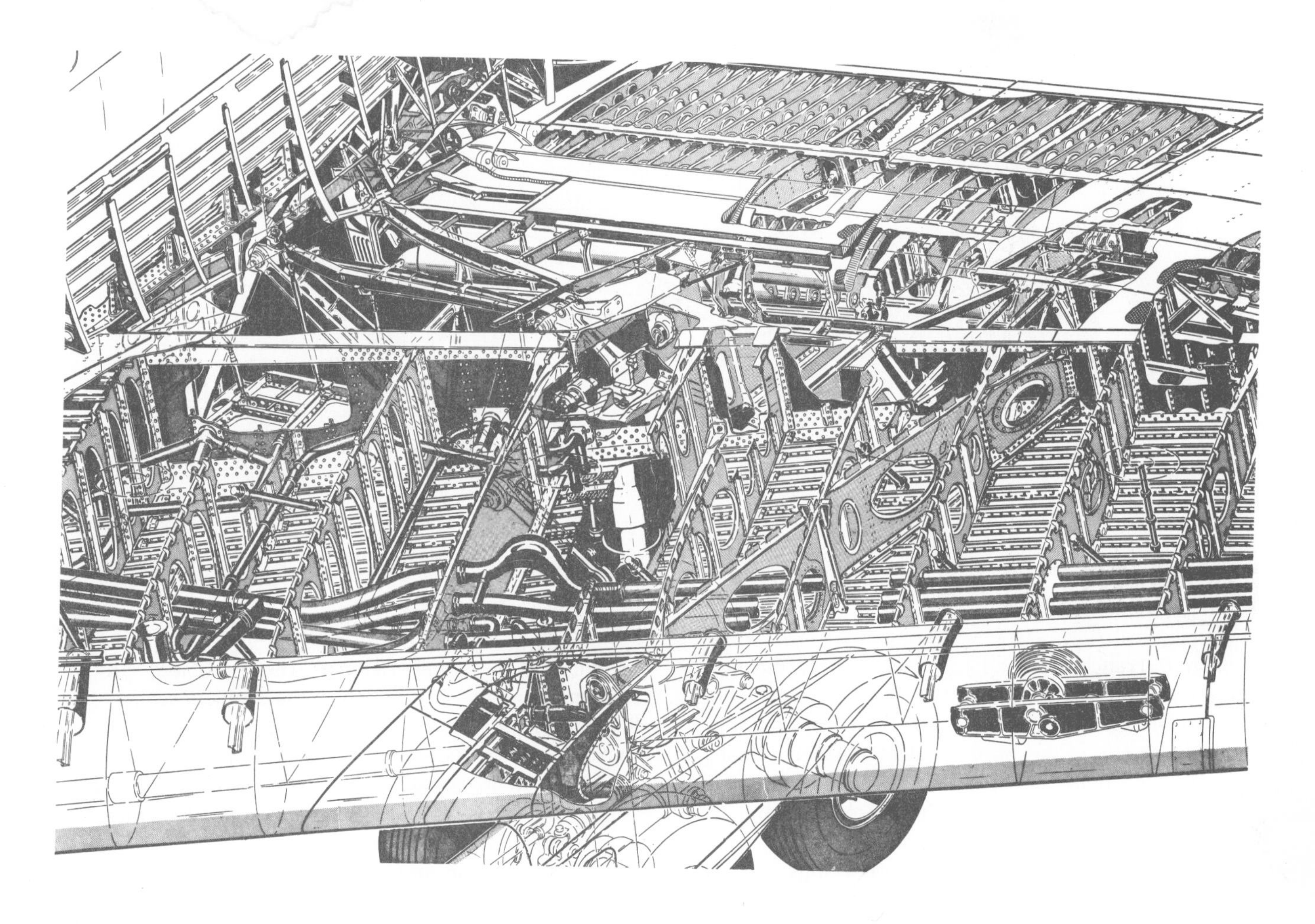

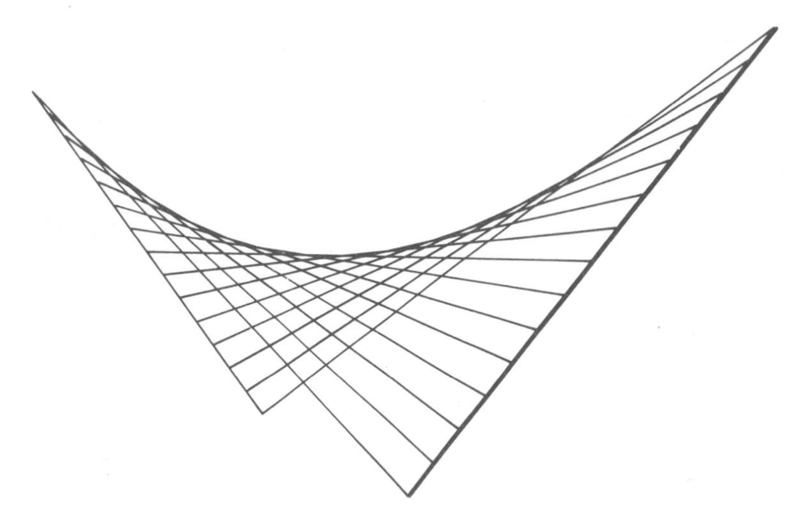

PREFACE

Thomas E. French and Carl L. Svensen wrote the first edition of *MECHANICAL DRAWING* to give students a thorough understanding of the basics of drafting. Through the use of this book, students learned to visualize in three dimensions, to develop and strengthen their technical imagination, to think precisely, and to read and write the language of industry. They also gained experience in making working drawings according to current commercial practices. Those skills have been carefully reinforced in subsequent editions.

Beginning with the eighth edition, some new concepts were introduced by two new authors. Many of these concepts were covered in new chapters, such as "Systems for Graphic Communications," "Basic Descriptive Geometry," and "Drafting Media and Reproduction." Also, the addition of four-color illustrations helped show procedural steps and methods of projection in some of the more complex concepts.

In the ninth edition, all chapters have been rewritten to simplify the language and to reduce the reading level. This will enable students of all ages and grade levels to more easily understand the subject matter. Difficult words and technical terms have been carefully defined and illustrated when first used. In addition, both text and illustrations have been revised to reflect a better ethnic, racial, and gender balance than was found in previous editions.

In certain chapters, some of the problems from the seventh edition of *MECHANICAL DRAWING* have been brought back by popular demand. These problems provide a greater range of difficulty in the problem sections, allowing the instructor greater flexibility in assigning practice work to students. A sufficient number of problems have been provided so that different student assignments can be given to several classes over a number of years without duplication.

Conversion to the metric system continues to be an important part of each edition. Careful planning has been done in the ninth edition to provide enough metrication to keep pace with the national trend. Enough flexibility is provided so that teachers can set the rate at which their programs move toward total metric conversion. Throughout the written text, the metric dimensions appear first, with the U.S. Customary equivalents following in parentheses. In most cases, however, a soft conversion is used, providing the nearest practical unit rather than a precise equivalent.

We wish to thank those users of previous editions whose suggestions have helped shape this edition. Their interest and help are greatly appreciated. Credit is also given to the many companies who contributed technical assistance as well as illustrations. Comments and suggestions are welcomed and appreciated.

Jay D. Helsel
Byron Urbanick

Chapter 1

Twentieth-Century Drafting

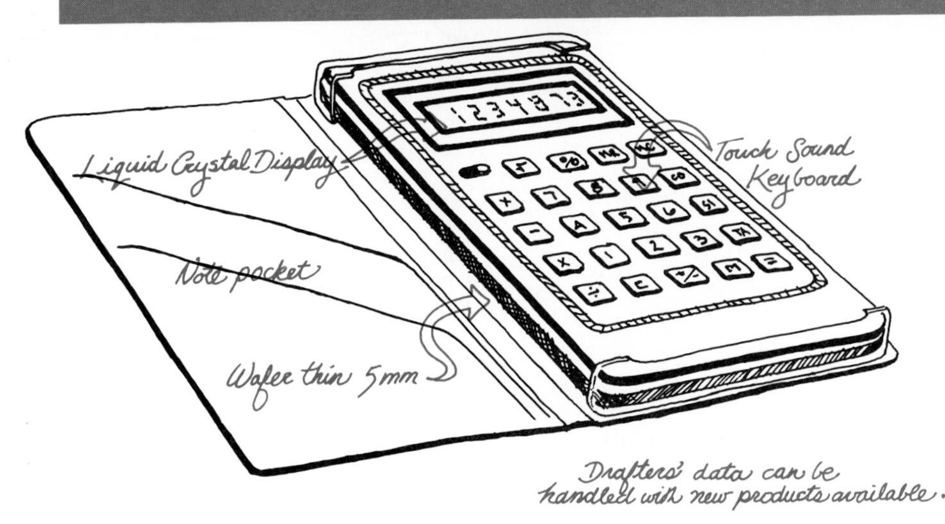

Fig. 1–1 **Modern devices are very convenient.** ***(Sketch by Jeff Podlensak.)***

DRAFTING IN A TECHNOLOGICAL WORLD

In our society, we buy, rather than make, most of the things we need to live. Most of the goods we need are made in factories. Twentieth-century technology has produced many useful things that are important parts of our daily lives. Each year, new discoveries and inventions help make our lives a little easier. There are modern machines for doing ordinary things like simple arithmetic (Fig. 1–1). There are machines for doing extraordinary things like space travel (Fig. 1–2).

For each of the manufactured products you see around you, a technical drawing was prepared at some point while it was being planned. That drawing was very accurate. It told someone how to make the product. If the thing has more than one part, a drawing was made of each part. The people who make these drawings of the products are called *drafters.*

How do drafters know how to draw the pictures from which goods can be made? They work with people who think up new *technology*—new products or new ways of making things. These people include engineers, architects, and designers. They use their own knowledge of drafting to sketch out new ideas that they have (Fig. 1–3). Drafters turn these sketches into detailed drawings. A knowledge of mathematics, science, and technology, as well as common sense and experience, are used by drafters to make technical drawings (Fig. 1–4). Anyone who likes putting mathematics and science to practical use

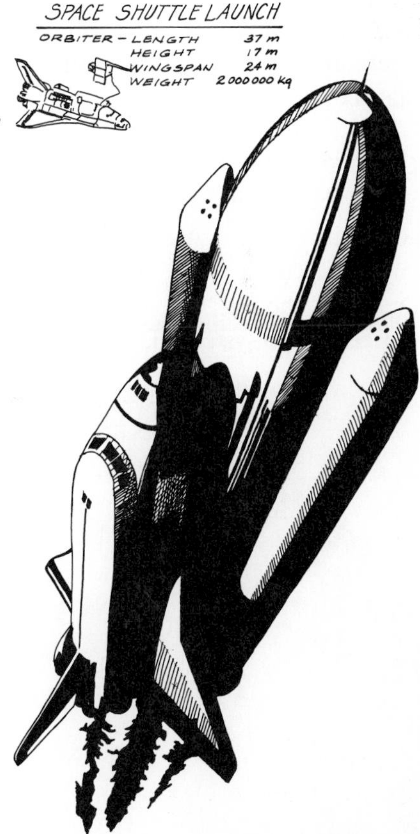

Fig. 1–2 **Space travel has provided many new career opportunities, including drafting.** ***(Sketch by Jeff Podlensak.)***

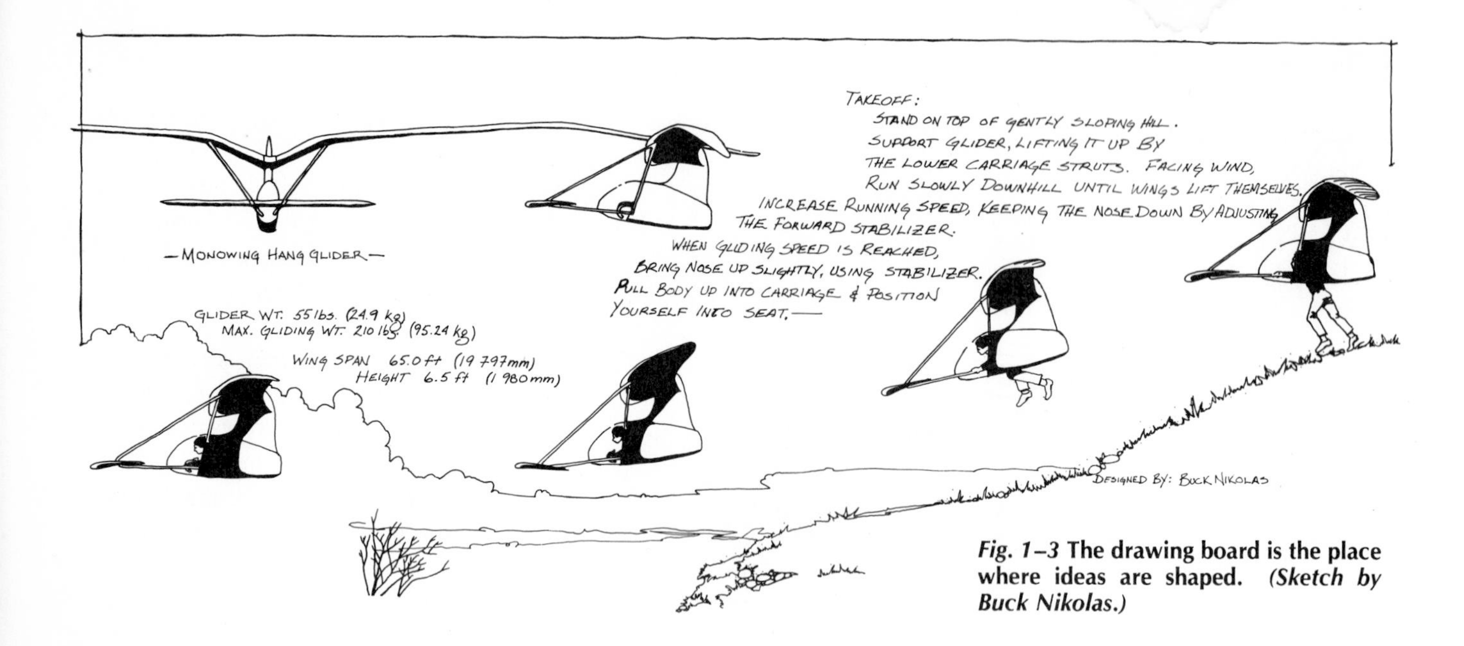

Fig. 1–3* The drawing board is the place where ideas are shaped. *(Sketch by Buck Nikolas.)

may find drafting a good career. There are other traits besides these that a drafter should have.

CAREER QUALIFICATIONS

It is important for drafters to be good with details. A drafter's work must be accurate, neat, and orderly. Drafters should be logical, resourceful, and systematic when they do detailed drawings. It is helpful if drafters are interested in machinery and how things work. It is good if they want to—and can—solve practical problems.

Drafting students will learn how to draw three-dimensional objects on paper, which, of course, has only two dimensions. They will learn how to read the *language of industry,* which is made up of lines and symbols. They will learn how to describe size and shape and how to use proportions and scales. After they leave school, professional drafters will always be learning more about new technology both in the industry they work for and in the drafting profession as a whole.

Fig. 1–4* Drafters must use their knowledge of mathematics and science. *(Photo by Toby Brown.)

As you will learn in Chapter 15, drafters work on a design team. Thus, they must be able to work well with others on the team and share ideas with them. Other members of the design team are the engineer, designer, and technician. Although you might start working toward your career by taking drafting courses in high school, you could become any one of the members of the design team. All team members should be able to do the following:

1. Solve technical problems by using the tools and techniques of drafting.

2. Make technical drawings that describe the design idea.
3. Read charts and get data from handbooks and technical reference books.
4. Write technical reports that are clear, concise, and accurate.

PATHS TO DRAFTING–RELATED CAREERS

A successful career in drafting, engineering, or designing begins with a strong high school background. The subjects most needed are mathematics, science, and English. Mathematical and scientific knowledge, combined with the communications skills of speaking, reading, and writing, help form a strong basis for a career.

Beyond high school, there are three main ways for a person to become a drafter, engineer, or designer (Fig. 1–5).

High School Followed by Apprenticeship Training

A good background in high school drafting may qualify you to be a trainee in an industrial drafting room. The broader your high school background, the better your chances of being accepted into an apprenticeship program.

In the training program, you might be called an *apprentice drafter,* a *junior drafter,* or a *detailer.* You would make simple detail drawings under the supervision of an experienced detailer. As you become more skilled, you would develop more difficult drawings.

Fig. 1–5 Three main career paths begin with high school drafting.

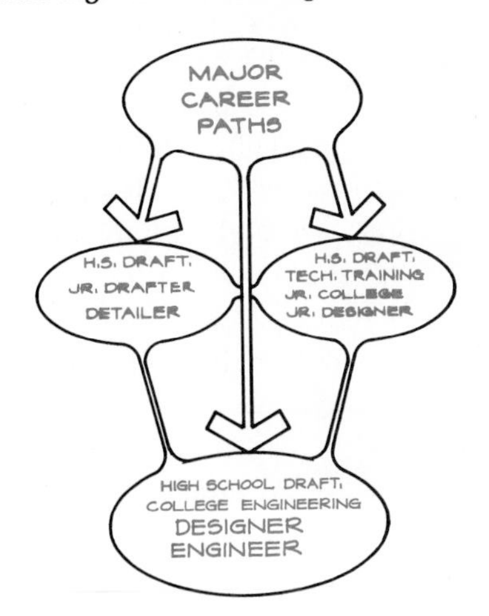

High School Followed by Technical Training

Beyond high school, there are good two-year courses that can train you to be a *drafting technician.* The technician assists the engineer. The training may include courses in surveying, mathematics, basic estimating, materials of industry, and production methods. You could advance from drafting technician to *senior detailer* or *junior designer.* As a senior detailer, you would have to use your own judgment when working on design drawings. You would also need to know a lot about manufacturing methods and procedures. Writing technical reports is often part of the job. These reports are usually submitted to the *chief drafter.*

High School Followed by College

Students planning to be engineers should carefully choose high school courses that will help them in college. You should have four years of mathematics and two or more years of science. All engineering students usually take the same courses for the first two years of college. At the end of the first two years, you would decide whether to become a mechanical, civil, electrical, or some other kind of engineer. After graduating from college, you would usually start in industry with the title of *engineer-in-training* or *engineer trainee.* After your training program, you would be given the title of *engineer.* You might then work for your company as an engineer in research, development, design, testing, production planning, construction operation, sales, service, or standards. No matter what field you choose, you will probably use drafting in your work. The person in charge of the *engineering graphics* (drafting) department is usually an engineer. You can get more information about careers in drafting, engineering, and related fields from these organizations:

American Institute for Design and Drafting
349 Price Road
Bartlesville, OK 74003

American Federation of Technical Engineers
1126 16th Street, N.W.
Washington, DC 20036

SPECIALIZED DRAFTING FIELDS

Just as engineers specialize because no one person can know all areas of engineering well, drafters also specialize. Some of the specialized drafting fields that you can choose are described below. These are not all the possible fields, but they are the most popular ones. Some chapters in this textbook will tell you about other drafting specialties and give you more details about

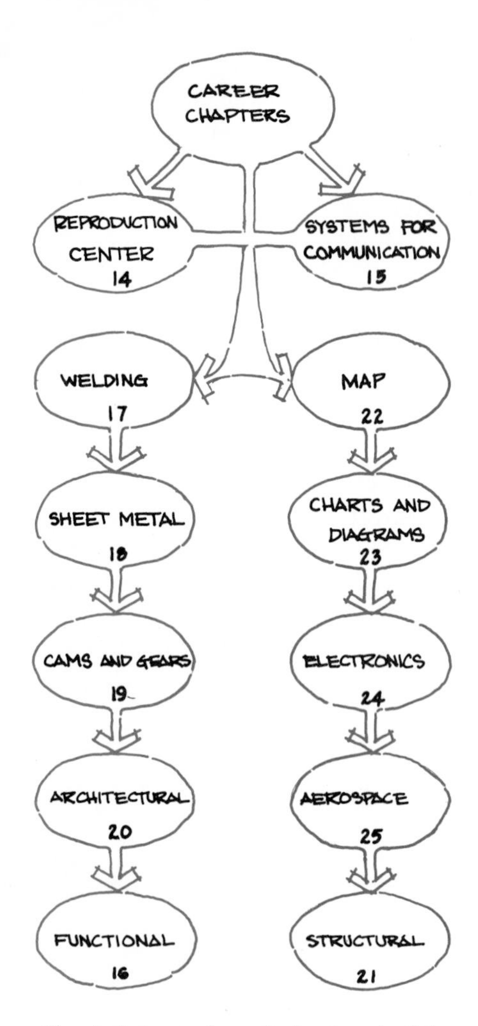

Fig. 1–6 **A number of chapters in this text offer career information.**

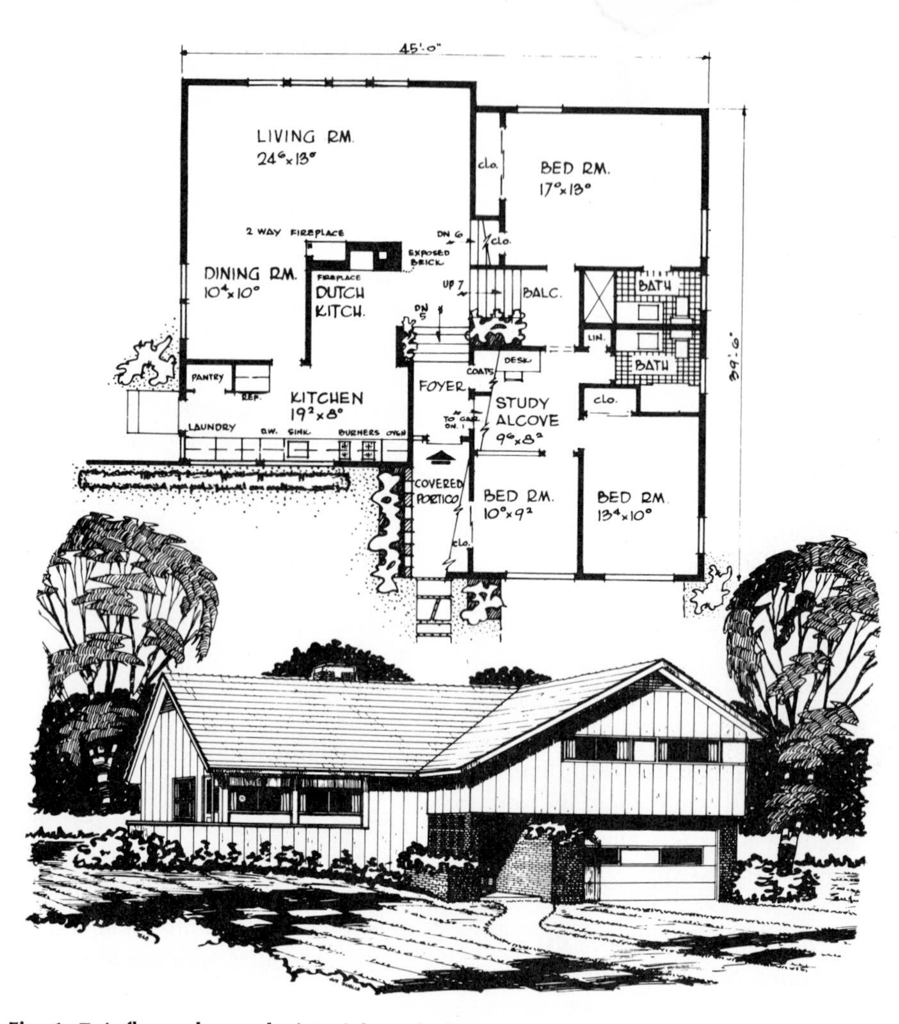

Fig. 1–7 **A floor plan and pictorial rendering.** ***(Herman H. York, architect, AIA, Jamaica, NY.)***

those described here. Figure 1–6 will tell you the name and number of the chapter that deals with a particular specialty.

Architectural Drafting

Architectural drafting has to do with drawing floor plans, wall-framing plans, roof-framing plans, outside elevations, inside elevations, and landscape plans. You sometimes draw small details to a large scale. Large structures are often drawn to a small scale. Figure 1–7 shows a floor plan and a pictorial, or picturelike, rendering of a house. These are typical drafting assignments. Sometimes, architects need to show what a building will look like by having a model made. The model will let the owner of the building see clearly how it will look on the site chosen (Fig. 1–8).

Aeronautical or Aerospace Drafting

The growing need for military and commercial aircraft and the

Fig. 1–8 **Checking an architectural model.** ***(Photo by Toby Brown.)***

space program have produced a great demand for aeronautical and aerospace drafters. As such a drafter, you would draw single-part details or subassembly drawings, which show how two or more parts are put together. You might also make major assembly drawings, installation drawings, and layouts. Figure 1–9 is an example of an aeronautical drawing.

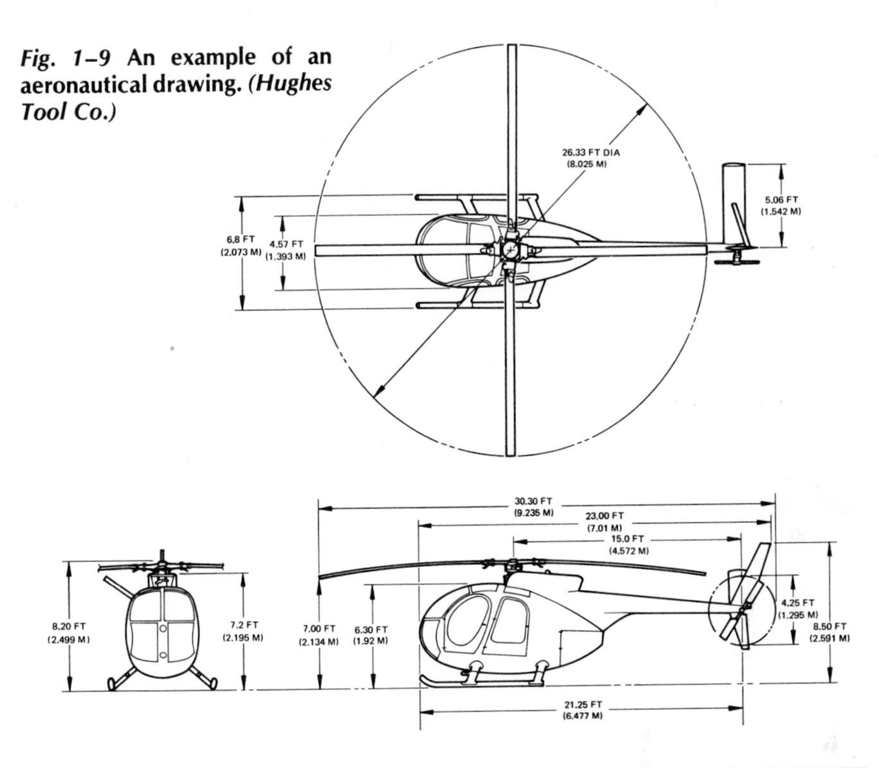

Fig. 1–9 **An example of an aeronautical drawing.** *(Hughes Tool Co.)*

Map Drafting

Map drafters work with civil engineers and surveyors. Governments need map drafters to keep records of property boundaries and road changes. Figure 1–10 is an example of the work map drafters do.

Fig. 1–10 **Map of Chicago lakefront.**

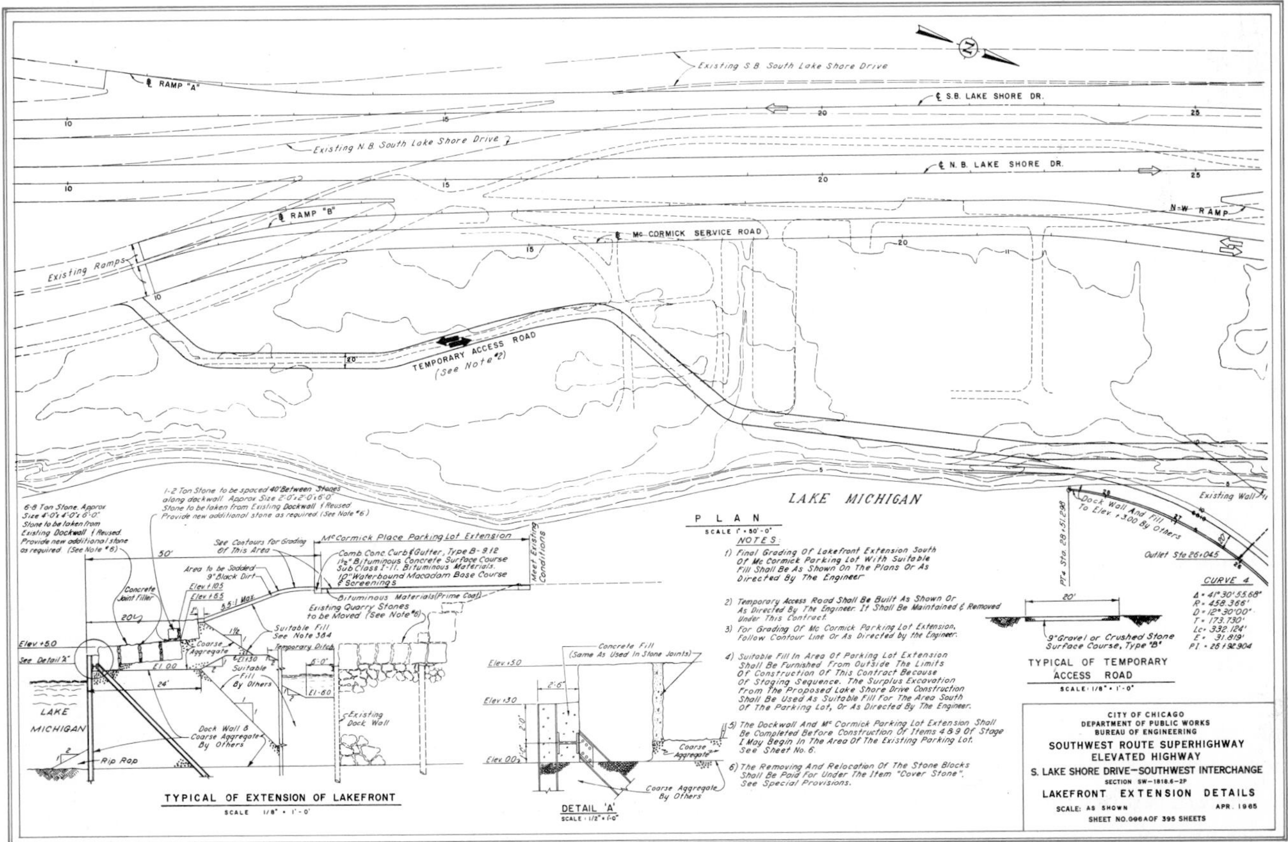

Machine Drafting

Machine drafters make detail and assembly drawings of machine parts and machines. They also specify parts made by other companies. All the information needed to make or assemble the products is found on drawings made by machine drafters. Figure 1–11 is an example of a machine drawing.

Electrical and Electronics Drafting

Electrical drafters make drawings that show where electric wires and fixtures are to be put in buildings. They also draw electrical fixtures and machines. Electronics drafters draw schematic diagrams of electronic devices such as radio, television, and stereo sets. The microcircuits found inside calculators, computers, and other electronic devices are also drawn by the electronics drafter. Figure 1–12 is an example of an electronic schematic drawing.

Technical Illustration

Technical illustrations can be simple sketches or very complex drawings. The technical illustrator generally has some artistic talent. This talent can be used even on technical drawings. Technical illustrations are usually pictorial drawings. They show how machines are put together. An exploded view, as shown in Fig. 1–13, is an example. Technical illustrations are found in owner's manuals and in technical manuals for service, repair, or operation. Catalogs and other advertising materials also have technical illustrations.

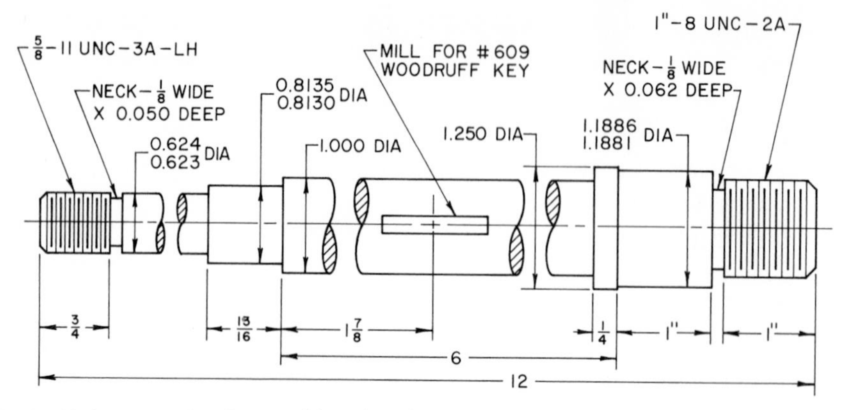

Fig. 1–11 **An example of a machine drawing.**

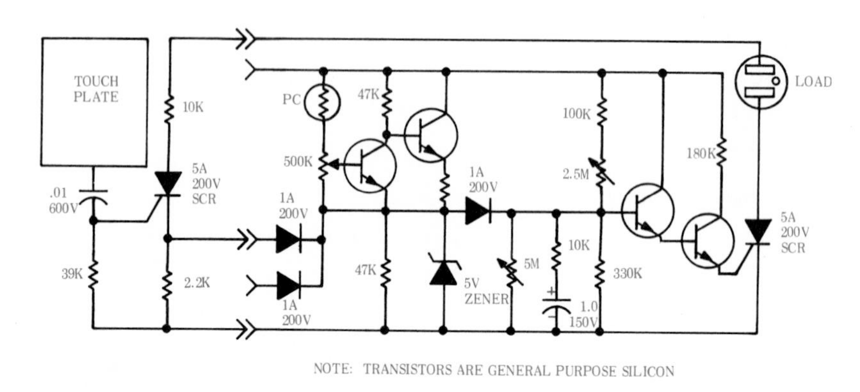

Fig. 1–12 **An example of an electronic schematic diagram.**

Fig. 1–13 **An exploded pictorial drawing.** ***(Industrial Division, Standard Precision, Inc.)***

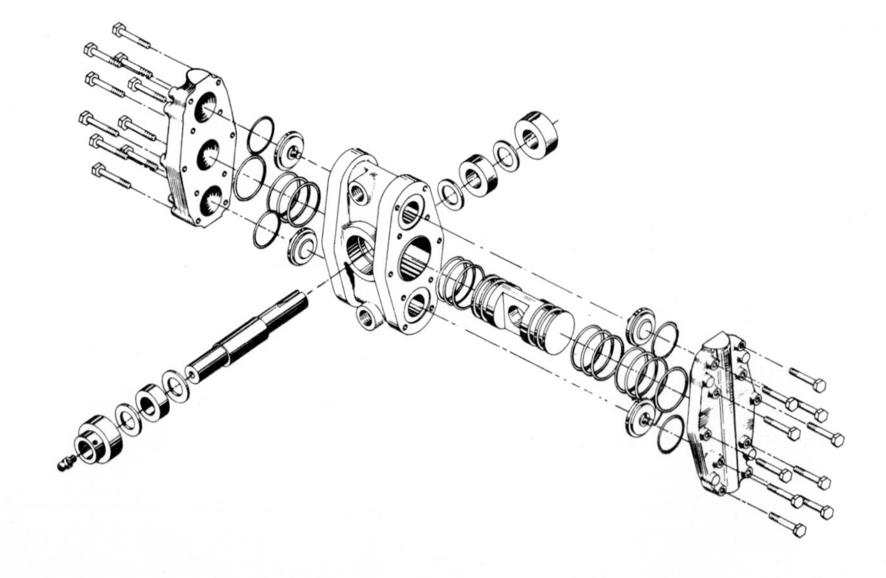

THE V-BLOCK IS TO BE MADE OF CAST IRON AND MACHINED ON ALL SURFACES. THE OVERALL SIZES ARE TWO AND ONE-HALF INCHES HIGH, THREE INCHES WIDE, AND SIX INCHES LONG. A V-SHAPED CUT HAVING AN INCLUDED ANGLE OF 90° IS TO BE MADE THROUGH THE ENTIRE LENGTH OF THE BLOCK. THE CUT IS TO BE MADE WITH THE BLOCK RESTING ON THE THREE INCH BY SIX INCH SURFACE. THE V-CUT IS TO BEGIN ONE-QUARTER INCH FROM THE OUTSIDE EDGES. AT THE BOTTOM OF THE V-CUT THERE IS TO BE A RELIEF SLOT ONE-EIGHTH INCH WIDE BY ONE-EIGHTH INCH DEEP.

Fig. 1–14 **A written description of a V-block.**

Each of these kinds of drawings is a way of getting information from one person to another. Each is a special form of *communication.*

COMMUNICATION

Each one of us spends much of our waking lives communicating. Mostly we talk or listen, but we also read and write, and sometimes we draw pictures. Drafting is a form of communication that is technical and very exact. Drafters the world over have agreed to use the same lines and symbols and the same methods for drawing. For this reason, drafting is called the *universal language of industry.* Drawings made in New York can be used by technicians in Chicago, London, Paris, or Tokyo.

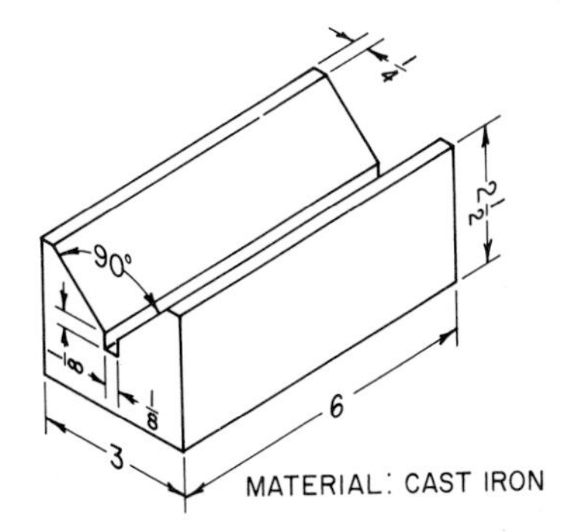

Fig. 1–15 **A technical drawing of a V-block.**

Drawings are used for communication in industry because they are the clearest way to tell someone what to make and how to make it. Think how hard it would be to tell a friend in words about the shape and size of all the parts in an automobile! Figure 1–14 is an example of a "word picture" of a simple tool, a V-block. Read it and see if you can tell what a V-block looks like. Could you make one from this description? Figure 1–15 is a drawing of this same V-block. See how much easier it is to know the shape and size from the drawing than it is from the written description. You know it would be easy to make one from this drawing. Such a drawing is an example of the fourth level of *graphic communication.* This is communication that is written or drawn.

FOUR LEVELS OF GRAPHIC COMMUNICATION

Graphic communication may be talked about as having four separate levels: (1) *creative communication,* (2) *technical communication,* (3) *market communication,* and (4) *construction communication.* Each of these levels has its place in the field of drafting (Fig. 1–16).

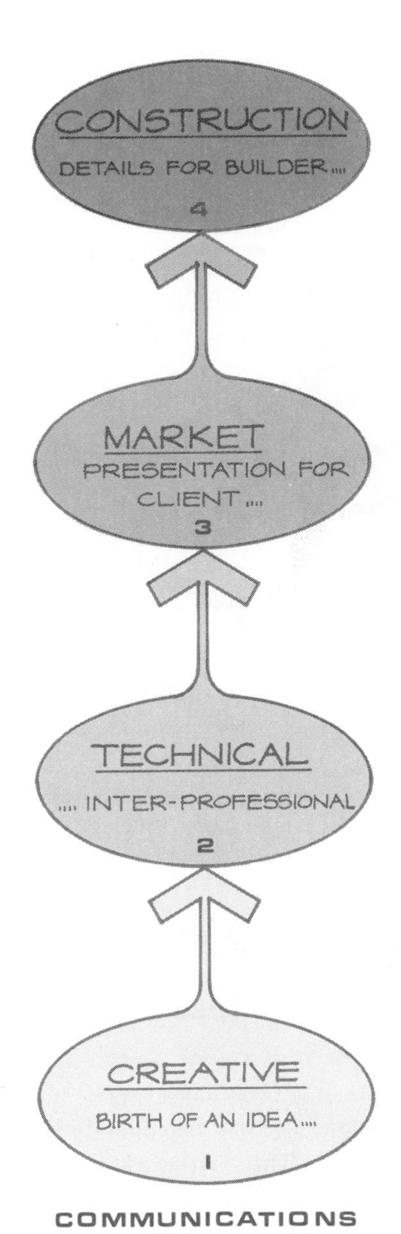

Fig. 1–16 **Four levels of graphic communication are used by designers and drafters.**

Level One:
Creative Communication

Graphic communication usually begins as an idea in the mind of a

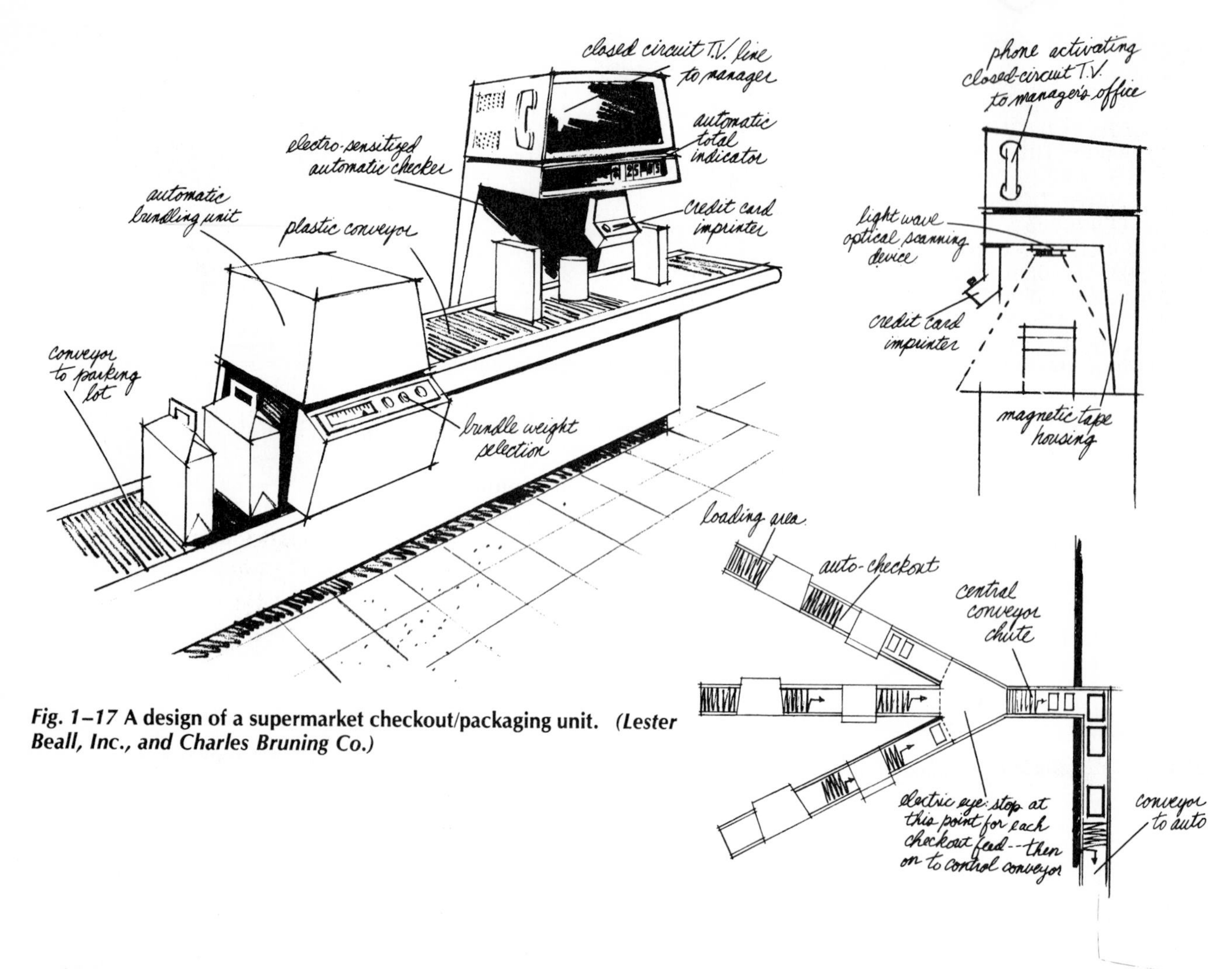

Fig. 1–17* A design of a supermarket checkout/packaging unit. *(Lester Beall, Inc., and Charles Bruning Co.)

designer. The designer's first sketches are the beginnings of a product. This first work may be thought of as the *birth of an idea.* The creative communication of sketching is important to people who need to express ideas quickly. Sketching can capture ideas for further study.

Level Two: Technical Communication

When the designer gives a sketch to the other members of the design team, it is ready for level two. Engineers and technicians or architects and their assistants study and change the original design to make it more practical. The design is refined, or improved, through the ideas of several people. Two (or more) heads are better than one! The result is a group of drawings that fully explain the design (Fig. 1–17).

Level Three: Market Communication

This level includes the evaluation of a design by a client or customer. This is especially used in architectural design. Before designs are made final, customers look at them to see if they like them. Clients must evaluate designs for style, form, and function. An example would be approving the designs of the floor plan (Fig. 1–18) and the outside (Fig. 1–19) of a house before actual construction work on the house begins.

Level Four: Construction Communication

This level includes all the details needed for manufacturing or construction. These drawings must be complete so that estimators can fig-

Fig. 1–18 **A client and architect study a floor plan.** ***(Photo by Toby Brown.)***

Fig. 1–19 **The design of a home is shown for approval.** ***(Inland Steel Co.)***

ure the exact costs of a project and factory superintendents can know exactly how a product is to be made. People who use the drafter's plans should not have to guess about details or ask the designer exactly what was meant.

Society needs students who can continue technical progress through drafting. If you come to thoroughly understand the first thirteen chapters of this book, you will have the basic skills to accept the challenge of the advanced chapters of this text.

Review

1. Drafting is often called the language of ________.

2. Who are the four members of the design team?

3. There are basically three paths to a career in drafting. Name them.

4. Name three specialized fields of technical drawing.

5. Name the four levels of graphic communication.

6. The designer's first sketches can be thought of as the ________.

7. Passing information from one person to another in a way that can be understood is ________.

8. What three high school subjects are needed for a drafting career?

Chapter 2

Sketching and Lettering

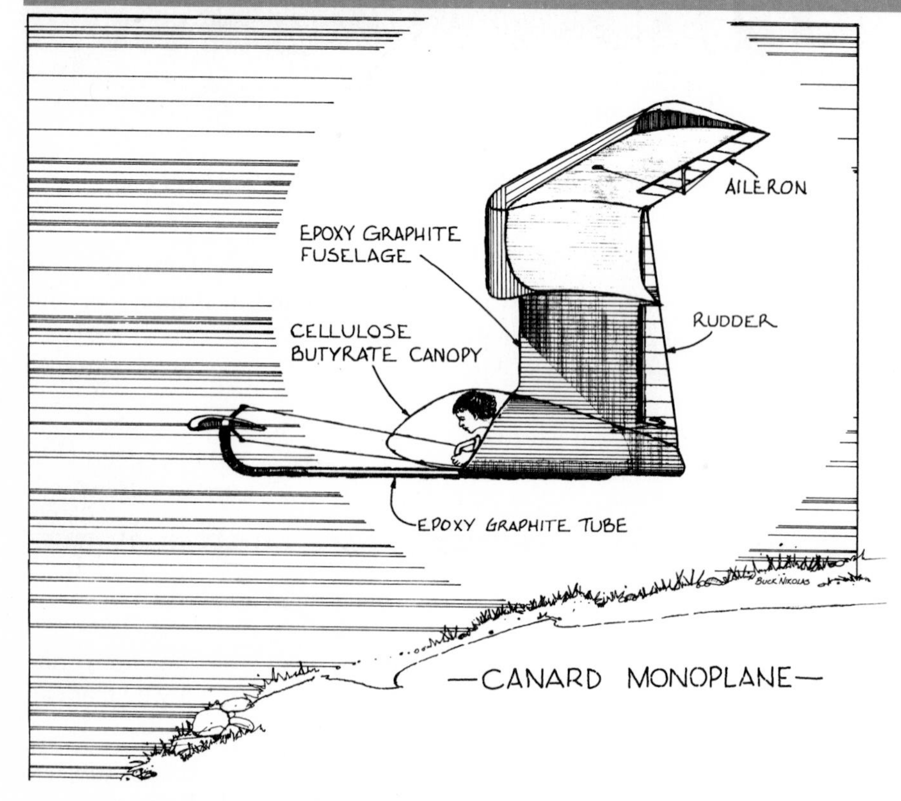

***Fig. 2–1* Sketching can describe new ideas.** *(Sketch by Buck Nikolas.)*

A LANGUAGE OF VISUAL SYMBOLS

When you see sketches bring ideas alive, you know the saying *progress begins on paper* is true. When words alone cannot describe new or futuristic forms, sketching is needed to show the thoughts that cannot be said (Fig. 2–1). The language of sketching has four basic *visual symbols* (things you can see). These are a point, a line, a plane, and a texture, or surface quality (Fig. 2–2). Any idea, no matter how simple or complicated or how plain or spectacular, can be sketched by using these four things. Sketching well often means using as few details as you can to show an idea fully.

REASONS FOR SKETCHING

There can be many reasons for doing technical sketching. The following are the most important:

1. To persuade people who make decisions about a project that an idea is good.
2. To develop a refined sketch of a proposed solution to a problem so that a client can respond to it (Fig. 2–3).
3. To show a complicated detail of a *multiview* (more-than-one-view) drawing enlarged or in a simple *pictorial* (picturelike) sketch.
4. To give design ideas to drafters so that they can do the detail drawings.
5. To develop a series of ideas for refining a new product or machine part.
6. To develop and analyze the best methods and materials for making a product.
7. To record permanently a design improvement on a project that already exists. The change may result from the need to repair a part that breaks over and over again. It may result from the discovery of an easier and cheaper way to make a part.
8. To show that there are many ways to look at or to solve a problem.
9. To spend less time in drawing. It is quicker to make a sketch,

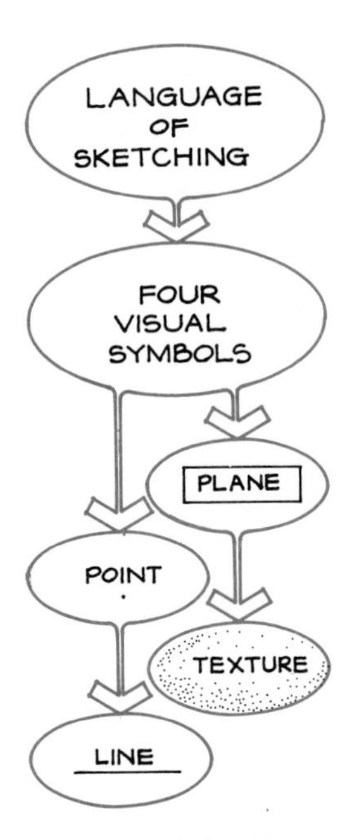

Fig. 2–2 **Elements of sketching are simple visual symbols.**

Fig. 2–3 **Sketching for the client.** *(A. W. Wendell and Sons, designer-contractor.)*

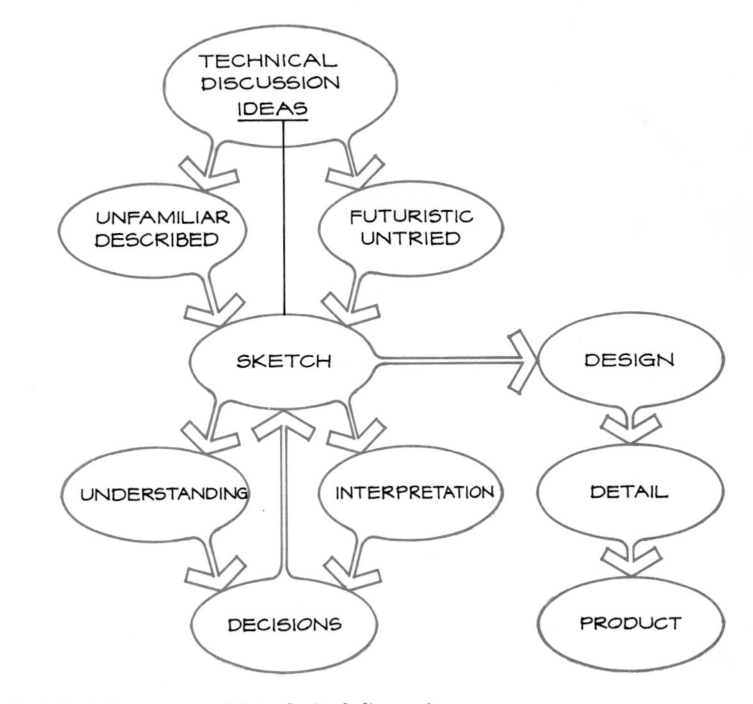

Fig. 2–4 **Sketches can assist technical discussions.**

which takes only a pencil and a sheet of paper, than to do a mechanical drawing.

SKETCHING AND TECHNICAL DISCUSSIONS

Freehand sketching is the simplest form of drawing. It is one of the quickest ways to express ideas. A sketch can help simplify a technical discussion. Designers, drafters, technicians, engineers, and architects will often explain complicated or unclear thoughts with a freehand sketch. Ideas imagined in the mind can be caught in sketches and thus held in simple lines for further study. The pencil is an instrument that aids clear thinking and creative communication. Figure 2–4 shows when sketching can become an important part of technical discussions and design decisions.

LETTERING

On most of the sketches you make, you will need to write notes and directions. To be sure that the notes and directions are clear, the drafter uses lettering instead of writing. Even on drawings made with instruments, notes are hand lettered. Good lettering is always needed by drafters throughout their careers. Lettering is one of the first things students of drafting study. This is because you do not learn how to do good lettering all at once. You learn it by practicing it little by little for a long time. Your lettering should get better with every sketch or drawing you do.

THE USE OF LETTERING

Simple freehand lettering, quickly made and perfectly *legible* (readable), is important to business, industry, and engineering. Single-stroke Gothic lettering (See Fig. 2–13) is used on technical drawings

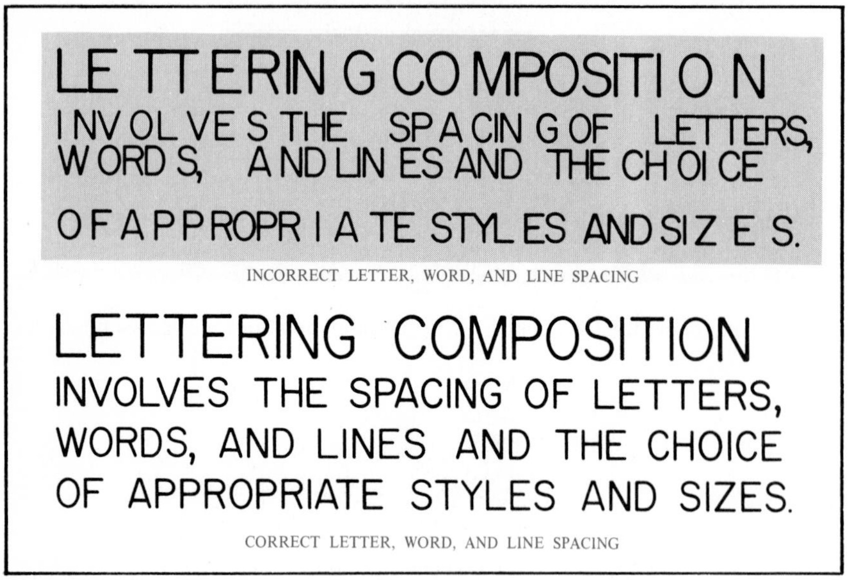

LE TT ERIN G CO MPOSITI O N
I NV OL VE S THE SP A CIN G OF LETTERS,
W ORD S, A ND LIN ES AND THE CH OI CE
O F A P P R OPR I A TE STYL ES AND SI Z E S.
INCORRECT LETTER, WORD, AND LINE SPACING
LETTERING COMPOSITION
INVOLVES THE SPACING OF LETTERS,
WORDS, AND LINES AND THE CHOICE
OF APPROPRIATE STYLES AND SIZES.
CORRECT LETTER, WORD, AND LINE SPACING

Fig. 2–5 **Single-stroke lettering and word spacing.**

GUIDELINES 2½
3 4 5
6 7 8
Lines 27
Lines 27
T-SQUARE
AEIO
Line 14
Line 14
T-SQUARE

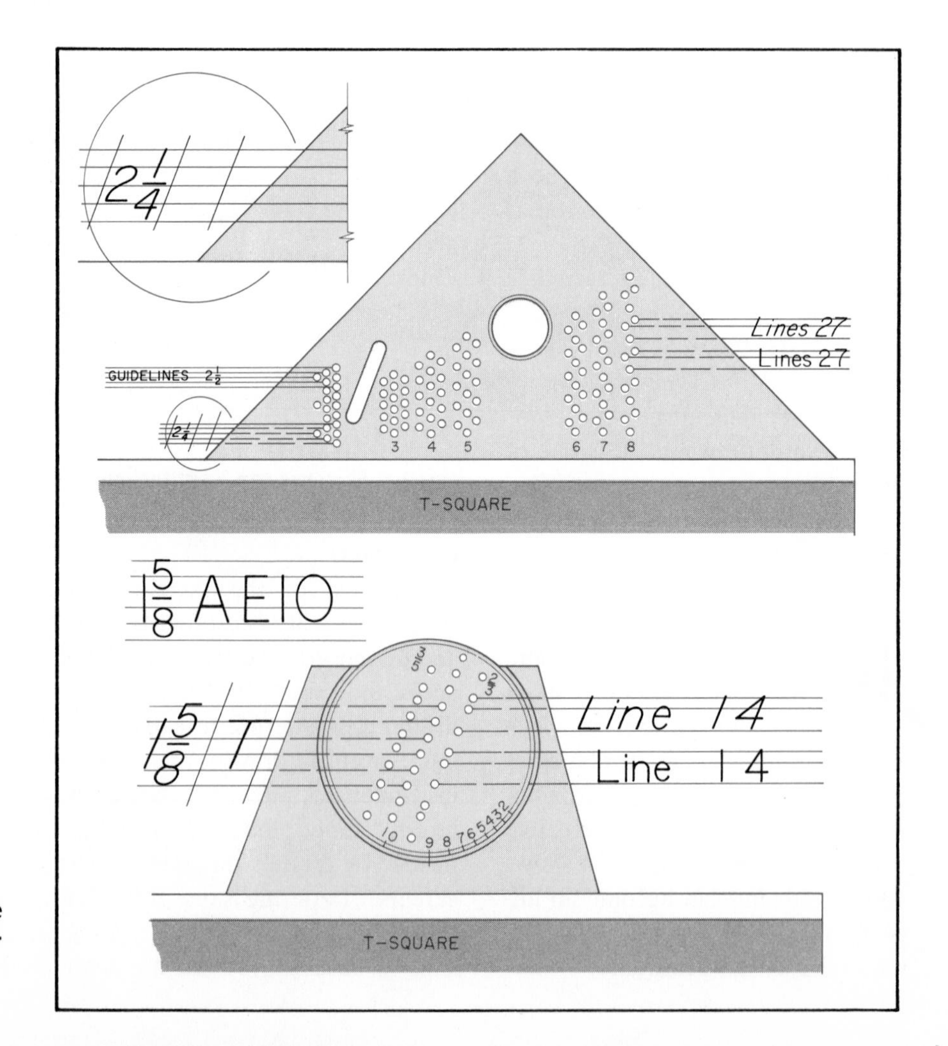

Fig. 2–6 **Lettering guidelines are to be evenly spaced by using a lettering triangle or an Ames lettering instrument.**

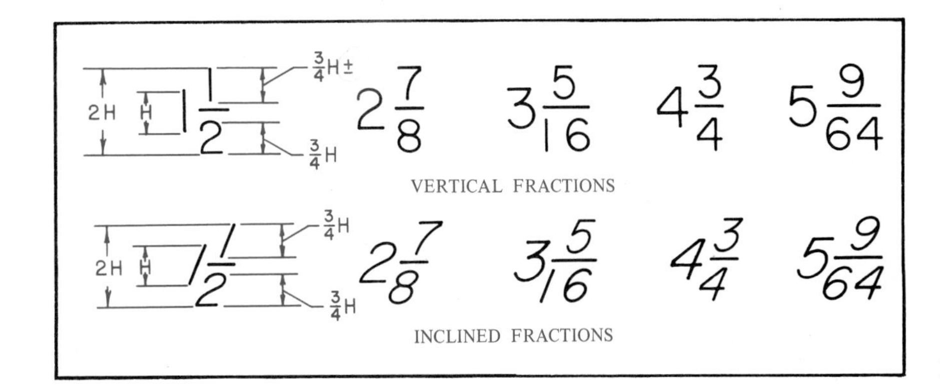

Fig. 2–7 **Guidelines for fractions.**

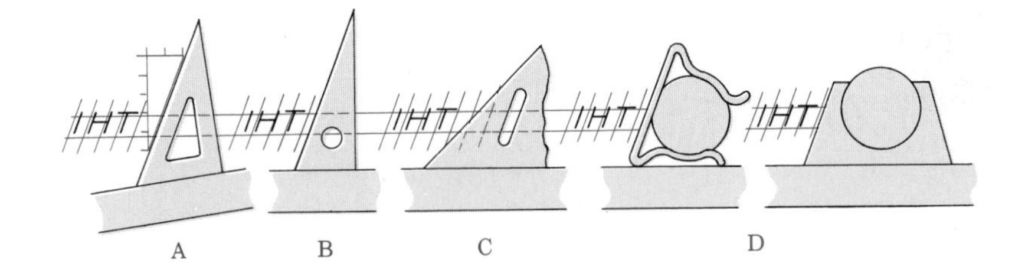

Fig. 2–8 **Various methods for laying out inclined guidelines.**

to tell the kinds of materials, sizes, distances, and amounts; to identify units; and to give other necessary information.

COMPOSITION

In lettering, *composition* means arranging words and lines with letters of the right style and size. Letters in words are not placed at equal distances from each other. They are placed so that the spaces between the letters *look equal.* The distance between words, called *word spacing,* should be about equal to the height of the letters. Figure 2–5 shows examples of proper and improper letter and word spacing.

GUIDELINES

In order to make letters the same height and in line, you must *rule* (draw with a straightedge) guidelines for the top and bottom of each line of letters. You draw guidelines lightly with a sharp pencil. The *clear distance* (open space) between lines of letters is from ½ to 1½ times the height of the letters. You can use a lettering triangle or an Ames lettering instrument to space guidelines accurately (Fig. 2–6). Guidelines are spaced 3 mm (⅛ in.) apart for regular letters and *numerals* (numbers) on a drawing. Guidelines for fractions are usually twice the height of whole numbers, as shown in Fig. 2–7.

FRACTIONS

Fractions are always made with a division line that is *horizontal* (side-to-side). The whole fraction is usually twice the height of regular numbers. The numbers in the fraction are about three-fourths the height of regular numbers. *Fraction numbers must never touch the division line* (Fig. 2–7).

SINGLE-STROKE INCLINED CAPITAL LETTERS AND NUMERALS

Inclined (slanted) letters and numerals should slope at an angle of about 67½° with the horizontal. Figure 2–8 shows some ways of laying out inclined guidelines. These may help you develop skill in hand lettering. Guidelines should be very thin and light so they will not detract from the letters.

Figure 2–9 shows single-stroke inclined letters and numerals on an inclined grid. Notice that the only difference between *vertical* (straight) and inclined letters is the slant.

SINGLE-STROKE VERTICAL CAPITAL LETTERS AND NUMERALS

The shapes and proportions of letters and numerals are shown in

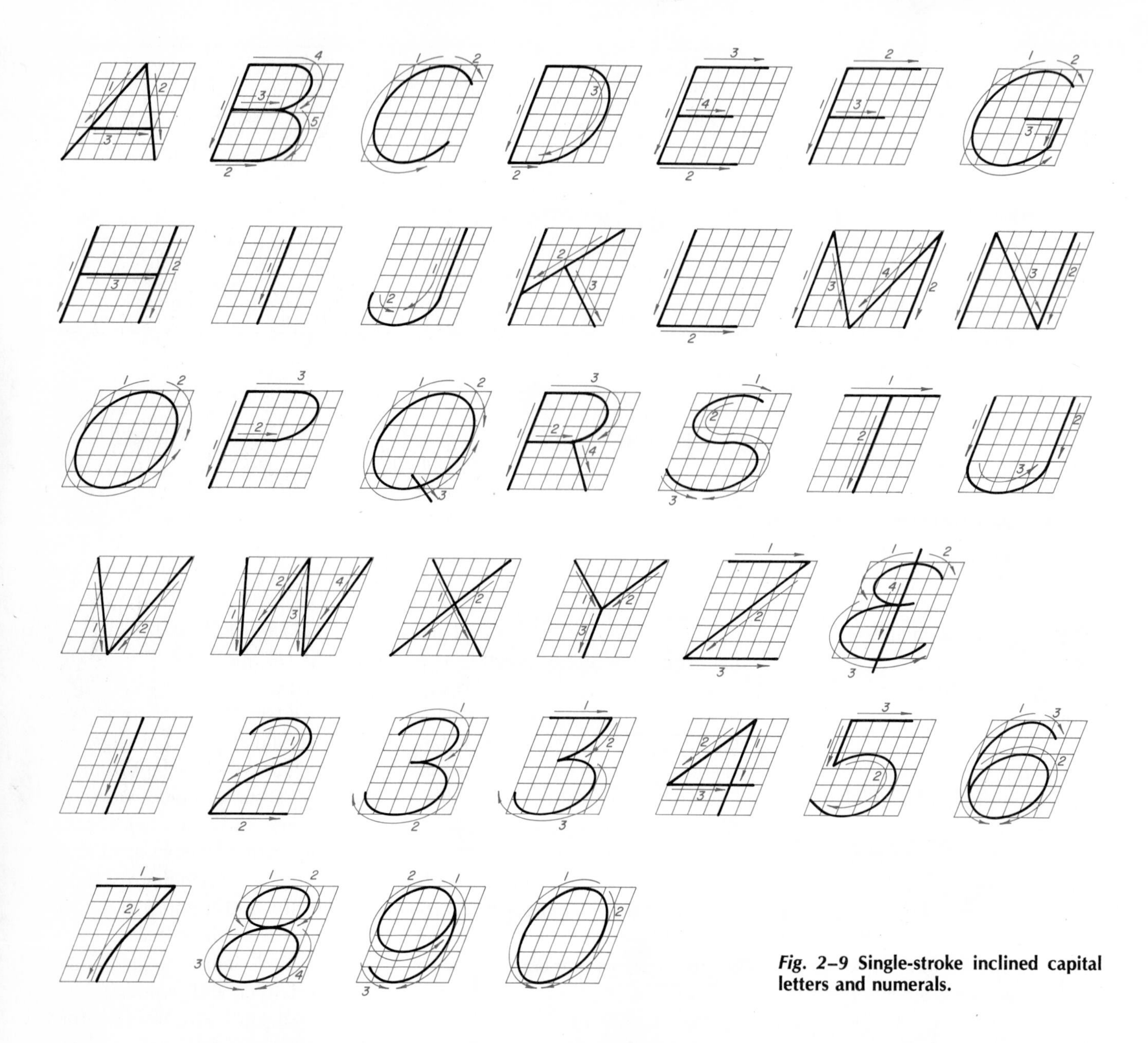

Fig. 2–9 **Single-stroke inclined capital letters and numerals.**

Fig. 2–10. You should study these characters carefully until you understand completely how each is made. Each character is shown in a square 6 units high. The squares are divided into unit squares. By following these, you can easily learn the right shapes, proportions, and strokes. You can vary your lettering to make it more individual. Figure 2–11 shows some of the possible variations.

VARIOUS STYLES OF LETTERING

There are various styles of letters, a few of which are shown in Fig. 2–12. Each style is right for a particular use. You should be careful in choosing one. For example, it is not practical to use fancy roman-style lettering for notes and dimensions on technical drawings. Such lettering would be more costly in time and money than the single-stroke styles in Fig. 2–13. Yet, it would serve the purpose of the drawing no better.

Fig. 2–10 **Single-stroke vertical capital letters and numerals.**

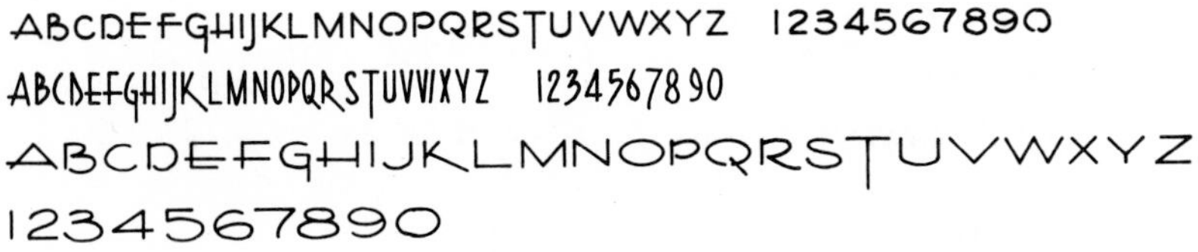

ABCDEFGHIJKLMNOPQRSTUVWXYZ 1234567890

ABCDEFGHIJKLMNOPQRSTUVWXYZ 1234567890

ABCDEFGHIJKLMNOPQRSTUVWXYZ

1234567890

Fig. 2–11 **Variations in lettering styles.**

The lettering style most commonly used on working drawings is *single-stroke commercial Gothic* (Fig. 2–13). This style is best because it is easy to read and easy to hand letter. It is made up of *uppercase* (capital) letters, *lowercase* (small) letters, and numerals. Nearly all companies now use only uppercase lettering. As a result, this book stresses uppercase lettering. Letters and numerals may be either vertical or inclined (Fig. 2–13). However, the same style should be followed throughout a set of drawings.

DISPLAY LETTERING

Display lettering is often used for title sheets, posters, and other places where large, easy-to-read lettering is needed. Although any style may be used for display, the drafter most often uses Gothic, the most legible of all styles. Figure 2–14 shows a block-style and a regular-style Gothic.

The block style is the easiest to make. It has no curved strokes and only takes a T-square and triangle. A light-line grid may be drawn on the sheet and later erased. If you are using a tracing *medium* (drawing material), you can put a grid sheet under the drawing sheet. You can block in the letters in pencil first. Then, if you want, you can trace them in ink. Letters can either be outlined or filled in, as shown in Fig. 2–14. The thickness of the strokes can vary from one-fifth to one-tenth the height of the letter. In Fig. 2–14, one-seventh is used. In Fig. 2–15, one-fifth through one-tenth are shown.

VERTICAL

lower case

INCLINED

COMPRESSED

EXTENDED

BOLD FACE

SLANT *Italics*

Old English

MODERN

OLD ROMAN

Fig. 2–12 **Various styles of lettering.**

Fig. 2–13 **Single-stroke Gothic letters, vertical and inclined.**

VERTICAL SINGLE - STROKE COMMERCIAL GOTHIC

ABCDEFGHIJKLMNOPQRSTUVWXYZ CAPITALS (UPPER CASE)

abcdefghijklmnopqrstuvwxyz LOWER CASE

1234567890 NUMERALS

INCLINED SINGLE - STROKE COMMERCIAL GOTHIC

ABCDEFGHIJKLMNOPQRSTUVWXYZ CAPITALS (UPPER CASE)

abcdefghijklmnopqrstuvwxyz LOWER CASE

1234567890 NUMERALS

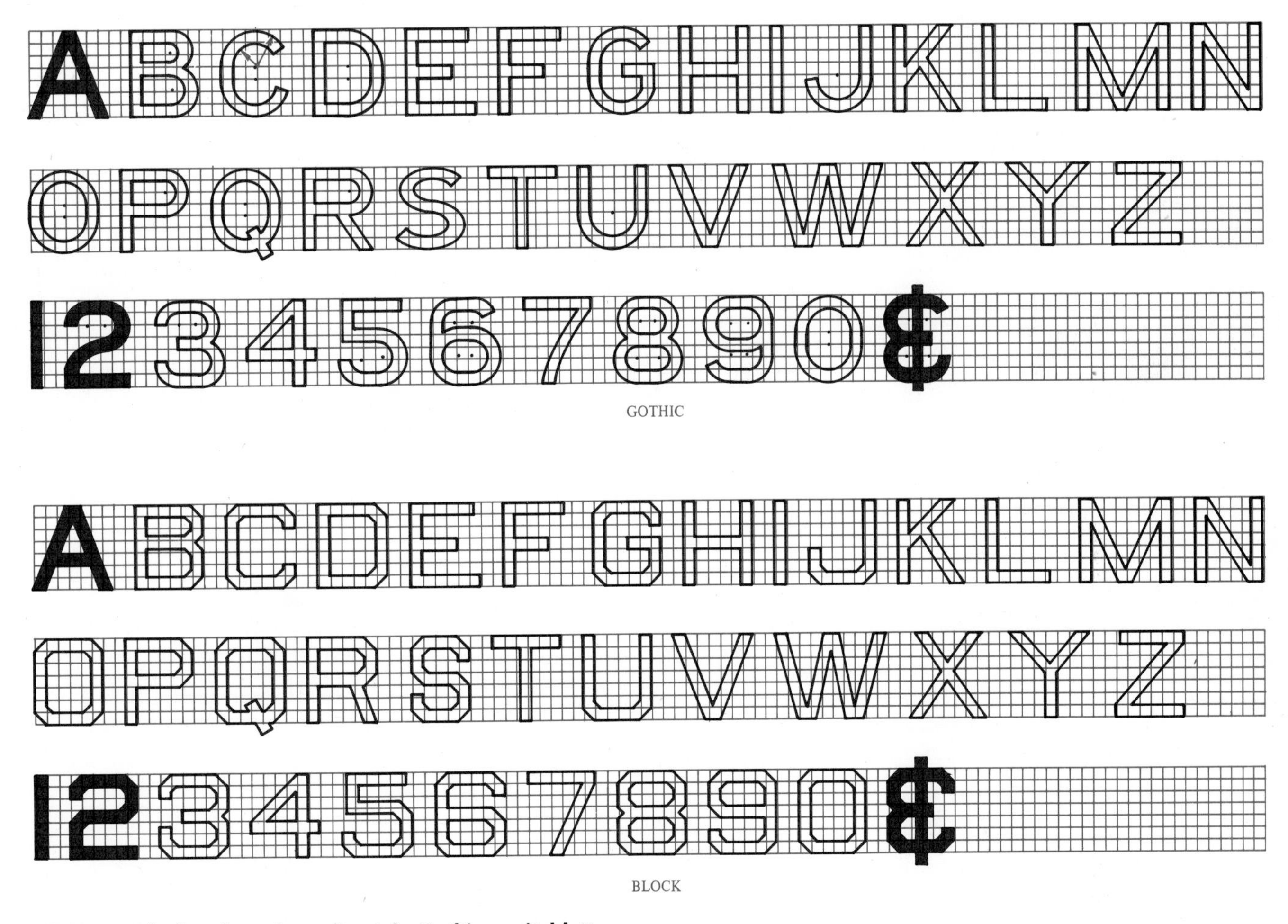

Fig. 2–14 **Block-style and regular-style Gothic capital letters.**

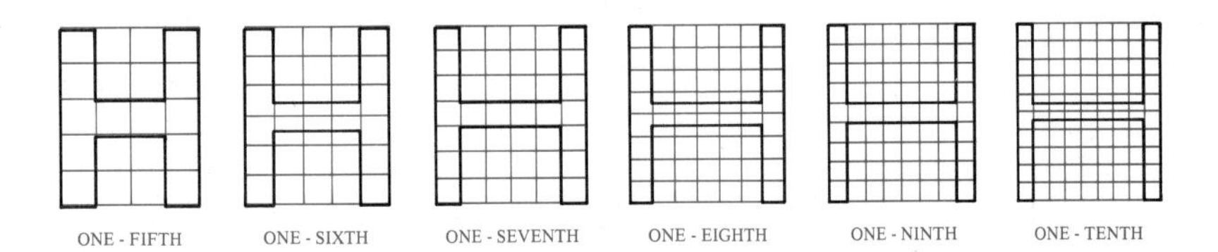

Fig. 2–15 **The thickness of strokes may vary.**

Regular Gothic capital letters are made in much the same way. The only difference is that you have to make curved strokes. You can do this with circle templates or a compass. You can also draw curves freehand.

You can change the size and proportion of letters by using the grid as shown in Fig. 2–16. Notice that letters can be made larger or smaller overall by using this method. In addition, they can be stretched or squeezed.

TYPES OF SKETCHES—ROUGH (UNREFINED) AND REFINED

Any image drawn on paper *freehand* (without a straightedge or other tools) may be called a *sketch.* The sketching may be *rough,* or *unrefined.* That is, the sketch may

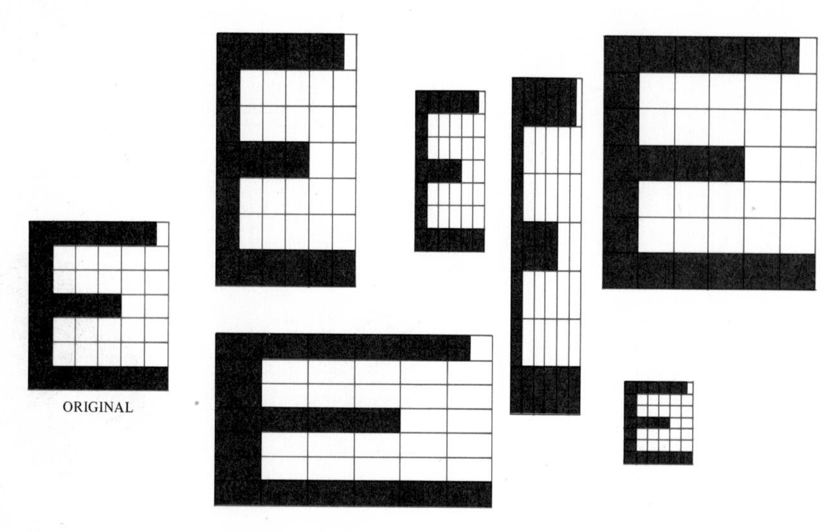

Fig. 2–16 **Grid method for changing the size and proportion of letters.**

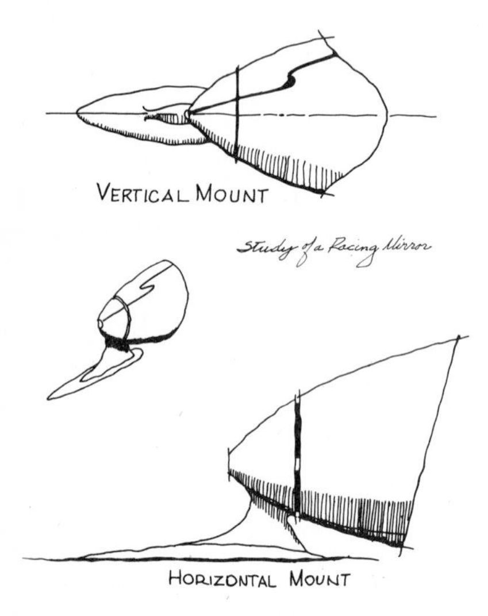

Fig. 2–17 **A two-positional automobile racing mirror.**

be drawn quickly with jagged lines. The rough sketch can quickly express thoughts. Figure 2–17 shows rough sketches that were used to develop *preliminary* (early) designs of a two-position automobile mirror. The sketches show several design choices. The other way of sketching is *refined.* That is, the sketch is neat and finished-looking. A refined sketch is carefully drawn. It shows good proportion and excellent line values. It may be more persuasive than an unrefined sketch. Many refined sketches are based on a rough sketch that has captured the general idea. Figure 2–18 shows the usual way sketching is used to communicate. A very good way to *refine* (improve) a sketch is to use an overlay.

THE OVERLAY

Sketches are often drawn on paper that can be seen through. This paper is called *translucent paper* or *tracing paper.* The best parts of a sketch may be quickly traced by putting a new piece of this paper, called the *overlay,* on top (Fig. 2–19). Thus, refining ideas means sketching over and over again on tracing paper until the design is right.

TWO USES FOR OVERLAYS

The overlay is used in two important ways that may seem very much alike. The first use is reshaping an idea. This might include refining the proportions of the parts of an object or changing its shape entirely. Secondly, an overlay can be used to refine, or improve, the drawing itself without really changing the design. These two things can be, and usually are, done at the same time (Fig. 2–20).

NATURE OF A SKETCH

A sketch is an important form of *graphic* (drawn) communication.

Fig. 2–18 **Sketching—rough to refined.**

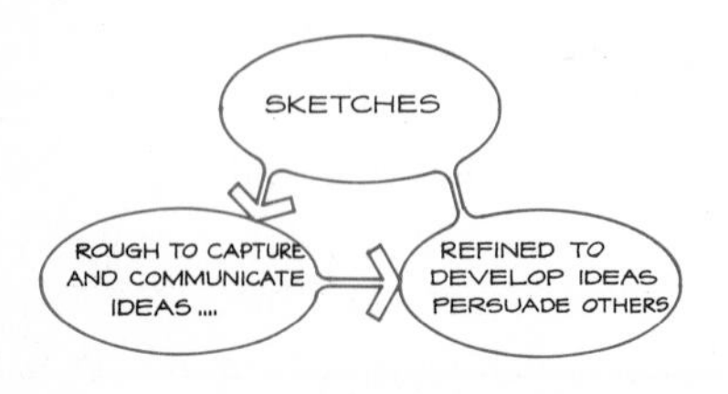

Fig. 2–19 **The overlay can speed up the design process.**

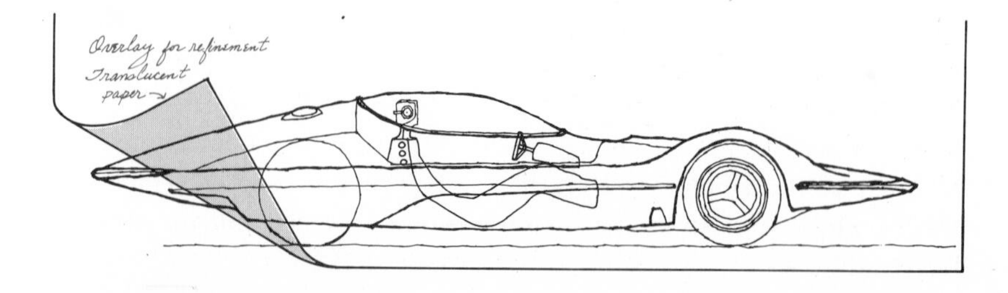

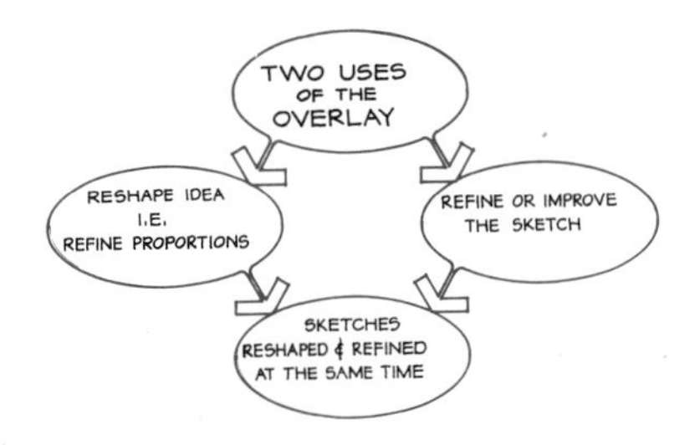

Fig. 2–20 **Two uses for the overlay process.**

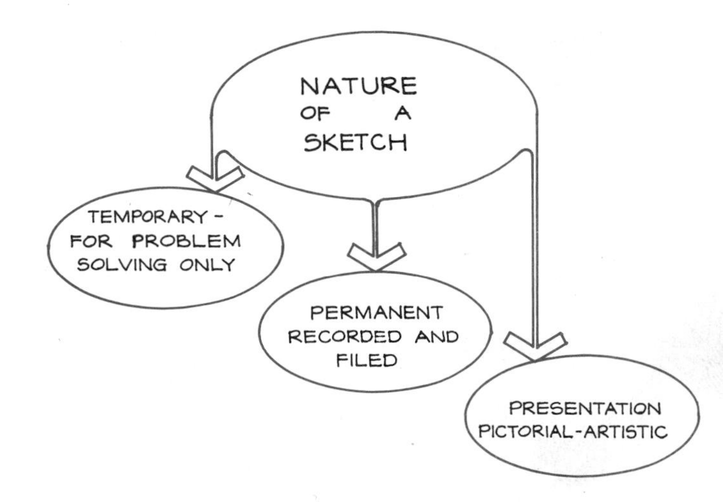

Fig. 2–21 **Sketches are classified in three ways.**

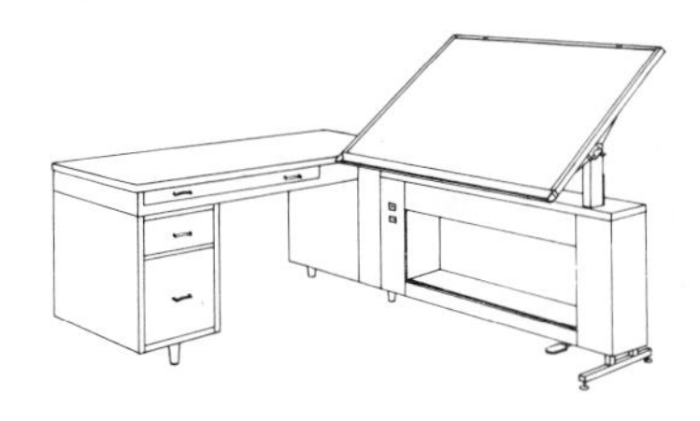

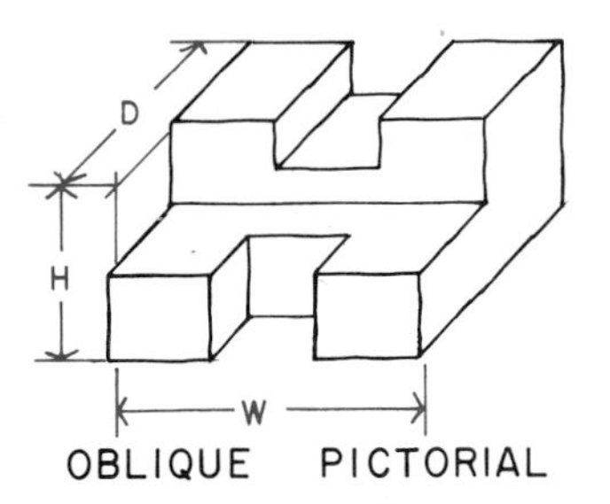

Fig. 2–22 **Comparing perspective and oblique pictorials.**

There are many levels of sketches for different uses. (Fig. 2–21). If you know the language of mechanical drawing well, you will know how to sketch your ideas quickly, clearly, and accurately.

Temporary

Many technical sketches have short lives. They are done merely to solve an immediate problem. Then they are thrown away. Other technical sketches are kept longer. It may take weeks or even months to study some sketches and make mechanical drawings from them. However, these sketches too may be thrown away some day.

Permanent

Sometimes the engineering department or the management of a company will include a sketch in a notice to other employees. Such a sketch is an important record and should be kept. Therefore, some sketches are filed as part of a company's permanent records.

Presentation

The sketch that is refined for presentation is generally *pictorial* (picturelike). It is used to convince a client or management to accept and approve the ideas presented. The pictorial sketch has a three-dimensional view that can be understood easily by nontechnical people. Such sketches are generally drawn so that they look glamorous, artistic, or eye-appealing.

VIEWS NEEDED FOR A SKETCH

In chapter 5, multiview projection will be discussed in full. However, you need to know some basic things about views and how they are placed in order to do sketches.

There are two types of drawings that you can sketch easily. One is called *pictorial* (picturelike) *drawing.* In this type of drawing, you show the width, height, and depth of an object in one view (Fig. 2–22). The other is called *multiview projection* or *orthographic projection.* In this type of drawing, you usually show an object in more than one view. You do this by drawing sides of the object and relating them to each other, as shown in Fig. 2–23.

ONE–VIEW SKETCH

If an object can be fully shown in only one view, this means that its shape can be described well in two *dimensions* (sizes). These are height and width. Things shown in one-view drawings generally have a depth or thickness that is *uniform* (the same throughout). This third dimension can be given in a note rather than being drawn. A typical one-view drawing is shown in Fig.

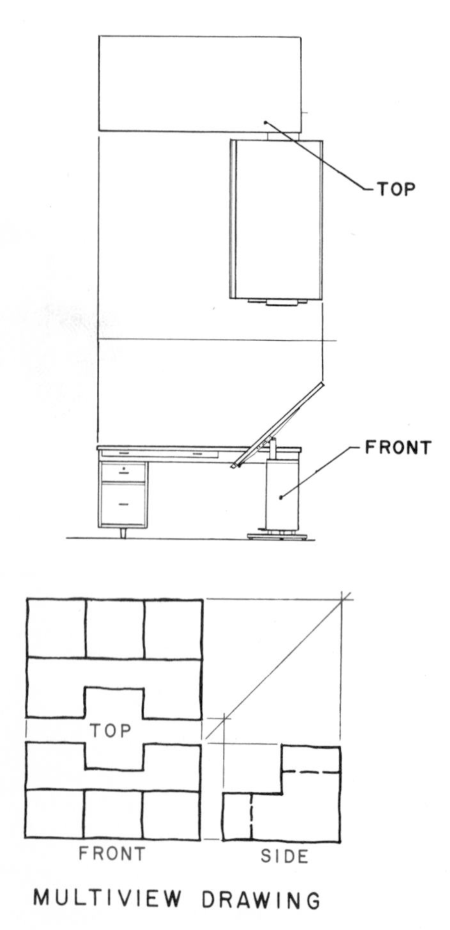

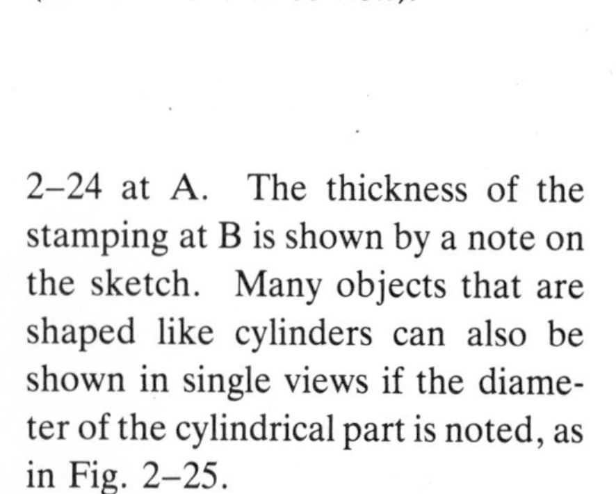

Fig. 2–23 **Typical multiview drawings (two-view and three-view).**

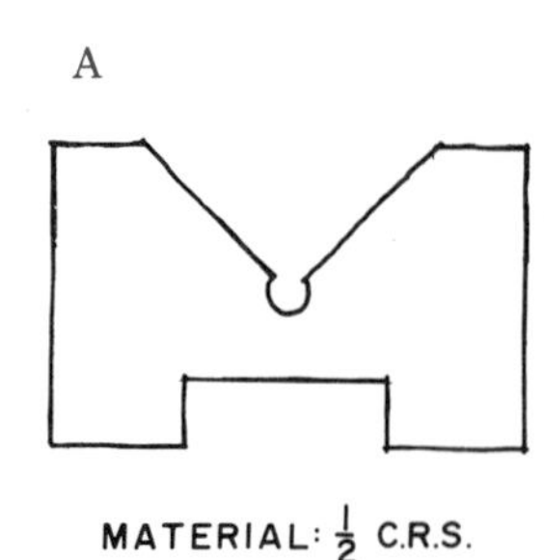

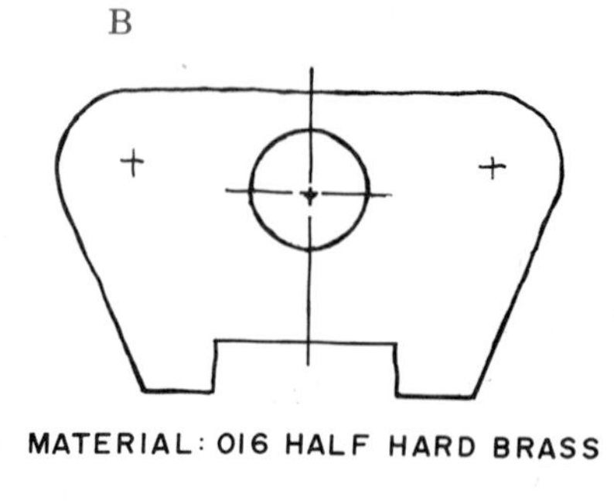

Fig. 2–24 **Typical one-view drawings.**

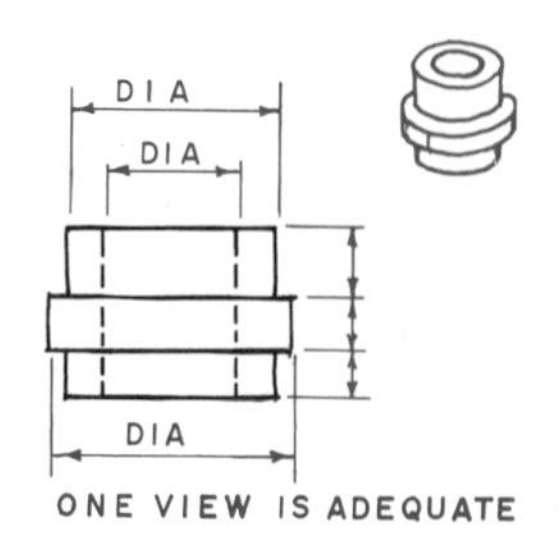

Fig. 2–25 **A cylindrical object may require only one view.**

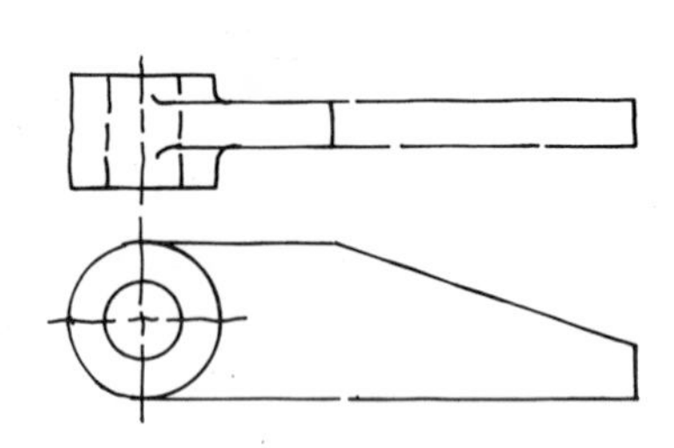

Fig. 2–26 **A two-view sketch.**

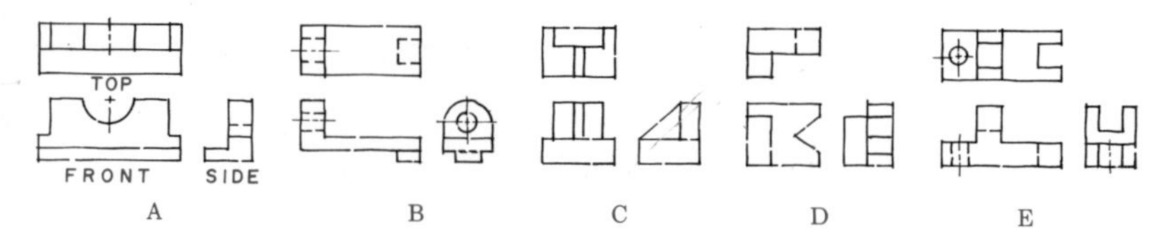

Fig. 2–27 **Select the two required views at A, B, C, D, and E.**

2–24 at A. The thickness of the stamping at B is shown by a note on the sketch. Many objects that are shaped like cylinders can also be shown in single views if the diameter of the cylindrical part is noted, as in Fig. 2–25.

TWO–VIEW SKETCHES

There are many objects that can be described in only two views, as shown in Fig. 2–26. If two views that describe the object well are carefully selected, this can help simplify the drawing. Figure 2–27 shows 5 three-view drawings in which the objects could have been described well with only two views. Which view is not needed at A, B, C, D, and E?

MULTIVIEW SKETCHES

A pictorial drawing shows how the object looks in three-dimensional form. Three directions are suggested for viewing the residence in Fig. 2–28. From in front, the width and height would show. This

Fig. 2–28 **A-frame residence in pictorial showing three-dimensional form.**

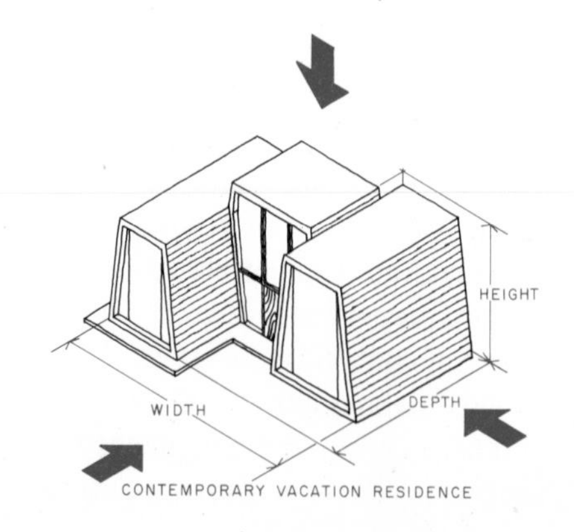

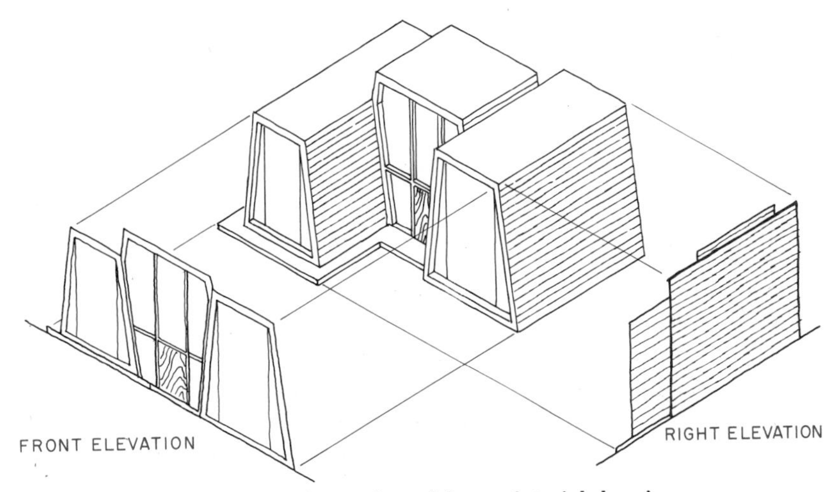

Fig. 2–29 **Two elevations can be projected from pictorial drawing.**

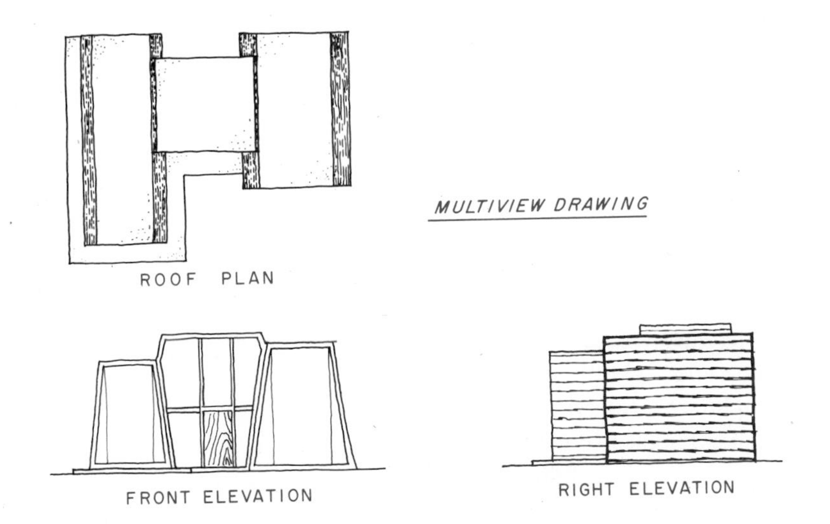

Fig. 2–30 **Three-view drawing of an A-frame residence.**

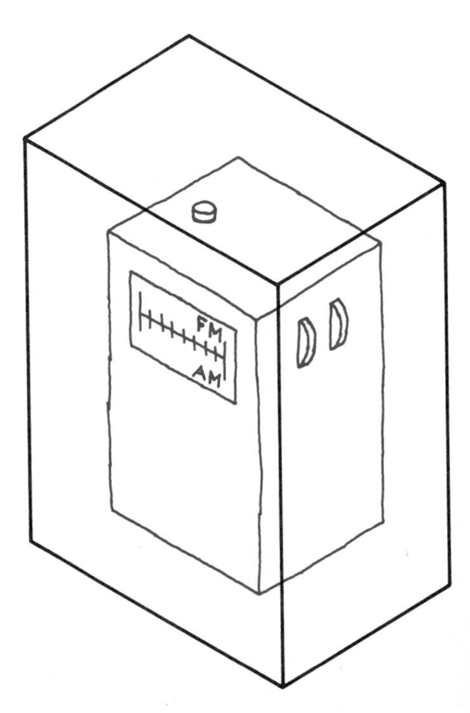

Fig. 2–31 **The portable radio can be viewed through a "glass box."**

is the front view. From the side, the depth and height would show. This is the right-side view. From above, the depth and width would show. This is the top view. However, a three-view pictorial does not fully describe the residence. There are lines and details that do not show completely (Fig. 2–29).

Multiview projection, also called *orthographic projection,* is the system that arranges views in relation to each other (Fig. 2–30).

THE GLASS BOX

The portable radio in Fig. 2–31 can be thought of as being inside a *transparent* (clear) glass box. By looking at each side of the radio through the glass box, you can see the six possible views of the object. When the glass box is opened up into one plane (Fig. 2–32), the views are placed as they would be drawn on paper.

MATERIALS FOR SKETCHING

The good things about sketching are that you need only a few materials and that you can do it anywhere. You are ready to sketch if you have a pencil, an eraser, and a pad of paper. If you find that you need more equipment than that, you are probably not as good a drafter as you could be.

Paper

You can use plain paper for sketching. If you need to refine the sketch, you should use tracing paper. You can control proportions

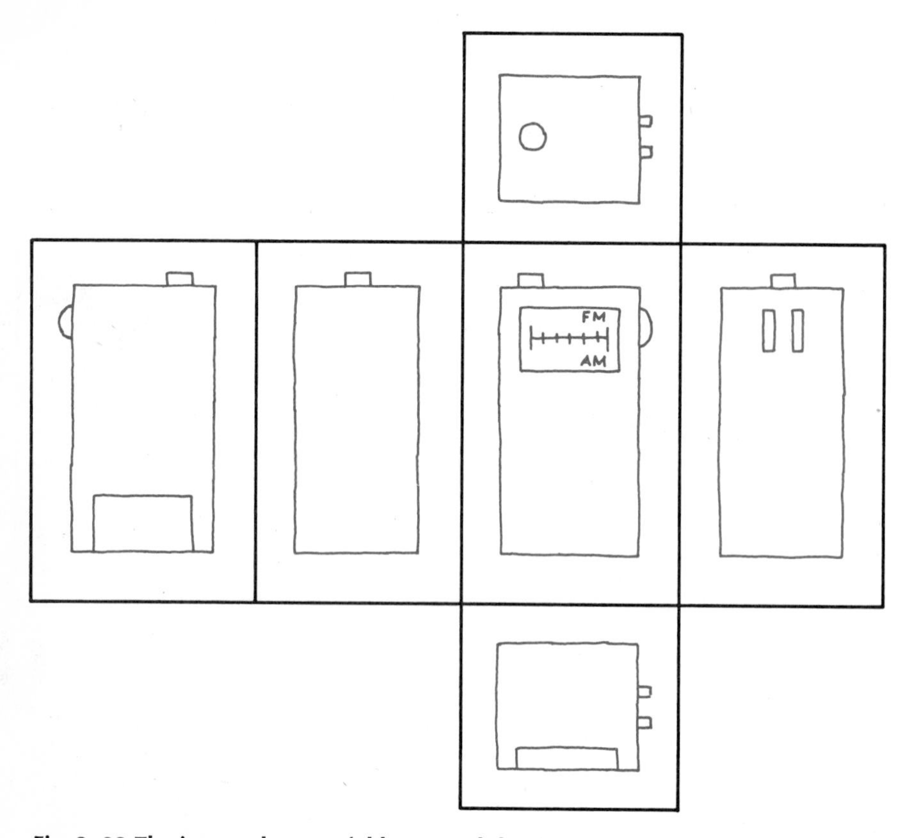

Fig. 2–32 **The image planes unfold to reveal the six sides of the radio.**

while sketching by using *cross-section paper,* also called *graph paper* or *squared paper.* Graph paper that is used most often has heavily ruled one-inch squares. The one-inch squares are then subdivided into lightly ruled one-tenth-, one-eighth-, one-quarter-, or one-half-inch squares. This paper is called ten to the inch, eight to the inch, and so on. Graph paper is also ruled in millimeters. There are many specially ruled graph papers for particular kinds of drawing such as isometric and perspective. These kinds of drawings are explained in Chapter 12.

You can sketch on any convenient size of paper. However, standard letter size (216 mm × 279 mm or 8½″ × 11″) is the best for making small sketches quickly. You can hold the paper on stiff cardboard or on a clipboard while you are working on it. If you use cross-section paper, you can put it under tracing paper to help guide line spacing.

Pencils and Erasers

Most drafters like to use soft pencils (grades F, H, or HB), properly sharpened. They also use an eraser that is good for soft leads, such as a plastic eraser or a kneaded-rubber eraser.

Use a drafter's pencil sharpener to remove the wood from the *plain* end of a pencil. This is done so that the grade mark (F, H, or HB) on the other end will show. Sharpen the lead to a point on a sandpaper block or on a file. If you are using a lead holder, use a lead pointer. Do not forget to adjust the grade mark in the window of the lead holder if it has one. Be careful to remove the needle point that the sharpener leaves by touching it gently on a piece of scrap paper. Then the pencil will not groove or tear the drawing paper.

Four types of points are used for sketching: sharp, near-sharp, near-dull, and dull (Fig. 2–33). The points should make lines of the following kinds:

Sharp point—a thin black line for center, dimension, and extension lines

Near-sharp point—visible or object lines

Near-dull point—cutting-plane and border lines

Dull point—construction lines

Lines drawn freehand have a natural look. They show freedom of movement because of their slight changes in direction. Hold the pencil far enough from the point that you can move your fingers easily and yet can put enough pressure on the point to make dense, black lines when you need to. You should draw light construction lines with very little pressure on the point. They should be light enough that they need not be erased.

STRAIGHT LINES

You can sketch lines in the following ways: (1) You can draw them continuously. (2) You can draw short dashes where the line should start and end. Then place the pencil point on the starting dash. Keeping your eye on the end

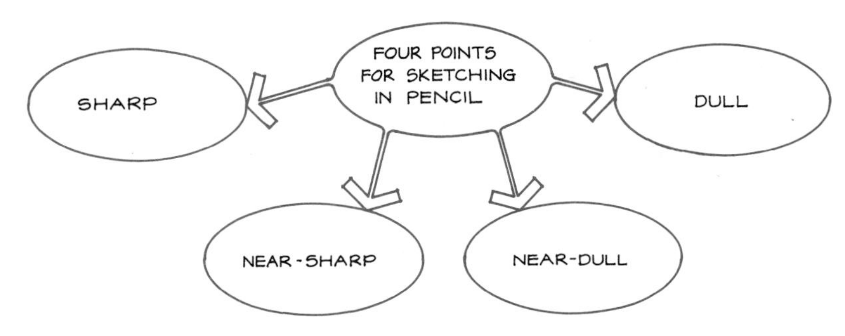

Fig. 2–33 **Four convenient pencil points for sketching.**

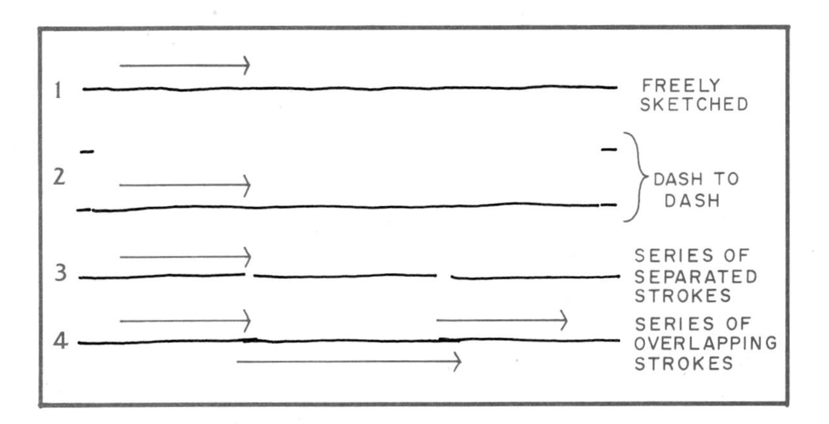

Fig. 2–34 **Some ways of sketching straight lines.**

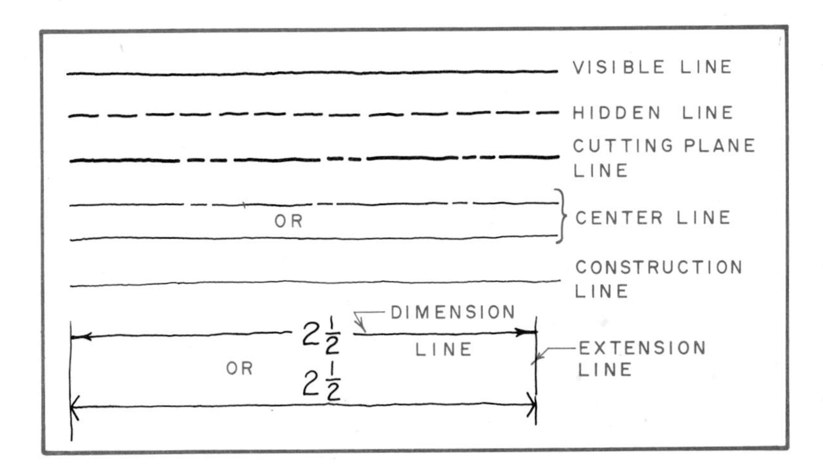

Fig. 2–35 **Types of lines used in sketching and the technique for drawing horizontal and vertical lines.**

dash, draw toward it. (3) You can draw a series of strokes that touch each other or are separated by very small spaces. (4) You can draw a series of overlapping strokes (Fig. 2–34).

Before you try to draw objects, practice sketching straight lines to improve your line technique. You draw horizontal lines from left to right. You draw vertical lines from the top down (Fig. 2–35).

SLANTED LINES AND SPECIFIC ANGLES

You can sketch slanted, or inclined, lines from left to right. You might find it easiest to turn the paper and draw an inclined line the same way you draw a horizontal one. When you are trying to sketch a specific angle, first draw a vertical line and a horizontal line to form a *right angle* (90°). Divide the right angle in half to form two 45° angles. Or, divide it in thirds to form three 30° angles. By starting with these simple angles, you can *estimate* (guess) other angles more exactly. Note the direction of the inclined lines drawn to form a desired angle in Fig. 2–36.

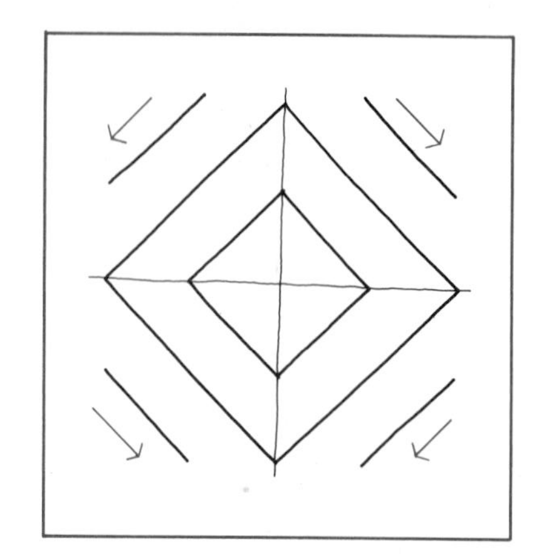

Fig. 2–36 **Sketching slanted lines and angles.**

PROPORTIONS FOR SKETCHING

You do not usually make sketches to *scale* (exact measure). Nonetheless, it is important to keep sketches in *proportion* (similar to exact measure). In preparing the layout, you look at the largest overall dimension, usually width, and estimate the size. Next, you determine the proportion of the height to the width. Then, as the front view with the width and height takes shape, you can compare the smaller details with the larger ones and fill them in too (Fig. 2–37).

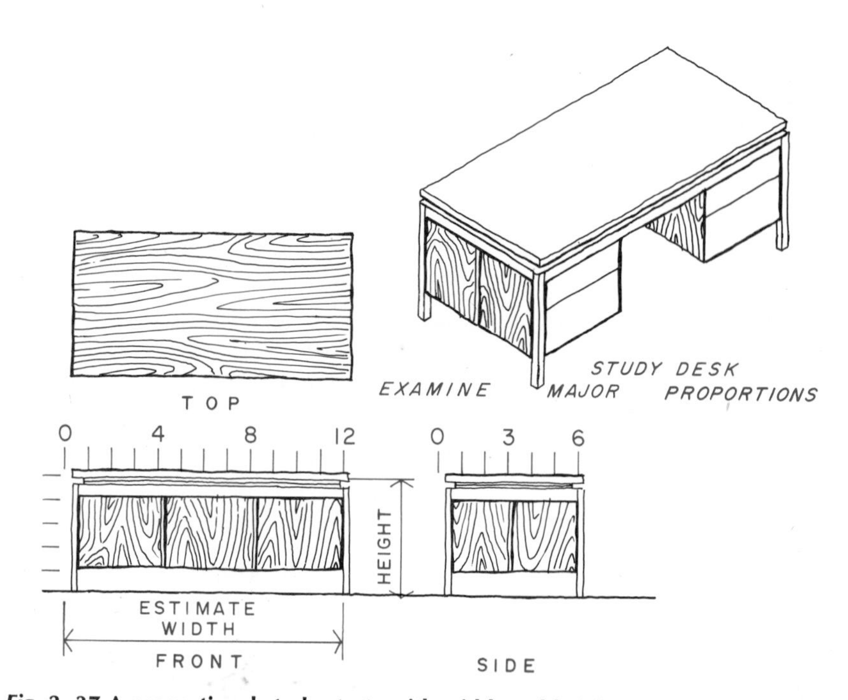

Fig. 2–37 **A proportional study starts with width and height.**

It is important that the design drafter, in sketching an object, have a good sense of how distances relate to each other. This will allow the drafter to show the width, height, and depth of an object in the right proportions.

For example, suppose that the design drafter plans a contemporary stereo cabinet to be four units wide. Each unit width in metric measure is about 400 mm, for a total of 1600 mm. The height of the cabinet is two units. Each unit height is about 400 mm, for a total of 800 mm. The depth of the cabinet is one unit. The one depth unit is 400 mm. Thus, the overall front-view proportions are a width of 1600 mm and a height of 800 mm. This is a proportion of two-to-one. The overall side-view proportions are a height of 800 mm and a depth of 400 mm. This is also a two-to-one proportion. If you were designing this cabinet in customary units, you might choose a 15 in. width, height, and depth unit. The width overall would be 60 in. The height would

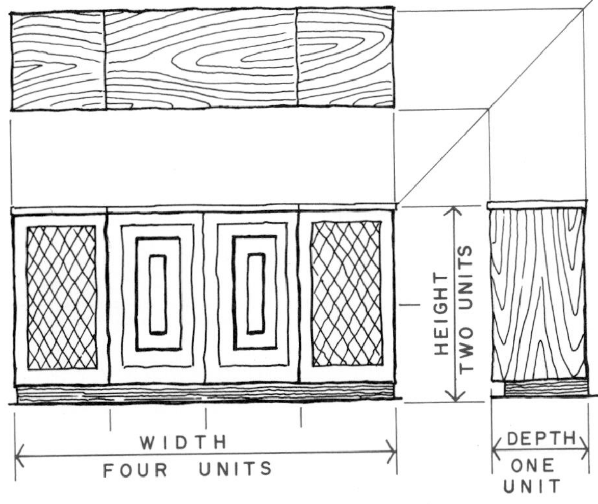

Fig. 2–38 **Sketching a contemporary stereo with proportional units.**

be 30 in., and the depth 15 in. The proportions are developed in two-to-one units in Fig. 2–38.

TECHNIQUE IN DEVELOPING PROPORTION

In order to sketch well, the design drafter must be able to divide a line in half by *eyeballing* (estimating by eye). The halves can be divided again to give fourths. Through practice, you can train your eye to work in at least two directions. Start by drawing a line of one unit. Increase it by one equal unit so that it is twice as long as at first. Practice adding an equal unit and dividing a unit equally in half. Practice developing units on parallel horizontal lines. Then develop them vertically. By learning to compare distances you can get better and better at estimating (Fig. 2–39).

Fig. 2–39 **Practice estimating (eyeballing) proportional units.**

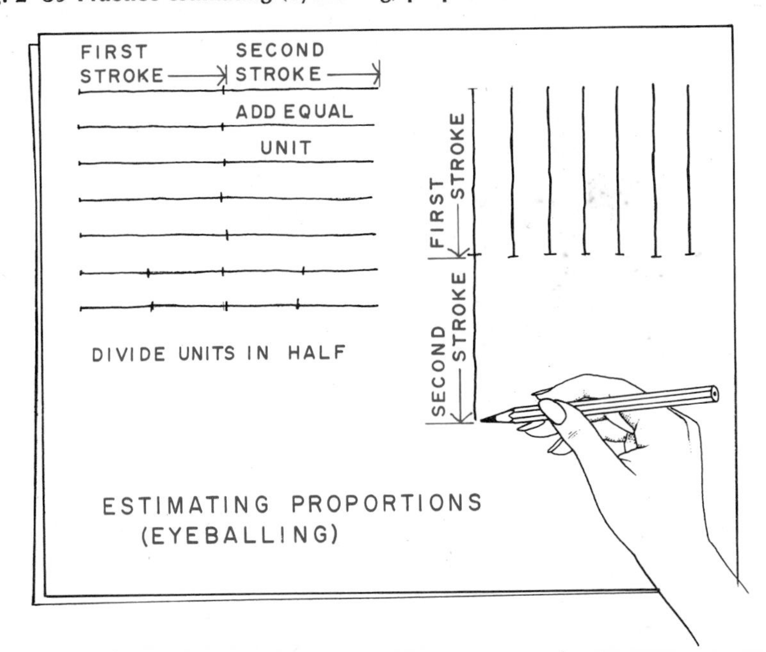

Design drafters can use scrap paper or a rigid card as a straightedge when they do not have *scales* (rulers) at hand. You can fold the paper in half to find the length of units on the marked edge, as in Fig. 2–40. In this way, you can draw lines of the same length several times in different directions. This will help you maintain the right proportions.

Figure 2–41 shows how to use *diagonal* (corner-to-corner) lines to multiply proportional units. At A, the square *ABCD* is sketched using diagonals. At B, a horizontal centerline is sketched through the *intersection* (crossing point) of the diagonals. The centerline is then extended. At C, the large diagonal lines *AL* and *BN* are sketched. The rectangle with a length three

Fig. 2–40 **Drafters develop techniques for blocking out sketches.**

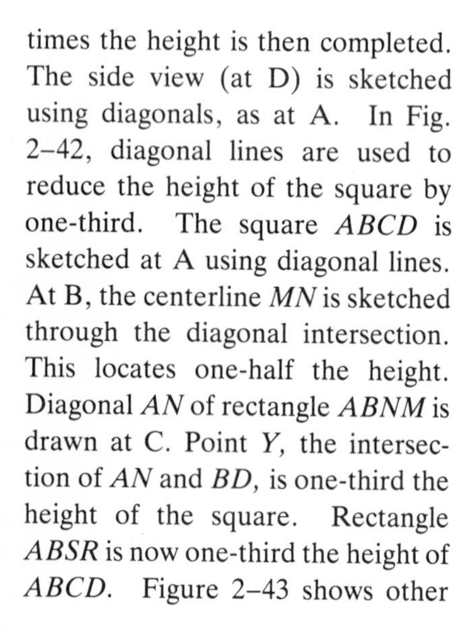

Fig. 2–41 **The diagonal is used to multiply proportions.**

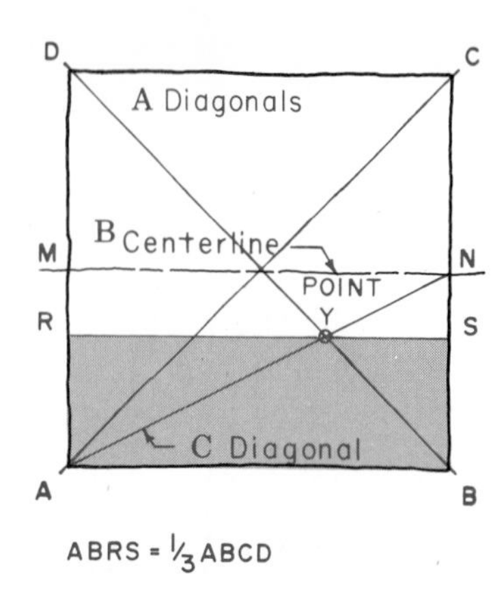

Fig. 2–42 **Diagonals are used to reduce proportional units.**

times the height is then completed. The side view (at D) is sketched using diagonals, as at A. In Fig. 2–42, diagonal lines are used to reduce the height of the square by one-third. The square *ABCD* is sketched at A using diagonal lines. At B, the centerline *MN* is sketched through the diagonal intersection. This locates one-half the height. Diagonal *AN* of rectangle *ABNM* is drawn at C. Point *Y*, the intersection of *AN* and *BD*, is one-third the height of the square. Rectangle *ABSR* is now one-third the height of *ABCD*. Figure 2–43 shows other ways of developing proportions for sketching.

EXAMPLE: SKETCH OF A CHAIR

The chair is shown in picture form in Fig. 2–44. Note that three views are enough to sketch all the important features. After you look at the chair carefully, estimate the width, height, and depth. Then, block in the major dimensions as shown. Large areas are first drawn in to the right proportion. Next, the smaller features are added in the right places. Never try to finish one view at a time. Block them all in, then work out the same details in each view. By sketching this way, you can develop views of an object that have the right proportions to each other.

Fig. 2–43 **Study the ways in which the diagonal line has divided the square.**

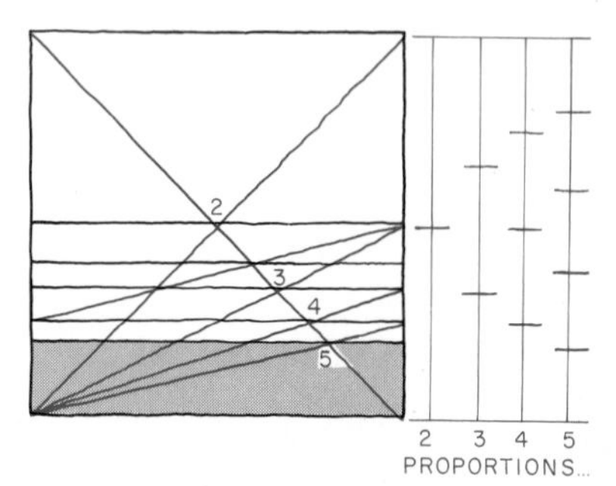

Fig. 2–44 **The development of a multiview drawing.**

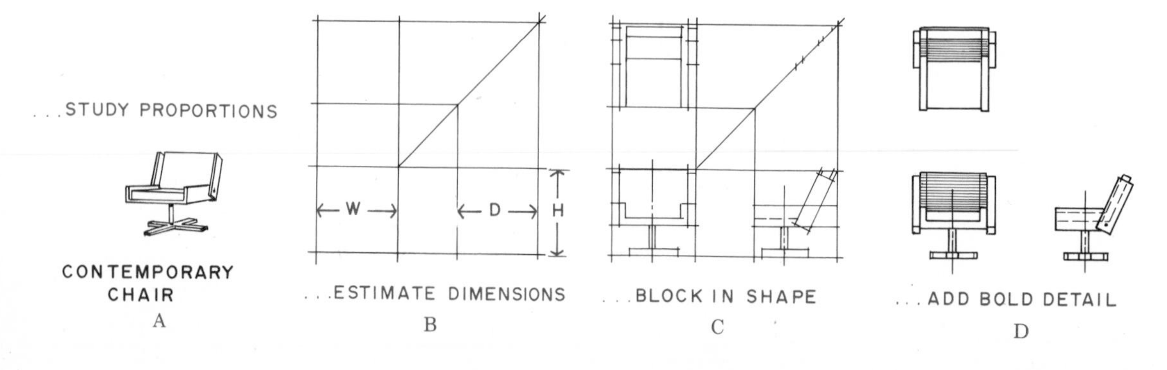

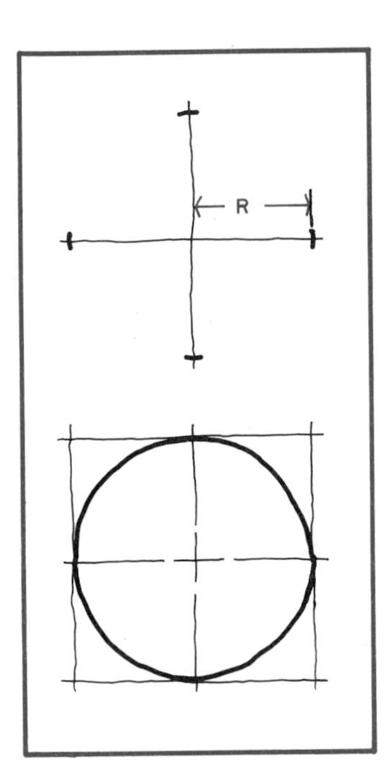

Fig. 2–45 **Mark off the radii and draw a square in which to sketch a circle.**

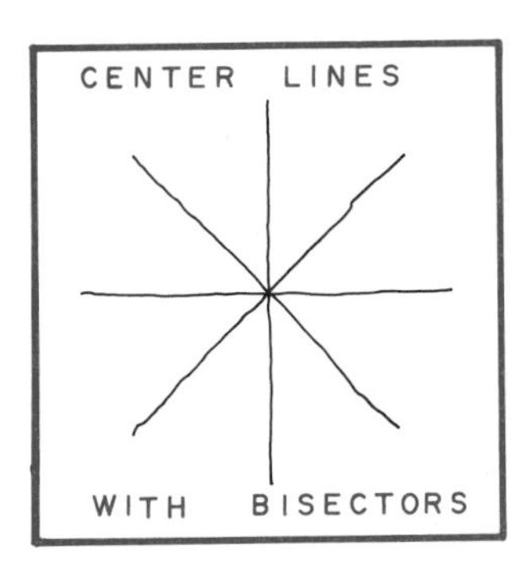

Fig. 2–46 **For a circle, draw centerlines with bisecting lines.**

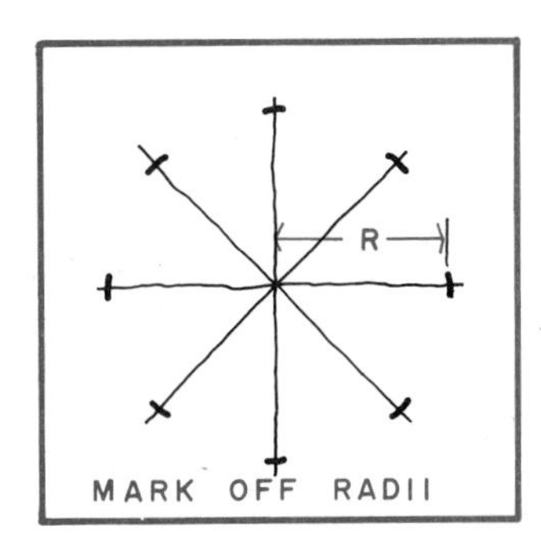

Fig. 2–47 **Mark off the estimated radii on all lines before sketching the circle.**

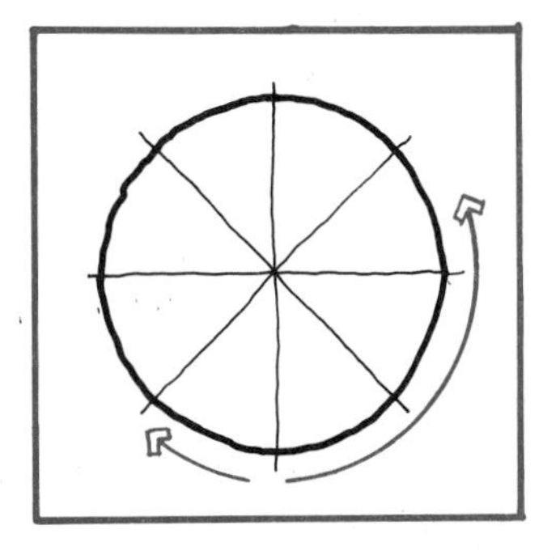

Fig. 2–48 **The bottom, or concave, side of a curve is the easiest to form.**

STEPS IN MAKING A SKETCH

You have to practice carefully to gain the skills needed to make good sketches. The steps in making a sketch are shown in Fig. 2–44.

1. Observe the chair at A.
2. Select the views needed to show all shapes.
3. Estimate the proportions carefully. Mark off major distances for width, height, and depth in all three views at B.
4. Block in the enclosing rectangles as at C.
5. Locate the details in each of the views. Block them in as at D.
6. Finish the sketch by darkening the object lines.
7. The blocking-in lines that you start with should be light enough that you do not have to erase them. You can easily draw the dashes for hidden lines over the light blocking-in lines.
8. Add the dimensions and notes that you need.

CIRCLES AND ARCS

There are two ways to sketch circles. For the first way, draw very light horizontal and vertical lines. Next, estimate the length of the *radius* (the distance from the center of the circle to its edge; plural, *radii*) and mark it off. Then, draw a square in which you can sketch the circle (Fig. 2–45). For the second way, first draw very light centerlines. Then draw *bisecting* (halving) lines through the center at convenient angles (Fig. 2–46). Next, mark off estimated radii of the same length on all lines (Fig. 2–47). The bottom of the curve is generally easier to form, so draw it first. Then turn your paper so the rest of the circle is on the bottom. Finish drawing it (Fig. 2–48). You can sketch *arcs* (parts of a circle); *tangent arcs* (parts of two circles that touch); and *concentric circles* (circles of different diameters that have the same center) in the same way you sketch circles (Fig. 2–49). Use light, straight, construction lines to block in the area of the figure.

For large circles and arcs and for *ellipses* (ovals), use a scrap of paper with the radius marked off along

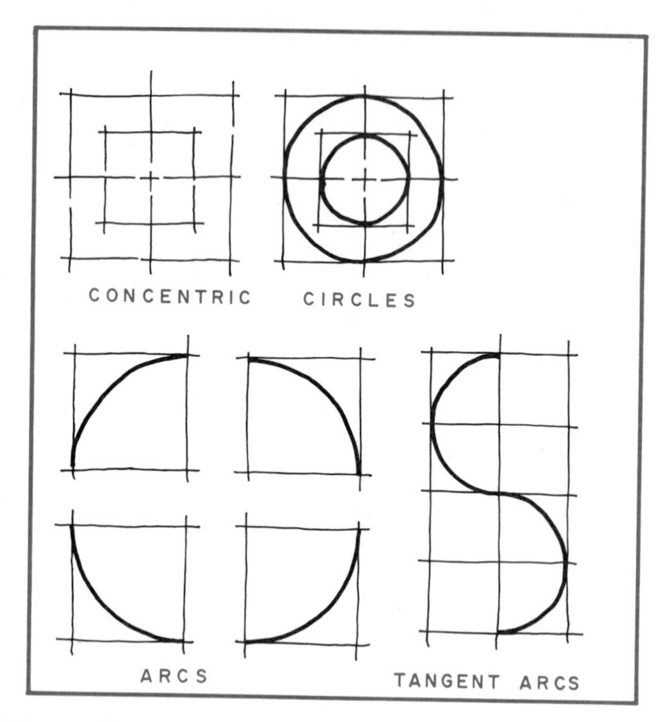

Fig. 2–49 **Arcs and concentric circles are controlled by sketching squares.**

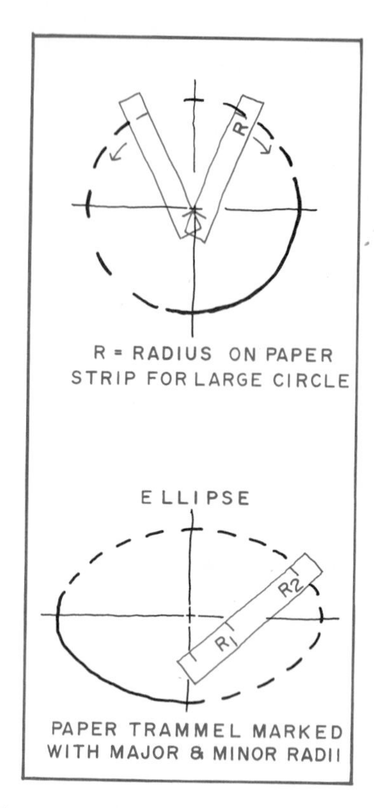

Fig. 2–50 **Large circles, large arcs, and ellipses are easily sketched with the aid of a strip of paper.**

one edge. Put one end of the marked-off radius on the center of the circle. Draw the arc by placing a pencil at the other end of the radius and turning the scrap paper (Fig. 2–50). Note in Fig. 2–50 that two radii are needed for an ellipse. To draw the ellipse, sketch both centerlines. Keep both radius points on the centerlines as you draw with the pencil at the other end.

You can also use your hand as a compass. To do this, you use your little finger as a *pivot* (turning point) at the center of the circle. You use your thumb and forefinger to hold the pencil rigidly at the radius you want. You turn the paper carefully under your hand, thereby drawing the circle (Fig. 2–51). You can also draw circles with two crossed pencils. Hold them rigidly with the two points as far apart as the length of the radius you want. Put one pencil point on the center. Hold it there firmly and turn the paper, drawing the circle with the other point (Fig. 2–52).

Fig. 2–51 **The hand as a compass, using the little finger as a pivot point.**

Fig. 2–52 **Two pencils can serve as a compass.**

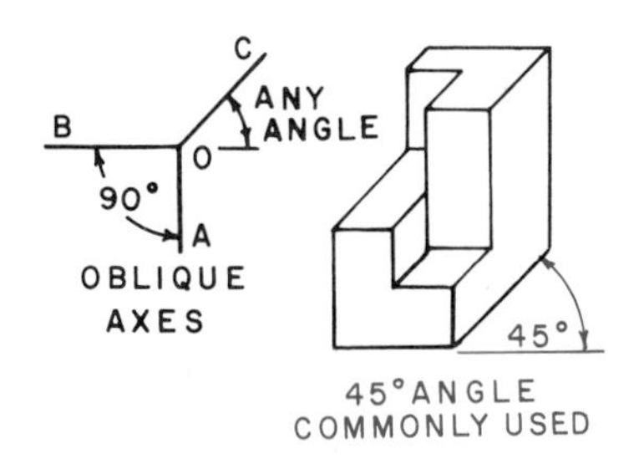

Fig. 2–53 **Oblique drawings always have one right-angle corner.**

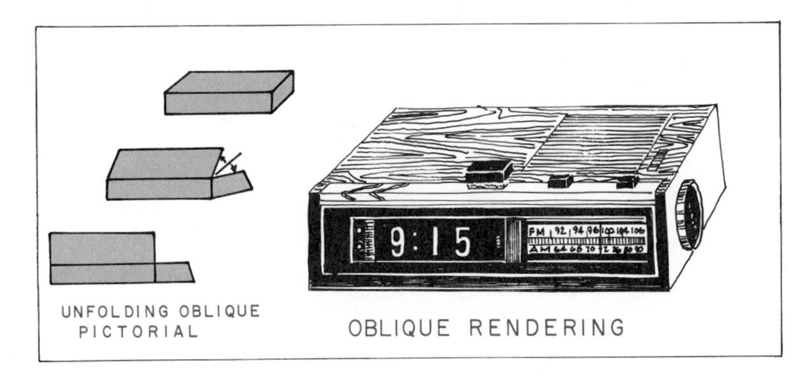

Fig. 2–54 **Oblique drawings can be used with rendering techniques.**

PICTORIAL DRAWING

There are several kinds of pictorial drawings. All types will be discussed in Chapter 12. For sketching, we will consider only two kinds of pictorial drawings, oblique and isometric. Making oblique and isometric sketches will help you learn how to "see" objects in your mind. You must be able to do this in order to draw *multiview projections* (three-view drawings). Pictorial drawings help people who are not trained to read multiview drawings understand basic shapes.

OBLIQUE SKETCHING

One of the easiest pictorial drawings to make is the *oblique* (inclined) sketch. In an oblique sketch, the depth of an object is drawn at any angle. Every object has three *dimensions* (sizes or directions): *width, height,* and *depth.* Each of these dimensions is called an *axis* (plural, *axes*). In oblique drawings, two of the axes are at *right angles* (90°) to each other. The third axis is drawn at any angle to the other two (Fig. 2–53). You may make any side of an object the front view. This view shows the width and height of the object. You usually make the side with the most detail the front view. In Fig. 2–54, a digital clock radio is used as an example. It is shown in an oblique pictorial view. The dial side has been made the front view. This is because it has the most detail and shows the width and height of the radio. The front view is sketched just as the clock radio would appear when you look at it.

OBLIQUE SKETCHING ON GRAPH PAPER

Graph paper is useful for oblique sketching. This is so because the front view of the oblique sketch is like the front view of a multiview sketch, as shown in Fig. 2–55 at A and B. If you develop the oblique pictorial drawing on graph paper from a multiview drawing on graph paper, you simply transfer the di-

Fig. 2–55 **Graph paper can assist in developing oblique pictorials.**

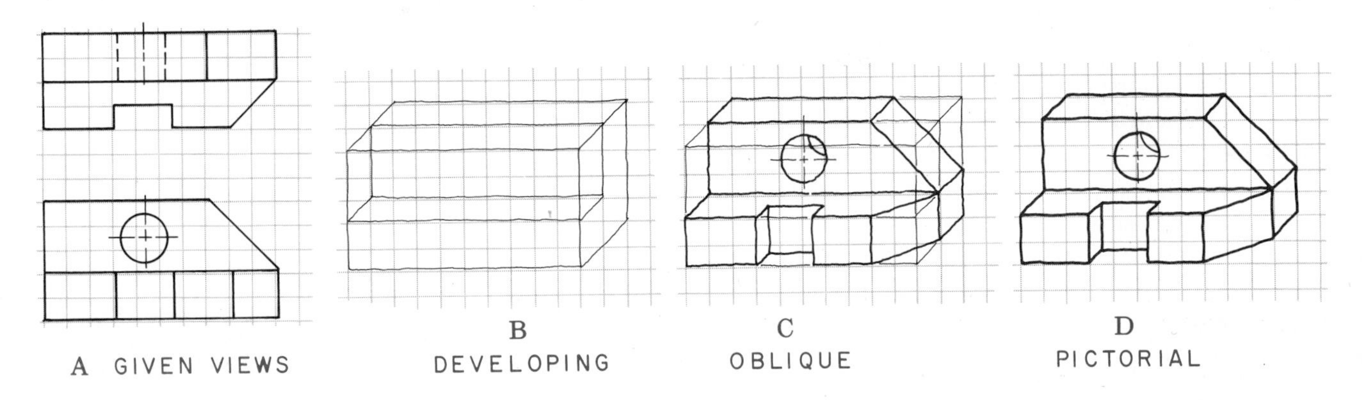

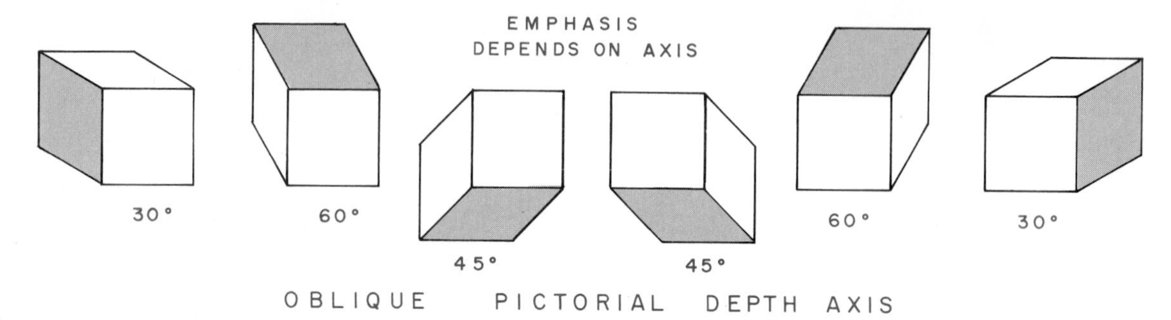

Fig. 2–56 **The many angles of oblique drawing.**

mensions from one to the other by counting the graph-paper squares:

1. Block in lightly the front face of the object by counting squares.
2. Sketch lightly the *receding* (going-away) axis by drawing a line diagonally through the squares. You find the depth by using half as many squares, as shown on the side view of Fig. 2–55 at A.
3. Sketch in any arcs and circles.
4. If you have drawn all the layout lightly, darken the final object lines.

OBLIQUE LAYOUT

To make a good oblique sketch, you should always follow a good layout procedure:

1. Block in lightly the front face of the object with the estimated units for the width and height.
2. Sketch in lightly the receding lines at any angle, for example, about 30° with the horizontal. You can choose an angle that will show as much as you want of the top and side. Thus, if you draw the third axis at a small angle like 30°, the side will show more clearly. On the other hand, if you choose 60°, the top will show more clearly (Fig. 2–56).
3. The receding axis may have the same proportioned units as the front axis. Or, you can reduce its size by up to one-half. When the depth dimension of a drawing is exactly one-half the true dimension, it is called a *cabinet sketch.* Using a full-depth dimension produces a *cavalier sketch* (Fig. 2–57).
4. If you have drawn all the layout lightly, darken the final object lines.

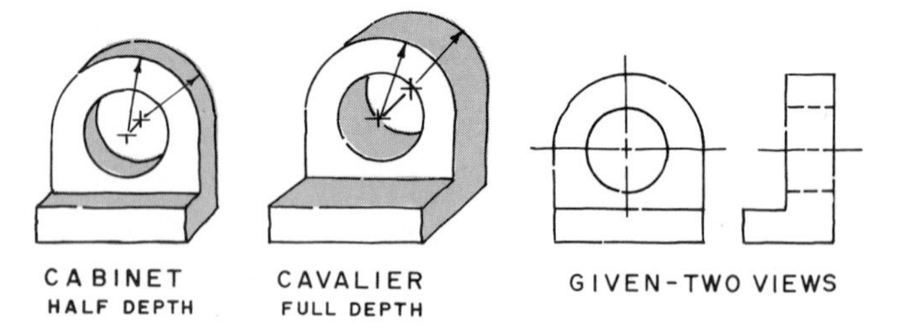

Fig. 2–57 **Oblique drawings can vary in depth.**

OBLIQUE CIRCLES

In oblique sketching, circles in the front view can be drawn in their true shape. However, oblique circles drawn in the top or side views appear *distorted* (not in their true shape). Indeed, you must draw an ellipse to show such circles. Ellipses are not good in oblique pictorial sketching. It is better to show the circular shapes of important parts in the front view.

ISOMETRIC AXES

An *isometric* (equal-measure) sketch is based on three lines called axes. These are used to show the three basic dimensions: width, height, and depth. A *cube* is an object with six equal square sides (Fig. 2–58). The isometric cube will have three equal sides. Thus, there are three equal angles at the *Y* axis at B (Fig. 2–58). The height *OA* is laid off on the vertical leg of the *Y* axis. The width *OB* is laid off to the left on a line 30° above the horizontal. The depth *OC* is laid off to the right on a line 30° above the horizontal. The 30° lines receding to the left and right can be lo-

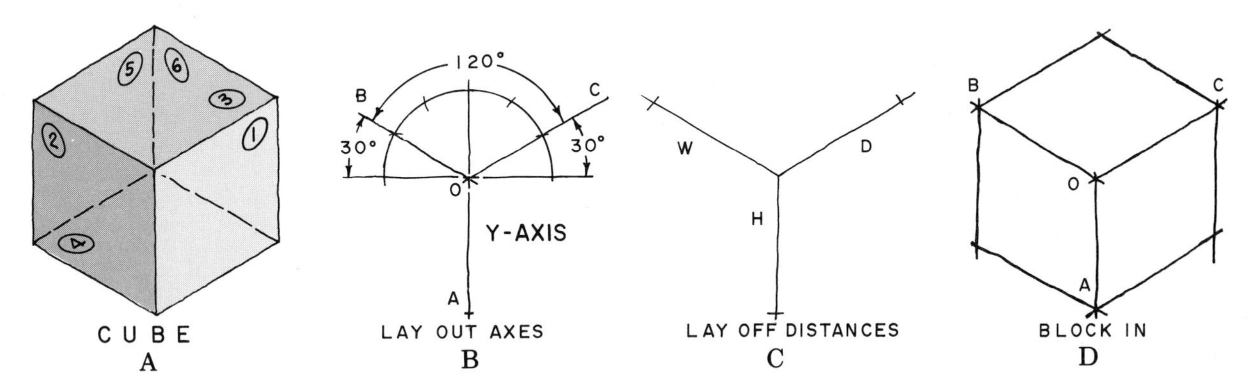

Fig. 2–58 **Sketching isometric angles and layout of *Y* axis.**

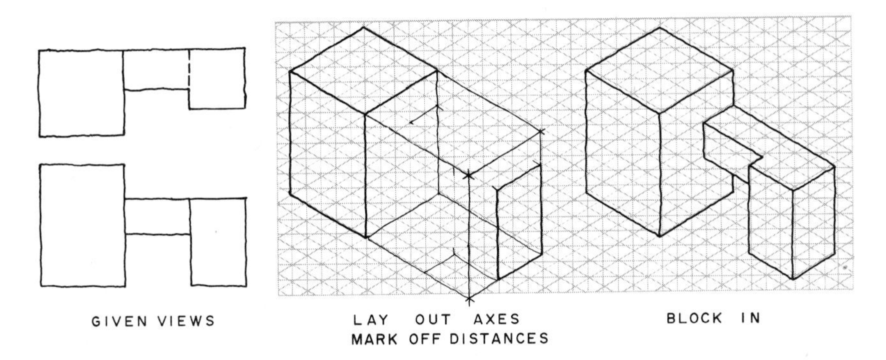

Fig. 2–59 **Isometric grid paper can be used for quick sketches.**

cated by estimating one-third of a right angle, as shown in Fig. 2–58. Lines parallel to the axes are called *isometric lines.* The estimated distances are laid off on them only as shown for the cube at C (Fig. 2–58).

The sketched lines for isometric axes tend to become steeper than 30° if you do not carefully prepare the layout. A better pictorial sketch will result when the angle is at 30° or a little less. Using special isometric graph paper with 30° ruling lets you make good pictorial sketches quickly and easily (Fig. 2–59). Figure 2–60 shows the steps in making an isometric sketch.

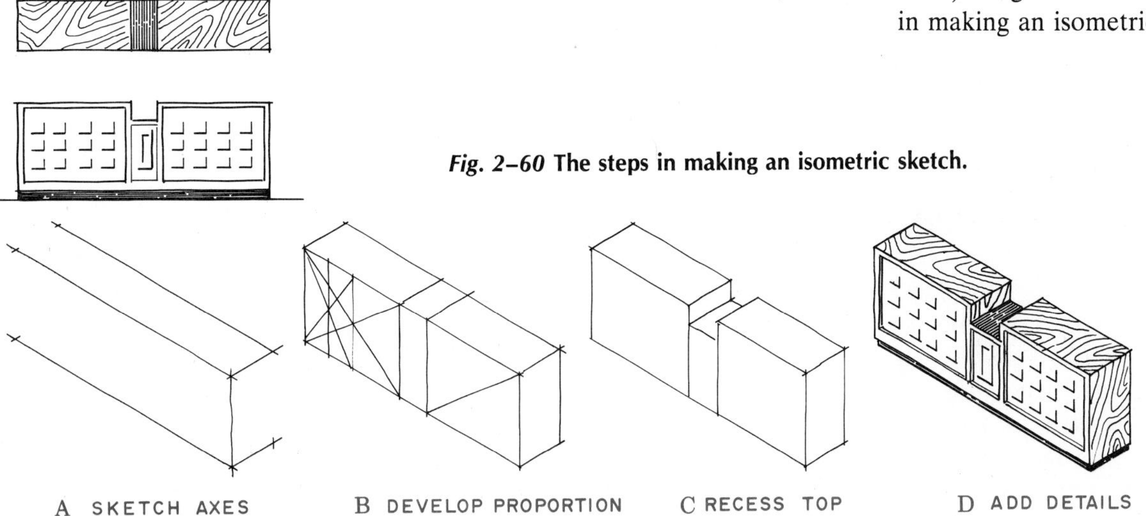

Fig. 2–60 **The steps in making an isometric sketch.**

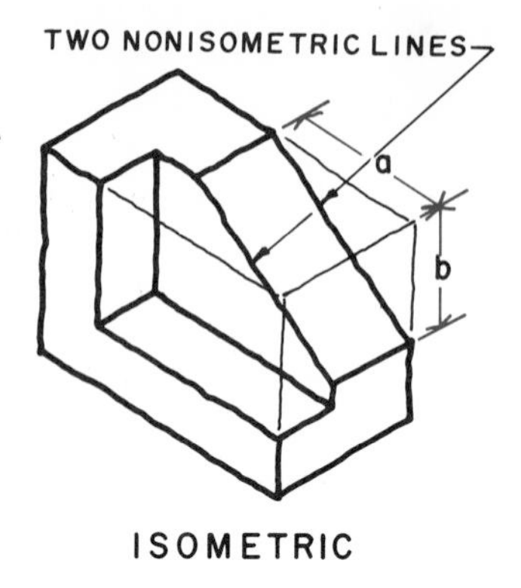

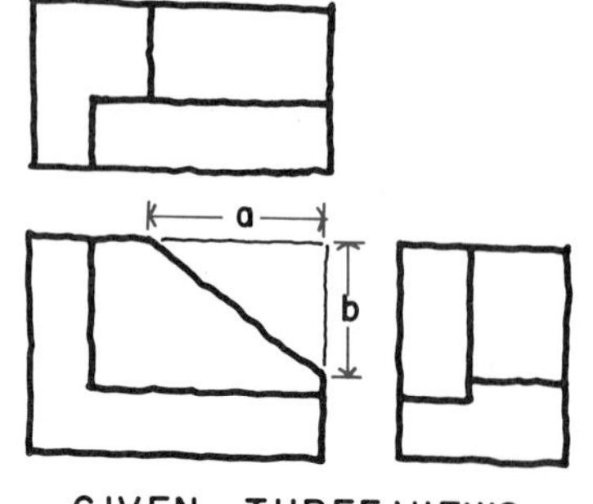

Fig. 2–61 **Identifying nonisometric lines that form an inclined plane.**

Fig. 2–62 **Development of two inclined planes in an isometric pictorial.**

NONISOMETRIC LINES ON ISOMETRIC SKETCHES

As we have seen, lines parallel to isometric axes are called isometric lines. Therefore, lines that are not parallel to the isometric axes are *nonisometric lines* (Fig. 2–61).

Any object may be sketched in a box, as suggested in Fig. 2–62. Note, however, that the objects in Figs. 2–63 and 2–64 have some lines that are not parallel to the isometric axes. Lines not parallel to the axes can be drawn by extending their ends to touch the blocked-in box. Locate points at the ends of slanted lines by estimating measurements along or parallel to isometric lines. Having located the ends of the non-isometric lines, you can sketch the lines from point to point. Lines that are parallel to each other will also show parallel on the sketch (Fig. 2–64). Note how the ends have been located on lines 1–2 and 1–3 in Fig. 2–64. Distances *a* and *b* are estimated and transferred from the figure at A to B. Any inclined line, plane, or specific angle must be found by locating two points of intersection on isometric lines.

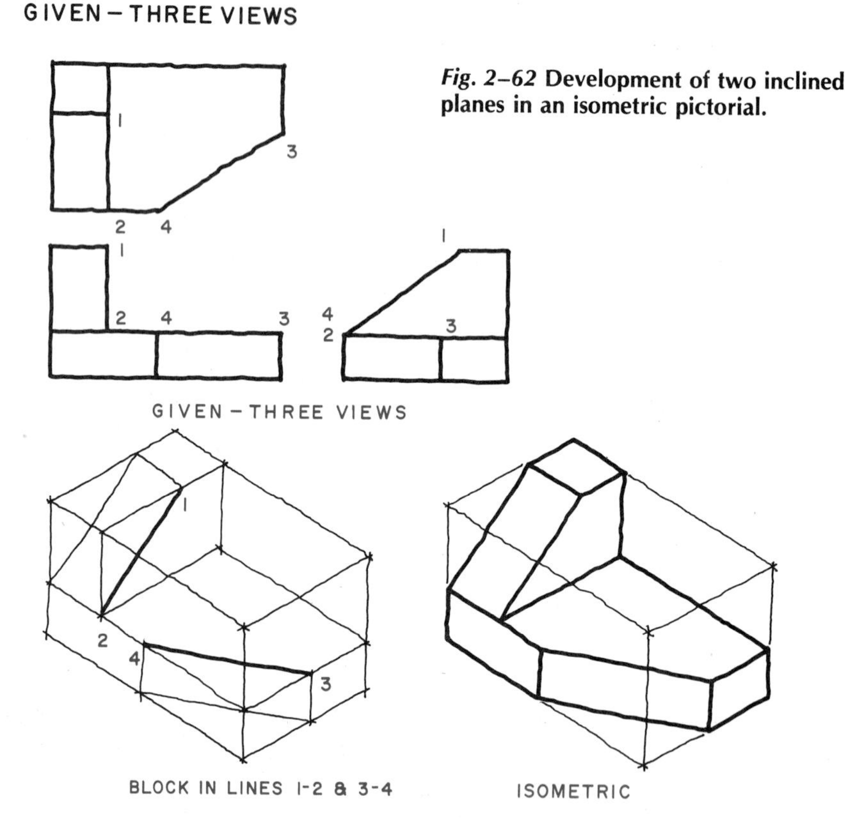

Fig. 2–63 **Development of an oblique plane in an isometric drawing.**

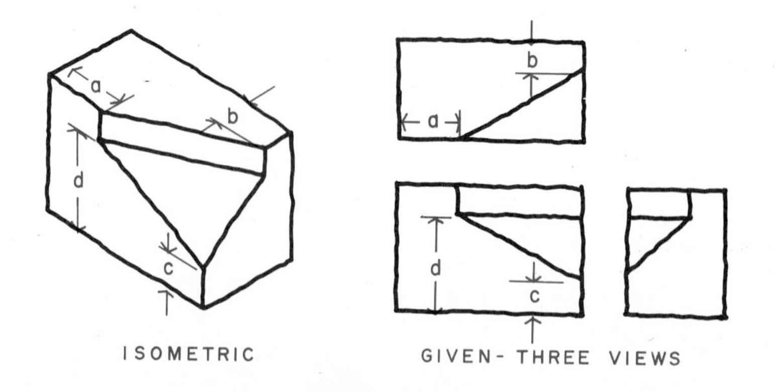

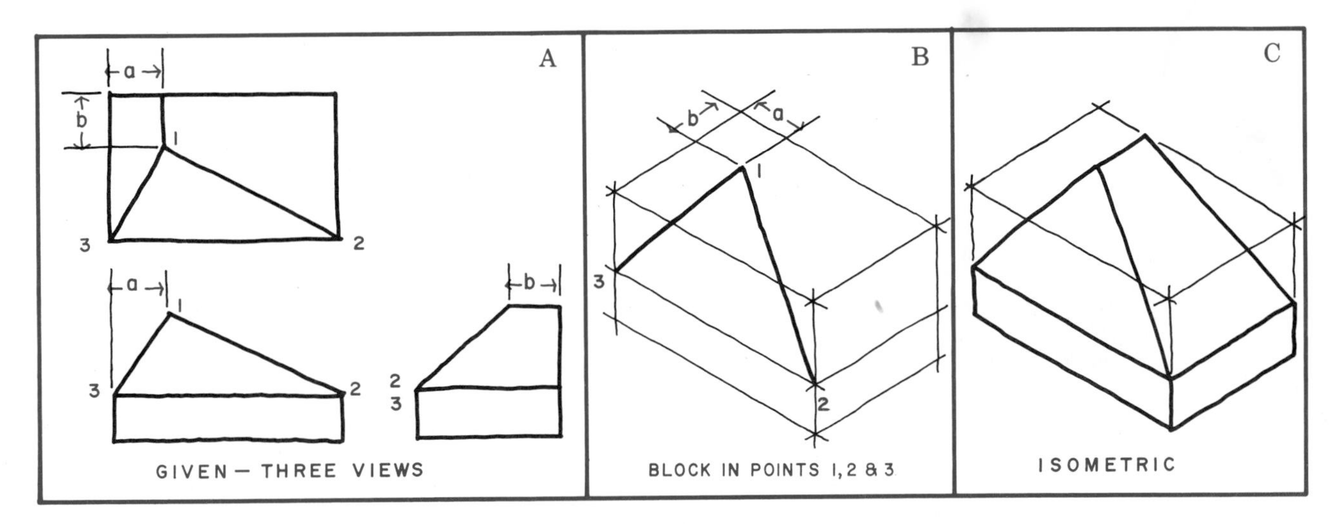

Fig. 2–64 **Development of three inclined surfaces in isometric drawing.**

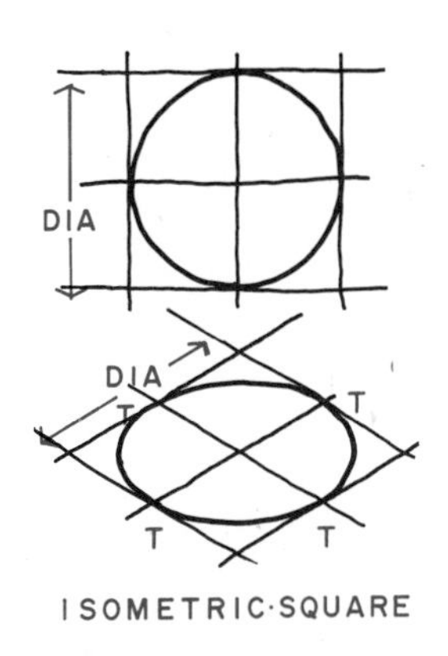

Fig. 2–65 **The isometric square with arcs tangent to form an ellipse.**

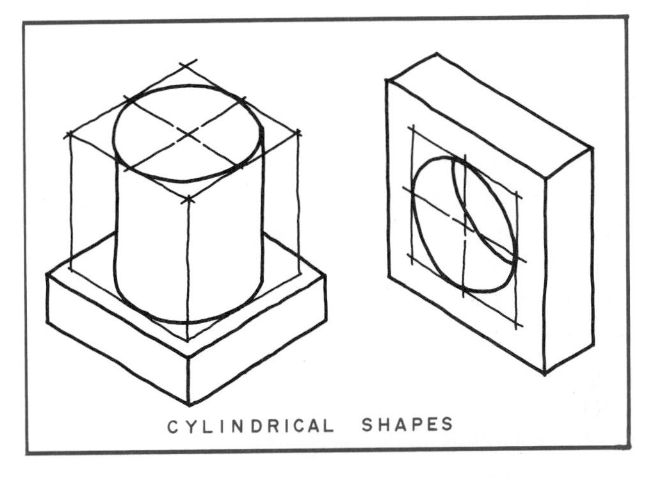

Fig. 2–67 **Isometric circles assist in describing cylindrical forms.**

Fig. 2–66 **Isometric circles (ellipses) on the front, top, and sides of a cube.**

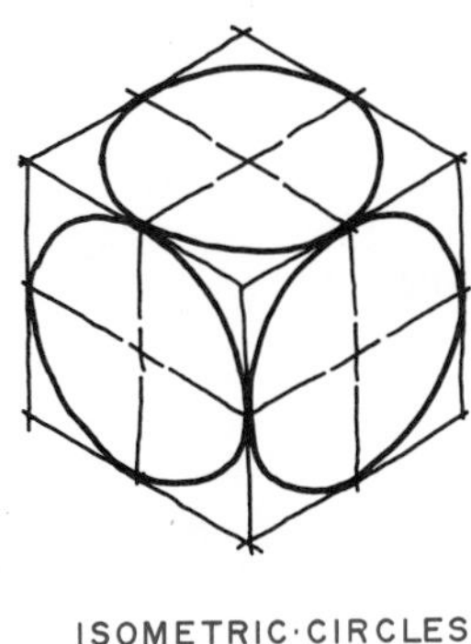

CIRCLES AND ARCS ON ISOMETRIC SKETCHES (ELLIPSES)

To sketch a circle in an isometric view (Fig. 2–65), you sketch an isometric square first. The small-end arcs are sketched *tangent to* (touching) the square. You then sketch the larger arcs tangent at points *T* to finish the ellipse. Note that the *major diameter* (long axis) is longer than the true diameter of the circle. The *minor diameter* (short axis) is shorter. Figure 2–65 shows an ellipse for a top view only.

Circles on the three faces of an isometric cube are sketched in Fig. 2–66. Some ways to block in *cylindrical* (cylinderlike) shapes are shown in Fig. 2–67. Some ways to block in *conical* (conelike) shapes are shown in Fig. 2–68.

Arcs developed in an isometric view are shown in Fig. 2–69. A semicircular opening and rounded corners appear in the front view.

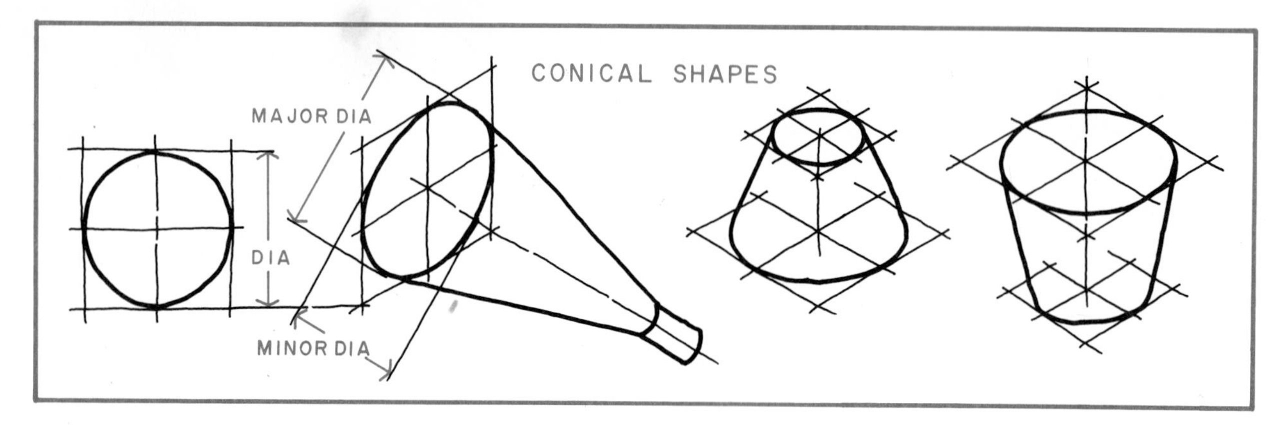

Fig. 2–68 **Isometric circles on conical shapes.**

The object is blocked in lightly at A, similar to the way isometric circles are plotted. Only partial circles or arcs are needed here. The outline of the object is darkened in at B. The rounded corners take up only a quarter of the full isometric circle that was plotted.

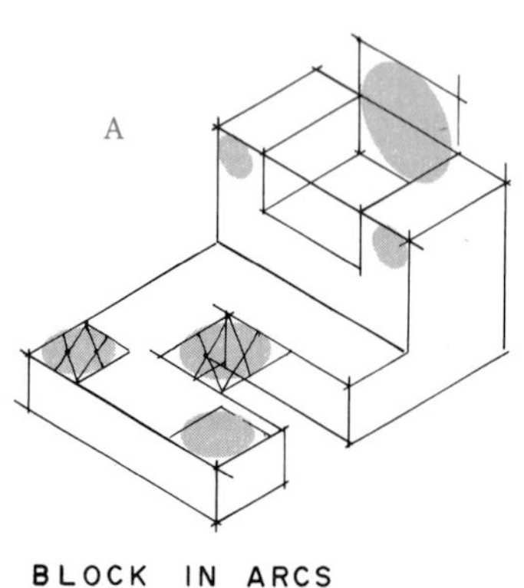

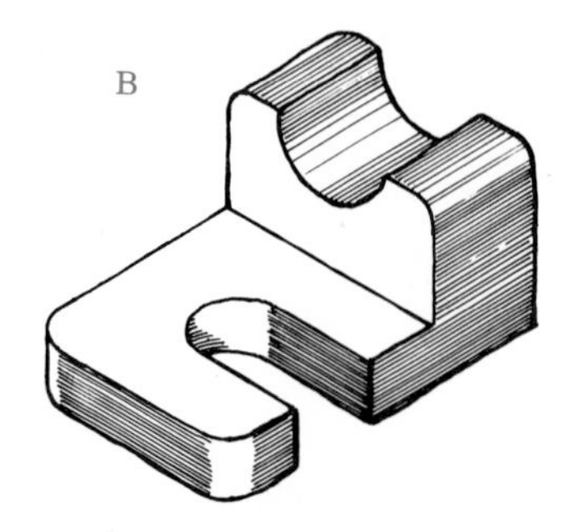

Fig. 2–69 **Arcs are blocked in similar to isometric circles.**

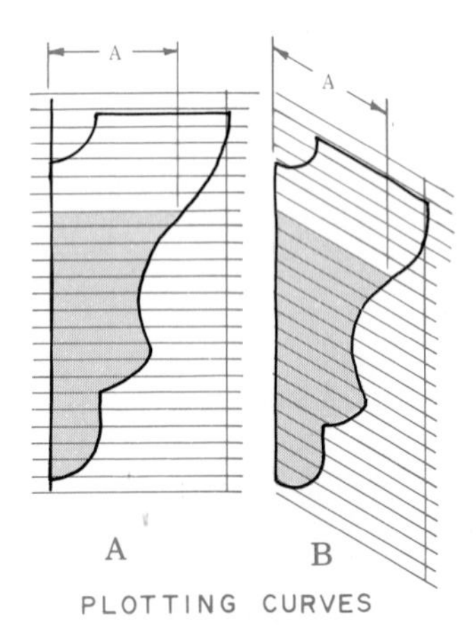

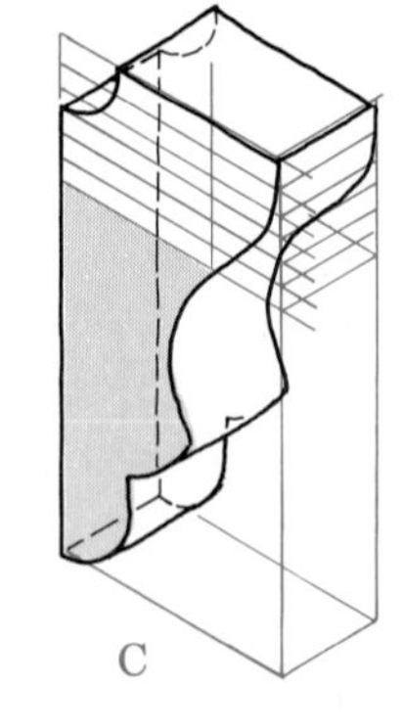

Fig. 2–70 **Irregular curves are plotted with a coordinated grid.**

SKETCHING CURVES AND ARCS

You can draw curves in pictorial sketches by plotting *coordinates* (points). To locate the curve, you generally transfer the coordinates from a multiview sketch. In Fig. 2–70 at A, coordinates are plotted that stand for the width from the left edge of the front view and the height from the top edge of the front view. The series of lines at B drawn parallel to the vertical and horizontal axes locate points of intersection as shown. Similar coordinates are plotted on the pictorial at C. The intersections serve as points for sketching the pictorial curve.

OTHER ISOMETRIC AXES

Figure 2–71 shows a regular *Y*-axis situation. You can draw other isometric axes in any position as long as there are 120° between the axes, as shown in Fig. 2–72. The reverse axis is shown for objects that can be seen better from the bottom. The position of the axes in Fig. 2–73 is good for long objects. In preparing a pictorial sketch that must clearly describe the shape of an object, you have to decide how to use the isometric axes.

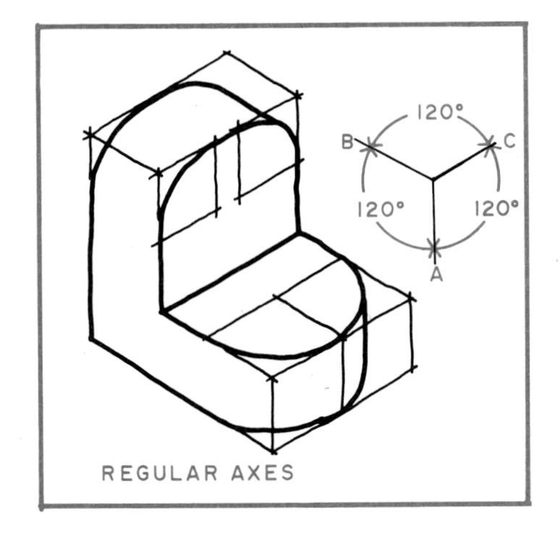

Fig. 2–71 **The regular isometric axes.**

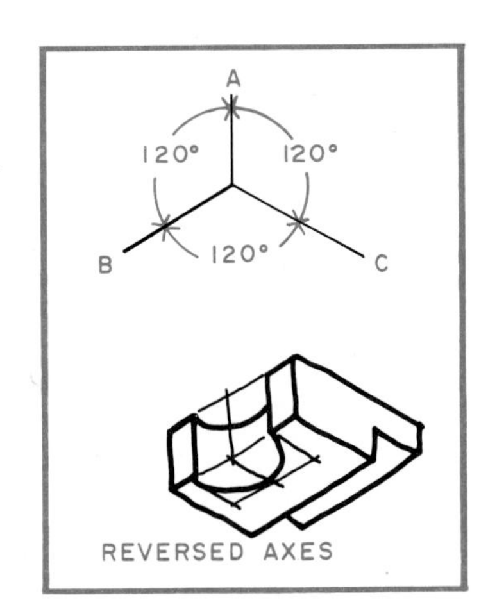

Fig. 2–72 **The isometric axes are reversed to emphasize the bottom view.**

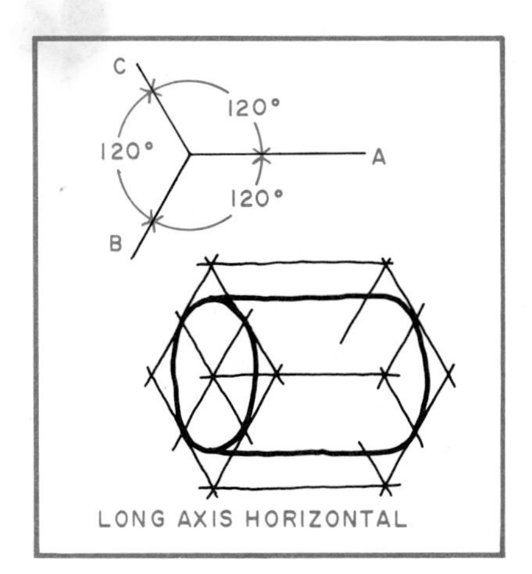

Fig. 2–73 **The horizontal isometric axis.**

Fig. 2–74 **The six views repeat geometric details.**

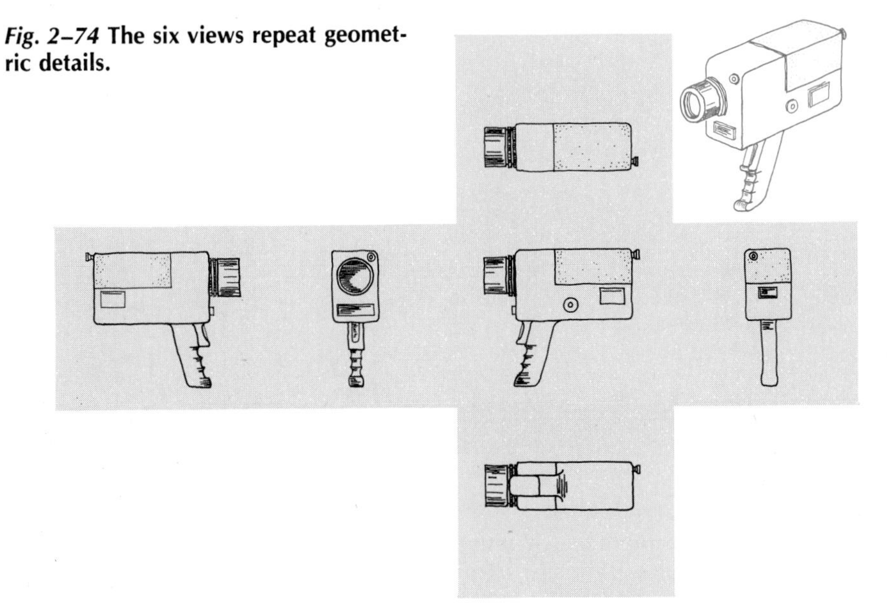

TOP
B
A
LEFT SIDE
FRONT
RIGHT SIDE

Fig. 2–75 **Selecting the left-side view for clarity.**

SIX VIEWS VS. THREE VIEWS

The purpose of a sketch is to give a description of the shape of an object.

Six views are shown for the movie camera in Fig. 2–74. Studying the pictorial sketch and the six views will help you see which views are needed. The top, front, and side views describe the shape best, as they have the fewest hidden lines. You should sketch only the views actually needed to show an object fully. Usually, fewer than six views need to be drawn. One, two, or three views are often enough.

TYPICAL STUDIES

Most three-dimensional objects can be shown fully with three views. The top, front, and right-side views as ordinarily drawn are shown in Fig. 2–75 at A. When a left-side view is drawn, it is placed as shown at Fig. 2–75 at B. In this example, the left-side view at B is better than the right-side view at A because it has fewer hidden lines. Because it is simpler, view B will show the object better.

Study the drawing in Fig. 2–76 to see how the views should be drawn for four different shapes:

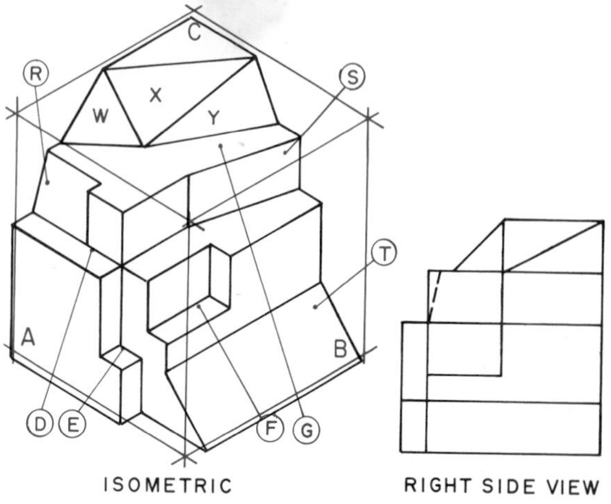

ISOMETRIC

RIGHT SIDE VIEW

Fig. 2–76 **Describe and identify the planes in the right-side view.**

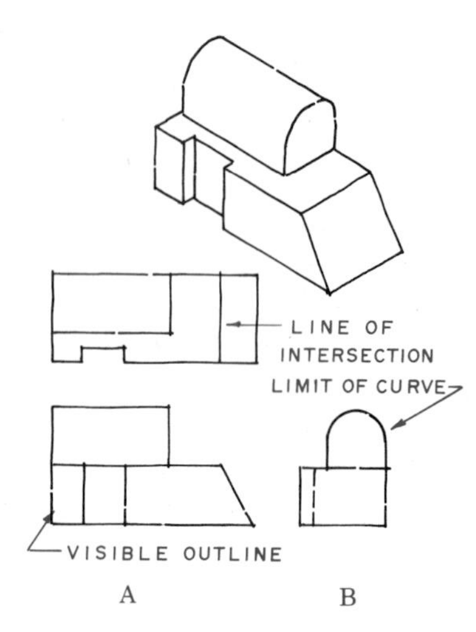

Fig. 2–77 **Visible line interpretation.**

1. The three principal planes A, B, and C
2. Stepped planes D, E, F, and G
3. Inclined planes R, S, and T
4. Slanted planes W, X, and Y

LINE INTERPRETATION

Sketched lines that are *visible* (seen) have three important roles:

1. The visible line forms the outline of the drawing.
2. The visible line shows how two surfaces *intersect* (meet), as shown in Fig. 2–77 at A.
3. The visible line shows the limit of a curved surface, as in Fig. 2–77 at B.

Sketched lines that are *invisible* or *hidden* are used to show parts of an object that you normally cannot see. The right ways to draw hidden lines are shown in Fig. 2–78. Invisible lines are shown by sketching dashes about 3 mm (1/8 in.) long with 1.5 mm (1/16 in.) spaces between. Hidden lines can help you show the shape of an object. When a visible line *coincides* (lines up) with a hidden line, only the visible line will show.

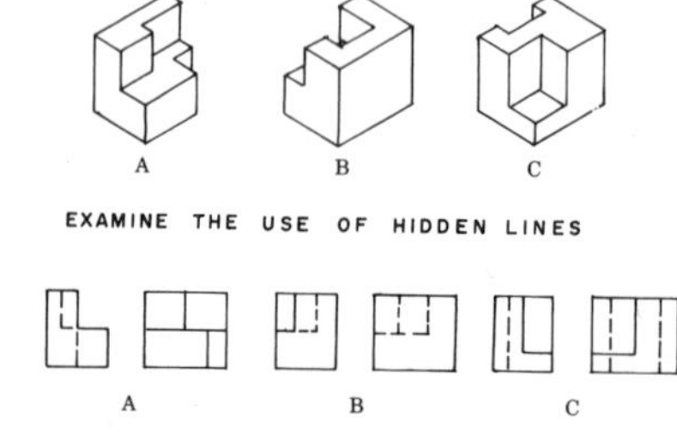

Fig. 2–78 **Hidden lines are used to describe the unseen.**

You must use centerlines when sketching objects that have symmetrical parts. An object is *symmetrical* when its shape is the same on both sides of a line drawn through its center. The centerline (symbol ℄) can help you develop proportions for cylinders and other curved surfaces. The centerline is shown in Fig. 2–79. The long dash of the centerline can be from 19 to 38 mm (3/4 to 1 1/2 in.) or more in length, depending on the size of the object. The short dashes are usually 3 mm (1/8 in.) long, with spaces on either side of about 1.5 mm (1/16 in.). When doing a sketch, make the centerline thinner than hidden and visible lines. When a centerline coincides with visible or invisible lines, omit it (Fig. 2–79).

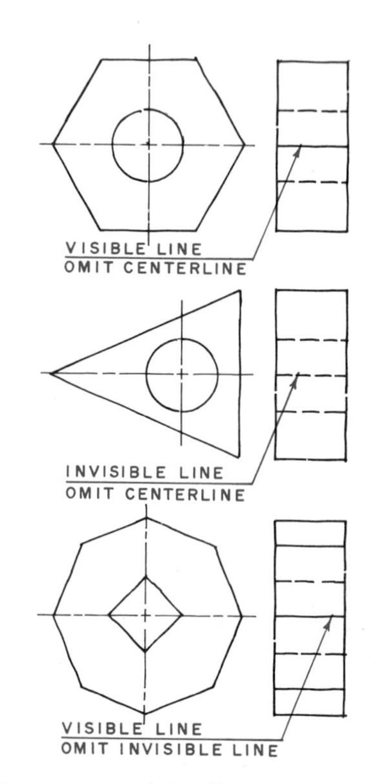

Fig. 2–79 **Coinciding lines.**

INTRODUCTION TO DIMENSIONING

A sketch must fully describe the object. Generally the sketch is made before the measurements of an object have been decided. After the needed dimensions are determined, they can be recorded on the sketch. Two types of dimensions are *specified* (given). Size dimensions (Figs. 2–80 at A) describe the overall geometric elements that give an object form. Location dimensions (Fig. 2–80 at B) relate these geometric elements to each other.

Definitions

A *dimension line* is used to show the direction of a dimension. Dimension lines have an *arrowhead* at each end to show where the dimension begins and ends. They have a break for the dimension numbers. Dimension lines generally stop at centerlines or extension lines. *Extension lines* are thin lines used to extend the shape of the object. An extension line starts 1.5 to 3 mm (1/16 to 1/8 in.) away from the object. It extends 3 mm (1/8 in.) beyond the arrowhead on the dimension line. Centerlines can serve as extension lines. *Leaders* are thin lines drawn from a note or dimension to the place where it applies. A leader starts with a 3 to 6 mm (1/8 to 1/4 in.) horizontal dash. It then angles off to the part featured, usually at 30°, 45°, or 60°. It ends with an arrowhead.

TYPICAL MULTIVIEW DIMENSIONING

Dimensions can be used to describe linear distances, angles, and notes. A typical dimensioned sketch is shown in Fig. 2–81. Chapter 6 discusses where to put dimensions in multiview drawings.

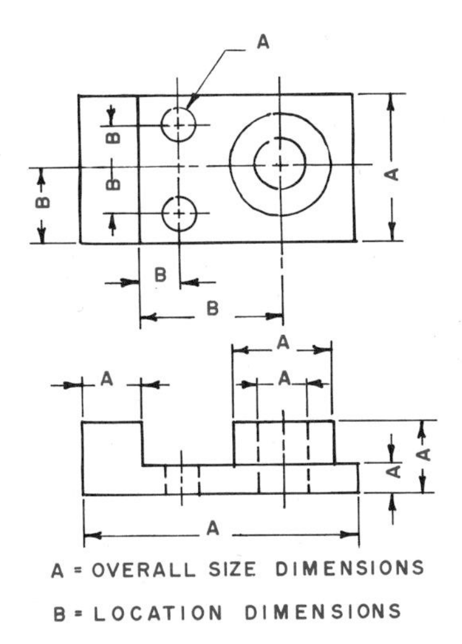

Fig. 2–80 **Size dimensions at A, location dimensions at B.**

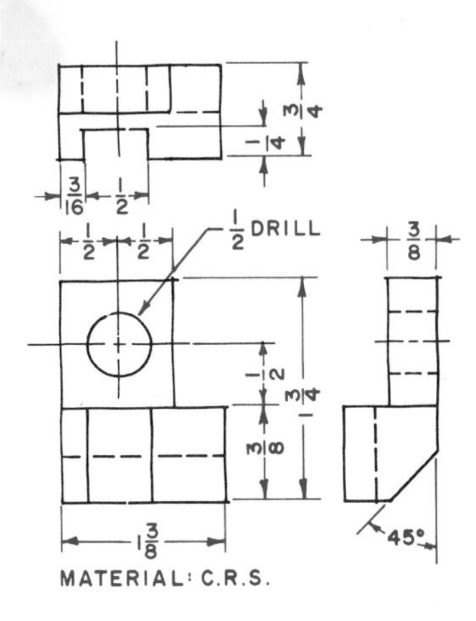

Fig. 2–81 **Multiview dimensioning.**

TYPICAL OBLIQUE DIMENSIONING

Oblique dimension lines are always parallel to the major axes. The lettering that goes with them is always vertical. The American National Standards Institute (ANSI) approves the aligned and the unidirectional systems that are shown in Fig. 2–82. Chapter 6 discusses where to put dimensions in oblique drawings.

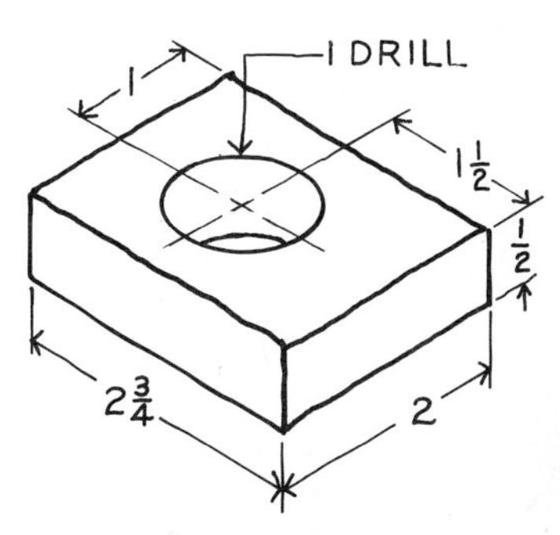

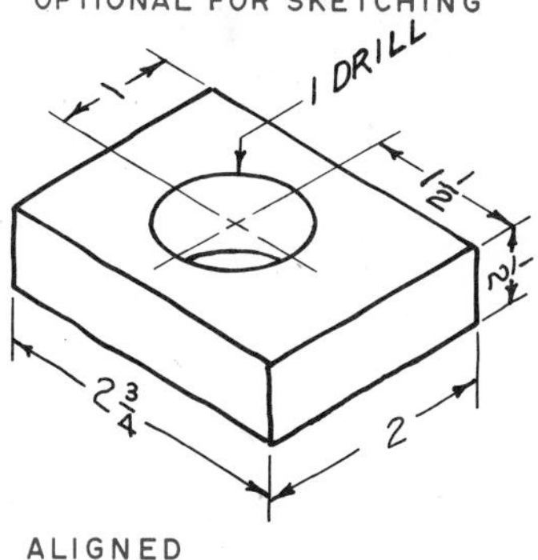

Fig. 2–82 **Dimensioning in pictorial.**

CONCLUSION

Drafting begins with an idea. The idea develops with a sketch. Many jobs depend on the ability of the drafter to put ideas on paper with sketches and notes. The successful drafter is able to sketch technical details accurately and to render pictorial sketches artistically. Engineers, architects, and designers need drafters and artists with these skills. Drafting is an exciting challenge for those with the skills and techniques that industry needs.

Review

1. What kind of paper is used in the overlay method?

2. Why do the drafter, designer, and engineer who must take part in highly technical discussions consider sketching a valuable skill?

3. Of all the reasons for sketching, which three seem most important?

4. Name the three terms that describe the nature of a sketch.

5. Why are the four visual symbols in the language of sketching essential in graphic communication?

6. What qualities would be good for the paper used for refining a sketch?

7. What is the term applied to estimating proportions? Why is estimating important to the drafter and designer?

8. List the uses for the four kinds of pencil points used in freehand sketching.

9. Straight lines can be developed easily by drawing vertical lines from the ________ down and horizontal lines from ________ to ________.

10. Name two advantages of pictorial sketching.

11. When preparing sketches of objects, what is the drafter's first important decision?

12. What is another name for multiview drawing?

13. Name two basic types of pictorial sketching.

14. What are the three principal planes of projection? What view appears on each plane?

15. When a visible line and a hidden line coincide, which line should be drawn? Which if the lines are a hidden line and a centerline?

Problems

Fig. 2–83

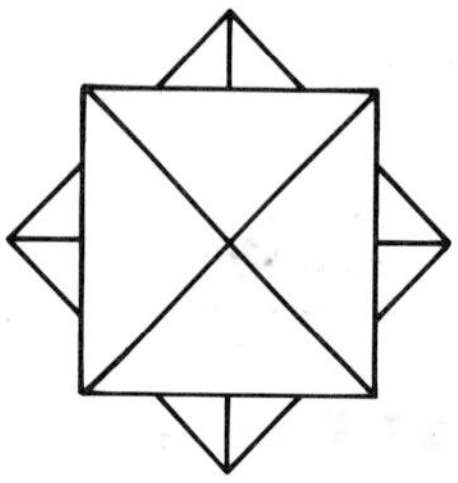

Fig. 2–84

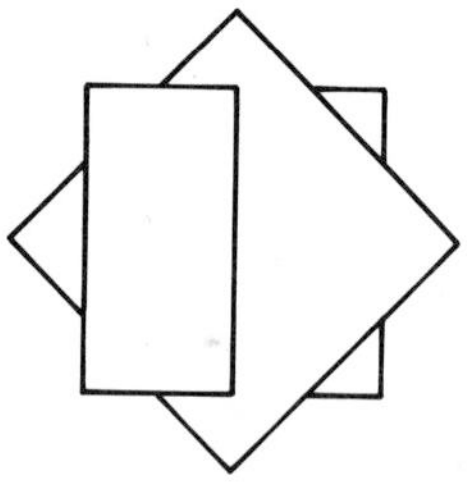

Fig. 2–85

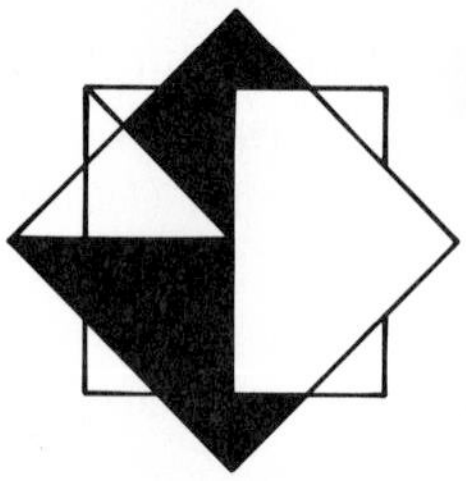

Fig. 2–86

Figs. 2–83 through 2–86 **Sketch in 2-in. overlapping squares as creative visual studies.**

Fig. 2–87

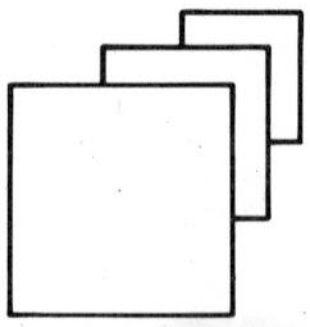

Fig. 2–88

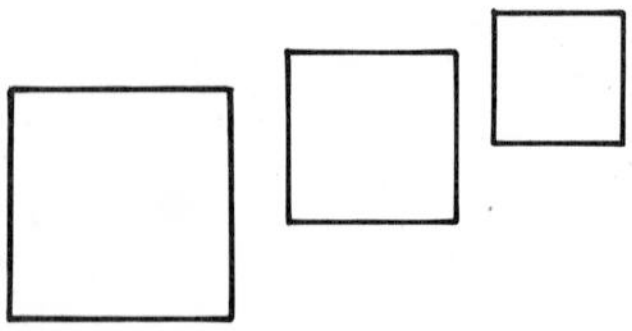

Fig. 2–89

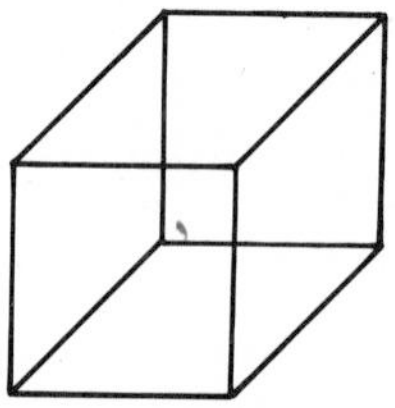

Fig. 2–90

Figs. 2–87 through 2–90 **Sketch the squares overlapping, diminishing, and as a transparent cube. Sizes are about 1½ in., 1⅛ in., and ⅞ in.**

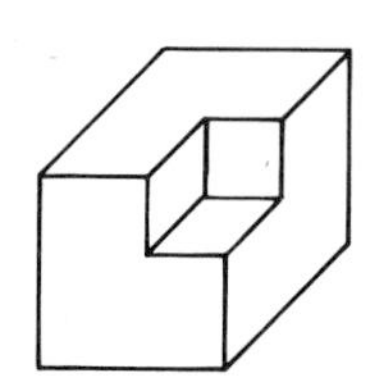

Fig. 2–91 **Sketch a cube with the alteration as shown, and observe the optical illusion.**

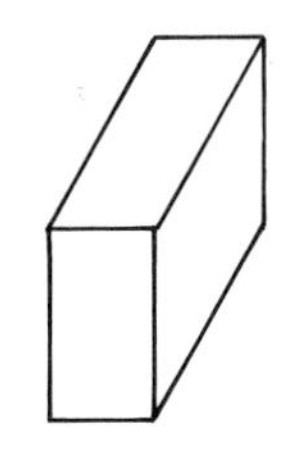

Fig. 2–92 **Sketch a rectangular solid with the approximate proportions shown.**

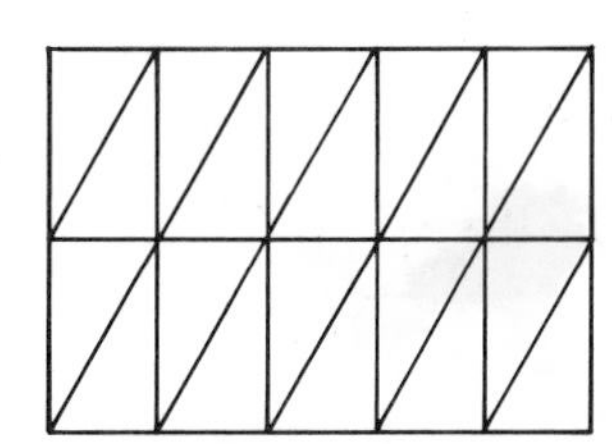

Fig. 2–93 **Sketch the apparent two-dimensional rectangular form using 6 diagonals. How many times will Fig. 2–92 be found in Fig. 2–93?**

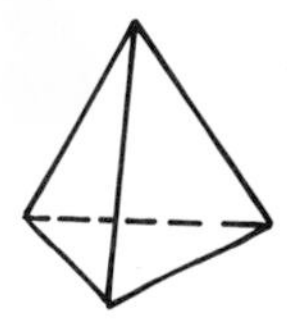

Fig. 2–97

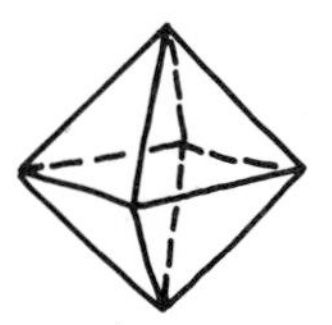

Fig. 2–98

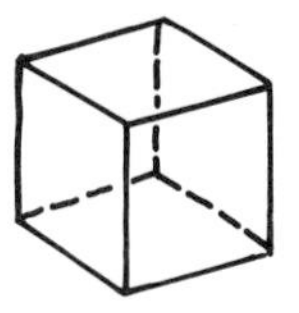

Fig. 2–99

Fig. 2–100

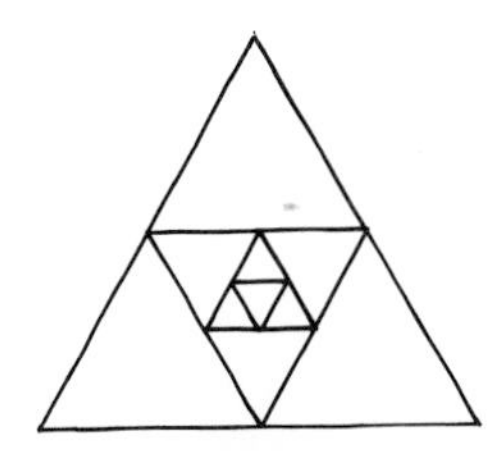

Fig. 2–94 **Sketch a 3-in. equilateral triangle with diminishing triangles at midpoints. Note the proportions.**

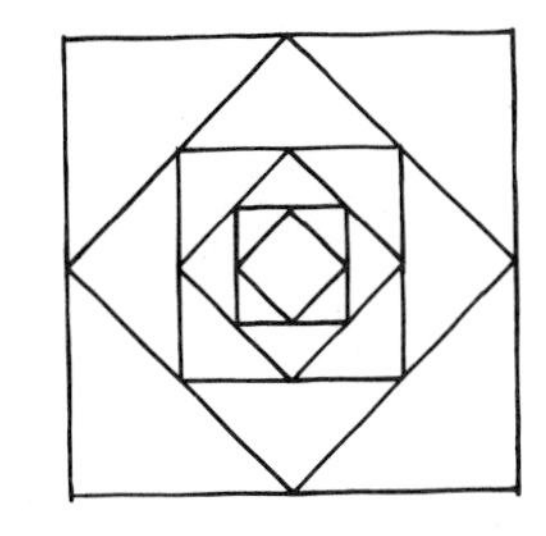

Fig. 2–95 **Sketch a 3-in. square with diminishing squares at midpoints. Note the proportions.**

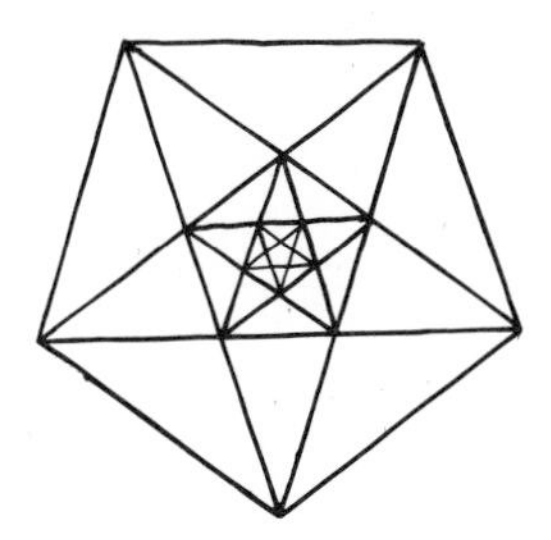

Fig. 2–96 **Sketch a pentagon using 2-in. sides with diminishing five-pointed stars. Note the proportions.**

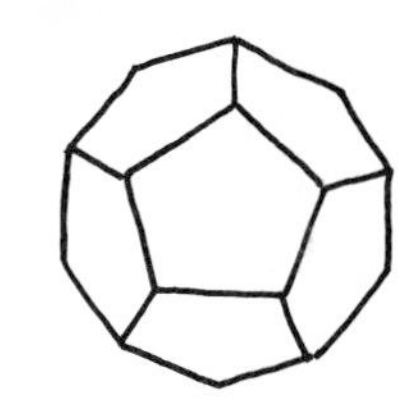

Fig. 2–101

Figs. 2–97 through 2–101 **Sketch the five basic solids.**

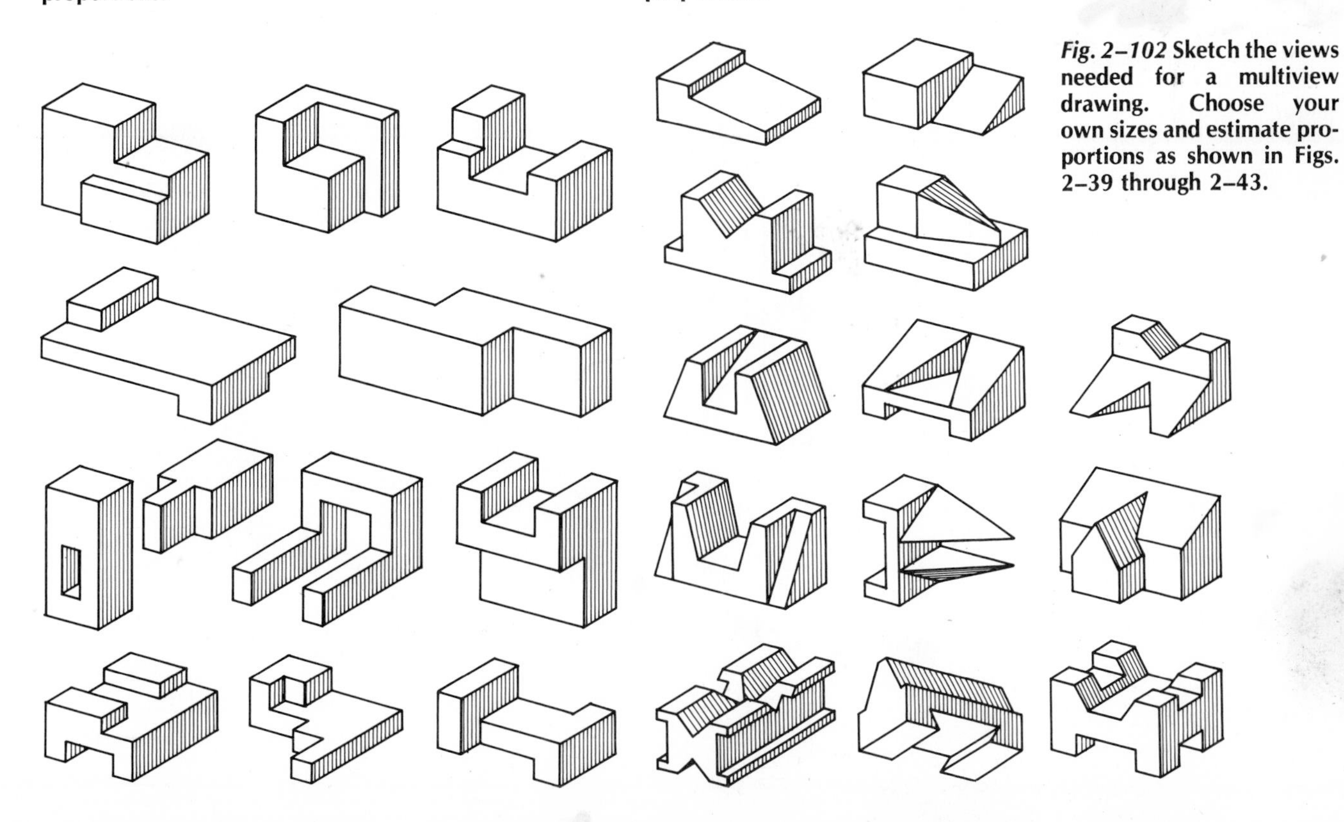

Fig. 2–102 **Sketch the views needed for a multiview drawing. Choose your own sizes and estimate proportions as shown in Figs. 2–39 through 2–43.**

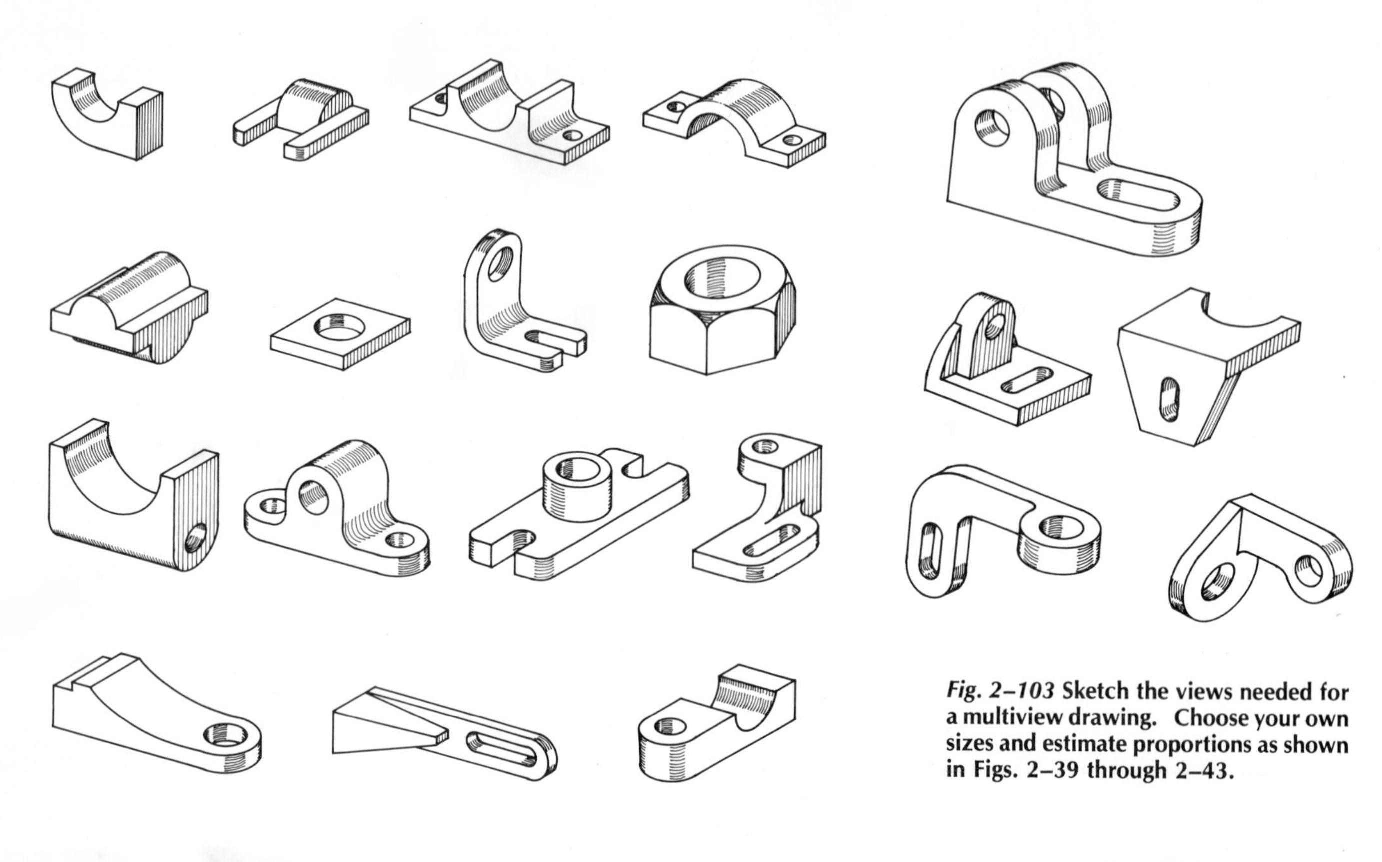

Fig. 2–103 **Sketch the views needed for a multiview drawing. Choose your own sizes and estimate proportions as shown in Figs. 2–39 through 2–43.**

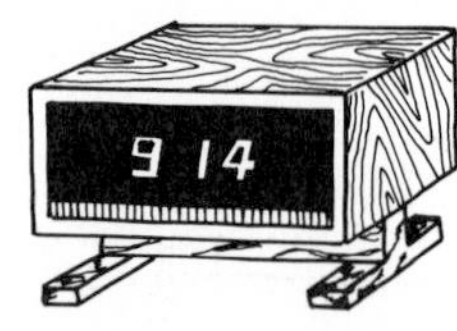

Fig. 2–105 **Prepare a multiview sketch of the digital-clock cabinet. Include dimensions and material specifications.**

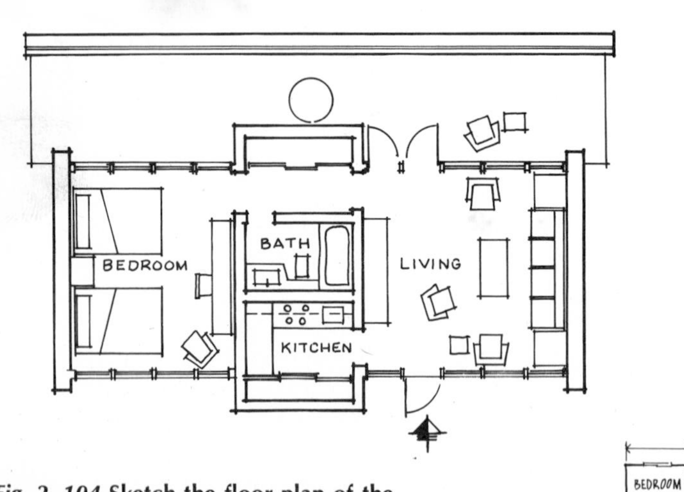

Fig. 2–104 **Sketch the floor plan of the vacation home, and prepare an elevation of your own design to match.**

Fig. 2–106 **Prepare a sketch of the ski lodge at $1/4$-in. scale. Add a side elevation as appropriate.**

29'-0
BEDROOM
BEDROOM
BATH
KITCHEN
LIVING
DINING
34'-0
8'-5
DECK
PLAN
1100 SQ.FT.

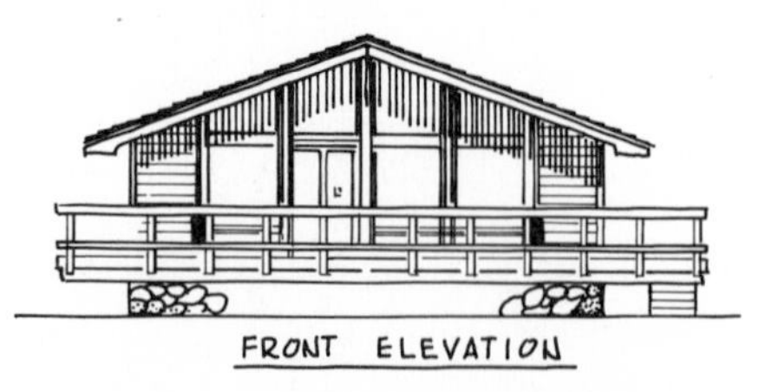

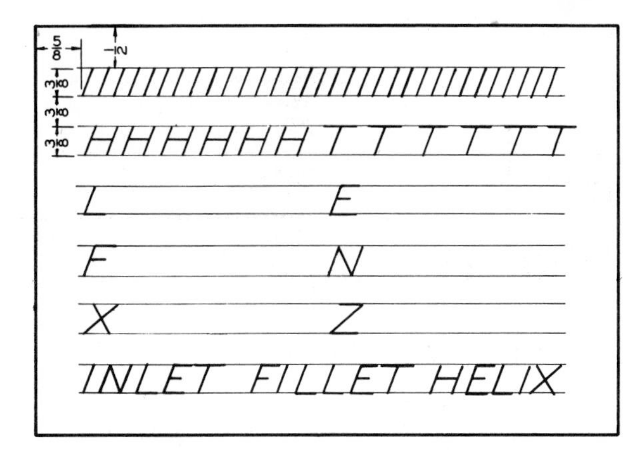

Fig. 2–107 3/8″ inclined capitals in pencil. Complete each line. Draw a few very light inclined guidelines.

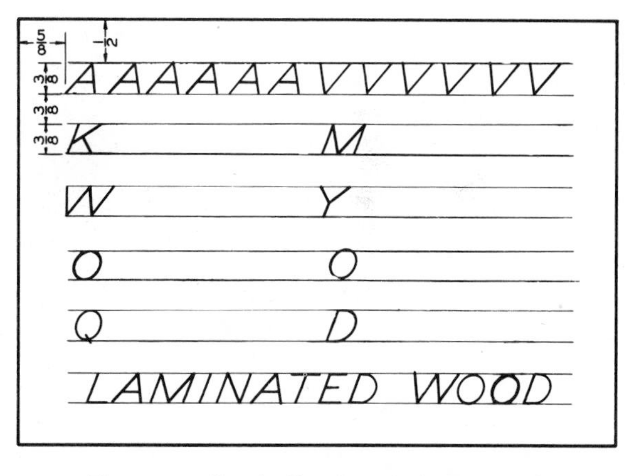

Fig. 2–108 3/8″ inclined capitals in pencil. Complete each line.

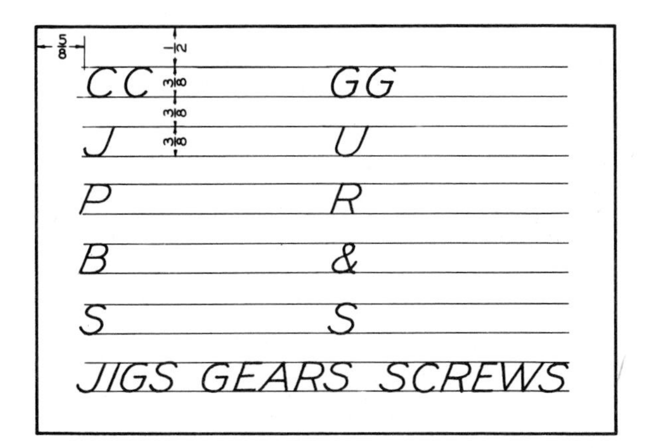

Fig. 2–109 3/8″ inclined capitals in pencil. Complete each line.

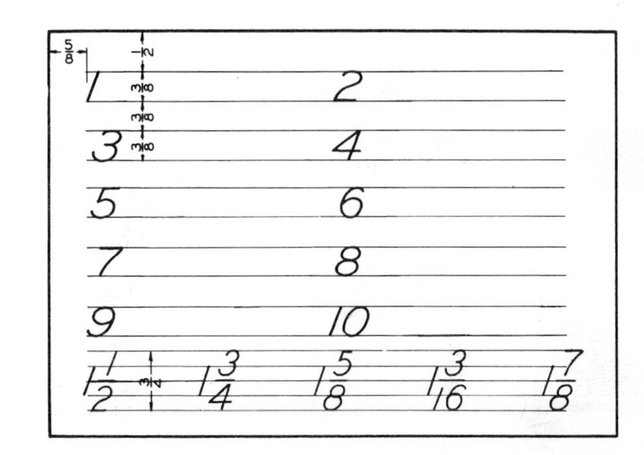

Fig. 2–110 3/8″ inclined numerals in pencil. Make total height of fractions two times height of numerals.

Fig. 2–111 Prepare a lettering sheet as you did for Fig. 2–107. Practice the same letters with vertical capitals.

Fig. 2–112 Prepare a lettering sheet as you did for Fig. 2–108. Practice the same letters with vertical capitals.

Fig. 2–113 Prepare a lettering sheet as you did for Fig. 2–109. Practice the same letters with vertical capitals.

Fig. 2–114 Prepare a lettering sheet as you did for Fig. 2–110. Practice the same numbers with vertical strokes.

Fig. 2–115 Prepare your name and your school name on a cover sheet for your drawings. Use block lettering as in Fig. 2–14.

Chapter 3

The Use and Care of Drafting Equipment

Fig. 3–1 **An industrial drafting room.** *(Phiz Mezey/DPI.)*

Fig. 3–2 **Student drafting kit.** *(AM Bruning International.)*

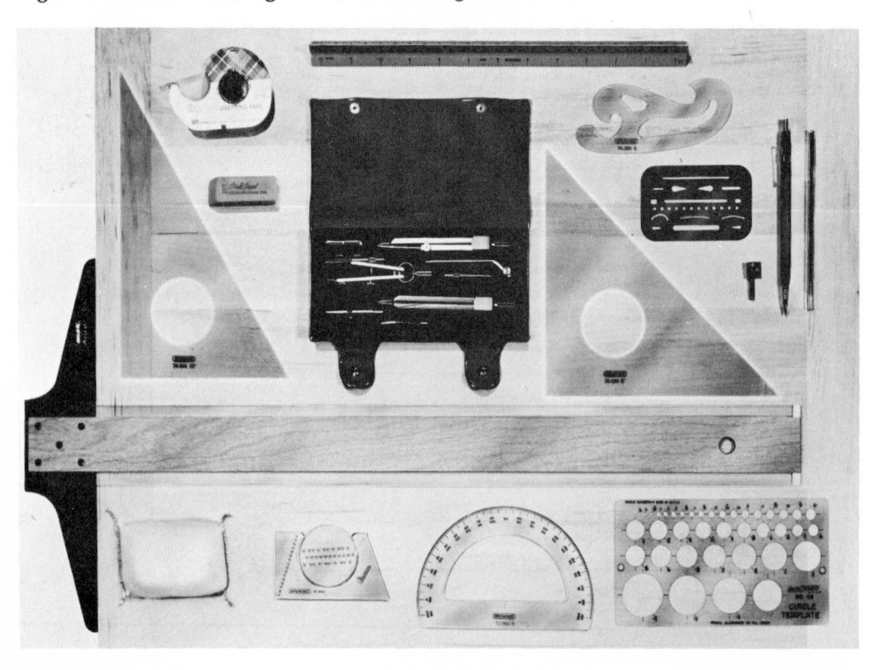

THE DRAFTING ROOM

The plans and directions that are needed for doing engineering work of all kinds are prepared in the drafting room (Fig. 3–1). There, the designs are worked out and checked. In Chapter 1, we learned that technical drawing is really a language. We saw why drawings are used in industrial, engineering, and scientific work. Sometimes, designers or scientists make freehand sketches to help them study new ideas or show ideas to other people. Usually, though, technical drawings must be more accurate than any sketch can be. Such accurate drawings are made with drafting *instruments* (tools). In learning to read and write the drafting language, you must learn which instruments to use on a particular problem. You need to know how to use drafting equipment skillfully, accurately, and quickly.

BASIC EQUIPMENT

Figure 3–2 shows the basic drafting equipment often found in a student drafting kit. This equipment and other common items are listed below.

Drafting board

T-square, or parallel-ruling straightedge, or drafting machine

Drawing sheets (paper, cloth, or film)

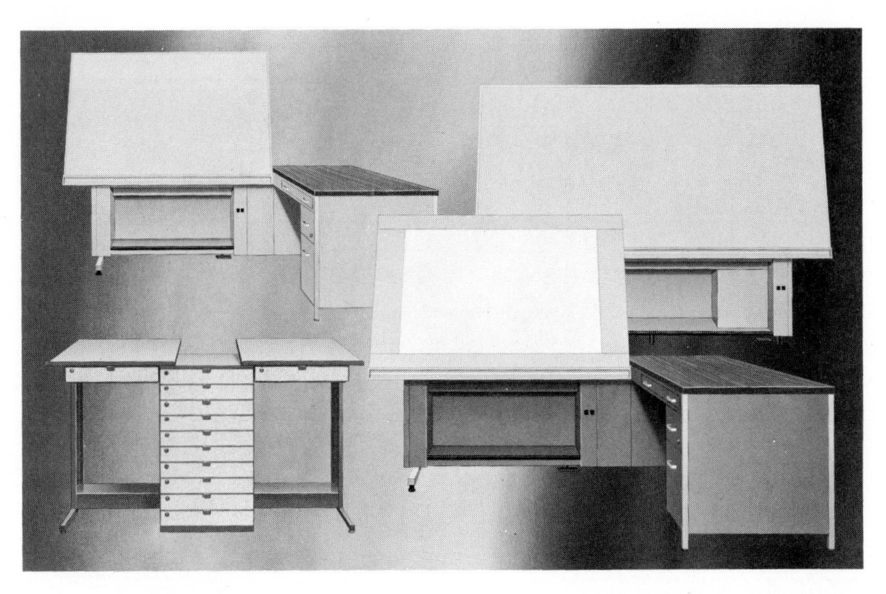

Fig. 3–3 **Drafting tables are available in a variety of sizes and styles.** ***(AM Bruning International.)***

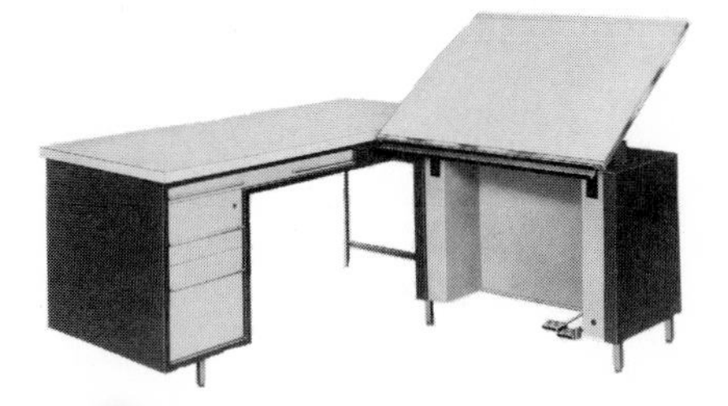

Fig. 3–4 **Modern combination desk and drafting table.** ***(Stacor Corp.)***

Drafting tape
Drafting pencils
Pencil sharpener
Eraser
Erasing shield
Triangles, 45° and 30°–60° (not required with drafting machine)
Architect's or engineer's scale
Irregular curve
Case instruments
Lettering instruments
Black drawing ink
Technical fountain pens
Brush or dust cloth
Protractor
Cleaning powder

Your teacher can tell you exactly what equipment will be needed for your course.

DRAWING TABLES AND DESKS

These items of equipment come in many different sizes and types (Fig. 3–3). One kind of table has a fixed top and a separate drawing board that you can move around. Another has a fixed top that is made to be used as a drawing board itself. A third has an adjustable top that holds a separate drawing board at the slope you want. Still another has an adjustable top that is made to be used as a drawing board.

Some tables are made to be used while you are standing up or sitting on a high stool. Other drawing tables are the same height as regular desks. There are also combined tables and desks that you can use either standing up or sitting down. The type that combines a drafting table, desk, and regular office chair is the most comfortable and efficient. This kind is replacing the high drawing table in drafting rooms and engineering offices (Fig. 3–4).

DRAWING BOARDS

The drawing sheet is attached to a drawing board (Fig. 3–5). Drawing boards used in school or at home usually measure 230 × 300 mm (9″ × 12″), 400 × 530 mm (16″ × 21″), or 460 × 600 mm (18″ × 24″). Boards used to make engineering or architectural drawings are typically larger and may be any size needed. Boards are generally made of soft pine or basswood. They are made

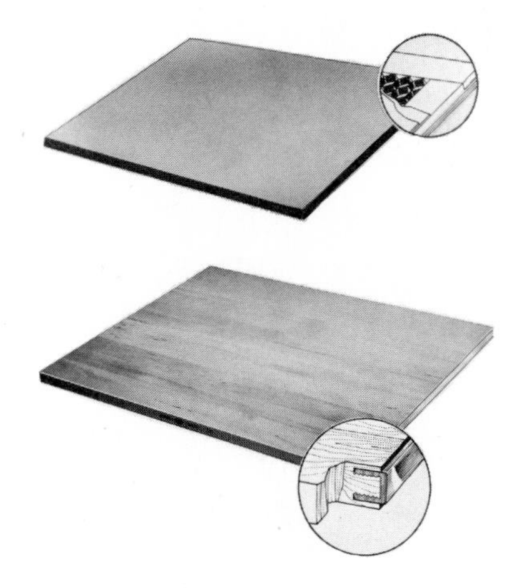

Fig. 3–5 **Drawing boards are available in many types.** ***(Teledyne Post.)***

so that they will stay flat and so that the guiding edge (or edges) will be straight. Hardwood or metal (steel or aluminum) strips are used on some boards to provide truer guiding edges.

T-SQUARES

T-squares (Fig. 3–6) are made of various materials. However, most have plastic-edged wood blades, or clear plastic blades, and heads of wood or plastic. Where great accuracy is needed, stainless steel or hard aluminum blades with metal heads are used. The blade (straightedge) must be very straight. It must be attached securely to the top surface of the T-square head.

You can easily find how accurate your T-square is, as shown in Fig. 3–7. First, draw a sharp line along the drawing edge of the T-square on a sheet of paper. Second, turn the drawing sheet around and line up the drawing edge of the T-square with the other side of the line. If the drawing edge and the pencil line do not match, the T-square edge is not accurate.

Some T-squares are made with an adjustable head that allows the blade to be lined up with the drawing. If the head of such a T-square has a protractor, the blade can be set at any angle. You find the size of a T-square by measuring along the blade from the contact surface of the head to the end of the blade.

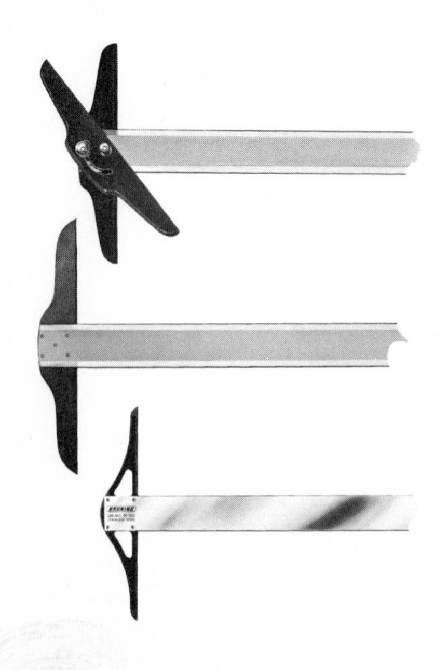

Fig. 3–6 T-squares are available in various styles and materials. *(AM Bruning International.)*

AMERICAN NATIONAL STANDARD TRIMMED SIZES OF DRAWING SHEETS

Three series of sizes are listed. One is based on the standard-size letter paper, 8½″ × 11″. A second is based on a sheet 9″ × 12″. The third is based on the ISO standard. The ISO standard is developed downward in size from a base sheet with an area of 1 square meter. Sheet sizes are based on a ratio of one to the square root of two ($1:\sqrt{2}$). Multiples of these sizes are used for larger sheets, as seen on this page.

U.S. CUSTOMARY SERIES		
Size	First series	Second series
A	8½″ × 11″	9″ × 12″
B	11″ × 17″	12″ × 18″
C	17″ × 22″	18″ × 24″
D	22″ × 34″	24″ × 36″
E	34″ × 44″	36″ × 48″

ISO STANDARD	
Size	Third series
A0	841 × 1189 mm
A1	594 × 841 mm
A2	420 × 594 mm
A3	297 × 420 mm
A4	210 × 297 mm

DRAWING SHEETS

Drawings are made on many different materials. Papers may be white or tinted cream or pale green. They are made in many thicknesses and qualities. Most drawings are made directly in pencil on tracing paper, vellum, tracing cloth, glass cloth, or film. When such materials are used, copies can be made by blueprinting or other reproduction methods (Chapter 14).

Vellum is tracing paper that has been treated to make it more *transparent* (clear). Tracing cloth is a finely woven cloth. It is treated to provide a good working surface and good transparency. Polyester films are widely used in industrial drafting rooms. They are very transparent, strong, and lasting. Drawing films are made with a *matte* (dull and rough) surface. They are suitable for both pencil and ink work.

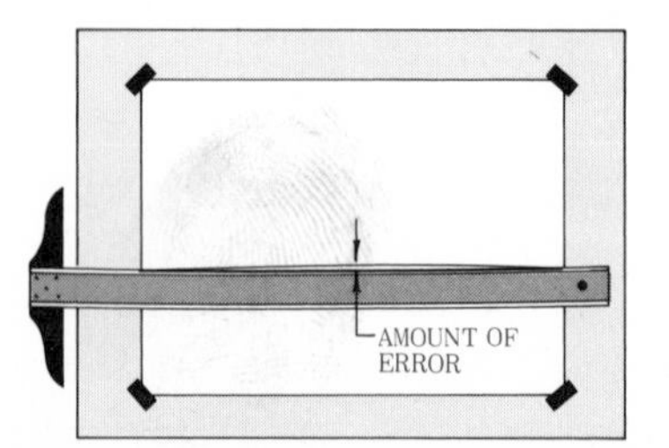

Fig. 3–7 Check to see that the T-square is accurate.

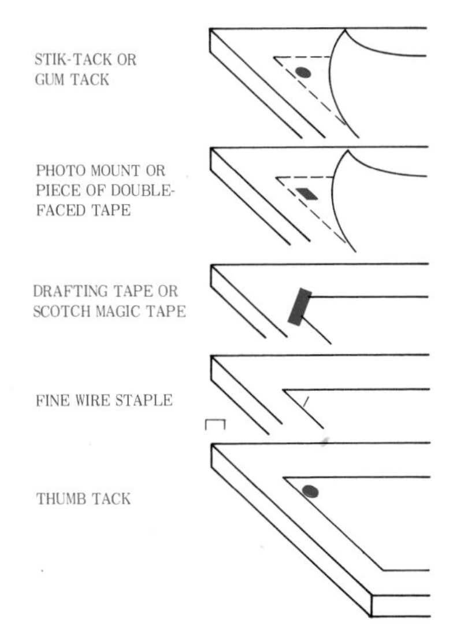

Fig. 3–8 **The drawing sheet can be fastened to the board in several ways.**

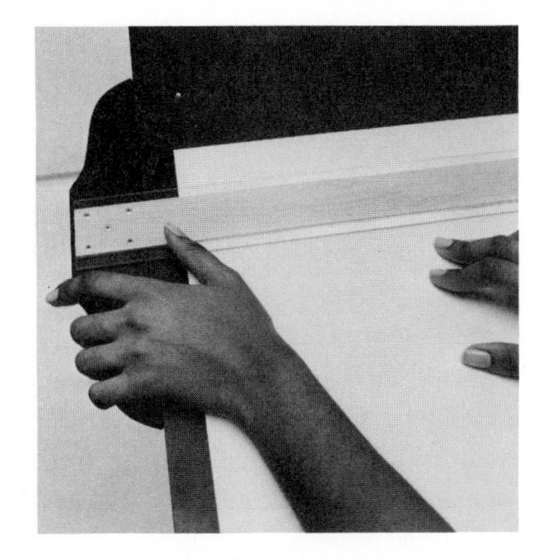

Fig. 3–9 **Fastening the drawing sheet to the board.** ***(R.J. Capece/McGraw-Hill.)***

FASTENING THE DRAWING SHEET TO THE BOARD

The sheet may be held in place on the board in several ways (Fig. 3–8). Drafting tape can be put across the corners of the sheet and, if needed, at other places. *Stik-tacks* (thin disks with adhesive on both sides) can be put on the board and under the sheet. These can be taken off and used again. Similar material comes in squares or as tape. These ways of attaching the sheet let you move the T-square and triangles freely over the whole sheet. In addition, they do not damage the corners or the edges of the sheet. They also can be used on composition boards or other boards with hard surfaces. As a result, most drafters prefer to use one of these methods. If you use thumbtacks, you should push them straight down until the heads are flat on the board. Wire staples can also be used, but they may damage the sheet. Thumbtacks and staples can be used only on soft pine or on other soft boards.

To fasten the paper or other drawing sheet (Fig. 3–9), place it on the drawing board with the left edge 25 mm (1 in.) or so away from the left edge of the board. (Left-handed students should work from the right edge.) Put the lower edge of the sheet at least 100 mm (4 in.) up from the bottom of the board so that you can work on it comfortably. Then line up the sheet with the T-square blade, as shown at A. Hold the sheet in position. Move the T-square down, as at B, keeping the head of the T-square against the edge of the board. Then fasten each corner of the sheet with drafting tape.

DRAWING PENCILS

Both regular wooden pencils and mechanical pencils are used for technical drawing. Four kinds of lead are now used in drawing pencils. Of course, pencil "leads" are not really made from lead, the metal. One pencil lead is made from *graphite,* a form of the element carbon. It also contains clay and *resins* (sticky substances). Graphite pencils have been used for over 200 years. They are still the most important kind. Drafting pencils of this type are usually made in 17 degrees of hardness, or grades. These grades are listed below.

6B (softest and blackest)
5B (extremely soft)
4B (extra soft)
3B (very soft)
2B (soft, plus)
B (soft)
HB (medium soft)
F (intermediate, between soft and hard)
H (medium hard)
2H (hard)
3H (hard, plus)
4H (very hard)
5H (extra hard)
6H (extra hard, plus)
7H (extremely hard)
8H (extremely hard, plus)
9H (hardest)

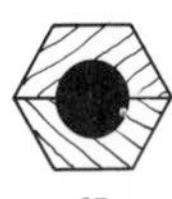

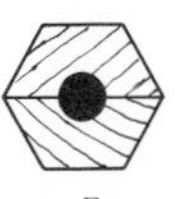

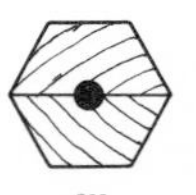

Graphite pencils are mostly used on paper or vellum. They may also be used on cloth.

Which grade of pencil you use depends on what kind of surface you are drawing on. It also depends on how *opaque* (dark) and how thick you want the line you are drawing to be. To lay out views on fairly hard-surfaced drawing paper, you usually use grades 4H and 6H. When you use tracing paper or cloth and draw finished views that are to be *reproduced* (copied by a machine), you use an H or 2H pencil. Grades HB, F, H, and 2H are sometimes used for sketching and lettering and for drawing arrowheads, symbols, border lines, and so on. The exact grade you use depends on the drawing and the surface. Very hard and very soft leads are seldom used in ordinary drafting.

Since film has come into use for drawings, new kinds of pencil lead have been developed. Three types are described below, based on information furnished by the Joseph Dixon Crucible Company. The first kind is a *plastic pencil.* This type is a black crayon. Its lead is *extruded* (pushed out) in a "plasticizing" process. Drawings made with this lead copy well on microfilm. The second type of lead is a "combination." It is part plastic and part graphite and is made by heating. This kind stays sharp, draws a good opaque line, does not smear easily, erases well, and microfilms well. It can be used on paper or cloth as well as on film. The third type is also a combination of plastic and graphite. This type is extruded, like the first kind, but not heated, like the second. The third type does not stay sharp. However, it draws a fairly opaque line, erases well, does not smear easily, and microfilms well. It is used mostly on film. These three types of lead are made in only five or six grades. Their grades are not the same as those used for regular graphite leads. The companies that make these pencils use different systems of letters and numbers to tell what kind of lead is in each pencil and how hard it is.

SHARPENING THE PENCIL

You sharpen a wooden pencil by cutting away the wood at a long slope, as at A in Fig. 3–10. Always sharpen the end opposite the grade mark. Be careful not to cut the lead. Leave it exposed for about 10 to 13 mm (3/8 to 1/2 in.). Then shape the lead to a long *conical* (cone-shaped) point. You do this by rubbing the lead back and forth on a sandpaper pad (or on a fine file), as at B, while turning it slowly to form the point, as at C or D. Some drafters prefer the flat point (or chisel point) shown at E. Keep the sandpaper pad at hand so that you can sharpen the point often.

Fig. 3–10 Sharpening the pencil properly is important. ***(R.J. Capece.)***

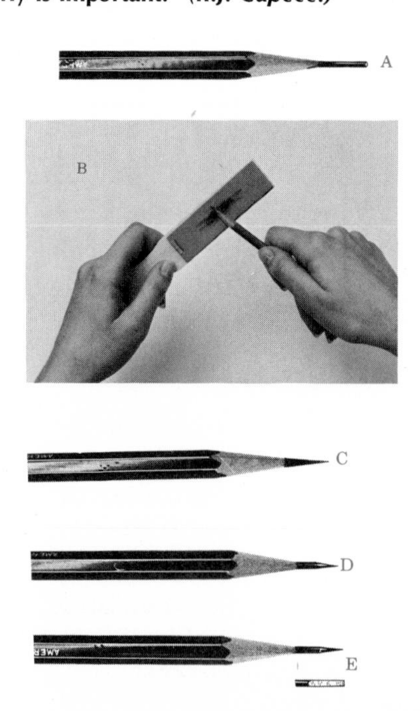

Mechanical sharpeners (Fig. 3–11) are made with special drafter's cutters that remove the wood as shown. Special pointers are made for shaping the lead, as in Fig. 3–12. Such devices may be hand-operated or electrically powered.

Mechanical pencils, also called lead holders, are widely used by drafters. They hold plain leads by means of a chuck that lets you adjust the exposed lead to any length you want. The lead is shaped as with wooden pencils. However, some refill pencils have a built-in sharpener that shapes the lead.

Pencil lines must be clean. They must be dark enough for the views to be seen when the standard lines

Fig. 3–11 A drafter's pencil sharpener cuts the wood, not the lead.

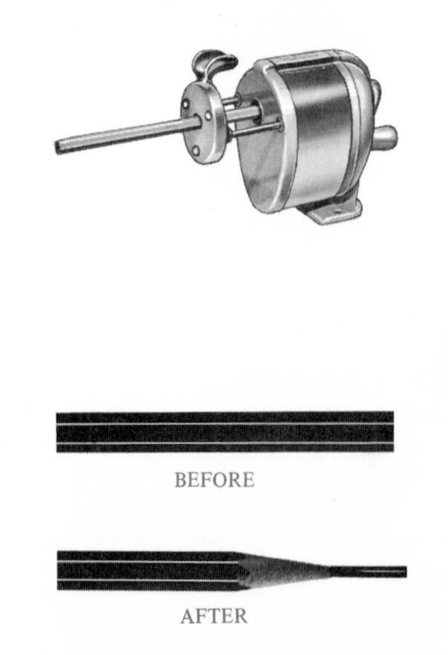

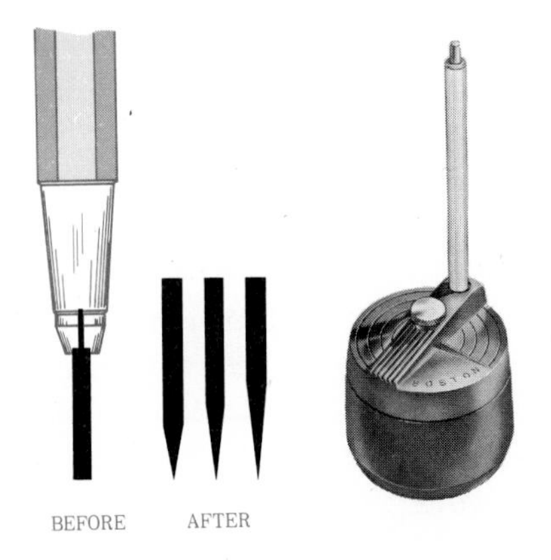

Fig. 3–12 **This lead pointer allows a choice of point shapes.** ***(Hunt Manufacturing Co.)***

in Fig. 3–13 are used. If you use too much pressure, you will groove the drawing surface. You can avoid this if you use the right grade of lead. Develop the habit of turning the pencil between your thumb and forefinger as you are drawing a line. This will help make the line uniform and keep the point from wearing down unevenly. *Never sharpen a pencil over the drawing board.* After you sharpen a pencil, wipe the lead with a cloth to remove the dust. Being careful in these ways will help keep your drawing clean and bright. This is very important when you have to use the original pencil drawing to make copies.

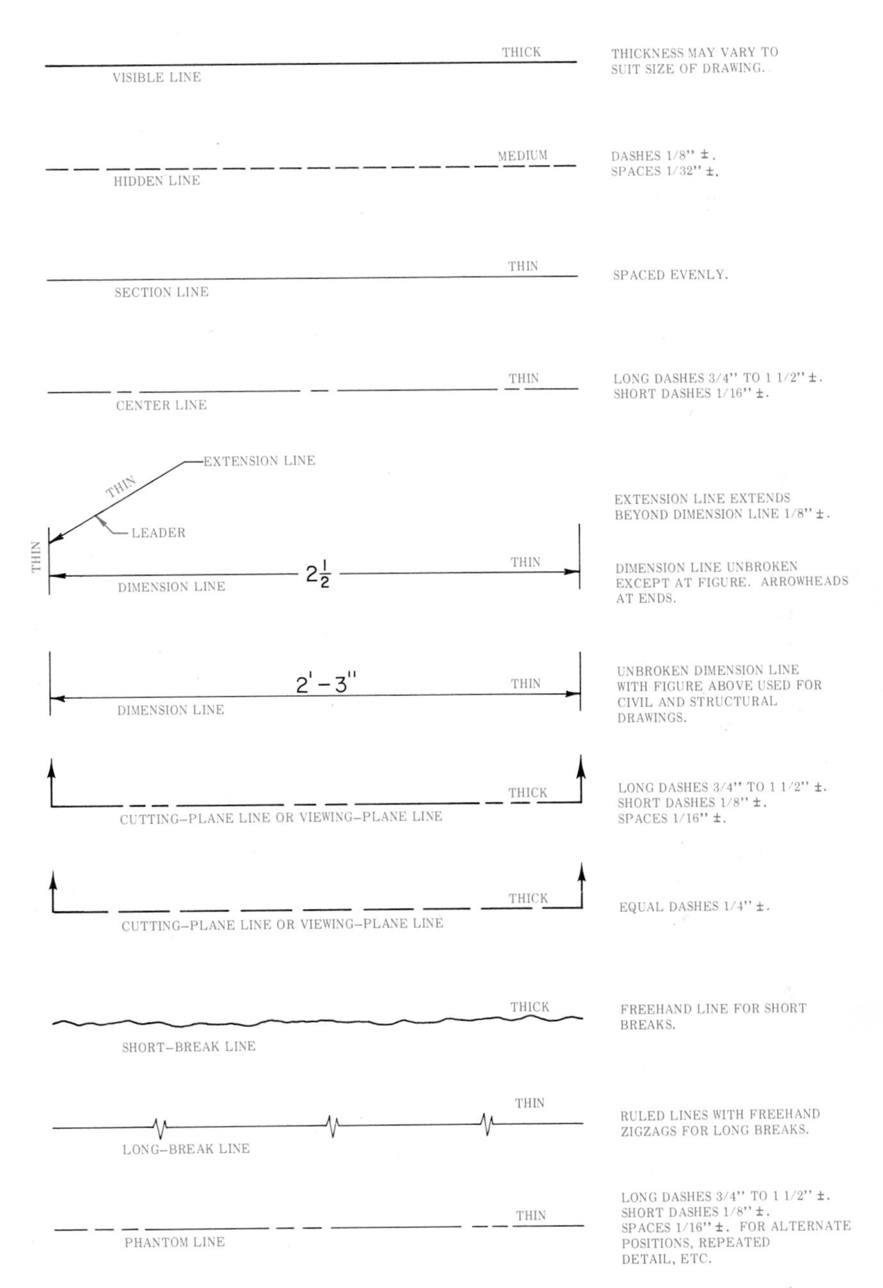

Fig. 3–13 **Alphabet of lines for pencil drawing. (See Chapter 13 for ink lines.)**

ALPHABET OF LINES

The different lines, or line symbols, used on drawings are a kind of graphical alphabet. The line symbols that the American National Standards Institute recommends for pencil drawings are shown in Fig. 3–13. Two widths of lines—thick and thin—are generally used. (Formerly, a medium line was used as a hidden line.) Drawings are much easier to read when there is good contrast between the different kinds of lines. Pencil lines should be uniformly sharp and black.

ERASERS AND ERASING SHIELDS

Soft erasers, such as the vinyl type, the Pink Pearl, or the Artgum, are used for cleaning soiled spots or light pencil marks from drawings. Rub-kleen, Ruby, or Emerald erasers are generally good for removing pencil or ink. Ink erasers contain grit. If they are used at all, they have to be used very carefully to keep from damaging the drawing surface. Electric erasing machines (Fig. 3–14) are in common use in drafting rooms.

Erasing shields, made of metal or plastic, have openings of different sizes and shapes. They are useful for protecting lines that are not to be erased (Fig. 3–15). Lines on paper or cloth are erased *along* the direction of the work. Lines on film are erased *across* the direction of the work. On film, you should use a vinyl eraser or other eraser without grit. Erasing on film is done very carefully by hand.

Fig. 3–14 **An electric erasing machine saves time.** ***(R.J. Capece/McGraw-Hill.)***

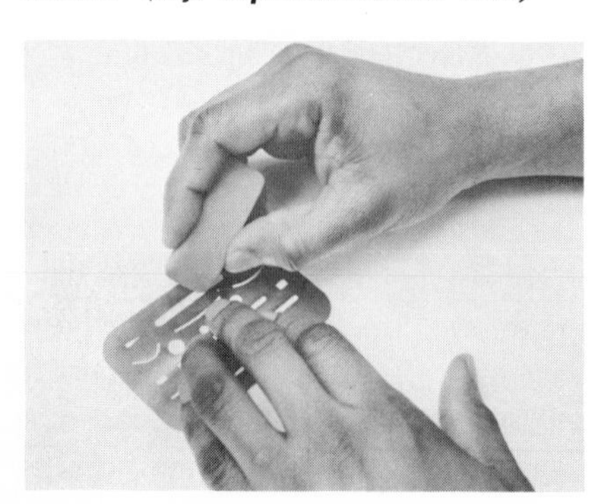

Fig. 3–15 **Erasing with the aid of a shield.** ***(R.J. Capece/McGraw-Hill.)***

TO DRAW HORIZONTAL LINES

The upper edge of the T-square blade is used as a guide for drawing *horizontal* (side-to-side) lines (Fig. 3–16). With your left hand, place the head of the T-square in contact with the left edge of the board. Keeping the T-square in contact, move it to the place you want to draw the line. Slide your left hand along the blade to hold it firmly against the drawing sheet. Hold the pencil about 25 mm (1 in.) from its point. Slant it in the direction in which you are drawing the line. (This direction should be left to right for right-handers and right to left for left-handers.) While drawing the line, turn the pencil slowly and slide your little finger along the blade of the T-square. This will help you control the pencil better. On film, you must keep the pencil at the same angle (55° to 65°) all along the line. You must also use less pressure than on paper or other material. Always keep the point of the lead a little distance away from the corner between the guiding edge and the drawing surface, as shown in Fig. 3–16. This will let you see where you are drawing the line. It will also help you keep from making a poor or smudged line. *Be careful to keep the line parallel to the guiding edge.*

TO DRAW VERTICAL LINES

You use a triangle and a T-square to draw *vertical* (up-and-down) lines (Fig. 3–17). Place the head of the T-square in contact with the left edge of the board. Keeping the T-square in contact, move it to a position below the start of the vertical line. Place a triangle against the T-square blade. Move the triangle to the place you want to draw the line. Keeping the vertical edge of the triangle toward the left, draw upward. Slant the pencil in the direction in which you are drawing the line. Be sure to keep this angle the same when you are drawing on film.

Fig. 3–16 **Drawing a horizontal line.** ***(R.J. Capece/McGraw-Hill.)***

Fig. 3–17 **Drawing a vertical line.** *(R.J. Capece/McGraw-Hill.)*

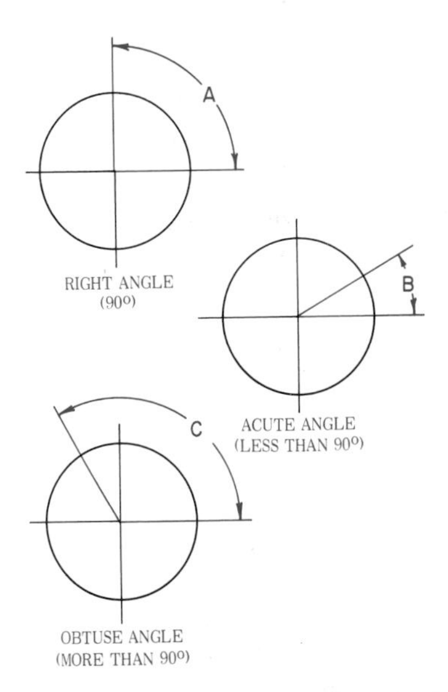

Fig. 3–18 **Angles are measured in degrees, minutes, and seconds.**

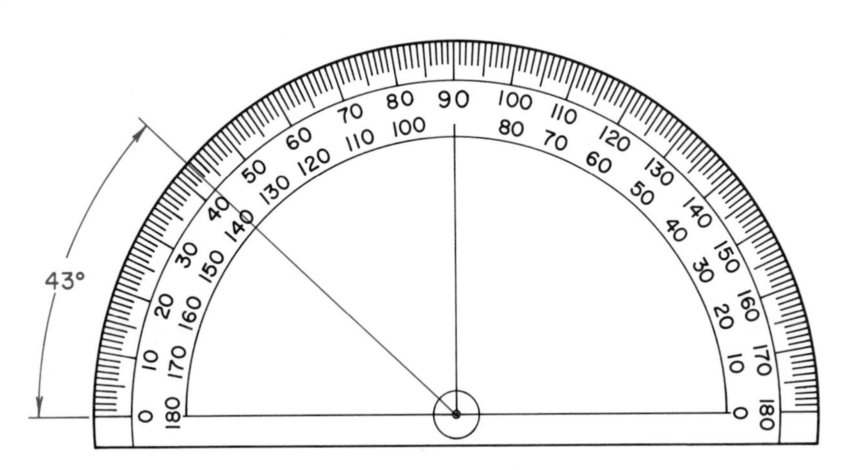

Fig. 3–19 **A protractor is used to lay out, or measure, angles.**

Keep the point of the lead far enough out from the guiding edge so that you can see where you are drawing the line. Be careful to keep the line parallel to the guiding edge.

ANGLES

An *angle* is formed when two straight lines meet at a point. The point where the lines meet is called the *vertex* of the angle. The lines are called the *sides* of the angle. Angles are measured in a unit called a *degree*. Degrees are marked by the symbol °. If the *circumference* (rim) of a circle is divided into 360 parts, each part is an angle of 1°. Then, one-fourth of a circle would be 360/4 = 90°, as at A in Fig. 3–18. A part of a circle is called an *arc*. Two lines drawn to the center of a circle from the ends of a 90° arc on that circle form a *right angle*. An *acute angle* (B) is less than 90°. An *obtuse angle* (C) is greater than 90°. For closer measurements, a degree is divided into 60 equal parts called *minutes* (′). Minutes in turn, are divided into 60 equal parts called *seconds* (″). Notice that the size of an angle does not depend on the length of its sides. Therefore, the number of degrees in a given angle will be the same, whatever the size of the circle and arc that define it.

An instrument used in measuring or laying out angles is called a *protractor*. A semicircular protractor is shown in Fig. 3–19, where an angle of 43° is measured. Drafting machines and adjustable-head T-squares have an adjustable protractor for laying out angles of any degree.

Angles of 45°, 30°, and 60° can be drawn directly, using the triangles. Angles varying by 15° can also be

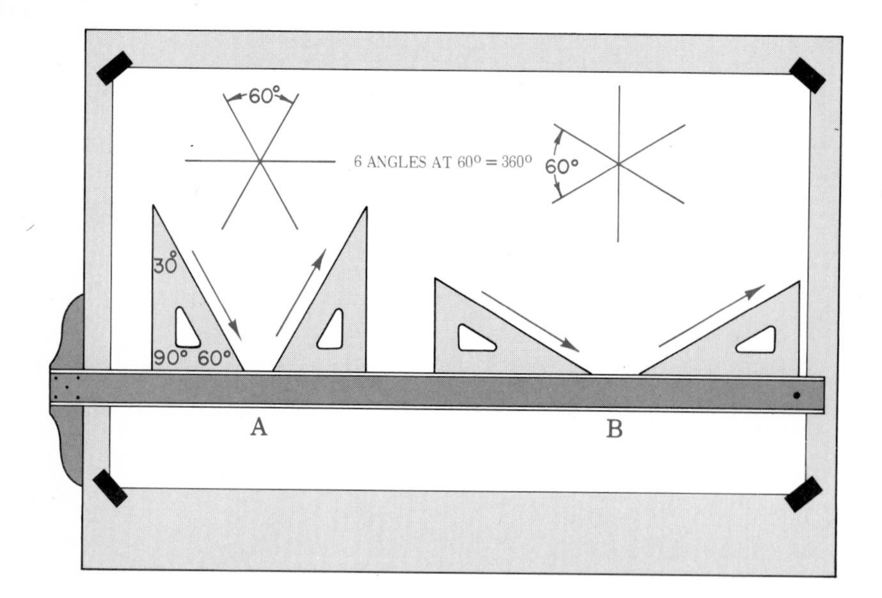

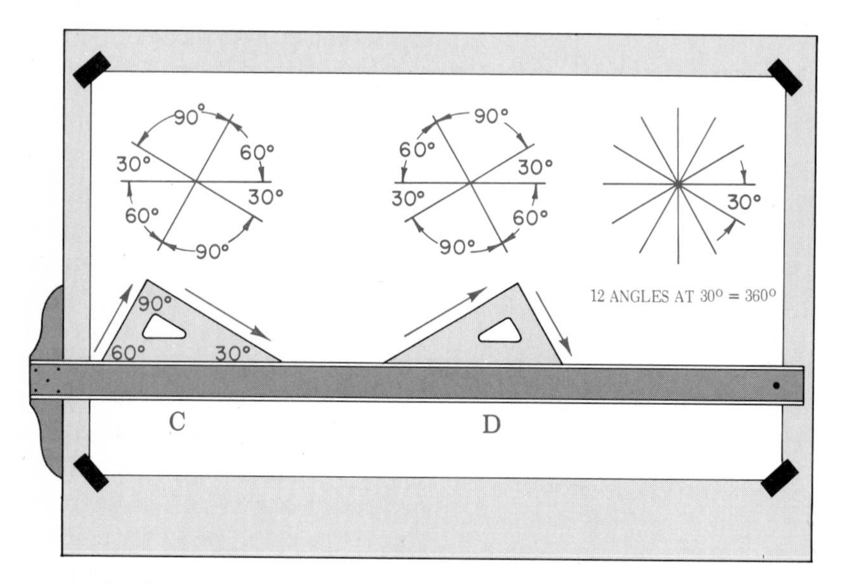

Fig. 3–20 **The 30°–60° triangle has angles of 30°, 60° and 90°.**

drawn with the triangles, as described in the following section.

TO DRAW LINES INCLINED 30°–60° AND 45°

The 30°–60° triangle (Fig. 3–20) has angles of 30°, 60°, and 90°. Lines inclined at 30° and 60° from horizontal or vertical lines are drawn with the triangle held against the T-square blade, as shown in Fig. 3–20, or against a horizontal straightedge. The 30°–60° triangle can be used to lay off equal angles, 6 at 60° or 12 at 30°, about a center and for many constructions.

The 45° triangle (Fig. 3–21) has two angles of 45° and one of 90°. Lines *inclined* (slanted) 45° from horizontal or vertical lines are drawn with the triangle held against the T-square blade, as shown at A, B, and C in Fig. 3–21, or against a horizontal straightedge. The 45° triangle can be used to lay off eight equal angles of 45° about a center and for many constructions.

TO DRAW LINES AT ANGLES VARYING BY 15°

The 45° and 30°–60° triangles, alone or together and combined with the T-square, can be used to draw angles increasing by 15° from the horizontal or vertical line. Some ways of placing the triangle to draw angles of 15° and 75° are suggested at A, B, C, and D in Fig. 3–22. Ways of getting the different angles are shown in Fig. 3–23. In Space O, the lines have been drawn for all the positions possible. All the angles are 15°. By placing the triangles as in Spaces B and M or in Spaces F and I, you get two angles of 30° and two angles of 15°. It is a good idea to try placing the triangles in all the different ways until you are used to them. (Of course, you can get any angle you want by using a protractor.)

TO DRAW PARALLEL LINES

You can draw *parallel* (side-by-side) horizontal lines with the T-square. You can draw parallel vertical lines by using a triangle along with the T-square. Parallel lines can be drawn at regular angles with the triangles, as suggested in Fig. 3–23. You can draw parallel lines in any place by using a triangle along with the T-square or another triangle, as shown in Fig. 3–24. Parallel lines can also be drawn directly with a drafting machine.

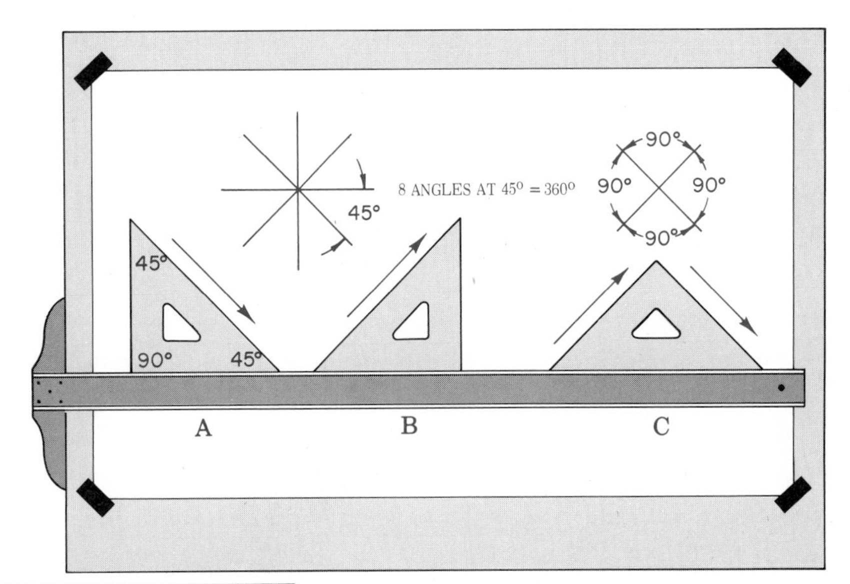

Fig. 3–21 **The 45° triangle has angles of 45° and 90°.**

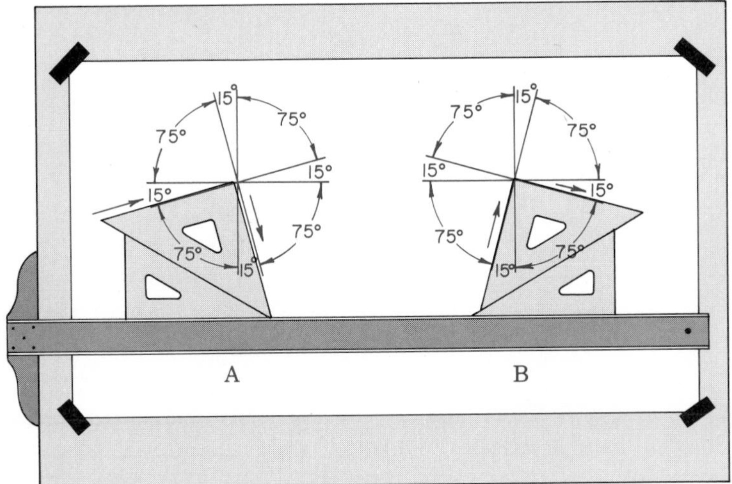

Fig. 3–22 **Drawing lines at 15° and 75° using the two triangles.**

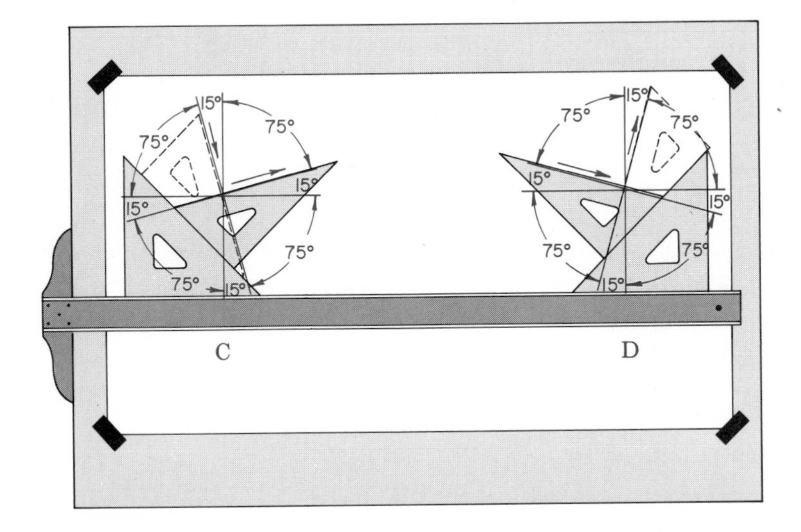

TO DRAW A LINE PARALLEL TO A GIVEN LINE

Place a triangle against the T-square blade and move them together until one edge of the triangle matches the given line (Fig. 3–24, Space A). Holding the T-square firmly, slide the triangle along its blade until it reaches the place you want. Then draw the parallel line (Space B). In Fig. 3–24, the *hypotenuse* (the edge opposite the right angle) of the triangle is used to draw the parallel line. However, other edges of the 30°–60° triangle can be used instead. In Space C, a triangle is used in place of the T-square. Parallel lines can also be drawn directly with a drafting machine.

TO DRAW PERPENDICULAR LINES

Lines at right angles (90°) with each other are *perpendicular*. A vertical line is perpendicular to a horizontal line. Perpendicular

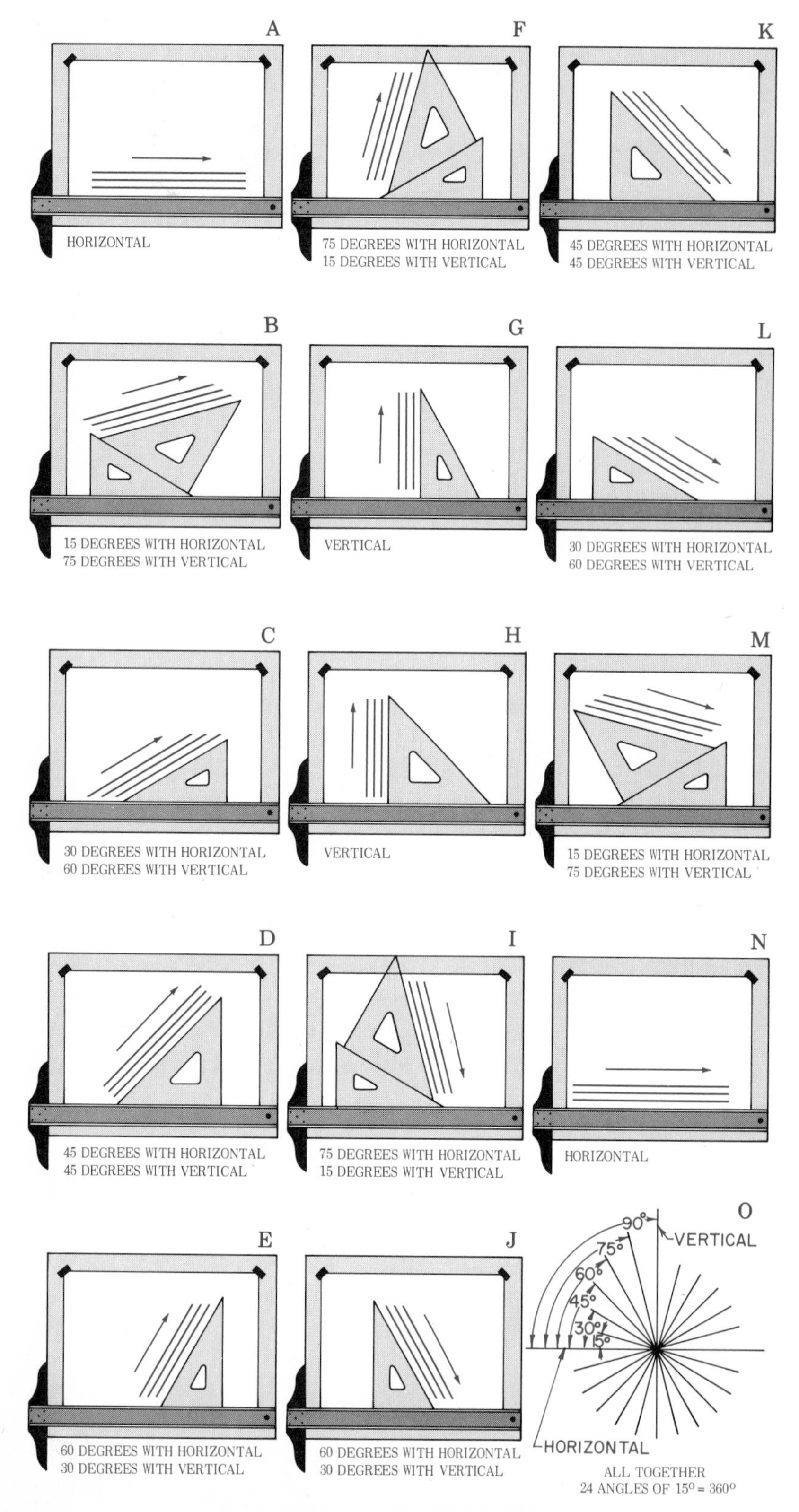

Fig. 3–23 **Drawing lines with the T-square and triangles.**

lines can be drawn using triangles and the T-square. They can also be drawn directly with a drafting machine.

TO DRAW A LINE PERPENDICULAR TO ANY LINE

Two methods can be used.

First method (Fig. 3–25). Place a triangle and the T-square together so that one edge of the triangle matches the given line, as in Space A. Hold the T-square firmly. Slide the triangle along the blade to

Fig. 3–24 **Drawing parallel lines.**

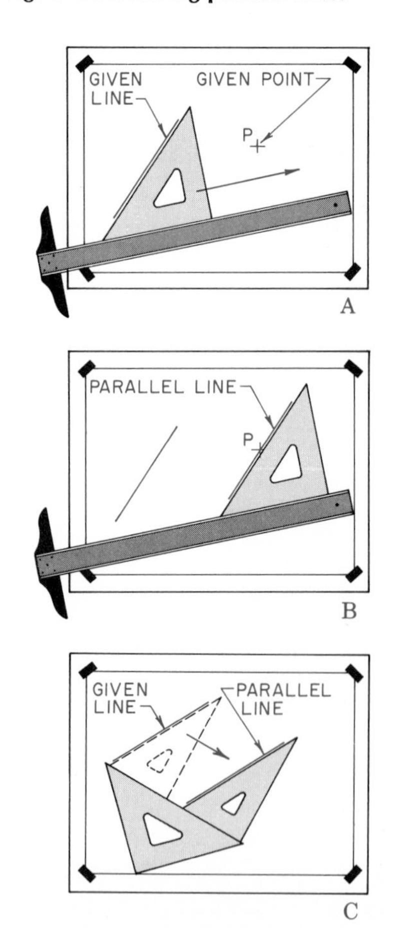

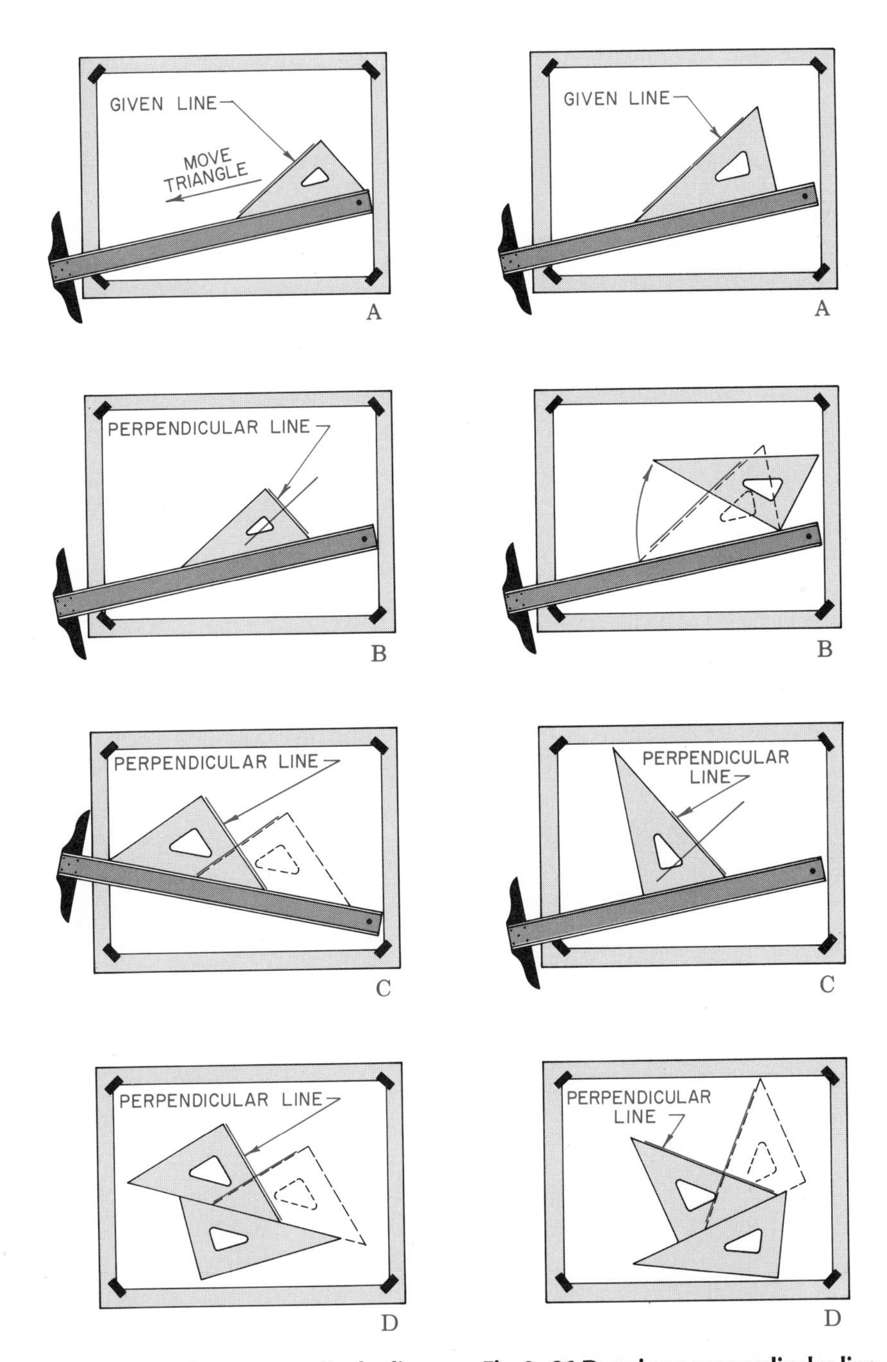

Fig. 3–25 **Drawing a perpendicular line, first method.**

Fig. 3–26 **Drawing a perpendicular line, second method.**

the desired position, as in Space B. Draw the perpendicular line. The 45° triangle can be used instead of the 30°–60° triangle, as in Space C. A triangle can be used instead of the T-square, as in Space D.

Second method (Fig. 3–26). Place a triangle and the T-square together so that the hypotenuse of the triangle matches the given line, as in Space A. Turn the triangle about its right-angled corner, as in Space B. Slide the triangle along the T-square blade until it is in the place you want, as in Space C. Now draw the perpendicular line. The 45° triangle can be used instead of the 30°–60° triangle. A triangle can be used instead of the T-square, as in Space D.

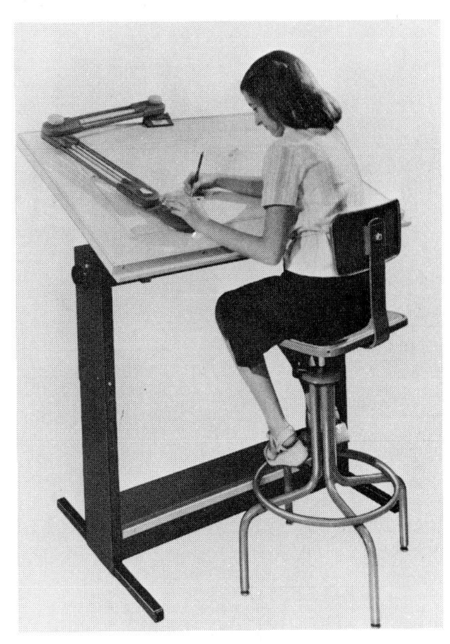

Fig. 3–27 **An arm-type drafting machine. *(Teledyne Post.)***

DRAFTING MACHINES

There are two kinds of drafting machines in wide use. One type, shown in Fig. 3–27, uses an anchor and two arms to hold a movable protractor head with two scales. The scales are ordinarily at right angles to each other. The arms allow the scales to be moved to any place on the drawing that is parallel to the starting position. Another type, Fig. 3–28, uses a horizontal

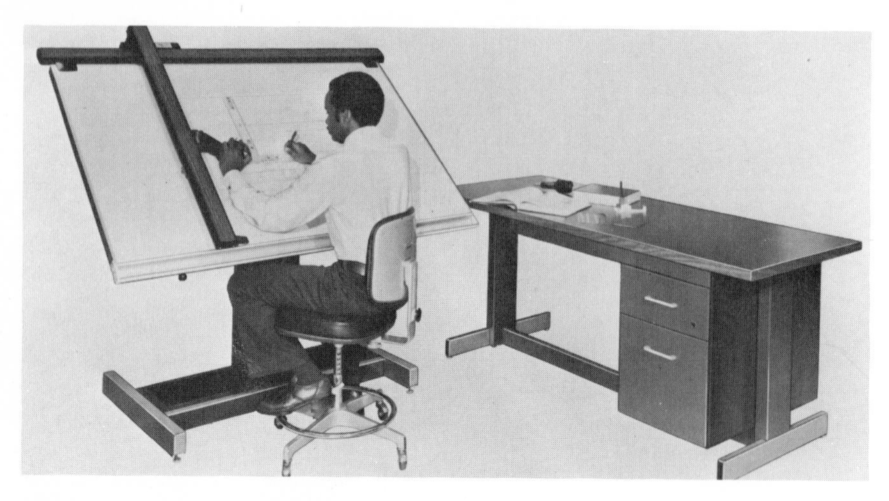

Fig. 3–28 **This track-type drafting machine is especially adapted for wide drawings, in addition to use for regular sizes of drawings.** ***(Teledyne Post.)***

guide rail at the top of the board and a moving arm rail at right angles to the top rail. An adjustable protractor head and two scales, ordinarily at right angles, move up and down on the arm. The scales may be moved to any place on the drawing that is parallel to the starting position. This type of drafting machine is easy to use on large boards or on boards placed vertically or at a steep angle.

Most industrial drafting departments now use drafting machines. Many school drafting departments teach drawing with drafting machines. Drafting machines combine the functions of the T-square, triangles, scales, and protractor. You can draw lines as long as you want, where you want, and at any angle you want just by moving the scale ruling edge to the right places. This lets you draw faster and with less work. You should learn how to use the drafting machine in all the possible ways and how to take care of it. If you do so, you will soon see how valuable a tool the drafting machine is.

Fig. 3–29 **A parallel-ruling straightedge is another convenient instrument used to save time.** ***(R.J. Capece/McGraw-Hill.)***

PARALLEL-RULING STRAIGHTEDGES

Many drafters find it very convenient to use parallel straightedges when working on large boards placed vertically or nearly so. A guide cord is clamped to the ends of the straightedge. The cord runs through a series of pulleys on the back of the board. This lets you move the straightedge up and down in parallel positions (Fig. 3–29).

SCALES

Scales are made in different shapes (Fig. 3–30). They are made of different materials, such as boxwood, plastic, white plastic on boxwood, and metal. They are also made with different divisions so that they can be used to make different kinds of drawings. The customary-inch system uses separate scales for architectural, mechanical engineering, and civil engineering drawings. Thus, customary-inch scales that are

Fig. 3–30 **Scales are made in various shapes.**

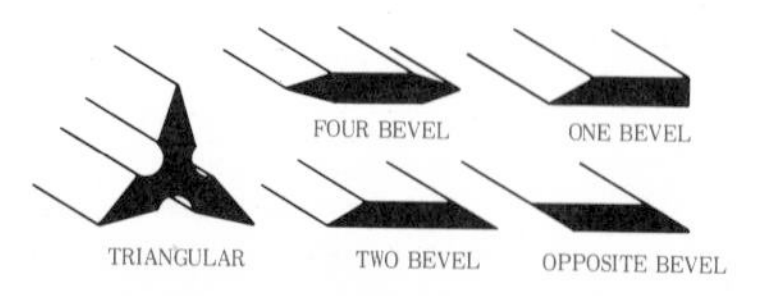

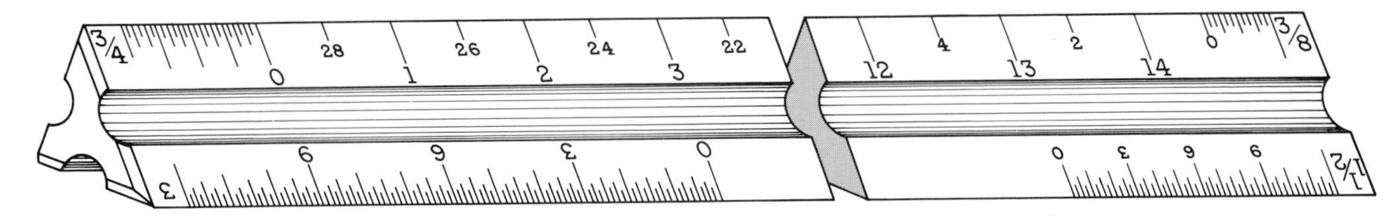

Fig. 3–31 **Architect's scale, open divided. The triangular form has many proportional scales.**

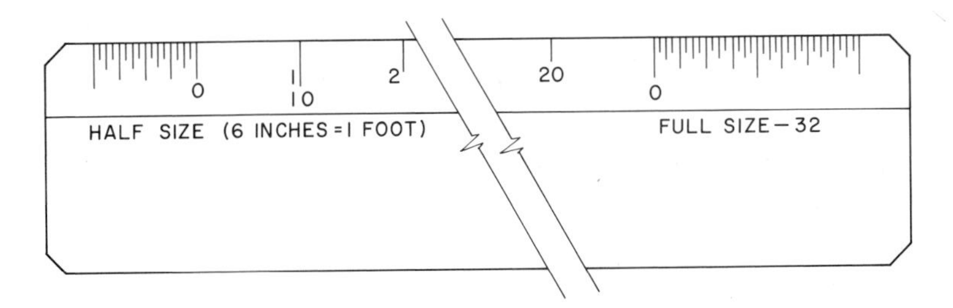

Fig. 3–32 **Mechanical engineer's scale, open divided.**

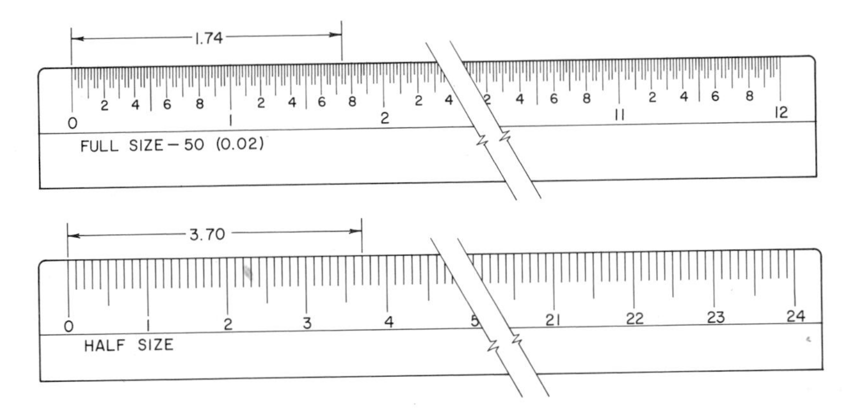

Fig. 3–33 **Civil engineer's scale, divided into decimals.**

commonly used include the architect's scale (Fig. 3–31), the mechanical engineer's scale (Fig. 3–32), and the civil engineer's scale (Fig. 3–33). By contrast, the proportional metric scale can be used on all types of engineering drawings. Both customary and metric scales can have any of the shapes shown in Fig. 3–30.

Scales are used for laying off distances and for making measurements. Measurements can be full size or in some exact proportion to full size. Scales can be *open-divided* (Figs. 3–31 and 3–32), with only the end units subdivided. Or they can be *full-divided* (Figs. 3–33 and 3–34), with subdivisions over their entire length.

THE ARCHITECT'S SCALE

The architect's scale (Fig. 3–31) is divided into proportional feet and inches. The triangular form shown is used in many schools and in some drafting offices because it has many scales on a single stick. However, many drafters prefer flat scales, especially when they do not have to change scales often. The symbol ′ is used for feet and ″ for inches. Thus, three feet four and one-half inches is written 3′–4½″. When all dimensions are in inches, the symbol is usually left out.

The usual proportional scales are listed on page 56.

Fig. 3–34 **Decimal-inch scales are used in drawing machine parts.**

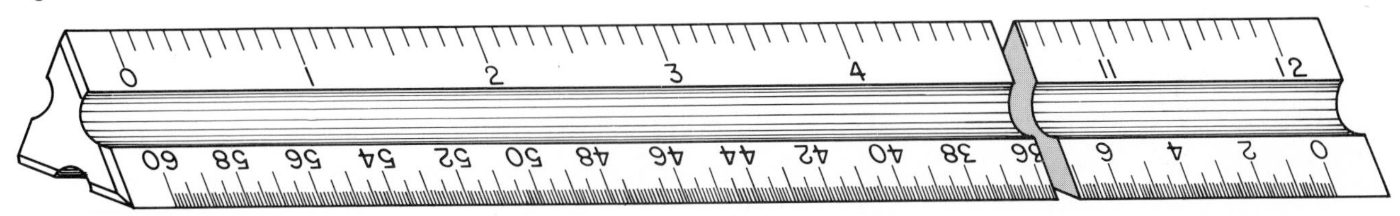

PROPORTION	GRADATIONS	RATIO
Full size,	12″ = 1′–0″	1:1
1/4 size,	3″ = 1′–0″	1:4
1/8 size,	1 1/2″ = 1′–0″	1:8
1/12 size,	1″ = 1′–0″	1:12
1/16 size,	3/4″ = 1′–0″	1:16
1/24 size,	1/2″ = 1′–0″	1:24
1/32 size,	3/8″ = 1′–0″	1:32
1/48 size,	1/4″ = 1′–0″	1:48
1/64 size,	3/16″ = 1′–0″	1:64
1/96 size,	1/8″ = 1′–0″	1:96
1/128 size,	3/32″ = 1′–0″	1:128

These scales are used in drawing buildings and in making mechanical, electrical, and other engineering drawings. They are also often used in drafting in general. The proportional scale to which the views are drawn should be given on the drawing. This is done in the title block if only one scale is used. If different parts of a drawing are in different scales, the scales are given near the views in this way:

Scale: 6″ = 1′–0″
Scale: 3″ = 1′–0″
Scale: 1 1/2″ = 1′–0″

THE MECHANICAL ENGINEER'S SCALE

The mechanical engineer's scale (Fig. 3–32) has inches and fractions of an inch divided to stand for inches. The usual divisions are the following:

Full size—1 in. divided into 32nds
Half size—1/2 in. divided into 16ths
Quarter size—1/4 in. divided into 8ths
Eighth size—1/8 in. divided into 4ths

These scales are used for drawing parts of machines or where larger reductions in scale are not needed. The proportional scale to which the views are drawn should be given on the drawing. This is done in the title block. If different parts of a drawing are in different scales, the scales are given near the views (as, for example, full size or 1″ = 1″; half size or 1/2″ = 1″).

The decimal-inch system uses a scale divided into *decimals* (parts divisible by ten) of an inch (Fig. 3–33). For full size, 1 in. is divided into 50 parts (each part of 1/50 = 0.02″). In this way, you can easily measure to hundredths of an inch by sight. Some usual divisions are given below:

Full size—1 in. divided into 50ths
Half size—1/2 in. divided into 10ths
Three-eighths size—3/8 in. divided into 10ths
One-quarter size—1/4 in. divided into 10ths

The decimal-inch system has been used in the automotive industry for many years. It is now in widespread use in other engineering fields, as well.

THE CIVIL ENGINEER'S SCALE

The civil engineer's scale (Fig. 3–34) has inches divided into decimals. The usual divisions follow:

10 parts to the inch
20 parts to the inch
30 parts to the inch
40 parts to the inch
50 parts to the inch
60 parts to the inch

With this scale, 1 in. may stand for feet, rods, miles, and so forth. It may also stand for quantities, time, or other units. The divisions may be single units or multiples of 10, 100, and so on. Thus, the 20-parts-to-an-inch scale may stand for 20, 200, or 2000 units. This scale is used for civil engineering work. This includes maps and drawings of roads and other public projects. It is also used where decimal divisions are needed. These uses include plotting data and drawing graphic charts.

The scale used should be given on the drawing or work as follows:

Scale: 1″ = 500 pounds
Scale: 1″ = 100 feet
Scale: 1″ = 500 miles
Scale: 1″ = 200 pounds

For some uses, a graphic scale is put on a map, drawing, or chart, as shown below.

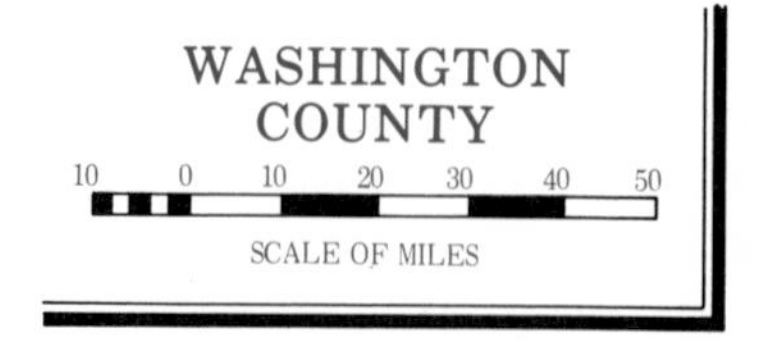

METRIC SCALES

Metric scales (Fig. 3–35) are divided into millimeters. The usual proportional scales in the metric system are listed as a ratio in the table on page 57.

ENLARGED	SAME SIZE	REDUCED
2000:1		
1000:1	1:1	
500:1		1:2
200:1		1:5
100:1		1:10
50:1		1:20
20:1		1:50
10:1		1:100
5:1		1:200
2:1		1:500
		1:1000
		1:2000

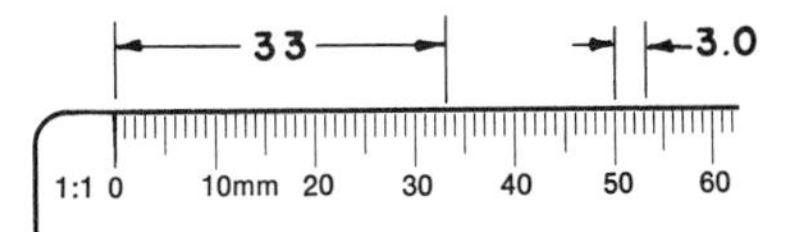

Fig. 3–35 **Metric scales are divided into millimeters.**

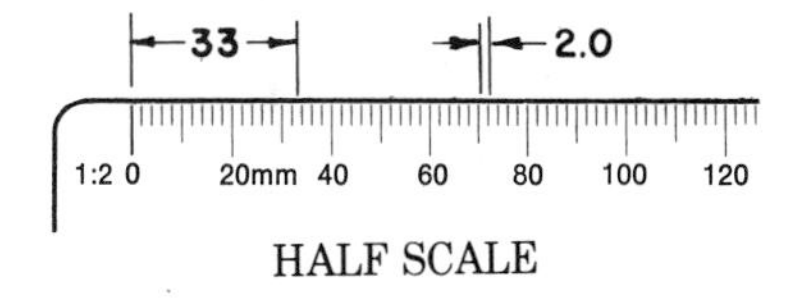

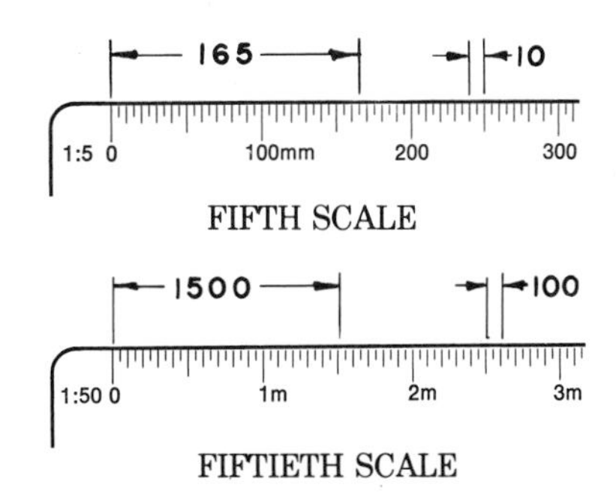

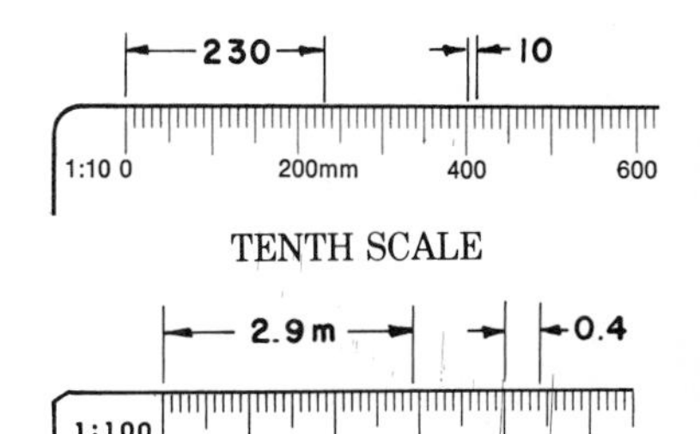

Fig. 3–36 **Metric scales for reduction.**

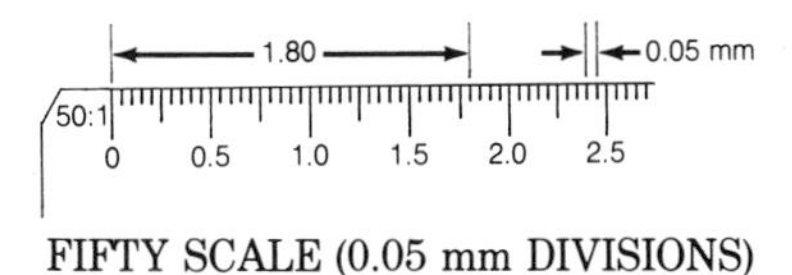

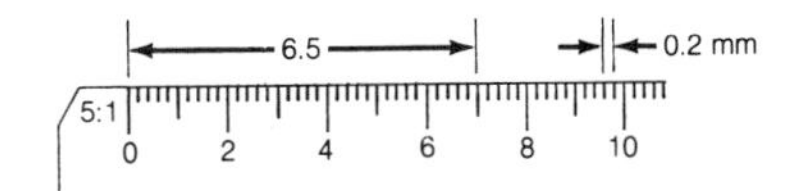

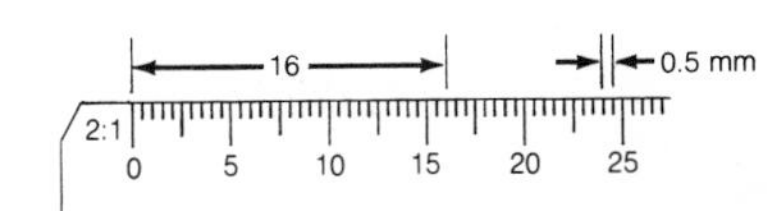

Fig. 3–37 **Metric scales for enlarged objects.**

Metric architectural proportions are used in drawing buildings and in making many mechanical, electrical, and other engineering drawings. They are also often used in drafting in general. The proportional scale to which the views are drawn should be given on the drawing. This is done in the title block if only one scale is used. If different parts are in different scales, the scales are given near the views in this way:

Scale: 1:2
Scale: 1:5
Scale: 1:10
Scale: 1:50
Scale: 1:100

To reduce the size of an object being drawn, a drafter uses one of the scales in Fig. 3–36. The proportional scales used to enlarge drawings of machine parts are shown in Fig. 3–37.

FULL-SIZE DRAWINGS

When an object is not too large for the paper on which you are drawing it, you can draw it full size (1:1). *To make a measurement,* put the scale on the paper in the direction in which you are measuring. Make a short, light dash next to the zero on the scale and another dash next to the division at the distance you want. Do not make a dot or punch a hole in the drawing sheet.

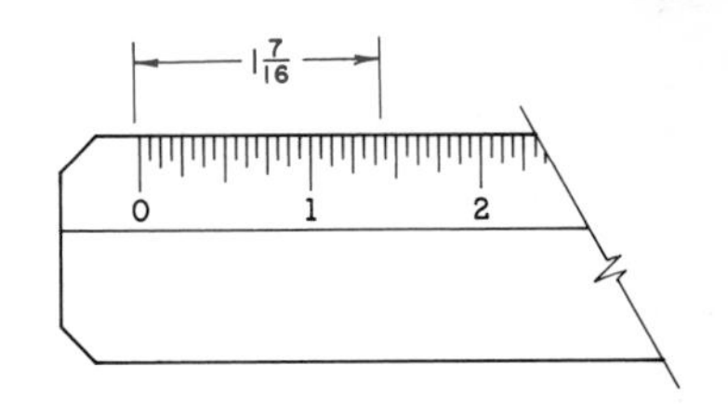

Fig. 3–38 **Making a measurement with the full-size scale.**

Figure 3–38 shows a distance of 1⁷⁄₁₆″ laid off on a scale.

DRAWING TO SCALE

If an object is large or has little detail, you can draw it in a *reduced* (smaller) *proportion.* In the customary system, the first reduction is to the scale of 6″ = 1′-0″, commonly called *half size.* You can use a full-size scale to draw a half size by letting each half inch stand for 1 in. and each 12 in. scale stand for a 24 in. scale. This is shown in Fig. 3–39, where 3⅝″ is laid off by three half inches and five-eighths of the next half inch. Always think full size. If you have a scale divided and marked for half size (Fig. 3–32), it would be easier to use.

If smaller views are needed, the next reduction that can be used is the scale of 3″ = 1′-0″, called *quarter size.* Find this scale on the architect's scale and look at it. The actual length of 3 in. represents 1 ft. divided into 12 parts. Each part stands for 1 in. and is further divided into eighths. Learn to think of the 12 parts as standing for real inches. Fig. 3–40 shows how to lay off the distance of 1′-3½″. Notice where the zero mark is. It is placed so that inches can be measured in one direction from it and feet in the other direction, as shown in the figure.

You can use a regular mechanical engineer's scale with a scale of ¼″ = 1″. For other reductions, the proportional scales listed in the sections on the architect's scale and the mechanical engineer's scale (see above) are used. For small parts, you can use enlarged scales, such as 24″ = 1′-0″ for double-size views. Very small parts can be drawn 4 or 8 times full size, or, for some purposes, 10, 20, or more times full size. The views of large parts and projects have to be drawn to a very small scale.

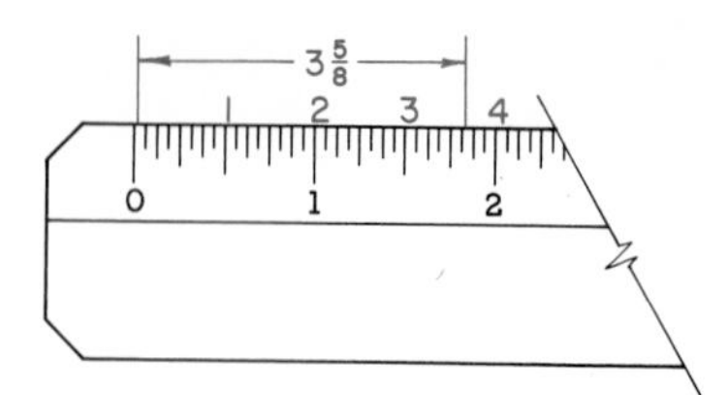

Fig. 3–39 **Measuring to half size on a full-size scale.**

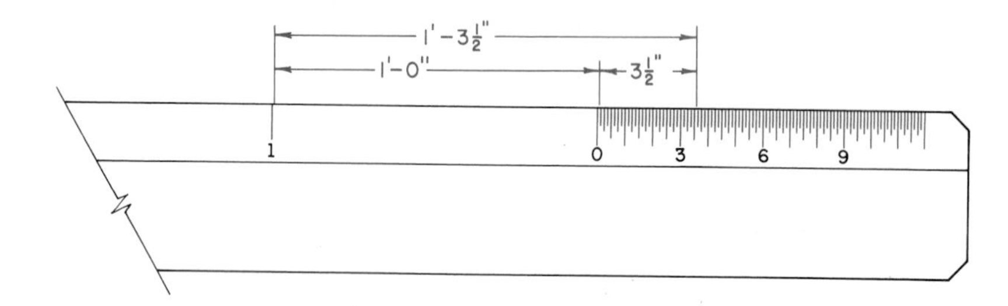

Fig. 3–40 **Reading the scale of 3″ = 1′-0″, called quarter size.**

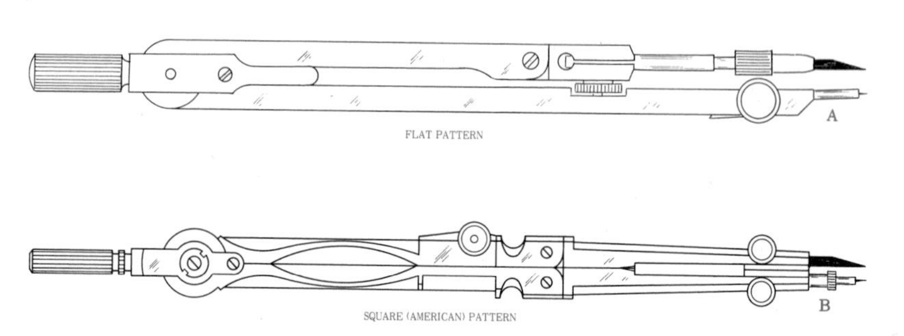

Fig. 3–41 **The two patterns of drawing instruments.**

When you draw to scale using the metric system, you can get reduced or enlarged proportions with the scales shown in Figs. 3–36 and 3–37.

CASE INSTRUMENTS

The two kinds of drafting instruments that are generally used are shown in Fig. 3–41. The flat pattern is at A and the square pattern at B. Some drafters prefer one pattern over the other. A set of instruments (Fig. 3–42) usually includes compasses with pen part, pencil part, lengthening bar, dividers, bow pen, bow pencil, bow dividers, and one or two ruling pens.

Large bow sets (Fig. 3–43) are favored by some drafters. They are known as *master,* or *giant,* bows and are made in several patterns. With large bows, 152 mm (6 in.) or longer, you can draw circles up to 330 mm (13 in.) in diameter or, with lengthening bars, up to 1016 mm (40 in.) in diameter. Large bow sets let you use one instrument in place of the regular compasses, dividers, and small bow instruments. Large bow instruments let you hold the *radius*

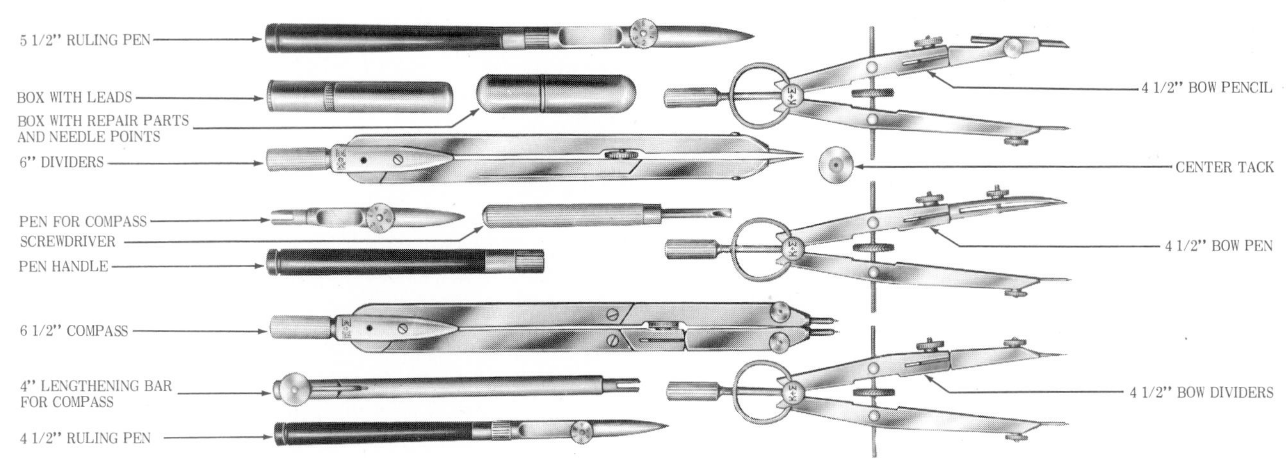

Fig. 3–42 **A three-bow set of drawing instruments.** ***(Keuffel & Esser Co.)***

(½ the diameter; plural, *radii*) securely at any distance you want, up to their largest possible radius.

THE DIVIDERS

Lines are divided and distances *transferred* (moved from one place to another) with dividers. You hold dividers in your right hand. They are adjusted as shown in Space A, Fig. 3–44.

To divide a line into three equal parts, adjust the points of the dividers until they seem to be about one-third the length of the line. Put one point on one end of the line and the other point on the line (Space B). Turn the dividers about the point that rests on the line, as in Space C. Then turn it in the alternate direction, as in Space D. If the last point falls short of the end of the line, increase the distance between the points of the dividers by an amount about one-third the distance *mn*. Then start at the beginning of the line again. You may have to do this a few times. If the last point overruns the end of the line, decrease the distance between the points by one-third the extra distance. For four, five, or more spaces, you follow the same rules, but you correct by one-fourth, one-fifth, and so forth, of the overrun or underrun. You divide an arc or the circumference of a circle in the same way.

THE COMPASS

Views on drawings are made up of straight lines and curved lines. Most of the curved lines are circles or parts of circles (arcs). You draw these with the compass (Fig. 3–45).

Fig. 3–43 **A large-bow set of drawing instruments.** ***(Keuffel & Esser Co.)***

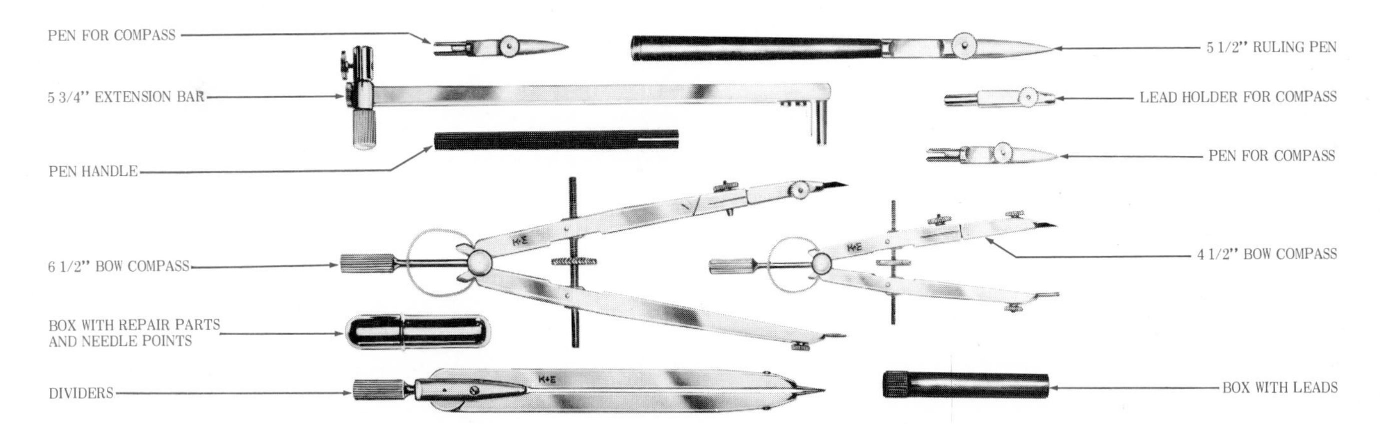

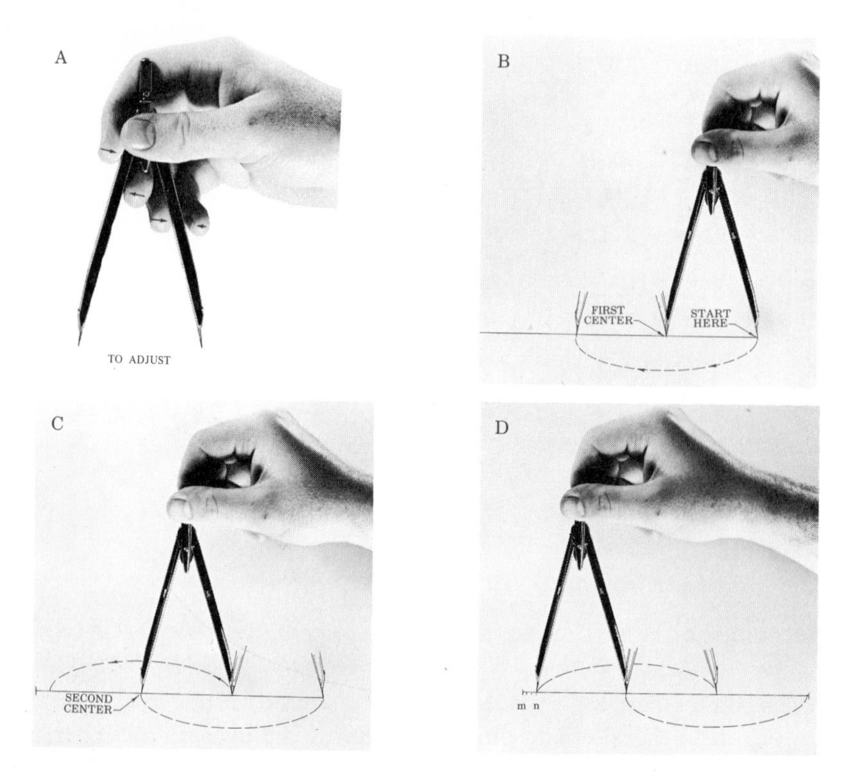

Fig. 3–44 The dividers are used to divide and transfer distances.

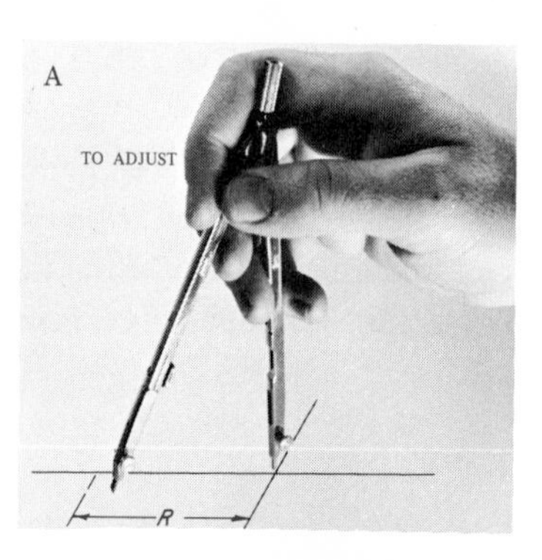

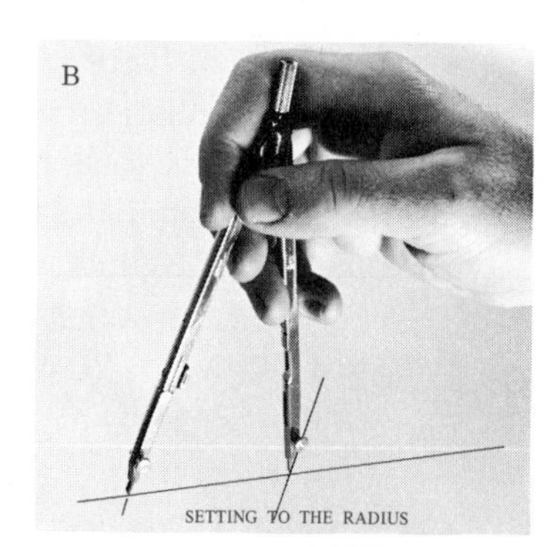

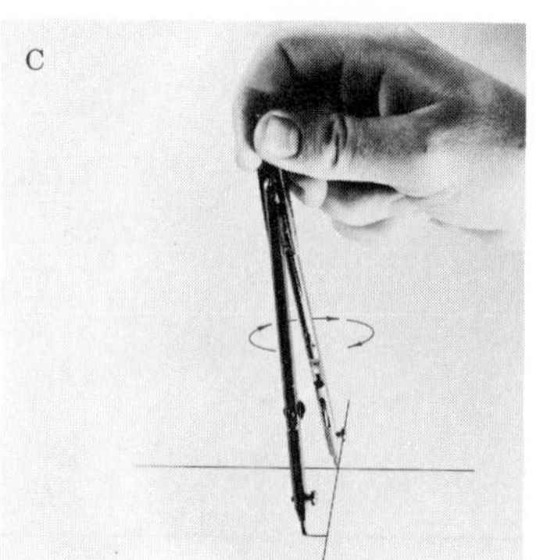

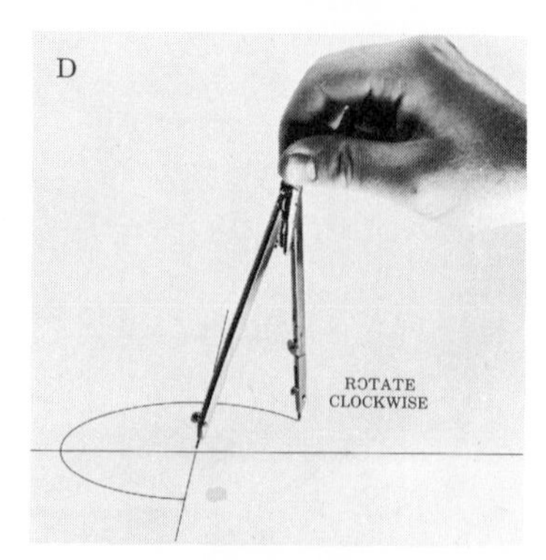

Fig. 3–45 The compass is used to draw circles and arcs.

You can leave the legs of the compass straight for radii under 50 mm (2 in.). For larger radii, you should make the legs perpendicular to the paper (Fig. 3–46). You can put in a lengthening bar (Fig. 3–47) when you need a very large radius, usually one over 200 mm (8 in.).

To get the compass ready for use, you should sharpen the lead as in Fig. 3–48, allowing it to extend about 10 mm (3/8 in.) Using a long *bevel* (slant) on the outside of the lead will keep the edge sharp when you increase the radius. Then adjust the shouldered end of the needle point until it extends a very little beyond the lead point, as shown in Fig. 3–48. You cannot use as much pressure on the lead in the compass as you can on a pencil. Therefore, you should use lead one or two degrees softer in the compass to get the same line *weight* (thickness and darkness).

When you are ready to use the compass, locate the center of the arc or circle you need by two *intersecting* (crossing) lines. Lay off the radius by a short, light dash (Fig. 3–45 in Space A). Open the compass by pinching it between your

Fig. 3–46 Adjusting the compass for large circles.

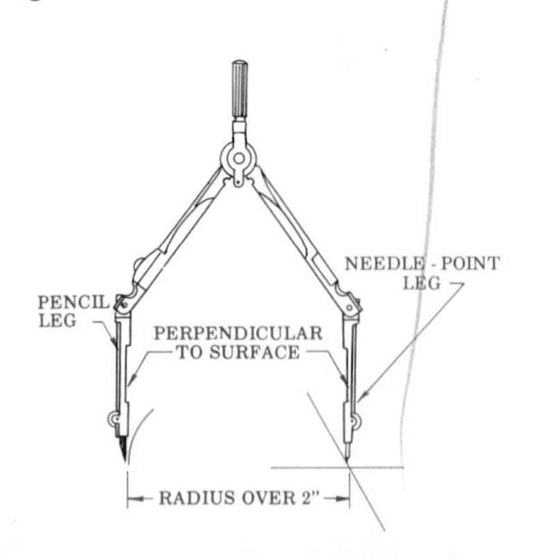

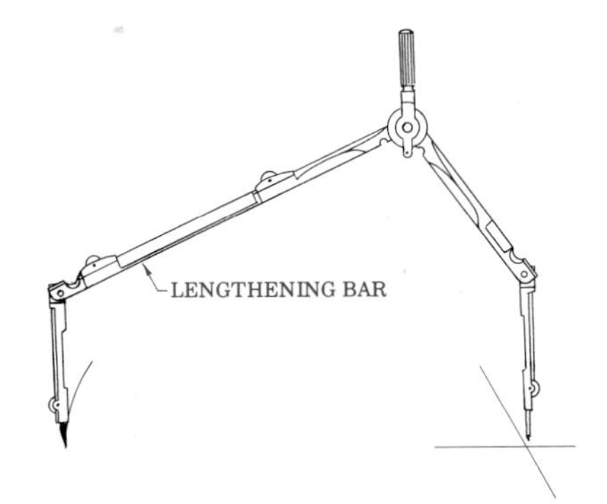

Fig. 3–47 **The lengthening bar is used in compasses for large radii.**

thumb and second finger (Space A). You set it to the right radius by putting the needle point at the center and moving the pencil leg with your first and second fingers (Space B). When the radius is set, raise your fingers to the handle (Space C). Turn the compass by twirling the handle between your thumb and finger. Start the arc near the lower side and turn clockwise (Space D). In doing this, slant the compass a little in the direction of the line. *Do not force* the needle point way into the paper. Use only enough pressure to hold the point in place. You can get long radii by using a lengthening bar in the compass to make the pencil leg longer. For extra-long radii, a beam compass is used (Fig. 3–53).

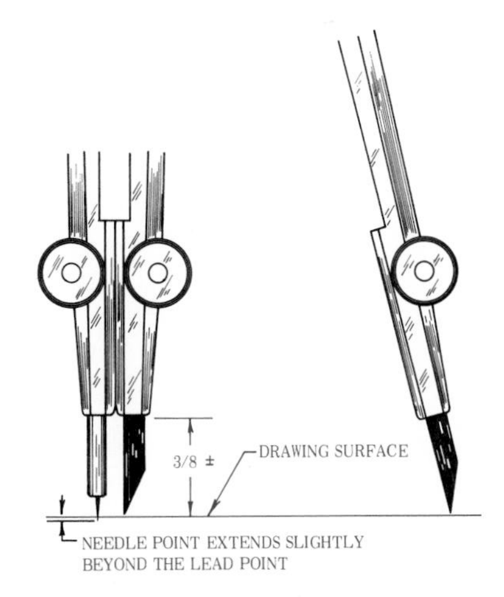

Fig. 3–48 **Adjusting the point and shaping the lead of the compass.**

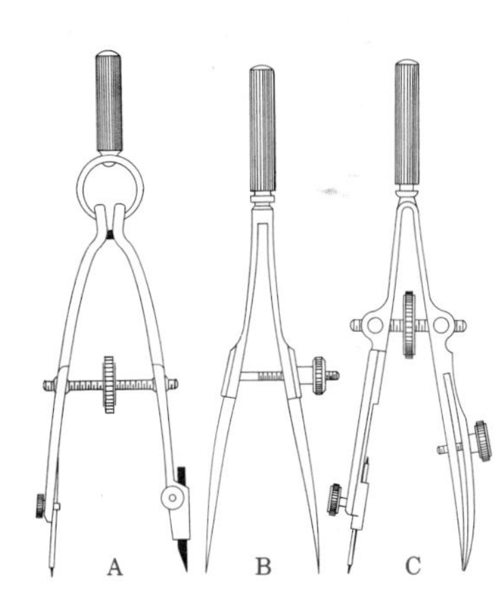

Fig. 3–49 **Bow instruments are used for drawing small circles and arcs.**

BOW INSTRUMENTS

A set of bow instruments (Fig. 3–49) is made up of the bow pencil (A), the bow dividers (B), and the bow pen (C). Any of them can be adjusted with either a center wheel, as at A and C, or a side wheel, as at B. They may be of the hook-spring type, as at A, or the fork-spring type, as at B and C. They are usually about 102 mm (4 in.) long.

You use the bow dividers for taking off (transferring) small distances. In addition, you use them for marking off a series of small distances. You also use them for dividing a line into small spaces. The bow pencil is used for drawing small circles. Whether you use instruments with center wheels or with side wheels is up to you. You sharpen and adjust the lead for the bow pencil as is shown at A in Fig. 3–50. The inside bevel holds an edge for small circles and arcs, as indicated at B. For larger radii, the outside bevel at C is better. Some drafters prefer a conical center point or an off-center point, as at D, E, and F. The bow instruments are easy to use and accurate for small distances or radii (less than 32 mm or 1¼ in.). They hold small distances better than the large instruments. You can make large adjust-

Fig. 3–50 **Adjusting the lead for the bow pencil.**

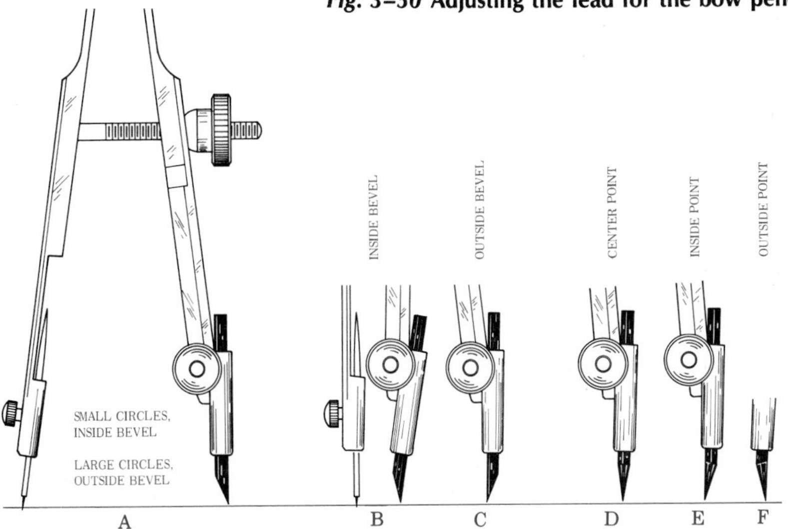

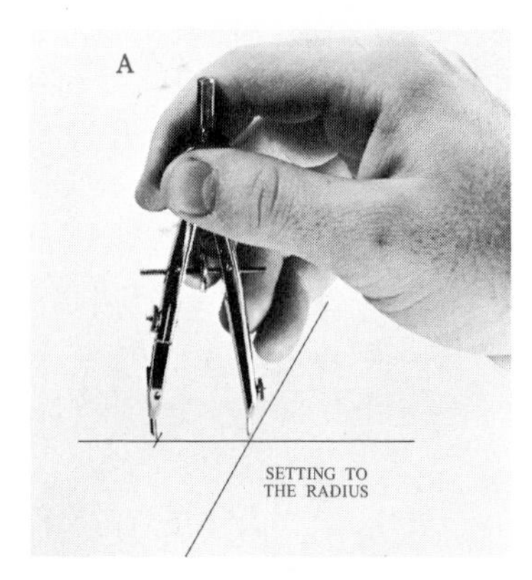

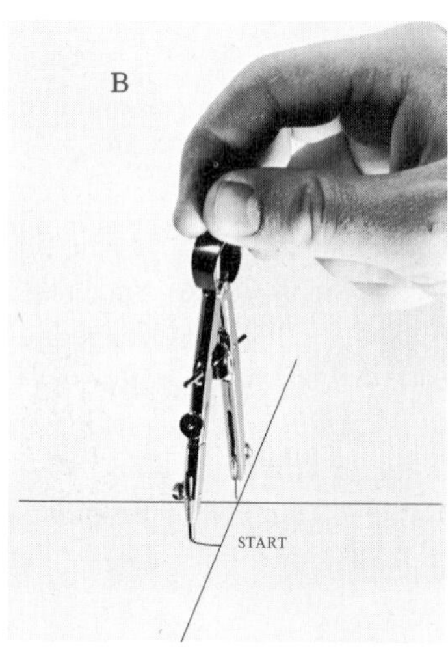

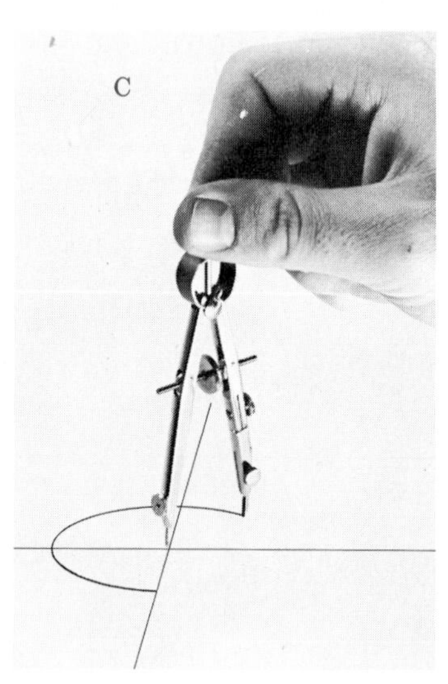

Fig. 3–51 Adjusting the radius for the bow pencil compass.

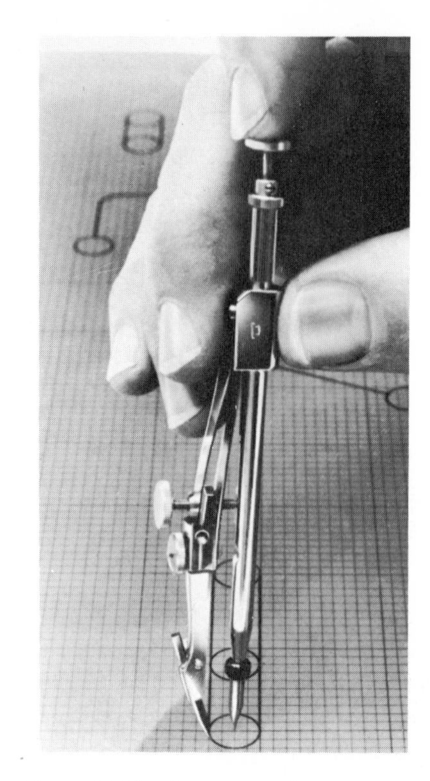

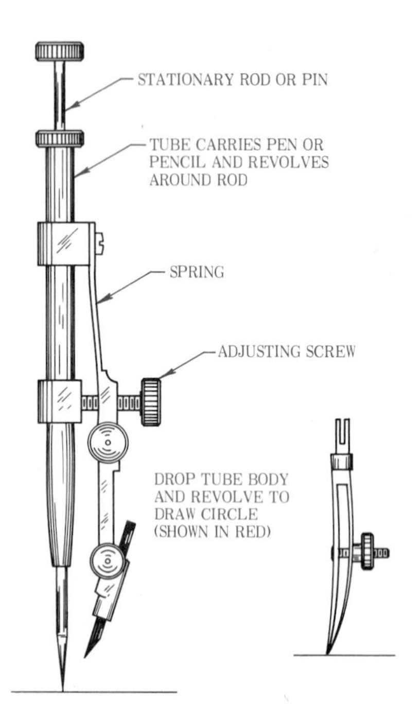

Fig. 3–52 **The drop-spring bow compass is used for drawing very small circles, especially where there are many to be drawn.** ***(Keuffel & Esser Co.)***

ments quickly with the side-wheel bows by pressing the fork and spinning the adjusting nut. Some center-wheel bows are built for making large, rapid adjustments. This is done by holding one leg in each hand and either pushing to close or pulling to open. You make small adjustments with the adjusting nut on both the side-wheel and the center-wheel bows. You use the bow pencil (Fig. 3–51) with one hand. Set the radius as in Space A. Start the circle near the lower part of the vertical center line (Space B). Turn clockwise, as in Space C.

DROP–SPRING BOW COMPASS

The drop-spring bow compass (Fig. 3–52) is good for drawing very small circles. It is very useful when you have to draw many small circles of the same size, such as when you are drawing rivets. You attach the marking point (pencil or pen) to a tube that slides on a pin. When you use the bow, the pin stays still while the pencil point turns around it. You set the radius with the spring screw. You hold the marking point up while you are putting the pin on the center. Then you drop it and turn it. The circles you draw will all be the size you set.

BEAM COMPASS

You use beam compasses (Fig. 3–53) to draw arcs or circles with large radii. The beam compass is made up of a bar (beam) on which movable holders for a pencil part and a needle part can be put and

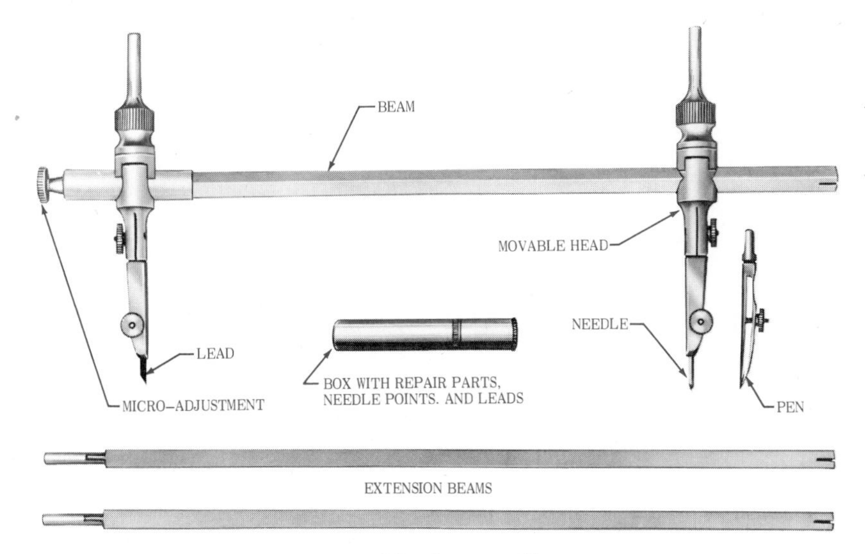

Fig. 3–53 **Beam compasses are used for large radii.**

Fig. 3–54 **Some irregular, or French, curves. They are made in a great variety of forms.** ***(Teledyne Post.)***

fixed as far apart as you want. You can also use a pen part in place of the pencil point. If you put a needle point in place of the pencil point you can use a beam compass as a divider or to set off long distances. The usual bar is about 330 mm (13 in.) long. However, by using a coupling to add extra length, you can draw circles of almost any size you want.

IRREGULAR CURVES

You use irregular, or French, curves (Fig. 3–54) to draw many noncircular curves (involutes, spirals, ellipses, and so forth). Irregular curves are also used for drawing curves on graphic charts. In addition, they can be used to plot motions and forces and to make some engineering and scientific graphs. Irregular curves are made of sheet plastic. They come in many different forms, some of which are shown. Sets are made for ellipses, parabolas, hyperbolas, and many special purposes.

To use an irregular curve (Fig. 3–55), find the points through which a curved line is to pass. Then set the path of the curve by drawing a light line, freehand, through the points. Adjust it as needed to make the curve smooth. Next,

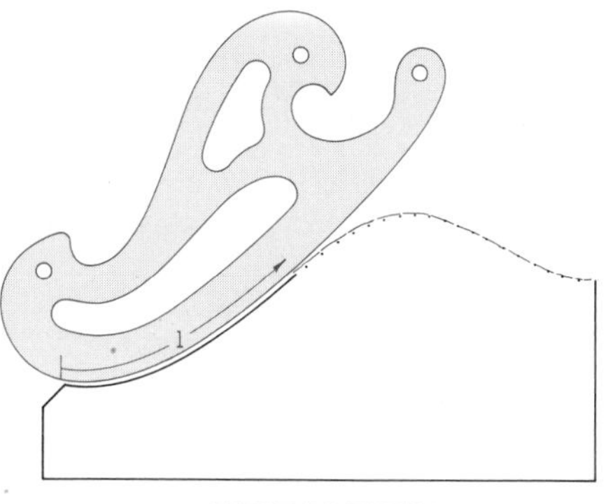

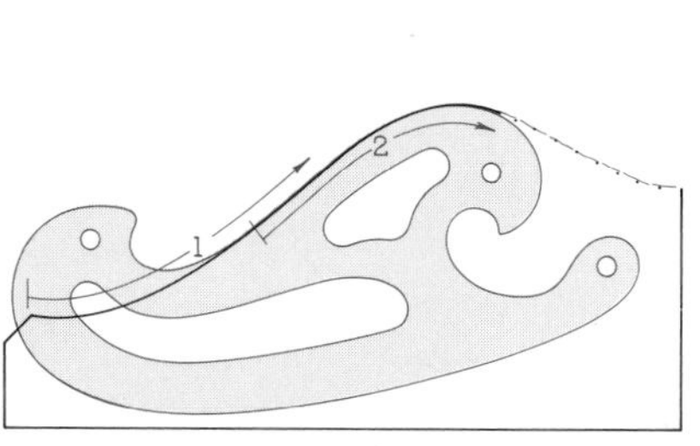

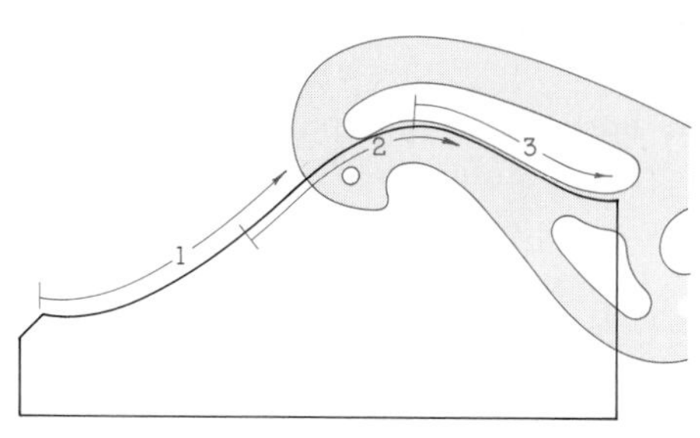

Fig. 3–55 **Steps in drawing a smooth curve.**

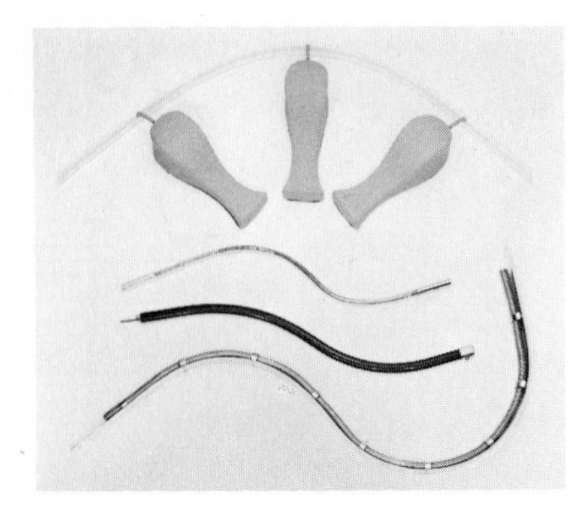

Fig. 3–56 **Flexible curves for plotting smooth curves.** ***(AM Bruning International.)***

match the irregular curve against a part of the curved line and draw part of the line. Move the irregular curve to match the next part, and so on. Each new position should fit enough of the part just drawn to make the line smooth. You must note whether the radius of the curved line is increasing or decreasing and place the irregular curve in the same way. Do not try to draw too much of the curve with one position. If the curved line is *symmetrical* (even) around an *axis* (center point), you can mark the position of the axis on the irregular curve with a pencil for one side. You then turn the irregular curve around to match and draw the other side. There are adjustable or flexible curves that you can use for certain special kinds of work (Fig. 3–56).

TEMPLATES

Templates (Fig. 3–57) are an important part of the equipment of engineers and professional drafters. They save a great deal of time in drawing shapes of details. These include bolt heads, nuts, and electrical, architectural, and plumbing symbols.

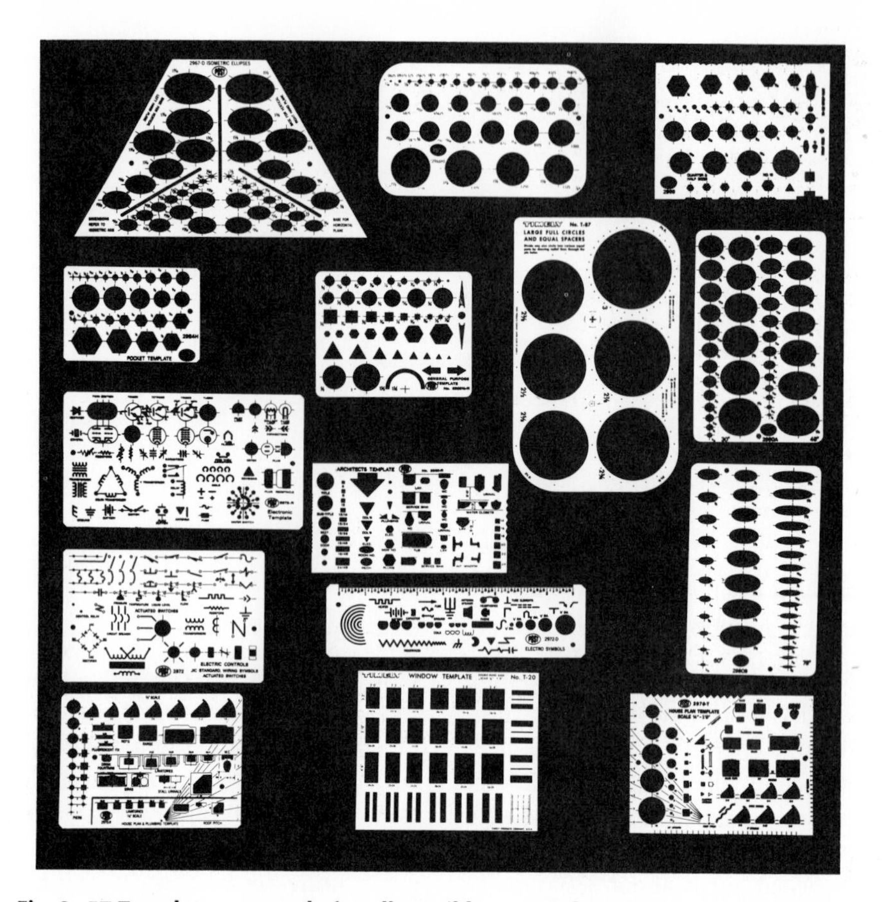

Fig. 3–57 **Templates are made for all possible uses and save a good deal of time.** ***(Teledyne Post.)***

Review

1. Explain how you can check the accuracy of a T-square.

2. Name three kinds of drawing sheets.

3. Drafting pencils are made in 17 degrees of hardness from ________ (softest and blackest) to ________ (hardest).

4. The shape of a drafting pencil point may be ________ or ________.

5. How many widths, or thicknesses, of lines are generally used in drafting?

6. You can draw angles in ________-degree intervals by using the 45° and 30°–60° triangles with the T-square.

7. Name three shapes of scales (measuring instruments) used in drafting.

8. What is the common advantage in using metric scales?

9. What is the difference between an A0 and an A4 size sheet?

10. What metric scale would you use for an architectural floor plan?

Problems

Fig. 3–58 Make a drawing of the template shown in Fig. 3–58. In this and several of the one-view drawings, the order of working is shown in progressive steps. These steps should be followed carefully, because they represent the drafter's procedure in making drawings. You should not consider the explanations as applying only to the particular problem. Instead, try to understand the system and apply it to all drawings.

1. On an 11″ × 17″ drawing sheet, lay out a working space 15″ wide and 10 1/2″ high.
2. Measure 3 1/8″ from left border line, and from this mark measure 8 3/4″ toward the right.
3. Lay the scale on the paper vertically near (or on) the left edge, make a mark 2 1/2″ up, and from this measure 5 1/2″ more. The sheet will appear as in Space 1.
4. Draw horizontal lines 1 and 2 with the T-square, and vertical lines 3 and 4 with T-square and triangle (Space 2).
5. Lay the scale along the bottom line of the figure with the measuring edge on the upper side and make marks 1 3/4″ apart. Then with the scale on line 3, with its measuring edge to the left, measure from the bottom line two vertical distances, 2 1/2″ and 1 1/2″ (Space 3).
6. Through the two marks draw light horizontal lines.
7. Draw the vertical lines with T-square and triangle by setting the pencil on the marks on the bottom line and starting and stopping the lines on the proper horizontal lines (Space 4).
8. Erase the lines not wanted and darken the lines of the figure to get the finished drawing (Space 5).

Fig. 3–58 Template.

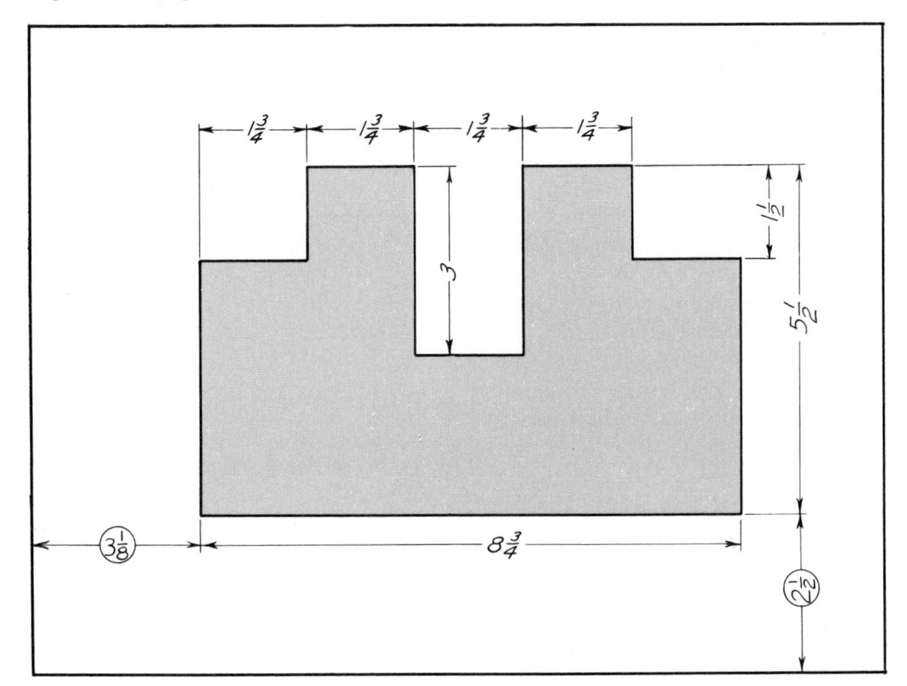

1

2

1

2

3

4

3

4

5

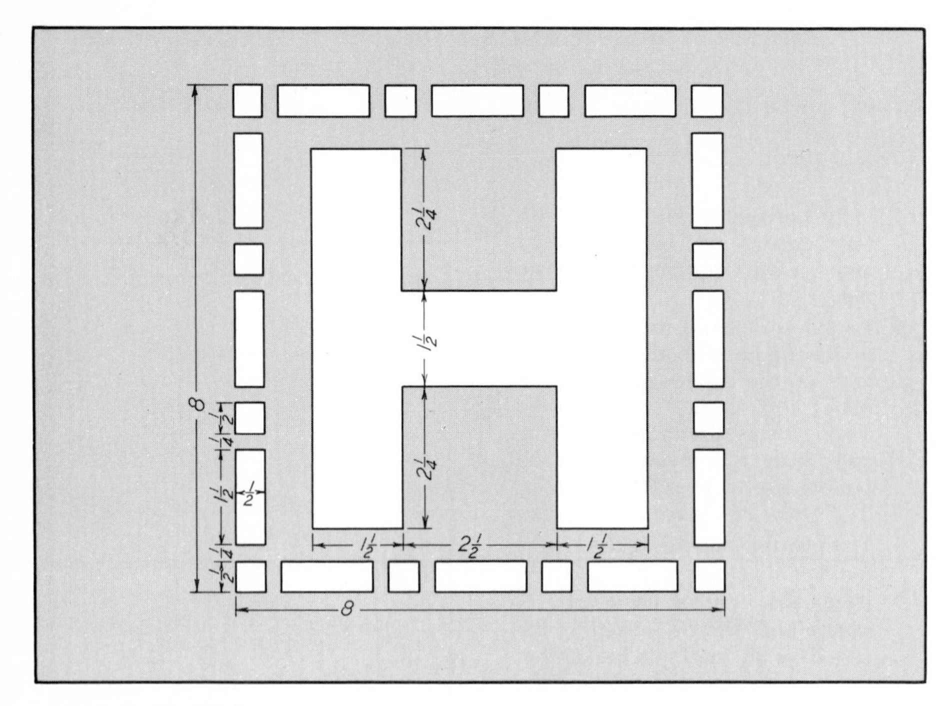

Fig. 3–59 Stencil.

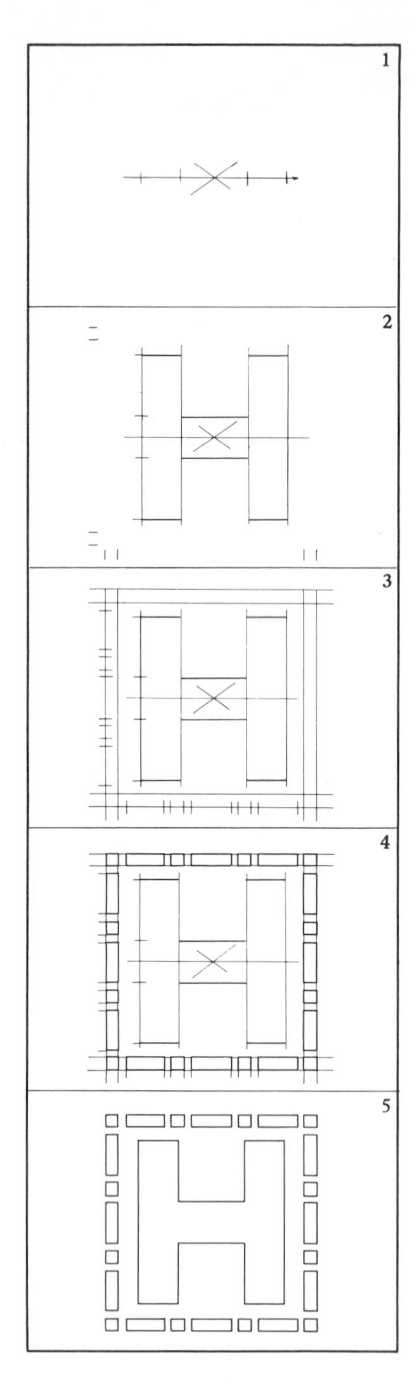

Fig. 3–59 Order of working for drawing the stencil.

Fig. 3–59 Make a drawing of the stencil shown in Fig. 3–59. This drawing gives practice in accurate measuring with the scale and making neat corners with short lines. The construction shown in the order of working should be drawn very lightly with a well-sharpened 3H or 4H pencil.

1. On an 11″ × 17″ drawing sheet, lay out a working space 15″ wide and 10½″ high. Find the center of this space by laying the T-square blade face down across the opposite corners and drawing short lines where the diagonals intersect (Space 1).
2. Through the center draw a horizontal centerline and on it measure and mark off points for the four vertical lines. The drawing will appear as in Space 1.
3. Draw the vertical lines lightly with T-square and triangle. On the first vertical line, at the extreme left, measure and mark off points for all horizontal lines. The drawing will now appear as it does in Space 2.
4. Draw the horizontal lines as finished lines. Measure points for the stencil border lines on the left side and bottom. The drawing will now appear as in Space 3.
5. Draw the border lines. On the lower and left-hand border lines, measure the points for the ties.
6. Complete the border by drawing the cross lines as finished lines and darkening the other lines as in Space 4.
7. Darken the vertical lines and finish as in Space 5.
8. Fill in the title block.

Figs. 3–60 through 3–71 **ASSIGNMENT: Make an instrument drawing of the figure assigned by your instructor. NOTE: Some dimensions have been deleted intentionally or placed incorrectly for instructional purposes. Include all centerlines. Do not add dimensions unless instructed to do so.**

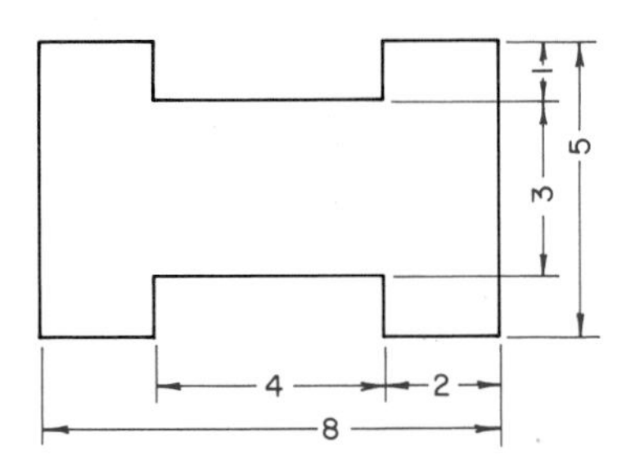

Fig. 3–60 **Sheet-metal pattern.**

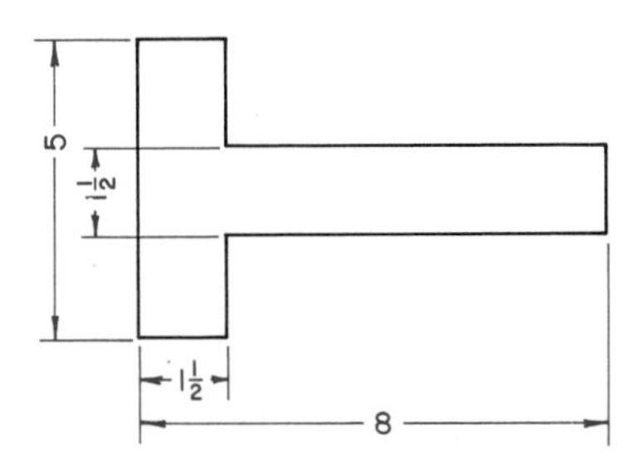

Fig. 3–61 **Template for letter T.**

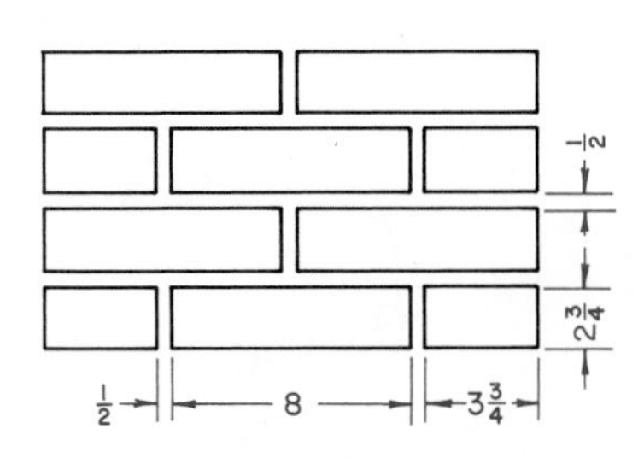

Fig. 3–62 **Brick pattern.**

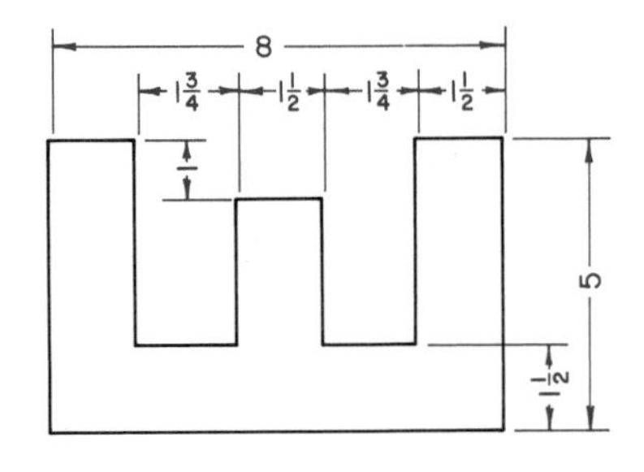

Fig. 3–63 **Template for letter E.**

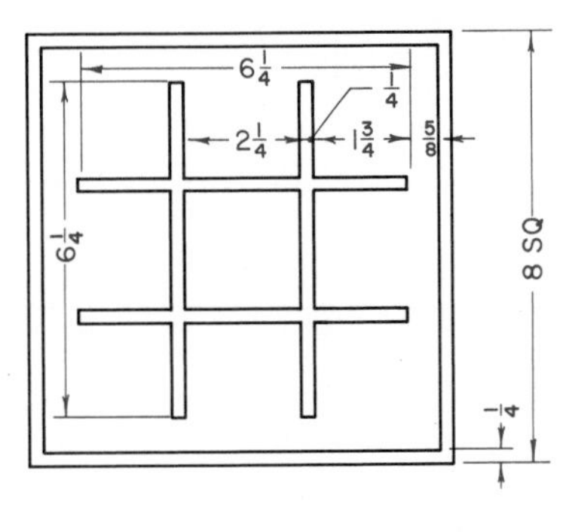

Fig. 3–64 **Tic-tac-toe board**

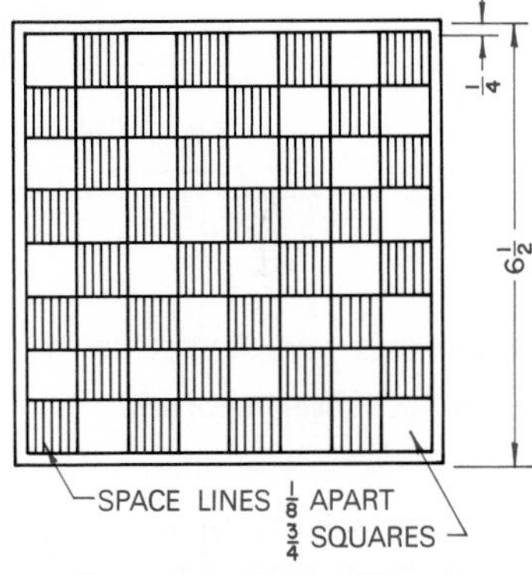

Fig. 3–65 **Checker board.**

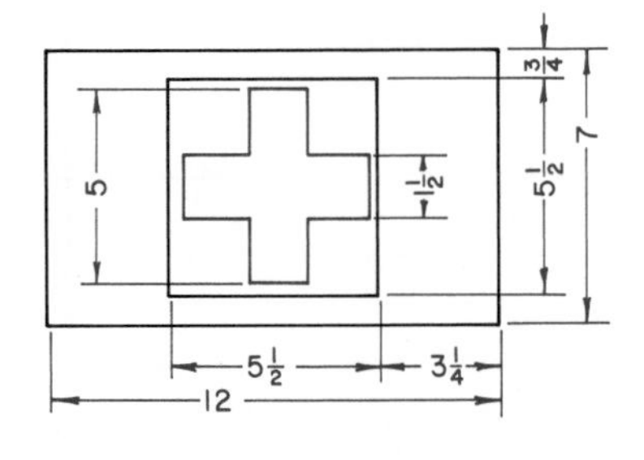

Fig. 3–66 **First-aid station sign.**

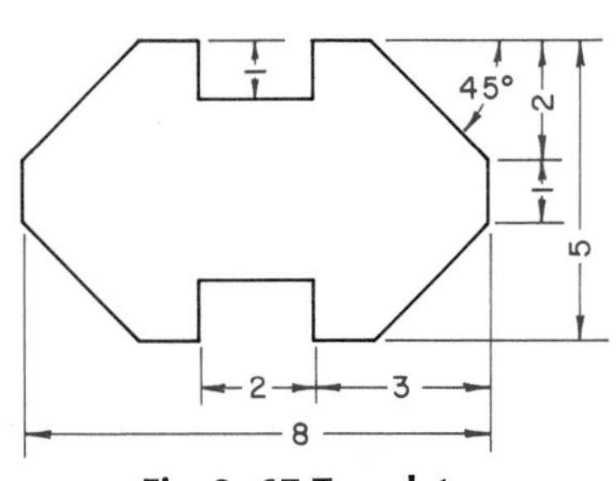

Fig. 3–67 **Template.**

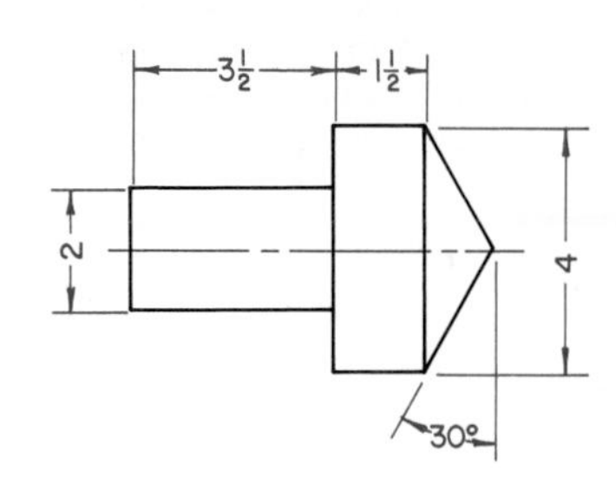

Fig. 3–68 **Pivot.**

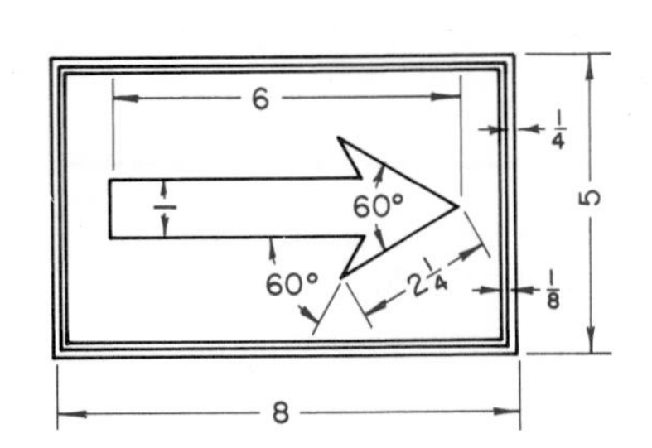

Fig. 3–69 **Direction sign.**

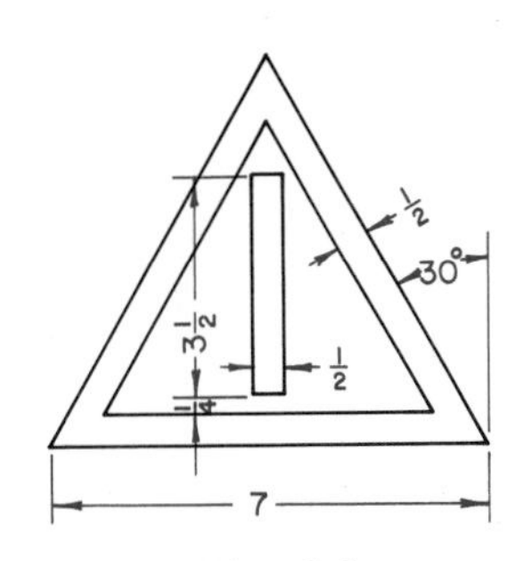

Fig. 3–70 **International danger road sign.**

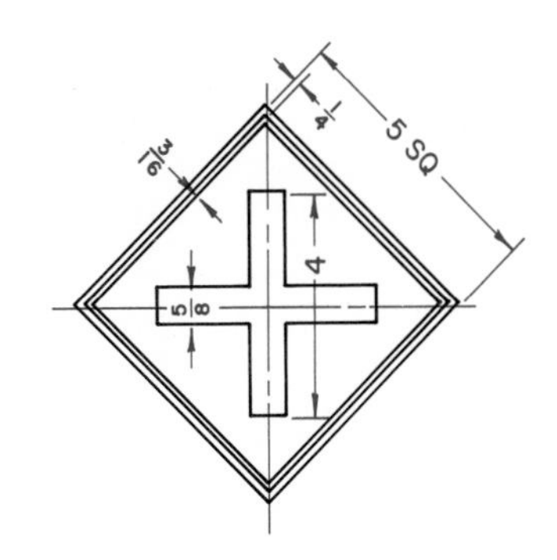

Fig. 3–71 **Highway warning sign.**

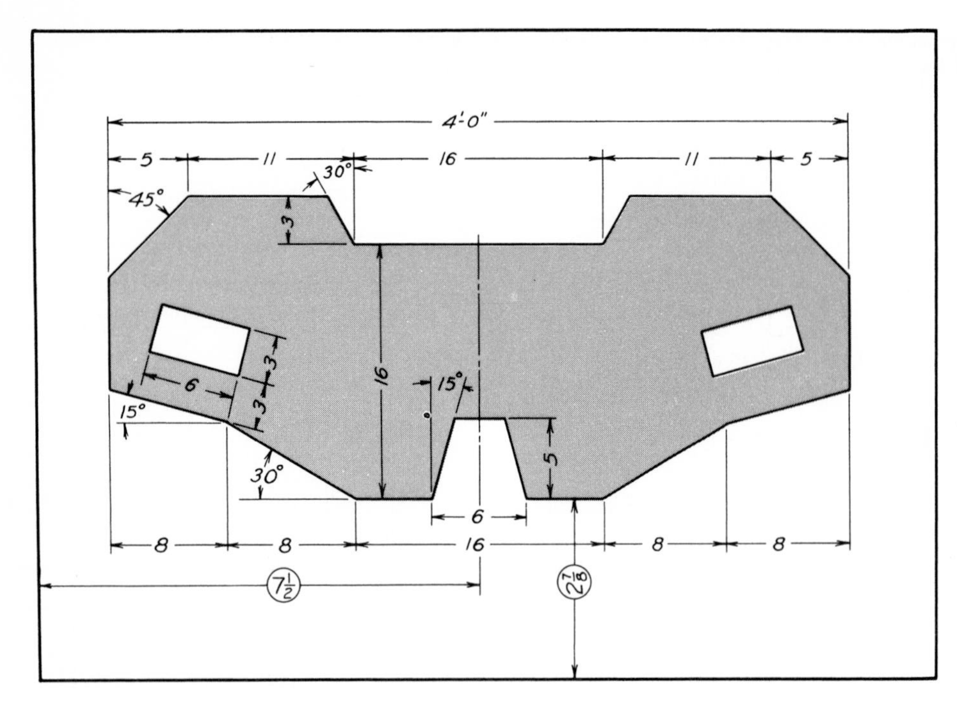

Fig. 3–72 **Shearing blank.**

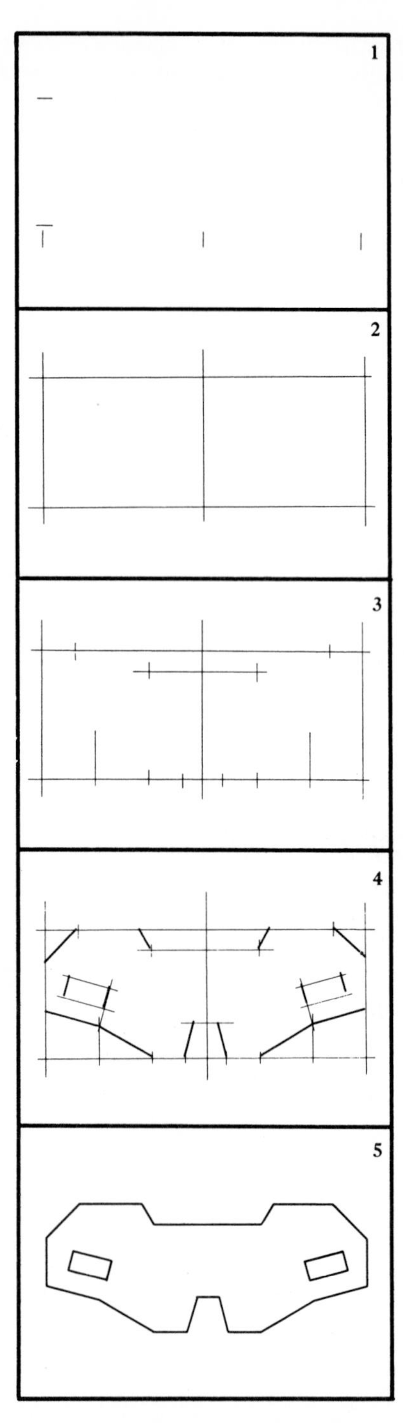

Fig. 3–72 **Order of working for drawing the shearing blank.**

Fig. 3–72 **Make a drawing of the shearing blank shown in Fig. 3–72. When a view has inclined lines, it should first be blocked in with square corners. Angles of 15°, 30°, 45°, 60°, and 75° are drawn with the triangles after locating one end of the line.**

1. **Locate vertical centerline and measure 2′ on each side (Space 1). Note that this drawing must be made to the scale of 3″ = 1′.**
2. **Locate vertical distances for top and bottom lines.**
3. **Draw main blocking-in lines as in Space 2.**
4. **Make measurements for starting points of inclined lines (Space 3).**
5. **Draw inclined lines with T-square and triangles (Space 4).**
6. **Finish as in Space 5.**
7. **Fill in the title block.**

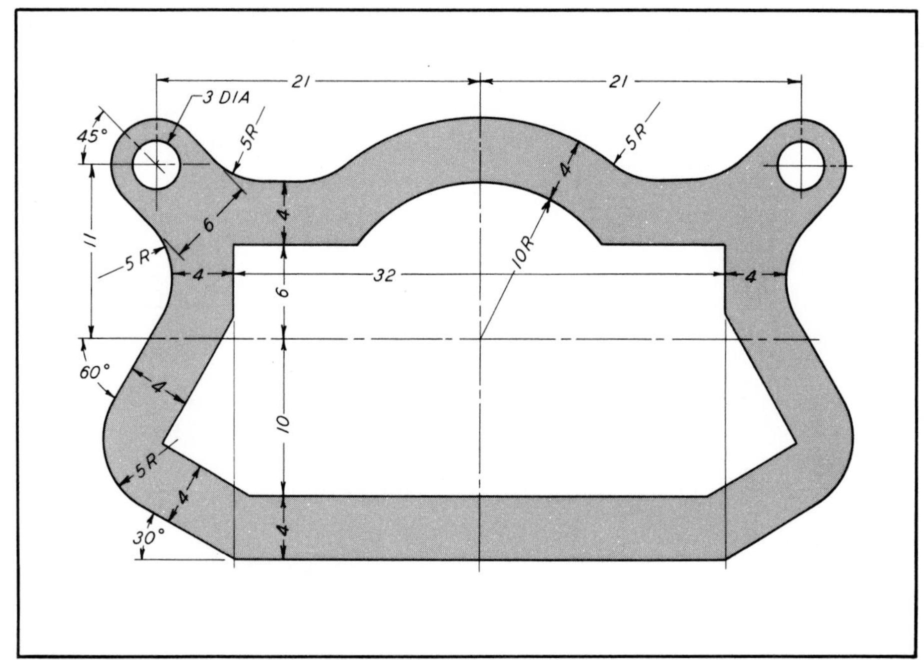

Fig. 3–73 Cushioning base.

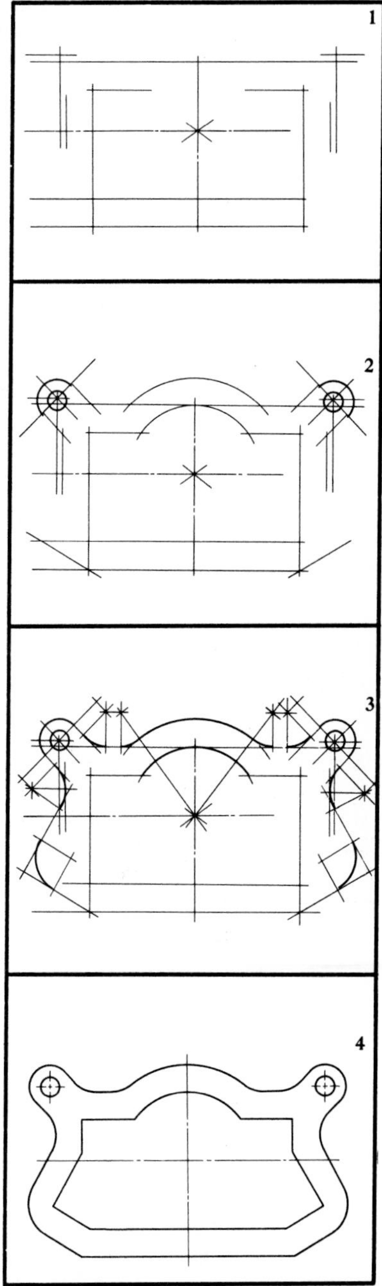

Fig. 3–73 Order of working for drawing the cushioning base.

Fig. 3–73 Make a drawing of the cushioning base (Fig. 3–73). The drawing is for practice with triangles, compasses, and scale. Centers of arcs and tangent points must be carefully located.

Order of working for drawing the cushioning base in Fig. 3–73:

1. Through the center of the working space, draw horizontal and vertical centerlines. Measure horizontal and vertical distances. This drawing must be made to the scale of 3″ = 1′. Then draw horizontal and vertical lines (Space 1).
2. Draw inclined lines with 45° and 30°-60° triangles. Then draw large arcs and two semicircles with tangents at 45° (Space 2).
3. Locate centers and tangent points for the 5″ radius tangent arcs. To do this, measure 5″ perpendicularly from each tangent line and draw lines parallel to the tangent lines. The intersection of these lines will be the required centers. To find the points of tangency, draw lines from the centers perpendicular to the tangent lines. To find the centers for the 5″ arcs tangent to the middle arc, proceed as follows: Increase the radius of the larger arc by 5″ and draw two short arcs cutting lines parallel to and 5″ above the top horizontal tangent line. These points will be the centers. Lines joining these centers with the center of the large arc will locate the points of tangency of the arcs (Space 3).

   Draw all the 5″ tangent arcs above the horizontal centerline. Locate points of tangency and draw the two 60° tangent lines. Locate centers and draw the 5″ tangent arcs below the centerline.
4. Complete the view by drawing the lines for the opening. Darken the lines and finish the drawing as shown in Fig. 3–73 (Space 4).

Figs. 3–74 through 3–85 **ASSIGNMENT: Make an instrument drawing of the figure assigned by your instructor. NOTE: Some dimensions have been deleted intentionally or placed incorrectly for instructional purposes. Include all centerlines. Do not add dimensions unless instructed to do so.**

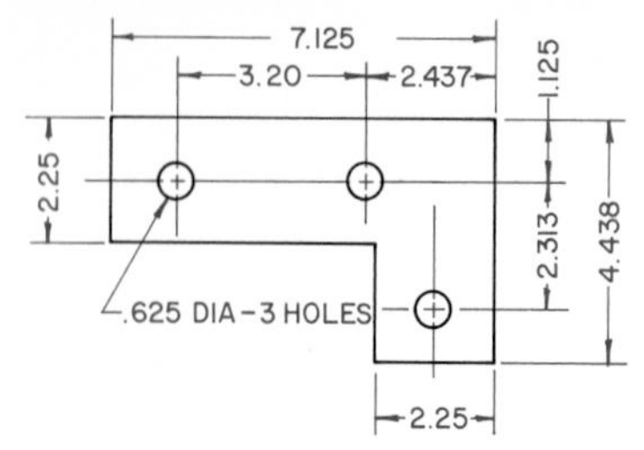

Fig. 3–74 **Angle bracket.**

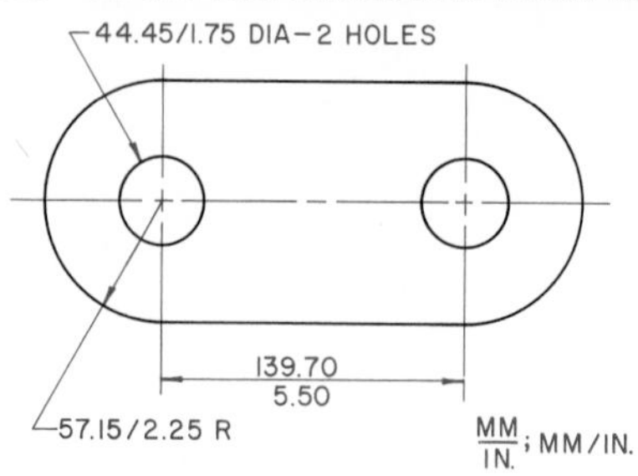

Fig. 3–75 **Gasket.**

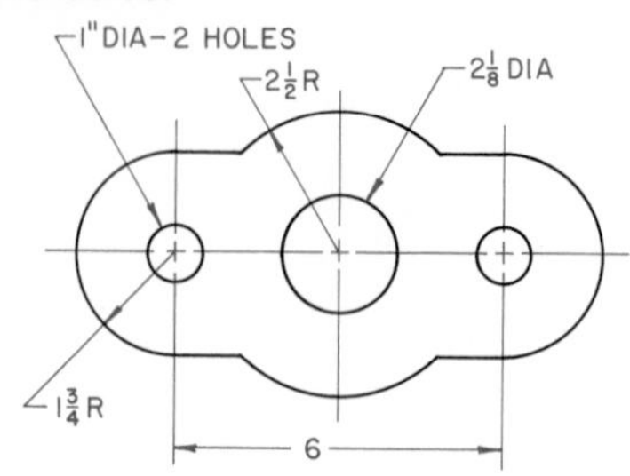

Fig. 3–76 **Armature support.**

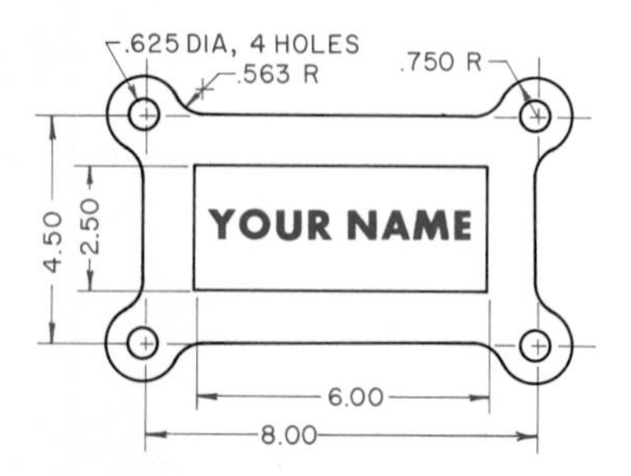

Fig. 3–77 **Identification plate.**

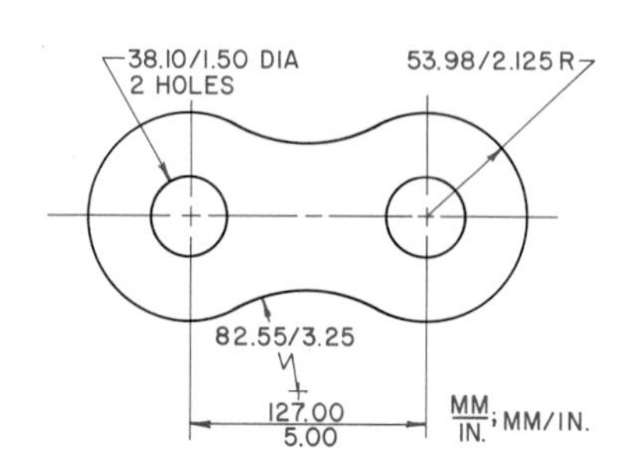

Fig. 3–78 **Bicycle chain link.**

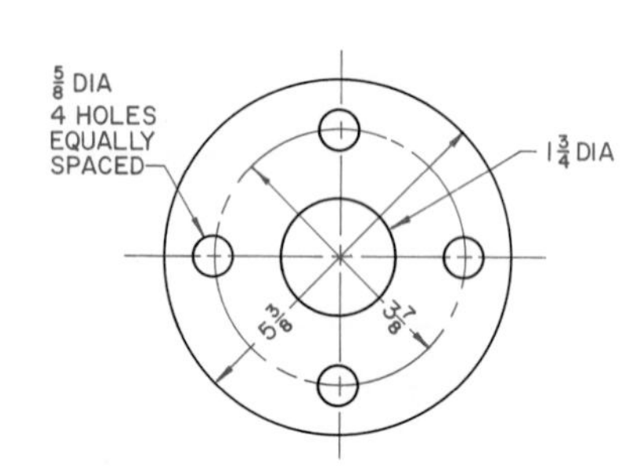

Fig. 3–79 **Round gasket.**

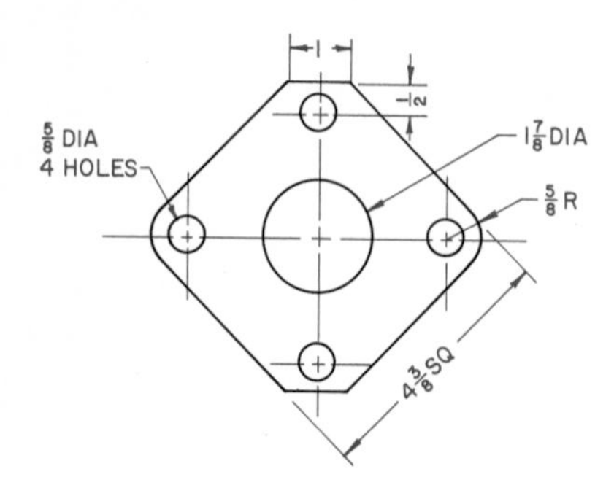

Fig. 3–80 **Bronze shim.**

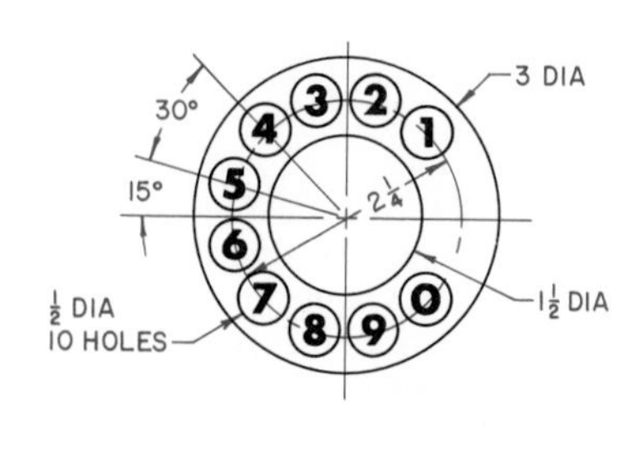

Fig. 3–81 **Telephone dial.**

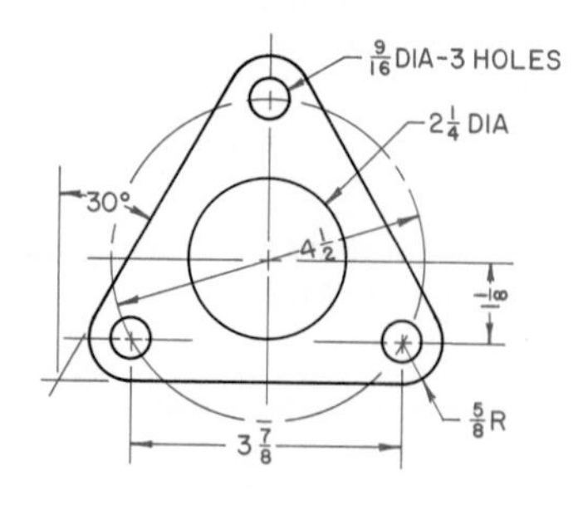

Fig. 3–82 **Base plate.**

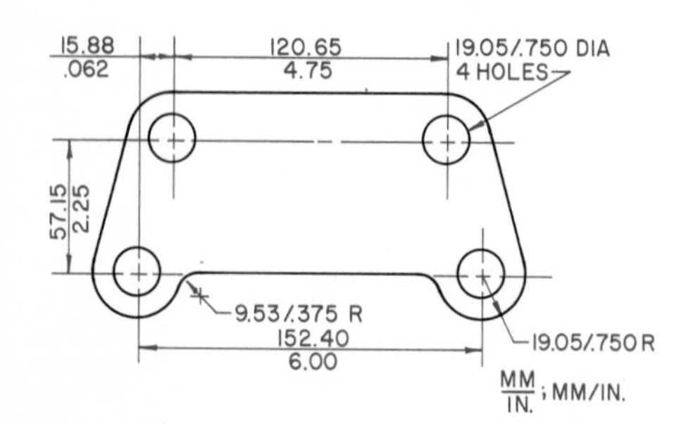

Fig. 3–83 **Cover plate.**

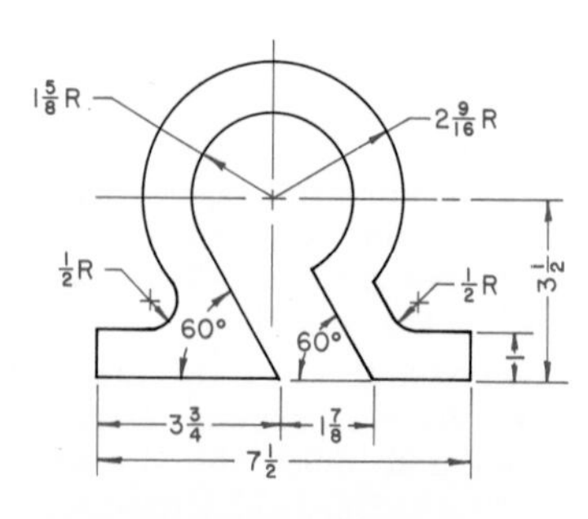

Fig. 3–84 **Housing.**

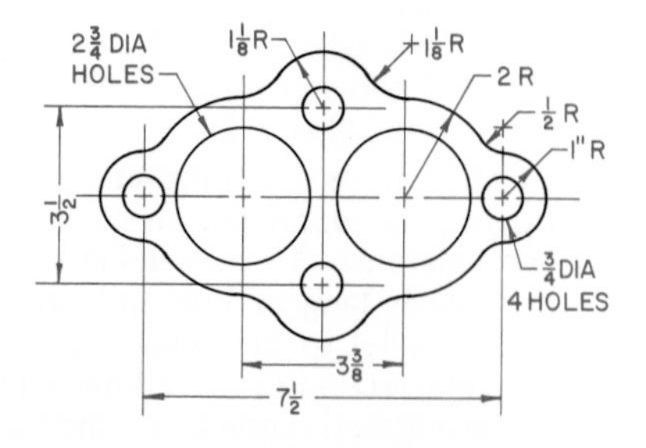

Fig. 3–85 **Carburetor gasket.**

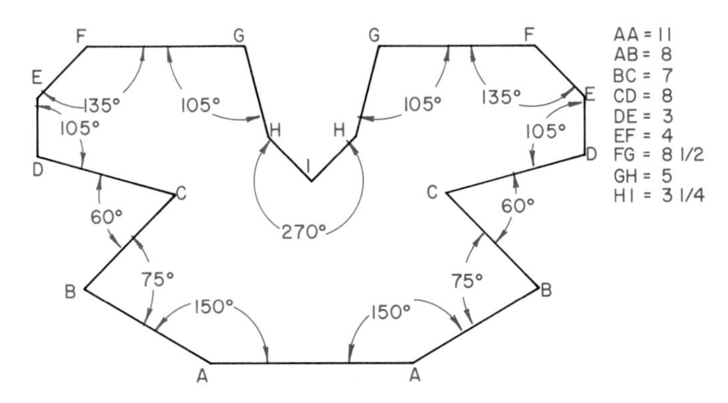

Fig. 3–86 **Irregular polygon. Construct as shown. Use a scale of 1/2″ = 1″. Begin by drawing line *AA* near the bottom of the sheet and centered horizontally. The length of each line is given at the right of the figure above. All angles may be drawn with the T-square and a combination of triangles.**

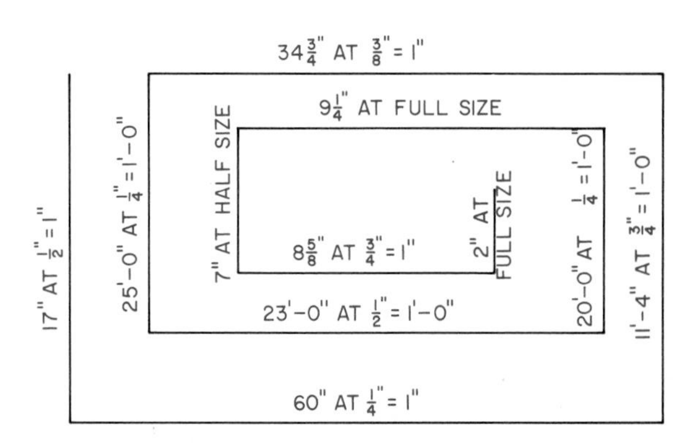

Fig. 3–87 **Scale drawing. Draw a figure similar to the one shown above. Draw lines to the lengths and at the scales indicated.**

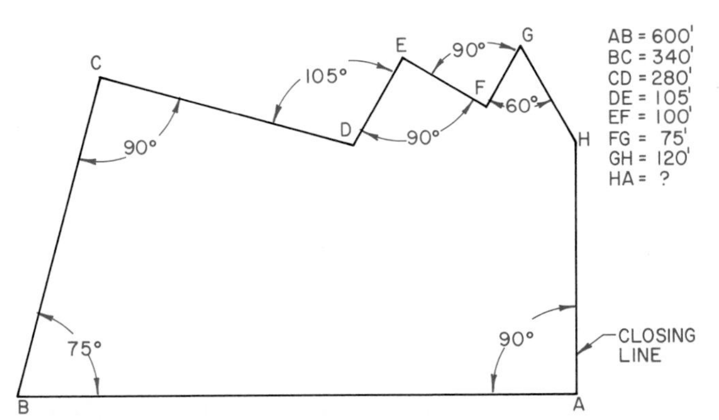

Fig. 3–88 **Civil engineer's scale. Draw a figure similar to the one shown above. Use a scale of 1″ = 40′-0″. Measure the length of the closing line to the nearest tenth of a foot and note it on your drawing.**

Fig. 3–89 **Measuring practice. Measure the lengths of lines *A* through *N* at full-size, 3/4″ = 1″, 1/2″ = 1″, 1″ = 40′-0″, 1″ = 1′-0″, etc., as assigned by your instructor.**

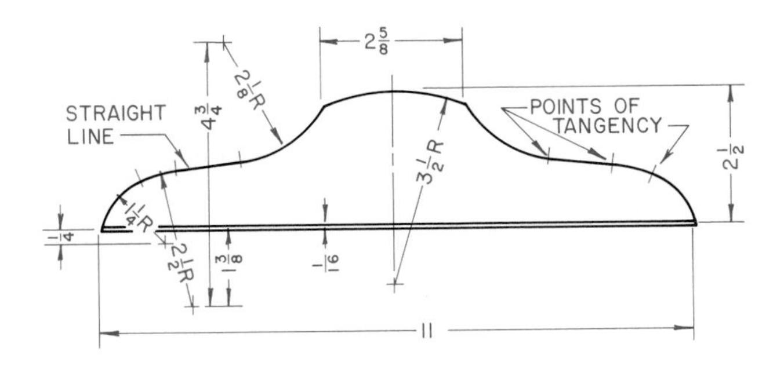

Fig. 3–90 **T-square head. Draw to a scale of 3/4″ = 1″. Be sure to locate points of tangency.**

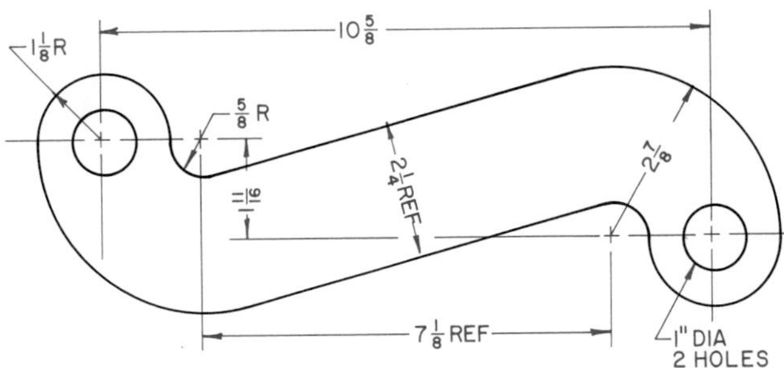

Fig. 3–91 **Offset bracket. Locate all center points before beginning to draw circles and arcs.**

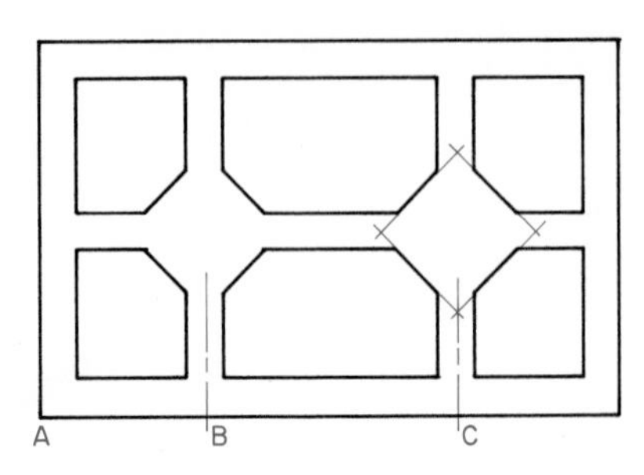

Fig. 3–92 **Grill plate. Scale: full size or as assigned. Make all ribs 1/2″ wide. The distance *A B* is 2 1/4″; *B C* is 3 1/2″. The diamond shapes are 1 1/2″ square.**

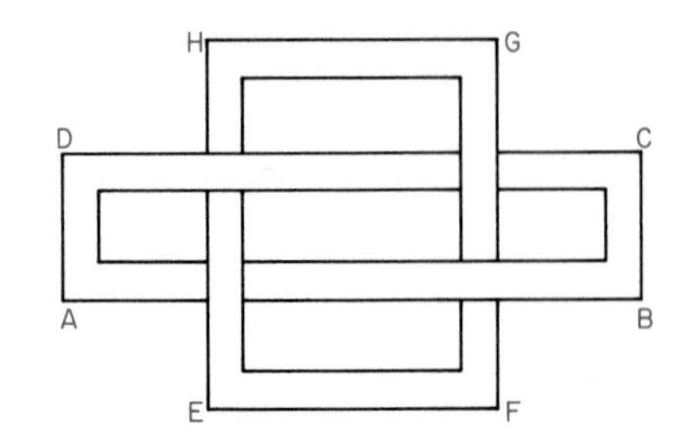

Fig. 3–93 **Inlay. Scale: full size or as assigned. Rectangle *EFGH* is 8″ wide × 10″ high. All ribs are 1″ wide. Each rectangle is centered on the other.**

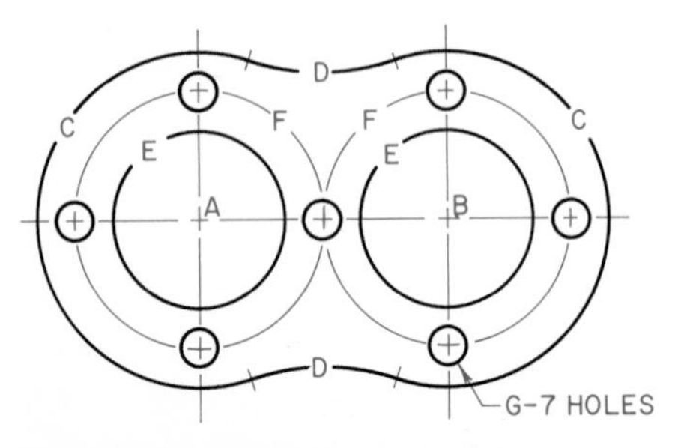

Fig. 3–94 **Head gasket. Scale: as assigned. Use metric scale. The distance between center points *A* and *B* is 45.50 mm. Radius of arc *C* is 30 mm. Radius of arc *D* is 43 mm. Diameter of hole *E* is 32 mm. Diameter of circle *F* is 45.50 mm. Holes labeled *G* are 7 mm DIA.**

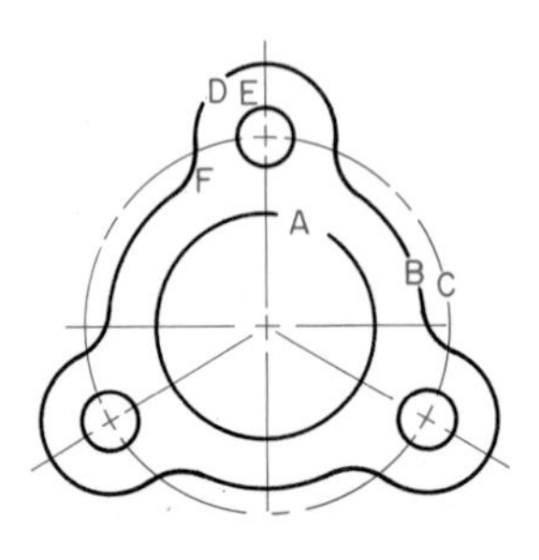

Fig. 3–95 **Gasket. Scale: full size (decimal inches). Hole *A* = 3.125″. Arc *B* = 2.250″R. Hole circle *C* = 5.250″. Arc *D* = 1.00″R. Hole *E* = 0.750″ DIA, 3 holes. Arc *F* = 0.750″R. Locate all tangent points before darkening lines.**

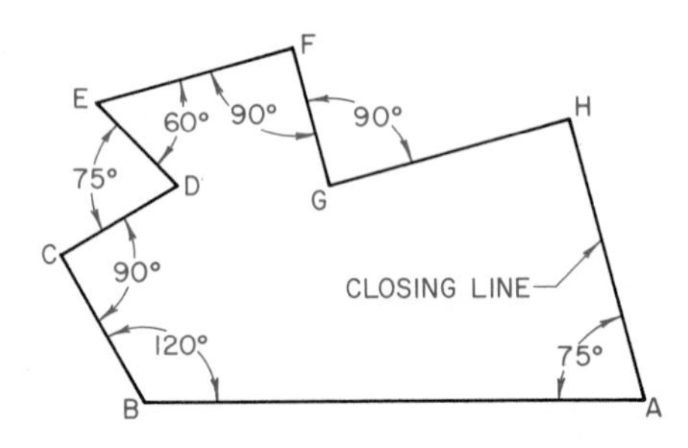

Fig. 3–96 **Metric measurement. Scale: as assigned. Draw horizontal line *AB* 180 mm long. Work clockwise around the layout. *Remember!* Angular dimensions are the same in both the customary and metric systems. *BC* = 60 mm. *CD* = 48 mm. *DE* = 42 mm. *EF* = 74 mm. *FG* = 50 mm. *GH* = 90 mm. Measure the closing line and measure and label the angle at *H*.**

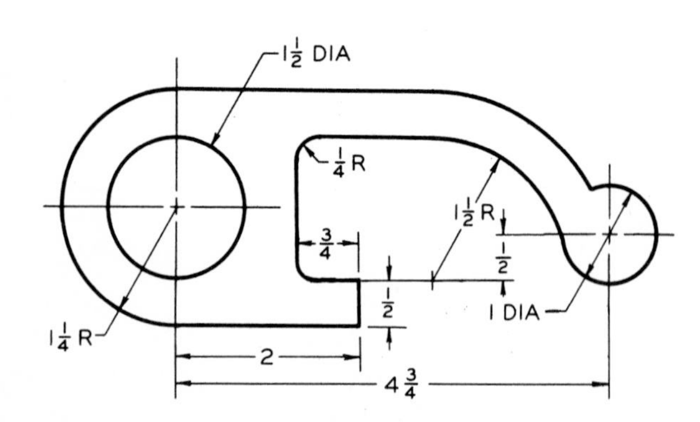

Fig. 3–97 **Draw the release lever. Scale: full size. Use compass and bow pencil. Make neat tangent joints.**

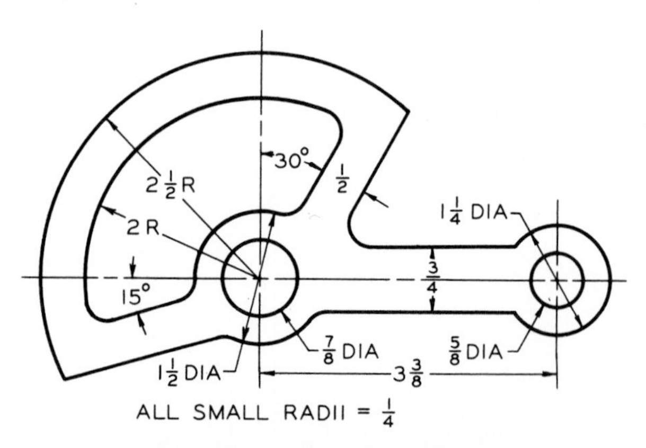

Fig. 3–98 **Draw the adjustable sector. Scale: full size. Make neat tangent joints. Use compass and bow pencil.**

Chapter 4

Geometry for Technical Drawing

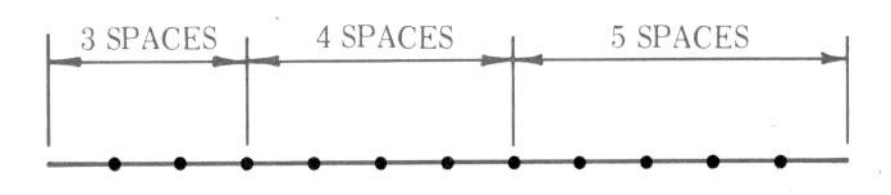

Fig. 4–1 **Rope used by rope stretchers.**

IMPORTANCE OF GEOMETRY

Geometry has always been important to people. It was used in ancient times for measuring land and making right-angle (90°) corners for buildings and other kinds of construction. The Egyptians used people called *rope stretchers* for this purpose. They used rope with marks or knots having 12 equal spaces. It was divided into 3-, 4-, and 5-space sections, as shown in Fig. 4–1. A square (right-angle) corner was made by stretching the rope and driving pegs into the ground at the 3-, 4-, and 5-space marks, as shown in Fig. 4–2. This was one way an ancient people used geometry. The 3-4-5 method of making a right angle is important in drafting geometry even today.

Fig. 4–2 **Rope stretched to make a right angle.**

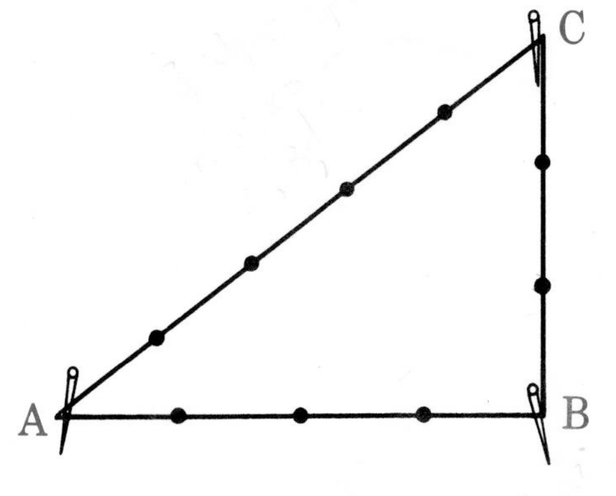

The use of the 3-4-5 triangle for making a right angle was proved by the mathematician Pythagoras in the sixth century B.C. This proof is called the Pythagorean theorem. The theorem can be shown graphically or mathematically. The Pythagorean theorem can be stated in this way: The sum of the square of each *adjacent* (next to) side of a right triangle equals the square of the *hypotenuse* (side opposite the right angle). The Pythagorean theorem can be stated in numbers in this way: $3^2 + 4^2 = 5^2$.

$$3 \times 3 = 9$$
$$\underline{4 \times 4 = 16}$$
$$5 \times 5 = 25$$
$$25 = 25$$

This method also works well for right triangles that have the same proportions. For example, in a right triangle with sides of 6, 8, and 10 units, the sum of the squares of the adjacent sides equals the square of the hypotenuse. The units may be millimeters, meters, inches, or fractions of an inch.

The *graphic* (drawn) proof of the Pythagorean theorem can be found in Fig. 4–3, the "dictionary of drafting geometry." The proof is also shown in Fig. 4–21.

Geometry is the study of the size and shape of things. The relationship of straight and curved lines in drawing shapes is also a part of geometry. Some geometric figures used in drafting include circles, squares, triangles, hexagons, and octagons. Many other shapes and lines are shown in Fig. 4–3. Study the "dictionary of drafting geometry" (Fig. 4–3) before beginning the geometric *constructions* (drawings) on the following pages. Be sure to refer back to the "dictionary of drafting geometry" as often as you need to in order to review geometric terms and figures.

Geometric constructions are made of individual lines and points drawn in proper relationship to one another. You need to accurately measure lines, angles, and the location of points to make an exact geometric construction.

The rules of geometric construction stated in this chapter are important to drafters, surveyors, engineers, architects, scientists, mathematicians, and designers. Geometric construction is important for making technical drawings. It is also needed for solving technical problems by the use of diagrams. Therefore, you will need to know most of the constructions explained in this chapter.

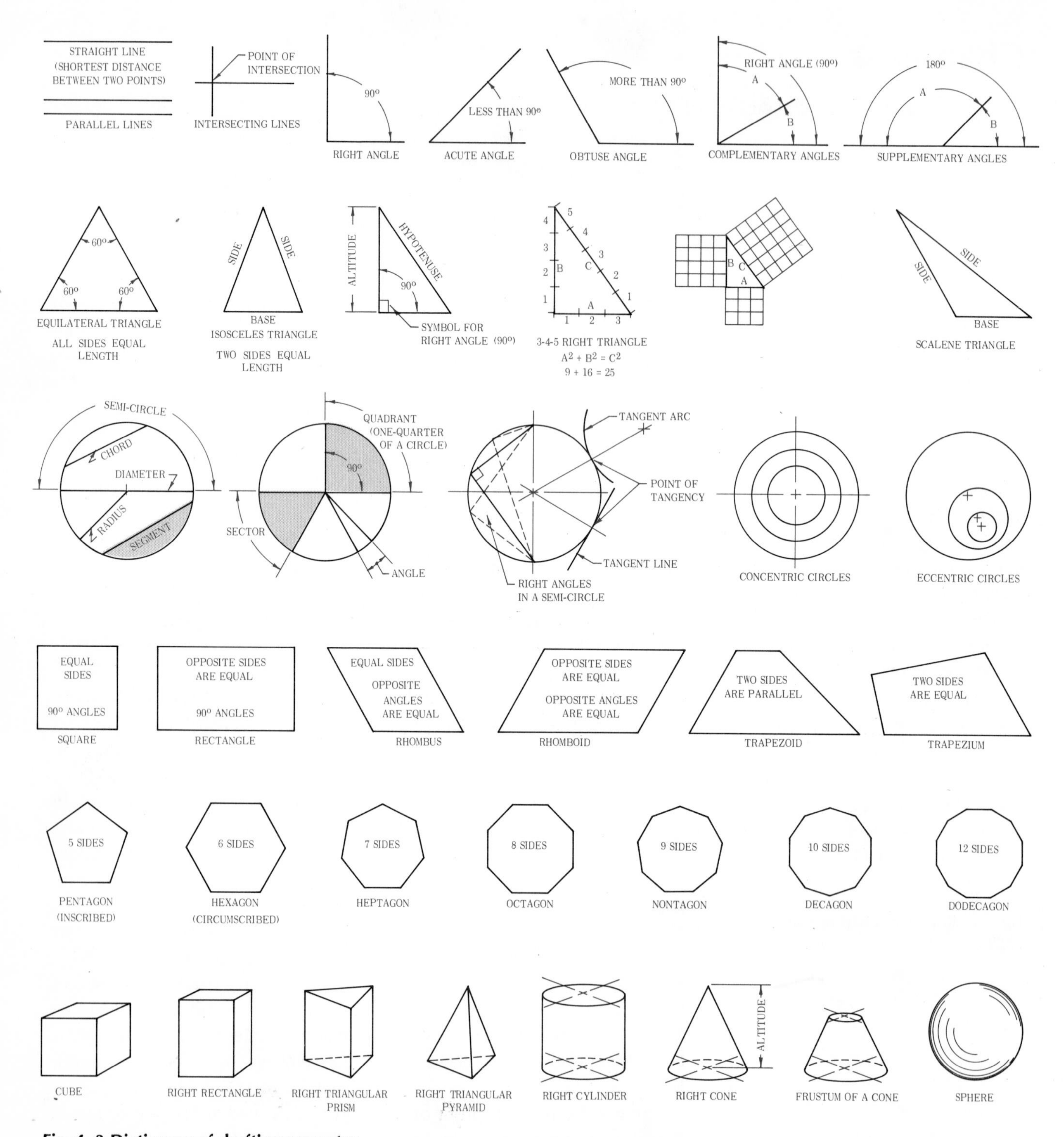

***Fig. 4–3* Dictionary of drafting geometry.**

PROBLEM: Bisect line AB or arc AB. NOTE: "Bisect" means to divide into two equal parts.

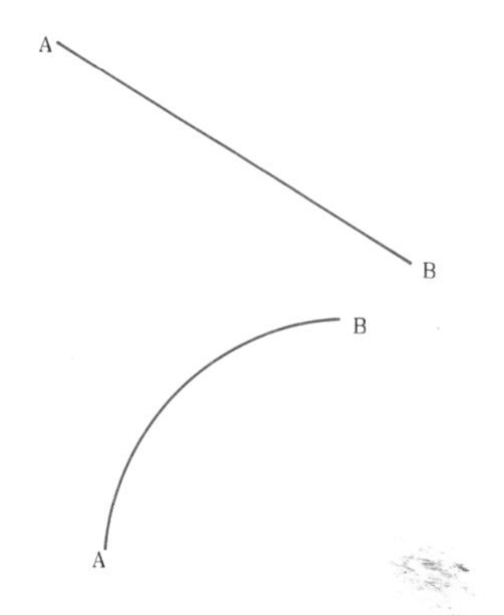

Given line *AB* or arc *AB*.

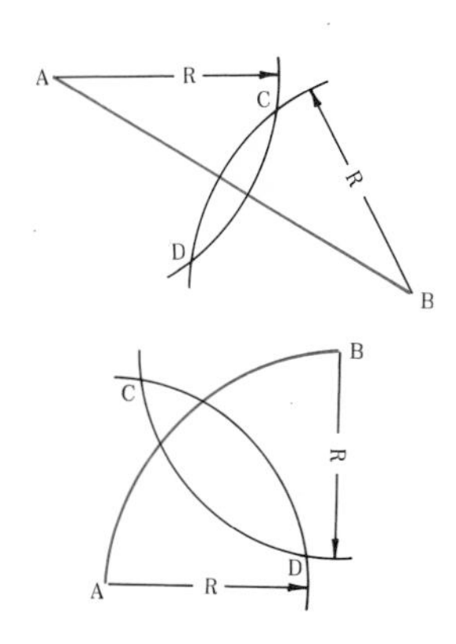

With *A* and *B* as centers and any radius *R* greater than one-half of *AB*, draw arcs to intersect (cut across) at *C* and *D*.

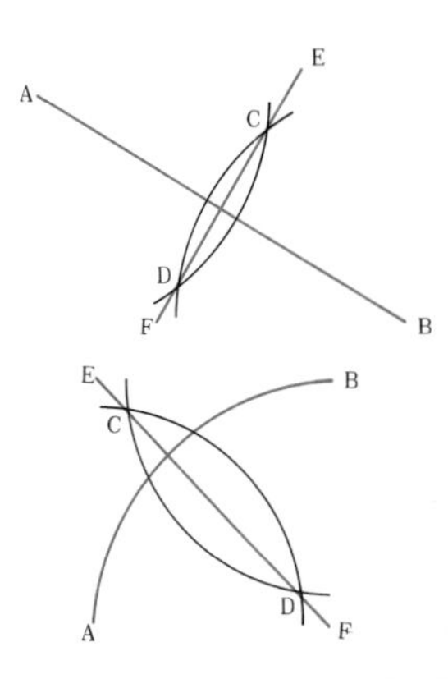

Draw line *EF* through intersections *C* and *D*.

Fig. 4–4 To bisect a straight line or arc.

PROBLEM: Divide line AB into eight equal parts.

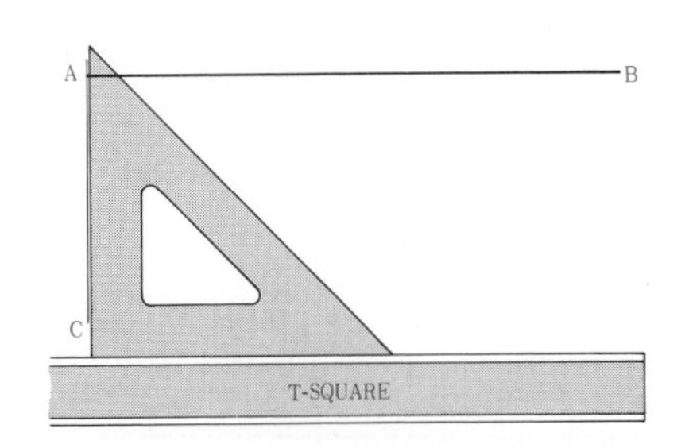

Draw a line of any length at *A* perpendicular (at right angles, 90°) to line *AB*.

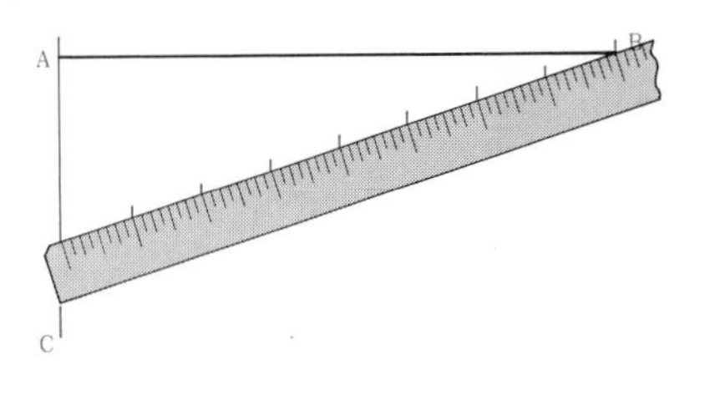

Place scale with zero at point *B* and adjust it along line *AC* until any eight equal divisions are included between points *B* and *C*. (In this case, eight inches.) Mark the divisions.

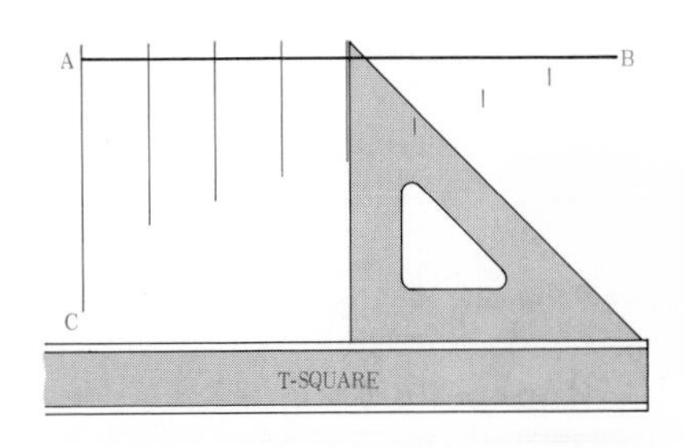

Draw lines parallel (side-by-side) to *AC* through the division marks to intersect line *AB*.

Fig. 4–5 To divide a straight line into any number of equal parts (first method).

PROBLEM: Divide line AB into five equal parts.

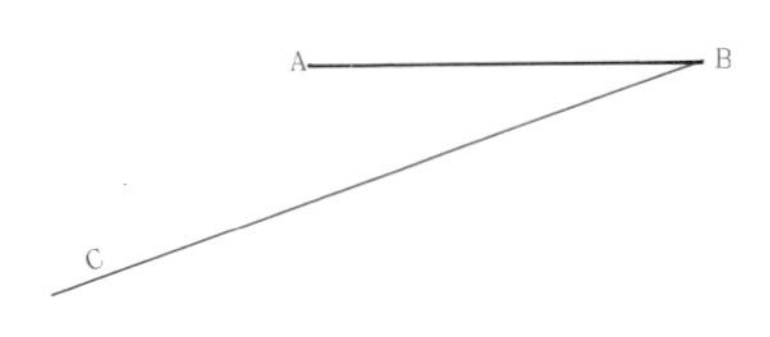

Draw line *BC* from point *B* at any convenient angle and of any length.

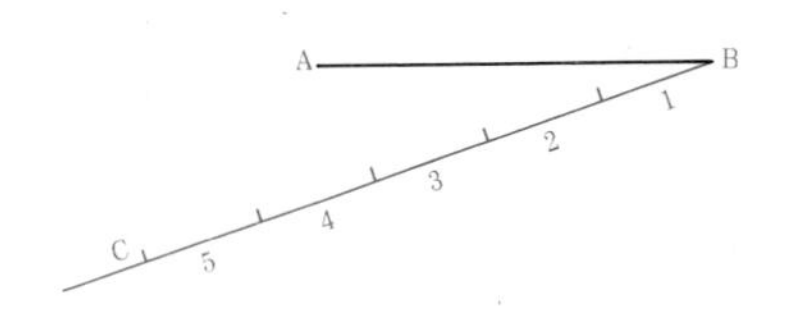

Use dividers or a scale to step off five equal spaces on line *BC* beginning at point *B*.

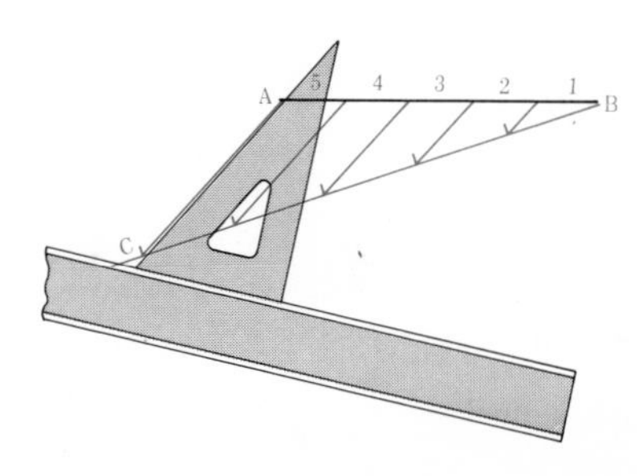

Draw a line connecting points *A* and *C*. Draw lines through each point on *BC* parallel to line *AC* to intersect line *AB*.

Fig. 4–6 To divide a straight line into any number of equal parts (second method).

PROBLEM: Draw a perpendicular line at point O on line AB.

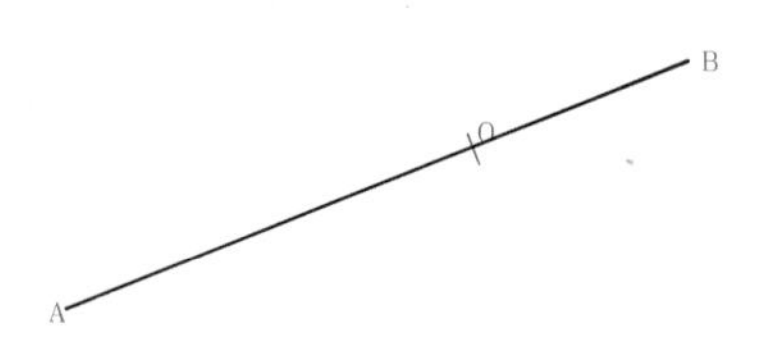

Given line *AB* and point *O*.

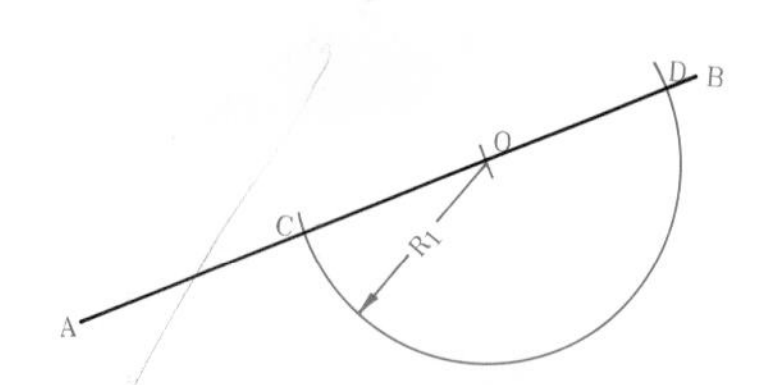

With *O* as the center and any convenient radius R_1, construct an arc cutting line *AB*, locating points *C* and *D*.

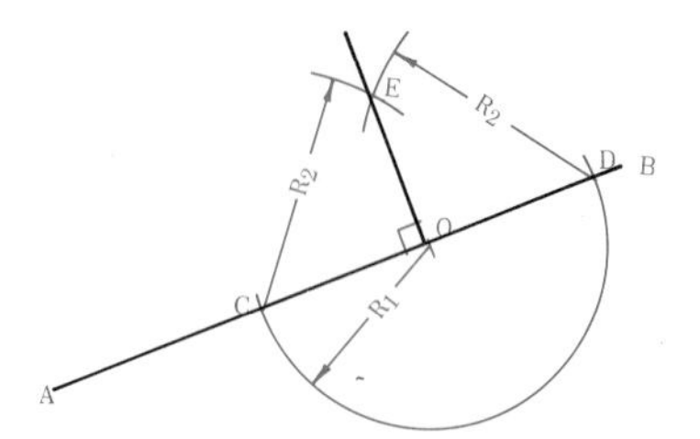

With *C* and *D* as centers and any radius R_2 greater than *OC*, draw arcs intersecting at *E*. Draw a line connecting points *E* and *O* to form the perpendicular.

Fig. 4–7 To draw a perpendicular line to a given line at a given point on the line (first method).

PROBLEM: Draw a perpendicular line at O near the end of line AB.

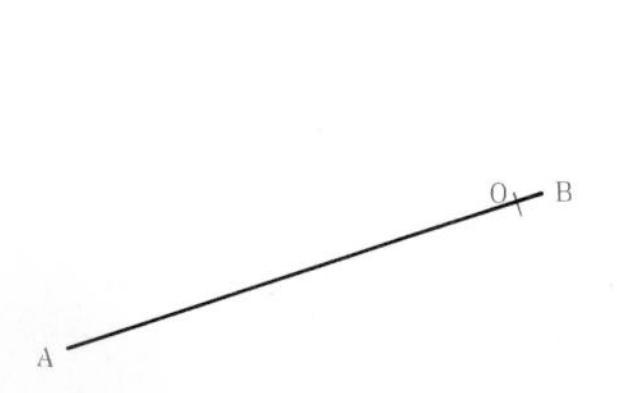

Given line AB and point *O*.

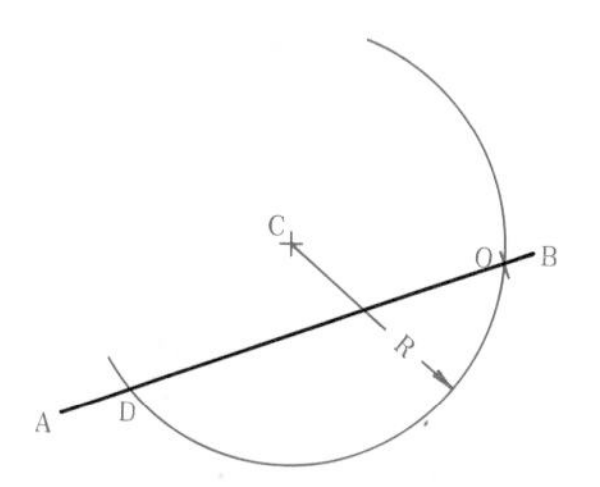

From any point *C* above line *AB*, construct an arc using *CO* as the radius and passing through line *AB* to locate point *D*.

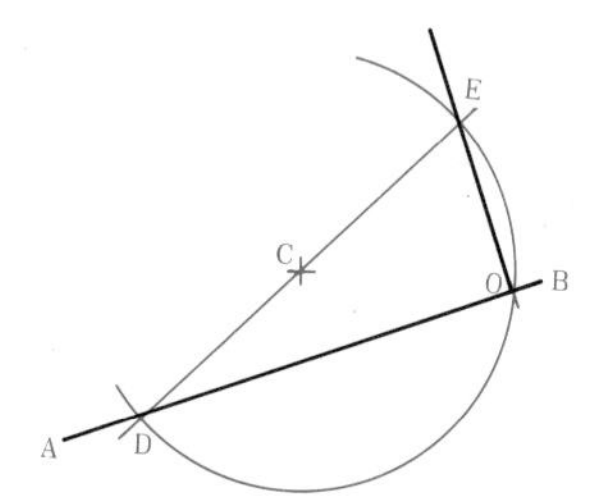

Draw a line through points *D* and *C*, extending it through the arc to locate point *E*. Connect points *E* and *O* to form the perpendicular line.

Fig. 4–8 To draw a perpendicular line to a given line at a given point on the line (second method).

PROBLEM: Draw a perpendicular line to line AB through point O.

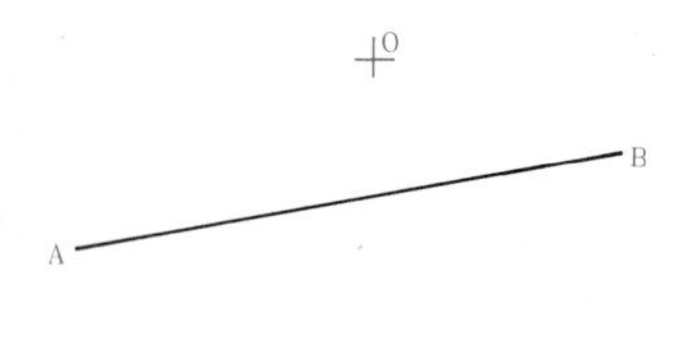

Given line AB and point *O*.

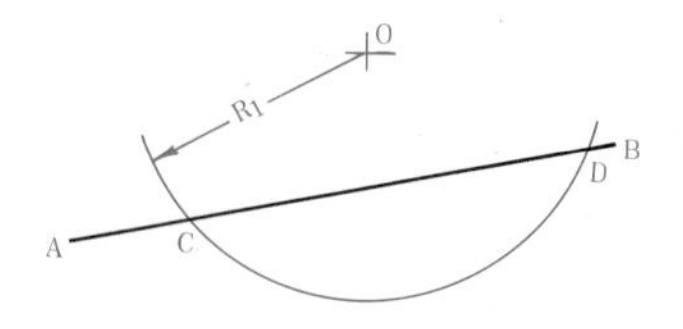

With *O* as the center, draw an arc with radius R_1 long enough to intersect line *AB* to locate points *C* and *D*.

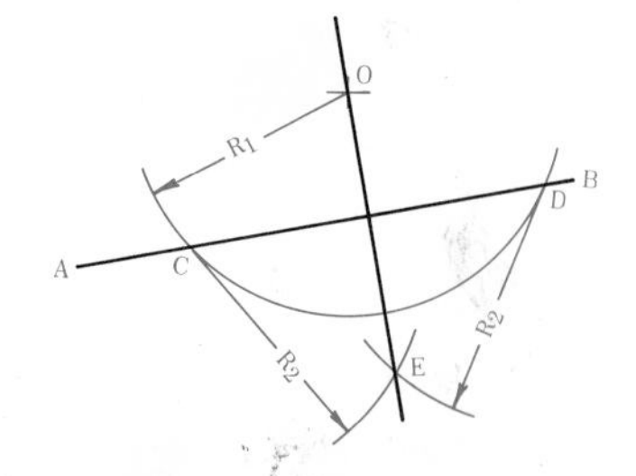

With *C* and *D* as centers and radius R_2 greater than one-half of *CD*, draw intersecting arcs to locate point *E*. A line drawn through points *O* and *E* is the perpendicular line.

Fig. 4–9 To draw a perpendicular line to a given line through a point outside the line (first method).

PROBLEM: Draw a perpendicular line to line AB through point O.

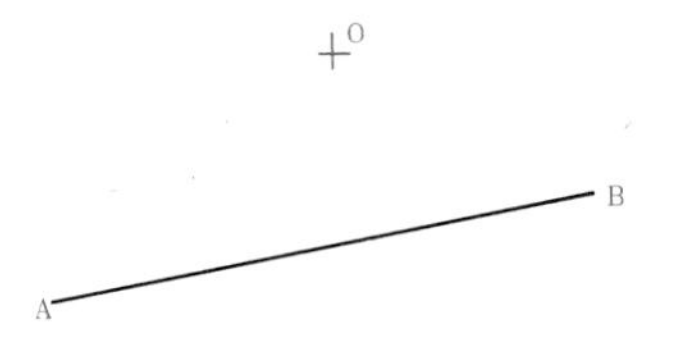

Given line AB and point O.

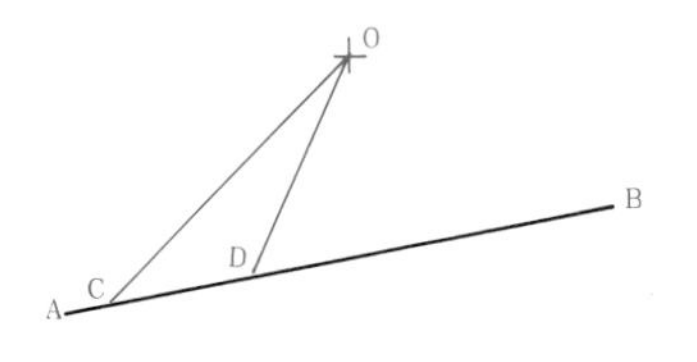

Draw lines from point *O* to any two points on line *AB*, locating points *C* and *D*.

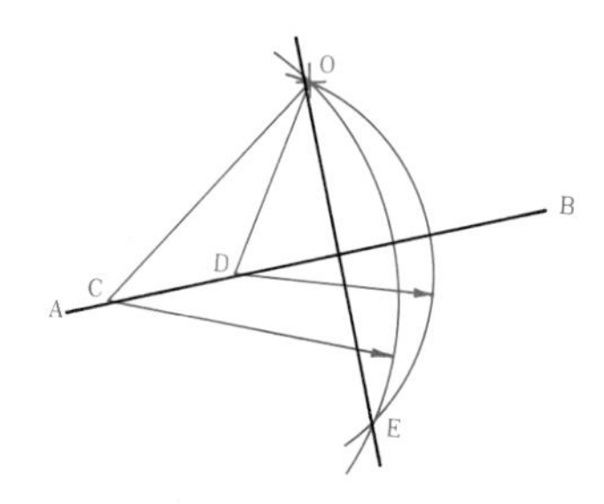

With *C* and *D* as centers and *CO* and *DO* as radii, draw arcs to intersect, locating point *E*. Connect points *O* and *E* to form the perpendicular line.

Fig. 4–10 To draw a perpendicular line to a given line through a point outside the line (second method).

PROBLEM: Draw a perpendicular line to line AB through point O.

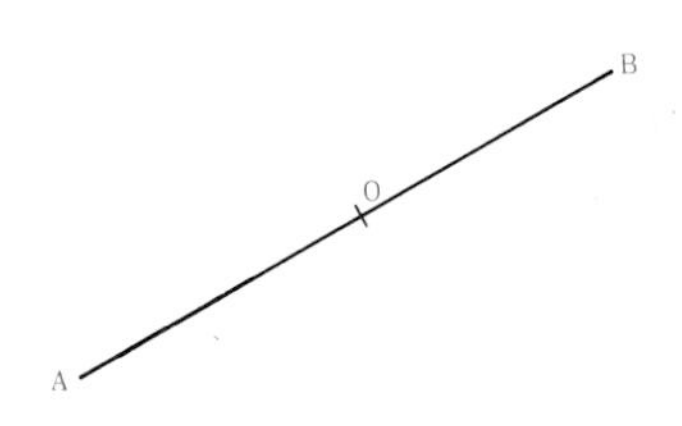

Given line *AB* and point *O*.

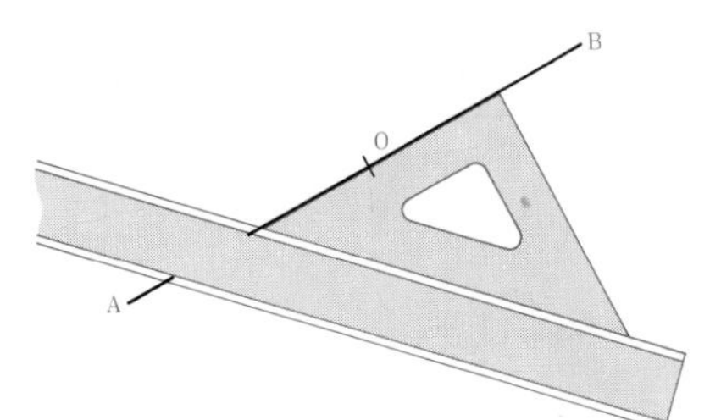

Place the T-square and triangle as shown.

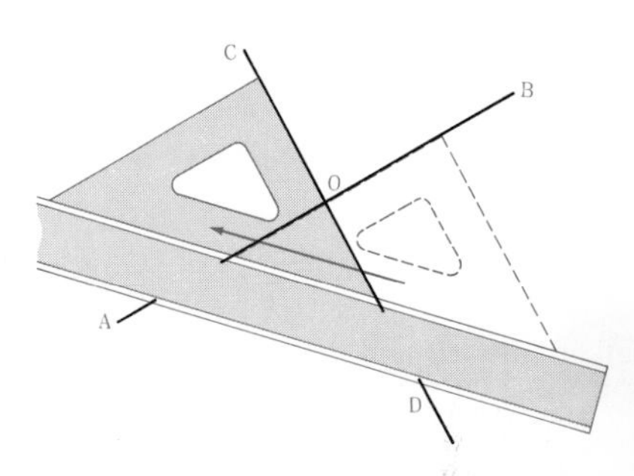

Slide the triangle along the T-square until the edge aligns with point *O* on line *AB*. Draw the perpendicular *CD*.

Fig. 4–11 To draw a perpendicular line to a given line through a point outside the line (third method).

PROBLEM: Draw a line parallel to line AB through point P. NOTE: "Parallel" lines are always the same distance apart.

Given line *AB* and point *P*.

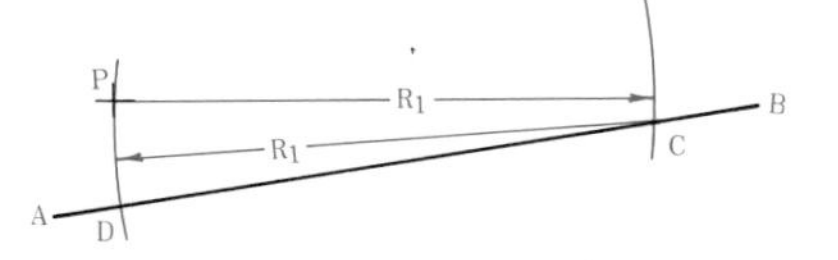

With point *P* as the center and any convenient radius R_1, draw an arc cutting line *AB* to locate point *C*. With point *C* as the center and the same radius R_1, draw an arc through point *P* and line *AB* to locate point *D*.

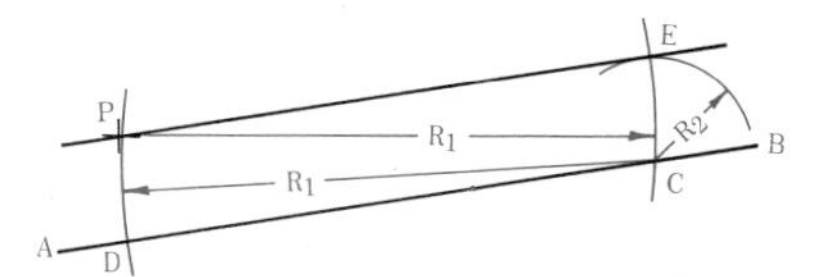

With *C* as the center and radius R_2 equal to chord *PD*, draw an arc to locate point *E*. Draw a line through points *P* and *E* that is parallel to *AB*.

Fig. 4–12 To construct (draw) a line parallel to a given line (first method).

PROBLEM: Construct (draw) a line parallel to AB at a distance you need from AB.

Given line *AB*.

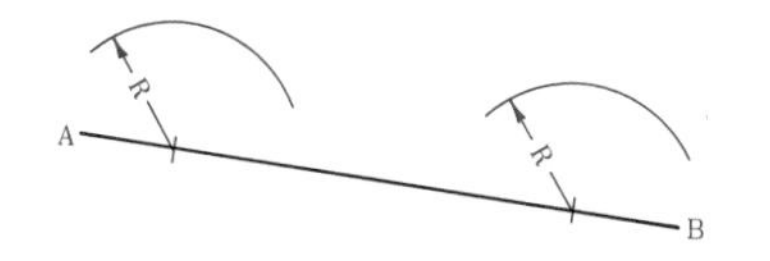

Draw two arcs with centers anywhere along line *AB*. The arcs should have a radius *R* equal to the distance you need between the parallel lines.

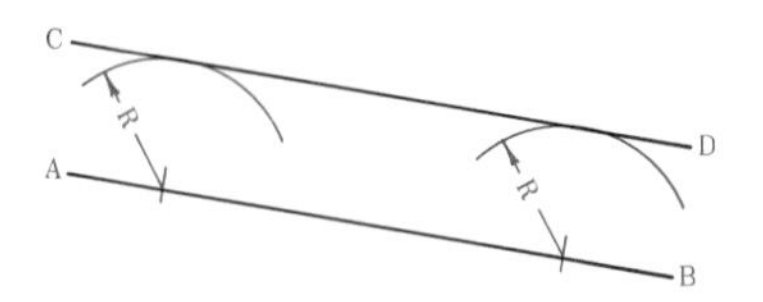

Draw the parallel line *CD* tangent (just touching) to the arcs.

Fig. 4–13 To construct a line parallel to a given line (second method)

PROBLEM: Construct a line parallel to AB through point P.

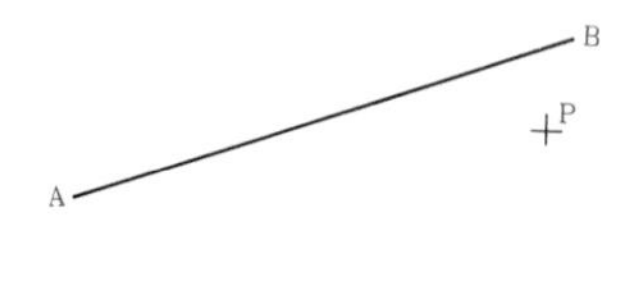

Given line *AB* and point *P*.

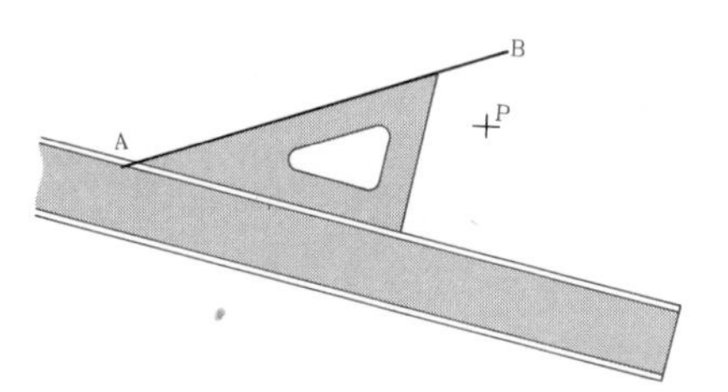

Place the T-square and triangle as shown.

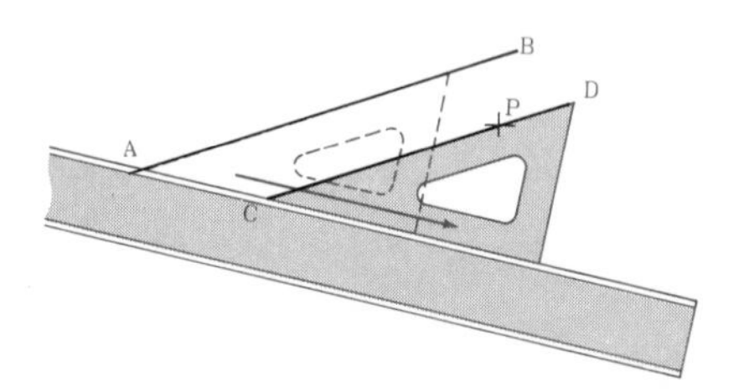

Slide the triangle until the edge aligns with point *P*. Draw the parallel *CD*.

Fig. 4–14 To construct a line parallel to a given line (third method).

PROBLEM: Construct a line to bisect angle AOB.

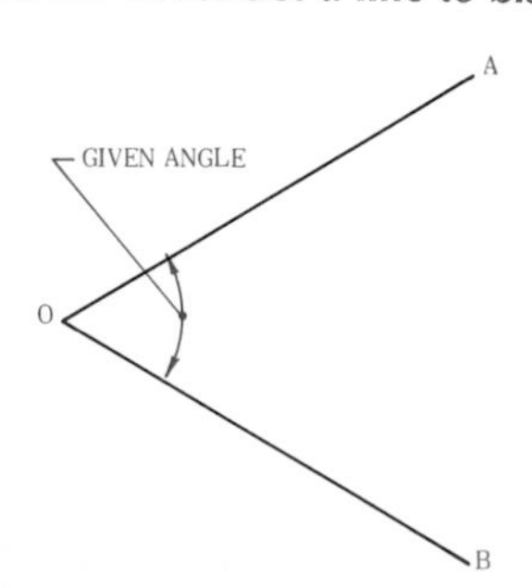

Given angle *AOB*.

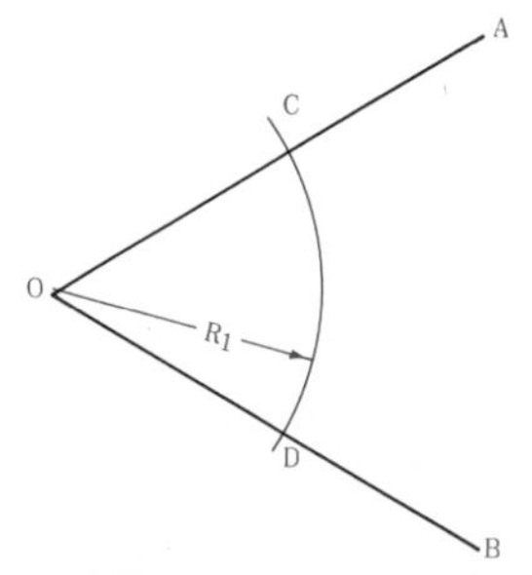

With point *O* as the center and any convenient radius R_1, draw an arc to intersect with *AO* and *OB* at *C* and *D*.

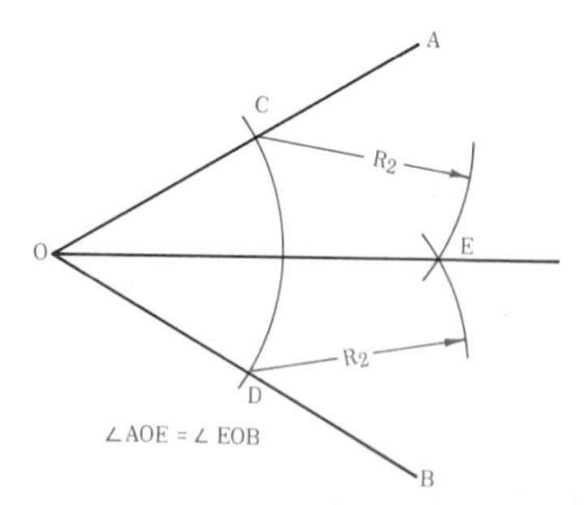

With *C* and *D* as centers and any radius R_2 greater than one-half of arc *CD*, draw arcs to intersect, locating point *E*. Draw a line through points *O* and *E* to bisect angle *AOB*.

Fig. 4–15 To bisect an angle.

PROBLEM: Construct angle AOB in a new location.

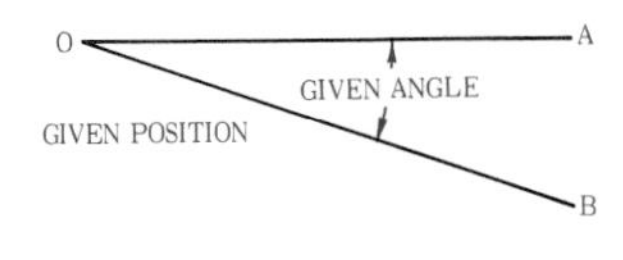

Given angle *AOB*.

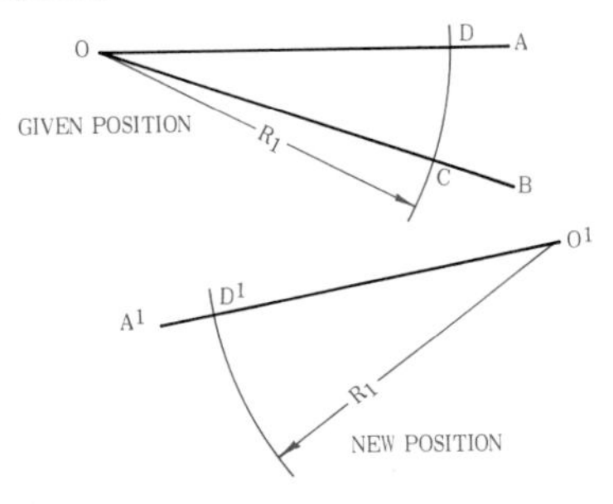

Draw one side O^1A^1 in the new position. With O and O^1 as centers and any convenient radius R_1, construct arcs to cut BO and AO at C and D and A^1O^1 at D^1.

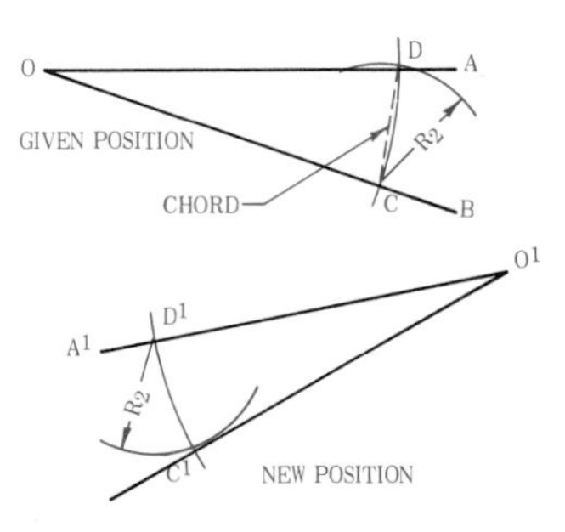

With D^1 as the center and radius R_2 equal to chord DC, draw an arc to locate point C^1 at the intersection of the two arcs. Draw a line through points O^1 and C^1 to complete the angle.

Fig. 4–16 To copy a given angle.

PROBLEM: Construct an isosceles triangle. NOTE: An "isosceles" triangle has two equal sides.

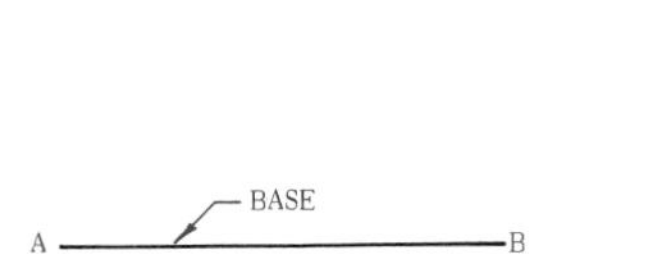

Given base line *AB*.

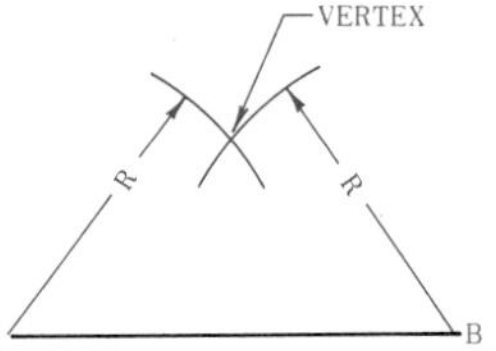

With points *A* and *B* as centers and a radius *R* equal to the length of the sides you want, draw intersecting arcs to locate the vertex (top point) of the triangle.

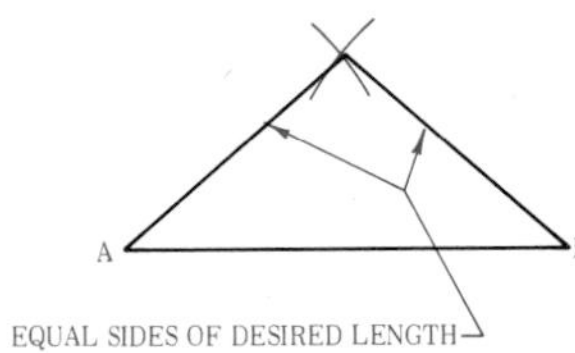

Draw lines through point *A* and the vertex and through point *B* and the vertex to complete the triangle.

Fig. 4–17 To draw an isosceles triangle.

PROBLEM: Construct an equilateral triangle. NOTE: An "equilateral" triangle has all sides equal and all angles equal.

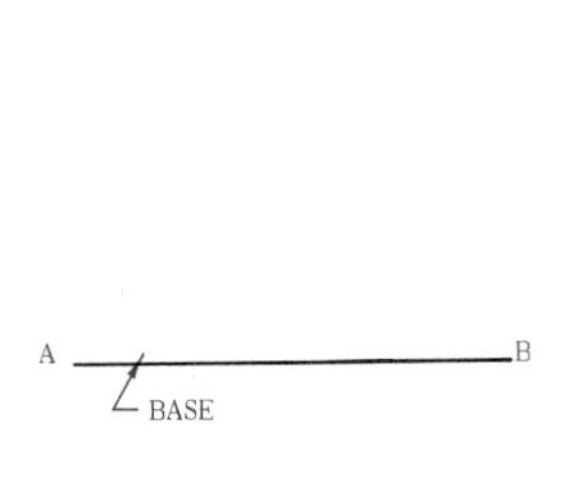

Given base line *AB*.

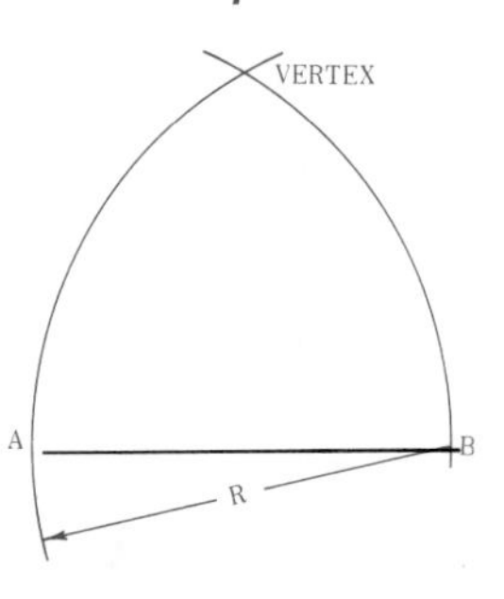

With points *A* and *B* as centers and a radius *R* equal to the length of line *AB*, draw intersecting arcs to locate the vertex.

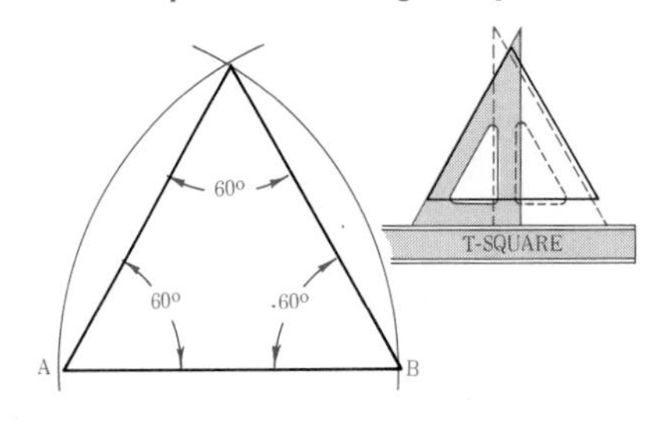

Draw lines through point *A* and the vertex and through point *B* and the vertex to complete the triangle. NOTE: An equilateral triangle may also be constructed by drawing 60° lines through the ends of the base line with the 30°–60° triangle, as shown at right.

Fig. 4–18 To draw an equilateral triangle.

PROBLEM: Construct a right triangle with two sides given. NOTE: A right triangle has one right angle (90°).

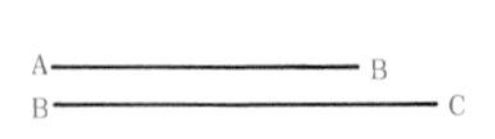

Given sides *AB* and *BC*.

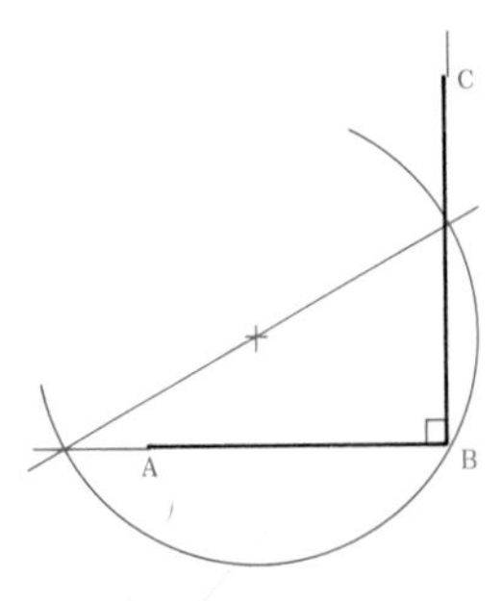

Draw side *AB* in the desired position. Draw a perpendicular line to *AB* at *B* equal to *BC*. NOTE: Use the method of Fig. 4–7 or the method of Fig. 4–11 to construct the perpendicular line.

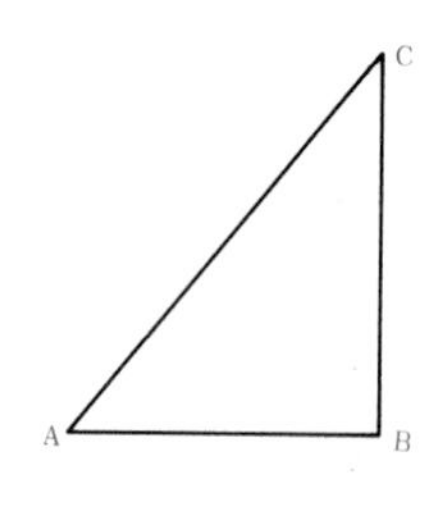

Draw a line connecting points *A* and *C* to complete the right triangle. NOTE: Line *AC* is called the *hypotenuse*. It is always the side opposite the 90° angle.

Fig. 4–19 To draw a right triangle with two sides given.

PROBLEM: Construct a right triangle with the hypotenuse and one side given.

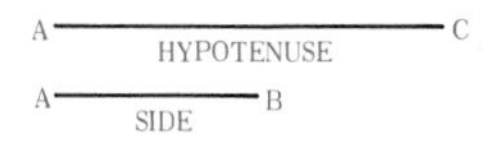

Given hypotenuse *AC* and side *AB*.

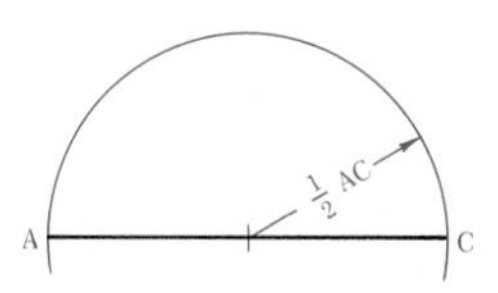

Draw the hypotenuse in the desired location. Draw a semicircle on *AC* using ½*AC* as the radius.

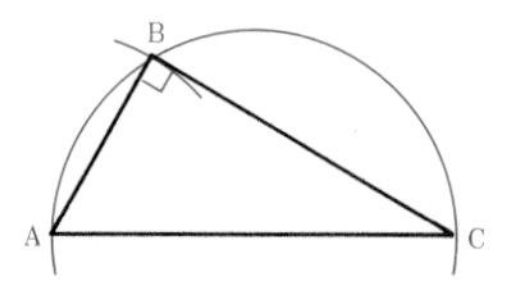

With point *A* as the center and a radius equal to side *AB*, draw an arc to cut the semicircle at *B*. Draw *AB* and then draw a line to connect *B* and *C* to complete the triangle.

Fig. 4–20 To draw a right triangle with the hypotenuse and one side given.

PROBLEM: Construct a right triangle on a base line three units long.

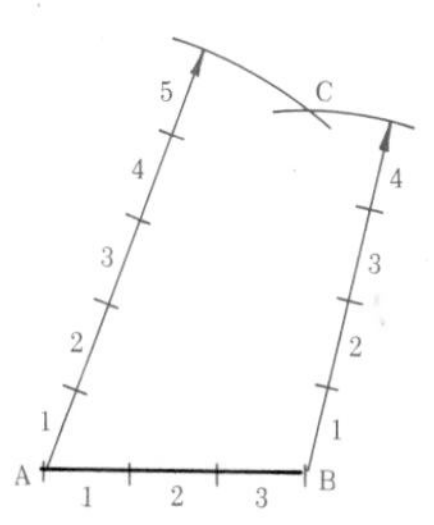

Given base line *AB* three units long.

With *A* and *B* as centers and radii four and five units long, draw intersecting arcs to locate point *C*.

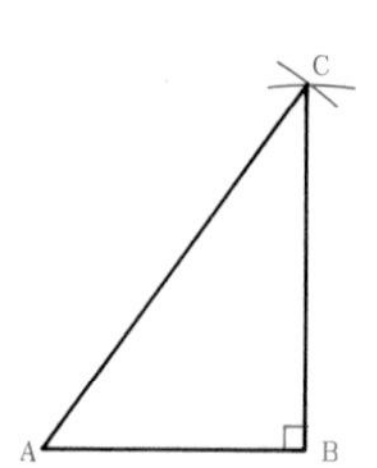

Draw lines *AC* and *BC* to complete the triangle.

Fig. 4–21 To draw a right triangle by the 3–4–5 method.

PROBLEM: Construct a triangle with three sides given.

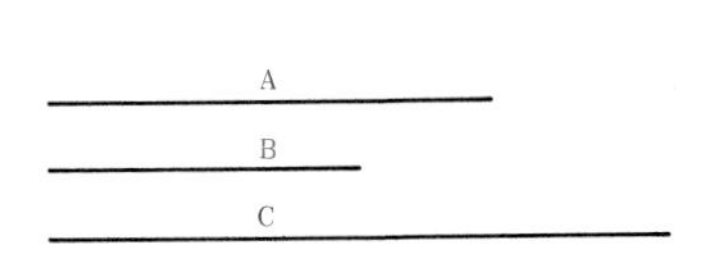

Given triangle sides *A, B,* and *C.*

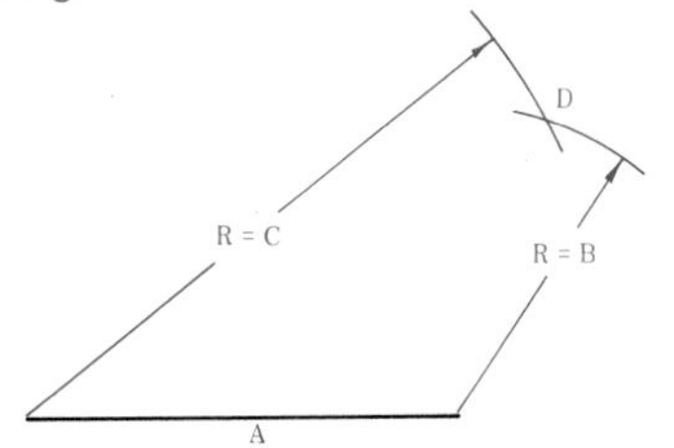

Draw base line *A* in the desired location. Construct arcs from the ends of line *A* with radii equal to lines *B* and *C* to locate point *D.*

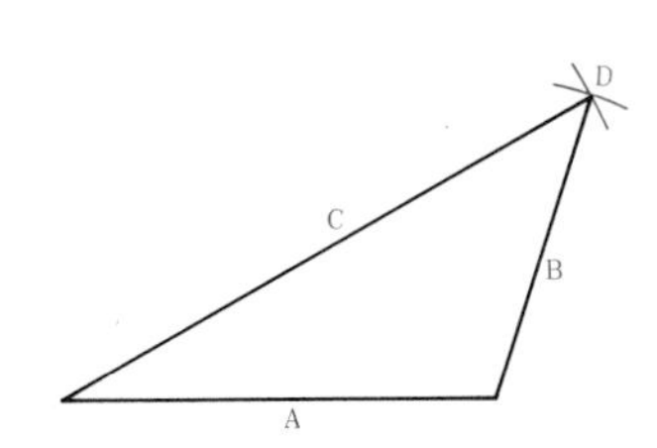

Connect the ends of line *A* with point *D* to complete the triangle.

Fig. 4–22 To draw a triangle with three sides given.

PROBLEM: Draw a square within a circle with corners tangent to (that touch but do not cross) the circle.

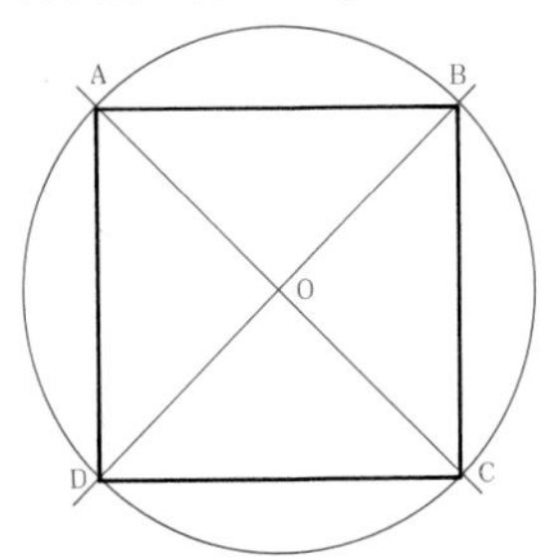

Given a circle with center point *O.* Construct 45° diagonals through the center point *O* to locate points *A, B, C,* and *D.* Connect *A* and *B, B* and *C, C* and *D,* and *D* and *A* to complete the square.

Fig. 4–23 To draw a square within a circle.

PROBLEM: Draw a square outside the circle with the circle tangent to the midpoints of the sides of the square.

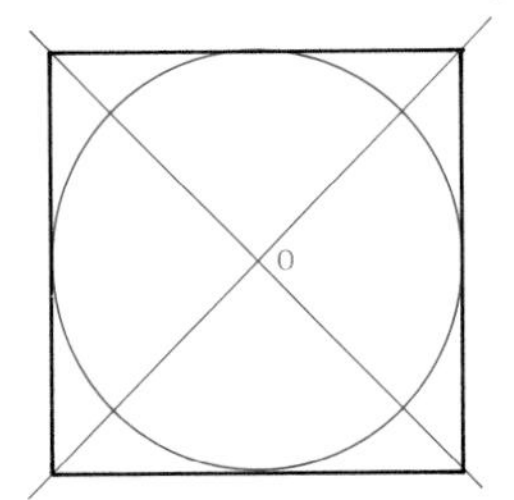

Given a circle with center point *O.* Construct 45° diagonals through the center point *O.* Draw sides tangent to the circle, intersecting at the 45° diagonals to complete the square.

Fig. 4–24 To draw a square about a circle.

PROBLEM: Construct a square.

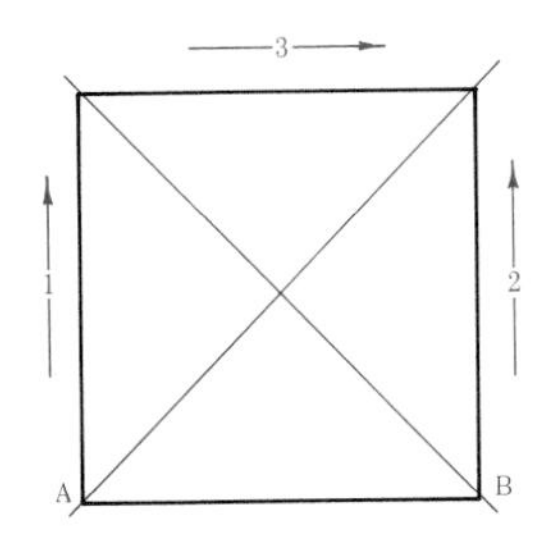

Given the length of the side *AB.* Construct 45° diagonals from ends of line *AB.* Complete the square by drawing the sides in the order shown by the numbered arrows.

Fig. 4–25 To draw a square.

PROBLEM: Construct a regular pentagon. NOTE: "Regular" means the object has equal sides and equal angles.

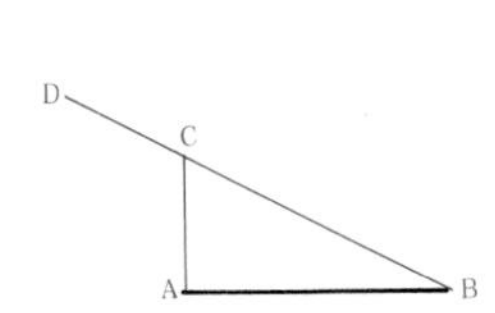

Given line *AB,* construct a perpendicular line *AC* equal to one-half of *AB.* Draw line *BC* and extend it to make line *CD* equal to *AC.*

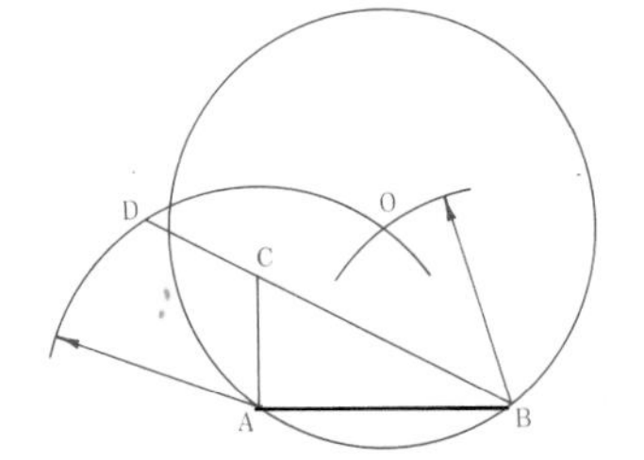

With radius *AD* and points *A* and *B* as centers, draw intersecting arcs to locate point *O.* With the same radius and *O* as the center, draw a circle.

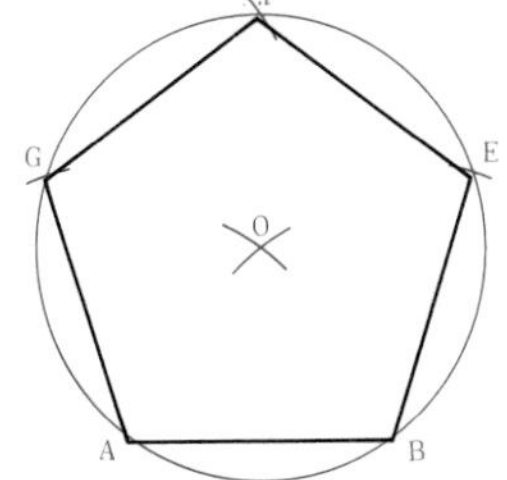

Step off *AB* as a chord to locate points *E, F,* and *G.* Connect the points to complete the pentagon.

Fig. 4–26 To construct a regular pentagon given one side.

PROBLEM: Draw a regular pentagon in a given circle.

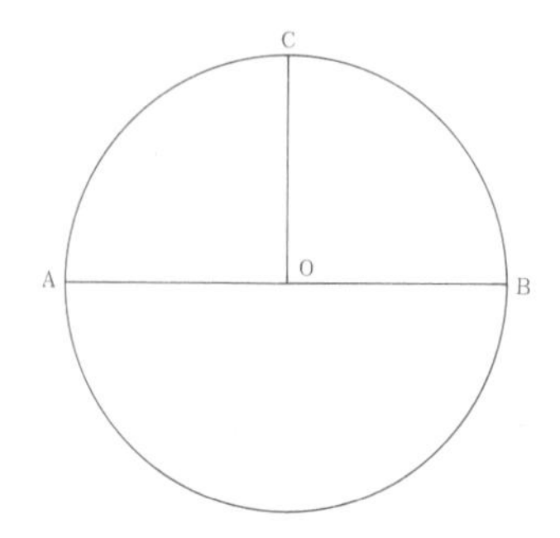

Given a circle with diameter *AB* and radius *OC*, draw a pentagon within the circle with points touching the circle.

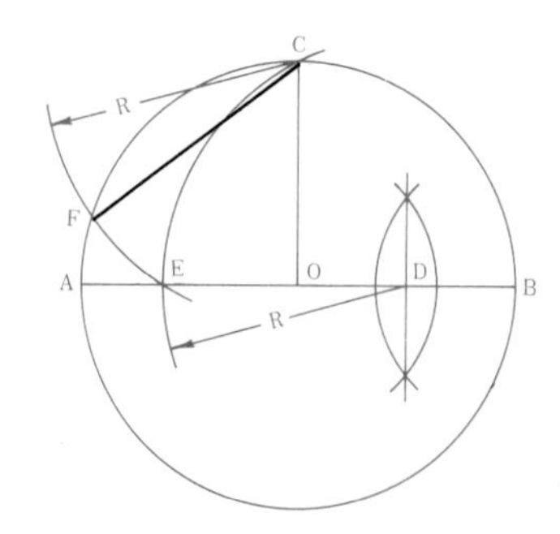

Bisect radius *OB* to locate point *D*. With *D* as center and radius *DC*, draw an arc to locate point *E*. With *C* as center and radius *CE*, draw an arc to locate point *F*. Chord *CF* is one side of the pentagon.

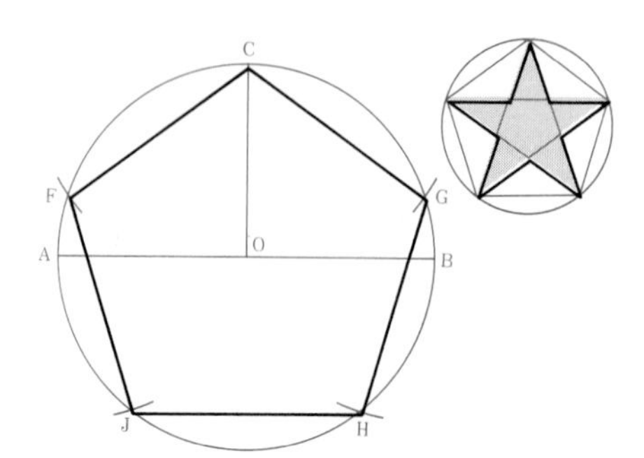

Step off chord *CF* around the circle to locate points *G*, *H*, and *J*. Draw the chords to complete the pentagon. NOTE: Another method for constructing a pentagon in a circle is to use dividers and locate the points by trial-and-error, as shown to the right.

Fig. 4–27 To draw a regular pentagon in a given circle.

PROBLEM: Construct a hexagon.

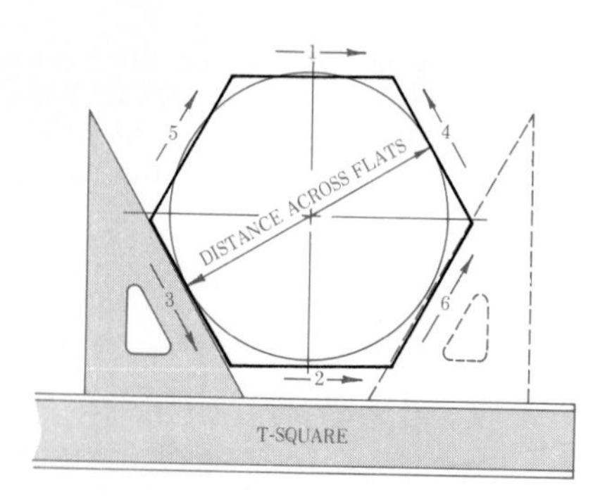

Given the distance across the flats of a regular hexagon. Draw centerlines and a circle with the diameter equal to the distance across the flats. With the T-square and 30°–60° triangle, draw the tangents in the order indicated.

Fig. 4–28 To draw a regular hexagon given the distance across the flats.

PROBLEM: Construct a hexagon.

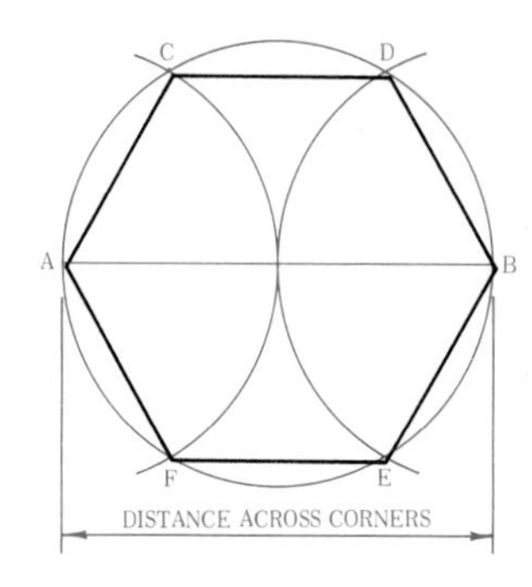

Given the distance *AB* across the corners. Draw a circle with *AB* as the diameter. With *A* and *B* as centers and the same radius, draw arcs to intersect the circle at points, *C*, *D*, *E*, and *F*. Connect the points to complete the hexagon.

Fig. 4–29 To draw a regular hexagon given the distance across the corners (first method).

PROBLEM: Construct a hexagon.

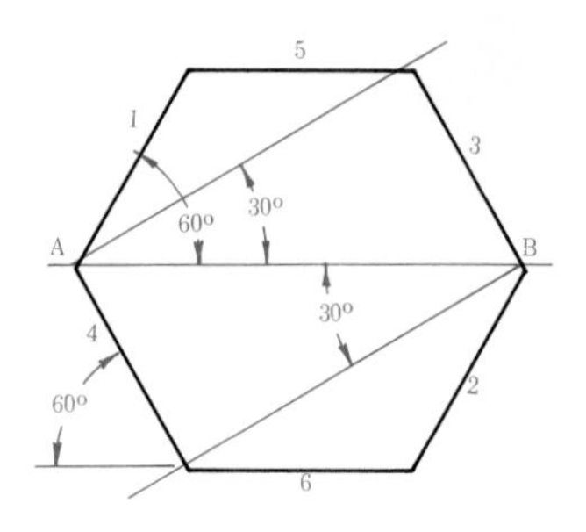

From points *A* and *B* draw lines of indefinite (any) length at 30° to line *AB*. With the T-square and 30°–60° triangle, draw the sides of the hexagon in the order indicated.

Fig. 4–30 To draw a regular hexagon given the distance across the corners (second method).

PROBLEM: Construct an octagon outside a circle that touches the midpoints of each of its sides.

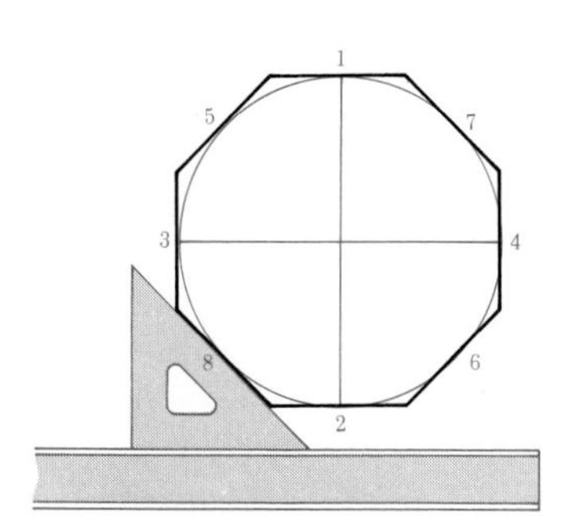

Given the distance across the flats, draw centerlines and a circle with the diameter equal to the distance across the flats. With the T-square and 45° triangle, draw lines tangent to the circle in the order indicated to complete the octagon.

Fig. 4–31 To draw a regular octagon about a circle.

PROBLEM: Construct an octagon within a circle so that the corners touch the circle.

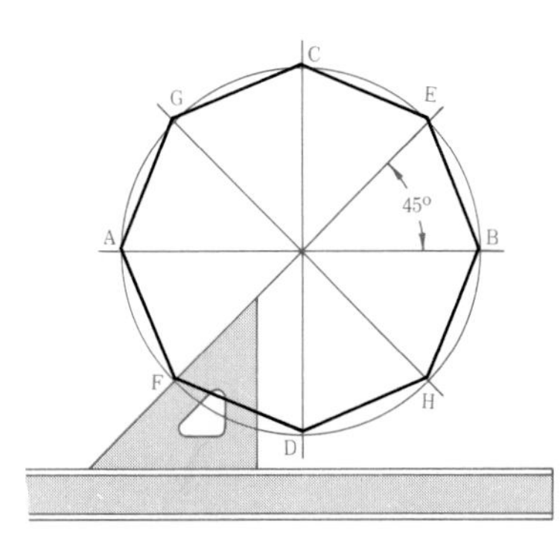

Given the distance across the corners, draw centerlines *AB* and *CD* and a circle with the diameter equal to the distance across the corners. With the T-square and 45° triangle, draw diagonals *EF* and *GH*. Connect the points to complete the octagon.

Fig. 4–32 To draw a regular octagon within a circle.

PROBLEM: Construct an octagon within a square.

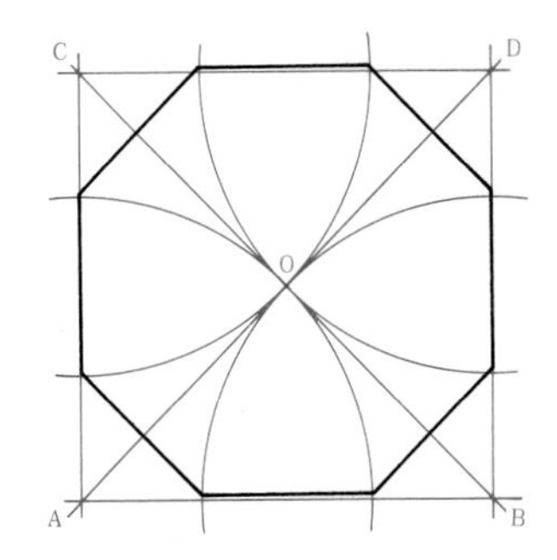

Given the distance across the flats, construct a square having the sides equal to *AB*. Draw diagonals *AD* and *BC* with their intersection at *O*. With *A, B, C,* and *D* as centers and radius $R = AO$, draw arcs to cut the sides of the square. Connect the points to complete the octagon.

Fig. 4–33 To draw a regular octagon in a square.

PROBLEM: Construct a circle through points A, B, and C.

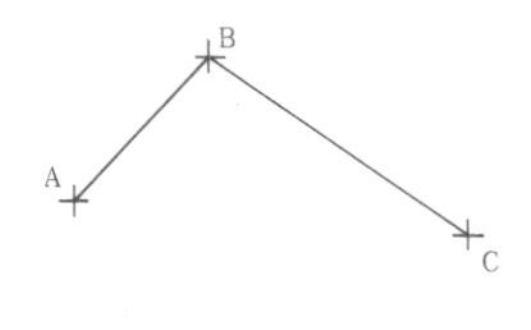

Given points *A, B,* and *C*. Draw lines *AB* and *BC*.

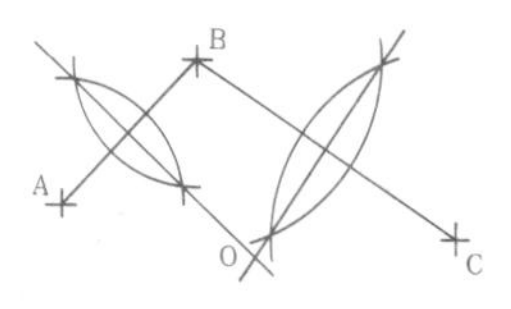

Draw perpendicular bisectors of *AB* and *BC* to intersect at point *O*.

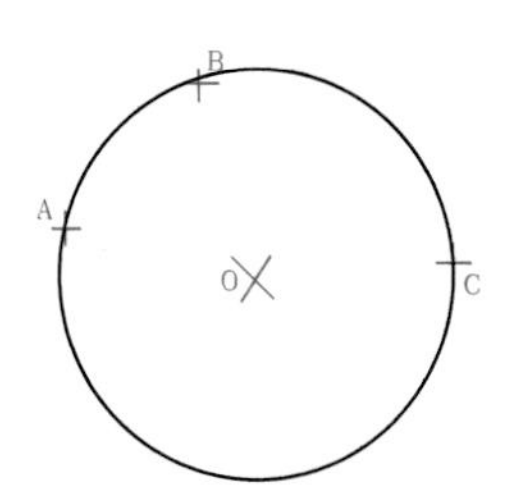

Draw the required circle with point *O* as the center, and radius $R = OA = OB = OC$.

Fig. 4–34 To construct a circle through any three points not in a straight line.

PROBLEM: Draw a tangent line through point P on the circle.

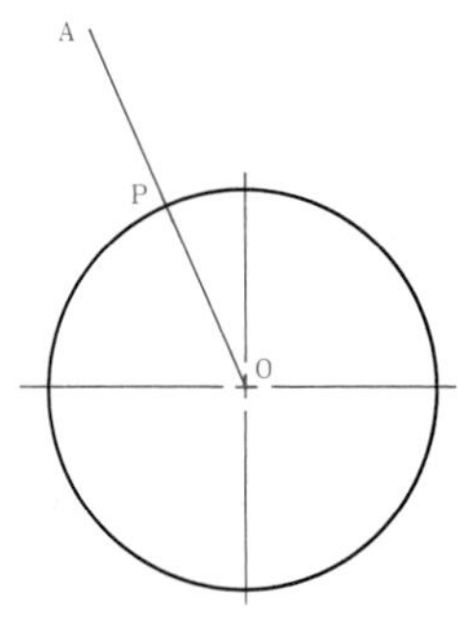

Given circle with center point O and tangent point P. Draw line OA from the center of the circle to extend beyond the circle through point P.

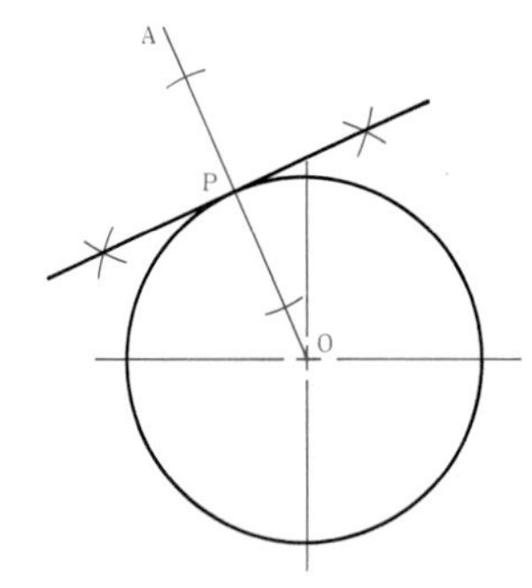

Draw a perpendicular line to OA at P. The perpendicular line is the tangent line.

Fig. 4–35 To draw a tangent to a circle at a given point P on the circle (first method).

PROBLEM: Draw a tangent line through point P on the circle.

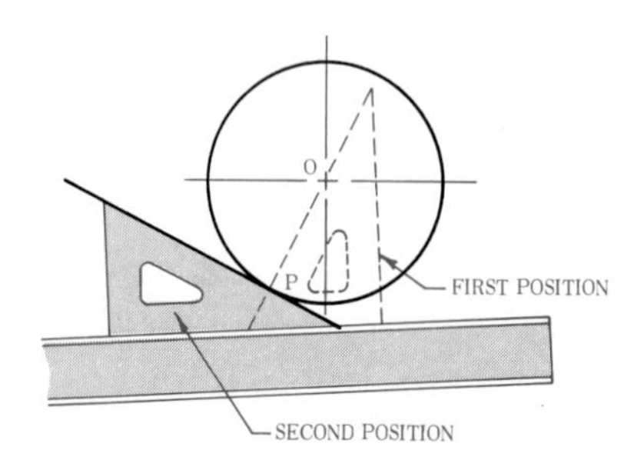

Given a circle with center point O and tangent point P. Place a T-square and triangle so that the hypotenuse of the triangle passes through points P and O (first position). Hold the T-square, turn the triangle to the second position at point P, and draw the tangent line.

Fig. 4–36 To draw a tangent to a circle at a given point P on the circle (second method).

PROBLEM: Draw a line from point P tangent to the circle.

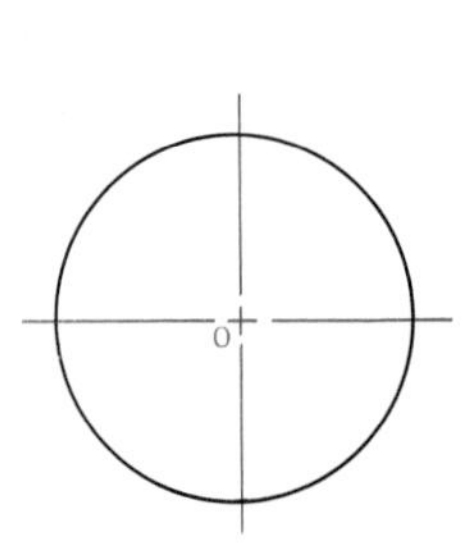

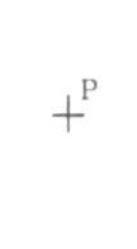

Given a circle with center point O and point P outside the circle.

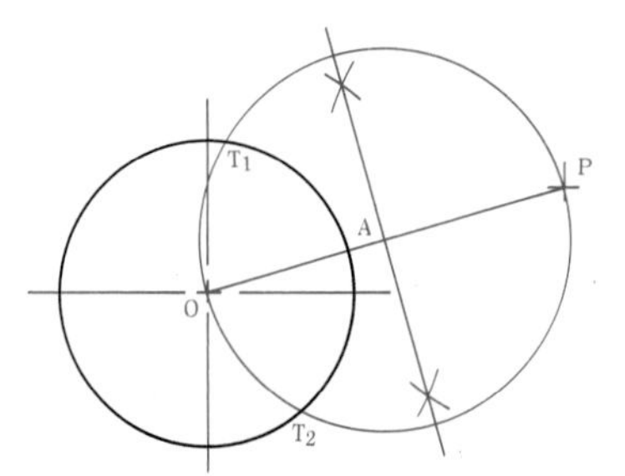

Draw line OP and bisect it to locate point A. Draw a circle with center A and radius $R = AP = AO$ to locate tangent points T_1 and T_2.

Draw PT_1 and PT_2. These lines are tangent to the circle.

Fig. 4–37 To draw a tangent to a circle from a point outside the circle.

PROBLEM: Construct an arc tangent to two straight lines.

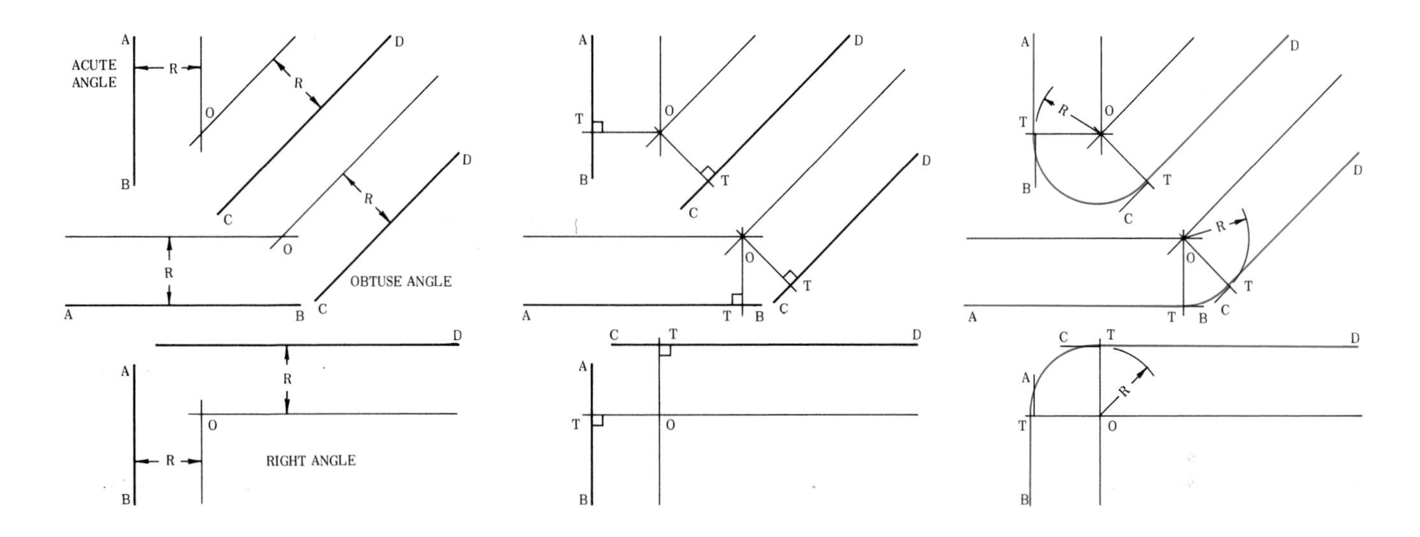

Given lines *AB* and *CD*, draw lines parallel to *AB* and *CD* at a distance *R* from them on the inside of the angle. The intersection *O* will be the center of the arc you need.

Draw perpendicular lines from *O* to *AB* and *CD* to locate the points of tangency *T*.

With *O* as the center and radius *R*, draw the needed arc.

Fig. 4–38 To construct an arc tangent to two straight lines at an acute angle, an obtuse angle, and a right angle.

PROBLEM: Draw a reverse, or ogee, curve between two straight lines.

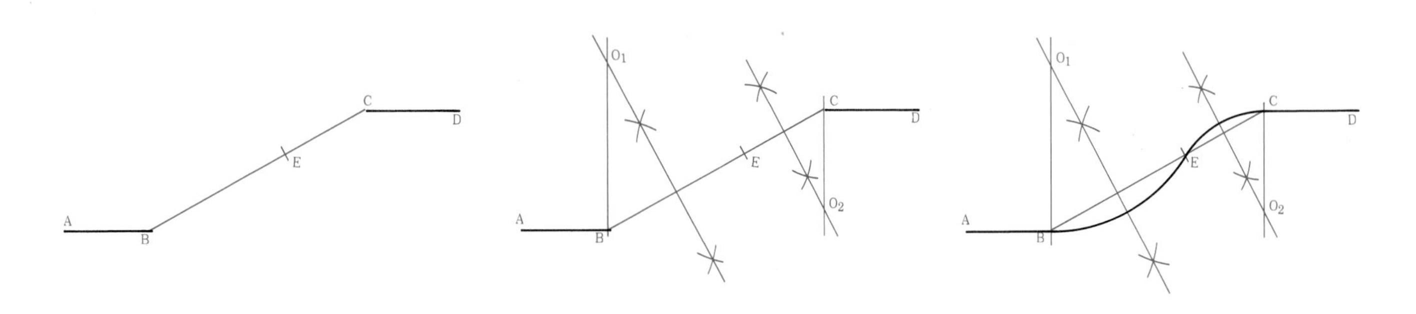

Given lines *AB* and *CD*, draw line *BC*. Select a point *E* on line *BC* through which the curve is to pass.

Draw perpendicular bisectors of *BE* and *EC*. Draw perpendicular lines to *AB* at *B* and to *CD* at *C*. They must cross the bisectors of *BE* and *EC* at O_1 and O_2, respectively.

Draw one arc with center O_1 and radius O_1E and the other with center O_2 and radius O_2E to complete the required curve.

Fig. 4–39 To draw a reverse, or ogee, curve.

PROBLEM: Construct an arc tangent to two given arcs.

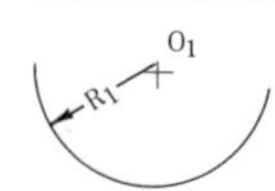

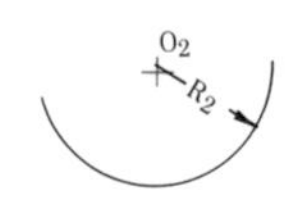

Given two arcs having radii R_1 and R_2 (radii may be equal or unequal) and radius R of the tangent arc.

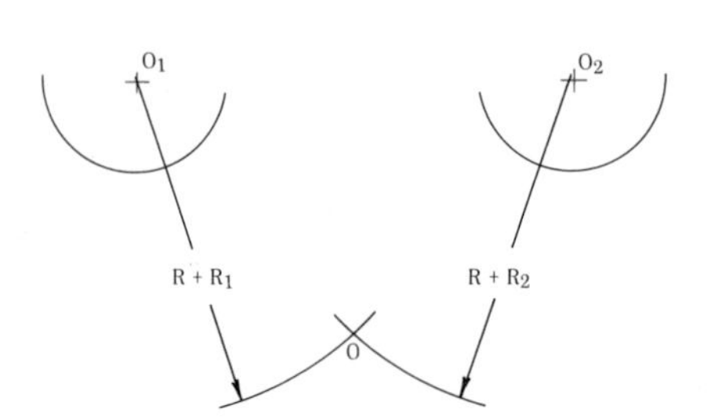

Draw an arc with center O_1 and radius $= R + R_1$. Draw an arc with center O_2 and radius $= R + R_2$. The intersection at O is the center of the tangent arc.

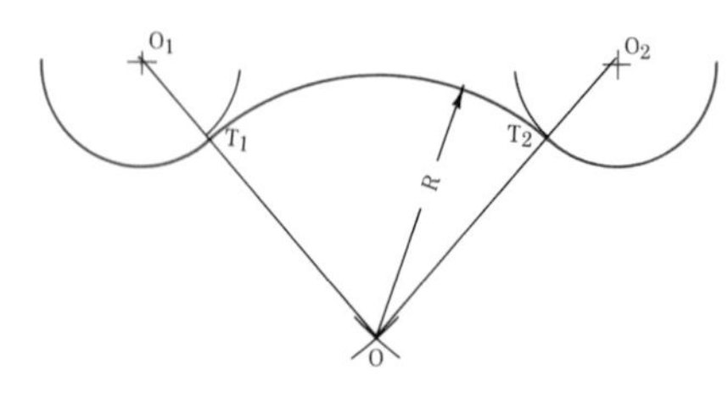

Draw lines O_1O and O_2O to locate tangent points T_1 and T_2. With point O as the center and radius R, draw the tangent arc needed.

Fig. 4–40 To draw an arc of a given radius tangent to two given arcs.

PROBLEM: Draw a tangent line to the exterior of two circles.

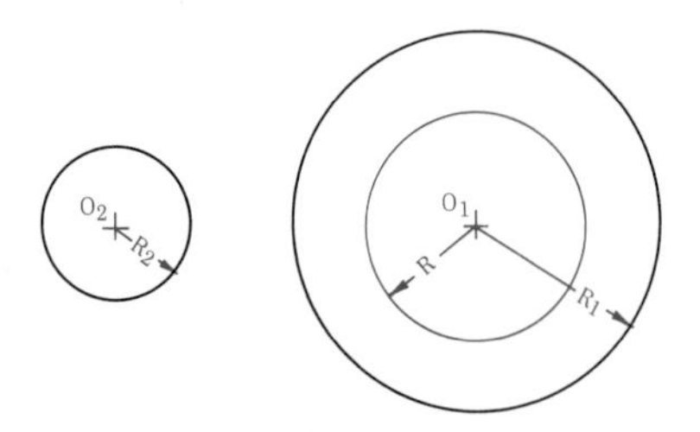

Given two circles with centers O_1 and O_2 and radii R_1 and R_2. $R = R_1 - R_2$. Using radius R and point O_1 as the center, draw a circle.

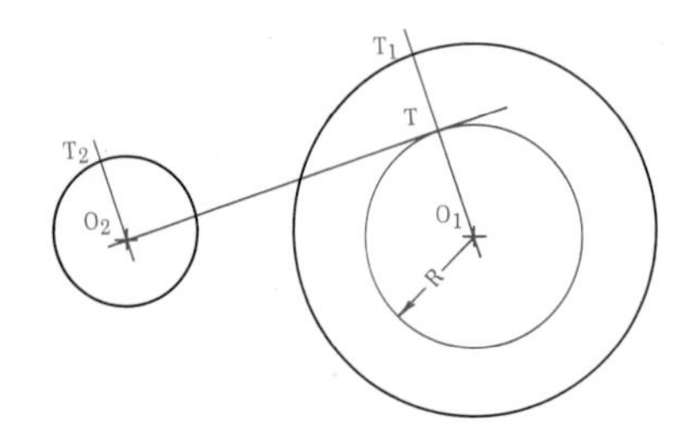

From center point O_2 draw a tangent O_2T to the circle of radius R. Draw radius O_1T. Extend it to locate tangent point T_1. Draw O_2T_2 parallel to O_1T_1.

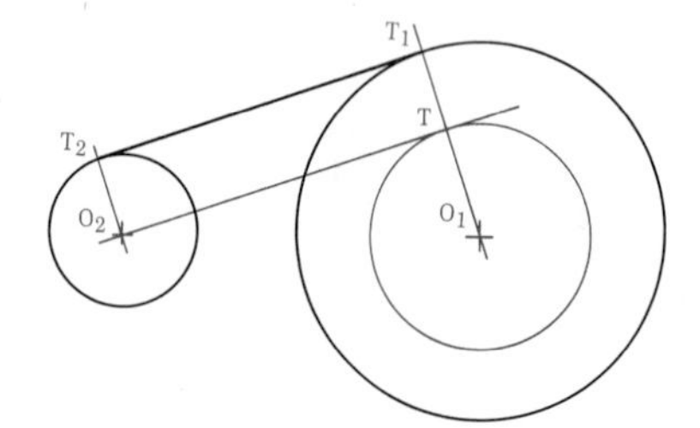

Draw the needed tangent T_1T_2 parallel to TO_2.

Fig. 4–41 To draw an exterior common tangent to two circles of unequal radii.

PROBLEM: Draw a tangent line to the interior of two circles.

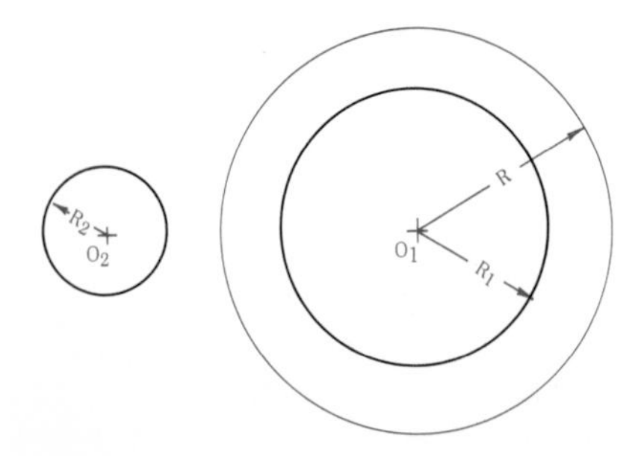

Given two circles with centers O_1 and O_2, and radii R_1 and R_2. $R = R_1 + R_2$. Using radius R and point O_1 as the center, draw a circle.

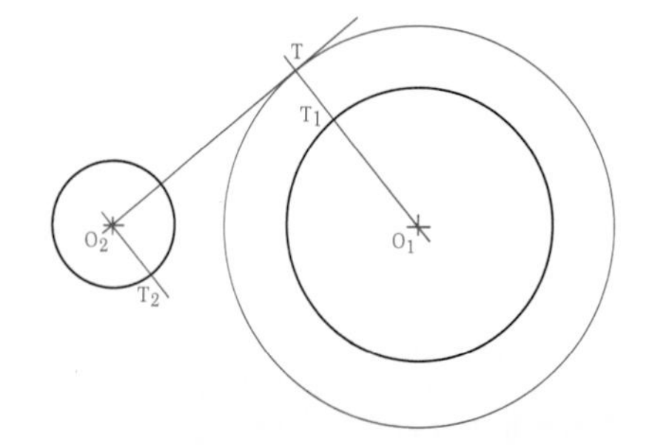

From center point O_2, draw a tangent O_2T to the circle of radius R. Draw radius O_1T to locate tangent point T_1. Draw O_2T_2 parallel to O_1T.

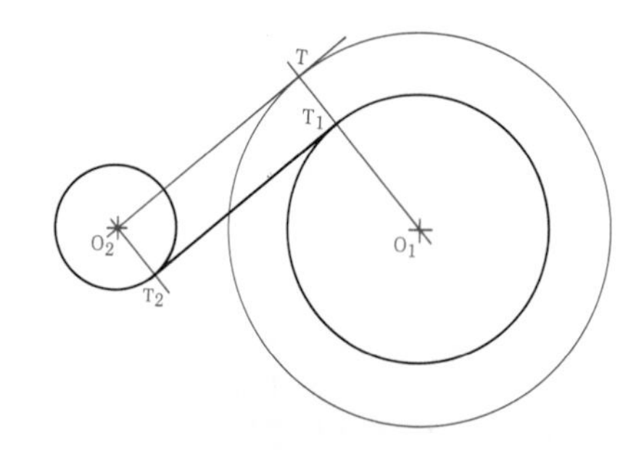

Draw the needed tangent T_1T_2 parallel to TO_2.

Fig. 4–42 To draw an interior common tangent to two circles of unequal radii.

PROBLEM: Construct an arc tangent to line AB and arc CD.

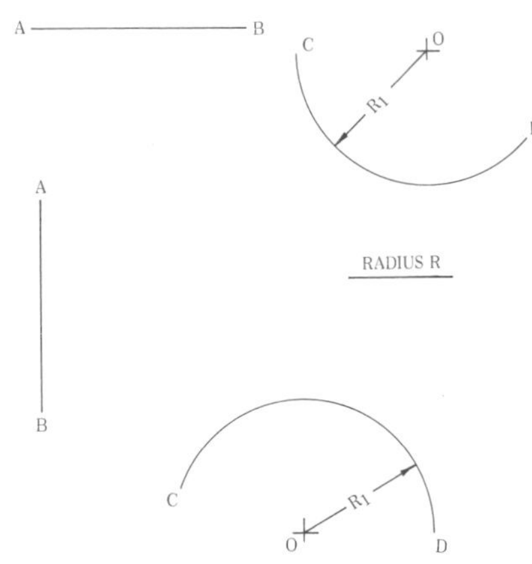

Given line *AB*, arc *CD*, and radius *R*.

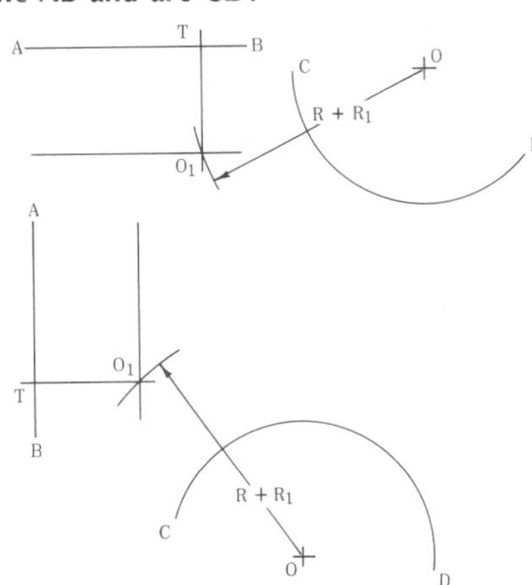

Draw a line parallel to *AB*, at distance *R*, toward arc *CD*. Use radius $R_1 + R$ to locate point O_1. A perpendicular line from O_1 to *AB* locates tangent point *T*.

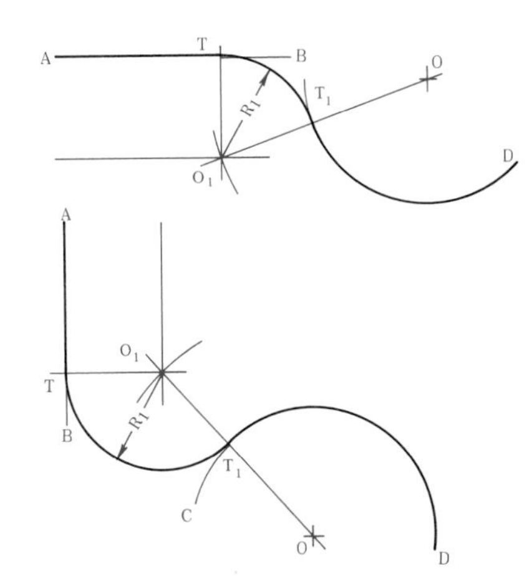

Draw a line from *O* to O_1 to locate tangent point T_1 on *CD*. With point O_1 as the center and radius *R*, draw the tangent arc.

Fig. 4–43 To draw an arc of given radius tangent to an arc and a straight line.

PROBLEM: Construct an ellipse (oval) by the pin-and-string method.

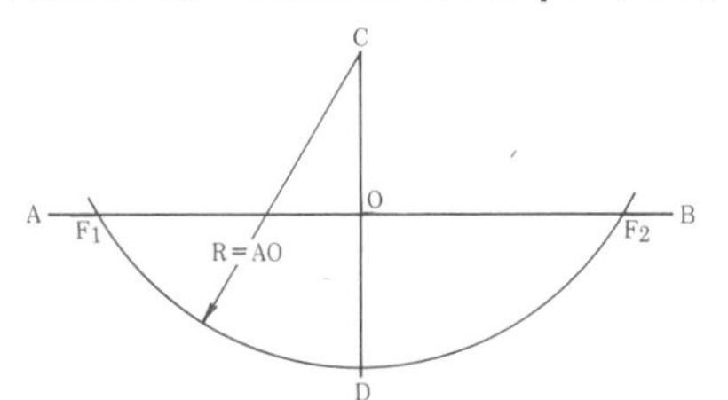

Given major axis *AB* and minor axis *CD* intersecting at *O*. With *C* as center and radius *R* = *AO*, an arc locates F_1 and F_2.

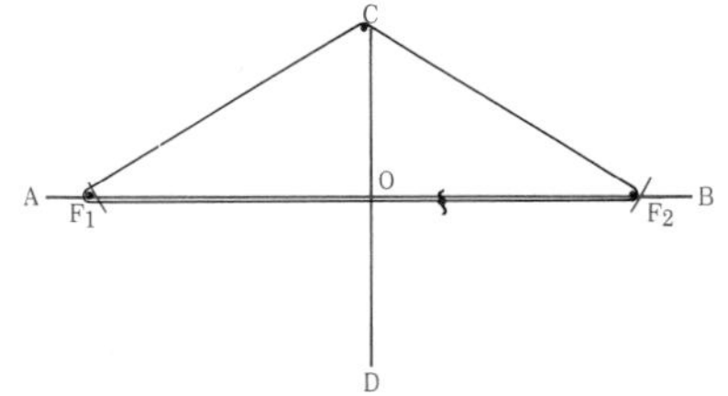

Place pins at points F_1, *C*, and F_2. Tie a string around the three pins and remove pin *C*.

Put the point of a pencil in the loop and draw the ellipse. Keep the string taut (tight) when moving the pencil.

Fig. 4–44 To draw an ellipse by the pin-and-string method.

PROBLEM: Construct an ellipse by the trammel method.

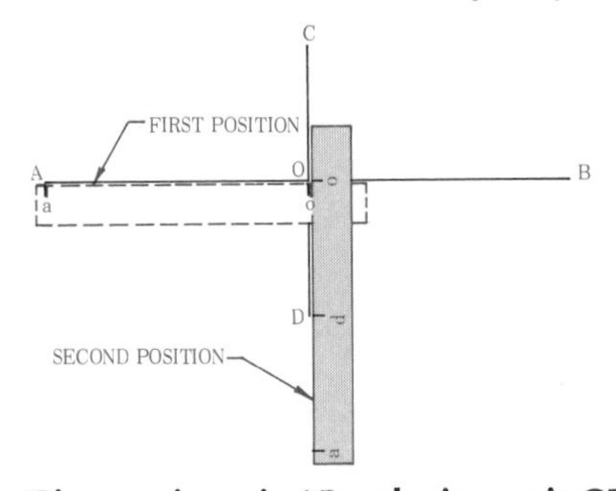

Given major axis *AB* and minor axis *CD* intersecting at point *O*. Cut a strip of paper or plastic (*trammel*). Mark off distances *AO* and *OD* on the trammel.

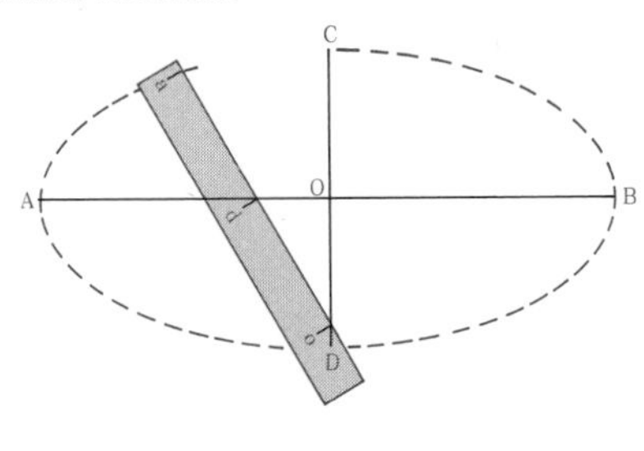

On the trammel, move point *o* along line *CD* (minor axis) and point *d* along line *AB* (major axis) and mark points at *a*.

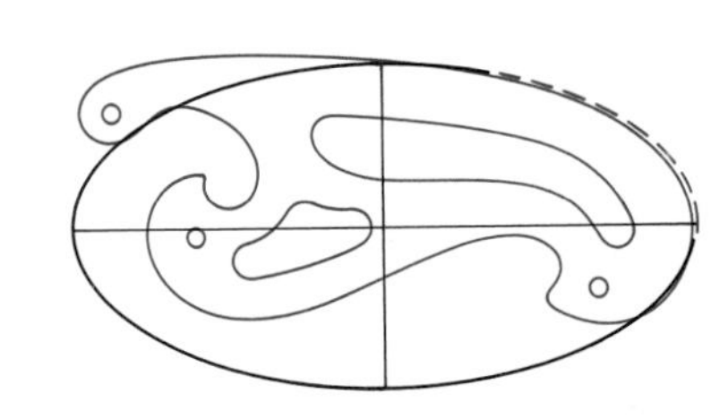

Use a French curve or flexible curve to connect the points to draw the ellipse.

Fig. 4–45 To draw an ellipse by the trammel method.

PROBLEM: Draw an ellipse using axes AB and CD.

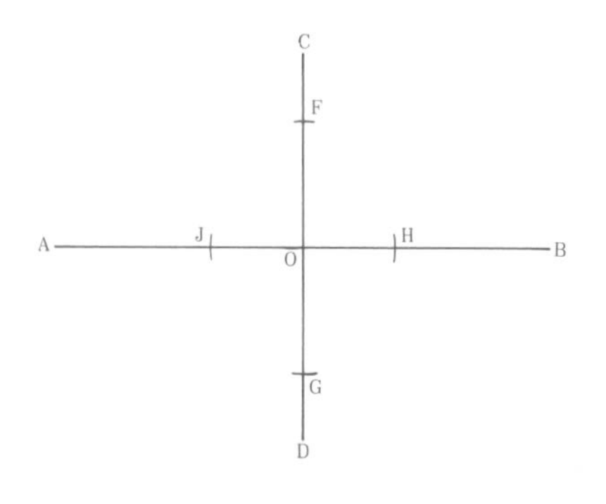

Given major axis *AB* and minor axis *CD* intersecting at point *O*. Lay off *OF* and *OG*, each equal to *AB* minus *CD*. Lay off *OJ* and *OH*, each equal to three-fourths of *OF*.

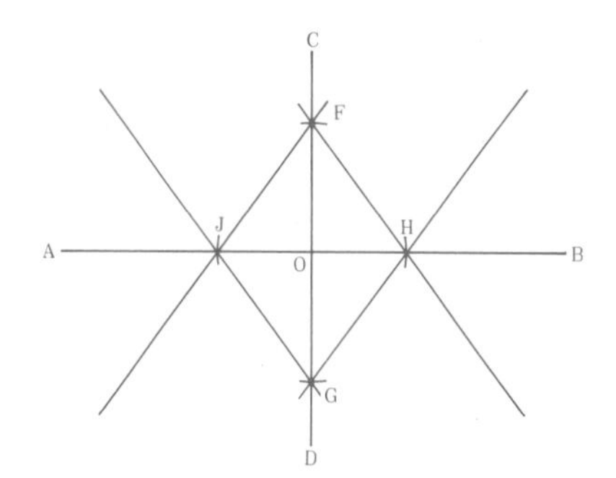

Draw and extend lines *GJ*, *GH*, *FJ*, and *FH*.

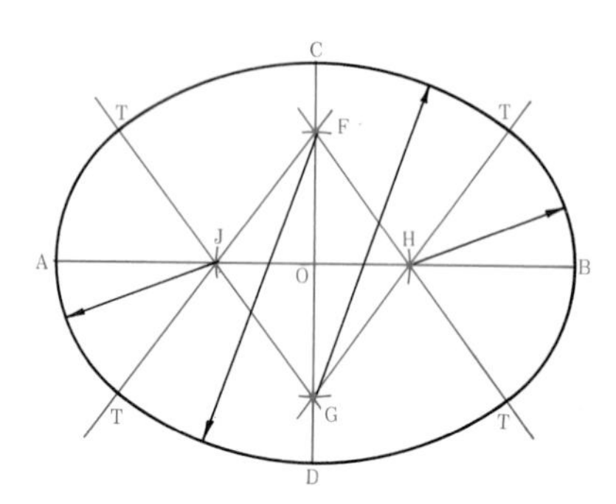

Draw arcs with centers *F* and *G* and radii *FD* and *GC* to the points of tangency. Draw arcs with centers *J* and *H* and radii *JA* an *HB* to complete the ellipse. The points of tangency are marked *T*.

Fig. 4–46 To draw an approximate ellipse when the minor axis is at least two-thirds the size of the major axis.

PROBLEM: Draw an ellipse using axes AB and CD.

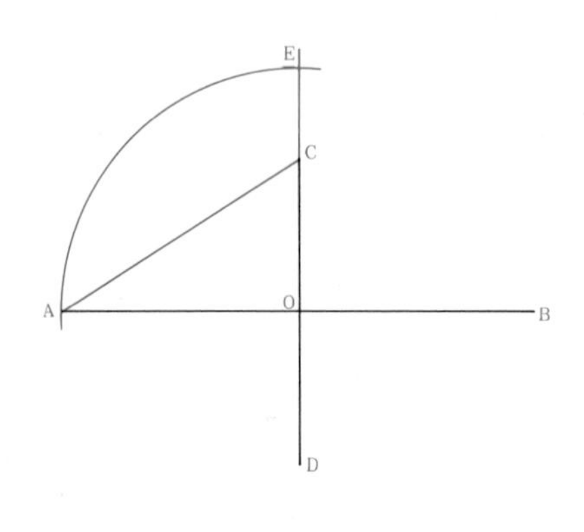

Given major axis *AB* and minor axis *CD* intersecting at point *O*. Draw line *AC*. Draw an arc with point *O* as the center and radius *OA* and extend line *CD* to locate point *E*.

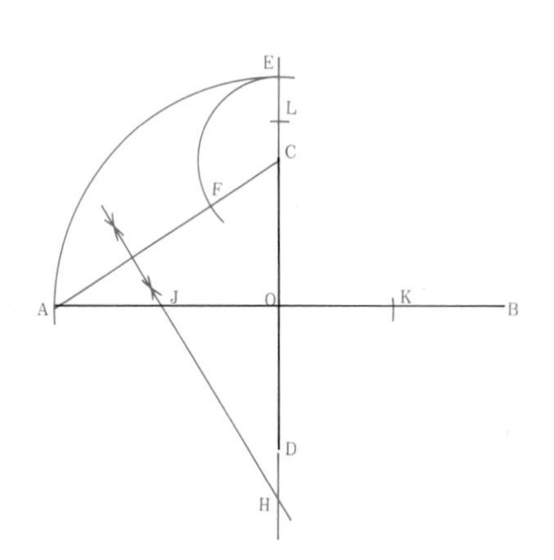

Draw an arc with point *C* as the center and radius *CE* to locate point *F*. Draw the perpendicular bisector of *AF* to locate points *J* and *H*. Locate points *L* and *K*. *OL* = *OH* and *OK* = *OJ*.

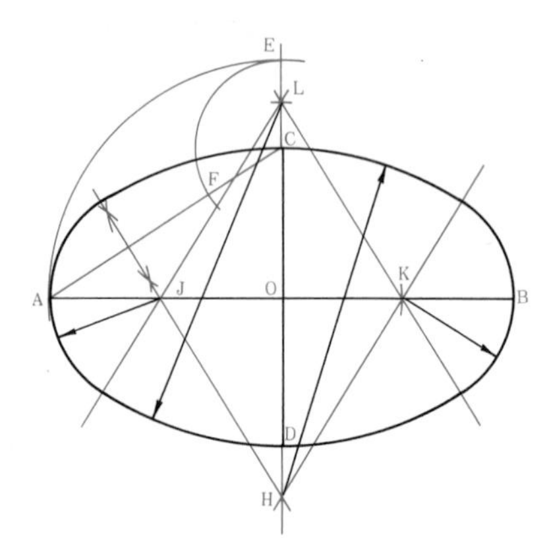

Draw arcs with *J* and *K* as centers and radii *JA* and *KB*. Draw arcs with *H* and *L* as centers and radii *HC* and *LD* to complete the ellipse.

Fig. 4–47 To draw an approximate ellipse when the minor axis is less than two-thirds of the major axis.

PROBLEM: Reduce or enlarge the drawing of the sailboat shown at A.

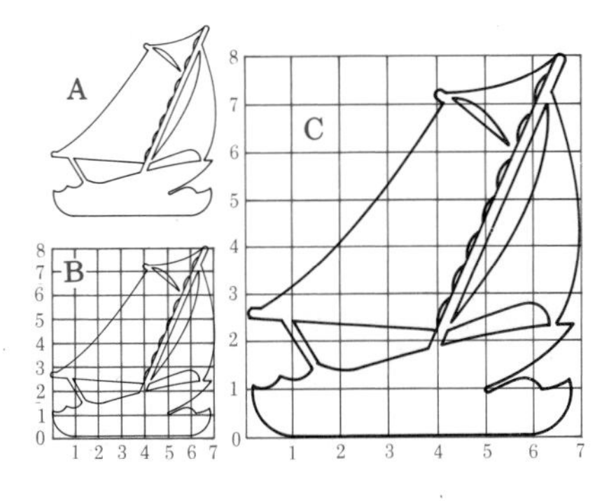

Lay out a grid over the drawing. Use squares of an approximate size. Draw a larger or smaller grid on a separate sheet of paper. The size of the grid depends upon the amount of enlargement or reduction needed. Use dots to mark key points on the enlarged or reduced grid corresponding to points on the original drawing. Connect the points to complete the new drawing.

Fig. 4–48 **To reduce or enlarge a drawing.**

PROBLEM: Change the proportion of the drawing shown at A.

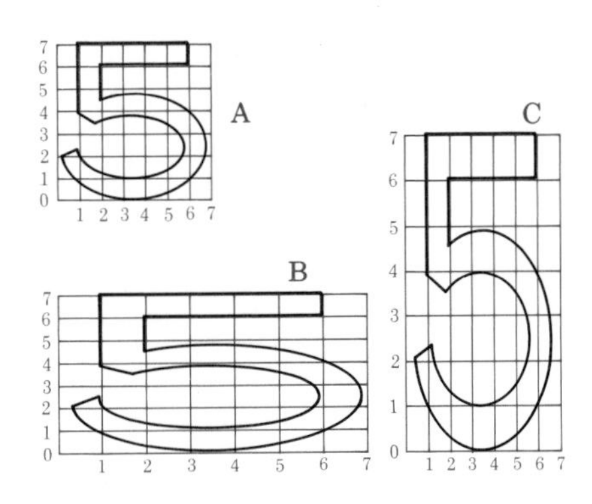

Draw a grid over the original drawing. Draw a grid on a separate sheet of paper in the needed proportion, as at *B* or *C*. Use dots to mark key points from the original drawing. Connect the points to complete the new drawing.

Fig. 4–49 **To change the proportion of a drawing.**

PROBLEM: Enlarge or reduce the original rectangle.

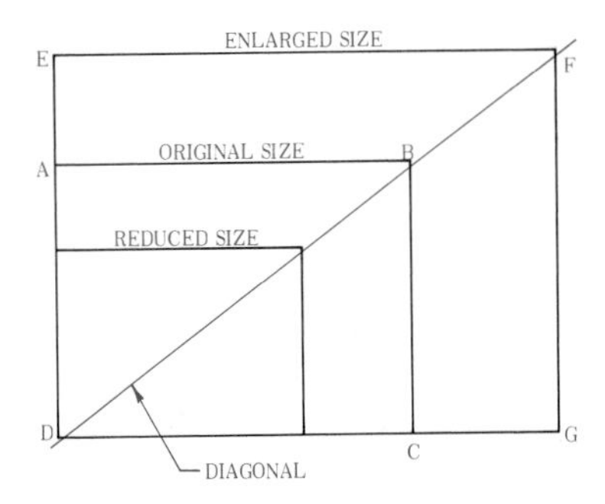

Draw a diagonal through corners *D* and *B*. Measure the width or height you need along line *DC* or *DA* (example: *DG*). Draw a perpendicular line from that point (*G*) to the diagonal. Draw a line perpendicular to *DE* intersecting at point *F*. Reductions are made in the same way.

Fig. 4–50 **To enlarge or reduce a square or rectangular area.**

Review

1. What does bisect mean?

2. What is another name for perpendicular?

3. The radius of a circle is what part of the diameter?

4. What name is given to the distance around a circle?

5. Describe an isosceles triangle.

6. Define an equilateral triangle.

7. What is an angle greater than 90° called?

8. What is an angle less than 90° called?

9. What is another name for a reverse curve?

10. Describe a right triangle.

11. A pentagon has how many sides?

12. Define a perpendicular bisector.

13. The shape of an ellipse is determined by two axes. Name them.

14. Name three ways of drawing an ellipse.

Problems

The problems for drafting geometry are designed to help the student develop accuracy in the use of instruments and to familiarize the student with the basic constructions that occur most frequently in drafting.

Suggestions

1. Make all construction (layout) lines thin and light.
2. Work accurately.
3. Do not erase construction lines.
4. Locate points by two short intersecting lines.
5. Show the exact length of lines by two short intersecting lines.

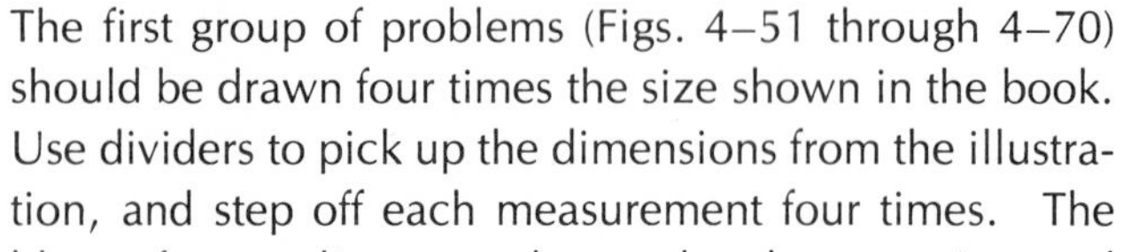

The first group of problems (Figs. 4–51 through 4–70) should be drawn four times the size shown in the book. Use dividers to pick up the dimensions from the illustration, and step off each measurement four times. The blue reference lines may be used to locate points and lines within the layout.

Nearly all the Problems for Chapter 3 may be used as Problems for Chapter 4 and vice versa.

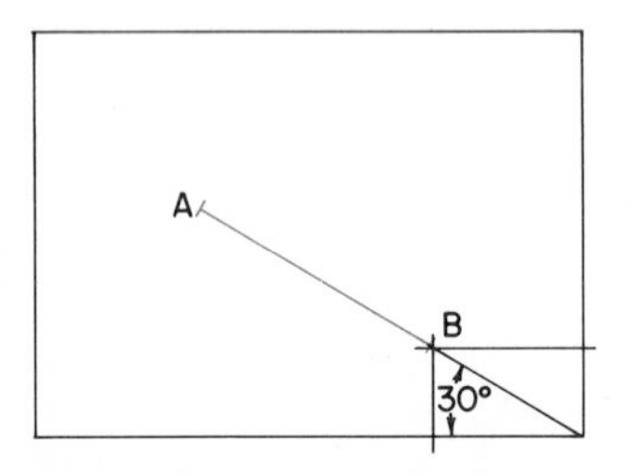

Fig. 4–51 **Bisect line *AB*.**

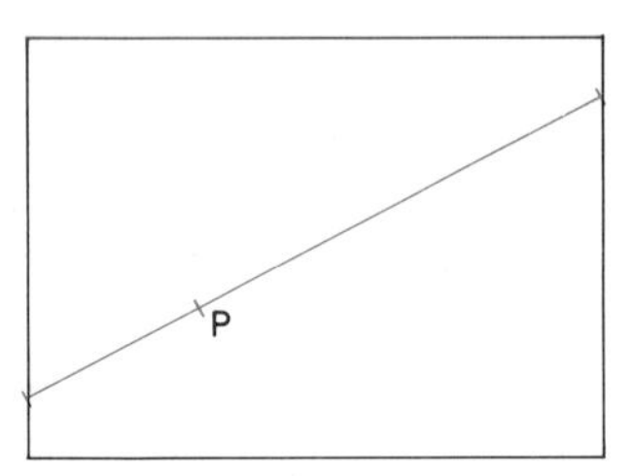

Fig. 4–52 **Construct a perpendicular at point *P*.**

Fig. 5–53 **Divide line *AB* into five equal parts.**

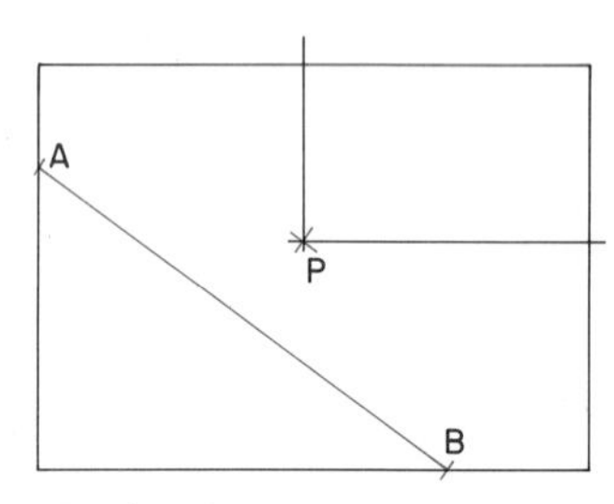

Fig. 4–54 **Construct line *CD* parallel to *AB* and equal in length to *AB* through *P*.**

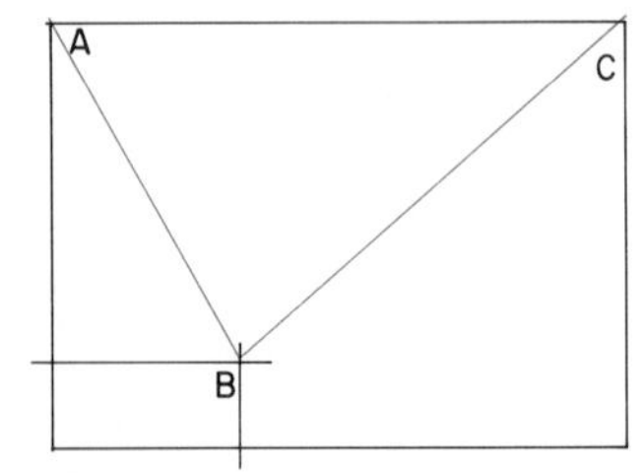

Fig. 4–55 **Bisect angle *ABC*.**

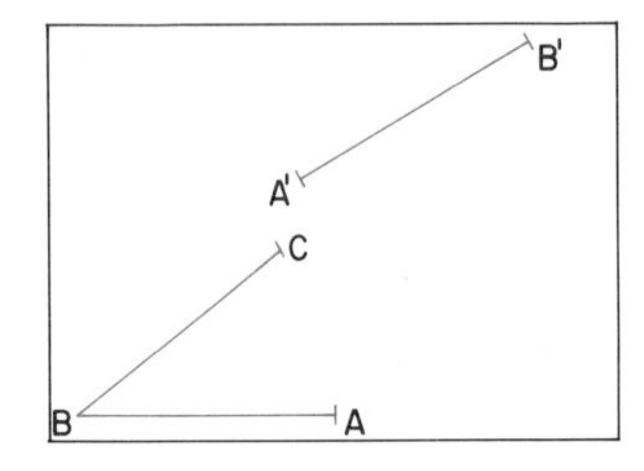

Fig. 4–56 **Copy angle *ABC* in a new location, beginning with *A′B′*.**

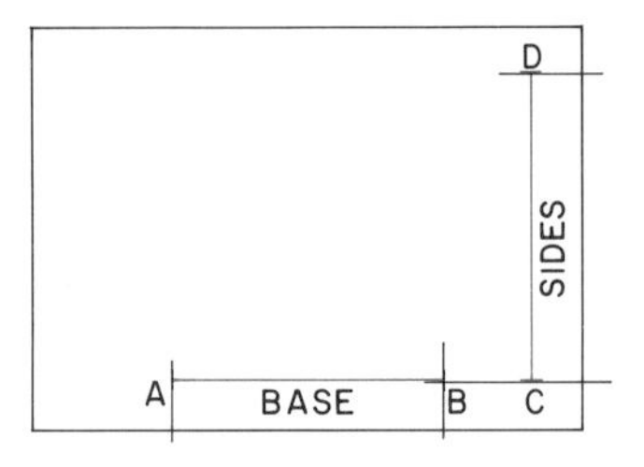

Fig. 4–57 **Construct an isosceles triangle on base *AB* with sides equal to *CD*.**

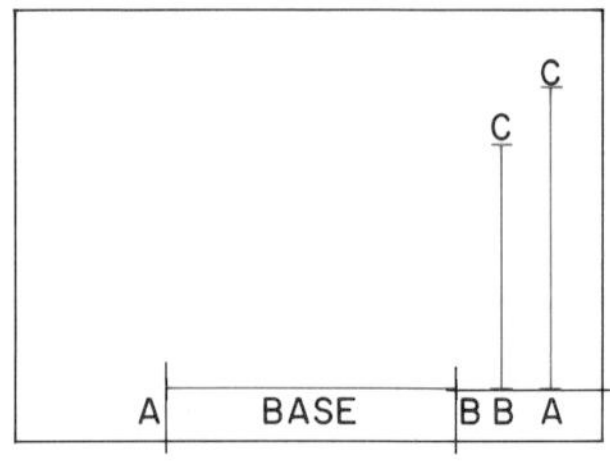

Fig. 4–58 **Construct a triangle on base *AB* with sides equal to *BC* and *AC*.**

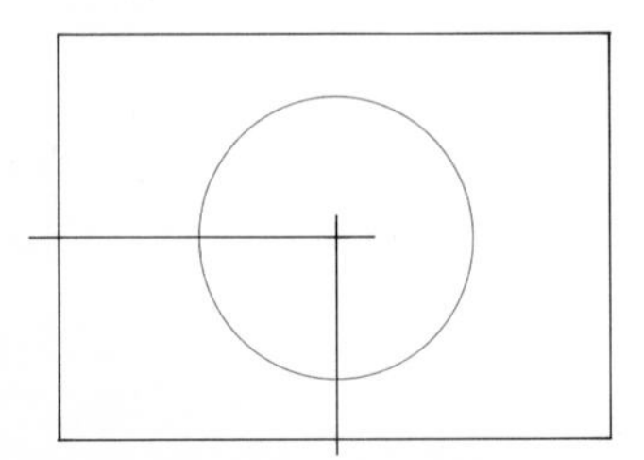

Fig. 4–59 **Construct a square in the circle with corners touching the circle.**

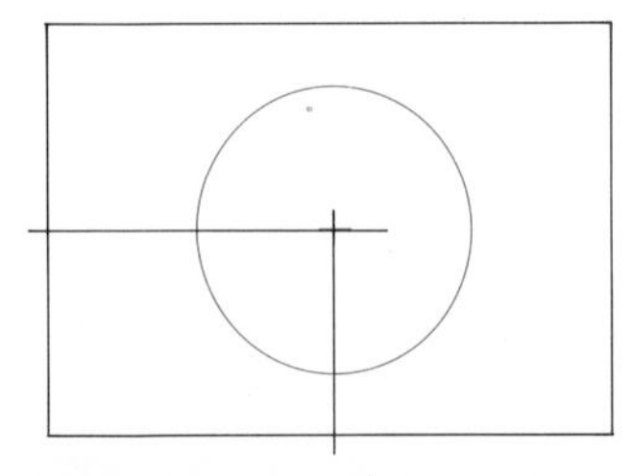

Fig. 4–60 **Construct a regular pentagon within the circle.**

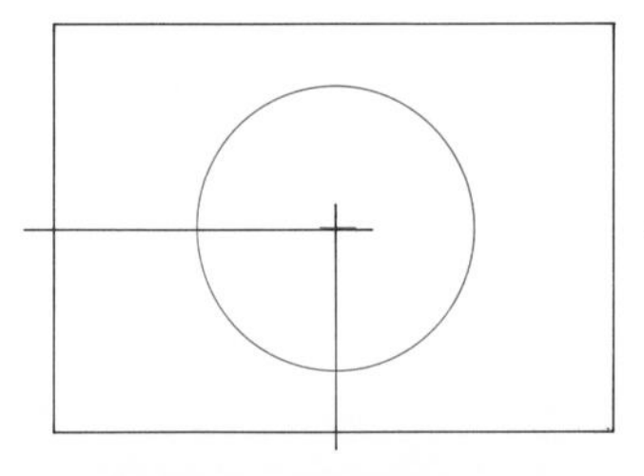

Fig. 4–61 **Construct a regular hexagon around the circle.**

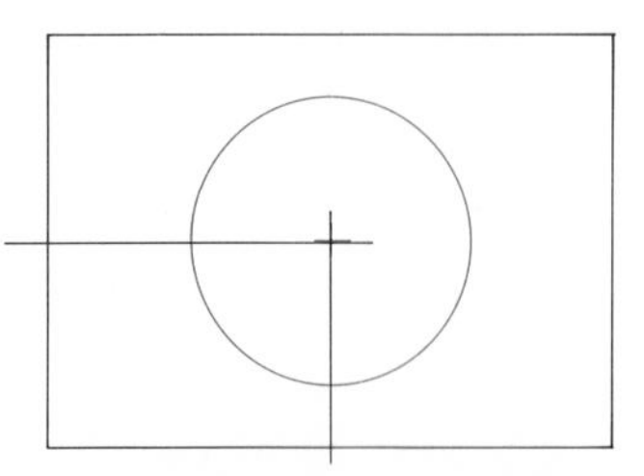

Fig. 4–62 **Construct a regular octagon around the circle.**

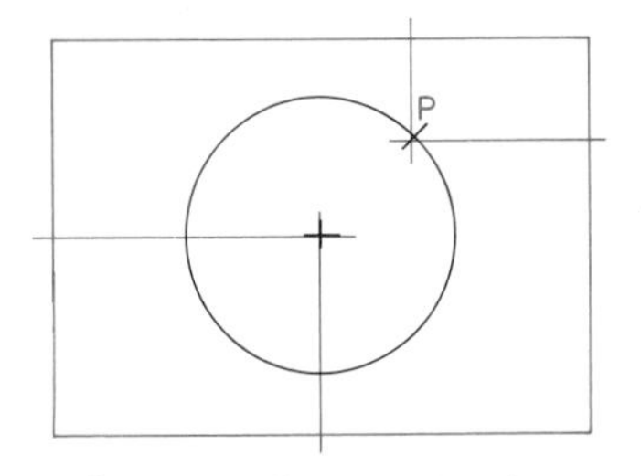

Fig. 4–63 Construct a tangent line through point *P* on the circle.

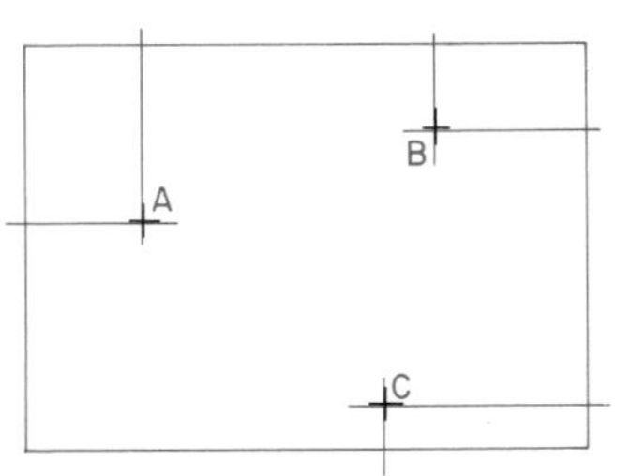

Fig. 4–64 Construct a circle through points *A, B,* and *C*.

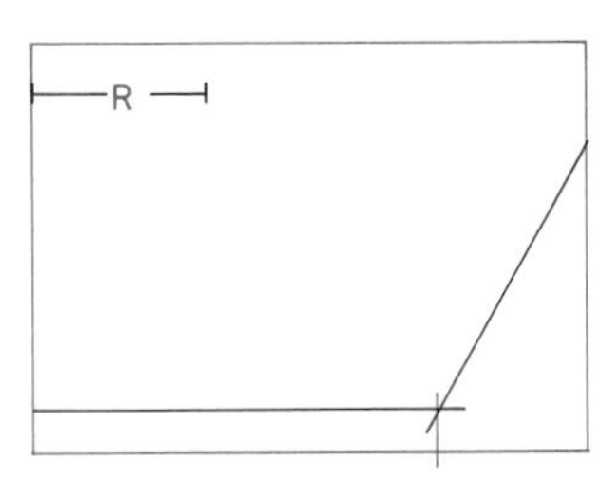

Fig. 4–65 Construct an arc having a radius *R* tangent to the two lines.

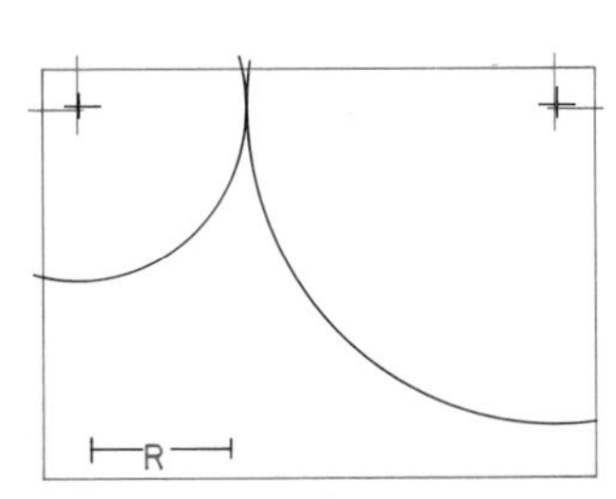

Fig. 4–66 Construct an arc having a radius *R* tangent to two given arcs.

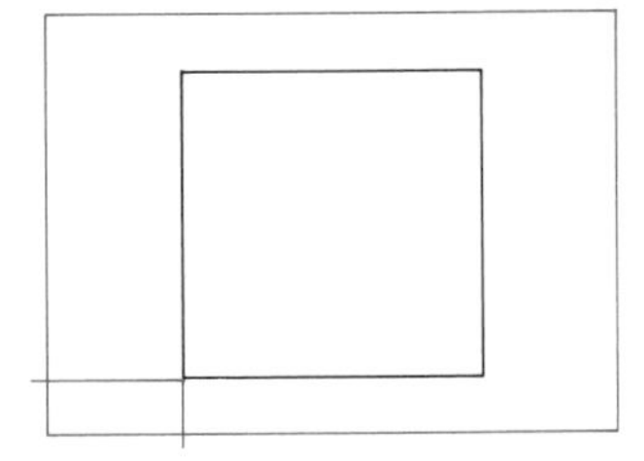

Fig. 4–67 Construct a regular octagon within the square.

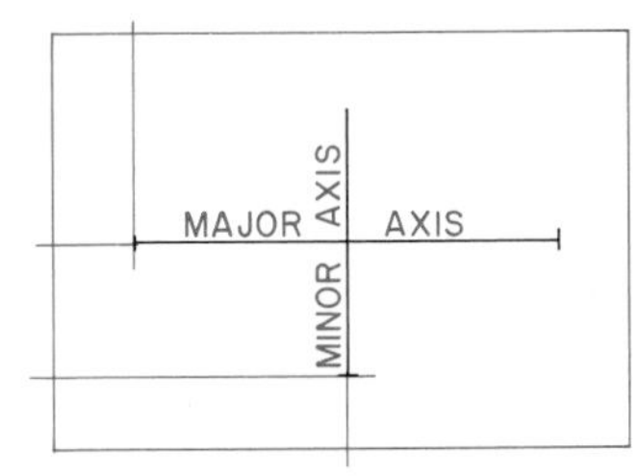

Fig. 4–68 Construct an ellipse on the major and minor axes. Use method assigned by the instructor.

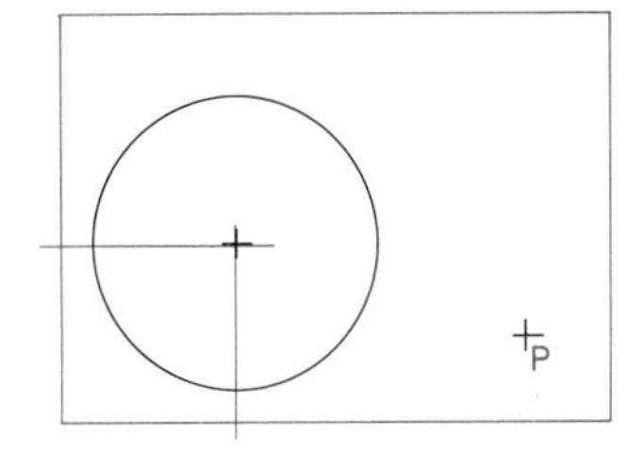

Fig. 4–69 Construct a line from point *P* tangent to the circle.

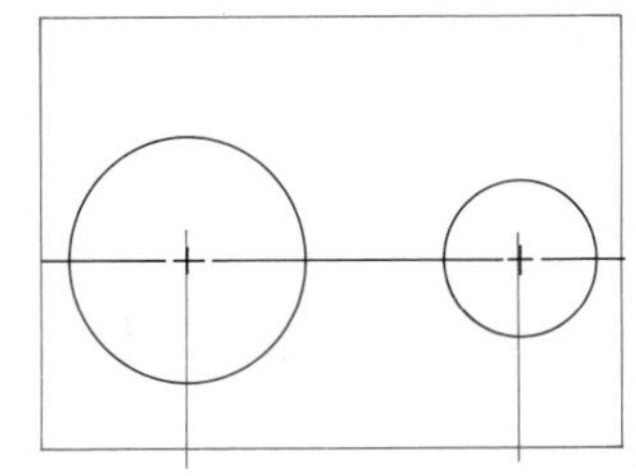

Fig. 4–70 Construct a tangent line to the exterior of the two circles.

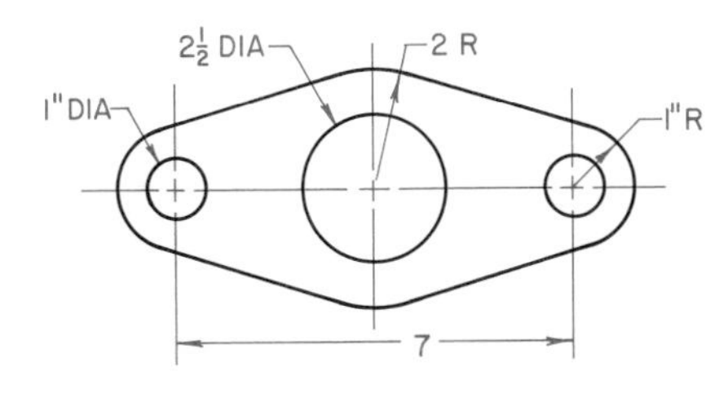

Fig. 4–71 Draw the view of the gasket full size or as assigned. Mark all points of tangency. Do not dimension.

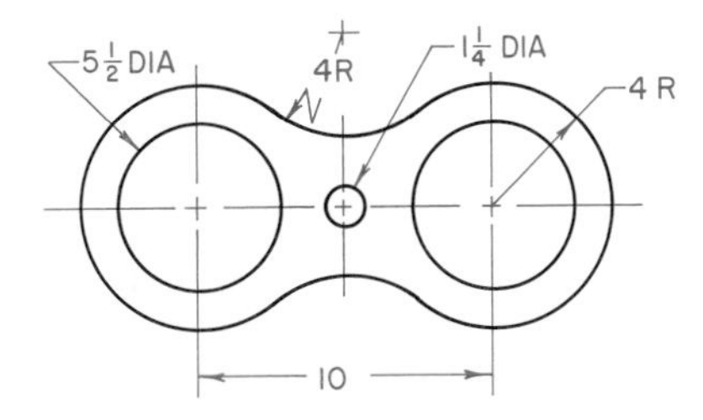

Fig. 4–72 Pipe support. Scale: as assigned. Locate and mark all centers and all points of tangency. Do not dimension.

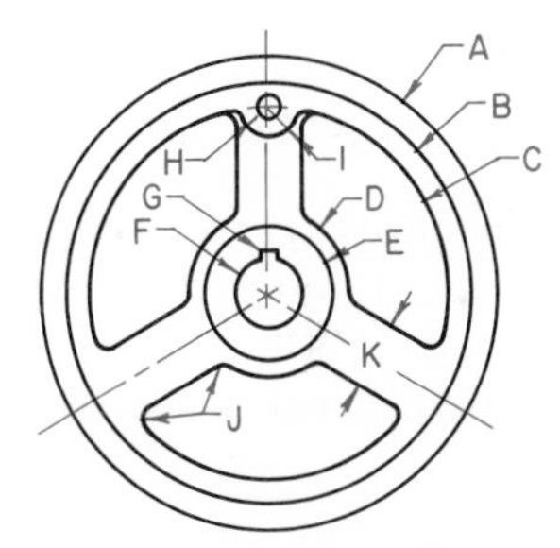

Fig. 4–73 Handwheel. Scale: as assigned. *A* = 7″ DIA, *B*, = 6$\frac{1}{8}$″ DIA, *C* = 2$\frac{3}{4}$″ R, *D* = 1$\frac{1}{4}$″ R, *E* = 2″ DIA, *F* = 1″ DIA, *G* (keyway) = $\frac{3}{16}$″ wide × $\frac{3}{32}$″ deep, *H* = $\frac{3}{8}$″ DIA, *I* = $\frac{3}{8}$″ R, *J* = $\frac{3}{16}$″ R, *K* = 1″.

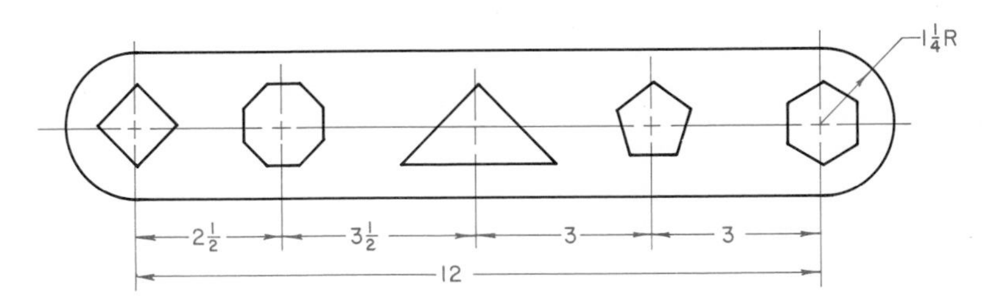

Fig. 4–74 Combination wrench. Scale: as assigned. Square: 1″; octagon: 1$\frac{3}{8}$″ across flats; isosceles triangle: 2$\frac{3}{4}$″ base, 2″ sides; pentagon: inscribed within 1$\frac{3}{8}$″-diameter circle; hexagon: 1$\frac{1}{4}$″ across flats. Use the method of your choice for constructing geometric shapes. Do not erase construction lines.

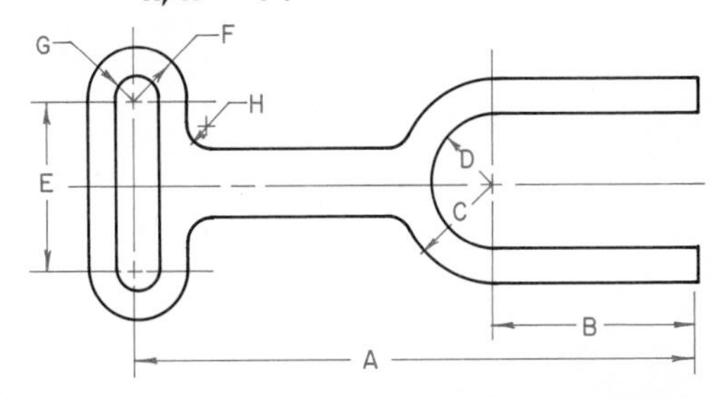

Fig. 4–75 Adjustable fork. Scale: as assigned. *A* = 220 mm, *B* = 80 mm, *C* = 40 mm, *D* = 26mm, *E* = 64 mm, *F* = 20 mm, *G* = 8 mm, *H* = 10 mm.

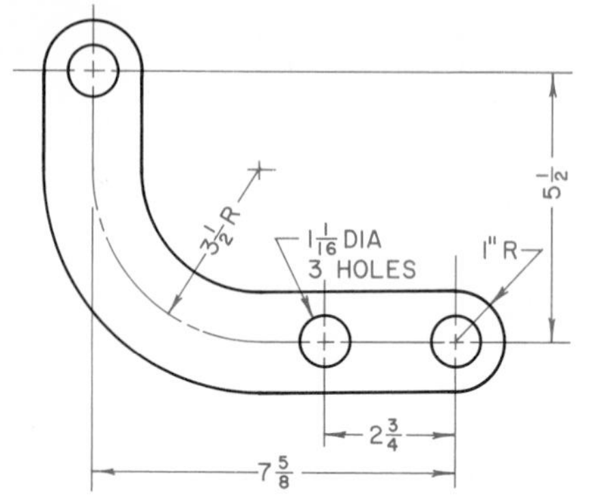

Fig. 4–76 **Rod support. Scale: as assigned by instructor.**

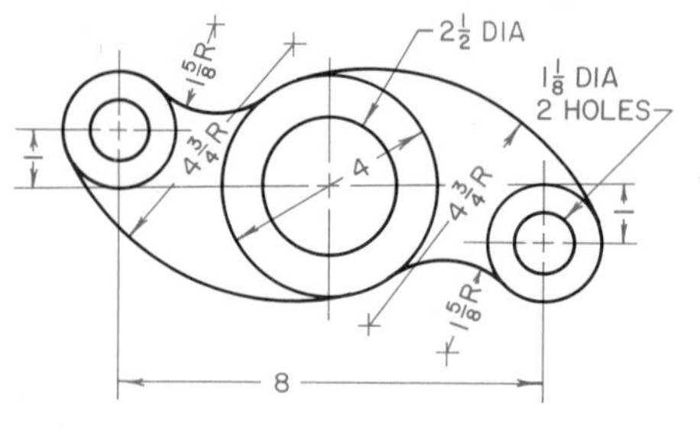

Fig. 4–77 **Rocker arm. Scale: full size or as assigned.**

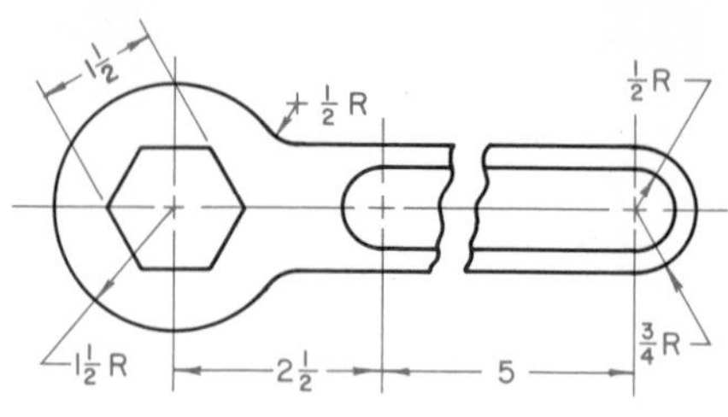

Fig. 4–78 **Hex wrench. Scale: as assigned. Mark all tangent points. Do not erase construction lines.**

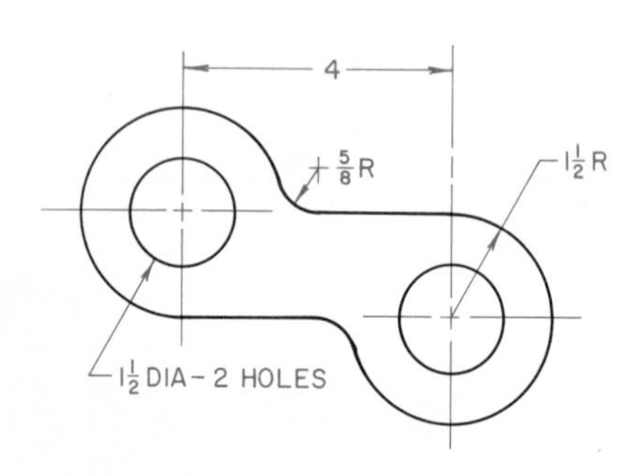

Fig. 4–79 **Offset link. Scale: full size or as assigned. Locate and mark all points of tangency.**

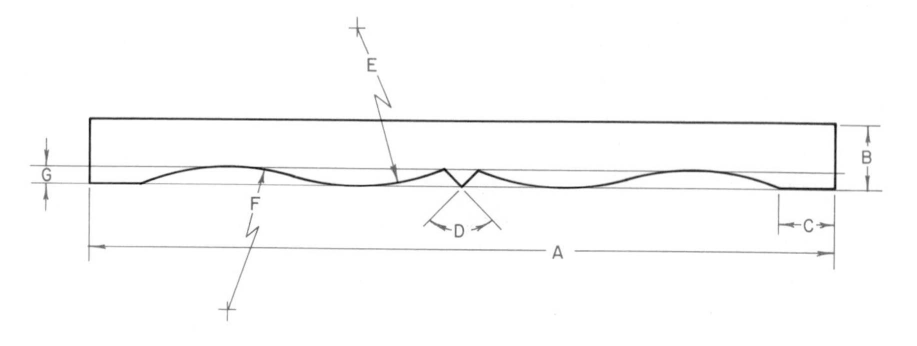

Fig. 4–80 **Valance board. Scale: 1″ = 1′–0″ or as assigned.** ***A*** **= 8′–0″,** ***B*** **= 0′–8″,** ***C*** **= 0′–7″,** ***D*** **= 90°,** ***E*** **= 2′–6″,** ***F*** **= 2′–6″,** ***G*** **= 0′–2″. Locate and mark points of tangency. Do not erase construction lines.**

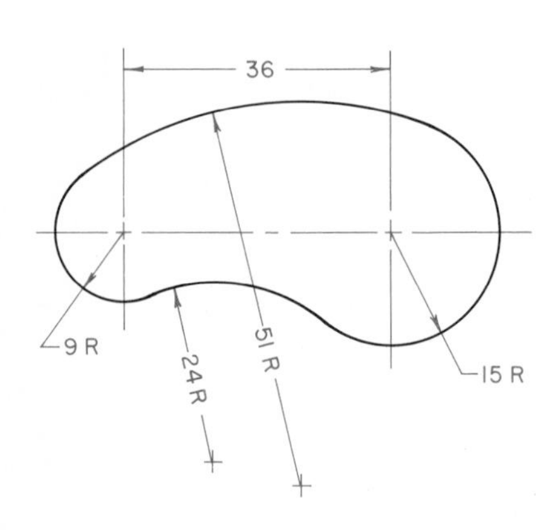

Fig. 4–81 **Kidney-shaped table top. Scale: full size or as assigned.**

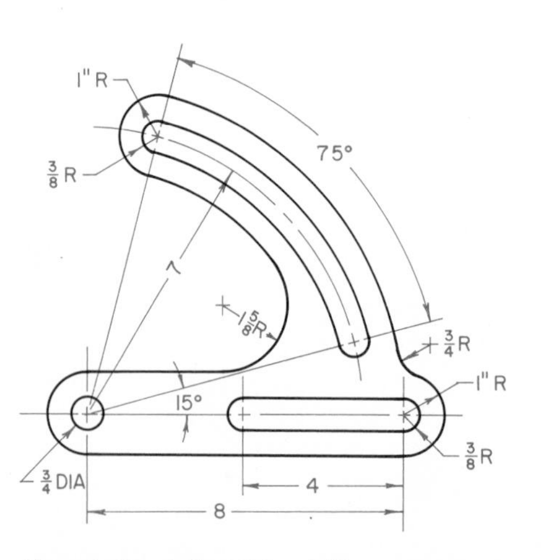

Fig. 4–82 **Adjustable table support. Scale: as assigned.**

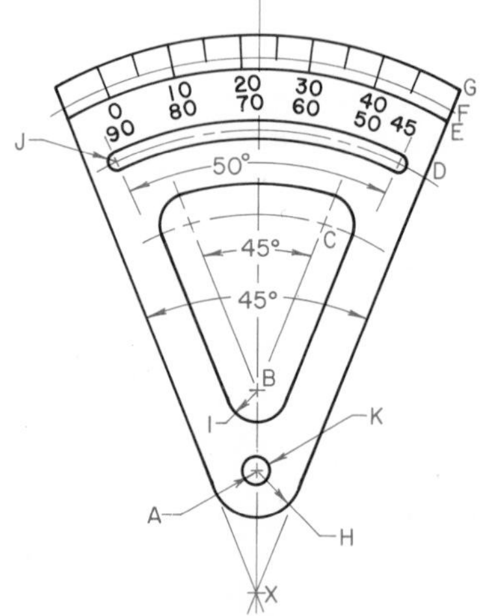

Fig. 4–83 **Tilt scale. Scale: as assigned.** ***AB*** **= 1¾,** ***AX*** **= 2⅝,** ***AC*** **= 5½,** ***AD*** **= 7¼,** ***AE*** **= 8½,** ***AF*** **= 8¾,** ***AG*** **= 9¼,** ***H*** **= 1″R** ***I*** **= ⅝ R,** ***J*** **= 3/16 R,** ***K*** **= ½ DIA.**

Chapter 5

Multiview Drawing

INTRODUCTION TO MULTIVIEW DRAWING

As you learned in Chapter 1, technical drawing is a way of communicating ideas. People communicate by verbal and written language and by *graphic* (pictorial) means. One of the graphic means is technical drawing. It is a language used and understood in all countries. When accurate *visual* (sight) understanding is necessary, technical drawing is the most exact method that can be used.

Technical drawing involves two things: (1) visualization and (2) implementation. *Visualization* is the ability to see clearly what a machine, device, or other object looks like in the mind's eye. *Implementation* is the drawing of the object that has been visualized. In other words, the designer, engineer, or drafter first visualizes the object and then explains it pictorially by a technical drawing. Thus, the idea is recorded in a form that can be used as a means of communication.

A technical drawing, properly made, will give a more accurate and clearer description of an object than a photograph or written explanation. Technical drawings made according to standard *principles* (rules) result in views that give an exact visual description of an object (Fig. 5–1).

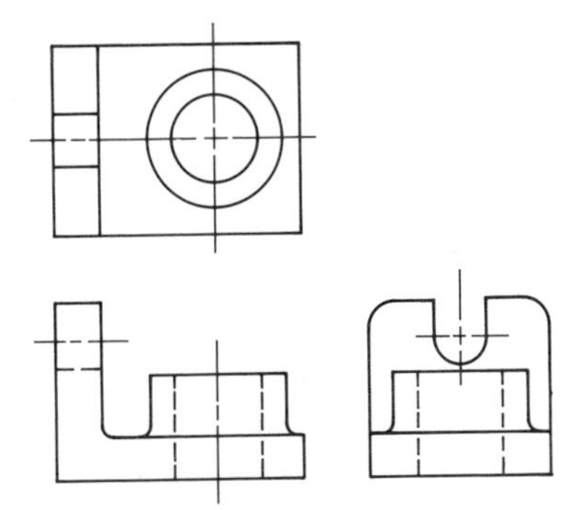

Fig. 5–1 **This three-view drawing gives an accurate description of the object.**

MULTIVIEW DRAWING

A photograph of a V-block is shown in Fig. 5–2. It shows the object as it appears to the eye. Notice that three sides of the V-block are shown in a single view. In Figure 5–3 the same photograph is shown with the three sides labeled according to their *relative* (related) positions. If all sides could be shown in a single photograph, it would also include a left-side view, a rear view, and a bottom view. Nearly all objects have six sides.

An object cannot be photographed before it has been built. Therefore it is necessary to use another kind of graphic representation. One possibility is to make a *pictorial drawing,* as shown in Fig. 5–4. It shows, just as a photograph, the way the object looks generally. However, it does not show the exact forms and relationships of the parts of the object. It shows the V-block as it appears, not as it really is. For example, the

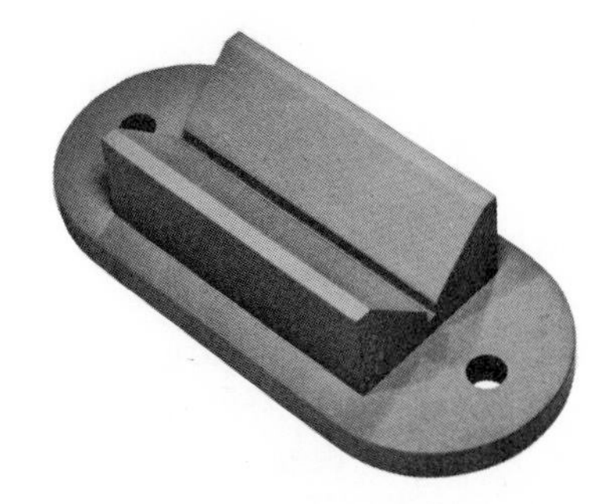

Fig. 5–2 **Photograph of a V-block.**

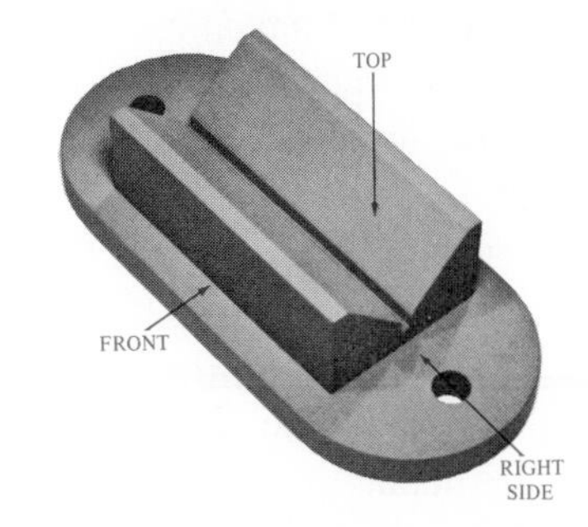

Fig. 5–3 **Photograph of a V-block with front, top, and side views labeled.**

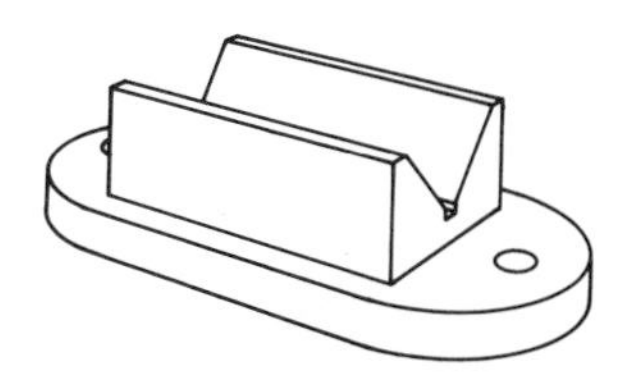

Fig. 5–4 **Pictorial drawing.**

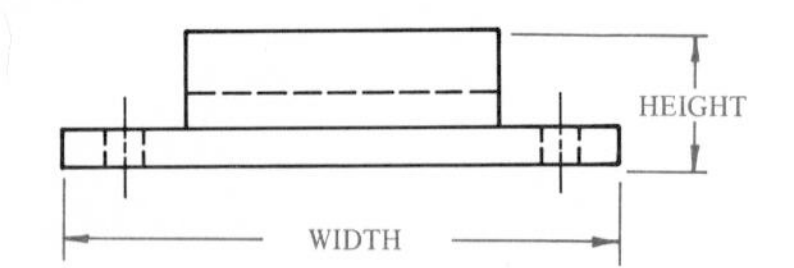

Fig. 5–5 **Front view of V-block.**

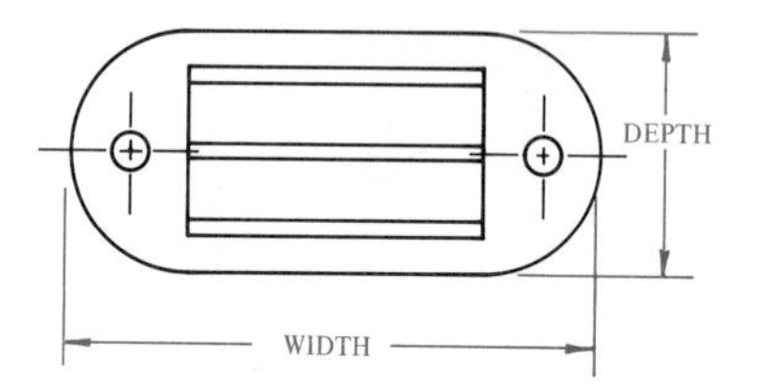

Fig. 5–6 **Top view of V-block.**

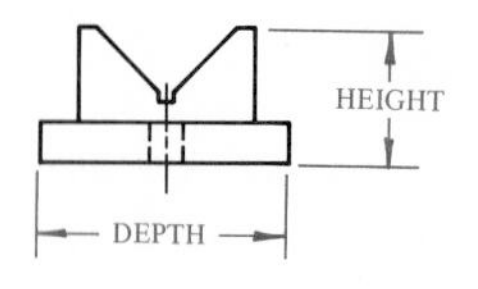

Fig. 5–7 **Right-side view of V-block.**

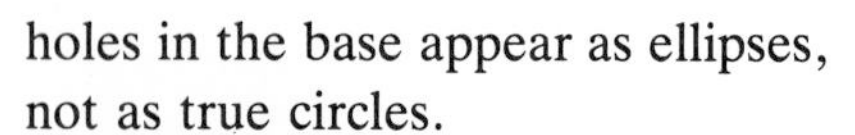

holes in the base appear as ellipses, not as true circles.

The problem, then, is to represent an object on a sheet of paper in a way that will describe its exact shape and proportions. This can be done by drawing views of the object as it is seen from different positions. These views are then arranged in a particular order.

In order to describe accurately the shape of each view, you must imagine a position directly in front of the object, then above it, and finally at the right side of it. This is where the ability to visualize is important. Figure 5–5 shows the exact shape of the V-block when viewed from the front. The dashed lines are used to show the outline of details behind the front surface (hidden details). Notice that this view shows the width and the height of the object.

Figure 5–6 is a top view of the V-block. It shows the width and the depth. Since the view is taken directly from above, the exact shape of the top is shown. Notice that the holes are true circles and that the rounded ends of the base are true radii. In the photograph and in the pictorial drawing, these appeared as elliptical shapes.

Figure 5–7 is a right-side view of the V-block. It shows the depth and the height. Notice that the shape of the V appears to be balanced in the drawing. It appears distorted or misshapen in the photograph and in the pictorial drawing.

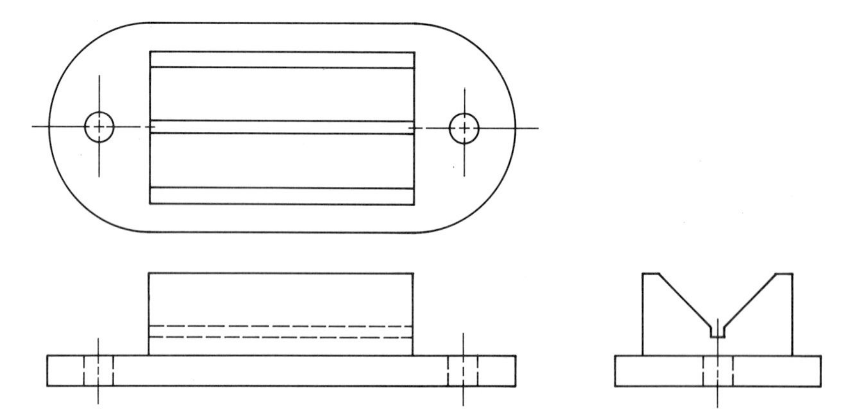

Fig. 5–8 **The relationship of the three normal views of the V-block.**

The Relationship of Views

Views must be placed in proper relationship to one another. Only in this way can technical drawings be read and understood properly. Figure 5–8 shows the V-block and how its three normal views have been *revolved* (turned) into their proper places. Notice that the top view is directly above the front view. The right-side view is directly to the right of the front view. Each of the normal views is where it logically belongs. When the normal views are placed in proper relationship to one another, the result is a *multiview drawing.* Multiview drawing is the exact representation of two or more views of an object on a *plane* (flat surface). These three views will usually give a complete description of the object. Figure 5–8 is an example of a multiview drawing.

Other Views

Most objects have six sides or six views. In most cases two or three views will completely describe the shape and size of all parts of an object. However, in some cases it may be necessary to show views other than the front, top, and right sides. In Fig. 5–9 dice are shown in the upper left-hand corner. Since the detail is different on each of the six sides, six views are needed for a complete graphic description. The six views are shown in their proper locations in the lower part of Fig. 5–9. Only in unusual cases are six views necessary.

ORTHOGRAPHIC PROJECTION

Earlier we said that multiview drawing is the exact representation of two or more views of an object on a plane (flat surface). These views are developed through the *principles* (rules) of orthographic projection. *Ortho-* means "straight or at right angles." *Graphic* means "written or drawn." *Projection* comes from two Latin words: *"pro,"* meaning "forward," and *"jacere,"* meaning "to throw." Thus, orthographic projection literally means "thrown forward, drawn at right angles." *Orthographic projection* is the method of representing the exact form of an object in two or more views on planes, usually at right angles to each other, by lines drawn *perpendicular* (at right angles) from the object to the planes.

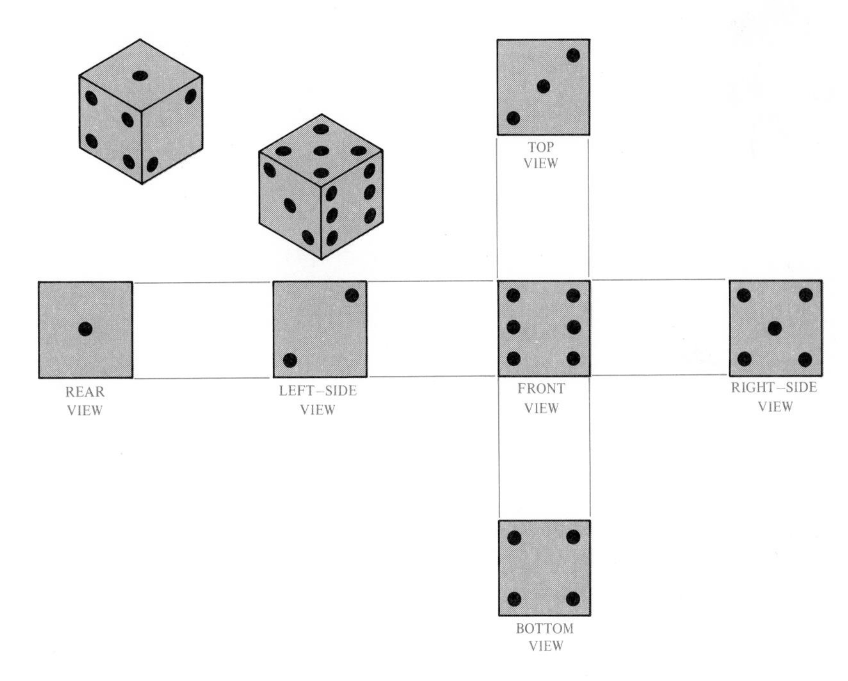

Fig. 5–9 **Pictorial drawing and six views of dice.**

Angles of Projection

Orthographic projection involves the use of three planes. They are the *vertical* (up and down) plane, the *horizontal* (side to side) plane, and the *profile* (side view) plane. These are shown in Fig. 5–10. In technical drawing, *a plane* is an imaginary flat surface that has no thickness. A view of an object is then projected and drawn upon this plane. Notice that the vertical and horizontal planes divide space into four *quadrants* (quarters of a circle). In orthographic projection, quadrants are usually called angles. Thus, we get the names *first-angle projection* and *third-angle projection.* First-angle projection is used in European countries. Third-angle projection is used in the United States and Canada. Second- and fourth-angle projections are not used in technical drawing.

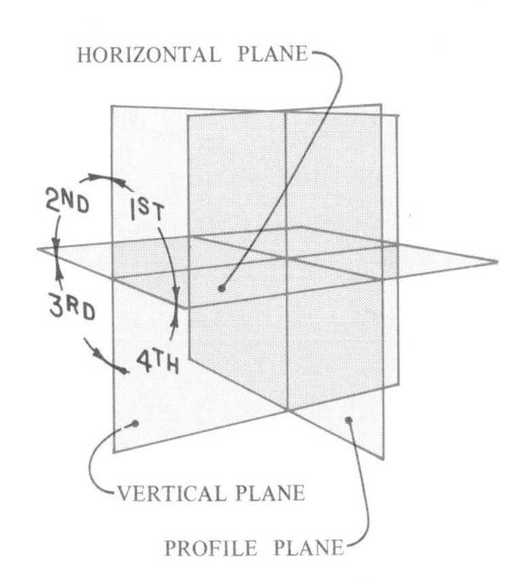

Fig. 5–10 **The three planes used in orthographic projection.**

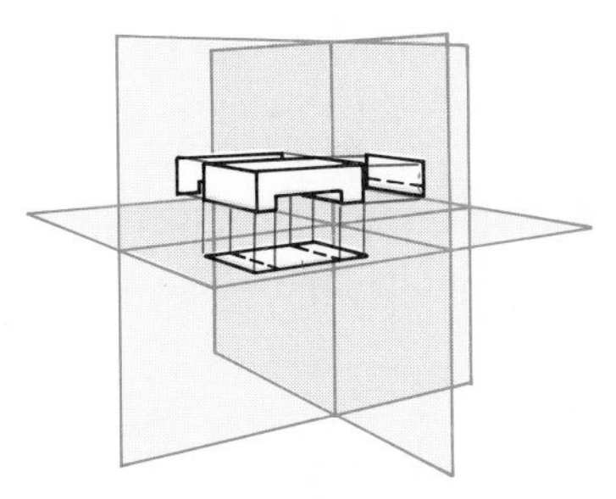

Fig. 5–11 **The position of the three planes used in first-angle projection.**

First-Angle Projection

Figure 5–11 shows an object within the three planes of the first quadrant for developing a three-view drawing in first-angle projection. The front view is projected to the vertical plane. The top view is projected to the horizontal plane. The *left-side view* is projected to the profile plane. The horizontal and profile planes are rotated into a single plane. *In this case the front view is above the top view.* The left-side view is to the right of the front view (Fig. 5–12).

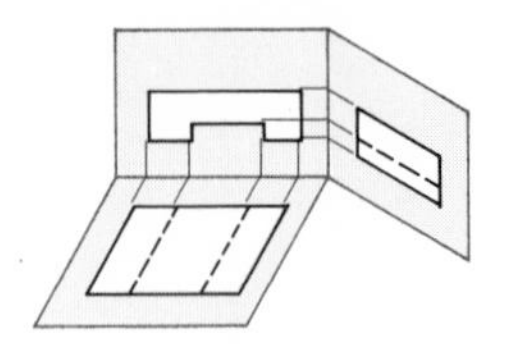

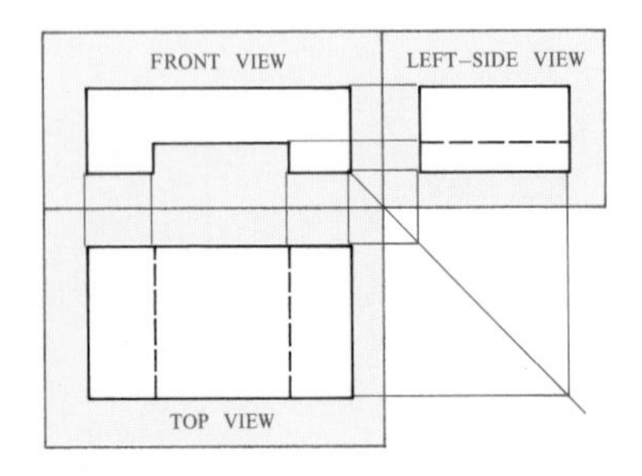

Fig. 5–12 **Three views in first-angle projection.**

Third-Angle Projection

Third-angle projection uses the same basic *principles* (rules) as first-angle. The main difference is in the *relative* (related) positions of the three planes. Figure 5–13 shows the same object placed within the third quadrant for developing a three-view drawing in third-angle projection. In this case the front view is projected to the vertical plane. The top view is projected to the horizontal plane. The *right-side view* is projected to the profile plane. The horizontal and profile planes are rotated into a single plane. *Thus, the top view is above the front view.* The right-side view is to the right side of the front view (Fig. 5–14).

THE GLASS BOX

The three views of an object have been developed by using imaginary *transparent* (see-through) planes. The views are projected onto these planes. We mentioned earlier that most objects have six sides. Therefore, six views may result. To explain the theory of projecting all six views, let us use an imaginary glass box.

Figure 5–15 shows the glass box partially opened with the six views labeled. When the box is fully opened up into one plane (Fig. 5–16), the views are in their relative positions as they would be if they had been drawn on paper. These views are arranged according to proper order for the six views. Notice that the rear view is located to the left of the left-side view.

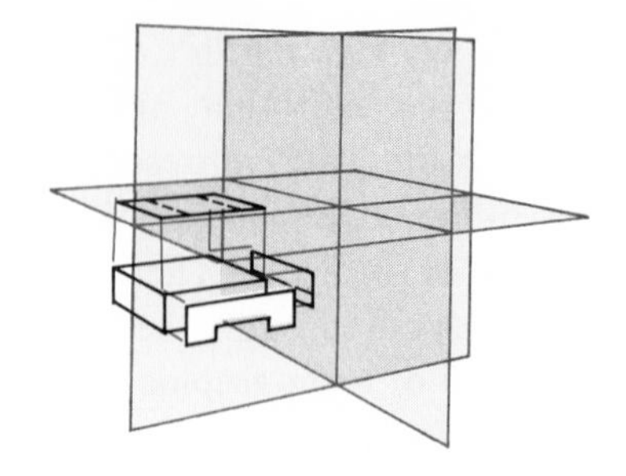

Fig. 5–13 **The position of the three planes used in third-angle projection.**

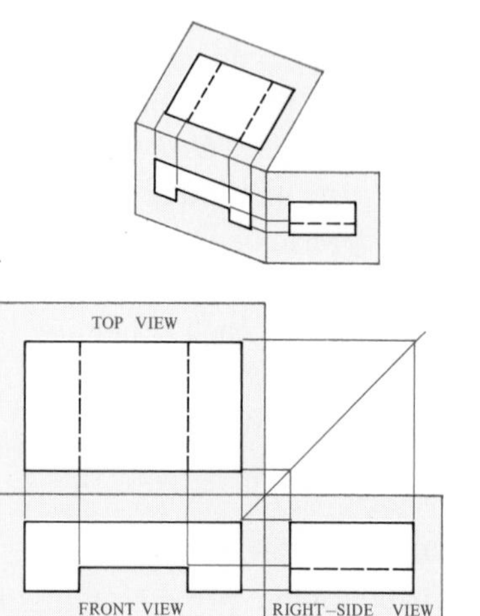

Fig. 5–14 **Three views in third-angle projection.**

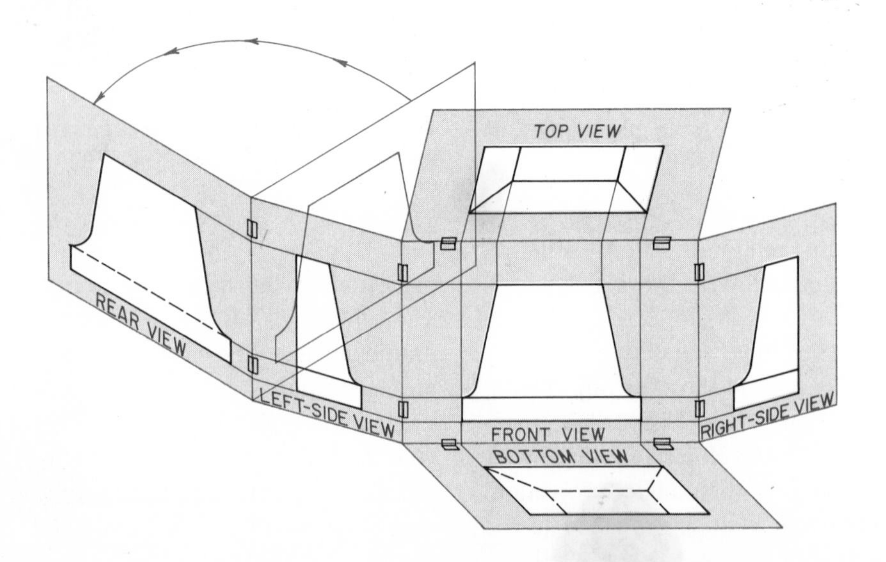

Fig. 5–15 **Opening the glass box.**

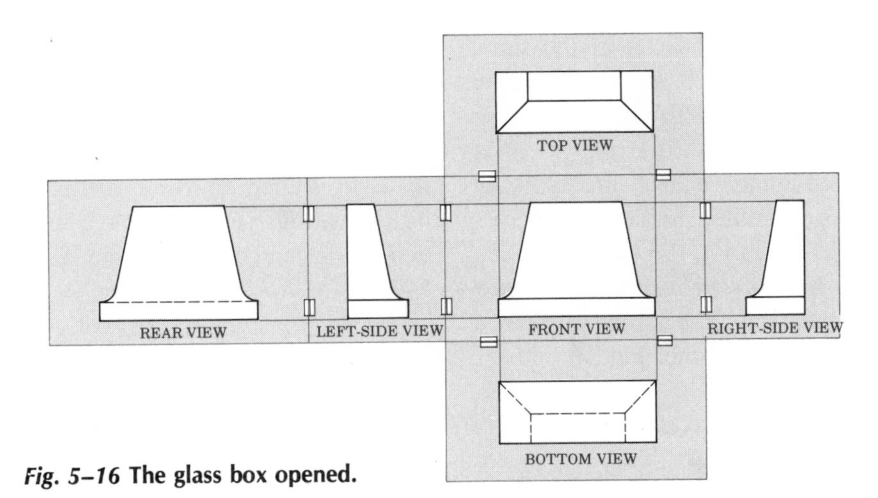

Fig. 5–16 **The glass box opened.**

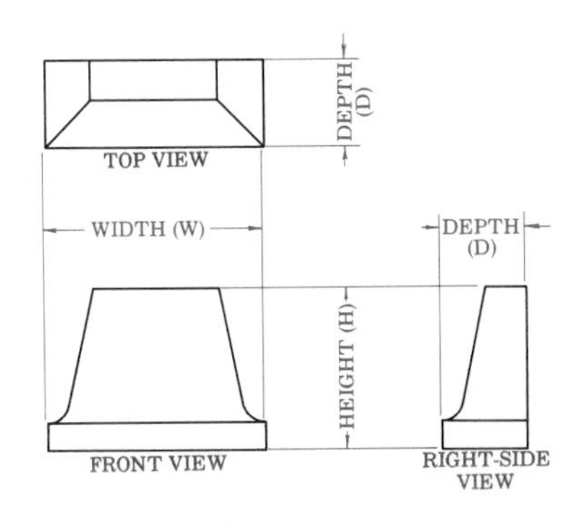

Fig. 5–17 **The front, top, and right-side views.**

Also notice that some views give the same information that is in other views. The views may also be mirror images of one another. Thus, it is not necessary to show all six views for a complete description of the object. The three normal views, as ordinarily drawn, are shown in Fig. 5–17.

HIDDEN LINES

It is necessary to describe every part of an object. Therefore, everything must be represented in each view, whether or not it can be seen. *Interior* (inside) features and *exterior* (outside) features are both projected in the same way. Parts that cannot be seen in the views are drawn with hidden lines that are made up of short dashes (Fig. 5–18). Notice that the first dash of a hidden line touches the line where it starts (Fig. 5–18 at A). If a hidden line is a continuation of a visible line, space is left between the visible line and the first dash of the hidden line, as at B. If the hidden lines show corners, the dashes touch at the corners, as at C.

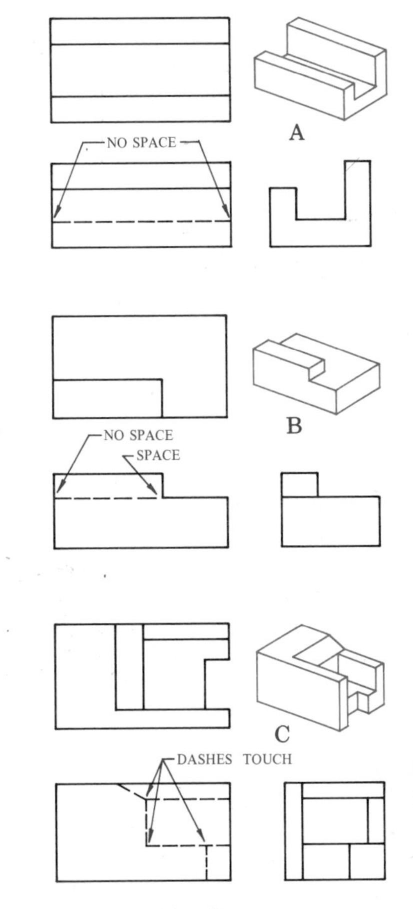

Fig. 5–18 **Hidden lines.**

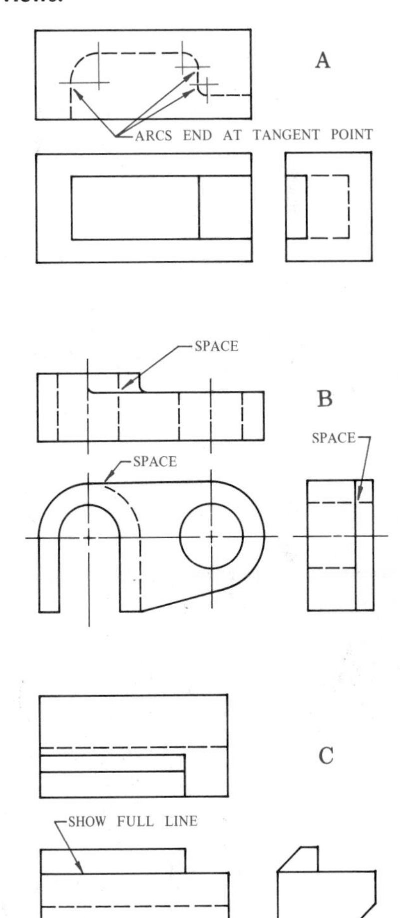

Fig. 5–19 **Hidden arcs.**

Dashes for hidden arcs (Fig. 5–19 at A) start and end at the tangent points. When a hidden arc is tangent to a visible line, a space is left, as at B. When a hidden line and a visible line project at the same place, show the visible line (Fig. 5–19 at C). When a centerline and

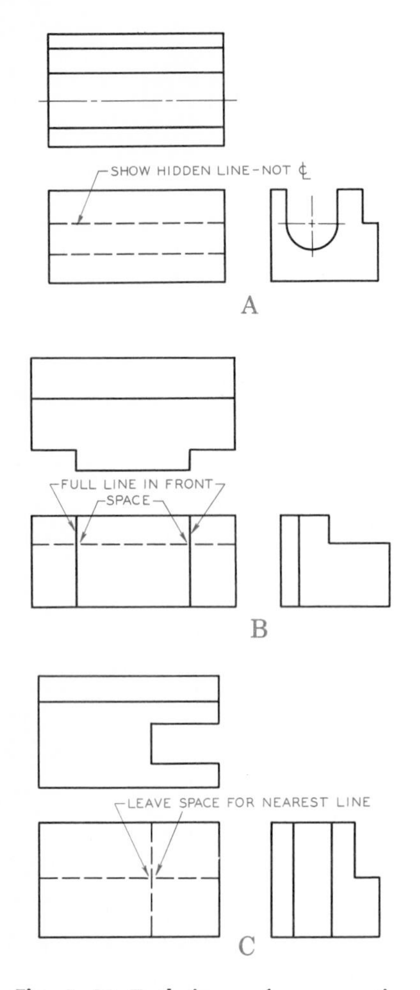

Fig. 5–20 **Technique of representing hidden and visible lines.**

a hidden line project at the same place (Fig. 5–20 at A), draw the hidden line. When a hidden line crosses a visible line (Fig. 5–20 at B), do not cross the visible line with a dash. When hidden lines cross (Fig.5–20 at C), the nearest hidden line has the "right of way." Draw the nearest hidden line through a space in the farther hidden line.

CENTERLINES

Centerlines are used to locate views and dimensions. (See the alphabet of lines, Fig. 3–13.) Primary centerlines, marked P in Fig. 5–21, locate the center on *symmetrical* (balanced) views where one part is a mirror image of another. Primary centerlines are used as major locating lines to help in making the views. They are also used as base lines for dimensioning. Secondary centerlines, marked S in Fig. 5–21, are used for drawing details of a part. Primary centerlines are the first lines to be drawn. The views are developed from them. Note that centerlines represent the axes (singular, axis; turning point) of cylinders in the side view. The centers of circles or arcs are located first so that measurements can be made from them to locate the lines on the various views. Show a hidden line instead of a centerline (Fig. 5–22).

CURVED SURFACES

Some curved surfaces, such as cylinders and cones, do not show as curves in all views. This is illustrated in Fig. 5–23. A cylinder with its *axis* (centerline) perpendicular to a plane will show as a circle on that plane. It will show as a rectangle on the other two planes. Three views of a cylinder in different positions are shown at B, C, and D. The holes may be thought of as *negative cylinders.* (In mathematics, "negative" means an amount less than zero. A hole is a "nothing" cylinder. However, it has size. Thus, in a sense, it is negative.) A cone appears as a circle in one view. It appears as a triangle in the others, as shown at E. One view of a frustum of a cone appears as two circles, as at F. In the top view, the conical surface is represented by the space between the two circles.

Cylinders, cones, and frustums of cones have single curved surfaces. They are represented by circles in

℄ = CENTERLINE
P = PRIMARY CENTERLINE
S = SECONDARY CENTERLINE

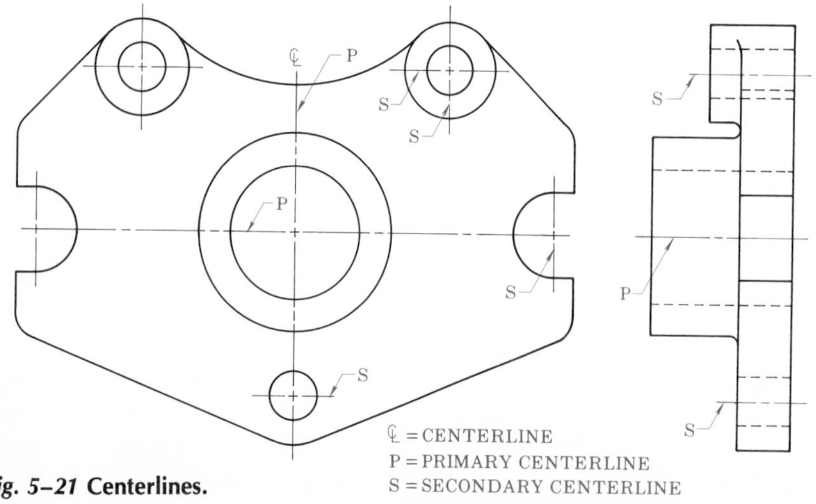

Fig. 5–21 **Centerlines.**

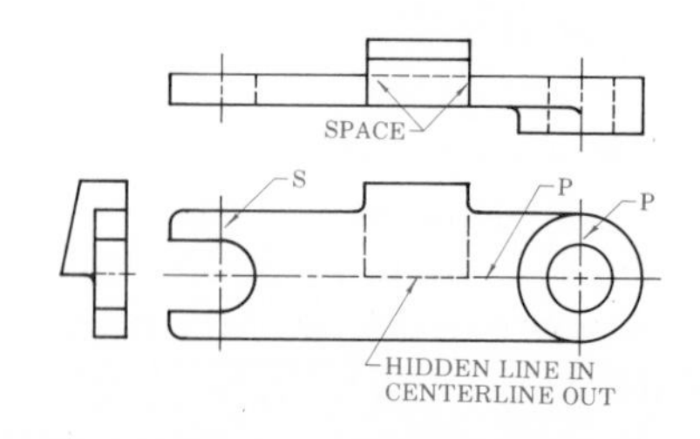

Fig. 5–22 **Centerlines and hidden lines.**

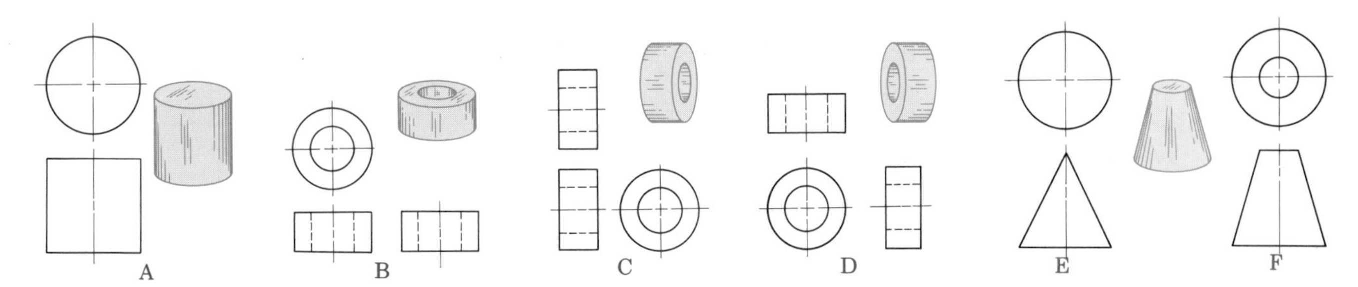

Fig. 5–23 **Curved surfaces. Cylinders and cones.**

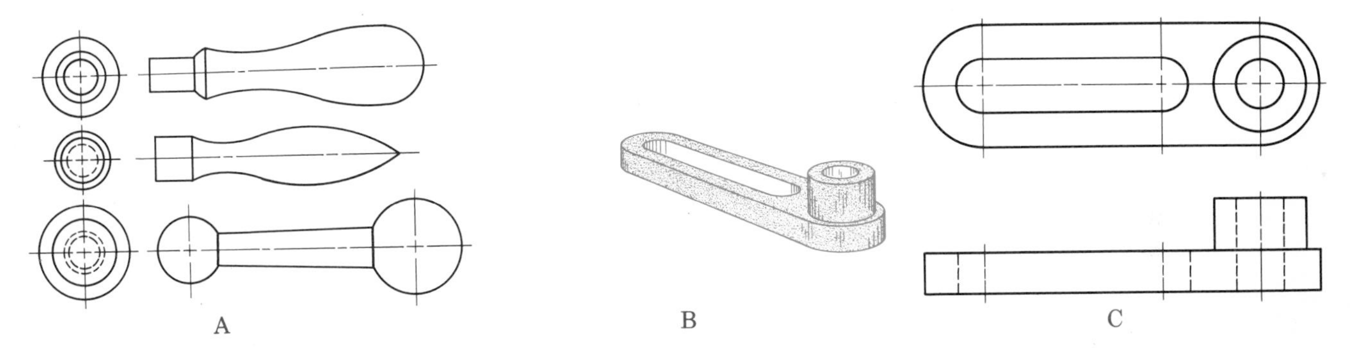

Fig. 5–24 **Curved surfaces.**

one view and straight lines in the other. The handles in Fig. 5–24 at A have double curved surfaces that are represented by curves in both views. The ball handle has spherical ends. Thus, both views of the ends are circles because a sphere appears as a circle when viewed from any direction. The slotted link in Fig. 5–24 at B and C is an example of tangent plane and curved surfaces. The rounded ends are tangent to the sides of the link and the ends of the slot are tangent to the sides. Therefore, the surfaces are smooth. There is no line of separation.

WHAT VIEWS TO DRAW

Six views are not always needed to describe most objects. Usually three views are sufficient. The general *characteristics* (different features) of an object will suggest the views that are required to describe its shape. Three properly selected views will describe most shapes. Sometimes, however, there are features that can be more clearly described by using more views or parts of extra views.

Most pieces can be recognized because they have a characteristic view. This is the first view to consider. Usually, it is the first view to draw. Next, consider the normal position of the part when it is in use. It is often desirable to draw the part in its normal position. However, it is not always necessary. For example, tall parts, such as vertical shafts, can be drawn in a horizontal position more easily. Views with the fewest hidden lines are easiest to read. They also take much less time to lay out and draw.

The practical purpose for drawing views is to describe the shape of something. Therefore, it is a waste of time to make more views than are necessary to describe an object. In fact, some objects require only one view in order to describe them adequately. Some things that can often be described in one view are the following: turned parts, such as the handles shown at A in Fig. 5–24; sheet material; plywood; plate material; and parts of uniform thickness, such as the latch and the stamping in Fig. 5–25 at A and B. For the handles in Fig. 5–24 at A, give the diameter in a note. For the latch or the stamping in Fig. 5–25 at A and B, give the thickness in a note. Parts, such as the bushing shown at D and the sleeve shown at E, are often shown in one view. Dimensions for the diameters,

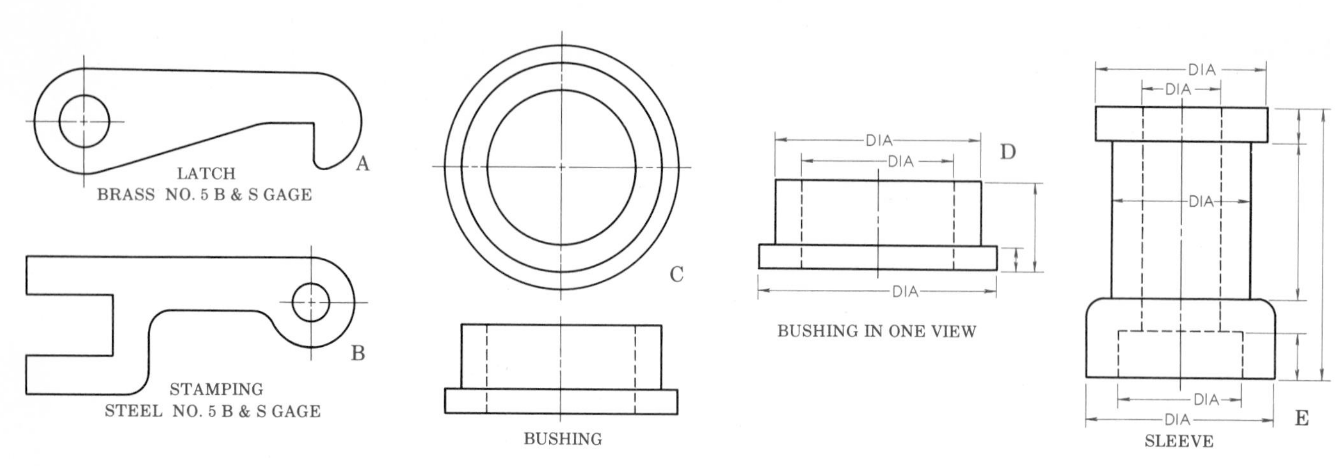

Fig. 5–25 **One-view drawings.**

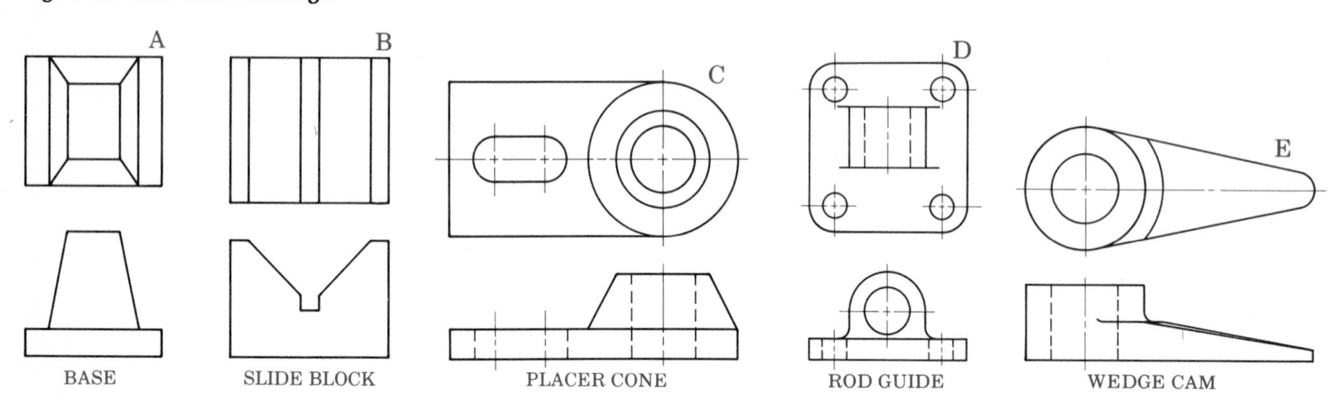

Fig. 5–26 **Two-view drawings.**

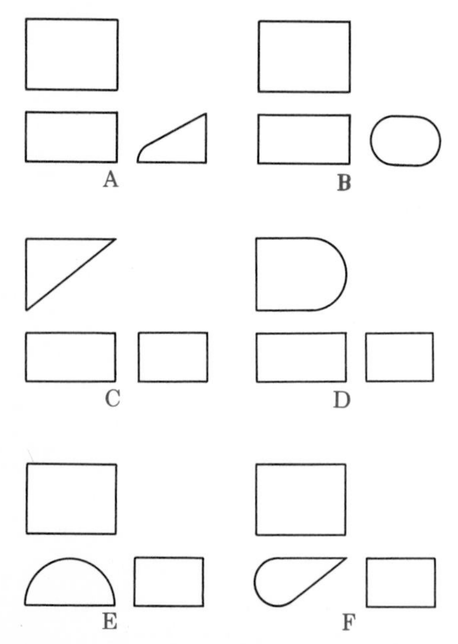

Fig. 5–27 **Selection of two views.**

marked Ø or DIA, are indicated. The two views of the bushing at C are not necessary, as shown at D.

Many things can be described in two views. These are shown in Fig. 5–26. When two views are used, they must be selected carefully so that they describe the shape of the object accurately. For the parts at A and B in Fig. 5–26, there would be no question about what views to make. For the part at A, the top view and either front or side view would be enough. A third view would add nothing to the description of the placer cone at C. There should be no question about the selection of views for the guide at D or for the wedge cam at E. Figure 5–27 shows some objects that can be described in two views. The top and front views at A and B are the same. Since the side views are necessary, the front and side views would be adequate. At C and D the top views are necessary. Therefore, the top and front views would be sufficient. At E and F the front views are necessary. Therefore, the front and top views or the front and side views would be sufficient.

Some things, such as the angle in Fig. 5–28, require three views. Three views are needed because it is necessary to describe shapes in each of the views.

Six views are shown for the sliding base in Fig. 5–29. As you look

Fig. 5–28 **Three-view drawings.**

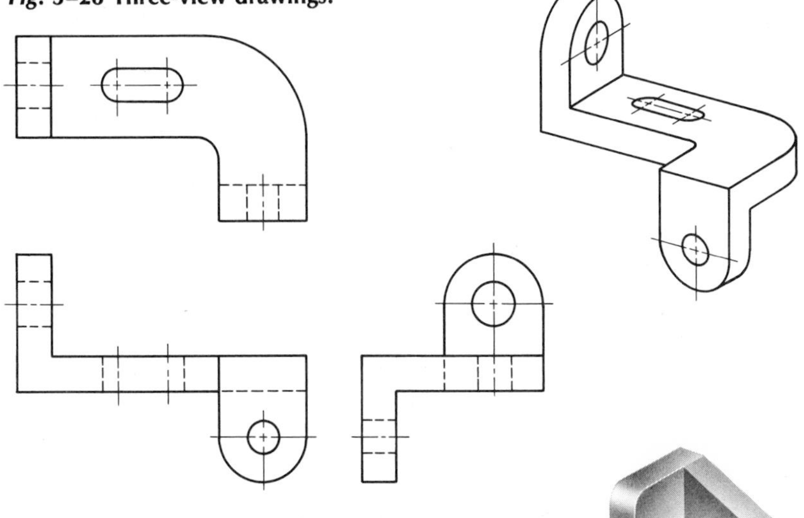

at the pictorial view you will see that the top, front, and right-side views will give the best shape description. These views also have the fewest hidden lines. The six views are shown here simply to illustrate and explain the selection of views. You would draw only the necessary views in real practice.

Careful thought about your "mind's-eye picture" of an object will help you decide which views best describe its shape.

SECOND POSITION OF THE SIDE VIEW

The proportions of an object or the size of the sheet sometimes makes you want to show the side view in the second position. This position is directly across from the top view, as shown in Fig. 5–30. This view can be made by revolving the side plane around its intersection with the top plane.

Fig. 5–29 **Choice of views.**

Fig. 5–30 **Second position of the side view.**

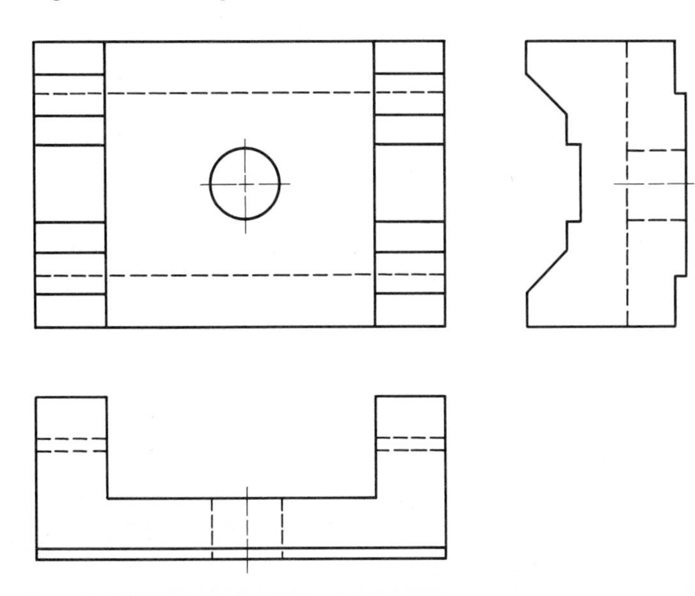

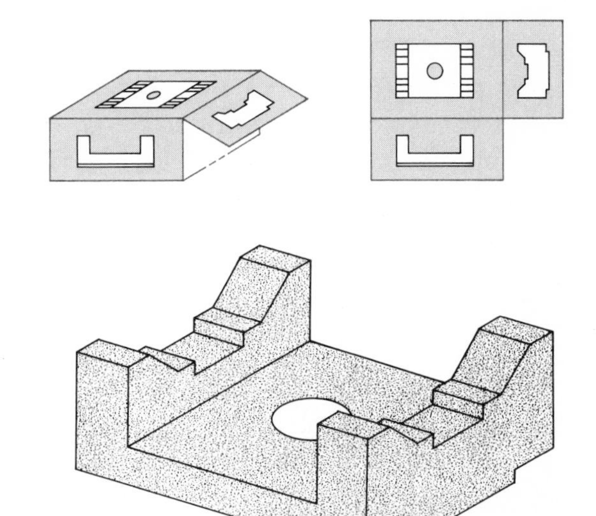

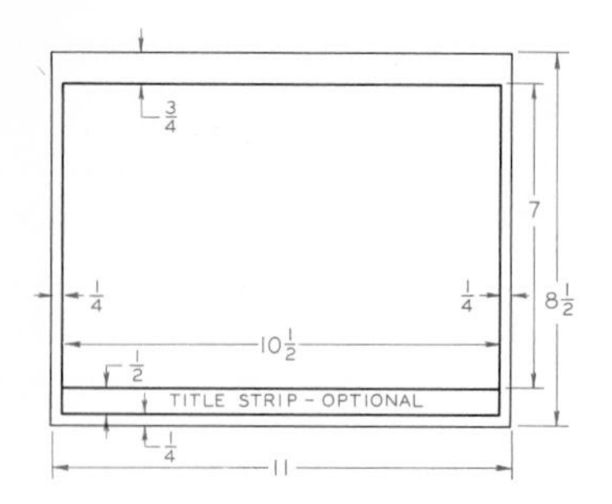

Fig. 5–31 **Sheet layout.**

PLACING VIEWS

The size of the drawing sheet should allow enough space for the number of views that you need to give a clear description of the part. Working space is suggested by the sheet layout in Fig. 5–31. Working space may also be specified by your instructor. The method for working out the positions of the views is the same for any space.

In Fig. 5–32 at A, a pictorial drawing of a slide stop is shown with its overall width, height, and depth dimensions. Some simple arithmetic is needed to place the three normal views properly. It may also be helpful to make a rough layout on scrap paper, as shown at B. This layout need not be made to scale.

A working space of 267 mm (10½″) × 178 mm (7″) is used to explain how to place the views of the slide stop in Fig. 5–32 at A. The overall dimensions are: width = W = 133 mm (5¼″); depth = D

Fig. 5–32 **Calculations for the placement of three views in the customary inch system.**

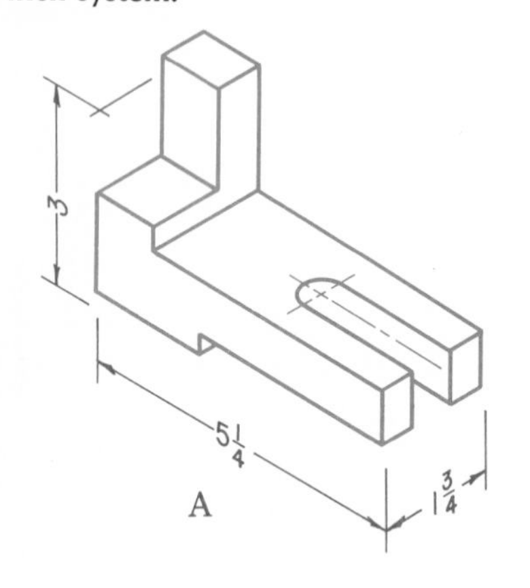

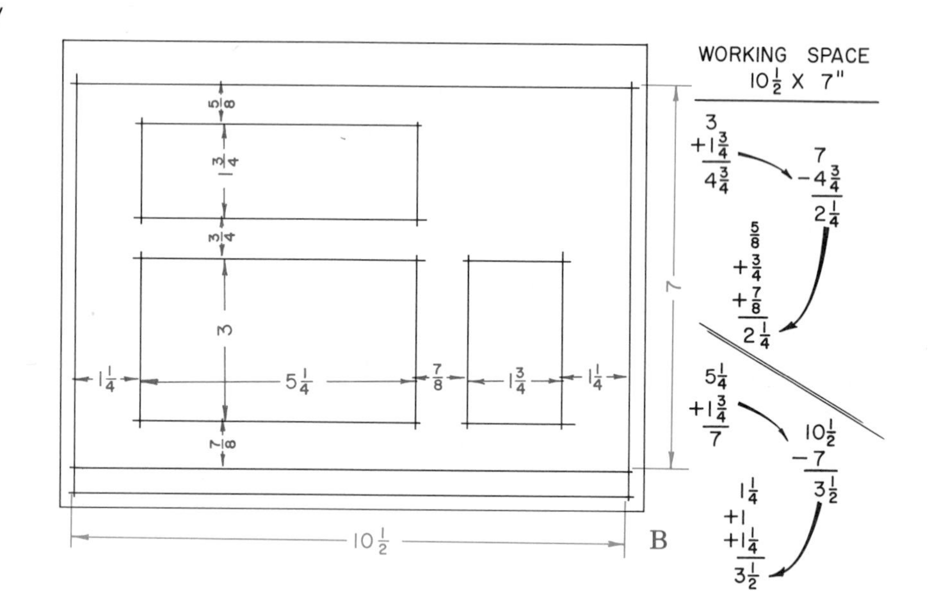

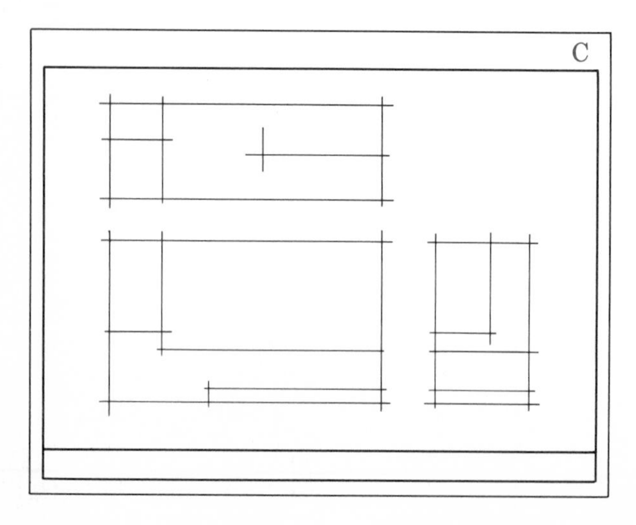

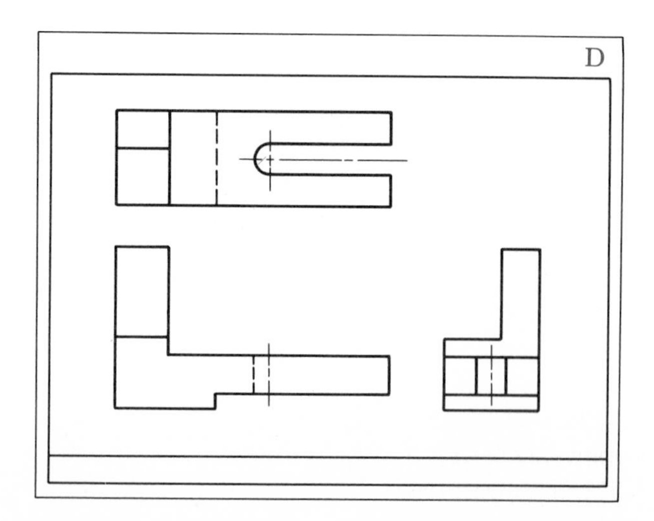

= 45 mm (1¾″); height = H = 76 mm (3″).

The width, depth, and height dimensions are given in red on the sketch (Fig. 5–32 at B). The dimensions in blue indicate the spacing at the top, bottom, side, and between views. Use the following procedure to determine spacing. Refer to Fig. 5–32.

1. Add the width, 133 mm (5¼ in.) and the depth, 45 mm (1¾ in.): 133 + 45 = 178 mm (5¼ + 1¾ = 7 in.). Subtract 178 mm (7 in.) from the width of the drawing space, 268 mm (10½ in.): 268 − 178 = 90 mm (10½ − 7 = 3½ in.). The remaining 90 mm (3½ in.) is the amount left for the space at the left, right, and between views. If a space of about 25 mm (1 in.) is used between the front and side views, it will allow 32 mm (1¼ in.) on the left and right sides. These spaces may be larger or smaller, depending upon the shapes of the views, the space available, and the space needed for dimensions and notes when added.
2. Next, add the height, 76 mm (3 in.) and the depth, 45 mm (1¾ in.): 76 + 45 = 121 mm (3 + 1¾ = 4¾ in.). Subtract 121 mm (4¾ in.) from the height of the drawing space, 178 mm (7 in.): 178 − 121 = 57 mm (7 − 4¾ = 2¼ in.). The remaining 57 mm (2¼ in.) is the amount left for the space at the bottom, top, and between views. If a space of 19 mm (¾ in.) is used between the front and top views, it will allow 38 mm (1½ in.) for spaces above and below the views. These could be 19 mm (¾ in.) each, but a better visual balance will result if 22 mm (⅞ in.) is used below and 16 mm (⅝ in.) above.

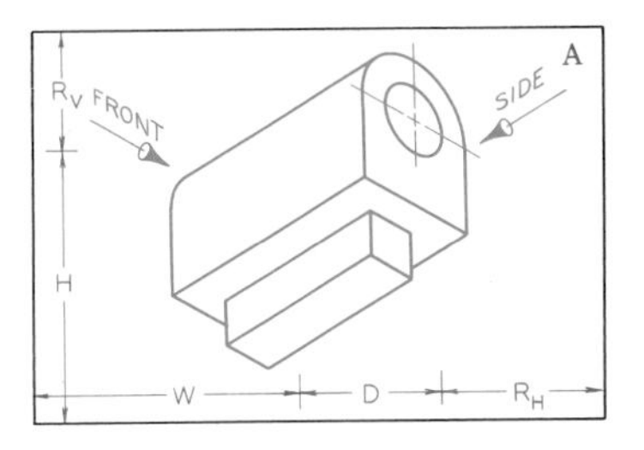

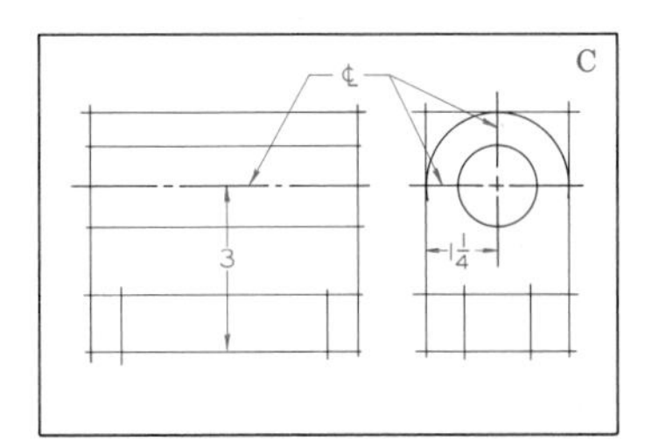

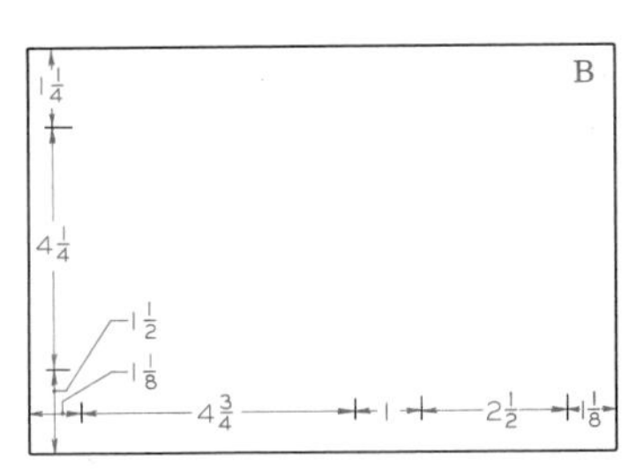

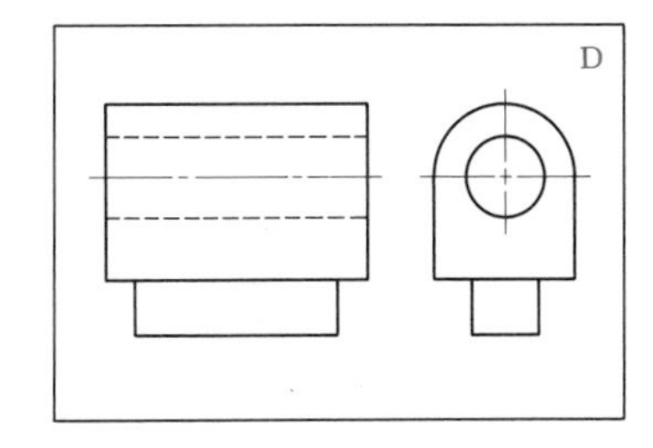

Fig. 5–33 **Calculations for the placement of two views using the customary inch system.**

After calculations are made, proceed with the layout on the final drawing sheet, as shown at C in Fig. 5–32. Notice that the views are blocked in with light construction lines until all details have been added. Figure 5–32 at D shows all necessary visible, hidden, and centerlines darkened.

Figures 5–33 and 5–34 show the same procedure being used for a two-view drawing. Regardless of the number of views, the basic procedure does not change. The views can be arranged as in Fig. 5–33 at D or as in Fig. 5–34 at D.

LOCATING MEASUREMENTS

After lines have been drawn to locate the views, make measurements for details. Then, draw the views (Fig. 5–35). Measurements made on one view can be transferred to another to save the time of making them again. This procedure will also ensure accuracy and correctness. Distances in the three directions, width *W*, height *H*, and depth *D*, can be easily transferred, as seen in Fig. 5–35.

1. Width, *W* (horizontal), measurements made on the front view can be located on the top view by drawing up from the front view. In the same way, measurements can be projected down from the top view to the front view.
2. Height, *H* (vertical), measurements on the front view can be located on the side view by drawing a light line across to the side view. Measurements can also be projected to the front view from the side view.
3. Depth, *D*, measurements show as vertical distances in the top view and as horizontal distances in the side view. Such measure-

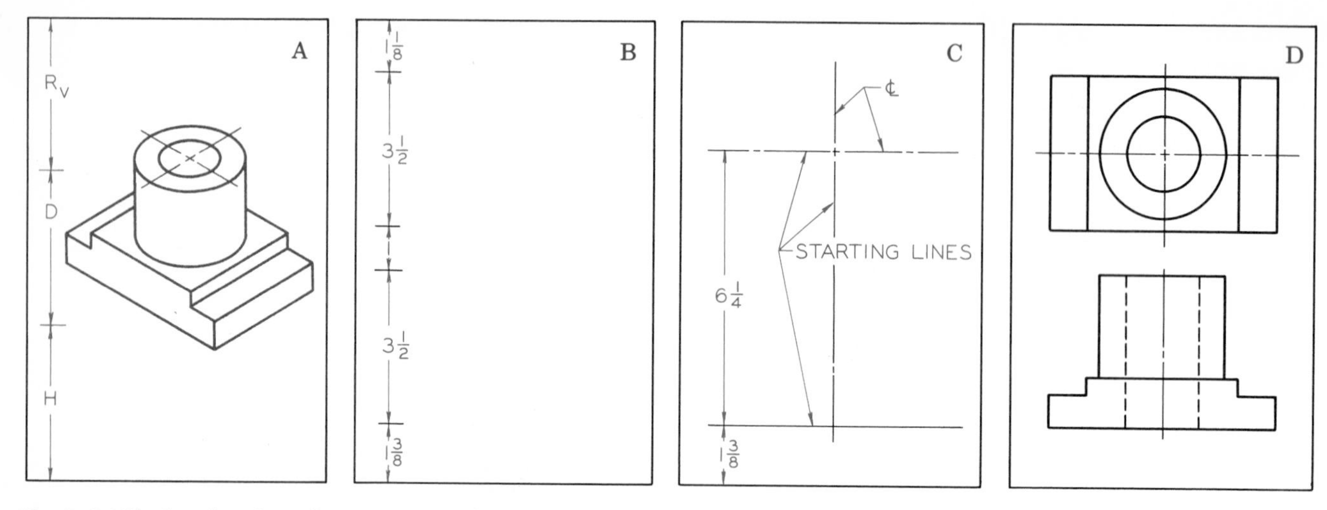

Fig. 5–34 **Placing the views for a two-view drawing.**

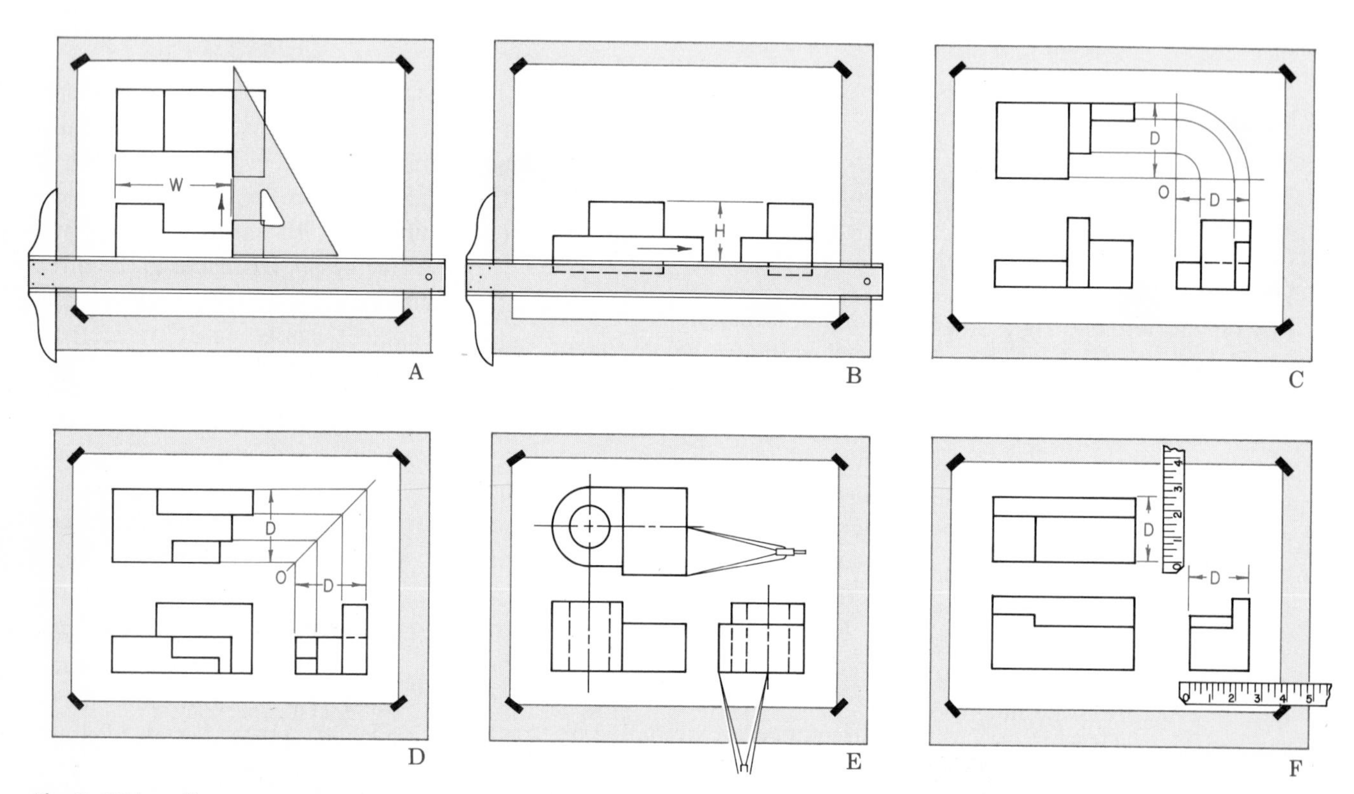

Fig. 5–35 **Locating measurements.**

ments can be taken from the top view to the side view by the following methods: by drawing arcs from a center *O* (at C), by using a 45° triangle through *O* (at D), by using the dividers (at E), or by using the scale as shown at F.

TO MAKE A DRAWING

You should follow a step-by-step method of working to be sure your drawing is accurate and easy to understand (Fig. 5–36). All views should be carried along together. Do not attempt to finish one view before starting the others. Use a hard pencil (4H or 6H) and light, thin lines for preliminary lines. Use a soft pencil (F, HB, or H) for final lines. The grade of pencil you use depends partly upon the surface of the paper, cloth, or film you use. The following order of working is suggested:

1. Consider the characteristic view (at A; the front view).
2. Determine the number of views (at A; three views needed).
3. Locate the views (at B).
4. Block in the views with light, thin lines (at C).
5. Lay off the principal measurements (at D).
6. Draw the principal lines (at E).
7. Lay off the measurements for details (at F; centers for arcs, circles, and triangular ribs).
8. Draw the circles and arcs (at G).
9. Draw any additional lines needed to complete the views.
10. Brighten the lines where necessary to make them sharp and black and of the proper thickness (at H).

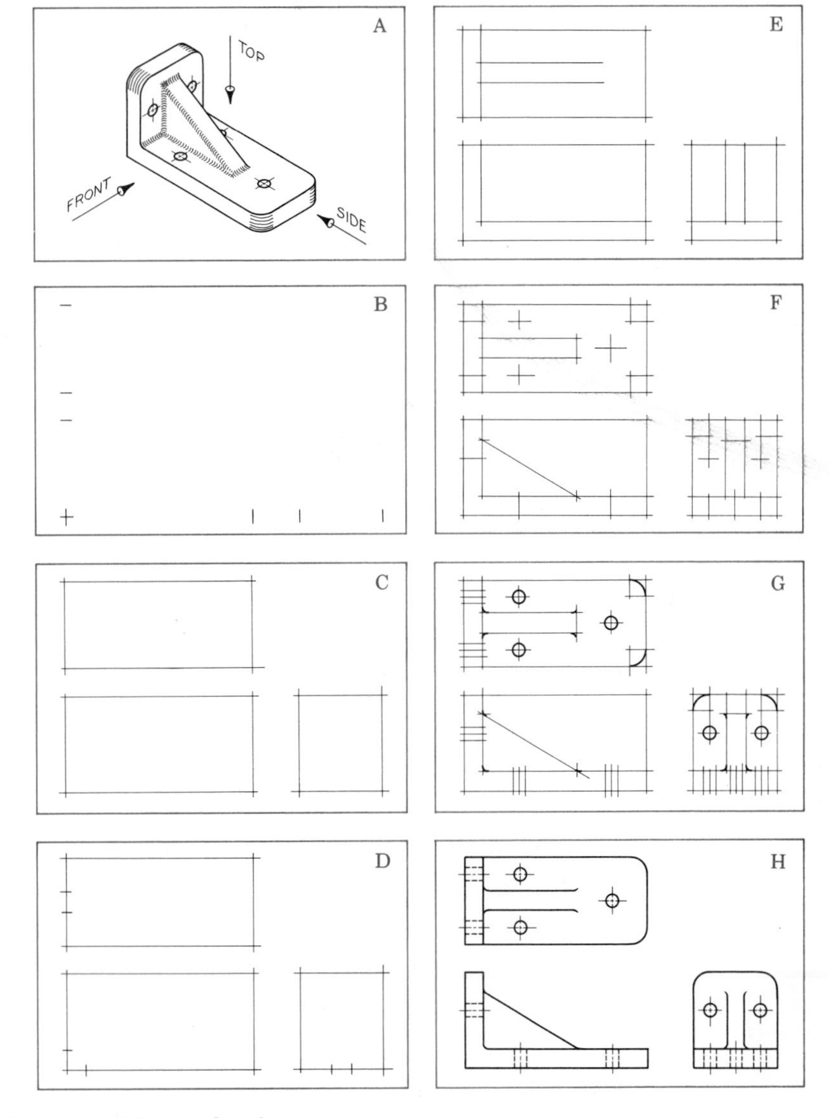

Fig. 5–36 **Making a drawing.**

Review

1. Describe multiview drawing.

2. Most objects have six sides or six views. Name them.

3. There are four angles of projection. Which two are used in technical drawing? Which one is used in the United States?

4. What type of line is used to represent *interior* (inside) details not seen on the outside of an object?

5. Name the two types of centerlines.

6. How many views are usually needed to completely describe an object having uniform thickness throughout?

7. How many views does a cylindrical object usually require?

8. If a centerline and a hidden line fall in the same place, which takes preference?

9. What determines the amount of space needed between views?

10. Which type of drawing gives the most accurate shape description of an object?

Problems

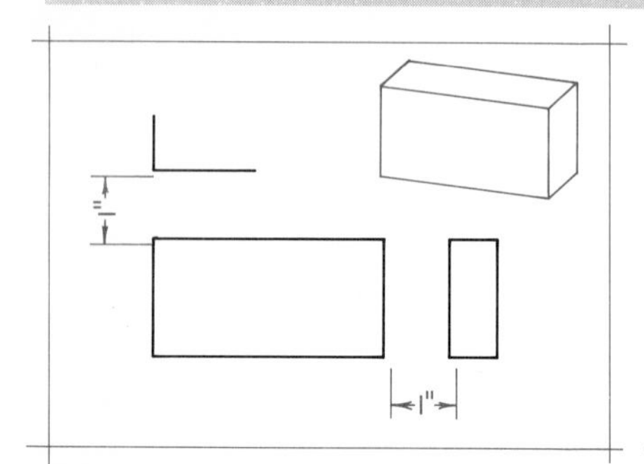

Fig. 5-37 Sanding block. A pictorial and two views are given. Draw full size the two views shown and complete the third (top) view. Do not draw the pictorial view. The block is 3/4″ thick, 1 3/4″ wide, and 3 1/2″ long.

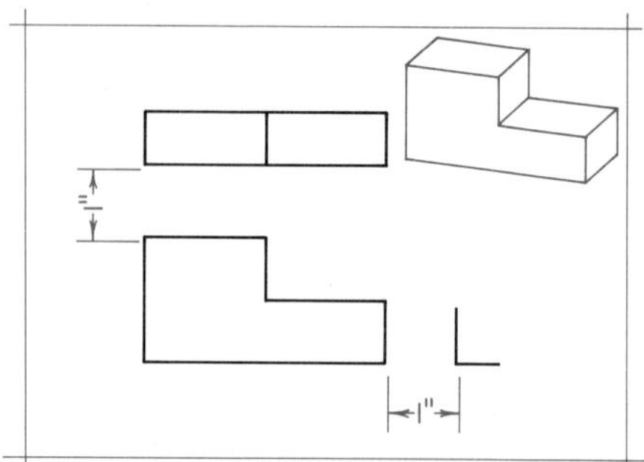

Fig. 5-38 Step block. Scale: full size. Draw the front and top views, but not the pictorial view. Complete the right-side view in its proper location. The step block is 3/4″ thick, 1 3/4″ wide, and 3 1/2″ long. The notch is 7/8″ × 1 3/4″.

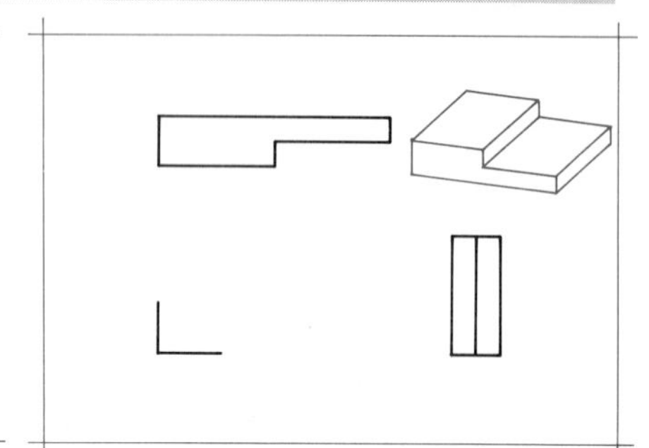

Fig. 5-39 Half lap. Scale: full size. Draw the top and right-side views, but not the pictorial. Complete the front view in its proper shape and location. The half lap is 3/4″ thick, 1 3/4″ wide, and 3 1/2″ long. The notch is 3/8″ × 1 3/4″.

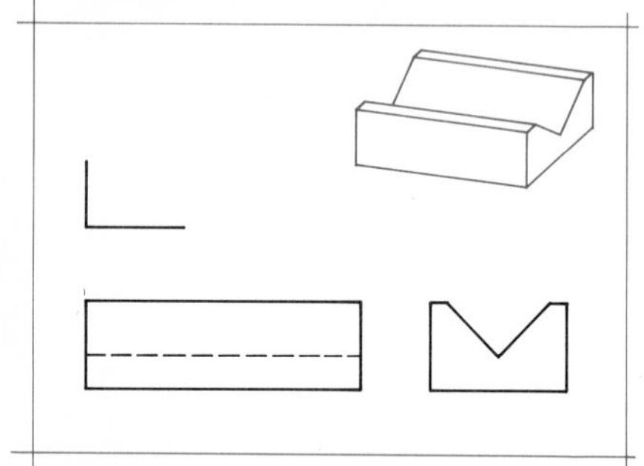

Fig. 5-40 V-block. Scale: full size. Draw the front and right-side views as shown. Complete the top view in its proper location. Do not draw the pictorial view. The overall sizes are 1 1/4″ high, 2″ wide, and 4″ long. The V-cut has a 90° included angle and is 3/4″ deep.

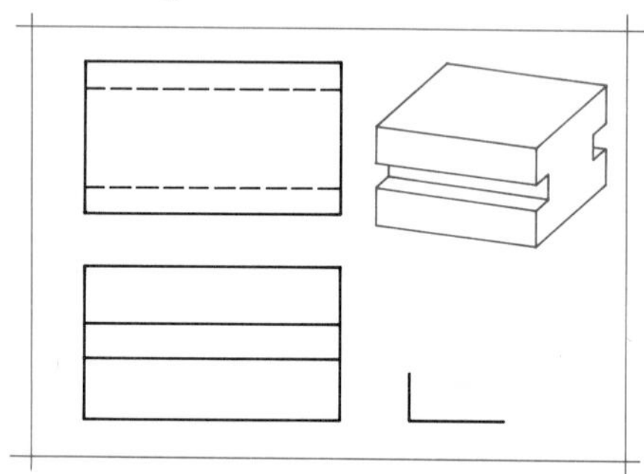

Fig. 5-41 Slide. Scale: full size. Draw the front and top views as shown. Complete the right-side view in its proper location. Do not draw the pictorial view. The overall sizes are 2 1/8″ square and 3 3/4″ long. The slots are 3/8″ deep and 1/2″ wide.

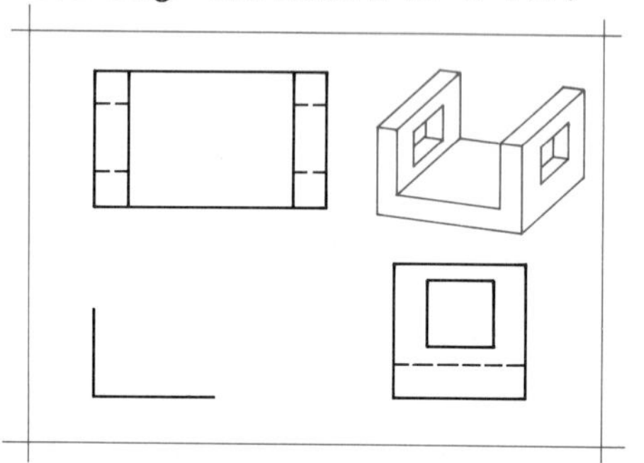

Fig. 4-42 Rod support. Scale: full size. Draw the top and right-side views, but not the pictorial. Complete the front view. The overall sizes are 2″ square by 3 1/2″ long. Bottom and ends are 1/2″ thick. The holes are 1″ square and are centered on the upper portions.

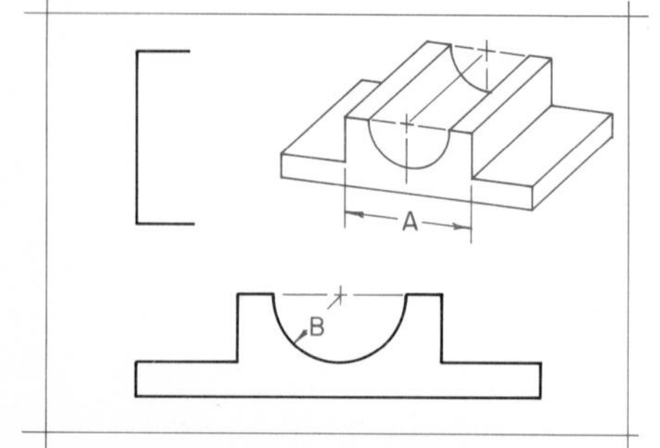

Fig. 5-43 Cradle. Scale: full size. Draw the front view, but not the pictorial. Complete the top view in the proper shape and location. Height = 2″, width = 2 1/2″, length = 6″. Base = 1/2″ thick. *A* = 3″, *B* = 1″R.

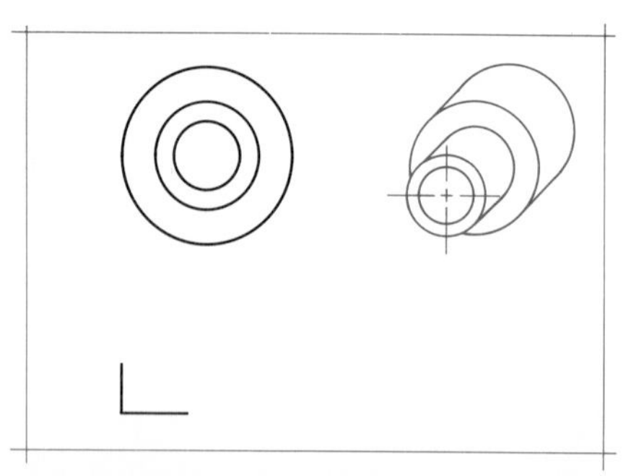

Fig. 5-44 Spacer. Scale: as assigned. Draw the top view, but not the pictorial. Complete the front view. Base = 2 1/2″ DIA × 1″ high. Top = 1 1/2″ DIA × 3/4″ high. Hole = 1″ DIA. A vertical sheet will permit a larger scale.

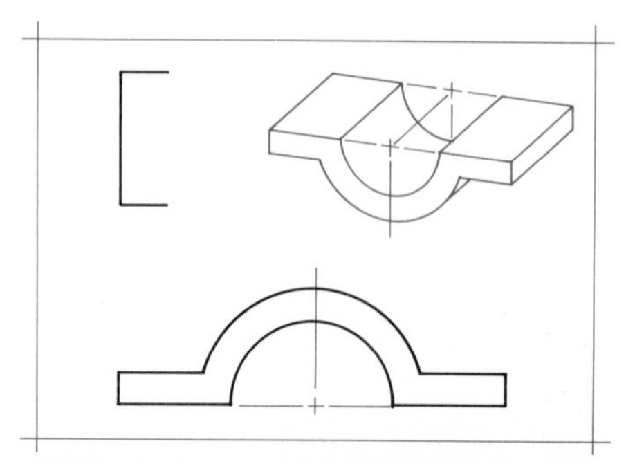

Fig. 5-45 Strap. Scale: full size. Draw the front view as shown. Complete the top view in the proper shape and location. Do not draw the pictorial view. Material is 1/2″ thick × 2″ wide. Inside radius is 1 1/4″. Overall length is 6″.

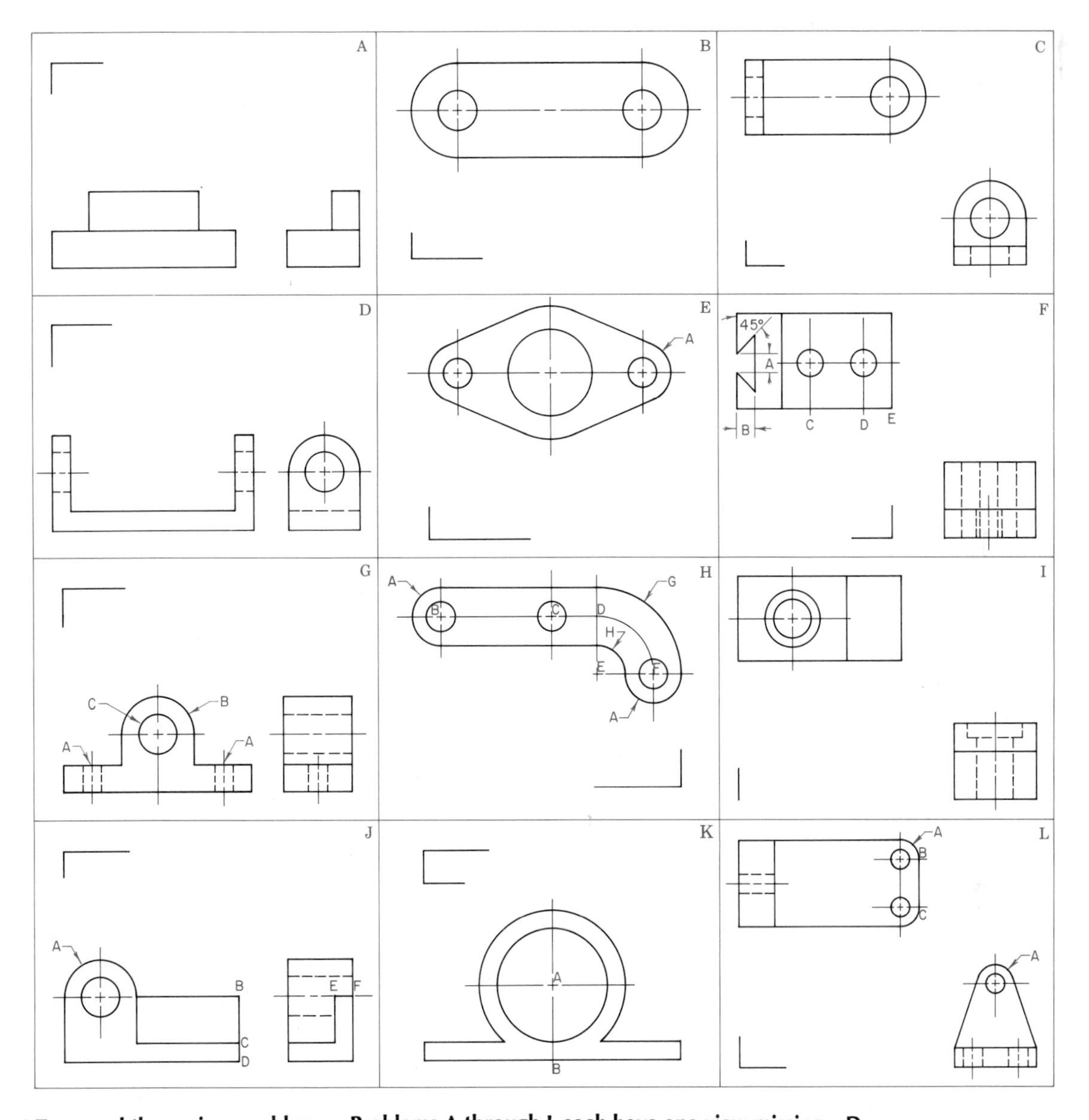

Fig. 5–46 **Two- and three-view problems. Problems A through L each have one view missing. Draw the view or views given and complete the remaining view in the proper shape and location. Scale: full size or as assigned. Do not dimension unless instructed to do so.**

A STOP: $L = 5$, $W = 2$, $H = 2$, base $= 1 \times 2 \times 5$, top $= \frac{3}{4} \times 1 \times 3$.

B LINK: $L = 7\frac{1}{2}$, $W = 2\frac{1}{2}$, thk $= 1\frac{1}{4}$, holes $= 1\frac{1}{16}$ DIA.

C ANGLE BRACKET: $L = 5$, $W = 2$, $H = 2\frac{1}{4}$, matl thk $= \frac{1}{2}$, holes = 1″ DIA.

D SADDLE: $L = 5\frac{1}{2}$, $W = 2$, $H = 2\frac{1}{2}$, matl thk $= \frac{1}{2}$. hole = 1″ DIA.

E SPACER: $L = 6\frac{1}{2}$, $W = 3\frac{1}{4}$, thk = 1″, holes $= 2\frac{3}{8}$ DIA, $\frac{3}{4}$ DIA, $A = \frac{3}{4}$R.

F DOVETAIL SLIDE: $L = 4\frac{1}{4}$, $W = 2\frac{1}{2}$, $H = 2$, base thk $= \frac{3}{4}$, upright thk $= 1\frac{1}{4}$, holes $= \frac{5}{8}$ DIA, $A = \frac{1}{2}$, $B = \frac{1}{2}$, $CD = 1\frac{1}{2}$, $DE = \frac{3}{4}$.

G ROD GUIDE: $L = 5\frac{1}{8}$, $W = 1\frac{7}{8}$, $H = 2\frac{1}{2}$, C = 1″ DIA, $A = \frac{1}{2}$ DIA, $3\frac{5}{8}$ apart, base thk $= \frac{3}{4}$, B = 1″R.

H HINGE PLATE: $A = \frac{3}{4}$R, $BC = 3$, $CD = 1\frac{1}{4}$, $DE = 1\frac{1}{2}$, $EF = 1\frac{1}{2}$, $G = 2\frac{1}{4}$R, $H = \frac{3}{4}$R. holes $\frac{3}{4}$ DIA, thk = 1″.

I OFFSET LUG: $L = 4\frac{1}{2}$, $W = 2\frac{1}{4}$, $H = 2$, notch $= \frac{3}{4} \times 1\frac{1}{2}$, hole = 1″ DIA, Cbore $= 1\frac{1}{2}$ DIA $\times \frac{3}{8}$ deep.

J PIN HOLDER: $L = 4\frac{3}{4}$, $W = 1\frac{3}{4}$, $H = 2\frac{3}{4}$, hole = 1″ DIA, A = 1″R, $BC = 1\frac{1}{4}$, $BD = 1\frac{3}{4}$, $EF = \frac{1}{2}$.

K RING: base $= \frac{1}{2} \times \frac{7}{8} \times 7$, ring = 4 OD, 3 ID, $AB = 2$.

L BRACKET: $L = 5$, $W = 2\frac{1}{4}$, $H = 2\frac{3}{4}$, base thk $= \frac{1}{2}$, upright = 1″, $A = \frac{1}{2}$R, $BC = 1\frac{1}{4}$, holes $= \frac{1}{2}$ DIA.

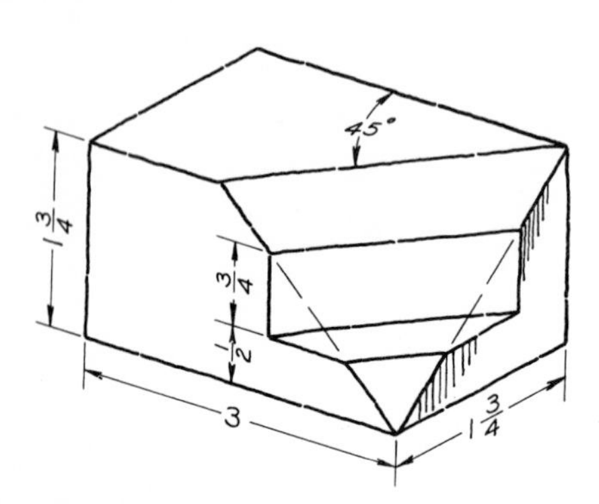

Fig. 5–47 **Draw three views of the corner lock.**

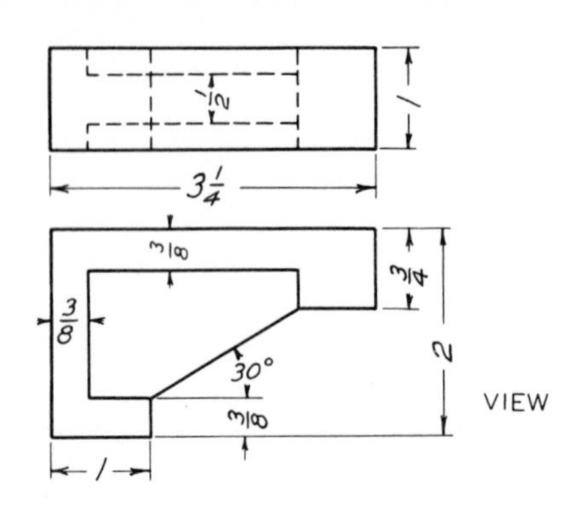

Fig. 5–48 **Draw three views of the bracket.**

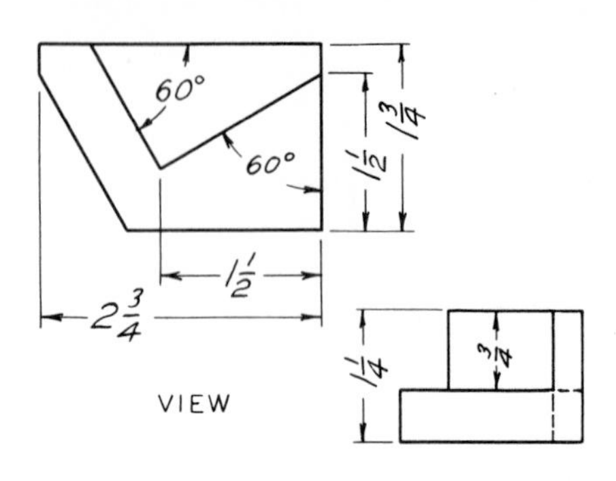

Fig. 5–49 **Draw three views of the locator.**

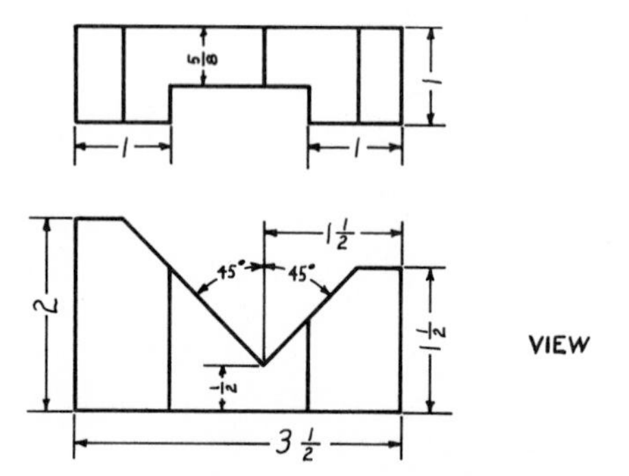

Fig. 5–50 **Draw three views of the V-slide.**

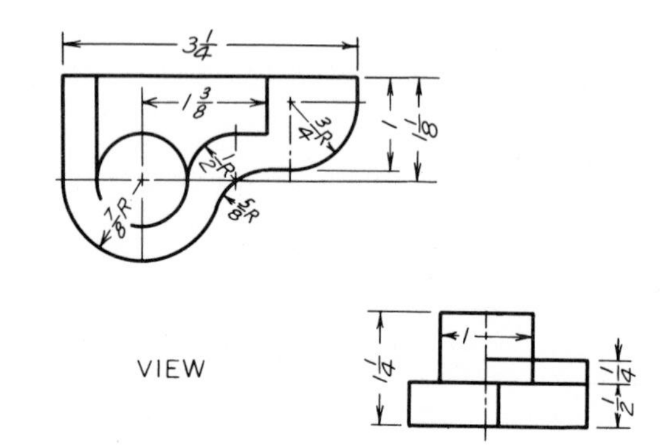

Fig. 5–51 **Draw three views of the pivot.**

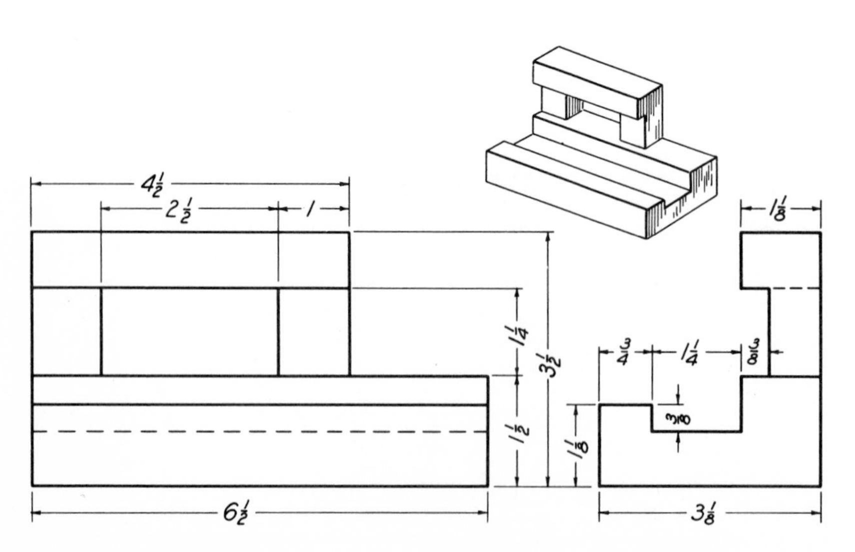

Fig. 5–52 **Draw three views of the support guide. Scale: full size.**

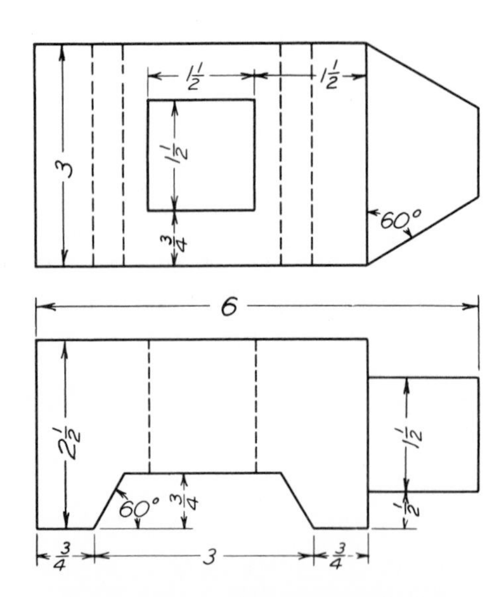

Fig. 5–53 **Draw three views of the locating support.**

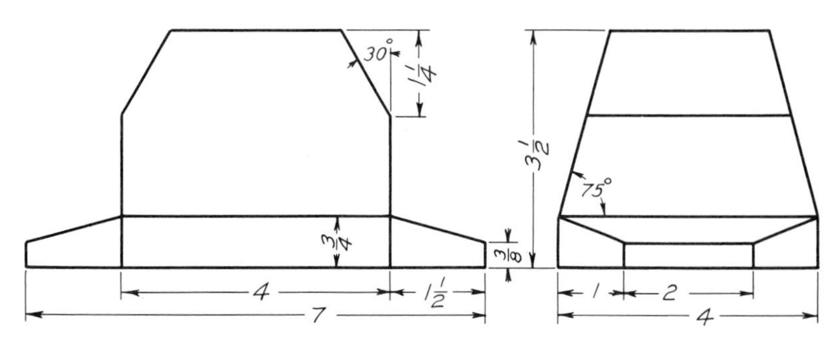

Fig. 5–55 **Draw three views of the base.**

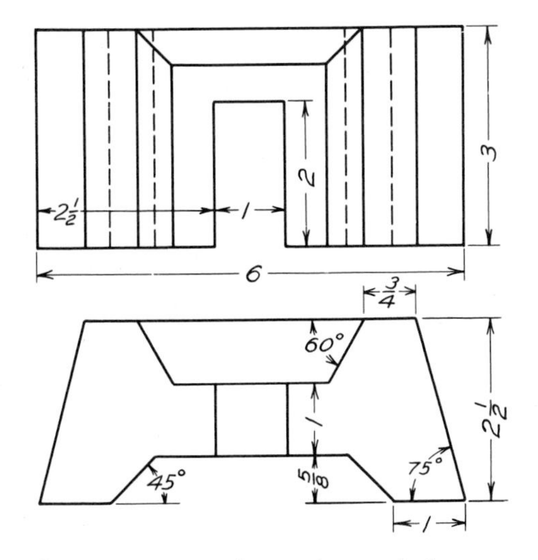

Fig. 5–54 **Draw three views of the base.**

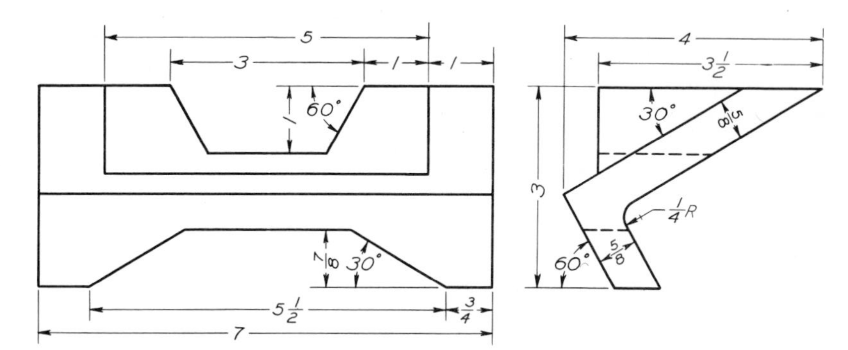

Fig. 5–56 **Draw three views of the separator.**

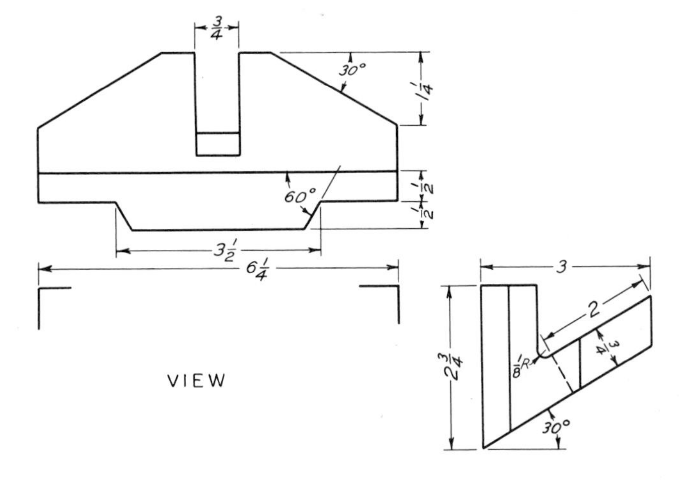

Fig. 5–57 **Draw three views of the secondary guide lug.**

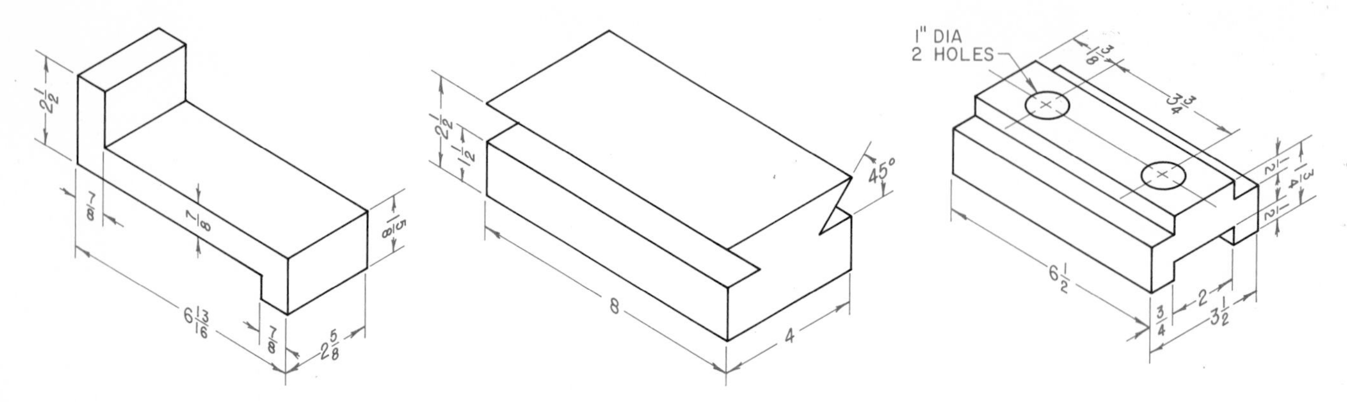

Fig. 5–58 **Stop.**

Fig. 5–59 **Dovetail slide.**

Fig. 5–60 **Slide.**

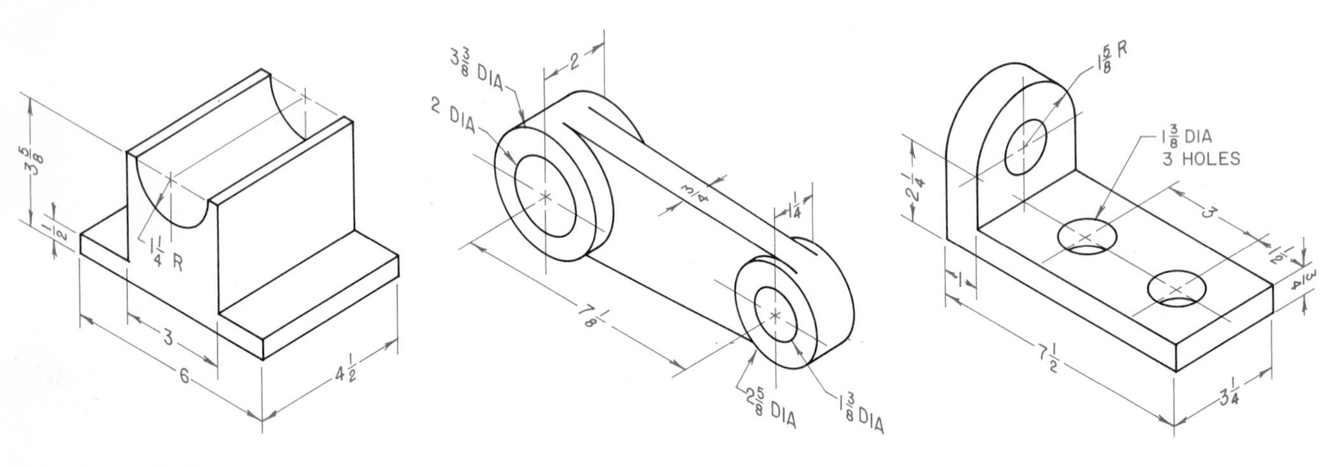

Fig. 5–61 **Cradle block.**

Fig. 5–62 **Pivot arm.**

Fig. 5–63 **Base.**

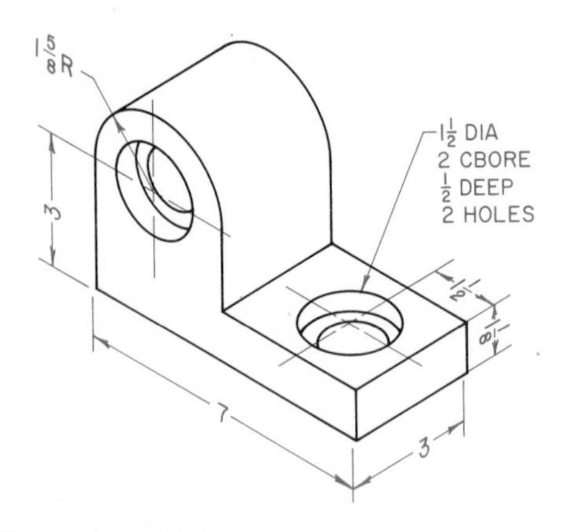

Fig. 5–64 **Shaft support.**

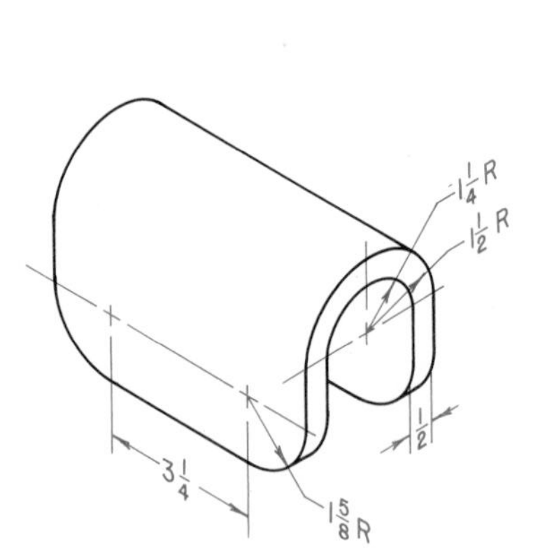

Fig. 5–65 **Edge protector.**

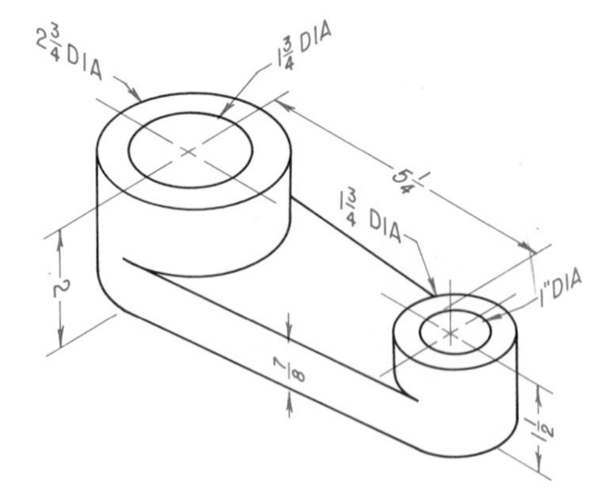

Fig. 5–66 **Swivel arm.**

ASSIGNMENT: **Make two- or three-view drawings of objects on this page with instruments. Scale: as assigned. Do not dimension unless instructed to do so. Include all centerlines.**

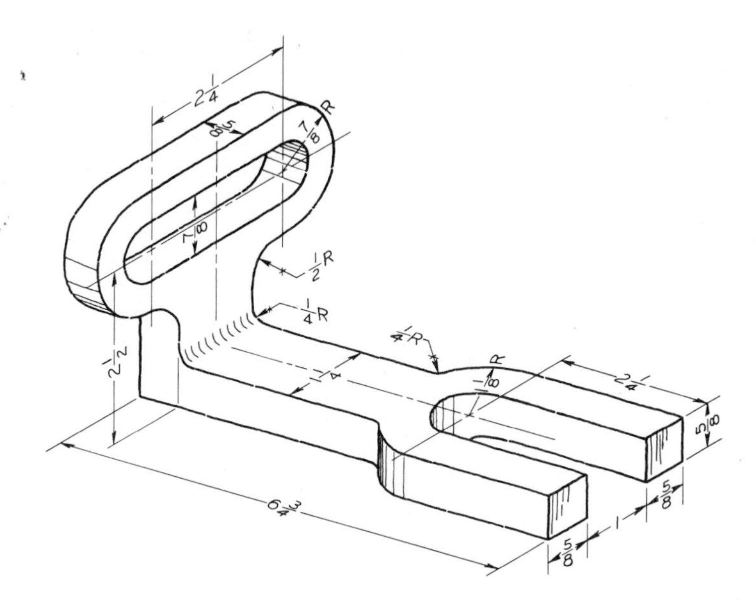

Fig. 5–67 **Draw three views of the adjustable fork.**

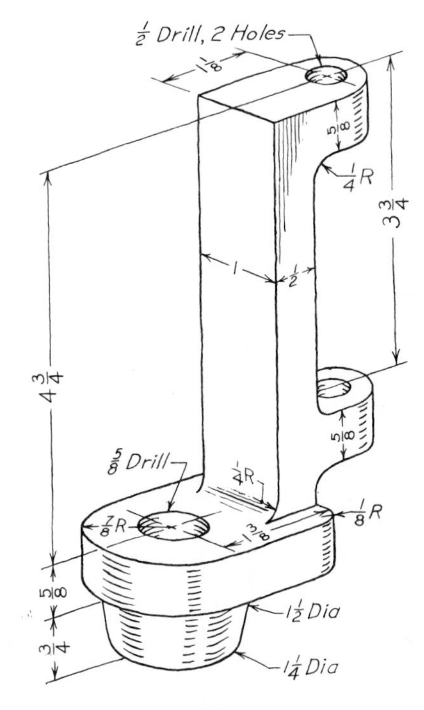

Fig. 5–68 **Draw three complete views of the vertical bracket.**

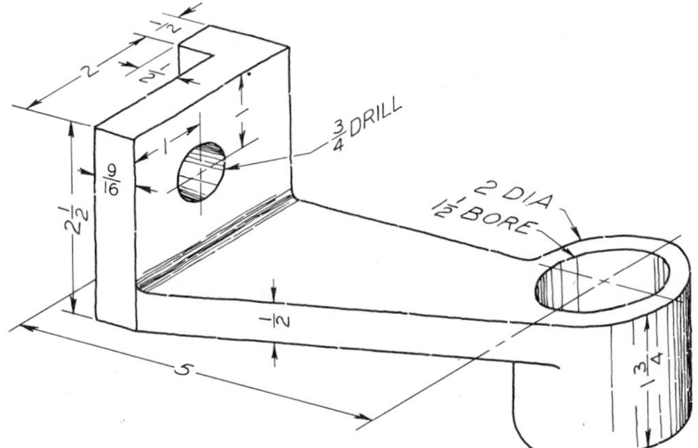

Fig. 5–69 **Draw three views of the bracket.**

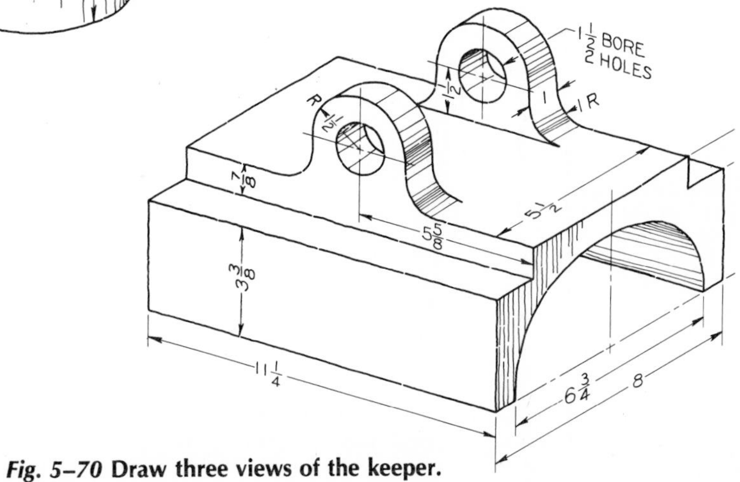

Fig. 5–70 **Draw three views of the keeper.**

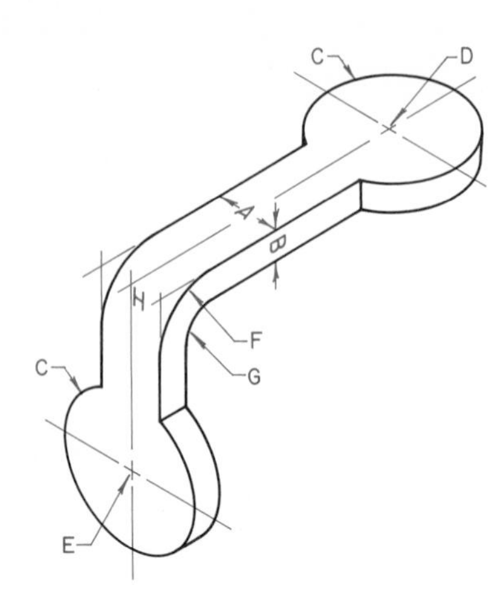

Fig. 5–71 Offset ring. $A = 1^1/_2''$, $B = {}^5/_8''$, $C = 1^5/_8''$R, $D = 1''$ DIA, $E = 1^1/_8''$ DIA, $F = 1^1/_2''$R, $G = {}^7/_8''$R, $EH = 4''$, $HD = 6^3/_4''$.

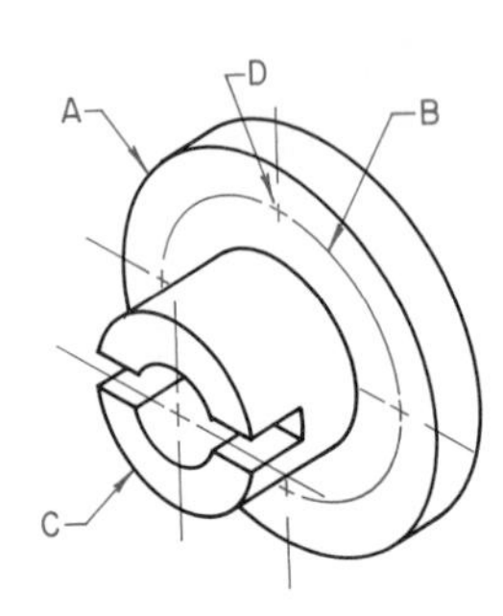

Fig. 5–72 Socket. Scale: as assigned. All dimensions are in millimeters (mm). A = 50.50 mm DIA × 7 mm thick, B = 38.0 mm, C = 25.25 mm DIA × 17 mm long with 13-mm-DIA hole through, Slots = 4.50 mm wide × 8.0 mm deep, D = 6 mm DIA, 4 holes equally spaced.

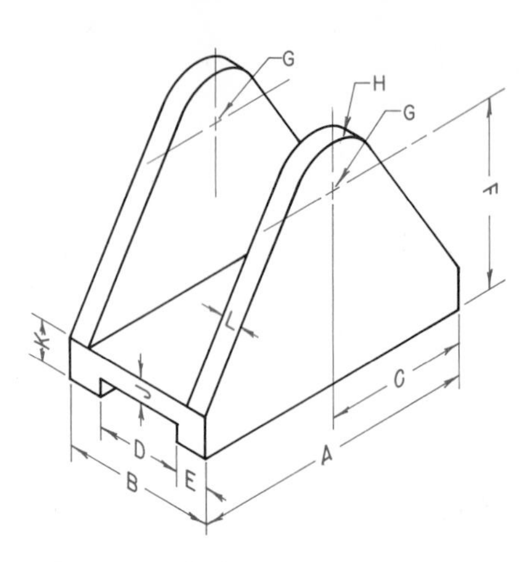

Fig. 5–73 Shaft guide. Scale: as assigned. $A = 7^1/_4''$, $B = 3^7/_8''$, $C = 3^5/_8''$, $D = 2^1/_4''$, $E = {}^{13}/_{16}''$, $F = 4^1/_2''$, $G = 1''$ DIA, 2 holes, $H = 1^1/_4''$R, $J = {}^1/_2''$, $K = 1''$, $L = {}^1/_2''$.

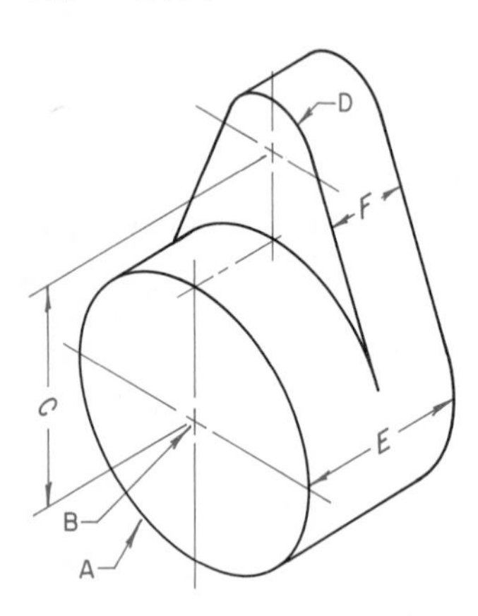

Fig. 5–74 Cam. $A = 2^5/_8''$DIA, $B = 1^1/_4''$ DIA, $1^3/_4''$ Cbore, $^1/_4''$ deep, both ends, $C = 2^1/_8''$, $D = {}^9/_{16}''$R, $E = 1^3/_4''$, $F = {}^7/_8''$.

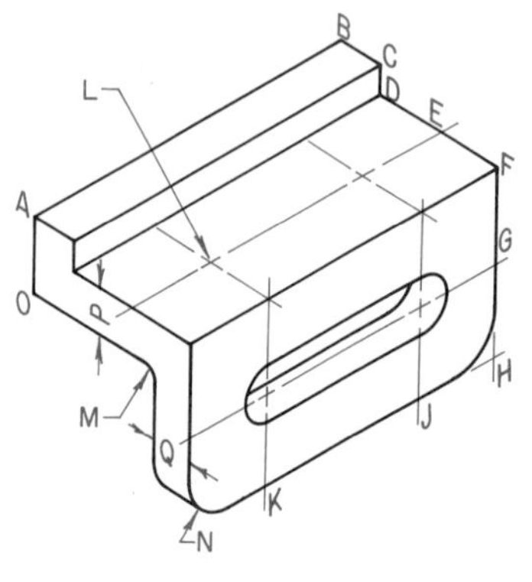

Fig. 5–75 Adjustable stop. Scale: as assigned. $AB = 10''$, $BC = 1^3/_8''$, $CD = {}^3/_4''$, $DE = 1^7/_8''$, $EF = 2''$, $FG = 2^1/_2''$, $GH = 2^1/_2''$, $FH = 5''$, $HJ = 2^1/_2''$, $JK = 5''$, $L = 1''$ DIA, 2 holes, $M = {}^1/_2''$R, $N = 1^1/_4''$R, $AO = 2''$, $P = 1^1/_4''$, $Q = 1^1/_4''$, Slot = $1^1/_2''$ wide.

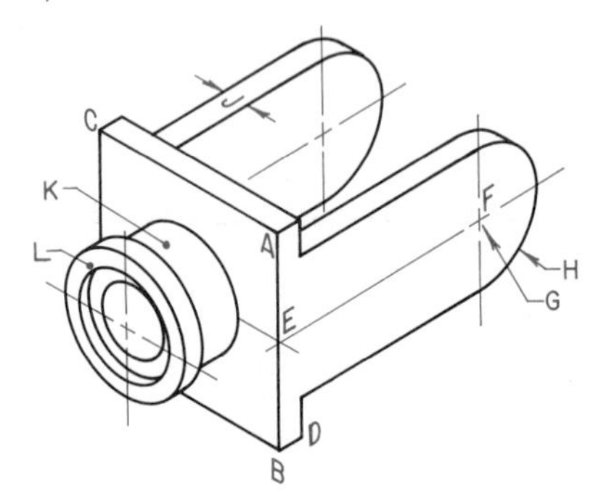

Fig. 5–76 Camera swivel base. Scale: as assigned. AB = 40 mm, AC = 38 mm, BD = 5.5 mm, BE = 20 mm, EF = 45 mm, H = 12 mm R, G = 10 mm DIA, J = 6 mm. Boss: K = 24 mm DIA × 9 mm long, L = 28 mm DIA × 4 mm long, hole = 12 mm DIA × 14 mm deep, Cbore = 18 mm DIA × 3 mm deep.

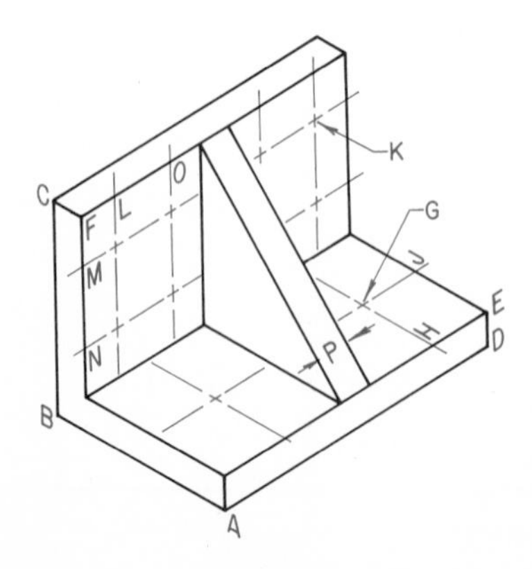

Fig. 5–78 Angle plate. Scale: as assigned. $AB = 6''$, $BC = 6^1/_2''$, $AD = 9^3/_4''$, $DE = 1''$, $CF = 1''$, $G = {}^3/_4''$ DIA, 2 holes, $EH = 2''$, $EJ = 2^1/_2''$, $FL = 1^1/_8''$, $LO = 2''$, $FM = 1^1/_2''$, $MN = 2^1/_2''$, $P = 1^1/_8''$, $K = {}^1/_2''$ DIA, 8 holes.

Fig. 5–77 Pipe support. Scale: as assigned. $AB = 8''$, $BC = 4''$, $AD = {}^3/_4''$, $E = 3^7/_8''$, $F = 2^3/_4''$ DIA × $2^1/_4''$ long, $G = 1^1/_8''$ DIA hole through, slots = $1''$ wide.

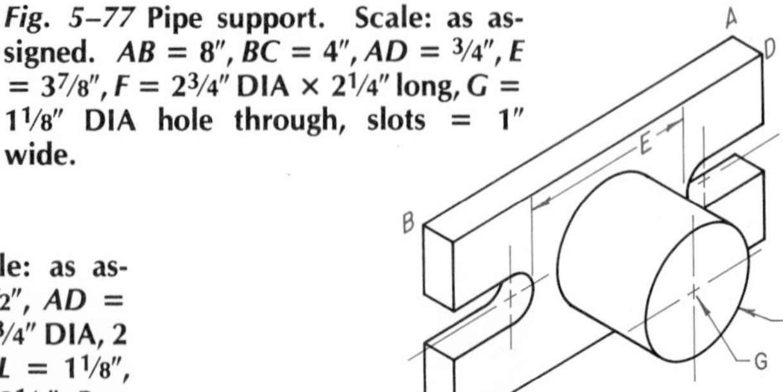

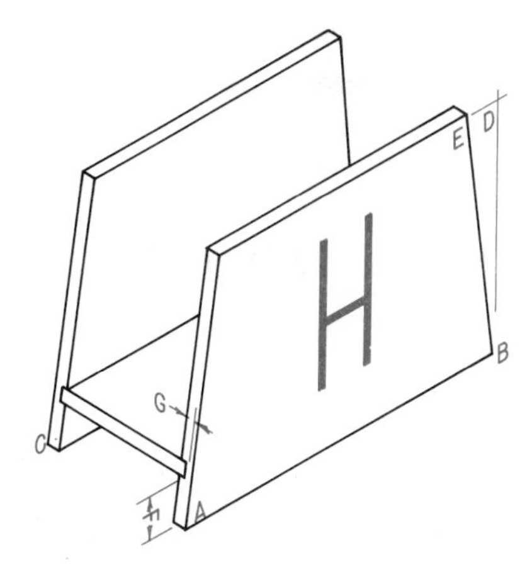

Fig. 5–79 Letter holder. Scale: full size or as assigned. Draw all necessary views. All material is 3/16″ thick (plastic or wood). *AB* is 4″, *AC* is 2″, *BD* is 3″, *DE* is 3/8″, *F* is 3/8″, *G* is 3/32″. Add a design to the front view. See Chapter 2 for layout of block-style lettering. Dimension only if instructed to do so.

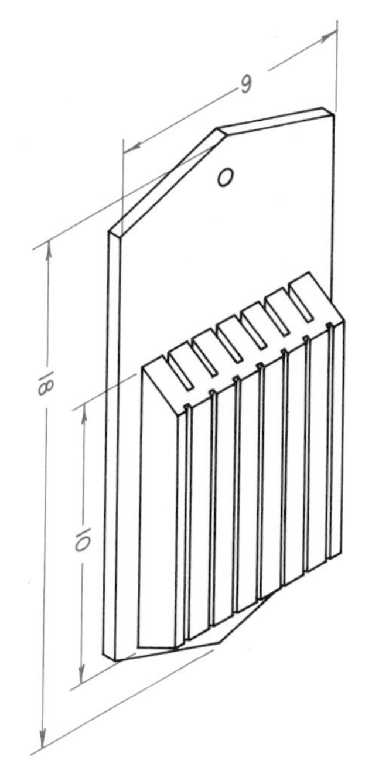

Fig. 5–80 Knife rack. Scale: half size or as assigned. Draw all necessary views. Back is 1/2″ × 9″ × 18″. Front is 1 1/2″ × 7″ × 10″ with 30° -angle bevels on each end. Slots for knife blades are 1/8″ wide × 1″ deep. Grooves on front are 1/8″ wide × 1/8″ deep. Estimate all sizes not given. Redesign as desired.

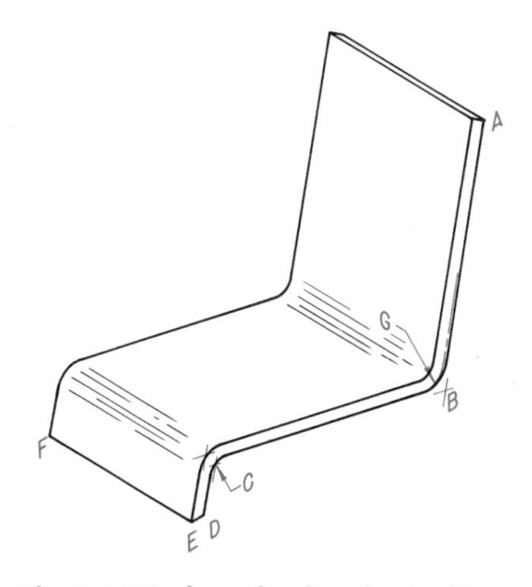

Fig. 5–81 Desk-top book rack. Scale: as assigned. Draw all necessary views. Material is laminated wood, plastic, or aluminum. *AB* is 8″, *BC* is 9 1/2″, *CD* is 2″, *DE* is 3/8″, *EF* is 6″, *G* is 1 1/4″R. All bends are 90°.

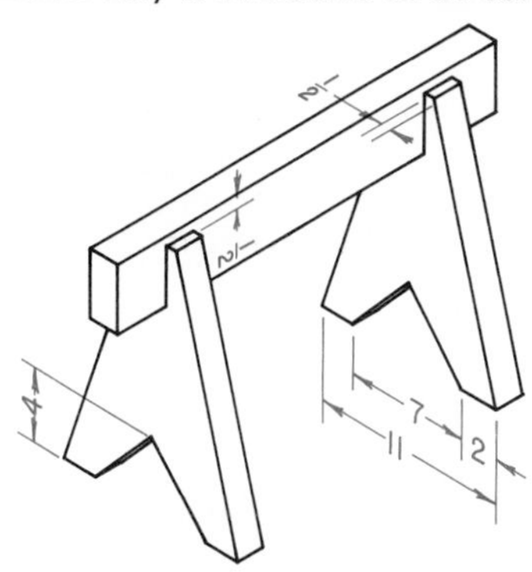

Fig. 5–82 Mini saw horse. Scale: 1/4″ = 1″ or as assigned. Top rail is 2″ × 4″ × 24″. Legs are cut from 2 × 12, 14 1/2″ long. Dimension only if instructed to.

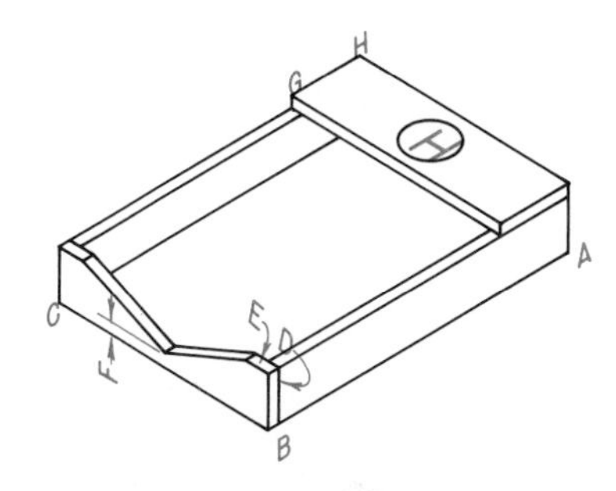

Fig. 5–83 Note-paper box. Scale: full size or as assigned. All stock is 1/4″ thick. *AB* is 6 5/8″, *BC* is 4 5/8″, *BD* is 1″, *DE* is 1/2″, *F* is 1/4″, *GH* is 1 1/2″. Draw all necessary views. Do not dimension unless instructed to do so. Initial inlay is optional. Redesign as desired.

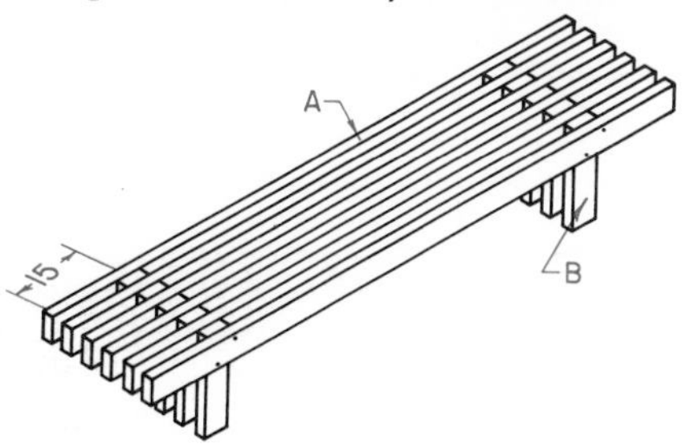

Fig. 5–84 Garden bench. Scale: 1″ = 1′-0″ or as assigned. *A* is 2 × 4 × 6′-0″, *B* is 2 × 4 × 1′-4″. Draw front, top, and right-side views of the bench. Use six or more top rails.

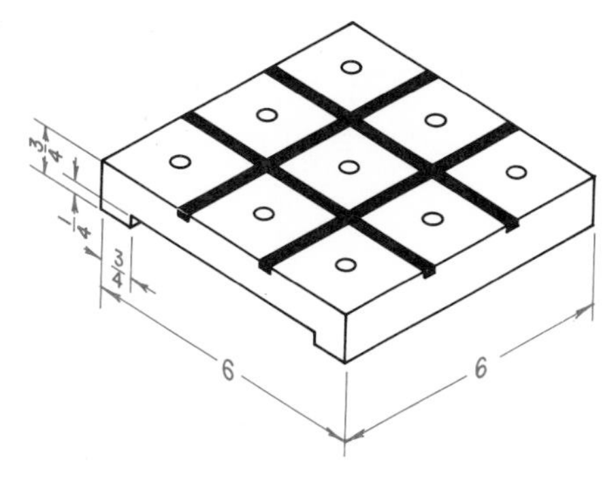

Fig. 5–85 Tic-tac-toe board. Scale: full size or as assigned. Material: hardwood with black plexiglas inlays. Inlays are 1/8″ × 1/4″ and are located 2″ OC. Holes are 5/16″ DIA × 1/4″ deep and are centered within squares. Game board is designed to use marbles.

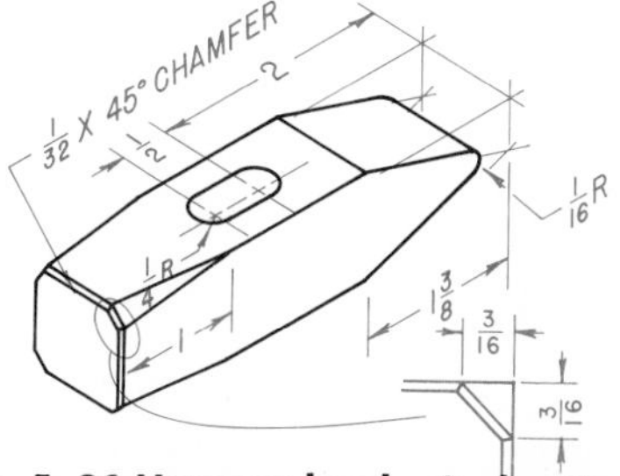

Fig. 5–86 Hammer head. Scale: 2″ = 1″ or as assigned. Draw all necessary views. Overall sizes are 7/8″ square × 3 1/2″ long. Do not dimension unless instructed to do so.

Chapter 6

Dimensioning

SIZE DESCRIPTION

To describe an object completely, you have to know two things. One is its shape. The other is its size. Most of this book deals with the ways of describing shape. In this chapter, however, you will learn how to show the size of the objects that you draw. It is very important that you clearly understand the rules and principles of size description. After all, a machinist in a factory cannot make parts correctly unless all the sizes on the drawing are accurate.

Another name for size description is *dimensioning. Dimensions* (sizes) are measured in either metric or U.S. customary units. Metric units, such as the millimeter, are used for most engineering drawings. Civil engineering drawings are mostly dimensioned in meters. Architects use both meters and millimeters, depending on the size of the item they are drawing. In the customary system, measurements are given in feet and inches, or in inches and fractions, or in decimal divisions of inches. Notes and symbols that show the kind of finish, materials, and other information needed to make a part are also part of dimensioning. So a complete *working drawing* includes shape description as well as measurements, notes, and symbols (Fig. 6–1).

Size description is an important part of a working drawing. Sometimes, all a drafter has to do is give nominal and ordinary sizes in millimeters or in fractions of an inch. On other drawings, a note is added stating that the dimensions can be plus or minus a specific amount, such as 0.4 mm (1/64 in.) or 0.8 mm (1/32 in.). For large castings, 1.6 mm (1/16 in.) or more would be close enough. Such a note may be placed on the drawing with the views. It can also be placed in the title block, usually in a space provided for it.

When dimensions must be exact, they are given in tenths, hundredths, or even thousandths of a

Fig. 6–1 **Dimensioning includes measurements, notes, and symbols.**

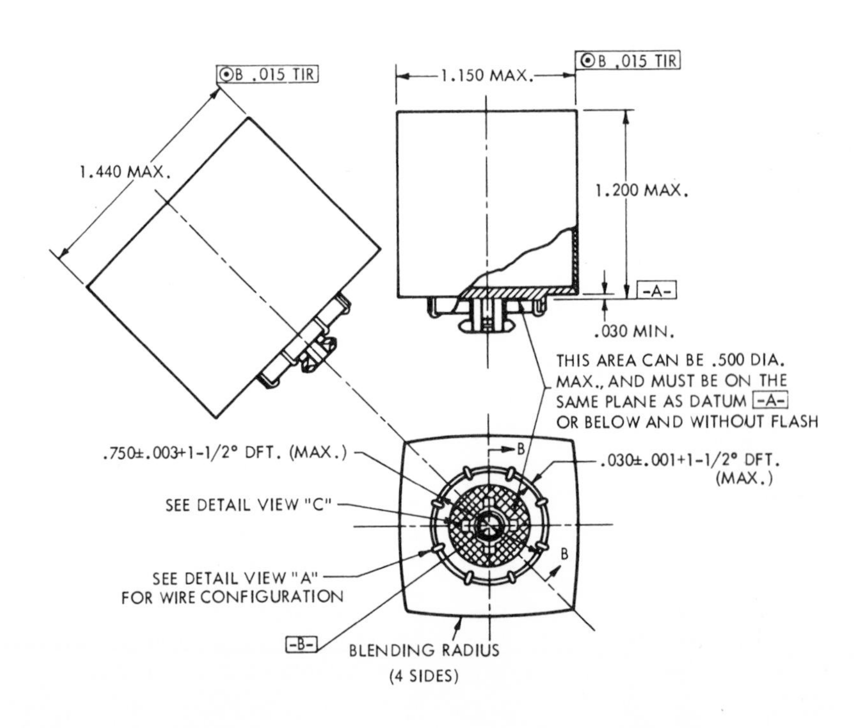

millimeter. If the customary system is being used, the measurements may be in hundredths, thousandths, or ten-thousandths of an inch.

DIMENSIONING

The views on drawings describe the shape of an object. In theory, size could be found by measuring the drawing and applying a scale. In reality, though, this would not be practical, even when the views are drawn to actual or full size. The measuring would simply take too much time. More important, it is impossible to measure a drawing accurately enough for many interchangeable parts that must fit closely together. To ensure accuracy and efficiency, size information is added to the drawing. This is done through a system of lines, symbols, and numerical values.

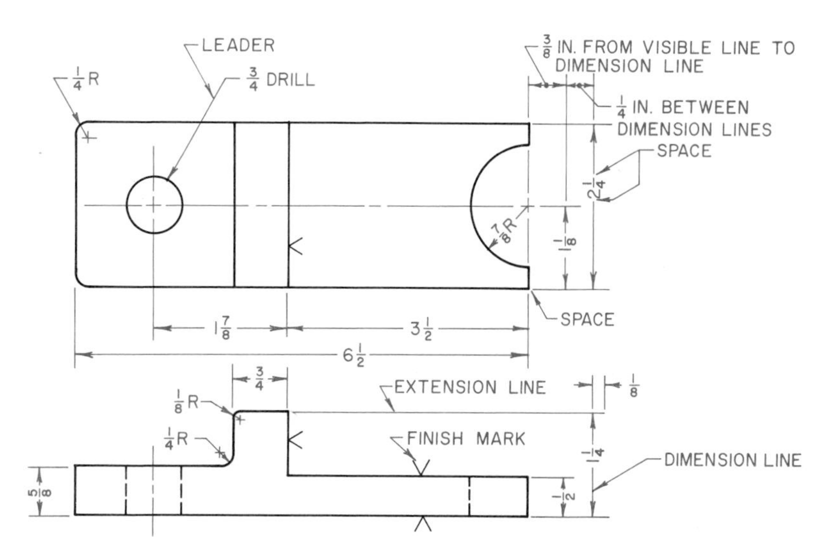

Fig. 6–2 **Dimensioning consists of lines and symbols, and placement techniques.**

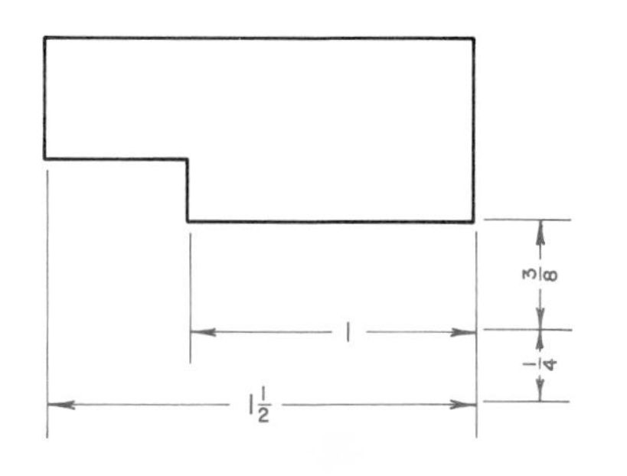

Fig. 6–3 **Dimension lines must be spaced to provide clearness.**

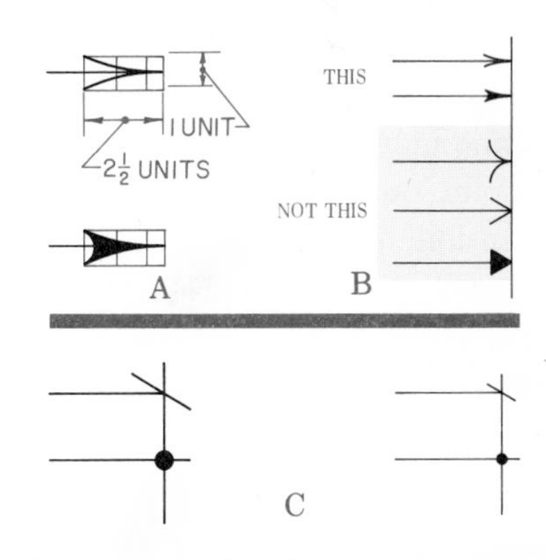

Fig. 6–4 **Arrowheads.**

LINES AND SYMBOLS

Lines and symbols are used on drawings to show where the dimensions apply (see Fig. 6–2). Professional and trade associations, engineering societies, and certain industries have agreed upon the symbols to be used. Therefore, the lines and symbols are recognized by the people who use the drawings. The latest standards information on drawings and symbols can be found in publications from the American National Standards Institute (ANSI), the Society of Automotive Engineers (SAE), and in the Military Standards.

To make a correct drawing, drafters must know these symbols as well as the principles of dimensioning. Drafters must also know the shop processes that are used to build or make the products they draw. Sometimes drafters include symbols in the drawing to show which processes are needed.

DIMENSION LINES

A dimension line is a thin line that shows where a measurement begins and where it ends (Fig. 6–3). It is also used to show the size of an angle. The dimension line should have a break in it for the dimension numbers. To keep the numbers from getting crowded, dimension lines should be at least 10 mm (3/8 in.) from the lines of the drawing. They should also be at least 6 mm (1/4 in.) from each other.

ARROWHEADS

Arrowheads are used at the ends of dimension lines. They show where a dimension begins and ends (Fig. 6–3). They are also used at the end of a leader (see Fig. 6–8) to show where a note or dimension applies to a drawing.

Arrowheads can be open or solid. Their shapes are shown enlarged in Fig. 6–4 at A and reduced to actual size at B. Draw arrowheads care-

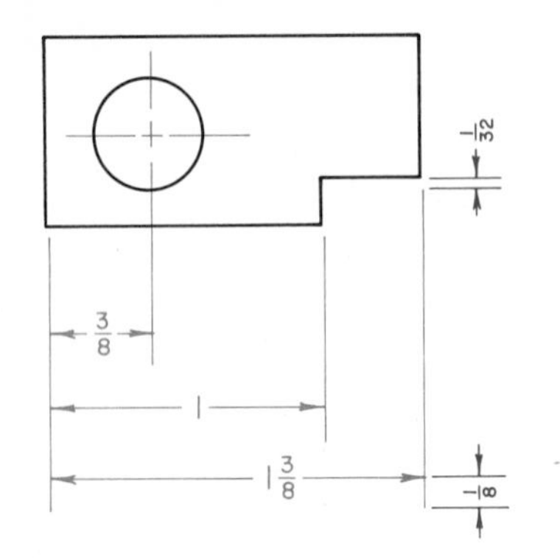

Fig. 6–5 **Extension (witness) lines. A centerline may be used as an extension line.**

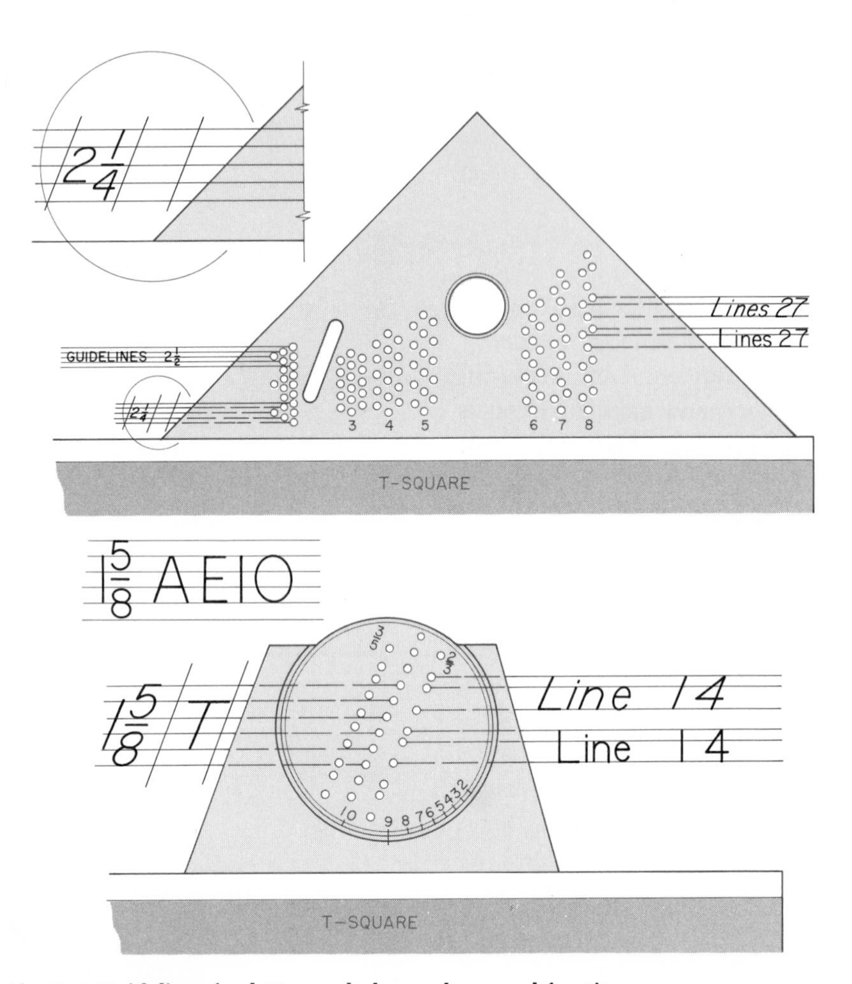

Fig. 6–6 **Guidelines for letters, whole numbers, and fractions.**

fully. In any one drawing, they should all be the same size. However, in a small space you may have to vary the size somewhat.

Some industries use other means to point out the end point of a dimension line or leader. Figure 6–4 at C shows examples of these. These methods do the same job as arrowheads. Nevertheless, the arrowheads shown in Fig. 6–4 at A and B are preferred.

EXTENSION LINES

Extension lines are thin lines that extend the lines of the views. They are used to show center points and to provide space for dimension lines (Fig. 6–5). Extension lines are not part of the views so they should not touch the outline. They should start after a visible space of about 1 to 1.5 mm (about 1/32 to 1/16 in.) and extend about 3 mm (about 1/8 in.) beyond the last dimension line.

NUMERALS AND NOTES

Numerals (numbers) and notes have to be easy to read. So they must be made carefully. Do not make them too large, however. They could overbalance the drawing. Capital letters are preferred on most drawings. They can be either *vertical* (straight up and down) or *inclined* (slanted). In general, make numerals about 3 mm (about 1/8 in.) high. When fractions are used, make the fractions about 6 mm (about 1/4 in.) and the fraction numerals about 2.5 mm (about 3/32 in.) high. Always make the fraction bar (division line) in line with the dimension, never at an angle.

Sometimes drawings are made to be microfilmed or to be reduced photographically and used at a smaller size. When this is the case, the numerals must be made larger and with heavier strokes so that they will be clear when reduced.

Light guidelines for figures and fractions may be drawn quickly and easily with a lettering triangle or an Ames lettering instrument (Fig. 6–6).

THE FINISH MARK

The finish mark, or surface-texture symbol, shows that a surface is to be machined (finished), as shown in Fig. 6–7. The old symbol form,

√, is still used occasionally, but it is being replaced. An even older symbol form, ✗, may be found on some very old drawings. The symbol √ is now in general use. The point of the symbol √ should touch the edge view of the surface to be finished. Modified forms of this symbol can be used to show that allowance for machining is needed, that a certain surface condition is needed, and for other conditions. These are described in ANSI B46.1, *Surface Texture.*

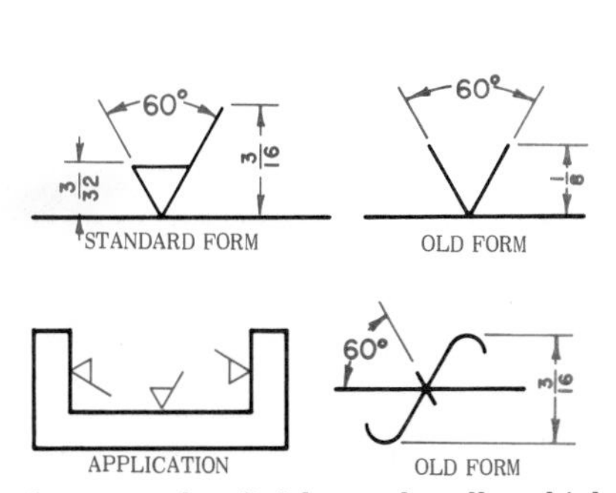

Fig. 6–7 The finish mark tells which surfaces are to be machined.

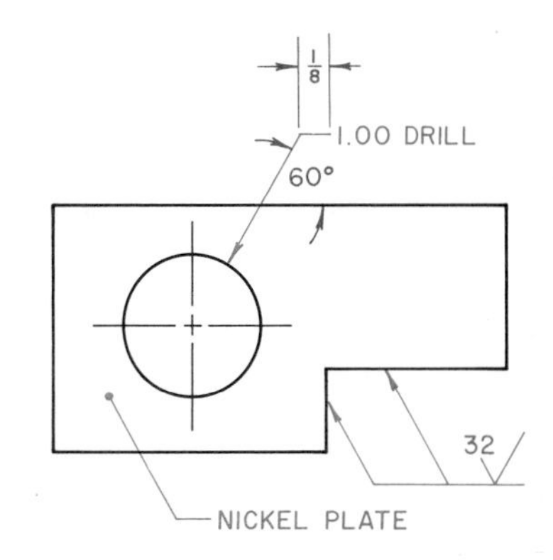

Fig. 6–8 Leaders point to the place where a note or dimension applies.

LEADERS

Leaders are thin lines drawn from a note or a dimension to the place where it applies (Fig. 6–8). Leaders are drawn at an angle to the horizontal. An angle of 60° is preferred, but 45°, 30°, or other angles may be used. A leader starts with a dash, or short horizontal line. This line should be about 3 mm (about 1/8 in.) long, but it may be longer if needed. It generally ends with an arrowhead. However, a dot or other symbol may be used for special identification.

When a number of leaders must be drawn close together, it is best to draw them parallel. A leader to a circle or arc should point to its center. Do not cross leaders. Do not draw long leaders. Do not draw leaders horizontally, vertically, at a small angle, or parallel to dimension, extension, or section lines.

SCALE OF A DRAWING

Scales used in making drawings are described in Chapter 3. The scale used should be given in or near the title. If a drawing has views of more than one part and different scales are used, the scale should be given close to the views. Usual scales are stated as full or full size, 1:1; half size, 1:2, and so forth. If enlarged views are used, the scale would be shown as two times full size, 2:1, five times full size, 5:1, and so forth.

The scales used on metric drawings are usually based on divisions of ten. Scales such as 100:1, 1:50, and 1:100 are examples.

UNITS AND PARTS OF UNITS

When the metric system is used, dimensions on drawings are given in millimeters, meters, and, for special applications, micrometers. When the customary system is used, measurements are given in feet and decimals of a foot, feet and inches, inches and fractions of an inch, or inches and decimals of an inch. The choice depends on the amount of accuracy that is needed. When customary dimensions are in inches, the inch symbol (″) is omitted. When feet and inches are used, the standard practice is to show the symbol for feet but *not* for inches: 7′–5, 7′–0, and so forth.

Feet and inches are used with common fractions, such as 1/2, 1/4, 1/8, and so on, when particular accuracy is not necessary. They are used, for example, when measurements do not have to be closer than ± 1/64 in.

Sometimes, parts must fit together with great accuracy. In that case, the machinist must work within specified limits. If the measurements are customary, the decimal inch is used. This is called decimal dimensioning. Such dimensions are used between finished surfaces, center distances, and pieces that must be held in a definite relationship to each other.

With both metric and customary measures, decimals to two places are used where limits of ± 0.01 mm or in. are close enough (Fig. 6–9 at A, B, and C). Decimals to three or more places are used where limits smaller than ± 0.01 mm or in. are required, as in Figs. 6–10A and B. For two-place decimals, fiftieths, such as 0.02, 0.04, 0.24, are preferred over decimals such as 0.03 and 0.05. Fiftieths can be divided by two and still end up as two-place decimals. This is useful when, for example, you divide a diameter by two to get a radius. The decimal point should be clear and placed on the bottom guideline in a space about the width of a 0.

Decimal dimensioning is used in some industries. It is the preferred method in drafting with customary measure. In recent years, a dual dimensioning system has been used in industries involved in international trade. This system uses both the decimal inch and the millimeter (Fig. 6–11). However, in many industries it is becoming more common to use the metric system alone.

PLACING DIMENSIONS FOR READING

There are two methods in use: the *aligned system* and the *unidirectional system.*

In the *aligned system* of dimensioning (Fig. 6–12) the dimensions are placed in line with the dimension lines. Horizontal dimensions always read from the bottom of the sheet. Vertical dimensions read from the right-hand side of the sheet. Inclined dimensions read in line with the inclined dimension line. If possible, all dimensions should be kept outside the shaded area in Fig. 6–13. The aligned system was once the only system in use.

In the *unidirectional system* of dimensioning (Fig. 6–14), all the dimensions read from the bottom of the sheet, no matter where they appear. In both systems, notes and dimensions with leaders should read from the bottom of the drawing.

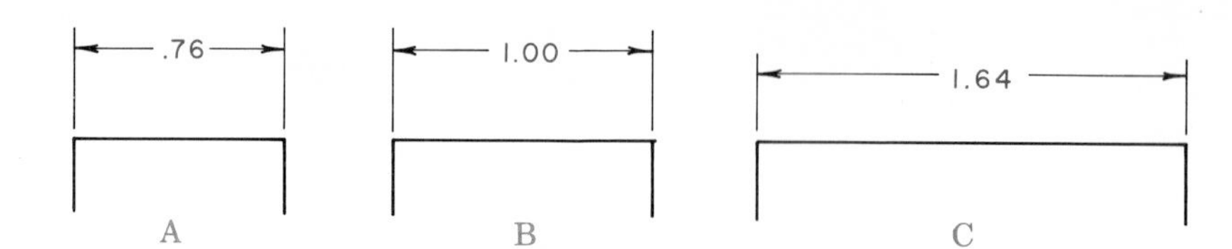

Fig. 6–9 **Decimal dimensions: two places.**

Fig. 6–10 **Decimal dimensions: three places.**

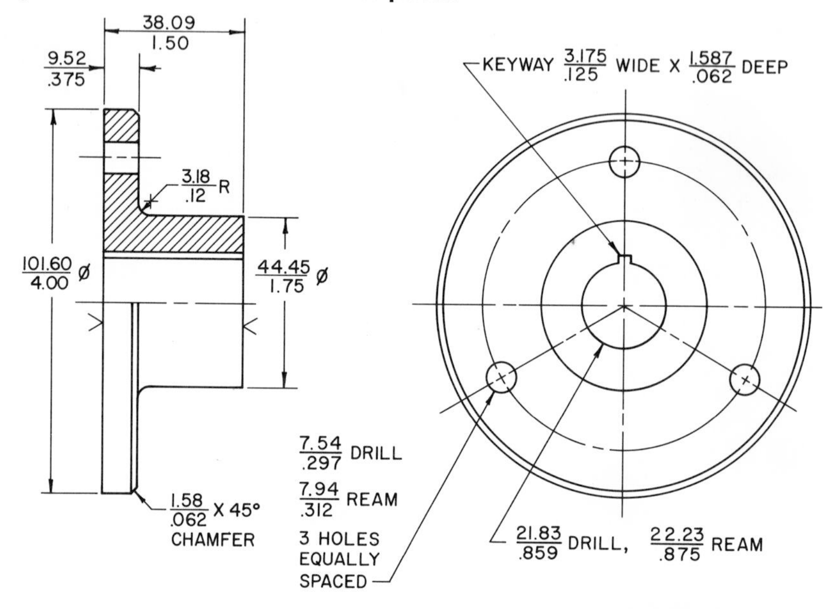

Fig. 6–11 **Typical dual-dimensioned drawing.**

MILLIMETER
INCH

THIRD ANGLE PROJECTION

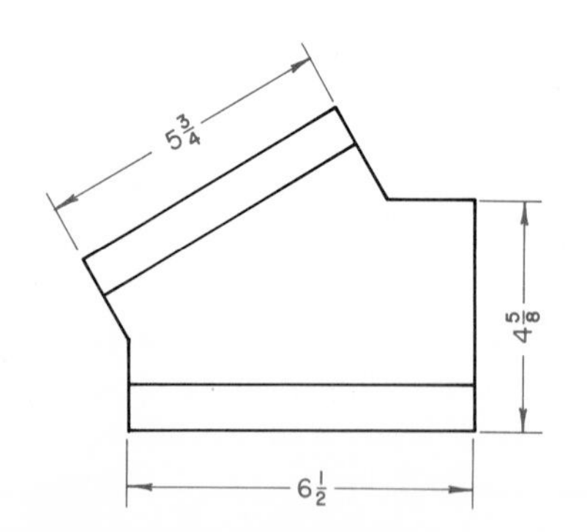

Fig. 6–12 **Dimensioning: the aligned system.**

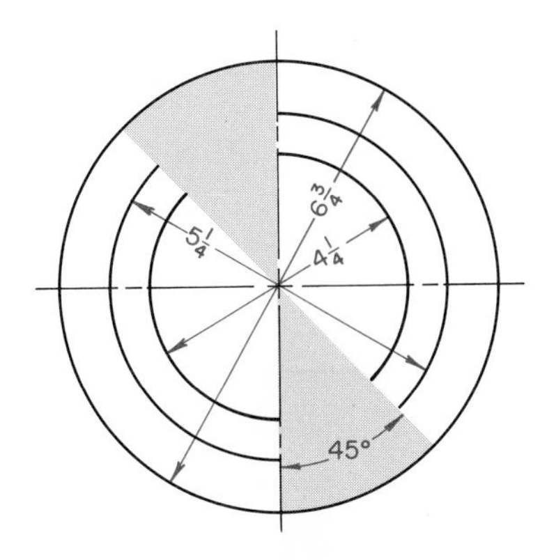

Fig. 6–13 **Avoid placing dimensions in the shaded area.**

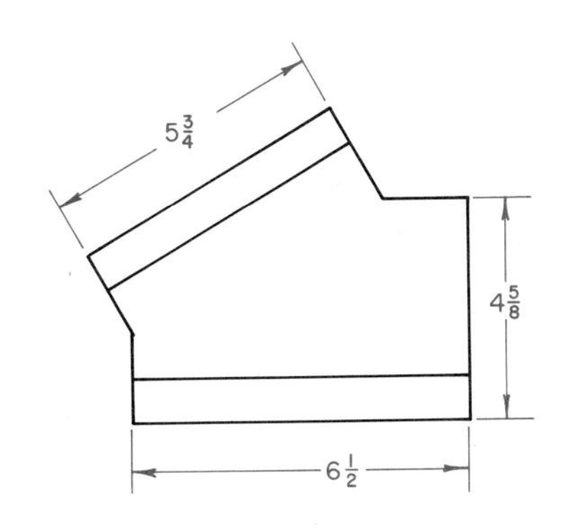

Fig. 6–14 **Dimensioning: unidirectional system.**

Automobile and aircraft companies first brought the unidirectional system into use. It is being adopted by other industries and is replacing the old aligned system. One of the reasons for its popularity is that there are special typing and dimensioning typewriters that write only unidirectionally. These machines save drafters a great deal of time.

THEORY OF DIMENSIONING

There are two basic kinds of dimensions: *size dimensions* and *location dimensions.* Size dimensions define each piece. Giving size dimensions is really a matter of giving the dimensions of a number of simple shapes. Every object is broken down into its geometric forms, such as prisms, cylinders, pyramids, cones, and so forth, or into parts of such shapes. This is shown in Fig. 6–15, where the bearing is separated into simple parts. A hole or hollow part has the same outlines as one of the geometric shapes. Think of such open spaces in an object as *negative* (not solid) shapes.

The idea of open spaces is especially valuable to certain industries. Drafters in the aircraft industry need to know the weights of parts. These weights are worked out from the volumes of the parts as solids. From these solids, the volumes of holes and hollow or open spaces (negative or minus shapes) are subtracted. To get the total weight, the result is then multiplied by the weight per cubic millimeter or per cubic inch of the material.

When the object being dimensioned has a number of pieces, the positions of each piece must also be

Fig. 6–15 **Parts can usually be broken down into basic geometric shapes for dimensioning.**

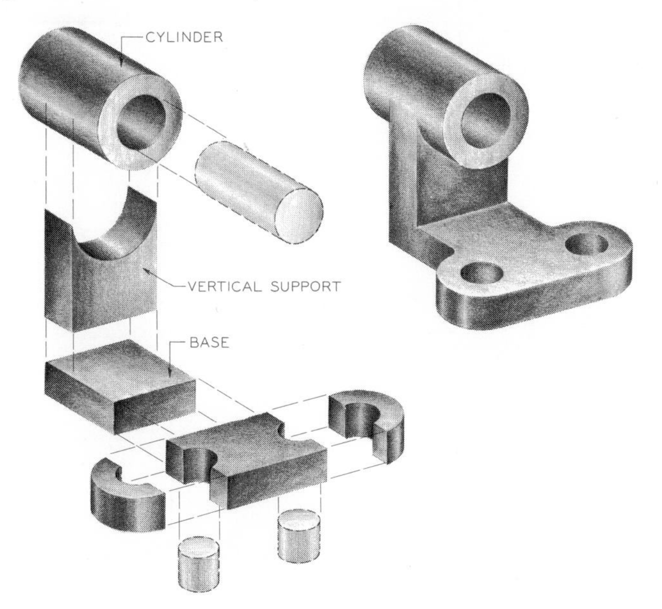

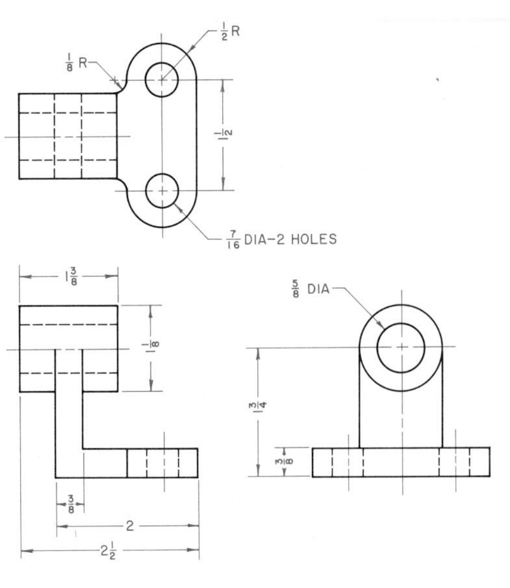

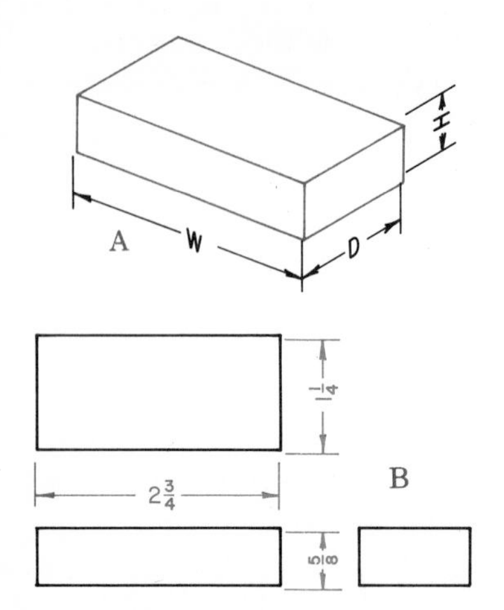

Fig. 6–16 The first shape.

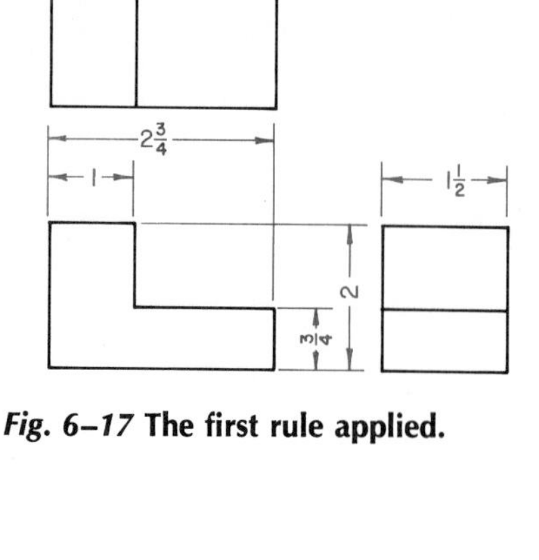

Fig. 6–17 **The first rule applied.**

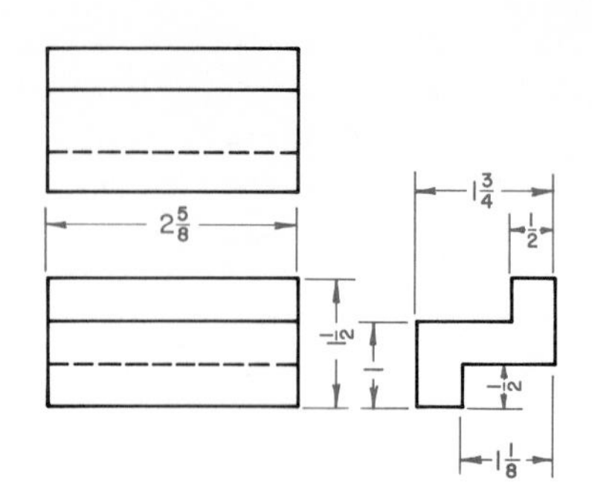

Fig. 6–18 **The first rule applied.**

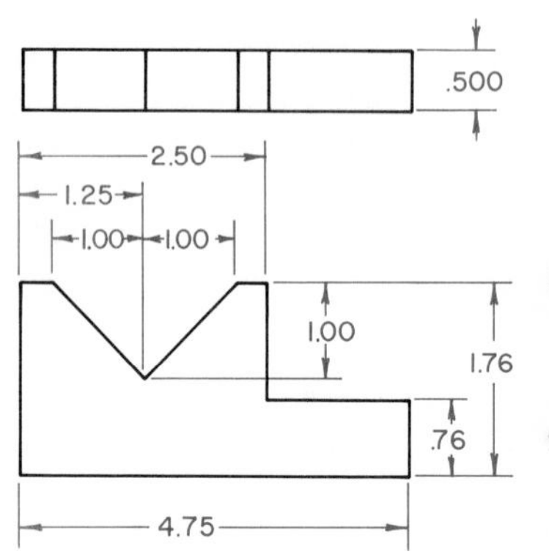

Fig. 6–19 **An irregular flat shape.**

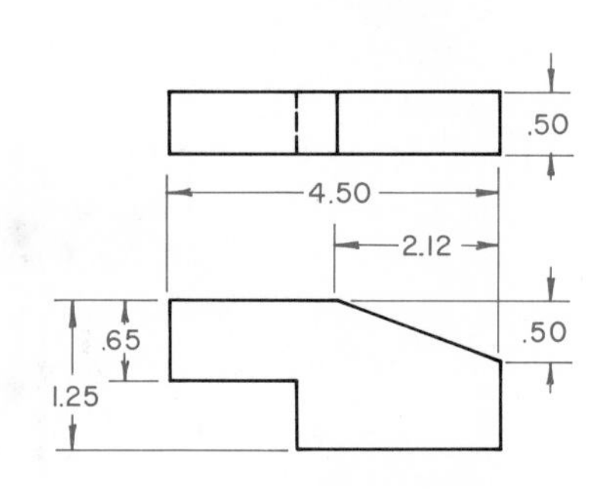

Fig. 6–20 **An irregular flat shape.**

given. These are given by location dimensions. Each piece is first considered separately and then in relation to the other pieces. When the size and location dimensions of each piece are given, the size description is complete. Dimensioning a whole machine, a piece of furniture, or a building is just a matter of following the same orderly pattern that is used for a single part.

SIZE DIMENSIONS

The first shape is the prism. For a rectangular prism (Fig. 6–16), the width *(W)*, the height *(H)*, and the depth *(D)* are needed. This basic shape may appear in a great many ways. A few of these are shown in Figs. 6–17 and 6–18. Flat pieces of irregular shape are dimensioned in a similar way (Figs. 6–19 and 6–20).

The rule for dimensioning prisms is as follows:

For any flat piece, give the thickness in the edge view and all other dimensions in the outline view.

Fig. 6–21 **Dimensioning prisms.**

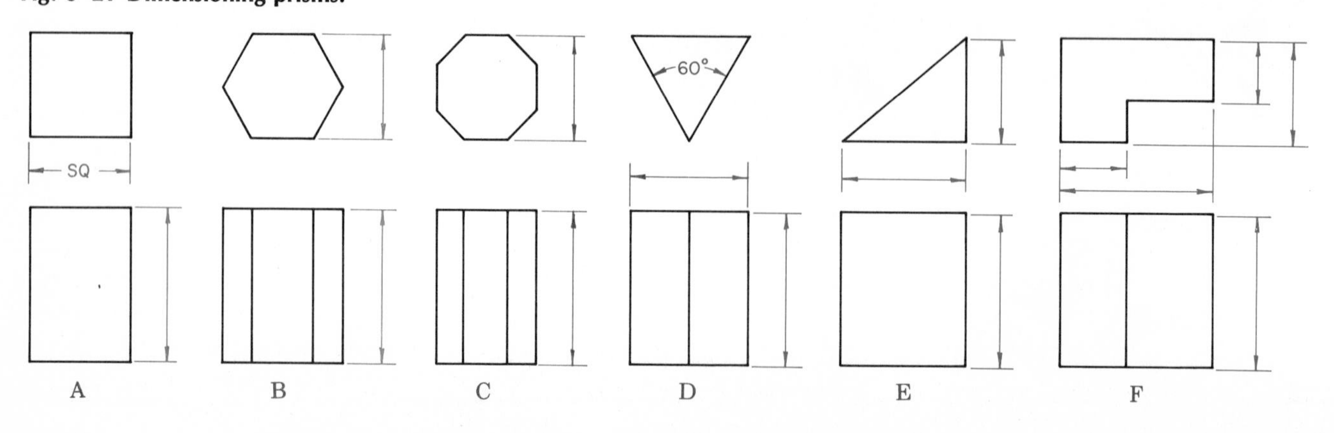

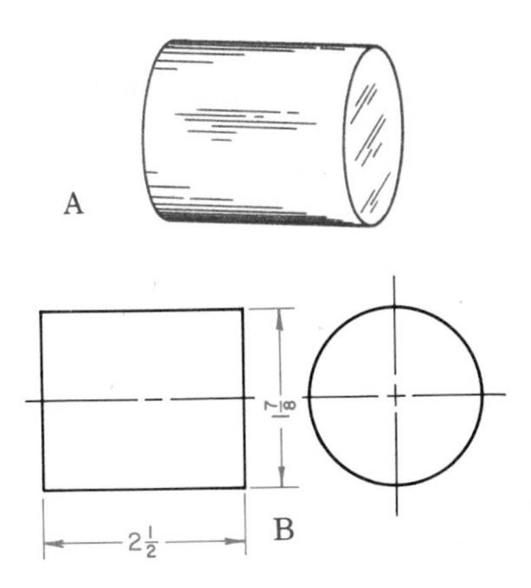

Fig. 6–22 **Dimensioning a cylinder: the second shape.**

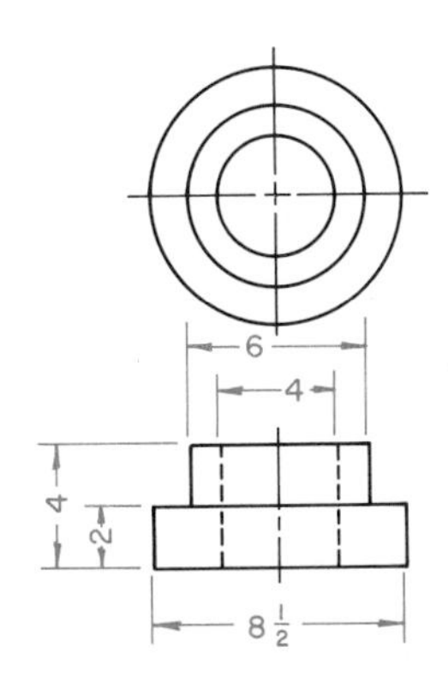

Fig. 6–23 **The second rule applied.**

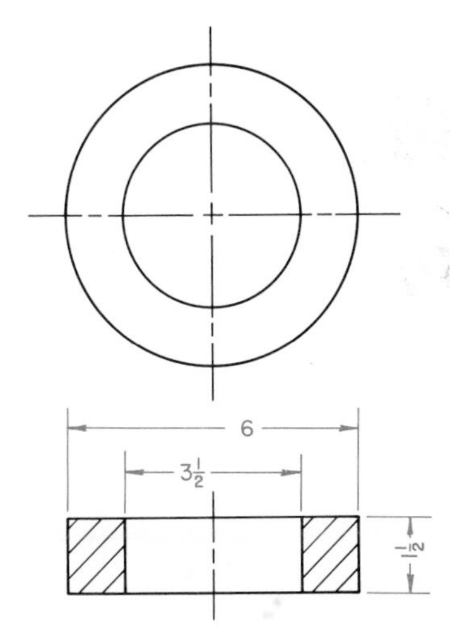

Fig. 6–24 **The second rule applied.**

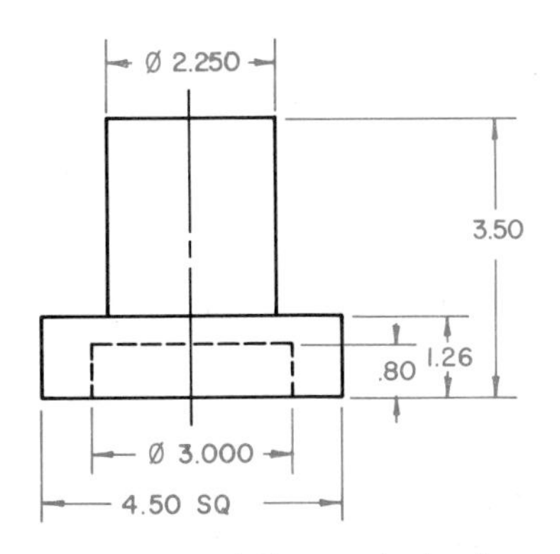

Fig. 6–25 **Use of Ø on a single view.**

The outline view is the one that shows the shape of the flat surface or surfaces. The front views in Figs. 6–19 and 6–20 are the outline views.

Other prisms are shown in Fig. 6–21. A square prism needs two dimensions. A hexagonal or an octagonal prism may also use two dimensions. A triangular prism may use three dimensions, and so on for other regular or irregular prisms.

The second shape is the *cylinder* (tube shaped). The cylinder needs two dimensions: the diameter and the length (Fig. 6–22). Three cylinders are dimensioned in Fig. 6–23. One of these is the hole. A washer or other hollow cylinder can be thought of as two cylinders of the same length (Fig. 6–24). The second rule applies to the second shape:

> *For cylindrical pieces, give the diameter and the length on the same view.*

When the circular view of a cylinder is not shown, the symbol Ø is placed with the diameter dimension (Fig. 6–25). The abbreviation DIA will be found on older drawings instead of the symbol.

Notes are generally used to give the sizes of holes. Such a note is usually placed on the outline view, especially when the method of forming the hole is also given (Fig. 6–26 at A). These notes show what operations are needed to form or finish the hole. For example, drilling, punching, reaming, lapping, tapping, countersinking, spot facing, and so forth, may be specified (Fig. 6–47). Either a dimension or a note can be used when a hole is to be formed by boring. When a hole in a casting is to be formed by a core, the word "core" is used in a note or with the dimension.

When parts of cylinders occur, such as fillets (Fig. 6–26 at B), rounds (Fig. 6–26 at C), and rounded corners, they are dimensioned in the views where the curves show. The radius dimension is given and is followed by the abbreviation R.

Some of the other shapes are the cone, the pyramid, and the sphere. The cone, the frustum, the square pyramid, and the sphere can be dimensioned in one view (Fig. 6–27). To dimension rectangular or other pyramids and parts of pyramids, two views are needed.

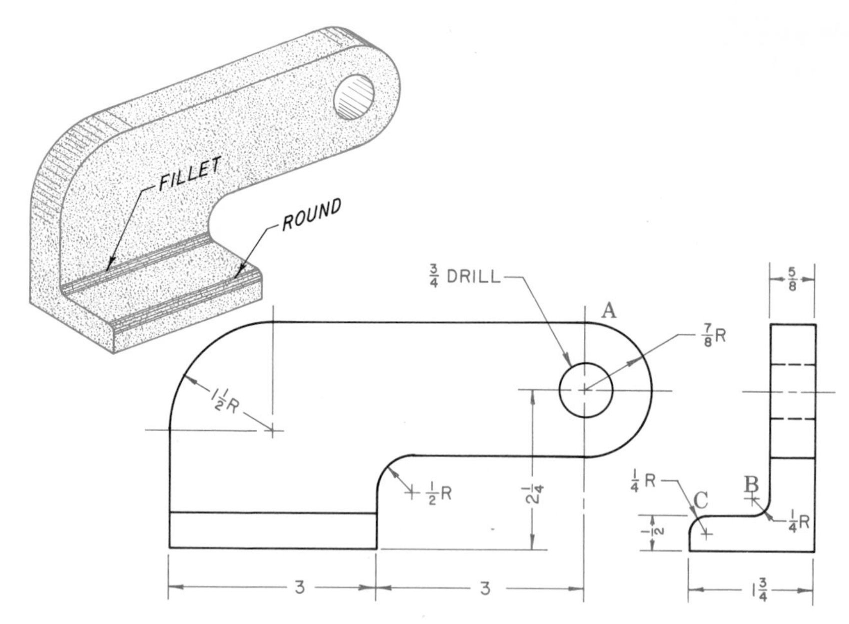

Fig. 6–26 **Fillets, rounds, and radii.**

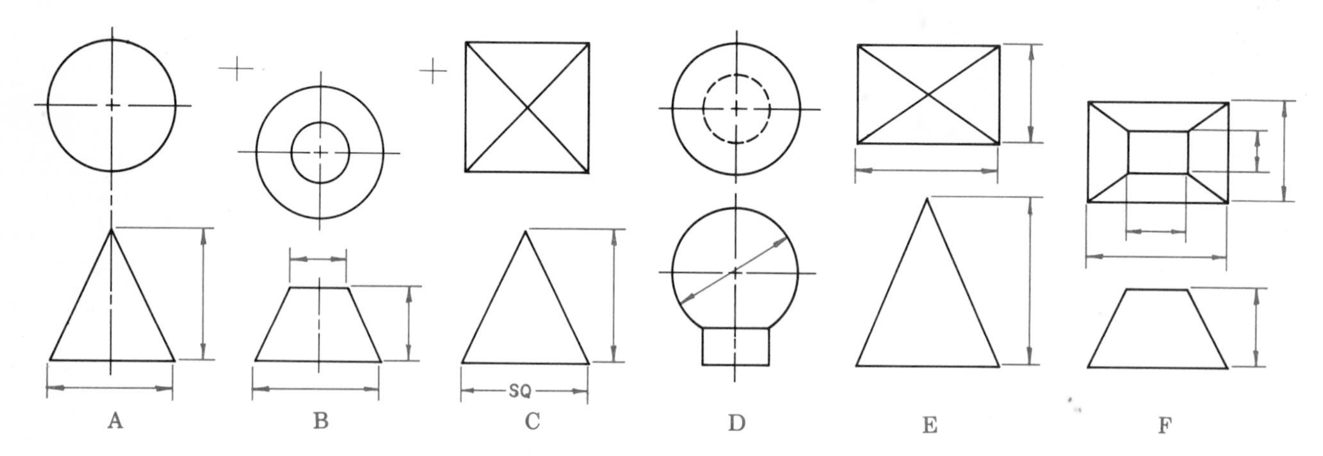

Fig. 6–27 **Dimensioning some elementary shapes.**

LOCATION DIMENSIONS

Location dimensions are used to show the relative positions of the basic shapes. They are also used to locate holes, surfaces, and other features. In general, location dimensions are needed in three mutually perpendicular directions: up and down, crossways, and forward and backward.

Finished surfaces and centerlines, or axes, are important for fixing the positions of parts by location dimensions. In fact, finished surfaces and axes are used to define positions. Prisms are located by surfaces, surfaces and axes, or axes. Cylinders are located by axes and bases. Three location dimensions are needed for both forms.

All the surfaces and axes must be studied together so that the parts will go together as accurately as necessary. That means that in order to include the right location dimensions and notes on a drawing, drafters must be familiar with the engineering practices needed for the manufacture, assembly, and use of a product.

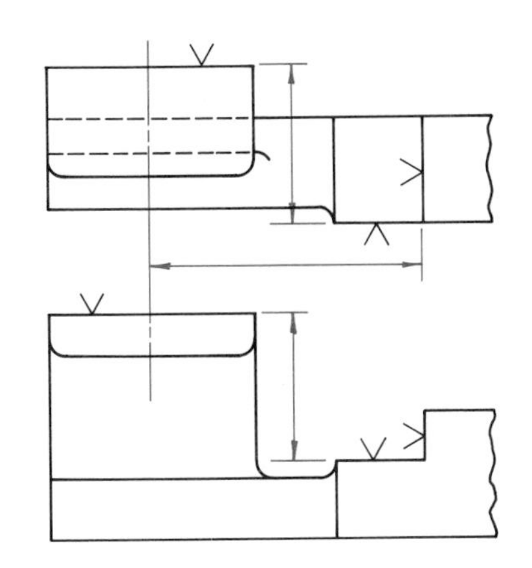

Fig. 6–28 **Locating dimensions for prisms.**

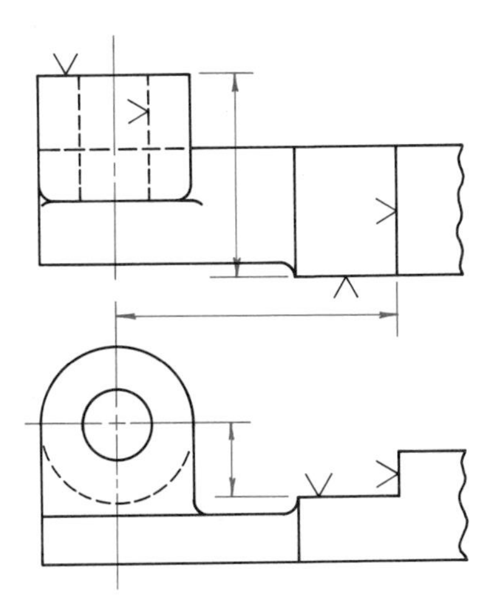

Fig. 6–29 **Locating dimensions for prisms and cylinders.**

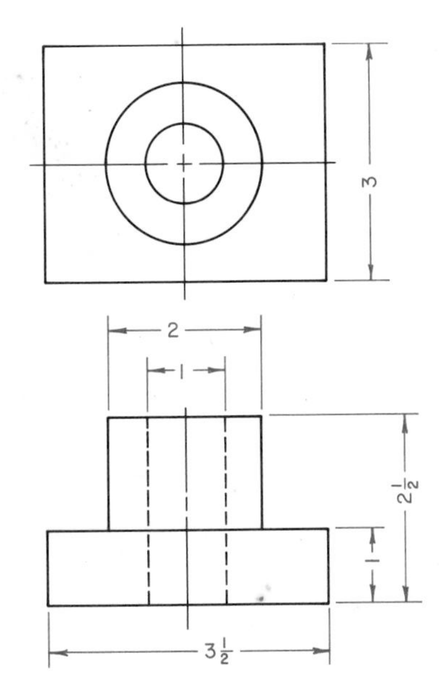

Fig. 6–30 **First and second shapes.**

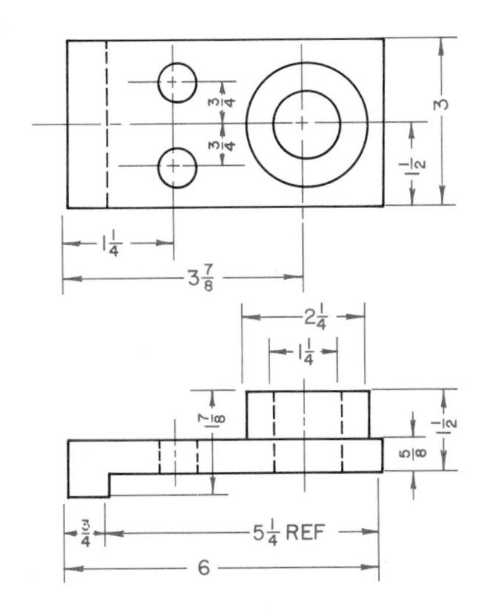

Fig. 6–31 **First and second shapes.**

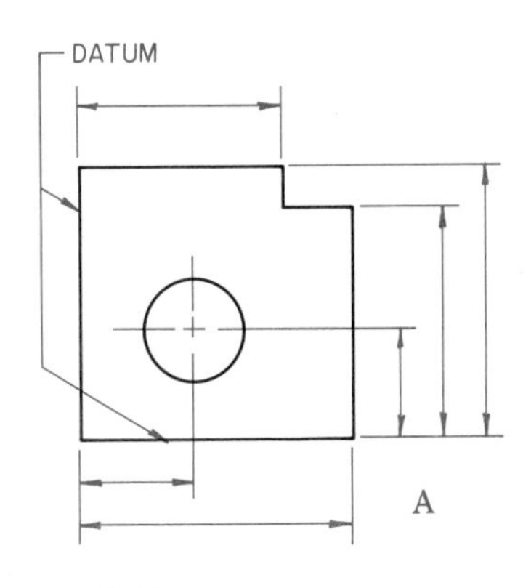

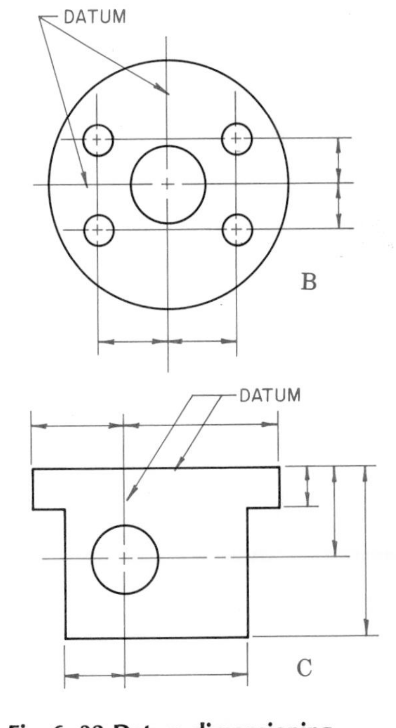

Fig. 6–32 **Datum dimensioning.**

There are two general rules for showing location dimensions:

Prism forms are located by the axes and the surfaces (Fig. 6–28). Three dimensions are needed.

Cylinder forms are located by the axis and the base (Fig. 6–29). Three dimensions are needed.

Combinations of prisms and cylinders are shown in Figs. 6–30 and 6–31. Dimensions marked *L* (Fig. 6–31) are location dimensions.

DATUM DIMENSIONS

Datums are points, lines, and surfaces that are assumed to be exact. Such datums are used to compute or locate other dimensions. Location dimensions are given from them. When positions are located from datums, the different features of a part are all located from the same datum.

Two surfaces, two centerlines, or a surface and a centerline are typical datums. In Fig. 6–32 at A, two surface datums are used. At B, two centerlines are used. At C, a surface and a centerline are used. A datum must be clear and visible while the part is being made. Mating parts should have the same datums, because they fit together.

GENERAL RULES

When adding dimensions to drawings, drafters follow certain practices. These methods have been found to be the best way to do things. So, in a way, they are rules for dimensioning.

1. Dimension lines should be spaced about 6 mm (about 1/4 in.) apart and about 10 mm (about 3/8 in.) from the view outline (See Fig. 6–3).
2. If the aligned system is used, dimensions must read in line with the dimension line and from the lower or right-hand side of the sheet (See Fig. 6–12).
3. If the unidirectional system is used, all dimensions must read from the bottom of the sheet (See Fig. 6–14).
4. On machine drawings, dimensions should be given in millimeters or in decimal inches, even if the drawings are of large objects such as airplanes or automobiles. Values are given to at least two digits, except when the value is 0. Single numbers from 1 to 9 use a decimal marker followed by a 0, for example, 3.0.
5. When all the dimensions are in millimeters or inches, the symbol is generally omitted. A note can be added to the drawing such as "All dimensions are in millimeters."
6. Very large areas on architectural and structural drawings may be dimensioned in meters. In that case, whole numbers stand for millimeters and decimalized dimensions stand for meters. However, this is true only on architectural and structural drawings. Customary measure is also used. Any measurement over 12 in. is given in feet and inches.
7. Sheet-metal drawings are usually dimensioned in millimeters or in inches.
8. Furniture and cabinet drawings are usually dimensioned in millimeters or in inches.
9. When using customary measure, feet and inches are shown as 7′–3. Where the dimension is in even feet, it is written 7′–0.
10. Dimensions should be positioned clearly. The same dimension is not repeated on different views.
11. Dimensions that are not needed should not be given. This is especially important for interchangeable manufacture where limits are used. Figure 6–33 at A shows unnecessary dimensions. They have been omitted in Fig. 6–33 at B.
12. Overall dimensions should be placed outside the smaller dimensions (Figs. 6–33 and 6–34). When the overall dimension is given, one of the smaller distances should not be dimensioned (Fig. 6–33 at B) unless it is needed for reference. Then REF should be added, as in Fig. 6–34.
13. On circular end parts, the center-to-center dimension is given instead of an overall dimension (Fig. 6–35).
14. When you have to put a dimension within a sectioned area, leave a clear space for the number (Fig. 6–36).
15. American National Standard practice is to avoid placing dimensions in the shaded area in Fig. 6–37 when the aligned system is used.

Fig. 6–33 **Omit unnecessary dimensions.**

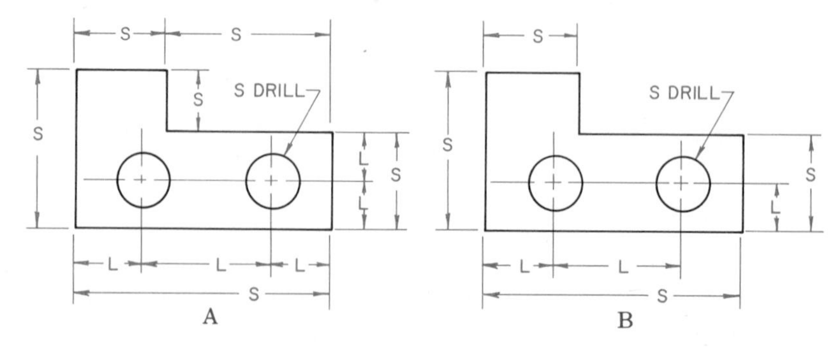

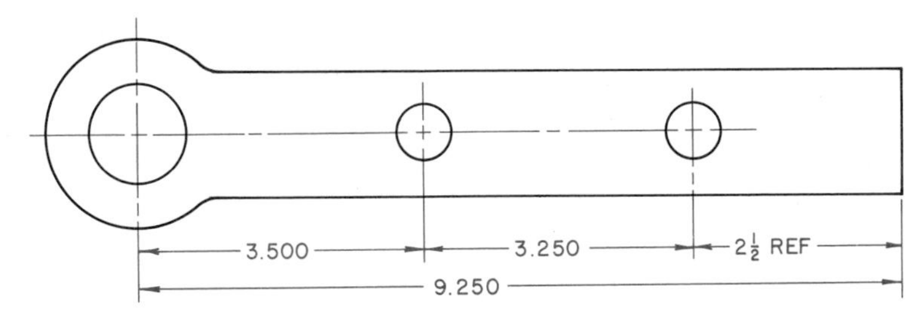

Fig. 6–34 **A dimension for reference should be indicated by REF.**

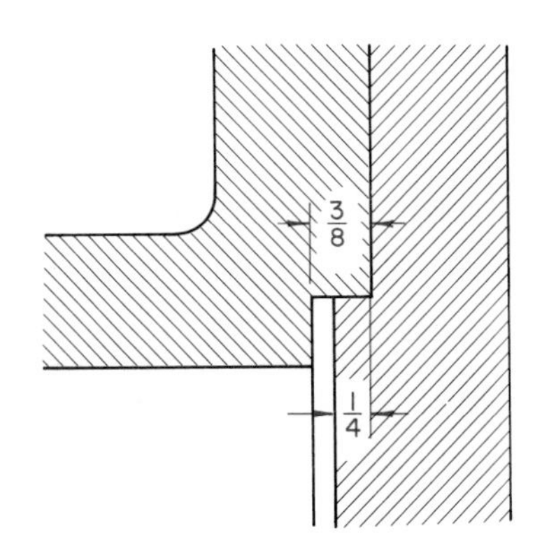

Fig. 6–36 **Dimensions within a sectioned area.**

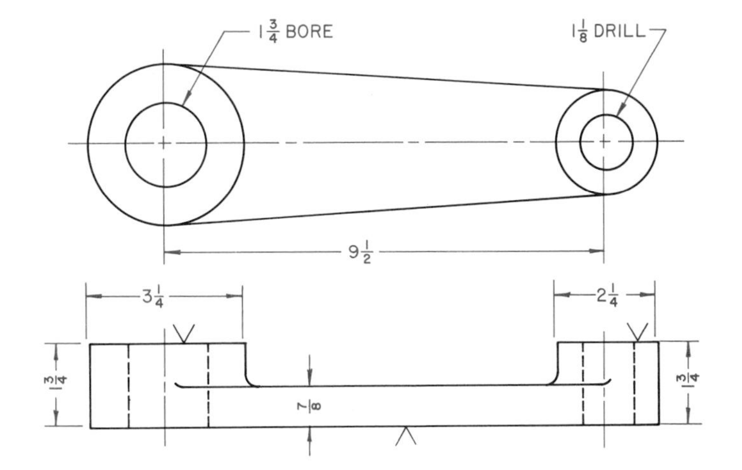

Fig. 6–35 **Center-to-center dimensions.**

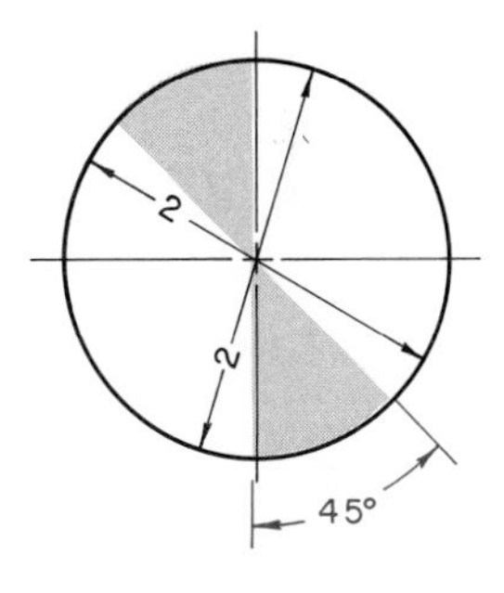

Fig. 6–37 **Avoid shaded area with aligned dimensions.**

16. Dimensions should be given from centerlines, finished surfaces, or datums where needed.
17. Never use a centerline or a line of the drawing as a dimension line.
18. Never have a dimension line that extends from a line of a view.
19. Never place a dimension where it is crossed by a line.
20. Always give the diameter of a circle, not the radius. The symbol Ø is used before the dimension, unless it is obviously a diameter.
21. The radius of an arc should always be given with the abbreviation R placed after the dimension.
22. In general, dimensions should not be placed inside the view outlines.
23. Extension lines should not cross each other or cross dimension lines if this can be done without making the drawing more complicated.
24. Do not dimension to hidden lines if possible.
25. Remember that there are no hard-and-fast rules or practices that are not subject to change under the special conditions or needs of a particular industry. However, when there is a variation of any rule, there must be a reason to justify it.

STANDARD DETAILS

The shape, the methods of manufacture, and the use of a part generally tell us which dimensions must be given and how accurate they must be. So a knowledge of manufacturing methods, patternmaking, foundry and machine-shop procedures, forging, welding, and so on, is very useful when you are choosing and placing dimensions. In fact,

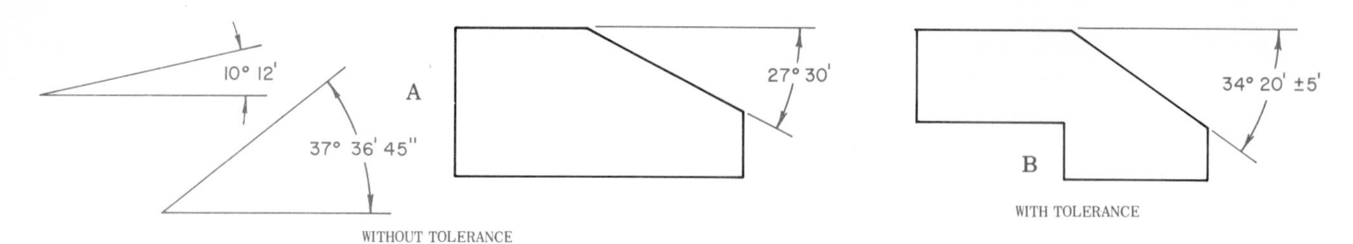

Fig. 6–38 **Dimensioning an angle.**

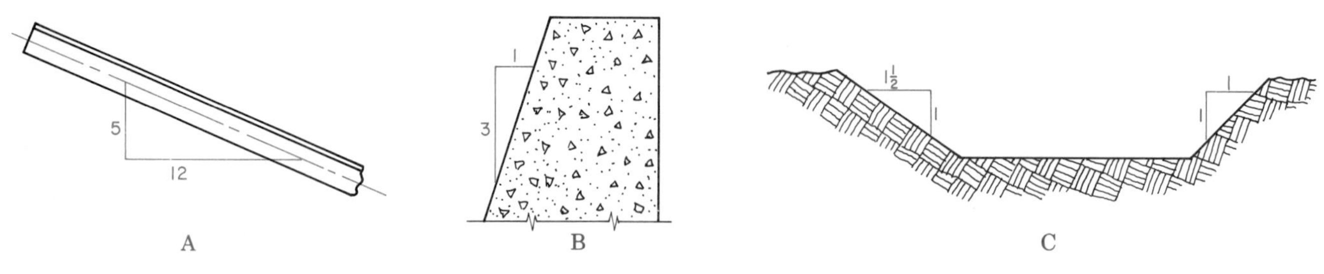

Fig. 6–39 **Dimensioning angles of slopes.**

most of the time you have to know these things. You must also consider whether one copy of a part is to be made. If many copies are to be made, quantity-production methods have to be used. In addition, there are purchased parts, identified by name or brand, that call for few, if any, dimensions. Some companies have their own standard parts for use in different machines or constructions. The dimensioning of these parts depends on how they are used and produced.

There are, however, certain more-or-less standard details or conditions. For these, there are suggested ways of dimensioning.

ANGLES AND CHAMFERS

Angles are usually dimensioned in degrees (°), minutes (′), and seconds (″) (Fig. 6–38 at A). However, decimalized angles are preferred. The abbreviation DEG may be used instead of the symbol ° when only degrees are given. Angular tolerance is usually bilateral (shown on both sides of the angles as plus or minus); for example, the tolerance is ± 0.08. If decimalized angles are not being used, the dimensions might be written ± ½° for degrees and ± 5′ for minutes (Fig. 6–38 at B). Angular tolerance is stated either on the drawing or in the title block. Angular measurements on structural drawings are given by run and rise, (Fig. 6–39 at A). A similar method is used for slopes, as at B and C. Here one side of the triangle is made equal to 1.

Two usual methods of dimensioning chamfers are shown in Fig. 6–40 at A and B.

Fig. 6–40 **Dimensioning chamfers.**

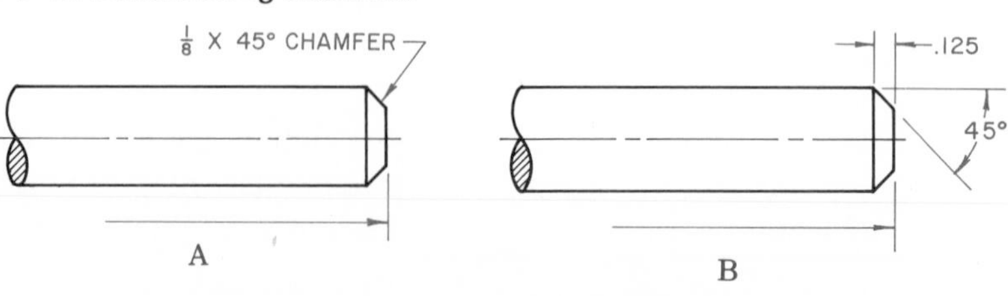

TAPERS

Tapers can be dimensioned giving the length, one diameter, and the taper as a ratio, as in Fig. 6–41 at A. Another method is shown at B. In this method, one diameter or width, the length, and either the American National Standard, or another standard taper number are given. For a

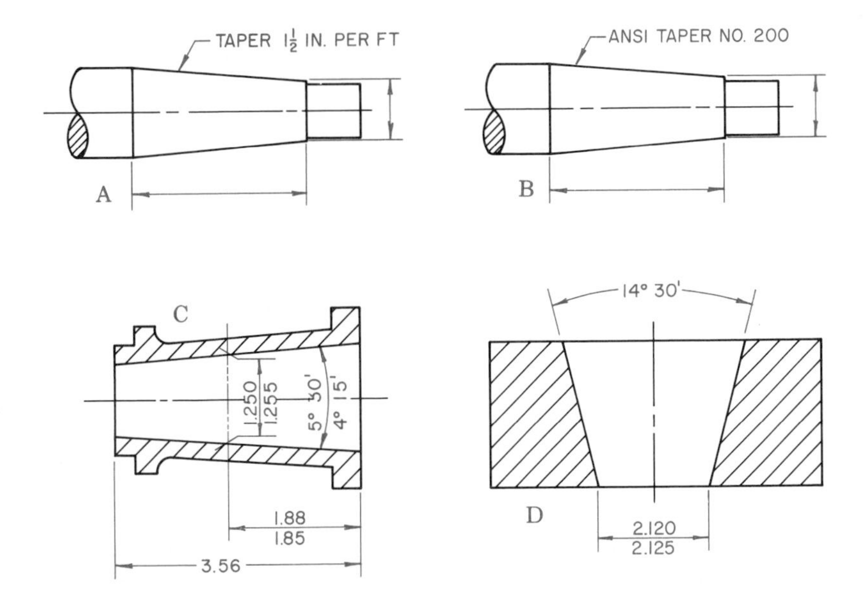

Fig. 6–41 **Dimensioning tapers.**

close fit, the taper is dimensioned as at C. The diameter is given at a located gage line. At D, one diameter and the angle are given.

DIMENSIONING CURVES

A curve made up of arcs of circles is dimensioned by the radii that have centers located by points of tangency (Fig. 6–42 at A and B). Noncircular or irregular curves (Fig. 6–43) can be dimensioned as at A. They can also be dimensioned from datum lines, as at B. A regular curve can be described and dimensioned by showing the construction or naming the curve, as at C. The basic dimensions must also be given.

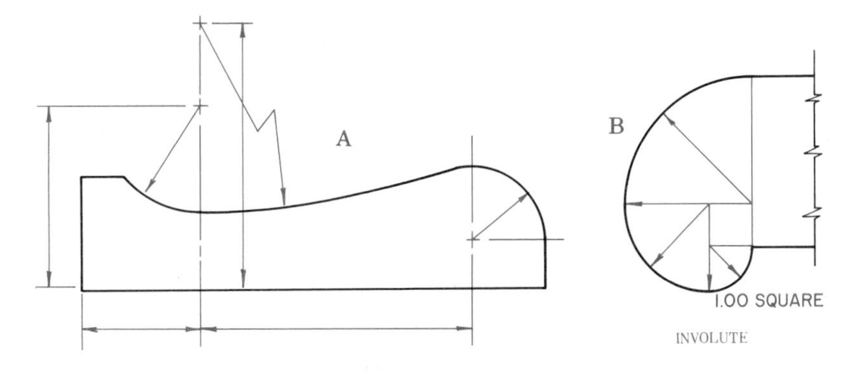

Fig. 6–42 **Dimensioning curves composed of circular arcs.**

DIMENSIONING A DETAIL DRAWING

A drawing for a single part that includes all the dimensions, notes, and information needed to make that part is called a *detail drawing.* The dimensioning should be done in the following order: (1) All the views of a drawing should be done before adding any dimensions or notes. (2) Think about the actual shape of the part and its characteris-

Fig. 6–43 **Dimensioning noncircular curves.**

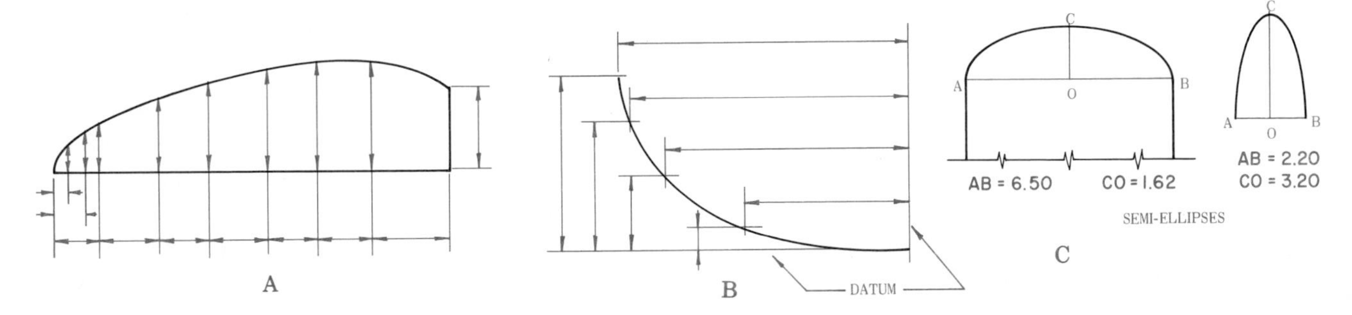

tic views. With this mind, draw all of the extension lines including the lengthening of any centerlines that may be needed. (3) Think about the size dimensions and the related location dimensions. Put on the dimension lines, leaders, and arrowheads. (4) After considering any changes, put in the dimensions and add any notes that may be needed.

DIMENSIONING AN ASSEMBLY DRAWING

When the parts of a machine are shown together in their relative positions, the drawing is called an *assembly drawing.* If an assembly drawing needs a complete description of size, the rules and methods of dimensioning apply.

Drawings of complete machines, constructions, and so on, are made for different uses. The dimensioning must show the information that a drawing is designed to supply.

1. If the drawing is only to show the appearance or arrangement of parts, the dimensions can be left off.
2. If a drawing is needed to tell the space a product requires, give overall dimensions.
3. If parts have to be located in relation to each other without giving all the detail dimensions, you would usually give center-to-center distances. You would also give the dimensions needed for putting the machine together or erecting it in position. *Photodrawings* (photographs to draw on), instead of regular drawings, can be made of a machine for the uses in paragraphs 1, 2, and 3.
4. In some industries, assembly drawings are completely dimensioned (Chapter 11). These *composite drawings* are used as both detail and assembly drawings.

For furniture and cabinetwork, sometimes only the major dimensions are given. For example, length, height, and sizes of stock may be given. The details of joints are left to the cabinetmaker or to the standard practice of the company. This is especially true if machinery is used and construction details are standardized.

NOTES FOR DIMENSIONS

Notes supply the information needed for some operations. Among these are the operations for making drilled holes (Fig. 6–44), reamed holes (Fig. 6–45), and counterboring or spot facing (Fig. 6–46). Notes that specify these and other dimensions and operations are shown in Fig. 6–47. In this figure, A is for a drilled hole, and B is for a hole to be drilled and reamed. C and D specify counterbore. E specifies countersink for a No. 24 flathead screw. F, G, and H are for countersunk and counterdrilled holes. I specifies a spot face to allow for a nut. J specifies a smooth-surface spot face.

Sometimes a hole is to be made in a piece after assembly with its mating piece. In that case, the words "at assembly" should be added to the note. Because such a hole is located when it is made, no dimensions are needed for its location. Some other dimensions with machining operations are suggested in Fig. 6–48.

Fig. 6–44 **Drilling a hole.**

Fig. 6–45 **Reaming a hole.**

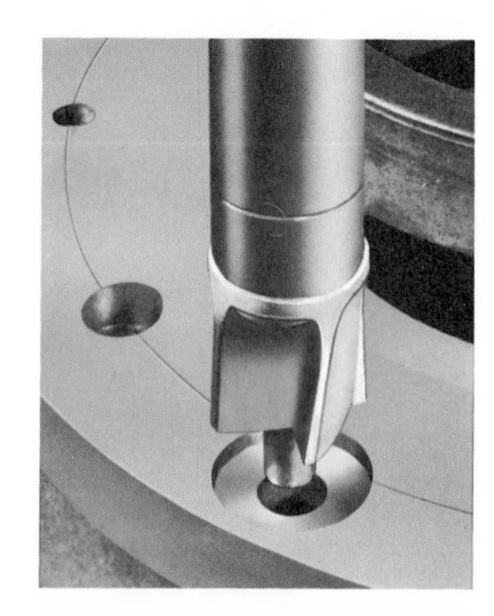

Fig. 6–46 **Counterboring to a specified depth. Spot facing is generally used to provide smooth spots.**

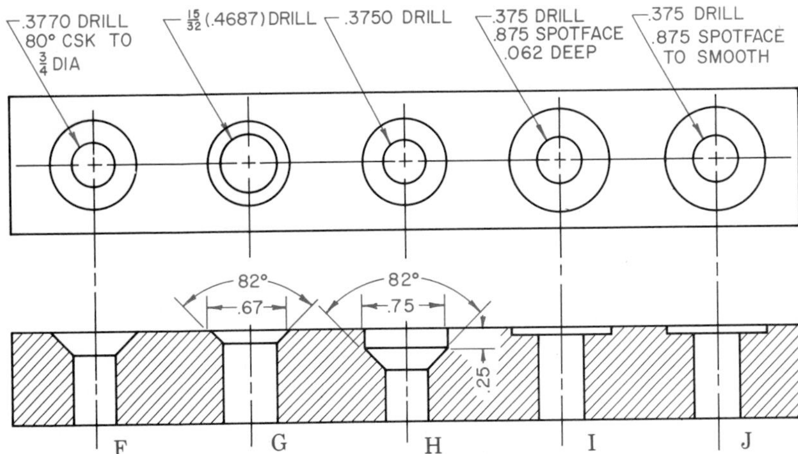

Fig. 6-47 **Dimensions for holes.**

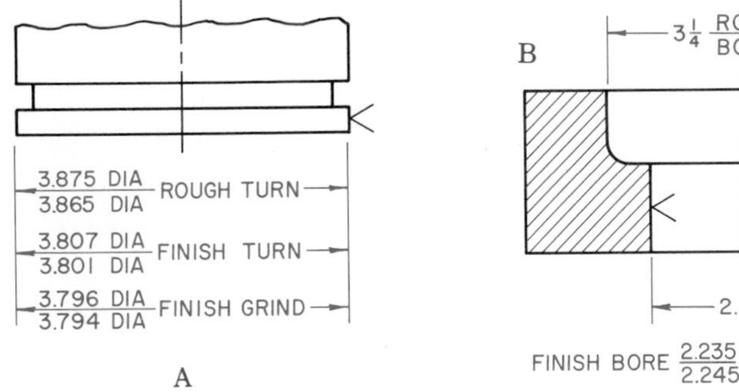

Fig. 6-48 **Operations with limits specified.**

ABBREVIATIONS

You already know many of the abbreviations used in dimensioning. A few of the more familiar examples from American National Standards are listed here:

Allowance	ALLOW
Alloy	ALY
Aluminum	AL
Babbitt	BAB
Bevel	BEV
Cast iron	CI
Centerline	CL or ℄
Chamfer	CHAM
Cold-rolled steel	CRS
Countersink	CSK
Degree	(°) DEG
Diameter	DIA or Ø
Dimension	DIM.
Inch	IN.
Key	K
Keyseat	KST
Keyway	KWY
Left	L
Left hand	LH
Limit	LIM
Material	MATL
Maximum	MAX
Millimeter	mm
National	NATL
Not to scale	NTS
Outside diameter	OD
Pattern	PATT
Radial	RAD
Radius	R
Reference	REF
Require	REQ
Revise	REV
Right hand	RH
Screw	SCR
Spherical	SPHER
Spot faced	SF
Square	SQ
Stock	STK
Surface	SUR
Tabulate	TAB
Thread	THD
Tolerance	TOL
United States Gage	USG
United States Standard	USS
Wrought Iron	WI

If you have a question about other abbreviations, look them up in the latest edition of *American National Standard Abbreviations for Use on Drawings,* ANSI Y1.1.

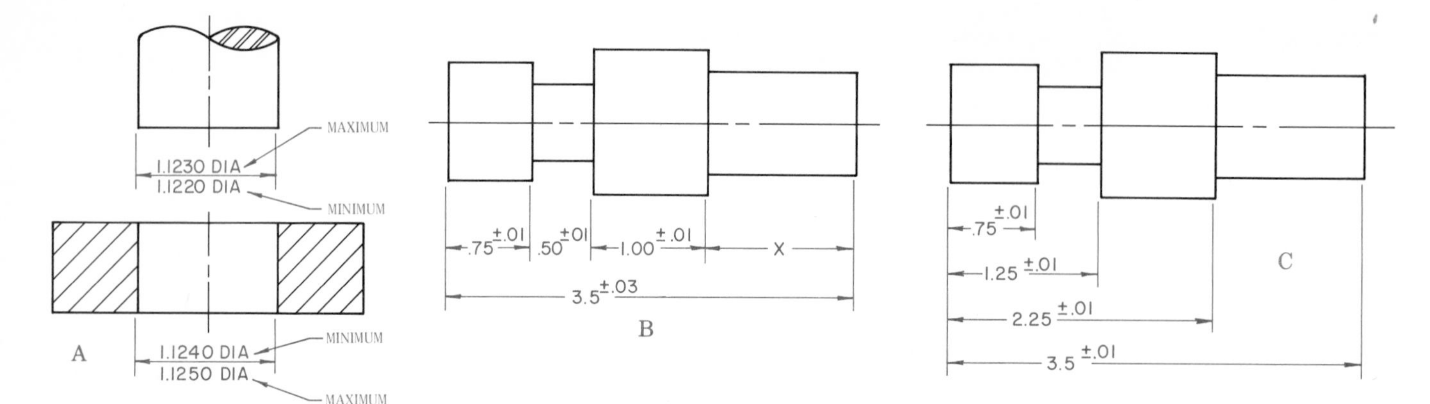

Fig. 6–49 **Limit dimensions.**

INTERCHANGEABLE MANUFACTURING

When many parts are made to be assembled with other parts, as on an assembly line, the parts have to be made so that they will all fit into place without any more machining or handwork. These parts are called mating parts. For mating parts to fit together, specified size allowances are necessary. For example, suppose two mating parts are a rod or shaft and the hole in which it turns. For these parts to work the way they are supposed to, the diameter of the rod would be limited. So would the diameter of the hole. The rod could not be too big or it would not turn. If it were too small, the rod would be too loose and not work properly.

LIMIT DIMENSIONING

Absolute accuracy cannot be expected. Instead, workers have to keep within a fixed limit of accuracy. They are given a number of tenths, hundredths, thousandths, or ten-thousandths of a millimeter or of an inch that the part is allowed to vary from the absolute measurements. This variation limit is called the *tolerance.* The tolerance may be stated in a note on the drawing or written in a space in the title block. An example would be "Dimension Tolerance 0.01 Unless Otherwise Specified." Limiting dimensions, or limits that give the maximum and minimum dimensions allowed, are also used to show the needed degree of accuracy. This is illustrated at A in Fig. 6–49. Note that the maximum limiting dimension is placed above the dimension line for the shaft (external dimension). But for the hole in the ring (internal dimension), the minimum limiting dimension is placed above the dimension line.

At B and C in Fig. 6–49, the basic sizes are given, and the plus-or-minus tolerance is shown. *Consecutive* dimensions (one after the other) are shown at B. In this case, the dimension *X* could have some variation. This dimension would not be given unless it was needed for reference. Then it would be followed by the abbreviation REF. *Progressive* dimensions (each starting at the same place) are shown at C. Here they are all given from a single surface. This kind of dimensioning is sometimes called baseline dimensioning.

Very accurate or limiting dimensions should not be called for unless they are truly needed. They greatly increase the cost of making a part. The detail drawing in Fig. 6–50 has limits for only two dimensions. All the others are nominal dimensions. The amount of variation in these parts depends on their use. In this case, the general note calls for a tolerance of ±0.250 mm (±0.010 in.).

PRECISION OR EXACTNESS

The latest edition of ANSI Y14.5 gives precise information on accurate measurement and position dimensioning. The following paragraphs are adapted from *American National Standard Drafting Practices* with the permission of the publisher, The American Society of Mechanical Engineers, 345 East 47th St., New York, NY, 10017.

EXPRESSING SIZE AND POSITION

Definitions relating to size. Size is a designation of magnitude. When a value is given to a dimension, it is called the size of that dimension. NOTE: The words "dimension" and "size" are both used to convey the meaning of magnitude.

Nominal size. The nominal size is used for general identification. Example: 13 mm (½ in.) pipe.

Basic size. The basic size is the size to which allowances and tolerances are added to get the limits of size.

Design size. The design size is the size to which tolerances are added to get the limits of size. When there is no allowance, the design size equals the basic size.

Actual size. An actual size is a measured size.

Limits of size. The limits of size (usually called "limits") are the maximum and minimum sizes.

Position. Dimensions that fix position usually call for more analysis than size dimensions. Linear and angular sizes locate features in relation to one another (point-to-point) or from a datum. Point-to-point distances may be enough for describing simple parts. If a part with more than one critical dimension must mate with another part, dimensions from a datum may be needed.

Locating round holes. Figures 6-51 through 6-56 show how to position round holes by giving distances, or distances and directions, to the hole centers. These methods can also be used to locate round pins and other features. Allowable variations for the positioning dimensions are shown by a tolerance with each distance or angle. Variations may also be shown by stating limits of dimensions or angles, or by true position expressions.

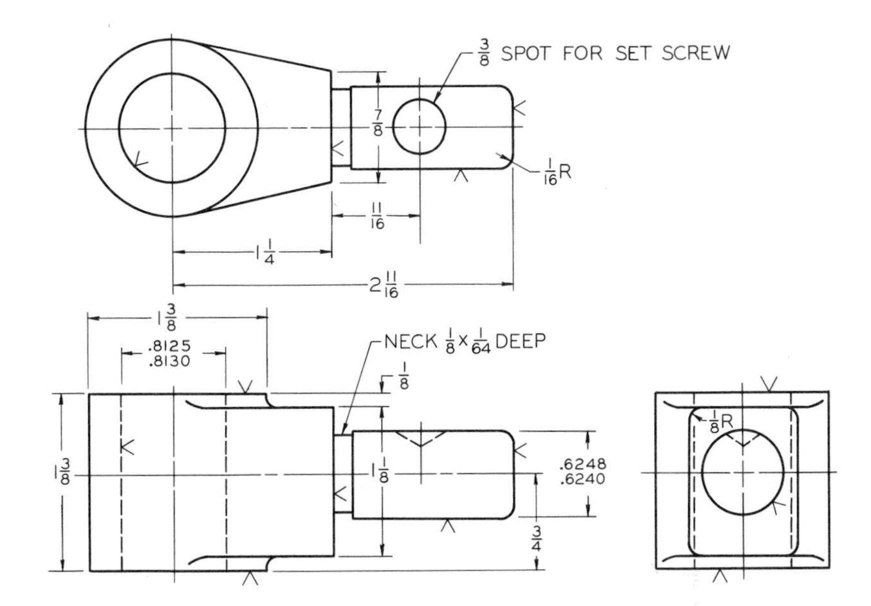

Fig. 6–50 A detail drawing with limits.

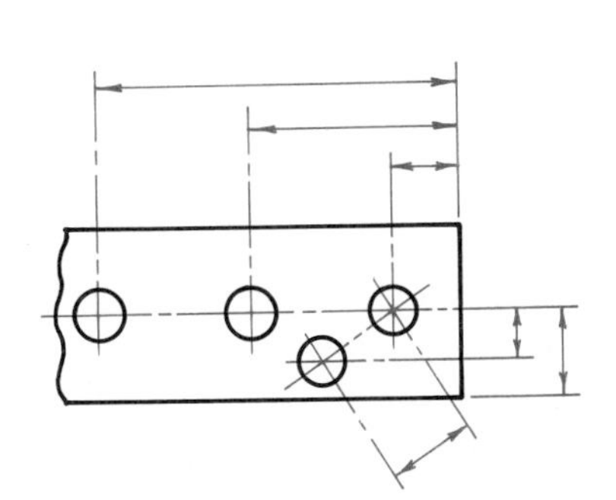

Fig. 6–51 Locating by linear distances.

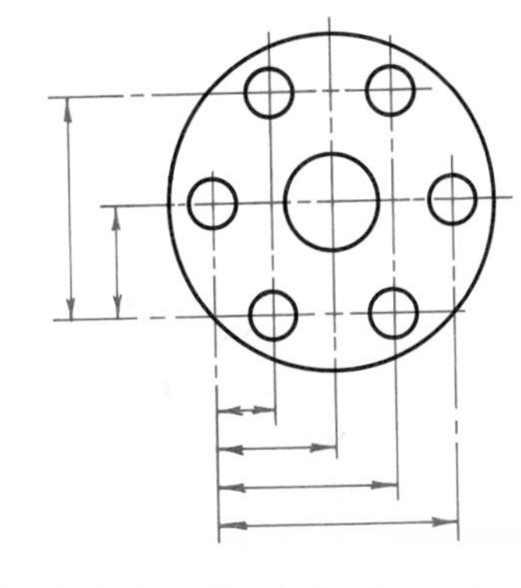

Fig. 6–52 Locating holes by rectangular coordinates. *(ANSI.)*

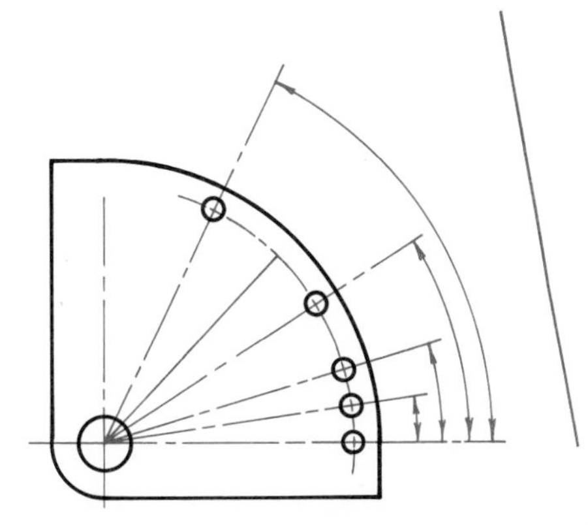

Fig. 6–53 Locating holes on a circle by polar coordinates. *(ANSI.)*

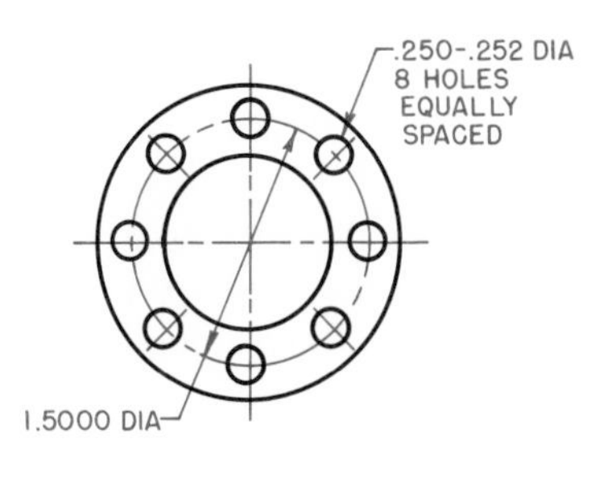

Fig. 6–54 Locating holes on a circle by radius or diameter and "equally spaced." *(ANSI.)*

.250 +.005 −.000
5 HOLES
EQUALLY SPACED

Fig. 6–55 **"Equally spaced" holes in a line.** *(ANSI.)*

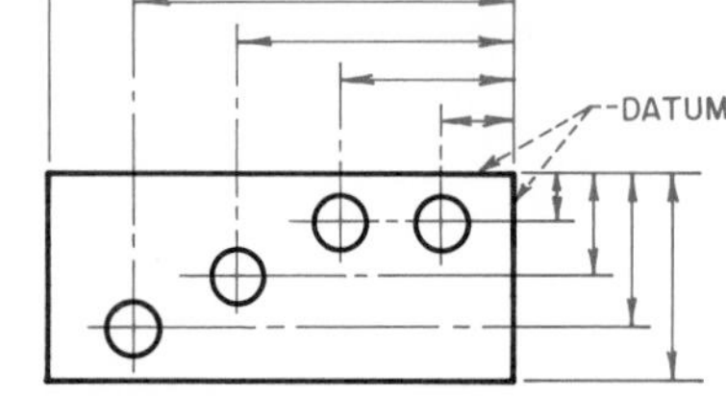

Fig. 6–56 **Dimensions for datum lines.** *(ANSI.)*

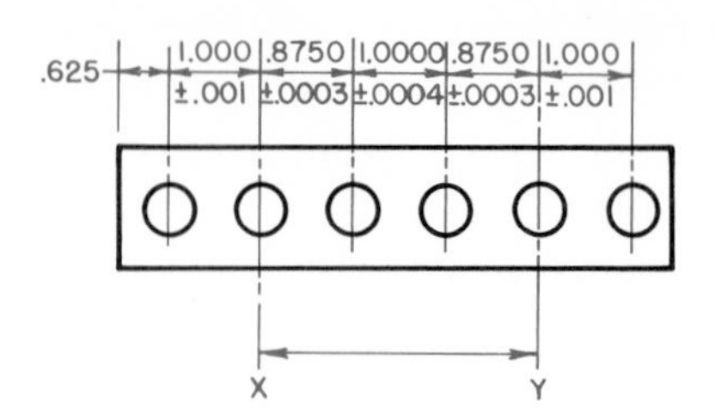

Fig. 6–57 **Point-to-point, or chain, dimensioning.** *(ANSI.)*

TOLERANCE

A tolerance is the total amount that a given dimension may vary. A tolerance should be expressed in the same form as its dimension. The tolerance of a decimal dimension should be expressed by a decimal to the same number of places. The tolerance of a dimension written as a common fraction should be expressed as a fraction. An exception is a close tolerance on an angle that can be expressed by a decimal representing a linear distance.

In a "chain" of dimensions with tolerances, the last dimension may have a tolerance equal to the sum of the tolerances between it and the first dimension. In other words, tolerances accumulate; that is, they are added together. The datum dimensioning method of Fig. 6–56 avoids overall accumulations. The tolerance on the distance between two features (first and second hole, for example) is equal to the tolerances on two dimensions from the datum added together. Where the distance between two points must be controlled closely, the distance between the two points should be dimensioned directly, with a tolerance. Figure 6–57 illustrates a series of "chain" dimensions where tolerances accumulate between points *X* and *Y*. Datum dimensions in Fig. 6–58 show the same accumulation with larger tolerances. Figure 6–59 shows how to avoid the accumulation without the use of extremely small tolerances.

Unilateral tolerance system. Unilateral tolerances allow variations in only one direction from a design size. This way of stating a tolerance is often helpful where a critical size is approached as material is removed during manufacture (see Fig. 6–60). For example, close-fitting holes and shafts are often given unilateral tolerances.

Bilateral tolerance system. Bilateral tolerances allow variations in both directions from a design size. Bilateral variations are usually given with locating dimensions. They are also used with any dimensions that can be allowed to vary in either direction. See Fig. 6–60.

LIMIT SYSTEM

A limit system shows only the largest and smallest dimensions allowed. See Figs. 6–61 and 6–62. The tolerance is the difference between the limits.

Expressing allowable variations. The amount of variation permitted when dimensioning a drawing can be given in several ways. The ways

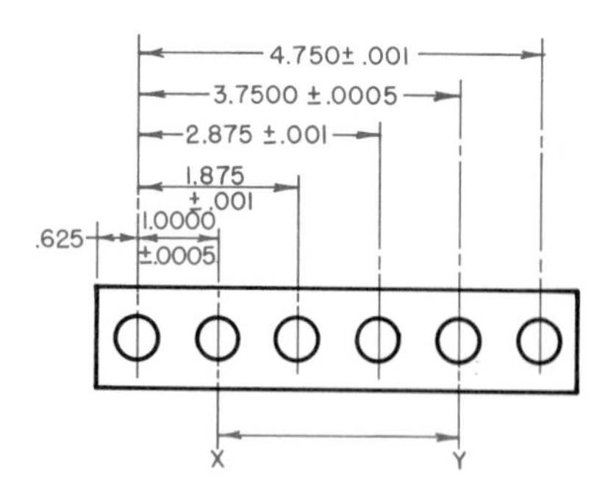

Fig. 6–58 **Datum dimensioning.** *(ANSI.)*

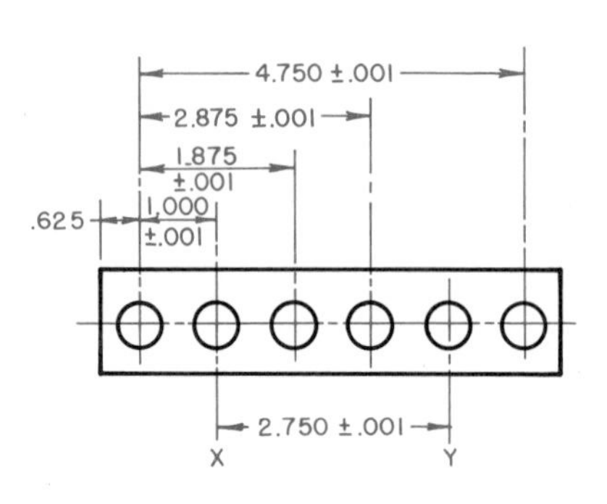

Fig. 6–59 **Dimensioning to prevent tolerance accumulation between *X* and *Y*.** *(ANSI.)*

Fig 6–60 **Giving a tolerance by a plus figure and a minus figure.** *(ANSI.)*

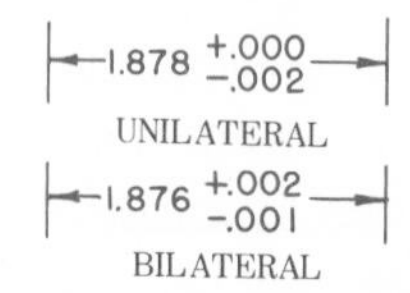

recommended in this book are as follows:

1. If the plus tolerance is different from the minus tolerance, two tolerance numbers are used, one plus and one minus. See Fig. 6–60. NOTE: Two tolerances in the same direction should never be called for.
2. When the plus tolerance is equal to the minus tolerance, use combined plus and minus sign followed by a single tolerance number. See Fig. 6–63.
3. The maximum and minimum limits of size are shown. The numerals should be placed in one of two ways. Do not use both on the same drawing.
   a. The high limit is always above the low limit where dimensions are given directly. The low limit always comes before the high limit where dimensions are given in note form. See Fig. 6–61.
   b. For location dimensions given directly (not by note), the high-limit number (maximum dimension) is placed above. The low-limit number (minimum dimension) is placed below. For size dimensions given directly, the number representing the maximum material condition is placed above. The number representing the minimum material condition is placed below. Where the limits are given in note form, the maximum number is first and the minimum number is second. See Fig. 6–62 for an example.

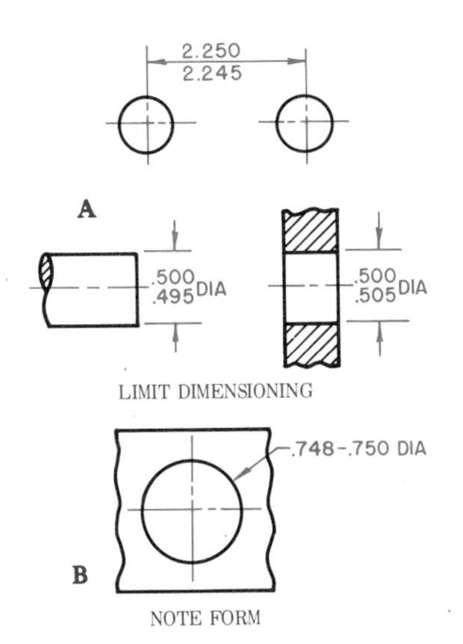

Fig. 6–61 **Specifying limits: first method. *(ANSI.)***

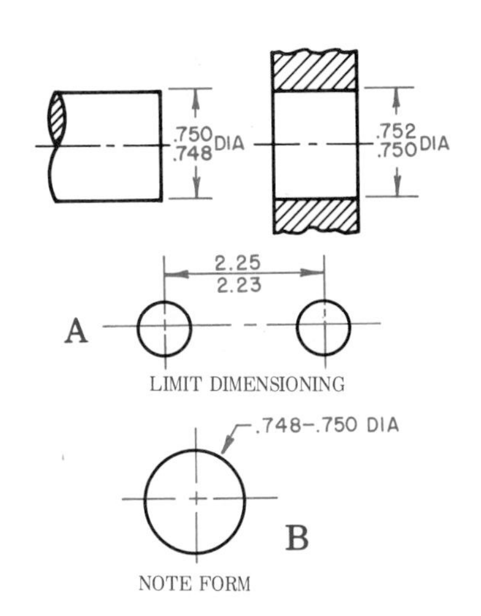

Fig. 6–62 **Specifying limits: second method. *(ANSI.)***

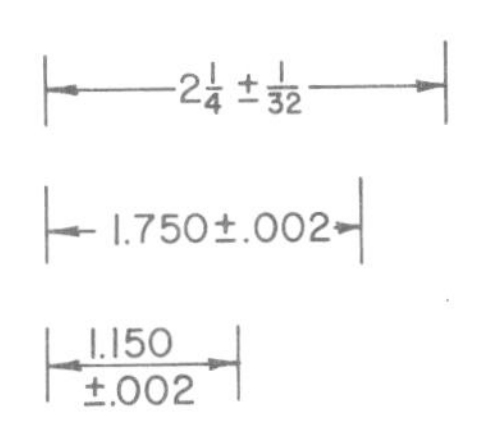

Fig. 6–63 **Using a combined plus and minus sign. *(ANSI.)***

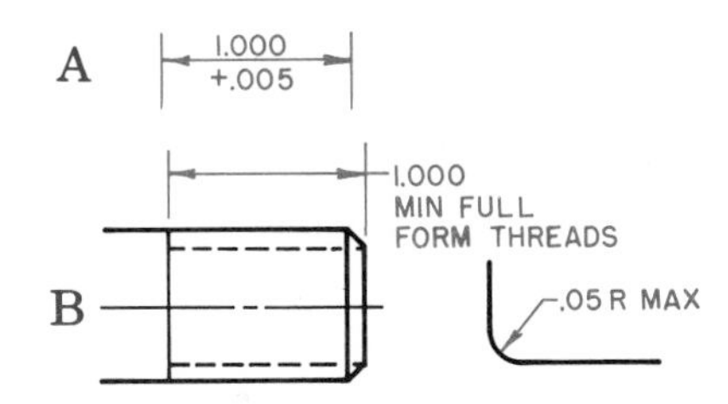

Fig. 6–64 **Expressing a single tolerance or limit. *(ANSI.)***

4. You do not always have to give both limits.
   a. A unilateral tolerance is sometimes given without stating that the tolerance in the other direction is zero. See Fig. 6–64 at A.
   b. MIN or MAX is often placed after a number when the other limit is not important. Depths of holes, lengths of threads, chamfers, etc., are often limited in this way. See Fig. 6–64 at B.
5. The above recommendations are for *linear* (in-line) tolerances, but the same forms are also used for angular tolerances. Angular tolerances may be given in degrees, minutes and seconds, or in decimals.

PLACING TOLERANCE AND LIMIT NUMERALS

You should place a tolerance numeral to the right of the dimension numeral and in line with it. You may also place the tolerance numeral below the dimension numeral with the dimension line between them. Figure 6–65 shows both arrangements.

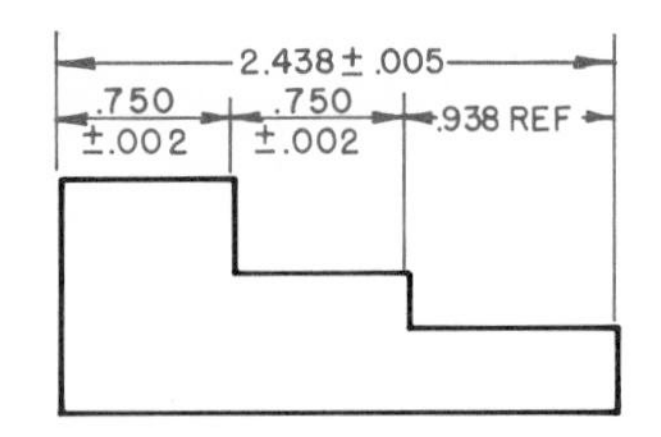

Fig. 6–65* Placing tolerance and limit numerals. Note the reference dimensions. *(ANSI.)

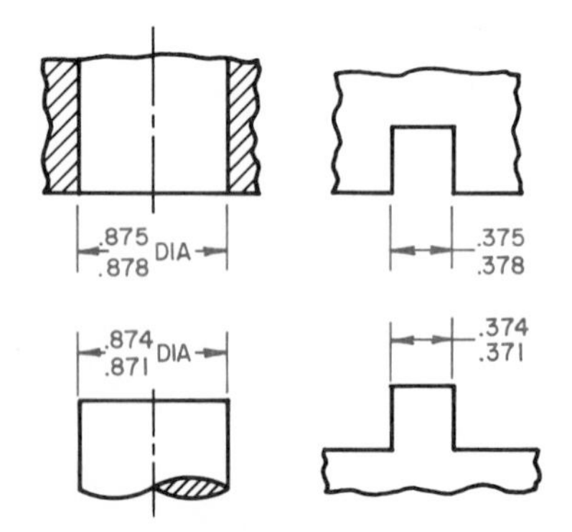

Fig. 6–66* Indicating dimensions or surfaces that are to fit closely. *(ANSI.)

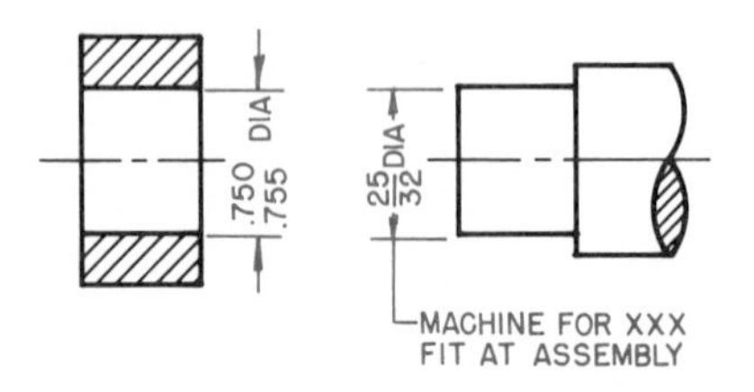

Fig. 6–67* Dimensioning noninterchangeable parts that are to fit closely. *(ANSI.)

Fig. 6–68* "Basic hole" fits. *(ANSI.)

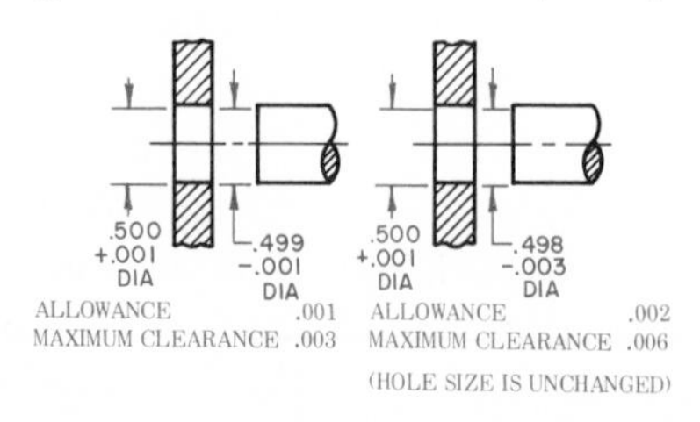

DIMENSIONING FOR FITS

The tolerances on the dimensions of interchangeable parts must allow these parts to fit together at assembly. See Fig. 6–66.

Figure 6–67 shows a way to dimension mating parts that do not need to be interchangeable. The size of one part does not need to be held to a close tolerance. It is to be made the proper size at assembly for the desired fit.

"Hole basis" and "shaft basis." You need to start calculating dimensions and tolerances of cylindrical parts that must fit well together from the minimum hole size or the maximum shaft size.

A basic hole system is one in which the design size of the hole is the basic size and the allowance is applied to the shaft.

A basic shaft system is one in which the design size of the shaft is the basic size and the allowance is applied to the hole.

NOTE: For further information on limits and fits, see ANSI B4.1–1967 or any later revision.

Basic hole system. To figure out the limits for a fit in the basic hole system, follow these steps: (1) Give the minimum hole size. (2) For a clearance fit, find the maximum shaft size by subtracting the desired allowance (minimum clearance) from the minimum hole size. For an interference fit, add the desired allowance (maximum interference). (3) Adjust the hole and shaft tolerances to get the desired maximum clearance or minimum interference. See Fig. 6–68. By using the basic hole system, tooling costs can often be kept down. This is possible because you can choose a size that can be produced by a standard tool (reamer, broach, etc.) or gaged with a standard plug gage.

Basic-shaft system. To figure out the limits for a fit in the basic shaft system follow these steps: (1) Give the maximum shaft size. (2) For a clearance fit, find the minimum hole size by adding the desired allowance (minimum clearance) to the maximum shaft size. Subtract for an interference fit. (3) Adjust the hole and shaft tolerances to get the desired maximum clearance or minimum interference. See Fig. 6–69. Use the basic shaft method only if there is a good reason for it, such as when a standard size shaft can be used.

Fig. 6–69* "Basic shaft" fits. *(ANSI.)

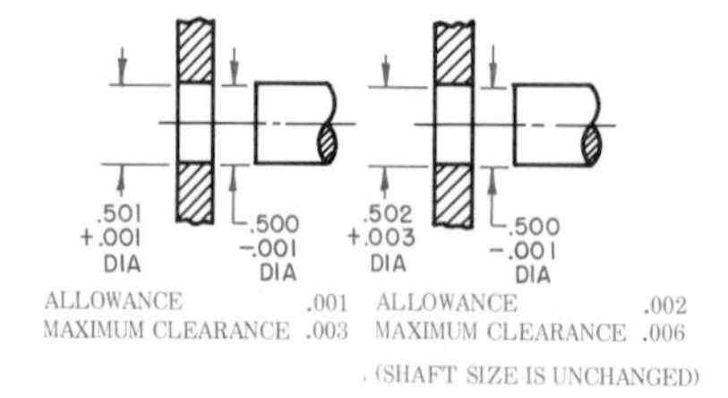

GEOMETRIC TOLERANCING

When you give geometric tolerances, you give the largest amount of variation in form (shape and size) and position (location) that a system of interchangeable parts can tolerate. Tolerances of position refer to the location of holes, slots, tabs, dovetails, and so on. Tolerances of form refer to such things as flatness, straightness, roundness, parallelism, perpendicularity, angularity, and so on.

SPECIFYING TOLERANCES OF POSITION AND FORM

Tolerances of position and form may be given in a note or written in symbol form (Fig. 6–70). Both methods are acceptable. However, most companies ask all drafters to use one or the other all the time.

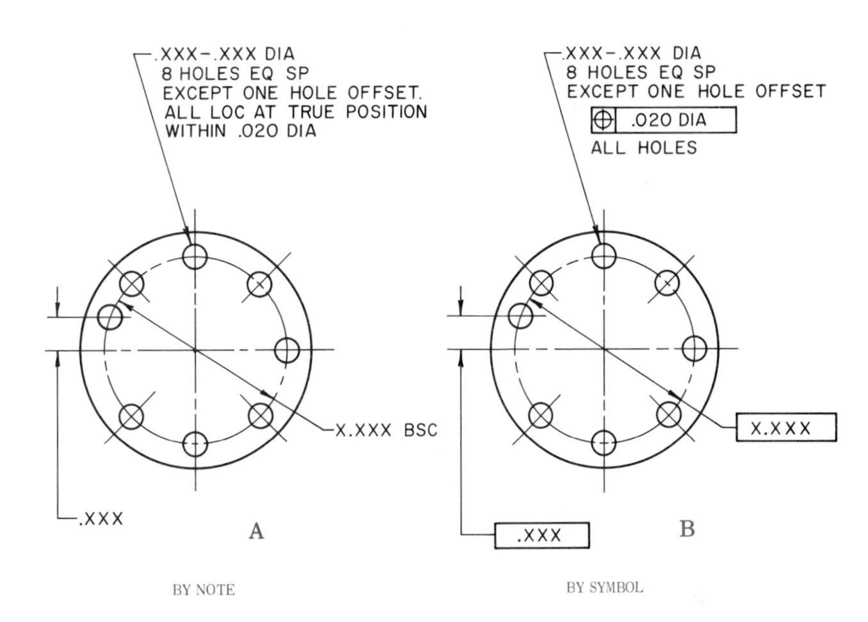

Fig. 6–70 **Tolerances may be specified in a note or by a symbol.**

GEOMETRIC CHARACTERISTIC SYMBOLS

The symbols that stand for geometric characteristics are shown in Fig. 6–71. These symbols are a drafter's shorthand way of writing form and position tolerances on drawings. These symbols can be used alone. They can also be used with other symbols and notes to give all the information needed for the shape or location of details.

DATUM

A datum is a reference that has been determined to be true and accurate. It may be a point, a line, a plane, a cylinder, and so on. Because its dimensions are exact, measurements can be made from it. Therefore, a datum is used to figure out the location of other features.

DATUM REFERENCE LETTERS

Datum features that need to be identified are given a reference letter. Any letter of the alphabet may be used except I, O, and Q. These letters can be confusing and may be misread. Single letters are used unless more than 23 are needed. After that, combinations, such as AA through AZ, may be used. Figure 6–72 shows a reference letter in a datum symbol. The box around the letter is about 8 mm (about 5/16 in.) high. Its length may vary according to its application on the drawing.

SYMBOLS FOR MMC AND RFS

The symbols Ⓜ and Ⓢ stand for *maximum material condition* (MMC) and *regardless of feature size* (RFS). MMC exists when a feature has the maximum amount of material. That would be, for example, when a hole is at its minimum diameter and a shaft is at its maximum diameter. RFS means that tolerance of position or form must be met no matter where the feature lies within its size tolerance. The symbols are used only in feature control symbols.

FEATURE CONTROL SYMBOLS

The feature control symbol is a frame that contains the geometric characteristic symbol followed by the tolerance. In some cases, the symbol Ⓜ or Ⓢ is also included.

GEOMETRIC CHARACTERISTIC SYMBOLS			
		CHARACTERISTIC	SYMBOL
FORM TOLERANCES	FOR SINGLE FEATURE	FLATNESS	⏥
		STRAIGHTNESS	—
		ROUNDNESS (CIRCULARITY)	○
		CYLINDRICITY	⌭
		PROFILE OF ANY LINE	⌒
		PROFILE OF ANY SURFACE	⌓
	FOR RELATED FEATURES	PARALLELISM	∥
		PERPENDICULARITY (SQUARENESS)	⊥
		ANGULARITY	∠
		RUNOUT	↗
POSITIONAL TOLERANCES		TRUE POSITION	⌖
		CONCENTRICITY	◎
		SYMMETRY	⌯

Fig. 6–71 **Geometric characteristic symbols. *(ANSI.)***

Fig. 6–72 **Datum-identifying symbol.**

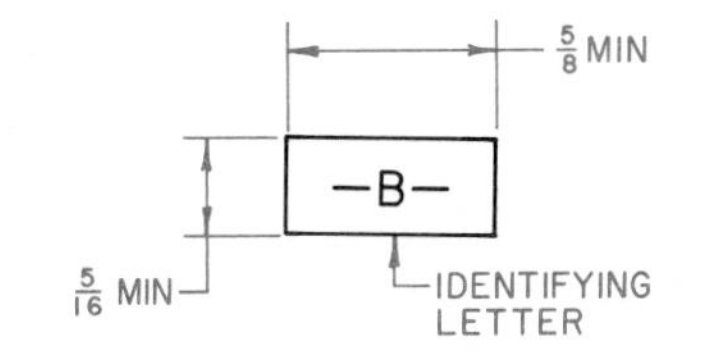

A vertical line separates the symbol from the tolerance, as shown in Fig. 6–73. The datum-identifying symbol may also be added, as shown in Fig. 6–74. ANSI Y14.5 (ISO R129; R406), *Dimensioning and Tolerancing for Engineering Drawings*, describes terms and symbols and their uses in detail.

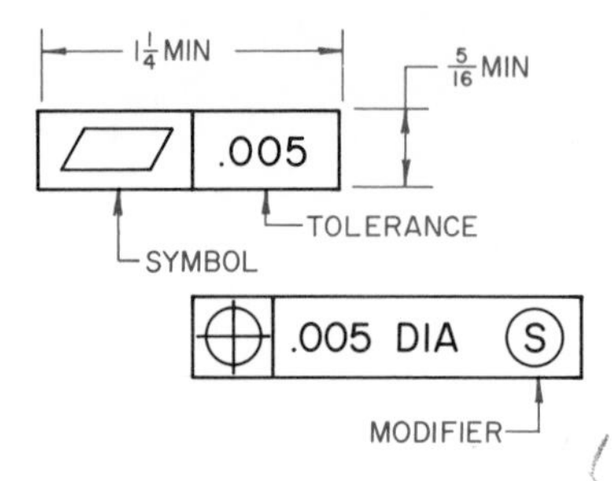

Fig. 6–73 **Feature-control symbols.**

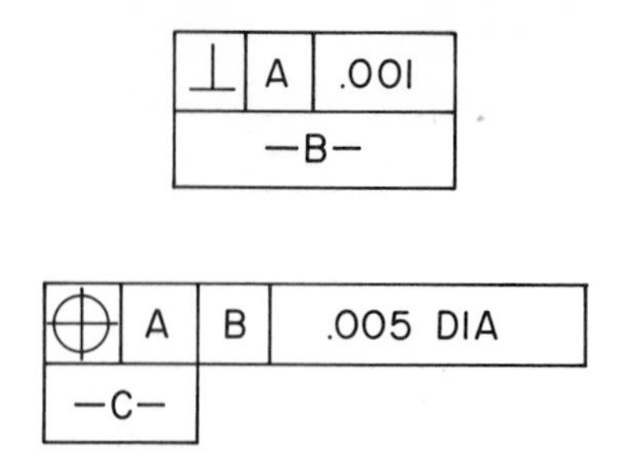

Fig. 6–74 **Combined feature-control and datum-identifying symbols.**

SURFACE TEXTURE

There is no such thing as a perfectly smooth surface. All surfaces have irregularities. Sometimes a drafter has to tell how much roughness and waviness the surface of a material can have and the lay direction of both. There are standards for these characteristics. There are also symbols that represent them.

Surface texture is discussed fully in the American National Standards ANSI B46.1. Use that text for study and as a reference. The following paragraphs about surface texture are adapted from *Surface Texture* (ANSI B46.1). They are included here with the permission of the publisher, The American Society of Mechanical Engineers, United Engineering Center, 345 East 47th St., New York, NY, 10017.

Surfaces, in general, are very complex in character. This standard deals only with the height, width, and direction of surface irregularities. These are of practical importance in specific applications.

CLASSIFICATION OF TERMS AND RATINGS RELATED TO SURFACES

The terms and ratings in this book have to do with surfaces made by various means. Among these are machining, abrading, extruding, casting, molding, forging, rolling, coating, plating, blasting, burnishing, and others.

Surface texture. Surface texture includes roughness, waviness, lay, and flaws. It includes repetitive or random differences from the nominal surface that forms the pattern of the surface.

Profile. The profile is the contour (shape) of a surface in a plane perpendicular to the surface. Sometimes an angle other than a perpendicular one is specified.

Measured profile. The measured profile is a representation of the profile obtained by instruments or other means. See Fig. 6–75.

Micrometer. A micrometer is one millionth of a meter (0.000 001 m). Micrometers may be abbreviated μm.

Microinch. A microinch is one millionth of an inch (0.000 001 in.). Microinches may be abbreviated μin.

Roughness. Roughness is the finer irregularities in the surface texture. Roughness usually includes irregularities caused by the production process. Among these are traverse feed marks and other irregularities within the limits of the roughness-width cutoff. See Fig. 6–76.

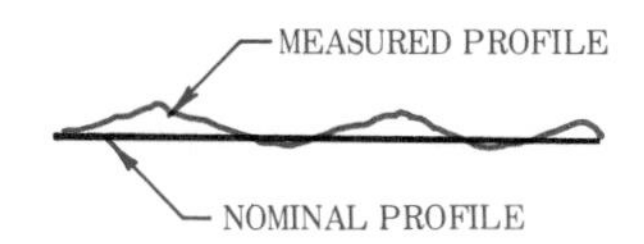

Fig. 6–75 **An enlarged profile shows that a surface is not as it appears.** ***(ANSI.)***

Roughness height. For the purpose of this book, roughness height is the arithmetical average deviation. It is expressed in micrometers or in microinches measured normal to the centerline. The preferred series of roughness height values is given in Table 6–1.

Roughness width. Roughness width is the distance between two peaks or ridges that make up the pattern of the roughness. Roughness width is given in millimeters or in inches.

Roughness-width cutoff. This is the greatest spacing of repetitive surface irregularities to be included in the measurement of average roughness height. Roughness-width cutoff is rated in millimeters or in inches. Standard values are given in Table 6–2. Roughness-width cutoff must always be greater than the roughness width in order to obtain the total roughness-height rating.

TABLE 6–1.	PREFERRED SERIES ROUGHNESS	
Roughness Values		
		Grade
50	2000	12
25	1000	11
12.5	500	10
6.3	250	9
3.2	125	8
1.6	63	7
0.8	32	6
0.4	16	5
0.2	8	4
0.1	4	3
0.05	2	2
0.025	1	1

TABLE 6–2.	STANDARD ROUGHNESS-WIDTH CUTOFF VALUES					
When no value is specified, the value 0.750 mm (0.030 in.) is assumed						
mm	0.075	0.250	0.750	2.500	7.500	25.000
in	0.003	0.010	0.030	0.100	0.300	1.000

Waviness. Waviness was covered by surface-texture standards. Geometric tolerancing now covers this surface condition under flatness. Flatness is a condition where all surface elements are in a single plane. Flatness tolerances are applied to surfaces to control variations in surface texture (see Fig. 6–76). Waviness was defined as the usually widely spaced component of surface texture. It generally was of wider spacing than the roughness-width cutoff. Waviness results from such factors as machine or work deflections, vibration, chatter, heat treatment or warping strains. Roughness may be thought of as superimposed on a "wavy" surface.

Waviness height. Waviness height was rated in inches as the peak to valley distance. The maximum waviness height is given in Table 6–13.

Waviness width. Waviness width was rated in inches as the spacing of successive wave peaks or successive wave valleys. When specified, the values were the maximum amounts permissible.

Lay. Lay is the direction of the predominant surface pattern. Ordinarily, it is determined by the production method used. Lay symbols are shown in Fig. 6–80.

Flaws. Flaws are irregularities that occur at one place or at relatively infrequent or widely varying intervals in a surface. Flaws include such defects as cracks, blow holes, checks, ridges, scratches, etc. The effect of flaws shall not be included in the roughness-height measurements, unless otherwise specified.

Contact area. Contact area is the amount of area of the surface needed to be in contact with its mating surface. Contact area should be distributed over the surface with approximate uniformity. Contact is specified as shown in Fig. 6–79.

DESIGNATION OF SURFACE CHARACTERISTICS

Where no surface control is specified, you can assume that the surface produced by the operation will be satisfactory. If the surface is critical, the quality of surface needed should be indicated.

Surface symbol. The symbol used to designate surface irregularities is the check mark with horizontal extension, as shown in Fig. 6–77. The point of the symbol must touch the line indicating which surface is meant. It may also touch the extension line, or a leader pointing to the surface. The long leg and extension is drawn to the right as the drawing is read. Where only roughness height is shown, the horizontal extension may be left off. Fig. 6–78 shows the typical use of the symbol on a drawing.

Where the symbol is used with a dimension, it affects all surfaces defined by the dimension. Areas of transition, such as chamfers and fillets, should usually be the same as the roughest finished area next to them.

Surface-roughness symbols always apply to the completed surface, unless you indicate otherwise. Drawings or specifications for plated or coated parts must indicate whether the surface-roughness sym-

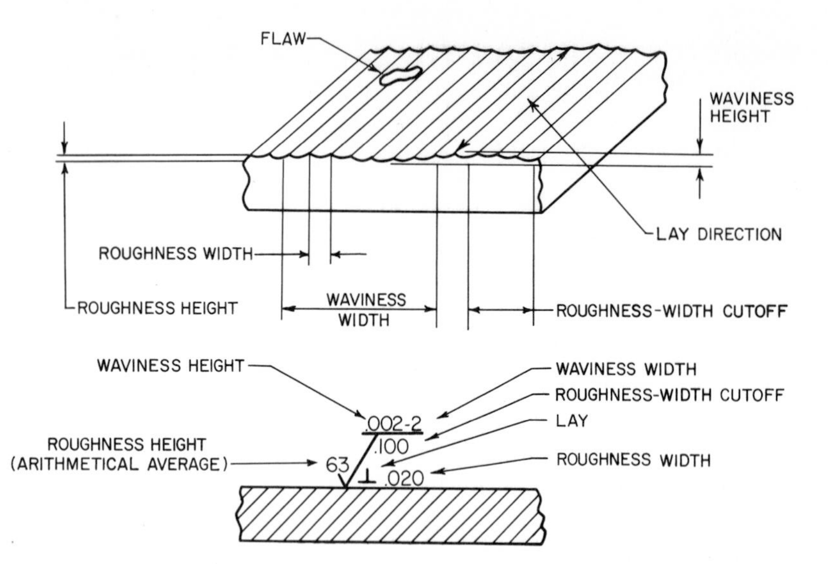

Fig. 6–76 **Relation of symbols to surface characteristics. Refer to Fig. 6–78.** *(ANSI.)*

Fig. 6–77 **The surface symbol.** *(ANSI.)*

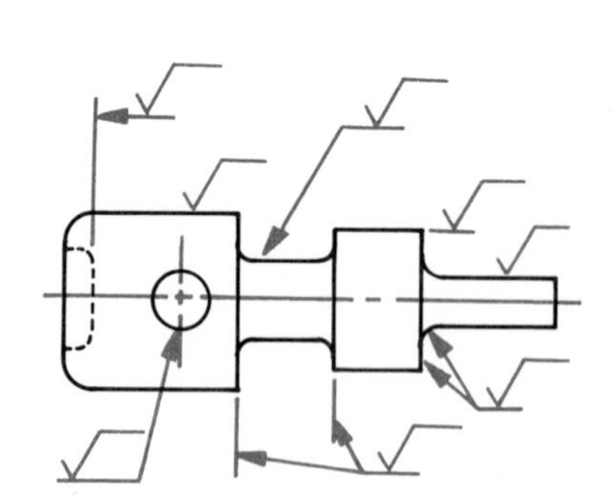

Fig. 6–78 **The surface symbol on a drawing.** *(ANSI.)*

bols apply before plating, after plating, or to both before and after plating.

Application of symbols and ratings. Fig. 6–79 shows the way roughness, waviness, and lay are called for on the surface symbol. Only those ratings needed to specify the desired surface needed to be shown in the symbol.

Symbols indicating direction of lay. Symbols for lay are shown in Fig. 6–80.

Roughness ratings usually apply in a direction that gives the maximum reading. This is normally across the lay.

This is the end of the material extracted and adjusted from ANSI *Surface Texture,* ANSI B46.1. It should give you an idea of the thought that may be given to surface quality. If you need more information, use the complete ISO and ANSI standard.

Fig. 6–79 **Applications of symbols and ratings.** *(ANSI.)*

A (a) Roughness height rating is centered above and between the two legs. The specification of only one rating shall indicate the maximum value and any lesser value shall be acceptable. A value is here applied to the symbol variations.	E (90%) Minimum requirements for contact or bearing area with a mating part or reference surface shall be indicated by a percentage value placed above the extension line as shown. Further requirements may be controlled by notes.
B (a1, a2) The specification of maximum value and minimum value roughness height ratings indicates the permissible range of value rating.	F (a ⊥) Lay designation is indicated by the lay symbol placed at the right of the long leg.
C (MILLED, a1, a2) If a final surface texture must be produced by a special production method, it is placed above the horizontal extension.	G (MILLED, 2.44, a ⊥) If it is necessary to indicate a sampling length, it is placed below the horizontal extension.
D (ZINC PLATED, a) Any indication as to treatment or coating is also placed above the horizontal extension. The numerical value of the roughness applies to the surface texture after treatment, unless stated otherwise.	H (MILLED, 2.44, a1, a2, ⊥ 0.05) Where required, maximum roughness width rating shall be placed at the right of the lay symbol. Any lesser rating shall be acceptable.

=	Lay parallel to the line representing the surface to which the symbol is applied.	 DIRECTION OF TOOL MARKS
⊥	Lay perpendicular to the line representing the surface to which the symbol is applied.	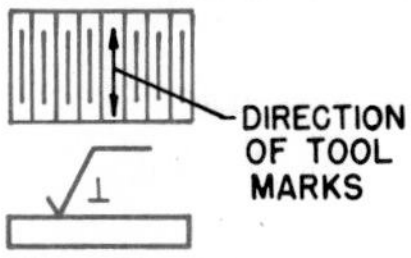 DIRECTION OF TOOL MARKS
X	Lay angular in both directions to line representing the surface to which symbol is applied.	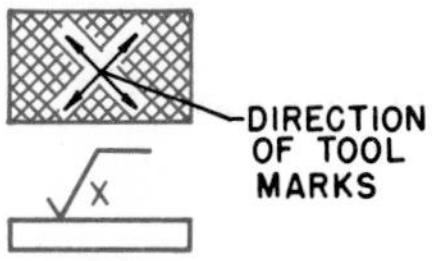 DIRECTION OF TOOL MARKS
M	Lay multidirectional.	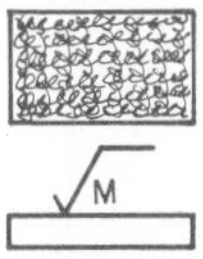
C	Lay approximately circular relative to the center of the surface to which the symbol is applied.	
R	Lay approximately radial relative to the center of the surface to which the symbol is applied.	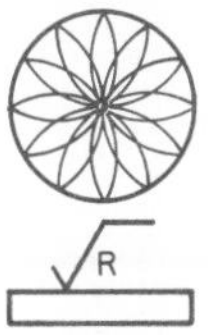

Fig. 6–80 **Lay symbols.** *(ANSI.)*

Review

1. What are the two things you must know about an object in order to do a complete graphic description of it?

2. What is another name for size description?

3. What unit of measure is used on metric drawings?

4. Name the method in which both U.S. customary units and metric units are given on the same drawing.

5. When placing dimensions on a drawing, there are two systems in common use. One is aligned. Name the other system.

6. There are two basic kinds of dimensions: (1) size dimensions and (2) ________ dimensions.

7. Draw the symbols for degrees, minutes, and seconds.

8. A complete circle is dimensioned as a diameter. An arc is dimensioned as a ________.

9. Name the abbreviations for the following: diameter, inch, centerline, millimeter, radius, tolerance, outside diameter, and countersink.

10. The total amount by which a given dimension may vary is called ________.

11. There are two kinds of tolerance: unilateral and ________.

12. Giving the maximum allowable variation in form and position is called ________.

Problems

Figures 6–81 through 6–93 offer a total of 54 dimensioning problems. Additional problems may be chosen from other chapters.

The problems in Figs. 6–80 through 6–83 are to be done as follows:

1. Take dimensions from the printed scale at the bottom of the page, using dividers.
2. Draw the complete views.
3. Add all necessary extension and dimension lines for size and location dimensions.
4. Add arrowheads, dimensions, and notes.

Figures 6–85 through 6–93 are more advanced. For these problems use the following procedure:

1. Determine the necessary views, and prepare a freehand sketch.
2. Dimension the sketch.
3. Decide on a scale and draw views mechanically.
4. Add all necessary dimensions and notes.

Practice in decimal-inch, metric, or dual dimensioning may be obtained by converting the fractional-inch dimensions for any of the problems. It is recommended that students gain some experience in all methods.

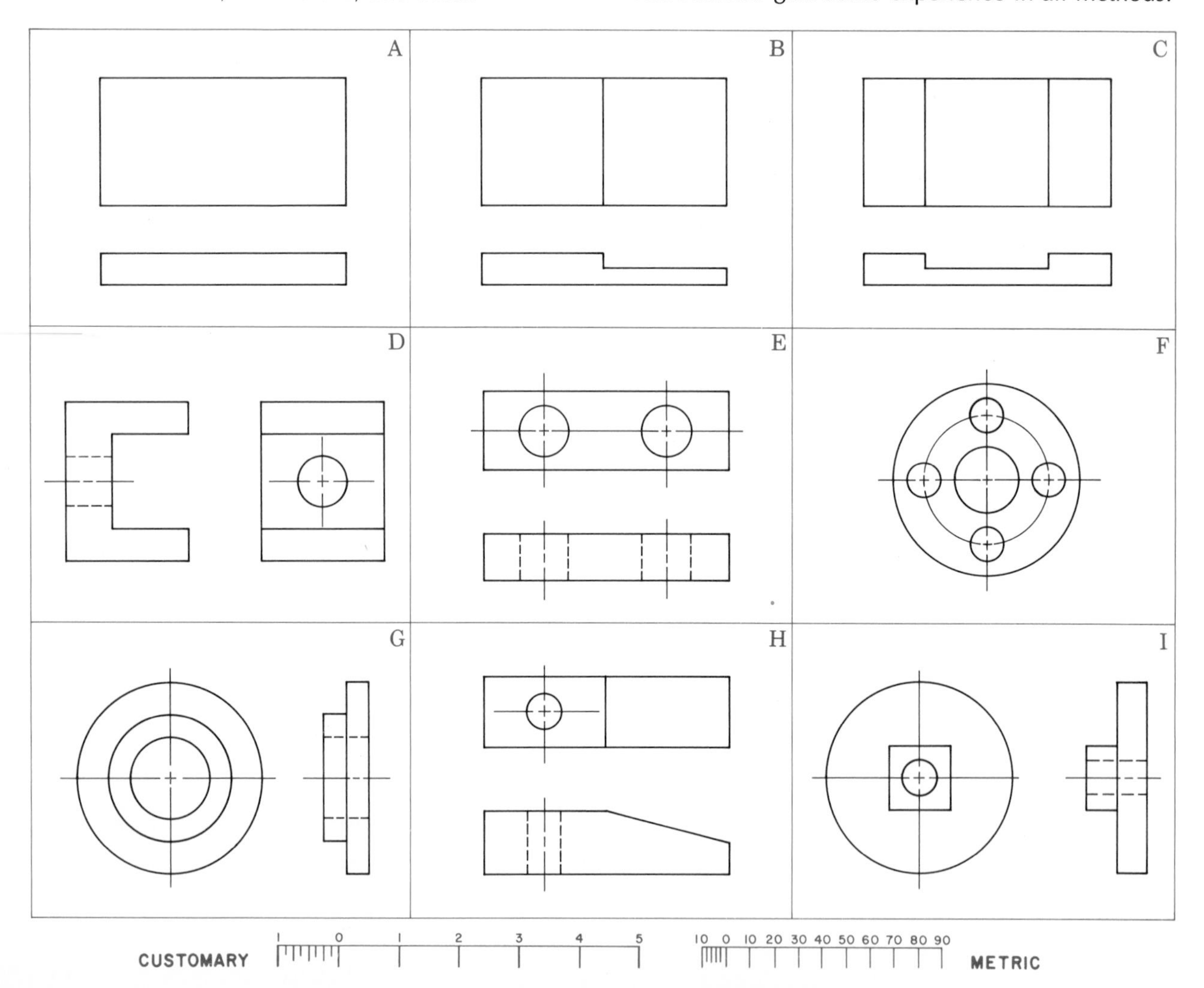

Fig. 6–81 **Problems for dimensioning practice. Take dimensions from the printed scale, using dividers. Draw the views as shown and add all necessary size and location dimensions.**

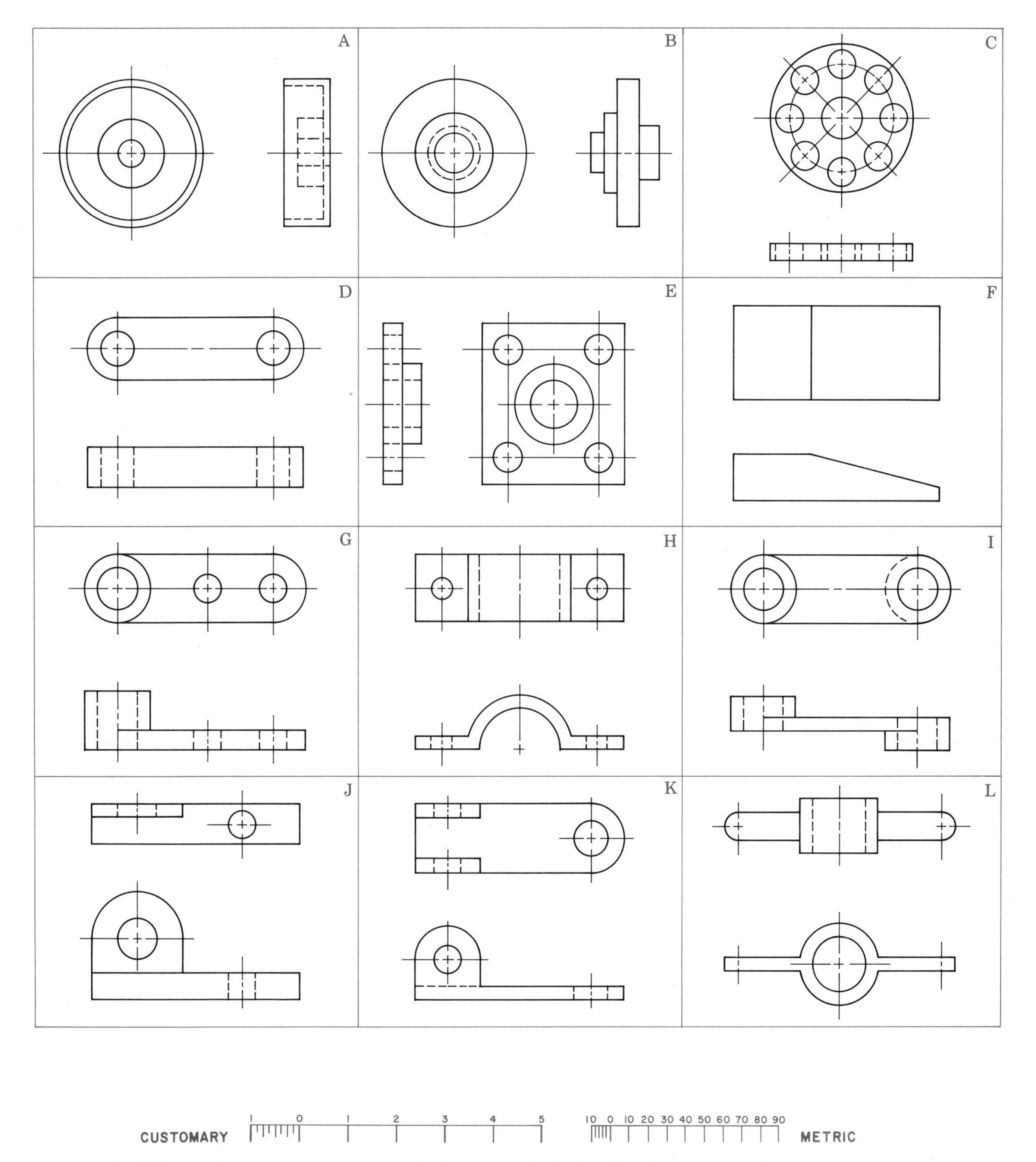

Fig. 6–82 **Problems for dimensioning practice. Take dimensions from the printed scale, using dividers. Draw the views as shown and add all necessary size and location dimensions.**

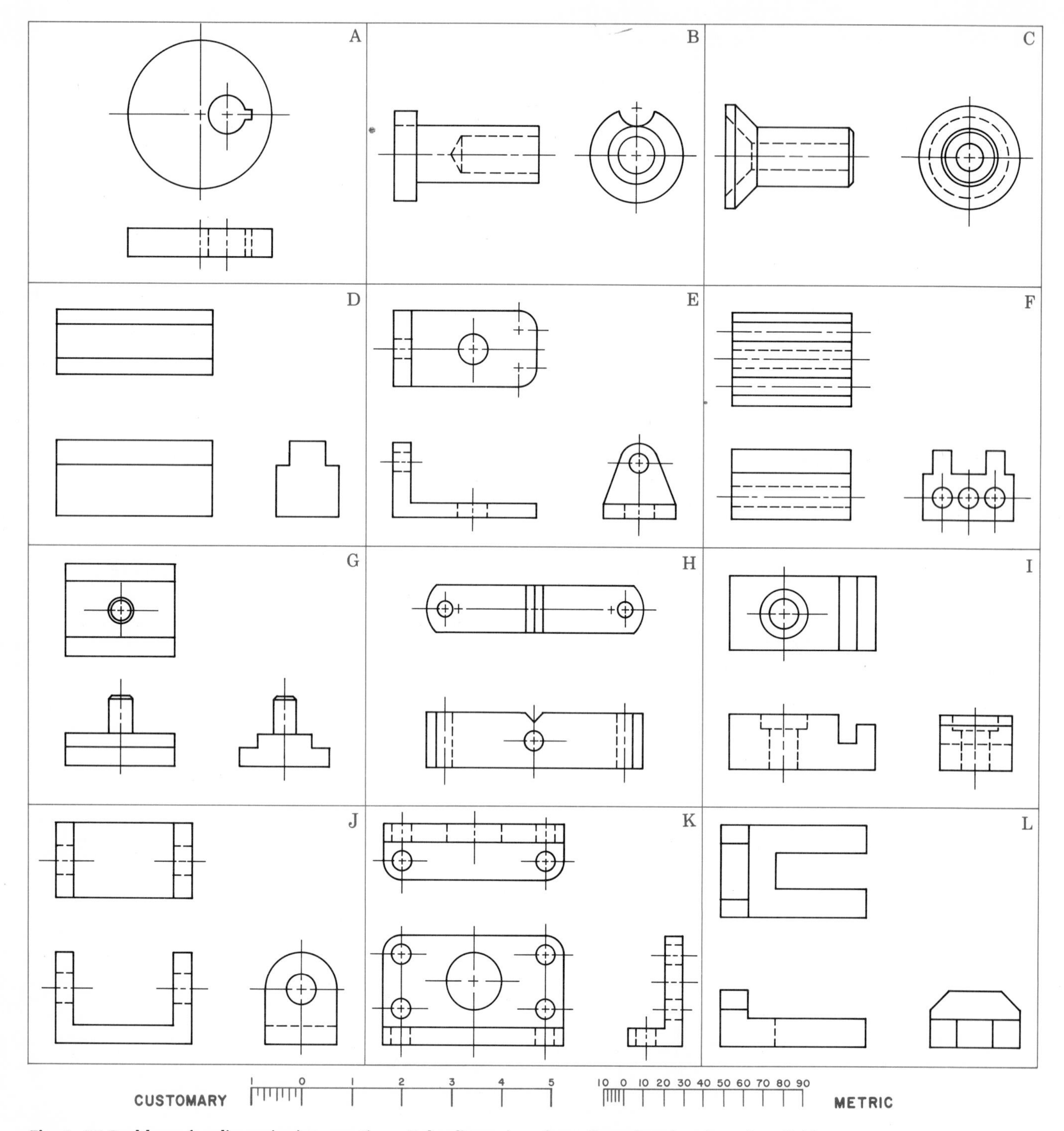

Fig. 6–83 **Problems for dimensioning practice. Take dimensions from the printed scale, using dividers. Draw the views as shown and add all necessary size and location dimensions.**

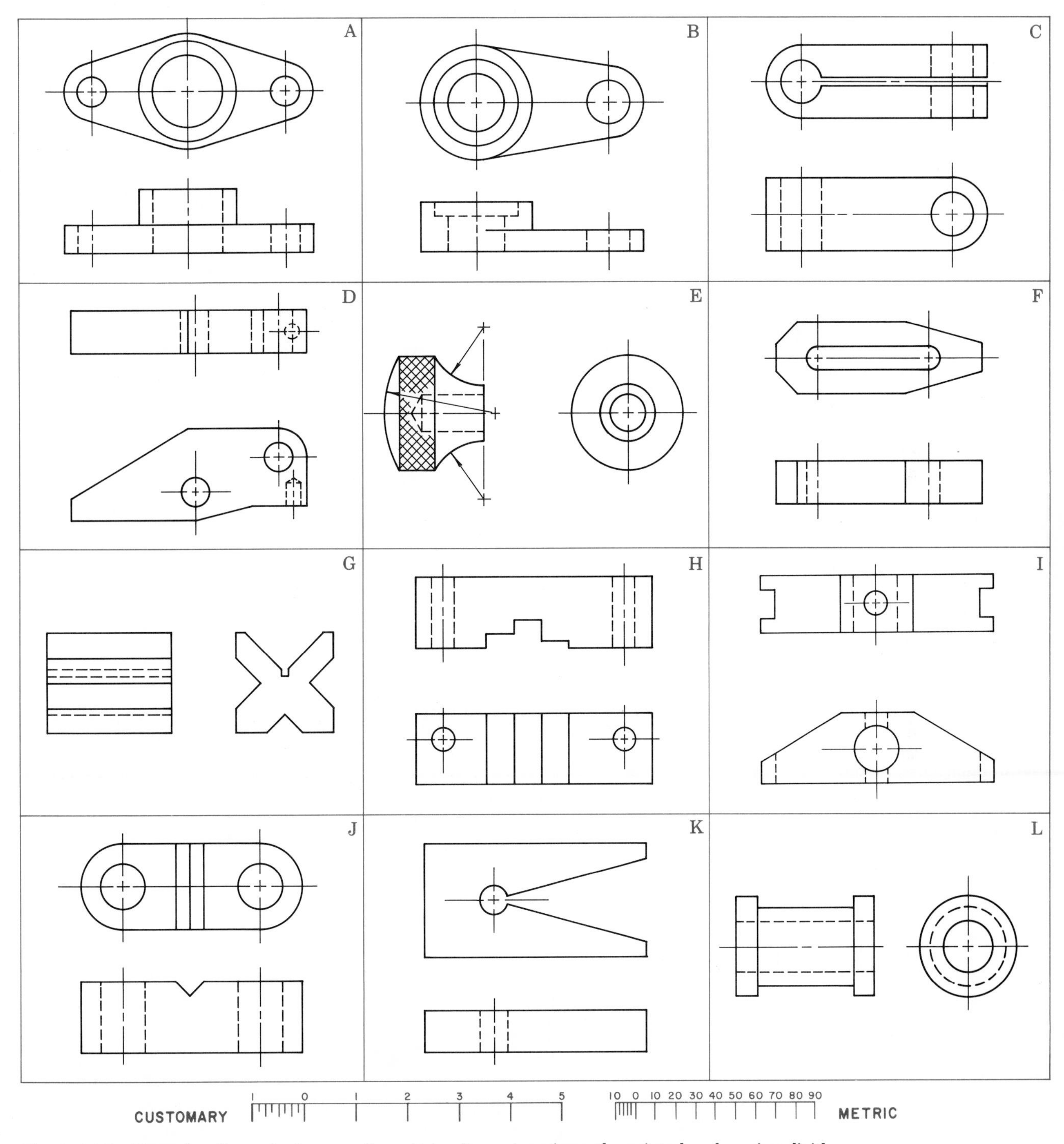

Fig. 6–84 **Problems for dimensioning practice. Take dimensions from the printed scale, using dividers. Draw the views as shown and add all necessary size and location dimensions.**

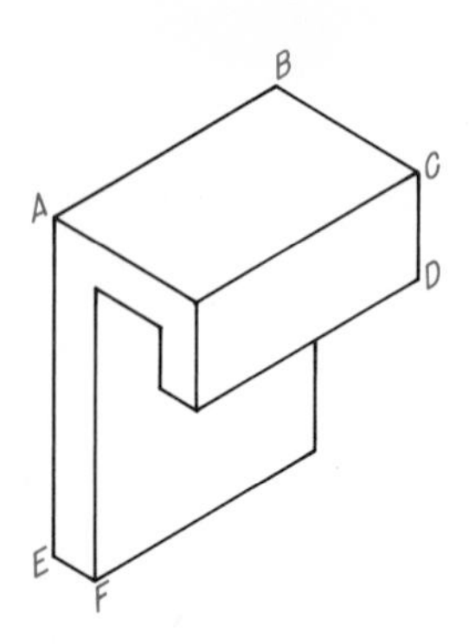

Fig. 6–85 Cut-off stop. Draw all necessary views and dimension. Scale: double size or as assigned. *AB* = 40 mm, *BC* = 26 mm, *CD* = 17 mm, *AE* = 53 mm, *EF* = 7.50 mm.

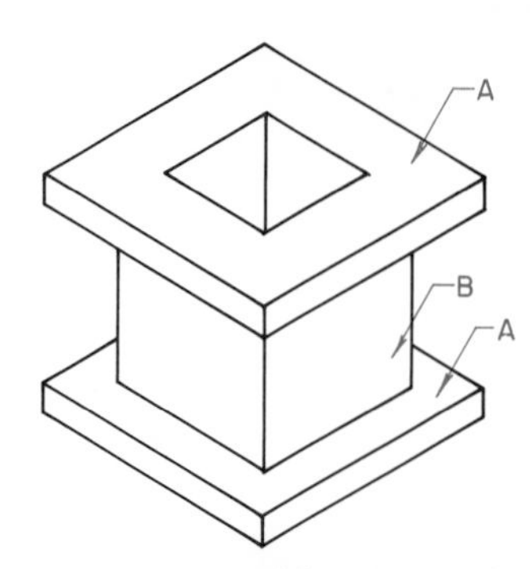

Fig. 6–86 Square guide. Draw all necessary views and dimension. Scale: full size or as assigned. *A* = 5 mm thick × 44 mm square, *B* = 30 mm square × 30 mm high, Hole = 20 mm square.

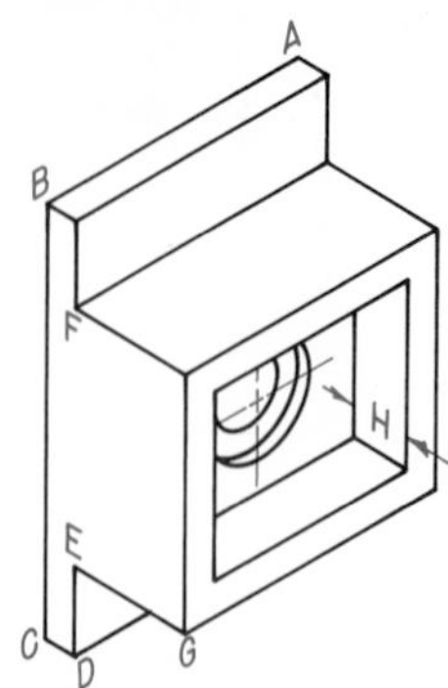

Fig. 6–87 Locator. Draw all necessary views and dimension. Scale: full size or as assigned. *AB* = 40 mm, *BC* = 60 mm, *CD* = 5 mm, *DE* = 12 mm, *EF* = 36 mm, *EG* = 18 mm, *H* = 8 mm, Hole = 10 mm DIA through, 18 mm Cbore, 2 mm deep.

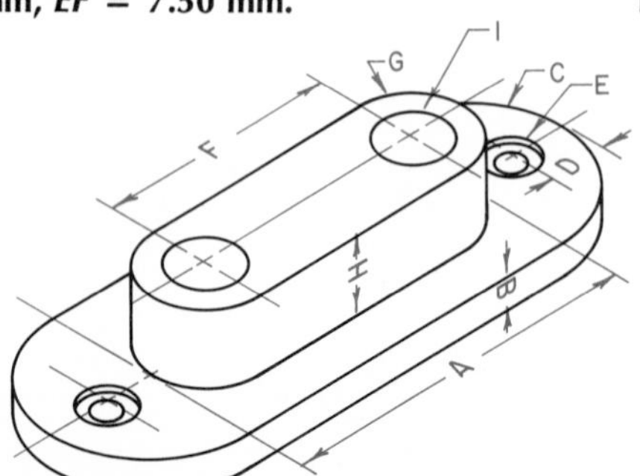

Fig. 6–88 Double-shaft support. Draw all necessary views and dimension. Scale: full size or as assigned. *A* = 67 mm, *B* = 7 mm, *C* = 21 mm R, *D* = 10 mm, *E* = 5 mm DIA through 10 mm Cbore, 2 mm deep, 2 holes, *F* = 43 mm, *G* = 12 mm R, *H* = 14 mm.

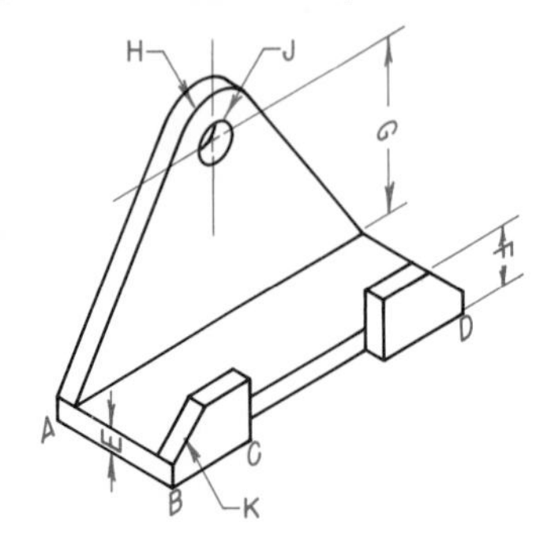

Fig. 6–89 Hanger. Draw all necessary views and dimension. Scale: $\frac{3}{4}$ size or as assigned. *AB* = $1\frac{1}{4}$, *BC* = $\frac{13}{16}$, *BD* = 3, *E* = $\frac{3}{16}$, *F* = $\frac{1}{2}$, *G* = $1\frac{9}{16}$, *H* = $\frac{7}{16}$ R, *J* = $\frac{3}{8}$DIA, *K* = 45°.

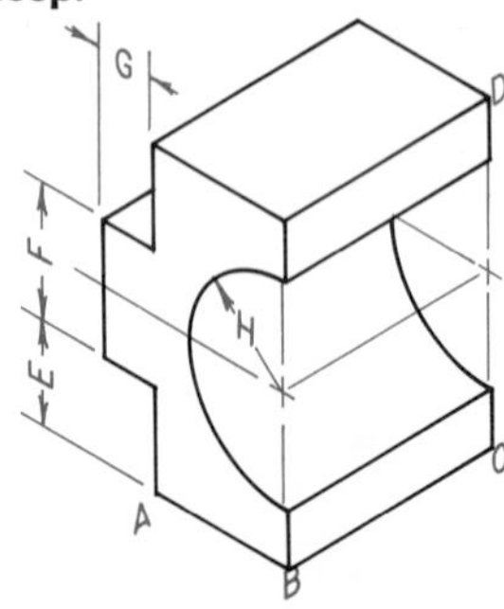

Fig. 6–90 Cradle slide. Draw all necessary views and dimensions. Scale: full size or as assigned. *AB* = $2\frac{3}{8}$, *BC* = $3\frac{9}{16}$, *CD* = $5\frac{1}{8}$, *E* = $1\frac{1}{2}$, *F* = $2\frac{1}{8}$, *G* = $\frac{7}{8}$, *H* = $1\frac{5}{8}$R.

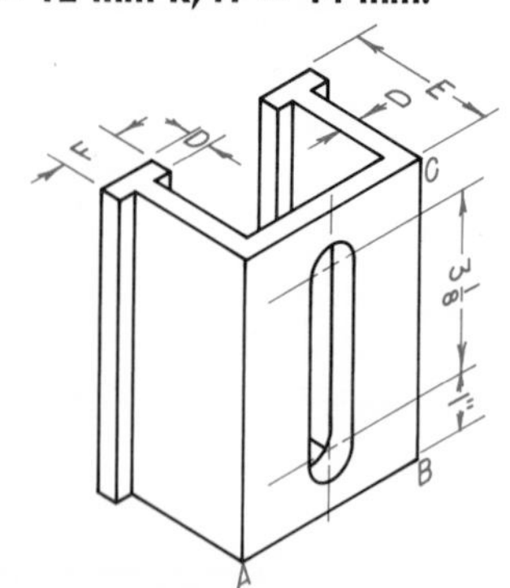

Fig. 6–91 Adjustable stop. Draw necessary views and dimension. Scale: $\frac{3}{4}$ size or as assigned. *AB* = $3\frac{5}{8}$, *BC* = $5\frac{1}{8}$, *D* = $\frac{5}{16}$, *E* = $2\frac{5}{8}$, *F* = 1, slot = $\frac{7}{8}$ wide.

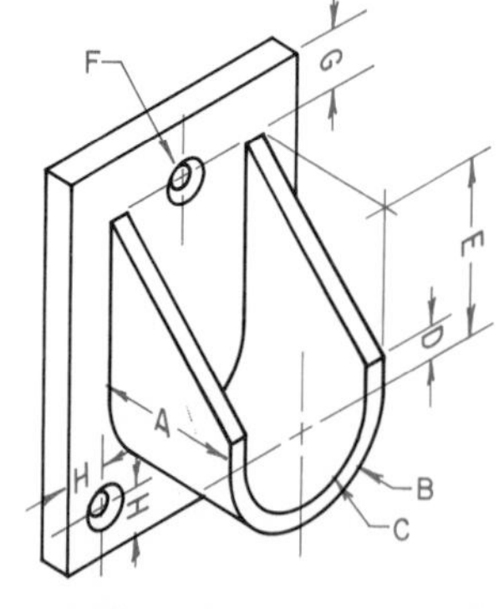

Fig. 6–92 Pipe support. Draw all necessary views and dimension. Scale: $\frac{1}{2}$ size or as assigned. Base plate = $\frac{1}{2}$ thick × $4\frac{1}{2}$ wide × $6\frac{1}{2}$ long, *A* = $2\frac{3}{8}$, *B* = $1\frac{1}{2}$R, *C* = $1\frac{1}{8}$R, *D* = $\frac{1}{2}$, *E* = 3, *F* = $\frac{3}{8}$ DIA hole through, CSK to $\frac{3}{4}$ DIA, 3 holes, *G* = 1, *H* = $\frac{3}{4}$.

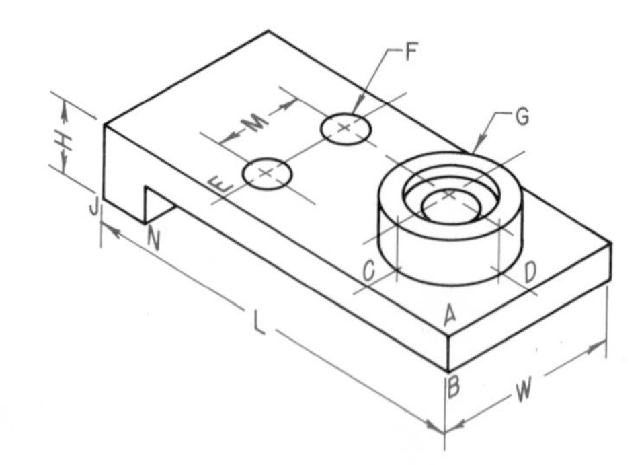

Fig. 6–93 Stop plate. Draw all necessary views and dimensions Scale: full size or as assigned. Overall sizes: L = $4\frac{1}{4}$, W = 2, H = $\frac{3}{4}$. *AB* = $\frac{3}{8}$, *AC* = 1, *AE* = $2\frac{3}{4}$, *AD* = 1, *JN* = $\frac{1}{2}$, *M* = 1, *F* = $\frac{7}{16}$DIA, 2 holes, *G* = Boss: $1\frac{1}{4}$DIA × $\frac{1}{2}$ high, $\frac{1}{2}$ DIA through, $\frac{7}{8}$ Cbore = $\frac{1}{8}$ deep.

Chapter 7

Auxiliary Views and Revolutions

INTRODUCTION TO AUXILIARY VIEWS

In Chapter 5, "Multiview Projections," we learned to describe an object with views on the three regular planes of projection. These are the top, or horizontal plane; the front, or vertical plane; and the side, or profile plane. With these planes, you can solve many graphic problems. However, to solve problems involving *inclined* (slanted) surfaces, you will need to draw views on *auxiliary* (additional) planes of projection. These are called *auxiliary views.* In this chapter you will learn to draw these views on planes that are parallel to the inclined surfaces (Fig. 7–1).

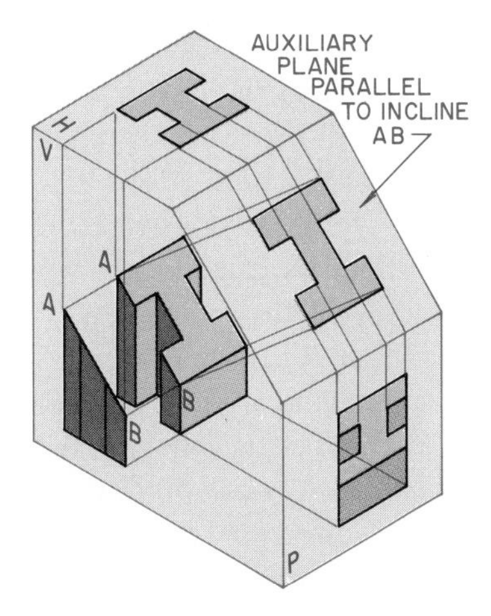

Fig. 7–1 **Pictorial study of a primary auxiliary view.**

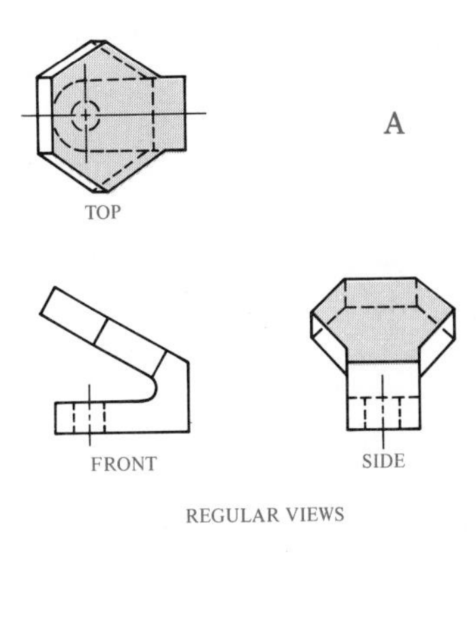

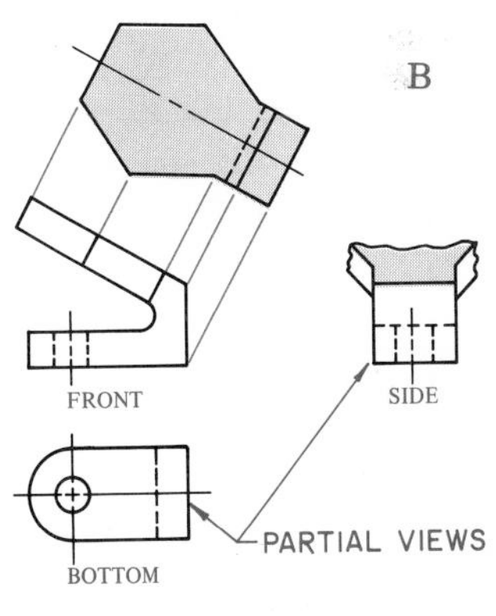

Fig. 7–2 **Compare the regular views at A with the auxiliary view at B.**

AUXILIARY VIEWS ARE "HELPER VIEWS"

When an object has inclined surfaces, these do not show up in true shape in regular views. For example, in Fig. 7–2 at A, neither the front view, top view, or side view shows the true size and shape of the object's inclined surface. However, a view on a plane parallel to the inclined surface, as at B, does show its true size and shape. This is an auxiliary view. It and the other views at B describe the object better than the views at A.

An auxiliary view is a projection on an auxiliary plane that is parallel to an inclined (slanting) surface. It is a view looking directly at the inclined surface in a direction perpendicular to it.

An anchor with a slanting surface is pictured in Fig. 7–3 at A. At B are three views of the same anchor on the regular planes. These views are hard to draw and to understand. Also, they show three circular features of the anchor as ellipses. At C, by contrast, the anchor is described completely in two views, one of which is an auxiliary view.

Auxiliary projections are important for describing the true geometric shapes of inclined surfaces. You also use them for dimensioning these shapes.

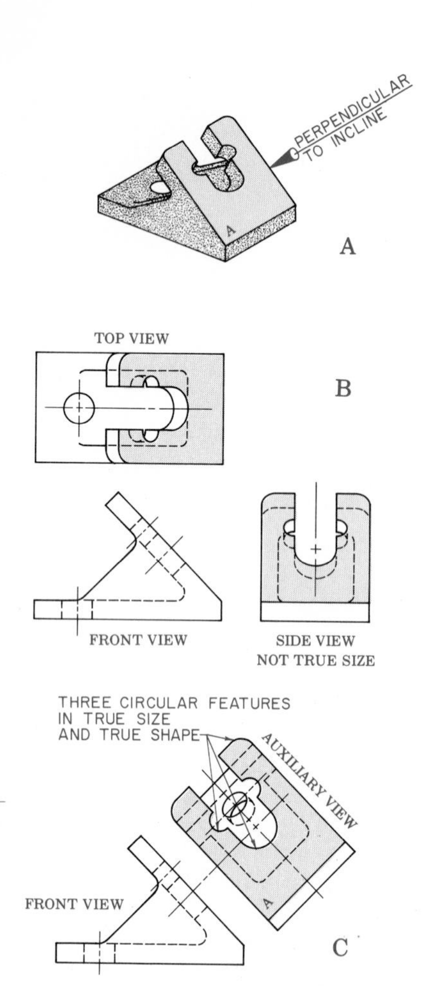

Fig. 7–3 **The pictorial view at A and the three-view drawing at B are difficult to draw.**

THE RELATIONSHIP OF AUXILIARY VIEWS TO REGULAR VIEWS

In Fig. 7–4, at A, a simple inclined wedge block is shown in the regular views. In none of these views does the slanted surface, called *A,* appear in its true shape. In the front view, all that shows is its edge line *MN.* In the side view, which is made by looking in the direction of arrow *Y,* surface *A* appears, but it is foreshortened. Surface *A* is also foreshortened in the top view. Line *MN* also appears in both views, but looking shorter than its true length, which shows only in the front view. To show surface *A* in its true shape, you need to imagine a plane parallel to it, as at B. This is called an *auxiliary plane.* A view of this plane from the direction of arrow *X,* which is perpendicular to it, will show the true size and shape of surface *A* at A^1. At C this *auxiliary view* has been *revolved* (turned) to align with the plane of the paper. By following this method, you can show the true size and shape of any inclined surface.

The Auxiliary Plane in Relation to the Regular Planes

Figure 7–5, at A, shows a picture of a wedge along with the regular front, top, and side planes. The planes have been drawn to look "hinged" together. At B, the planes have been "unfolded," or revolved, to make a regular technical drawing. Auxiliary views are generally projected from one of these planes. At C, an auxiliary plane has been drawn hinged to the front and side planes. Note that the hinge line *XY* is parallel to the slanted surface of the wedge. At D, the auxiliary plane has been unfolded, or revolved, to align with the plane of the paper. The top view has been left out in this drawing, and the side view also could be, since the wedge is completely described by the front and auxiliary views.

KINDS OF AUXILIARY VIEWS

Auxiliary views are classified according to which of the three regular planes they are developed from. There are three primary auxiliary

Fig. 7–4 **Basic relationship of the auxiliary view to three-view drawing.**

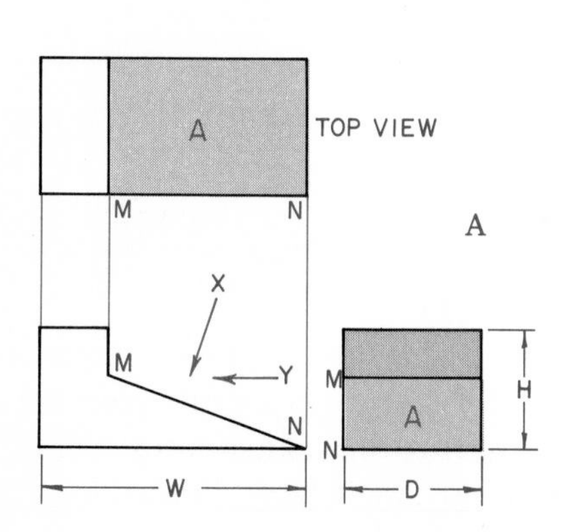

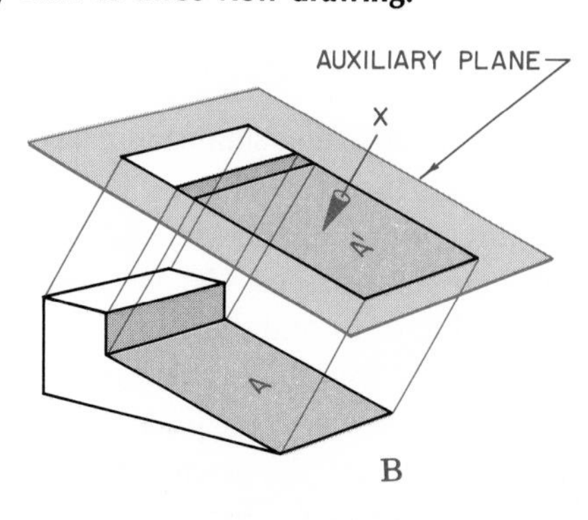

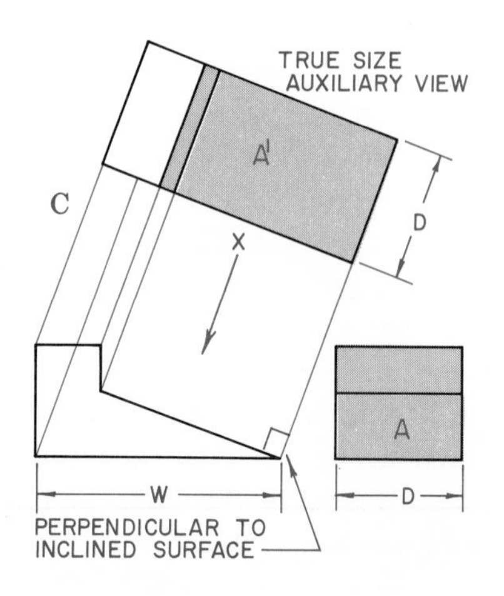

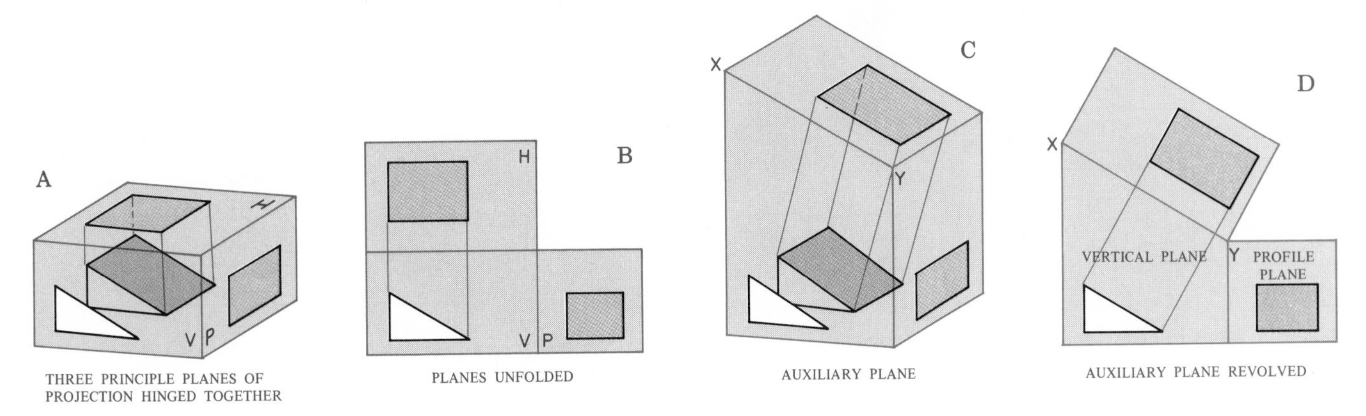

Fig. 7–5 **Basic relationship of the auxiliary plane to the regular planes.**

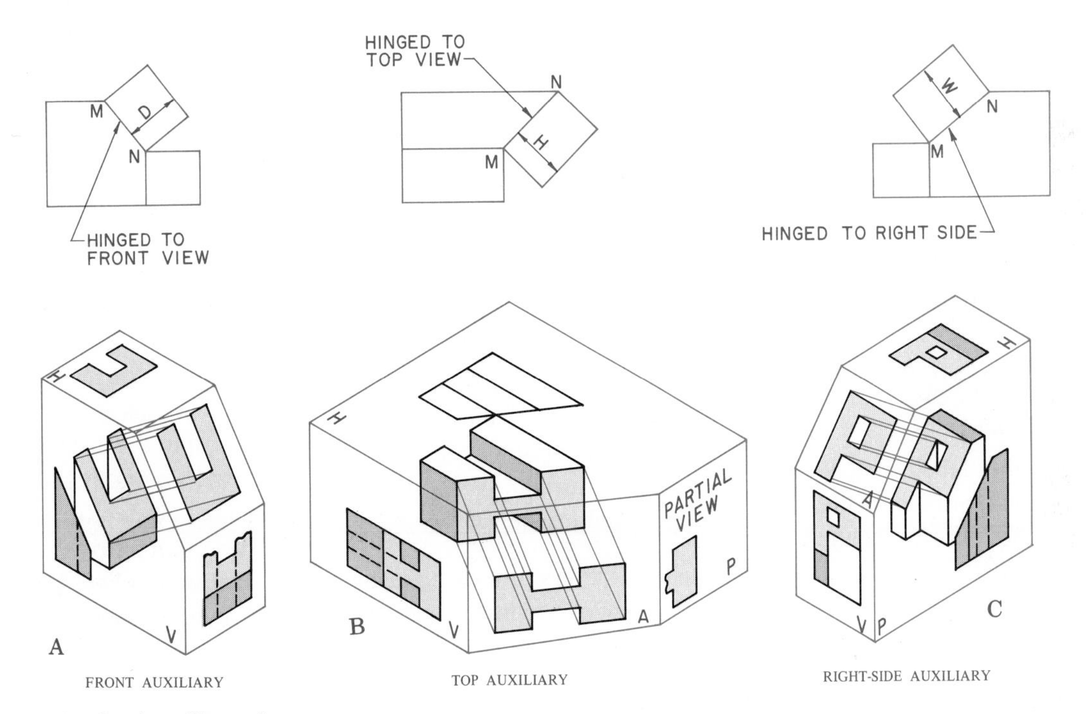

Fig. 7–6 **Three kinds of auxiliary views.**

views (Fig. 7–6). Each is developed by projecting from one of the three regular views, using as a *primary reference* a dimension—height, width, or depth—obtained from another regular view.

The first primary auxiliary view is the front auxiliary view, shown at A. It is "hinged" on the front view, and its primary reference is the depth. The second auxiliary view is the top auxiliary view, shown at B. It is hinged on the top view and its primary reference is the height. The third auxiliary view is the right-side auxiliary view, shown at C. It is hinged to the side view and its primary reference is the width. The three principal planes always show the auxiliary plane as an inclined line *(MN)*. This line is considered the edge view of the auxiliary plane.

CONSTRUCTING AN AUXILIARY VIEW

To construct any primary auxiliary view, use the following steps (Fig. 7–7). NOTE: The method shown is for a front auxiliary view.

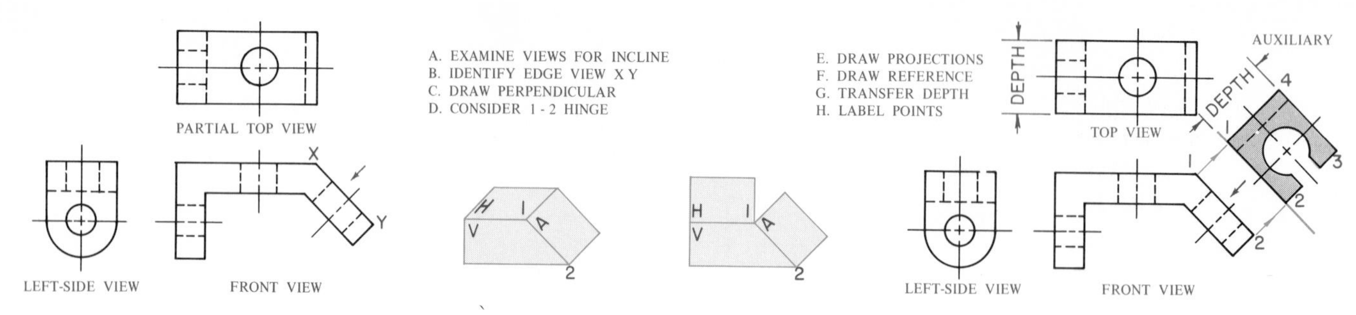

Fig. 7–7 **Steps in constructing an auxiliary view.**

A. Examine the views that are given for an inclined surface.

B. Find the line that is considered the edge view of the inclined plane.

C. In the front view, draw a light construction line at right angles to the inclined surface. This is the line of sight.

D. Think of the auxiliary plane as being attached by hinges to the front (vertical) plane from which it is developed.

E. From all points labeled on the front view, draw projection lines at right angles to the inclined surface (parallel to the line of sight).

F. Draw a reference line parallel to the edge view of the inclined surface and at a convenient distance from it.

G. Transfer the depth dimension, which in this case is the primary reference, to the reference line as shown.

H. Project the labeled points and connect them in sequence to form the auxiliary view. The points used to identify the shape are for solving difficult problems (instructional purposes). You would not normally leave them on the final drawing.

Figure 7–8 shows how to make an auxiliary view of a symmetrical object. At A, the object is shown in a pictorial view. In this case, you use a center plane as a reference plane, as at B (center plane construction). The edge view of this plane appears as a centerline, line *XY*, on the top view. Number the points on the top view. Then transfer these numbers to the edge view of the inclined surface on the front view, as shown. Parallel to this edge view and at a convenient distance from it, draw the line *X′Y′*, as at C. Now, in the top view, find the distances from the numbered points to the centerline. These are the depth measurement. Transfer them onto the corresponding construction lines that you have just drawn, measuring them off on

Fig. 7–8 **To draw an auxiliary view using the center plane reference.**

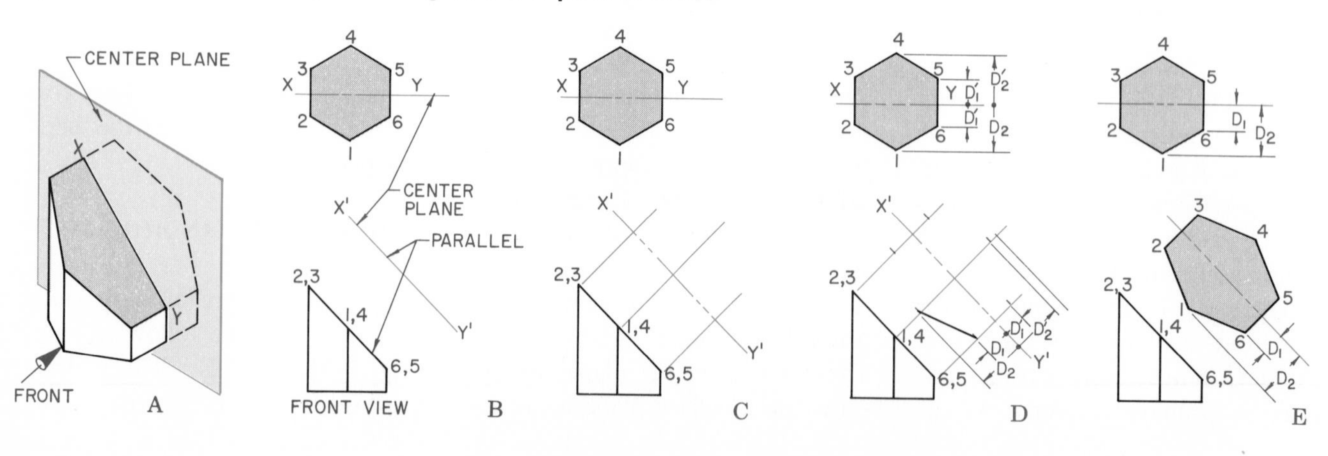

either side of line $X'Y'$, as shown at D. The result will be a set of points on the construction lines. Connect and number these points, as at E, and you will have the front auxiliary view of the inclined surface. You could also, if you wished, project the rest of the object from the center reference plane.

Figure 7–9 shows how to draw a front auxiliary view of a nonsymmetrical object by using a vertical reference plane. First, place the object on reference planes, as shown. These planes are located strictly for convenience in taking reference measurements. The vertical plane can be in front or in back of the object. In this case it is in back. The construction is similar to that in Fig. 7–8, except that the depth measurements D_1, D_2, and D_3 are laid off in front of the vertical plane. The drawing shows the entire object projected onto the front auxiliary plane.

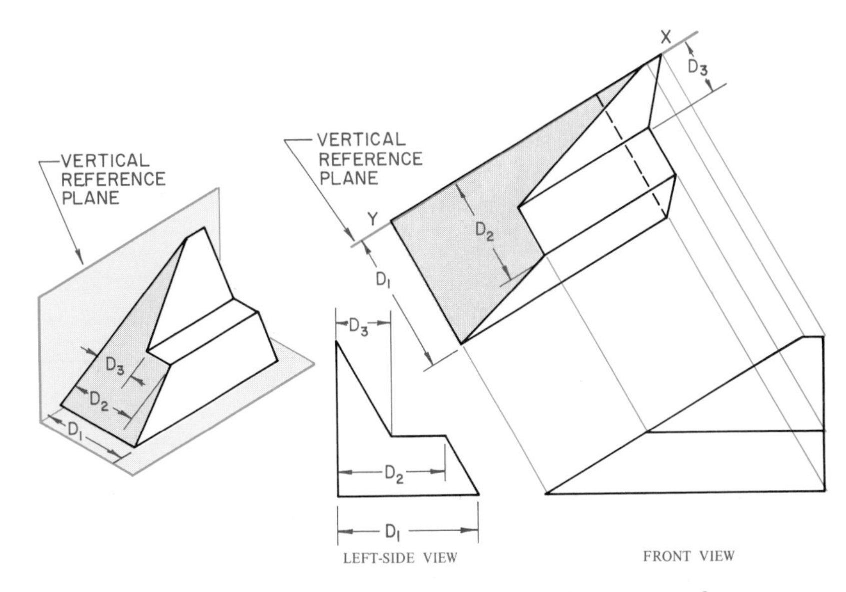

Fig. 7–9 **To draw a front auxiliary view using a vertical reference plane.**

Figure 7–10 shows how to draw a top auxiliary view by using a horizontal reference plane. The object shown at A is a molding cut at a 30° angle. First, imagine a reference plane XY under the molding, as shown. Then find points 1 to 6 in the top and left-side views. In the top view, find the edge line of the slanted surface. Draw reference line $X'Y'$ parallel to it and a convenient distance away. Then, from every point in the top view, project a line out to line $X'Y'$ and at right angles to it. Now, in the side view, find height measurements for the various numbered points by measuring up from XY. Lay off these same measurements up from $X'Y'$ along the lines leading to the corresponding points in the top view. Locate more points on the curve as needed in order to draw it more accurately. The result will be a top auxiliary view, with its base on line $X'Y'$.

Fig. 7–10 **To draw a top auxiliary view with a horizontal reference plane.**

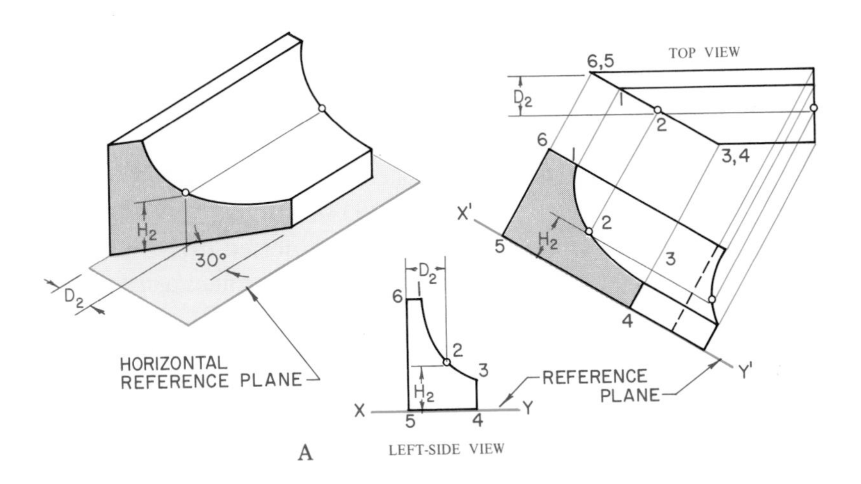

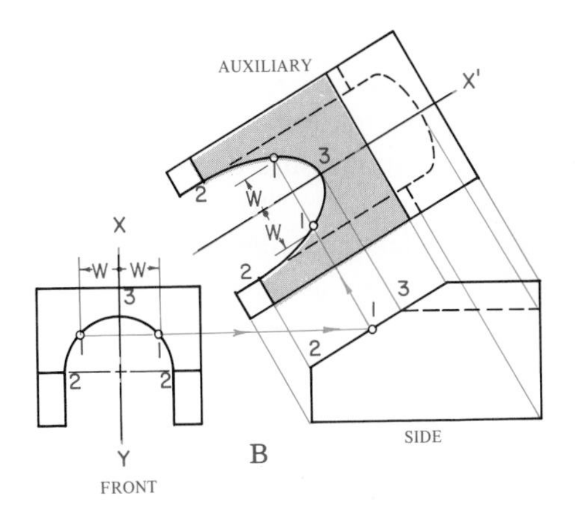

CURVES ON AUXILIARY VIEWS

You draw an auxiliary view of a curved line by locating a number of points along that line. A drawing of a simple curve is shown in Fig. 7–10 at B. Figure 7–11 shows how to make an auxiliary view of the curved cut surface of a cylinder. The cylinder is shown in a horizontal position. It has been cut at an angle, so that the true shape of the slanting cut surface is an ellipse.

This auxiliary view is a front auxiliary view with the depth as its primary reference. To draw it, begin by locating the vertical centerline *XY* in the side view. This line will serve as a center reference plane. Next, locate a number of points along the rim of the side view. The more points you locate, the more accurate your curve will be. Then, project lines from these points over to the edge view of the cut surface in the front view. Now, parallel to this edge view and at a convenient distance from it, draw the new centerline *X′Y′*. From the points you have located on the edge view, project lines out to line *X′Y′* and perpendicular to it. Continue these lines beyond *X′Y′*. Finally, find your depth measurements in the side view by measuring off the distances D_1, D_2, etc. between the centerline *XY* and the points you located along the rim. Take these distances and measure them off on either side of *X′Y′*. Draw a smooth curve through the points marked to form the ellipse, as shown.

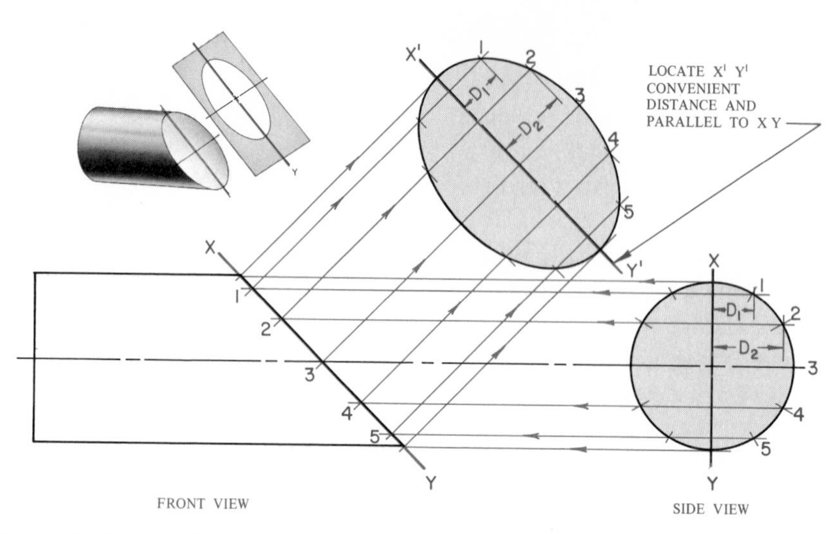

Fig. 7–11 **The auxiliary-view curve (ellipse) of the cut surface of a cylinder.**

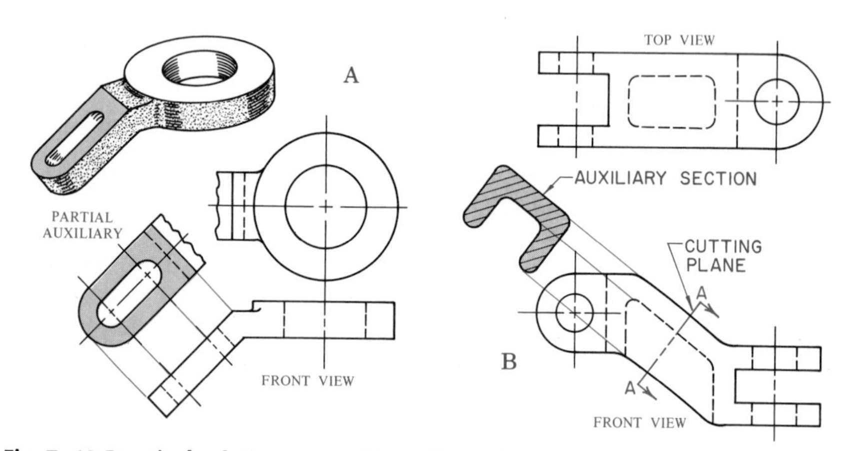

Fig. 7–12 **Practical solutions—partial auxiliary views and auxiliary sections.**

PARTIAL AUXILIARY VIEWS

Figure 7–12 at A shows a partial auxiliary view. If you use break lines and centerlines properly, you can leave out complex curves while still describing the object completely, as shown. You can draw a half view if the object is symmetrical in a way that is simple and clearly understood.

AUXILIARY SECTIONS

Another practical use of auxiliary views is the auxiliary section (Fig. 7–12 at B). In this case, the slanted surface you are drawing is one made by a plane that cuts through the object. (See Chapter 9, "Sectional Views.") You can locate the surface in *cross section* (crosshatched) by using cutting-plane line *AA*.

SECONDARY AUXILIARY VIEWS

A view projected from a primary auxiliary view is called a *secondary auxiliary* (Fig. 7–13). You must use such a view if you wish to show the true size and shape of a surface that is *oblique* (inclined to all three of the regular planes).

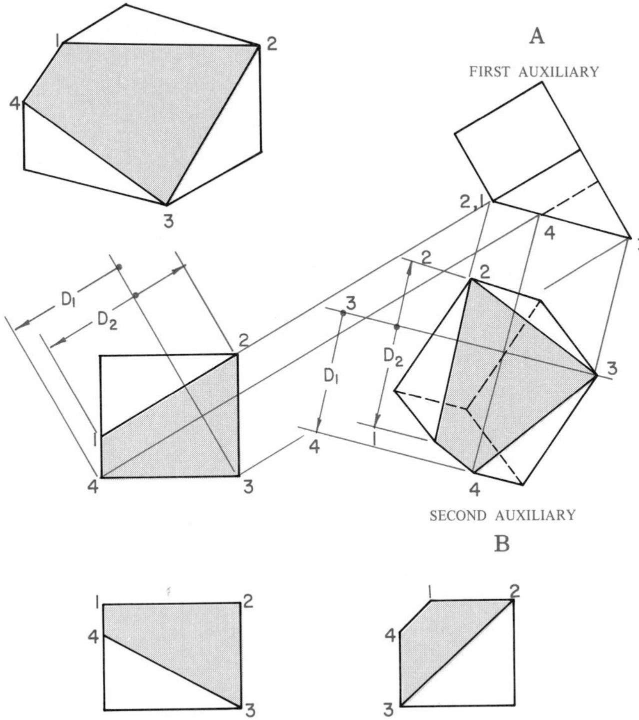

Fig. 7–13 **Secondary auxiliary views assist in finding the true shape of a surface.**

In Fig. 7–13, surface 1-2-3-4 is inclined to the three regular planes. At A, a first auxiliary view has been drawn. It is on a plane perpendicular to the inclined surface. Note that, in this view, points 1, 2, 3, and 4 appear as a line or edge view of the plane. At B, a secondary auxiliary view has been drawn from the first. It is on a plane parallel to the surface 1-2-3-4. This view shows the true shape of the surface.

Figure 7–14 shows another example. In this case, an octahedron (eight triangles making a regular solid) is shown in three views. Triangle surface 0-1-2 is inclined to all three. At A, a first auxiliary view has been drawn. It is on a plane perpendicular to triangle surface 0-1-2. Note that line 1-2 in the top view appears as point 1-2 in this auxiliary view and that the triangle now appears as an edge line 0′-1-2. At B, a secondary auxiliary view has been drawn. It is on a plane parallel to the edge view of triangle surface 0′-1-2 in the first auxiliary view. This secondary auxiliary view shows the true shape of triangle 0-1-2.

INTRODUCTION TO REVOLUTIONS

When the true size and shape of an inclined surface do not show in a drawing, one solution is, as we have seen, to make an auxiliary view. Another, however, is to keep using the regular reference planes while

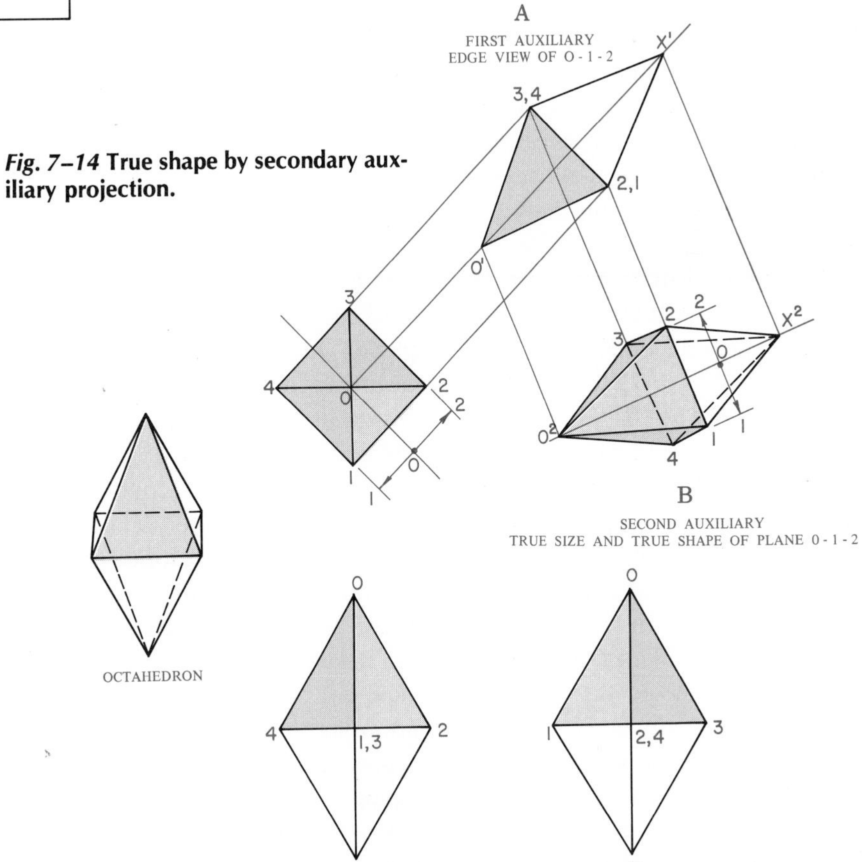

Fig. 7–14 **True shape by secondary auxiliary projection.**

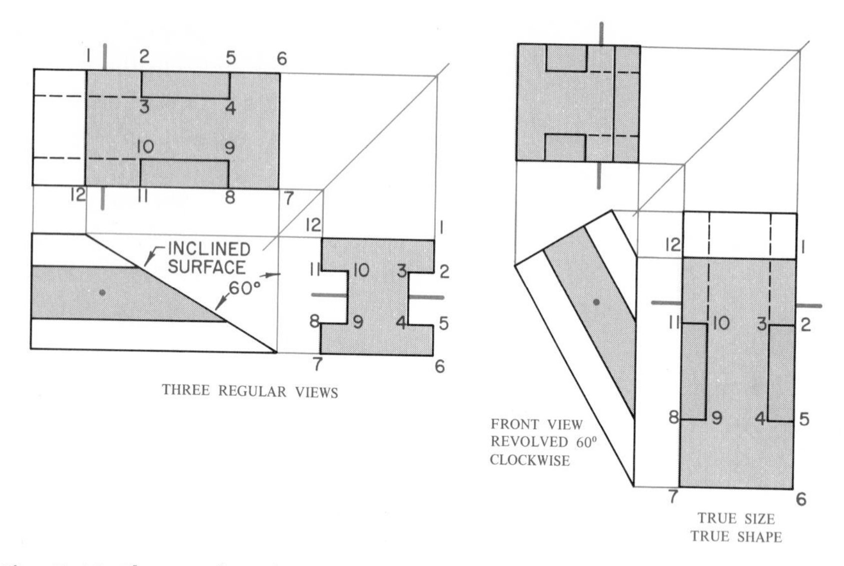

Fig. 7–15 **The regular planes remain in order in revolution. The object is revolved.**

imagining that the object has been *revolved* (turned). (See Fig. 7–15.) Remember, in auxiliary views, you set up new reference planes to look at objects from new directions. Understanding *revolutions* (ways of revolving objects) should help you better understand auxiliary views. Both methods will be examined for spatial problems in Chapter 8, "Basic Descriptive Geometry."

THE AXIS OF REVOLUTION

An easy way to picture an object being revolved is to imagine that a shaft or axis has been passed through it. Imagine, also, that this axis is perpendicular to one of the principal planes. In Fig. 7–16, the three principal planes are shown with an axis passing through each one and through the object beyond.

An object can be revolved to the right (clockwise) or to the left (counterclockwise) about an axis perpendicular to either the vertical or the horizontal plane. The object can be revolved forward (counterclockwise) or backward (clockwise) about an axis perpendicular to the profile plane.

SINGLE REVOLUTION

As we have seen, an axis of revolution can be perpendicular to the vertical, horizontal, or profile plane. In Fig. 7–17 at A, the usual front and top views of an object are shown in Space 1. Space 2 shows the same views of the object after it has been revolved 45° counterclockwise about an axis perpendicular to the vertical plane. Notice that the front view is the same in size and shape as in Space 1, except that it has a new position. The new top view has been made by projecting up from the new front view and across from the old top view in Space 1. Note that the depth remains the same from one top view to the other.

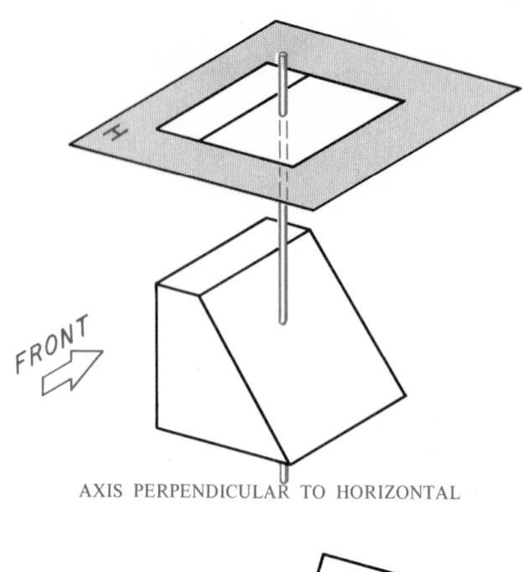

AXIS PERPENDICULAR TO HORIZONTAL

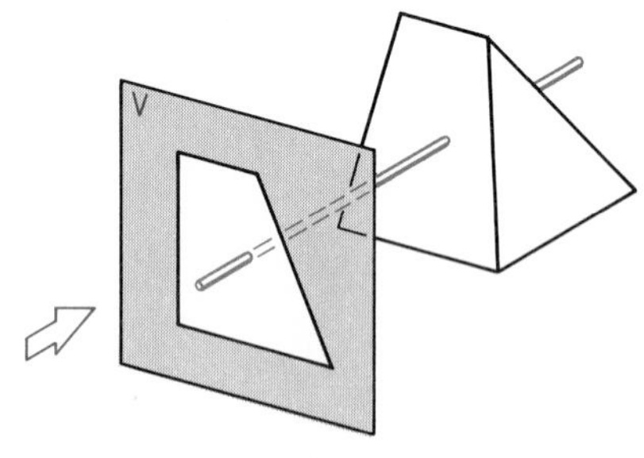

AXIS PERPENDICULAR TO VERTICAL

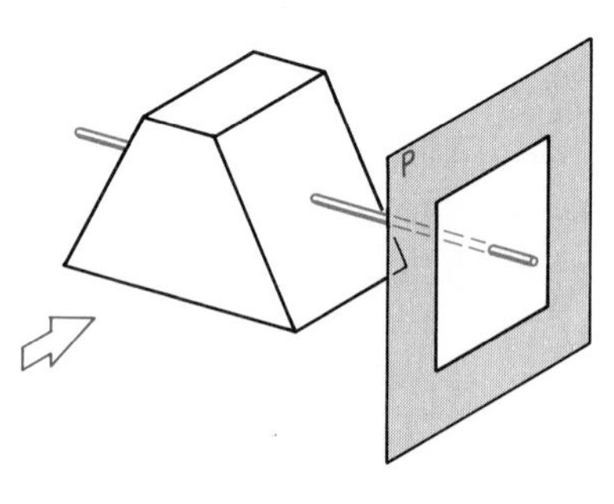

AXIS PERPENDICULAR TO PROFILE

Fig. 7–16 **Three positions for the axis of revolution.**

In Fig. 7–17 at B, a second object is shown in Space 1 in the usual top and front views. Space 2 shows the same views of the object after it has been revolved 60° clockwise about an axis perpendicular to the horizontal plane. The new top view is the same in size and shape as in Space 1. The new front view has been made by projecting down from

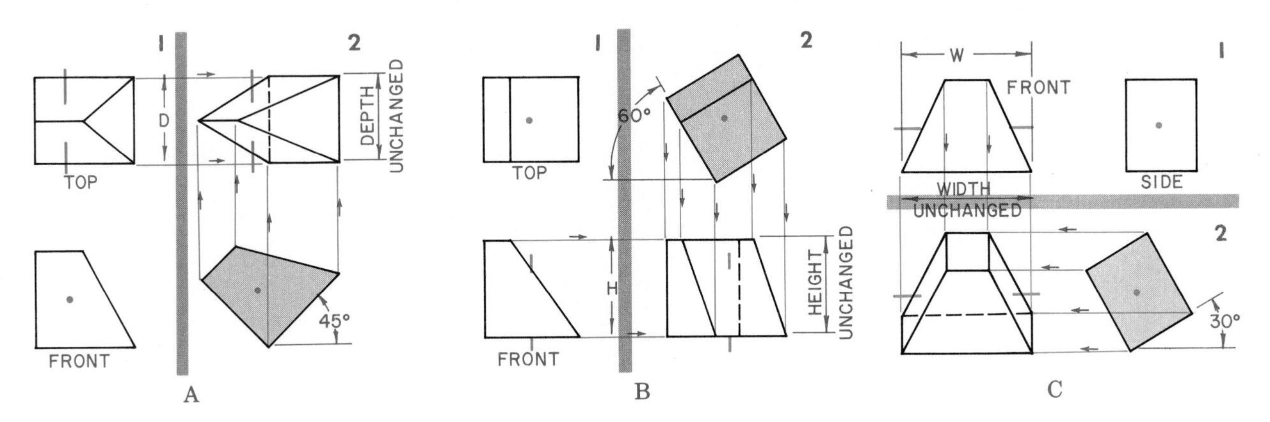

Fig. 7–17 **Single revolution about each of the three axes.**

the new top view and across from the old front view of Space 1. Note that the height remains the same from the original front view to the revolved front view.

In Fig. 7–17 at C, a third object is shown in Space 1 in the usual front and side views. Space 2 shows the same views of the object after it has been revolved forward (counterclockwise) 30° about an axis perpendicular to the profile plane. The new front view has been made by projecting across from the new side view and down from the old front view in Space 1. Note that the width remains the same from one front view to the other. Revolution can be clockwise, as at Fig. 7–17 at B, or it can be counterclockwise, as at A and C.

THE RULE OF REVOLUTION

The rule of revolution has two parts. (1) *The view that is perpendicular to the axis of revolution stays the same except in position.* (This is true because the axis is perpendicular to the plane on which it is projected.) (2) *Distances parallel to the axis of revolution stay the same.* (This is true because they are parallel to the plane or planes on which they are projected.) Figure 7–18 illustrates the two parts of the rule of revolution.

REVOLUTION ABOUT AN AXIS PERPENDICULAR TO THE VERTICAL PLANE

Figure 7–19 shows how to draw a primary revolution. At A, an imaginary axis *AX* is passed horizontally through a truncated right octagonal prism. In the front view, it shows as a dot. In the top view, and later in the side view, it shows as a line. At B, the prism has been revolved clockwise about the axis into a new position. You can see that the new front view has the same size and shape as the old. Just its position has changed. However, the side view now shows the true size and shape of the truncated

Fig. 7–18 **The rule of revolution.**

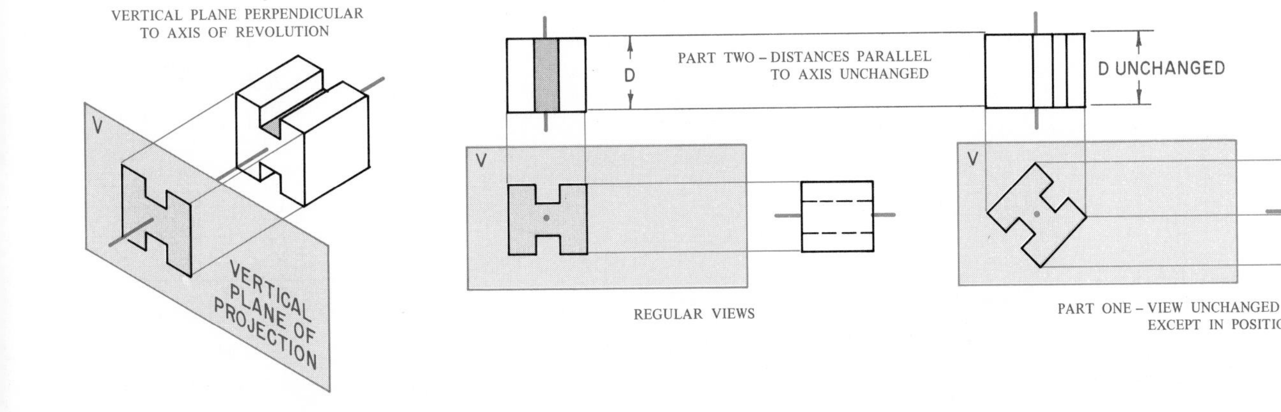

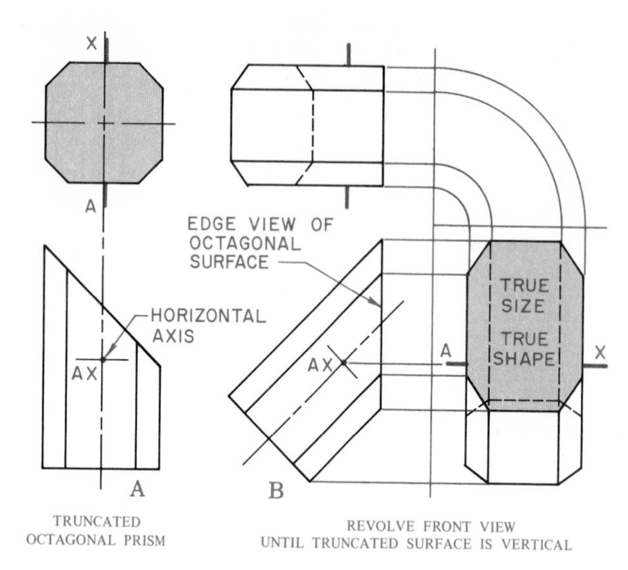

Fig. 7–19 **Revolution about an axis perpendicular to the vertical plane.**

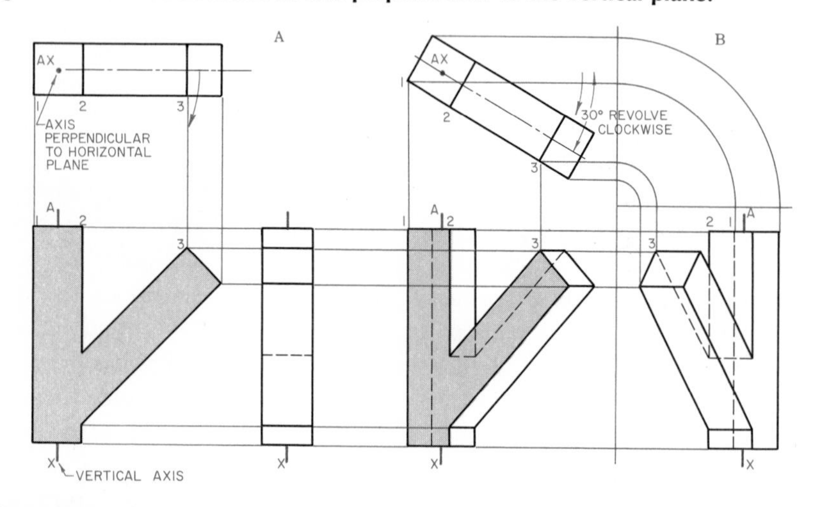

Fig. 7–20 **Revolution about an axis perpendicular to the horizontal plane (clockwise).**

Fig. 7–21 **Revolution about an axis perpendicular to the horizontal plane (counterclockwise).**

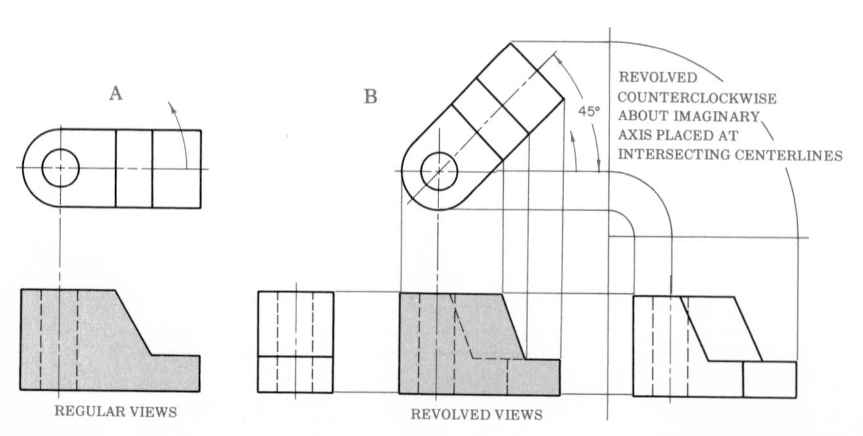

surface. This side view is made by projecting across from the new front view and by transferring the depth from the top view.

REVOLUTION ABOUT AN AXIS PERPENDICULAR TO THE HORIZONTAL PLANE

Figure 7–20 shows how to draw an object that is revolved clockwise about an imaginary vertical axis *AX*. At A, the three regular views are given. At B, the top view has been revolved 30° clockwise about the axis *AX*. Since this is a vertical axis, revolution does not change the height of the object. Therefore, points from the old vertical plane at A can be projected to make the new one at B. The new side view is made from the front and top views in the usual way. Figure 7–21 shows how to draw an object revolved counterclockwise through 45°.

Fig. 7–22 **Practical use of revolutions.**

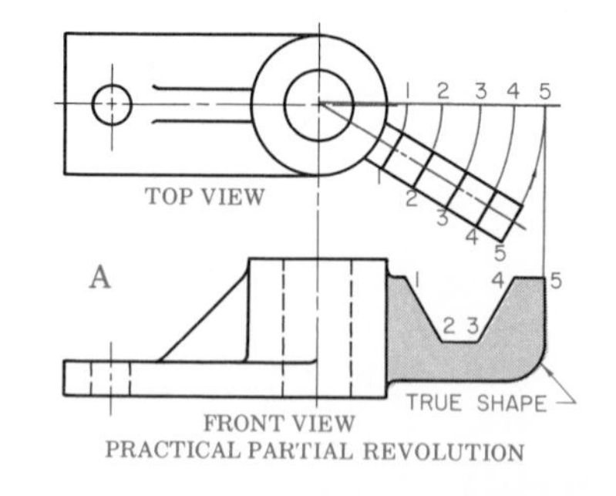

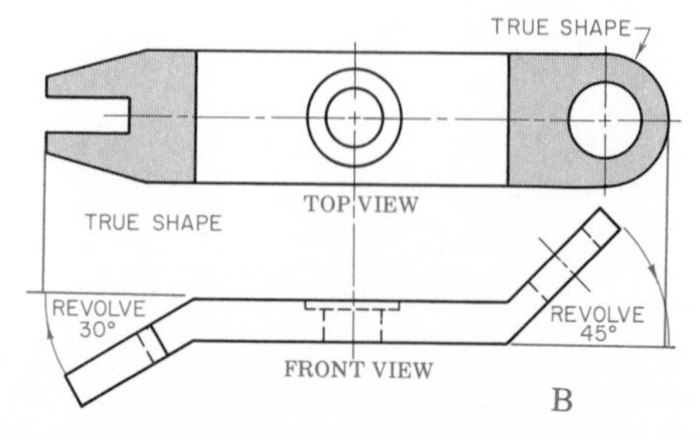

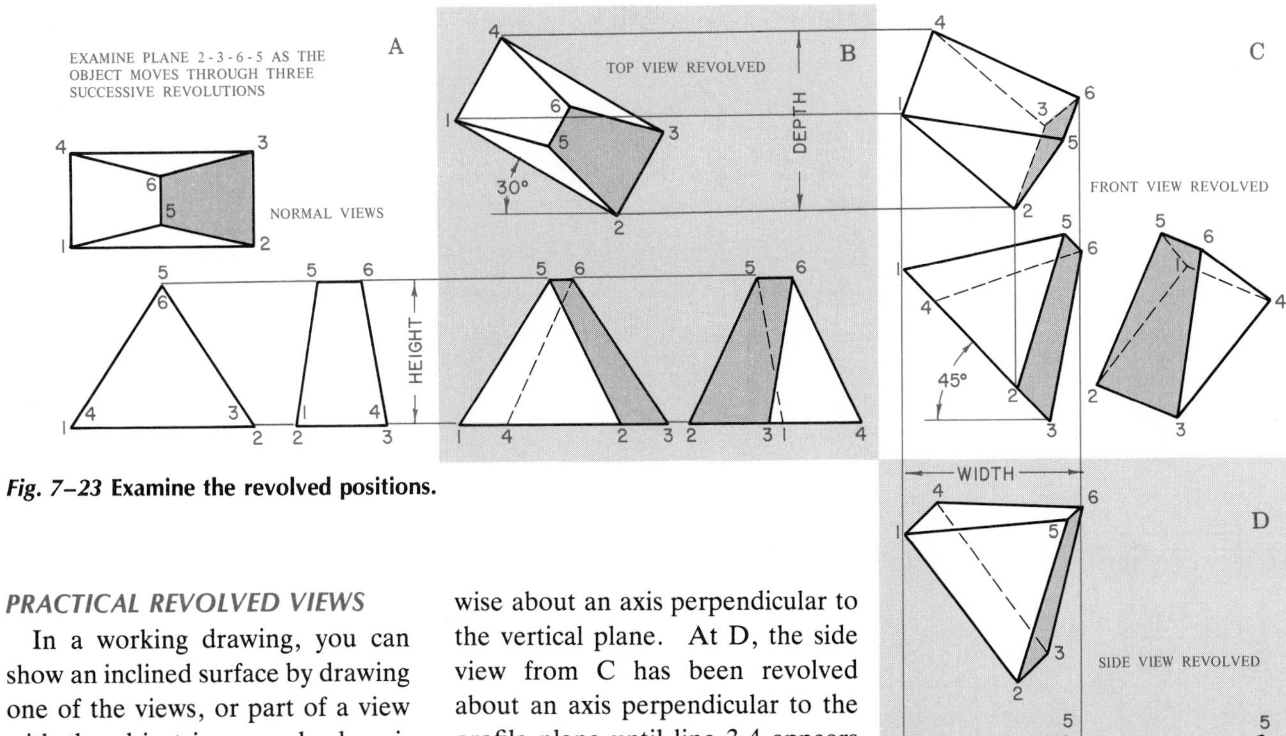

Fig. 7–23 **Examine the revolved positions.**

PRACTICAL REVOLVED VIEWS

In a working drawing, you can show an inclined surface by drawing one of the views, or part of a view with the object in a revolved position. In Fig. 7–22 at A, the top view shows the angle of a V-shaped part. In the front view, the part is revolved to show its true shape. In Fig. 7–22 at B, the front view shows the angles at which surfaces of a part are inclined. The inclined surfaces are then revolved in the front view. Next, their dimensions are transferred to the top view. There, the surfaces appear in true size and shape.

SUCCESSIVE REVOLUTIONS

After an object is revolved about an axis perpendicular to one plane, it can be revolved again about an axis perpendicular to another plane. This process is shown in Fig. 7–23. At A, an object is shown in the normal views. At B, the object has been revolved 30° clockwise about an axis perpendicular to the horizontal plane. At C, the front view from B has been revolved 45° clockwise about an axis perpendicular to the vertical plane. At D, the side view from C has been revolved about an axis perpendicular to the profile plane until line 3-4 appears as a horizontal line. In all, then, three revolutions occurred. There has been one in each of the three principal planes of projection.

AUXILIARY VIEWS AND REVOLVED VIEWS

You can show the true size of an inclined surface either by an auxiliary view (Fig. 7–24 at A) or a revolved view (Fig. 7–24 at B and C). In a revolved view, the inclined surface is turned until it is parallel to one of the principal planes. The revolved view at B and C is similar to the auxiliary at A.

In the auxiliary view, it is as if the observer has changed position to look at the object from a new direction. Conversely, in the revolved view, it is as if the object has changed position. Both revolutions and auxiliaries help you visualize things better. They also work equally well in solving problems.

TRUE SHAPE OF AN OBLIQUE PLANE BY SUCCESSIVE REVOLUTIONS

A surface shows its true shape when it is parallel to a plane. In Fig. 7–25 at D, an object is pictured on which surface 1-2-3-4 is an oblique plane. It is oblique because it is inclined to all three of the normal planes. At A, the object is drawn in its normal position. At B, it has been revolved about an axis perpendicular to the horizontal plane until surface 1-2-3-4 is perpendicular to the vertical plane. Now, in the front view, all you see of this surface is its edge line. At C, the object has been revolved about an axis perpendicular to the front plane

Fig. 7–24 **True size shown by an auxiliary view and a revolved view.**

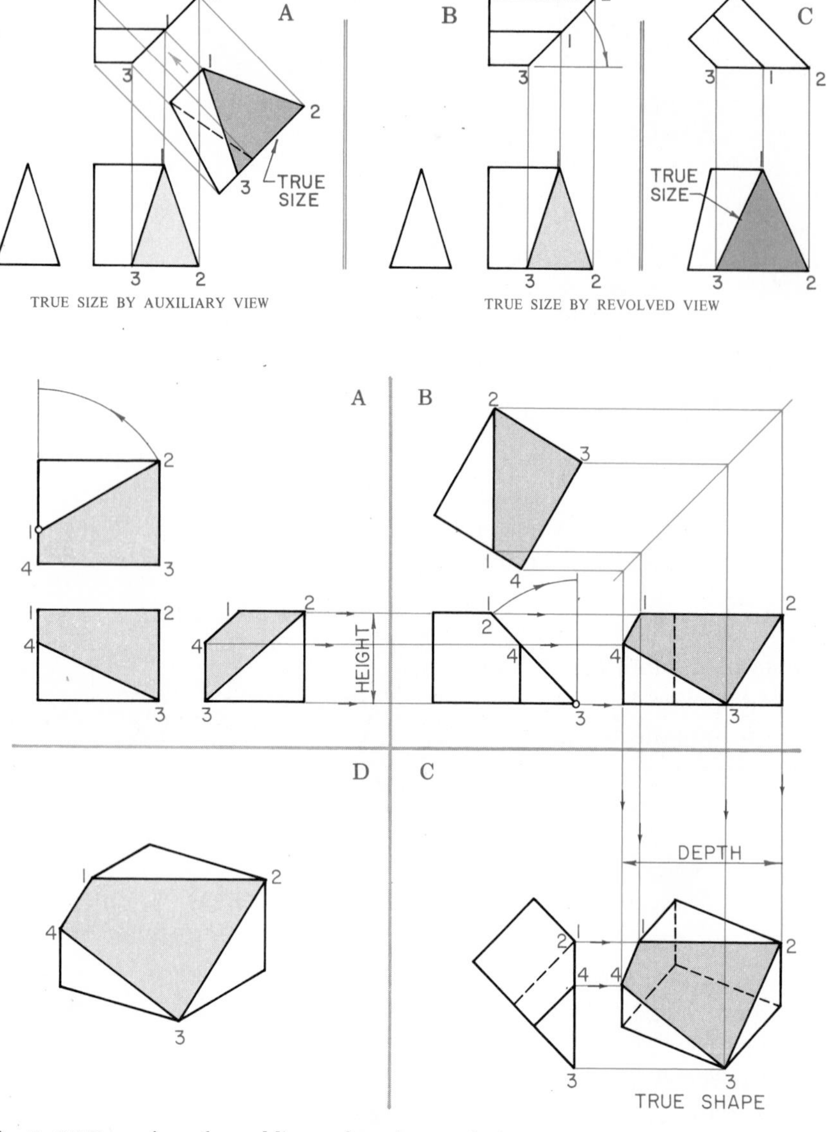

Fig. 7–25 **True size of an oblique plane by revolution.**

until surface 1-2-3-4 is parallel to the profile plane. There it shows its true shape.

TRUE LENGTH OF A LINE

Since an auxiliary view shows the true size and shape of an inclined surface, you can also use it to find the true length of a line. In Fig. 7–26 at A, the line *OA* does not show its true length in the top, front, or side view because it is inclined to all three of these planes of projection. At B, however, in the auxiliary plane, it does show its true length *(TL)*. This is because the auxiliary plane is parallel to the surface *OAB*.

Figure 7–26 at C shows another way to show the *TL* of *OA*. This is to revolve the object about an axis perpendicular to the vertical plane until surface *OAB* is parallel to the profile plane. The side view will then show the true size of *OAB* and also the *TL* of *OA*. A shorter method of showing the *TL* of *OA* is to revolve only the surface *OAB*, as shown at D.

At E, the object is revolved in the top view until line *OA* in that view is horizontal. The front view now shows *OA* in its true length because this line is now parallel to the vertical plane.

At F and G, still another method is shown. In this case, instead of the whole object being revolved, just line *OA* is turned in the top view until it is horizontal at OA^1. The point A^1 then can be projected to the front view. There, OA^1 will be in true length.

You can revolve a line in any view to make it parallel to any one of the three principal planes. Projecting the line on the plane to which it is parallel will show its true length. In Fig. 7–26 at H, the line has been revolved parallel to the horizontal plane. The true length then shows the top view.

INDUSTRIAL APPLICATIONS

The following illustrations show how a drafter uses revolution. Figure 7–27 shows a tractor with various lift positions. Figure 7–28 shows a product designer's use of a revolved position.

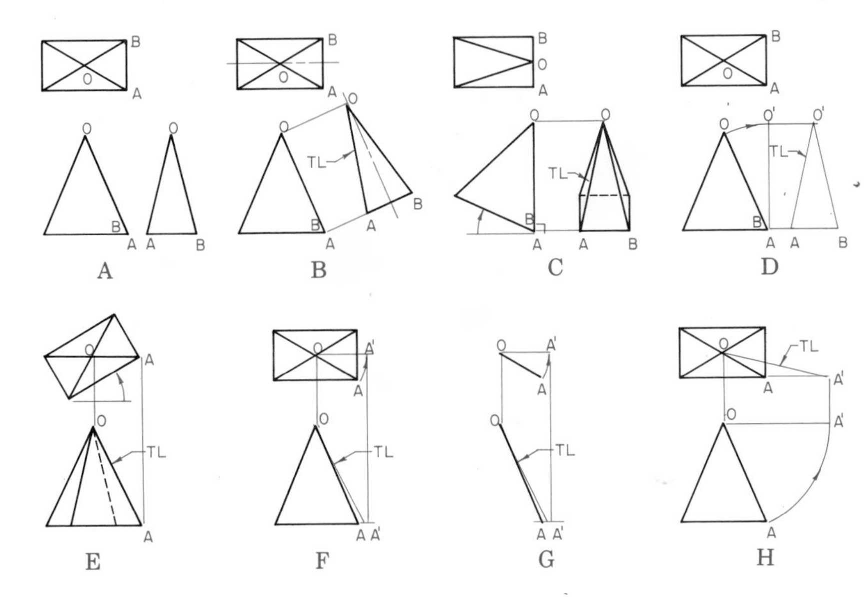

Fig. 7–26 **Typical true-length problems examined.**

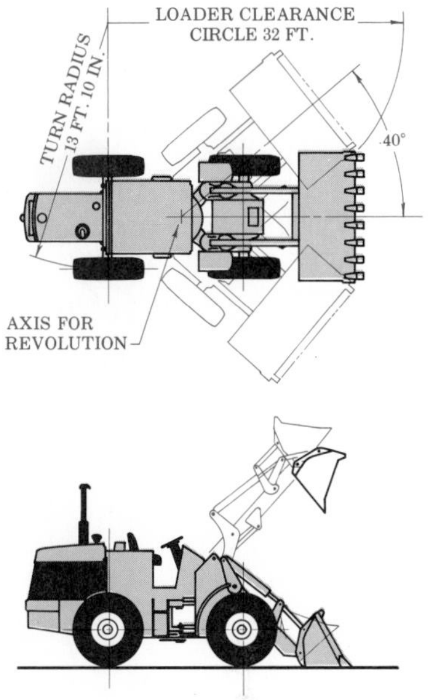

Fig. 7–27 **The profile of a tractor shows several positions of the loader. The plan view shows additional revolutions of the tractor.**

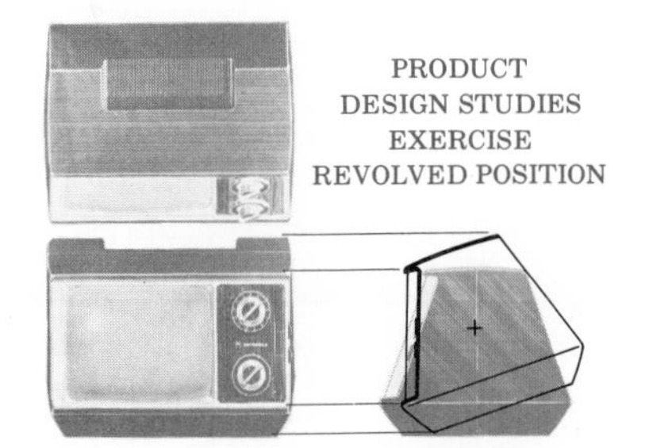

Fig. 7–28 **Note that the television profile has been revolved to show the screen in true shape.** ***(Motorola, Inc.)***

Review

1. Why do you need auxiliary views?

2. How do you place the auxiliary plane in relation to the inclined surface?

3. What are the two chief reasons for drawing auxiliary projections?

4. Name the three primary auxiliary projections.

5. List the steps in drawing an auxiliary view.

6. What is a reference-plane construction?

7. Can you plot curved lines on auxiliary views?

8. Is doing partial auxiliary views an accepted drawing practice in industry?

9. What auxiliary projection do you use to find the true size of an oblique surface?

10. What is the basic reason for revolving the view of an object?

11. What is the axis of revolution?

12. In your own words, describe the first rule of revolution.

13. Name the three basic single revolutions.

14. What is the chief use of successive revolutions?

15. Can you use both auxiliary views and revolved views to find the true lengths of inclined and oblique lines?

Problems

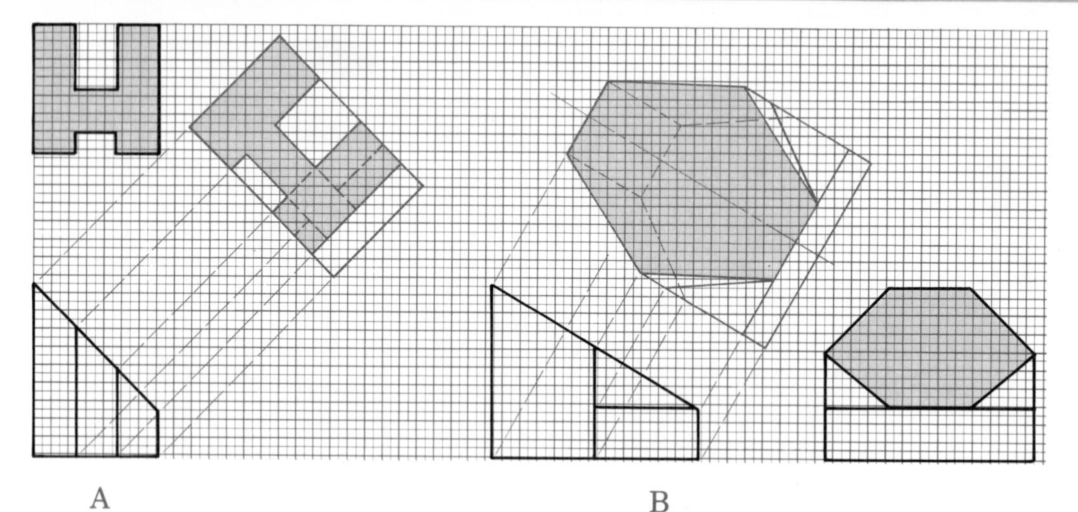

Fig. 7–29 **Problem 1: Draw the front, top, and side view of each figure, A and B. Complete the front auxiliary projection. Problem 2: Change the angle of the inclined surface at A to 30° and at B to 45°. Scale: each square = 1/4″.**

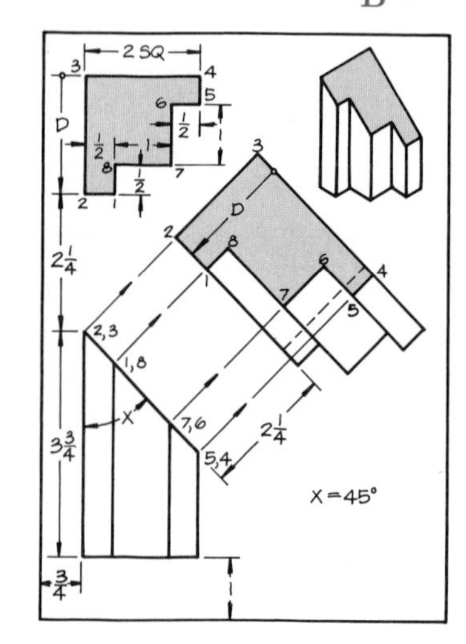

Fig. 7–30

in.	mm
1/2	13
3/4	19
1	25
1-1/2	38
1-3/4	44
2	50
3-3/4	95

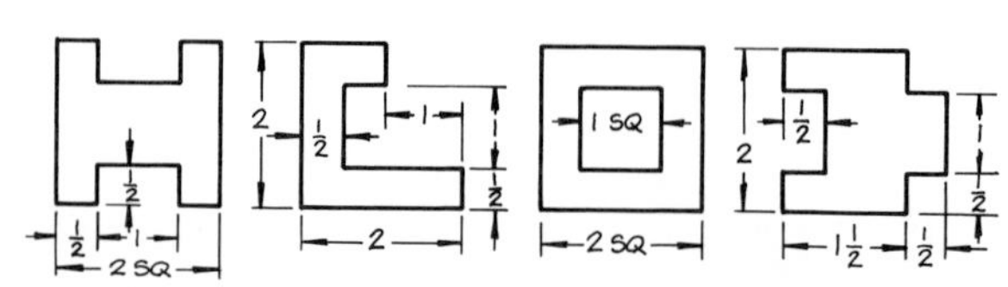

Fig. 7–31 *Fig. 7–32* *Fig. 7–33* *Fig. 7–34*

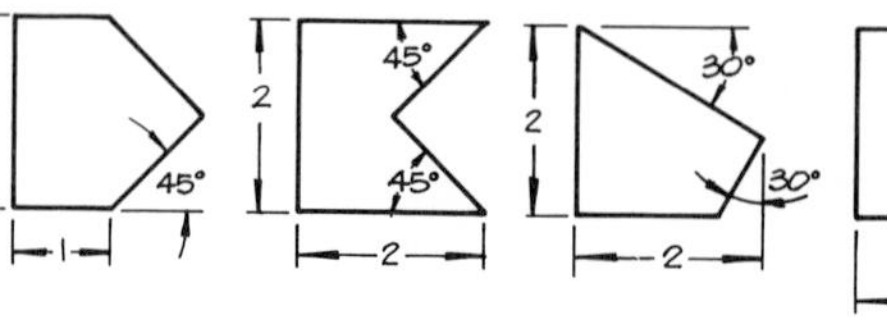

Fig. 7–35 *Fig. 7–36* *Fig. 7–37* *Fig. 7–38*

Fig. 7–39 *Fig. 7–40* *Fig. 7–41* *Fig. 7–42*

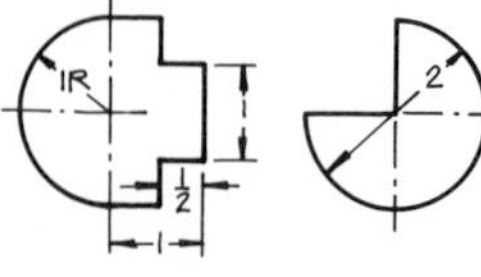

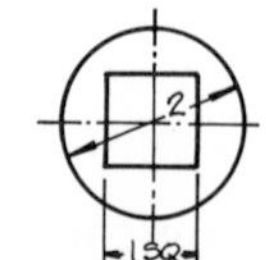

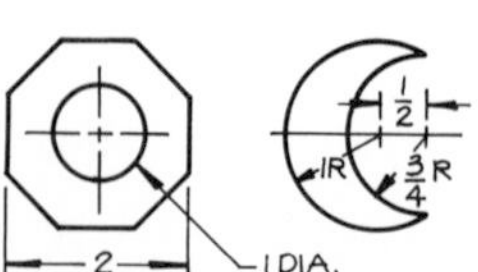

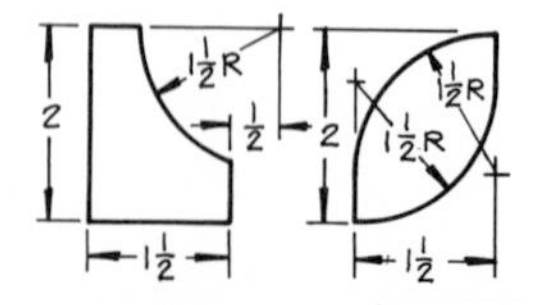

Fig. 7–43 *Fig. 7–44* *Fig. 7–45* *Fig. 7–46* *Fig. 7–47* *Fig. 7–48* *Fig. 7–49*

Figs. 7–30 through 7–49 **Each of Figs. 7–31 through 7–49 shows a top view. For each figure, draw the top and front views and either the complete auxiliary view or the cut face only, as directed by the instructor. Figure 7–30 has been worked as an example. The angle *X* may be 45° or 60° as assigned. The total height of the front view is 3¾″ for all Figs. 7–30 through 7–49.**

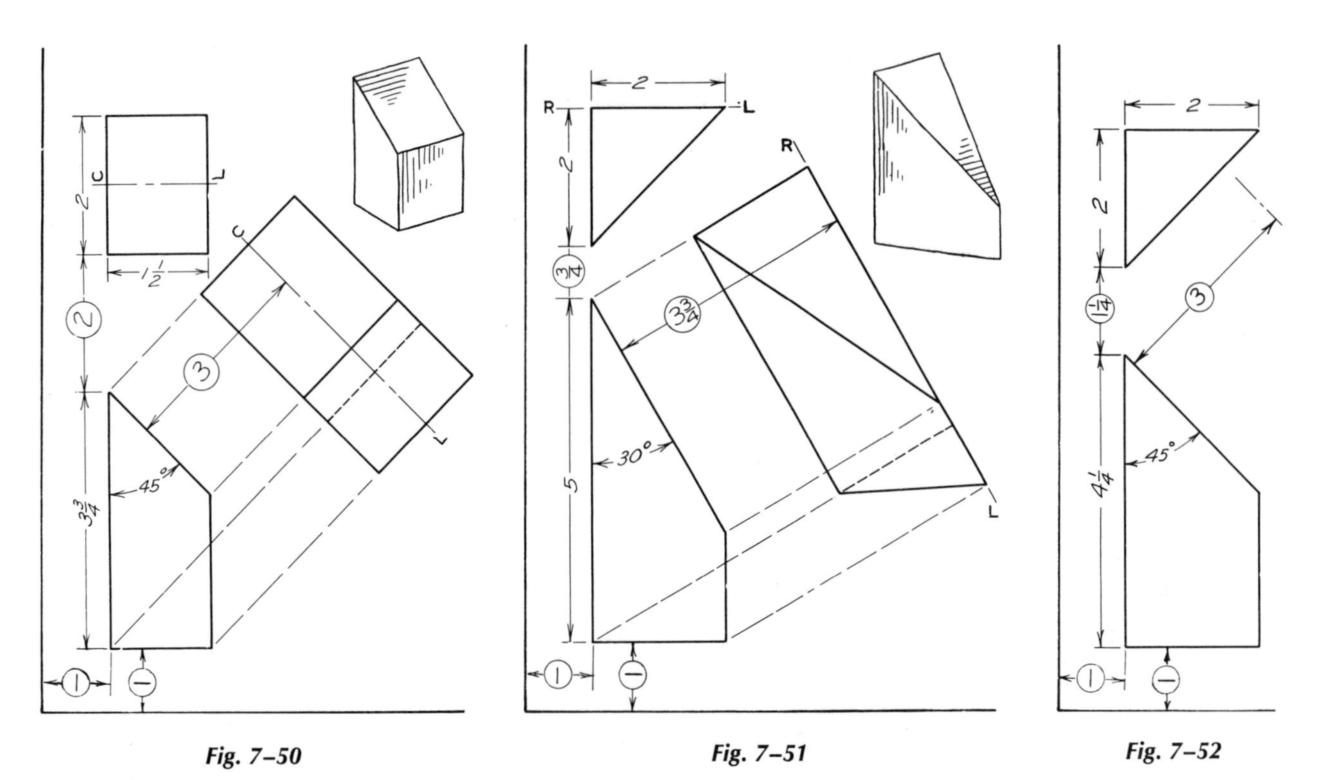

Fig. 7–50 Fig. 7–51 Fig. 7–52

Figs. 7–50 through 7–52 **Draw the front, top, and complete auxiliary views. The solutions are given for Figs. 7–50 and 7–51. Do not copy the solutions. Instead, cover them. Then work the problems and compare your solutions.**

Figs. 7–53 through 7–56 **Draw the front, top, and complete auxiliary views.**

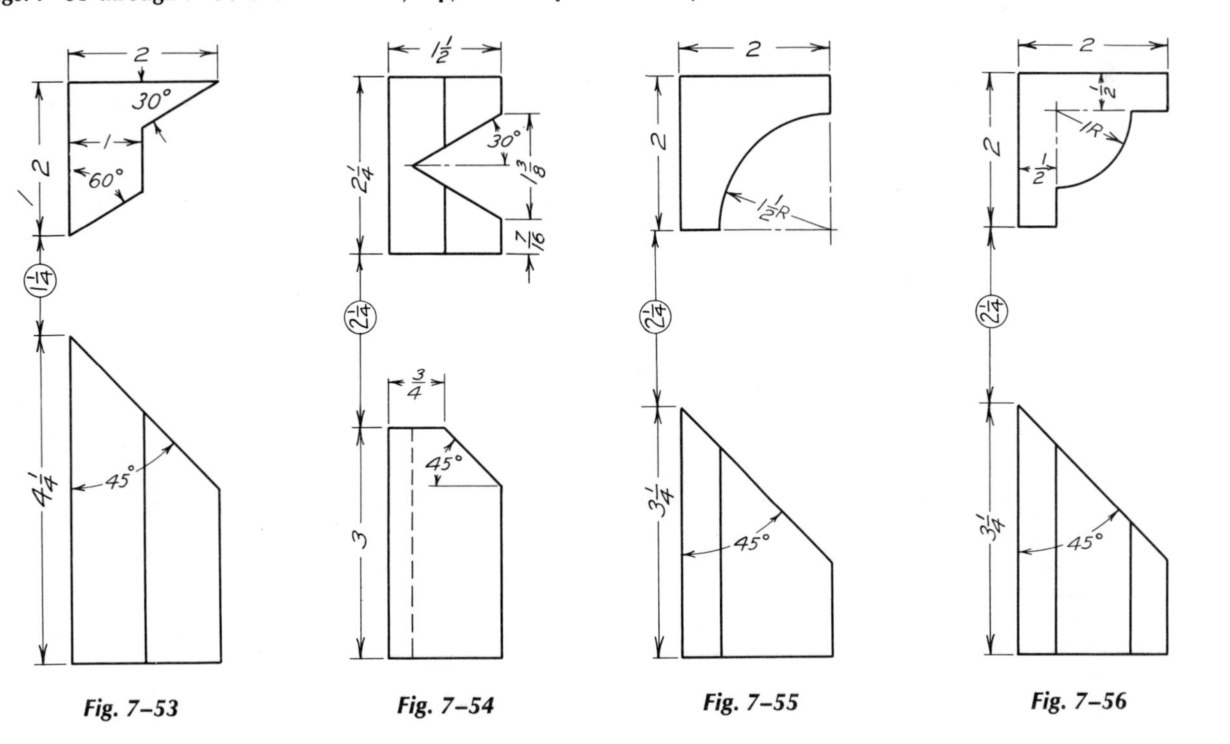

Fig. 7–53 Fig. 7–54 Fig. 7–55 Fig. 7–56

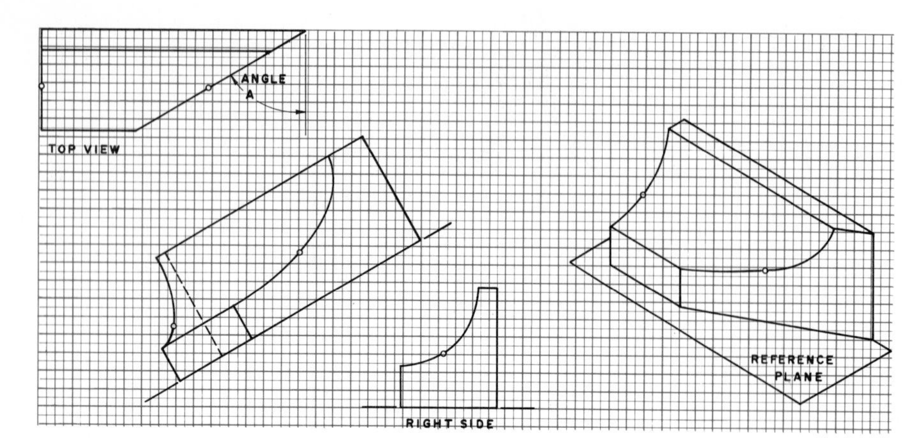

Fig. 7–57 Develop the three views, changing the angle *A* to 45°. Complete the top auxiliary projection. Scale: each square = 1/4″.

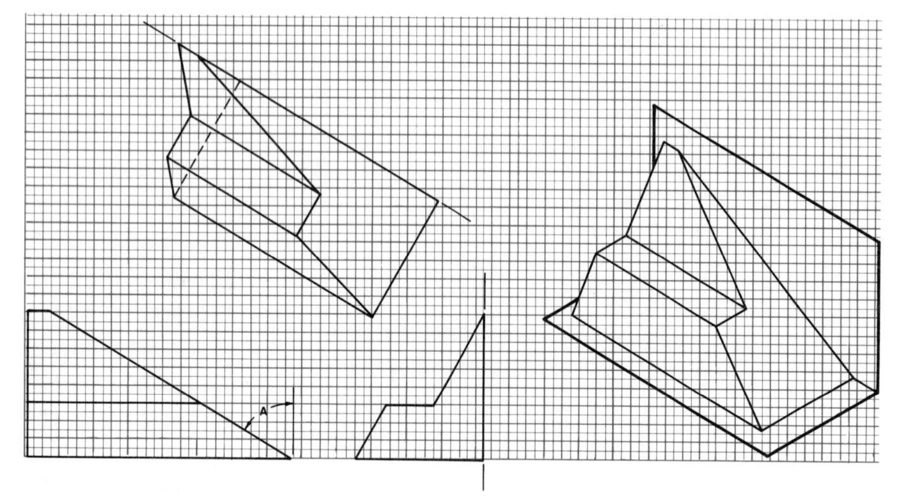

Fig. 7–58 Determine the necessary views, letting angle *A* = 60°. Complete the front auxiliary view. Scale: each square = 1/4″.

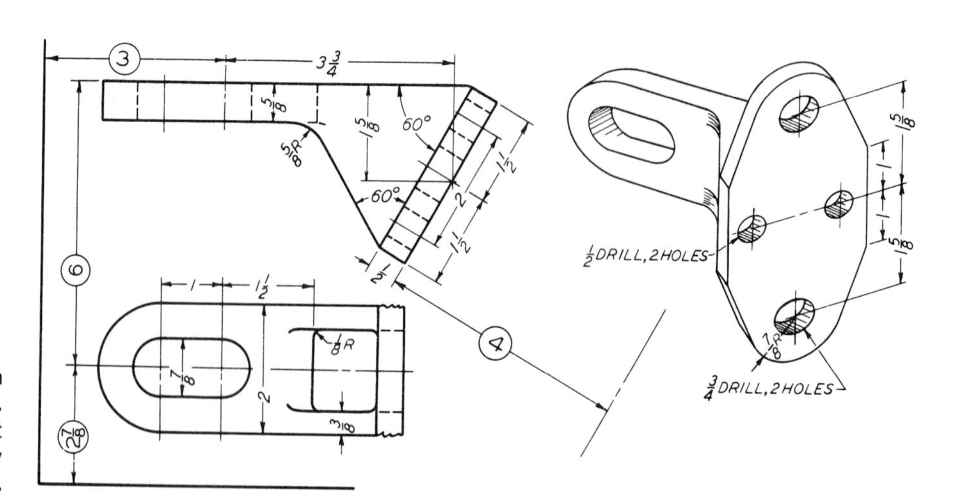

Figs. 7–59 A picture and a layout for an angle plate are shown in the figure. Draw the top view and the part front view as shown. Draw a part auxiliary view where indicated on the layout. Note that this is an auxiliary elevation, as it is made on a plane perpendicular to the horizontal plane.

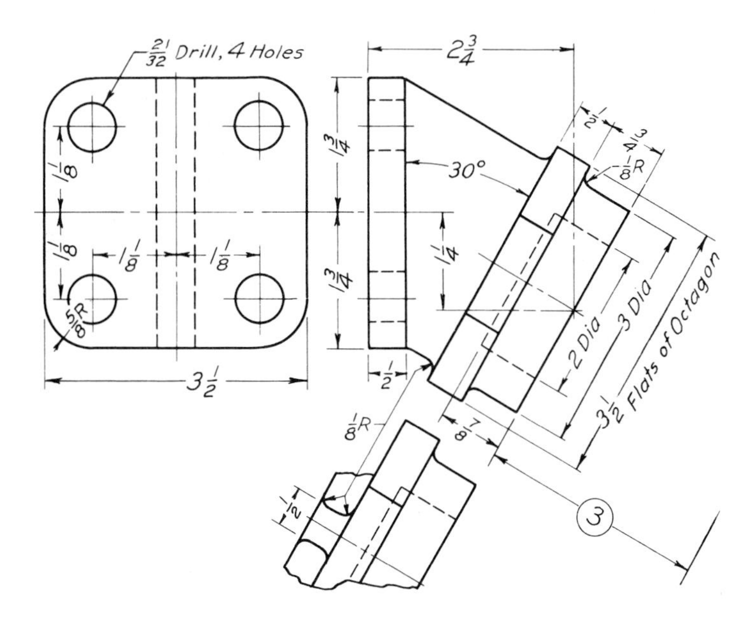

Fig. 7–60 A part front view, a right-side view, and a part auxiliary view of an angle cap are shown on the layout in the figure. Draw the views given and another part auxiliary view where indicated on the layout. Note that this last auxiliary view is a rear auxiliary view and that the one shown on the layout is a front auxiliary view. Dimension if required by the instructor.

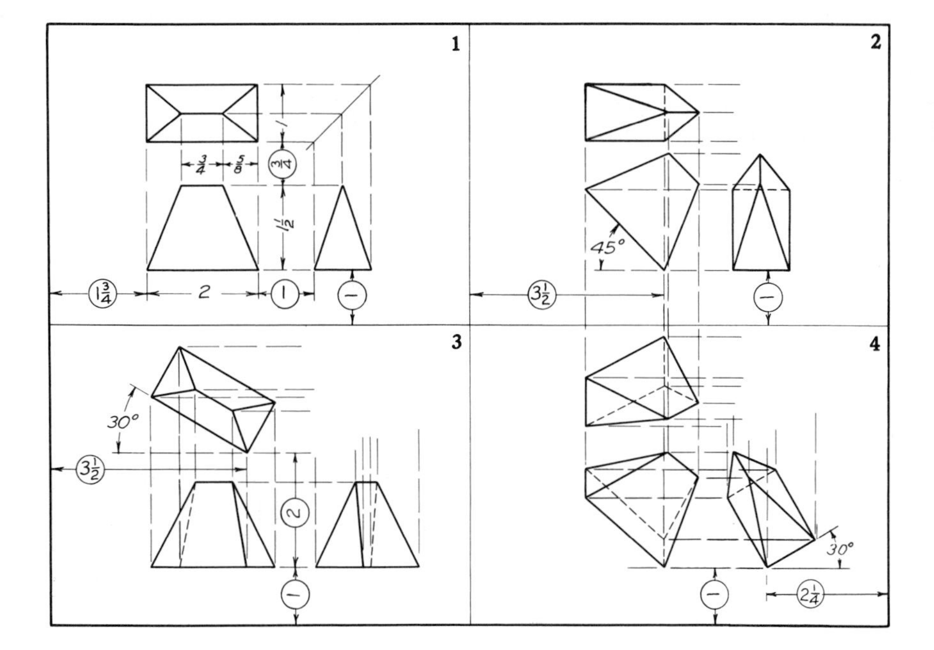

Fig. 7–61 The figure shows the completed problem. It is given for comparison and is not to be copied. In Space 1 is a three-view drawing of a block in its simplest position. In Space 2 (upper right) the block is shown after being revolved from the position in Space 1, through 45°, about an axis perpendicular to the frontal plane. The front view was drawn first, copying the front view of Space 1. The top view was obtained by projecting up from the front view and across from the top view of Space 1.

In Space 3 (lower left) the block has been revolved from position 1 through 30° about an axis perpendicular to the horizontal plane. The top view was drawn first, copied from the top of Space 1. In Space 4, the block has been tilted forward from position 2 about an axis perpendicular to the side plane. The side view was drawn first, copied from the side view of Space 2. The widths of front and top view were projected from the front view of Space 2.

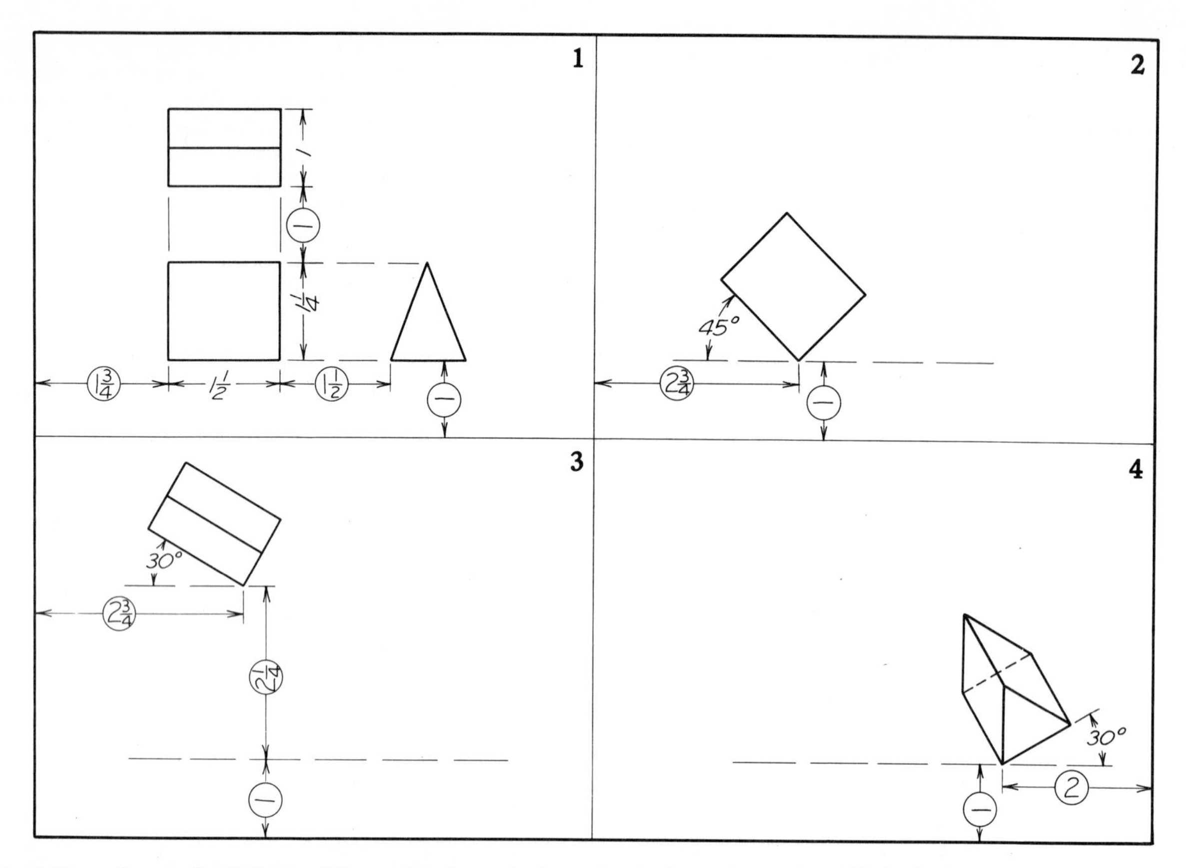

Fig. 7–62 **Draw the revolved views of the wedge shown in Space 1. In Space 2, revolve 45° clockwise about an axis perpendicular to the frontal plane and draw three views. In Space 3, revolve 30° forward from the position of Space 2. In Space 4, revolve from the position of Space 1, 30° counter-clockwise about an axis perpendicular to the horizontal plane.**

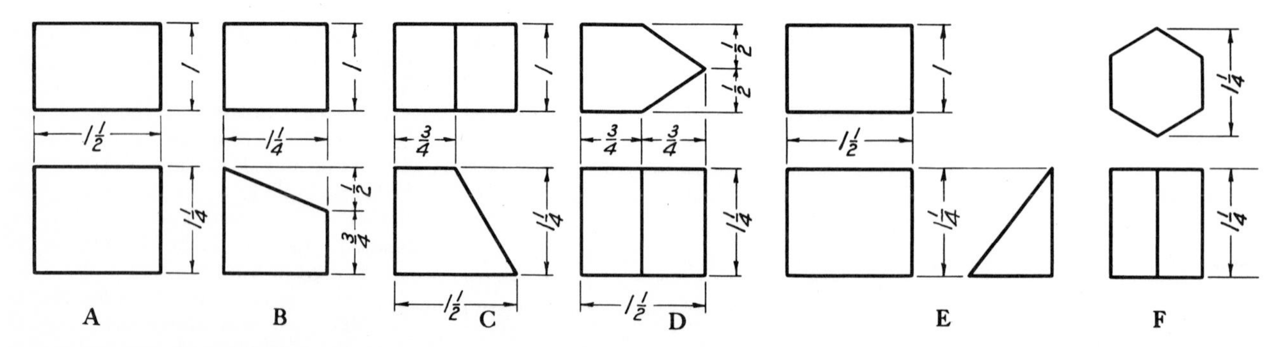

Fig. 7–63 **Follow the directions for Fig. 7–62 for the object assigned.**

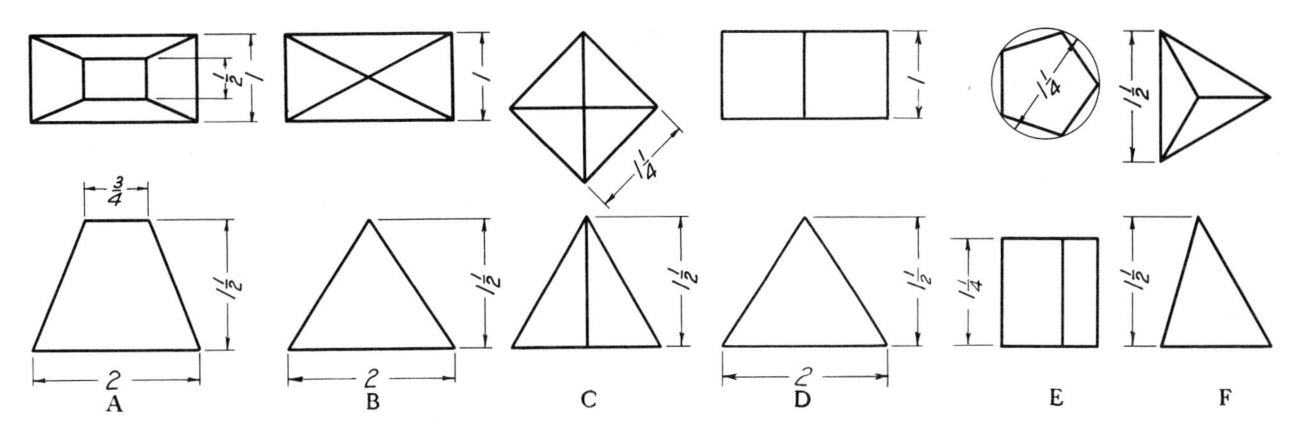

Fig. 7–64 **Follow the directions for Fig. 7–62 for the object assigned.**

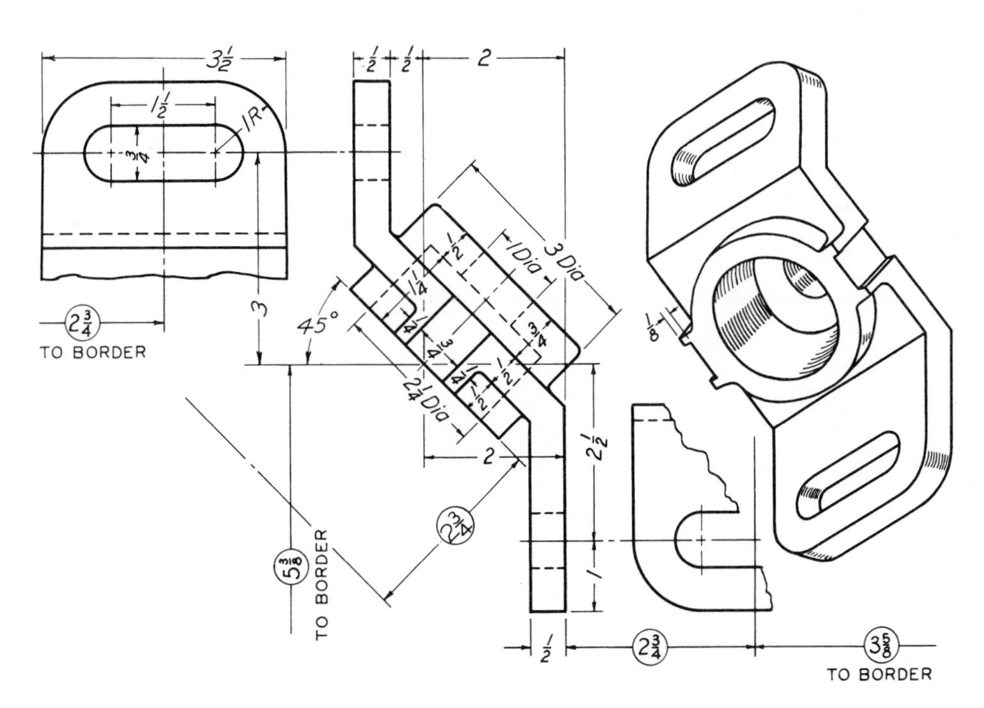

Fig. 7–65 **A picture and a layout for an inclined stop are shown. The complete view in the middle of the space is the right-side view. Draw this side view. Draw the part front view shown in the upper left-hand part of the space. Draw a part rear view on both sides of the vertical centerline in the lower right-hand part of the space. Draw an auxiliary view of the inclined part. Note that this will be a front auxiliary view.**

Chapter 8

Basic Descriptive Geometry

Fig. 8–1 **Geometric space-frame structure. Franklin Park Mall, Toledo, Ohio.** *(Unistrut Corp.)*

GRAPHICS AND MATHEMATICS

The designer who works with an engineering team can help solve problems by producing drawings made up of geometric elements. *Geometric elements* are points, lines, and planes defined according to the rules of geometry. Every structure has a three-dimensional form made up of geometric elements (Fig. 8–1). In order to draw three-dimensional forms, you must understand how points, lines, and planes relate to each other in space to form a certain shape. Problems that you might think need mathematical solutions can often be solved through drawings that make manufacturing and construction possible. *Basic descriptive geometry* is one of the ways a designer thinks about and solves problems. In the eighteenth century a French mathematician, Gaspard Monge, developed a drawing system for solving *spacial* (space) problems related to military structures. Descriptive geometry was brought to the U.S. Military Academy at West Point by Claude Crozet in 1816. The Mongean method was changed over the years. However, its basic principles are still taught in engineering schools throughout the world. In studying descriptive geometry, you develop a reasoning ability that lets you solve problems through drawing.

In this chapter you will learn a way of drawing that lets you analyze all geometric elements. Learning to see geometric elements will let you describe a structure of any shape. Most structures designed by people have been shaped like a rectangle. This is because it is easy to plan a structure with this shape. Figure 8–2 shows the basic geomet-

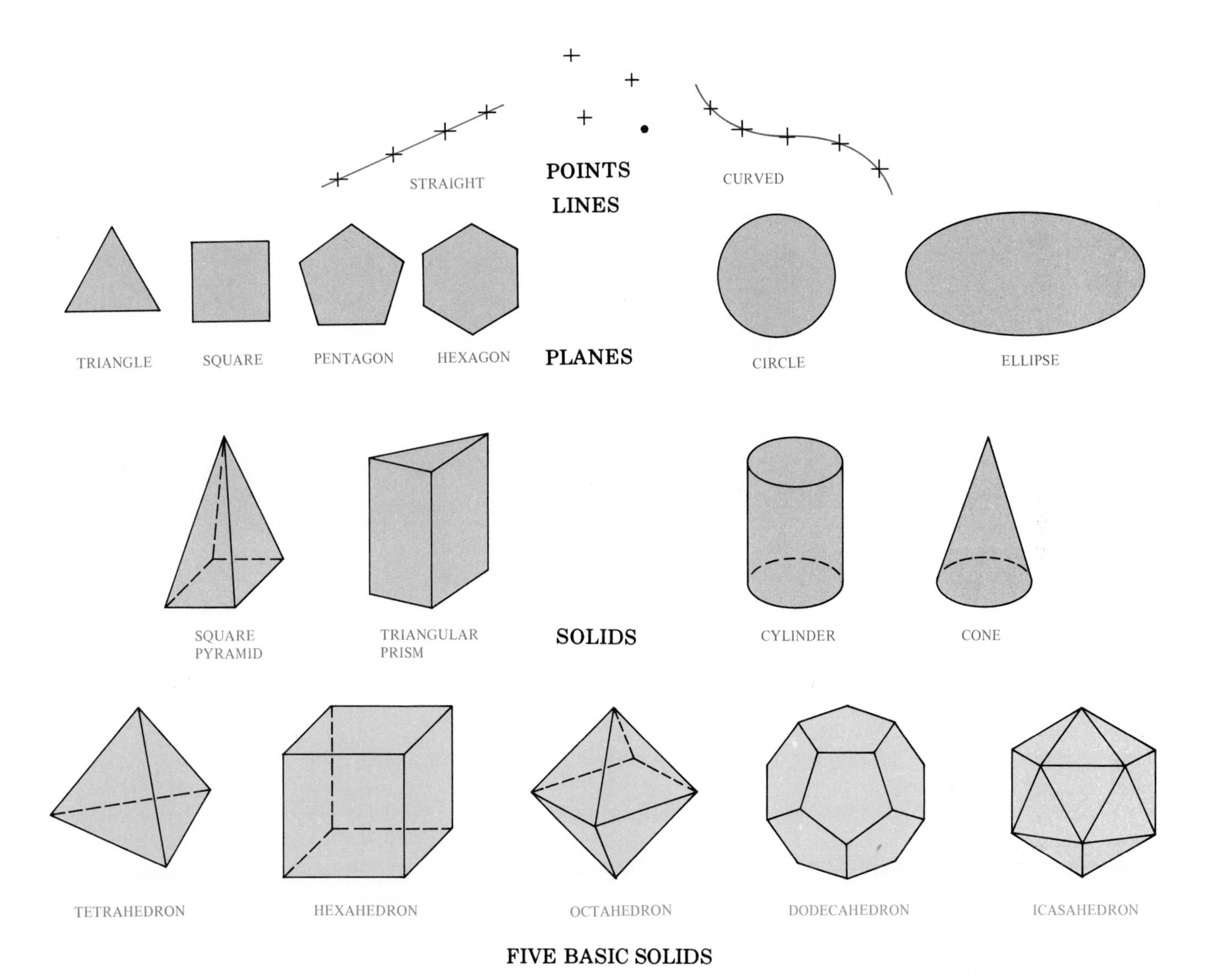

***Fig. 8–2* Basic geometric elements and shapes.**

ric elements. It also shows some of the common geometric features found in engineering designs.

POINTS

A point can be thought of as having an actual physical existence. On a drawing, you can locate a point with a small dot or a small cross. Normally, you identify a point with two or more projections. In Fig. 8–3, at A, the small cross for point number 1 is shown in the front, top, and right-side views. At B, in Fig. 8–3, the regular reference planes are shown in a pictorial view with point 1 projected to all three planes. The reference planes are shown again in Fig. 8–4. Notice that when the three planes are unfolded, a flat two-dimensional surface with fold lines is formed. The fold lines are labeled as shown to indicate that *V* stands for the vertical view, *H* stands for the horizontal, or top, view. *P* stands for the profile, or right-side, view. Points may also stand for the intersection of two lines or the corners on an object.

FIXED POINTS

Figure 8–5 shows a group of points. Points are related to each other by distance and direction. These are measured on the coordi-

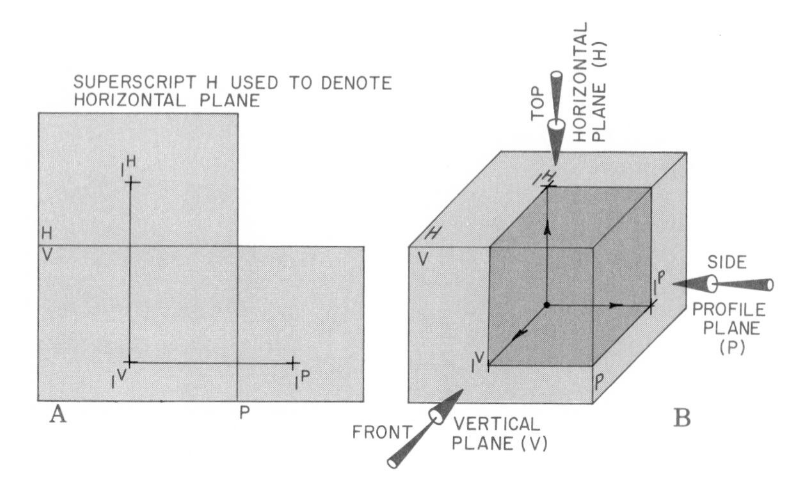

Fig. 8–3 **Locating and identifying a point in space.**

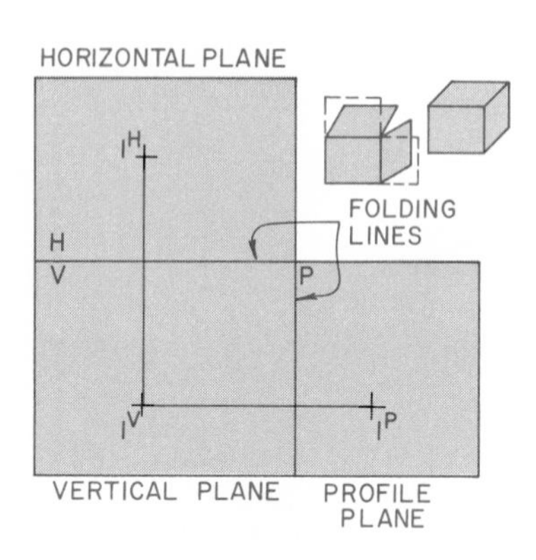

Fig. 8–4 **Points identified on unfolded reference planes.**

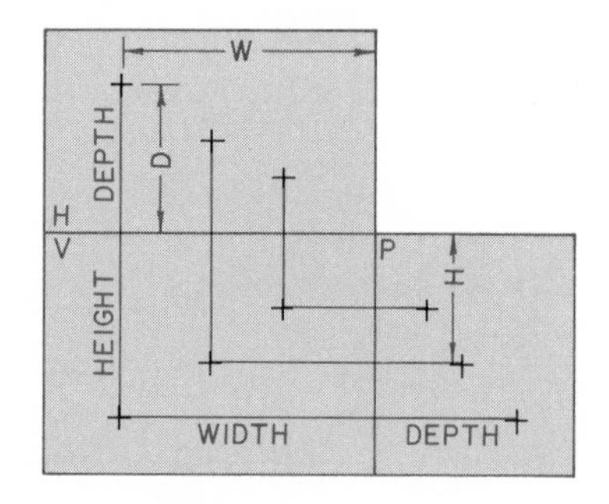

Fig. 8–5 **Points are related to coordinated reference planes.**

nated reference planes. The vertical height relation can be seen in the front and side views. The relative width dimensions can be seen in the front and top views. The relative depth dimensions can be seen in the top and side views. Note that the three basic dimensions are indicated by *H, W,* and *D* (Fig. 8–5).

LINES

If a point moves away from a fixed place, its path forms a line. A line has location, direction, and length. A straight line moves in only one direction. It is easy to draw circular and straight lines. However, you must plot irregular curves very carefully. A straight line can be determined by two points. Or, it can be determined by one point and a fixed direction.

THE BASIC LINES

There are three kinds of straight lines. The kind of line depends on how it relates to the coordinating reference planes.

1. *Normal Lines.* A normal line is one that is *perpendicular* (at right angles) to one of the three reference planes. It will project on that plane as a point (Fig. 8–6 at A, B, and C). If a normal line is parallel to the other two reference planes, as shown in Fig. 8–6 at D, E, and F, it is shown true length (TL), as noted.
2. *Inclined Lines.* An inclined line is slanted in one of the three main reference planes but is parallel to one other. Inclined lines are shown in Fig. 8–7 at A, B, and C. An inclined line is *foreshortened* (not in true size) in two planes. It is shown true length (TL), as noted.
3. *Oblique Lines.* An oblique line appears inclined in all three reference planes (Fig. 8–8). It makes an angle with all three planes. In other words, it is not perpendicular or parallel to any of the three planes. The true length is not shown in any of these views. Also, the angles of direction cannot be measured on the main reference planes.

AUXILIARY REFERENCE PLANE—TRUE LENGTH OF AN OBLIQUE LINE.

A normal line and an inclined line each project parallel to at least one of the main planes of projection. Therefore, a line parallel to a plane of projection shows true length in that projection. An oblique line is not parallel to any of the three main reference planes. Thus, in order to

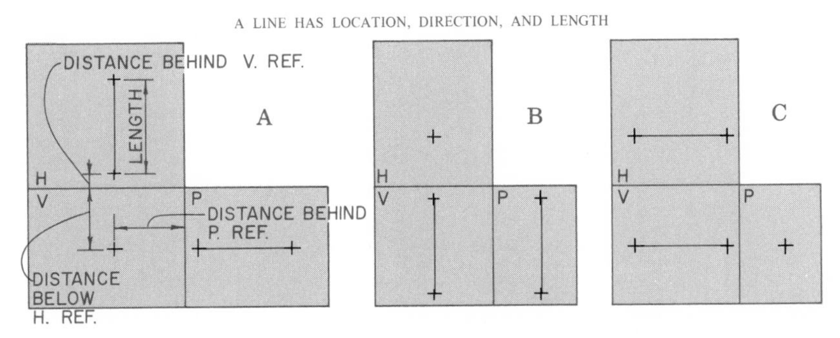

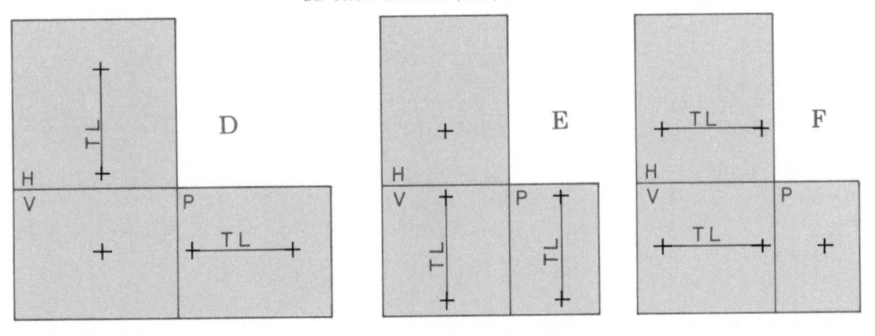

Fig. 8–6 **Normal lines in true length are parallel to two reference planes.**

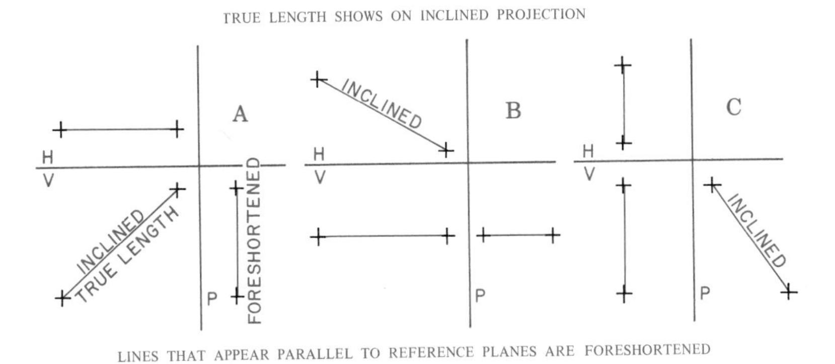

Fig. 8–7 **Inclined lines will be parallel to one reference plane.**

show an oblique line true length, you must place an auxiliary reference plane parallel to the oblique line in any plane. This is shown in Fig. 8–9 at A, B, and C. The auxiliary and regular planes of projection will have the same relationship as any two main planes. First, they must always be perpendicular to each other (see Fig. 8–9 at D). Secondly, they must be measured in relation to the previous plane on which they are related. The true length (TL) is obtained as noted.

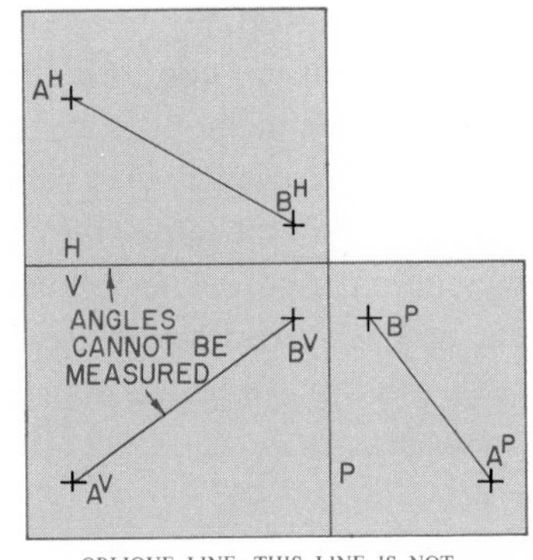

Fig. 8–8 **Oblique lines appear inclined in all projections.**

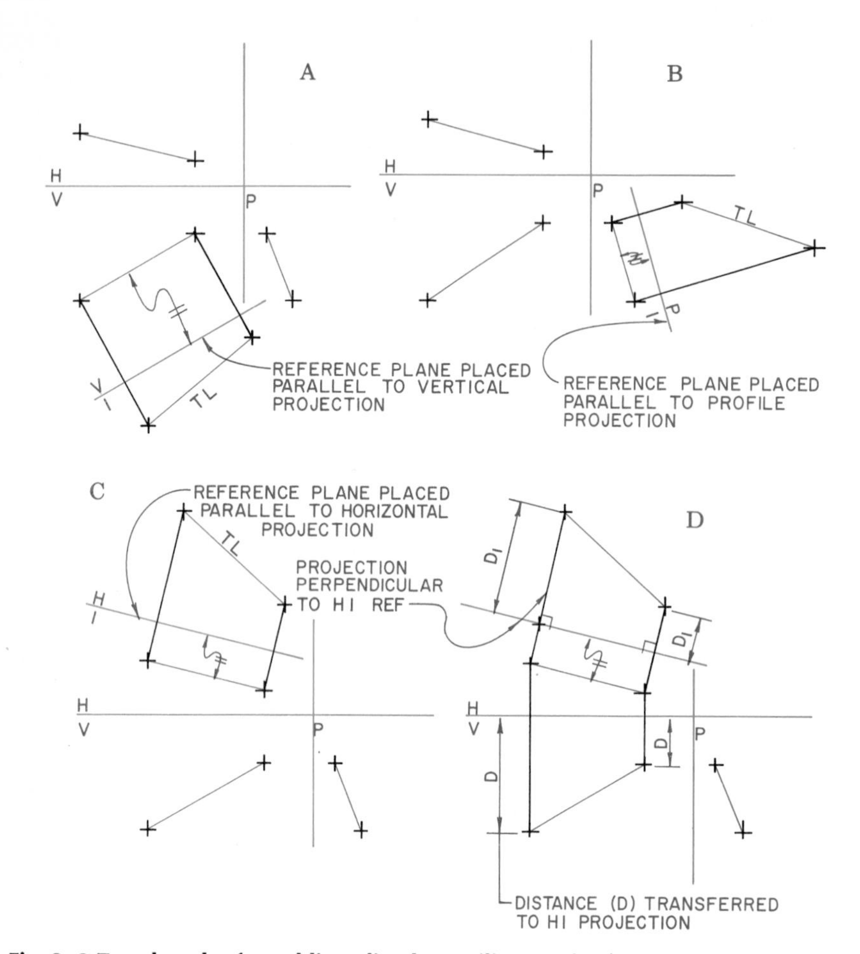

***Fig. 8–9* True length of an oblique line by auxiliary projection.**

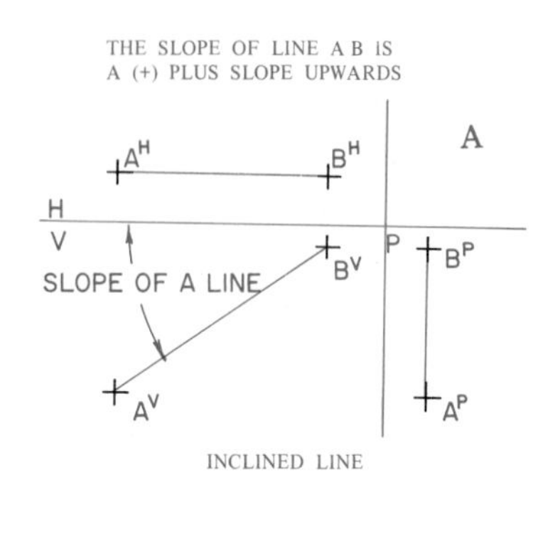

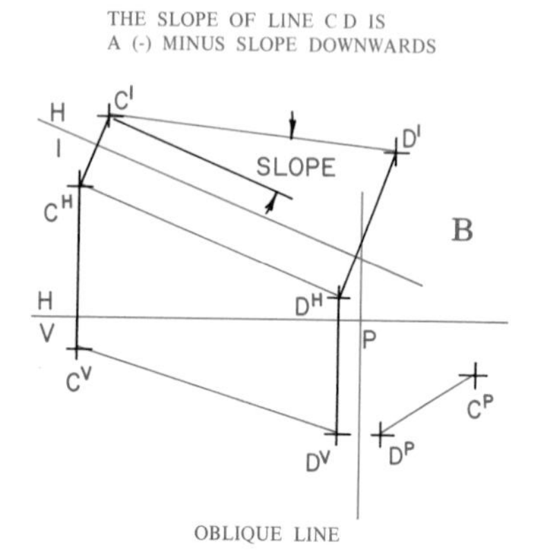

***Fig. 8–10* Slope of a line in the vertical (elevation) projection.**

LINE TERMINOLOGY

It may seem to you that lines drawn on paper mean little and are worth little. However, they reflect real things. In that sense, they are always being used. The following terms related to lines are used in mining, geology, engineering, and navigation.

Slope

A line that makes an angle with the horizontal plane has a *slope* measured in degrees. In Fig. 8–10 at A, slope is shown in the front view when the line is in true length. At B, slope is found for an oblique line in true length in an auxiliary projection perpendicular to a horizontal reference.

Bearing

The angle a line makes in the top view with a north-south line is called its *bearing*. The north-south line is generally vertical. North is at top. Therefore, right is east and left is west. The measurement should be made in the horizontal projection. It should be dimensioned in degrees, as shown in Fig. 8–11 at both A and B.

Azimuth

A measurement that defines the direction of a line off due north is the *azimuth*. The azimuth is always measured off the north-south line in the horizontal plane with clockwise dimensioning, as shown in Fig. 8–11 at C.

Grade

Incline, or *grade,* is measured as a percentage. Figure 8–12 shows the scale for constructing a highway with a +12% grade. The grade rises 3.6 m (12 feet) in every 30 m (100 feet) of horizontal distance.

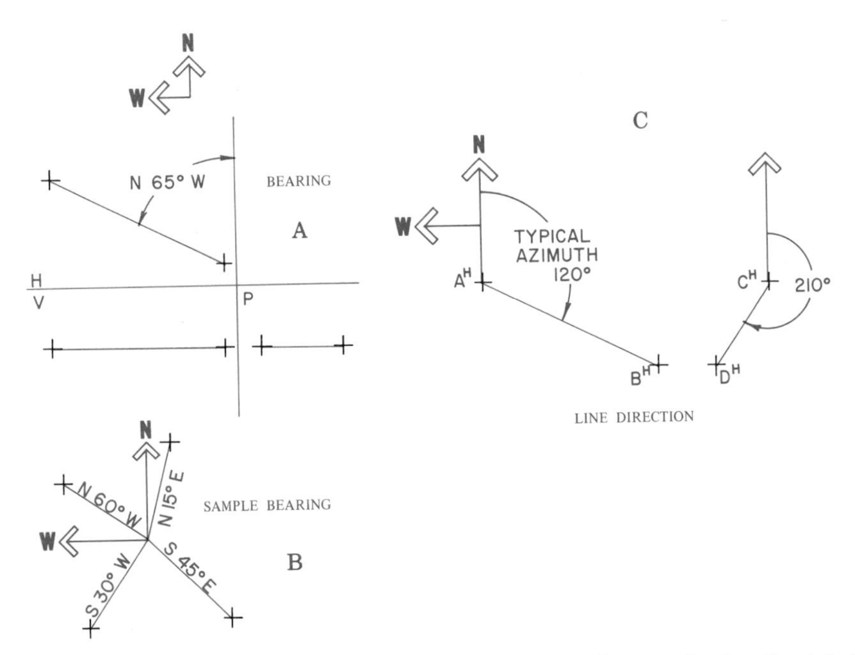

Fig. 8–11 **Bearing in the horizontal projection, bearing readings, and azimuth related to due north.**

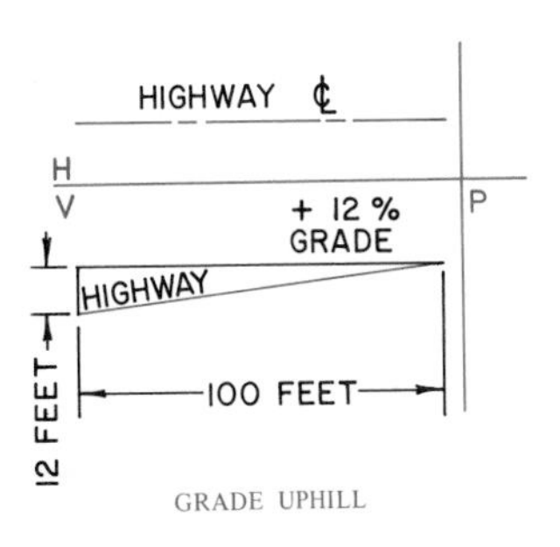

Fig. 8–12 **Grade is measured in the vertical projection.**

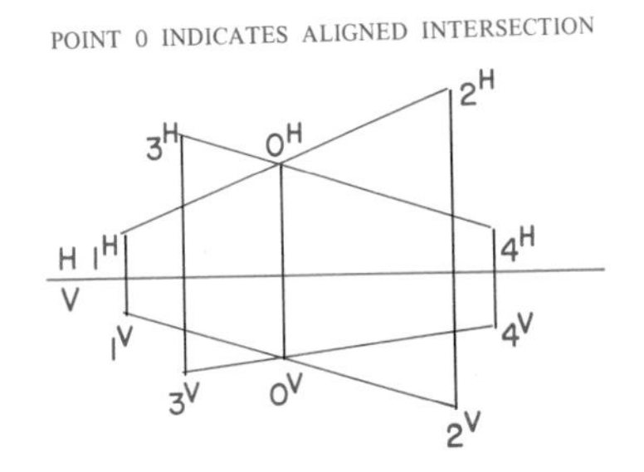

Fig. 8–13 **Intersecting points.**

Fig. 8–14 **Lines in space examined for intersection.**

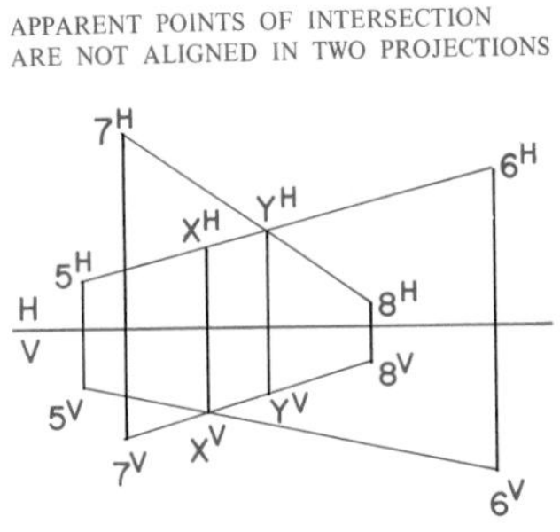

APPARENT POINTS OF INTERSECTION ARE NOT ALIGNED IN TWO PROJECTIONS

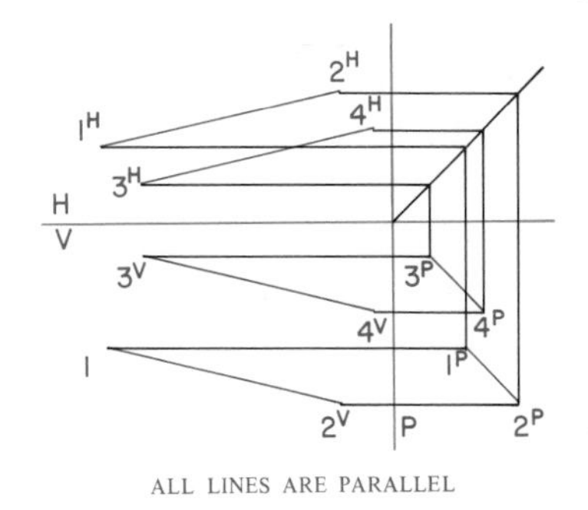

Fig. 8–15 **Lines are parallel when all three of the reference projections are parallel.**

LINES IN SPACE

If two lines intersect, they will have at least one point in common. Figure 8–13 shows the alignment needed to check the point of intersection of two straight lines. In Fig. 8–14, the points of intersection in the H and V projection are not aligned. Thus, the intersection is incomplete. How would the intersection look if completed? (Note that the cross symbol for locating points will no longer be used in the figures.)

Figure 8–15 shows the relationship of parallel lines in a three-view study. All line projections are parallel if they appear parallel in all three reference planes. Note that the lines in Fig. 8–16 seem parallel in the front and top views but are not parallel in the side view.

Perpendicular lines are examined in Fig. 8–17 at A and B. To find out if two lines are perpendicular, and thus have a right angle between them, you must first find the true length of one line. This lets you know if the angle between the lines is actually a right angle. Note that

Fig. 8–16 **Lines in space examined for parallel relationship.**

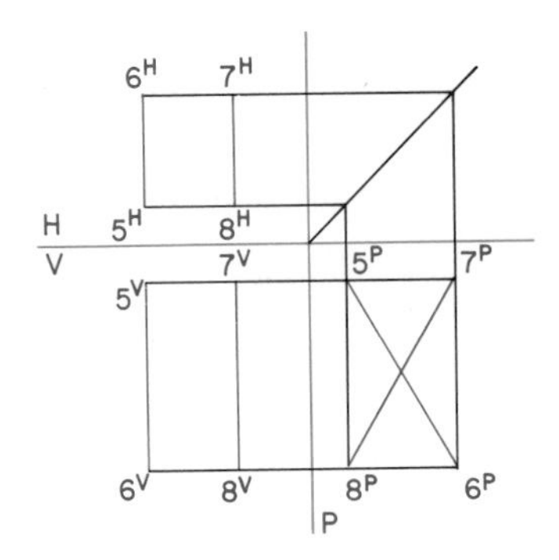

the TL at A indicates that one line is parallel to a main plane of projection. The oblique lines at B are examined in an auxiliary projection to find true length and the right angle.

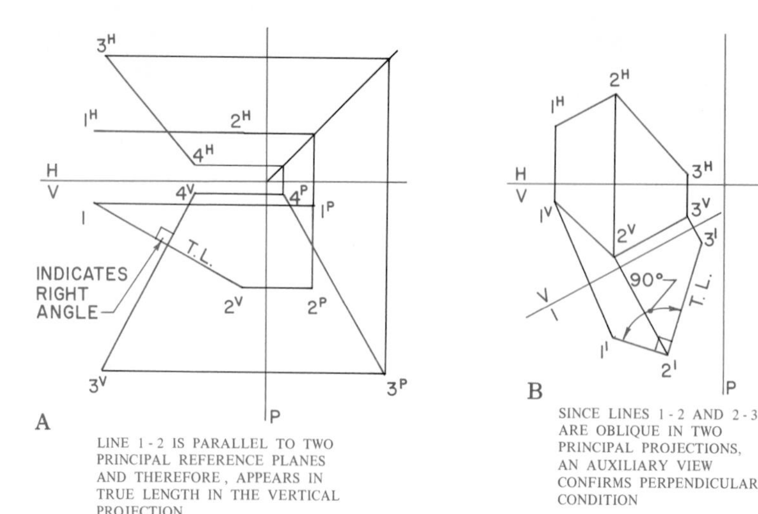

Fig. 8–17 **Perpendicular lines.**

PLANES

If a line moves away from a fixed place, its path forms a plane. In drawings, planes are thought of as having no thickness. They can also be extended as far as you like. A plane can be shown or determined by intersecting lines, two parallel lines, a line and a point, three points, or a triangle.

BASIC PLANES

There are three basic planes. What kind a plane is depends on how it relates to the three main reference planes. Plane one is perpendicular to two of the reference planes and parallel to the third. Plane two is perpendicular to one reference plane and inclined to the other two. Plane three is inclined to all three reference planes.

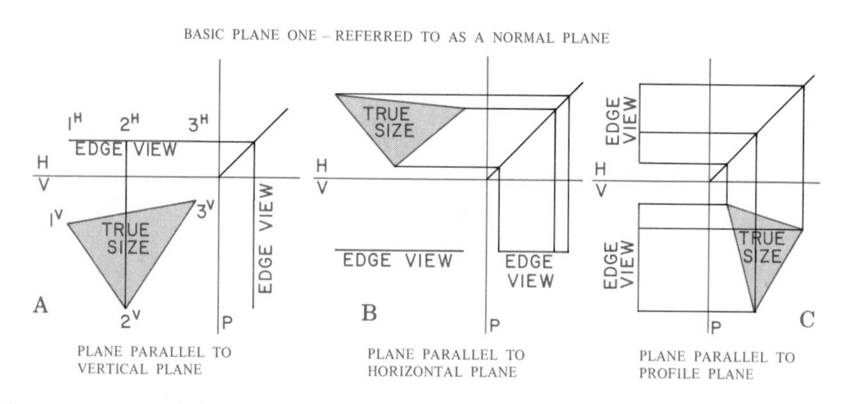

Fig. 8–18 **Normal planes, three examples.**

Basic Plane One

Sometimes this plane is called a *normal plane.* Three similar examples are shown in Fig. 8–18 at A, B, and C. Two of the main reference planes in each example show the edgewise view (a line) of the plane. At A, the plane is parallel to the vertical reference plane and perpendicular to the horizontal and profile planes. At B, the plane is parallel to the horizontal reference plane and perpendicular to the vertical and profile planes. At C, the plane is parallel to the profile reference plane and perpendicular to the vertical and horizontal planes.

Basic Plane Two

This type of plane is also called the *inclined plane.* Three similar examples are shown in Fig. 8–19 at A, B, and C. In one of the main reference planes, the plane shows as a line (edge view). Thus, it is perpendicular to that plane. The other two reference planes show the plane as a foreshortened surface. At A, the inclined plane is perpendicular to the vertical reference plane. It is inclined to the horizontal and profile planes, where it is foreshortened. At B, the inclined plane is perpendicular to the horizontal reference plane, where it shows as a line. The other two reference planes show the plane

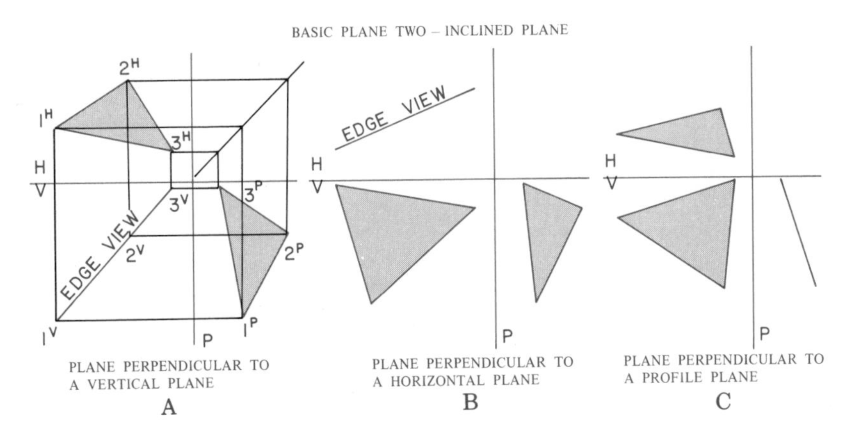

Fig. 8–19 **Inclined planes, three examples.**

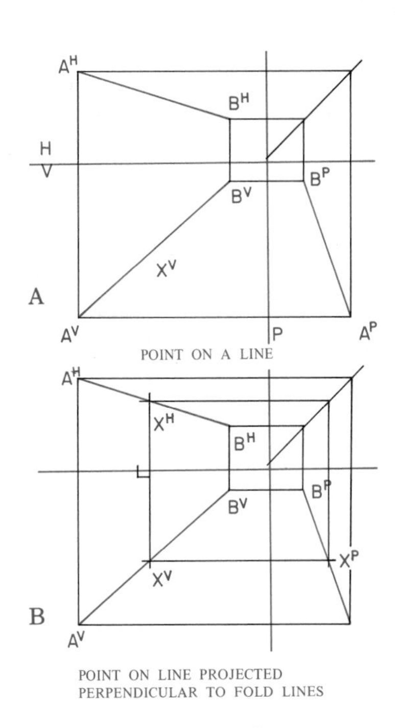

Fig. 8–21 **A point on a line.**

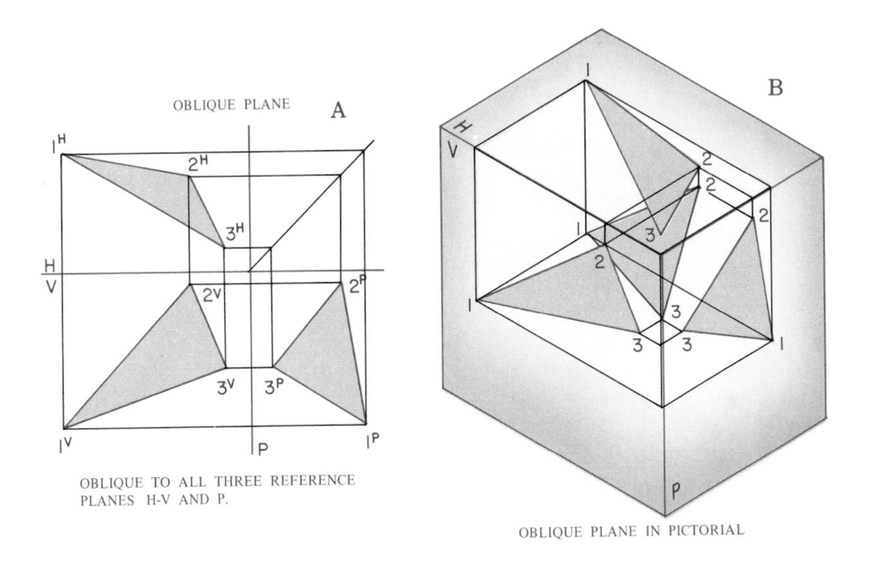

Fig. 8–20 **Oblique plane in three-view projection and pictorial.**

foreshortened. At C, the inclined plane is perpendicular to the profile reference plane, where it shows as a line. The plane shows as a foreshortened surface in the other two reference planes.

Basic Plane Three

This third type of plane is also called an *oblique plane.* An example is shown in Fig. 8–20 at A. The oblique plane is not perpendicular to any of the three main reference planes. It therefore cannot be parallel to any one of the three planes. Thus, it shows as a foreshortened plane in each of the three regular views. Figure 8–20 at B shows the same oblique plane in a pictorial rendering.

A POINT ON A LINE

In Fig. 8–21 at A, the line AB on the vertical plane has a point X. To place the point on the line in the other two reference planes, you must project construction lines perpendicular to the folding lines, as at B. The construction lines are projected across to $A^H B^H$ and $A^P B^P$ to locate point X in the horizontal and profile projections. Straight lines can be extended to new points on either end as needed to solve problems (Fig. 8–22). A point may seem to be on a line in one view. However, another view may show that it is really in front, on top, or in back of the line, as in Fig. 8–23 at A, B, and C.

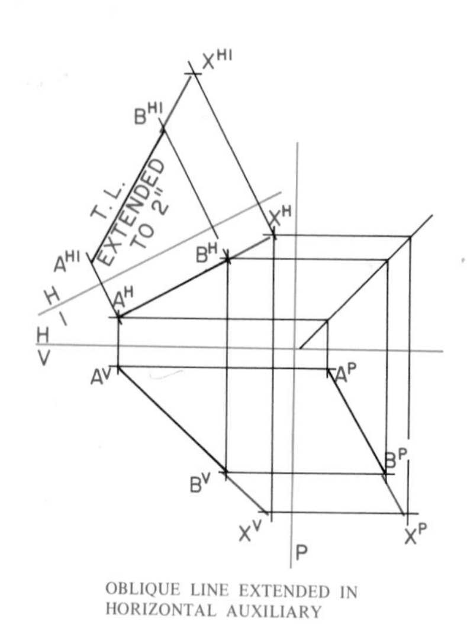

Fig. 8–22 **Straight lines may be extended.**

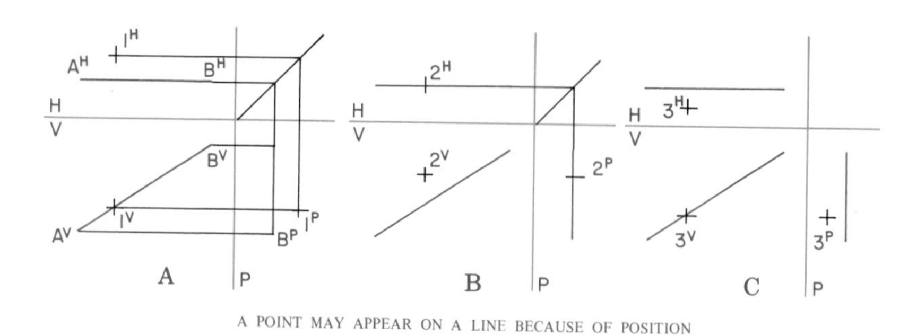

Fig. 8–23 **Point-line relationship.**

A LINE IN A PLANE

A line lies in a plane if it intersects two lines of a plane. It also lies in a plane if it intersects one line and is parallel to another line of that plane. Figure 8–24 at A, B, C, and D shows how lines can be added to planes. At A, the line *RS* must be a part of the plane *ABC,* since *R* is on line *AB* and *S* is on *BC*. You know that line *RS* is an oblique line because it is not parallel to a main plane of projection and is clearly not perpendicular to the reference planes.

At B, a horizontal line *MN* is constructed in the vertical projection of plane *ABC*. This line is called a *level line.* Projecting *MN* to the right lines in the horizontal projection shows that it is an inclined line. The top view will show the true length (TL).

At C, a line *XY* is constructed parallel to the HV reference line in

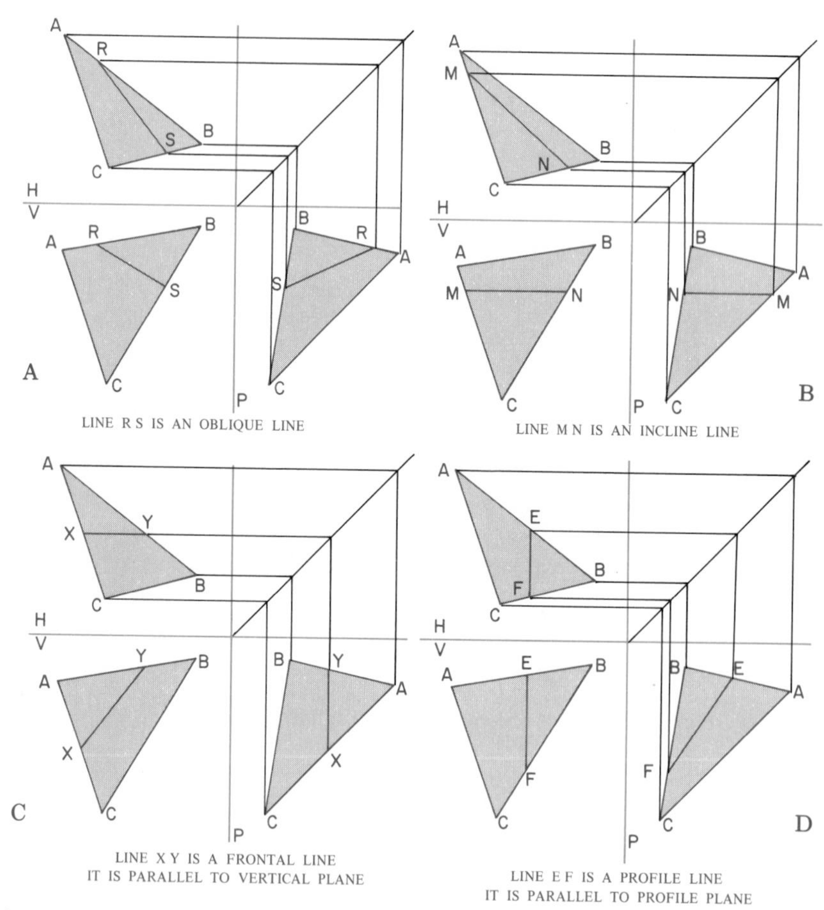

Fig. 8–24 **A line in a plane.**

the horizontal reference plane. Projected into the vertical plane, it shows as an inclined line. It will be in true length (TL). This line is called a *frontal line,* since it is parallel to the vertical plane.

At D, a vertical line *EF* is constructed within plane *ABC*. It is parallel to the profile reference plane. The line *EF* projected to the profile reference shows in true length. The line EF is called a *profile line.*

LOCATING A POINT IN A PLANE

A point can be located in a plane by adding a line containing the point to the plane. Figure 8–25 at A shows that a point O appears within the plane ABC. At Fig. 8–25 at B, the line AX containing point O is projected. The line AX at Fig. 8–25 at C is projected to ABC in the horizontal reference plane. Point O is located on the line by drawing a vertical projection to line AX in the horizontal reference plane.

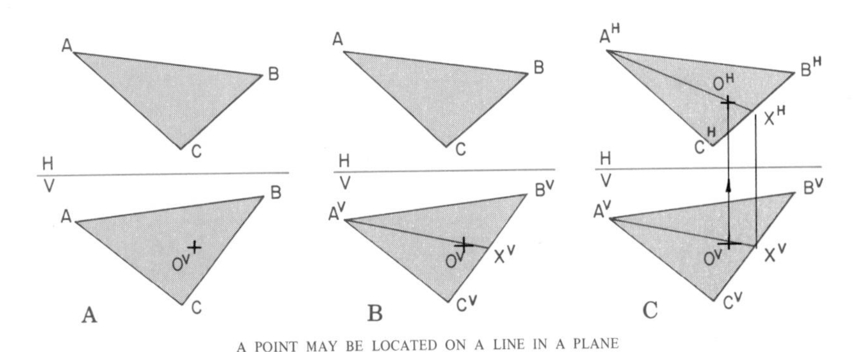

***Fig. 8–25* Locating a point in a plane.**

A POINT VIEW OF A LINE

If a line is perpendicular to a reference plane, it will project as a point on that plane. In Fig. 8–26 at A, the line AB is parallel to two main reference planes. It therefore shows as a point in the third vertical reference plane. At B and C, the same conditions exist. The line projects as a point in the horizontal (B) and profile (C) planes.

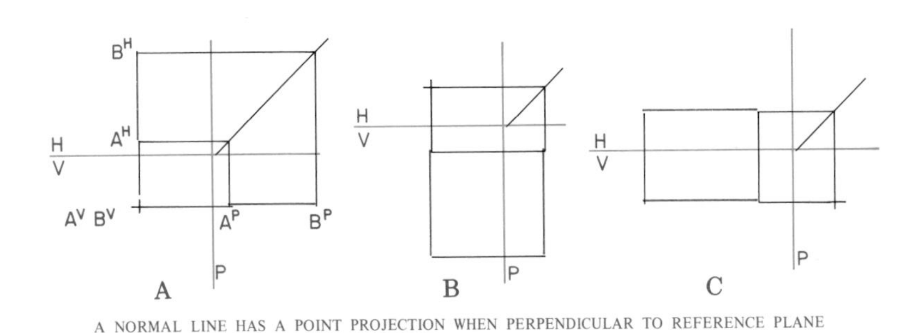

***Fig. 8–26* A point view of a line.**

When a line is parallel to only one main reference plane and inclined to the other two, as in Fig. 8–27, the point is projected by auxiliary projection. As shown in Fig. 8–28, a reference plane is placed perpendicular to the inclined line at a distance you choose. It is labeled H/1. The distance D is transferred as shown for a vertical or a horizontal auxiliary projection.

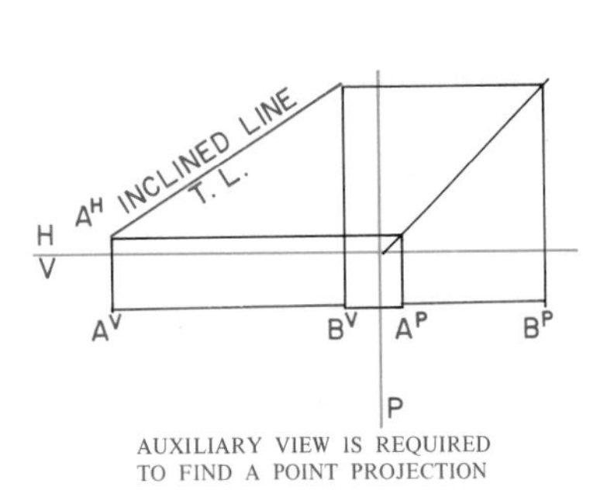

***Fig. 8–27* Point projection obtained by auxiliary projection.**

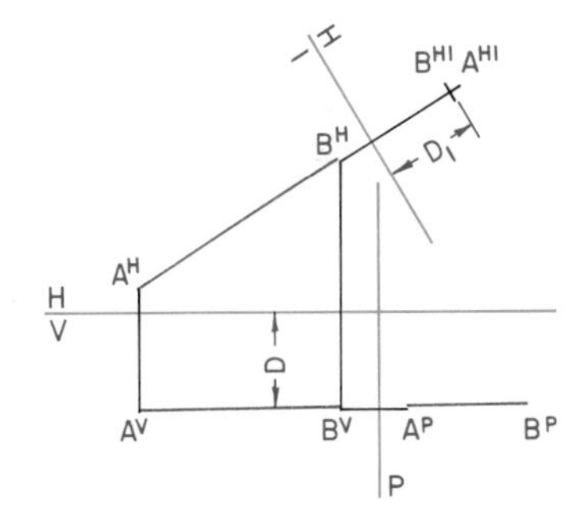

***Fig. 8–28* Transferring a line to the auxiliary projection.**

If a line seems inclined in all three reference planes (an oblique line), the point can be projected by using two auxiliary projections. As shown in Fig. 8–29 at A, set up the first auxiliary reference plane parallel to the oblique inclined line. Then find the true length. The second auxiliary reference plane is placed perpendicular to the true-length line of the first auxiliary. In Fig. 8–29 at B, the point projection is located by transferring the distance X.

PARALLEL LINES

Point projection is one way to show the true distance between two parallel lines. In Fig. 8–30, the

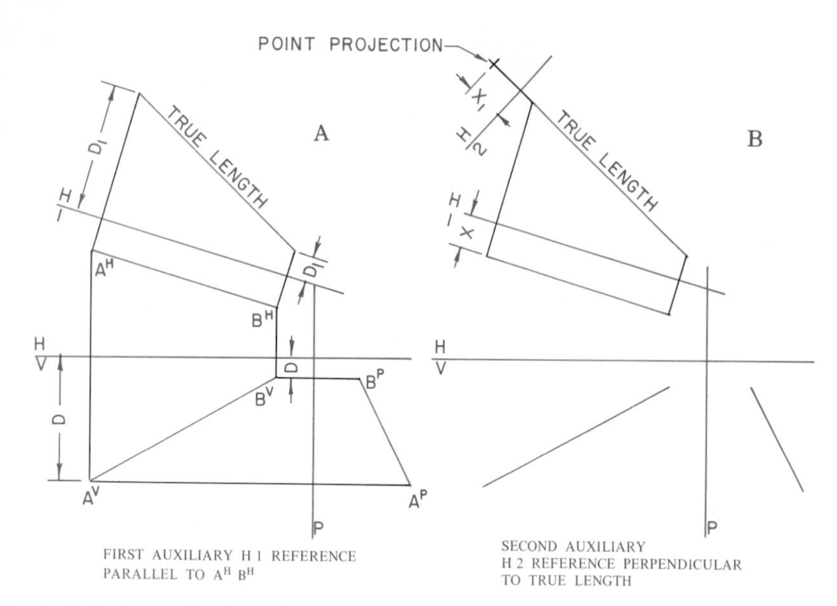

Fig. 8–29 **Point projection of an oblique line.**

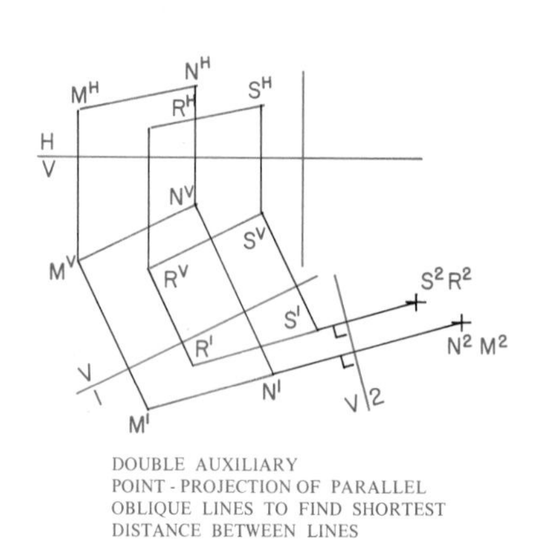

Fig. 8–30 **Distance between parallel lines.**

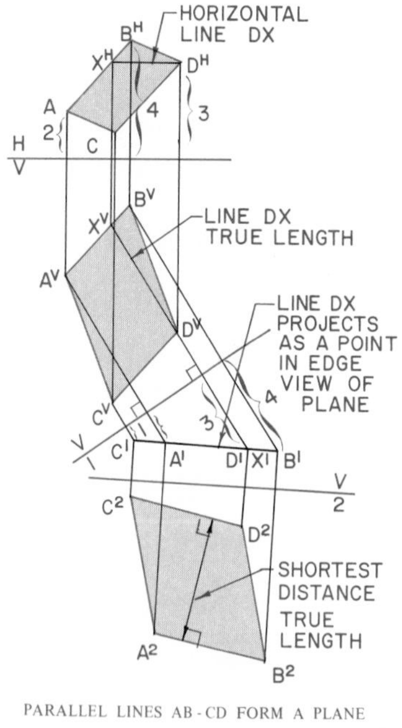

PARALLEL LINES AB - CD FORM A PLANE

Fig. 8–31 **Distance between lines forming a plane.**

parallel lines *MN* and *RS* are considered oblique. Two auxiliary projections are needed to find the point projections. The first auxiliary reference plane V/1 is parallel to *MN* and *RS*. In it, lines *MN* and *RS* are shown true length. The second auxiliary reference plane V/2 is perpendicular to the true-length lines in the first auxiliary. The distance between the point projections of the lines is a true distance.

A second way of finding the shortest distance between two parallel lines is shown in Fig. 8–31. Lines *AB* and *CD* are thought of as parts of a plane. Connect the points *A, B, C,* and *D* to form a plane. Draw a horizontal line in the top view *DX*. Project the point *X* into the vertical view. Draw the line *DX* in the vertical plane. Draw the first reference plane V/1 perpendicular to *DX* in the vertical view. The edge view of the plane *ABCD* is found by transferring distances 1, 2, 3, and 4, as shown. In the second auxiliary V/2, the true lengths of *AB* and *CD* show. The plane formed is in true size. The true distance between the lines is measured perpendicularly from *AB* to *CD,* as shown.

POINT–LINE RELATIONS

To find the shortest distance from a point to a line, you project the line as a point. Point *A* and oblique line *CD* in Fig. 8–32 are projected into the first auxiliary projection H/1. In H/1, the true length of *CD* is labeled (TL). The second auxiliary reference plane H/2 is placed perpendicular to line *CD*. Line *CD* is projected as a point in this plane. As shown, the distance between points in this projection is true length.

SHORTEST DISTANCE BETWEEN SKEW LINES

Skew lines *AB* and *CD* in Fig. 8–33 are not parallel and do not intersect. They are basically oblique. They show inclined in all

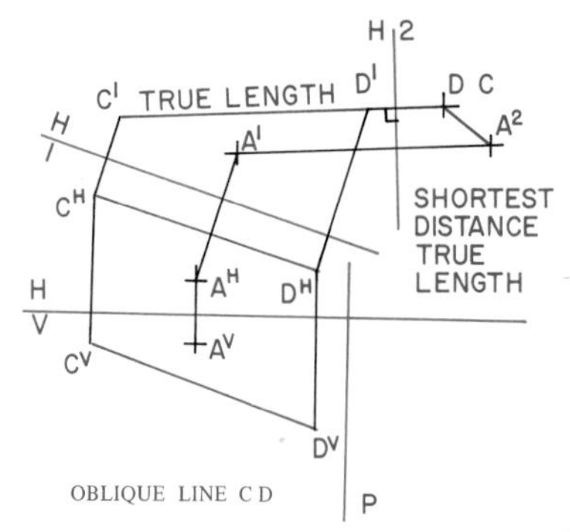

Fig. 8–32 **Distance from a point to a line.**

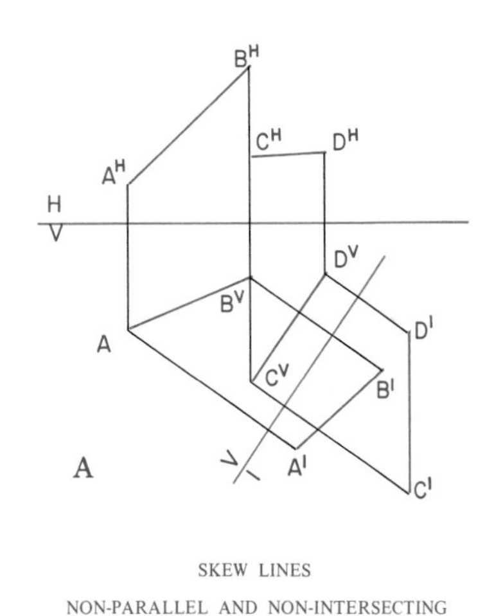

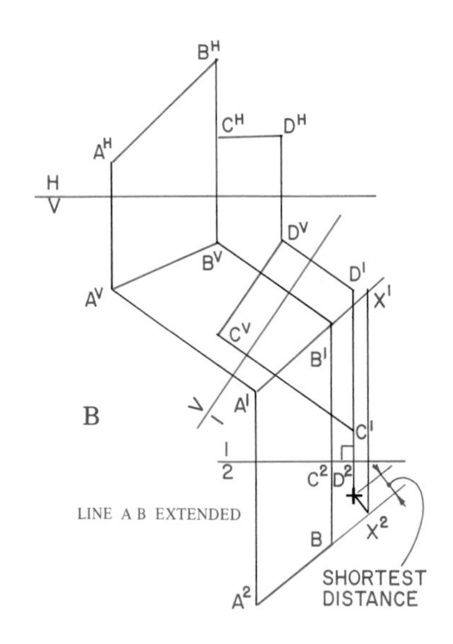

Fig. 8–33 **Distance between skew lines.**

main views. The shortest distance is a perpendicular line between the point view of one of the lines and the other line as shown in Fig. 8–33 at B. First, you find the true length of one of the lines, *CD,* in the first auxiliary. You do this by placing a V/1 reference line parallel to line *CD*. The second auxiliary reference 1/2 is placed perpendicular to the true length of line *CD*. The point projection of line *CD* is found as shown. Extend line *AB* as shown. Construct a perpendicular line from the point projection of *CD* to line *AB*. The perpendicular line intersects *AB* extended at point *X*. You can transfer the intersecting projection back to the first auxiliary, as shown on the extension of line *AB*.

THE TRUE SIZE OF AN INCLINED PLANE

In Fig. 8–34, the plane *ABC* shows as an edge view in the top view. The auxiliary reference plane H/1 is placed conveniently parallel to the edge view. Perpendicular projections are made. The distances *XY* and *Z* are transferred as shown. This lets you find the true size of the plane in the first auxiliary projection.

THE TRUE SIZE OF AN OBLIQUE PLANE

If the plane *ABC* in Fig. 8–35 at A is projected on a plane perpendicular to any line in the figure, it will show an edge view in the first auxiliary. In the top view, a line *BX* is drawn parallel to the reference plane. Reference line V/1 is placed

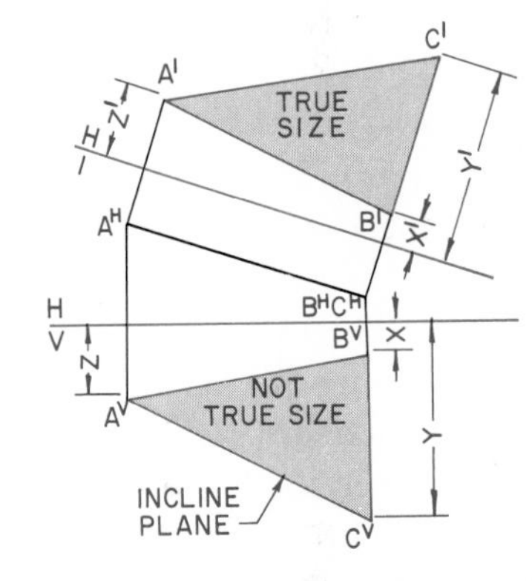

Fig. 8–34 **True size of an inclined plane.**

Fig. 8–35 **True size of an oblique plane.**

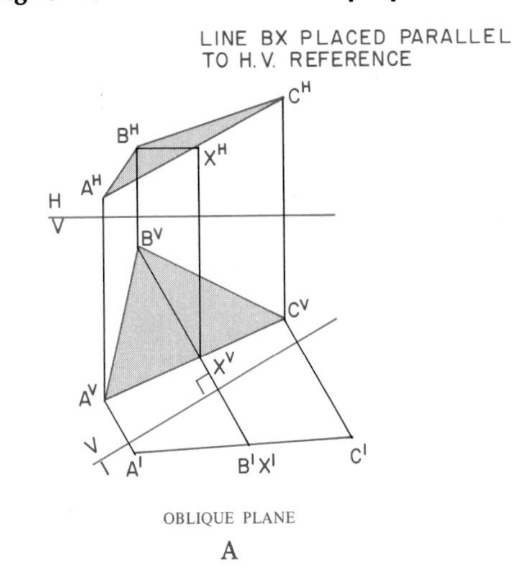

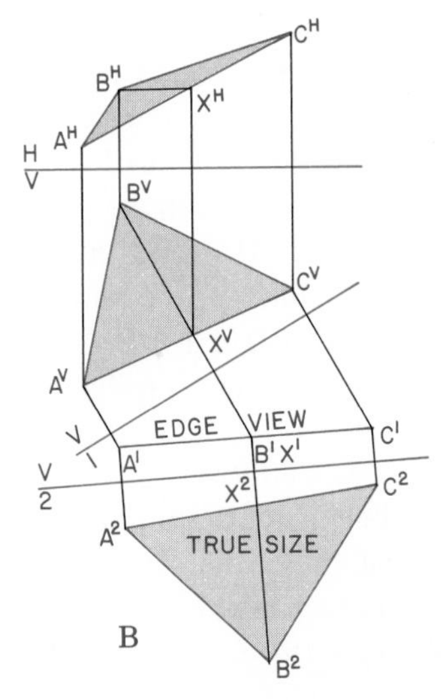

perpendicular to the front view of *BX*. The front view of *BX* is now projected into a point projection in the first auxiliary. The point projection is in the edge view of plane *ABC,* as shown. In Fig. 8–35 at B, the second auxiliary reference line V/2 is placed parallel to the edge view. The projection of plane *ABC* in the secondary auxiliary shows the true size.

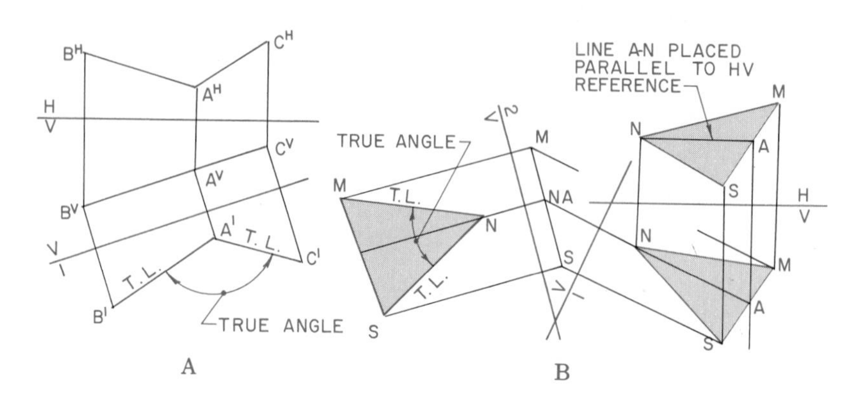

Fig. 8–36 True angle between lines.

TRUE ANGLES BETWEEN LINES

If two lines show in true length, the angle between them will appear in its true value. In Fig. 8–36 at A, the two lines show as an inclined plane. This is because the vertical view shows that lines *AB* and *AC* *coincide* (lie in a single line). The V/1 auxiliary reference is placed parallel to the two lines in the vertical view. The two lines are drawn true length in the auxiliary view. The first auxiliary shows the true angle between the lines.

The oblique condition of lines *MN* and *NS* does not show in an edge-wise view. Fig. 8–36 at B shows how you can solve the problem with two auxiliary planes. The first reference plane is perpendicular to the plane formed by the lines *NA*. The second reference plane is parallel to the first auxiliary view. That is, it is parallel to the edge view of the lines *MN* and *NS.* The second auxiliary view shows *MN* and *NS* in true length. It also shows the true angle between them.

PIERCING POINTS

If a line is not in, or parallel to, a plane, it must intersect the plane. The point of intersection, which is common to the plane and the line, is called a *piercing point.* The line can be thought of as piercing the plane.

EDGE–VIEW SYSTEM

The edge view of a plane contains all the points in the plane. Therefore, a line crossing the edge view will show the point where the line pierces the plane. If a line lies in a plane or is parallel to it, it cannot intersect the plane. In Fig. 8–37, the straight line is neither in the plane nor parallel to the plane. It intersects the plane at a point common to both. The edge view of plane *ABC* is shown in the vertical plane. The line *RS* in the horizontal plane pierces the plane at point *P* when projected to the vertical plane. If you look at line *RS* closely in the vertical projection, you will see that element *A* of the triangle is lower than point *R* of the piercing line. Therefore, the dashed portion of line *RS* is invisible.

You can solve the problem of the piercing point of a line *MN* intersecting an oblique plane *ABC* with the edge-view system, as shown in Fig. 8–38. The first auxiliary will determine the edge view of plane

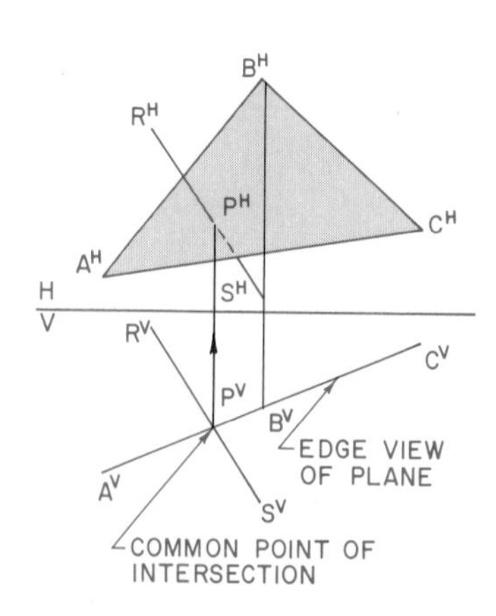

Fig. 8–37 Piercing-point-edge-view system.

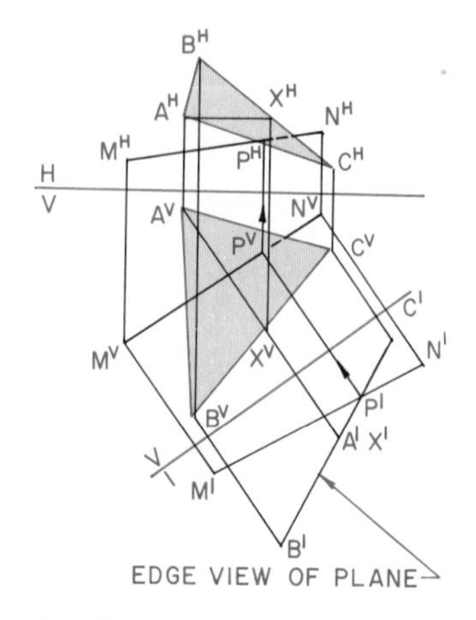

Fig. 8–38 Piercing point—a line with an oblique plane.

ABC. The piercing point *P* of line *MN* can then be carried back to the vertical and horizontal with the projections shown by arrows.

CUTTING-PLANE SYSTEM

When a line *RS* intersects an oblique plane *ABC*, a cutting plane containing the line will determine other working points, as shown in Fig. 8–39. The cutting plane seems to intersect the triangle in the front view at points 3 and 4. Points 3 and 4 are projected to the top view on lines *AC* and *BC*. When the points are connected across the plane, the piercing point *P* is found. Point *P* is projected to the front view. Since you now know what parts of line *RS* are not visible, you can add hidden lines. Figure 8–40 shows the cutting plane in the horizontal projection intersecting line *AB* at point 5 and line *AC* at point 6. Can the visibility of line *MN* be determined by inspection?

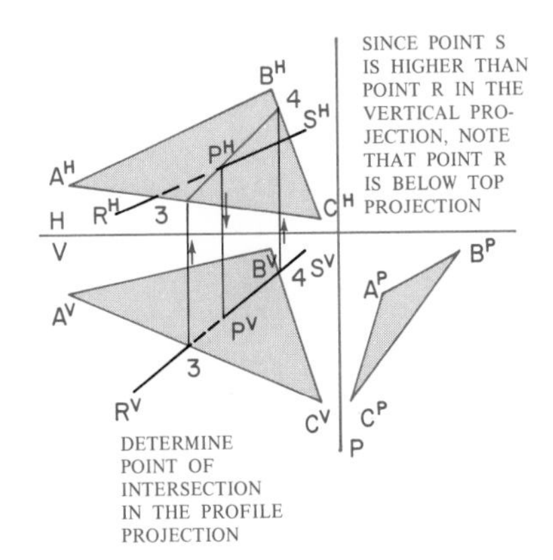

Fig. 8–39 **Piercing-point-cutting-plane system.**

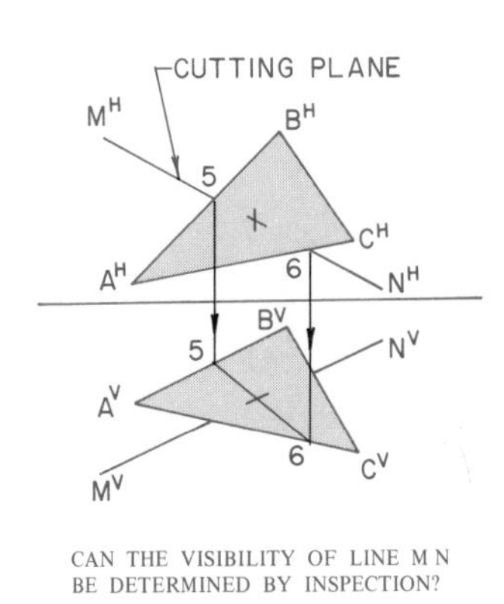

Fig. 8–40 **Piercing point developed.**

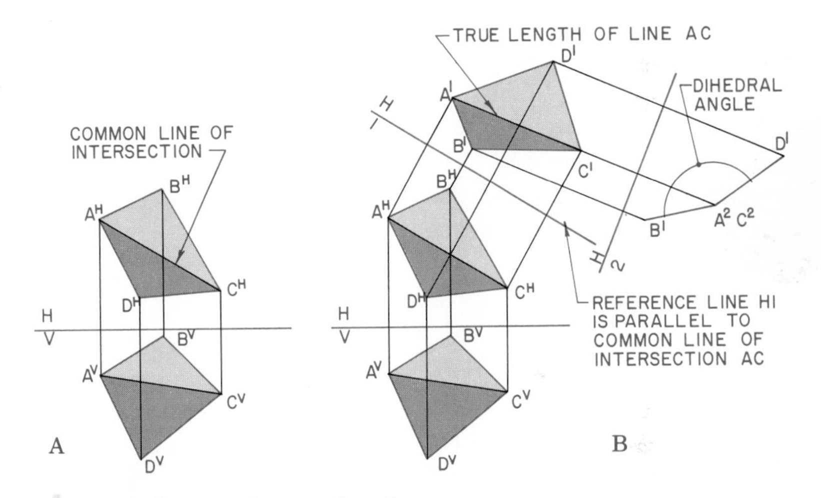

Fig. 8–41 **Angle between intersecting planes.**

ANGLE BETWEEN INTERSECTING PLANES

A dihedral angle is formed when planes intersect. Two planes that intersect have a straight line in common. The dihedral angle formed can only be measured perpendicular to the line of intersection. When a point view of the line of intersection is found, the planes are shown as edges. The angle between the planes will be true in size. In Fig. 8–41 at A, the planes *ABC* and *ACD* have a common line of intersection, *AC*. The first auxiliary projection H/1 in Fig. 8–41, at B, is taken in the horizontal plane of the line *AC*. The H/1 auxiliary lets *AC* be drawn in true length (TL). In the second auxiliary, the point projection of the true-length line *AC* also shows the two given planes as edges. The true angle is measured in the second auxiliary, as shown in Fig. 8–41 at B.

ANGLE BETWEEN A LINE AND A PLANE

A view that shows a plane in the edge view along with a true-length line will also show the true angle in between.

Plane Method

In Fig. 8–42, oblique line *XY* intersects the oblique plane *ABC*. You first find the edge view of the oblique plane *ABC*. The first auxiliary is found by placing H/1 perpendicular to a true-length line in the horizontal projection. Note that line *XY* is not shown true length in this first auxiliary view. After you project the edge view in H/1, you place the reference plane H/2 parallel to plane *ABC*. In the

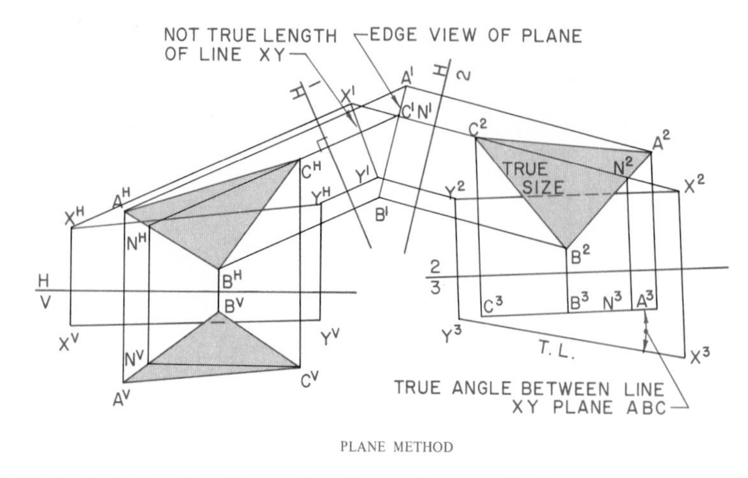

Fig. 8–42 **Angle between a line and a plane (plane method).**

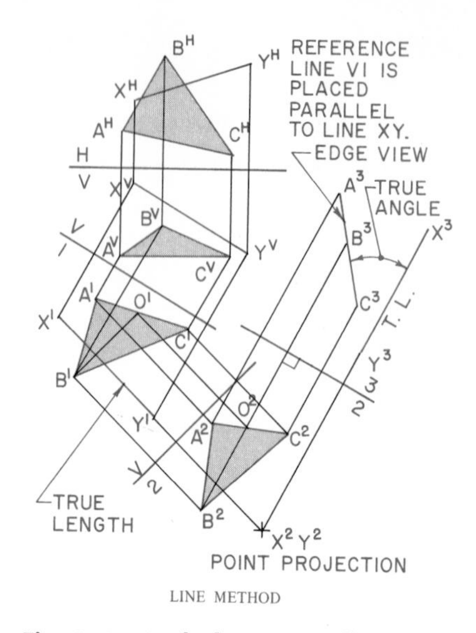

Fig. 8–43 **Angle between a line and a plane (line method).**

second auxiliary, the true size of plane *ABC* is plotted. In the third auxiliary, 2/3, the reference line is placed parallel to line *XY*. The new edge view of the plane and the true-length line form the true angle.

Line Method

In Fig. 8–43, oblique plane *ABC* and the oblique line *XY* intersect in the top view. The first auxiliary V/1 is placed parallel to the line *XY*. The true length of *XY* is thus found. In the second auxiliary, a reference V/2 is placed perpendicular to the true length of line *XY*. The point projection of *XY* is thus found. Finally, in the third auxiliary, the 2/3 reference line is set perpendicular to the true length of B^2O^2. Plane *ABC* is thereby shown as an edge view. The intersection of the line and plane shows the true size of the angle.

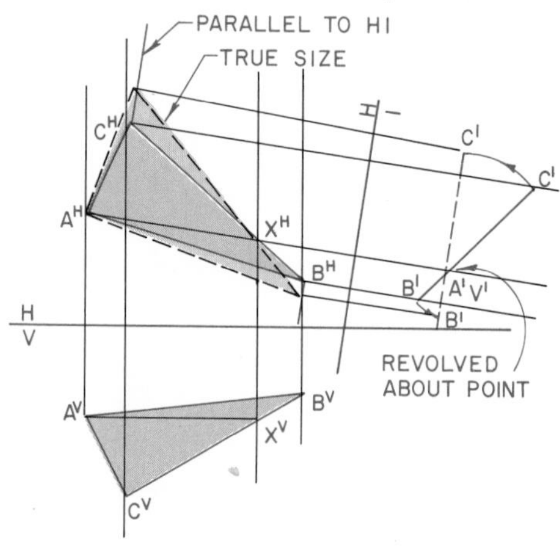

Fig. 8–44 **True size of an oblique plane by revolution.**

REVOLUTION

We read about the rule of revolution in Chapter 7. You can often solve basic problems by changing the position of an object so that the new view can show needed information. Descriptive geometry problems can be solved by revolving an object.

TRUE SIZE OF AN OBLIQUE PLANE BY REVOLUTION

In Fig. 8–44, it is clear that the plane *ABC* is inclined to all main planes. A line *AX* is placed in the plane *ABC* parallel to the horizontal reference line. Line *AX* projected to the top view appears as an inclined line in true length. The first auxiliary projected in the horizontal has H/1 reference perpendicular to line *AX*. The point view of *AX* is found in the first auxiliary. Line *ABC* appears as an edge view, with *X* within. At point *AX*, the edge view $B'A'C'$ is revolved (dashed line) so as to appear parallel to reference line H/1. Projecting points *B* and *C* to the horizontal projection allows the true size of plane *ABC* to be drawn in the top view, as shown.

Almost all problems in descriptive geometry can be worked out by using auxiliary planes. You can solve problems (Fig. 8–45) by knowing how to find

1. the true length of a line,
2. the point projection of a line,
3. the edge view of a plane, and
4. the true size of a plane figure.

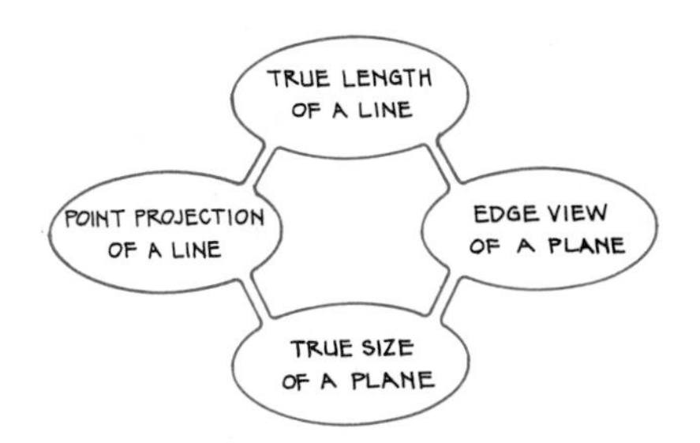

Fig. 8–45 **Four basic approaches for solving problems.**

Review

1. Any given shape may be formed by which three basic geometric elements?

2. What is the basic shape of most structures designed by people? Why is this form common?

3. Do you believe that basic descriptive geometry is one of the designer's ways of thinking and of solving problems? Why?

4. How is a point located in a drawing?

5. What are the three characteristics of a line?

6. Name the three basic lines used in descriptive geometry.

7. How does a normal line relate to the three main planes of projection?

8. How does an inclined line relate to the three main planes of projection?

9. In how many of the main planes of projection will an inclined line show in true length?

10. How does an oblique line relate to the three main planes of projection?

11. How many auxiliary projections are needed to find the point projection of an oblique line?

12. Name the three basic planes.

13. What are the major characteristics of a plane?

14. How can you find if two lines really intersect?

15. What is the difference between an inclined and an oblique plane?

16. How can a line be added to a plane in space?

17. How can you determine the distance between two parallel inclined lines?

18. How is the auxiliary reference plane placed in finding the true size of an inclined line?

19. How can you find the true angle between intersecting lines?

Problems

Figures 8–46 through 8–70 **Using the 3 mm (1/8″) grid, determine the relationship of lines and points to the HVP planes and lay out the following problems. Letter all points 3 mm (1/8″) high, using the proper superscript.**

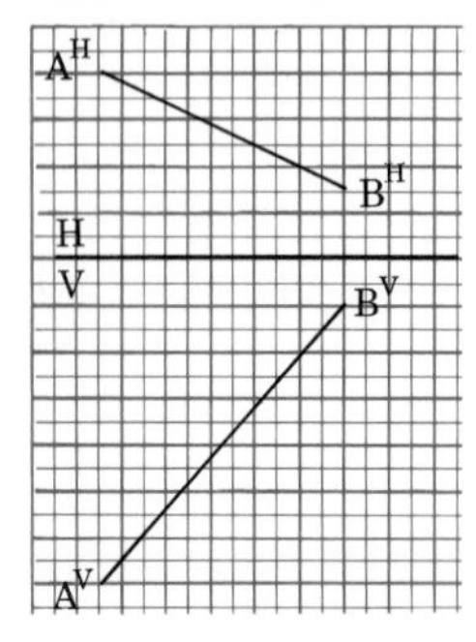

Fig. 8–46 **Find the true length of line *AB*.**

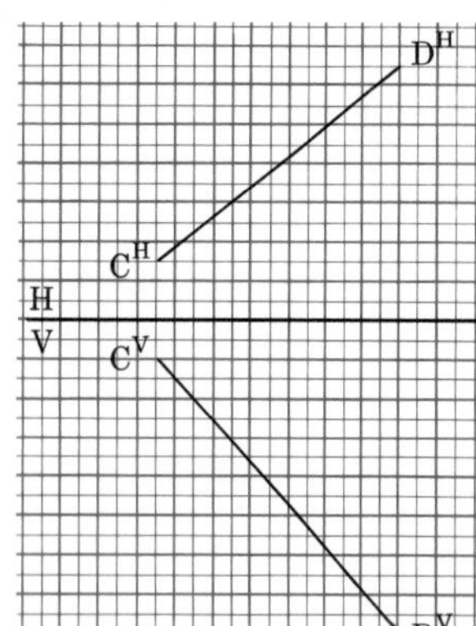

Fig. 8–47 **Determine the true length and slope of line *CD*.**

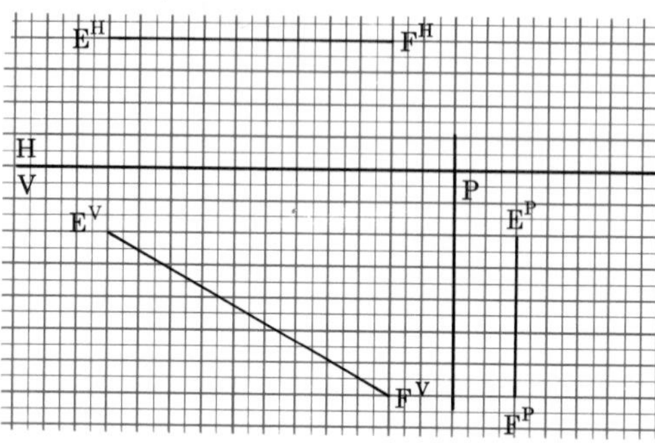

Fig. 8–48 **Line *EF* is the centerline of a pipeline. Scale: 1:500 (1″ = 40′-0″). Locate line *X* 6 m (20′) below *E* on all three projections. Determine the grade of line *EF*. Determine the true distance from *E* to *X*.**

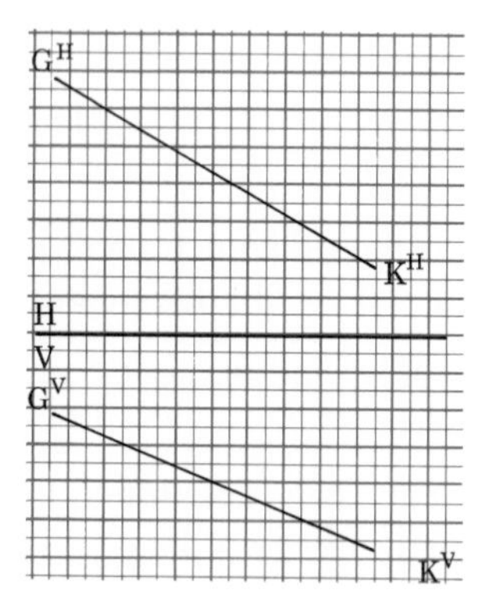

Fig. 8–49 **Determine the point projection of line *GK*.**

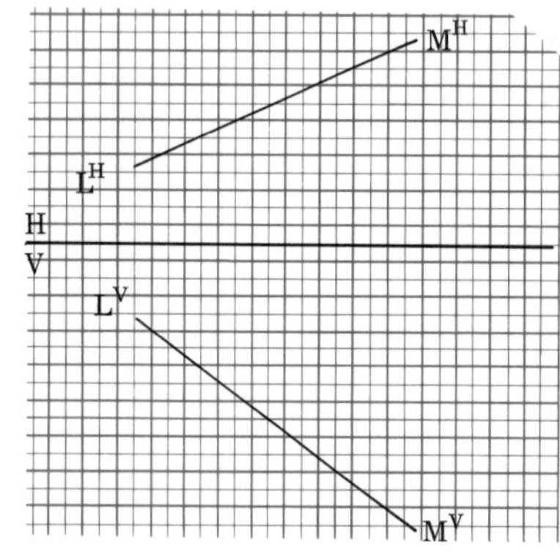

Fig. 8–50 **Determine the angle *LM* makes with the vertical plane. What is the bearing?**

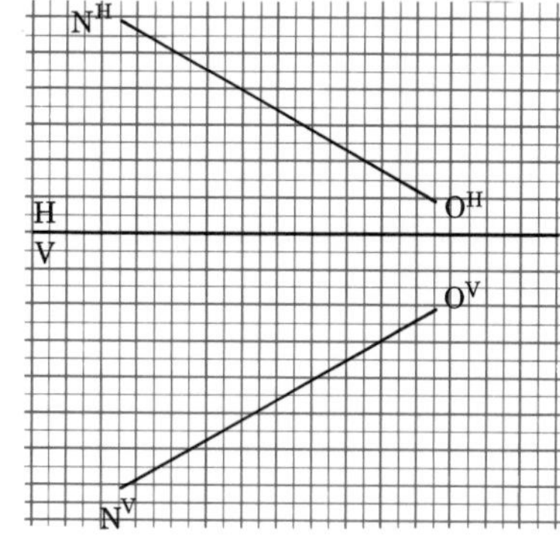

Fig. 8–51 **Determine the slope of line *NO*. Extend *NO* to measure 56 mm (2 1/4″) long. Draw all three views.**

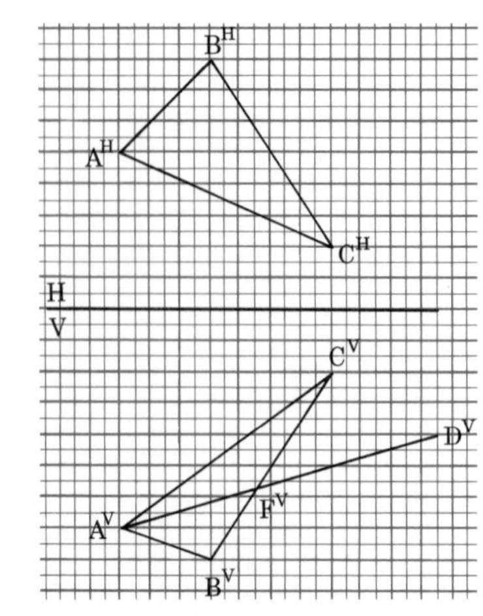

Fig. 8–52 **Locate point *D* in the plan (horizontal) projection. Determine the length of line *AD*.**

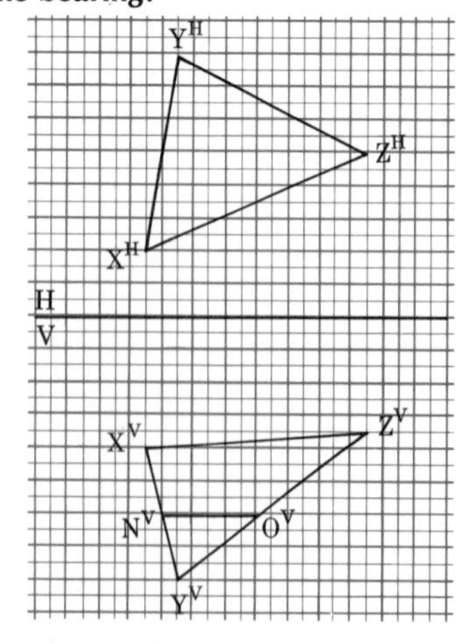

Fig. 8–53 **What is the bearing of line *NO* located on plane *XYZ*? Determine the true size of plane *XYZ*.**

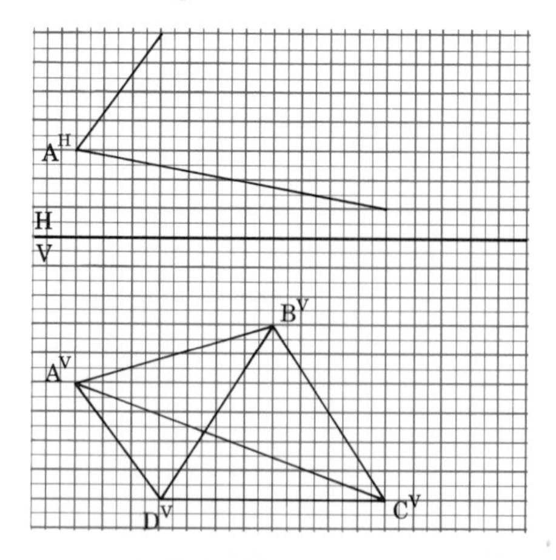

Fig. 8–54 **Complete the plan view of plane *ABCD* and develop a side view.**

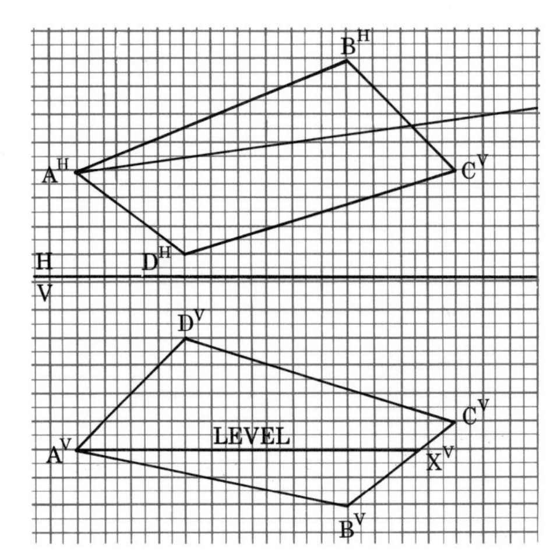

Fig. 8–55 **Find the edge view of plane *ABCD* and determine the angle that it makes with the horizontal plane.**

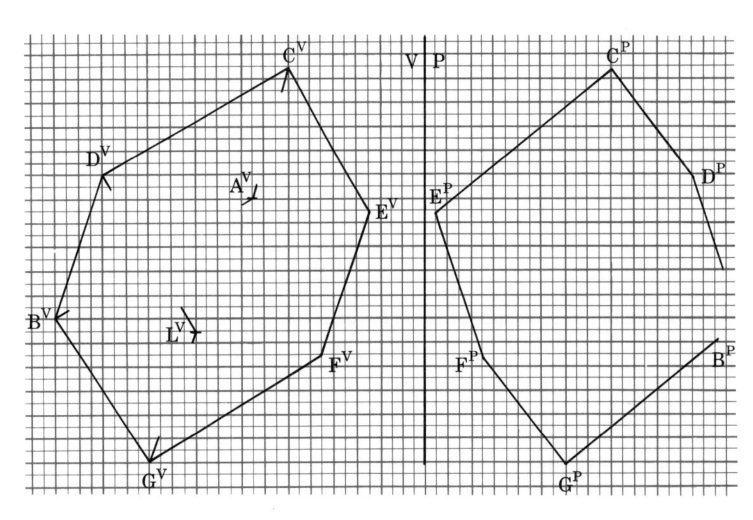

Fig. 8–56 **Determine the visibility of the edges relating to points *A* and *L* of the parallelepiped. Alter the location of points *A, G,* and *L* to develop the parallelepiped.**

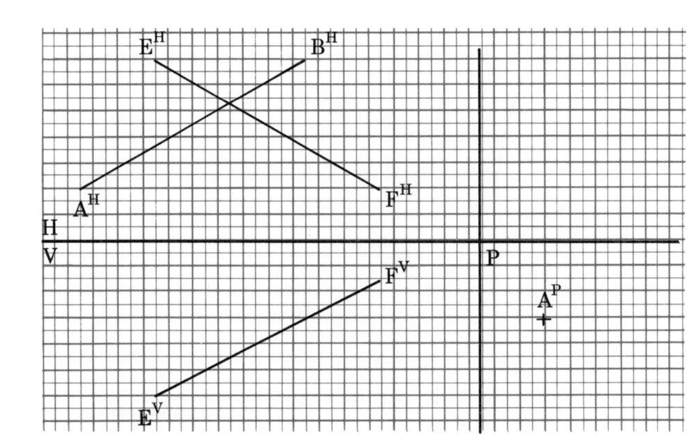

Fig. 8–57 **Complete the three views showing the intersection of lines *AB* and *EF*.**

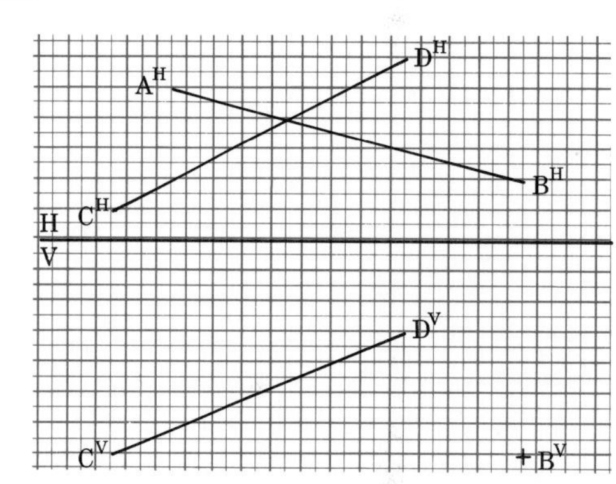

Fig. 8–58 **Draw the front view of line *AB* which intersects line *CD*. What is the distance from *C* to *A*?**

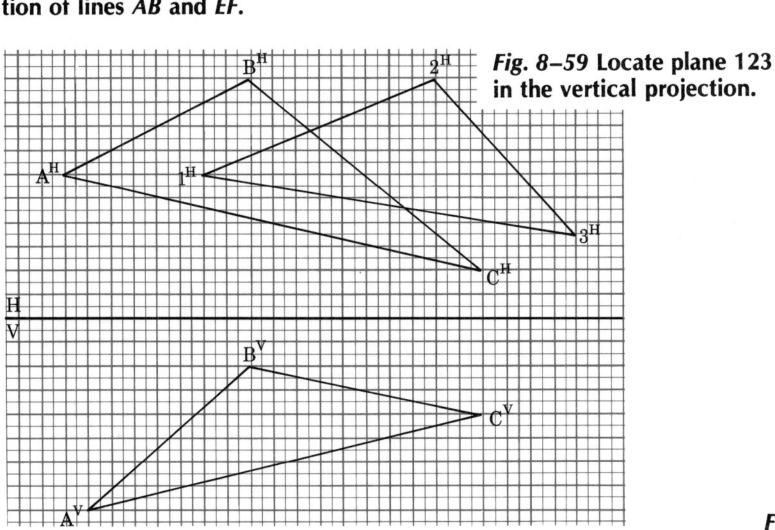

Fig. 8–59 **Locate plane 123 in the vertical projection.**

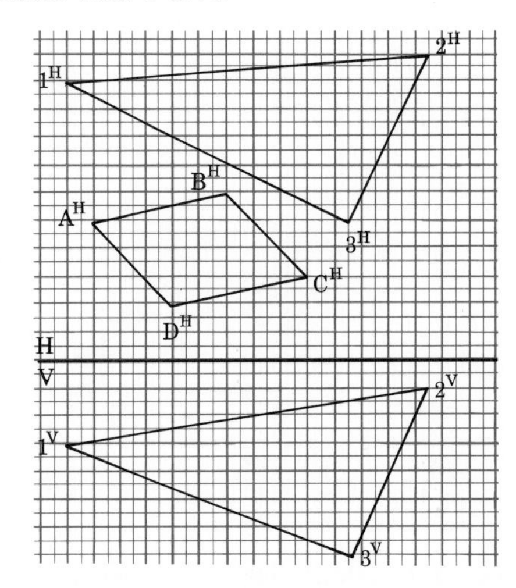

Fig. 8–60 **Construct plane *ABCD* parallel to plane 123.**

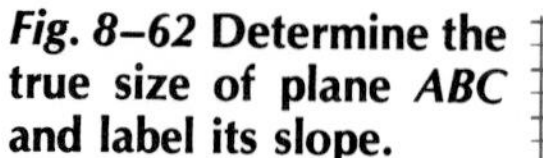

Fig. 8–61 Determine the true size of oblique plane ABC. Draw line XY parallel to plane ABC in the plan view.

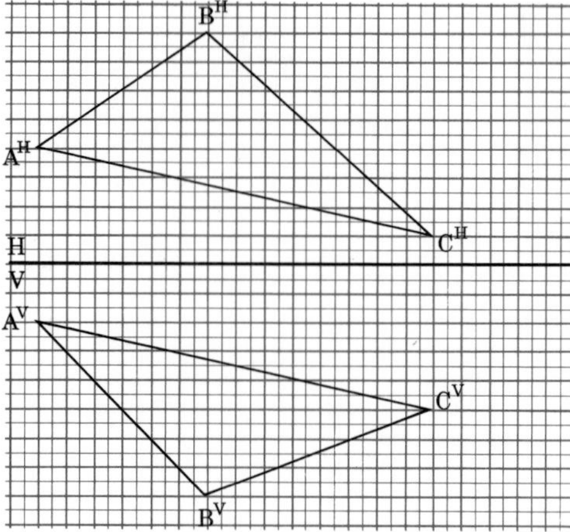

Fig. 8–62 Determine the true size of plane ABC and label its slope.

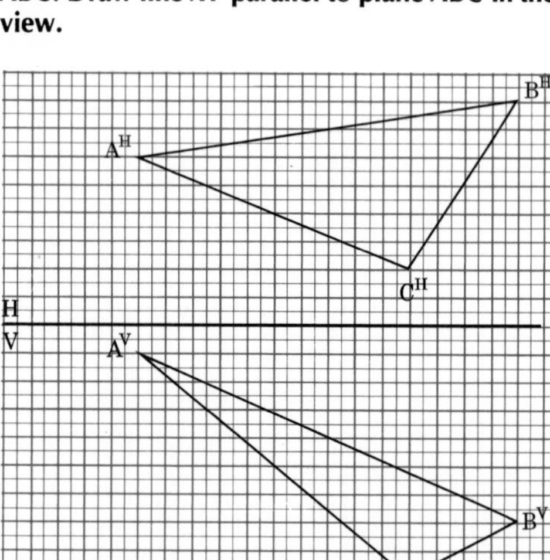

Fig. 8–63 Draw the true size of plane ABC and dimension the three angles.

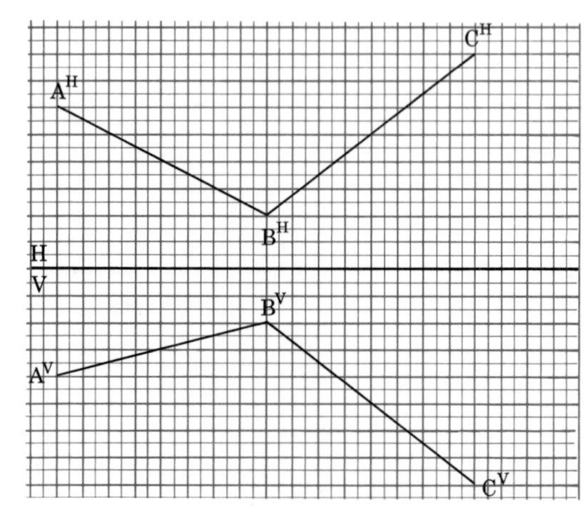

Fig. 8–64 Find the true angle between lines AB and BC.

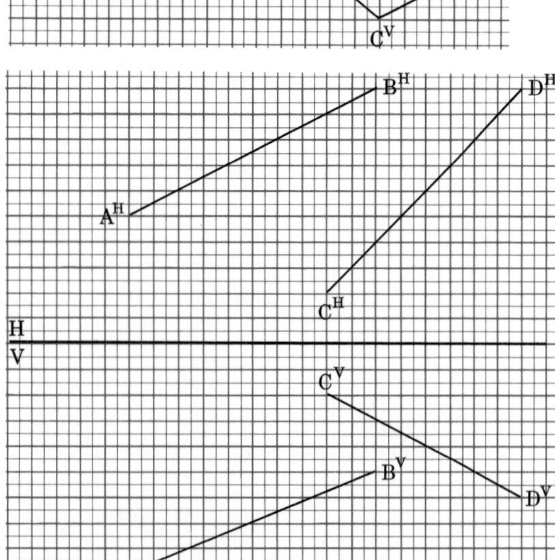

Fig. 8–65 Determine and locate the shortest distance between skew lines AB and CD.

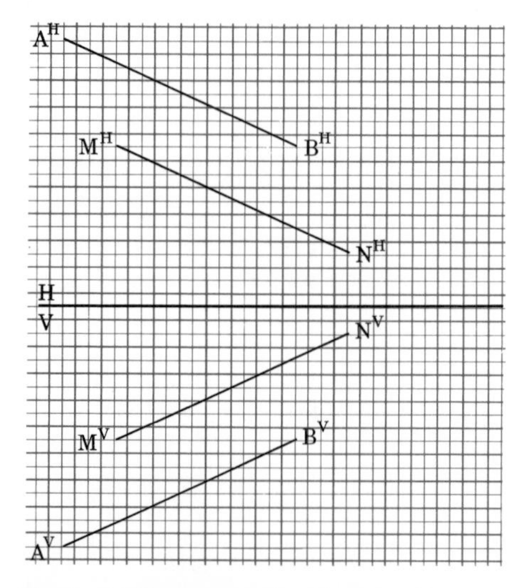

Fig. 8–66 Determine the shortest distance between parallel lines AB and MN.

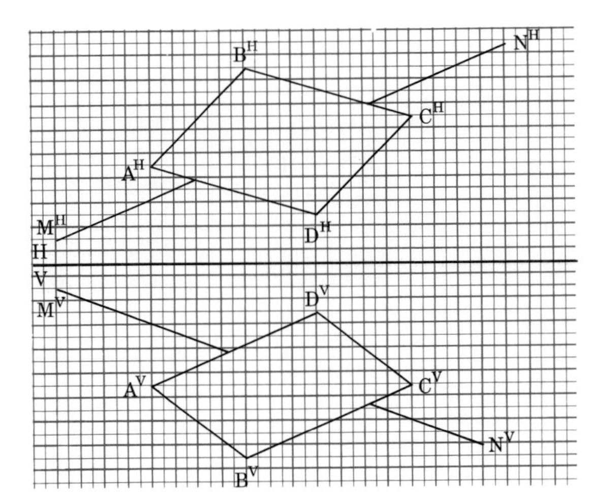

Fig. 8–67 **Determine if line *MN* pierces plane *ABCD*.**

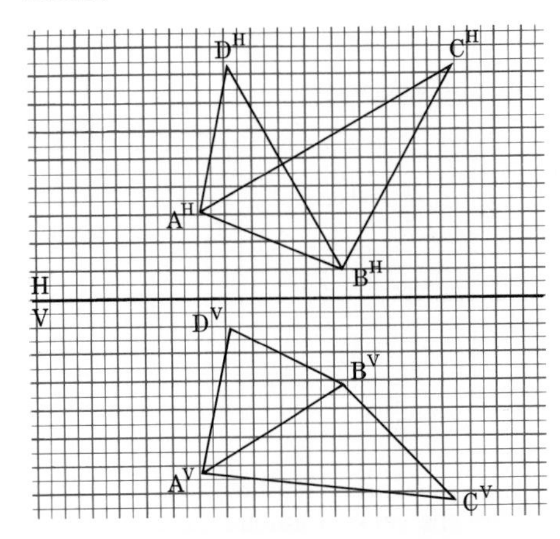

Fig. 8–69 **Determine the visibility and angle between planes *ABC* and *ABD*.**

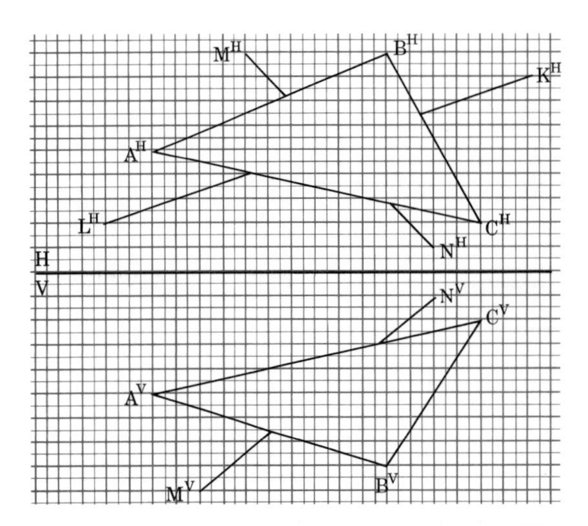

Fig. 8–68 **Determine if line *MN* pierces plane *ABC*. Locate line *KL* so that it pierces the center of plane *ABC*.**

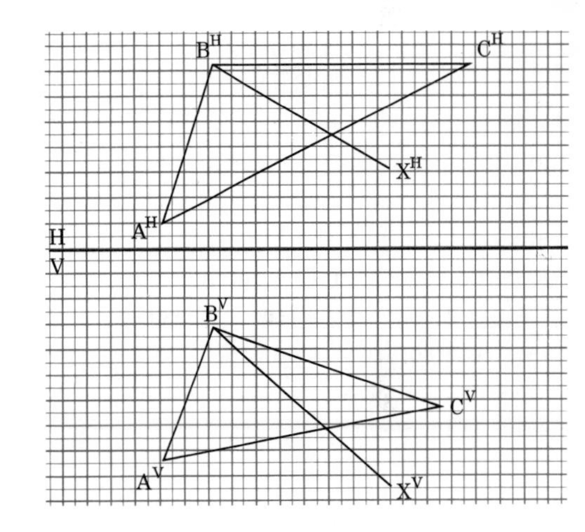

Fig. 8–70 **Determine the angle between line *BX* and plane *ABC*.**

Fig. 8–71 **After examining Figs. 8–37 through 8–40 on piercing points, create a problem in which an oblique plane is pierced by one inclined line and one oblique line. Show construction projection using the cutting-plane system.**

Fig. 8–72 **Study Figs. 8–41 through 8–43 carefully. Create two intersecting oblique planes, and determine the angle formed at the intersection.**

Chapter 9

Sectional Views and Conventions

SECTIONAL VIEWS

Your technical drawings must show all parts of an object, including the insides and other parts that you cannot see. In Chapter 5, we learned that you can draw these hidden parts with *hidden lines* made with short dashes. But this method works well only if the hidden part has a fairly simple shape. If the shape is complicated, dashed lines may show it poorly. They can also be confusing, as shown in Fig. 9–1. Instead of using dashed lines, then, you can draw a special view called a *section* or *sectional view.* A sectional view shows an object as if part of it were cut away to expose the insides (Fig. 9–2).

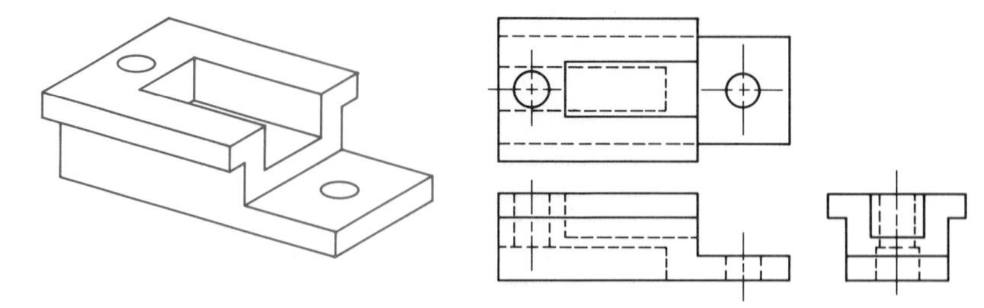

Fig. 9–1 **Pictorial view of object and the three normal views.**

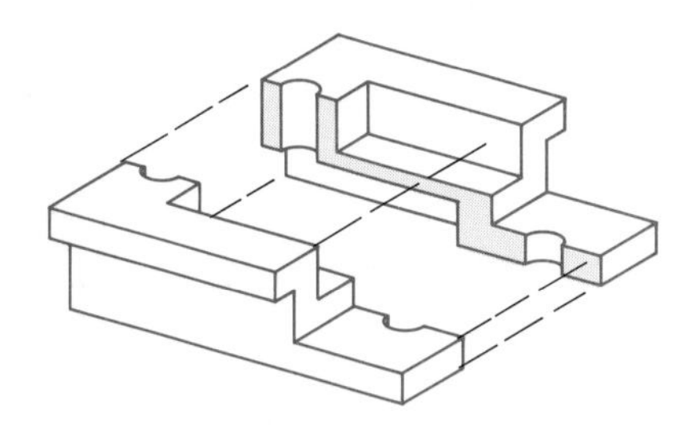

Fig. 9–2 **Object cut to show inside details. The front of the object has been removed.**

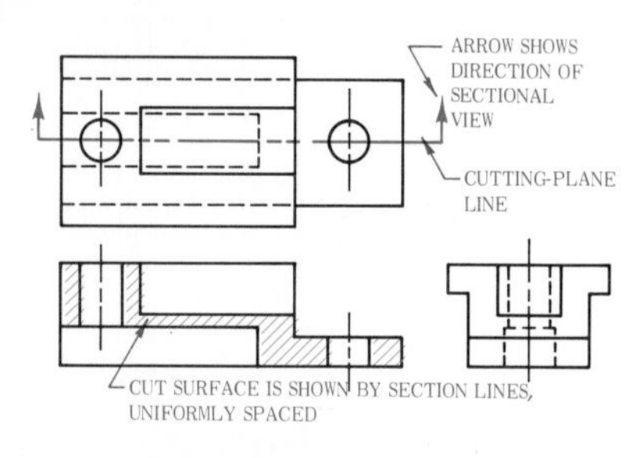

Fig. 9–3 **Three views with cutting-plane line and section lining.**

To draw a sectional view, imagine that a wide-blade knife has cut through the object. Call this knife a *cutting plane.* Then imagine that everything in front of the plane has been taken away, so that you can see the cut surface and whatever is inside (Fig. 9–2). On a normal view, you can show where a cutting plane will pass by drawing a special line, the *cutting-plane line* (Fig. 9–3). On a sectional view, you show the cut surface by marking it with evenly spaced thin lines. This is called *section lining.* It is also called *crosshatching.* The basic section lining pattern described above is not the only pattern. However, it is the one that is used in most cases. You can use it for objects made of any material, so it is called the general-purpose symbol. You use it especially when you do not have to show more than one kind of material, such as in a drawing of a single object. However, if you wish, you can also use special section lining patterns (also called symbols) to show what materials are used. The American National Standards provide for many symbols to stand for different materials (Fig. 9–4). Under this system, the general-purpose symbol can also mean that an object is made of cast iron. These special symbols are most useful in a drawing showing several objects made of different materials. You would use them, for example, in an *assembly drawing* (a drawing showing how different parts fit together). You can not, however, show what materials to use just with these symbols. You must also *specify* (call for) exactly the materials needed in a note, a list of materials, or some other way.

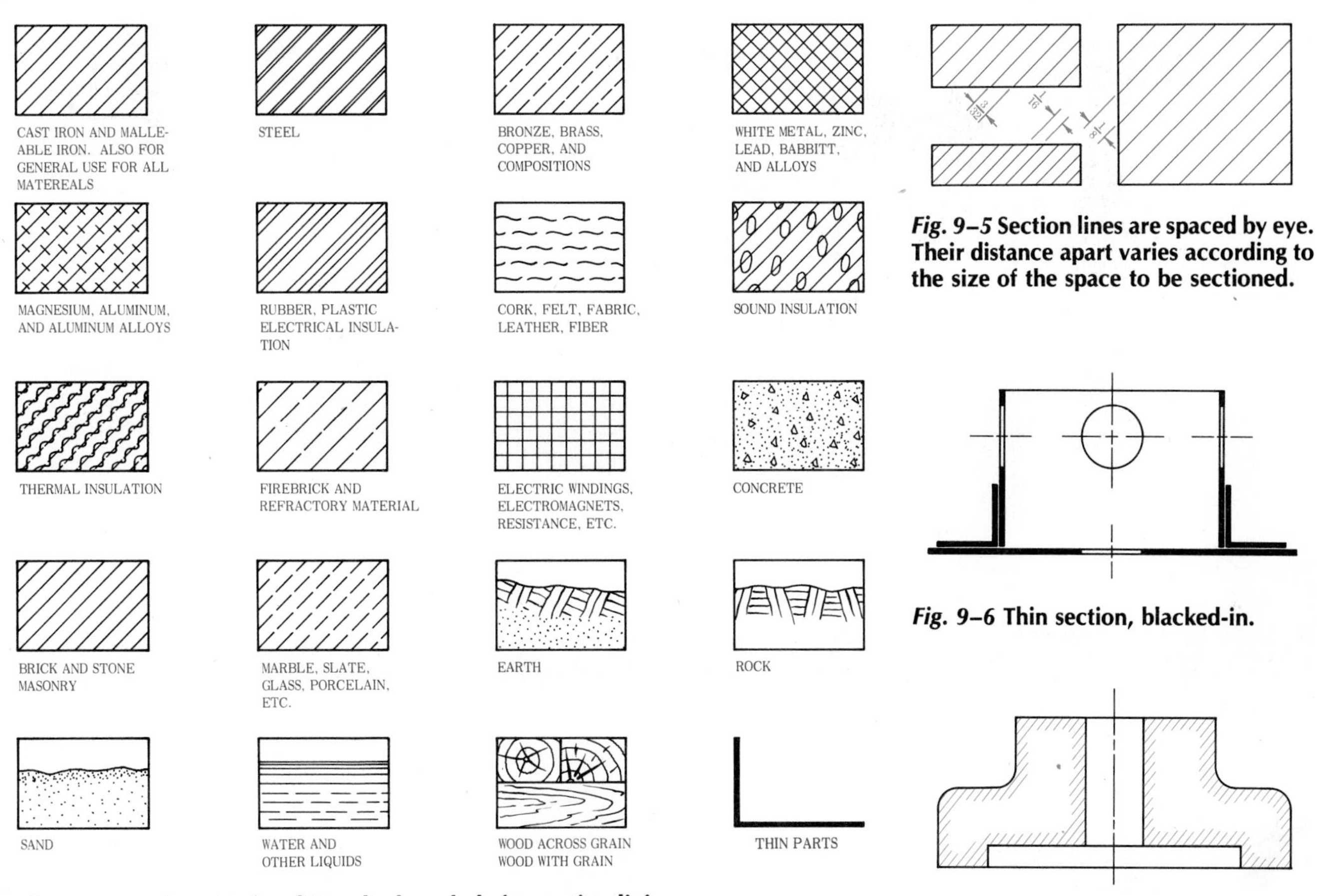

Fig. 9–4 **American National Standard symbols for section lining.**

Fig. 9–5 **Section lines are spaced by eye. Their distance apart varies according to the size of the space to be sectioned.**

Fig. 9–6 **Thin section, blacked-in.**

Fig. 9–7 **Outline sectioning.**

SPACING OF SECTION LINES

You can draw your section lines close together or far apart, depending upon how much space you have to fill (Fig. 9–5). According to the American National Standards, section lines can be anywhere from about 1.0 mm (1/32 in.) to 3.0 mm (1/8 in.) apart. However, they must be evenly spaced. They should also usually be slanted at a 45° angle. Choose the wider spacing when you need to show the sectioned area clearly. Your drawing will be neater if you *do not* space your section lines too close together. You will also save time. In most cases, the lines will look best spaced about 2 mm (3/32 in.) apart. You need not measure them off, just space them by eye. If the area to cover is large, space the lines more widely, up to 3 mm (1/8 in.) or more apart. If the area is small, space the lines more closely, down to 1.5 mm (1/16 in.) or less apart. If the area is very small, as for thin plates, sheets, and structural shapes, you can use *blacked-in* (solid black) sections. These are shown in Fig. 9–6. Note the white space between the parts.

When you are drawing a large sectioned area, one way to save time is to use outline sectioning. This method is shown in Fig. 9–7. Drafters who use it often draw the section lines freehand and spaced widely apart. You can also gray the sectioned area (Fig. 9–8), gray only along its outline (Fig. 9–9), or rub pencil dust over it. Apply a *fixative* (sealing substance) to prevent smudging.

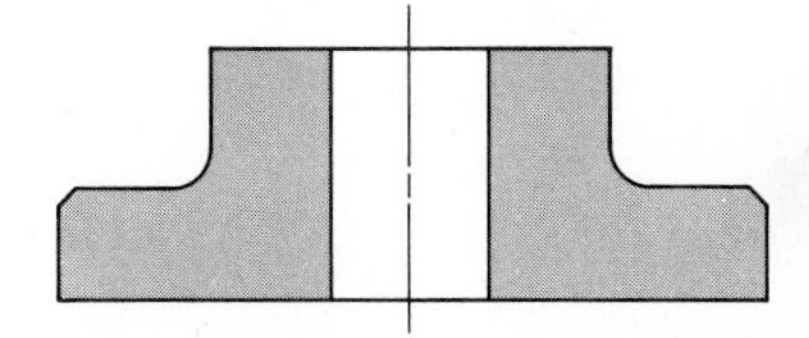

Fig. 9–8 **Cut surface may be grayed.**

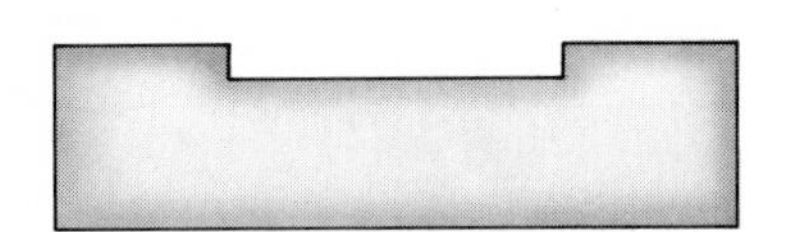

Fig. 9–9 **Cut surface may have grayed outline.**

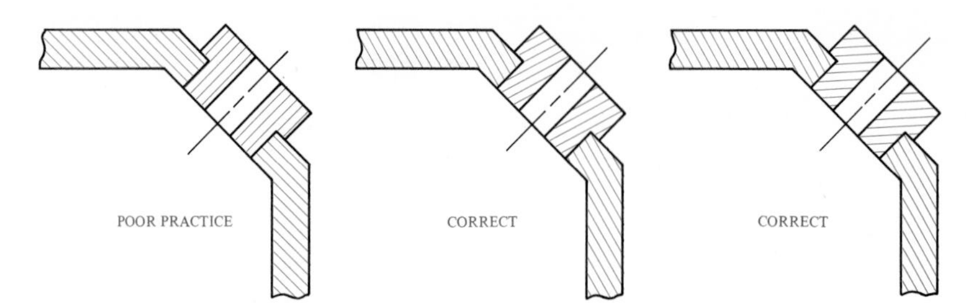

Fig. 9–10 **Do not draw section lines parallel or perpendicular to a main line of the view.**

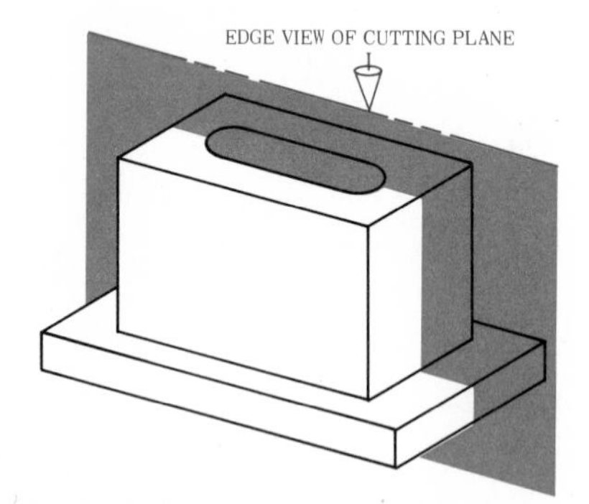

Fig. 9–11 **The cutting-plane line represents the edge view of the cutting plane.**

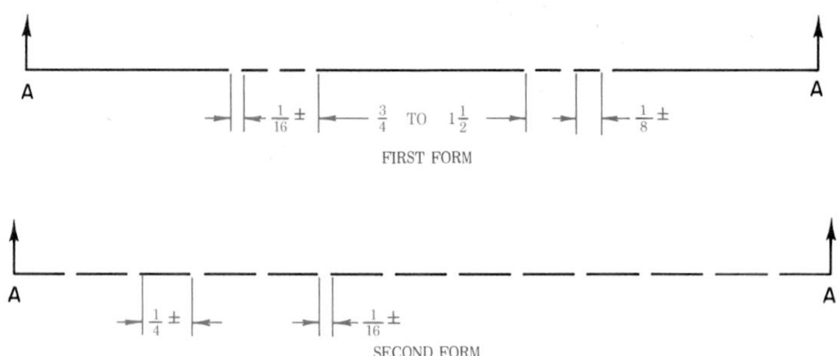

Fig. 9–12 **American National Standard cutting-plane lines.**

Never draw your section lines parallel to or at right angles to an important visible line (Fig. 9–10). You may, however, draw them at any other suitable angle and space them at any width. You can use section lines with different angles and spacing to identify different sectioned parts.

THE CUTTING–PLANE LINE

The cutting-plane line represents the cutting plane as viewed from an edge (Fig. 9–11). You can draw this line in either of two ways approved for American National Standards (Fig. 9–12). The first form is better for most uses. The second shows up well on complicated drawings. At each end of your line, draw a short arrow to show the direction for looking at the section. Make the arrows at right angles to the line. Place bold capital letters at the corners as shown, if needed for reference to the section.

You do not need a cutting-plane line when it is clear that the section is taken along an object's main centerline or at some other clearly seen place (Fig. 9–13).

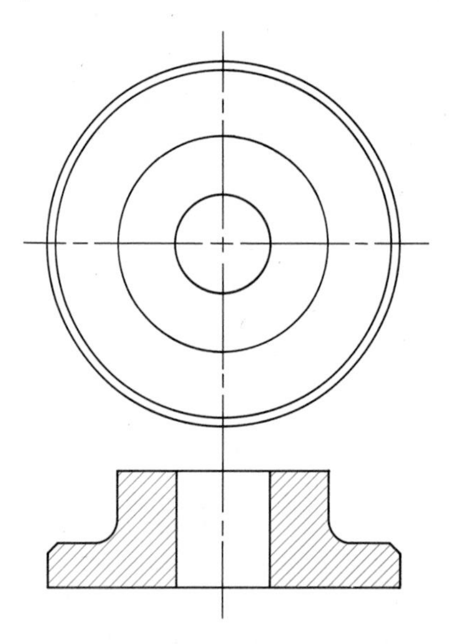

Fig. 9–13 **A centerline may be used to represent a cutting-plane line.**

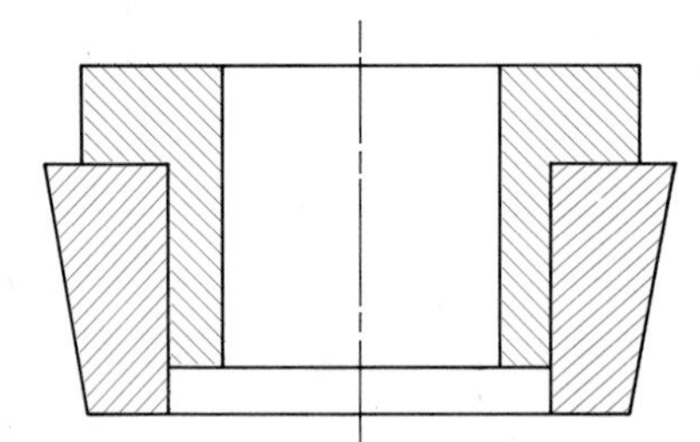

Fig. 9–14 **Two pieces. Section lines in two directions.**

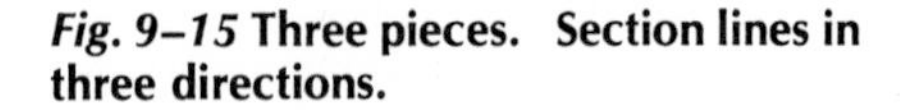

Fig. 9–15 **Three pieces. Section lines in three directions.**

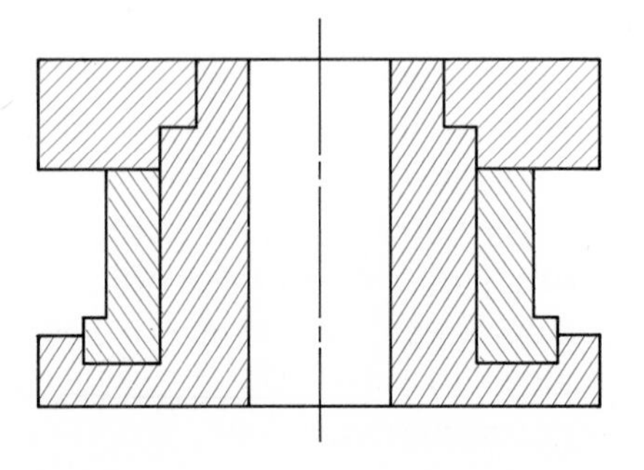

Fig. 9–16 **Full sectional view.**

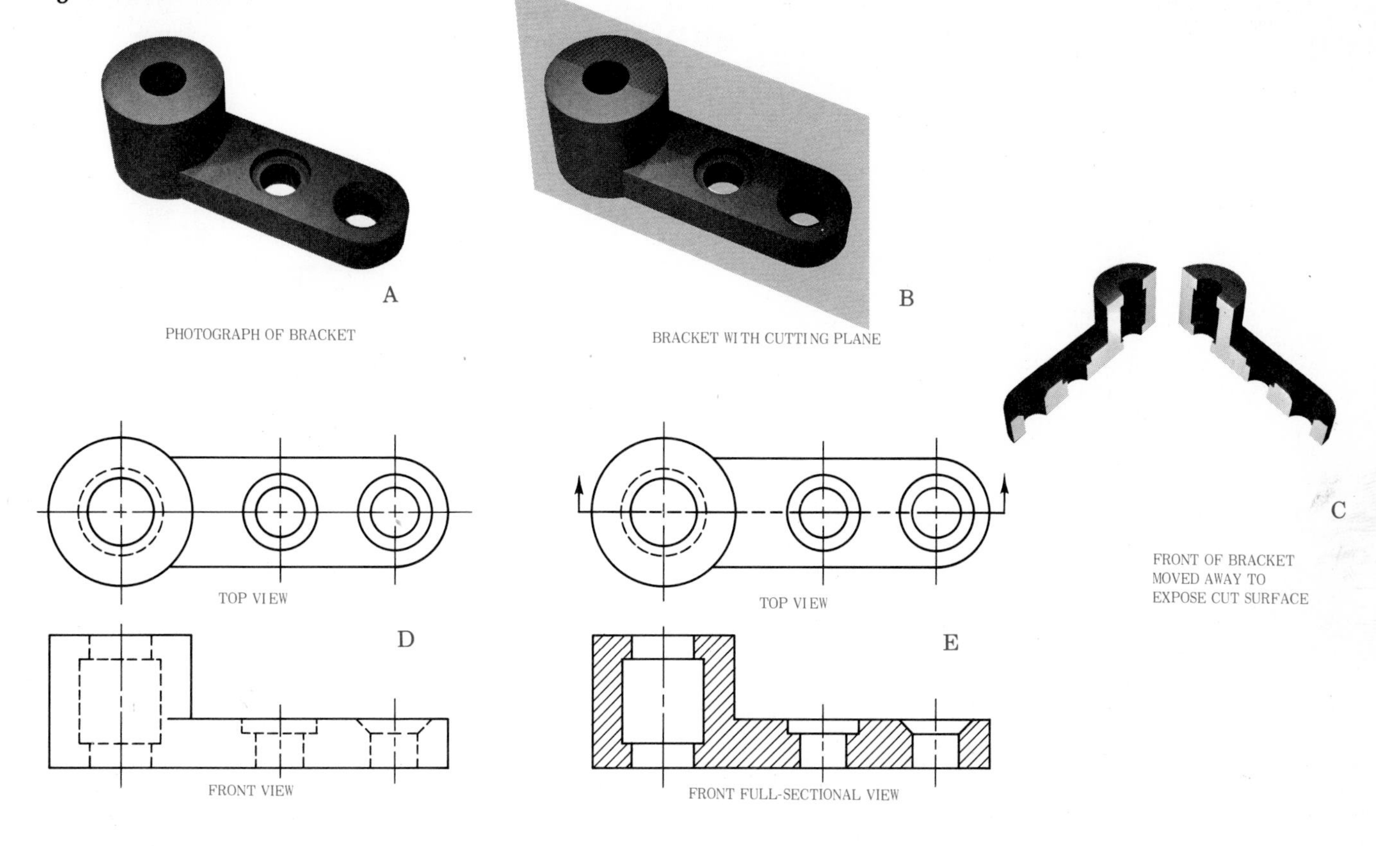

SECTIONS THROUGH ASSEMBLED PIECES

If your drawing shows more than one piece in section, draw the section lines in a different direction on each piece (Figs. 9–14 and 9–15). Remember, however, that any piece can, in turn, show several cut surfaces. Make sure that all the cut surfaces of any one piece have section lines in the same direction, as in Fig. 9–15.

FULL SECTIONS

A full sectional view shows an object as if it were cut completely apart from one end or side to another, as in Fig. 9–16. Such views are usually just called *sections*. The two most common types of full sections are *vertical* and *profile* (Figs. 9–17 and 9–18).

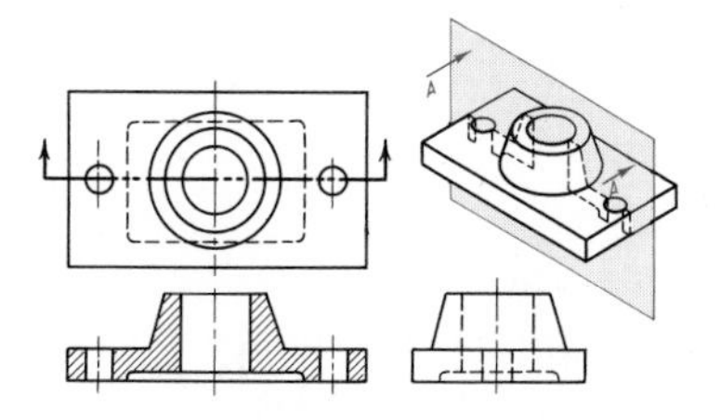

Fig. 9–17 **Vertical section.**

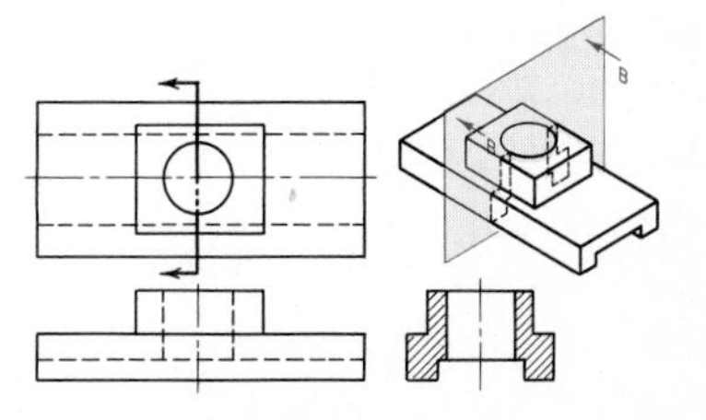

Fig. 9–18 **Profile section.**

OFFSET SECTIONS

In sections, the cutting plane is usually taken straight through the object. But it can also be *offset* (shifted) at one or more places in order to show some detail or to miss some part. An offset section is shown in Fig. 9–19. Here a cutting plane through the center is offset to pass through the two bolt holes. If the plane were not offset, the bolt holes would not show in the sectional view. You indicate an offset by drawing it on the cutting-plane line. No indication is given on the sectional view. If reference letters

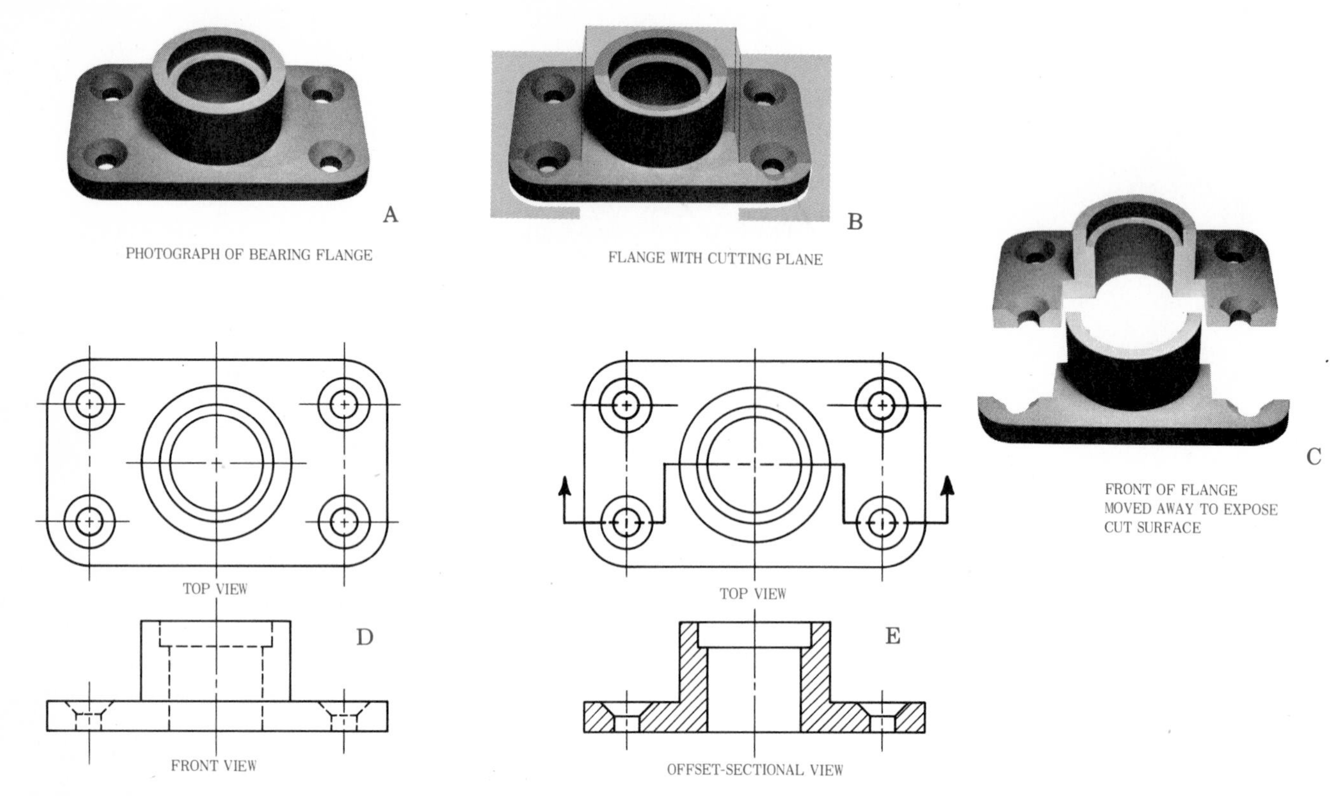

***Fig. 9–19* An offset section.**

are needed on the cutting-plane line, place them at the ends, opposite the arrowheads. Use capital letters.

HALF SECTIONS

A *half section* is one-half of a full sectional view. Remember, a full-sectional view makes an object look as if half of it has been cut away. So, a half-sectional view looks as if one-quarter of the original object has been cut away. Imagine that two cutting planes at right angles to each other slice through the object to cut away one-quarter of it, as in Fig. 9–20. The half-sectional view shows one-half of the front view in section. In the lower drawing at E, the right-hand side is the half-sectional view. The left-hand side shows the *exterior* (outside) of the object. This arrangement can be very useful when you are drawing a *symmetrical* object (one whose halves are mirror images of each other), because it lets you show both the inside and the outside in one view. Use a centerline where the exterior and half-sectional views meet because the object is not actually cut. Also, in the top view, show the complete object because no part is actually removed. If you wish to show the dimensions, you will have to draw some, or all, of the hidden lines in the exterior view. If you must show the direction for looking at the section, use only one arrow, as at E. In the top view, at E, the cutting-plane line could have been left out, since there is no doubt about where the section is taken.

HIDDEN AND VISIBLE LINES ON SECTIONAL VIEWS

Do not draw hidden lines on sectional views, unless they are needed for dimensioning or for clearly describing the shape. In Fig. 9–21 at A, a hub is clearly described using no hidden lines. Compare it with the view at B.

On *sectional assembly drawings* (sectional views of how parts fit together), you should leave out most hidden lines. This will keep

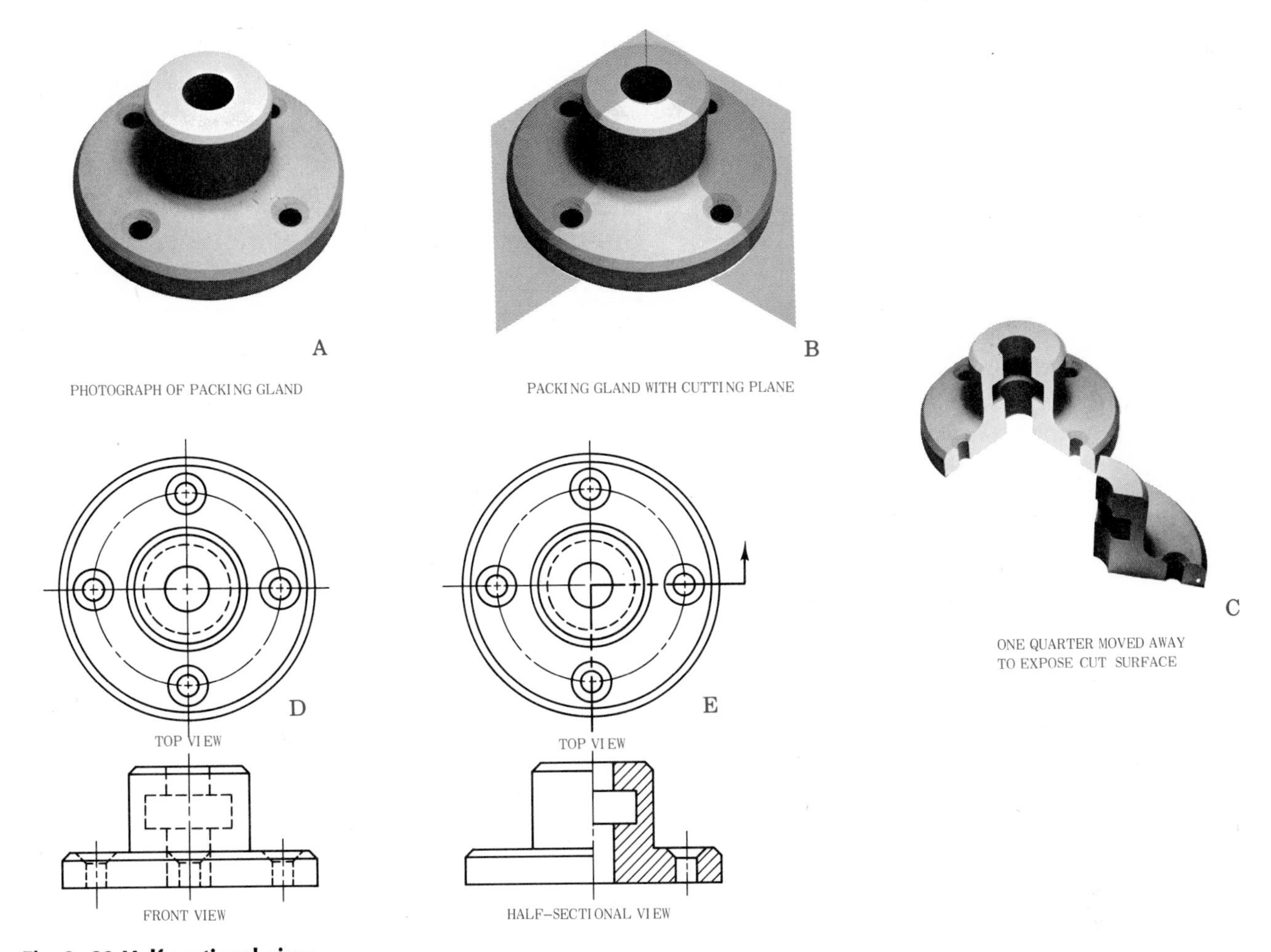

Fig. 9-20 **Half-sectional view.**

Fig. 9-21 **Omit hidden lines when not needed for clearness or dimensioning.**

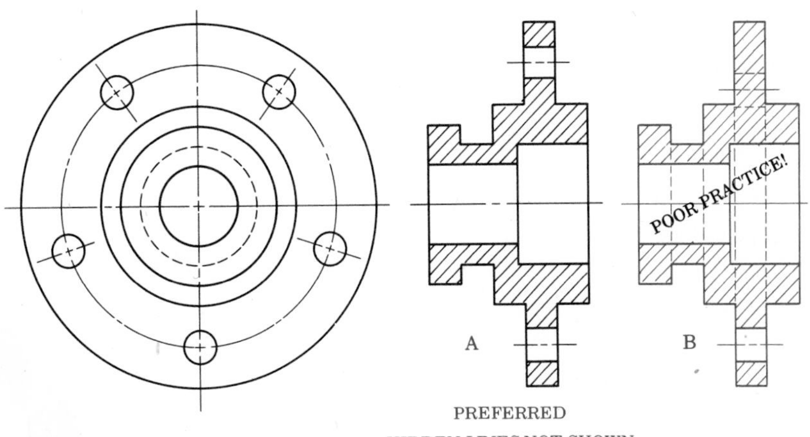

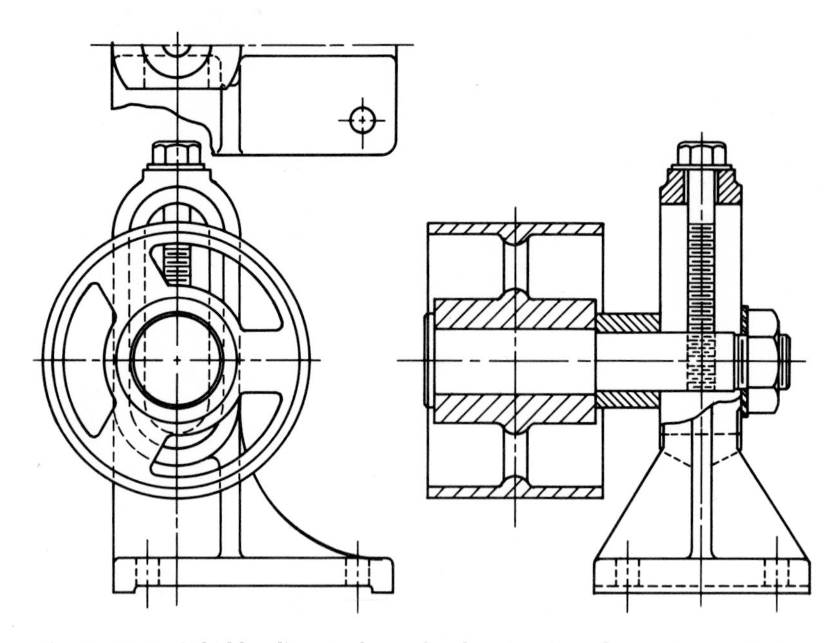

Fig. 9–22 **Omit hidden lines to keep the drawing from becoming confusing.**

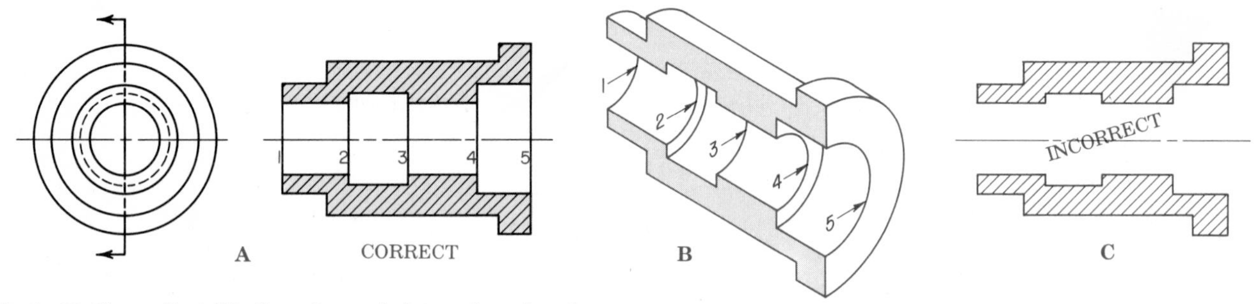

Fig. 9–23 **Show all visible lines beyond the sectioned surface.**

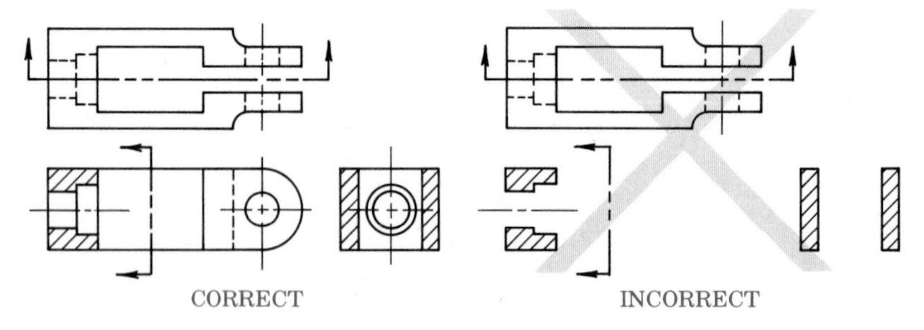

Fig. 9–24 **Correct and incorrect uses of visible lines beyond the plane of the section.**

the drawing from becoming cluttered and hard to read (Fig. 9–22). Sometimes a good way to avoid using hidden lines is to draw a half section or part section.

Normally, in a sectional view, you include all the lines that would be visible on or beyond the plane of the section. In Fig. 9–23, for example, the section drawing at A correctly includes the numbered lines, which match the lines on the drawing at B. A drawing without these lines, as at C, would have no value. *Never* draw sections that way. Figure 9–24 is another example of how wrong it is to leave out lines visible beyond the plane of the section.

BROKEN–OUT SECTIONS

A view with a *broken-out section* shows an object as it would look if a portion of it were cut partly away

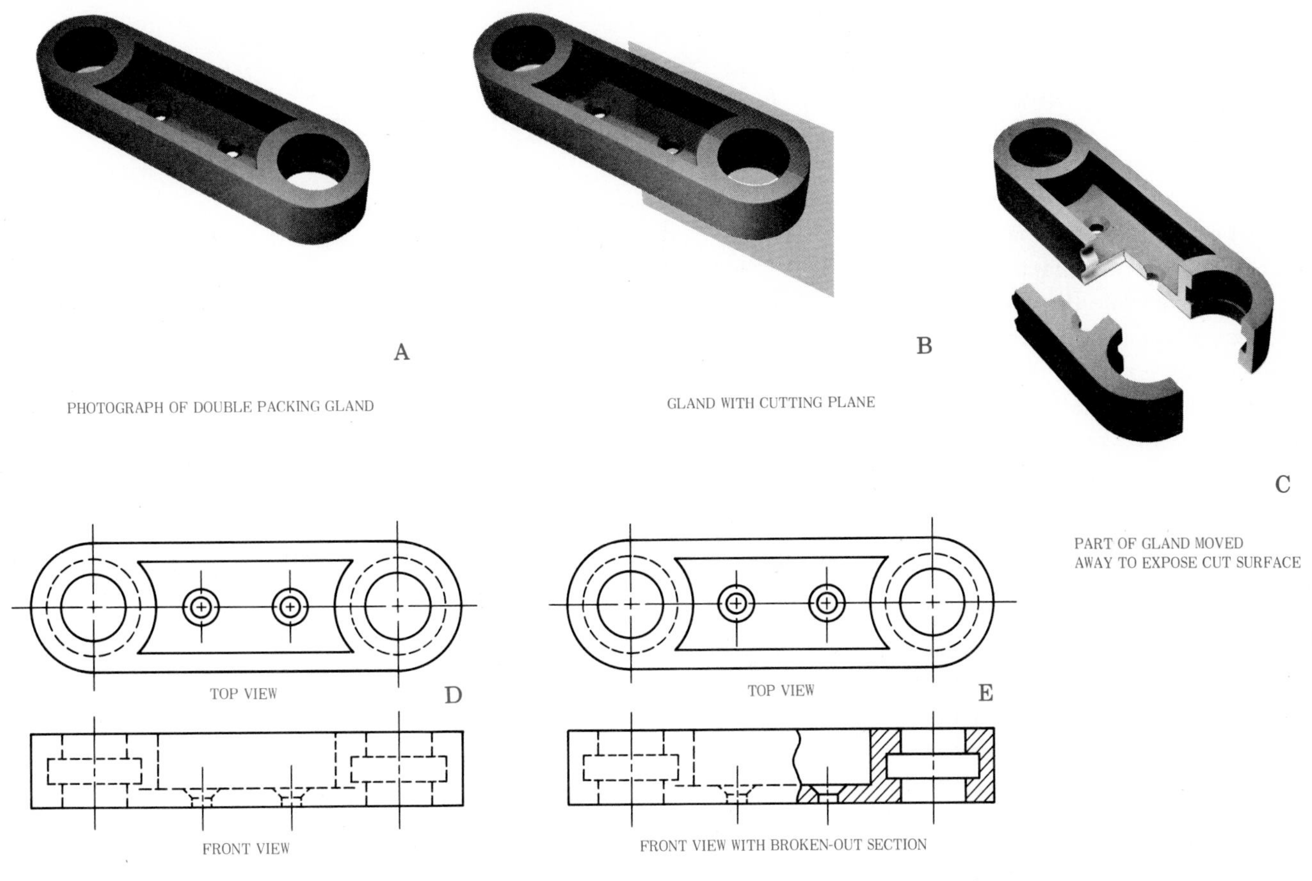

Fig. 9–25 **Broken-out section.**

from the rest by a cutting plane and then "broken off" to reveal the cut surface and insides (Fig. 9–25). You draw this view when you want to show some inside detail without drawing a full or half section.

Note that a broken-out section is bounded by a *break line.* You can draw break lines freehand. Make them the same thickness as visible lines. Figure 9–26 shows two more examples of broken-out sections.

REVOLVED SECTIONS

Think of a cutting plane passing through a part of an object, as in Fig. 9–27. Now think of that cut

Fig. 9–26 **Two additional examples of broken-out sections.**

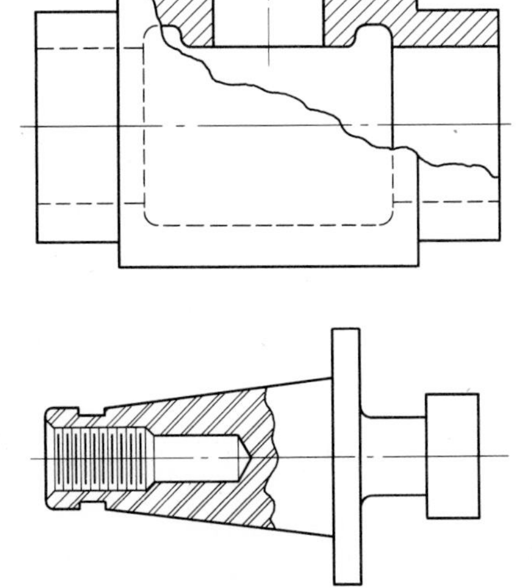

Fig. 9–27 **Cutting plane in position for revolved section.**

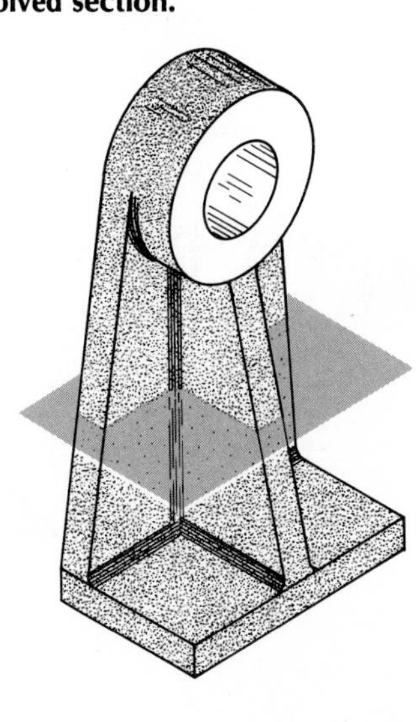

surface as *revolved* (turned) 90°, so that you can see its shape clearly (Fig. 9–28). The result is a *revolved section.* It can also be called a *rotated section.*

You can draw revolved sections, as in Fig. 9–29, if the part is long and thin and its shape in cross section is the same throughout. In such cases, you can shorten the view and give the length of the part by a dimension. This lets you draw a large part with a revolved sectional view in a short space.

REMOVED SECTIONS

Sometimes you will have to take a sectional view from its normal place on the view and move it somewhere else on the drawing sheet. When you do this, the result will be a *removed sectional view.* It can also be called a *removed section.* Remember, however, the removed

Fig. 9–28 **Revolved section.**

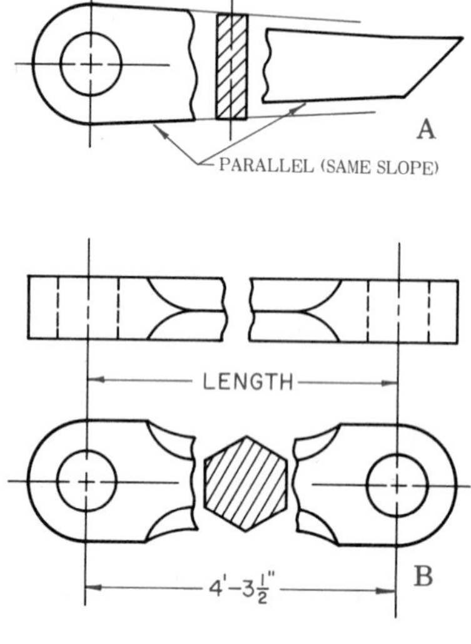

Fig. 9–29 **Revolved sections in long parts.**

Fig. 9–30 **Correct and incorrect positions of removed sections.**

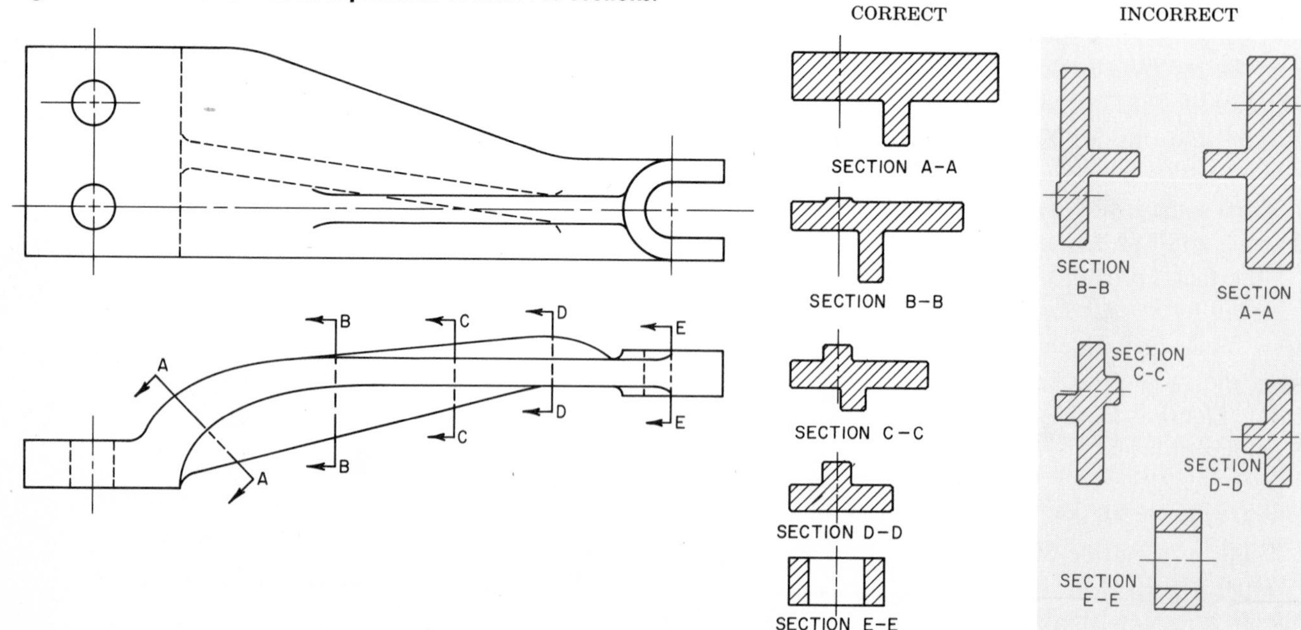

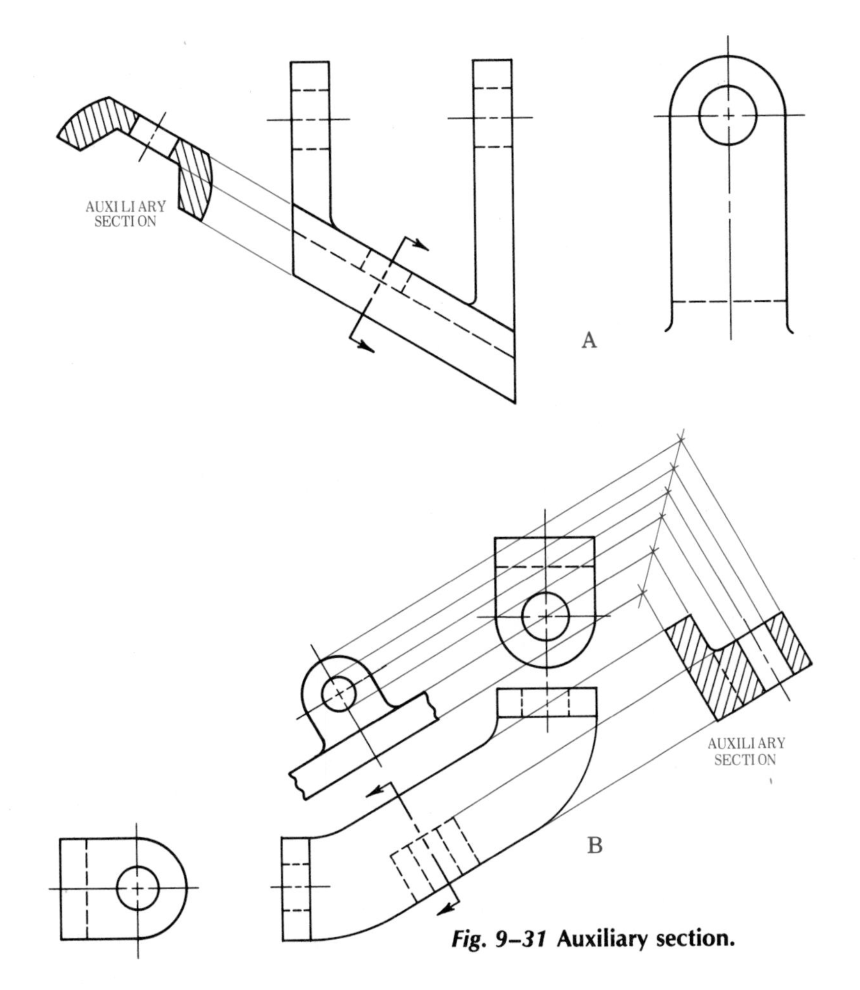

Fig. 9–31 **Auxiliary section.**

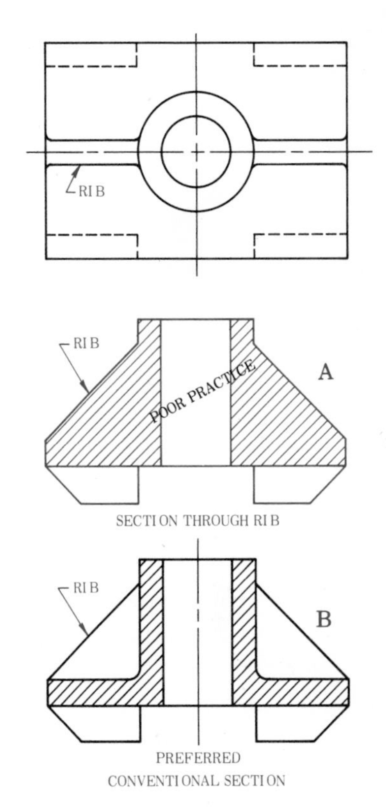

Fig. 9–32 **Ribs in section.**

section will be easier to understand if you show it looking just as it would if it were in its normal place on the view. In other words, do not rotate it in any direction. Fig. 9–30 shows right and wrong ways to position removed sections. Use bold letters to identify a removed section and its corresponding cutting plane on the regular view. Place the letters as shown in Fig. 9–30.

A removed section can be a *sliced section* (the same as a revolved section). Or it can show some additional detail visible beyond the cutting plane. You can draw it to a larger scale if you must show details clearly or to fit in dimensions.

Besides removed sectional views, you can also draw removed views of the *exterior* (outside) of an object. You can make these to the same scale or to a larger one. You can also make them complete pictures or just use them to show certain features.

AUXILIARY SECTIONS

When a cutting plane is passed through an object at an angle, as in Fig. 9–31 at A, the resulting sectional view, taken at the angle of that plane, is called an *auxiliary section.* You draw it just like you draw any other auxiliary view (see Chapter 7).

Usually, on working drawings, all you show in the auxiliary section is the cut surface. However, if you need to, you can also show any or all parts beyond the auxiliary cutting plane. In Fig. 9–31 at B, the auxiliary section contains one hidden line. There are also three incomplete views.

RIBS AND WEBS IN SECTION

Ribs and webs are thin, flat parts of an object used to brace or strengthen another part of the object. When a cutting plane passes through a rib or web *parallel to* (in line with) the flat side, do not draw section lining for that part, as in Fig. 9–32 at B. Instead, think of the plane passing just in front of the rib. A true section, as at A, would give the idea of a very heavy, solid piece.

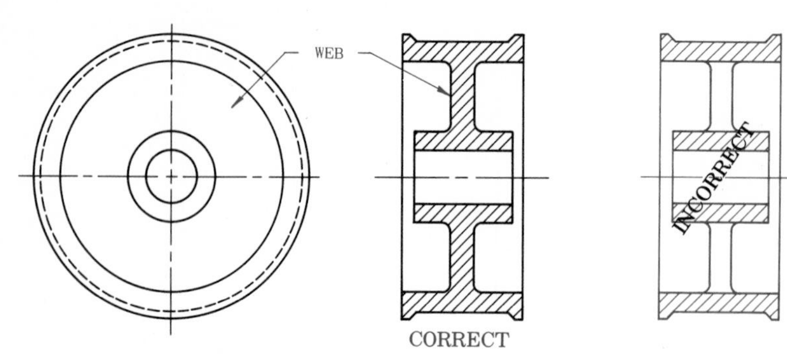

Fig. 9–33 **Web in section.**

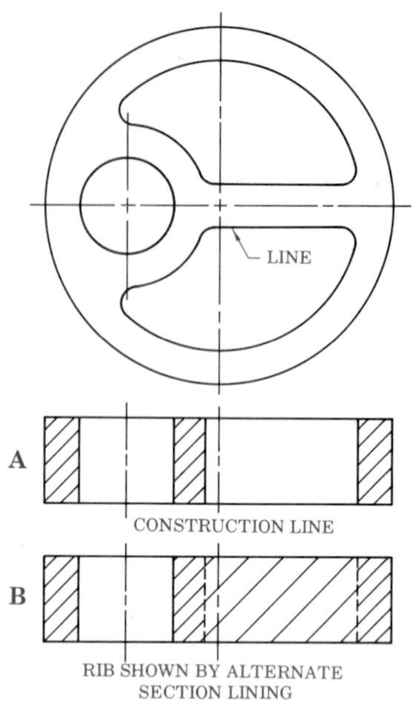

Fig. 9–34 **Alternate, or wide, section lining.**

This would not be a true description of the part.

If a cutting plane passes through a rib, web, or other thin, flat part at right angles to the flat side, draw in the section lining for that part. Figure 9–33 is an example.

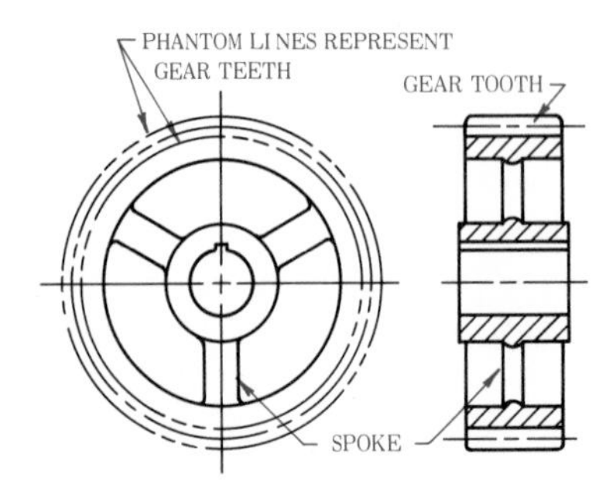

Fig. 9–35 **Spokes and gear teeth not sectioned.**

ALTERNATE SECTION LINING

Alternate (or wide) section lining is a sectioning pattern made by leaving out every other section line. You can use it to indicate a rib or other flat part in a sectional view when that part otherwise would not show up clearly. In Fig. 9–34 at A, an eccentric piece (circular shapes not using the same centerlines) is drawn in section. A rib is visible in the top view, but in the sectional view it is not indicated by any section lining. As we saw above, this is standard practice with a flat part like a rib. But there are no visible lines to indicate the rib either, because its top and bottom are both even with the surfaces they join. In fact, without the top view, you might not know that the rib was there. A drawing of an eccentric piece without a rib would look exactly the same. The problem is solved, however, at B. Here, alternate section lining is used with hidden lines to show the extent of the rib.

Alternate section lines are useful to show ribs and other thin, flat pieces in one-view drawings of parts or in assembly drawings.

OTHER PARTS USUALLY NOT SECTIONED

Do not draw section lining on spokes and gear teeth when the cutting plane passes through them. Leave them as shown in Fig. 9–35. Do not draw section lining either on shafts, bolts, pins, rivets, or similar items when the cutting plane passes through them *lengthwise* (through the axis). Leave them in full, as shown in Fig. 9–36. These objects are left in full because they have no inside details. Also, sectioning might give a wrong idea of the part. A drawing showing them in full is easier to read. It also takes less time to draw. However, when such parts are cut *across* the axis, you should section them (Fig. 9–37). See the sectional assembly in Fig. 9–38 for names and drawings of a number of other items that are not sectioned.

PHANTOM (HIDDEN) SECTIONS

You use a *phantom,* or *hidden,* section when you want to show in one view both the inside and the outside of an object that is not completely symmetrical. Figure 9–39 shows an object with a circular boss on one side. The object is not symmetrical, so you cannot show

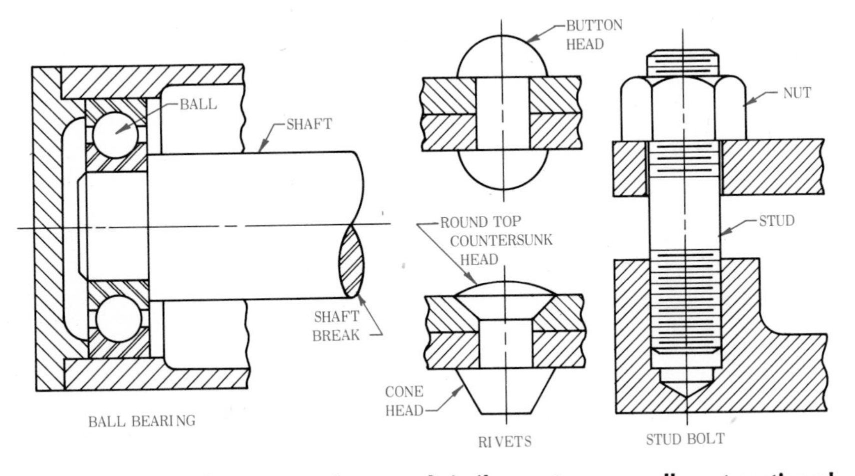

Fig. 9–36 **Shafts, bolts, screws, rivets, and similar parts are usually not sectioned.**

Fig. 9–37 **A cross section.**

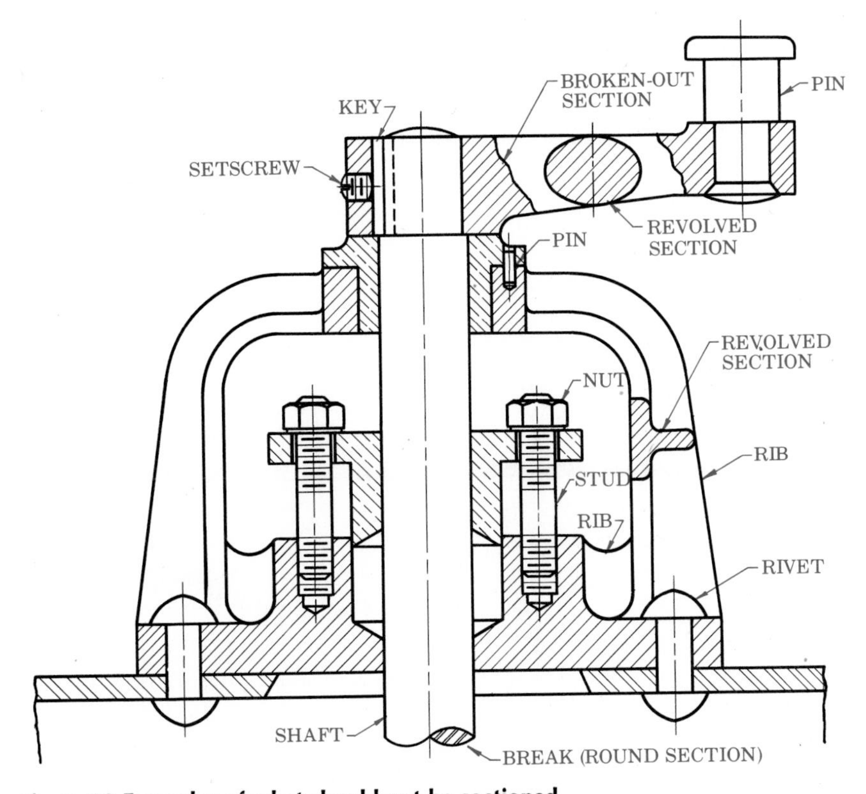

Fig. 9–38 **Examples of what should not be sectioned.**

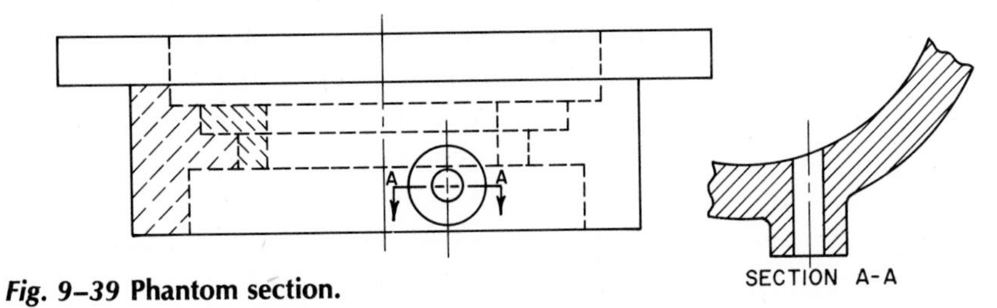

Fig. 9–39 **Phantom section.**

the inside with a half section. You must use the phantom section instead. A partial phantom section can sometimes be better than a broken-out section when you want to show a detail in section on an exterior view.

ROTATED FEATURES IN SECTION

A section or *elevation* (side, front, or rear view) of a symmetrical piece can sometimes be hard to read if drawn in true projection. It can also be hard to draw. If you are drawing such a view, follow the example in Fig. 9–40. In that example, a symmetrical piece with ribs and lugs is drawn twice in section. The first section drawing is a true projection. But it does not show the true shape of the ribs and lugs. In the second section, the ribs and lugs have been *rotated* (turned) on the vertical axis until they appear as mirror images of each other on either side of the centerline. Now you can see their true shape. This is the right way to draw this kind of object. Note that only the parts that extend all the way around the vertical axis are drawn with section lining. In Fig. 9–41, the lugs are rotated to show true shape. Note that they are not drawn with section lining.

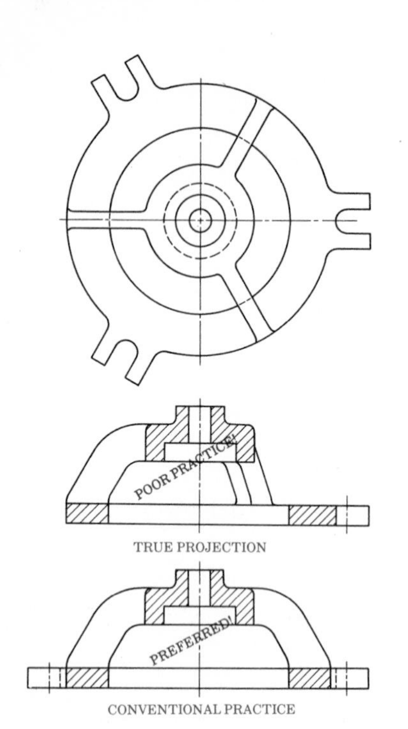

Fig. 9–40 Some features should be rotated to show true shape.

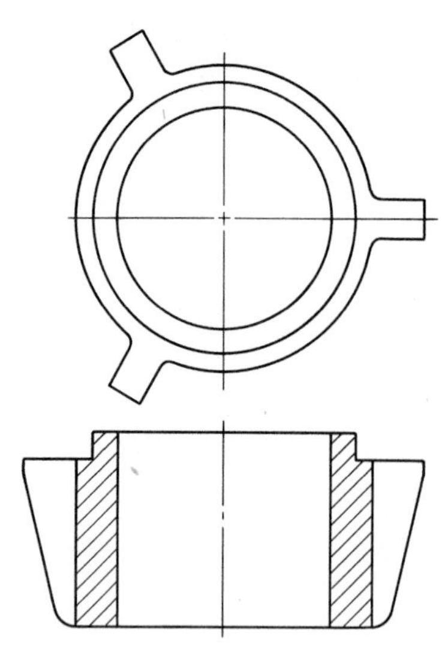

Fig. 9–41 Do not section lugs.

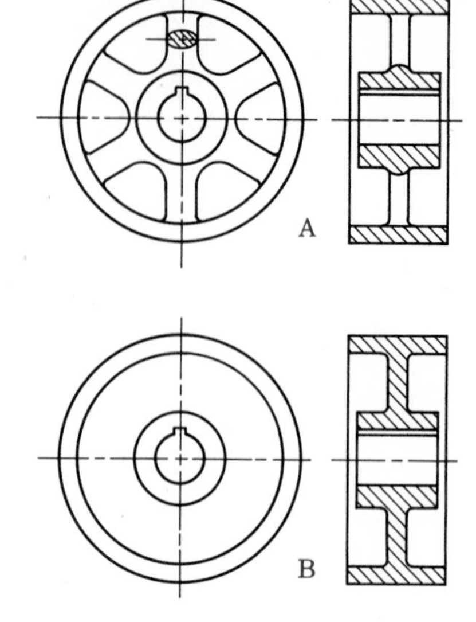

Fig. 9–42 Section through spokes.

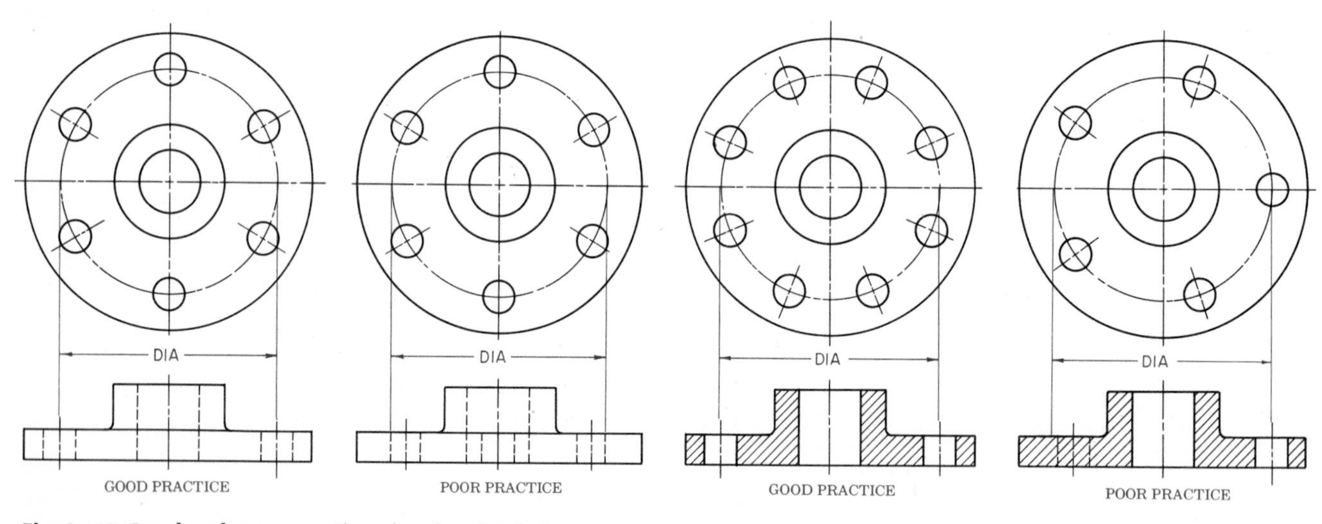

Fig. 9–43 Good and poor practices for showing holes.

When a section passes through spokes, do not draw section lining on the spokes. Leave them as in the section drawing in Fig. 9–42, at A. Compare this drawing with the section drawing for a solid web (Fig. 9–42 at B). Notice that it is the section lining that shows that the web is made solid rather than made with spokes.

When you draw a section or elevation of a part with holes arranged in a circle, follow the good practice examples in Fig. 9–43. In these examples, the holes have been rotated for the section drawing until two of them lie squarely on the cutting plane. These views then show the true distance of the holes from the center, when a true projection would not.

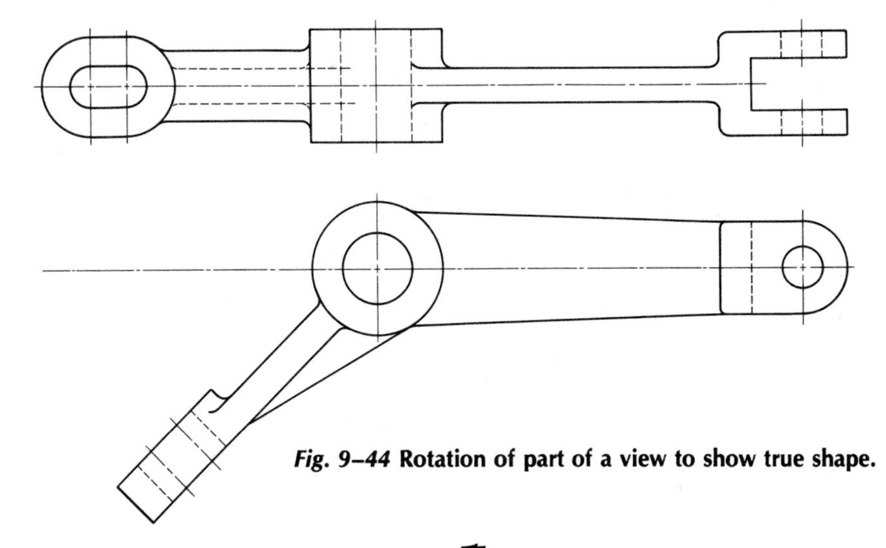

Fig. 9–44 **Rotation of part of a view to show true shape.**

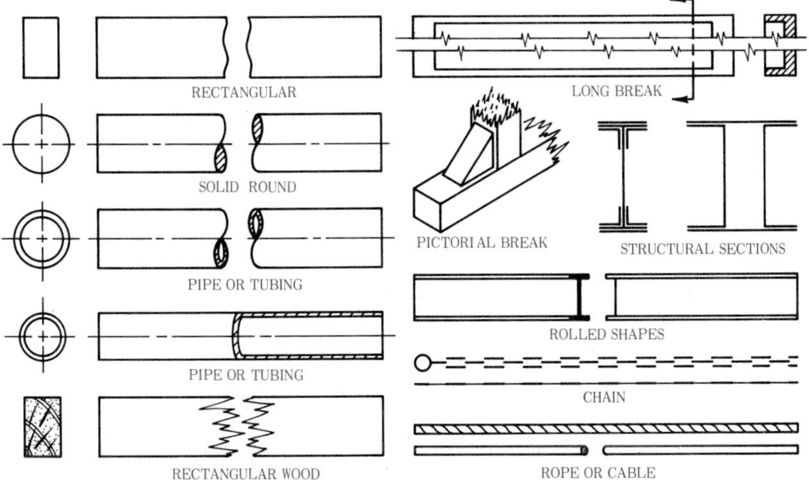

Fig. 9–45 **Conventional breaks and symbols.**

Fig. 9–46 **Drawing the break symbols for cylinders and pipes.**

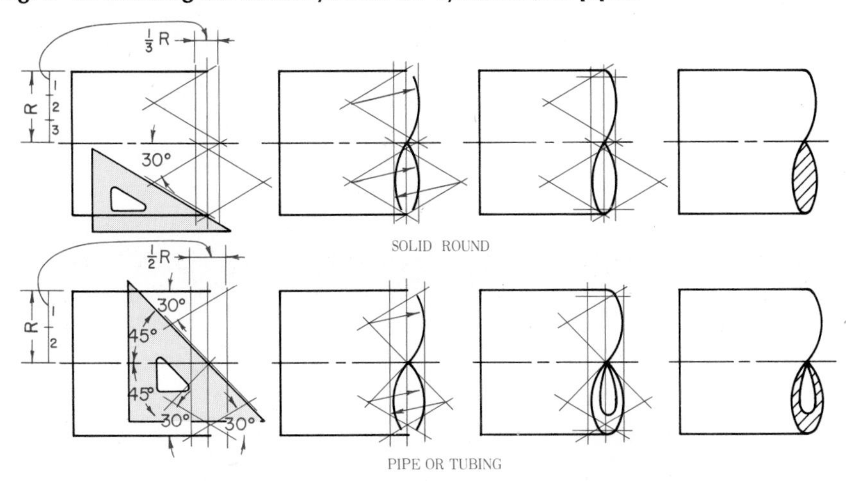

Rotating features in drawings are very useful when you want to show true conditions or distances that would not show in a true projection. Moreover, for some objects, you need to rotate in only part of the view. This is the case with the bent lever in Fig. 9–44.

CONVENTIONAL BREAKS AND SYMBOLS

Conventional breaks and symbols are used to make an object in a drawing easier to draw and easier to understand. Figure 9–45 shows the methods used to draw long, evenly shaped parts and to *break out* (shorten) the drawing of parts. Using a break lets you draw a view to a larger scale. Since your break shows how the part looks in cross section, you usually do not have to draw an end view. Give the length by a dimension. Usually, you can draw the symbols for conventional breaks freehand. However, on larger drawings, you might need to use tools to give a neat appearance. Figure 9–46 shows how to draw the break for cylinders and pipes.

INTERSECTIONS IN SECTION

An *intersection* is a point where two parts join together (Fig. 9–47). Drawing a true projection of an intersection is difficult and takes too much time. Also, such a line drawing is of little or no use to a blueprint reader. Instead, you should draw approximated and preferred sections such as those shown in Fig. 9–48.

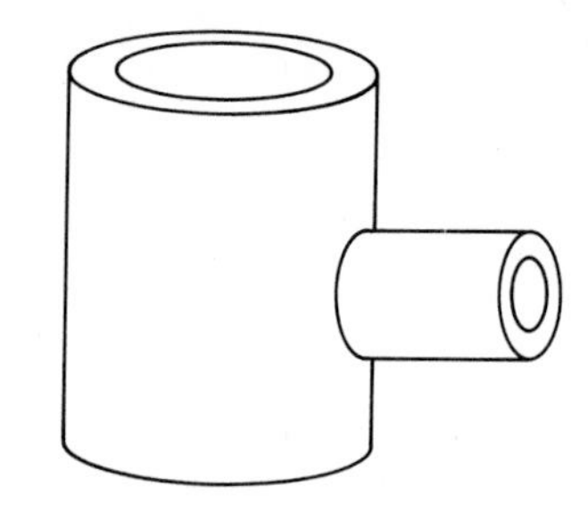

Fig. 9–47 **Intersecting parts.**

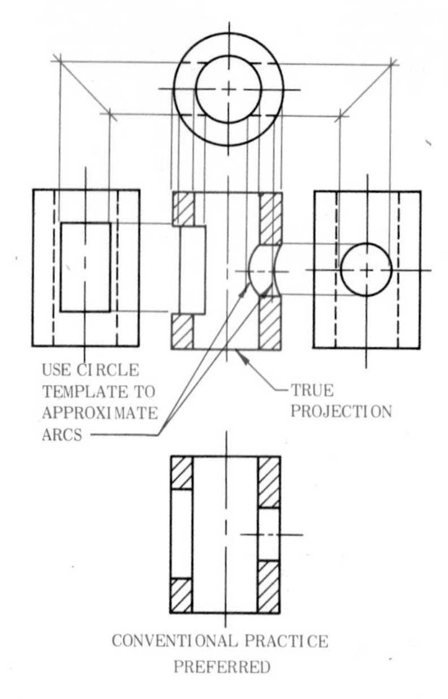

Fig. 9–48 **Approximated and preferred sections.**

Review

1. You use a sectional view to show ________ details.

2. Where a cutting plane passes on a normal view is shown by a line called a ________.

3. To show the cut surface in a sectional view, you use _____.

4. Name four types of sectional views.

5. Thin, flat parts of an object used to brace or strengthen another part of the object are called ________ or ________.

6. When a rib (or similar flat feature) does not show clearly in a sectional view, you should use ________ section lining.

7. To draw a sectional view, think of an imaginary ________ that passes through the object to expose the insides.

8. The section lining symbol for cast iron can also be called the ________ symbol.

9. When you draw an object as if it were cut completely apart, you are drawing a ________ sectional view.

10. If you imagine that two cutting planes at right angles to each other slice through an object to cut away one quarter of it, you will have a ________ sectional view.

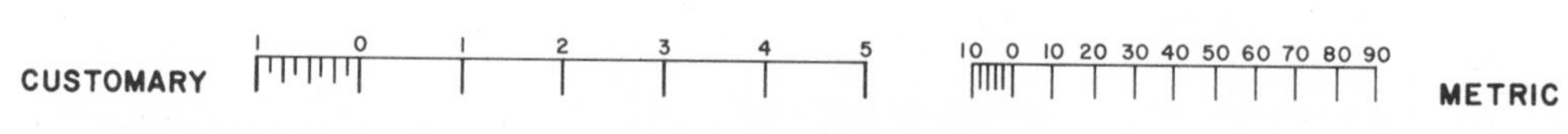

Scales for Fig. 9–49.

Problems

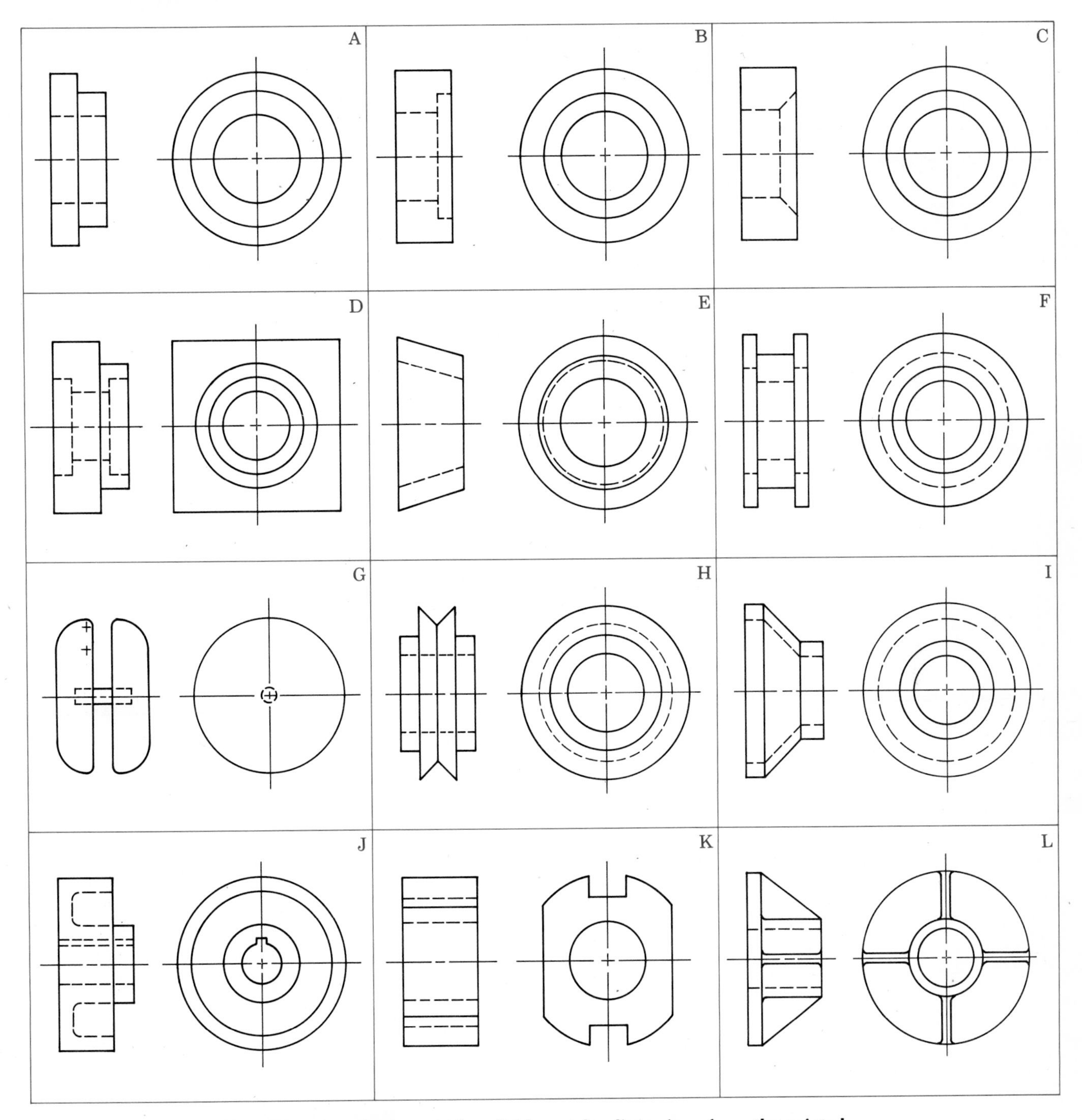

Fig. 9–49 **Problems for practice in sectioning. Using dividers, take dimensions from the printed scales on page 198. Draw both views. Make full- or half-sectional view as assigned. Add dimensions if required. Estimate the size of radii.**

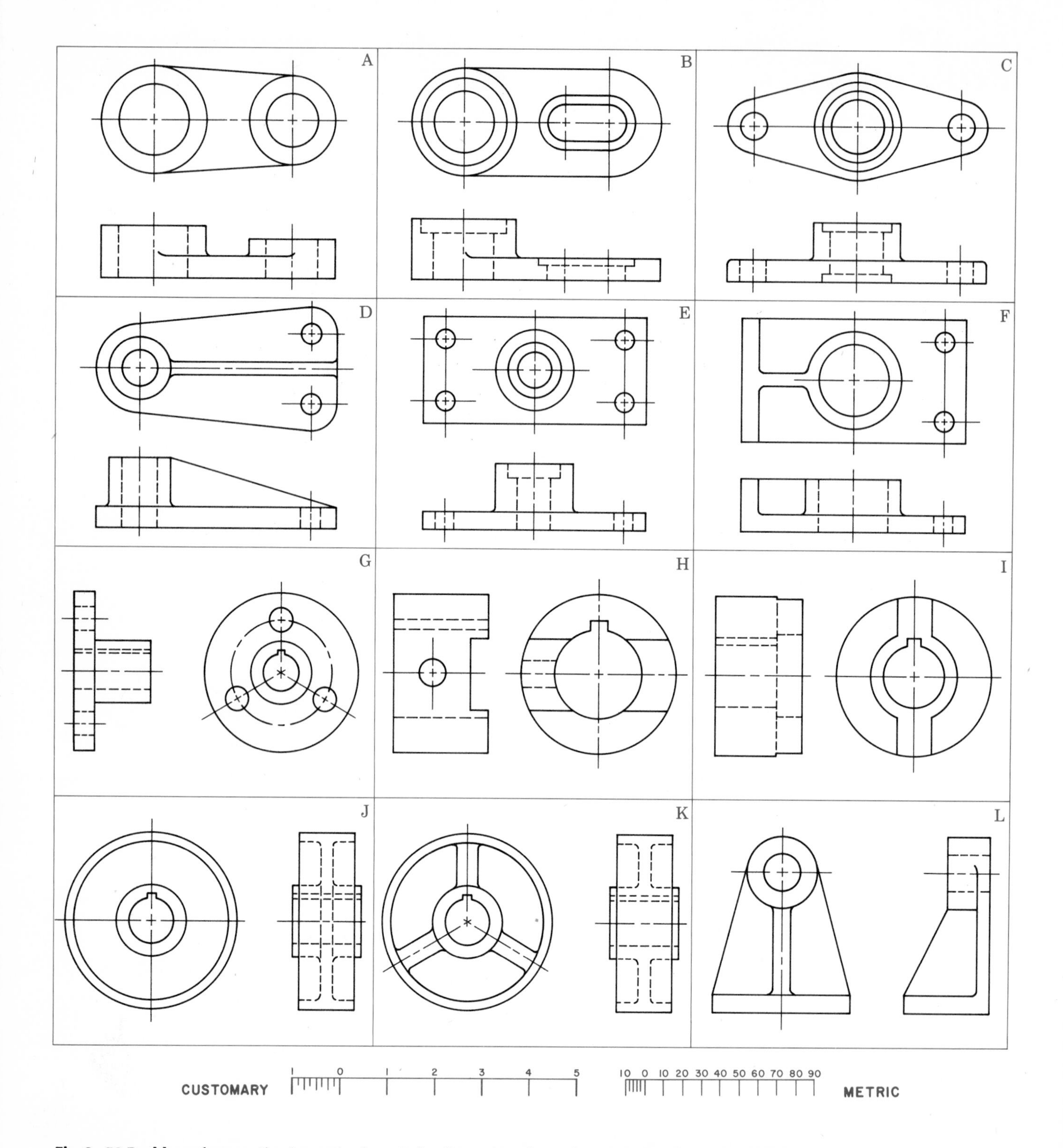

Fig. 9–50 **Problems for practice in sectioning. Take dimensions from the printed scale, using dividers. Draw both views. Make full- or half-sectional view as assigned. Add dimensions if required. Estimate the size of radii.**

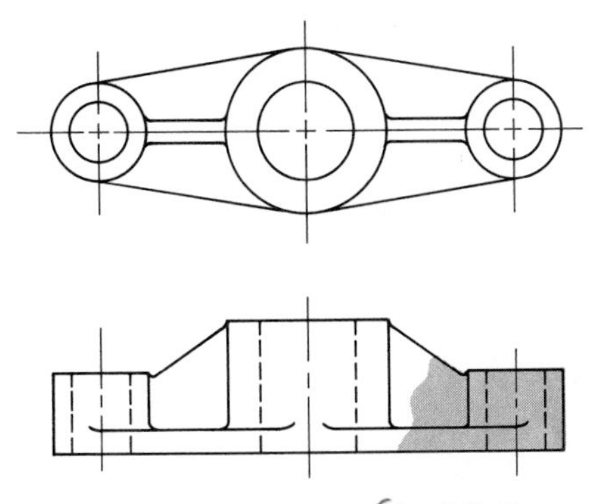

Fig. 9–51 **Rod guide. Scale: full size or as assigned. Make top view and broken-out section as indicated by the colored screen.**

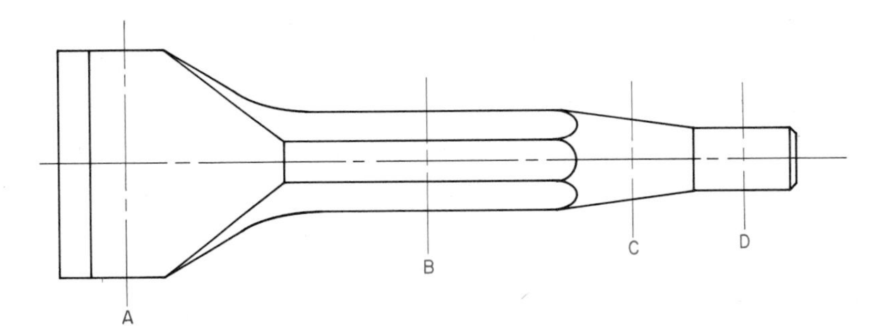

Fig. 9–52 **Chisel. Scale: full size or as assigned. Make revolved or removed sections on colored centerlines. A is a 1/2″ × 4 1/2″ rectangle, B is a 2″ (across flats) octagon, C is a circular cross section, D is a circular cross section.**

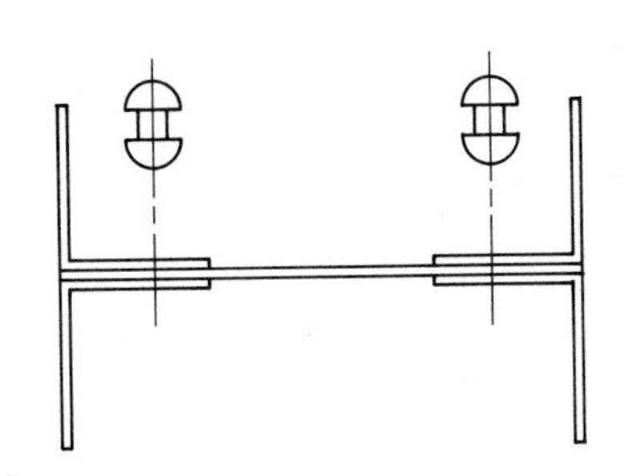

Fig. 9–53 **Structural joint. Scale: 3/4 size or as assigned. Make full-sectional view of joint with rivets moved into their proper position on the centerlines.**

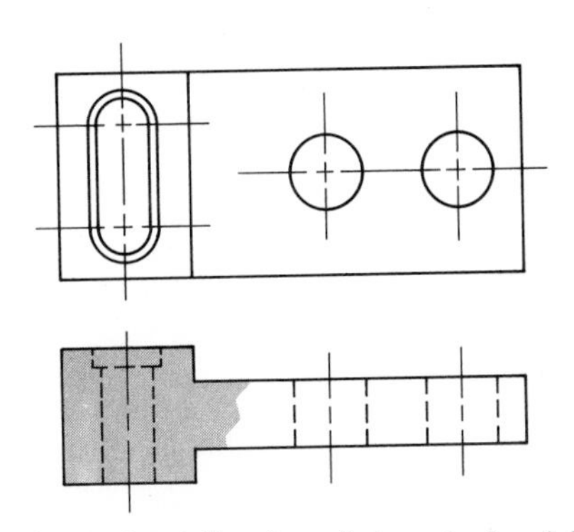

Fig. 9–54 **Adjusting plate. Scale: full size or as assigned. Draw front and top views. Make broken-out section as indicated by the colored screen.**

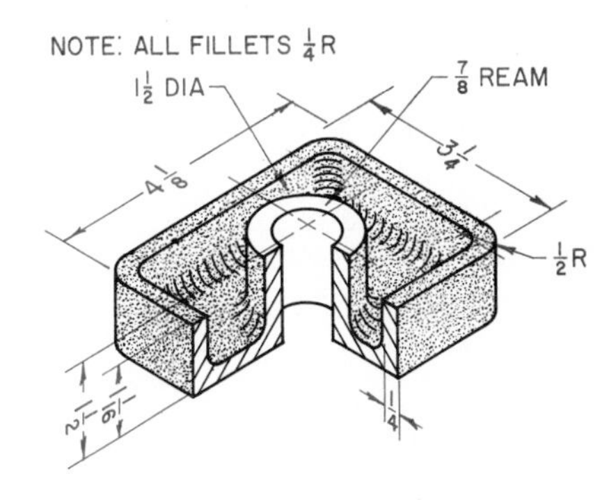

Fig. 9–55 **Shaft base. Scale: 3/4 size or as assigned. Material: cast iron.**

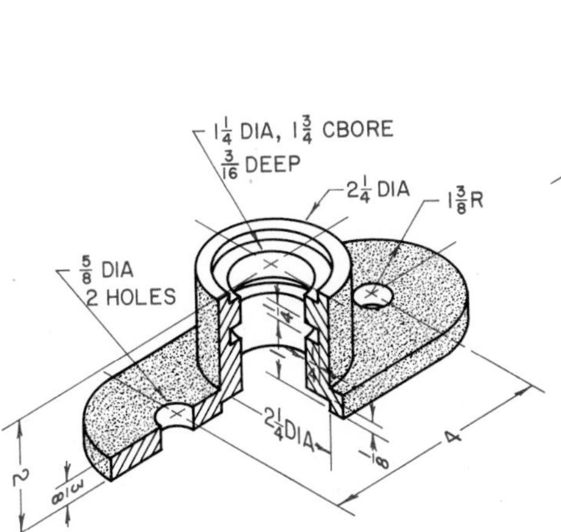

Fig. 9–56 **Lever bracket. Scale: full size or as assigned. Material: cast iron.**

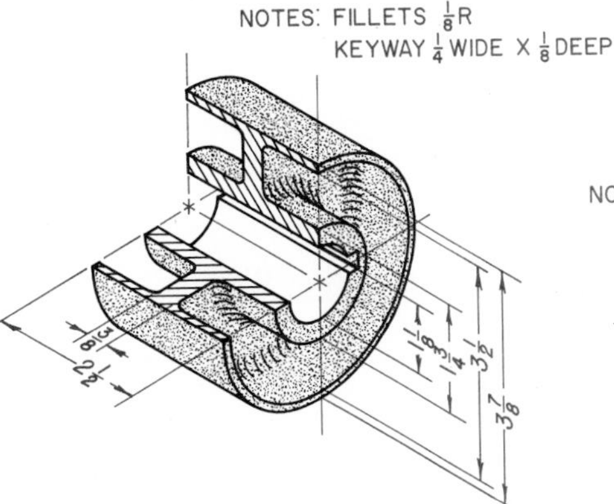

Fig. 9–57 **Idler pulley. Scale: full size or as assigned. Material: cast iron.**

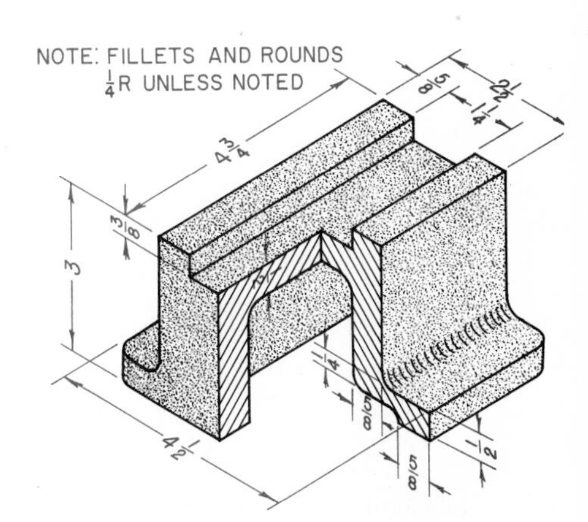

Fig. 9–58 **Rest. Scale: 3/4 size or as assigned. Material: cast aluminum.**

Fig. 9–59 **Make a two-view drawing of the hood bearing. Show the right-hand view in section.**

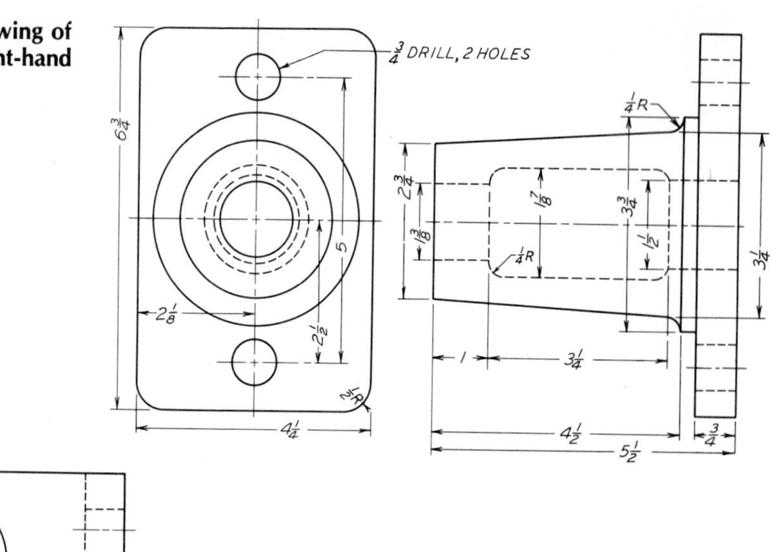

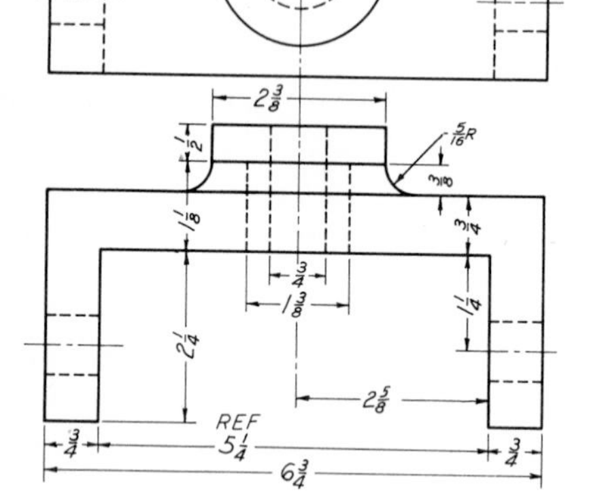

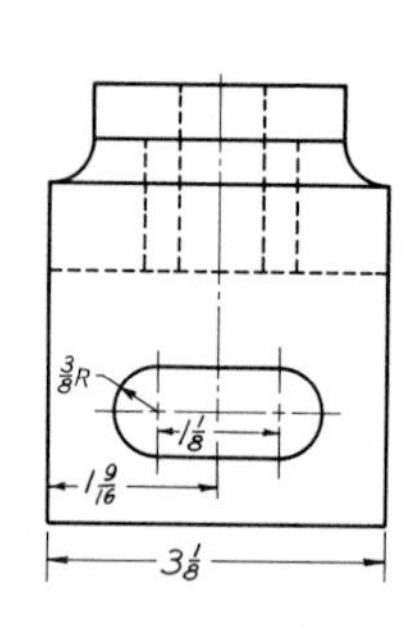

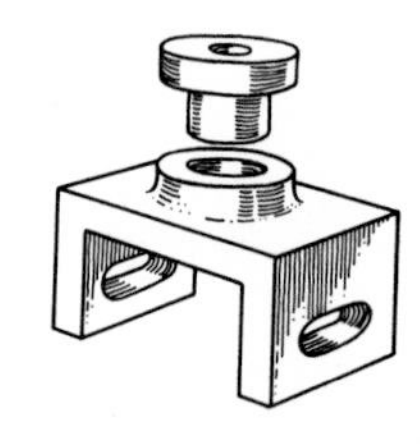

Fig. 9–60 **Draw three views of the yoke, the front view in section. There are two pieces: the yoke and the bushing. Do not copy the picture.**

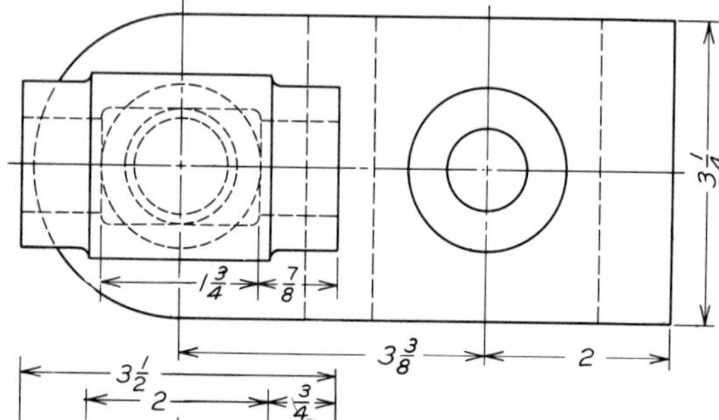

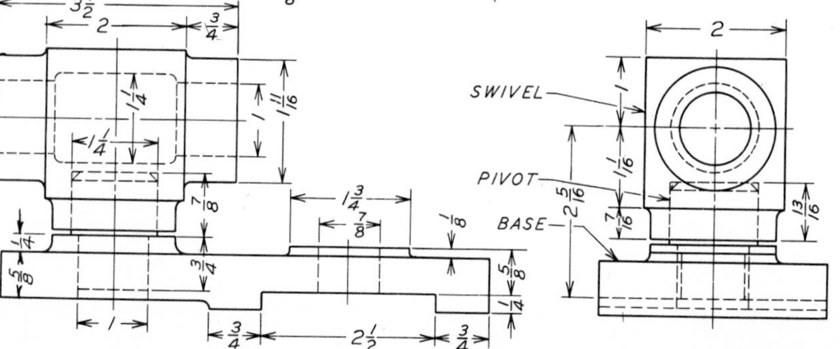

Fig. 9–61 **Draw three views of the swivel base, the front view in section.**

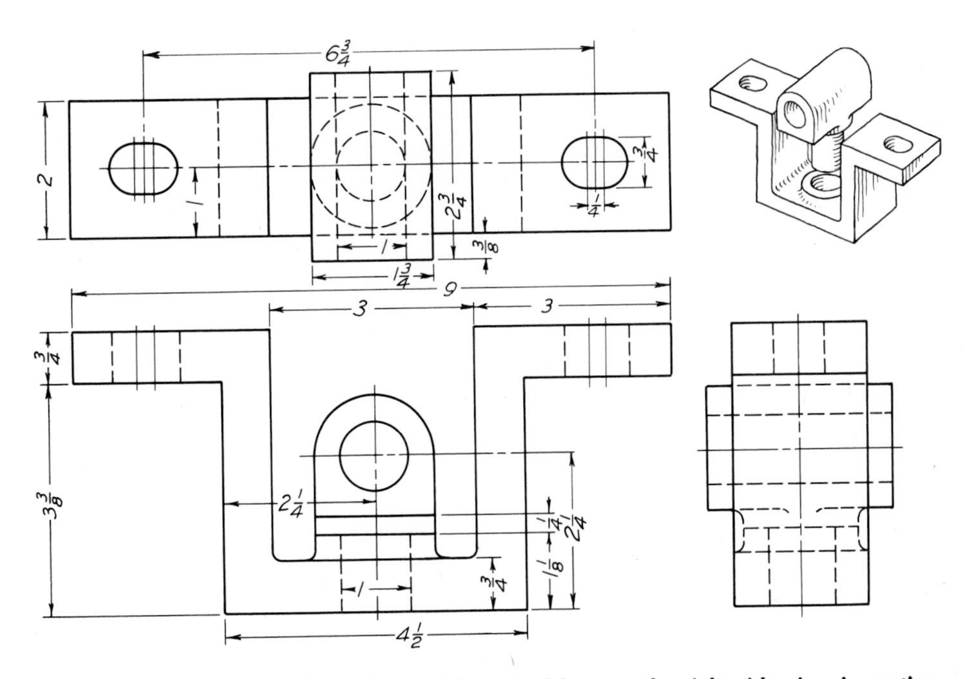

Fig. 9–62 **Draw three views of the swivel hanger, the right-side view in section. There are two pieces: the hanger and the bearing.**

Fig. 9–63 **Draw three views of the thrust bearing, the right-hand view in section. There are three parts: the shaft, the hub, and the base.**

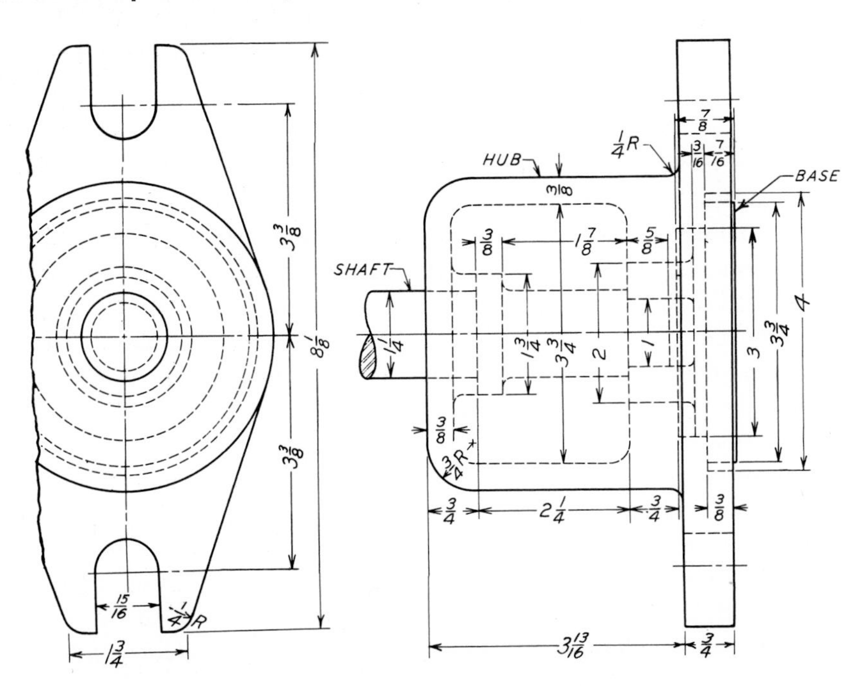

Chapter 10

Fasteners

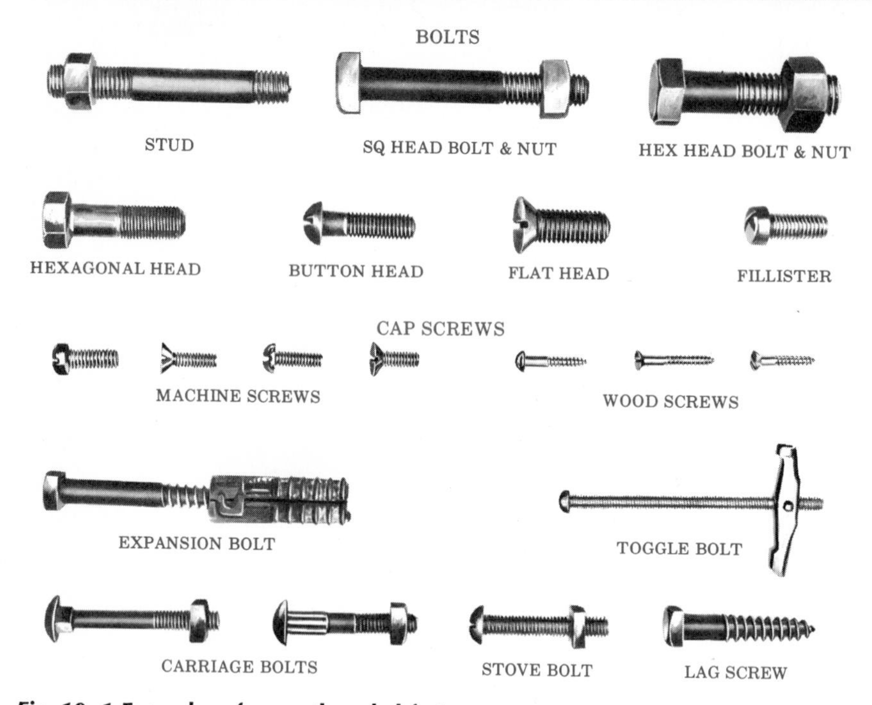

Fig. 10–1 **Examples of some threaded fasteners.**

THE FUNCTION OF FASTENERS

A fastener is any kind of device or method for holding parts together. Screws, bolts and nuts, rivets, welding, brazing, soldering, adhesives, collars, clutches, and keys are all fasteners. Each of these fastens in a different way. Each can also fasten parts permanently or so that they can later be taken apart again or adjusted.

SCREWS AND SCREW THREADS

The principle of the screw thread has been known for so long that no one knows who discovered it. Archimedes (287–212 B.C.), a Greek mathematician, put the screw to practical use. He used it in designing a screw conveyor to raise water. Similar devices are still used today to move flour and sugar in commercial bakeries, to raise wheat in grain elevators, to move coal in stokers, and for many other purposes.

Screws and other fasteners have so many uses and have become so important that engineers, drafters, and technicians must become familiar with their different forms (Fig. 10–1). They must also be able to draw and *specify* (call for) each type correctly.

THE TRUE SHAPE OF A SCREW THREAD

All screw threads are shaped basically as a *helix,* or *helical curve.* Technically, a *helix* is the curving path that a point would follow if it were to travel in an even spiral around a cylinder and *parallel* to (in line with) the axis of that cylinder. In simpler terms, if a wire is wrapped around a cylinder in evenly spaced coils, it forms helical curves. A coil spring is another example of helical curves.

Another way to visualize the shape of screw threads is to cut out a right triangle in paper and wrap it around a cylinder, as shown in Fig. 10–2 at A. If the triangle's base is the same length as the cylinder's circumference, its hypotenuse, wrapped around the cylinder, will form one turn of a helix. The triangle's altitude will be the pitch of the helix. A right triangle and the projections of the corresponding helix are shown in Fig. 10–2 at B.

To draw the projections of a helix, follow the method shown in Fig. 10–3 in spaces A and B. First, draw two projections of a cylinder as in space A. Lay off the *pitch*

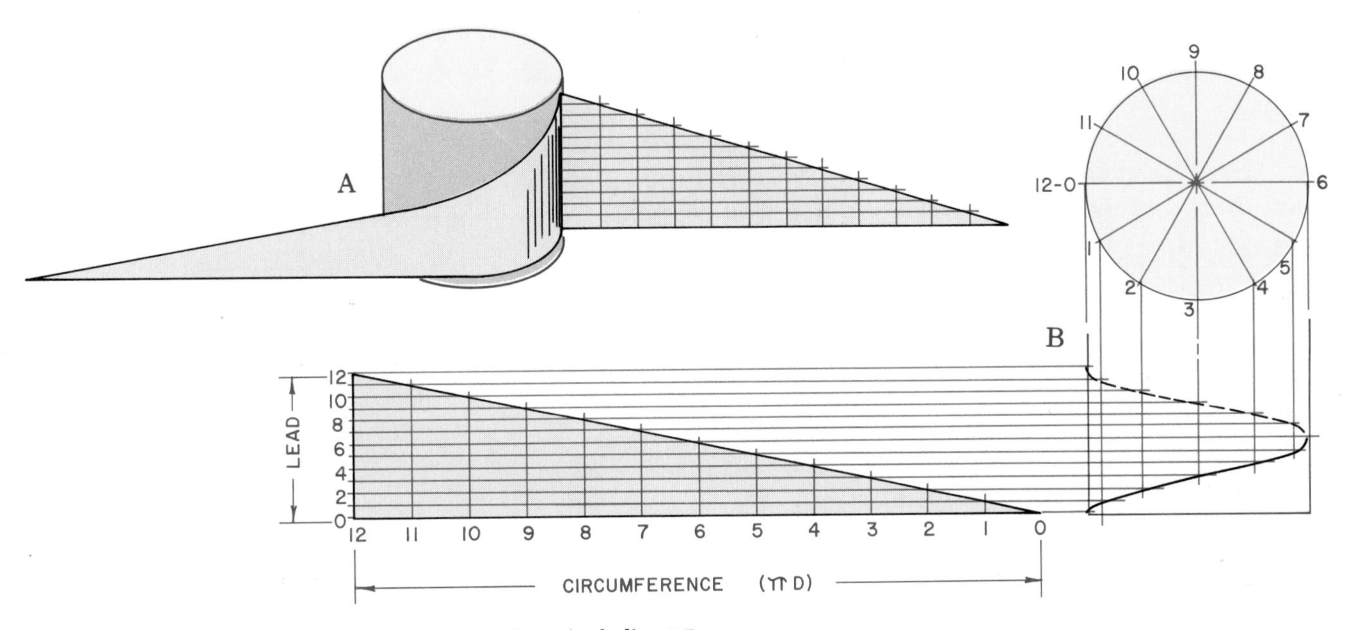

Fig. 10–2 **Picture of a helix at A and a projection of a helix at B.**

(the distance from a point on the thread form to the corresponding point on the next form). Divide the circumference into a number of equal parts. Divide the pitch into the same number of equal parts. From each division point on the circumference, draw lines parallel to the axis. From each division point on the pitch, draw lines at right angles to the axis. Then, draw a smooth curve through the points where these lines meet, as in space B. This will give the projection of the helix.

The application of the helix is shown in space C, the actual projection of a square thread. However, such drawings are seldom made, since they take too much time. Also, they are no more useful than conventional representations.

SCREW-THREAD STANDARDS

The first screws were made for one purpose. They were made without any thought of how anyone else might make one of the same diameter. Later, as industry developed, goods were produced in quantity by using *interchangeable* parts (parts in standard sizes that can be substituted for each other). The need then arose for standards for screws and screw threads.

Screw-thread standards in the United States were developed from a system that William Sellers presented to the Franklin Institute in Philadelphia in 1864. Screw-thread standards in England came from a paper presented to the Institution of Civil Engineers in 1841 by Sir Joseph Whitworth. These two standards were not interchangeable.

More and better screw-thread standards have been drawn up as industrial production has grown more complex. In 1948, standardization committees of Canada, Great Britain, and the United States agreed on the Unified Thread

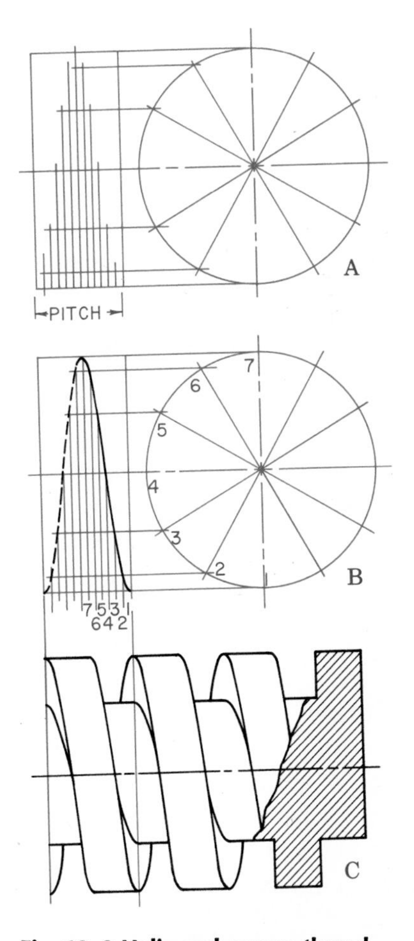

Fig. 10–3 **Helix and square thread.**

Standards. These standards are now the basic American National Standards. They are listed in the *American National Standard Unified Screw Threads for Screws, Bolts, Nuts and Other Threaded Parts* (ANSI B1.1–1960) and in Handbook H–28, *Federal Screw Thread Specifications.*

Since 1948, the different thread systems have been brought increasingly into line with each other. In 1968, the International Organization for Standardization (ISO) adopted the Unified Standard as its inch screw-thread system standard. The organization already had a metric screw-thread system standard. The two systems are alike in some ways.

These similarities are very important as the metric system of measurement comes into worldwide use. To help bring other screw-thread standards into line with this system, the American National Standards Institute has published the *American National Standard Unified Screw Threads—Metric Translation* (ANSI B1.1a–1968).

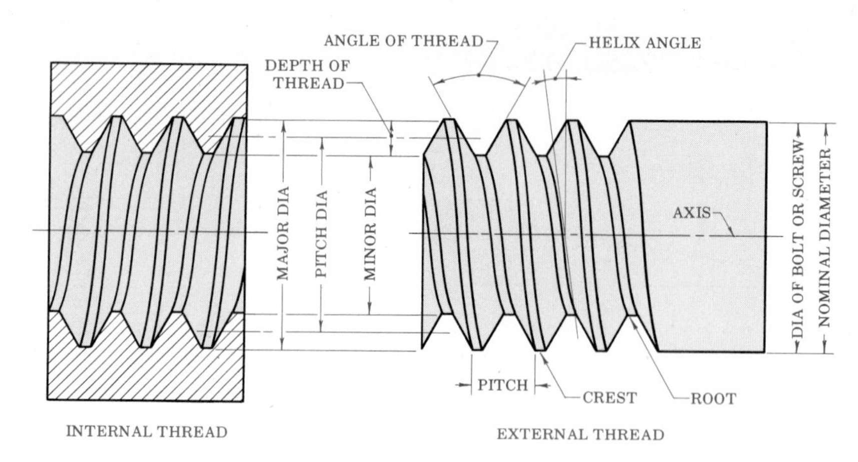

Fig. 10–4 **Screw-thread terms.**

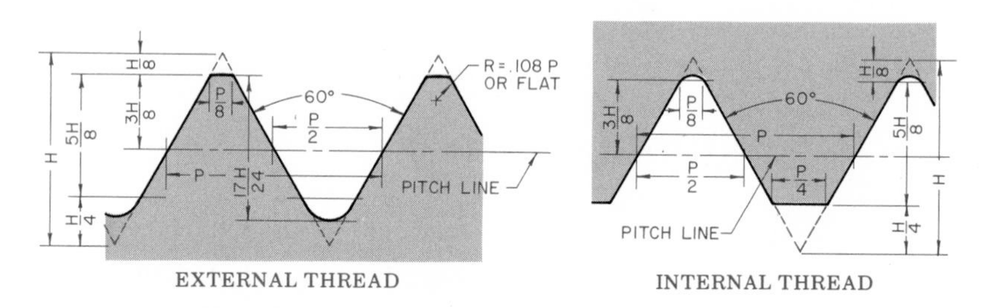

Fig. 10–5 **Unified screw-thread terms.**

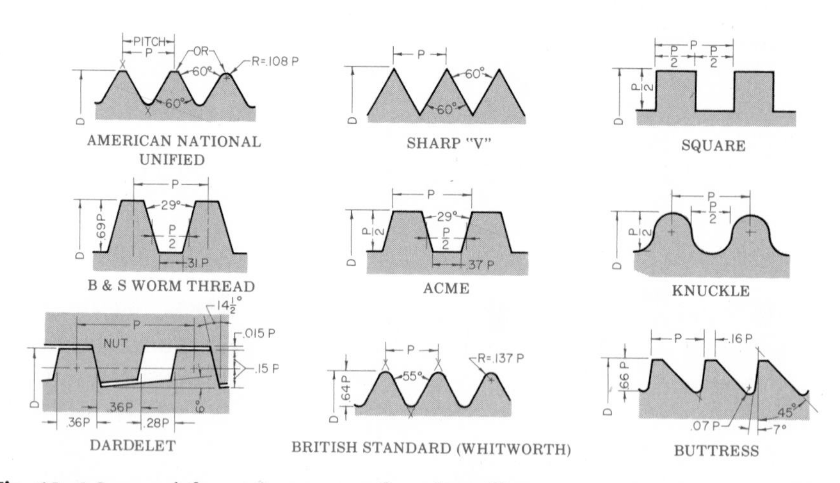

Fig. 10–6 **Some of the various screw-thread profiles.**

SCREW–THREAD TERMS

Figure 10–4 shows the main terms used to describe screw-threads. The Unified and American (National) screw-thread profile shown in Fig. 10–5 is the form used for fastening in general. Other forms of threads are used to meet special fastening needs. Some of these threads are shown in Fig. 10–6. The sharp V is seldom used today. The square thread and similar forms (worm thread and acme thread) are made especially to transmit motion or power along the line of the screw's axis. The knuckle thread is the thread used in most electric-light sockets. It is also a "cast" thread. The Dardelet thread automatically locks a screw in place. It was designed by a French military officer. The former British Standard (Whitworth) has rounded crests and roots. Its profile forms 55° angles. The former United States Standard had flat crests and roots and 60° angles. The buttress

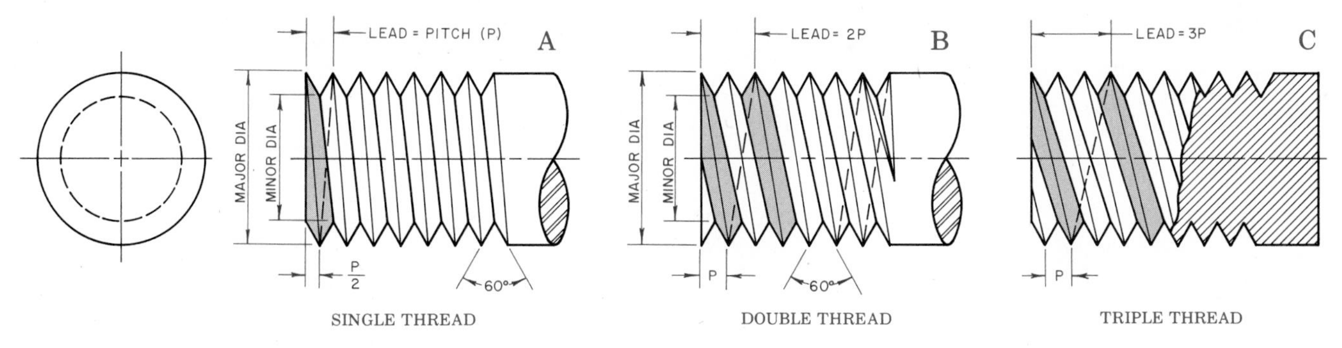

Fig. 10–7 **Single, double, and triple threads.**

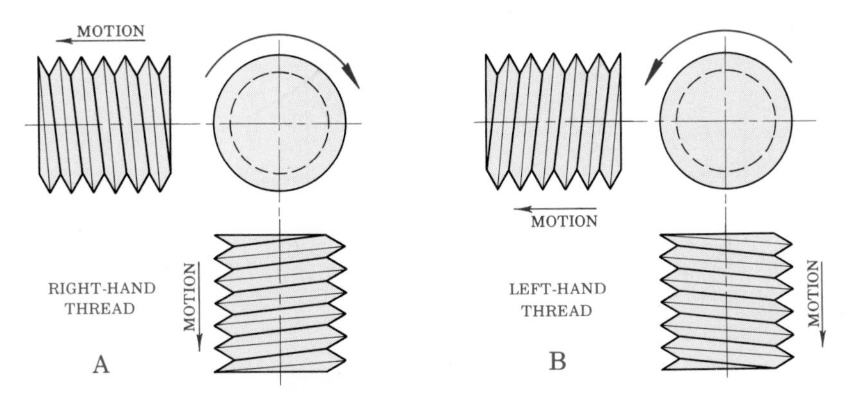

Fig. 10–8 **A right-hand screw thread at A and left-hand screw thread at B.**

thread takes pressure in one direction only: against the surface at an angle of 7°. On this thread, one face is not slanted. Instead, it is at right angles to the screw's axis.

DEFINITION OF A SCREW THREAD

A screw thread is a helical ridge on the external or internal surface of a cylinder. A screw thread is also made in the form of a conical spiral on the external or internal surface of a cone or frustum of a cone.

The *pitch* of a thread is the distance from one point on the thread form to the corresponding point on the next form. Pitch is measured parallel to the thread's axis (Fig. 10–7). The *lead* (say "leed") L is the distance along this same axis that the threaded part moves against a fixed mating part when given one full turn. It is the distance a screw enters a threaded hole in one turn.

SINGLE AND MULTIPLE THREADS

Most screws have single threads (Fig. 10–7 at A). A screw has a single thread unless it is marked otherwise. A single thread is a single ridge in the form of a helix. If you give a single-thread screw one full turn, the distance it will advance into the nut (lead) will equal the pitch of the thread.

A double thread (Fig. 10–7 at B) is two helical ridges side by side. The lead, in this case, is twice the pitch. A triple thread (Fig. 10–7 at C) is three ridges side by side. The lead for this thread is three times the pitch.

Multiple threads are used where parts must screw together quickly. For example, technical pen caps and toothpaste tube caps, have multiple threads.

RIGHT- AND LEFT-HAND THREADS

A *right-hand thread* screws in when you turn it clockwise as you view it from the outside end (Fig. 10–8 at A). A *left-hand thread* screws in when you turn it counterclockwise (Fig. 10–8 at B). Threads are always right hand unless marked with the initials *LH,* meaning a left-hand thread. Some devices, such as the turnbuckle (Fig. 10–9), have both right- and left-hand threads. Others, such as bicycle pedals, can have either. A left-hand pedal has left-hand threads; a right-hand pedal, right-hand threads.

REPRESENTATIONS OF SCREW THREADS ON DRAWINGS

When you draw screw-threads, you use special symbols. These are the same whether the threads are

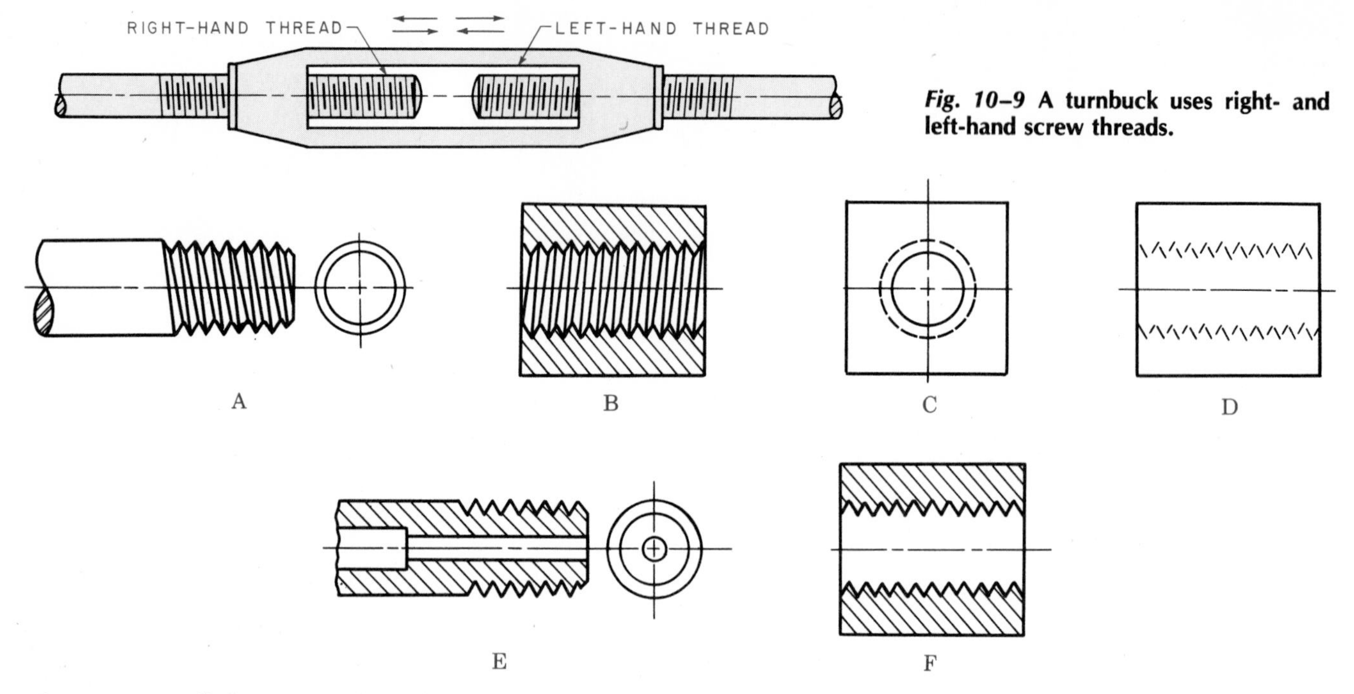

Fig. 10–9 **A turnbuck uses right- and left-hand screw threads.**

Fig. 10–10 **Detailed representations of screw threads.**

coarse or fine, or right-hand or left-hand. You use notes to give the necessary information.

Under ANSI rules, you can draw screw-threads in three ways:

1. A *detailed representation* approximates the real look of threads (Fig. 10–10). For this kind of drawing, you do not have to draw the pitch exactly to scale. Instead, just make it about the right size for the drawing. Draw the helixes as straight lines. Draw the threads as sharp Vs. In general, you do not usually use the detailed representation in working drawings (see Chapter 11), except where you need it for clearness. You also do not usually use it if the screw is less than 25 mm (1 in.) in diameter.
2. A *schematic representation* shows the threads with symbols rather than as they really look. For this kind of drawing, you leave out the Vs (Fig. 10–11). You also need not draw the pitch to scale. Just make it about the right size for the drawing. Then draw the crest and root lines accordingly. Space them by eye to make them look good. These lines can be at right angles to the axis or slanted to show the helix angle (see Fig. 10–17). The American National Standards calls for crest lines to be thin and root lines to be thick. However, you may make all your lines the

Fig. 10–11 **Schematic representations of screw threads.**

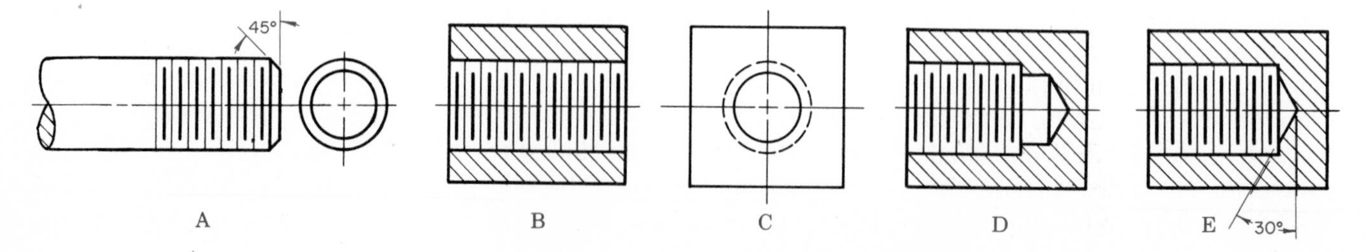

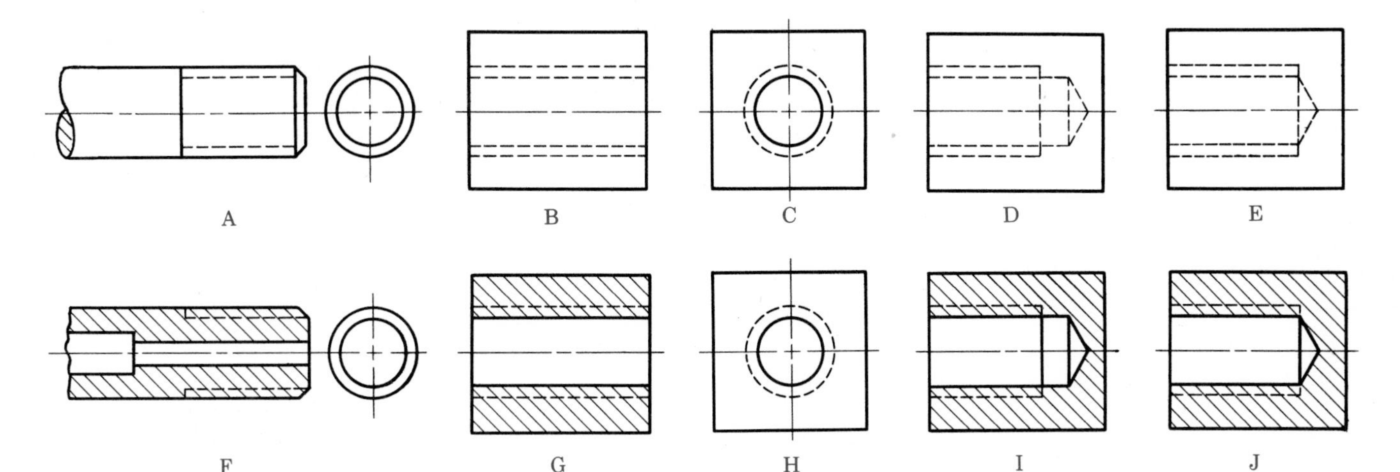

Fig. 10–12 **Simplified representations of screw threads.**

same thickness to save time. This is especially so on regular pencil working drawings.

3. A *simplified representation* is much like a schematic representation. In this case, however, you draw the crest and root lines as dotted lines, except where either of them would show as a visible solid line (Fig. 10–12 at A, C, F, G, H, I, and J). You save time by using the simplified representation because it leaves out useless details.

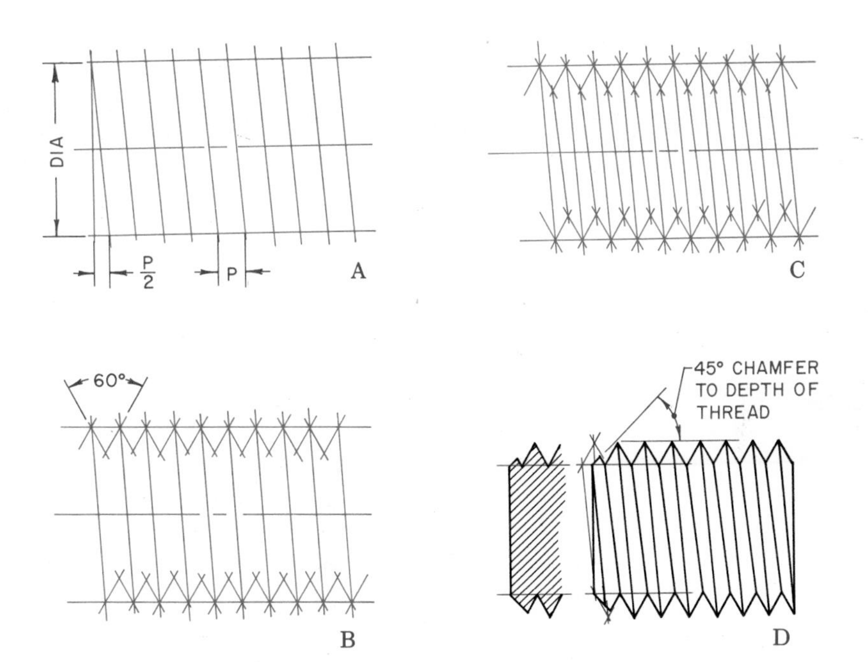

Fig. 10–13 **To draw the detailed representation of screw threads.**

TO DRAW THE DETAILED REPRESENTATION OF SCREW THREADS

For a detailed representation, you draw the screw threads with the sharp-V profile. Use straight lines to represent the helixes of the crest and root lines. To draw the V-form thread, follow the steps shown in Fig. 10–13. First, lay off the pitch *P*, as at A. You do not have to draw the pitch to scale. Just make it about the right size for the drawing. Also lay off the half pitch *P*/2 at the end of the thread, as shown. Construct a right triangle with a base of *P*/2 and an altitude equal to the outside diameter of the screw. Adjust your triangle, or the ruling arm of your drafting machine, to the slope of this right triangle. Now draw all the crest lines with this slope. Next, use your 30°–60° triangle (or your drafting machine ruling arm set at a 30° angle) to draw one side of the *V* for the threads, as at B. Then reverse the triangle, or ruling arm, and complete the Vs. Now set the triangle, or the ruling arm of the drafting machine, to the slope of the root lines. Draw them as shown at C. Notice that the root lines do not *parallel* (have the same slope as) the crest lines. This is because the root diameter is less

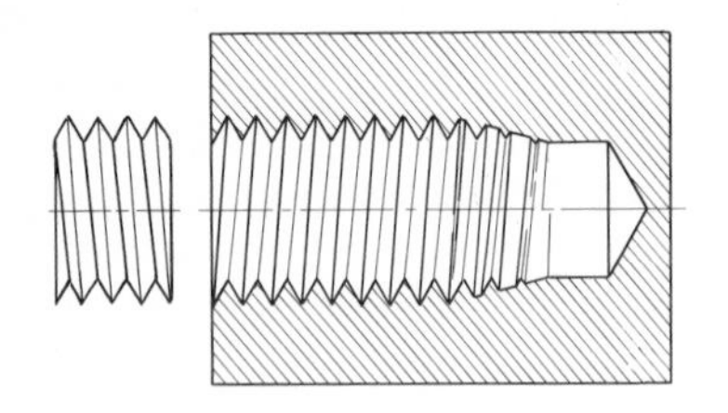

Fig. 10–14 **Internal threads in section (threaded hole).**

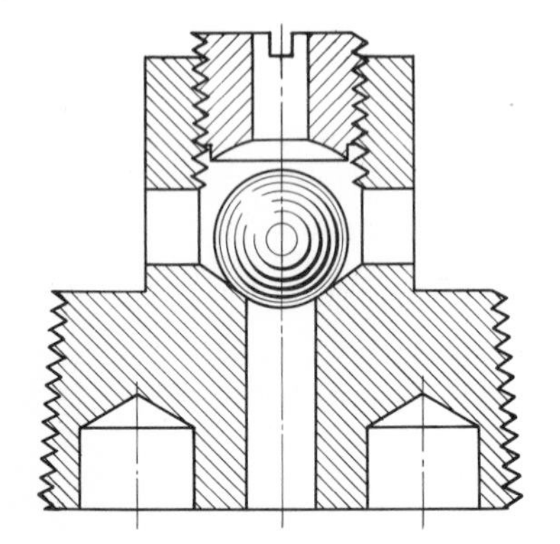

Fig. 10–15 **Threads in section on assembled pieces.**

than the major diameter. Finally, draw 45° chamfer lines using the construction shown in red at D.

A hole with an internal right-hand thread is shown as a sectional view in Fig. 10–14. Notice that the thread-line slope is in the opposite direction from the *external* right-hand thread lines on a mating screw. This is because the internal thread lines must match the far side of the screw.

The realistic, or V-form thread representation can make a drawing clearer where two or more threaded pieces are shown in section (Fig. 10–15).

TO DRAW THE SCHEMATIC REPRESENTATION OF SCREW THREADS

Follow the steps in Fig. 10–16. First, lay off the outside diameter of the screw thread as at A. Then follow the steps shown at B to lay off the thread depth and the chamfer. Next, draw thin crest lines at right angles to the axis, as at C. Then draw thick root lines parallel to the crest lines, as at D.

You may wish to draw your crest and root lines at a slope (Fig. 10–17 at A). If you do so, give each thread a slope of half the pitch. You may also make your crest and root lines the same width on pencil drawings (Fig. 10–17 at B). Finally, you need not lay off the pitch to scale. Just space it to look good.

TO DRAW THE SIMPLIFIED REPRESENTATION OF SCREW THREADS

Follow the steps in Fig. 10–18. First, lay off the outside diameter of the screw as at A. Next, follow the steps shown at B to lay off the screw-thread depth and the chamfer. Then draw the chamfer and a line to show the length of the thread, as at C. Finally, draw dash lines for the threads, as at D to complete the drawing.

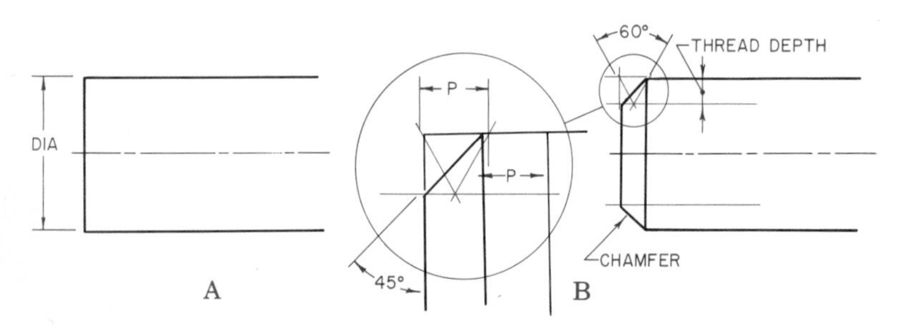

Fig. 10–16 **To draw the schematic representation of screw threads.**

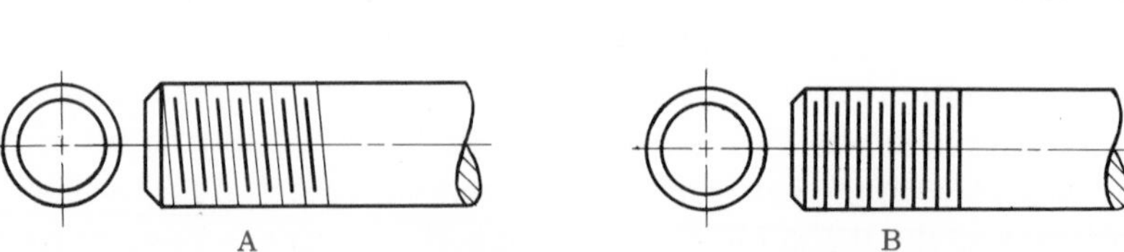

Fig. 10–17 **Slope-line representation at A and uniform-width lines at B.**

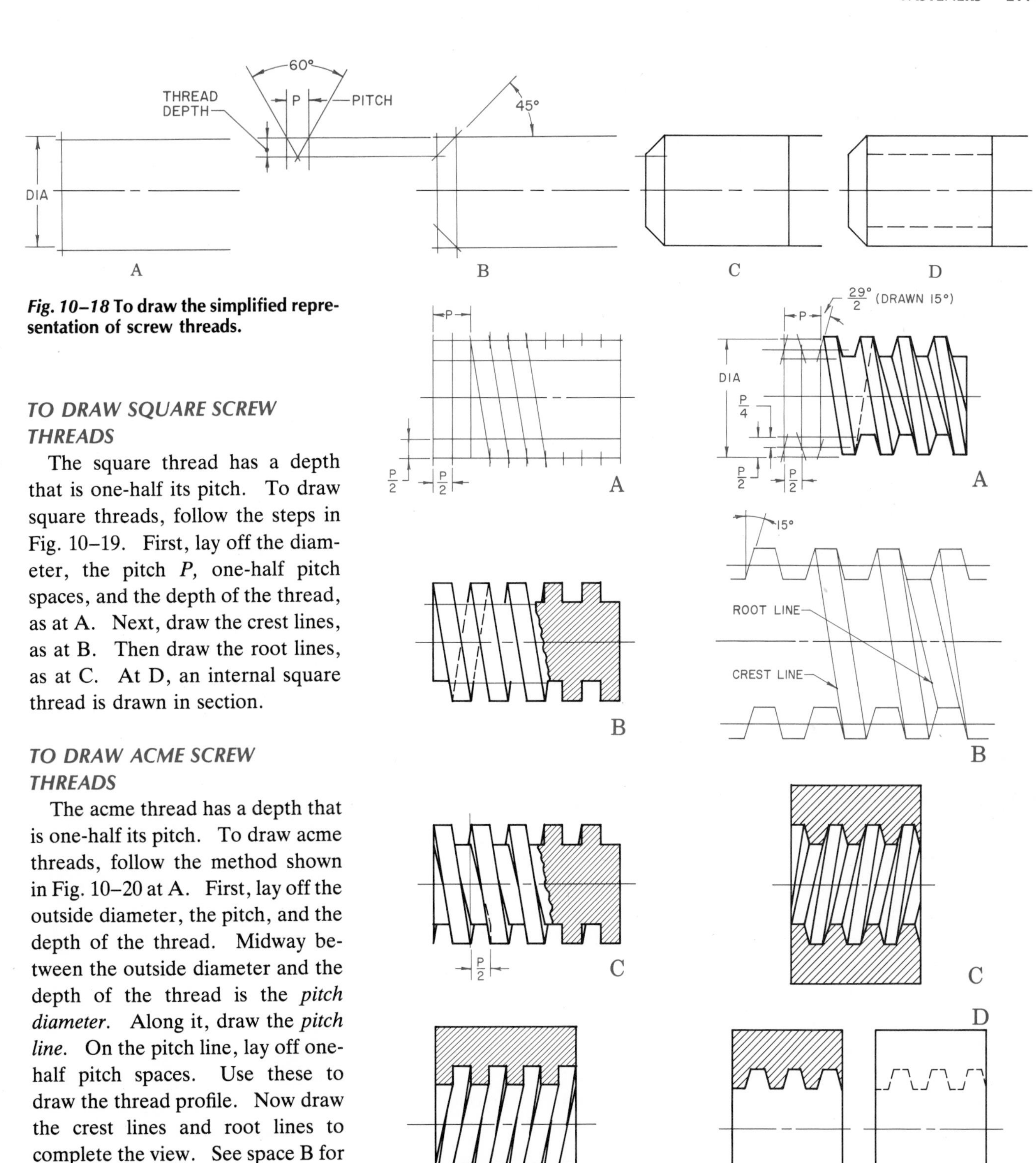

Fig. 10–18 **To draw the simplified representation of screw threads.**

Fig. 10–19 **To draw square threads.**

Fig. 10–20 **To draw acme threads.**

TO DRAW SQUARE SCREW THREADS

The square thread has a depth that is one-half its pitch. To draw square threads, follow the steps in Fig. 10–19. First, lay off the diameter, the pitch *P,* one-half pitch spaces, and the depth of the thread, as at A. Next, draw the crest lines, as at B. Then draw the root lines, as at C. At D, an internal square thread is drawn in section.

TO DRAW ACME SCREW THREADS

The acme thread has a depth that is one-half its pitch. To draw acme threads, follow the method shown in Fig. 10–20 at A. First, lay off the outside diameter, the pitch, and the depth of the thread. Midway between the outside diameter and the depth of the thread is the *pitch diameter.* Along it, draw the *pitch line.* On the pitch line, lay off one-half pitch spaces. Use these to draw the thread profile. Now draw the crest lines and root lines to complete the view. See space B for an enlarged view.

You can draw an internal acme thread in section, as shown at C. You can also draw it in other ways, including with dashes for the hidden lines or the outline in section, as shown at D.

THREAD SERIES FOR UNIFIED AND AMERICAN NATIONAL STANDARD SCREW THREADS

Screws of the same diameter are made with different pitches (number of threads per inch) for different uses. The various combinations of diameter and pitch have been grouped in *screw-thread series.* These series are listed in ANSI B1.1. Each is denoted by letter symbols, as follows:

Coarse-thread Series (UNC or NC). In this series the pitch for each diameter is relatively large. This series is used for engineering in general.

Fine-thread Series (UNF or NF). In this series, the pitch for each diameter is smaller (there are more threads per inch) than in the coarse-thread series. This series is used where a finer thread is needed, as in making automobiles and airplanes.

Extra-fine-thread Series (UNEF or NEF). In this series, the pitch is even smaller than in the fine-thread series. This series is used where the thread depth must be very small, as on aircraft gear or thin-walled tubes.

There are also several constant-pitch-thread series. They are indicated by the letter symbol, UN. These series have 4, 6, 8, 12, 16, 20, 28, or 32 threads per inch. They offer a variety of pitch-diameter combinations that can be used where the coarse, fine, and extra-fine series are not suitable. However, when selecting a constant-pitch series, the first choices should be the 8-, 12-, or 16-thread series. Constant-pitch threads are mainly used on parts that must be screwed together and unscrewed so often that thread repairs may be needed.

Eight-thread Series (8UN or 8N). This series uses 8 threads per inch for all diameters.

Twelve-thread Series (12UN or 12N). This series uses 12 threads per inch for all diameters.

Sixteen-thread Series (16 UN or 16N). This series uses 16 threads per inch for all diameters.

Special Threads (UNS, UN, or NS). These are nonstandard, or special, combinations of diameter and pitch.

American National Standard thread series have been largely replaced by the Unified Standard. However, you should still know what they and their letter symbols are. These series and their symbols are listed below.

Coarse-thread Series (NC). This series is used for screws, bolts, and nuts produced in quantity, and also for fastening in general.

Fine-thread Series (NF). This series has a smaller pitch for each diameter than NC. It is used where the coarse series is not suitable.

Extra-fine thread Series (NEF). This series is used when an even smaller pitch than NF is needed, as on thin-walled tubes.

The symbol NS denotes special threads.

Eight-thread Series (8N). This series is a constant-pitch series for large diameters.

Twelve-thread Series (12N). This series is a constant-pitch series that has a medium-fine pitch for large diameters.

Sixteen-thread Series (16N) is a constant-pitch series that has an even smaller pitch per diameter.

Notice that the American system has only three constant-pitch series, while the Unified system has eight, including all three from the American system. The three in the American system are used most.

CLASSES OF FITS FOR UNIFIED AND AMERICAN NATIONAL SCREW THREADS

Screw threads are also divided into *screw-thread classes* based on their *tolerances* (amount of size different from exact size) and *allowances* (how loosely or tightly they fit their mating parts). You can get exactly the screw thread you need by choosing both a series and a class. In brief, the classes for Unified threads are: Classes 1A, 2A, and 3A for external threads only; Classes 1B, 2B, and 3B for internal threads only.

Classes 1A and 1B have a large allowance (loose fit). They are used on parts that must be put together quickly and easily.

Classes 2A and 2B are the thread standards most used for general purposes, such as for bolts, screws, nuts, and similar threaded items.

Classes 3A and 3B are stricter standards for fit and tolerance than the others. They are used where thread size must be more exact.

Classes 2 and 3 are American National Standard. You will find them described, with tables of dimensions, in Appendix 1 of ANSI B1.1.

SCREW–THREAD SPECIFICATIONS

You *specify* (call for) a particular screw thread by telling its diameter (nominal, or major, diameter), number of threads per inch, length of thread, initial letters of the series, and class of fit. Any thread you specify will be assumed to be both single and right hand unless you say otherwise. If you mean the thread to be left hand, include the letters LH after the class symbol. If you mean it to be double or triple, include "double" or "triple."

Some examples using fractional sizes follow:

1¼—7UNC—1A (1¼ diameter, 7 threads per inch, Unified threads, coarse threads, class 1, external)

0.750—10UNC—2A (0.750 diameter, 10 threads per inch, Unified threads, coarse threads, class 2, external) See Fig. 10–21.

⅞—14UNF—2B (⅞ diameter, 14 threads per inch, Unified threads, fine threads, class 2, internal)

Fig. 10–21* Customary specifications. *(McGraw-Hill Ryerson Ltd.)

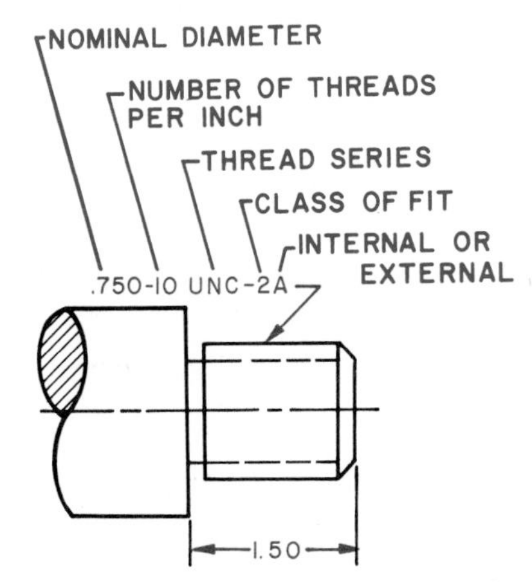

CUSTOMARY INCH THREAD CALLOUT

1⅝—18UNEF—3B—LH (1⅝ diameter, 18 threads per inch, Unified threads, extra-fine threads, class 3, internal, left-hand)

You specify tapped (threaded) holes by a note giving the diameter of the tap drill (27/64″); depth of hole (1⅜″); thread information (½ diameter, American National threads, Class 2); and length of thread (1″), as:

27/64 DRILL × 1⅜ DEEP
½—13NC—2 × 1 DEEP

THREAD DESIGNATION: METRIC THREADS

You *designate* (specify) an ISO metric screw thread by giving its nominal size (basic major diameter) and pitch, both expressed in millimeters. Include an "M" to denote that the thread is an ISO metric screw thread. Place the "M" before the nominal size. Use "x" to separate the nominal size from the pitch. For the coarse thread series only, you do not show the pitch unless you are also giving the length of the thread. If you are giving this length, separate it from the other designations with an "x." For external threads, you may give the length of thread as a dimension.

For example, you would designate a 10 mm diameter, 1.25 pitch, fine-thread series as M 10 × 1.25. A 10 mm diameter, 1.5 pitch, coarse-thread series would, however, be designated as M 10. Remember that for coarse threads, you do not give the pitch unless you are also giving the length of the thread. If, in this last example, the length were 25 mm, and if you had to call for it on the drawing, you would designate the thread as M 10 × 1.5 × 25.

To designate an ISO metric screw thread fully, you must do even more. In addition to the basic designation, you must also give the designation for the thread's tolerance class. This designation is separated from the basic designation by a dash. It consists of two sets of symbols. The first denotes the pitch diameter tolerance. The second, which follows immediately, denotes the diameter tolerance. Each of these two sets of symbols is composed of a number denoting the grade tolerance, followed by a letter (capital for internal threads and lowercase for external threads) denoting the tolerance position. If the symbols for the pitch diameter tolerance happen to be the same as those for the crest diameter tolerances, you can leave one set out (See Fig. 10–22).

THREADED FASTENINGS

Fasteners are made in many forms for different uses. The following sections will tell you how to identify and draw the threaded fasteners most commonly used on machines, constructions, and engineering projects. These include: square- and hexagonal-head bolts, square and hexagonal nuts, studs, machine screws, cap screws and setscrews, etc. You will find dimensions for drawing bolts, nuts, and some of the other generally used threaded fasteners in Appendix A.

Certain bolt and nut dimensions have been designated as Unified

Standard for use in the United States, Great Britain, and Canada. For complete metric information, consult the latest American National Standard or ISO standard.

STANDARD SQUARE AND HEXAGON BOLTS AND NUTS

These are so important that you should know the main terms used (Fig. 10–23). You should also be able to draw the necessary views. You will learn both in the sections that follow.

In general, bolts and nuts are either regular or heavy. They are also either square or hexagon. Regular bolts and nuts are used for the general run of work. Heavy bolts and nuts are somewhat larger. They are used where the bearing surface or the hole in the part being held must be larger. The types of bolts and nuts made in regular sizes include: square bolts and nuts, hexagon bolts and nuts, and semifinished hexagon bolts and nuts. Types made in heavy sizes includes hexagon bolts and nuts, semifinished hexagon bolts and nuts, finished hexagon bolts, and square nuts.

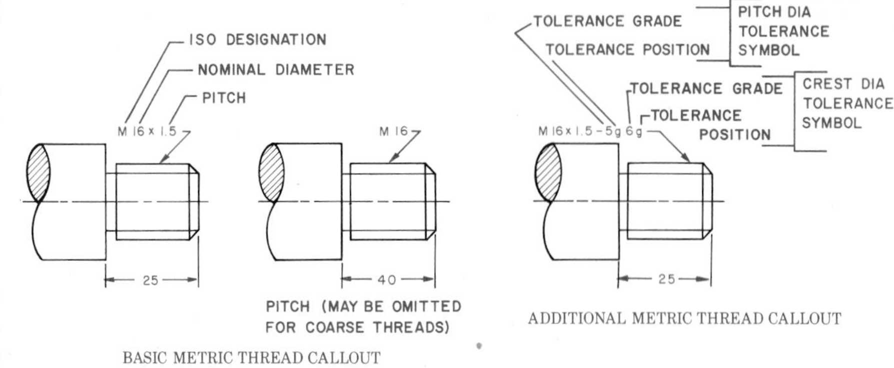

Fig. 10–22 **Metric thread specifications.** ***(McGraw-Hill Ryerson Ltd.)***

Regular bolts and nuts are not finished on any surface. Semifinished bolts and nuts are processed to have a flat bearing surface. "Finished bolts and nuts" are so called only because of the quality of their manufacture and the closeness of their tolerance. They do not have completely machined surfaces. Semifinished boltheads and nuts (Fig. 10–24) have on the bearing surface either a washer face or a face with chamfered corners. Each face has a diameter equal to the distance *across the flats* (between opposite sides of the bolthead or nut). The thickness of the washer face is about 0.4 mm (1/64 in.).

REGULAR BOLTHEADS AND NUTS

When you draw regular boltheads and nuts, figure the dimensions from the proportions given in Figs. 10–25 and 10–26 or from Appendix A. You can draw the chamfer angle at 30° for either the hexagon or the square forms. (The standard for the square form is actually 25°.) Find the radii for the chamfer arcs

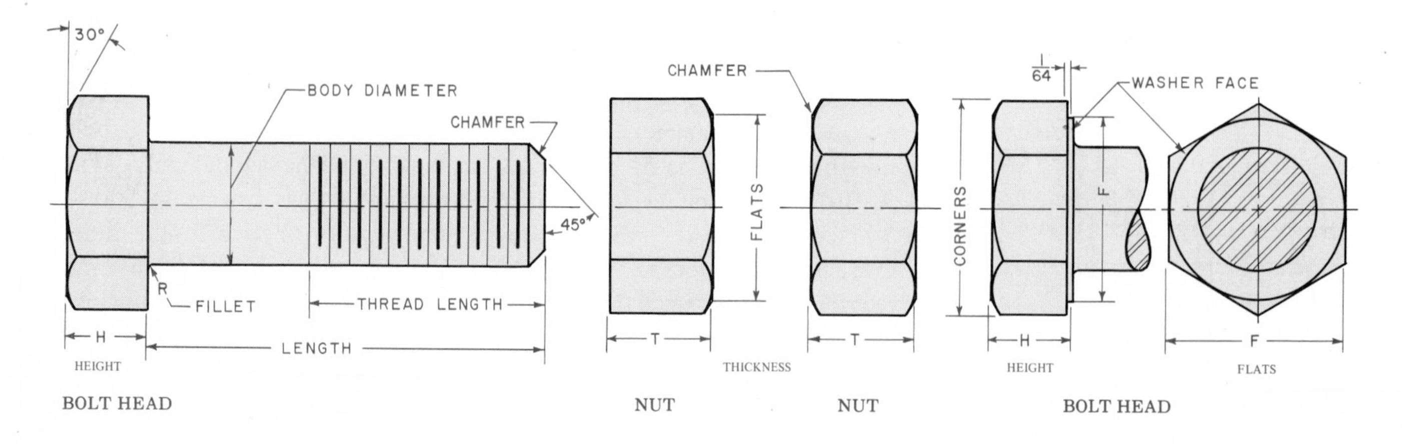

Fig. 10–23 **Bolt and nut terms.**

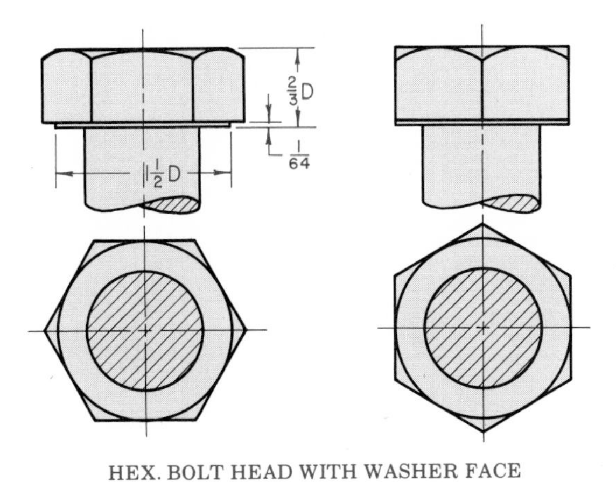

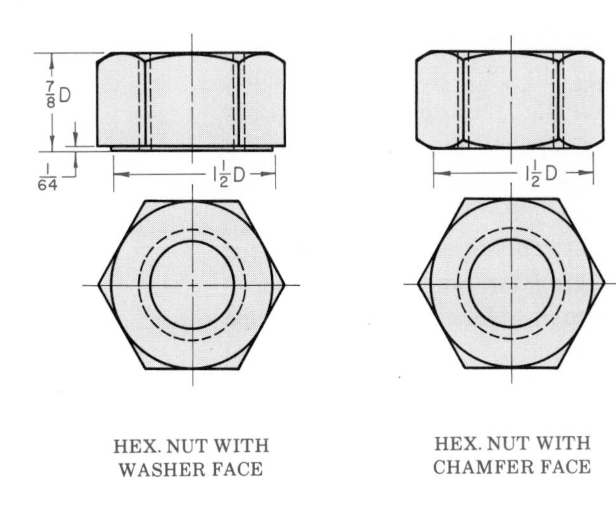

Fig. 10–24 **Semifinished boltheads and nuts.**

HEXAGON

Fig. 10–25 **Regular hexagon bolthead and nut.**

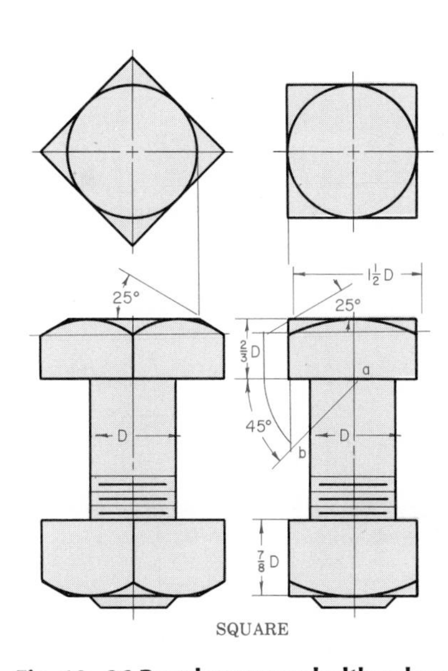

Fig. 10–26 **Regular square bolthead and nut.**

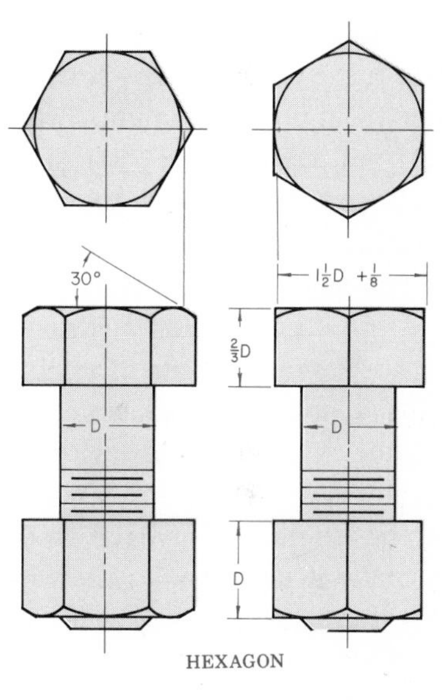

Fig. 10–27 **Heavy hexagon bolthead and nut.**

by trial. Note that you can find one-half the distance across corners, *ab*, by the construction shown.

HEAVY HEXAGON BOLTHEADS AND NUTS

When you draw heavy hexagon boltheads and nuts, figure the dimensions from the proportions given in Fig. 10–27 or from Appendix A. See also Appendix V-1962 of ANSI B18.2–1960 and B18.2.1–1965.

A new standard covers hexagon structural bolts. These bolts are made of high-strength steel. They are used for structural-steel joints. You will find the dimensions for drawing these high-strength steel bolts in Appendix A.

TO DRAW A REGULAR SQUARE BOLTHEAD ACROSS FLATS

(Square boltheads are not made in heavy sizes.) To draw a regular square bolthead across the flats, use the following proportions. If D is the major diameter of the bolt, and W is the width of the bolthead across the flats, then $W = 1.5D$. The height of the head, $H = 0.67D$.

Now, follow the method shown in Fig. 10–28. Draw centerlines, then start the top view as in space A. Draw a chamfer circle with a diameter equal to the distance across the flats. Then draw a square about this circle. Below the square, start the front view, as in space B. Draw a horizontal line repesenting the bearing surface of the head. Lay off the height of the head, *H*. Then draw the top line of the head. To get the sides of the head, project lines down from the top view. Finally, to draw the chamfer, begin by drawing line *ox*, as in space A. Revolve the line to make line *oy*, as shown. Now, run a line from point *y* straight down to the front view (space B). From point *a* in the front view, draw a 30° chamfer line *ab* out to meet the line from point *y*. This forms the chamfer depth. From point *b*, extend a line horizontally across to establish points *c* and *d*, as shown. Now draw the chamfer arc through points *c*, *e*, and *d*, using radius *R*. You can find the length of *R* by trial, or you can make it equal to *W*, the width across the flats. Complete the view, as in space D.

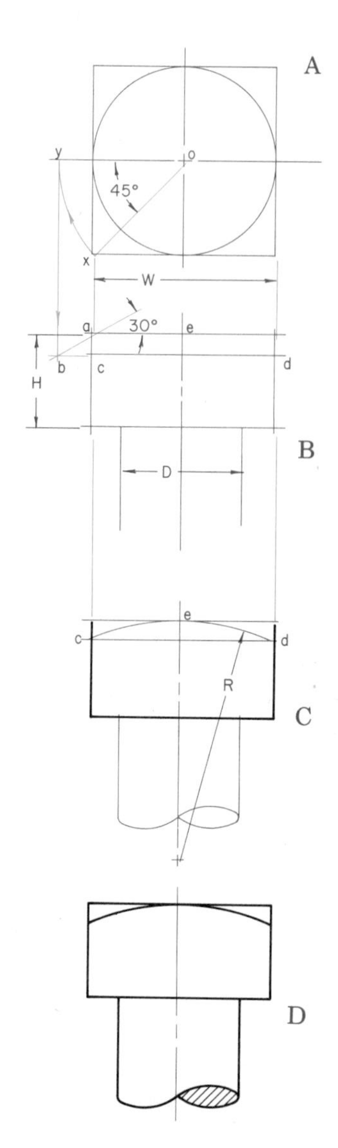

Fig. 10–28 To draw a regular square bolthead across flats.

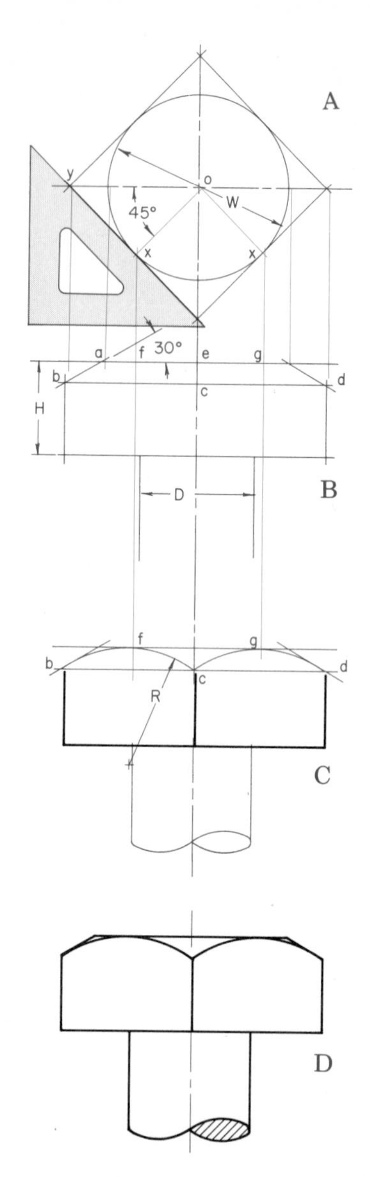

Fig. 10–29 To draw a regular square bolthead across corners.

TO DRAW A REGULAR SQUARE BOLTHEAD ACROSS CORNERS

(Square boltheads are not made in heavy sizes.) To draw a regular square bolthead across the corners, use the following proportions. If *D* is the major diameter of the bolt, and *W* is the width across the flats, then $W = 1.5D$. The height of the head, $H = 0.67D$. Begin your drawing, as in space A of Fig. 10–29, by drawing centerlines. Start the top view by drawing a chamfer circle with a diameter equal to *W*, the distance across the flats. Then, using a 45° triangle, draw a square about this circle, as shown. Next, start the side view, as in space B. Draw a horizontal line to represent the bearing surface of the head, *H*. Then draw the top line of the head. To get the two sides of the head, drop lines down from the corners of the square in the top view. Next, project lines down from the diameter of the chamfer circle in the top view to meet the top line in the front view. From this point, draw 30° chamfer lines to meet the lines at the sides of the head.

Now, find the two tangent points *x* shown in space A. Project these points to the top view to get points *f* and g, as shown in space B and C. Also draw line bcd. Then draw the chamfer arcs bfc and cgd using radius R. You can find radius R easily by trial, or you can figure that $R = yo = \frac{1}{2}$ distance across corners. Now, complete the view as in space D. You can use the same method to draw a nut except that $T = 0.88\ D$ for regular nuts, and $T = D$ for heavy nuts.

You usually draw boltheads and nuts across corners on all views of design drawings, no matter what the projection. You do this to show the largest *clearance* (space needed for turning) that the bolt or nut must have. You also do it to keep hexagon heads and nuts from being confused with square heads and nuts.

TO DRAW A HEXAGON BOLTHEAD ACROSS CORNERS

Use the same proportions as for the preceding drawings. Start the top view as in Fig. 10–30, space A. Draw a chamfer circle with a diameter of W, the distance across the flats. About this circle, draw a hexagon as indicated by the lines 1, 2, 3, 4, 5, 6. Begin the front view as in space B. Draw a horizontal line representing the bearing surface or undersurface of the head. Lay off the height of the head. Now draw the top line. Then project the edges from the corners of the top view. Also project the chamfer points. With these guides, draw in the chamfer line.

Draw line $abcd$, as in space C, to locate the chamfer intersections. Now, draw the arc bc, using radius R_1. You can find this radius by trial. Complete the view as in space D by drawing arcs ab and cd, using radius R_2. You can find this radius, too, by trial.

To draw a hexagon bolthead across flats, proceed as shown in Fig. 10–31. You can draw hexagon nuts in the same way, but note the difference between the height of the head and the thickness of the nut.

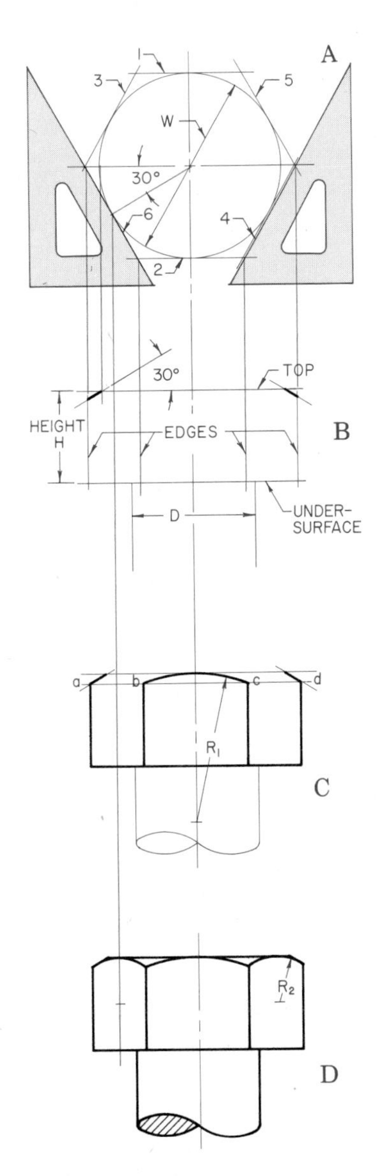

Fig. 10–30 **To draw a regular hexagon bolthead across corners.**

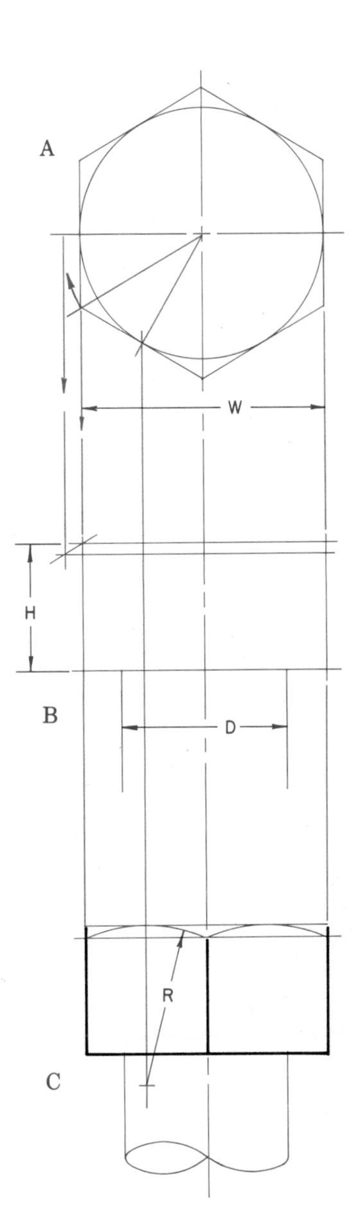

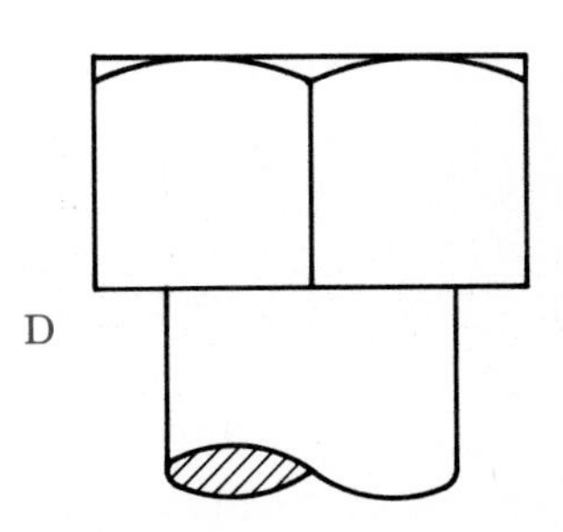

Fig. 10–31 **To draw a regular hexagon bolthead across flats.**

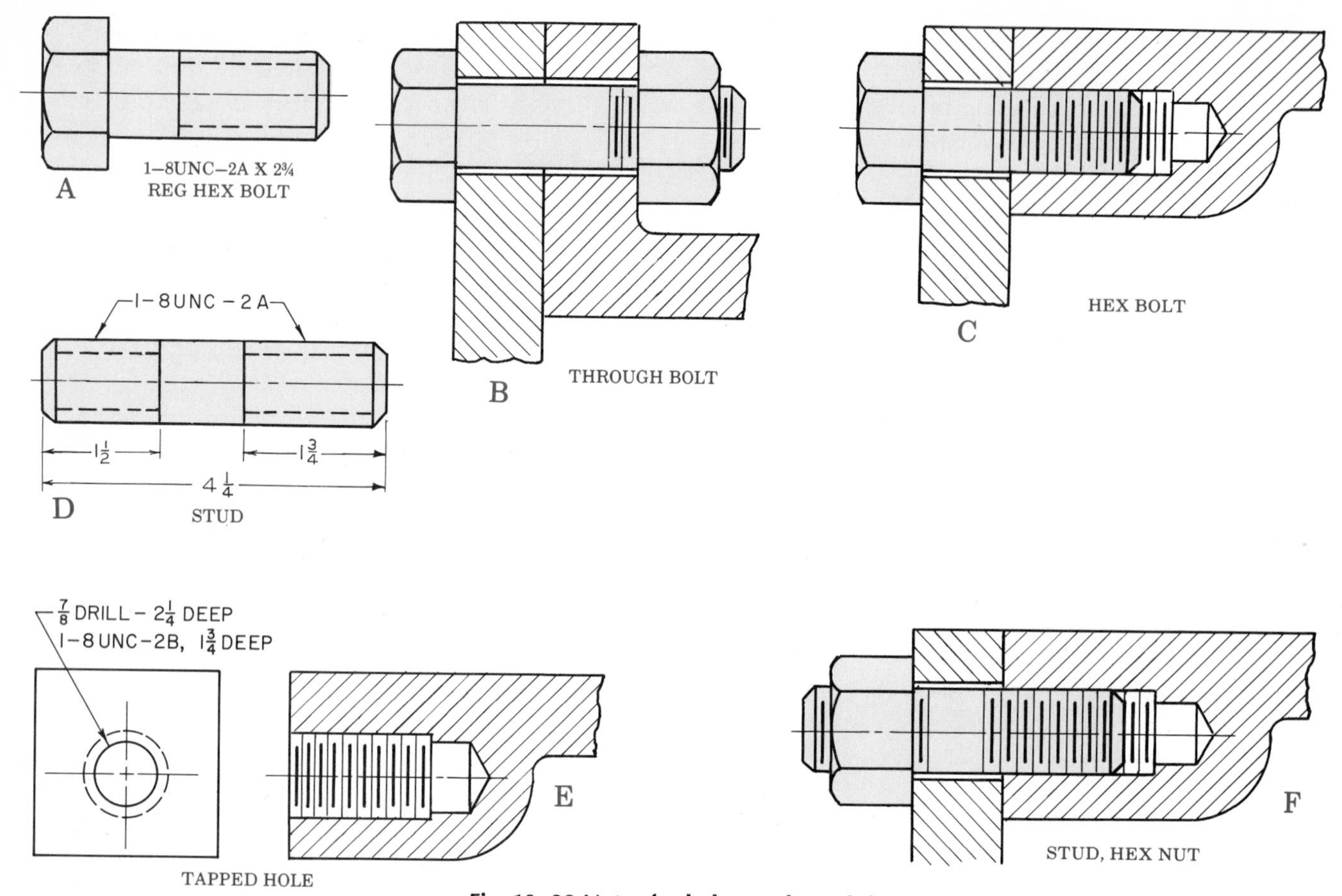

Fig. 10–32 **Notes for bolts, studs, and threaded holes.**

BOLTHEADS AND NUTS

Boltheads and nuts come in standard sizes. As a result, you often don't have to dimension them on drawings. Instead, you can give the needed information in a note, as in Fig. 10–32 at A. The note specifies a 1-in. diameter, 8 threads per inch, Unified coarse-thread series, Class 2A fit, 2¾ in. long, regular hex-head bolt.

A bolt may hold one part to another by passing through both of them and using a nut, as at B. A bolt may also pass through the first part and screw into a threaded hole in the second, as at C.

A stud or stud bolt (Fig. 10–32 at D) has threads on both ends. It is used where a bolt is not suitable and for parts that must be removed often. You dimension the length of thread from each end as shown. A tapped (threaded) hole is dimensioned as at E. One way to use a stud is to screw it into a threaded hole permanently. The hole in another part is put over the stud and a nut is screwed onto the end, as at F. In certain cases, a stud can be passed through two parts and a nut placed on each end.

LOCK NUTS

Lock nuts and various devices are used to keep nuts, or bolts and screws, from working loose. Many special devices are available.

Some forms of lock nuts are shown in Fig. 10–33. There are many special forms made to provide positive locking.

A self-locking fastener, the Jay Lock (Fig. 10–34), consists of an epoxy chemical locking agent that combines with a hardening agent. When a bolt or screw is screwed into it, it provides a strong, vibration-proof bond.

CAP SCREWS

A cap screw fastens two parts together by passing through a clearance hole in one and screwing into a

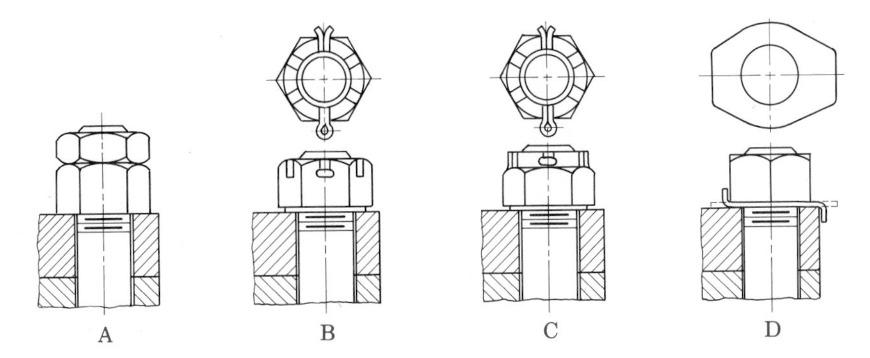

Fig. 10–33 **Locking threaded fastenings.**

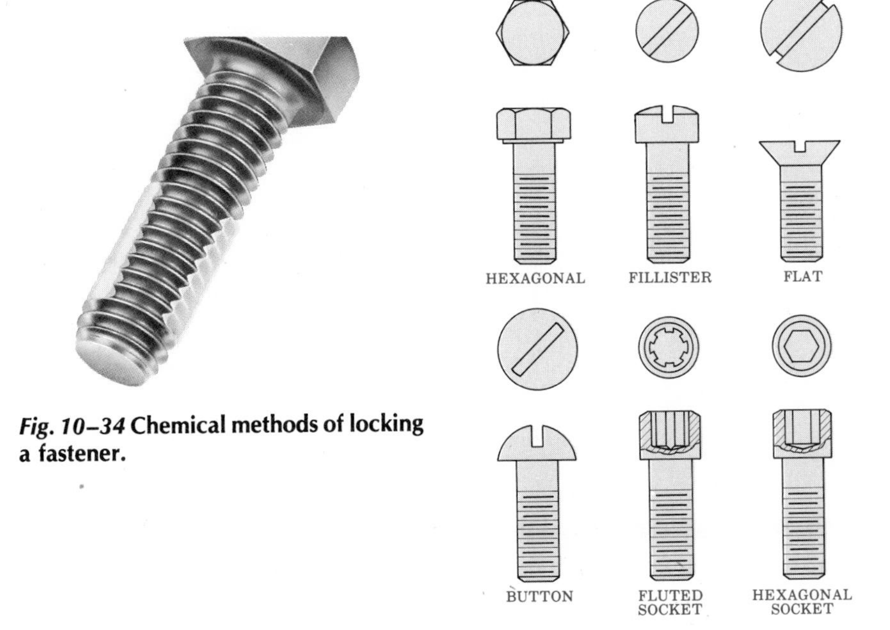

Fig. 10–34 **Chemical methods of locking a fastener.**

Fig. 10–35 **Cap screws.**

Fig. 10–36 **Machine screws.**

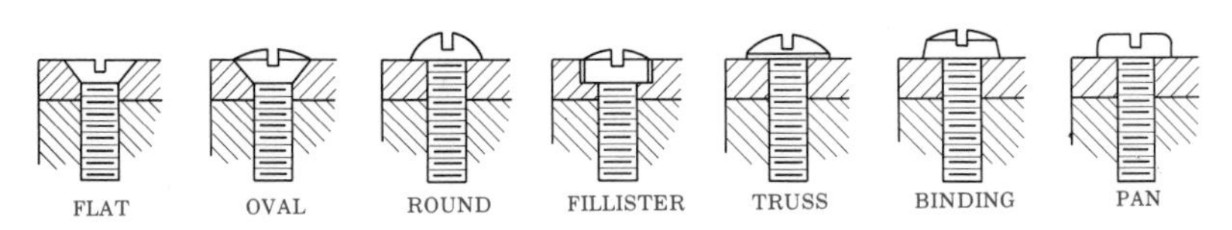

tapped hole in the other (Fig. 10–35). In most cases, you do not have to show the clearance hole on your drawing. Cap screws have a naturally bright finish. This is in keeping with the machined parts with which they are used. Coarse, fine, or 8 threads may be used on cap screws. Socket-head cap screws may have Class 3A threads. Class 2A threads will be found on other head styles. You will find dimensions for drawing various sizes of American National Standard cap screws in Appendix A.

MACHINE SCREWS

Machine screws are used where the fastener must have a small diameter (Fig. 10–36). Sizes below 6 mm (1/4 in.) in diameter have been specified by number. Machine screws may screw into a tapped hole or extend through a clearance hole and into a nut. The nuts used are square nuts. The finish on machine screws is bright. The ends are flat, as shown. Machine screws up to 50 mm (2 in.) long are threaded full length. Coarse or fine threads and Class 2 threads may be used on machine screws.

SETSCREWS

Setscrews are used to hold two parts together in a desired position. They do so by screwing through a threaded hole in one part and *bearing* (pushing) against the other (Fig. 10–37). There are two general types: square head and headless. Square-head setscrews can cause accidents when used on *rotating* (turning) parts. They may violate safety codes. Headless setscrews can have either a slot or a socket. Any of the points shown can be used on any setscrew.

WOOD SCREWS

Wood screws are made of steel, brass, or aluminum. They are finished in various ways (Fig. 10–38). Steel screws can be bright (natural finish), blued, galvanized, or copper plated. Both steel and brass screws are sometimes nickel plated. Round-head screws are set with the

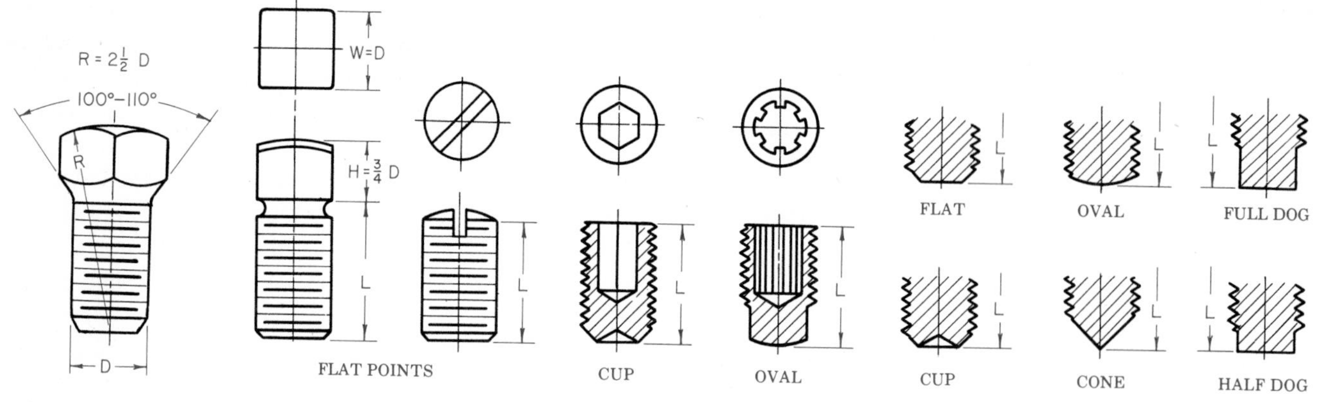

Fig. 10–37 **Setscrews.**

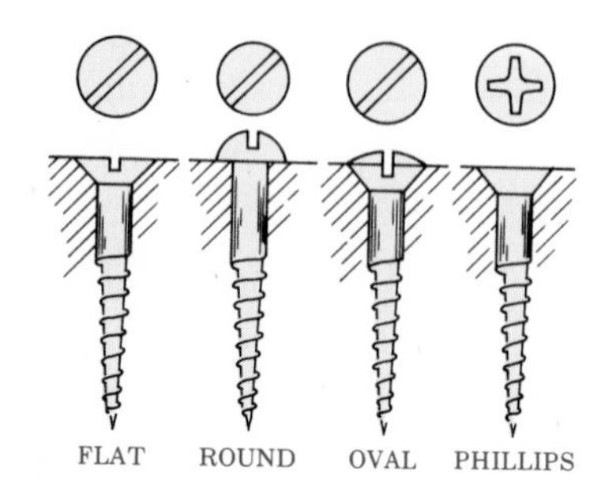

Fig. 10–38 **Wood screws.**

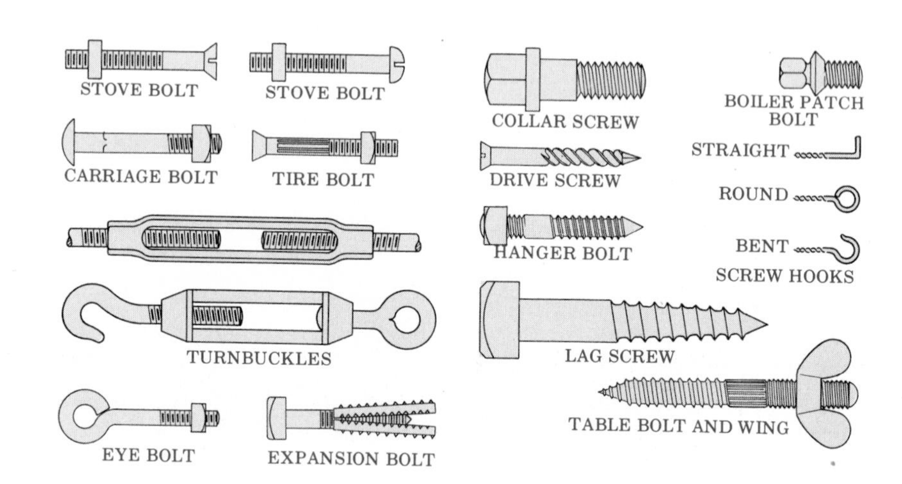

Fig. 10–39 **Miscellaneous threaded fastenings.**

head above the wood. Flat-head screws are set flush, or countersunk. Wood screws may be drawn as shown. You specify them by number, length, style of head, and finish. For flat-head screws, the length is measured overall. For round-head screws, it is measured from under the head to the point. For oval-head screws, it is measured from the largest diameter of the countersink to the point. You will find sizes and dimensions listed in Appendix A.

SOME MISCELLANEOUS THREADED FASTENINGS

These are shown in Fig. 10–39. The names denote the ways in which they are used. Screw hooks and screw eyes are specified by diameter and overall length.

A lag screw, or lag bolt, is used to fasten machinery to wood supports. It is also used in heavy wood constructions when a regular bolt cannot be used. It is similar to a regular bolt but has wood-screw threads. You specify a lag bolt by its diameter and the length from under the head to the point. The head of the lag bolt has the same proportions as a regular bolthead.

MATERIALS FOR THREADED FASTENERS

Threaded fasteners are usually made from steel, brass, bronze, aluminum, cast iron, wood, and nylon. Nylon screws and bolts are made in various bright colors, such as red, blue-green, yellow, white, etc.

KEYS

Keys are used to secure pulleys, gears, cranks, and similar parts to a shaft (Fig. 10–40). Keys are made in different forms for different uses.

They range from the saddle key, for light duty, to special forms, such as two square keys, for heavy duty. The common sunk key can have a breadth of about one-fourth the shaft diameter. Its thickness can vary from five-eighths the breadth to the full breadth. The Woodruff key is often used in machine-tool work. It is made in standard sizes. You specify it by number (see Appendix A). Special forms of pins have been developed to replace keys for some uses. These pins need only a drilled hole instead of the machining that is needed to make keys.

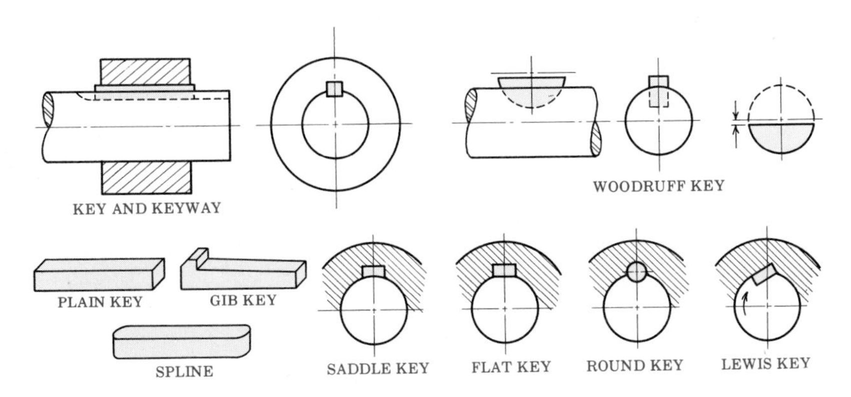

Fig. 10–40 **Keys.**

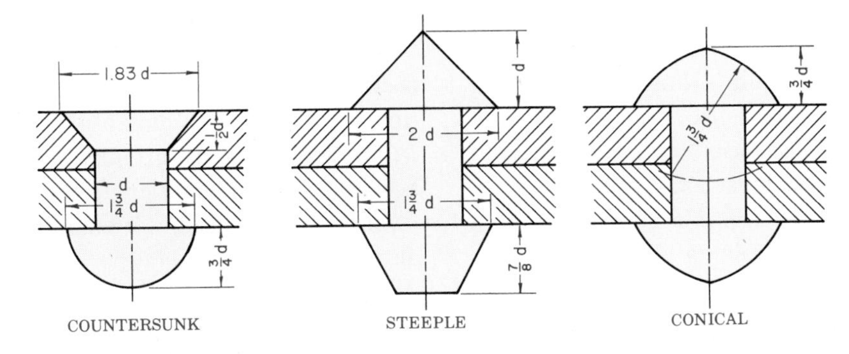

Fig. 10–41 **Large rivets.**

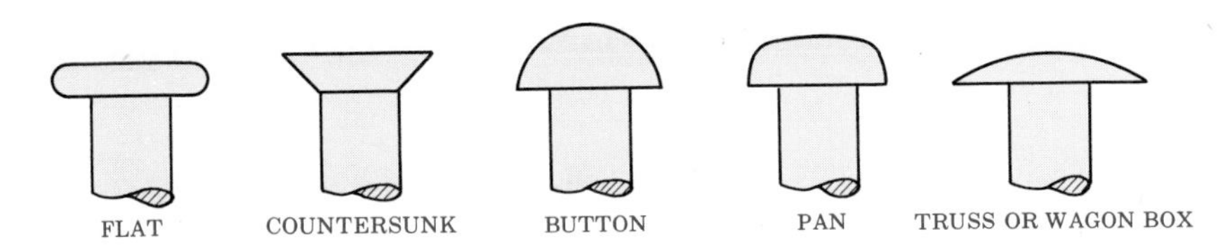

Fig. 10–42 **Small rivets.**

Fig. 10–43 **Explosive rivet.** *(Explosives Department, E.I. du Pont de Nemours & Co.)*

RIVETS

Rivets are rods of metal with a preformed head on one end. They are used to fasten permanently sheet-metal plates, structural steel shapes, boilers, tanks, and many other items. The rivet is first heated red hot. Then it is placed through the parts to be joined. It is held there in place while a head is formed on the projecting end. The rivet is then said to have been "driven."

Large rivets (Fig. 10–41) have nominal diameters ranging in size from 12 mm to 45 mm (1/2 to 1 3/4 in.). Small rivets (Fig. 10–42) range from 2 or 3 to 11 mm (1/16 or 3/32 to 7/16 in.) in diameter.

Some rivets are made especially for use where one side of the plates cannot be reached or where the space is too small to use a regular rivet. These are called *blind* rivets. One type is the du Pont explosive rivet (Fig. 10–43). It has a small explosive charge in a cavity. After

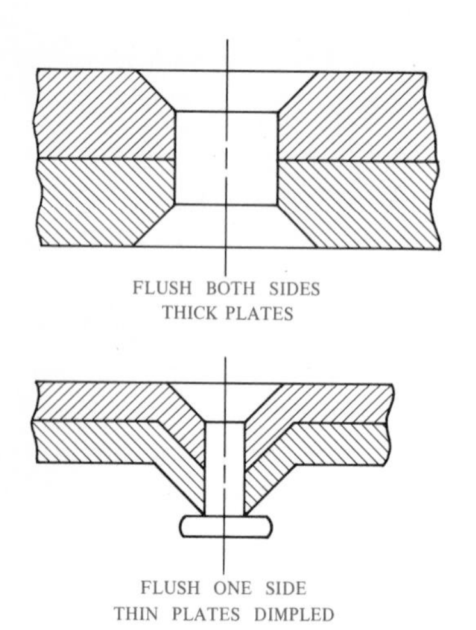

Fig. 10–44 **Flush rivets.**

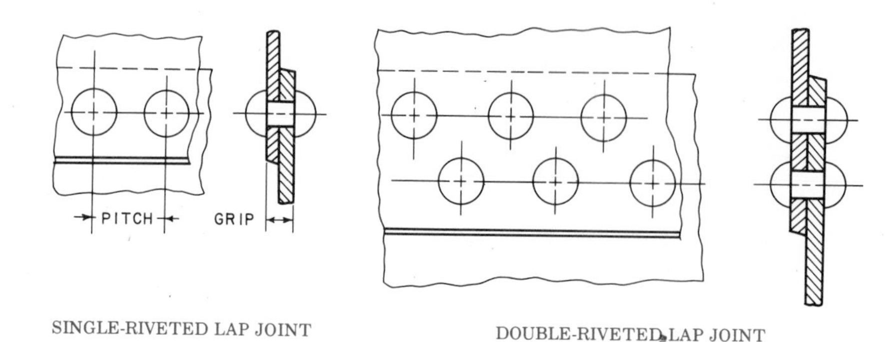

Fig. 10–45 **Riveted joints.**

the rivet is inserted, the charge is exploded, forming a head. This rivet is thus excellent for blind riveting, since the head can be formed inside places that are closed or impossible to reach.

Sometimes plates that are riveted together need to clear surfaces. This requires flush riveting (Fig. 10–44) on one or both sides. These rivets are used on airplanes, automobiles, spacecraft, etc.

Riveted joints (Fig. 10–45) are used for joining plates. They may have lap or butt joints. They may also have single or multiple riveting. (See Appendix A for American National Standard rivet dimensions.)

For some uses, as in tanks, steel buildings, etc., high-strength structural bolts are used. Welding is also in wide use.

Review

1. Name five types of fasteners.

2. The first known use of the principle of the screw was ________.

3. The true shape of a screw thread is based on a ________ curve.

4. One complete turn will advance a threaded part into the nut a distance equal to the ________ on a single thread.

5. What type of thread is used to screw parts together quickly?

6. Left-hand threads are indicated by the initials ________.

7. Name the three types of thread representation.

8. What is the difference in designating metric fine- and coarse-thread series?

9. The three most common Unified thread series are coarse, ________ and ________.

10. Name the classes of fits for Unified screw threads.

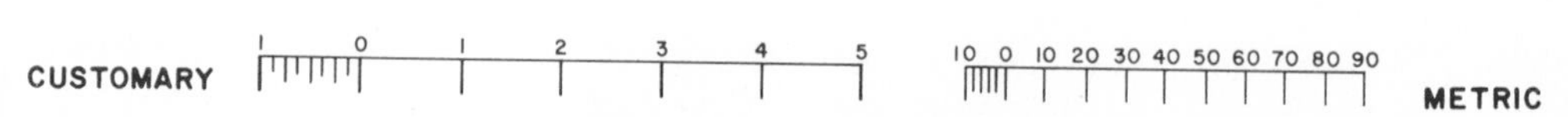

Scale for Figs. 10–50 and 10–51.

Problems

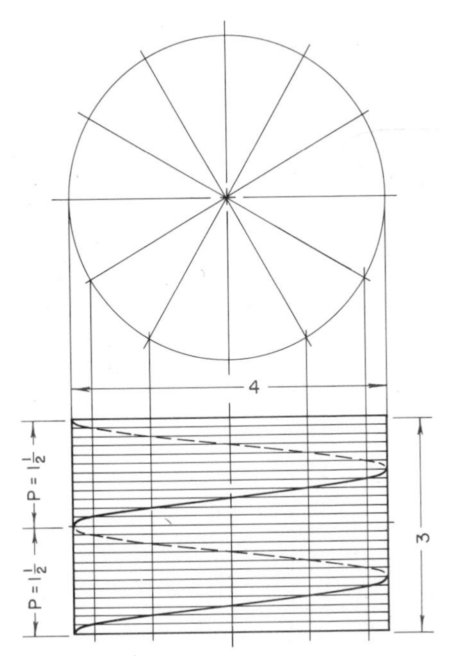

Fig. 10–46 Draw two complete turns of a right-hand helix as shown above. Use dimensions indicated and work full size.

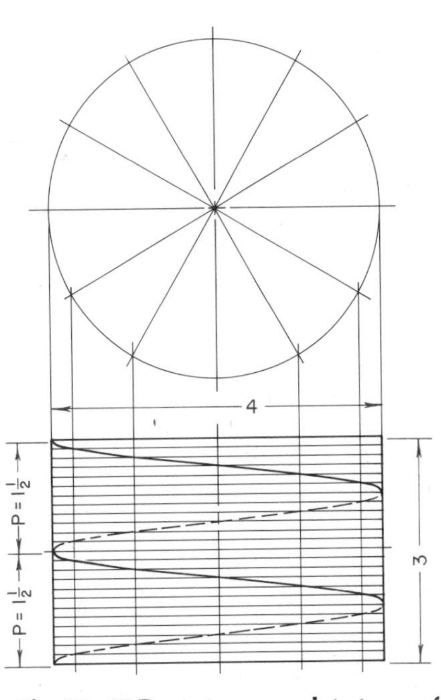

Fig. 10–47 Draw two complete turns of a left-hand helix as shown above. Use dimensions indicated and work full size.

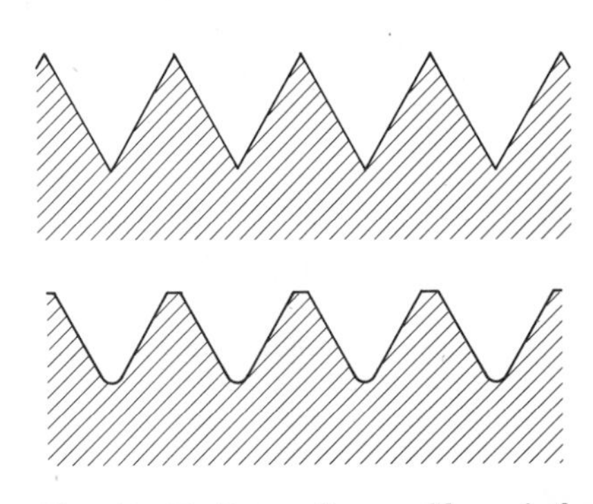

Fig. 10–48 Draw the profiles of the sharp "V" and the American National Unified thread. Letter the name of each under it. Pitch = 1″.

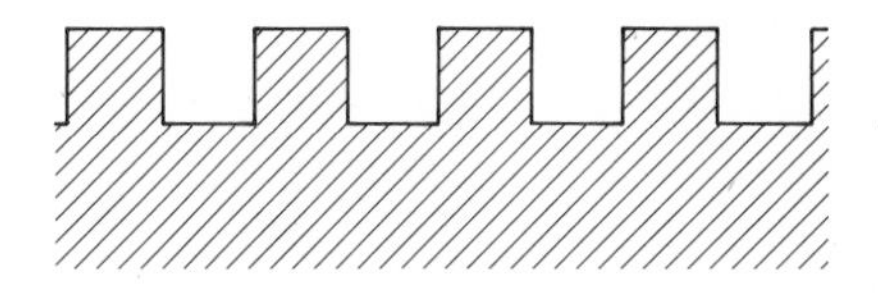

Fig. 10–49 Draw the profile of the square thread. Letter the name under it. Pitch = 1″.

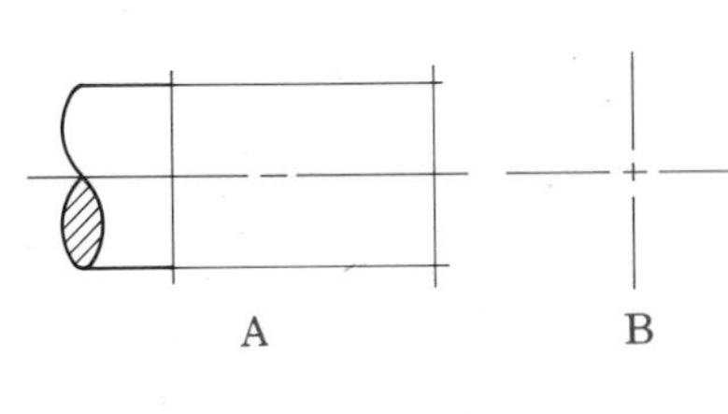

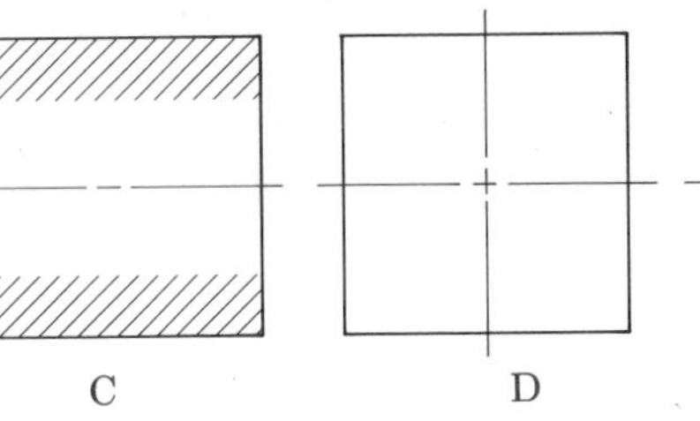

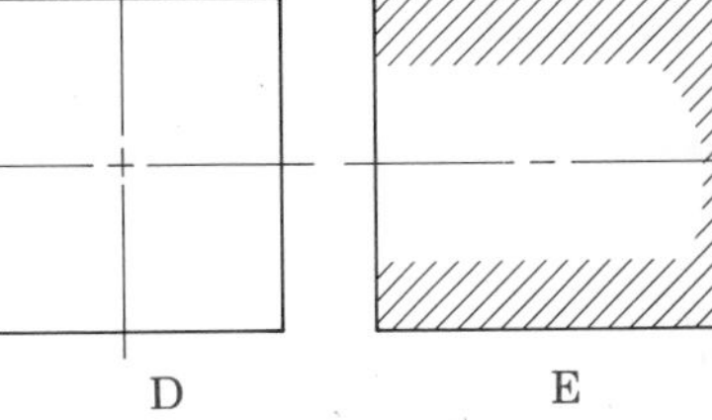

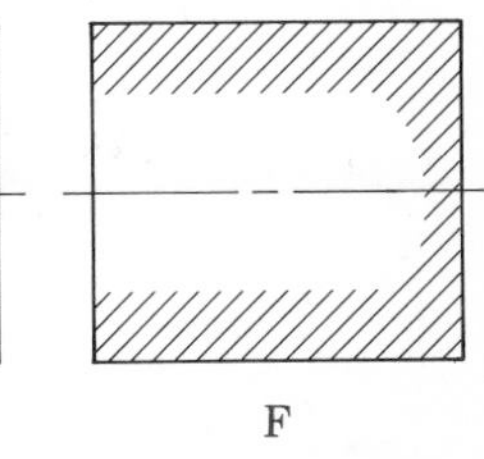

Fig. 10–50 Schematic drawing of screw threads. Take dimensions from scale on page 222. Draw the views as follows: A = schematic drawing showing 1″-8UNC-2A threads; B = end view of A; C = schematic drawing of section through 1″-8UNC-2B (internal) threads; D = right-side view of C; E = schematic drawing of section through 7/8 drill × 1½ deep, 1″-8UNC-2B × 1⅛ deep; F = schematic drawing of section through 7/8 drill × 1½ deep, 1″-8UNC-2B × 1½ deep.

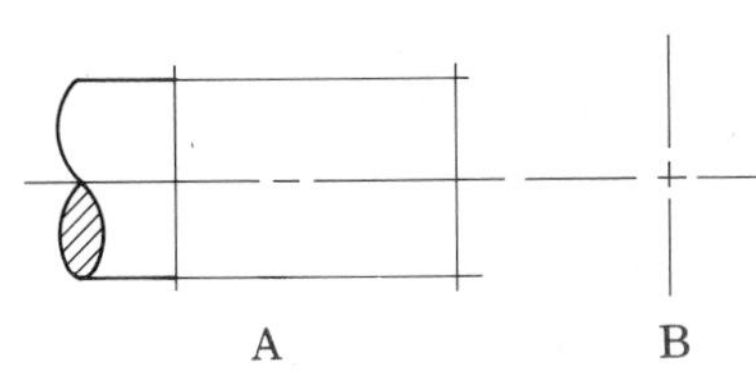

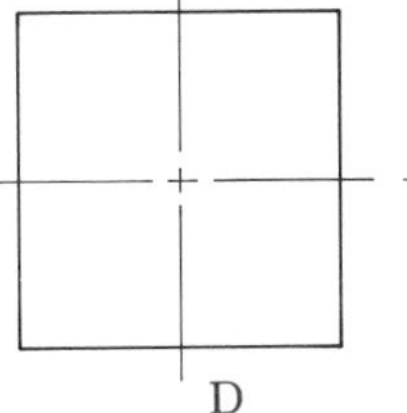

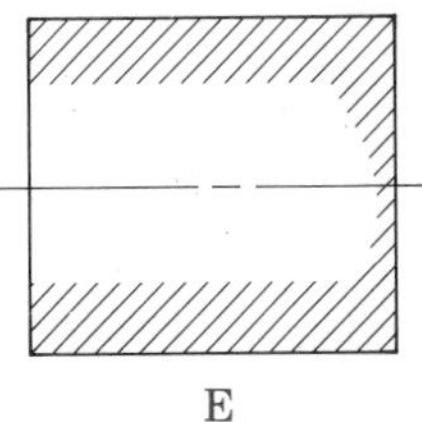

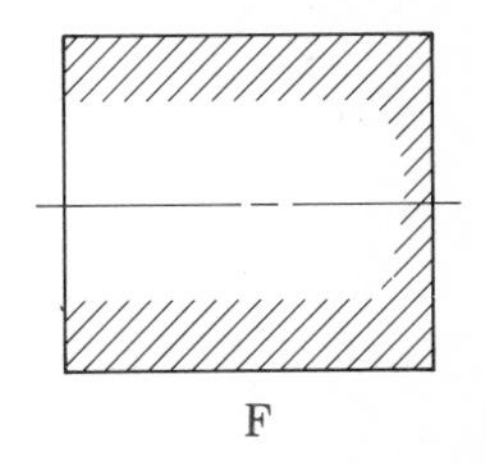

Fig. 10–51 Simplified drawing of screw threads. Take dimensions from scale on page 222. Draw the views as follows: A = simplified drawing showing 1″-8UNC-2A threads; B = end view of A; C = simplified drawing of section through 1″-8UNC-2B (internal) threads; D = right-side view of C; E = simplified drawing of section through 7/8 drill × 1½ deep, 1″-8UNC-2B × 1⅛ deep; F = simplified drawing of section through 7/8 drill × 1½ deep, 1″-8UNC-2B × 1½ deep.

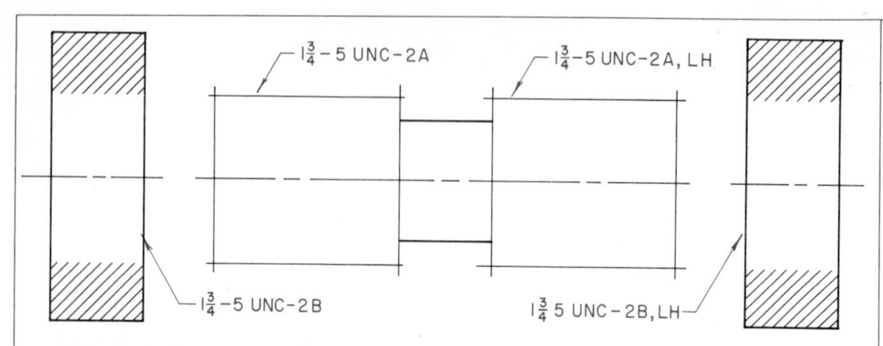

Fig. 10–52 **Detailed representation of screw threads. Take dimensions from the printed scale at the bottom of the page, using dividers. Draw the views as shown and complete each according to the specifications noted on each. Use detailed thread representation.**

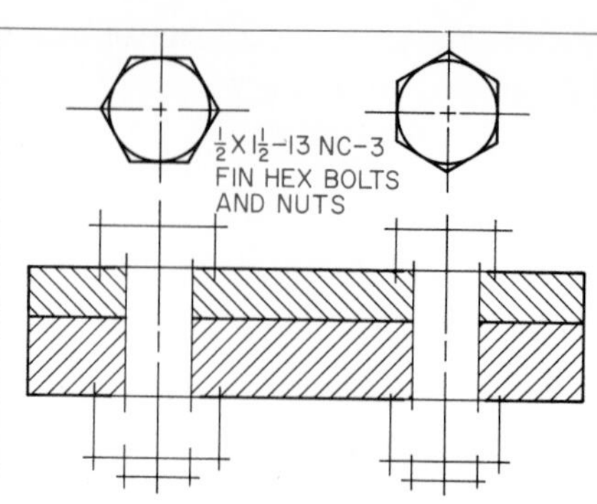

Fig. 10–53 **Regular hexagon bolt and nut. Draw the views and complete the bolts and nuts in the sectional view. See Appendix A for bolt and nut detail sizes.**

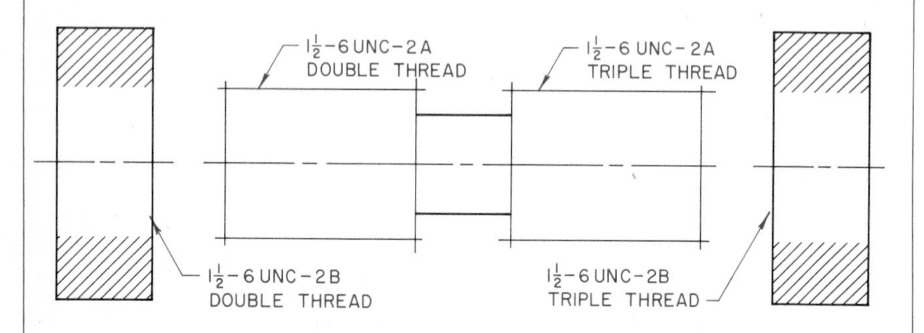

Fig. 10–54 **Double and triple threads. Take dimensions from the printed scale at the bottom of the page, using dividers. Draw the views as shown and complete each according to the specifications noted on each. Use detailed thread representation.**

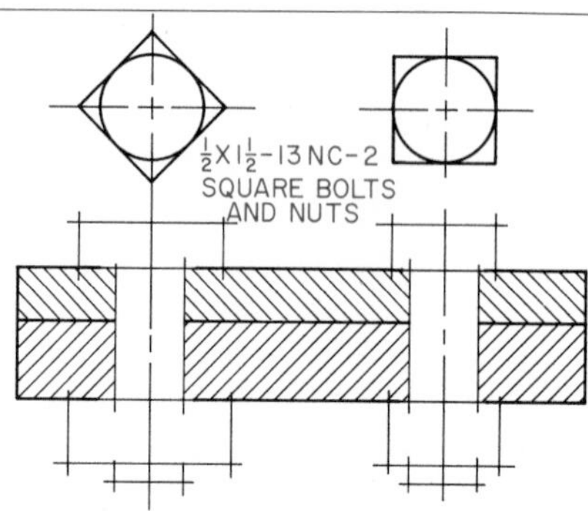

Fig. 10–55 **Regular square bolt and nut. Draw the views and complete the bolts and nuts in the sectional view. See bolt and nut detail sizes.**

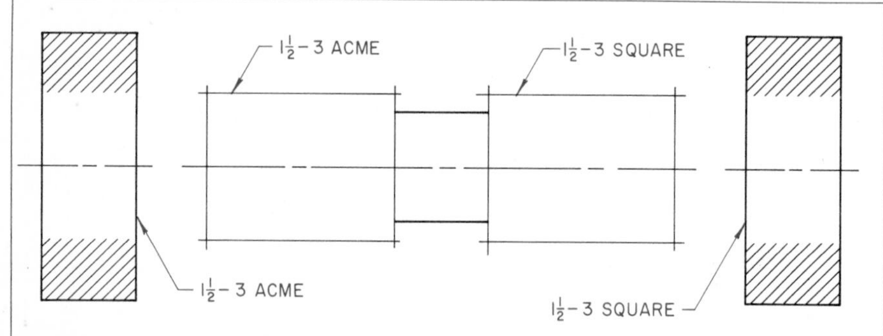

Fig. 10–56 **Acme and square threads. Take dimensions from the printed scale at the bottom of the page, using dividers. Draw the views as shown and complete each according to the specifications noted on each. Use detailed thread representation.**

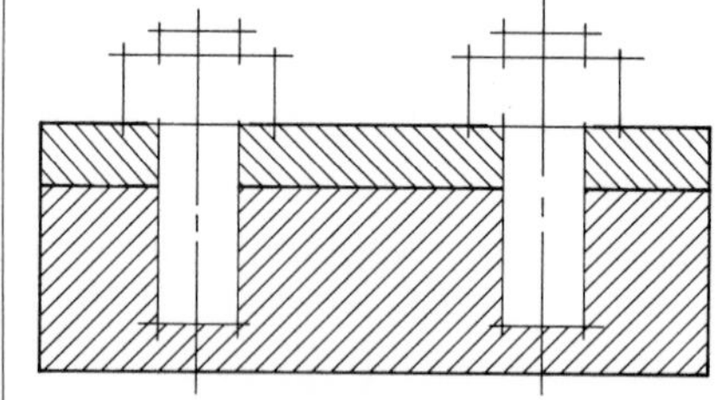

Fig. 10–57 **Studs. Draw the view as shown and complete the 1/2" = 1 3/4" studs and regular semifinished hexagon nuts. Check Appendix A for specific nut sizes. Other dimensions may be taken from the printed scale at the bottom of the page. Use schematic thread representation.**

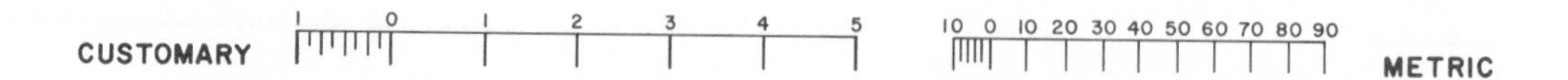

Fig.	Bolt DIA	A	B	C	D	E	F	G	H	I
10–58	1/4	1/4	5/16	5/8	1/2	2	3/4	3/8	9/32	3/32
10–59	5/16	5/16	3/8	5/8	5/8	2	3/4	3/8	11/32	3/32
10–60	3/8	3/8	7/16	5/8	5/8	2 1/4	3/4	3/8	13/32	1/8
10–61	7/16	7/16	1/2	3/4	5/8	2 1/2	1	1/2	15/32	1/8
10–62	1/2	1/2	9/16	7/8	5/8	2 3/4	1 1/4	5/8	9/16	1/8
10–63	9/16	9/16	5/8	1	3/4	3	1 1/2	3/4	5/8	1/8
10–64	5/8	5/8	11/16	1 1/8	3/4	3 1/4	1 3/4	3/4	11/16	1/8
10–65	3/4	3/4	13/16	1 1/4	7/8	3 1/2	2	7/8	13/16	3/16
10–66	7/8	7/8	1	1 3/8	1	3 3/4	2 1/4	1	15/16	1/4
10–67	1	1	1 1/8	1 1/2	1 1/8	4	2 1/2	1 1/8	1 1/8	1/4
10–68	1 1/8	1 1/8	1 1/4	1 5/8	1 1/4	4 1/4	2 3/4	1 1/4	1 1/4	1/4

Figs. 10–58 to 10–68 **Take all dimensions from the table for the problem assigned and draw the flange and head plate as shown. On the colored centerlines, draw American National Standard bolts and nuts (hex or square) as assigned. Place bolthead at the left and show bolthead across flats; nut across corners.**

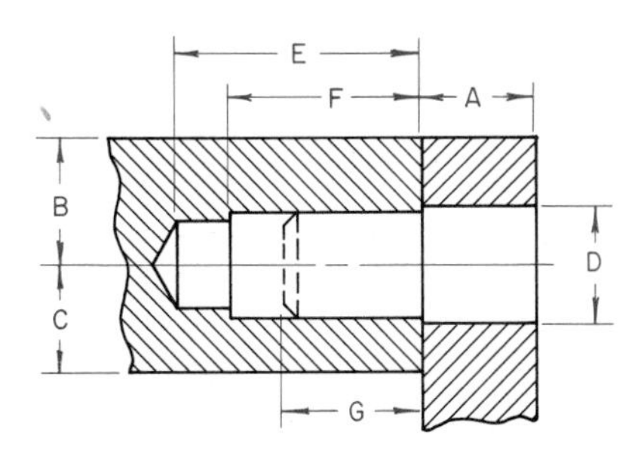

Fig.	Stud DIA	Nut	A	B	C	D	E	F	G
10–69	3/4	Hex	13/16	7/8	3/4	13/16	1 3/4	1 3/8	1
10–70	7/8	Sq	15/16	1 1/4	7/8	15/16	2	1 9/16	1 1/8
10–71	1	Hex	1 1/8	1 1/8	1	1 1/8	2 1/4	1 3/4	1 1/2
10–72	1 1/8	Sq	1 1/4	1 1/2	1 1/8	1 1/4	2 3/4	2 1/8	1 1/2

Figs. 10–69 to 10–72 **On the centerline shown, draw a stud with hexagon or square nut, across flats or corners, as directed by the instructor. Take dimensions from the table.**

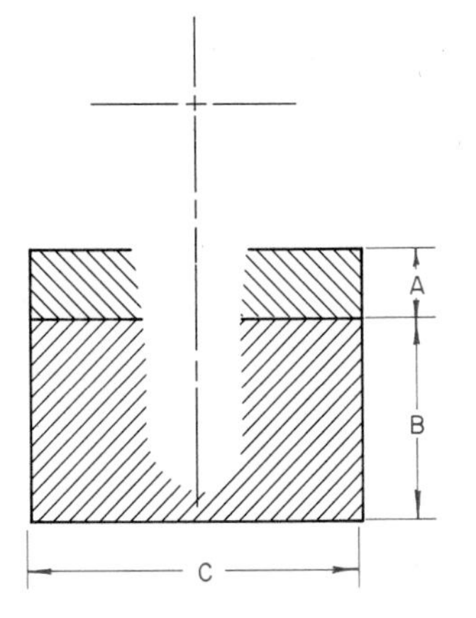

Fig.	Bolt DIA	Head Style	A	B	C
10–73	3/8	Button	1/2	1 1/2	2 1/2
10–74	1/2	Button	3/4	1 3/4	3
10–75	5/8	Button	1	2	3 1/2
10–76	3/8	Flat	1/2	1 1/2	2 1/2
10–77	1/2	Flat	3/4	1 3/4	3
10–78	5/8	Flat	1	2	3 1/2
10–79	3/8	Fillister	1/2	1 1/2	2 1/2
10–80	1/2	Fillister	3/4	1 3/4	3
10–81	5/8	Fillister	1	2	3 1/2

Figs. 10–73 to 10–81 **Take dimensions from the table for the problem assigned and draw the figure shown at the left. Refer to Appendix A for sizes and draw the assigned style of head and size of cap screw. Also, draw a top view of the screw head.**

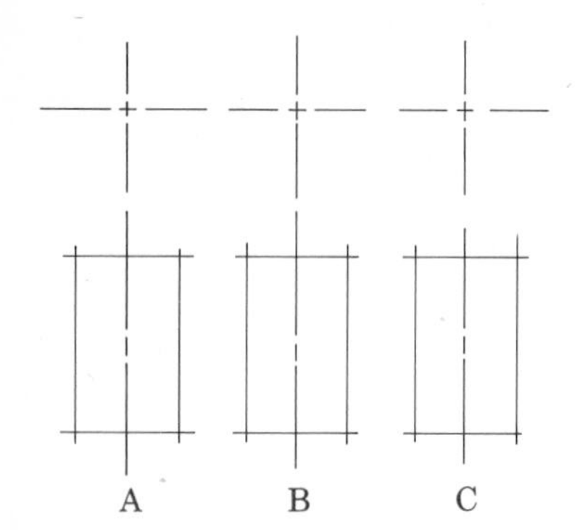

Fig. 10–82 Setscrews. Draw three setscrews: A = 3/4 DIA × 1 1/4 long, square head, flat point. B = 3/4 DIA × 1 1/4 long, slotted head, oval point. C = 3/4 DIA × 1 1/4 long, socket head, cup point. Use schematic thread representation. Do not section.

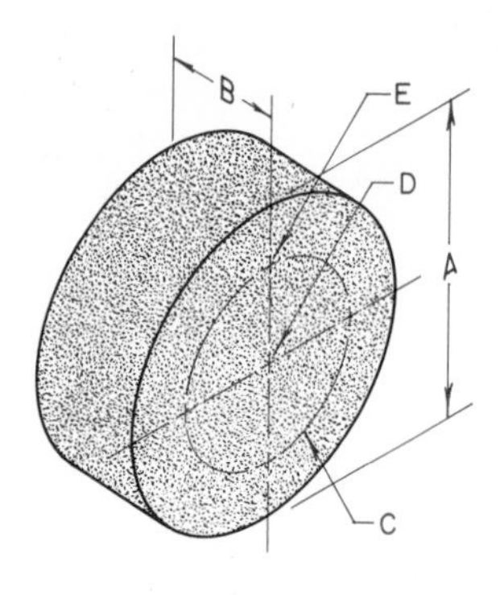

Fig. 10–83 Spacer. Draw two views of the spacer. Use schematic representation to show the threaded holes. A = 4″ DIA, B = 1 1/2″, C = 2 1/2″ DIA, D = 1″-8UNC-2B (through), E = 3/8-16UNC-3B (through). Add notes and all necessary dimensions.

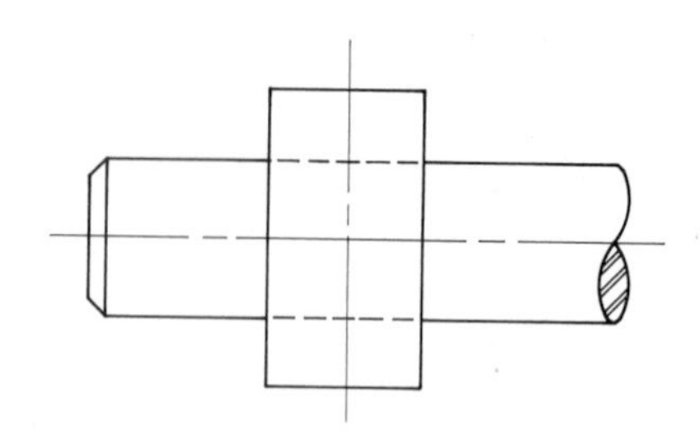

Fig. 10–84 Draw the 1″ DIA × 3 1/2″ long shaft and 1 7/8″ DIA × 1″ collar as shown. Draw the collar in full section. Add a No. 4 × 2″ American National Standard taper pin on the colored centerline. Estimate sizes not given. Materials: shaft–steel; collar–cast iron.

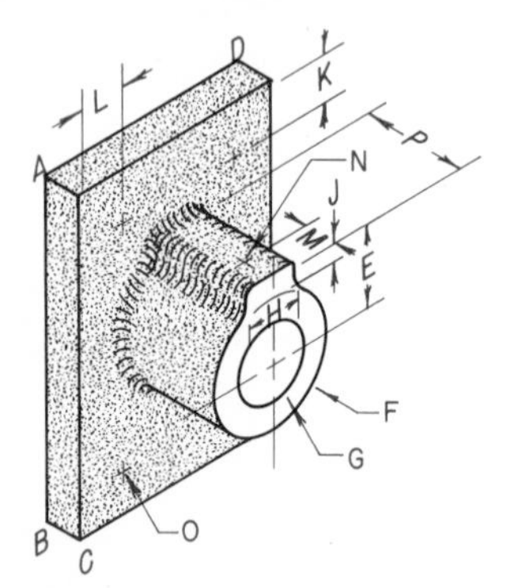

Fig. 10–85 Shaft support. Draw necessary views. At *N*, draw a 5/16″ setscrew (square head, flat point). At *O*, draw 3/8″ coarse threads (simplified representation). All fillets and rounds = 1/8 R. *AB* = 4 1/2, *BC* = 1/2, *AD* = 3, *E* = 1 1/2, *F* = 7/8 R, *G* = 1″ DIA, *H* = 3/4, *J* = 1/4, *K* = 5/8, *L* = 5/8, *M* = 1/2, *P* = 1 1/2.

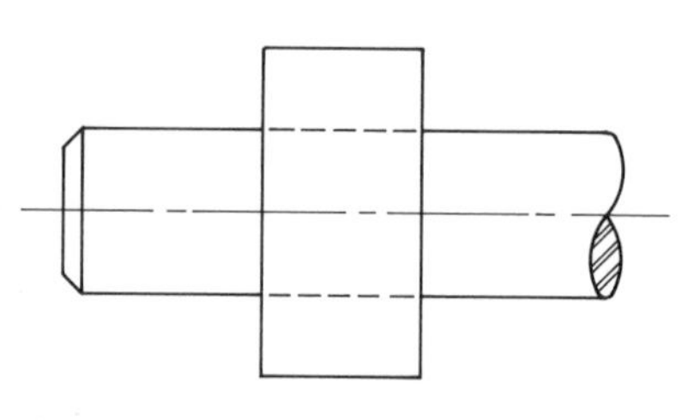

Fig. 10–86 Draw the 1″ DIA × 3 1/2″ long shaft and 2″ DIA × 1″ collar as shown. Draw the collar in half section and add an ANSI No. 608 Woodruff key at the top of the shaft. Estimate sizes not given. Materials: shaft—steel; collar—cast aluminum.

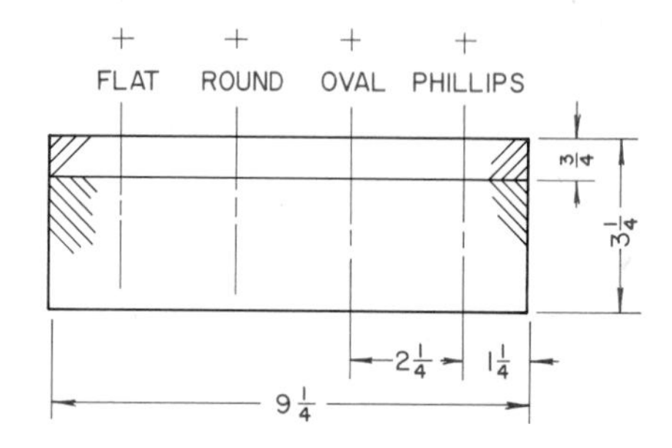

Fig. 10–87 Wood screws. Draw the view shown above. Add 2 1/2″ #12 wood screws on the colored centerlines. Show the four head types as indicated and draw a top view of each on the center mark above the view.

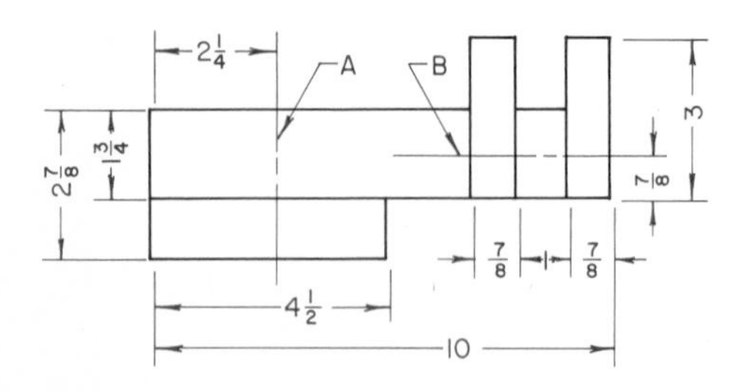

Fig. 10–88 Draw the view shown. On centerline at A, draw 1/2″ × 2 1/2″ fillister-head cap screw (head at top). At B, draw a 3/8″ × 4″ flat-head cap screw. Show view in section. Material: steel. Use schematic thread representation.

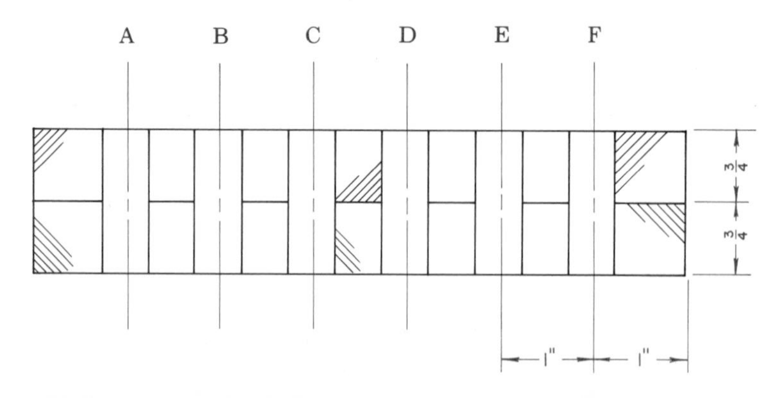

Fig. 10–89 Rivets. Draw the figure shown above. Overall sizes = 1 1/2″ × 7″. Refer to Chapter 10 and Appendix A and draw the heads (top and bottom) for 1/2″ DIA rivets. Do not dimension. A = button head; B = high button head; C = cone head; D = flat-top countersunk head; E = round-top countersunk head; F = pan head.

Chapter 11

Working Drawings

DEFINITIONS

A *working drawing* tells all that needs to be known for making a single part or a complete machine or structure. It tells precisely what the shape and size should be. It also tells exactly what kinds of material should be used, how the finishing should be done, and what accuracy is needed. A pictorial drawing and a working drawing of a simple machine part are shown in Fig. 11–1. A working drawing can be a *detail drawing,* an *assembly drawing,* or an *assembly working drawing.* This chapter will tell you all about each of these terms.

WORKING DRAWINGS

Working drawings are usually multiview drawings with complete dimensions and notes added. When you prepare a working drawing, make sure it includes all the information needed. Nothing must be left to guess.

A good working drawing must follow the style and practices of the office or industry where it is made. Most industries follow the standard recommended by the American National Standards Institute (ANSI). That way, plans can be easily read and understood from one industry to another. However, in some cases, shortcuts are used to save time. You will learn about these in Chapter 16.

No matter what special styles and practices you use in working drawings, there are certain rules you must follow. You must use proper

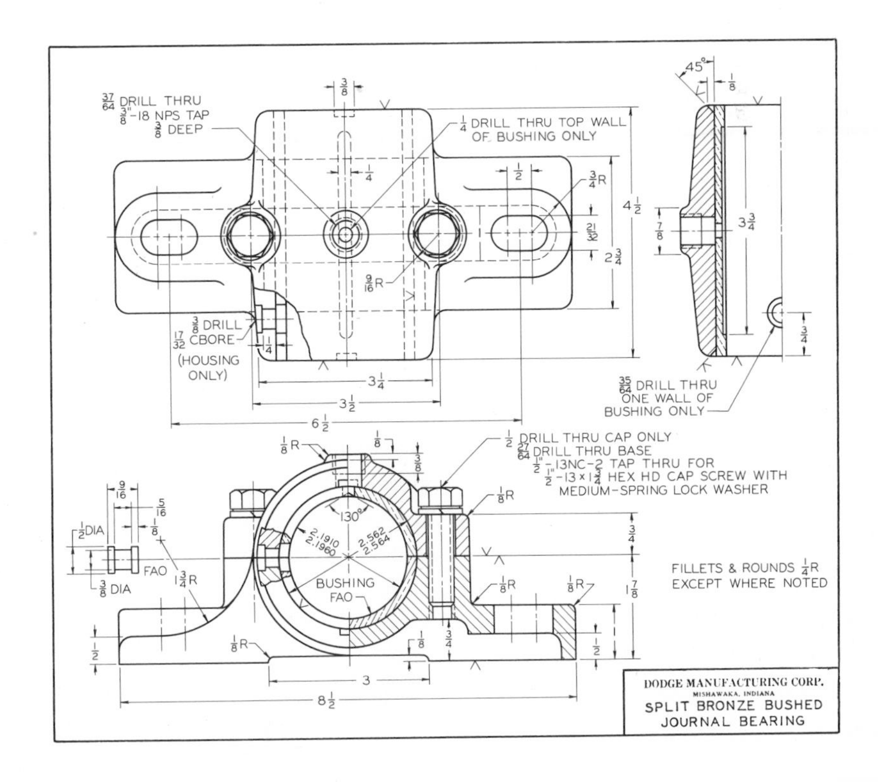

Fig. 11–1* Pictorial drawing and a working drawing of a split bronze-bushed journal bearing.** ***(Dodge Manufacturing Corp.)

line technique to make the contrast sharp. You must also make your numbers easy to read and your letters alike in style. The terms and abbreviations you use must be the standard ones. When you finish a working drawing, check it thoroughly to make sure that you have made no mistakes and done the best job possible. Do this before submitting the drawing to your supervisor or checker for approval.

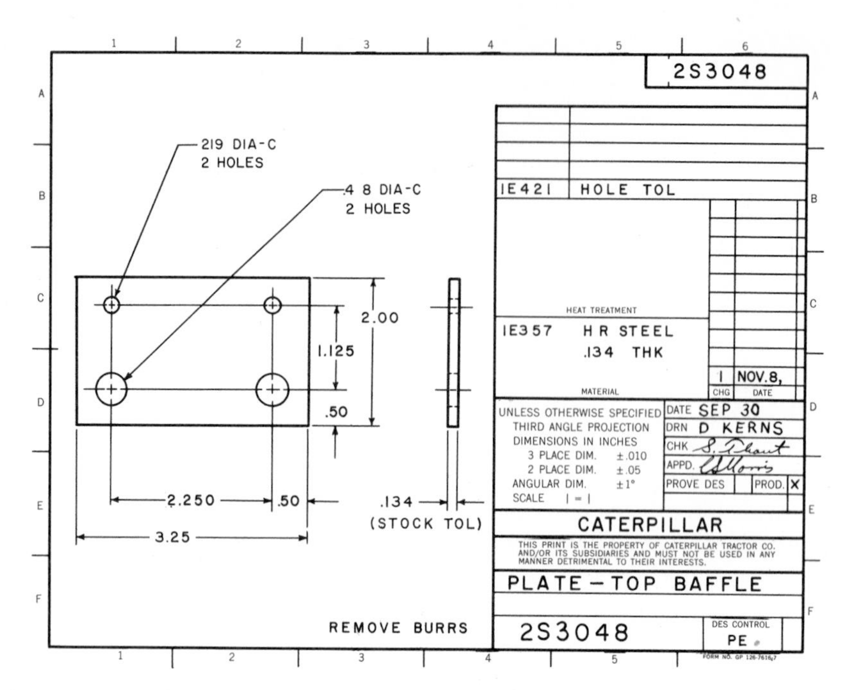

Fig. 11–2 **A working drawing of a simple detail.** ***(Caterpillar Tractor Co.)***

MAKING DRAWINGS FOR INDUSTRIAL USE

In the language of drawing, you describe an object by telling its shape and its size. All drawings follow this principle, whether they are for steam or gas engines, machines, buildings, airplanes, automobiles, missiles, or satellites.

A student's drawing can look rough and unfinished next to a "real drawing" made by a drafter or engineer. The drafter's skill in giving a finished look comes from a thorough knowledge of engineering drafting and its use in industry. In this chapter, you will learn the right order of going about the work and some of the procedures that drafters usually follow. You must become thoroughly familiar with these practices. Otherwise, your drawings will not have the style and good form that is demanded by today's industry.

Most of the drawings done for industry are done in pencil. If you work in pencil, you must have good technique. Your work must be neat, you must make few erasures, and your lines must be dense and sharp.

There is also, however, a growing need for high-quality drawings in ink. This is because today many drawings are put through modern micro-reproduction, storage, and retrieval systems. Their lines must therefore be thick and sharp, and their lettering must be large and neat.

DETAIL DRAWINGS

A drawing of a single piece that gives all the information needed for making it is called a *detail drawing.* An example is the drawing of a simple part in Fig. 11–2. A detail working drawing must be a full and exact description of the piece. It should show carefully selected views and include well-placed dimensions (Fig. 11–3).

When a large number of machines are to be produced, it is usual to make a detail drawing of each part on a separate sheet. This is done especially when some of the parts may be used on other machines. In some industries, however, it is usual to detail several parts of a machine on the same sheet. Sometimes a separate detail drawing is made for each of several workers involved, such as the patternmaker, machinist, or welder. Such a drawing shows only the dimensions and information needed by the worker for whom it is made. Figure 11–4 shows an index-plunger operating handle made by the Hartford Special Machinery Company. The picture shows the piece as forged and also after it has been machined. Figure 11–5 shows a working drawing of the same piece. Notice how the parts to be removed after all work is done are shown.

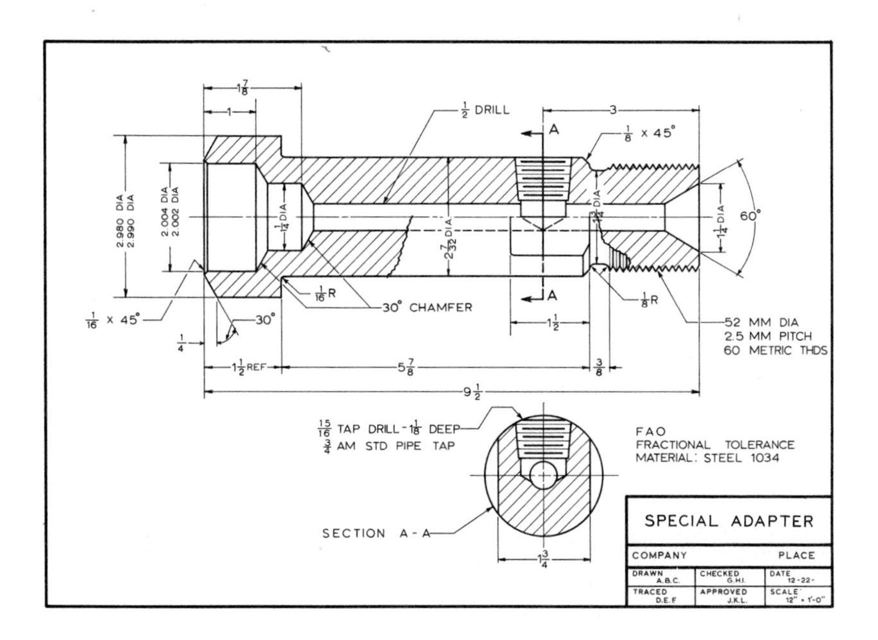

Fig. 11–3 **A one-view working drawing. The view and the extra section provide a complete description.**

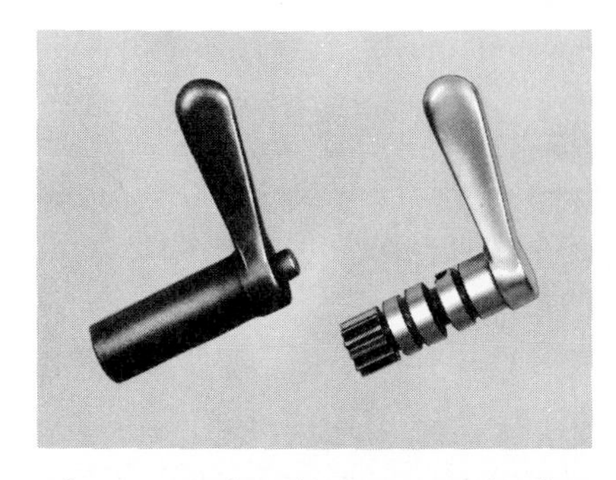

Fig. 11–4 **Index-plunger operating handle—forging and finished part.** ***(The Hartford Special Machinery Co.)***

Fig. 11–5 **Working drawing of part shown in Fig. 11–4.** ***(The Hartford Special Machinery Co.)***

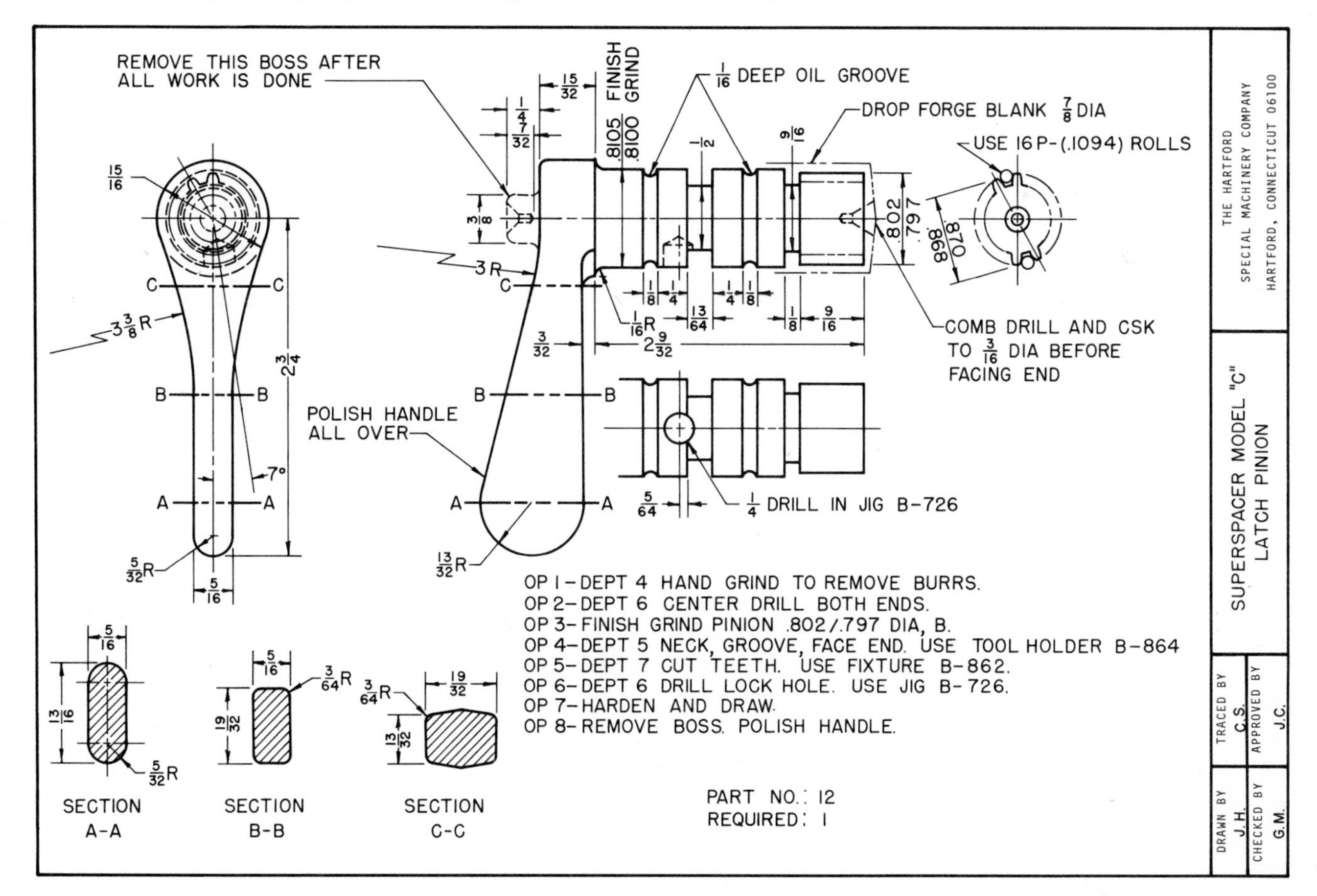

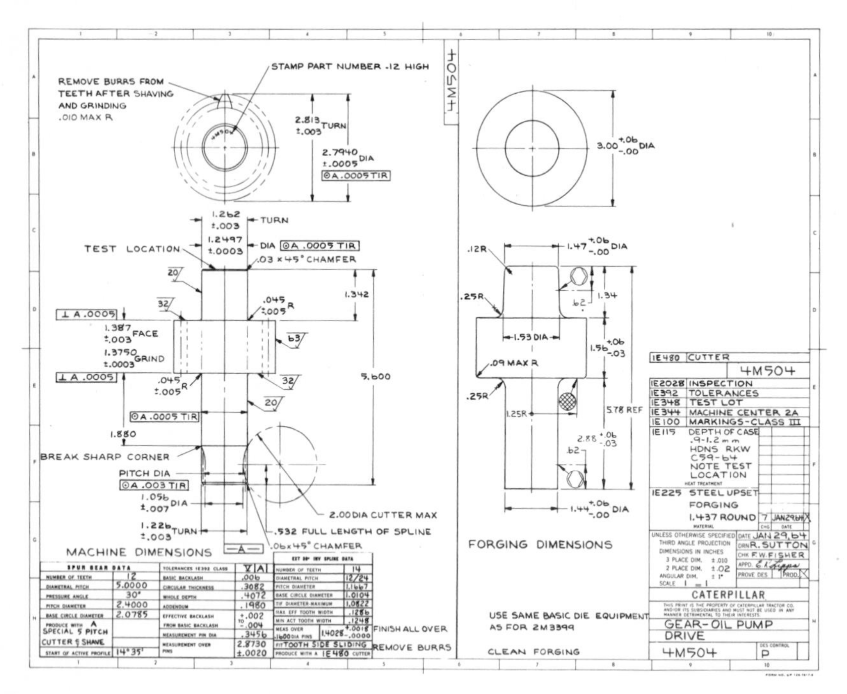

Fig. 11–6 **A two-part detail drawing showing separate information for forging and machining.** *(Caterpillar Tractor Co.)*

Fig. 11–7 **Tabulated (tabular) drawing.**

BUSHING					
PART NO.	A	B	C	D	E
CB 1	1.500	2.000	1.000	0.500	0.375
CB 2	1.625	2.125	1.125	0.625	0.437
CB 3	1.750	2.250	1.250	0.750	0.500
CB 4	1.875	2.375	1.375	0.875	0.562
CB 5	2.000	2.500	1.500	1.000	0.625

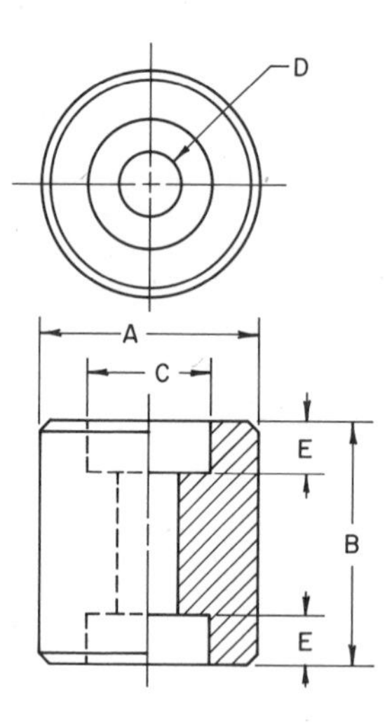

Notice also the detailed list of machine operations. Figure 11–6 shows a *combination drawing* (two-part detail drawing) for an oil-pump drive gear. The right-hand half gives the dimensions for the forging. The left-hand half gives the machining dimensions and notes. A detail drawing can also contain calculated data.

Standard detail drawings are often made for standard parts that come in a range of sizes. When such parts are used often, *tabulated,* or *tabular drawings* are made (Fig. 11–7). In this kind of drawing, the different dimensions are identified by letters. A table placed on the drawing tells what each dimension is for different sizes of the part. Either all or some of the dimensions can be given in this way. A similar kind of drawing is one in which the views are drawn with blank spaces for dimensions and notes (Fig. 11–8 at A). These are then filled in as needed with the required information (as at B). The views will not be to scale, except perhaps for one size.

ASSEMBLY DRAWINGS

A drawing of a fully assembled construction is called an *assembly drawing.* Such drawings vary greatly in how completely they show detail and dimensioning. Their special value is that they show how the parts fit together, the look of the construction as a whole, the dimensions needed for installation, the space the construction needs, the foundation, the electrical or water connections, and so forth. When an assembly drawing gives complete information, it can be used as a working drawing. It is then called an *assembly working drawing.* This kind of drawing can be made only when there is little or no complex detail. An example is shown in Fig. 11–9. You can often show furniture and other wood construction in assembly working drawings by adding enlarged details or partial views as needed (Fig. 11–10).

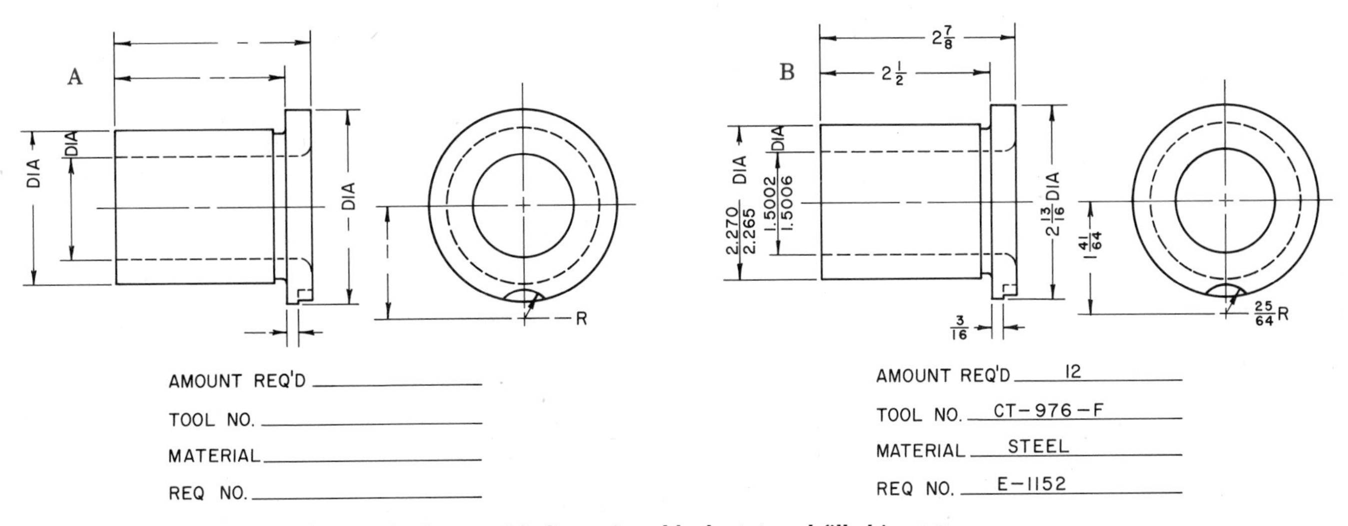

Fig. 11–8 **Detail drawing of a standard part with dimensions blank at A and filled in at B.**

Fig. 11–9 **An assembly working drawing for a belt tightener.**

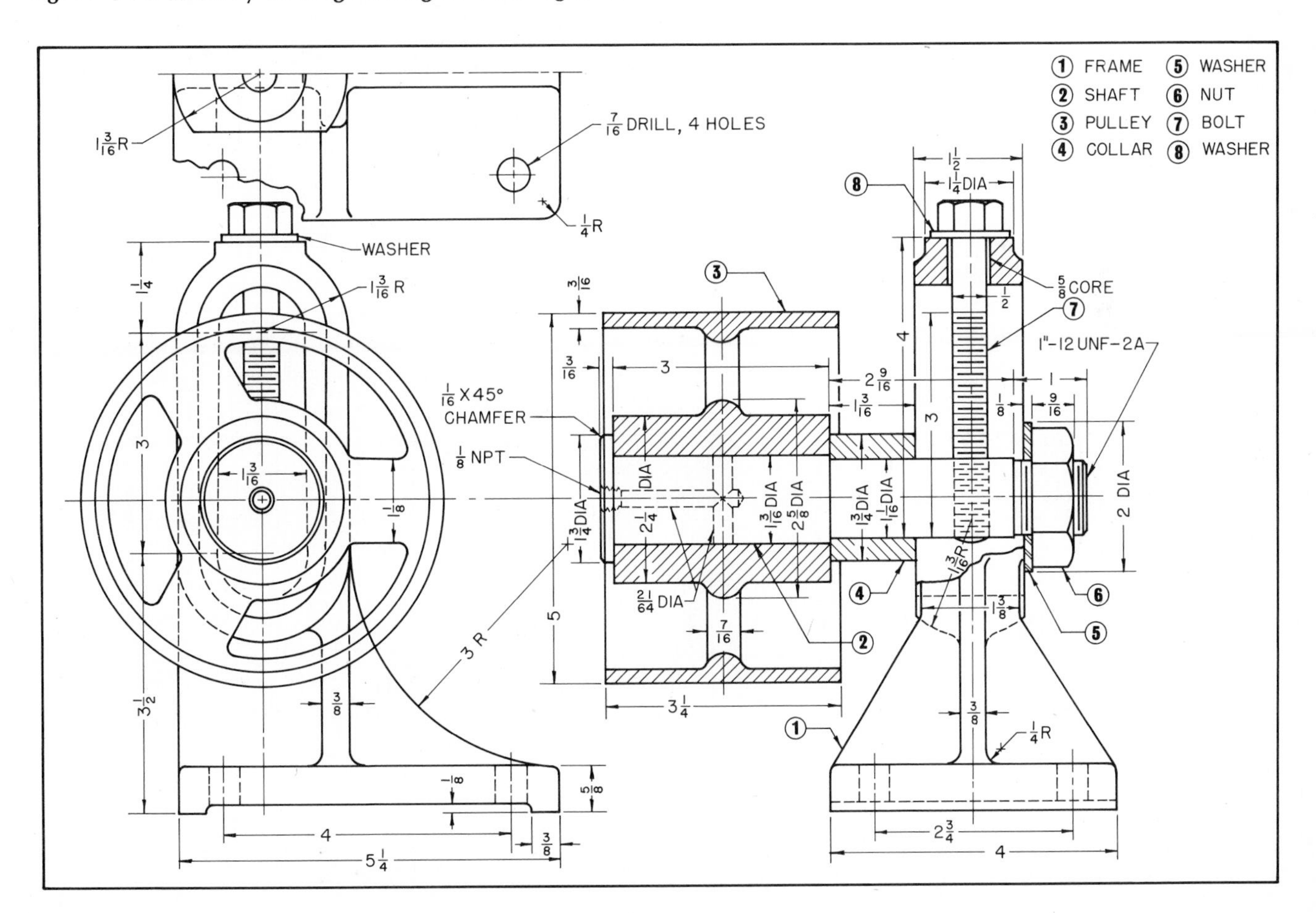

PICTORIAL RENDERING OF STEREO CABINET

Fig. 11–10 **An assembly working drawing with enlarged details and partial views.**

You generally make assembly drawings of machines to a small scale. Choose dimensions to tell overall distances, important center-to-center distances, and local dimensions. You can leave out all, or almost all, hidden lines. Also, if you are drawing to a very small scale, you can leave out unnecessary detail. This has been done, for

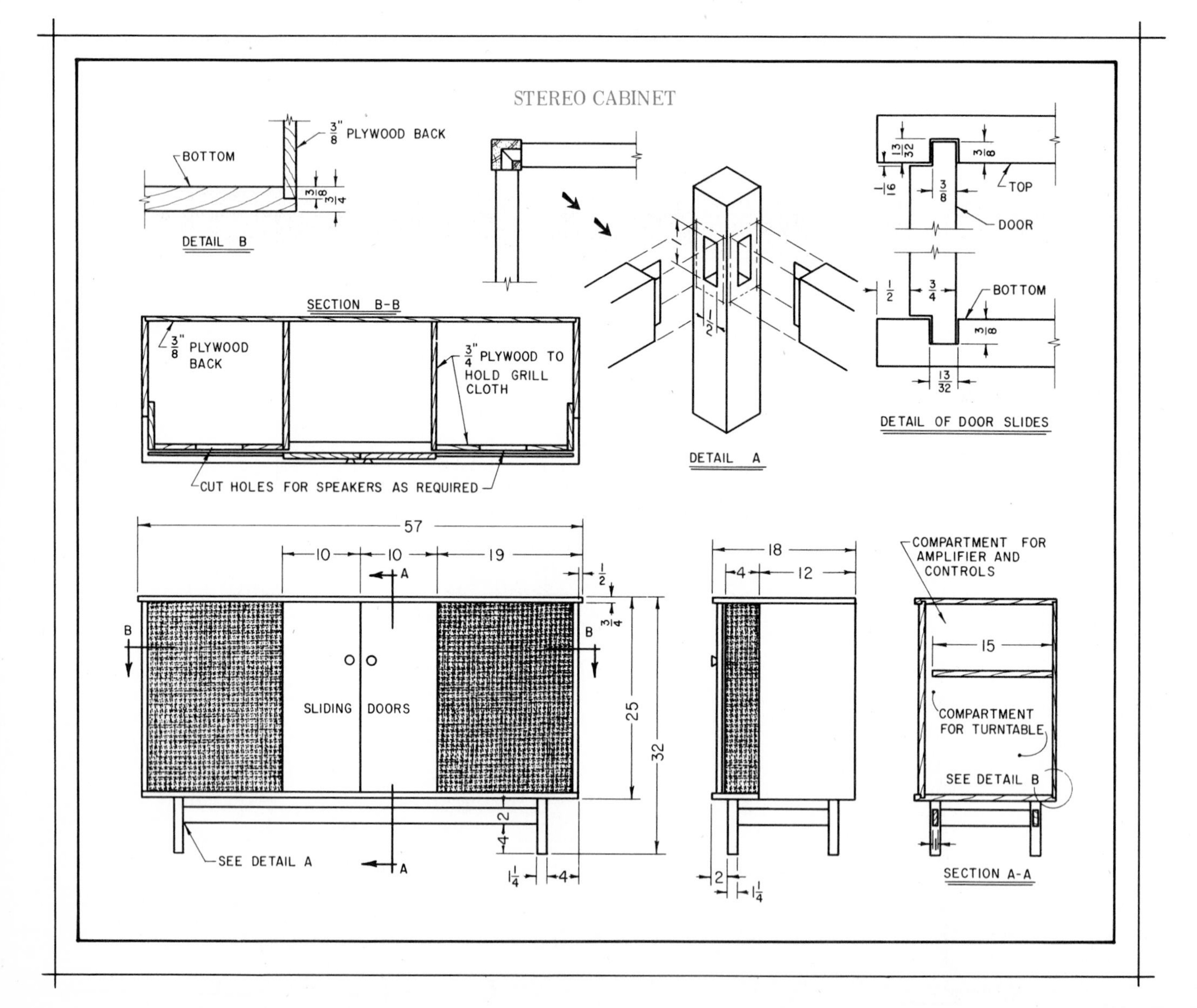

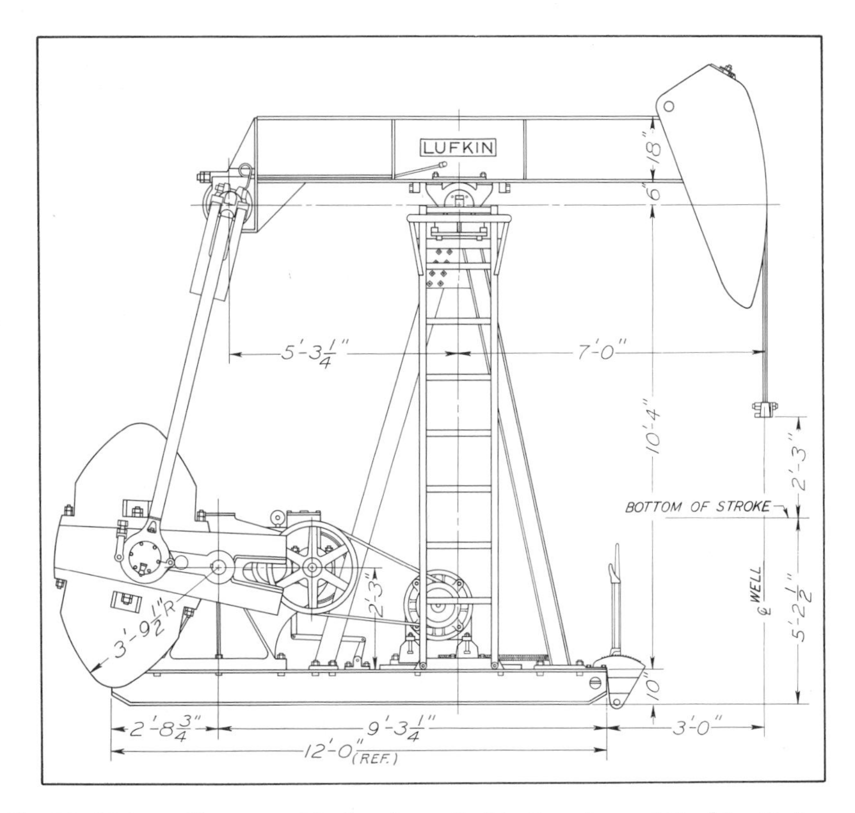

Fig. 11–11 **An outline assembly drawing.** ***(Lufkin Foundry and Machine Co.)***

Fig. 11–12 **A reference assembly drawing.** ***(Link-Belt Co.)***

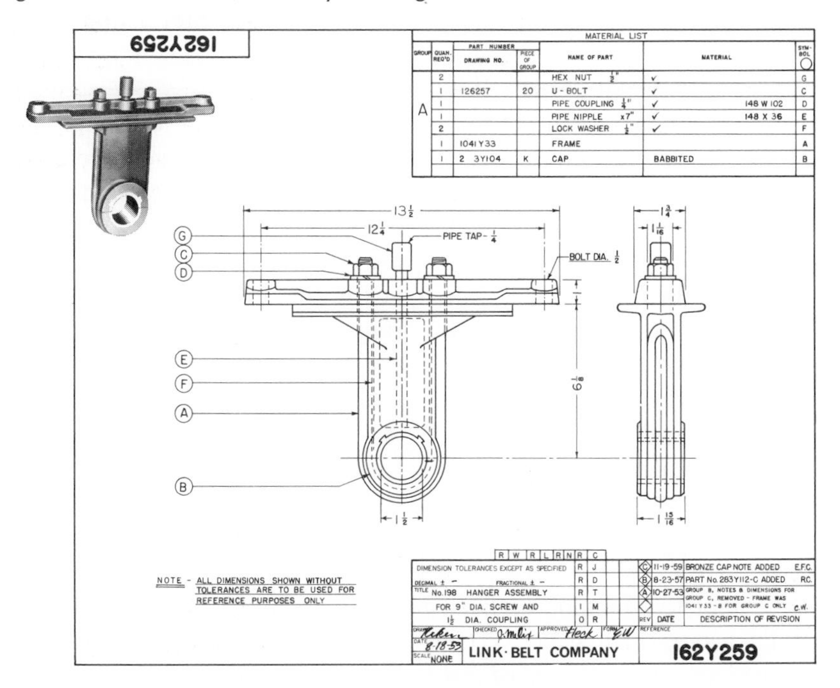

example, in Fig. 11–11. You can give either exterior or sectional views. When your main aim is just to show the general look of the construction, you need draw only one or two views. Because some assembled constructions are so big, you may have to draw the different views on separate sheets. Use the same scale on all sheets.

A special assembly drawing is the *reference assembly drawing* (Fig. 11–12). It is made for reference to identify parts to be assembled. Note the tabular list in the upper right-hand corner. Note also what dimensions the drafter has chosen to show.

Many other kinds of assembly drawings are made for special purposes. These include part assemblies for groups of parts, drawings for use in assembling or erecting a machine, drawings to give directions for upkeep and use, and so forth. A most important kind of assembly drawing is the *design layout.* It is from this kind of drawing that the detail drawings are made.

CHOICE OF VIEWS

Your drawings will be easier to use if you choose the views properly. To completely describe an object, you generally need at least two views. Although a drawing is not a picture, the views you choose should always be those that are easiest to read. Each view must have a part in the description. Otherwise, it is not needed and you should not draw it. In some cases, you need show only one view, as long as the shape and size are standard. Also, you may be able to

RIVETS EXC. NOTED	CENTRAL TEXAS IRON WORKS
HOLES EXC. NOTED	WACO TEXAS ABILENE
PAINT:	BUILDING
BLUE PRINT RECORD	LOCATION
NO. / DATE / ISSUED FOR	CUSTOMER
	ARCHITECT
	MADE BY / CHK. BY / SHEET NO.
	DATE / REVISIONS: / ORDER NO.

A

LESLIE LITTLEBEAR COMPANY					
414 SO. DETROIT – TULSA, OKLA.					
FOR:					
DRAWN BY: Chris Davis	DATE 4-3-79	CHECKED:	DATE	DWG. NO.	REV.
SCALE: NONE		APPROVED:	DATE		△

B

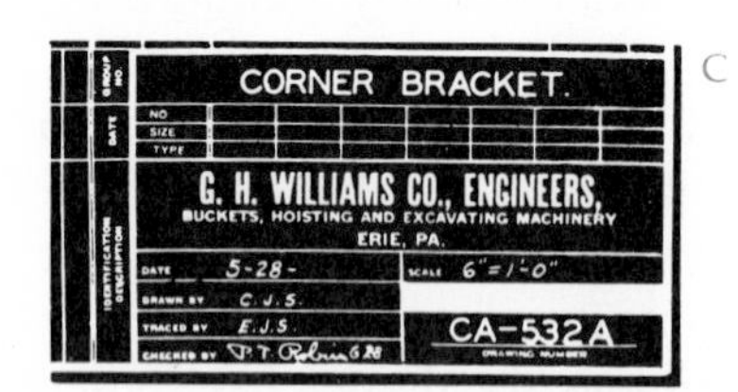

C

D

UNIT				NAME OF PIECE	
DR.	DATE	SYMBOL OF MACHINES USED ON	SUPERSEDES DRAW.	STOCK CASTING DROP FORGING	
DR.					
TR.		THE LODGE & SHIPLEY MACHINE TOOL CO.	SUPERSEDED BY DR.	MATERIAL	PIECE NO.
TR. CH.		FORM 795 CINCINNATI, OHIO. U. S. A.			

Fig. 11–13 **Titles. Boxed titles at A, B, and C. A strip title at D.**

give information in a note rather than draw a second view. For complex pieces, you may have to give more than three views. Some of these may be partial views, auxiliary views, or sectional views. If you are wondering what views to give, think of why you are making the drawing. A drawing is judged on how clearly and precisely it gives the information needed for its purpose.

CHOICE OF SCALE

You choose the scale for a detail drawing according to three factors:

1. How big the drawing must be to show details clearly.
2. How big it must be to carry all dimensions without crowding.
3. How big the paper is.

In most cases the best choice is a full-size scale. Other scales commonly used are half, quarter, and eighth. You should avoid such scales as 1:6 (2″ = 1′), 1:3 (4″ = 1′), and 1:1.33 (9″ = 1′). If a part is very small, you can sometimes draw it to an enlarged scale, perhaps twice full size or more.

When you draw a number of details on one sheet, you should, if possible, make them all to the same scale. If you use different scales, note them near each drawing. It is often useful to draw a detail, or part detail, to a larger scale on the main drawing. This will save you the work of making separate detail drawings. For general assembly drawings, choose a scale that will show the details you want and look good on the size of paper you are using. For sheet-metal pattern drawings for practical use, always use a full-size scale. You can make small-scale layouts of such drawings. These are used in building practice models.

For complete assemblies, you generally use a small scale. The scale is often fixed by the size of the paper the company has chosen for assemblies. For part assemblies, choose a scale to suit the purpose of the drawing. This might be to show how parts fit together, to identify the parts, to explain an operation, or to give other information.

TITLES

Every sketch or drawing must have some kind of title. However, the form, completeness, and location of this title can vary. On working drawings, you can place it in a box in the lower right-hand corner (Fig. 11–13 at A, B, and C). Or you can include it in a record strip running across the bottom or end of the sheet (Fig. 11–13 at D) as far as needed.

The title of a sketch gives as much as is needed of the following information:

1. The name of the construction, machine, or project.

2. The name of the part or parts shown, or simple details.
3. Manufacturer, company, or firm name, and address.
4. The date, usually the date of completion.
5. The scale or scales.
6. Heat treatment, working tolerances, and so forth.
7. Numbers of the drawing: of the shop order or of the customer's order, according to the system used.
8. Drafting-room record: names or initials, with dates, of drafter, tracer, and checker, and approval of chief drafter, engineer, and so forth.
9. A revision block or space for recording the changes, when needed, should be placed above or at the left of the title block.

A form for a basic layout of a title block is shown in Fig. 11–14. It is taken from American National Standards Drafting Practices, *Section 1, Size and Format* (ANSI Y14.1–1975) with the permission of the publisher, The American Society of Mechanical Engineering, 345 East 47th Street, New York, NY. The arrangement, size, and content may vary.

In large drafting rooms, the title is generally printed on the paper, cloth, or film, leaving spaces to be filled in. However, many firms use separate printed adhesive titles.

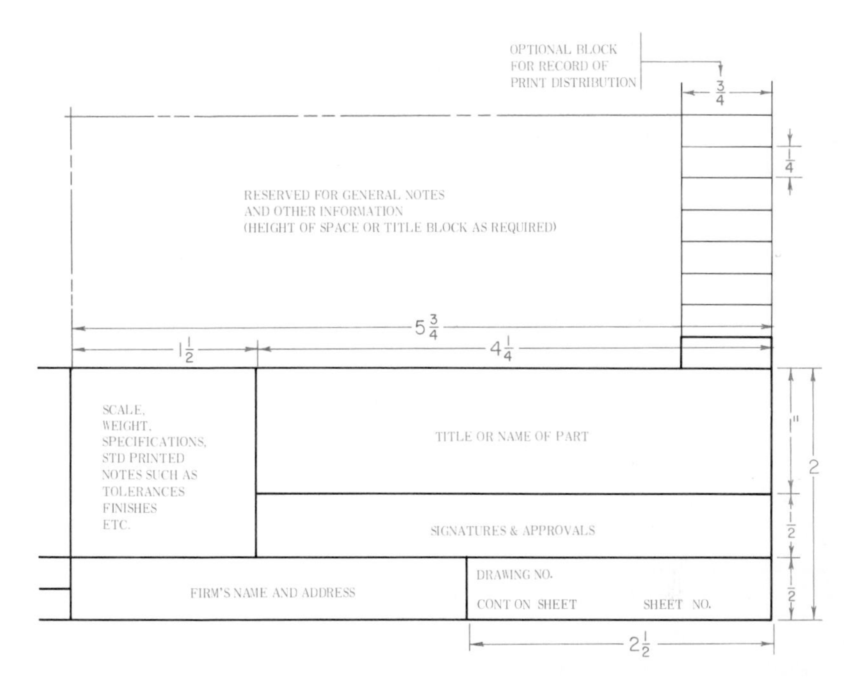

Fig. 11–14 **Basic layout for a title block.** ***(From American National Standards Institute ANSI Y14.1–1975.)***

BILL OF MATERIAL FOR IDLER PULLEY			
NAME	REQ.	MAT'L	NOTES
IDLER PULLEY	1	C.I.	
IDLER PULLEY FRAME	1	C.I.	
IDLER PULLEY BUSHING	1	BRO.	
IDLER PULLEY SHAFT	1	C.R.S.	
$\frac{5}{8}$ SAE HEX NUT	1		$\frac{3}{8}$ HIGH PURCHASED
WOODRUFF KEY 405	1		PURCHASED
$\frac{1}{8}$ OILER	1		PURCHASED

Fig. 11–15 **A bill of material.**

LIST OF MATERIALS OR PARTS LIST

You must, or should, include on most drawings a list of parts, the materials of which they are made, identification numbers, or other information. You need this information, especially on assembly drawings of various kinds. You also need it on detail drawings where a number of parts are shown on the same sheet.

You can sometimes give the names of parts, material, number required, part numbers, and so forth in notes near the views of each part. It is better, however, to place just the part numbers near the views, link them to the views with leaders, and then collect all the other information in tabulated lists. Such a list is called a *list of material, bill of material,* or *parts list* (Fig. 11–15). You can sometimes place it above the title. However, American National Standards recommends placing it in the upper right corner of the sheet. You can also sometimes write or type it on a separate sheet. You would title this sheet "Parts List for Drawing No. 00" to identify it. The American National Standards (ANSI Y14.1) recommends the form shown in Fig. 11–16. You can vary the column widths as needed.

GROUPING AND PLACING PARTS

When a number of details are used for a single machine, you often group them on a single sheet or set of sheets. A convenient method is to group the forging details together, the casting details together, the brass details together, and so on for other materials. In general, you should show parts in the position that they will occupy in the assembled machine. That way, related parts will appear near each other. Do not, however, draw long pieces, such as shafts and bolts, this way. Draw them with their long dimensions parallel to the long dimension of the sheet (Fig. 11–17).

NOTES AND SPECIFICATIONS

Information that you cannot make clear in a drawing must be given in lettered notes and symbols. For example, trade information that is generally understood by those on the job is often given in this way. You should use such notes for the following items: the number needed, the material, the kind of finish, the kind of fit, the method of machining, the kinds of screw threads, the kinds of bolts and nuts, the sizes of wire, and the thickness of sheet metal.

The materials in general use are wood, plastic, cast iron, wrought iron, steel, brass, aluminum, and various alloys. All parts to go together must be of the proper size so that they will fit. Pieces may be left rough, partly finished, or completely finished. Wood used in furniture is shaped with woodworking tools and machines. Many metals, such as cast iron, brass, aluminum, and so forth, are made into shapes by molding, casting, and machining. First, a wooden pattern of the required shape and size is made. Then, this pattern is placed in sand to make an impression, or mold, into which the molten metal is poured. Wrought iron and steel are made into shapes by rolling or forging in the rolling mill or blacksmith shop. Some kinds of steel may be cast.

There are many interesting ways of forming metals for special purposes. There are also many special alloys that this book does not have room to describe. However, you will learn much by observing the different shapes of machine parts and the many materials of which they are made.

After a part is cast or forged, it must be machined on all surfaces that are to fit other surfaces. Round surfaces are generally formed on a lathe. Flat surfaces

Fig. 11–16 **Recommended form for a list of materials.** *(From American National Standards Institute ANSI Y14.1–1975.)*

LIST OF MATERIAL								
GROUP NO. AND QUANTITY						PART NO.	NAME	DRAWING NO. OR DESCRIPTION

.25
.3
1.8
.03
.4
1.4
7.6

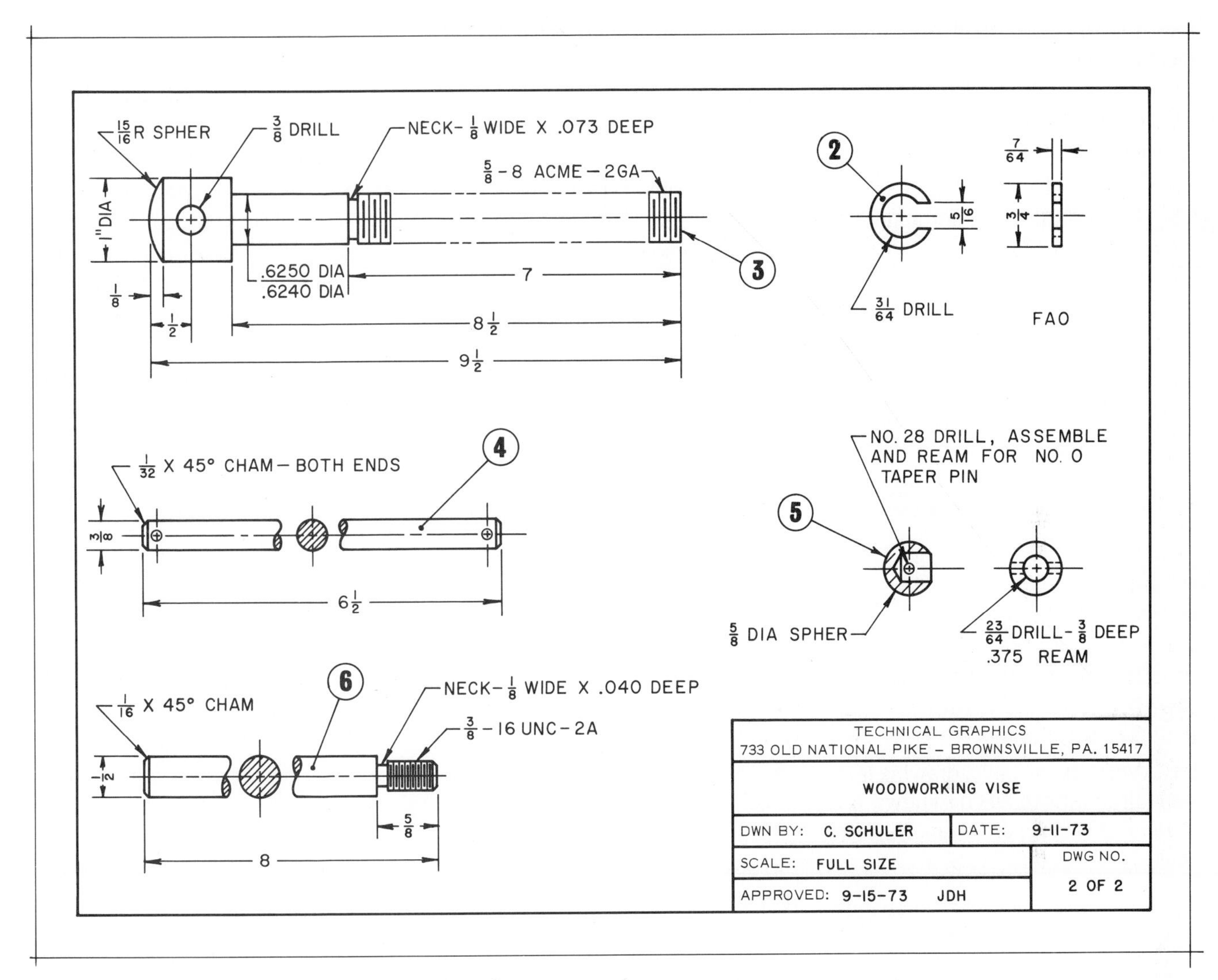

***Fig. 11–17* Several machine parts drawn on one sheet.**

are finished or smoothed on a planer, milling machine, or shaper. Holes are made with drill presses, boring mills, or lathes. Extra metal is allowed for surfaces that are to be finished. To specify such surfaces, place a V-symbol on the lines that represent their edges. If the entire piece is to be finished, write a note such as "Finish All Over" or "FAO." No other mark is needed.

You specify the kinds of machining, finish, or other treatment in notes. Such notes would read, for example, "spot face," "grind," "polish," "knurl," "core," "drill," "ream," "countersink," "counterbore," "harden," "caseharden," "blue," or "temper." Often you also have to add other notes for special directions, to explain the assembly, for example, or the order of doing work.

CHECKING A DRAWING

After you finish a drawing, it must be looked over very carefully before it is used. This is called *checking the drawing.* It is very important work. A drawing you have made should be checked by someone other than you. A person who has not worked on the drawing will be better able to spot errors.

To make a thorough check, there is a set order of procedure that you must follow. You must make all of the following checks:

1. See that the views completely describe the shape of each piece.
2. See that there are no unnecessary views.
3. See that the scale is large enough to show all detail clearly.
4. See that all views are to scale and that the right dimensions are given.
5. See that there will be no parts that will interfere with each other during assembly or operation and that necessary clearance space is provided around all parts that need it.
6. See that enough dimensions are given to define the sizes of all parts completely, and that no unnecessary or duplicate dimensions are given.
7. See that all necessary location or positioning dimensions are given with necessary precision.
8. See that necessary tolerances, limits and fits, and other precision information is given.
9. See that the kind of material and the number needed of each part are specified.
10. See that the kind of finish is specified, that all finished surfaces are marked, and that a finished surface is not called for where one is not needed.
11. See that standard parts and stock items, such as bolts, screws, pins, keys or other fastenings, handles, catches, etc., are used where suitable.
12. See that all necessary explanatory notes are given and that they are properly placed.

Each drafter must inspect his or her own work for errors or omissions before having it checked.

Review

1. What do you call a drawing that tells all that needs to be known for making a single part or a complete machine?

2. Are most industrial drawings done in pencil or in ink?

3. What do you call a drawing of a single part that gives all information necessary for making it?

4. What is a drawing of a completely assembled construction called?

5. Name three factors that help determine the scale to be used on a drawing.

6. Name two types of title blocks.

7. What scale do you usually use for sheet-metal pattern drawings?

8. What is another name for a parts list?

9. If an entire part is to be finished, what general note do you use?

10. What symbol do you use to identify specific surfaces that are to be finished?

Problems

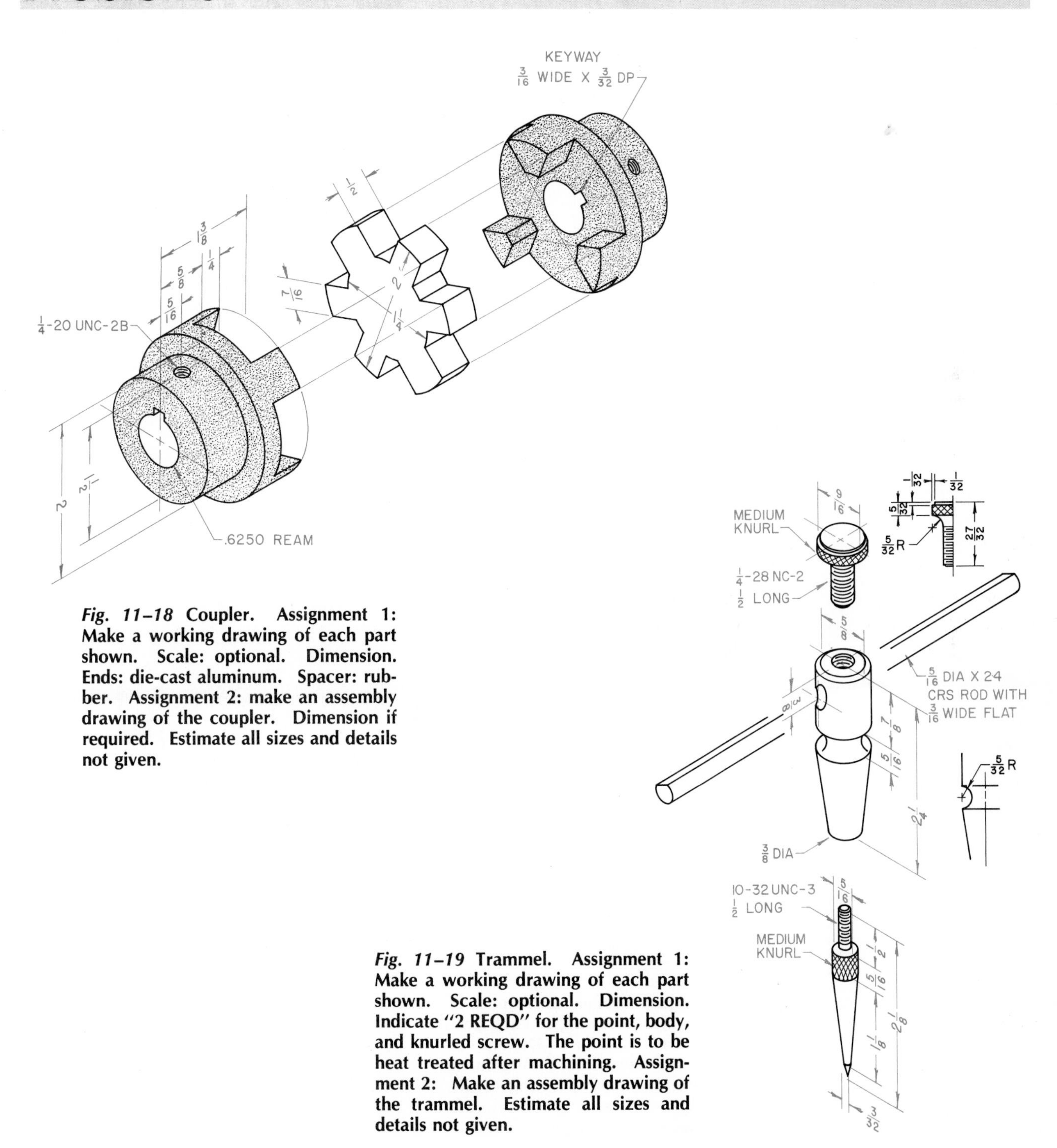

Fig. 11–18 Coupler. Assignment 1: Make a working drawing of each part shown. Scale: optional. Dimension. Ends: die-cast aluminum. Spacer: rubber. Assignment 2: make an assembly drawing of the coupler. Dimension if required. Estimate all sizes and details not given.

Fig. 11–19 Trammel. Assignment 1: Make a working drawing of each part shown. Scale: optional. Dimension. Indicate "2 REQD" for the point, body, and knurled screw. The point is to be heat treated after machining. Assignment 2: Make an assembly drawing of the trammel. Estimate all sizes and details not given.

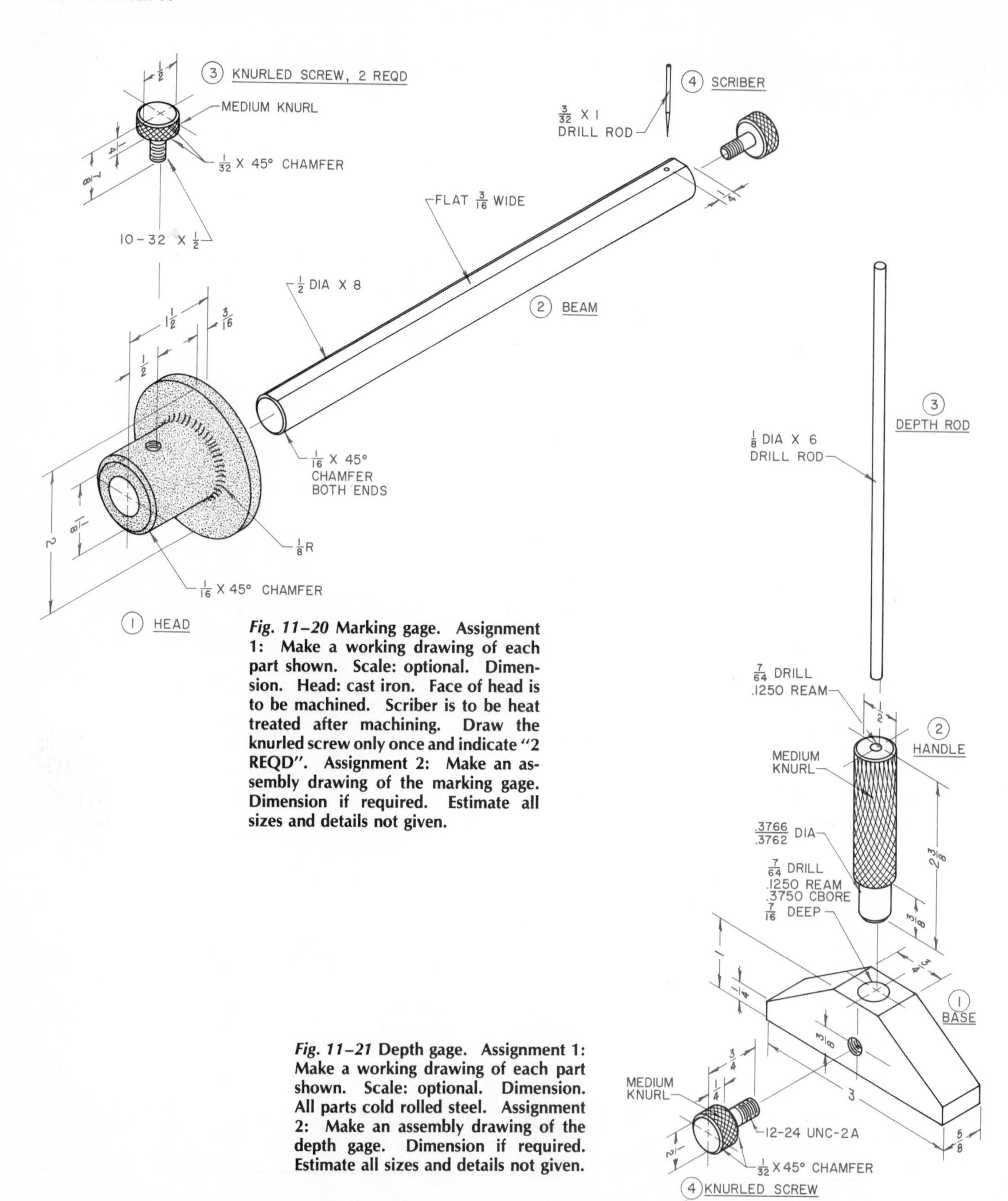

Fig. 11–20 **Marking gage. Assignment 1: Make a working drawing of each part shown. Scale: optional. Dimension. Head: cast iron. Face of head is to be machined. Scriber is to be heat treated after machining. Draw the knurled screw only once and indicate "2 REQD". Assignment 2: Make an assembly drawing of the marking gage. Dimension if required. Estimate all sizes and details not given.**

Fig. 11–21 **Depth gage. Assignment 1: Make a working drawing of each part shown. Scale: optional. Dimension. All parts cold rolled steel. Assignment 2: Make an assembly drawing of the depth gage. Dimension if required. Estimate all sizes and details not given.**

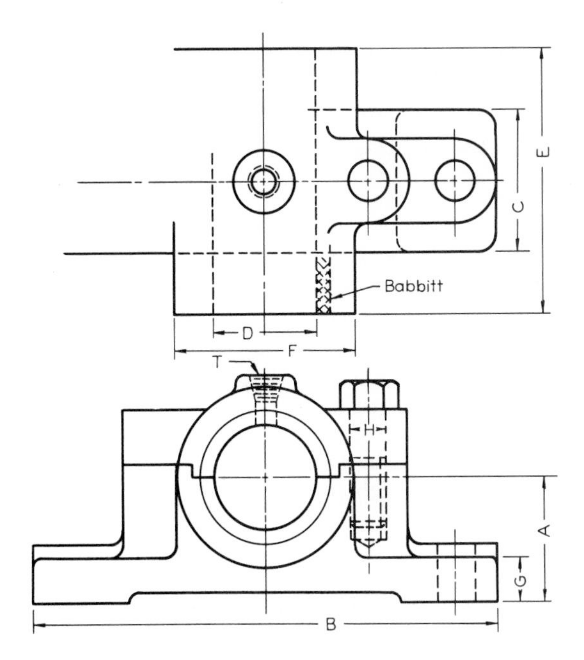

Fig. 11–22 **Make a complete working drawing of a babbitted bearing, with front and side views as half sections. Missing dimensions to be supplied by the student. For dimensions see accompanying table.**

Size	D	A	B	C	E	F	G	H	T
w	1	1¼	4¾	1⅜	2½	1¾	½	⅜	⅛ pipe
x	1½	1¾	6¾	2	3¾	2⅝	⅝	½	⅛ pipe
y	2	2¼	8¾	2½	5	3½	¾	½	¼ pipe
z	2½	2¾	10¾	3⅛	6¼	4⅜	⅞	⅝	¼ pipe

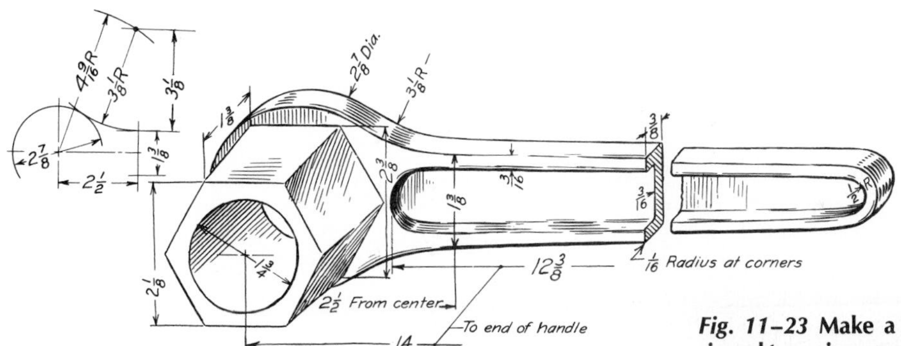

Fig. 11–23 **Make a completely dimensioned two-view working drawing of the plug wrench. Show a revolved section through the handle. Consider the choice of views, scale, and placing of views before starting the drawing.**

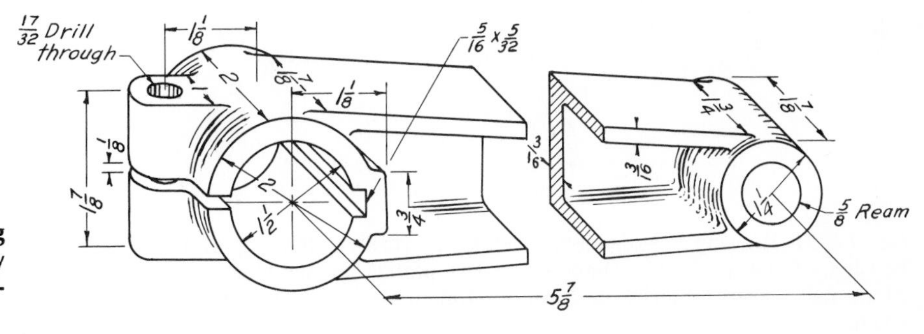

Fig. 11–24 **Make a two-view working drawing of the lever with all necessary dimensions. Use your judgment regarding finished surfaces.**

Figures 11–25 through 11–30 **refer to illustrations in the chapter text.**

Fig. 11–25 **See Fig. 11–1, a pictorial and a working drawing of a journal bearing. Prepare a detail drawing of the journal base. Scale: full size.**

Fig. 11–26 **See Fig. 11–3, a working drawing of a special adapter. Prepare a detailed section as shown and one exterior view with simplified threads. Scale: full size.**

Fig. 11–27 **See Fig. 11–5, a working drawing of a latch pinion. Prepare detailed drawings as shown. Include finishing notes. Scale: full size.**

Fig. 11–28 **See Fig. 11–9, a working drawing of a belt tightener. Prepare three exterior views. Scale: full size.**

Fig. 11–29 **See Fig. 11–9. Prepare detail drawings of the frame, shaft, and pulley. Scale: full size.**

Fig. 11–30 **See Fig. 11–10, a working drawing of a stereo cabinet. Prepare a bill of material and three views with dimensions in millimeters. Scale: full size.**

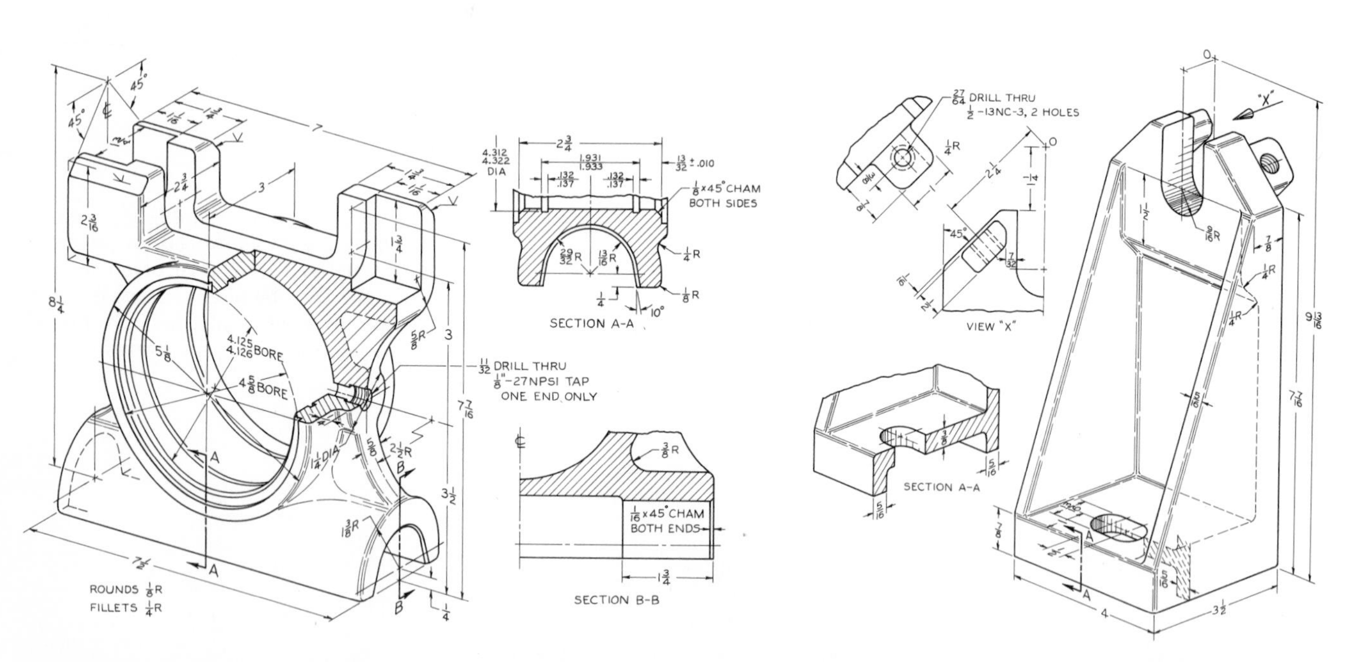

Fig. 11–31 **Housing. Assignment: Make a working drawing of the housing. Scale: optional. Dimension. Use partial and sectional views where necessary. Material: cast iron. Estimate all sizes and details not given.**

Fig. 11–32 **End base. Assignment: Make a working drawing of the end base. Scale: optional. Dimension. Use partial and sectional views where necessary. Material: cast iron. Estimate all sizes and details not given.**

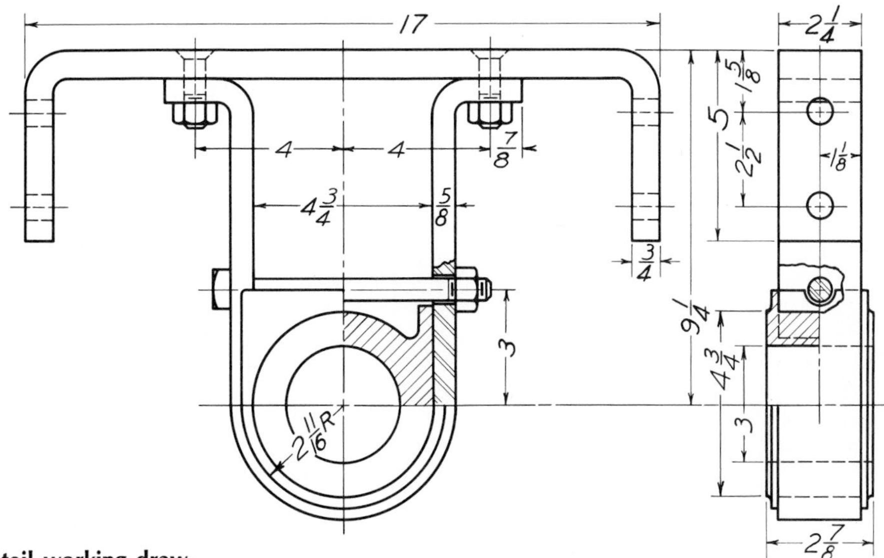

Fig. 11–33 **Make detail working drawings of the hung bearing, showing each piece fully dimensioned. All bolts are 5/8″ in diameter.**

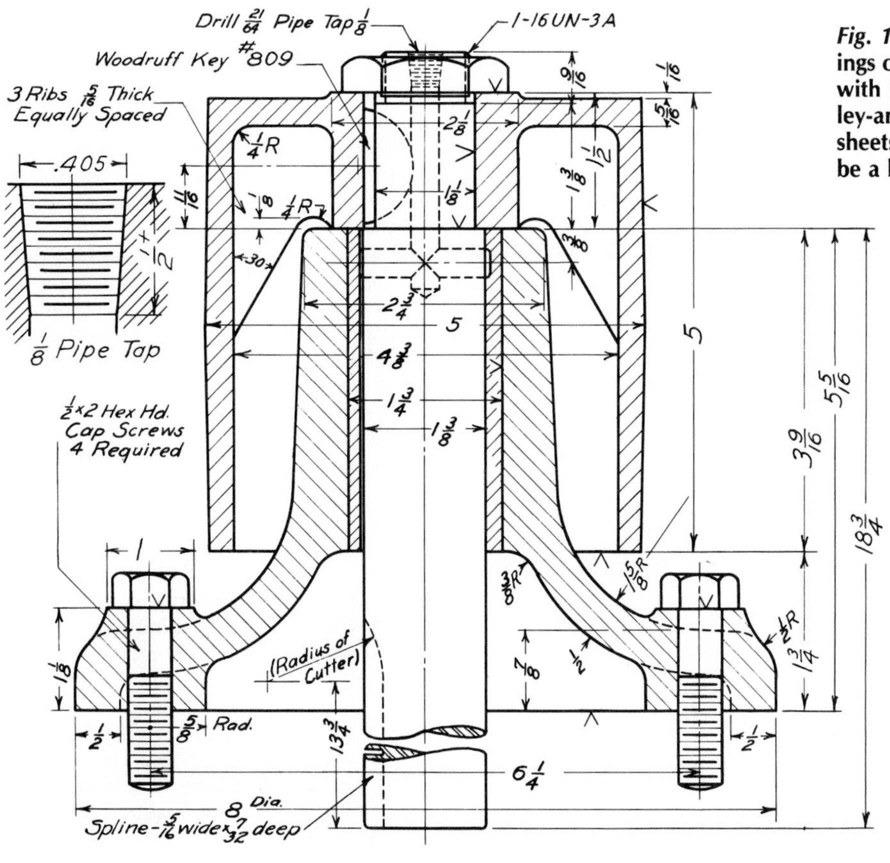

Fig. 11–34 **Make working detail drawings of base, pulley, bushing, and shaft, with bill of material for complete pulley-and-stand unit. Full size, three sheets. If necessary, the top view may be a half plan.**

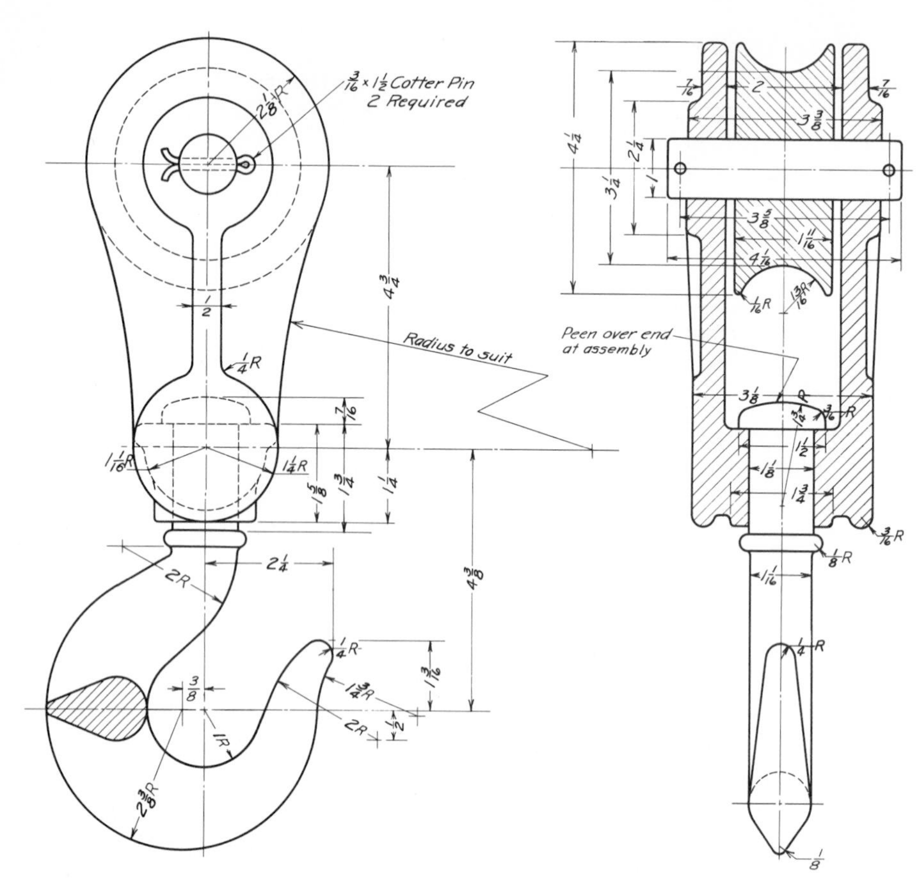

Fig. 11–35 Make detail drawings of the crane hook.

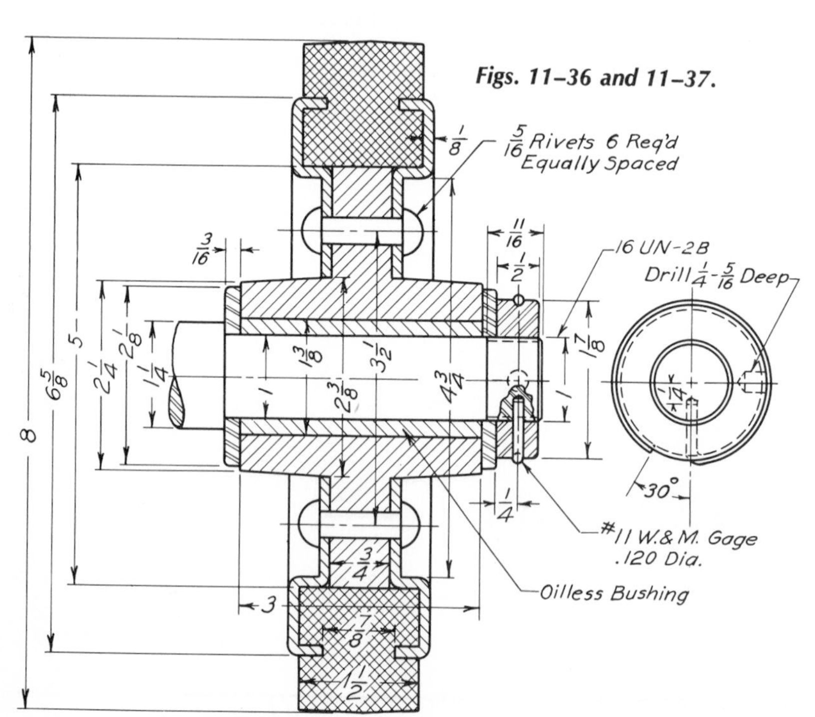

Figs. 11–36 and 11–37.

Fig. 11–36 Make a front view and section of the cushion wheel, full size. This type of wheel is used on warehouse or platform trucks to reduce noise and vibration.

Fig. 11–37 Make a complete set of detail drawings, full size, with bill of material, for the cushion wheel. Three sheets will be needed. Rivets are purchased and, therefore, would not be detailed but would be specified in the bill of material.

Figs. 11–38 through 11–40 A jig is a device used to hold a machine part (called the *work* or *production*) while it is being machined, or produced, so that all the parts will be alike within specified limits of accuracy. Note the production shown in the upper left corner.

Fig. 11–38 Make a detail working drawing of the jig body.

Fig. 11–39 Make a complete set of detail drawings for the jig with bill of material. Use as many sheets as necessary or use larger sheets.

Fig. 11–40 Make a complete assembly drawing of the jig, three views. Give only such dimensions as are necessary for putting the parts together and using the jig.

Figs. 11–38 through 11–40

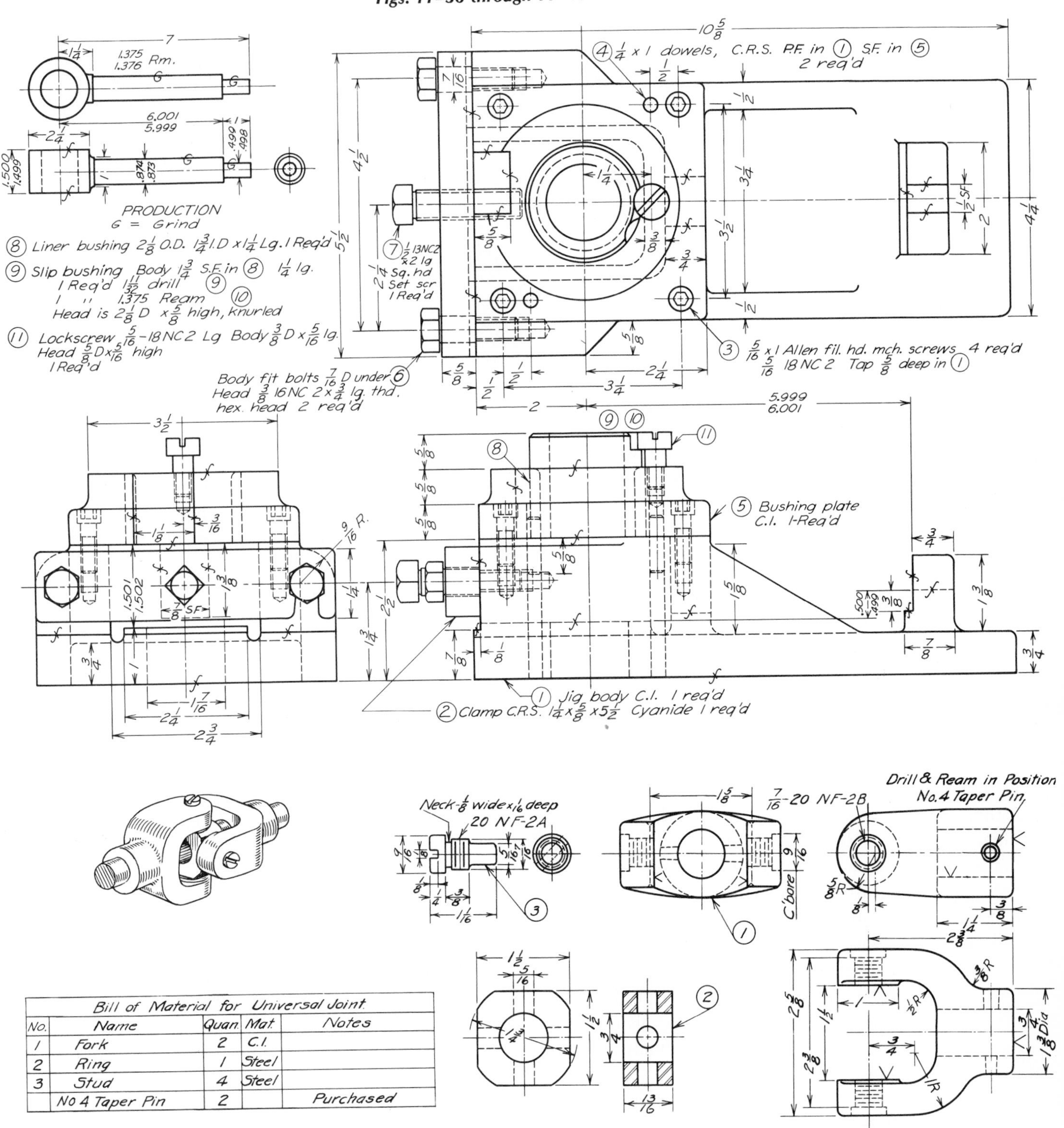

Bill of Material for Universal Joint				
No.	Name	Quan.	Mat.	Notes
1	Fork	2	C.I.	
2	Ring	1	Steel	
3	Stud	4	Steel	
	No 4 Taper Pin	2		Purchased

Fig. 11–41 Make a two-view assembly drawing of the universal joint in section.

Chapter 12

Pictorial Drawing

Fig. 12–1 **Pictorial drawings show objects as they appear.** *(Inland Steel Urban Development Corp.)*

USES OF PICTORIAL DRAWING

Pictorial drawing is an essential part of the graphic language. It is very important in engineering, architecture, science, electronics, technical illustration, and in many other professions. It appears in all kinds of technical literature as well as in catalogs and assembly, service, and operating manuals. Architects use pictorial drawing to show what the finished building will look like (Fig. 12–1). Advertising agencies use pictorial drawings to display a new product.

Pictorial drawing is often used in exploded views on production and assembly drawings. These views are made to illustrate parts lists (Fig. 12–2), to explain the operation of machines, apparatus, and equipment, and for many other commercial and technical purposes. In addition, most people use some form of pictorial sketches to help convey ideas that are hard to describe in words.

TYPES OF PICTORIAL VIEWS

Figure 12–3 shows different kinds of pictorial drawings. The first is a perspective view. It shows an object (in this case a stereo cabinet) as it actually looks to the eye. Drafters who are familiar with technical subjects often make perspective sketches. The other views are made with either of two pictorial projection methods: isometric or oblique. The isometric and oblique views are easier to draw than the perspective view, but they do not look as good. All three drawing methods allow the drafter to show three sides of an object in one view. Their real advantage is that the important lines can be measured directly.

Cabinet drawing is a special kind of oblique drawing. In a cabinet drawing, distances on the receding axes are reduced by one-half. In other words, a cabinet drawing is drawn only half as deep as a normal oblique view.

AXONOMETRIC PROJECTION

Isometric projection is one form of *axonometric* projection. Other forms are *dimetric* and *trimetric* projection. All three are made according to the same theory; the difference is the angle of projection (Fig. 12–4). In isometric projection, the axes form three equal angles of 120° on the plane of projection. Only one scale is needed for measurements along each axis. Isometric drawings are the easiest type of axonometric drawing to make. In dimetric projection, only two of the angles are equal, and two special foreshortened scales are needed to make measurements. In trimetric projection, all three angles are dif-

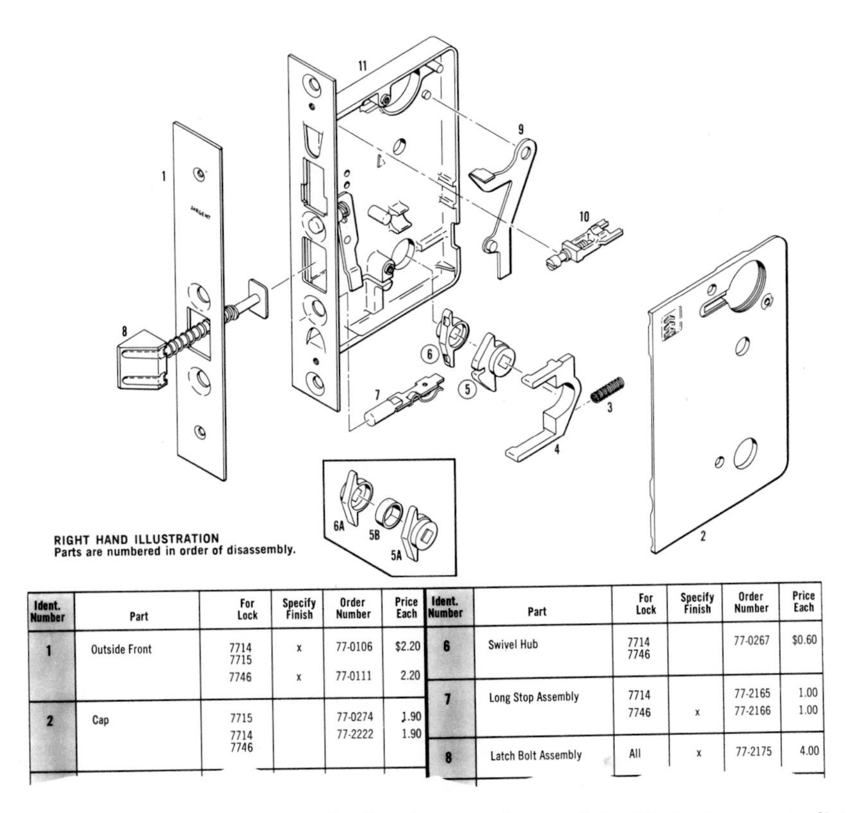

Ident. Number	Part	For Lock	Specify Finish	Order Number	Price Each	Ident. Number	Part	For Lock	Specify Finish	Order Number	Price Each
1	Outside Front	7714 7715	x	77-0106	$2.20	6	Swivel Hub	7714 7746		77-0267	$0.60
		7746	x	77-0111	2.20	7	Long Stop Assembly	7714		77-2165	1.00
2	Cap	7715		77-0274	1.90			7746	x	77-2166	1.00
		7714 7746		77-2222	1.90	8	Latch Bolt Assembly	All	x	77-2175	4.00

Fig. 12–2 **An exploded assembly drawing may be used to illustrate a parts list. *(Sargent & Co.)***

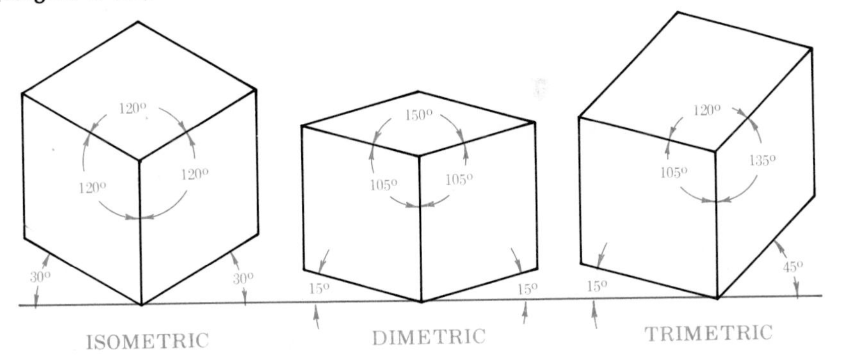

Fig. 12–4 **The three types of axonometric projections.**

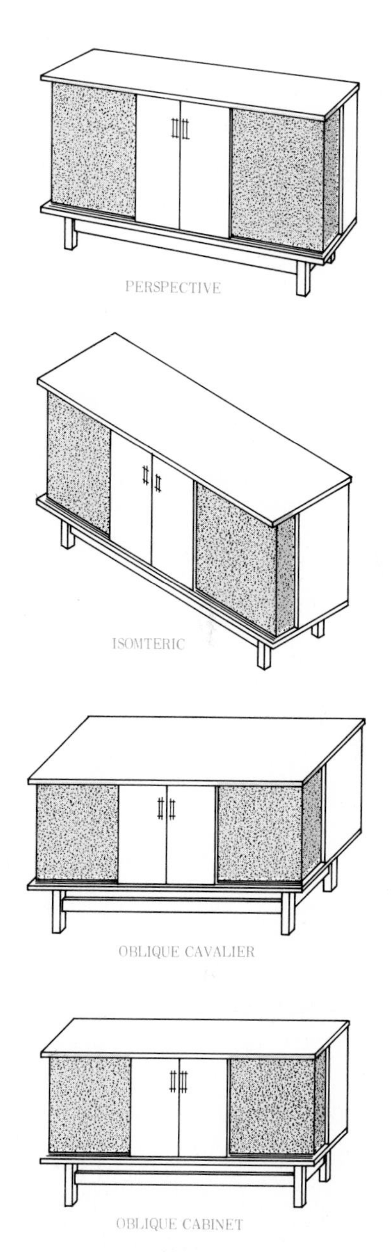

Fig. 12–3 **Stereo cabinet drawn in various types of pictorial views.**

ferent, and three special foreshortened scales are needed.

Dimetric and trimetric drawings are complicated to draw. Therefore, drafters use them less often than other types of pictorial drawing. This chapter will not deal with dimetric and trimetric drawings. It will cover only the isometric, oblique, and perspective types of pictorial drawing.

ISOMETRIC PROJECTION AND ISOMETRIC DRAWING

Isometric projection and isometric drawing are not the same thing, even though many people think they are. In this section, we will learn what the differences are. You probably will not have to make pictorial drawings using isometric projection, but it is a good idea for you to understand the theory behind it.

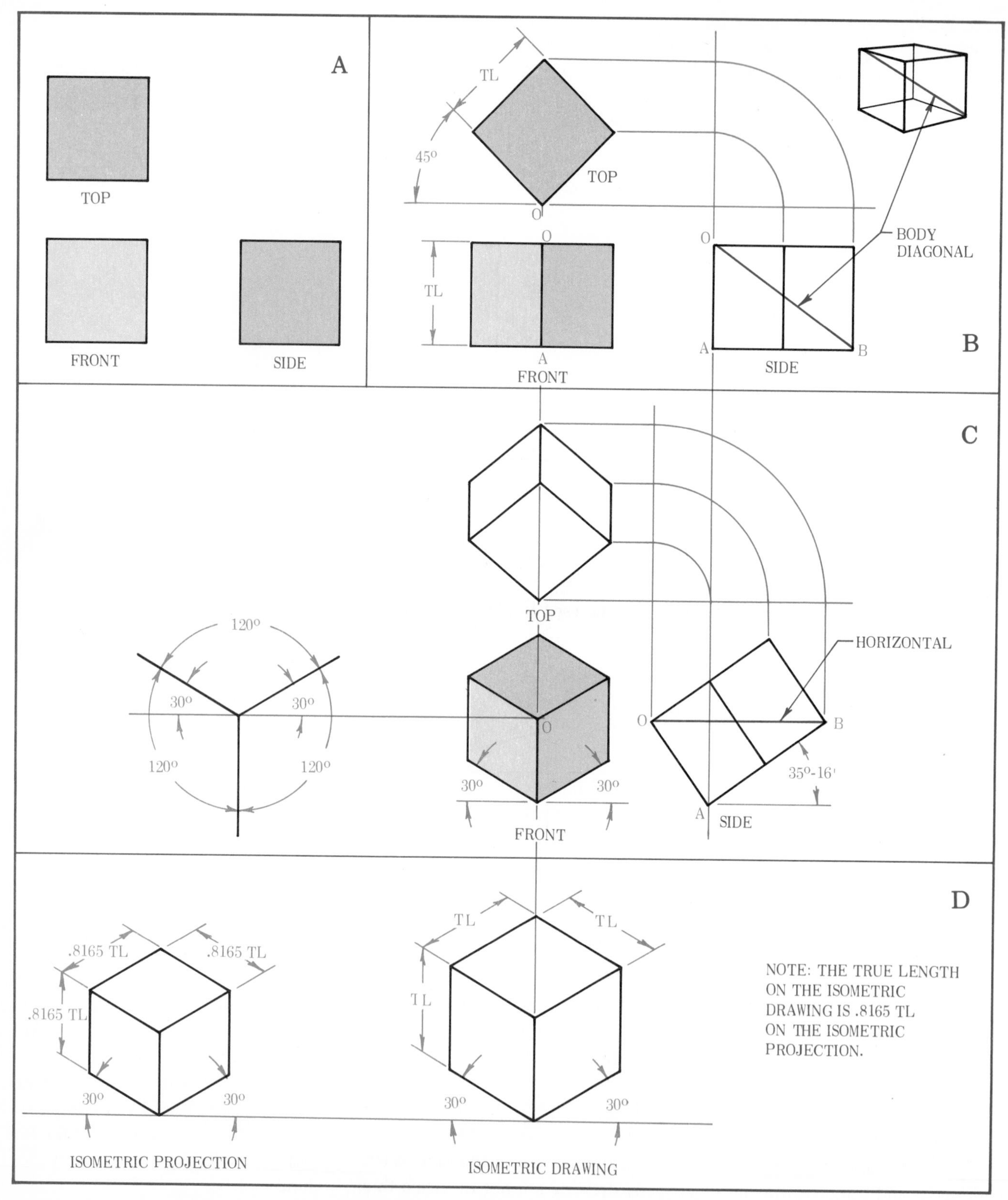

Fig. 12–5 **Isometric projection and isometric drawing.**

One way you can make an isometric projection is by a method called *revolution* (drawing an object as if it were revolved or turned). Figure 12–5 shows a cube in the three normal views of a multiview drawing. In Fig. 12–5B, each of the three views has been revolved 45°. Notice that the front and side views now show as two equal rectangles. On the side view, a diagonal is drawn from point *O* to point *B*. This is called the *body diagonal*. It is the longest straight line that can be drawn in a cube.

In Fig. 12–5C, the cube is revolved upward until the body diagonal is horizontal, as shown in the side view. Notice that the cube was revolved 35°16′ to achieve this. The front view now forms an isometric projection. Its lower edges form an angle of 30° to the horizontal.

Since the front view (isometric projection) of the revolved cube is made by projection, its lines are foreshortened. The actual difference is 0.8165 to 1 in. In other words, 1 in. on the cube in Fig. 12–5A has been reduced to 0.8165 in. on the isometric projection in Fig. 12–5C.

Figure 12–5D shows an isometric drawing and an isometric projection of the same cube. In the isometric drawing, all edges are drawn their true length instead of the shortened length. The drawing shows the shape of the cube just as well as the projection. Its advantage is that it is easier to draw, because all its measurements can be made with a regular scale. It can also be drawn at once, without projecting from other views or without using a special scale.

Isometric Drawing

A multiview drawing of a filler block is shown in Fig. 12–6 at A. To make an isometric drawing of the block, begin with the three lines shown at B. These lines are the *isometric axes.* They will represent three edges of the block. Draw them to form three equal angles, each of 120° (3 × 120° = 360°). Draw axis line *OA* vertically. Draw axes *OB* and *OC* with the 30°–60° triangle. The point at which the lines meet represents the upper front corner *O* of the block as shown at C.

Measure off the width *W*, the depth *D*, and the height *H* of the block on the three axes. Then draw lines parallel to the axes to make the isometric drawing of each block. To locate the rectangular hole shown at D, lay off 1 in. along *OC* to *c*. Then from *c* lay off 2 in. to c^1. Through *c* and c^1 draw lines parallel to *OB*. In like manner, locate *b* and b^1 on axis *OB* and draw lines parallel to *OC*. Draw a vertical line from corner 3. NOTE: *The dimensions, letters, and numerals are for instructional purposes. You would not normally put them on your drawing.* Darken all necessary lines to complete the drawing (Fig. 12–6E).

You make a pictorial drawing, in general, to show how something looks. Hidden edges (lines) are not "part of the picture"; therefore, you normally leave them out. However, you might include them in some particular case when you want to indicate a certain feature for explanation.

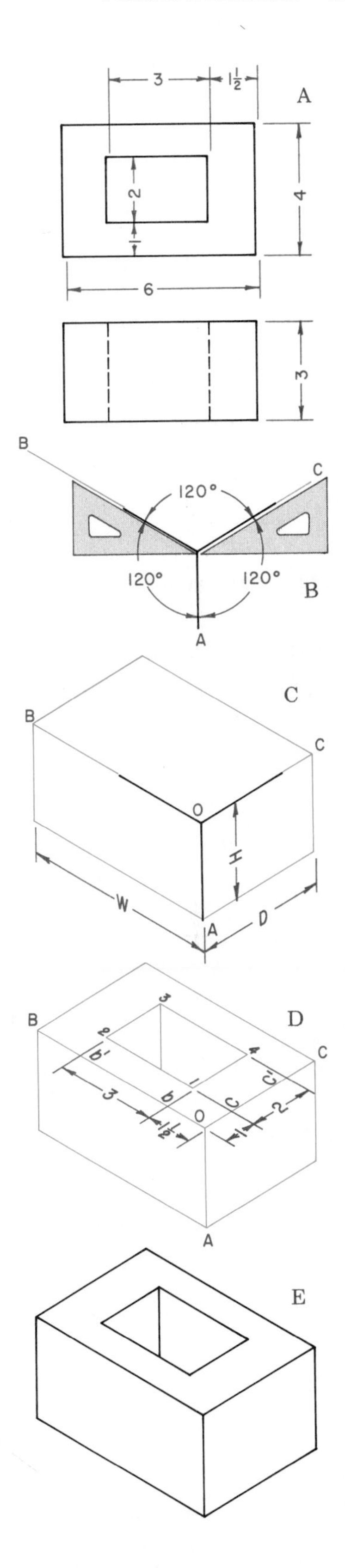

Fig. 12–6 Steps in making an isometric drawing.

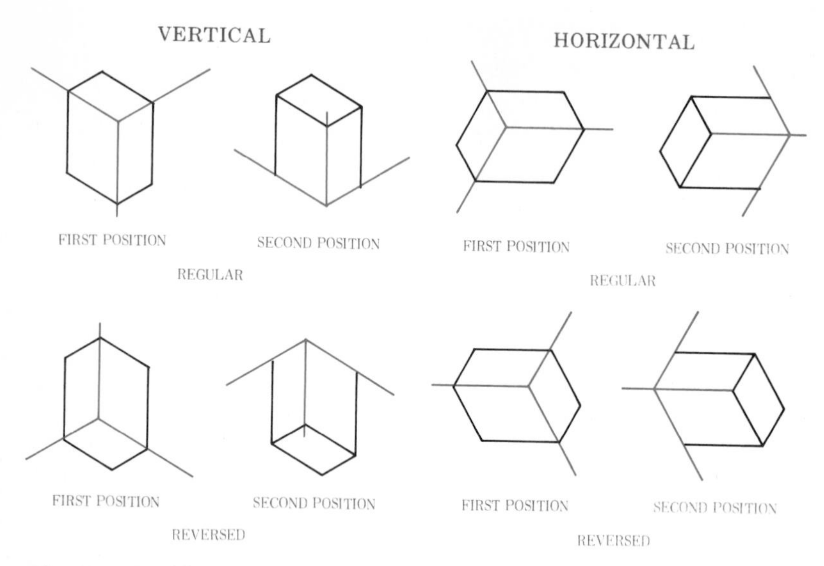

Fig. 12–7 **Positions for isometric axes.**

Position for the Isometric Axes

Isometric axes can be arranged in different ways, provided they remain at 120° angles from each other. Several positions are shown and identified in Fig. 12–7. You will see how to apply them later in this chapter.

The arrangement of the axes in Fig. 12–6 is called the first position. Here the axes represent the three edges of the cube that meet at the upper front corner. Often it is more convenient to place the axes in the second position. There they represent the edges that meet at the lower front corner (Fig. 12–8).

Any line of an object parallel to one of the edges of a cube is drawn parallel to an isometric axis. Such a line is called an *isometric line.* An important rule of isometric drawing is this: *Measurements can be made only on isometric lines.*

Nonisometric Lines

Lines that are not parallel to any of the isometric axes are called *nonisometric lines* (Fig. 12–9). Such lines do not show in their true length and cannot be measured. To draw them, you must first locate their two ends and then connect the points. Angles on isometric drawings also do not show in their true size. Therefore, they cannot be measured in degrees.

Figure 12–10 shows how to locate and draw nonisometric lines in an isometric drawing. The method is called the *box method.* The lines in

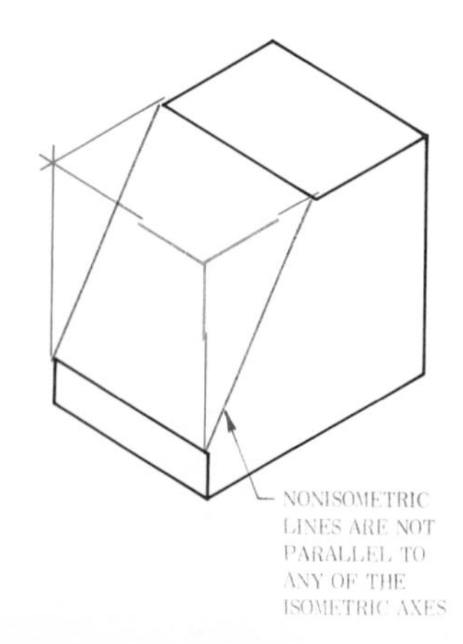

Fig. 12–9 **Nonisometric lines.**

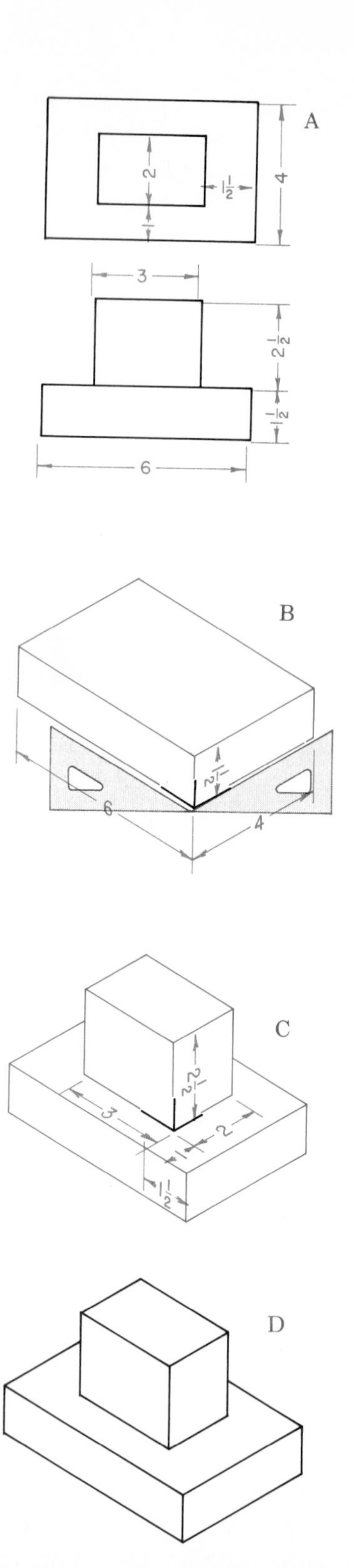

Fig. 12–8 **Isometric drawing with the axes in the second position. See also Fig. 12–6.**

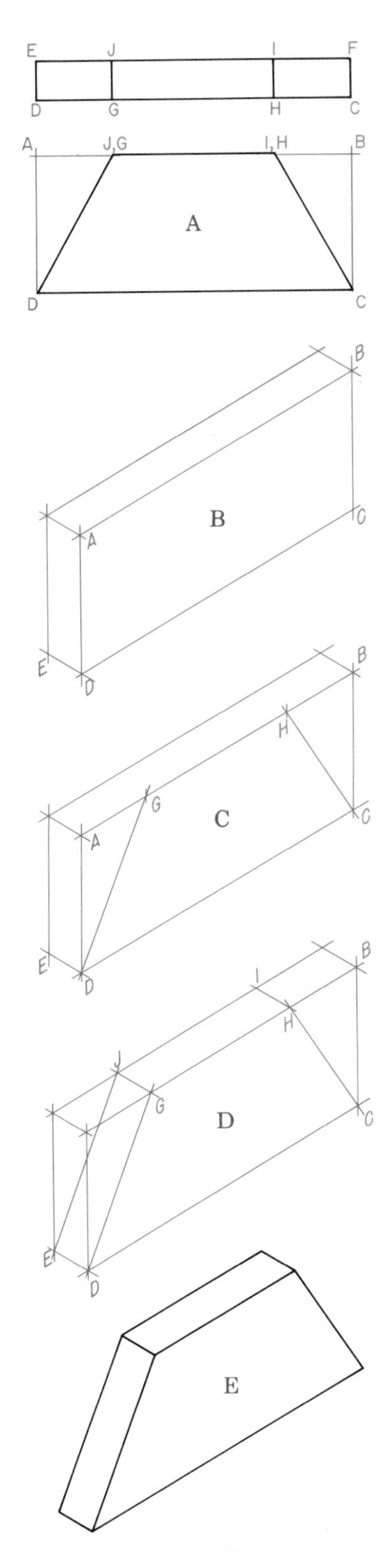

Fig. 12–10 **Drawing nonisometric lines.**

question are the slanted sides of the packing block shown in a multiview drawing at A. NOTE: *The colored lines are for instructional purposes only. You would not normally put them on your finished drawing.* To make an isometric drawing of the block, use the following procedure:

1. Block-in the overall sizes of the packing block to make the isometric box figure as shown at B.
2. Use dividers or a scale to transfer the distances *AG* and *HB* from the multiview drawing to the isometric figure. Lay these distances off along line *AB* to locate points *G* and *H*. You can then draw the lines connecting point *D* with *G* and *C* with *H*. This is shown in Fig. 12–10C.
3. Complete the layout by drawing *GJ* and *HI* and by connecting points *E* and *J* to form a third nonisometric line. Erase the construction lines to complete the drawing (Fig. 12–10E).

Angles in Isometric

Angles in some isometric drawings can be measured with a protractor. To draw an angle such as the 40° angle shown in Fig. 12–11A, use the following procedure:

1. Make *AO* and *AB* any convenient length. Draw *AB* perpendicular to *AO* at any convenient place.
2. Transfer *AO* and *AB* to the isometric cube (Fig. 12–11B). Lay off *AO* along the base of the cube. Draw *AB* parallel to the vertical axis.
3. Connect points *O* and *B* to complete the isometric angle. If you check this angle with a protractor, you will find that it does not measure 40°.

Follow the same steps to construct the angle on the top of the isometric cube. You can use this method to lay out any angle on any isometric plane.

Figure 12–12 is a multiview drawing of an object with four oblique

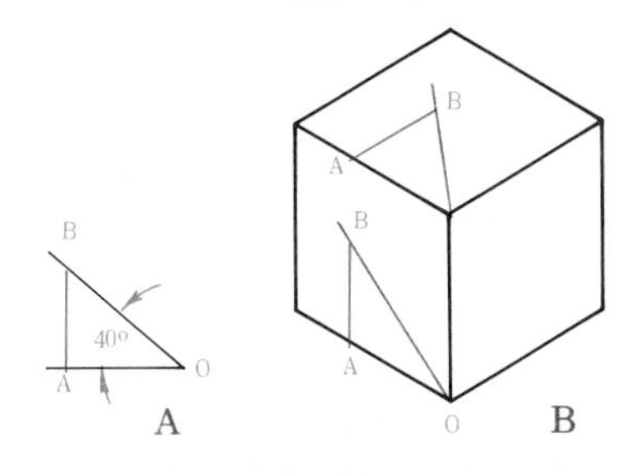

Fig. 12–11 **Constructing angles in isometric drawing.**

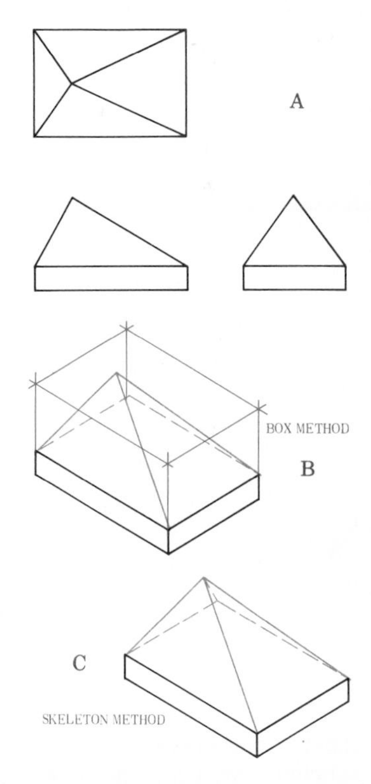

Fig. 12–12 **Drawing oblique surfaces in isometric.**

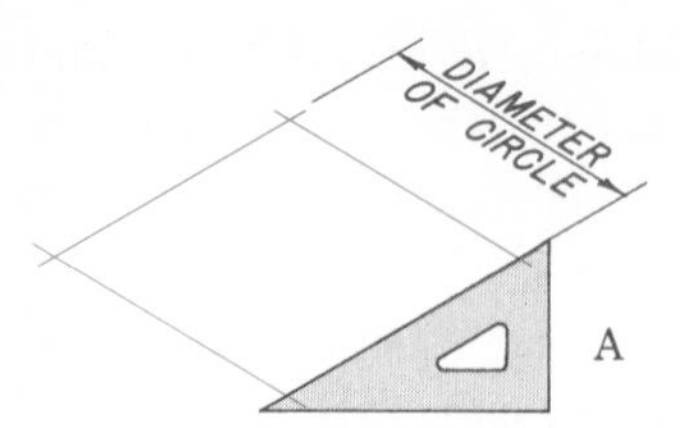

Draw an isometric square with the sides equal to the diameter of the circle.

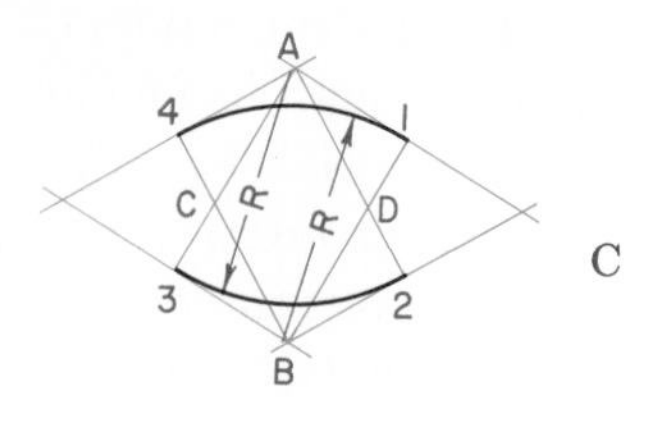

With *A* and *B* as centers and a radius equal to *A*2, draw arcs as shown.

Use a 30°-60° triangle to locate points *A*, *B*, *C*, *D*, and 1, 2, 3, 4.

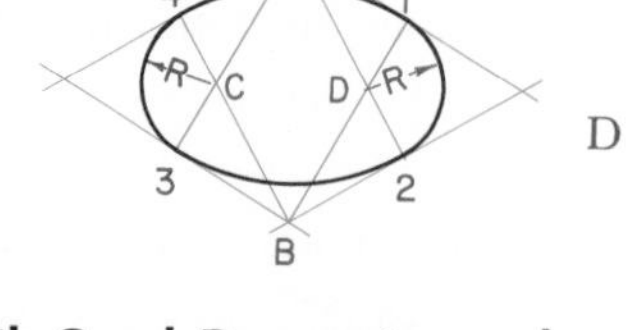

With *C* and *D* as centers and a radius equal to *C*4, draw arcs to complete the isometric circle (ellipse).

Fig. 12–13 **Steps in drawing an isometric circle.**

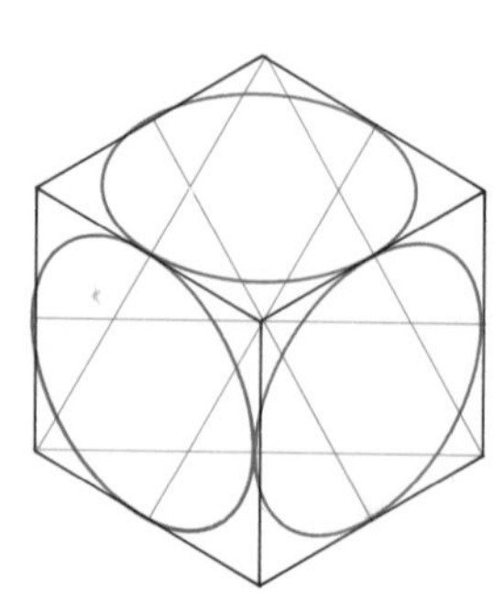

Fig. 12–14 **Isometric circles on a cube.**

surfaces. You can make an isometric view of this object by either the box or the skeleton method, as shown at B and C.

Isometric Circles

In isometric drawing, circles appear as ellipses. Since it takes a long time to plot a true ellipse, you can draw a four-centered approximation. The way to do this is shown and explained in Fig. 12–13. Figure 12–14 shows isometric circles drawn on three surfaces of a cube.

Figure 12–15 shows how to make an isometric drawing of a cylinder. Notice that the radii for the arcs at the bottom match those at the top.

You draw quarter rounds in isometric the same way you draw quarters of a circle. This is illustrated in Fig. 12–16. Notice that in each case the radius is measured along the tangent lines from the corner.

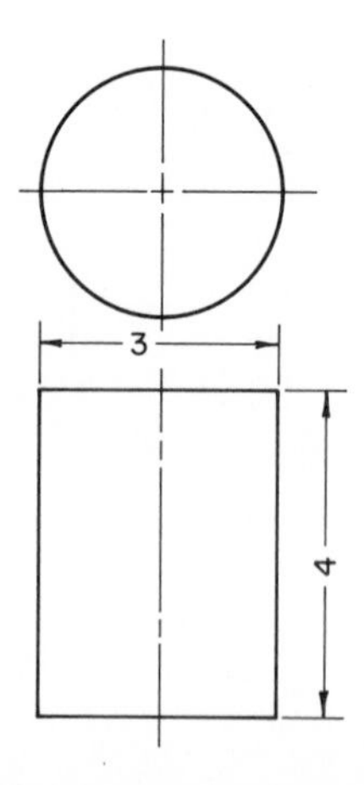

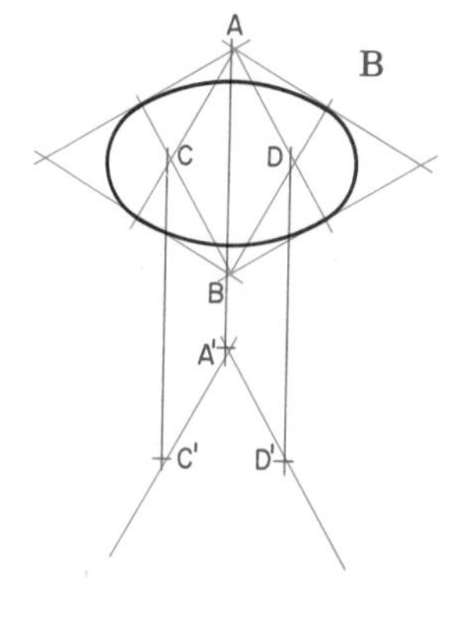

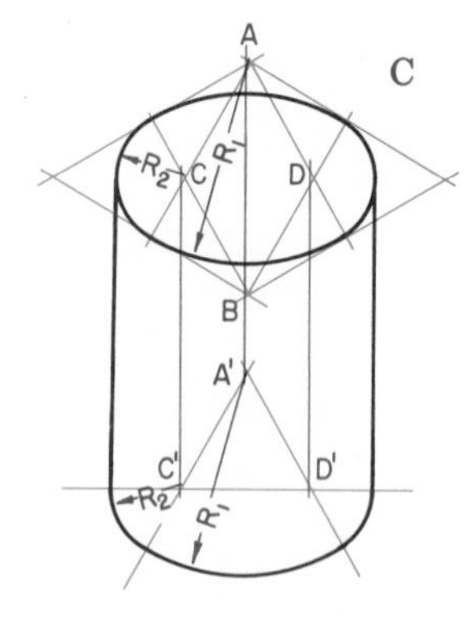

Fig. 12–15 **Steps in drawing an isometric cylinder. Multiview drawing of a cylinder. Draw an ellipse as described in Fig. 12–13. Drop centers *A*, *C*, and *D* a distance equal to the height of the cylinder (4 in. in this case). Draw lines *A′C′* and *A′D′*. A line through *C′D′* will locate the points of tangency.**

Draw the arcs using the same radii as in the ellipse at the top. Draw the vertical lines to complete the cylinder.

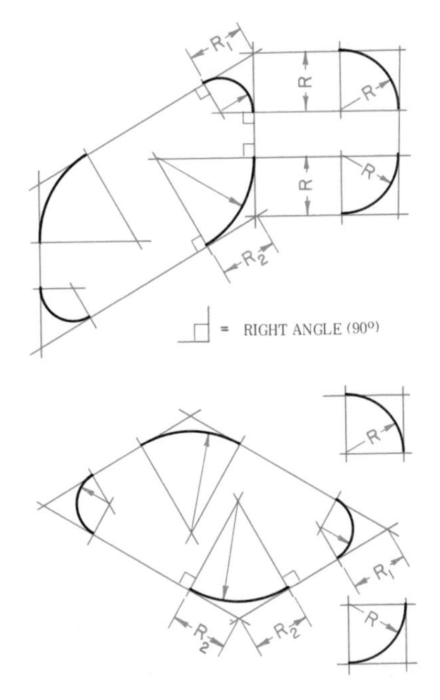

Fig. 12–16 **Drawing quarter rounds in isometric.**

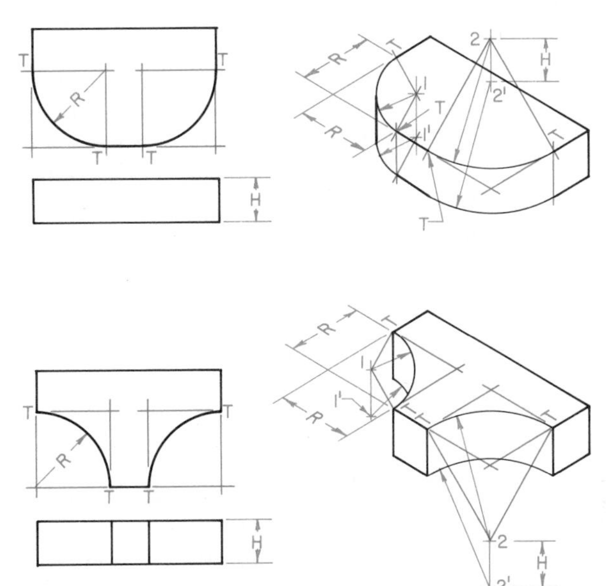

Fig. 12–17 **Construction for outside and inside arcs.**

Then the actual perpendiculars are drawn to locate the centers for the isometric arcs. Observe that r_1 and r_2 are found in the same way as the radii of an isometric circle.

When an arc is more or less than a quarter circle, you can sometimes plot it by drawing all or part of a complete isometric circle and using as much of the circle as you need.

Figure 12–17 shows how to draw outside and inside corner arcs. Note the tangent points, *T*, and centers *1* and *1′* and *2* and *2′*.

Irregular Curves in Isometric

Irregular curves in isometric cannot be drawn using the four-center method. You must first plot points and then connect them using a French curve, as shown in Fig. 12–18.

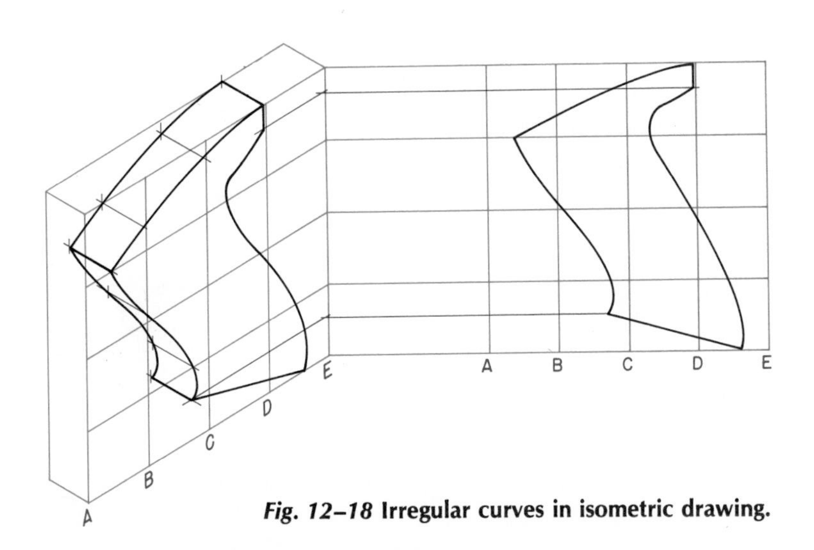

Fig. 12–18 **Irregular curves in isometric drawing.**

Isometric Templates

You will find isometric templates in a variety of forms. They are convenient and will save you time when you have to make many isometric drawings. Many of them have openings for drawing ellipses as well as 60° and 90° guiding edges. Simple homemade guides (Fig. 12–19) are convenient for straight-line work in isometric. Ellipse templates are very convenient for drawing true ellipses (Fig. 12–20). If you use these templates, your drawings will look better and you will not have to spend time plotting approx-

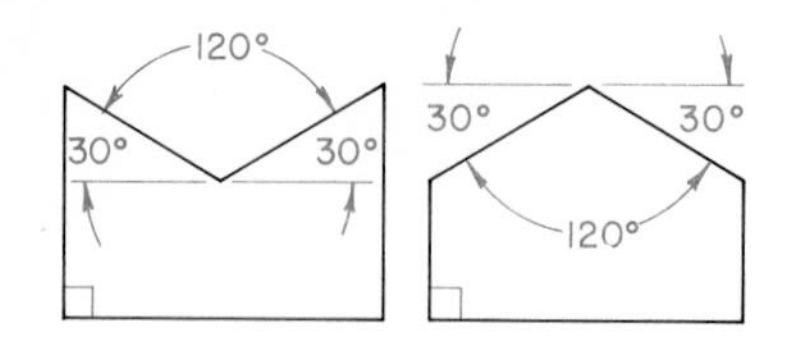

Fig. 12–19 **Simple isometric templates.**

imate ellipses. See Chapter 3 for information on how to use and care for templates.

Making an Isometric Drawing

Figure 12–21 shows how to make an isometric drawing of a guide. The guide is shown in a multiview drawing at *A*. Study the size, shape, and relationship of views before you proceed.

1. Draw the axes *AB, AC,* and *AD* in the second position (Fig. 12–21B). If you do not recall this position, refer to Fig. 12–7, which shows the positions of the isometric axes. Next, lay off the length, width, and thickness measurements given at A; that is, measure from *A* the length 76 mm (3 in.) on *AB*. Measure from *A* the width 50 mm (2 in.) on *AC*. Measure from *A* the thickness 16 mm (5/8 in.) on *AD*. Through the points you have found, draw isometric lines parallel to the axes. This "blocking-in" will produce an isometric view of the base.
2. Block-in the upright part in the same way, using the measurements of 50 mm (2 in.) and 19 mm (3/4 in.) given in the top view at *A*.
3. Find the center of the hole and draw centerlines as shown. Block-in a 19-mm (3/4-in.) isometric square and draw the hole as an approximate ellipse. To make the two quarter rounds, measure the 12-mm (1/2-in.) radius along the tangent lines from both upper corners (Fig. 12–21C). Draw real perpendiculars to find the centers of the quarter circles. See Fig. 12–16 for more information on drawing isometric quarter rounds.
4. Darken all necessary lines. Erase all construction lines to complete the isometric drawing as in Fig. 12–21D.

Isometric Sections

Isometric drawings are generally outside views. Sometimes, however, you need to draw a sectional view. To do so, you take a section on an *isometric plane* (a plane parallel to one of the faces of the cube).

Figure 12–22 shows isometric full sections taken on a different plane for each of three objects. Note the construction lines showing the parts that have been cut away. Isometric

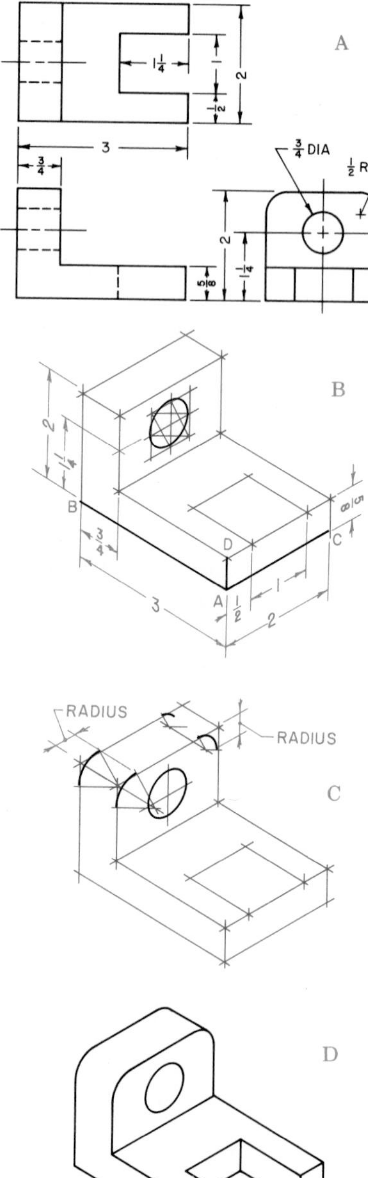

Fig. 12–21 **Steps in making an isometric drawing.**

Fig. 12–20 **Ellipse templates.**

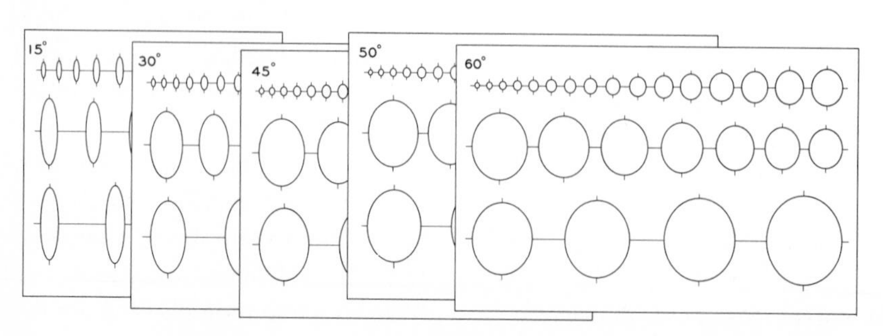

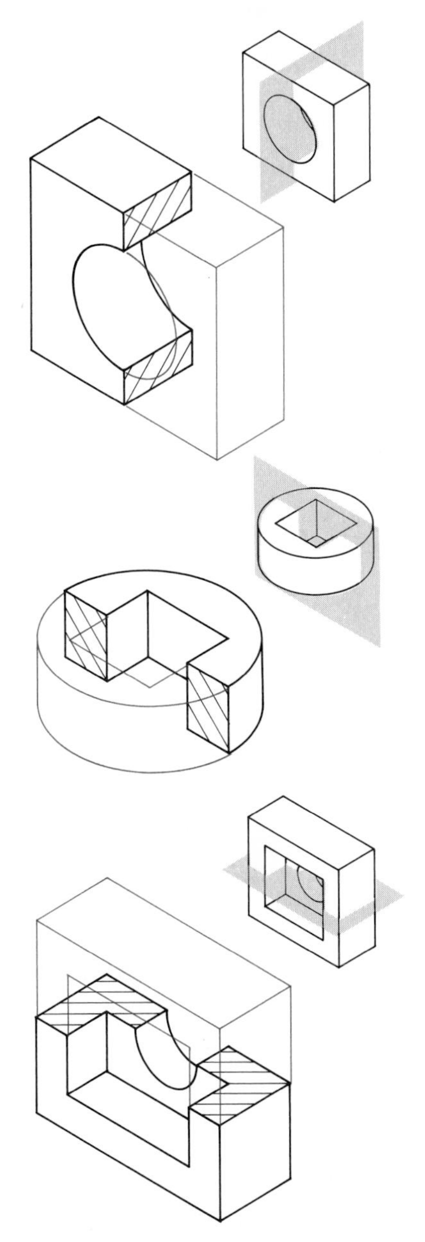

Fig. 12–22 **Examples of isometric full sections.**

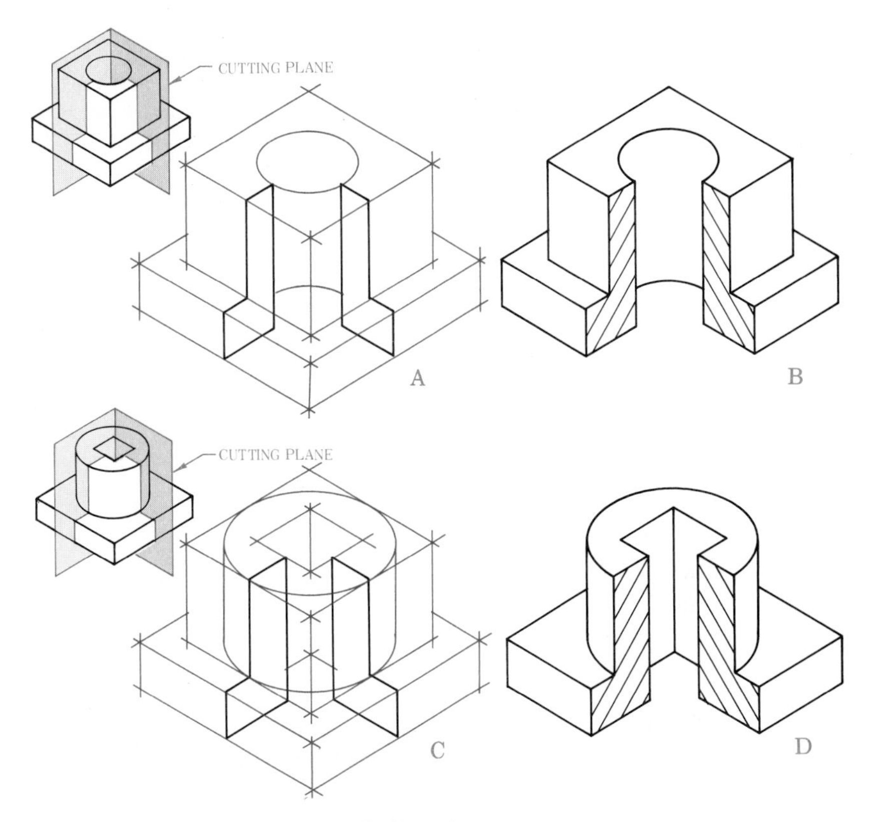

Fig. 12–23 **Examples of isometric half sections.**

half sections are illustrated in Fig. 12–23. The construction lines at A and C are for the complete outside views of the original objects. Notice the outlines of the cut surfaces in both views. There are two ways to make the sectional views shown at B and D. In one, the cut method, you draw the complete outside view plus the isometric cutting plane. Then you erase that part of the view that the cutting plane has cut away. In the other method, you first draw the section on the isometric cutting plane. Then you work from it to complete the view.

Reversed Axes

Sometimes you will want to draw an object as if it were being viewed from below. You do this in isometric drawing by reversing the position of the axes. Follow the example in Fig. 12–24. Consider how an object appears in a regular multiview drawing, as at A. Then begin the isometric view, as at B, by drawing the axes in reversed position. Complete the view, as at C, with dimensions taken from the multiview drawing. Darken the lines to finish the drawing.

Long Axis Horizontal

When you draw long pieces in isometric, you can make the long axis horizontal. Then you complete the picture just like any other isometric view. A drawing of this kind is illustrated in Fig. 12–25. At A, a long object is shown in a multiview drawing. At B is the start of an isometric view, with the axes shown by heavy black lines. At C the view is completed with dimensions taken from the drawing at A. Remember that in isometric drawing you draw circles first as isomet-

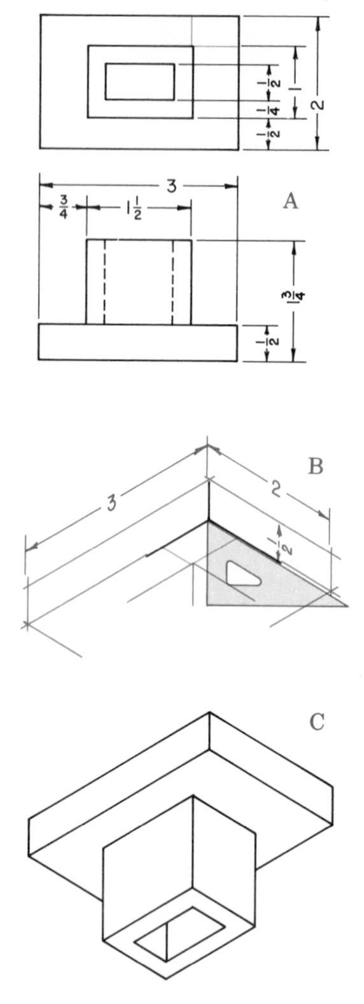

Fig. 12–24 **Steps in making an isometric drawing with reversed axes.**

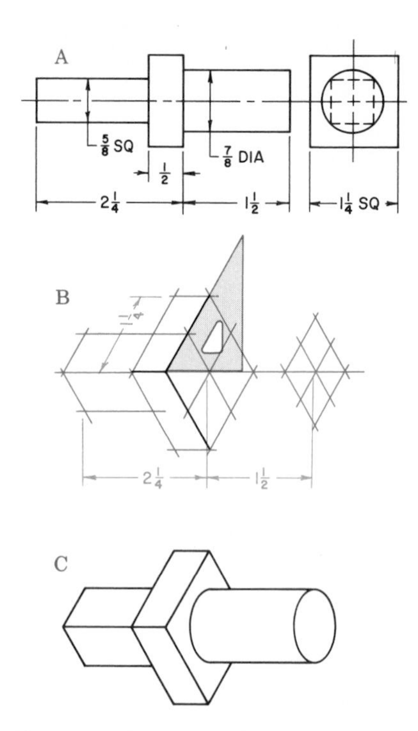

Fig. 12–25 **Steps in making an isometric drawing with the long axis horizontal.**

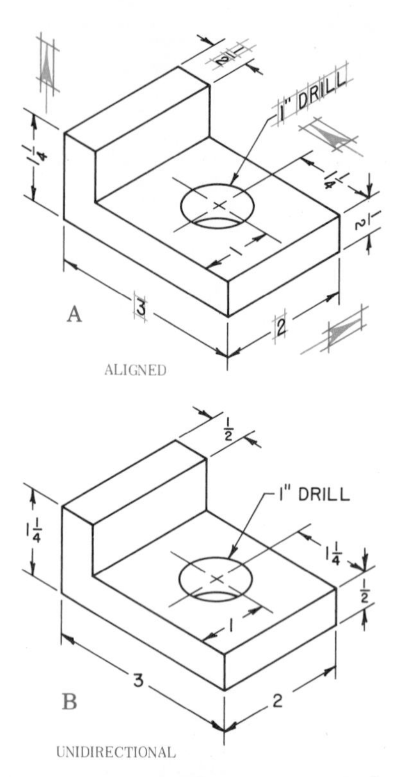

Fig. 12–26 **Two methods of dimensioning isometric views.**

ric squares, then complete them by the four-center method or with an ellipse template.

Dimensioning Isometric Drawings

There are two general ways to put dimensions on isometric drawings. The older method is to place them in the isometric planes, or extensions of them, and to adjust the letters, numerals, and arrowheads to isometric shapes, as shown in Fig. 12–26 at A. The newer unidirectional system, shown at B, is simpler. In this system, numerals and lettering are read from the bottom of the sheet. However, since isometric drawings are usually not used as working drawings, you seldom dimension them at all.

OBLIQUE PROJECTION AND OBLIQUE DRAWING

Oblique projection, like isometric, is a way of showing depth. A main difference, however, is that whereas isometric shows an object as if viewed from on edge, oblique shows it as if viewed face on. That is, one side of the object is seen squarely, with no distortion, because it is parallel to the *picture plane* (the plane on which the view is drawn). Depth is shown, as in isometric, by *projectors* (lines representing receding edges of the object). These lines are drawn at an angle other than 90° from the picture plane, to make the receding planes visible in the front view. And, as in isometric, lines on these receding planes that are actually parallel to each other are drawn parallel. Figure 12–27 shows how an oblique projection is developed. You probably will never have to develop an oblique projection in this way, but it is a good idea to understand the theory behind it.

Usually, no distinction is made between oblique projection and oblique drawing. This is another difference from isometric. In oblique, if the receding lines are drawn full length, the projection is called *cavalier.* If they are drawn one-half size, the projection is called *cabinet.* Many drafters use three-quarter size. This is sometimes called *normal,* or *general, oblique.* Generally, however, all three types are simply called oblique drawings.

Because oblique drawing can show one face of an object without distortion, it has a distinct advantage over isometric. It is especially useful for showing objects with irregular outlines.

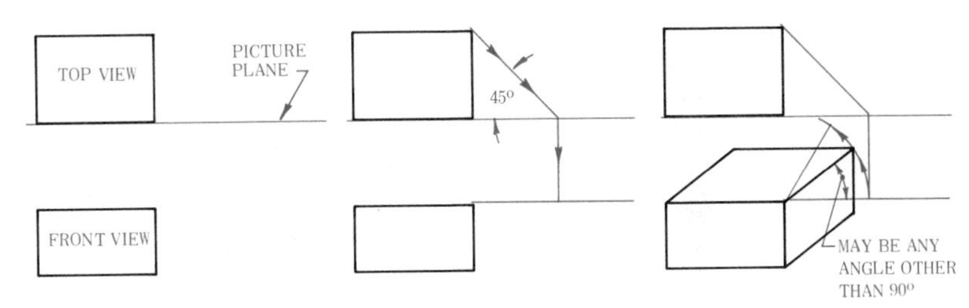

Fig. 12–27 **Oblique projection.**

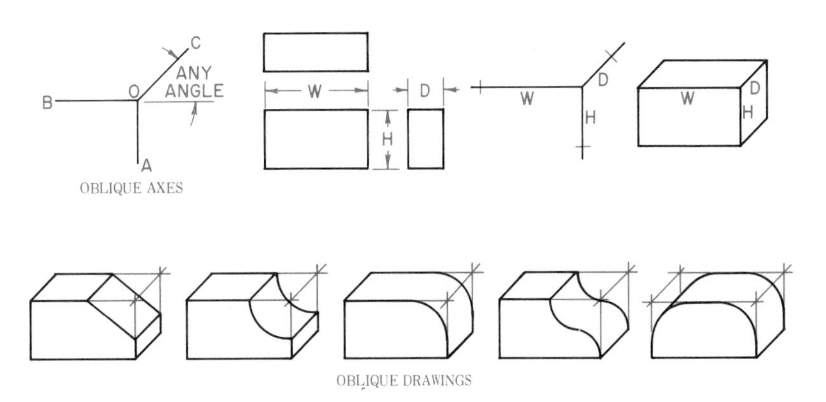

Fig. 12–28 **The oblique axes and oblique drawings.**

Oblique Drawing

You plot oblique drawings in the same way as isometric drawings, that is, on three axes. However, in oblique, two axes are parallel to the picture plane, rather than just one, as in isometric. These two axes always make right angles with each other (Fig. 12–28).

The methods and rules of isometric drawing also apply to oblique drawing. Oblique also has some special rules. The first is this:

Place the object so that the irregular outline or contour faces the front (Fig. 12–29A).

The second rule is this:

Place the object so that the longest dimension is parallel to the picture plane (Fig. 12–29B).

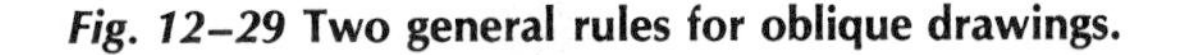

Fig. 12–29 **Two general rules for oblique drawings.**

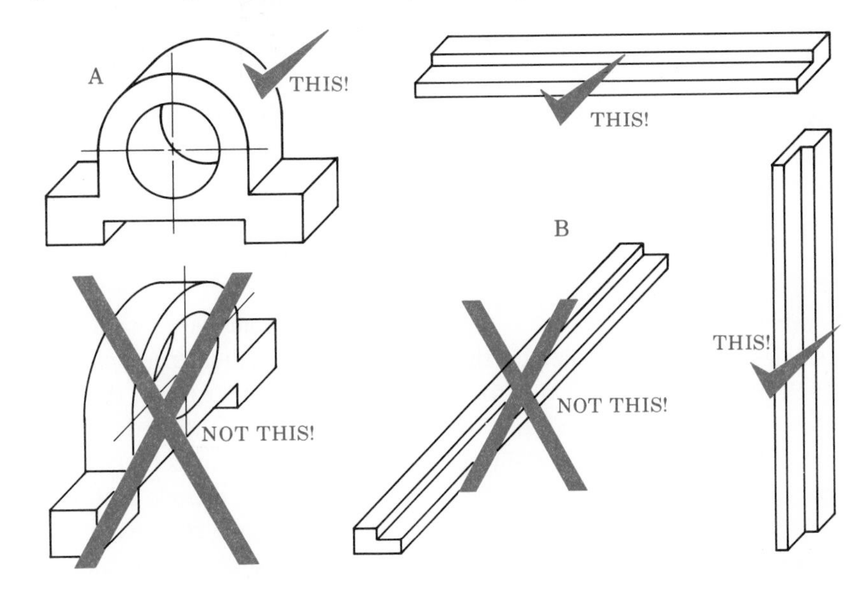

Positions for the Oblique Axes

Figure 12–30 shows several positions for oblique axes. In all cases,

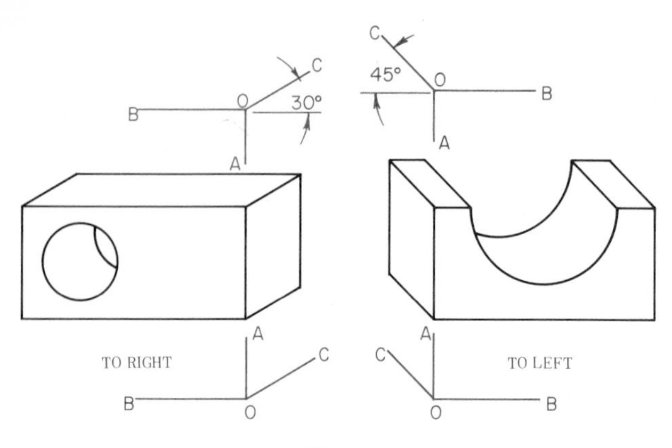

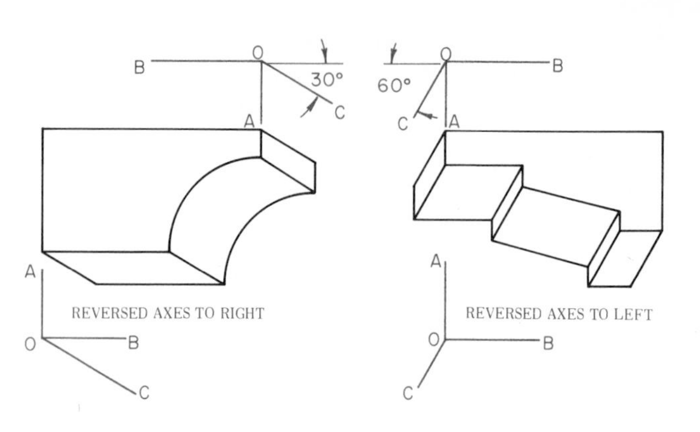

Fig. 12–30 **Positions for oblique axes.**

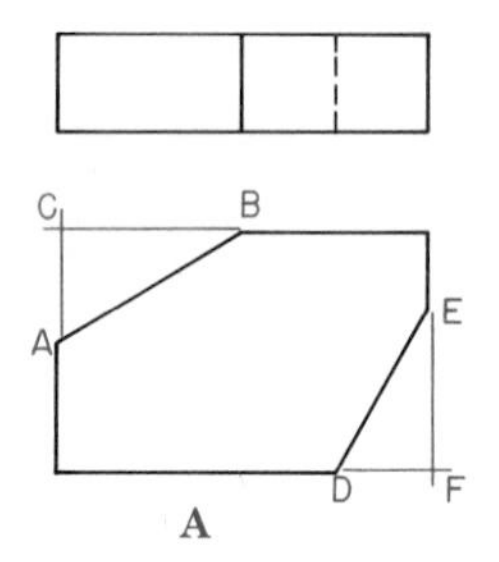

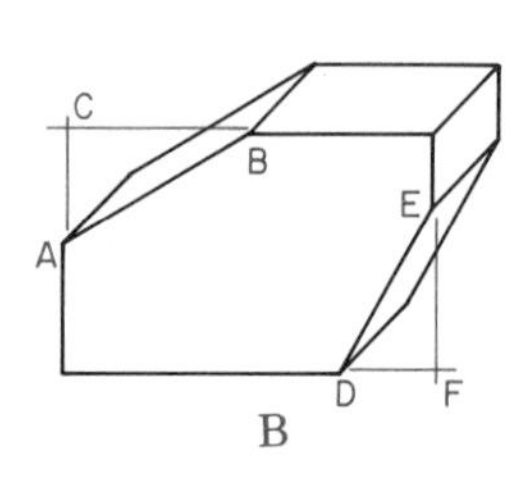

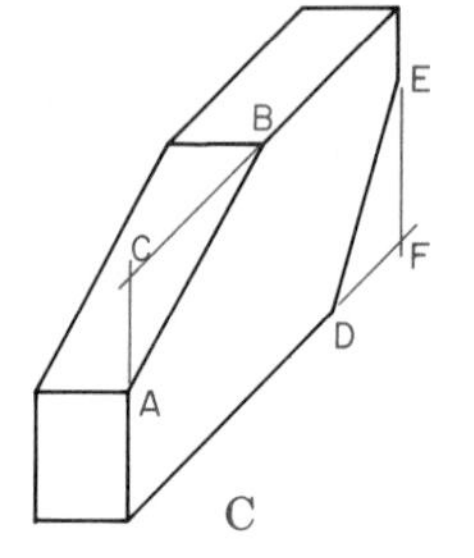

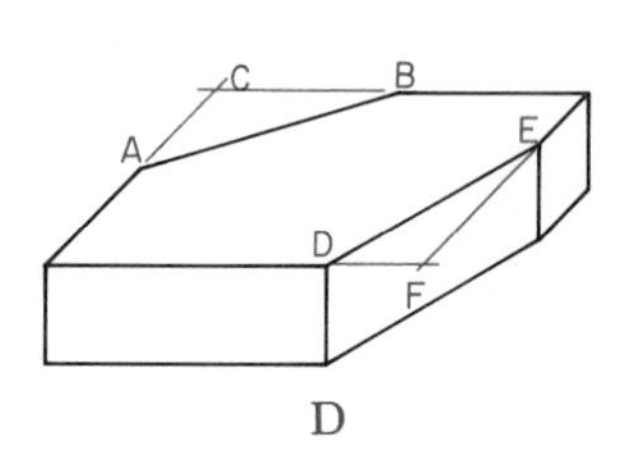

Fig. 12–31 **Angles on oblique drawings.**

two of the axes, *AO* and *OB*, are drawn at right angles. The oblique axis, *OC*, can be at any angle to the right, left, up, or down, as illustrated. The best way to draw an object is usually at the angle from which it would normally be viewed.

Angles and Inclined Surfaces on Oblique Drawings

Angles that are parallel to the picture plane show in their true size. Other angles can be laid off by locating both ends of the slanting line.

Figure 12–31 shows a plate with the corners cut off at angles. At B, C, and D, the plate is shown in oblique drawings. At B, the angles are parallel to the picture plane. At C, they are parallel to the profile plane. In each case, the angle is laid off by measurements parallel to one of the oblique axes. These measurements are shown by the construction lines.

Oblique Circles

On the front face, circles and curves show in their true shape (Fig. 12–32). On other faces, they show as ellipses. You can draw these ellipses by the four-center method. In Fig. 12–32A, a circle is shown as you would draw it on a front plane, a side plane, and a top plane.

Figure 12–32B is an oblique drawing with some arcs in a horizontal plane. Figure 12–32C is an oblique drawing with some arcs in a profile plane.

When you draw oblique circles by the four-center method, your results will be satisfactory for some purposes, but not pleasing. Ellipse templates, when available, give much better results. If you use a template, you should first block-in the oblique circle as an oblique square. This will show you where to place the ellipse. Blocking-in the circle first also helps you choose the proper size and shape of the ellipse. If you do not have a template, you can plot the ellipse as shown in Fig. 12–33.

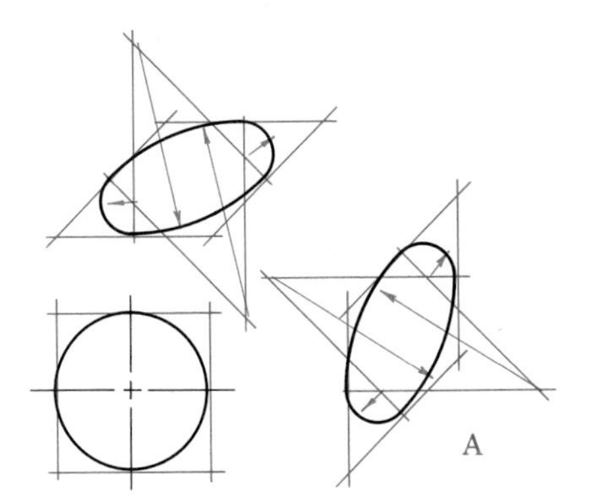

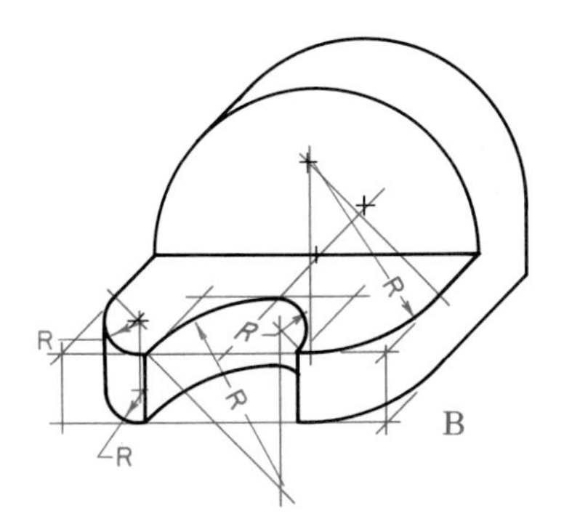

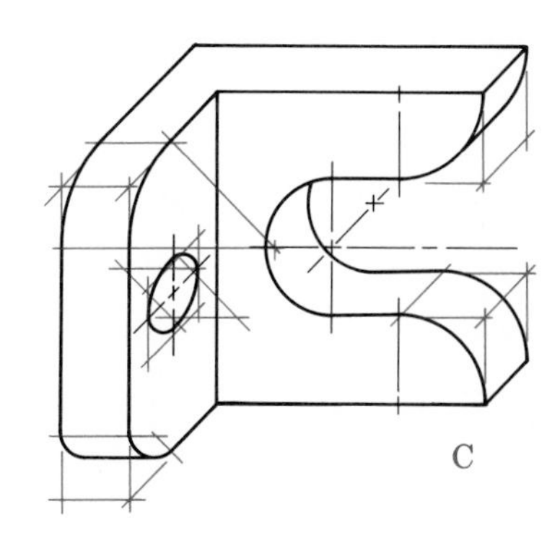

Fig. 12–32 **Circles parallel to the picture plane are true circles; on other planes, ellipses.**

Fig. 12–33 **To plot oblique circles.**

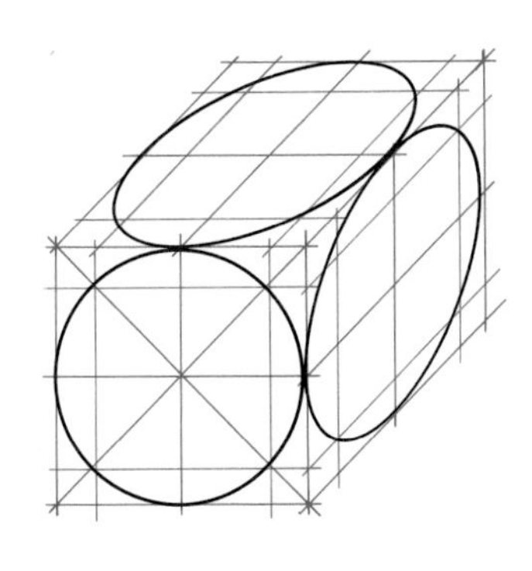

To Make an Oblique Drawing

Figure 12–34 shows the steps in making an oblique drawing. Notice that the drawing can show everything but the two small circles in true shape.

Oblique Sections

Oblique drawings are generally outside views. Sometimes, however, you need to draw a sectional view. To do so, take a section on a plane parallel to one of the faces of an oblique cube. Figure 12–35 shows an oblique full section and an oblique half section. Note the construction lines indicating the parts that have been cut away.

Fig. 12–34 **Steps in making an oblique drawing.**

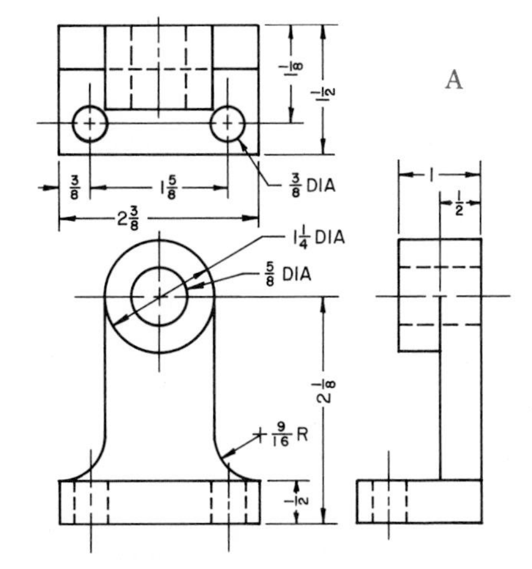

Multiview drawing of the object to be drawn in oblique.

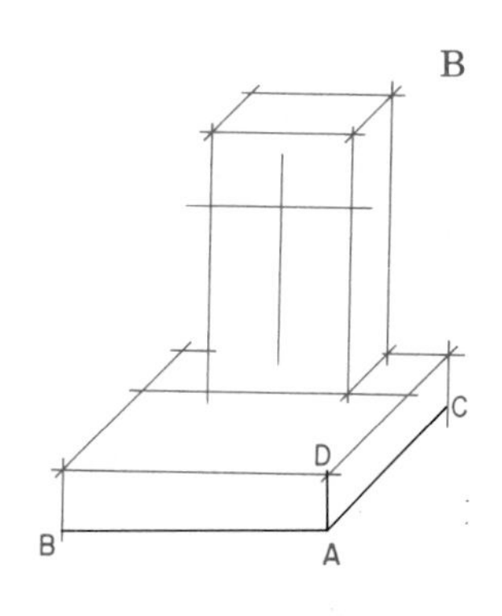

Draw the axes *AB, AC,* and *AD* for the base in second position, and on them measure the length, width, and thickness of the base. Draw the base. On it, block-in the upright, omitting the projecting boss as shown.

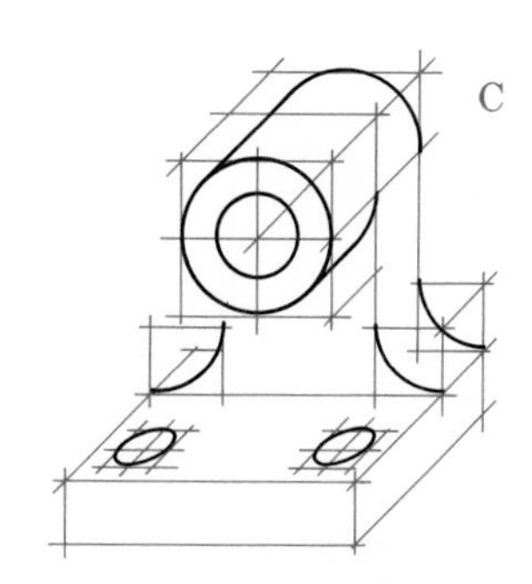

Block-in the boss and find the centers of all circles and arcs. Draw the circles and arcs.

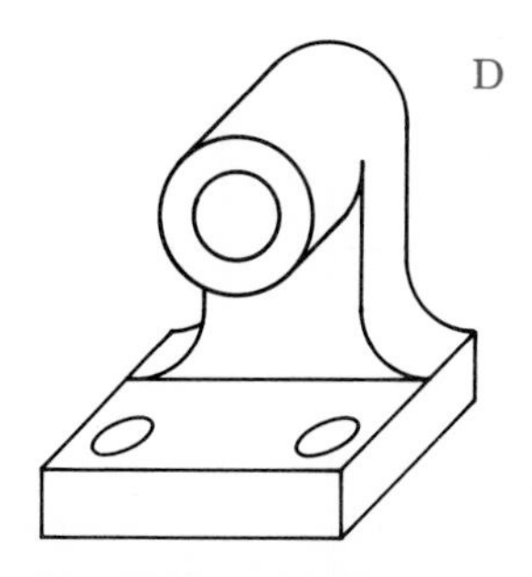

Darken all necessary lines, and erase construction lines to complete the drawing.

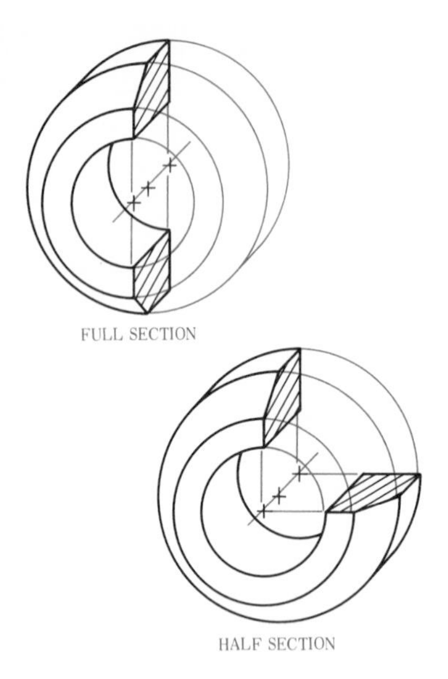

Fig. 12–35 **Oblique full and half sections.**

Cabinet Drawings

A cabinet drawing is an oblique drawing in which distances parallel to the oblique (receding) axis are drawn one-half size. Figure 12–36 shows a bookcase drawn in cavalier, normal oblique, and cabinet drawings. Cabinet drawings are so named because they are most often used in the furniture industry.

PERSPECTIVE DRAWING

Perspective drawing (Fig. 12–37) is a three-dimensional representation of an object as it looks to the eye from a particular point. Of all pictorial drawings, perspective drawings look the most like photographs. The distinctive feature of perspective drawing is that in perspective, lines on the receding planes that are actually parallel are not drawn parallel, as they are in isometric and oblique drawing. Instead, they are drawn as if they are *converging* (coming together).

Definition of Terms

Figure 12–38 illustrates perspective terms. A card appears on the plane at the right. The sight lines leading from points on this card and converging at the eye of the observer are called *visual rays.* The *picture plane* (PP) is the plane on which the card at right is drawn. The *station point* (SP) is where the observer is when looking at the card. A horizontal plane passes through the observer's eye. Where it meets the picture plane, it forms the *horizon line* (HL). Where the ground plane on which the observer stands meets the picture plane, it forms the *ground line* (GL). The *center of vision* (CV) is the point at which the *line of sight* (LS) (visual ray from the eye perpendicular to the picture plane) pierces the picture plane.

Figure 12–39 shows how in perspective drawing the *projectors* (receding axes) converge. The point at which they meet is called the *vanishing point* (VP).

Figure 12–39 also shows how the observer's eye level affects the perspective view. This eye level can be anywhere on, above, or below the ground. If the object is seen from above, the view is an *aerial,* or *birds's-eye, view.* If the object is seen from underneath, the view is a

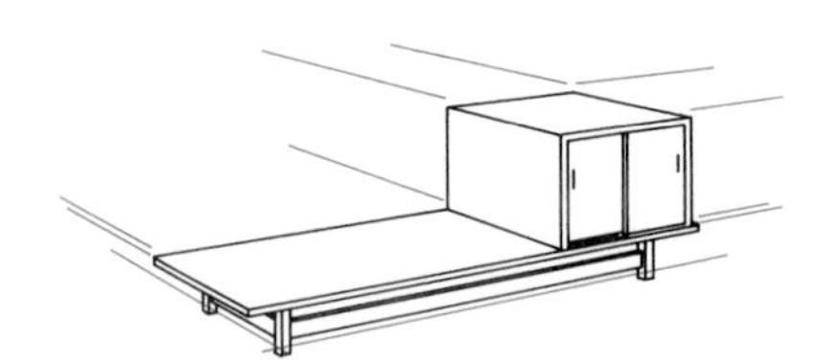

Fig. 12–37 **Perspective drawing of a music center.**

Fig. 12–36 **Three types of oblique drawings.**

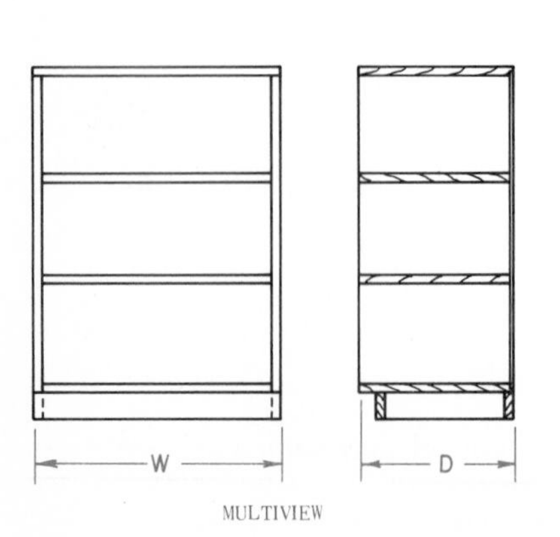

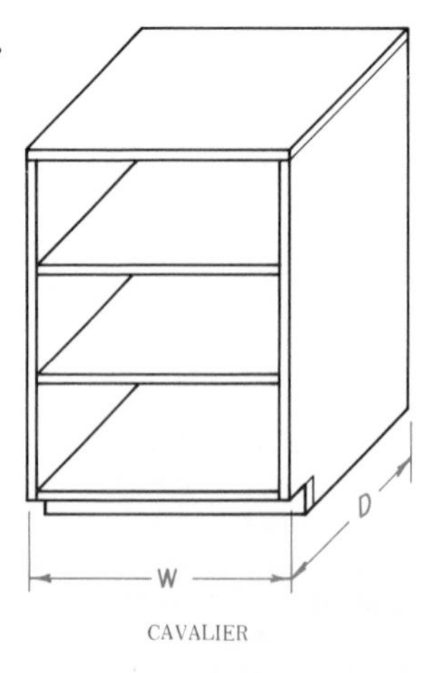

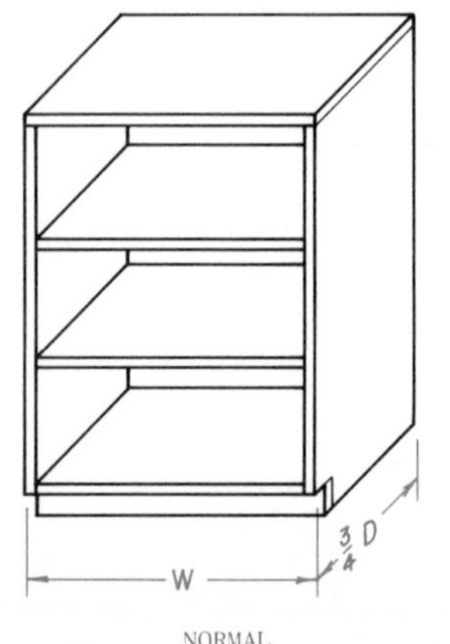

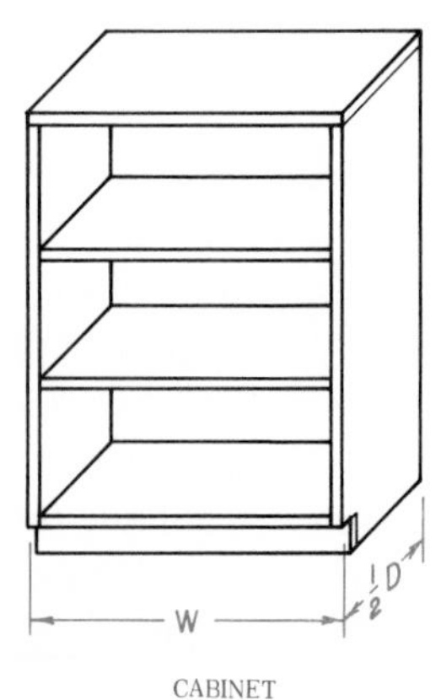

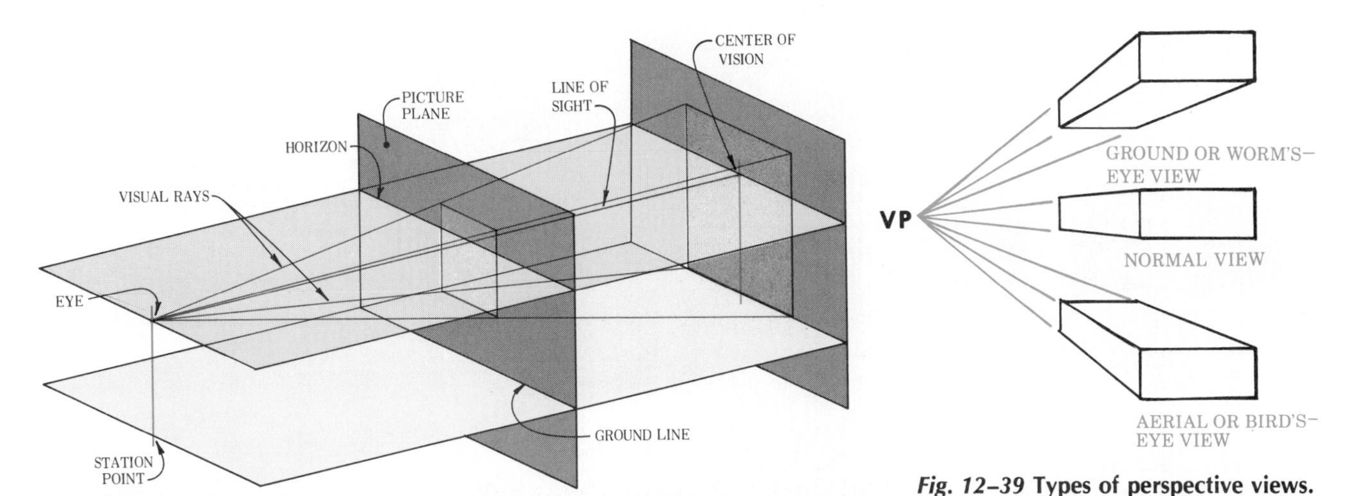

Fig. 12–38 **Some perspective terms.**

Fig. 12–39 **Types of perspective views.**

ground, or *worm's-eye, view.* If the object is seen face on, so that the line of sight is directly on it rather than above or below, the view is a *normal view.* The view in Fig. 12–38 was this kind of view.

Factors that Affect Appearance

Two factors affect how an object looks in perspective: its *distance* from the viewer and its *position* (angle) in relation to the viewer.

How Distance Affects the View

You know that the size of an object seems to change as you move toward or away from it. The farther away from the object you go, the smaller it looks. The closer you get, the larger it seems to grow. Figure 12–40 shows a graphic explanation of this effect of distance. An object is placed against a scale at a normal reading distance from the viewer. In that position, it looks to be the size indicated by the scale. However, if it is moved back from the scale to a point twice as far away from the viewer, it looks only half as large. Notice that each time the distance is doubled, the object looks only half as large as before.

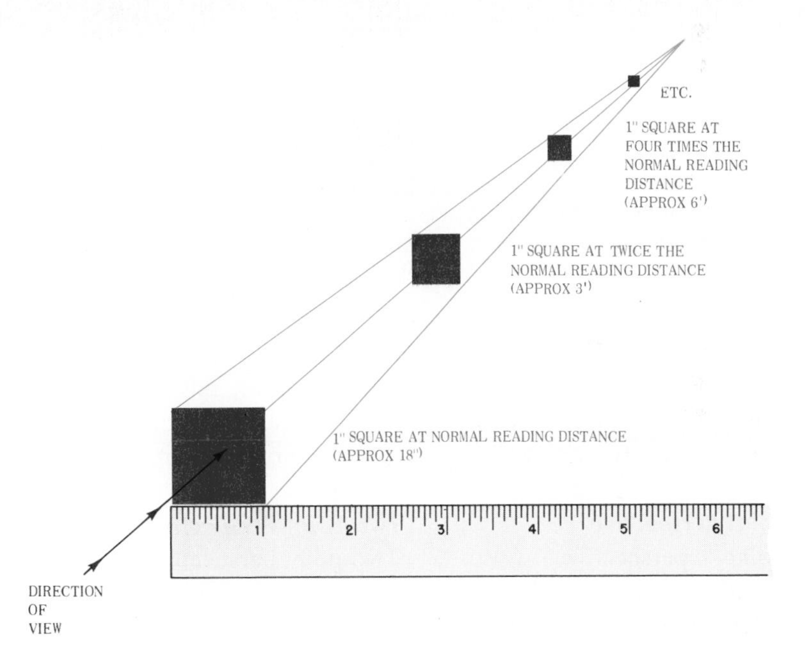

Fig. 12–40 **The size of an object appears half as large when the distance from the observer is doubled.**

How Position Affects the View

The shape of an object also seems to change when you see it from different positions (angles). This is illustrated in Fig. 12–41. If you look at a square face on, the top and bottom edges are parallel. But if the square is rotated so that you see it at an angle, these edges seem to converge. The square also appears to grow narrower. This foreshort-

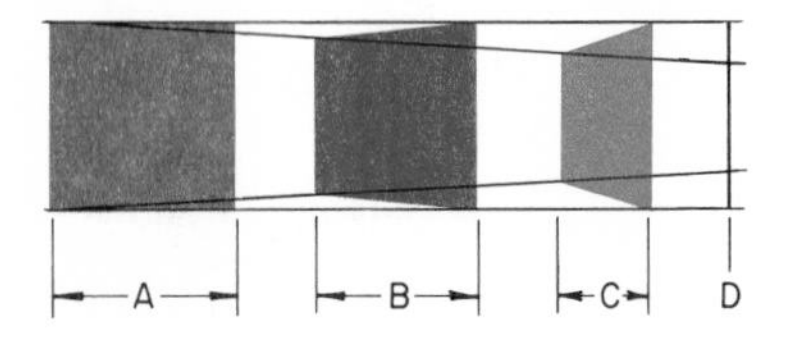

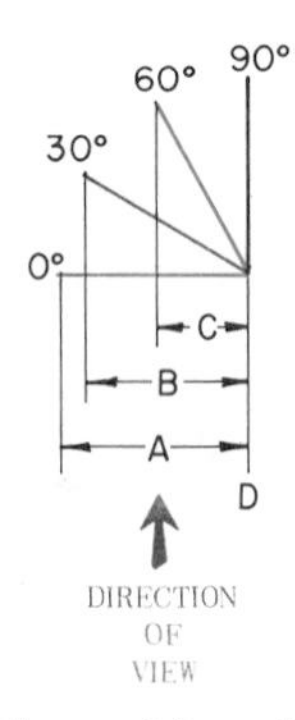

Fig. 12–41 **The position of the object in relationship to the observer affects its appearance.**

Fig. 12–42 **The lines of the porch rail, deck, roof, and building's side appear to converge at a distance.** ***(Designed by Howard A. Friedman & Associates.)***

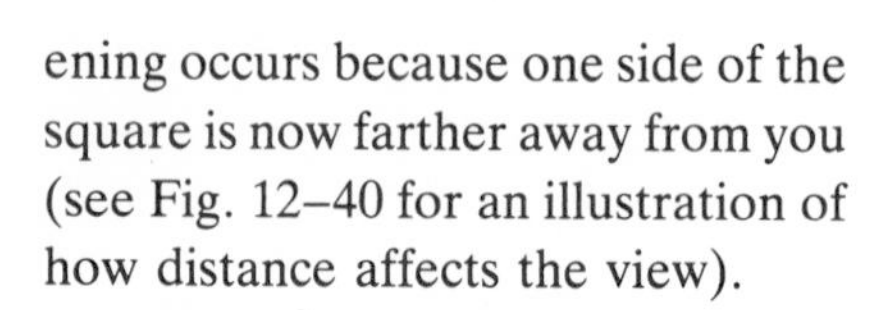

ening occurs because one side of the square is now farther away from you (see Fig. 12–40 for an illustration of how distance affects the view).

One-Point Perspective

One-point perspective (also called *parallel perspective*) means *one vanishing point.* Figure 12–42 shows how one-point perspective looks. Figure 12–43 shows an object in multiview and isometric drawings. Figure 12–44 shows how to draw the same object in a one-point bird's-eye view perspective. In Figs. 12–45 and 12–46, the object is drawn in one-point perspective in the other positions. Notice that in all three cases, one face of the object is placed on the picture plane (thus the name parallel perspective). Therefore, this face appears in true size and shape. True-scale measurements can be made on it.

Two-Point Perspective

Two-point perspective means *two vanishing points.* It is also called *angular perspective,* since none of the faces are drawn parallel to the picture plane. Figure 12–47 shows how two-point perspective looks.

Figure 12–48 shows an object in multiview and isometric drawings. Figure 12–49 shows how to draw this same object in two-point bird's-eye view perspective. Figures 12–50 and 12–51 show the object drawn in two-point perspective in the other positions.

Fig. 12–43 **Multiview and isometric drawings of an object to be drawn in single-point perspective.**

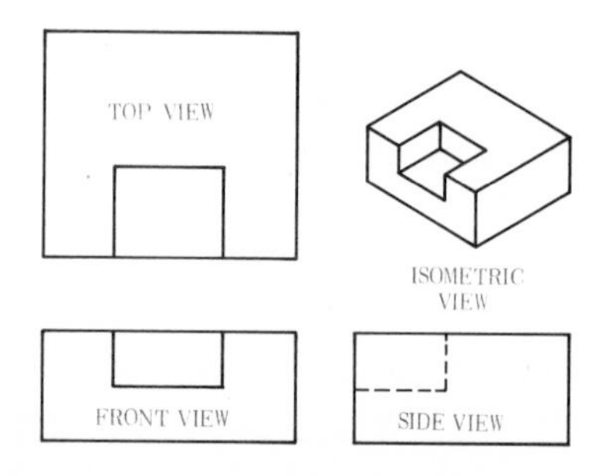

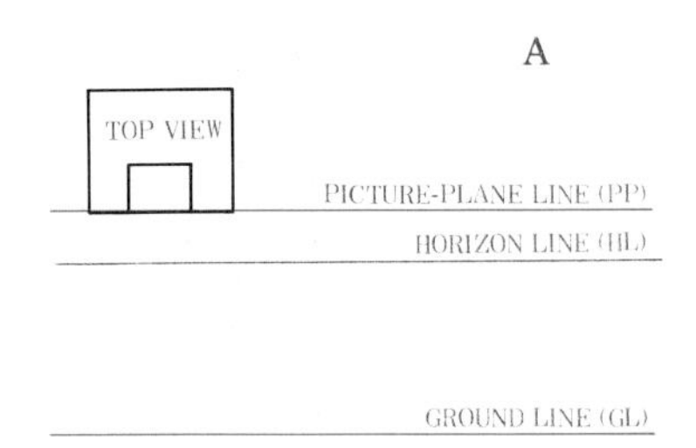

Decide on the scale to be used and draw the top view near the top of the drawing sheet. A more interesting view is obtained if the top view is drawn slightly to the right or to the left of center. Draw an edge (top) view of the picture plane (PP) through the front edge of the top view. Draw the horizon line (HL). The location will depend upon whether you want the object to be viewed from above, on, or below eye level. Draw the ground line. Its location in relation to the horizon line will determine *how far* above or below eye level the object will be viewed.

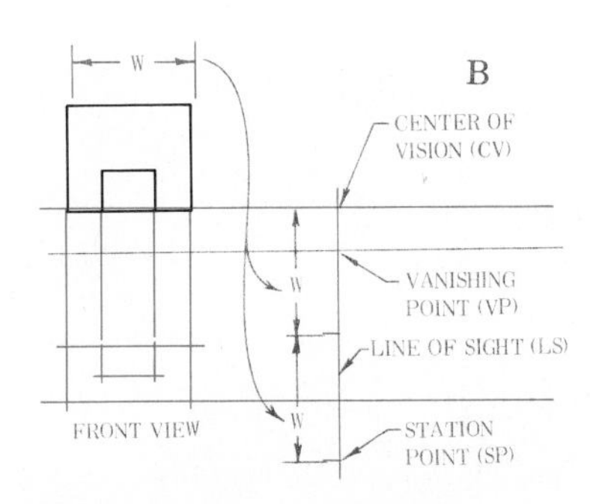

Locate the station point (SP). (a) Draw a vertical line (line of sight) from the picture plane toward the bottom of the sheet. Draw the line slightly to the right or to the left of the top view. (b) Set your dividers at a distance equal to the width (*W*) of the top view. (c) Begin at the center of vision on the picture plane and step off two to three times the width (*W*) of the top view, along the line of sight, to locate the station point (SP). Project downward from the top view to establish the width of the front view on the ground line. Complete the front view.

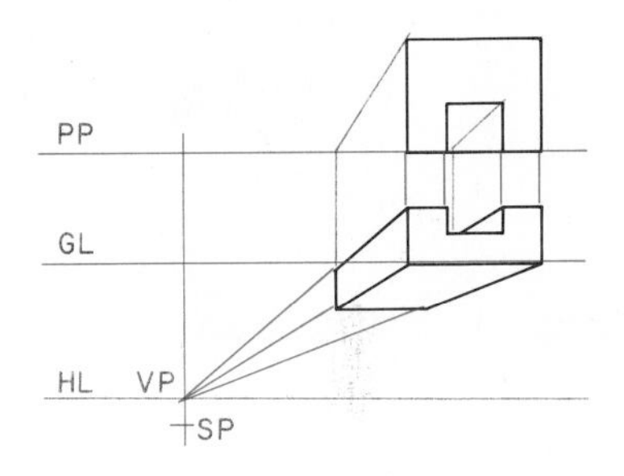

Fig. 12–45 Single-point perspective, worm's-eye view.

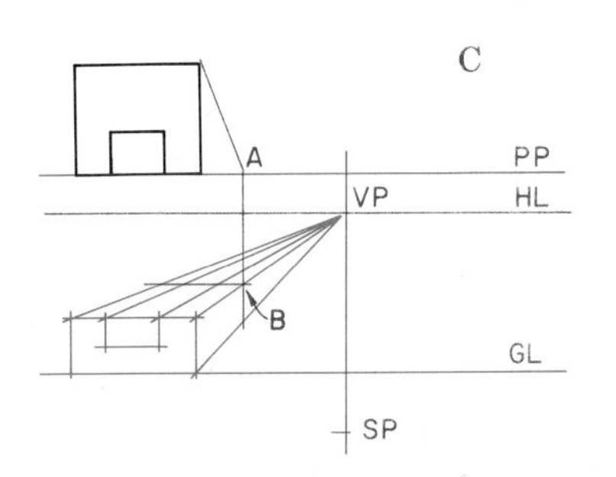

The vanishing point (VP) is the intersection of the line of sight (LS) and the horizon line (HL). Project lines from points on the front view to the vanishing point. Establish depth dimensions in the following way: (a) Project a line from the back corner of the top view to the station point. (b) At point *A* on the PP, drop a vertical line to the perspective view to establish the back vertical edge. (c) Draw a horizontal line through point *B* to establish the back top edge.

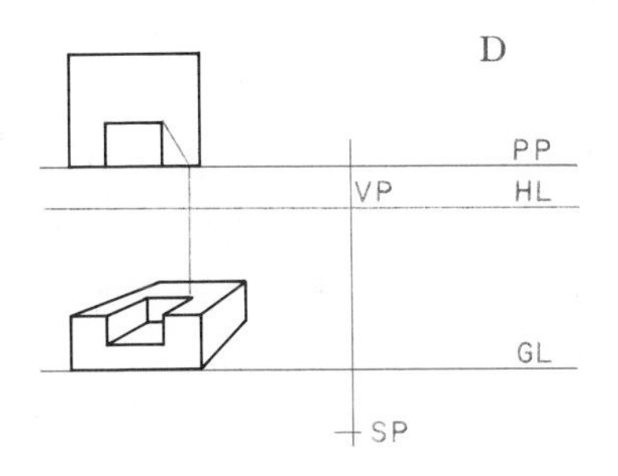

Proceed as in the previous step to lay out the slot detail. Darken all necessary lines, and erase construction lines as desired to complete the drawing.

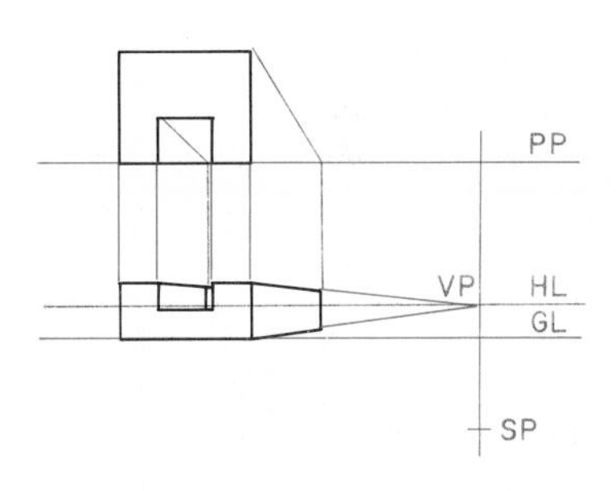

Fig. 12–46 Single-point perspective, normal view.

Fig. 12–44 Procedure for making a single-point, or parallel, perspective drawing (bird's-eye view).

Fig. 12–47 **When a building is viewed at an angle, two sides can be seen. The top and ground lines of each side appear to converge toward points. This is the effect in two-point, or angular, perspective.** ***(AM Bruning International.)***

Fig. 12–48 **Multiview and isometric drawings of an object to be drawn in two-point perspective.**

Inclined Surfaces

You plot inclined surfaces in perspective by finding the ends of inclined lines and connecting them. This method of drawing is shown in Fig. 12–52.

Circles and Arcs in Perspective

Figure 12–53 shows how to make a perspective view of an object with a cylindrical surface. Notice that points are first located on the front and top views. They are then projected to the perspective view. Where the projection lines meet, a path is formed. Along the path, the drafter draws the perspective arc, using a French curve or an ellipse template.

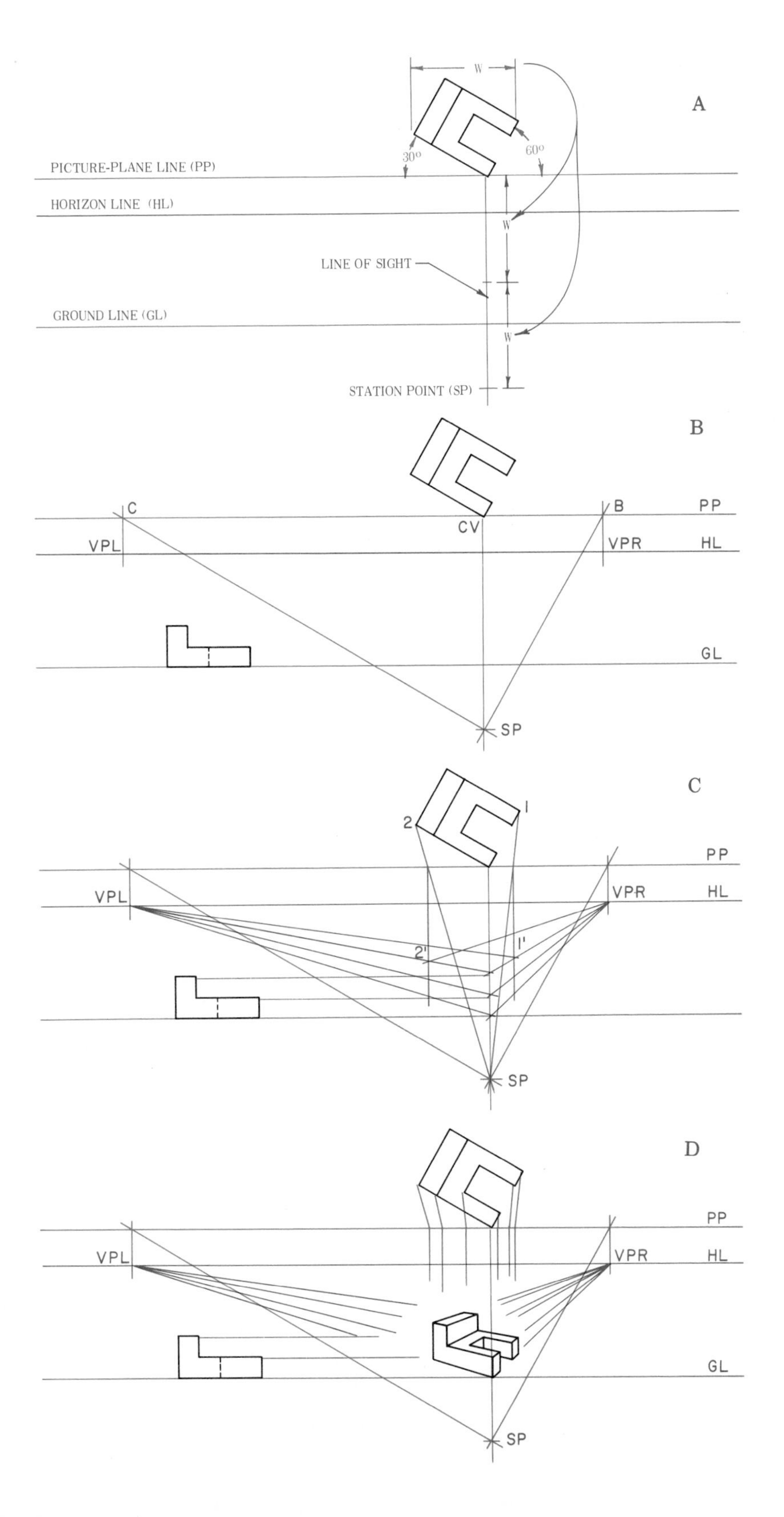

Fig. 12–49 **Procedure for making a two-point perspective drawing (bird's-eye view). Draw an edge view of the picture plane (PP). Allow enough space at the top of the sheet for the top view. Draw the top view with one corner touching the PP. In this case, the front and side of the top view form angles of 30° and 60°, respectively. Other angles may be used, but 30° and 60° seem to give the best appearance on the finished perspective drawing. The side with the most detail is usually placed along the smaller angle for a better view. Draw the horizon line (HL) and the ground line (GL). (Follow the procedure given in Fig. 12–44.) Draw a vertical line (line of sight) from the center of vision (CV) toward the bottom of the sheet to locate the station point. (Follow the procedure given in Fig. 12–44.)**

Draw line *SP-B* parallel to the end of the top view and line *SP-C* parallel to the front of the top view. (Use a 30°-60° triangle.) Drop vertical lines from the picture plane (PP) to the horizon line (HL) to locate vanishing point left (VPL) and vanishing point right (VPR). Draw the front or side view of the object on the ground line as shown.

Begin to block-in the perspective view by projecting vertical dimensions from the front view to the line of sight (also called measuring line) and then to the vanishing points. Finish blocking in the view as follows: (a) Project lines from points 1 and 2 on the top view to the station point. (b) Where these lines cross the picture plane (PP), drop vertical lines to the perspective view to establish the length and width dimensions. (c) Project point 1′ to VPL and 2′ to VPR.

Add detail by following the procedure described in the previous two main steps. Darken all necessary lines and erase construction lines as desired.

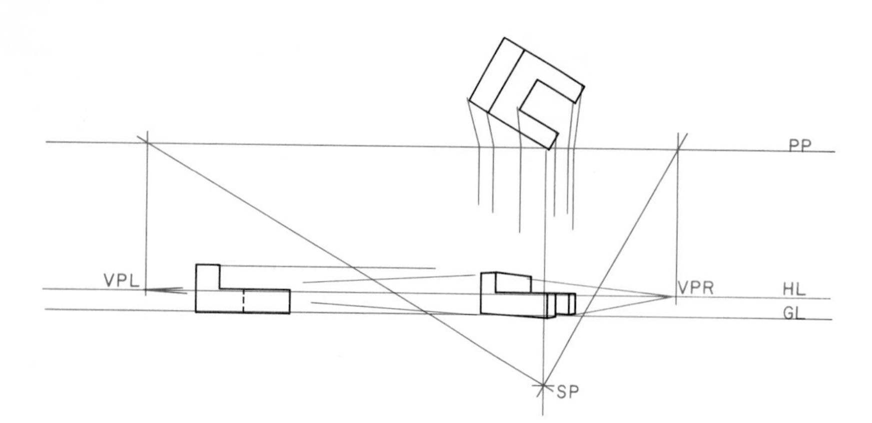

Fig. 12–50 **Two-point perspective, normal view.**

Fig. 12–51 **Two-point perspective, worm's-eye view.**

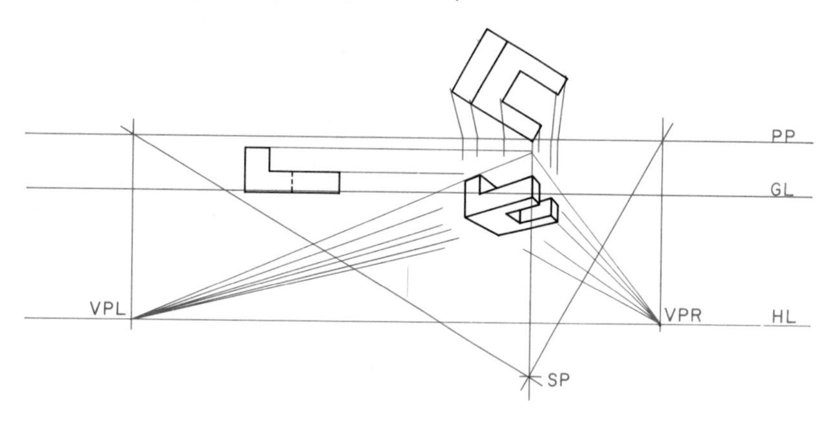

Fig. 12–52 **Two-point perspective with an inclined surface.**

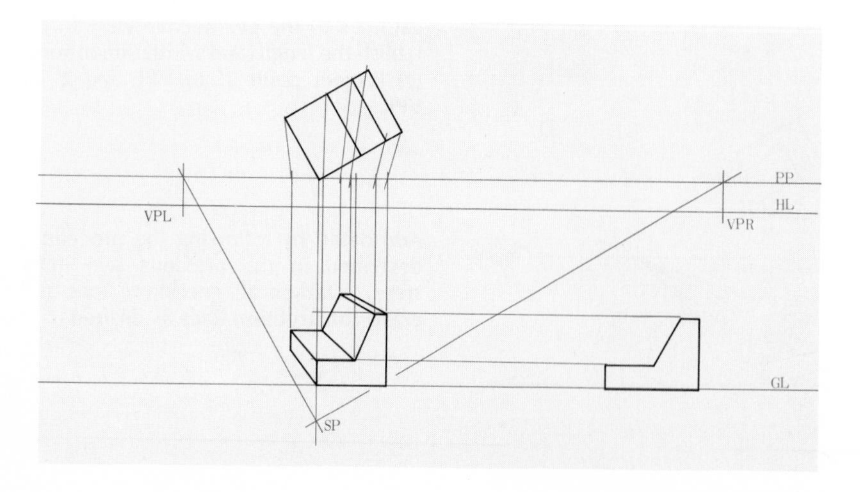

Perspective Drawing Shortcuts

Perspective drawing can take a lot of time. This is because you must do so much layout work before you can start the actual perspective view. Also, you often need a large drawing surface in order to locate distant points. However, you can offset these disadvantages by using various shortcuts. These are described below.

Perspective Grids

One of the simpler shortcuts is the perspective grid. Examples are shown in Fig. 12–54. There are many advantages in using grids. But there is one major disadvantage: a grid cannot show a variety of views. It is limited to one type of view based on one set of points and one view location. However, for the work done in some industrial drafting rooms, this may be all that is needed.

Perspective grids can be bought. The drafter can also make them. This is only practical, however, if the drafter has a number of perspective drawings to make in a special style.

Perspective Drawing Boards

The Klok perspective board (Fig. 12–55) is another timesaving device for making accurate perspective drawings. It includes a special T-square with the top edge of the blade centered on the head for drawing converging lines. It also includes perspective scales. These eliminate the need for extensive projection or calculation.

You can make your own perspective drawing board by using a regular drawing board and some thin

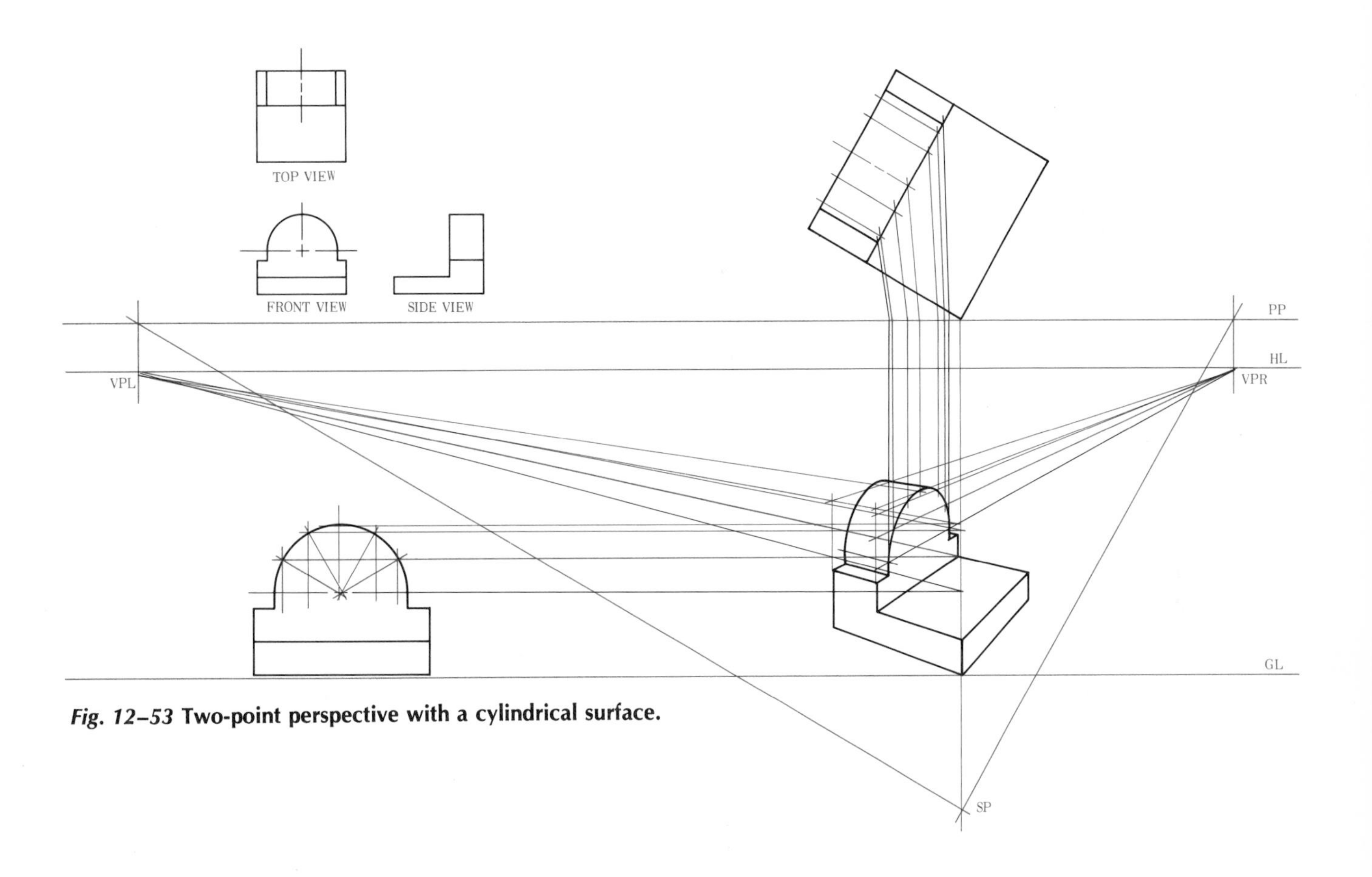

Fig. 12–53 **Two-point perspective with a cylindrical surface.**

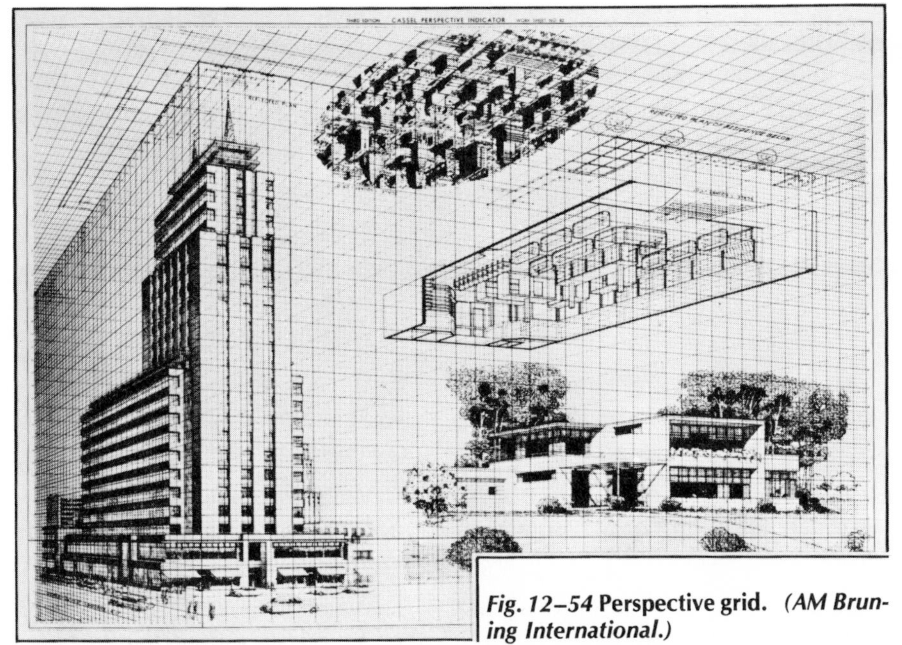

Fig. 12–54 **Perspective grid.** ***(AM Bruning International.)***

pieces of wood, cardboard, or hardboard, and by making or modifying a T-square (Fig. 12–56). Like the perspective grid, this device is practical only if you have to make many drawings of about the same size and style, such as perspective drawings of houses.

The way to make a perspective board is as follows:

1. Follow the procedure in Fig. 12–49 for locating two vanishing points. Use the top view of an object similar in size to the ones that will be drawn on the finished board. Make your layout on a large sheet of paper in order to locate distant vanishing points.

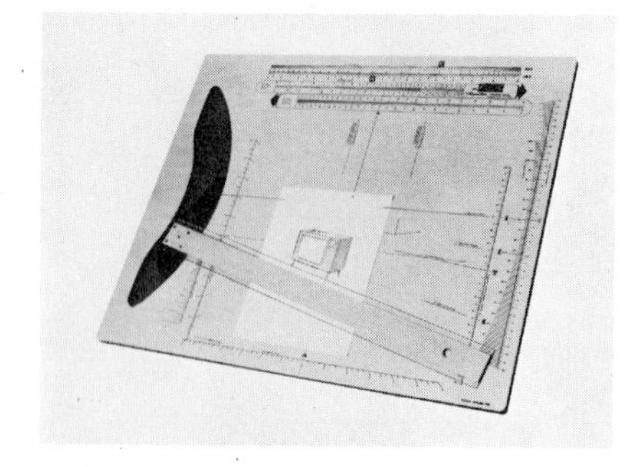

Fig. 12–55 **Klok perspective drawing board.** ***(Modulux Division, United States Gypsum Co.)***

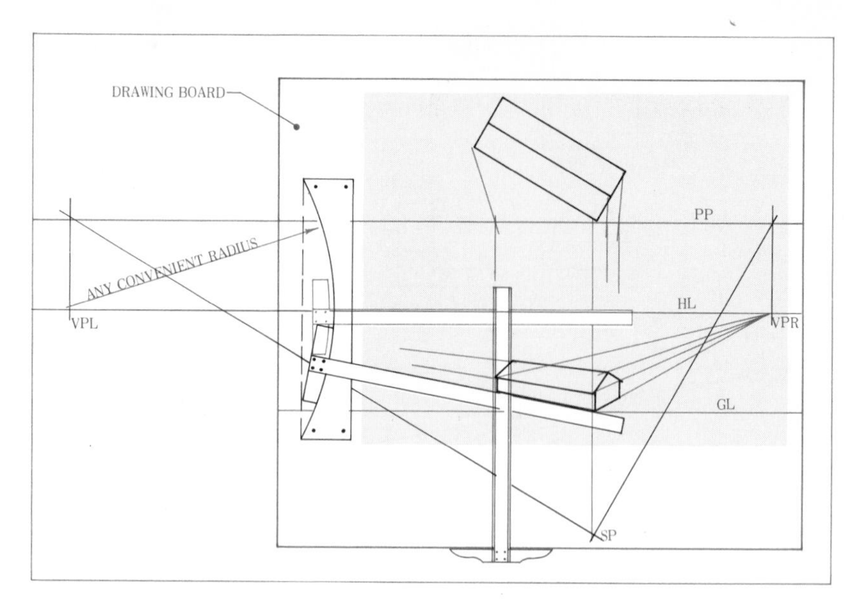

Fig. 12–56 **A homemade perspective drawing board.**

2. Place a drawing board within the layout, as shown in Fig. 12–56. For guides, use heavy cardboard or thin wood. Fasten the guide material in place. Then strike the arcs, using the vanishing points as centers and any convenient radius.
3. Cut the arcs with a sharp knife to form the guides. Draw the PP, HL, GL, LS, and SP on the board. You can now remove the board from the layout sheet.
4. Construct the T-square as shown. Notice that the top edge of the blade falls on the centerline of the heads. You must use thin material for the head as well as for the blade.

To use the perspective board:

1. Place a sheet of tracing paper in position between the guides (Fig. 12–56). Tracing paper will allow you to use the lines drawn on the board without redrawing them.
2. Draw the top view of the desired object in its proper position.
3. Proceed as in Fig. 12–49. Draw the lines that project toward VPR with the T-square head against the left guide. For those that project toward VPL, use the right-hand guide. Draw the vertical lines with the T-square head against the bottom edge of the board.

Review

1. Three types of axonometric projection are ________, ________, and ________.

2. Is an isometric projection larger or smaller than an isometric drawing?

3. Name the three most common types of pictorial drawing.

4. Since it takes a long time to draw a true ellipse, what shortcut can you use?

5. What do you call a line that is not parallel to any of the isometric axes?

6. Describe how you draw an irregular curve in pictorial drawings.

7. What is the difference between cavalier- and cabinet-type oblique drawings?

8. What are the two most common types of sectional views used in pictorial drawing?

9. What do you call the position of the observer in perspective?

Problems

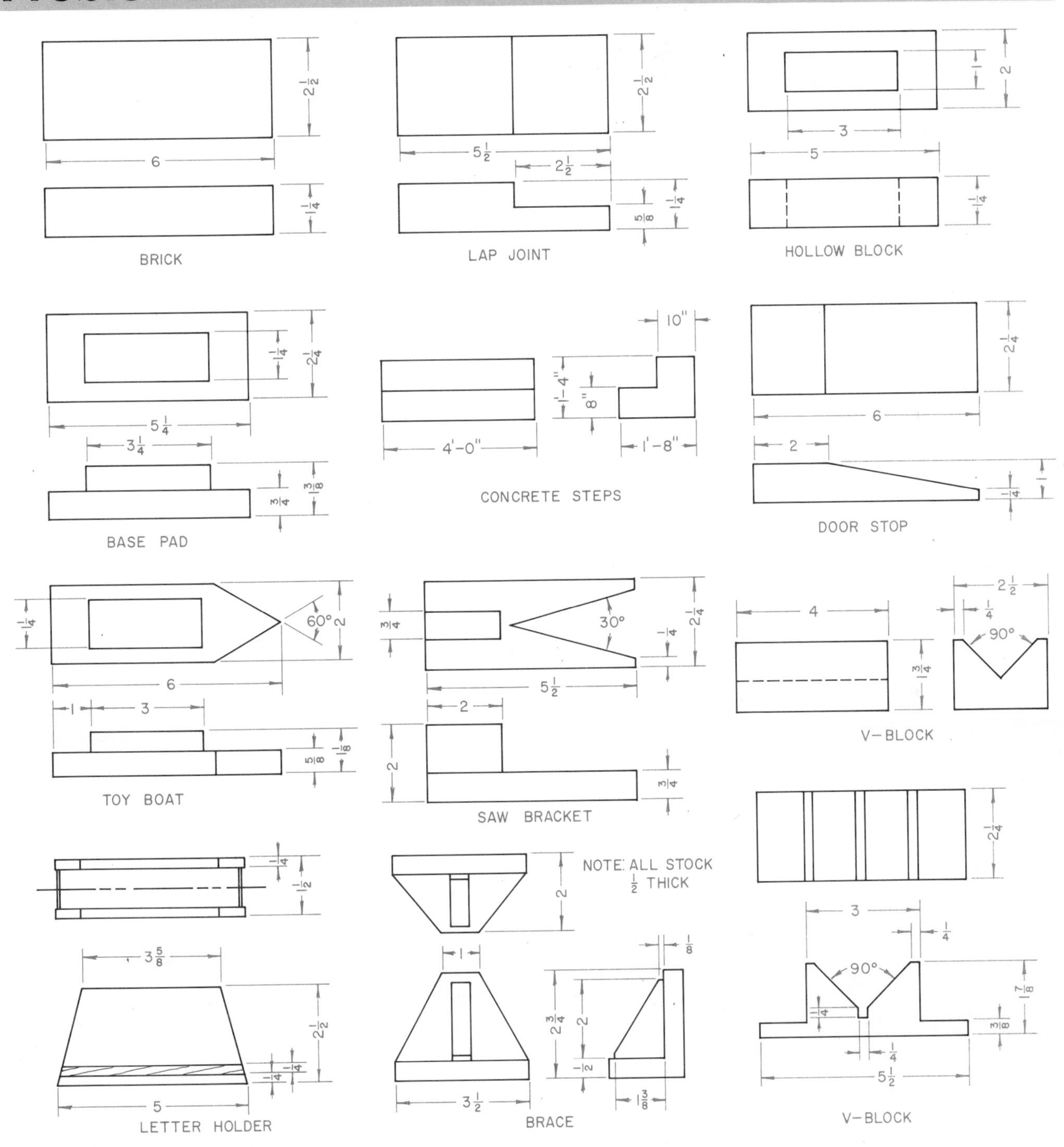

Fig. 12–57 **Isometric drawing problems. Scale: optional. Assignment 1: Make an isometric drawing of the object assigned. Assignment 2: Make an isometric half- or full-sectional view as assigned. NOTE: These problems may also be used for oblique and perspective drawing practice.**

Fig. 12-58

Fig. 12-59

LAYOUT

Fig. 12-60

Fig. 12-58 Make an isometric drawing of the plate.

Fig. 12-59 Make an isometric drawing of the notched block. Start at the corner indicated by thick lines.

Fig. 12-60 Make an isometric drawing of a babbitted stop.

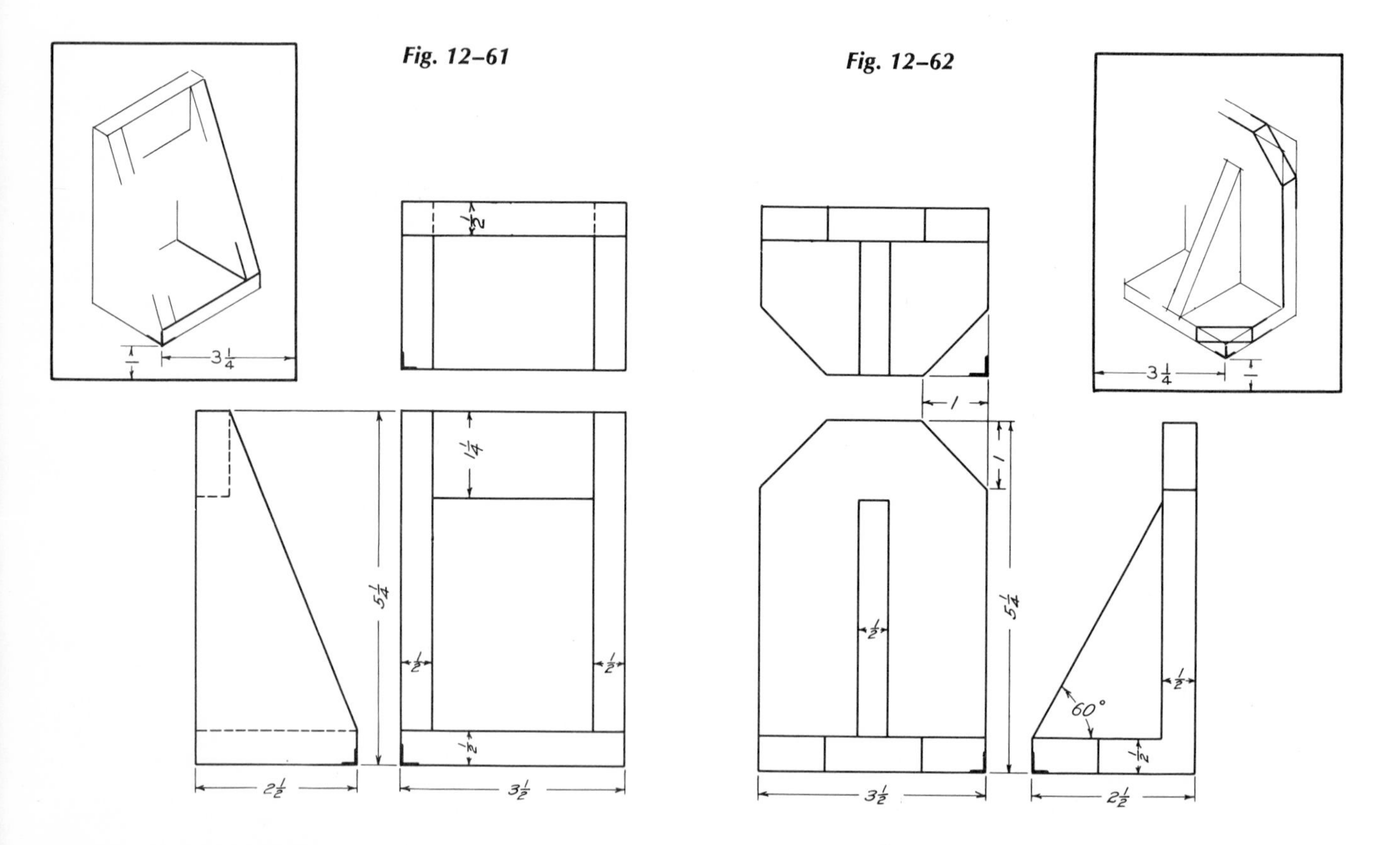

Fig. 12-61 Make an isometric drawing of the stirrup. The drawing is started on the layout at the upper left. Note the thick starting lines.

Fig. 12-62 Make an isometric drawing of the brace. The drawing is started on the layout at the upper right. Note the thick starting lines.

Fig. 12–63

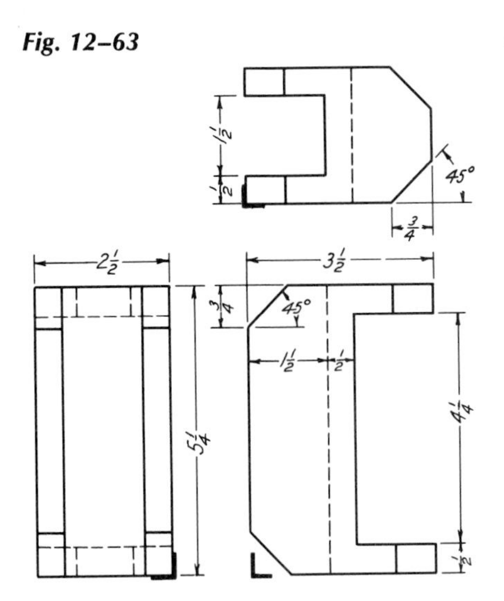

Fig. 12–63 **Make an isometric drawing of the cross slide. Use the layout of Fig. 12–61.**

Fig. 12–64

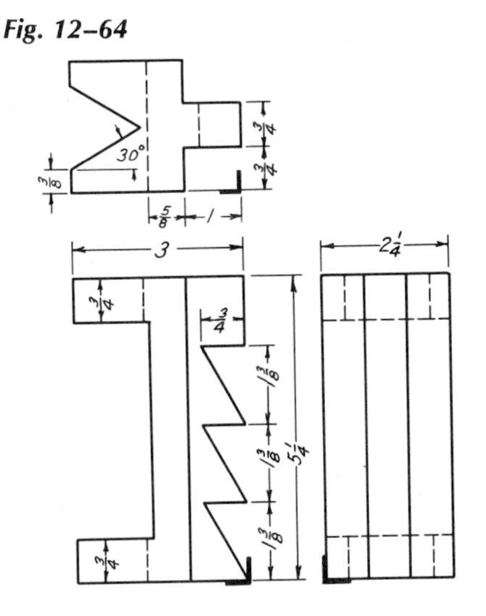

Fig. 12–64 **Make an isometric drawing of the ratchet. Use the layout of Fig. 12–61.**

Fig. 12–65

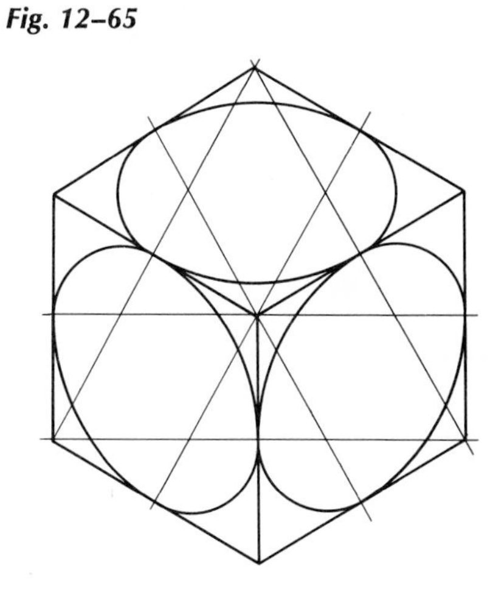

Fig. 12–65 **Make an isometric drawing of a 3″ cube with an isometric circle on each visible face.**

Fig. 12–66

Fig. 12–67

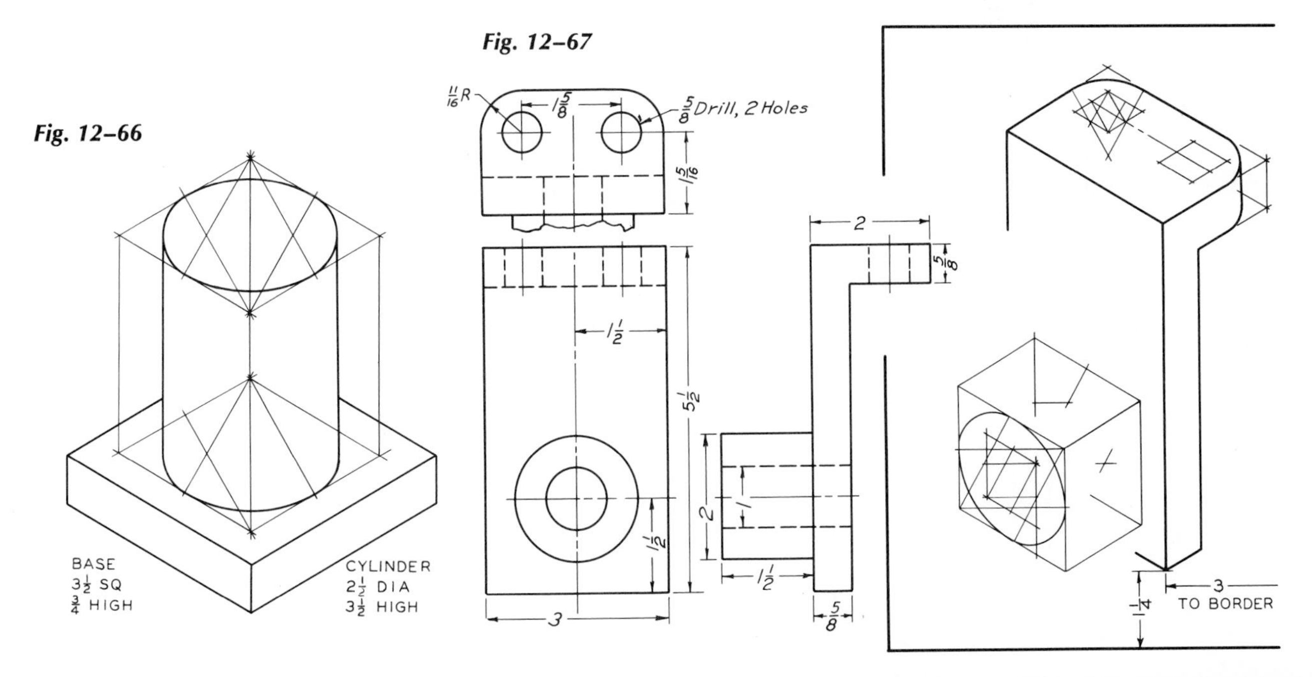

Fig. 12–66 **Make an isometric drawing of a cylinder resting on a square plinth (square base).**

Fig. 12–67 **Draw the three views given and an isometric drawing of the hung bearing. Most of the construction is indicated on the layout. Make the drawing as though all corners were square, and then construct the curves. Use two small sheets or one large sheet.**

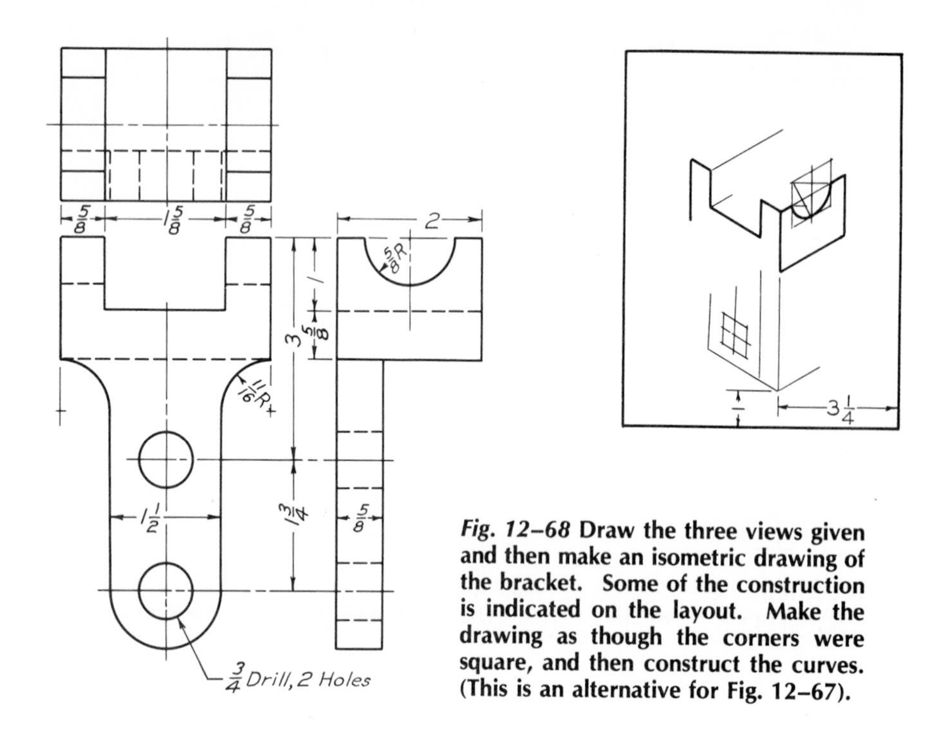

Fig. 12–68 **Draw the three views given and then make an isometric drawing of the bracket. Some of the construction is indicated on the layout. Make the drawing as though the corners were square, and then construct the curves. (This is an alternative for Fig. 12–67).**

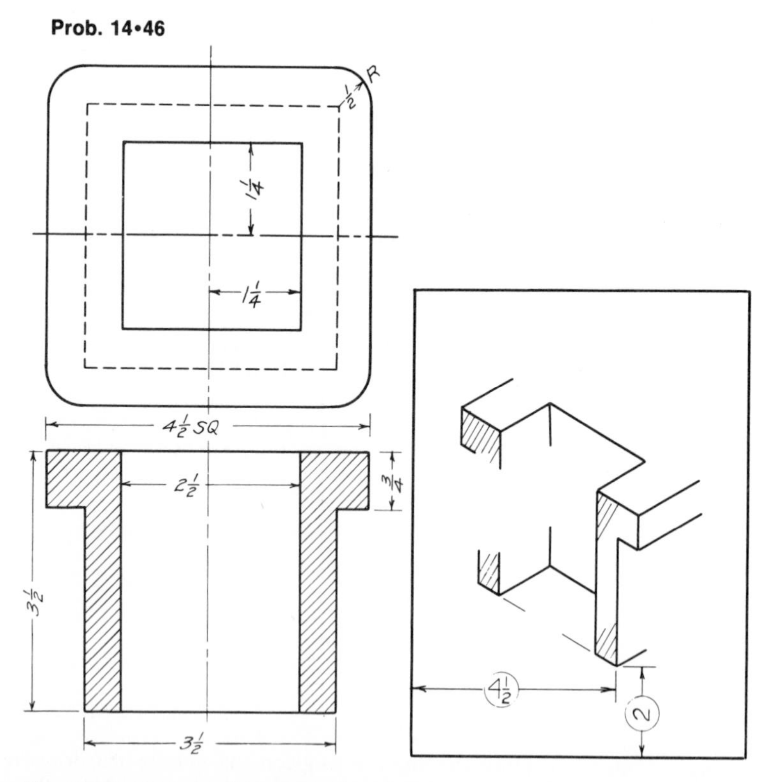

Fig. 12–69 **Make an isometric drawing in section of the post-socket. See the layout.**

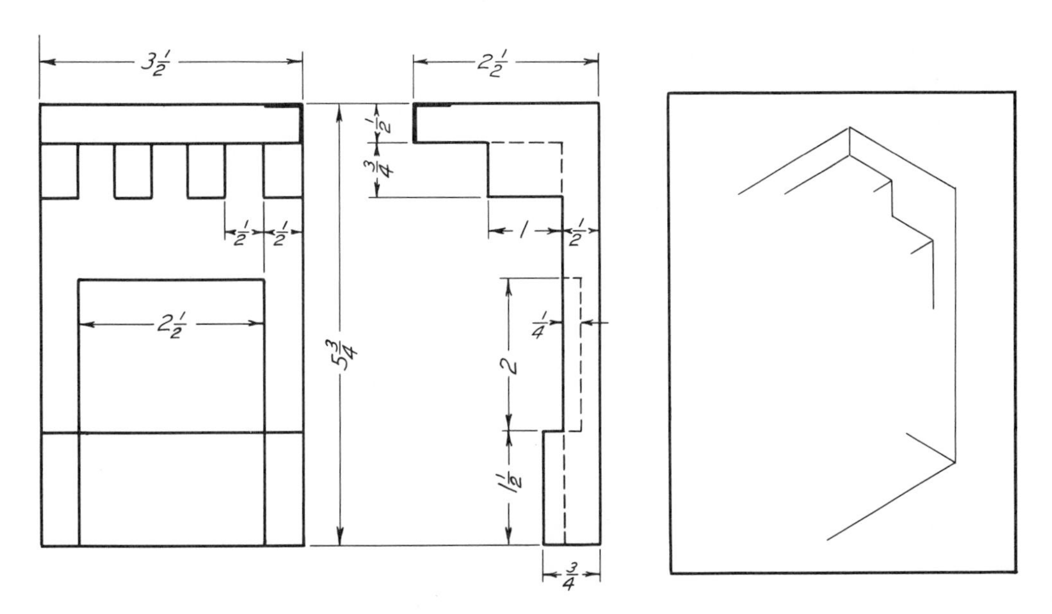

Fig. 12–70 Make an isometric drawing of the tablet. Use reversed axes.

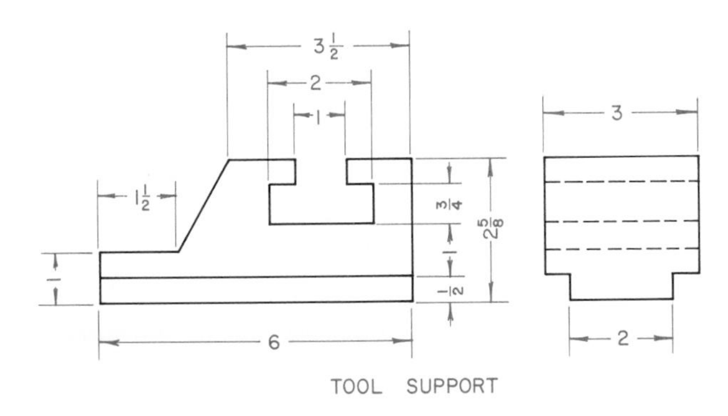

Fig. 12–71 Make a one-point-perspective or a two-point-perspective drawing of the tool support. Use light, thin lines for the construction lines and do not erase them. Brighten the finished perspective drawing lines. Use any suitable scale. The locations of the *PP, HL, GL,* etc., are optional.

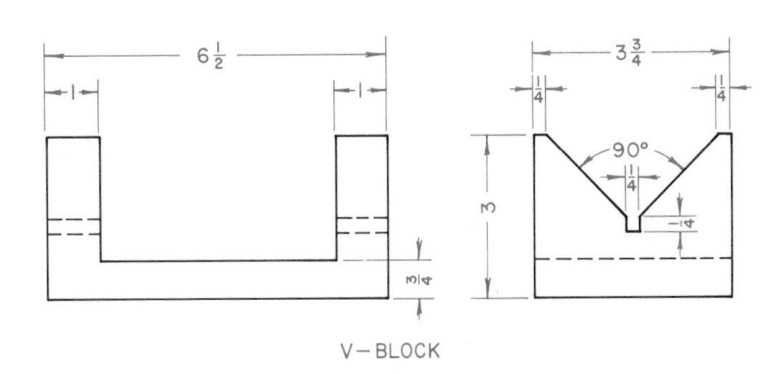

Fig. 12–72 Make a one-point-perspective or a two-point-perspective drawing of the V-block. Use light, thin lines for the construction lines and do not erase them. Brighten the finished perspective drawing lines. Use any suitable scale. The locations of the *PP, HL, GL,* etc., are optional.

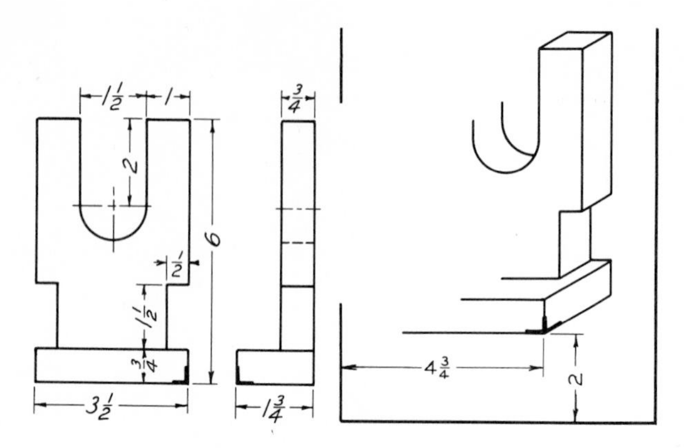

Fig. 12–73 **Make an oblique drawing of the angle support.**

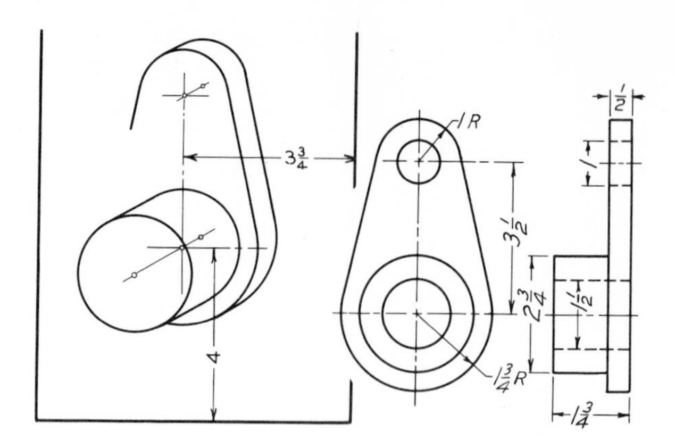

Fig. 12–74 **Make an oblique drawing of the crank.**

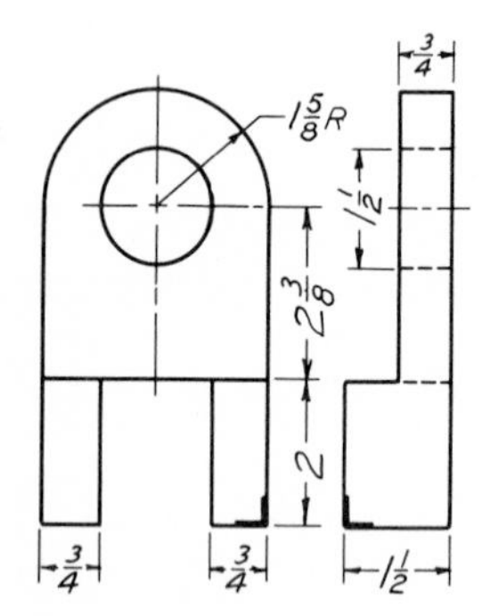

Fig. 12–75 **Make an oblique drawing of the forked guide.**

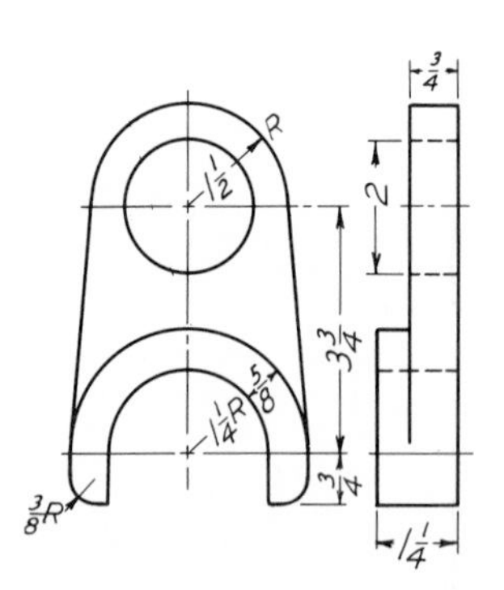

Fig. 12–76 **Make an oblique drawing of the guide link.**

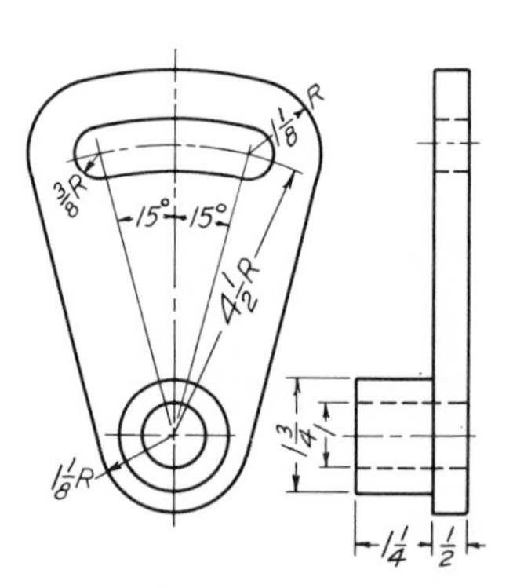

Fig. 12–77 **Make an oblique drawing of the slotted sector.**

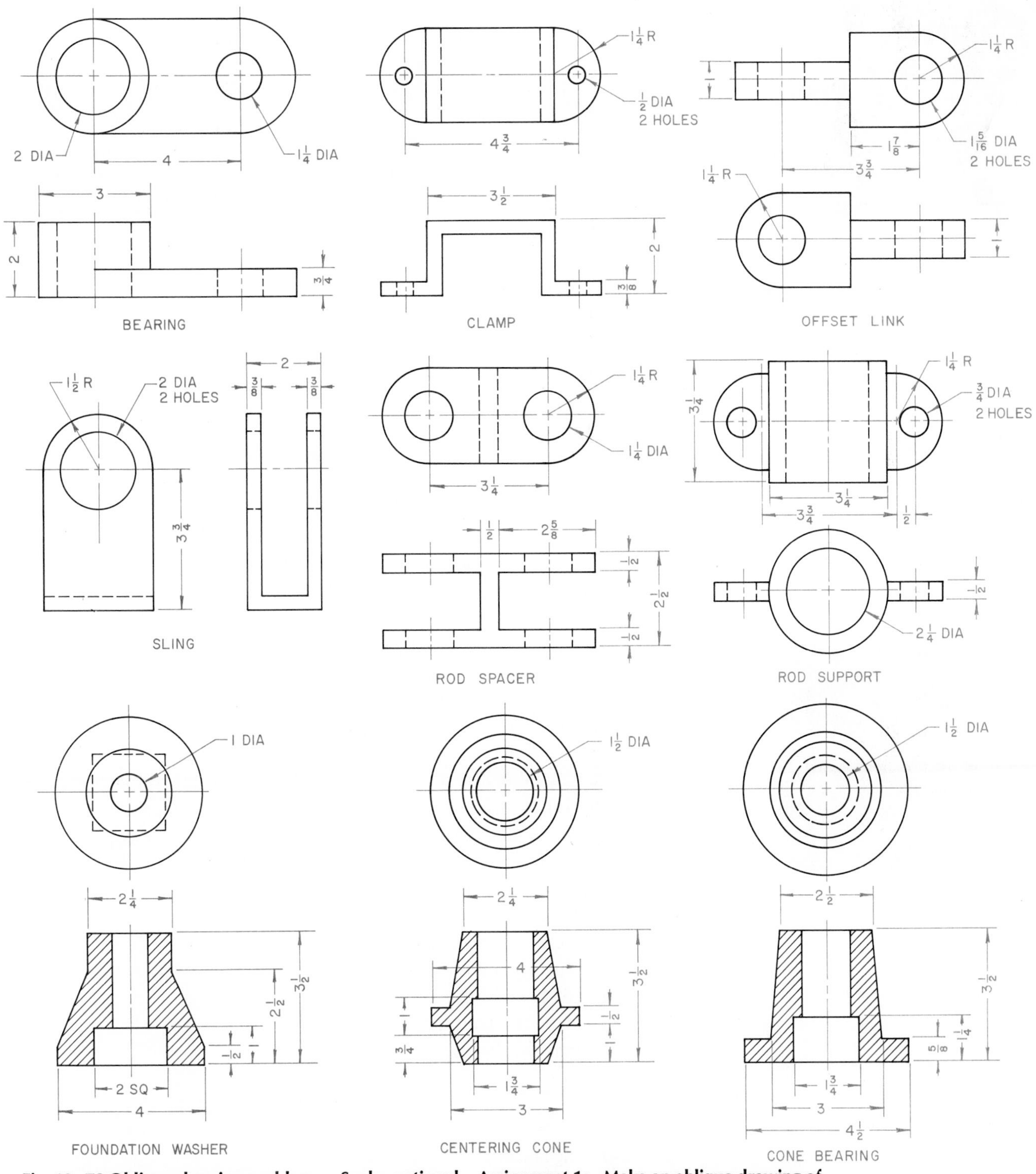

Fig. 12–78 **Oblique drawing problems. Scale: optional. Assignment 1: Make an oblique drawing of the object assigned. Assignment 2: Make an oblique half- or full-sectional view as assigned. Do not dimension unless instructed to do so. NOTE: These problems may also be used for isometric and perspective drawing practice.**

Chapter 13

Inking

NATURE AND PURPOSE OF INKED DRAWINGS

Most technical drawings are made with pencil on a good-quality tracing paper (vellum) or on drafting film. However, ink is often used to make high-quality tracings that can be copied. As the use of ink drawings increases, better inking techniques and equipment are being developed. For example, the technical pen with cartridge ink supply, and all its accessories and attachments has made inking more efficient. It has also improved the quality of the finished drawing.

Microfilming and similar copying methods (also called micrographics) have made new inking methods and high-quality ink drawings necessary (see Chapters 14 and 15). When technical drawings are reduced or enlarged photographically, some *definition* (detail and sharpness) is lost. A high-quality original always produces a high-quality reproduction (Fig. 13–1). If the original drawing is of poor quality, the reproduction may not be usable.

From original drawing

To microfilm reduction

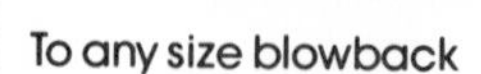

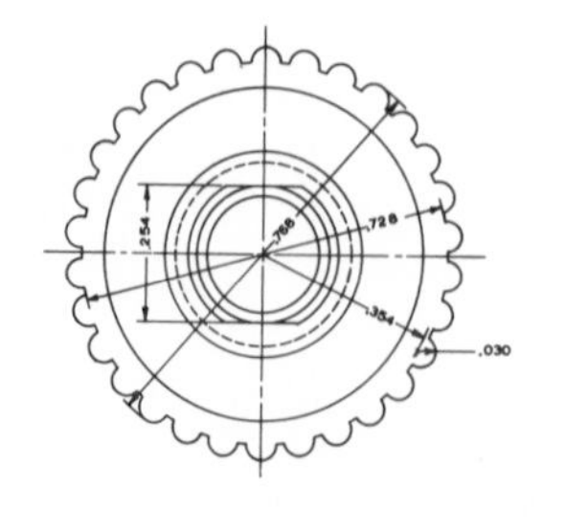

Fig. 13–1 **High-quality original ink drawings produce high-quality copies. *(Koh-I-Noor Rapidograph, Inc.)***

DRAWING INK

Ink used for technical drawings is also called India ink. It must have some special characteristics. For example, it must flow freely and yet dry fast. It must *adhere* (stick) well and must not chip, crack, peel, or smear after drying (Fig. 13–2). Also, it must be completely opaque in order to produce good uniform line tone.

Both waterproof and nonwaterproof ink can be used on high-quality paper, cloth, polyester film, and illustration board. The waterproof ink is best suited for mechanical drawings. Nonwaterproof ink is best for fine-line pictorial drawings. Both types of ink produce opaque lines that reproduce well when they are photographed or copied by diazo machines (see Chapter 14).

Fig. 13–2 **Qualities of ink for technical drawings.**

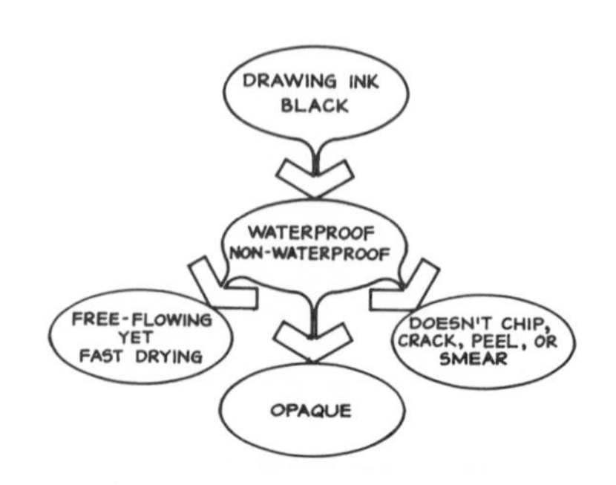

SPECIAL INKS
ACID BASE TO ETCH ACETATE
COLOR OPAQUE
BALL POINT COLOR

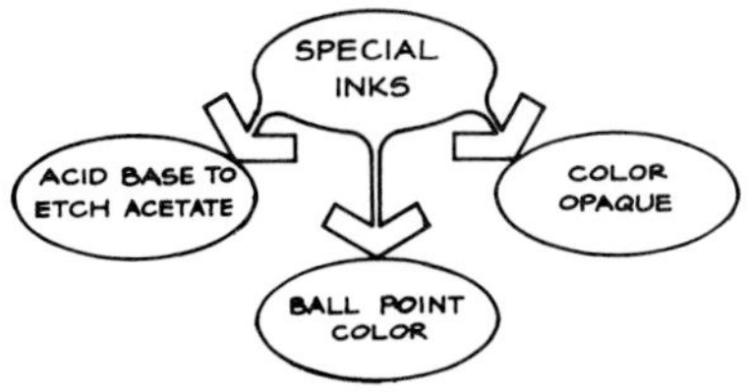

Fig. 13–3 **Special inks for new methods of presentation.**

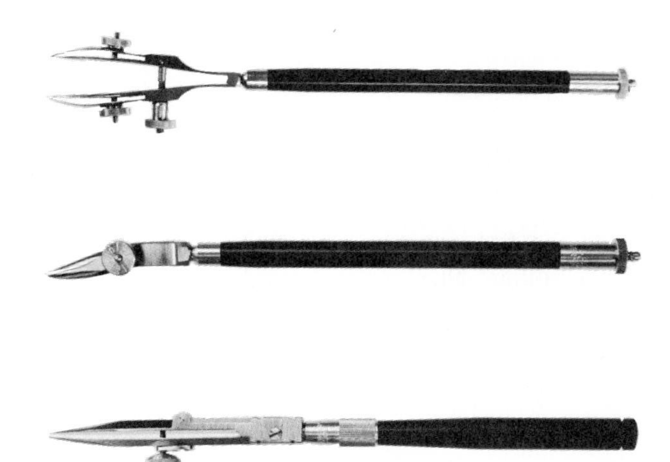

Fig. 13–5 **Special ruling pens.** ***(Teledyne Post.)***

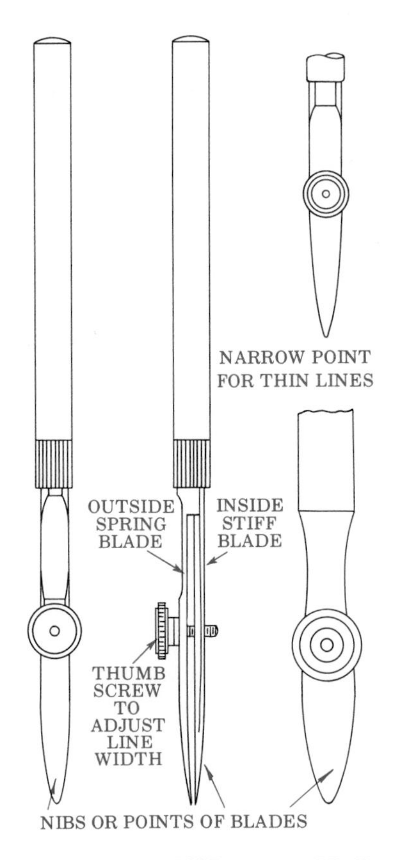

Fig. 13–4 **The all-purpose ruling pen.**

Acid-base ink is used on high-gloss acetate film. It *etches* (dissolves) the glossy surface of the film. Etching usually cannot be corrected. However, some companies have developed special techniques for making corrections and revisions. Other special inks come in colors. They include waterproof inks for ball-point pens that are used on computer or tape controlled plotting machines. Different kinds of ink are made to be used on different kinds of surfaces (Fig. 13–3).

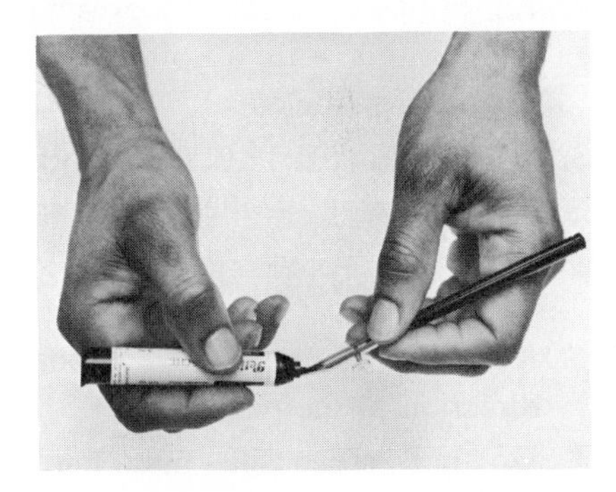

Fig. 13–6 **Filling the pen from a cartridge tube.** ***(R. J. Capece/McGraw-Hill.)***

BASIC INKING INSTRUMENTS

The ruling pen (Fig. 13–4) has blades that you can adjust to draw lines of different widths. It can be used to draw both straight and curved lines. Ruling pens are sometimes called *all-purpose pens.* The special ruling pens shown in Fig. 13–5 are used to draw double lines, contour lines, and thick (heavy) lines. They can be filled from a cartridge tube (Fig. 13–6), a squeeze bottle, or a dropper cap (Fig. 13–7).

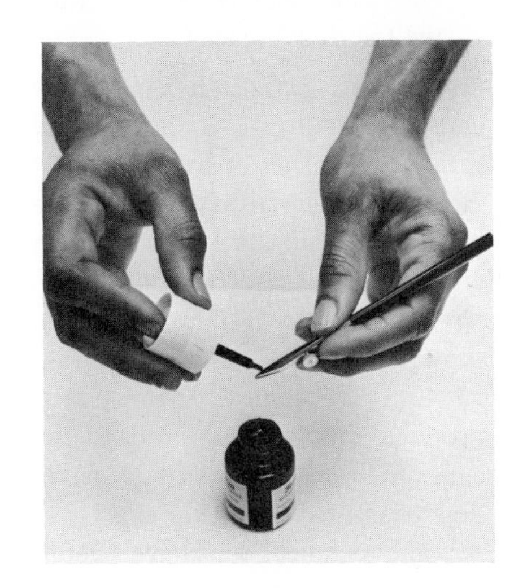

Fig. 13–7 **Filling the pen from a dropper cap.** ***(R. J. Capece/McGraw-Hill.)***

The simple rules listed below will help you produce high-quality ink drawings. Be sure to study them carefully before you try to use the ruling pen.

1. Do not hold the pen over the drawing when you are filling it.
2. Do not fill the pen too full; about 3 to 6 mm (1/8 to 1/4 in.) is usually enough.
3. Never dip the pen into the ink bottle or allow ink to get on the outside of the blades.
4. Keep blades clean by wiping them often with a soft towel, tissue, or pen cleaner.
5. While inking a line, keep both *nibs* (points) of the blades in contact with the drawing surface.
6. Keep the blades slanted in the direction of the pen stroke.
7. Do not press the nibs hard against the straightedge. Pressing too hard will make line widths uneven.
8. Do not press the nibs too hard against the drawing surface. The blades may get dull or damage the drawing surface.

Faulty lines can result from many causes. Figure 13–8 shows some mistakes that can be avoided.

Technical pens have made the inking of technical drawings much easier (Fig. 13–9). Technical pens have points of different sizes to draw different line widths (Fig. 13–10). Technical pen points produce uniform line widths because their design provides a steady flow of ink that does not clog.

Some technical pens have a refillable cartridge for storing ink (Fig. 13–11). Others have a cartridge that is used once and then replaced. Still others have a piston-action filling system that can easily be refilled from a standard drawing-ink bottle (Fig. 13–12).

Technical pen points are made of different materials for use on differ-

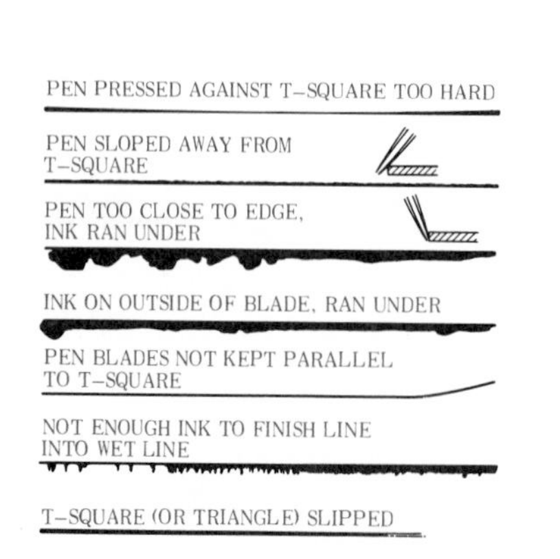

Fig. 13–8 **Causes of faulty lines from a ruling pen.**

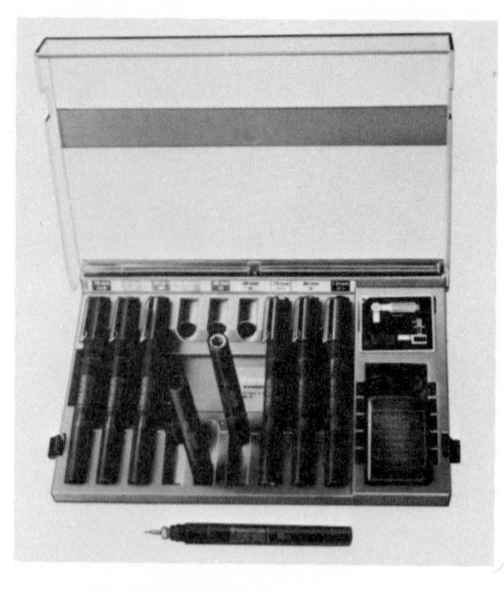

Fig. 13–9 **Technical fountain pens. *(J. S. Staedtler, Inc.)***

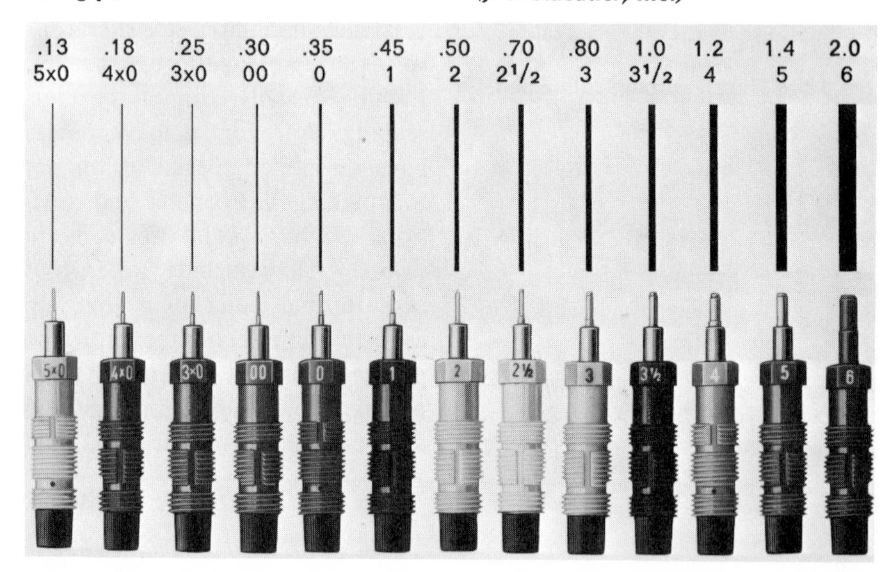

Fig. 13–10 **The range of lines available from technical pen points. *(J. S. Staedtler, Inc.)***

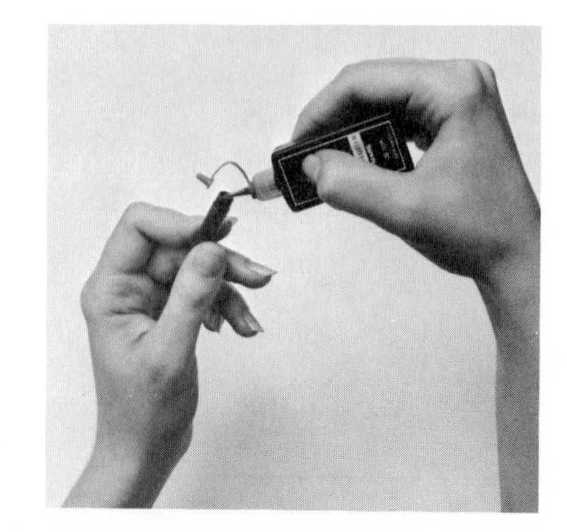

Fig. 13–11 **Refillable cartridge. *(R. J. Capece/McGraw-Hill.)***

Fig. 13–12 **Piston-action filling system.** ***(R. J. Capece/McGraw-Hill.)***

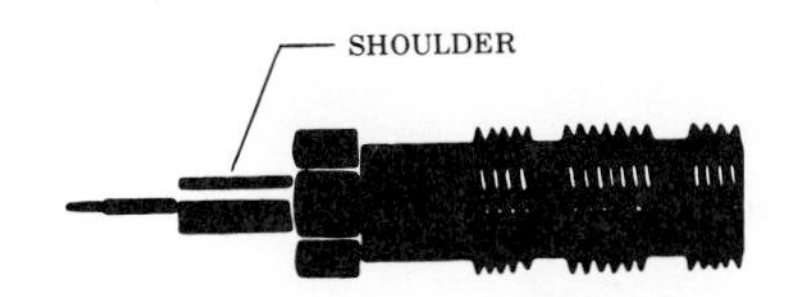

Fig. 13–13 **The offset of the barrel shoulder moves easily along all guiding edges.** ***(J. S. Staedtler, Inc.)***

Fig. 13–14 **The technical pen in use. Note that the shoulder keeps the barrel of the pen above the curve. (R. J. Capece.)**

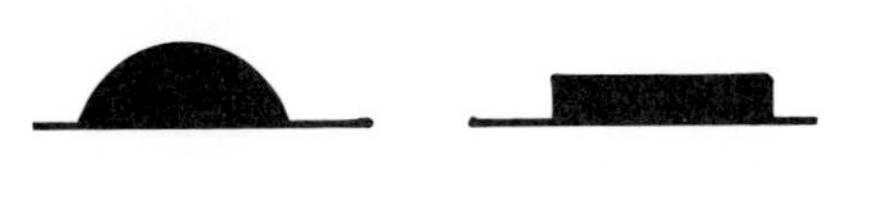

Fig. 13–15 **Section view of typical ink flow from a ruling pen and a technical pen.**

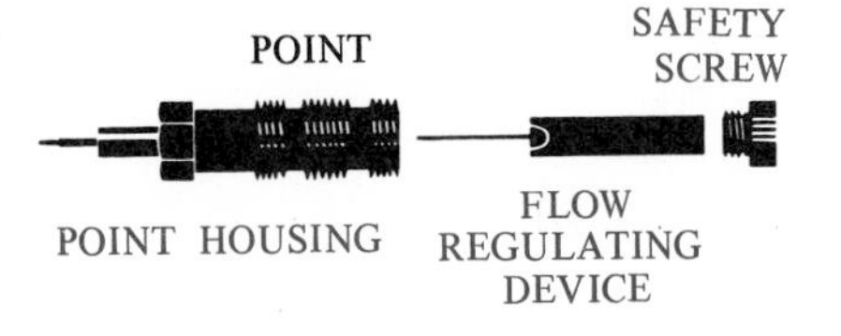

Fig. 13–16 **Clicking sound from cleaning wire assures smooth flow of ink.** ***(J. S. Staedtler, Inc.)***

ent surfaces. There are three main kinds of points:

1. Hardened stainless steel points, used on all surfaces.
2. Self-polishing jewel points, used on very *abrasive* (rough) surfaces, such as drafting film.
3. Wear-free tungsten carbide points, used on drafting film and with high-speed automated drafting equipment.

Technical pen points have a shoulder, as shown in Fig. 13–13, to prevent smudging. The shoulder is barrel-shaped for use with curved or straight guiding instruments.

The technical pen has some definite advantages over the ruling pen. They are the following:

1. It is easy to produce lines of uniform width (no line adjustments are required).
2. The barrel-shaped shoulder prevents smearing, because point and shoulder are offset (Fig. 13–14).
3. The cartridge is easy to load ink. Also, it needs refilling less often than a ruling pen.
4. The ink laid down by a technical pen can dry more uniformly because it flows more evenly (Fig. 13–15).
5. The technical pen does not often need to be cleaned, because it has a weighted cleaning wire inside the capillary tube of the point (Fig. 13–16).

Fig. 13–17 **Humidified container for technical pens.** ***(Keuffel & Esser Co.)***

The technical pen can be stored in a humidified container that prevents the ink from drying out (Fig. 13–17). In some technical pens, a *hermetic* (airtight) seal forms in the cap when the cap is tightened with a sharp twist. If you are careful in filling and cleaning your pens, your inking will be better.

INKING STRAIGHT LINES

Figure 13–18 shows the correct position for drawing lines with a technical pen. Note the direction

Fig. 13–18 **The position of the technical pen is important when drawing lines.** ***(Teledyne Post.)***

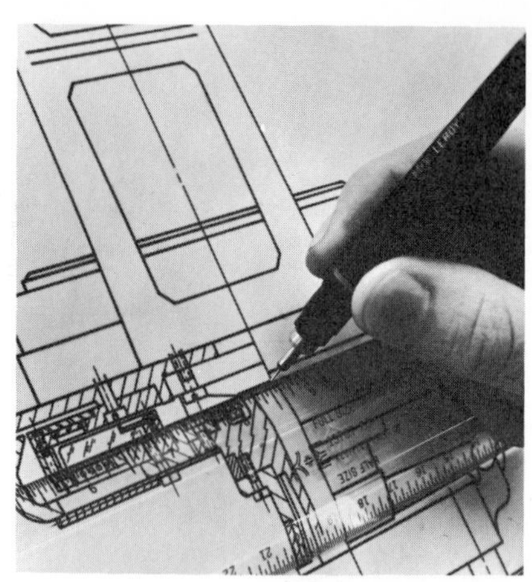

Fig. 13–19 **Correct position of the technical pen when drawing lines.** ***(Keuffel & Esser Co.)***

of the stroke and the angle of the ruling pen. Hold the technical pen in a nearly vertical position to get the most uniform line (Fig. 13–19).

MAKING CIRCLES AND ARCS

You can use templates and compasses with technical pens (Fig. 13–20). However, most templates cannot be used easily with ruling pens. Many compasses have ruling-pen attachments (Fig. 13–21) that are used for inking arcs and circles. Others have technical-pen adaptors, as shown in Fig. 13–22. To use a compass for inking, remove the pencil leg from the compass and replace it with the ruling-pen leg or the technical pen. Adjust the needle point carefully, as shown in Fig. 13–23, so that the legs are perpendicular to the drawing surface. Always draw the circle in one stroke. The compass can be slanted a little when you are using a ruling pen, but hold it vertically when using a technical pen (Fig. 13–24). The lengthening bar, or

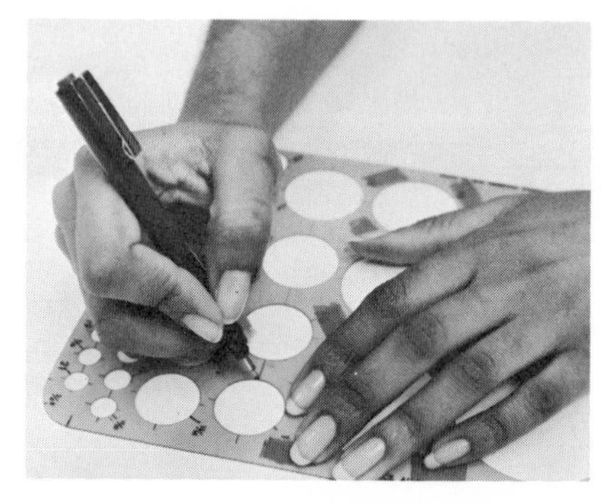

Fig. 13–20 **Technical pens are easily used with templates. (R. J. Capece.)**

Fig. 13–21 **Compass and ruling-pen attachment.** ***(Teledyne Post.)***

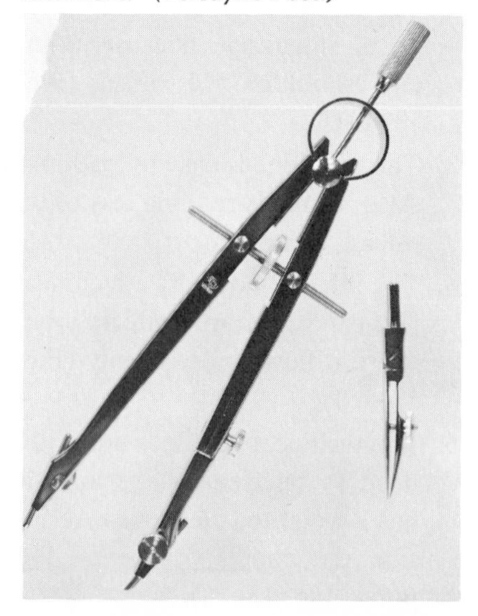

Fig. 13–22 **Technical-pen adaptors for compasses. The universal adaptor allows the pen to be positioned vertically.** ***(J. S. Staedtler, Inc.)***

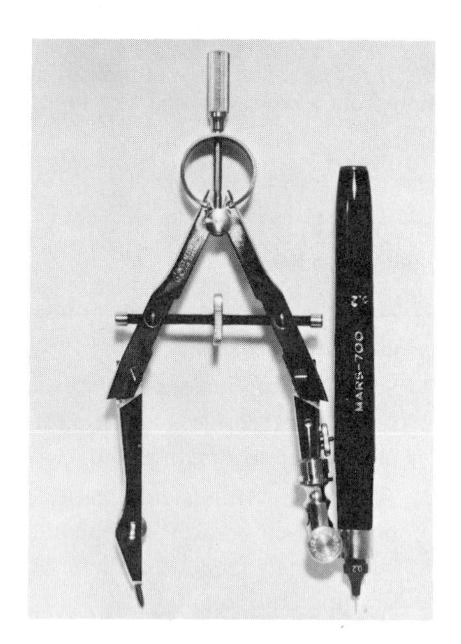

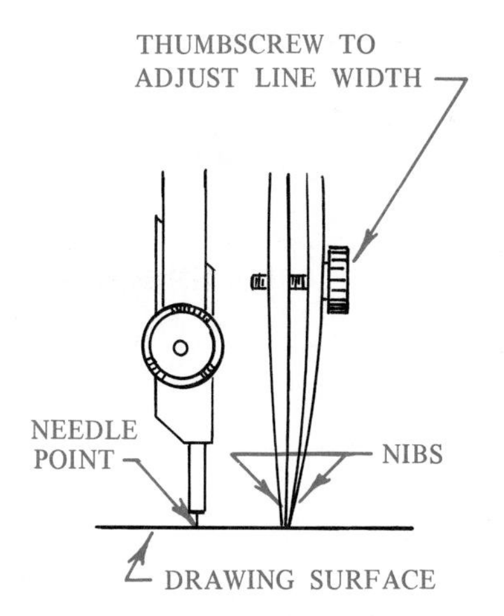

Fig. 13–23 **Needle point of compass adjusted so that the ruling nibs and point are perpendicular to the drawing surface.**

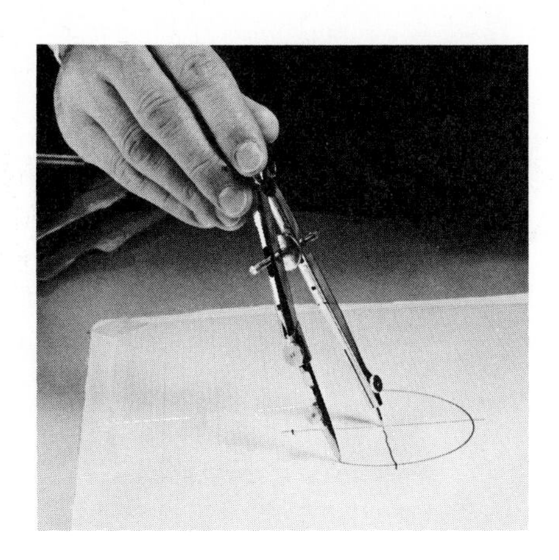

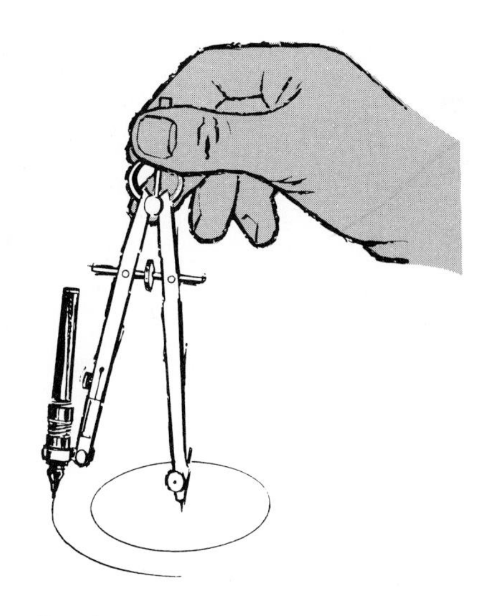

Fig. 13–24 **Inking with a compass.** ***(Teledyne Post.)***

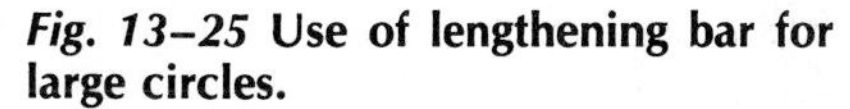

Fig. 13–25 **Use of lengthening bar for large circles.**

Fig. 13–26 **Fixed and adjustable irregular curves.**

beam compass, can be used for large circles, as shown in Fig. 13–25.

INKING CURVED LINES

Irregular (French) curves can be used to guide the pen when you are inking curves that are not circular arcs. You can use either fixed curved templates or irregular adjustable curves (Fig. 13–26). The following steps will help you in inking curves.

1. Locate the tangent points that change the direction of the curve or join it to a straight line (Fig. 13–27 at A).
2. Draw the curve lightly in pencil. Note where it changes from a smaller to a larger radius.
3. Ink over the middle of the pencil line (Fig. 13–27 at B).

LETTERING GUIDES AND EQUIPMENT

The lettering set shown in Fig. 13–28 has three basic tools for ink-

Fig. 13–27 **Ink curves to point of tangency first (A). Ink lines must be centered over pencil lines (B).**

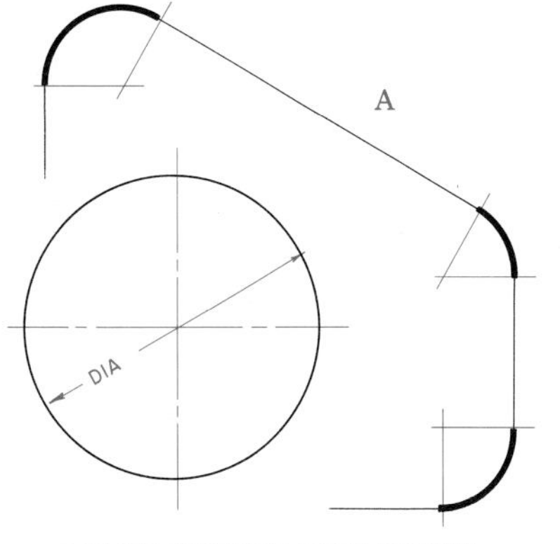

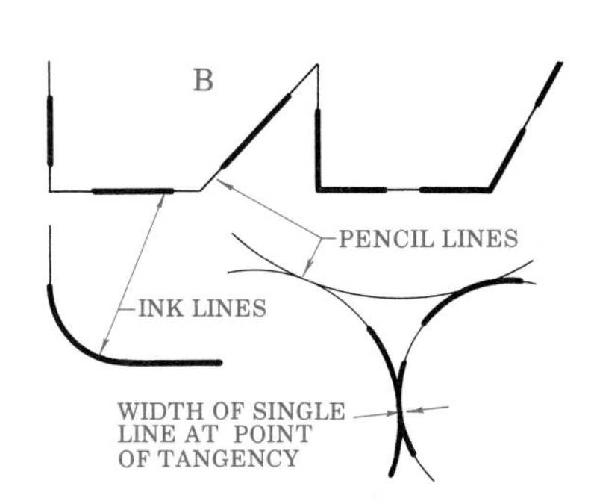

Fig. 13–28 **One of the drafter's most useful tools is a lettering set.** ***(Keuffel & Esser Co.)***

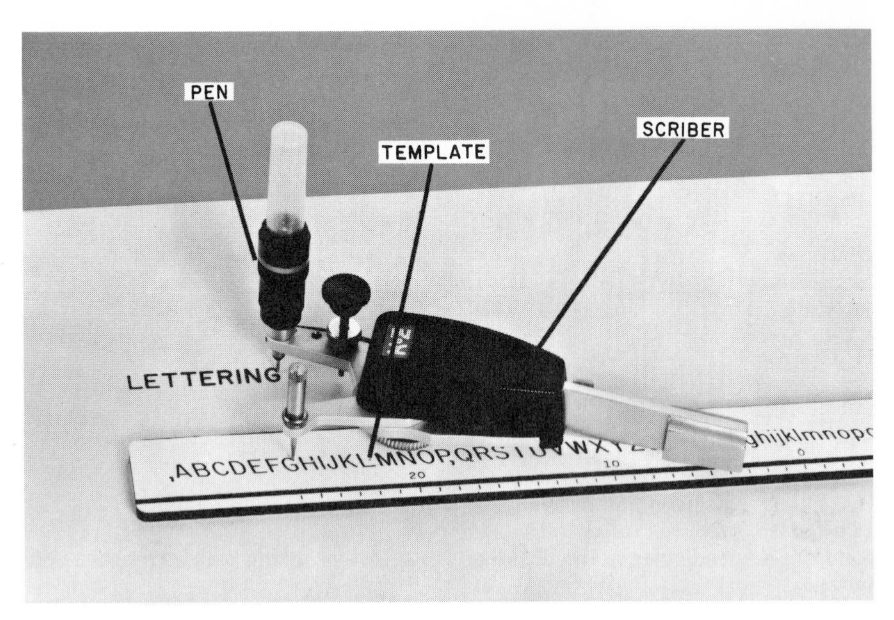

Fig. 13–29 **The three basic parts of a lettering set are the pen, the template, and the scriber.** ***(Keuffel & Esser Co.)***

ing: a scriber, a set of lettering templates, and a set of technical pens. A scriber has a tailpin to follow the horizontal groove in the bottom of the template, a tracer pin to follow the engraved letter, and a barrel slot. A technical pen is set in the slot. When you move the tracer pin along the letter, the pen copies the letter above the template (Fig. 13–29). Some scribers (Fig. 13–30) can be adjusted to make letters of different heights or letters slanted at different angles. Lettering templates also come in many styles and sizes of lettering, from about 1.5 to 50 mm (1/16 to 2 in.) high. Some templates are used to draw symbols instead of letters (Fig. 13–31). The width of the pen point you use depends on the height of the letters and the size of the symbols.

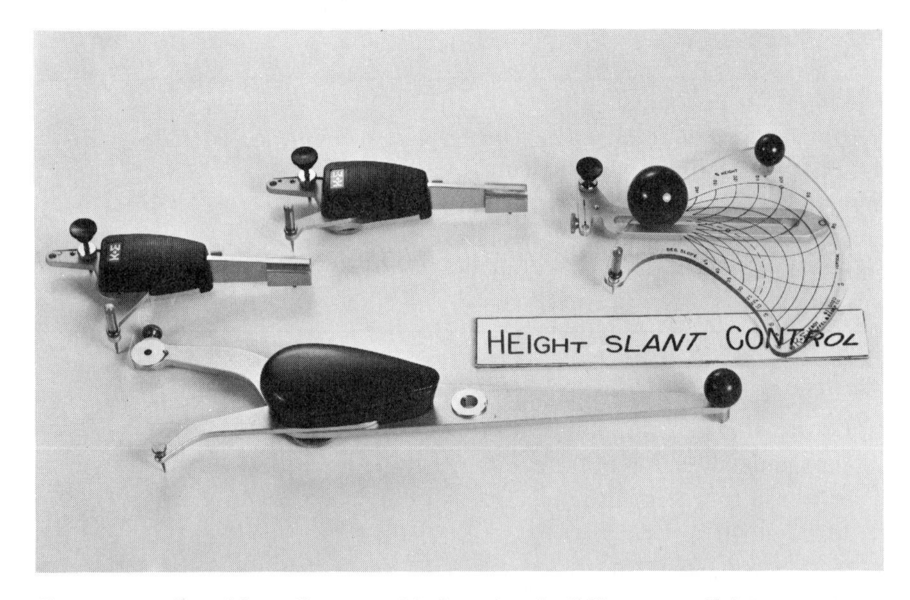

Fig. 13–30 **Adjustable scribers provide lettering flexibility.** ***(Keuffel & Esser Co.)***

ERASING TECHNIQUES

The ink used on polyester drafting film is waterproof. However, you can easily remove ink from the film by rubbing it with a moistened plastic eraser. Do not use any pressure in rubbing. The polyester film does not absorb ink, and therefore all the ink dries on top of its highly finished surface (Fig. 13–32). You can remove ink from other surfaces, such as tracing vellum or illustration board, with regular ink erasers. But be very careful. Press lightly with strokes in the direction of the line to remove ink caked on the surface. Too much pressure damages the surface and makes it hard to revise the drawing. The tooth surface shown in Fig. 13–33 holds the ink more firmly and makes it harder to erase. Different erasing techniques work better on different

Fig. 13–31 **Templates are available in various styles of lettering and kinds of symbols. *(Keuffel & Esser Co.)***

Fig. 13–32 **The results of good (left) and poor (right) ink adhesion.**

Fig. 13–33 **Section showing ink in contact with drawing surface.**

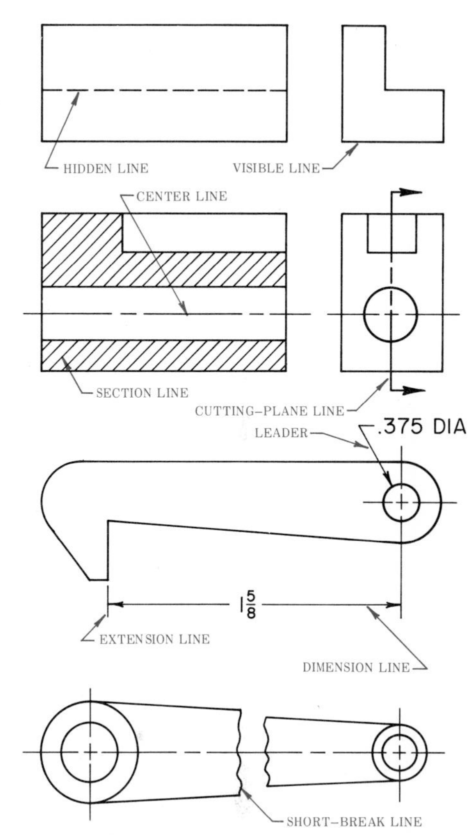

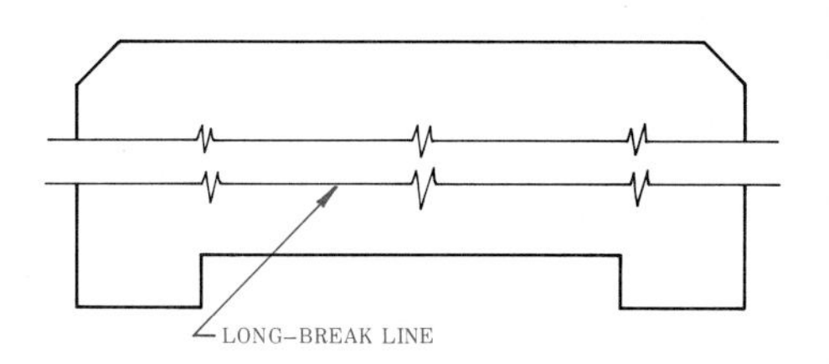

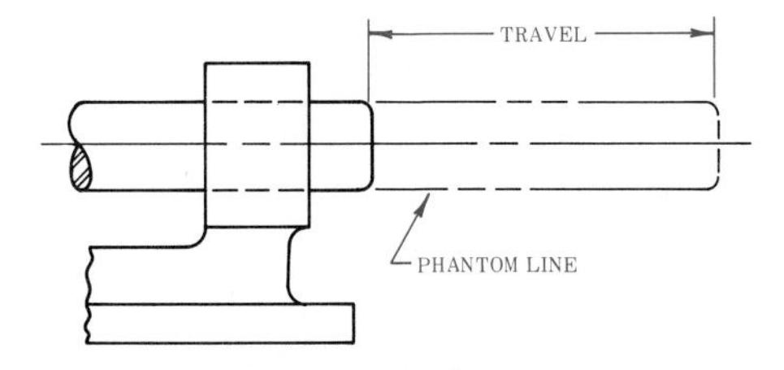

Fig. 13–34 **Alphabet of lines.**

surfaces. You should try different techniques to find the best one for each surface.

THE ALPHABET OF LINES

Line symbols, the different kinds of lines used to make drawings, are a kind of graphic alphabet. The *American National Drafting Standards Manual* recommends two widths of lines: thick and thin. Figure 13–34 shows the different line symbols and what they mean. Each kind of line has a definite meaning and must not be used for anything else. Detail drawings should have fairly wide (thick) outlines. The center and dimension lines and hidden lines should be thin. In this way, the drawing will have contrast and be easy to read. If all the lines have the same width, the drawing looks flat and is hard to read.

ORDER OF INKING

Smooth joints and tangents, sharp corners, and neat fillets make a drawing look better and make it easier to read. To do good inking, you need careful practice and a definite order of working.

The usual order of inking or tracing is shown in Fig. 13–35. First, you ink the arcs, centered over the

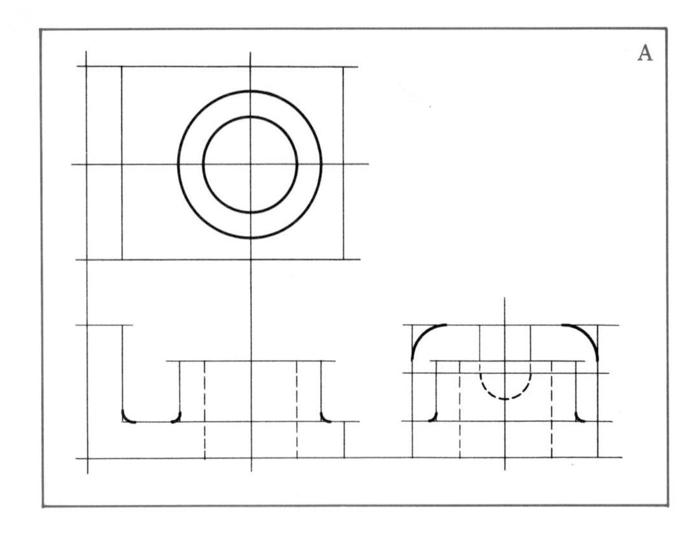

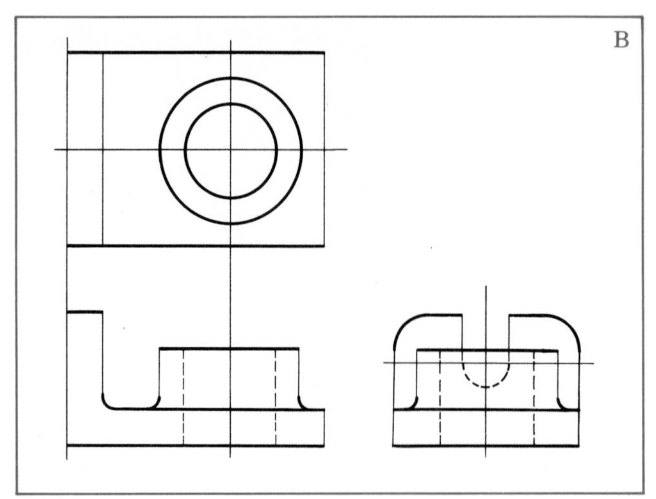

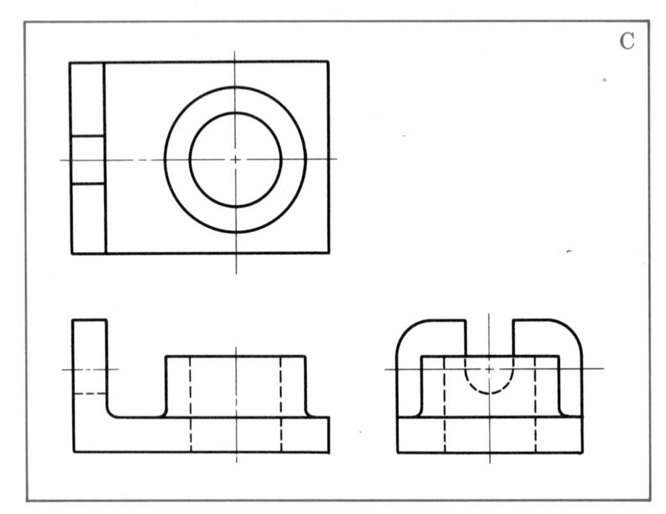

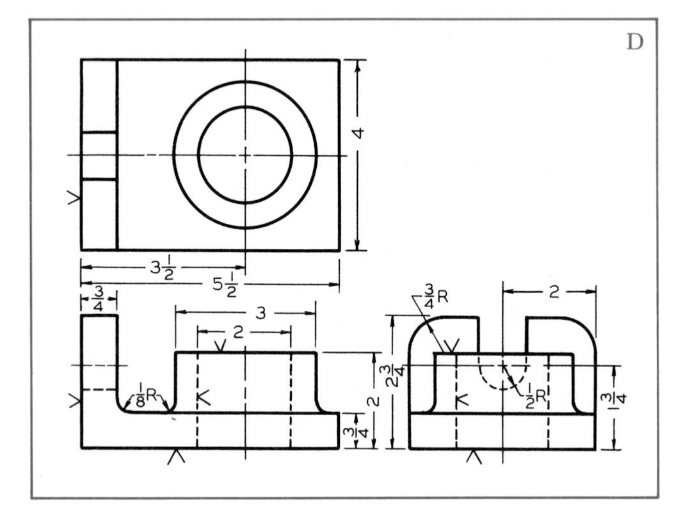

Fig. 13–35 **Order of inking or tracing a drawing.**

pencil lines, as in space A. Ink the horizontal lines next, as in space B. Complete the drawing with the vertical lines, as in space C. Then add the dimension lines, arrowheads, finish marks, and so on, and fill in the dimensions, as in space D. The order of inking is:

1. Ink main centerlines.
2. Ink small circles and arcs.
3. Ink large circles and arcs.
4. Ink hidden circles and arcs.
5. Ink irregular curves.
6. Ink horizontal full lines.
7. Ink vertical full lines.
8. Ink slanted full lines.
9. Ink hidden lines.
10. Ink centerlines.
11. Ink extension and dimension lines.
12. Ink arrowheads and figures.
13. Ink section lines.
14. Letter notes and titles.
15. Ink border lines.
16. Check drawing carefully.

ULTRASONIC CLEANER

The technical pen and the ruling pen sometimes need cleaning. The ultrasonic cleaner is very useful because it works so fast. The cleaning usually takes only 20 to 30 seconds, and the pen is ready to use again right away. To use an ultrasonic cleaner, just dip the pen point into a well containing cleaning fluid. From 60 000 to 80 000 cycles per second of sound energy act to loosen the dried ink on the point (Fig. 13–36). The normal high range of

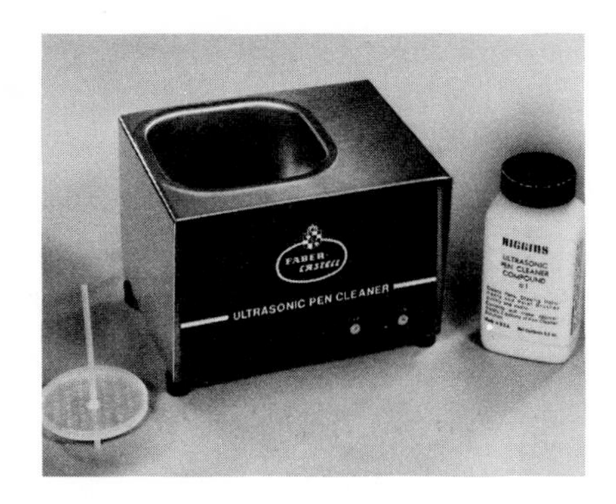

Fig. 13–36 **An ultrasonic cleaner.** ***(Faber-Castell.)***

human hearing is about 15 000 cycles per second.

To clean a technical pen thoroughly, you may have to take it apart. Remove the cap and ink cartridge and soak the point in a special cleaning fluid. After you remove the dried ink from the point, flush the parts in a stream of cold water. Dry all the parts before you use the pen again. If the parts are not dry, the ink will be *diluted* (mixed with water) and lines will not be dark enough. If you are careful in handling and storing your inking equipment, it will work well and need very little repair.

Review

1. Name the two kinds of ink pens used in drafting.

2. What is the best position in which to hold a technical pen?

3. Name two ways to ink circles and arcs.

4. Name the instruments used to guide the pen when you are drawing curved lines that are not arcs.

5. How do you erase India ink from drafting film?

6. How many widths of lines are recommended for use on ink drawings? Name them.

7. What kind of ink is used on acetate film?

8. Waterproof inks are used for mechanical drawings; nonwaterproof inks are used for ________.

9. Do not fill a ruling pen too full; about ________ to ________ mm is usually enough.

10. Metric line widths are based on a ________ progression of the ________.

Problems

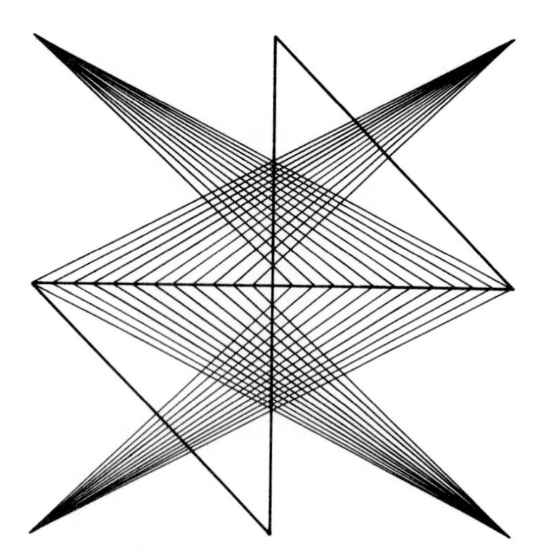

Fig. 13–37 Using light pencil construction lines, lay off a 6″ square and divide it equally into four 3″ squares. On the horizontal midline, mark off 24 1/4″ units. Proceed to ink in the lines to construct the given design.

Fig. 13–38 Construct in ink the intersecting units that have an illusion of depth. The axis to the left is 20°, and to the right 25°. The depth of a unit is 2″ to the left and 2 3/4″ to the right. The height is 2″.

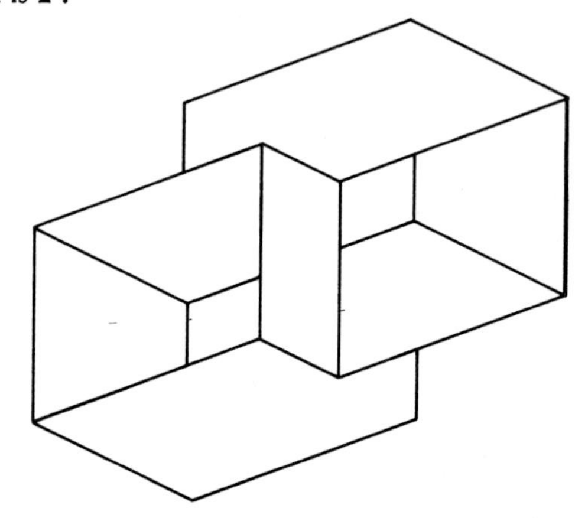

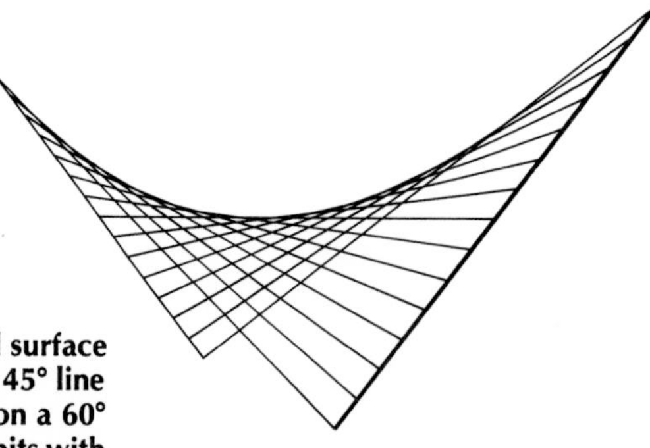

Fig. 13–39 Develop the warped surface shown. Place 15 1/4″ units on a 45° line to the left. Place 15 3/8″ units on a 60° line to the right. Connect the units with straight lines, as shown, to form the curve.

Fig. 13–40 See Fig. 13–35. Prepare an inked drawing of the figure. Scale: 2:1.

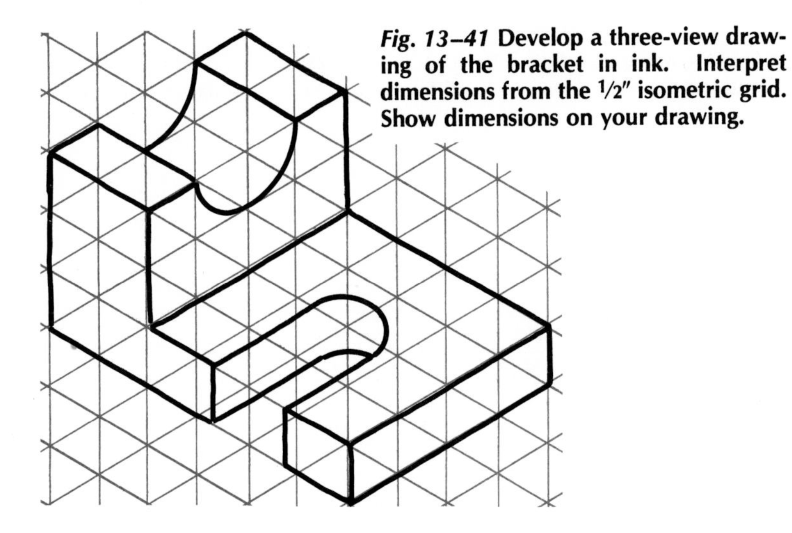

Fig. 13–41 **Develop a three-view drawing of the bracket in ink. Interpret dimensions from the 1/2" isometric grid. Show dimensions on your drawing.**

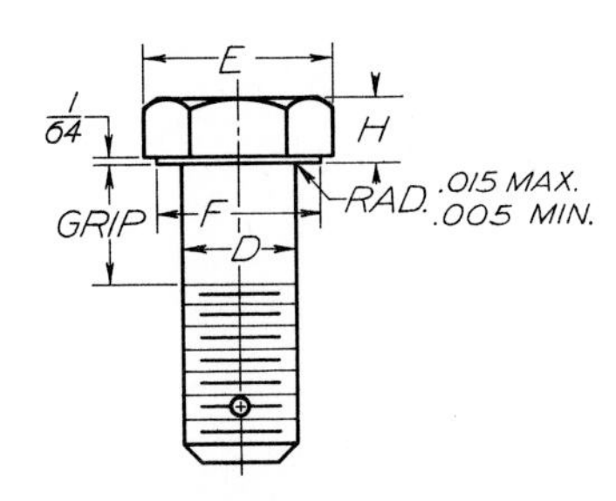

Fig. 13–42 **Develop two views of the aircraft bolt. D = 1", E = 1.65625, F = 1.4375, H = 1/2D. Threads per inch = 14. Ink all lines.**

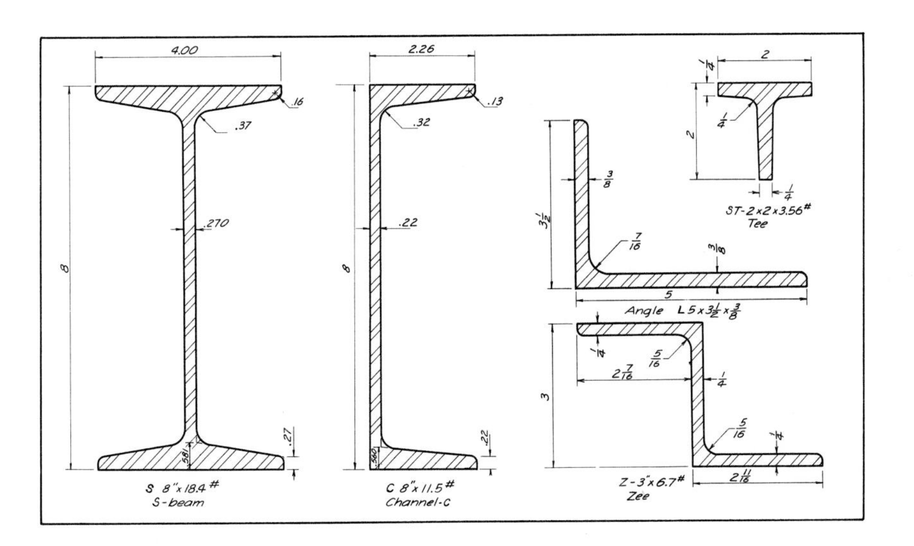

Fig. 13–43 **Prepare ink drawings of the five structural shapes at full scale on a B-size sheet.**

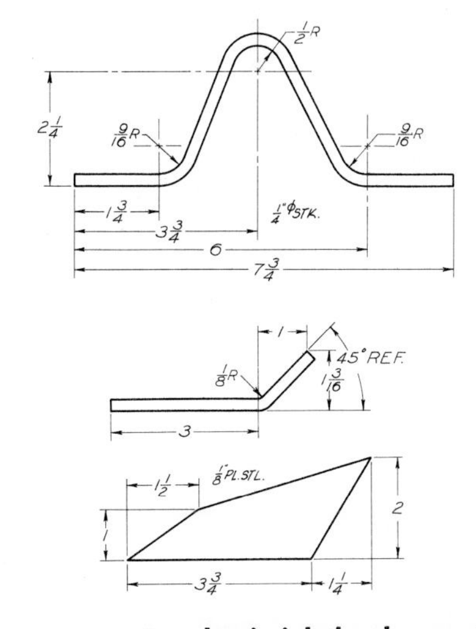

Fig. 13–44 **Develop in ink the shapes of the 1/4" round barstock and the sheet stock shown. Dimension the drawings.**

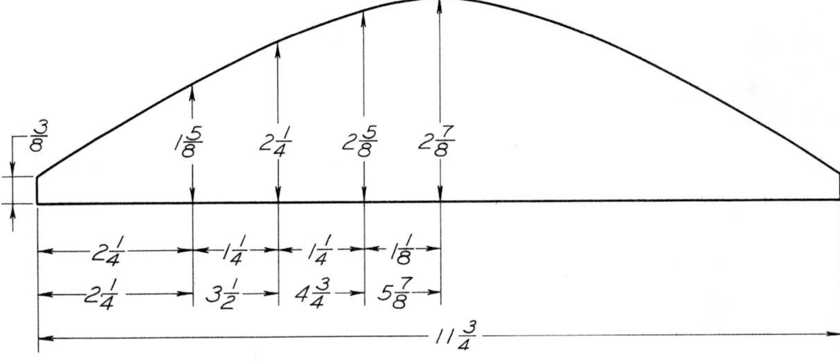

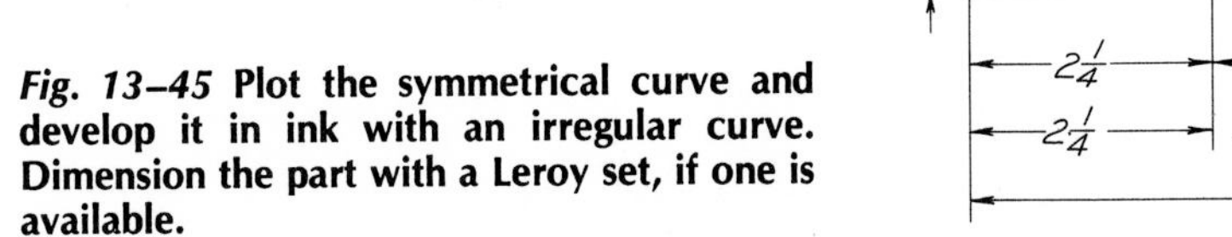

Fig. 13–45 **Plot the symmetrical curve and develop it in ink with an irregular curve. Dimension the part with a Leroy set, if one is available.**

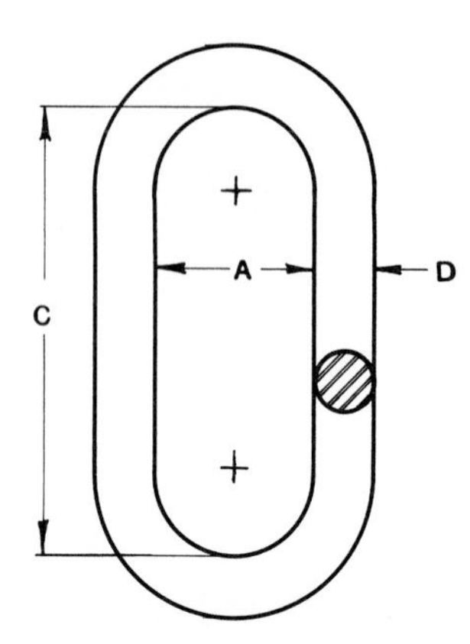

Fig. 13–46 **Prepare one view of the wire rope short link with revolved section. Use dimensions selected by the instructor. Dimension your drawing. *(United States Steel.)***

Wire Rope Link

A	C	D
1	3	3/8
1 1/4	4	1/2
1 1/2	4	5/8
1 3/4	5	3/4
2	5	7/8

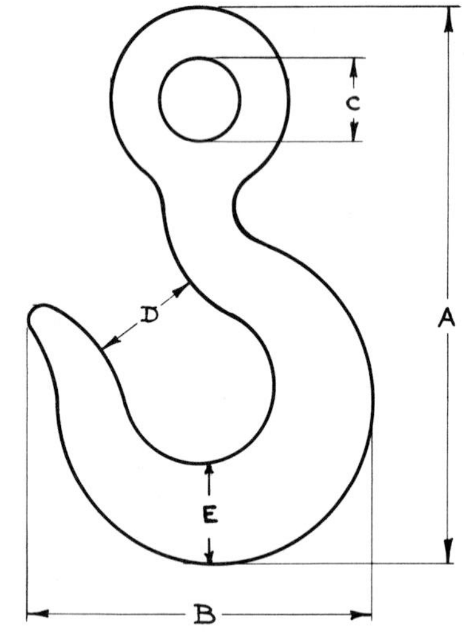

Fig. 13–47 **Prepare a wire rope hook using the dimensions selected by the instructor. Determine radii necessary for smooth tangencies. Dimension your drawing. *(United States Steel.)***

Wire Rope Hook

A	B	C	D	E
4 15/16	3 3/16	7/8	1 1/16	27/32
5 13/32	3 16/32	1	1 1/8	29/32
6 1/4	4 3/32	1 1/8	1 1/4	1 1/8
6 7/8	4 17/32	1 1/4	1 3/8	1 5/16
7 5/8	4 7/8	1 3/8	1 1/2	1 3/8
8 19/32	5 3/4	1 1/2	1 11/16	1 9/16
9 1/2	6 3/8	1 5/8	1 7/8	1 11/16

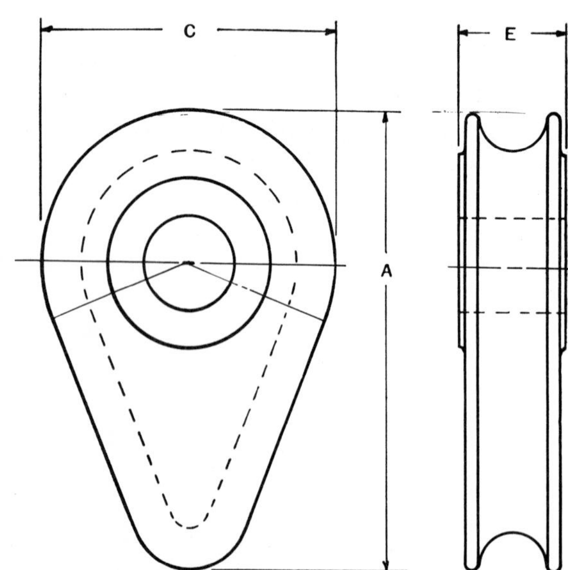

Fig. 13–48 **Prepare a two-view drawing of the solid cast-steel thimble with dimensions selected by the instructor. Select suitable radii for fillets and rounds. Note the points of tangency to determine the tapered shape. Dimension your drawing. *(United States Steel.)***

Cast-Steel Thimble

Wire Rope DIA	A	C	E
1/2	3 1/16	2 1/8	7/8
5/8	5 1/16	3 3/8	1 1/4
3/4	5 1/16	3 3/8	1 1/4
7/8	6 9/16	4 1/2	1 1/2
1	6 9/16	4 1/2	1 1/2

Chapter 14

Drafting Media and Reproduction

DRAFTING MEDIA

The graphic communications usually prepared in the drafting room today are made on paper, illustration board, cloth, or film. The basic *media* (materials to draw on; singular, *medium*) have been greatly improved in recent years. This is because industry has demanded an ever increasing standard of quality. These materials are generally used in the four levels of communication (Fig. 14–1) described in Chapter 1 as follows:

Level one: creative communication. Sketch paper and transparent paper are used for refining the ideas captured on the original layout. Grid paper is used for controlling proportions and developing schematic layout diagrams.

Level two: technical communication. Sketches are refined on tracing paper. Charts and diagrams are made on grid paper, or special graph paper is used for pictorial drawings.

Level three: market communication. Illustration paper or board is used for good ink and color renderings of pictorial or cutaway views. These highly developed artistic sketches explain the details of the project to the client.

Level four: construction or manufacturing communication. These drawings are made on high-grade tracing paper, vellum, cloth, or polyester film. Paper and film that are *ageproof* (will not discolor or fall apart) are required for permanent records.

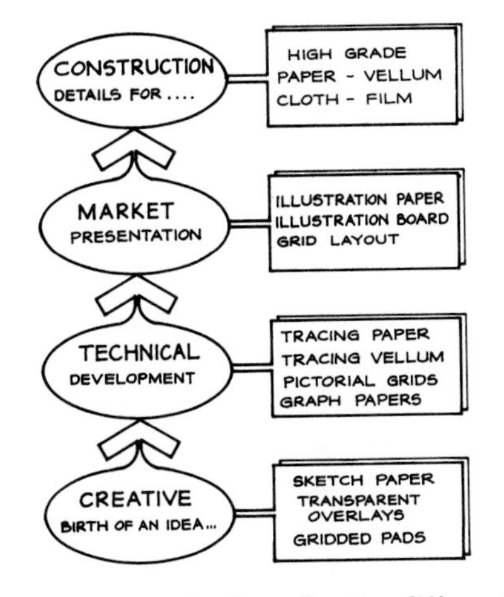

Fig. 14–1 **Typical media for different levels of communication.**

The engineering design team needs the basic media with their differing capabilities for specific uses.

Paper

A wide range of opaque and translucent paper is available. Designers consider opaque drawing paper more limited than translucent paper for reproduction. High-grade opaque paper can be used for black and white and for colored pictorial renderings that later can be photographed for reproduction. The opaque paper can be lined for drawing graphs or diagrams or for plotting mathematical solutions.

Fig. 14–2 **Translucent media for good reproduction.** ***(Teledyne Post.)***

Translucent tracing papers are used for developing original detail drawings. This medium must have a good surface quality. It must take pencil well and withstand repeated erasing and also a lot of handling by the engineering team (Fig. 14–2).

Translucent Tracing Papers

Natural tracing papers are made of rag. These are strong, but are not really transparent enough to be used all the time. The natural translucent paper without rag should be used only for sketching because it does not last very long. Advances have been made in paper-making to produce thin tracing paper that is reasonably transparent and retains pencil and ink well.

Transparentized Tracing Papers

These treated papers are sometimes called *vellum*. They are ideal for quick reproduction. This paper has rag in it and is thick. It is treated with resins to make it transparent. It makes sharp prints quickly. Good erasability, long life, and high strength are other qualities of transparentized papers. These papers are also considered ageproof. Nevertheless, they still have one disadvantage. They show fold marks.

Cloth

Cloth, or linen as it used to be called, is of higher quality than paper. Today it is an expensive medium that is excellent for pencil and ink. It is made of cotton fiber sized with starch. It is durable and erasable, and it reproduces well. Cloth used to expand or shrink a little, depending on the moisture in the air. Moisture-resistant cloths have been developed, making this medium competitive with others. Cloth is now stable enough to meet the needs of the engineering design team (Fig. 14–3).

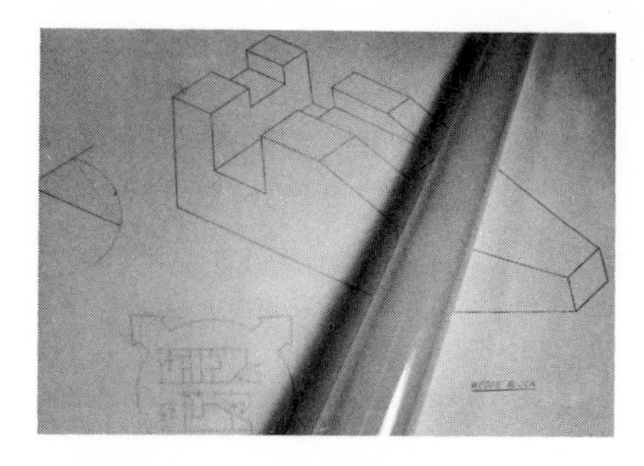

Fig. 14–3* Tracing cloth is translucent and highly durable. *(AM Bruning International.)

Film

The finest medium developed for the engineering team in recent years is polyester film. It will not shrink, and it is transparent, ageproof, and waterproof. Pencil or ink produce a clear image on the surface of film. That surface is stable even after repeated erasing. Film is classified by thickness and according to whether the surface is *matte* (not glossy) on one side or both. Special films such as acetate tracing films and adhesive transparent film are also available.

In part, better results are being obtained from the various media because better lead, ink, and erasers have become available. Each new or improved medium has a surface that is *workable* (erasable and reproducible).

Fade-out Grid and Lined Stock

High grades of tracing paper and film are available with a grid pattern. This pattern is very useful in drawing. It does not show up on copies (Fig. 14–4). The patterns generally have ten squares, eight squares, or four squares to the inch. There are also millimeter grids.

Fig. 14–4* Preprinted grid lines do not show on reproduction copies. *(Keuffel & Esser Co.)

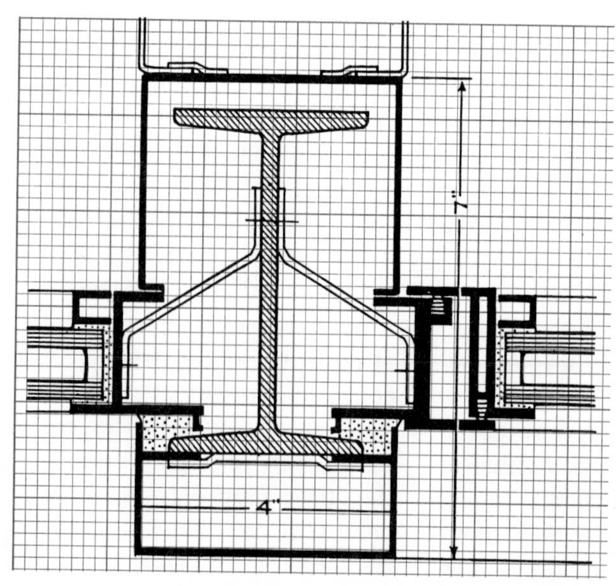

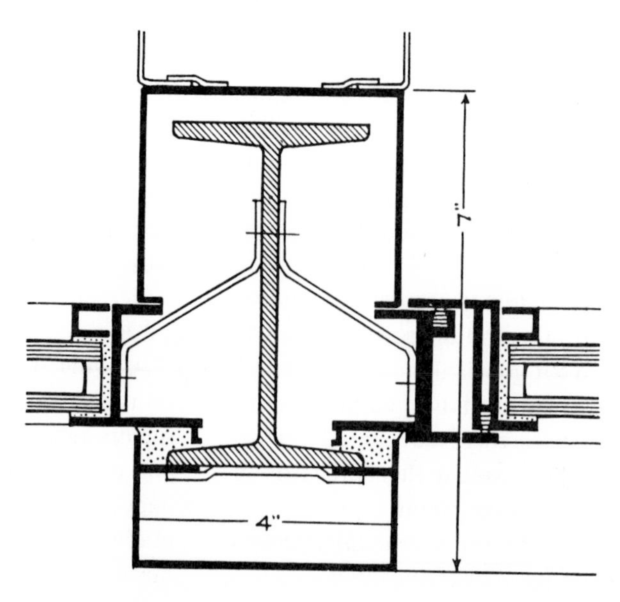

This paper, called cross-section paper, is available in rolls, sheets, and pads. There is generally a choice of colored lines, though light blue is most commonly used. Special kinds of cross-section paper are made for orthographic and isometric drawings.

Illustration Board

Artists' drawing and illustration boards are used in preparing drawings for the client's approval. The surface quality of these boards varies. The high surface (glossy) is smooth and is used for ink renderings. The medium and regular surfaces are preferred for pencil and color-wash drawings.

REPRODUCTION

Once a drawing is completed, many copies of it are often needed for various purposes. The design and engineering team need copies they can refer to or use to develop or revise drawings. The factory needs copies to make and inspect parts. Subcontractors, which are other companies that supply parts, also need copies to make and inspect those parts. The purchasing department needs copies to order standard parts not made in the shop. These might include bolts, nuts, washers, gaskets, and the like. If the item is *assembled* (put together) in a place other than the shop, additional copies are needed for assembly.

Fast, accurate, and economical methods of *reproduction* (making copies) have been available for many years. It is important to choose the right reproduction process for a particular purpose. In making a choice, you need to consider such factors as these:

1. Input of the originals: What is the size, weight, color, and opacity of the medium used?
2. Output quality: How readable must the copies be? Are they to be transparent or opaque? Must they be workable?
3. Size of reproduction: Are the copies to be enlarged, reduced, or the same size? Are they to be folded by machine?
4. Speed of reproduction: How many copies of each original are needed? What is the speed of the copying machine?
5. Color of reproduction: Is the process blue line, blueprint, or other?
6. Cost of reproduction: What are the costs of materials, operations, and overhead?

The equipment available today for copying engineering drawings includes blueprint, diazo, electrostatic, thermographic, and photographic (Fig. 14–5).

Blueprint

This is an inexpensive process that makes a copy the same size as the original on light-sensitive contact photographic paper. The copy is like a negative, in that it has white lines on a blue background. Blueprint is the oldest copying process. Prints have to be washed in water after being exposed to light. They are then treated with a potash solution to make the blue darker, and rewashed. Blueprinting is a continuous machine process. The only hand work that it involves is trimming the copies with shears.

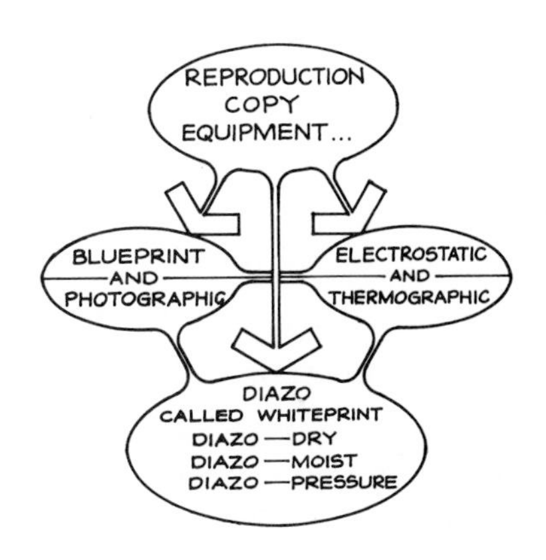

Fig. 14–5 **The kinds of reproduction processes.**

Diazo

Sometimes the diazo process is called *whiteprint.* The diazo print has the opposite (positive) look to the blueprint. That is, it has dark lines on a white background. It is made from a drawing having dark lines on a translucent medium. Diazo printing is available in a few colors, including blue line, black line, and red line. The paper, cloth, or film stock has a light-sensitive coating. This coating is developed by either a dry process or a semimoist process.

Diazo Dry Process

The print paper in the diazo dry process has a dye coating. The paper is passed through a light source while in contact with the original translucent drawing (Fig. 14–6). After this exposure, the dye coating on the print paper remains only where the light did not pass through the original drawing. The print paper is then passed through ammonia *vapor* (gas). The remaining dye is thereby developed into a line image. The original drawing itself does not pass through the developer.

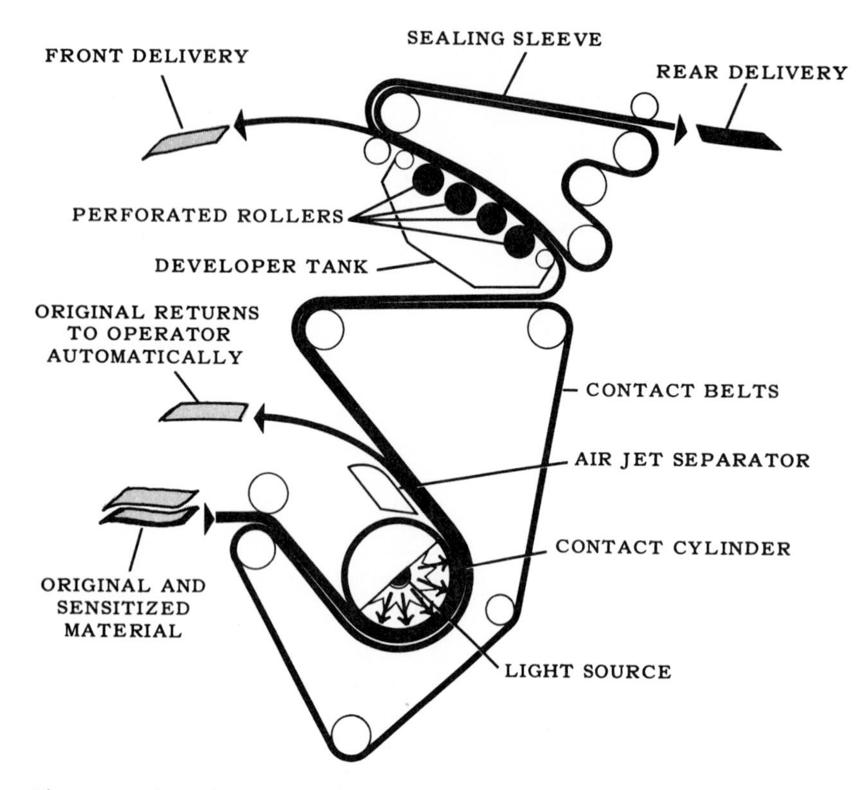

Fig. 14–6 **Diazo dry-process flow diagram.** ***(AM Bruning International.)***

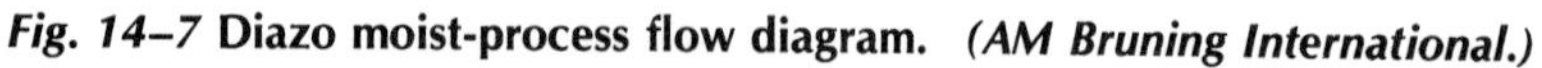

Fig. 14–7 **Diazo moist-process flow diagram.** ***(AM Bruning International.)***

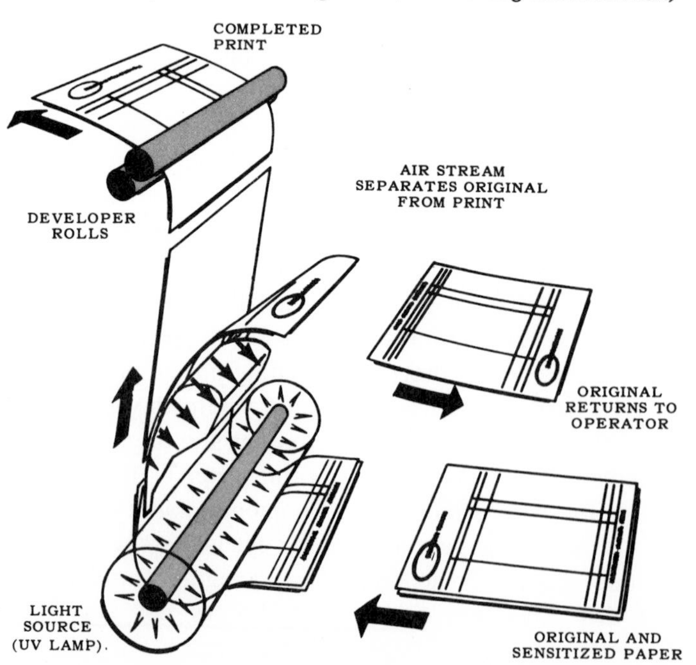

Diazo Moist Process

The print paper in the diazo moist process has a light-sensitive coating. This coating is developed with a liquid that only makes the paper damp. The drawing and print paper are exposed to light, while in contact, much as in the dry process. However, in the development phase, the print paper is fed through fine-grooved rollers where the developing liquid brings out the image. The original drawing does not pass through the developer (Fig. 14–7).

Diazo prints are made on paper, cloth, and film. Developing can be fast, medium, or slow, depending on the coating on the print material. The speed must be adjusted to make prints with good contrast. The cloth and films are available in matte finish and are workable on either side. These copies are of such high quality that they can be used in place of originals or as new originals for design changes.

A new pressure diazo process has been developed that makes dry, odorless whiteprints (Fig. 14–8).

Electrostatic Reproduction

Xerography is a dry process. It uses an electrostatic force to deposit dry powder. The powder makes positive, black images on copy paper (Fig. 14–9). The electrostatic process can produce copies that are enlarged, reduced, or the same size.

Xerographic machines have an electrostatically charged plate coated with the metal selenium. Where light strikes the charged surface, the charge *dissipates* (goes away). But the surface of the plate stays charged where no light falls. The blank spaces on an original cause the charge to dissipate. The lines and lettering on an original let the charge remain.

An image is produced on the surface of the exposed plate by a *toner,* or developer, of magnetic powder. This powder sticks to the places on the plate where the image on the original has let the charge remain. The powder image is passed from the plate surface to copy paper by an electrical *discharge* (jump). The powder is *fused* (joined) to the paper by heat.

Drawings can be sent from city to city by electronic transmission. Originals, such as engineering drawings, can be changed by machine into electronic signals. These signals can be *transmitted* (sent) to other places. There, the signals can be changed by machine back into xerographic copies of the original. This process can be controlled by a computer. The signals can be transmitted by microwave radio or by telephone wires.

The newest reproduction processes are electrostatic. These processes produce copies ranging from translucencies to photographs.

Thermographic Reproduction

Thermofax is a trade name of the company that first made thermographic copying machines. This kind of copying is based on the fact that a dark surface *absorbs* (soaks up) more heat than a light surface.

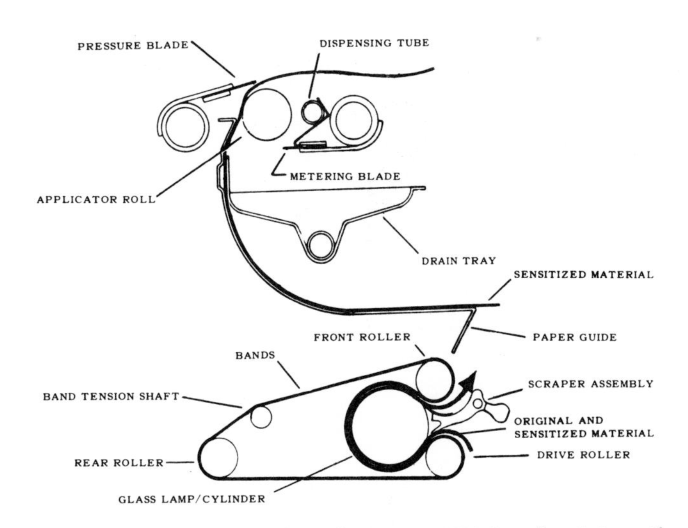

Fig. 14–8* Diazo pressure-process flow diagram.** ***(AM Bruning International.)

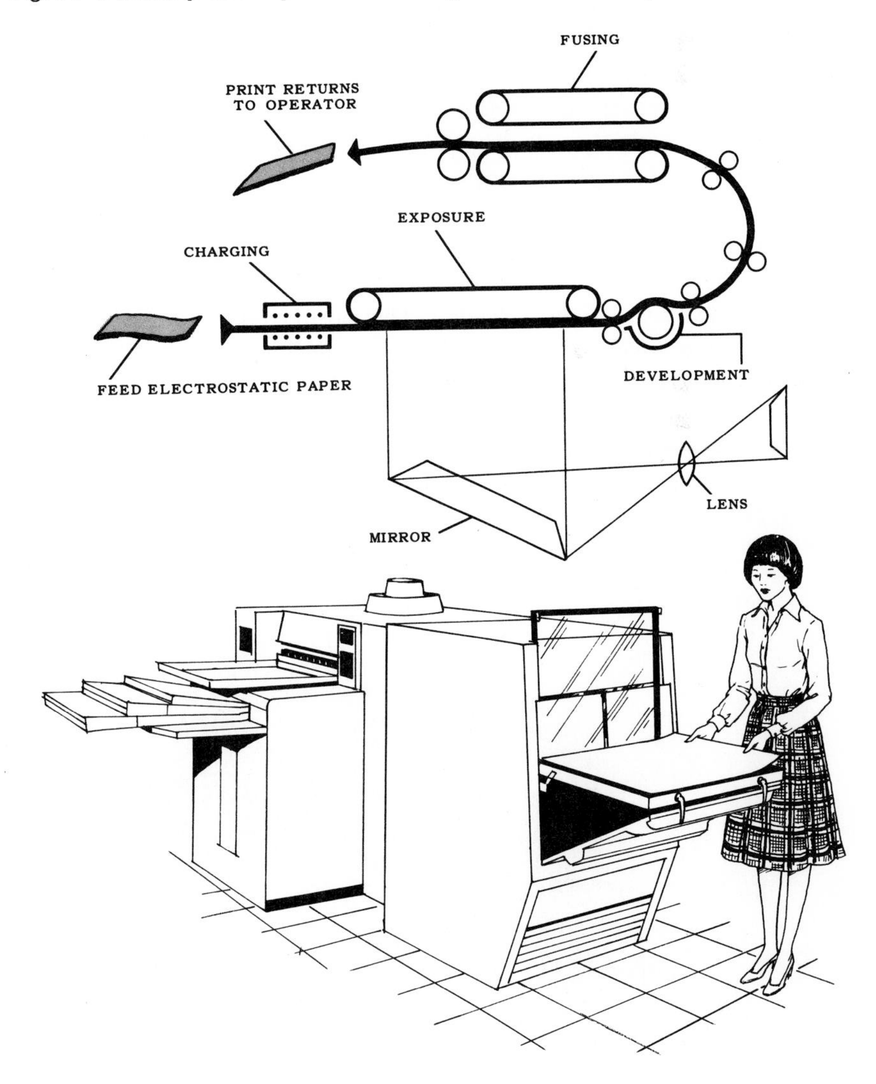

Fig. 14–9* Electrostatic reproduction machine and flow diagram.** ***(AM Bruning International.)

The print paper is affected by heat instead of light. Thus the original engineering drawing does not have to be on translucent paper. However, the markings on the original must be carbon or metallic. Such markings are heated by *infrared light,* a kind of light that you cannot see. There is no developer in this process. Since the thermographic process is dry, copies can be colored or translucent.

Photographic Reproduction

The companies that make film have found ways to help reproduce engineering drawings. Copy cameras (Fig. 14–10) make images on light-sensitive material, such as film that is coated with the chemical silver halide. The images are made permanent with chemical solutions and other ways of fixing an image. Ways of reducing and enlarging images are made easy. High-contrast films are used to give lines uniform widths and blackness (Fig. 14–11).

Film makes better prints than the commonly used diazo process. The film has a matte finish. Changes can be made with pen and ink or pencil. Film is most often used to make *intermediates* (copies used as originals), which you will learn about in Chapter 15.

Microfilm

This form of film is very good for keeping engineering records. Microfilm is excellent for handling engineering information. Records can be kept on different kinds of microfilm (Fig. 14–12). Roll film, jacketed film, micro-opaque microfilm, and aperture cards are all commonly used for engineering records. Roll film comes in widths of 16 mm, 35 mm, 70 mm, and 105 mm. There are scanning machines for studying and copying from the microfilm.

Jacketed film is best for records that are used and changed often. The positive micro-opaque film is an opaque card stock. It is used with readers that can magnify it.

Microfiche is a sheet of transparent film that has many very small images in rows. Some of the sizes used for storage are 75 × 125 mm (3 × 5 in.), 100 × 150 mm (4 × 6 in.), and 125 × 200 mm (5 × 8 in.). Copies can be printed after the right image is chosen on a screen reader. Copies can be made in different sizes.

Aperture Card

An aperture card is a standard-size data file card. It has a rectangular slot that can hold a framed microfilm (Fig. 14–13). The most common card is 83 × 187 mm (3¼ × 7⅜ in.). Data is punched in the card except in the area near the film. This data card can give engineering information about the film and make it easy to store. Cards can be sorted, filed, and retrieved by machine. Images can be reproduced in the size needed.

DRAFTING MATERIALS AND SYSTEMS

On the engineering team, it is the drafter who chooses the right drawing medium for the level of commu-

Fig. 14–10 Copy camera for photographic reproduction. ***(Keuffel & Esser.)***

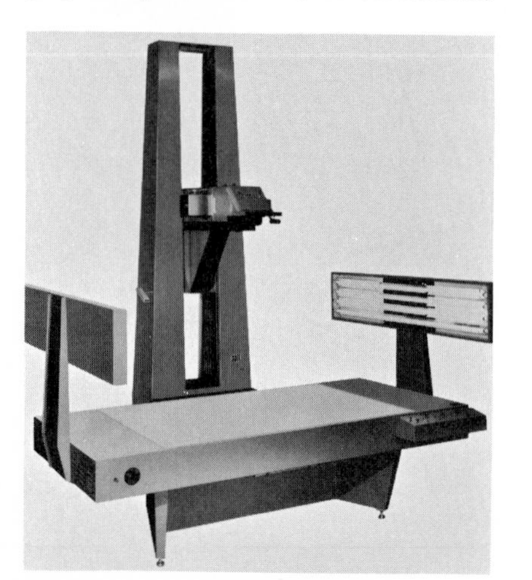

Fig. 14–11 This rendering has been photographically reproduced. While the size has been reduced, clarity of line has not been lost. ***(E. Zavada.)***

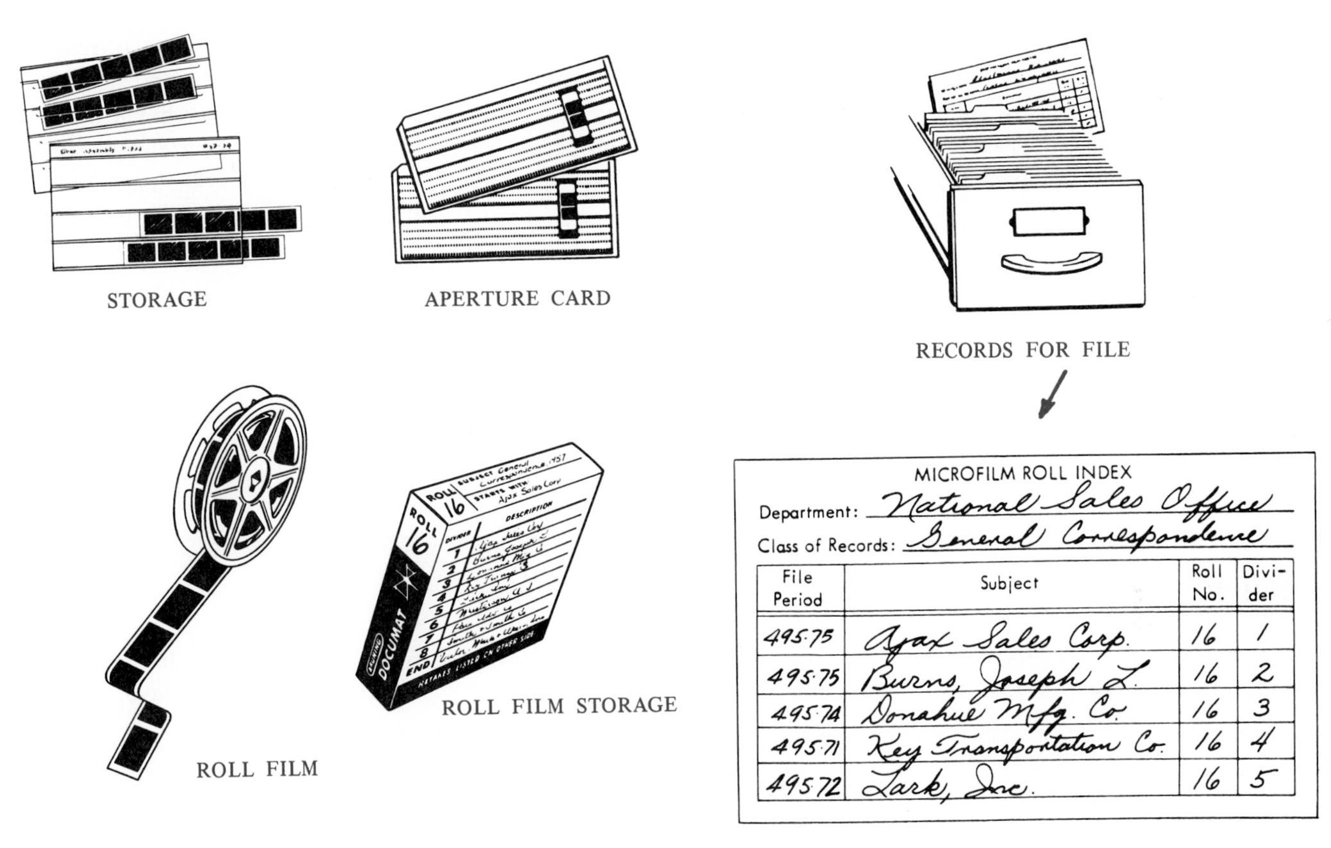

Fig. 14–12 **A microfilm system includes these elements.** ***(AM Bruning International.)***

Fig. 14–13 **A microfilm aperture card.** ***(Keuffel & Esser Co.)***

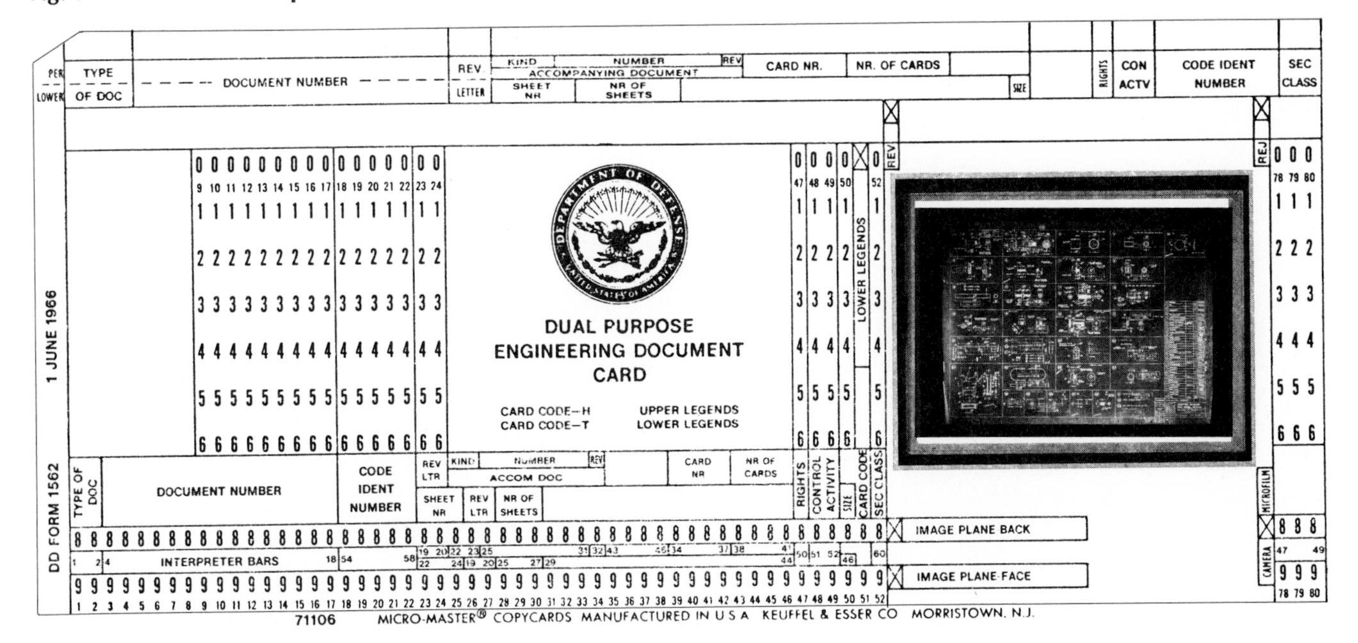

nication needed to do the job. The technologies of the older media have been improved. New media have been invented. These advances have usually involved complicated new reproduction equipment that makes many copies. These machines have greatly lessened the amount of clerical work that takes place in the modern engineering office. All these changes give drafters wider and better choices than ever. Thus, producing graphic communications has been made easier. Rapid change is likely to continue in this area. Drafters will always have to be learning how to use new tools of their trade.

Review

1. Name the best media for each of the four levels of communication.

2. What are the best uses for opaque paper?

3. Name five good qualities of vellum and one disadvantage for its use in drafting.

4. Tracing cloth (has or doesn't have) some qualities that are better than those of paper.

5. Name some advantages of using drafting film.

6. List several things that affect the selection of a reproduction process.

7. What is the difference between the copy made on a blueprint machine and the one made on a diazo machine?

8. List several capabilities of electrostatic copiers that other reproduction processes do not have.

9. Microfilm is not a good storage system for original drawings because the image is too small to see properly. True or false?

10. List reasons why a large company might put a microfilm storage and retrieval system in its engineering department.

Chapter 15

Systems for Graphic Communication

ORGANIZING GRAPHIC COMMUNICATION

All the drawings and reports done by the engineering team are usually processed in the drafting room. Two systems are used in processing design projects: written and *graphic* (drawn).

System One: Written Communication

Even graphic forms of communication are generally controlled by a written communication system. This system is made up of technical reports. There are five kinds of technical reports:

1. *Design report,* which includes a statement of the needs of the client and an early *analysis* (study) of the project by the designers.
2. *Technical report,* which is a design analysis of all the things that go into researching, designing, and developing the project.
3. *Market report* (also called *client report*), which is a formal report that includes recommendations. This report comes out before it is decided whether or not to go ahead with manufacturing or construction on the project.
4. *Contract report,* which includes all the details of manufacturing or construction. This report is done after it has been agreed to start work on the project.
5. *Progress report,* which is made once in a while to check on the quality of the project and take another look at the specifications in the contract. This report usually appears at certain chosen dates or when certain stages of the project have been finished.

Fig. 15–1 **System one: written communication.**

These reports are necessary to make the work of a design team on a project a success. They add to and support the graphic communication needed to design, start, finish, and sell a project (Fig. 15–1).

System Two: Graphic Communication

The four levels of graphic communication described in Chapter 1 are used over and over again as design teams work on design projects (Fig. 15–2).

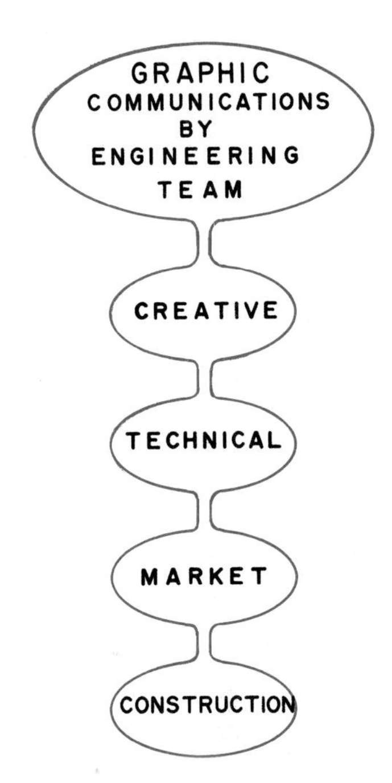

Fig. 15–2 **System two: graphic communication.**

Level one: creative communication. This is also called *the birth of an idea.* It usually takes the form of the designer's sketches showing new ideas to be studied.

Level two: technical communication. Several technical people put together what they know about materials and ways of doing things to suggest how to solve problems. These solutions are usually shown *graphically* (in drawings) because they are often difficult to put into words. Technical graphic communications include sketches, schematic diagrams, and graphs.

Level three: market communication. The technical study is refined and illustrated so the client can judge the form, function, and style of the design. Pictorials, cutaways, and exploded views can help the client make a decision about the design.

Level four: construction communication. Detailed drawings are made for contracts, estimating, construction, or manufacturing. They must be full and accurate. Written specifications often go with detailed drawings.

Which levels of graphic communication are produced by the design team depends partly on how the team is put together. Large companies organize design teams to solve major problems. Their solutions help improve our technology and raise our standard of living. All the consumer products, as well as many services, result from the work of the engineering design team. The scientist, engineer, technician, designer, and drafter are members of this team.

THE ENGINEERING TEAM

Two good examples of companies that need engineering teams would be a large architectural design firm and an automobile maker. The managers of large firms must use the design team to meet the needs of clients or future customers. The work of the team must be organized so that problems can be solved as quickly and cheaply as possible. From the early study to the final working drawing, all work is done according to a set schedule. This is true whether the product is a skyscraper or a sports car (Fig. 15–3). The head of a design project may make a flowchart to help in managing the design process.

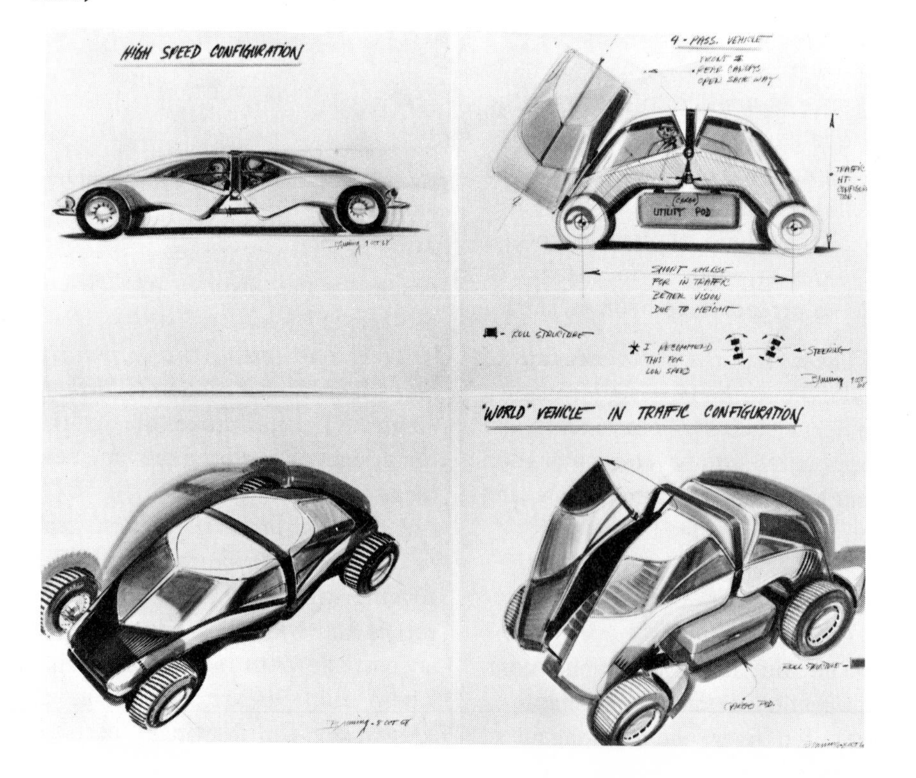

Fig. 15–3 A preliminary graphic study. ***(General Motors Corporation Design Staff.)***

ORGANIZING THE LEVELS OF DRAFTING WORK

The organization chart in Fig. 15–4 shows a set way of processing work through the drafting room. At all levels, the design process rests on the principles of mechanical drawing. This basic technical skill is used in every design project. The drafter can help the design team at all four levels of communication.

Level one: The layout drafter helps the designer by working closely with the design sketches, changing them into mechanical drawings.

Level two: The experienced senior drafter helps study and judge sketches, layouts, and diagrams as design decisions and recommendations are made.

Level three: The drafter-illustrator generally assists in the presentation studies. This drafter specializes in making pictorial views and renderings for the client to study and approve.

Level four: The senior and junior detail drafters prepare the finished construction and manufacturing drawings based on the layout work. They are supervised by the chief drafter. The final details are closely checked, often

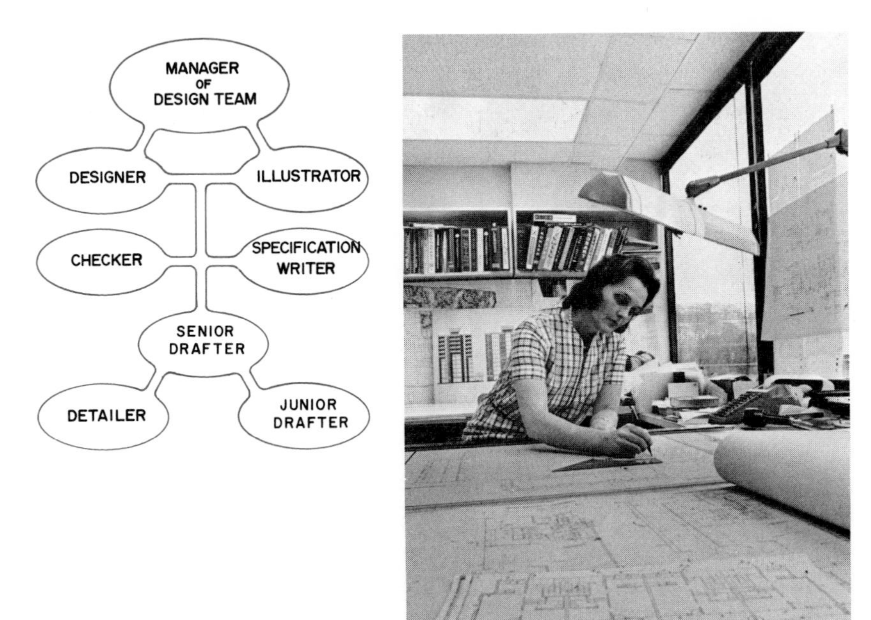

Fig. 15–4 **The design team manager and levels of the design team.** *(Mimi Forsyth/Monkmeyer.)*

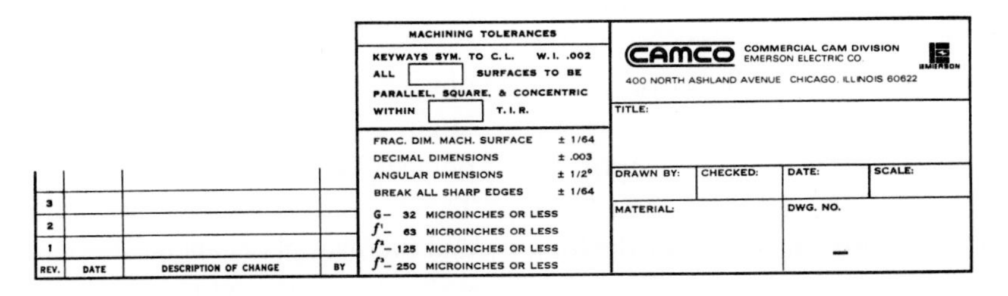

Fig. 15–5 **The basic title block.** *(Camco.)*

by a senior detailer who has the job of drawing checker.

SYSTEM FOR DRAFTING PROCEDURES

As the design team develops a project that uses drafting services, a record is kept of the progress under a contract number or job number. This number must be used on all reports and drawings. The number is a part of the information given in the title block of a drawing.

TITLE BLOCKS

The basic title block used in industrial drawings is shown in Fig. 15–5. The drawing-number block and the drawing title are needed to catalog and cross-index the drawing. The blocks giving the date, scale, signatures, initials, tolerance information, material specification, and revisions are needed to read the drawing and manufacture the product. All drawings, regardless of size, must have a title block.

REVISION BLOCK

Changes on a drawing are recorded in the revision block. It is usually found in the lower right-hand corner of the drawing sheet. Letters or numbers are used to identify the revised part of the drawing. These letters or numbers are needed to find out what change was made and who made it.

BASIC SHEET SIZES

Title blocks and revision blocks may differ a little in size and in where they are put. Their exact size and place depend on how big the drawing sheet is. Design teams in all industries have been using standard-size sheets. These come in multiples of 8½″ by 11″ and 9″ by 12″. The sizes are shown in Fig. 15–6. Both size ranges are identified by the letters A, B, C, D, and E.

The ISO "A" series is developed from a base sheet of paper with an area of 1 square meter (1 m^2). The ratio of the sides of all the sheets of paper derived from the base is $1{:}\sqrt{2}$. Therefore, the size of the base sheet is 841 mm × 1189 mm. Sizes are designated with the letter A followed by a number. This number tells how many times the base sheet has been divided. The number A1 means that the A0 sheet has been halved once. The number A2 means that it has been halved twice (Fig. 15–7). If sheets larger than A0 are required, they are designated by a number in front of the A0 designation. The next size larger than A0 would be designated 2A0, or 1189 mm × 1682 mm. The 4A0 sheet measures 1682 mm × 2375 mm.

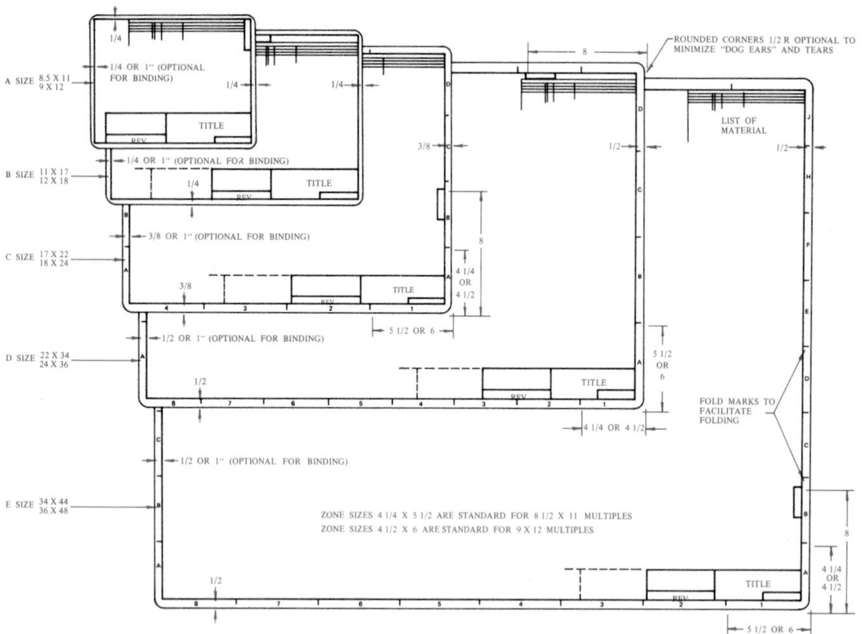

Fig. 15–6* Basic drawing sheet sizes. *(ANSI.)

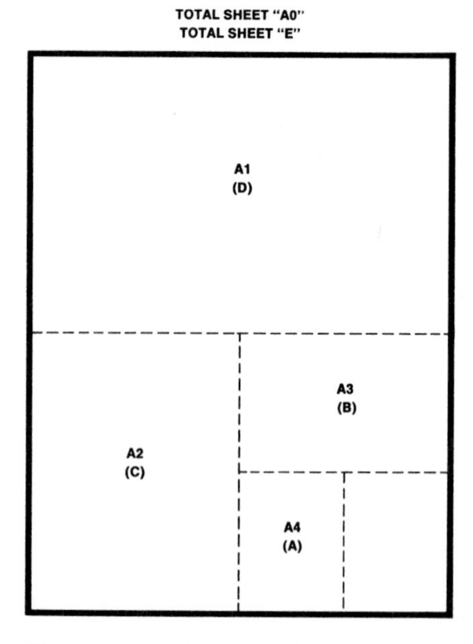

***Fig. 15–7* Metric paper sizes.**

ISO Series	"A" Paper Sizes (in millimeters)
A0	841 × 1189
A1	594 × 841
A2	420 × 594
A3	297 × 420
A4	210 × 297
A5	148 × 210
A6	105 × 148
A7	74 × 105
A8	52 × 74
A9	37 × 52
A10	26 × 37

CHECKING REPORT

Checking working drawings for accuracy is an important part of the drafting procedures system. The checking is usually done by an experienced detailer who reports to the chief drafter. The checker uses a print to look over the work, marking errors for correction or marking approval. This job is important because all the detail work must be accurate and use standard drafting practices.

PROGRESS REPORT

Drafting departments record the progress of the team detailing a project. Each drafter may have to report progress either daily or weekly. The report will record the job contract number and the number of hours spent working on each drawing. Engineering departments are generally given a certain time to finish a project. Therefore, the reports are important for keeping the job within the set time and for dividing the drawing work among the drafters.

DRAFTING STANDARDS

A major industry or a large company will generally set up a system for processing construction or manufacturing drawings. Some of the large manufacturing companies publish standard drafting practices. These help their own employees and their suppliers and customers as well. International Harvester, Caterpillar Tractor, and International Business Machines are just a few of the major companies that have published standards.

Early in the twentieth century, five engineering societies in the United States formed an organization to coordinate the development of national standards for engineering practices. The American Standards Association (ASA) has served United States industry for many years. In 1966, the ASA was reorganized as the United States of America Standards Institute and used the symbol USASI. This leading organization has been called the American National Standards Institute, Incoporated (ANSI), since an expansion in 1969. The institute has approved over 4000 standards. There are 16 standards and 2 supplements that make up the *American National Standard Drafting Manual.* Each of the standards was developed by national committees to give designers and drafters preferred drafting practices. The Y14 series of standards is of particular interest to designers, architects, and drafters. A chart listing series titles appears below.

The ANSI drafting standards are needed because of the many kinds of industrial work done across the country. Managements consider standards important in making communications the best possible. The Society for Automotive Engineers (SAE) has also set standards that are useful to the design and drafting profession. The great number of drawings done in the drafting room each day are not only standardized according to national practice, but they are also stored, filed, or controlled according to specific industrial standards.

CARE AND CONTROL OF DRAWINGS

In most companies, a system has been developed for cataloging and storing original drawings. Large companies keep the original drawings for a contract together under a contract number or job number. This makes the drawings easily available for copying or revising.

Some companies have a *closed catalog system* as part of the standards department. In this system, the number of people who can handle original drawings is tightly controlled. One person is responsible for keeping the drawings safe. That person may have a staff who file and copy the documents. Other companies favor an *open file system* within the engineering department. There, the original drawings are always available to anyone in the department. The open file system is generally not good because easy access to the original drawings can lead to mishandling. A company with such a system might have to go to the needless expense of having to retrace or redraw originals.

Some rules for storing originals efficiently and safely follow:

1. Develop a closed catalog system that requires expert handling.
2. If original drawings have to be revised a lot, handle them carefully. Work from a copy if you can. It is best to use an erasable intermediate copy. This is a print of the original drawing made on an erasable drawing paper. This paper is different from ordinary blueprint paper in that other prints can be made from it. (Prints, of course, cannot be made from other blueprints.)
3. The original drawing should be used only for making prints. It

AMERICAN NATIONAL STANDARD DRAFTING MANUAL

Title of Standard	Y14 Number
Drawing Sheet Size and Format..........................	Y14.1 —1975
Line Conventions and Lettering..........................	Y14.2M —1979
Multiview and Sectional-View Drawings.................	Y14.3 —1975
Pictorial Drawing..	Y14.4 —1957
Dimensioning and Tolerancing...........................	Y14.5 —1973
Screw-Thread Representation............................	Y14.6 —1978
Gears, Splines, and Serrations..........................	Y14.7 —1958
Gear-Drawing Standards—Part 1........................	Y14.7.1 —1971
Forgings..	Y14.9 —1958
Metal Stampings...	Y14.10 —1959
Plastics...	Y14.11 —1958
Mechanical Assemblies..................................	Y14.14 —1961
Electrical and Electronics Diagrams, including supplements Y14.15a—1970 (R1973) and Y14.15b—1973....	Y14.15 —1966 (R1973)
Fluid Power Diagrams..................................	Y14.17 —1966 (R1974)
Dictionary of Terms for Computer-Aided Preparation of Product Definition Data..............................	Y14.26 —1975
Chassis Frames—Passenger Car and Light Truck—Ground Vehicle Practices....................................	Y14.32.1—1974

should never be used for desk reference.

4. Original drawings should be stored flat. They should not be folded because creases will show up as lines on blueprints.
5. Cleanliness in the drafting room is essential. Vellum and paper stock can absorb moisture and thereby pick up stains. Therefore, clean, dry hands and clean working surfaces are necessary.

A good drafting-room system provides easy, but controlled, access to the drafting files.

Fig. 15–8 **Vertical file storage with a capacity of 3000 to 5000 drawings.** ***(Ulrick Plan File.)***

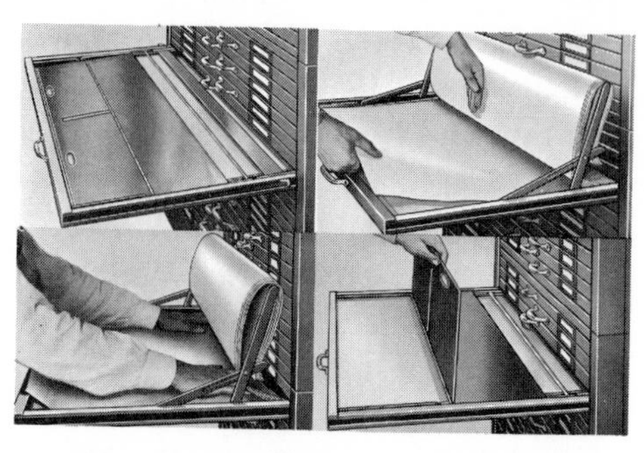

Fig. 15–9 **Horizontal file storage.** ***(AM Bruning International.)***

TYPICAL DRAWING STORAGE

Original drawings are stored in vertical files, as in Fig. 15–8. They can also be stored in horizontal file drawers, as in Fig. 15–9. The vertical files for drawings are as convenient as ordinary files for letters. All important documents should be flat and smooth. They should be protected from water and dust. Many drawings of sizes A (8½″ × 11″), B (12″ × 18″), C (18″ × 24″), as well as small ISO sheets, can be conveniently filed in vertical files.

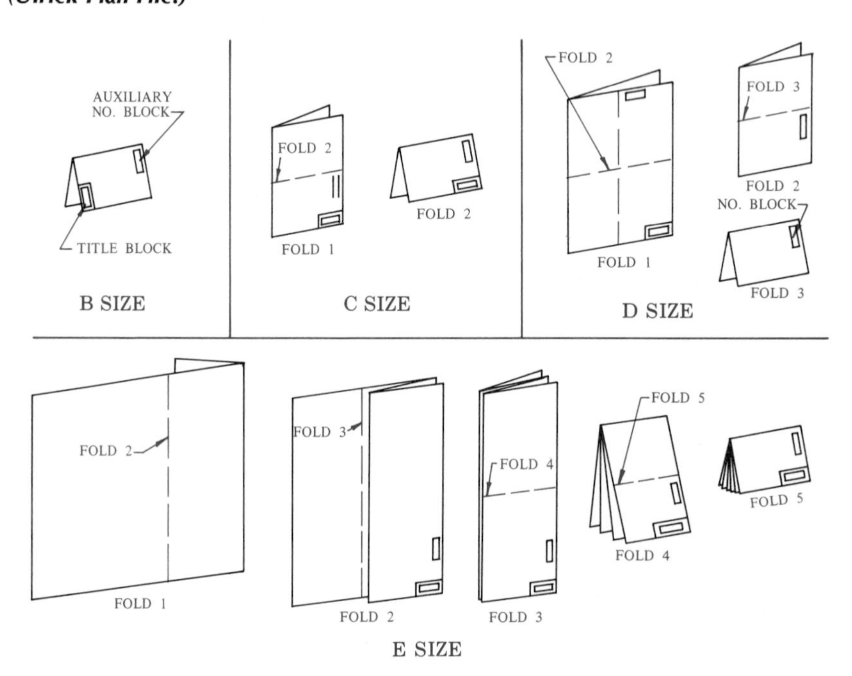

Fig. 15–10 **ANSI print-folding system.**

Folding the Prints

Original drawings should not be folded. However, prints filed for reference are folded. The folded print is filed with the creased edge on top. This is done so that other prints will not be caught in an open fold. Figure 15–10 shows the ANSI system for folding B-, C-, D-, and E-size prints. A number of systems are used to store prints of current work in the drafting room so they can be referred to quickly. Figure 15–11 shows a compact quick-access filing system, sometimes called a roll filing system. This system can be moved around.

Storing folded prints can be avoided by microfilming them. Drawings must be of a high quality to be good for microfilming. The many reference details and supply catalogs for parts and equipment can also be microfilmed. In large companies, reader-printers like the one in Fig. 15–12 are used to retrieve design information.

Microfilming and Data Control

Storing drawings usually takes up a lot of space and is difficult and expensive. Microfilming was brought into the drafting room to make storing and copying drawings easier. The National Microfilm Association (NMA) sets standards for

Fig. 15–11 **A quick-access filing system.** ***(Plan Hold Corp.)***

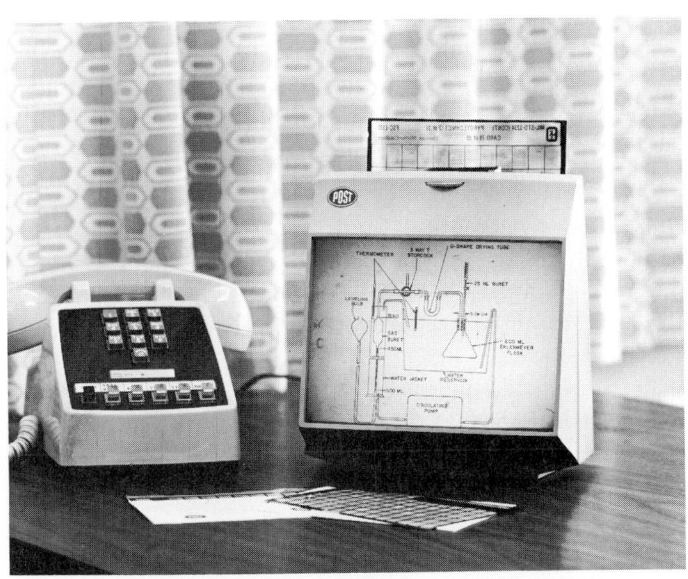

Fig. 15–12 **Microfilm reader.** ***(Teledyne Post.)***

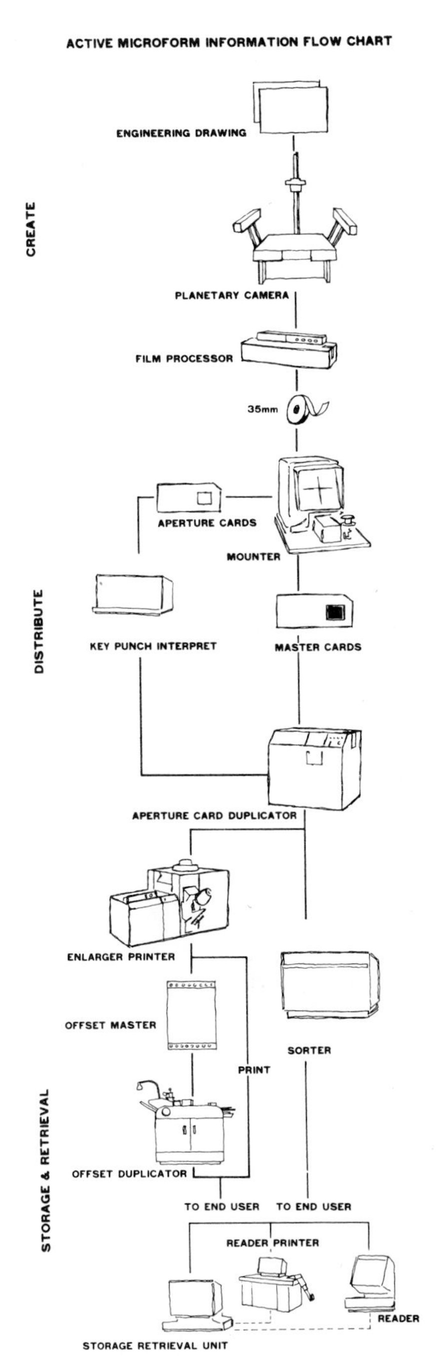

Fig. 15–13 **The microfilm process.** ***(AM Bruning International.)***

microfilming in the United States. This association has worked with the American National Standards Institute in setting its microfilming standards.

MICROFILMING PROCESS

A number of photographic systems are used in microfilming. However, 16mm and 35mm are the most popular film sizes. Film also comes in 70mm and 105mm. The film must be high contrast, high resolution, and fine-grained to be good for copying drawings. Only a small storage space is needed for microfilm. The data card that holds the microfilm can be taken out of storage by machine. The card may be used for printing a copy or for studying in a microfilm reader-printer. Figure 15–13 shows the flow of the microfilm process. Copies from microfilm can be full size, half size, or original intermediates depending on the film or paper used. Half sizes are popular because they are easy to handle and cut paper costs in half.

MICROFILMING TECHNIQUES FOR QUALITY REPRODUCTION

Microfilm images and copies from microfilm can be good only if the drafting in the original drawing is good. Microfilming standards have helped make drafting better. Good drafting has line work that is clear and clean. It has well-selected and well-positioned views. It has dimensions that can be read easily. The National Microfilm Association published Monograph No. 3, *Modern Drafting Techniques for Quality Microreproductions,* which has guidelines for preparing drawings for microfilming.

The major concerns in preparing high-quality original drawings for microfilming are listed as follows:

1. Selecting standard-size drawing media of high-quality vellum or film with good surface texture.
2. Selecting opaque pencil leads or ink that are right for the medium, that give good line contrasts, and yet that can be erased easily.
3. Using lettering that can be easily read. An example is Microfont, a letter style set by NMA. The rules for height and spacing of letters are very important, as shown in Fig. 15–14.
4. Making the drafting easy to use by spacing the views well and showing only as much detail as is needed.

These microfilming standards are important to industrial progress. Good camera work is wasted on poor drafting. Microfilm copies cannot be any better than the original drawings. Several systems have been developed that make secondary originals cheaply.

SYSTEMS OF PREPARING INTERMEDIATES

An *intermediate* is a copy of a drawing that is used in place of the original. It is often called a *secondary original.* The designer and drafter can change the intermediate while leaving the original alone. Using intermediates avoids tracing from, or redrawing on, originals when changes are needed. The seven methods that follow are only some of the ways that time can be saved on professional drafting jobs.

1. *Scissors drafting.* The unwanted sections of an intermediate are removed with scissors, knife, or razor blade (Fig. 15–15). This is called *editing.* A new print or new intermediate is then made. More changes can then be drawn on this new original.
2. *Correction fluids.* Unwanted data can be removed from an intermediate by putting on correction fluid. New data can then be drawn in the cleaned area. Prints are made from the corrected intermediates (Fig. 15–16).
3. *Erasable intermediates.* This kind of copy can be erased with ease. Thus, corrections can be made quickly without fluid. Redrawing can be done equally well with pencil or ink. The original remains the same (Fig. 15–17).
4. *Block-out, or masking.* This system is best when there is little time for reworking a drawing. The area to be blocked out is covered with opaque tape. A print is then made in which the

Fig. 15–14 Microfont, a lettering style developed by NMA. *(National Microfilm Association.)*

Unit	Comments	Not This	Possible Error
B	Upper part small, but not too small.	B	8
H	Bar above center line.	H	—
M	Center portion extends below center line. Slight slant on uprights.	M or M	—
S	Lower part large, ends open. Slight angle on center bar.	S	8
T	Horizontal bar shall be full width of letter "E."	T	7
U	Full width.	V	V
V	Sharp point.	V	U
Z	All lines straight.	Z	2
1	Full height and heavy enough to be identified. No serifs.	1	7
2	Upper section curved with open hook. Bottom line straight.	2	8 or Z
3	Upper portion same as lower. Never flat on top.	3 or 3	8 or 5
4	Body open, ends extended.	4	7 or 9
5	Body large, curve dropped to keep large opening. Top fairly wide.	5 or 5 or 5	6 or 3 or S
6	Large body, stem curved and open.	6	8
8	Lower part larger than upper, full and round to avoid blur.	8 or 8	B
9	Large body, stem curved but open.	9	8

MICROFONT
ABCDEFGHIJKLMNOPQR
STUVWXYZ 1234567890

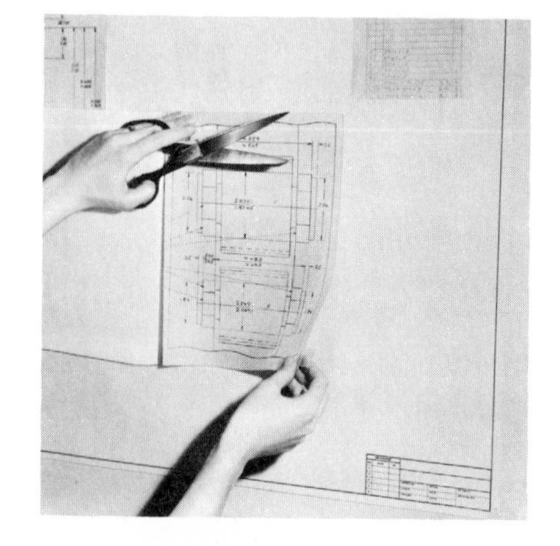

Fig. 15–15 Scissors drafting is used in the intermediate systems. *(Teledyne Post.)*

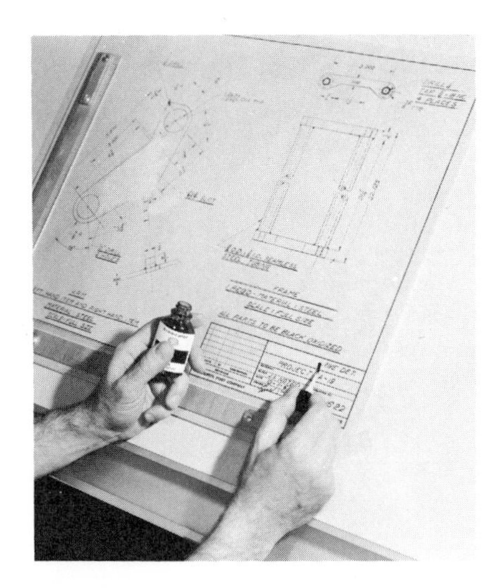

Fig. 15–16 **Using correction fluids on an intermediate.** ***(Teledyne Post.)***

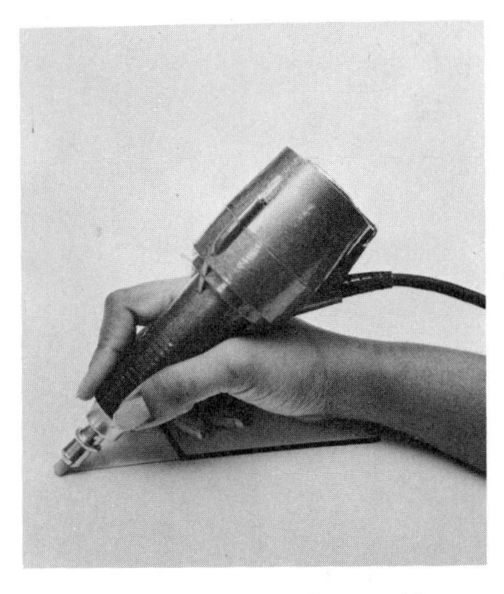

Fig. 15–17 **Using an erasing machine on an intermediate.** ***(R. J. Capece.)***

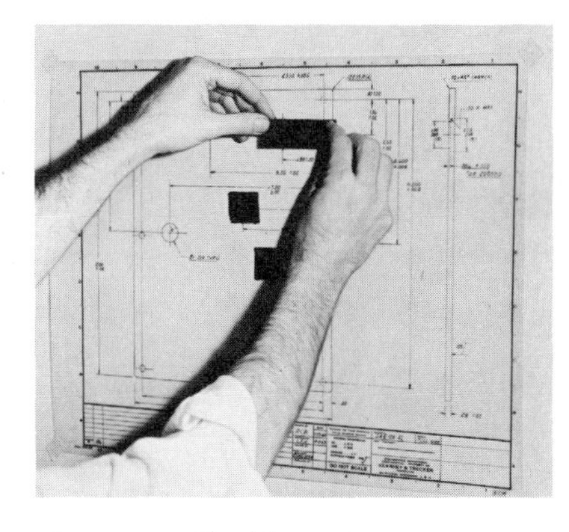

Fig. 15–18 **Blocking out an area on an intermediate.** ***(Teledyne Post.)***

masked areas produce blank spots. Changes are then entered in these areas (Fig. 15–18).

5. *Transparent tape.* Added data can be put on paper, cloth, or film and then taped in place on an intermediate. Dimensions, notes, or other data symbols can be added in this way. The notes can be typewritten on transparent press-on material, which is then put in place (Fig. 15–19).
6. *Composite grouped intermediates.* When several drawings are to be combined, a composite grouping should be cut from intermediates and taped in place. A final intermediate is then made (Fig. 15–20).
7. *Composite overlays.* Translucent original drawings can be combined into a composite intermediate by placing the originals in the desired places. The composite is then photographed. Photo positives or negatives can then be drawn on. The intermediate master is rerun for a second copy.

The drafter often must decide which intermediate system gives the best-quality secondary original for the least amount of time and materials. The quality of the copies made by the system chosen must meet the needs of the job. Along with intermediate systems, the computer has become a working tool of the design team.

COMPUTER-AIDED GRAPHICS SYSTEMS

The automated drafting process is a plotting system that operates at a high speed. It makes drawings from data in a computer (Fig. 15–21). The computer can produce data from punched cards, from punched tape, or from an *alphanumeric* (letters-and-numbers) keyboard. Computer-aided drafting (CAD) can change *XY* coordinates into accurate multiview drawings. Using mathematical formulas and stored programs, CAD can also

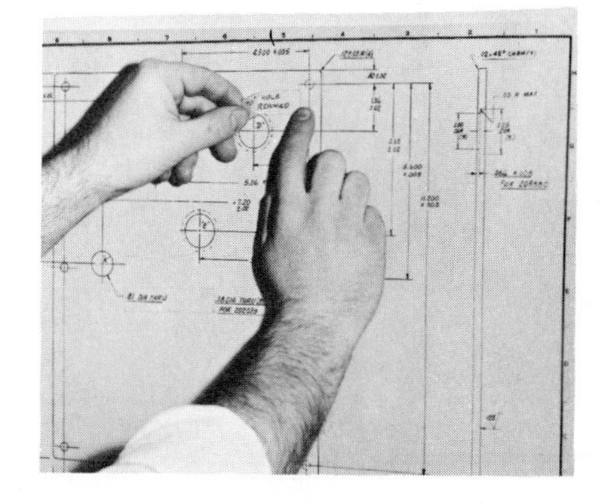

Fig. 15–19 **Using transparent tape on an intermediate.** ***(Teledyne Post.)***

Fig. 15–20 **Making a composite intermediate.** ***(Teledyne Post.)***

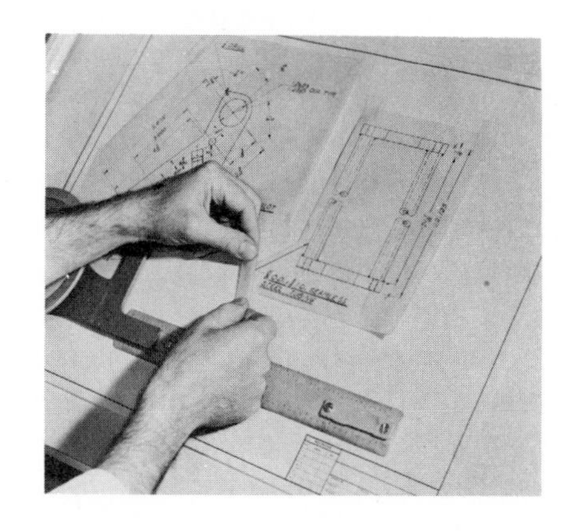

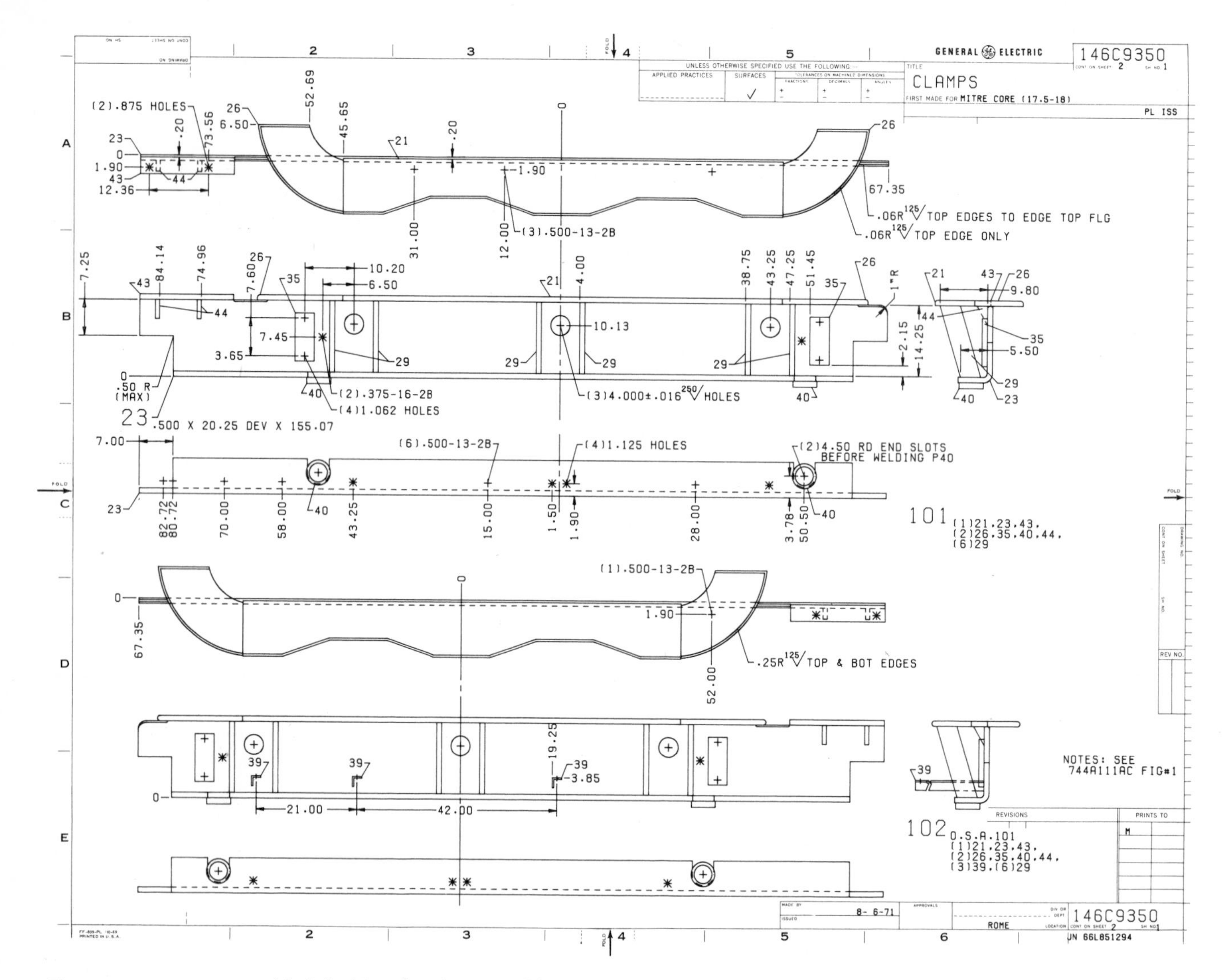

Fig. 15–21 **A computer-aided drafting drawing.** *(California Computer Products, Inc.)*

make pictorial drawings with one-, two-, or three-point perspectives to any height or angle. Computer-aided drafting uses a programmed drafting language as input. This drafting language is made up of letters and numbers. These letters and numbers are put into the system to produce data in numbers only, or *digital* data. (Most computers use a *binary,* or two-part, number system based on 0 and 1.) The drawing is made by a digital plotter (Fig. 15–22). The plotter may be either a drum or a flatbed type. The drawings can be plotted on tracing paper, graph paper, vellum, or film. The drum plotter is a precision instrument. It has a turning drum and a moving *stylus* (drawing head) on a crossbar. The flatbed plotter has a moving beam on a carriage above the flatbed. The plotters can take ball-point pens, capillary pens, or scribers. Figure 15–23 shows a pictorial drawing made on an automated drafting machine.

Drafters will not lose work because of the progress in computer-aided drafting systems. However, they may need a good education in science and mathematics to keep up with the changes. Figure 15–24 shows the cathode-ray tube (CRT) and the graphic plotting of images on the screen.

The systems discussed in this chapter are helping the design team in its work. As new systems constantly appear, each member of the design team will be affected by the technological change.

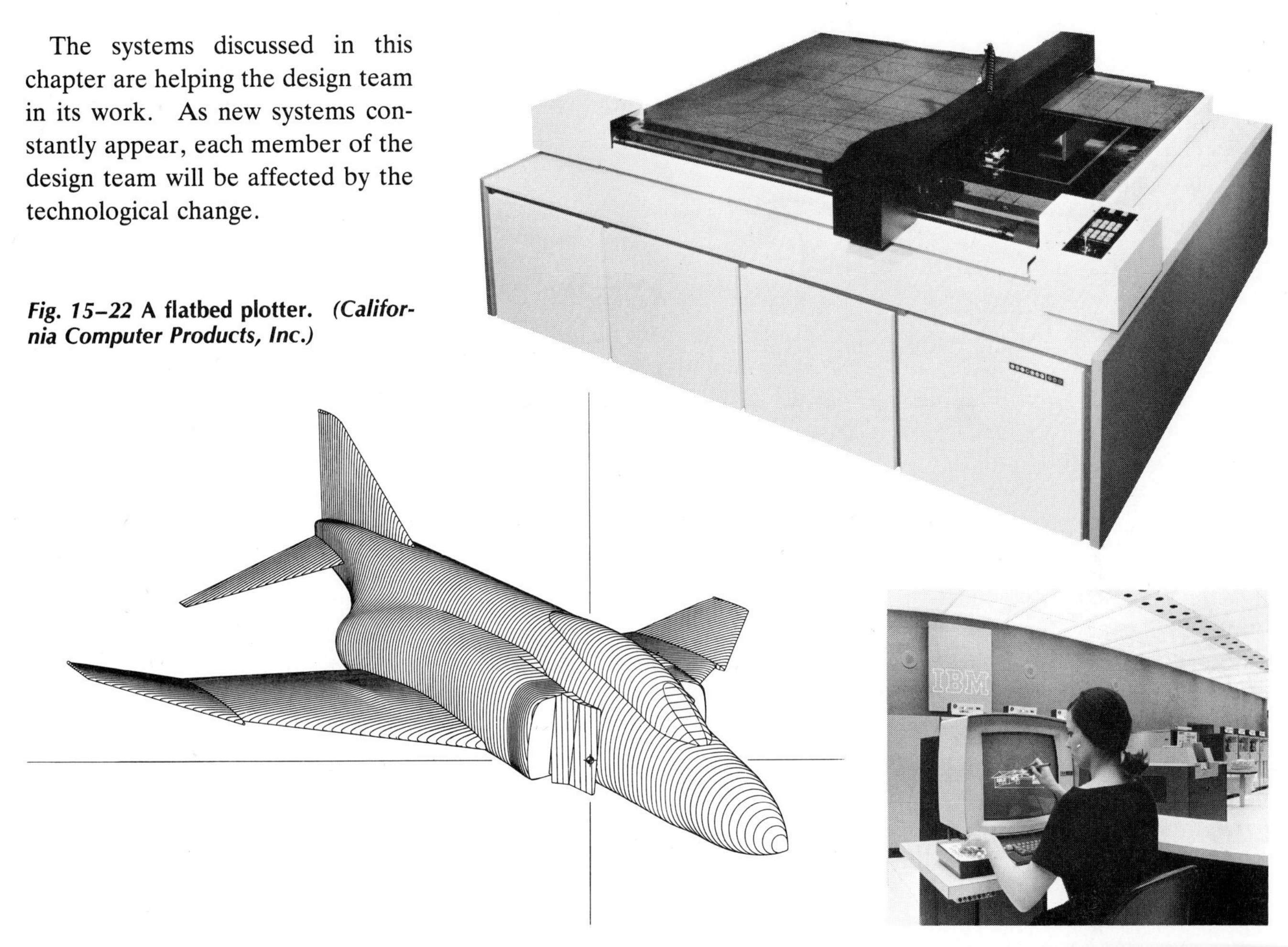

Fig. 15–22 **A flatbed plotter.** ***(California Computer Products, Inc.)***

Fig. 15–23 **A pictorial drawing done through computer-aided drafting.** ***(McDonnell Douglas Corp.)***

Fig. 15–24 **A cathode-ray tube.** ***(IBM.)***

Review

1. Why are drafting standards important?

2. If a drafter's job is to make drawings, why should he or she be required to provide information for reports?

3. List several reasons why a drafter should not check her or his own drawing.

4. Why is the proper care and control of drawings important?

5. List advantages and disadvantages of using intermediates.

6. The use of computers in drafting will reduce the number of jobs available for drafters. Do you agree or disagree? Why?

7. What is the main purpose of the American National Standards Institute?

8. How is the ISO sheet size expanded beyond an A0 sheet?

9. Which of the four levels of graphic communication is related to client approval?

10. What figure in Chapter 15 would relate to level one? Why?

Chapter 16

Functional Drafting

THE MEANING OF FUNCTIONAL DRAFTING

The term *functional drafting* is not often used by drafters. However, drafters have been doing functional drafting for many years but calling it *simplified drafting.* This term is often confused with the term *functional drafting.* Functional drafting is a general name for many different ways of drafting, including simplified drafting. In other words, simplified drafting is one kind of functional drafting, but there are other kinds.

Drafting has been called a way to communicate through a graphic language. Every technical drawing has a job to do. Your drawing should do its job in the best possible way. It must be clear. It must have only one meaning. It must be easy to read and understand. It should not have too much or too little information. To get your message across, your drawing must have only the lines, views, notes, and dimensions needed to be clear and complete. *Your technical drawing must not leave anything to chance.*

Definition

Functional drafting is a way of drawing that is clear and direct. A functional drawing can be easily read and understood by a user. Functional drawing is a way of drawing that shows exactly what is meant to be shown. It uses the fewest possible views, lines, and details to show all that is needed.

Functional means practical, useful, and *efficient* (not wasteful). It means being exactly suited to a purpose. Therefore, when you are making a functional drawing, you must keep in mind the person who will be using it.

CLASSES OF FUNCTIONAL DRAWING

There are three types of functional drawings.

Class 1: In-company or local working drawings. Drawings of a company's product used only within the company can be made simpler because the other people working there are familiar with the product. Very little detail and information is needed. Such drawings are functional within the company. However, you must remember that these drawings would not be very useful in another company for an unrelated product.

Class 2: For a field of industry or engineering. Some fields, such as the machine tool, aeronautical, automotive, and electrical fields, can use simpler drawings common to that industry. These industries have developed easier drawing methods that serve their own needs. Some of these methods serve more than one industry. After years of use, they become general methods of drawing.

Class 3: General functional drawing. All functional drawings are based on a graphic language understood by all drafters. The *American National Standards Institute* (ANSI) decides how to develop functional drawings for general use and for certain kinds of engineering use.

SIMPLIFIED DRAFTING

To save time and make technical drawing more functional, simplified drafting is used. Drafting shortcuts have come about over many years. Early technical drawings were very detailed works of art. They usually had more views than were needed. In other words, drawings took a long time to make. They were hard to read and understand because they had so many details.

As time went on, a larger number of technical drawings were needed. They also had to be more complex. For these reasons, a simpler way of drawing was needed. Drafters stopped using shading, unnecessary views, certain details, and special treatments. Today, drawings are

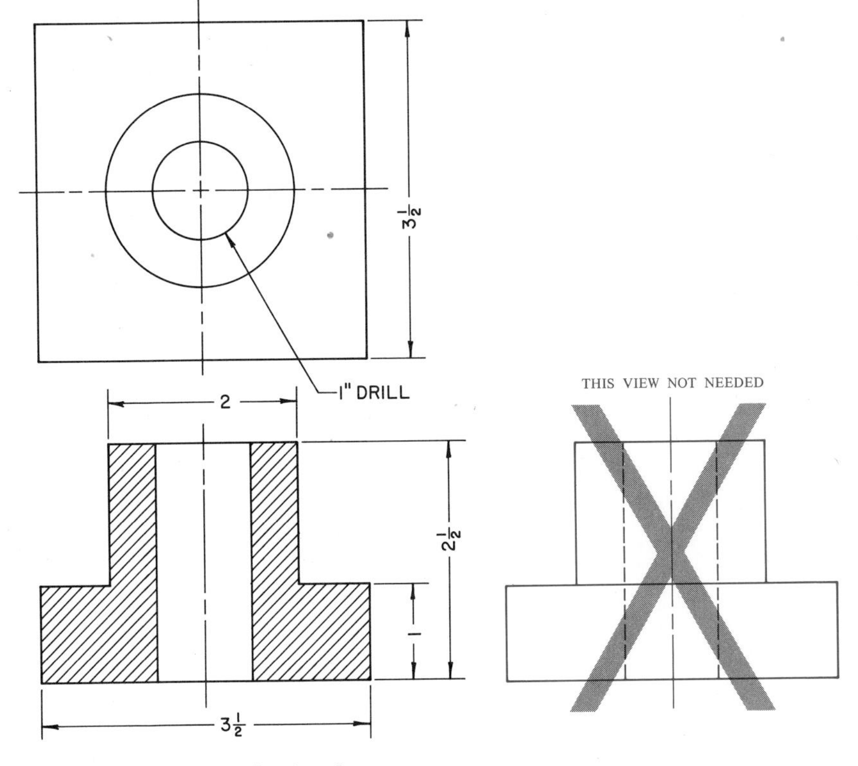

Fig. 16–1 **The right-side view is unnecessary.**

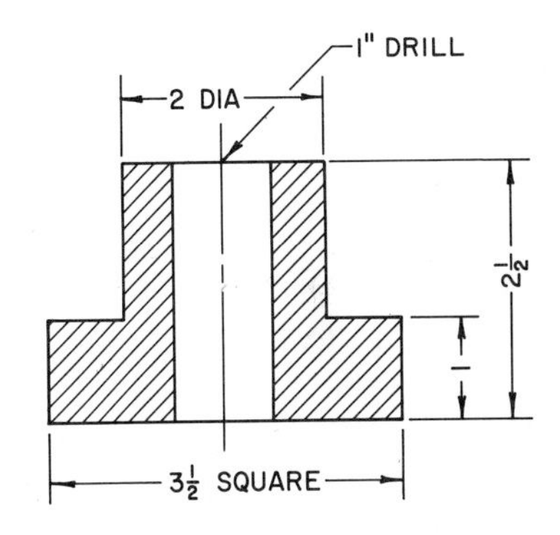

Fig. 16–2 **A complete description can be given in one view.**

easier to read and understand. Objects are drawn more simply, and symbols are used for familiar features. The two main reasons for making drawings simpler are to make them more functional and to save drawing time.

Unnecessary Views

Figure 16–1 is a three-view drawing of a flange. The front and right-side views have the same shape. Each shows exactly the same detail. It is clear, then, that one view is not needed. In other words, the front and top views show as much about the flange as do all three views. Figure 16–2 shows the same flange with another view eliminated. Notice how the part has been dimensioned. Now you can see its size in a single view.

One view is usually enough to describe parts that have the same thickness. Figure 16–3 shows three views of a spacer block at A. Either the top or the right view can be eliminated since their shapes are the same (Fig. 16–3 at B). Since the thickness is the same throughout, the top view can also be eliminated. The thickness can be given in a dimension, as shown in Fig. 16–3 at C. It can also be given in a note, as in Fig. 16–3 at D. The same drawing could be made even simpler, as shown in Fig. 16–3 at E.

The circular view of many cylindrical parts does not help describe them. In many cases, the principal view and note make a more functional drawing (Fig. 16–4). Place your hand or a piece of paper over the circular view. Notice that DIA, the abbreviation for diameter, quickly tells as much as the circular view. Views that are not needed waste the user's time.

The same is more true where two views are better than three views. Do not draw a third view if it does not tell anything more than two views do. Notice that the top view in Fig. 16–5 does not describe the part any more clearly.

Unnecessary Detail

You can take out a great deal of detail by using notes to describe the size and shape of holes that are to be

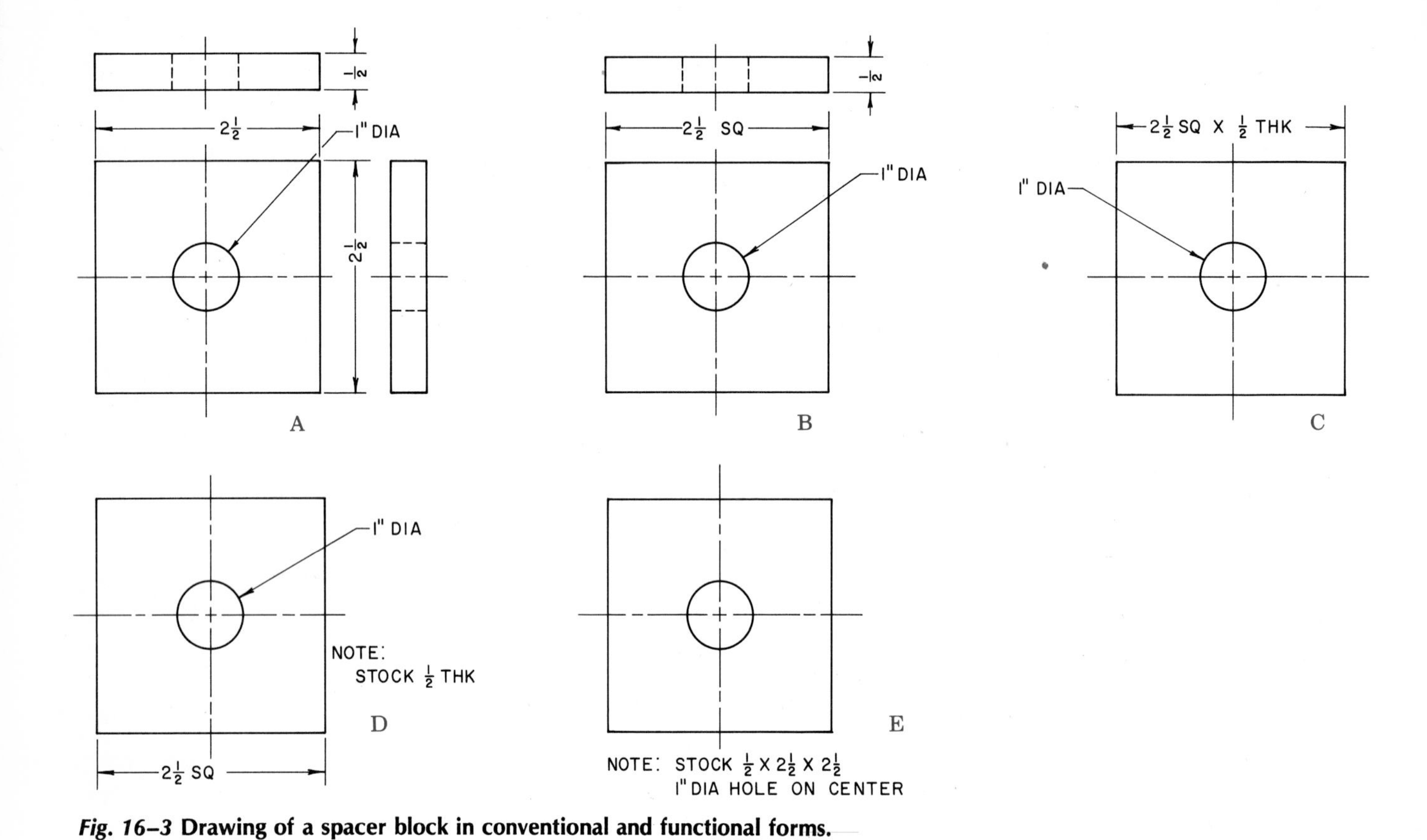

Fig. 16–3 **Drawing of a spacer block in conventional and functional forms.**

Fig. 16–4 **The circular view is not necessary. Omit it.**

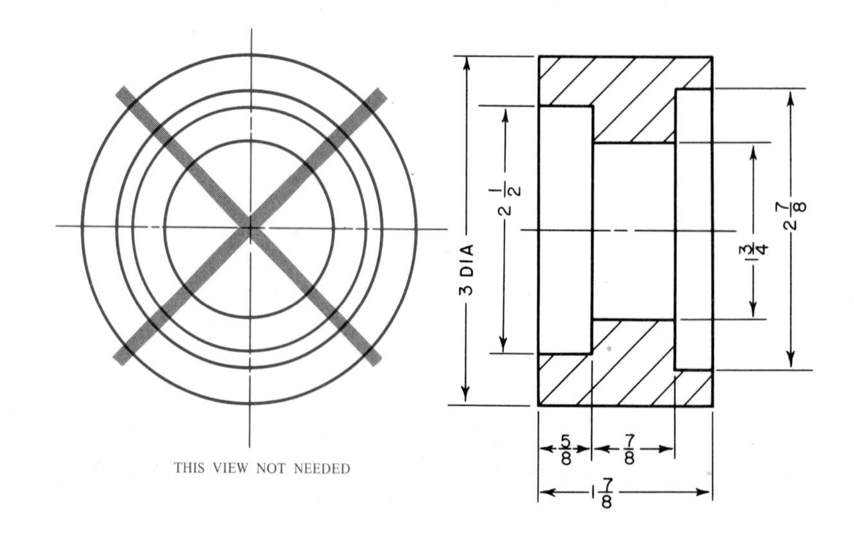

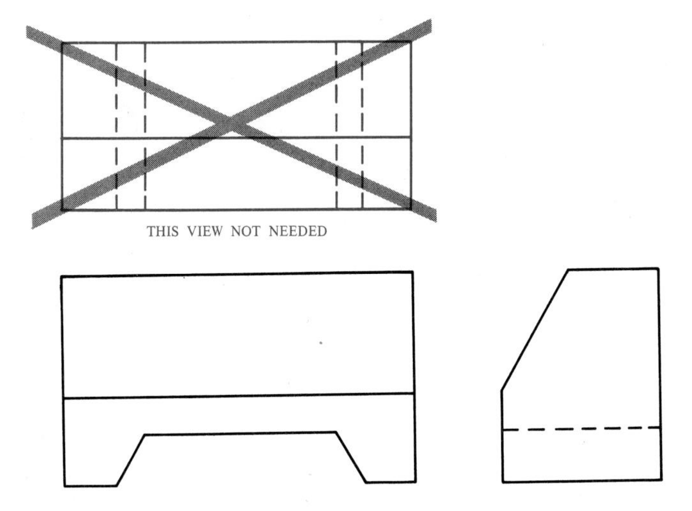

Fig. 16–5 **In this case, two views are better than three.**

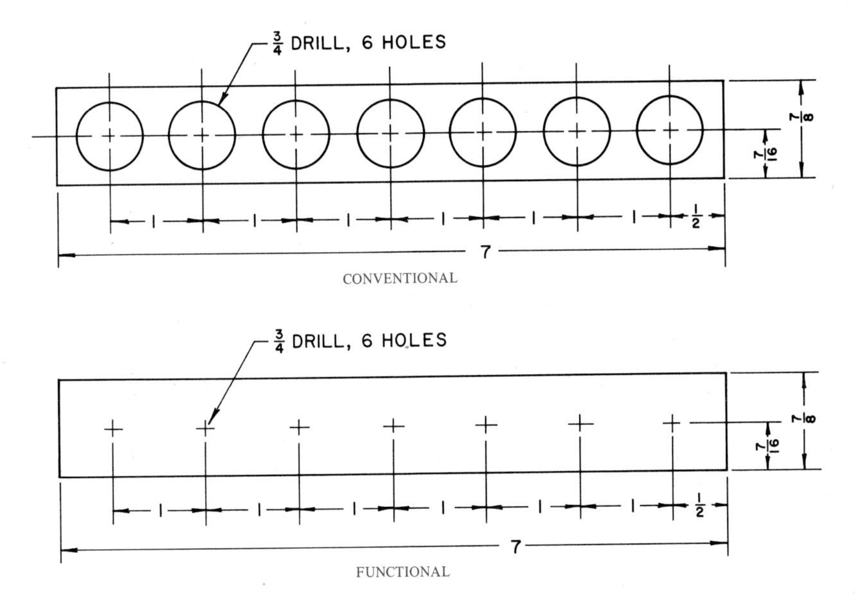

Fig. 16–6 **Holes need not be drawn.**

drilled, reamed, countersunk, and so on. You can show where the holes should be by using centerlines. You can give information about the holes in a note (Fig. 16-6). Using notes on a complex drawing saves time.

Sometimes you need to draw a detailed object more than once. To save time, you can draw it in detail once and show where else it goes with a note (Fig. 16–7 at A). Shown here is a simplified screw-thread symbol (not standard). You can draw it quickly and keep an outside thread, B, separate from an inside thread, C. See Chapter 10 for other simplified thread forms.

Other Simplified Ways to Draw

Other ways of making drawings easier to draw and read and so more functional are widely used. You learned about them in other chapters of this book. For example, you learned in Chapter 9 about the wide spacing for section lining, or just short lines or shading around the outline. Examples of these are shown in Fig. 16–8. Using half views for symmetrical objects also saves time (Fig. 16–9).

Base-line Dimensioning

The base-line system of dimensioning also makes drafting faster and easier (Fig. 16–10). In this case, dimension lines and arrowheads are taken out. All dimensions are taken from the base lines (marked 0).

Arrowheads

Using many arrowheads takes time. They may make your drawing no clearer. Many drafters either use no arrowhead at all or they use a dot (Fig. 16–11).

Freehand Sketches

In many cases, you can use simple freehand sketches as working drawings. You can sketch in any of the simpler ways to draw you learned about earlier in this chapter. Sketches should be used when you are drawing simpler parts or when you do not need an exact scale.

In-Company Standards

Many companies have their own ways to make drawings simpler. A few examples from Brown & Sharpe Manufacturing Company are shown

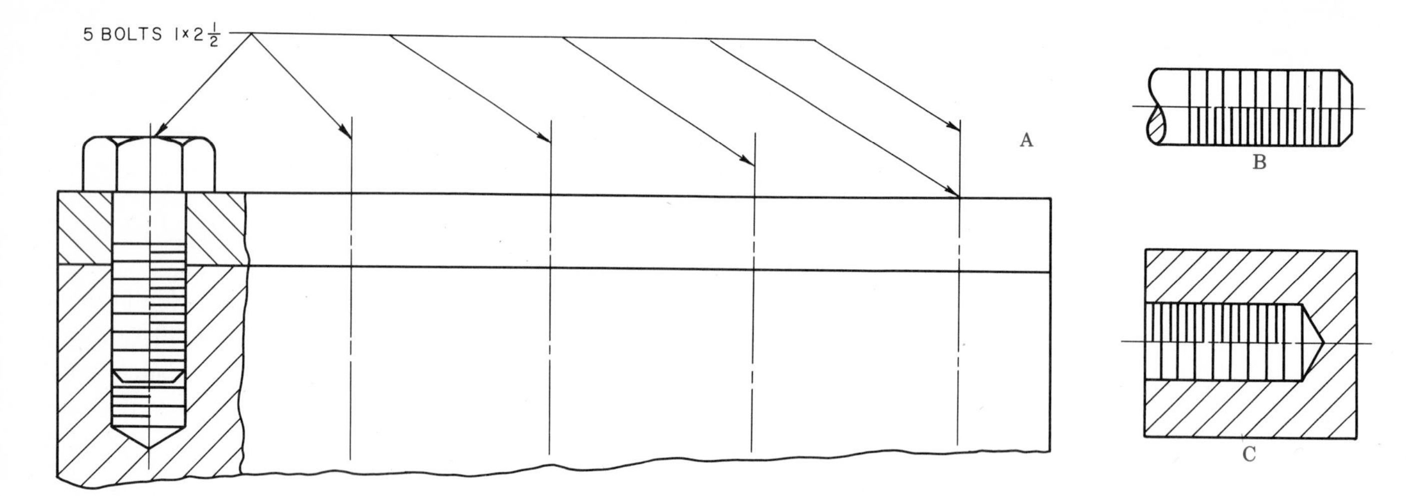

Fig. 16–7 **Repeated details need not be drawn.**

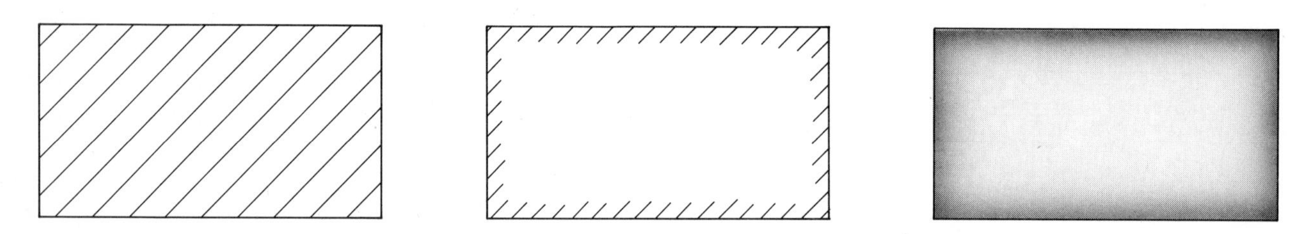

Fig. 16–8 **Functional section-lining techniques.**

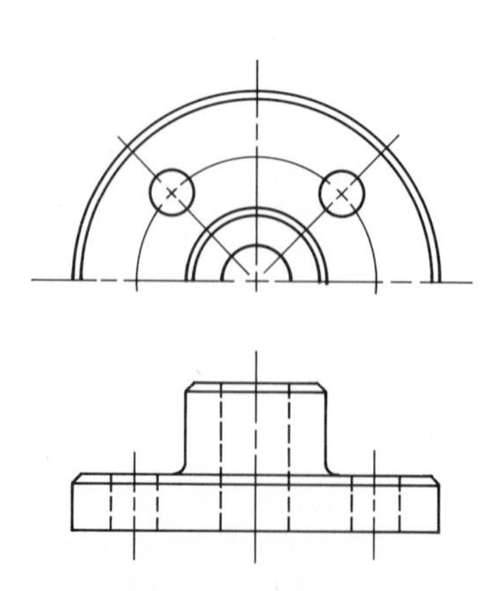

Fig. 16–9 **The use of half views saves time and space.**

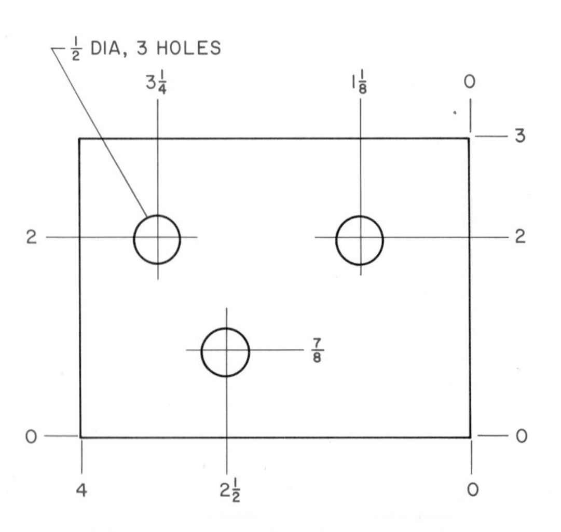

Fig. 16–10 **Base-line dimensioning.**

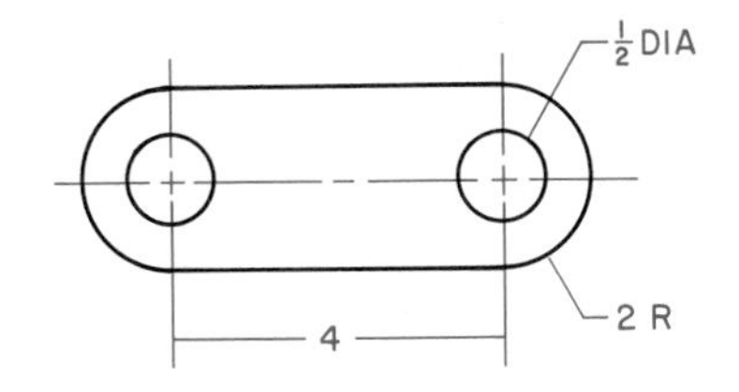

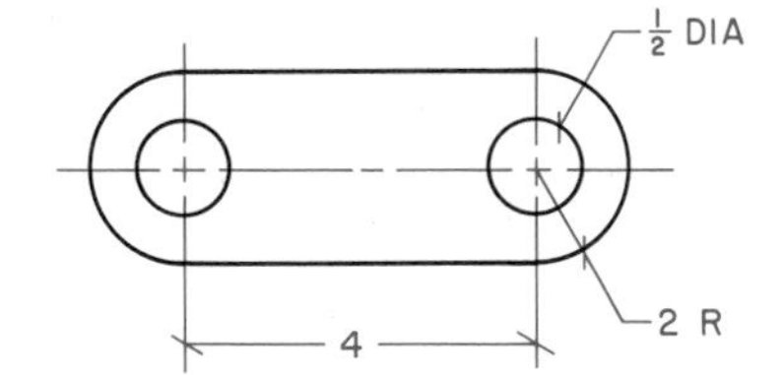

Fig. 16–11 **Substitutes for arrowheads.**

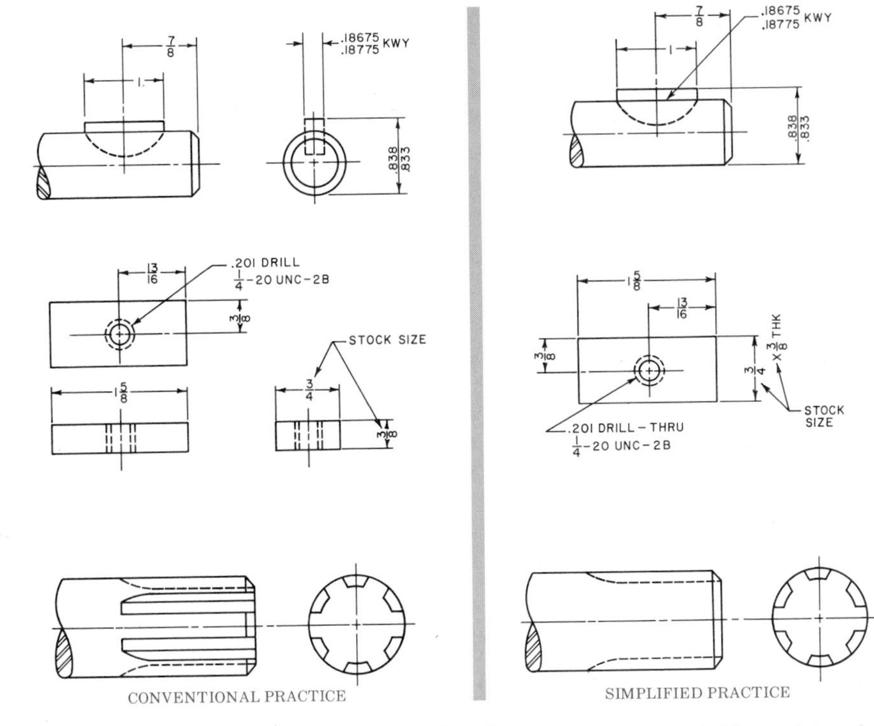

Fig. 16–12 **Samples of in-company simplified practices.** ***(Brown & Sharpe Manufacturing Co.)***

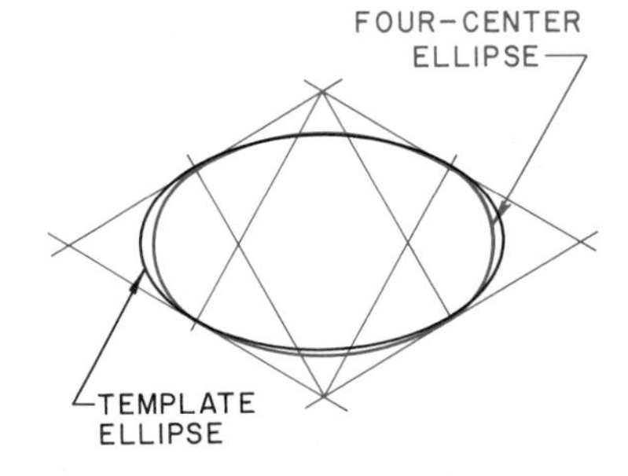

Fig. 16–13 **Templates save time and may be more accurate than other drawing methods.**

in Fig. 16–12. This figure shows how in-company standards can improve a drawing. Some of these simpler ways are general, so they are used by most drafters. To make a good functional drawing, you must know when, where, and how to use these simpler ways. This takes experience and thought.

Timesaving Symbols

Almost all fields of industry and engineering have symbols that everyone uses in that industry. For example, many drafters use special symbols in architectural and structural drafting (Chapters 20 and 21), electrical and electronics drafting (Chapter 24), aerospace drafting (Chapter 25), welding drafting (Chapter 17), surface development (Chapter 18), and so on. Symbols are used so much by industry that they have become *standard* (the same within each industry) through the American National Standards Institute. You can receive manuals about the symbols used in any of the fields mentioned above by writing to the American National Standards Institute, Inc., 1430 Broadway, New York, NY 10018.

TEMPLATES

Many different kinds of drafting *templates* (patterns) make the drafter's job easier and save time. Such detailed objects as circles, squares, ellipses, ovals, electrical and electronics symbols, architectural symbols, and many others can often be drawn faster and better with templates. For example, Fig. 16–13 shows a four-center *approximate* (not quite exact) ellipse in red. You can also see an ellipse of the same size (in black) drawn with a template. Notice that the template ellipse is a *true* (perfect) ellipse.

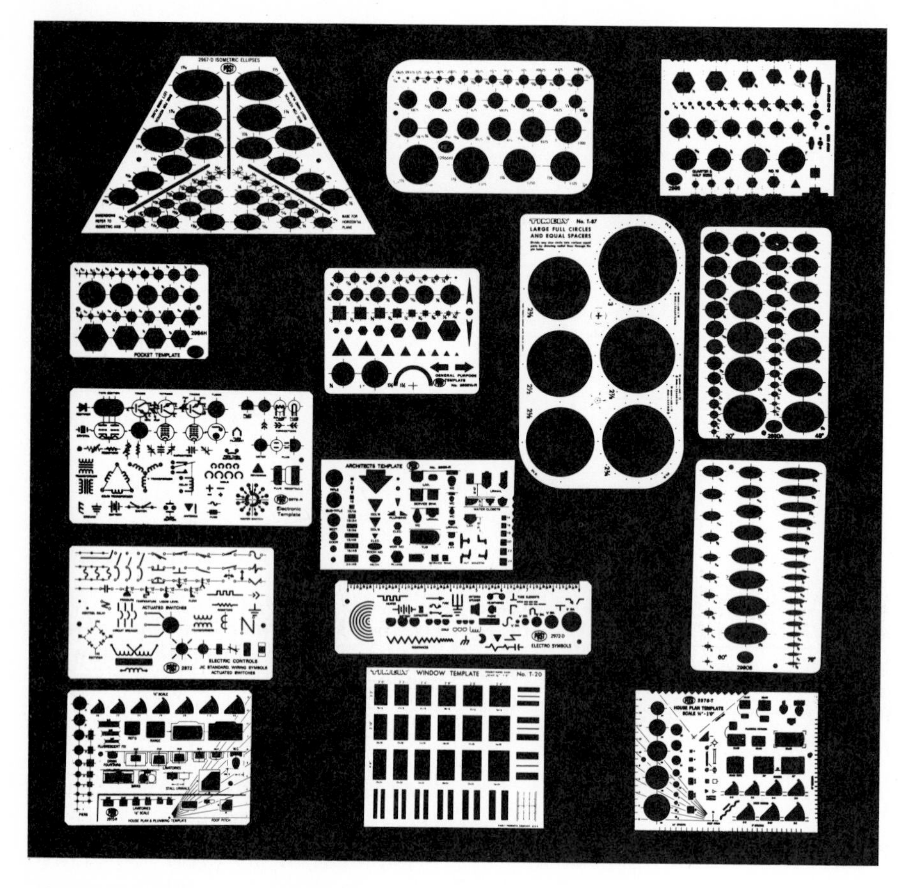

Fig. 16–14 **Examples of various types of templates.** ***(Rapidesign, Inc.)***

Fig. 16–15 **Leroy electronic symbol templates.** ***(R. J. Capece/McGraw-Hill.)***

The four-center ellipse is less than true.

Figure 16–14 shows many common drafting templates. Most templates are made of thin plastic sheets. Many of them have reference marks and other helpful information. Using the reference marks and information shown on the templates saves even more time and makes your drawing more exact.

Also, you can use many symbol templates with standard Leroy lettering equipment.

Figure 16–15 shows an electronics template being used. Figure 16–16 shows an electronics schematic drawing made with the Leroy template. You can see that the values of the different parts have been typed and added on.

PRESSURE–SENSITIVE OVERLAYS

Pressure-sensitive overlays are printed or plain. The sheets are transparent or translucent. The backs of the sheets are adhesive so they will stick to the drawing. A throw-away carrier sheet against the adhesive back keeps the sheets from sticking together. To use a pressure-sensitive overlay, peel the sheet with the image you want away from the carrier sheet. Place the overlay where you want it on the drawing. Rub the overlay briskly to make it stick (Fig. 16–17).

Overlay sheets come with many different standard symbols or patterns (Fig. 16–18) and in blank sheets. A *matte* (dull and rough) surface on the blank sheet accepts typed letters as well as pencil or ink lines. You can use this material for making corrections on drawings. You can also add lists of materials

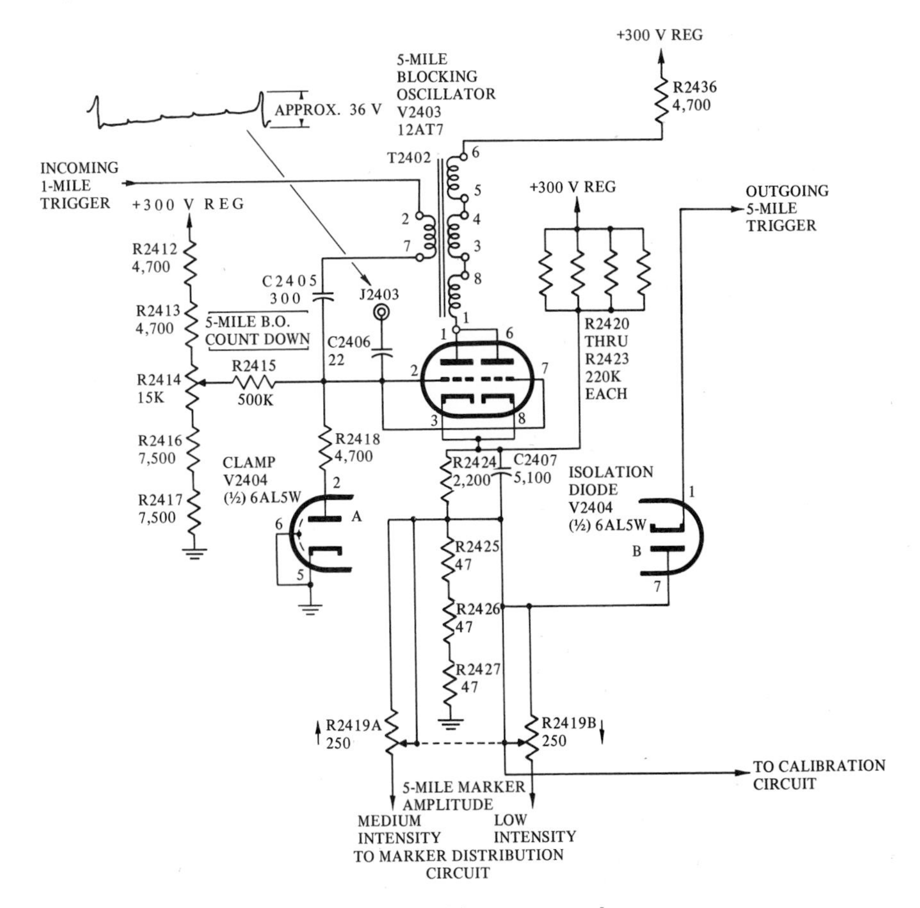

Fig. 16–16 **Schematic drawing made with Leroy template.**

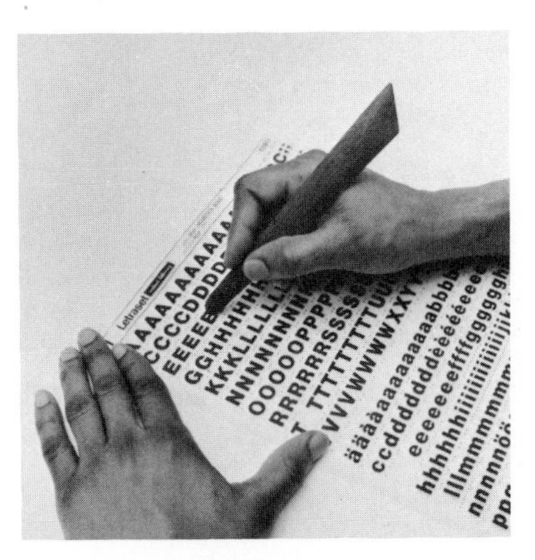

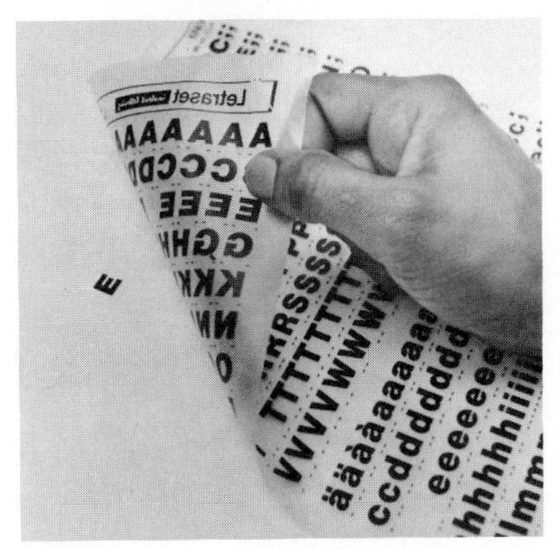

Fig. 16–17 **The application of pressure-sensitive material.** ***(R. J. Capece.)***

Fig. 16–18 **Examples of standard printed symbols and patterns available on pressure-sensitive overlays.** ***(Graphic Products Corp.)***

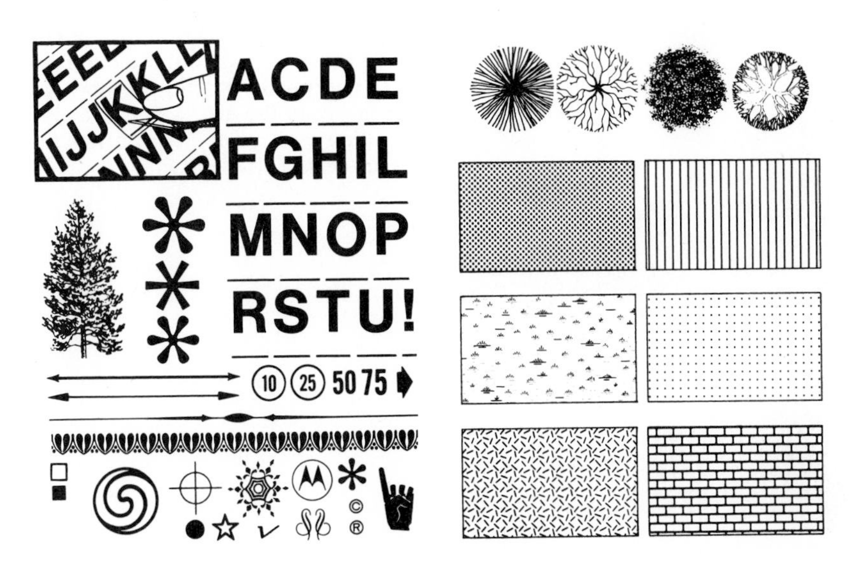

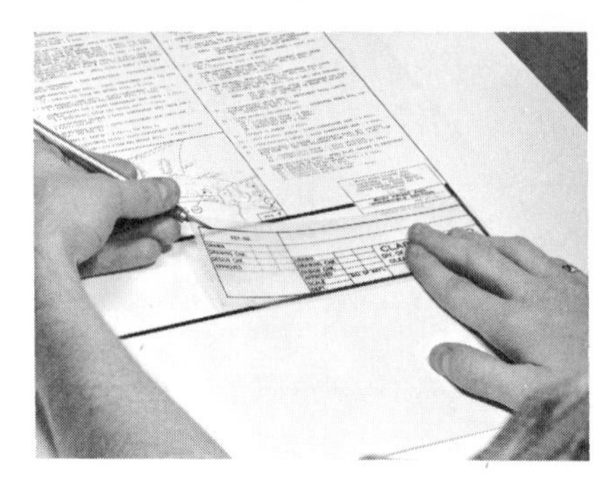

Fig. 16–19 Type or draw an image on the sheet, cut out, and position on the drawing. *(L: R. J. Capece/McGraw-Hill, R: Para-Tone, Inc.)*

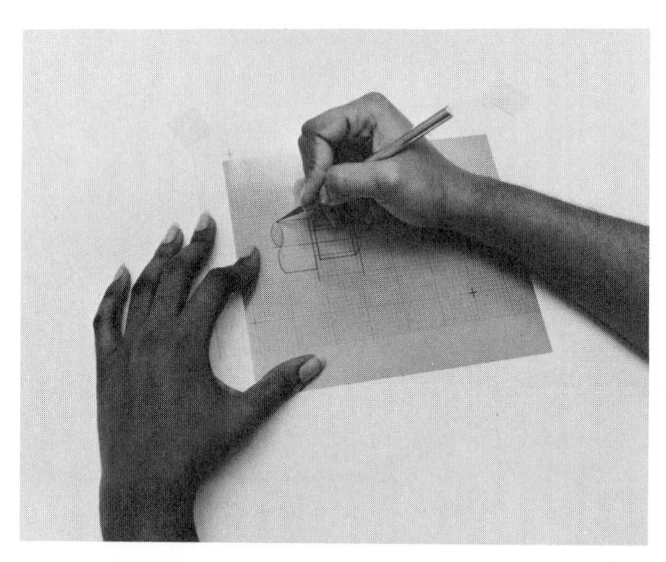

A

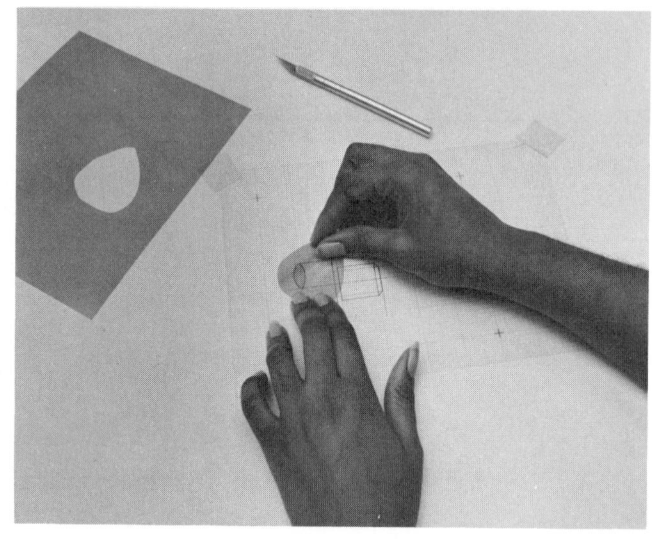

B

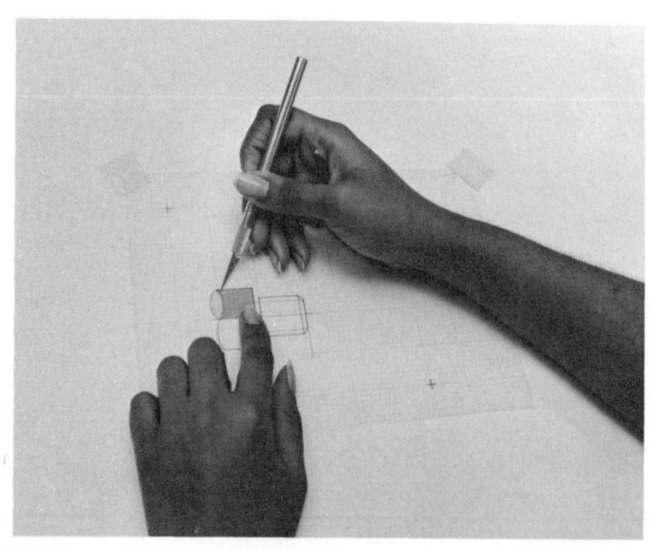

C

Fig. 16–20 Method of applying cut-out-type pressure-sensitive overlays. A. Cut a section slightly larger than needed. B. Position the overlay over the part to be covered, and press it firmly to the art. C. Use a sharp knife to cut the exact outline you need. Trim away the extra material. *(R. J. Capece/McGraw-Hill.)*

needed or detailed notes. These can often be typed faster than they can be hand lettered (Fig. 16–19).

Two basic kinds of pressure-sensitive overlays are available: cut-out and transfer. To apply a cut-out overlay, position the image you want in the right place on the drawing. *Burnish* (rub) the image area. Cut around it to remove the part you do not want (Fig. 16–20). The transfer pressure-sensitive overlay is different. To apply a transfer overlay, remove the carrier sheet from the translucent image sheet. Place the area of the image sheet you want transferred in the right place on the drawing. Then rub the image you want transferred over the top surface of the transfer sheet. Use a burnishing stick or other dull instrument. Lift the transfer sheet. You will see that the image remains on the drawing (Fig. 16–21). After you have transferred the image, place the carrier sheet over the image and burnish it again.

Pressure-sensitive overlays come not only in sheets. They also come in different patterns and colors of tape (Fig. 16–22).

You can use these tapes for borders, standard lines, charts and graphs, and many other drawing needs. You can apply them by hand or with a tool called a tape-pen (Fig. 16–23).

PHOTO DRAWING

Another type of functional drafting is *photo drawing.* In photo drawing, photographs are used as an engineering shortcut. Drafters generally use photo drawings in one of two ways. First, they use them as a quick and inexpensive way of

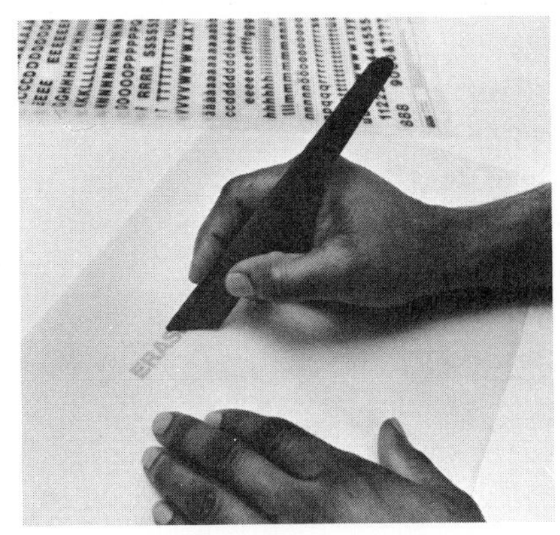

Fig. 16–21 **Application of transfer-type pressure-sensitive overlay.** *(R. J. Capece.)*

Fig. 16–22 **Some examples of pressure-sensitive tapes. These are available in a variety of colors and patterns.** *(Chartpak, A Division of Avery Products Corp.)*

Fig. 16–23 **Pressure-sensitive tape may be applied with a tape-pen or by hand.** *(Chartpak, A Division of Avery Products Corp.)*

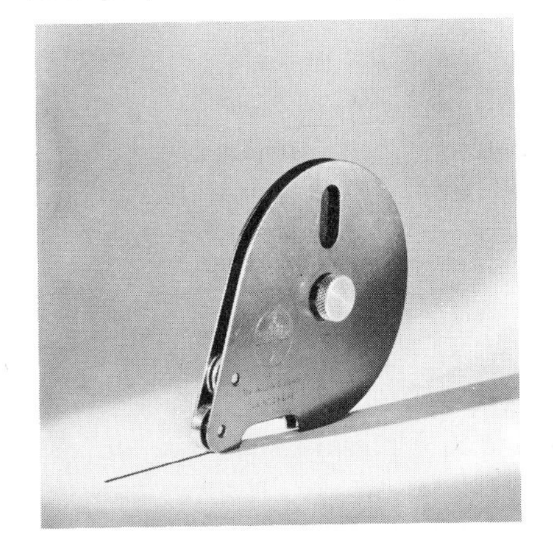

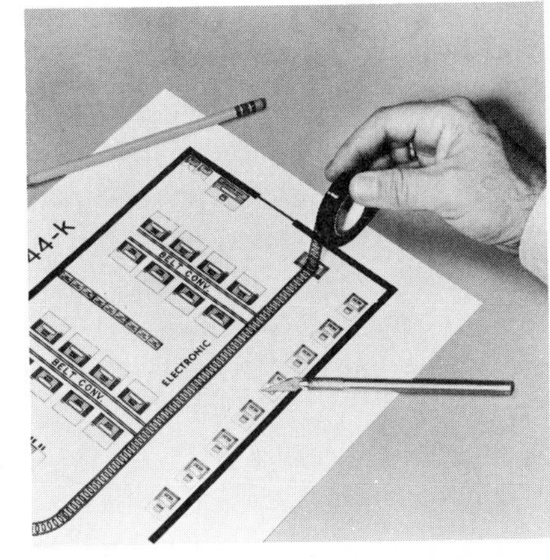

showing changes to a machine or device that has already been built. Second, they use a photo drawing when developing a product from the model or mock-up stage to the final product.

Sometimes changes are made in a device already built. Perhaps the least expensive way to record the changes is to make a photograph and mark the changes on it (Fig. 16–24).

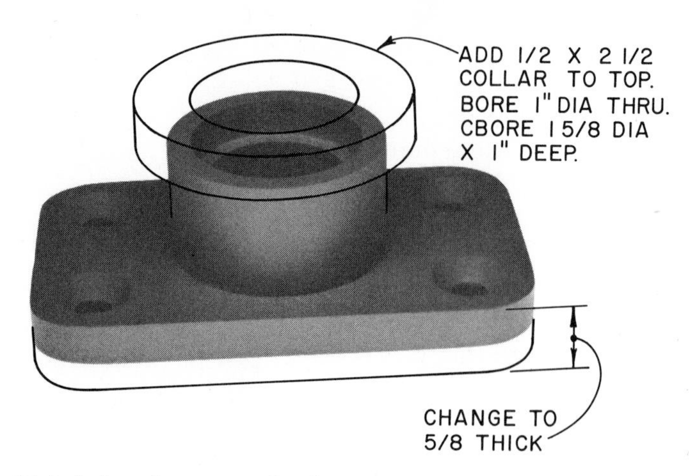

Fig. 16–24 **A design change may be shown on a photograph.**

You can use a regular photograph (paper print) if you need only one copy. However, more than one photograph is often needed for filing and distribution. In this case, a special print, called a *reproducible intermediate,* is made.

A reproducible intermediate is any *medium* (material) from which prints can be made. A reproducible intermediate is usually made by a photographic process on a transparent or translucent film. You can add lines with a soft-lead pencil or drawing ink. The part of the image you do not want can be removed with a chemical bleach or with a wet eraser. After you have made the changes, as many regular photographs can be made as are needed.

Engineers and other designers sometimes work with scale models and mock-ups during the early design stage. Sometimes they use photographs to save time. Small changes in the shape and size of parts can be made on a photograph. In this way, there is no need to build several design models.

SCISSORS DRAFTING

Scissors drafting is a way of preparing a finished working drawing by combining two or more drawings already made. In other words, drafters may have to draw some parts of an object. However, they may add other parts by cutting and pasting parts from a drawing already made. Scissors drafting can save a great deal of time.

A good thing about scissors drafting is that you do not have to use only transparent or translucent materials. You can use catalog illustrations or pictures from magazines, books, or other printed material. However, you must have the proper copying equipment. To use scissors draftings, simply tape or paste a picture or drawing to a sheet of white paper that is the right size. Then make a copy with any copying equipment that uses reflected light. Sometimes you may want to combine drawings made on translucent material (tracing paper, cloth, or film). To make copies of these drawings, use methods such as blueprinting or the diazo method.

You can make the *paste-up* (the sheet on which the drawings are pasted) in many different ways. Rubber cement, paste, or other adhesive may be used to fasten an *opaque* sheet (one that you cannot see through) to the drawing. A reflected-light copier is used with opaque materials. For this reason, the paste or other adhesive will not affect the final copy. If you attach the drawings on transparent or translucent material, you should use a transparent tape to fasten them together. To do this, place the drawing of the part you want added under the main drawing so that it is in the place it should be on the main drawing. Cut through both sheets with a knife or razor blade. Remove the top cutout. Tape the inserted piece in place with transparent tape. It is usually better to place the tape under the drawing. You can also remove details in this way. Cut out the unneeded detail. Then insert a blank sheet and tape it in place.

LETTERING

Functional lettering is a part of functional drawing. When deciding what ways of lettering you will use, keep in mind what is available and why you need the lettering. Freehand lettering is most widely used. However, templates, guides and scribers, printed and typed adhesives, and special typewriters are used also to help make functional drawings. To save time, you can type notes and parts lists.

Review

1. There are two basic kinds of pressure-sensitive overlays. Name them.

2. Name the functional drafting method that involves the use of photographs.

3. What is the drafting term used when a drawing is prepared by cutting and pasting two or more drawings already made?

4. A way of drawing that shows exactly what is meant to be shown using the fewest views, lines, and details that will show all that is needed in a way that the user can both easily read and understand is called ________.

5. Removing the views, details, and arrowheads that are not needed are methods used in ________ drafting.

6. Some companies have their own standards for simplified drawings. What are these called?

7. What is a reproducible intermediate?

8. Describe two uses for photo drawing.

9. Name the two main reasons for making drawings simpler.

10. Are at least two views of a cylindrical part always needed?

Problems

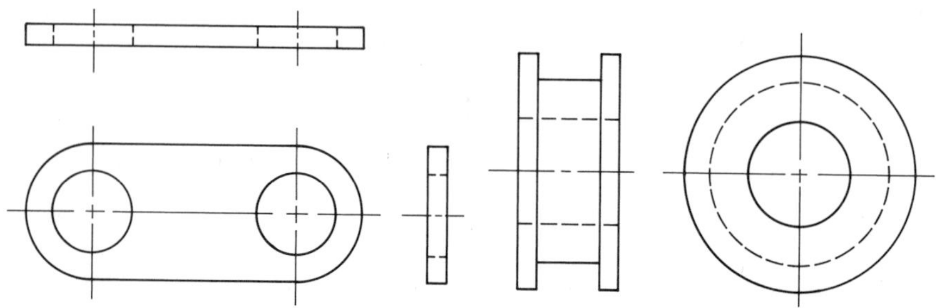

Fig. 16–25 Link. Take dimensions from the printed scale at the bottom of page 321, using dividers. Make a working drawing of the link, using functional drafting techniques. Show only essential views and dimensions.

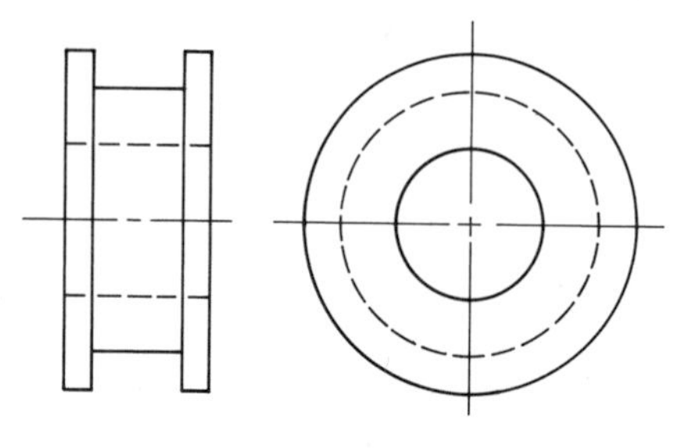

Fig. 16–26 Spool. Take dimensions from the printed scale at the bottom of page 321, using dividers. Make a working drawing of the spool, using functional drafting techniques. Show only essential views and dimensions.

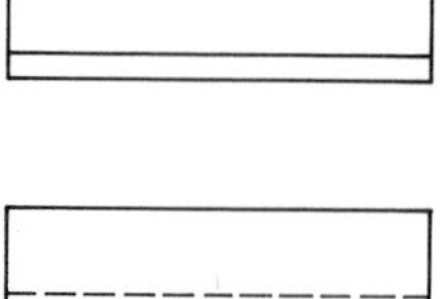

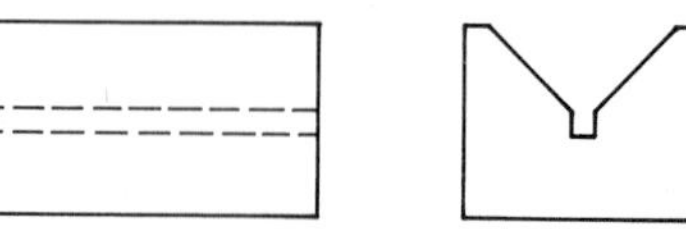

Fig. 16–27 V-block. Take dimensions from the printed scale at the bottom of page 321, using dividers. Make a working drawing of the V-block, using functional drafting techniques. Eliminate unnecessary views. Dimension.

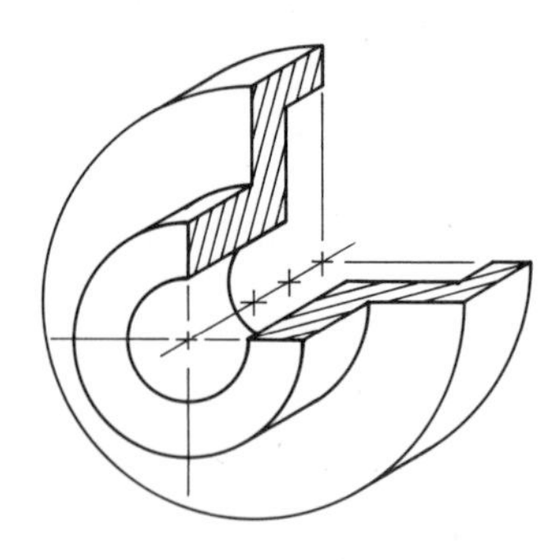

Fig. 16–28 Retainer. Take dimensions from the printed scale at the bottom of page 321, using dividers. Make a working drawing of the retainer, using functional drafting techniques. Show only essential views and dimensions.

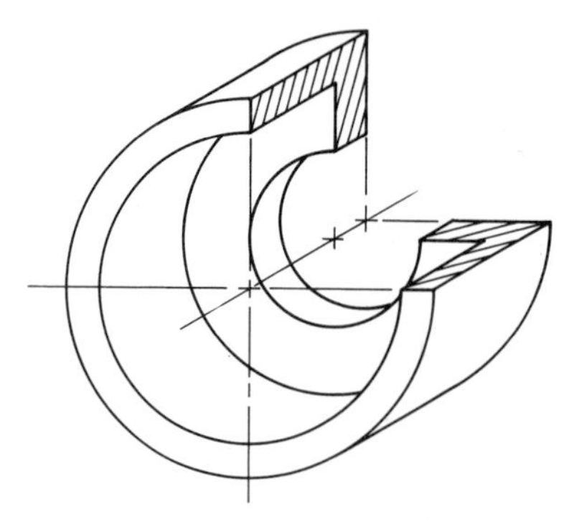

Fig. 16–29 Cap. Take dimensions from the printed scale at the bottom of page 321, using dividers. Make a working drawing of the cap, using functional drafting techniques. Show only essential views and dimensions.

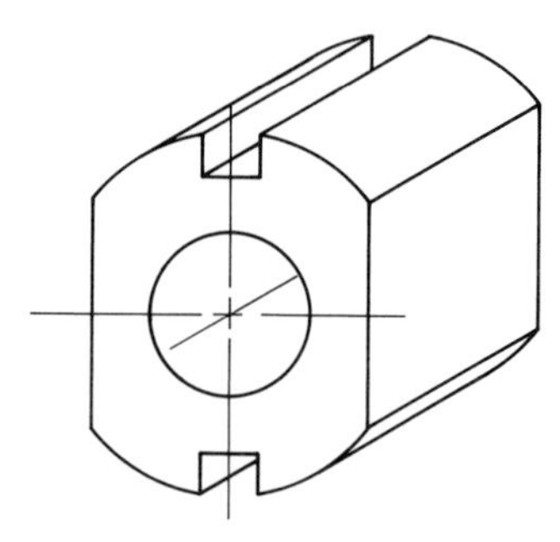

Fig. 16–30 Clutch block. Take dimensions from the bottom of page 321, using dividers. Make a working drawing of the clutch block, using functional drafting techniques. Show only essential views and dimensions.

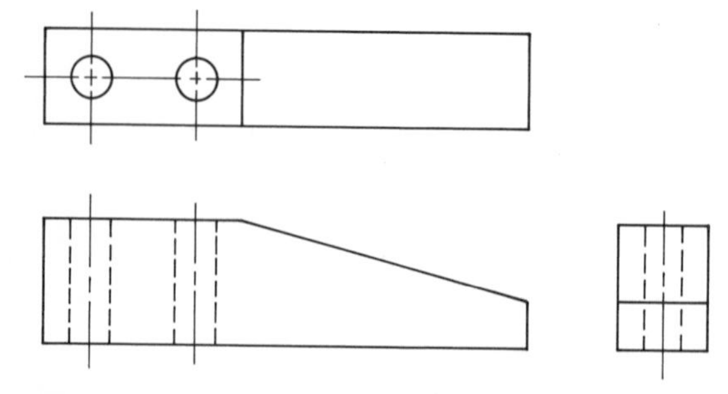

Fig. 16–31 Clamp jaw. Take dimensions from page 321. Make a working drawing of the clamp jaw, using functional drafting techniques. Show only essential views and dimensions.

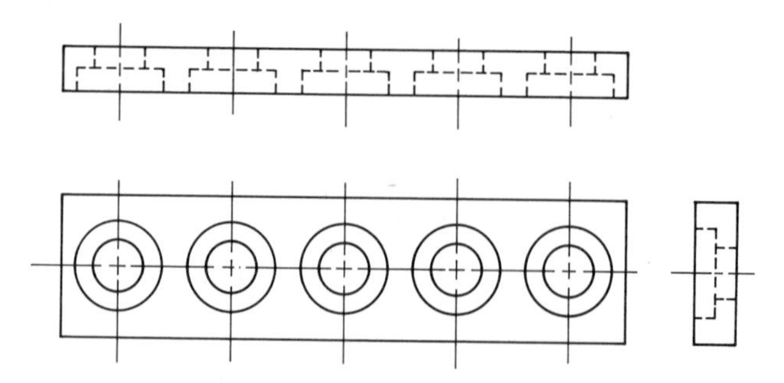

Fig. 16–32 Tie block. Take dimensions from page 321. Make a working drawing of the tie block, using functional drafting techniques. Show only essential views and dimensions.

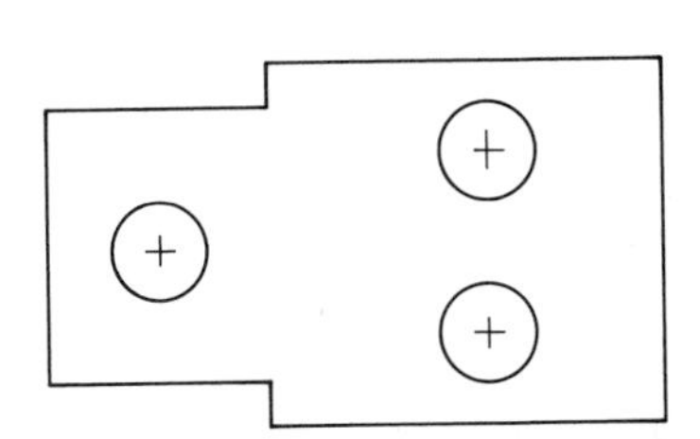

Fig. 16–33 **Base plate. Take dimensions from the printed scale at the bottom of the page, using dividers. Make a working drawing of the base plate, using functional drafting techniques. Use base-line dimensioning. Indicate hole sizes in a note.**

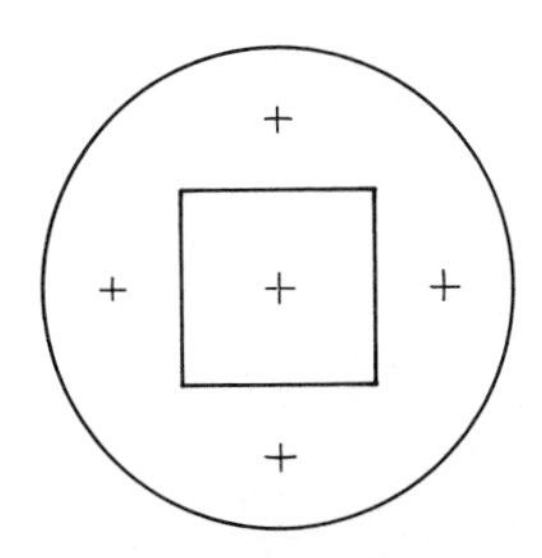

Fig. 16–34 **Gasket. Take dimensions from the printed scale at the bottom of the page, using dividers. Make a working drawing of the gasket, using functional drafting techniques. Use baseline dimensioning. Indicate hole sizes in a note.**

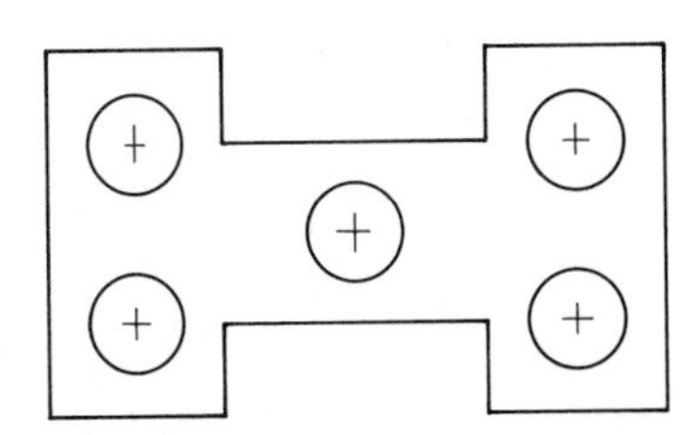

Fig. 16–35 **Spacer. Take dimensions from the printed scale at the bottom of the page, using dividers. Make a working drawing of the spacer, using functional drafting techniques. Use baseline dimensioning. Indicate hole sizes in a note.**

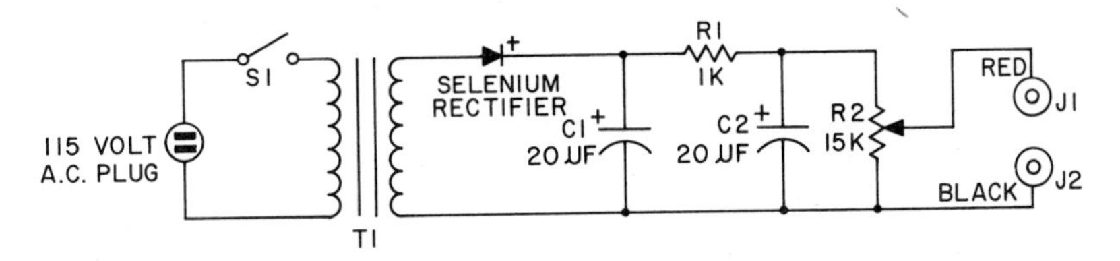

Fig. 16–36 **Schematic drawing. Use an electronic schematic drawing symbols template to make a finished drawing of the schematic shown. Use a lettering template for all lettering. Estimate all sizes not given on templates.**

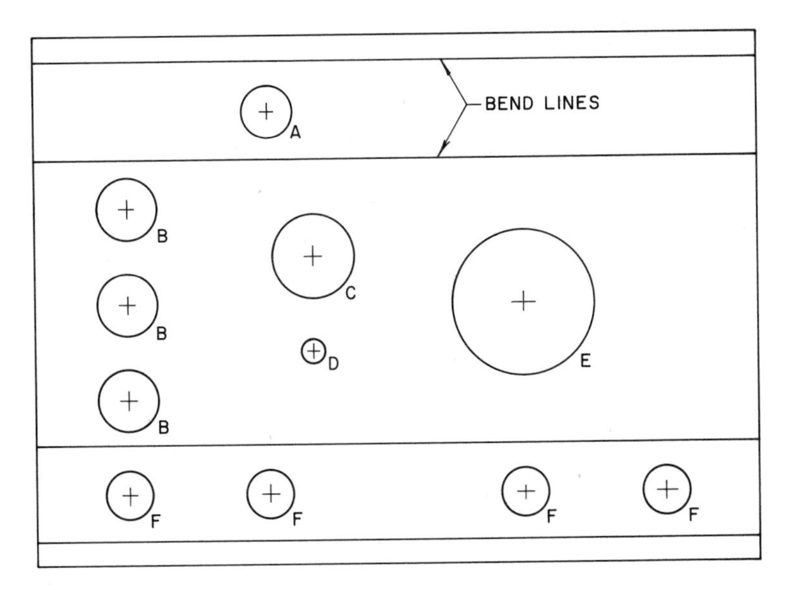

HOLE	DIA	QTY
A		
B		
C		
D		
E		
F		

Fig. 16–37 **Chassis layout. Take dimensions from the printed scale at the bottom of the page, using dividers. Make a working drawing of the chassis, using functional drafting techniques. Use base-line dimensioning. Indicate hole sizes and quantities by completing the chart to the right of the chassis layout.**

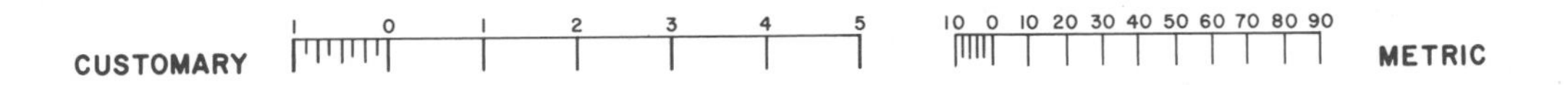

Chapter 17

Welding Drafting

JOINING METALS

Welding is a way of joining metal parts together. The art of welding is very old. In prehistoric times, it was already used to make rings, bracelets, and other jewelry. Today it is very important in industry. Standard shaped steel pieces, such as plates and bars, are welded together to make machine bases, frames, and other parts. Buildings are also put together by welding. Welded steel parts are lighter and stronger than parts made by forging or casting. Figure 17–1 shows a pulley housing made by casting. Compare it with the similar part made by welding shown in Fig. 17–2.

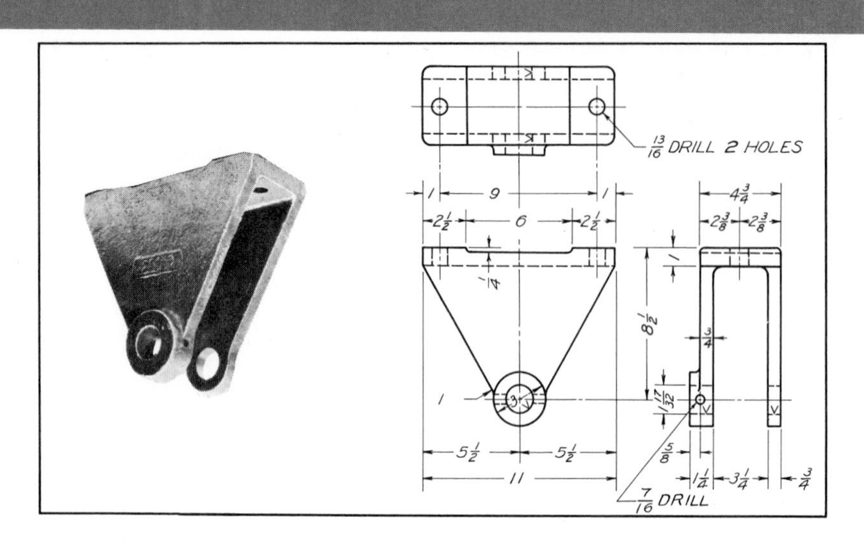

Fig. 17–1 **Pulley housing made by casting.** ***(Wellman Engineering Co. and The Lincoln Electric Co.)***

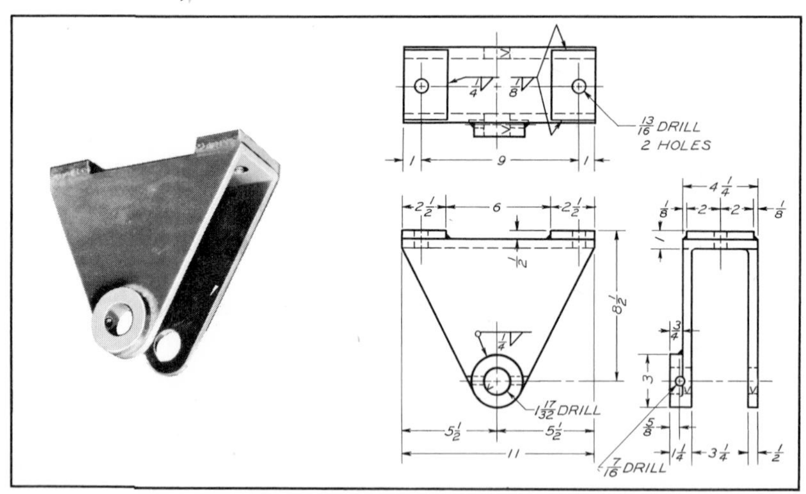

Fig. 17–2 **Pulley housing made by welding.** ***(Wellman Engineering Co. and The Lincoln Electric Co.)***

Welding has become a major assembly method in industries that use steel, aluminum, and magnesium to build cars, airplanes, ships, or buildings (Fig. 17–3). In a single truck (Fig. 17–4), there are hundreds of welds. The basic framework of the truck is put together by welding. Welding makes products stronger and more long-lasting. It also makes them look better. A drafter who is drawing parts to be welded works with a design engineer who knows what kinds of welding to use with different metals. Standard drafting symbols for welding have been established by the American Welding Society (Appendix Table A–28).

Welding joins materials by either heat or pressure or a combination of the two. There are two basic kinds of welding. Fusion welding uses only heat. Resistance welding combines heat and pressure.

Fig. 17–3 **Welding building components for Walt Disney World.** ***(United States Steel.)***

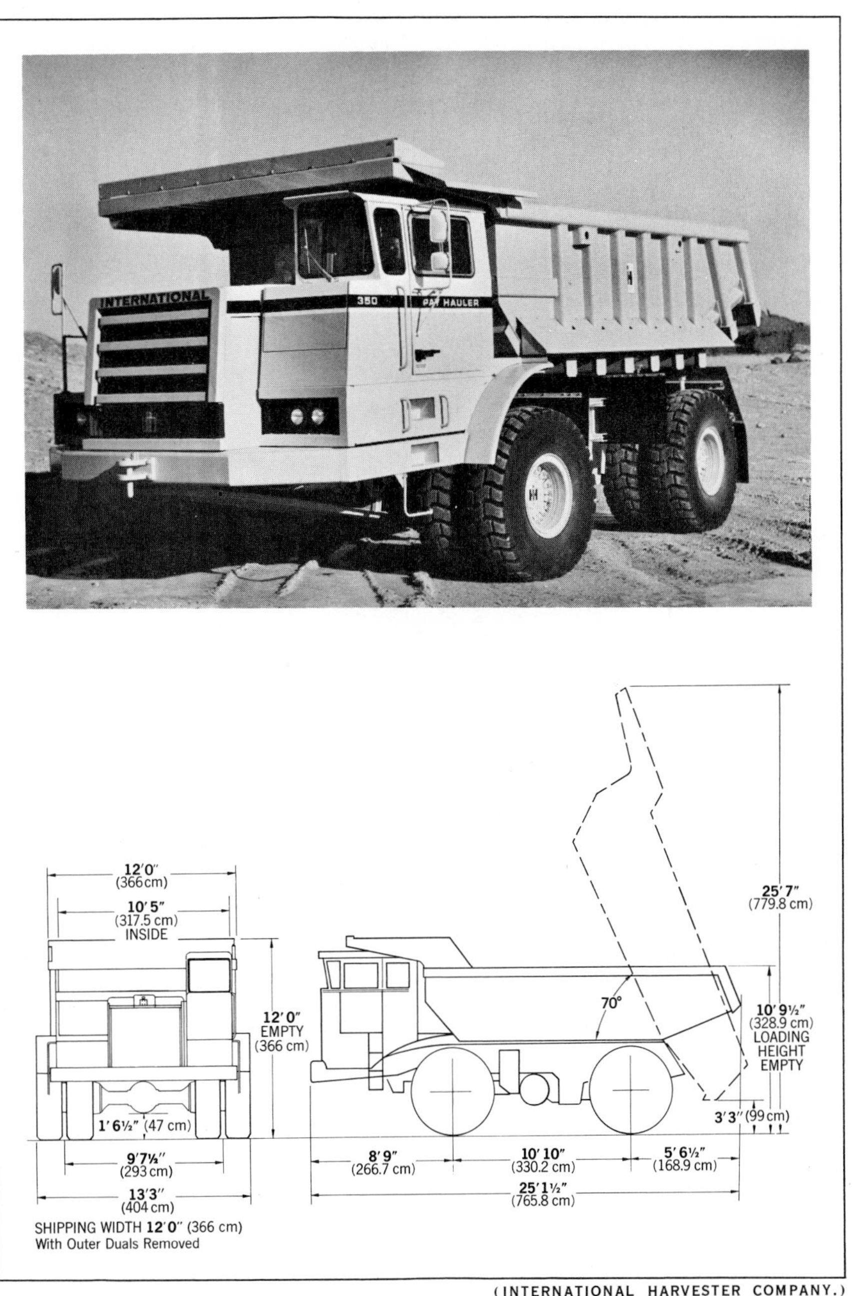

(INTERNATIONAL HARVESTER COMPANY.)

Fig. 17–4 **Items such as trucks have hundreds of welds.**

Resistance welding was developed in 1857 by James Prescott Joule. Industry began using it after the 1880s when electric power became available in large quantities. One kind of fusion welding, arc welding, was first performed by De Meritens in France in 1881. Another kind, gas welding, was effectively developed in 1885, when two gases—oxygen, from liquid air, and acetylene, from calcium carbide—were brought into use.

WELDING PROCESSES

Welding processes include fusion, gas, arc, thermit, gas and shielded arc, and resistance welding. Soldering and brazing, although called by their separate names, are also forms of welding.

Fusion Welding

Fusion welding is done with welding materials in the form of a wire or rod. This material is heated with a gas flame or a carbon arc. When it melts, it fills in a joint and combines with the metal being welded.

Gas Welding

Gas welding is fusion welding done with a flame of oxygen and acetylene, air and acetylene, or any other combination of gases that produces enough heat (Fig. 17–5). Burning oxygen and acetylene gives temperatures of between 2760° and 3595°C (5000° and 6500°F.)

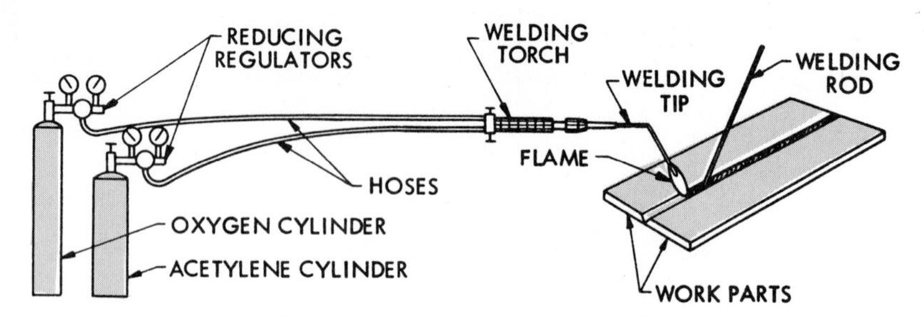

Fig. 17–5 **The gas-welding process.** ***(General Motors Corp.)***

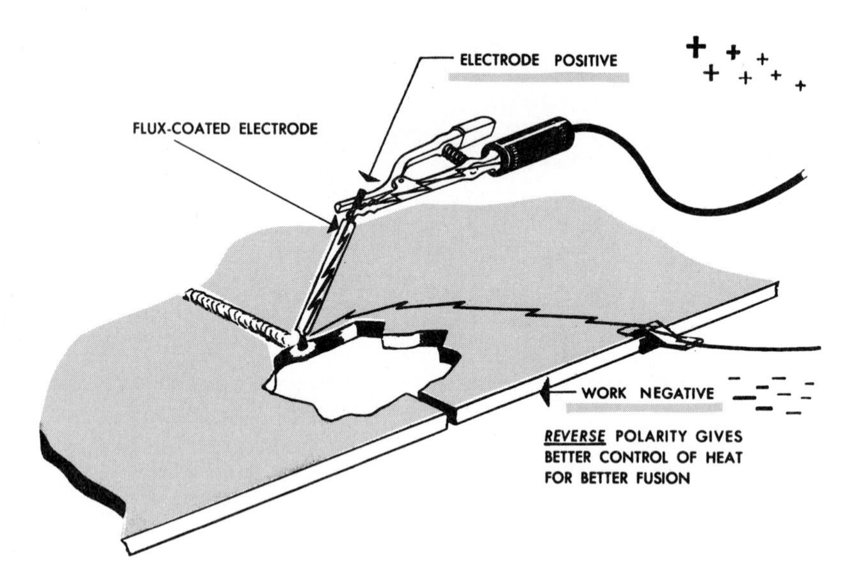

Fig. 17–6 **The arc-welding process.** ***(Republic Steel Corp.)***

Arc Welding

Arc welding is done by forming an electric arc between the *work* (part to be welded) and an electrode (Fig. 17–6). This arc is made with a direct current (dc) generator or other power source. It causes intense heat to develop at the tip of the electrode. This heat is then used to melt a spot on the work and on a rod of welding filler material, so that both can be fused together.

Thermit Welding

Thermit welding is based on the natural chemical reaction of aluminum with oxygen. A mixture, or charge, made of finely divided aluminum and iron oxide is ignited by a small amount of special ignition powder. The charge burns rapidly, producing a very high temperature. This melts the metal, which then flows into molds and fuses mating parts.

Gas and Shielded Arc Welding

Aluminum, magnesium, low-alloy steels, carbon steels, stainless steel, copper, nickel, Monel, and titanium are some of the metals that can be welded with this kind of process. There are two forms of gas and shielded arc welding. They are *Tungsten-Inert-Gas* (TIG) and *Metallic-Inert-Gas* (MIG). In TIG welding, the electrode that provides the arc for welding is made of tungsten. Since it provides only the heat for fusion, some other material must be used with it for filler. In MIG welding, the electrode contains a consumable metallic rod. It provides both the filler material and the arc for fusion.

Resistance Welding

This kind of welding is done with heat and pressure. It is a good way to fuse thin metals. To join two metal pieces, an electric current is passed through the points to be welded. At those points, the resistance to the charge heats the metal to a plastic state. Pressure is then applied to complete the weld. When current and pressure are confined to a small area between electrodes, the resulting weld is called a spot weld.

WELDING DRAWINGS

Standard drafting symbols for welding have been established by the American Welding Society. They let the drafter give the type, size, location, and all other specifications for a weld. There are many combinations available. Weld symbols are shown in Fig. 17–7. Also shown is the way in which other welding information is given along with these symbols. Every drafting room should have a copy of the latest edition of the American Welding Society Standard Welding Symbols. (See Appendix A for Table A–28)

BASIC WELDED JOINTS

There are five basic kinds of joints used in welding (Fig. 17–8). Each can have many variations. The many combinations and varieties of joints are needed because welding is used in so many different situations. To choose the right joint, one must be familiar with the materials and their conditions and have practical welding experience. Choosing the right weld depends on the type of material, the tools to be used, and the cost of preparation.

BASIC ARC AND GAS WELD SYMBOLS							
TYPE OF WELD							
Bead	Fillet	Plug or slot	Groove				
			Square	V	Bevel	U	J
⌒	◺	⏢	‖	V	⁄	U	⊔

SUPPLEMENTARY SYMBOLS			
Weld all around	Field weld	Contour	
		Flush	Convex
○	●	—	⌒

Finish symbol
Contour symbol
Root opening, depth of filling for plug and slot welds
Reference line
Size: size or strength for resistance welds
Specification or detail reference
Tail (may be omitted when reference is not used)
Basic weld symbol or detail reference
Groove angle: included angle of countersink for plug welds
Length of weld
Pitch (center-to-center spacing) of welds
Weld all around symbol
Field weld symbol
Arrow connecting reference line to arrow side of joint, to grooved member, or both
Number of spot or projection welds
F
A
R
S
T
L-P
(N)
(Both sides)
(Other side)
(Arrow side)

Fig. 17–7 **Location of welding information on welding symbols.**

Fig. 17–8 **Five basic types of joints: A, butt joint; B, lap joint; C, corner joint; D, edge joint; E, T-joint.**

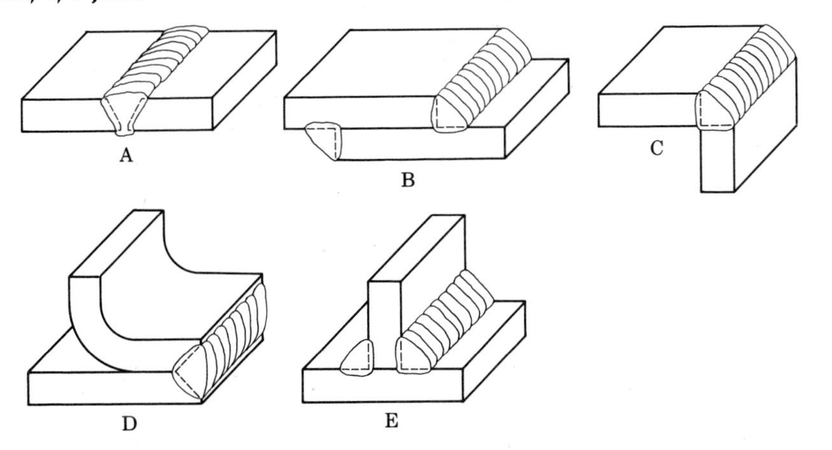

BASIC TYPES OF WELDS

Figure 17–9 shows the basic types of grooved welds. Note that they may be single or double in form. These basic types of welds are shown as applied to a butt joint. They can also be applied to all the other basic types of joints. Typical dimensions for a butt joint with a V-grooved weld are shown in Fig. 17–10. The dimensions for a T-joint with a bevel-grooved weld are shown in Fig. 17–11. Typical dimensions for a U-grooved weld on a butt joint are given in Fig. 17–12. Figure 17–13 shows the typical dimensions for a J-grooved weld on a T-joint.

SYMBOLS FOR BASIC ARC AND GAS WELDS

Figure 17–14 once again shows the basic and supplementary welding symbols, as in Fig. 17–7. You can use these symbols to describe any desired weld. By combining them, you can describe the most simple or the most complicated joints. Figure 17–7 showed the standard way in which other welding information is given along with welding symbols. The notes in the illustration told how to place symbols and data in relation to the reference line. Note that when you draw a fillet, bevel, or J-grooved weld symbol, you always place the

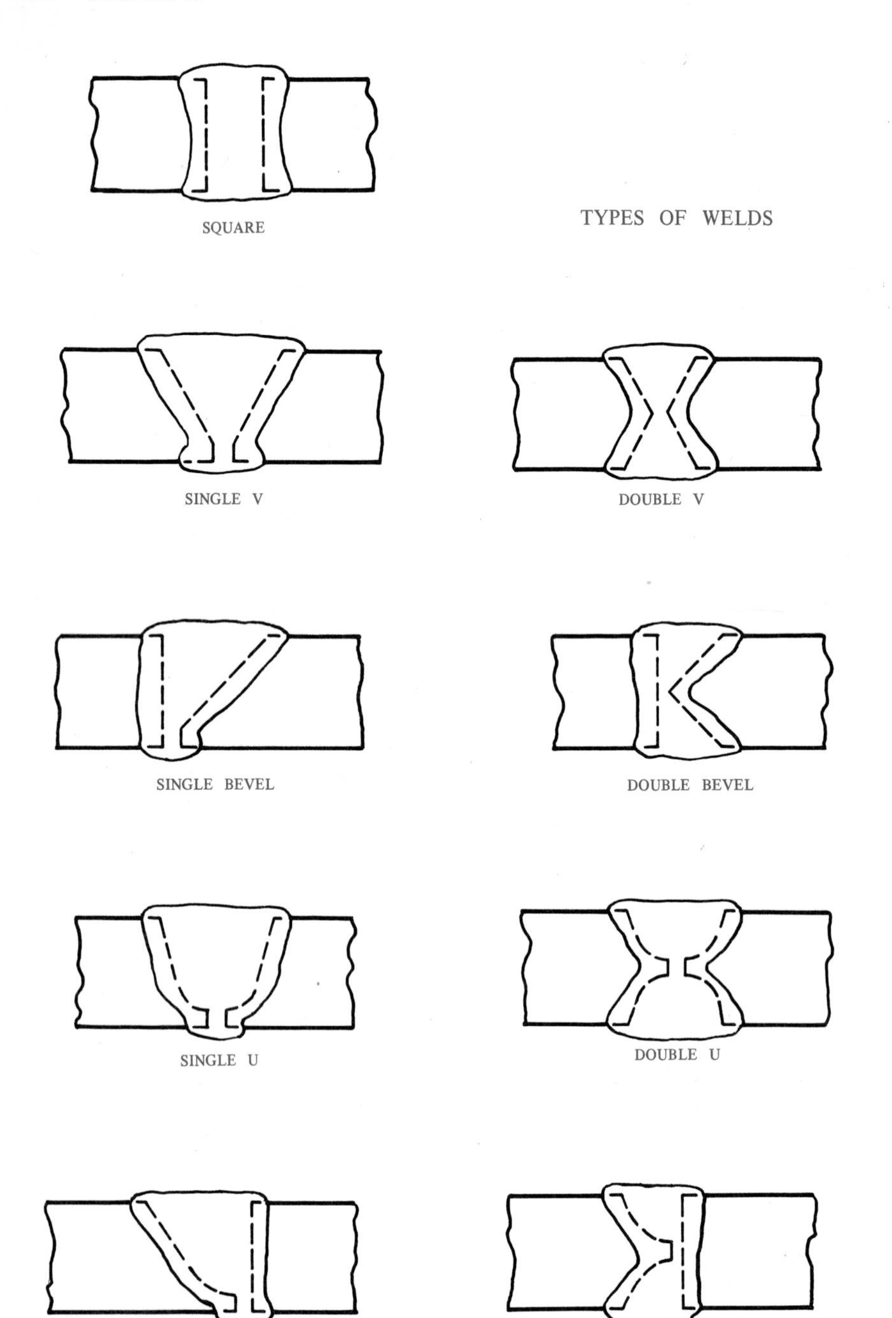

Fig. 17–9 **Basic types of grooved welds as applied to a butt joint.**

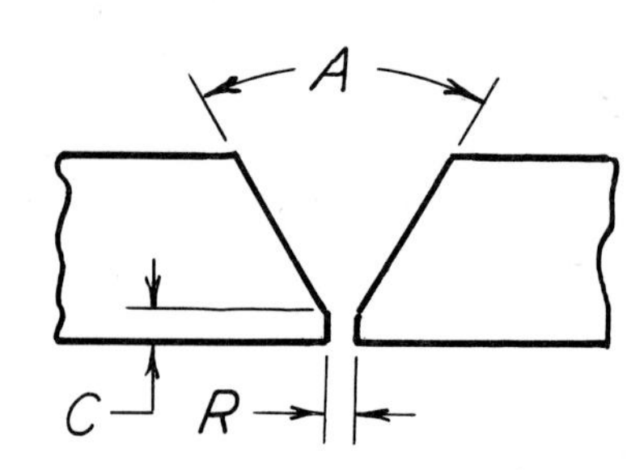

Fig. 17–10 **Dimensions for a V-grooved weld.** $A = 60°$ **min.,** $C = 0$ **to** $1/8$ **in.,** $R = 1/8$ **to** $1/4$ **in.**

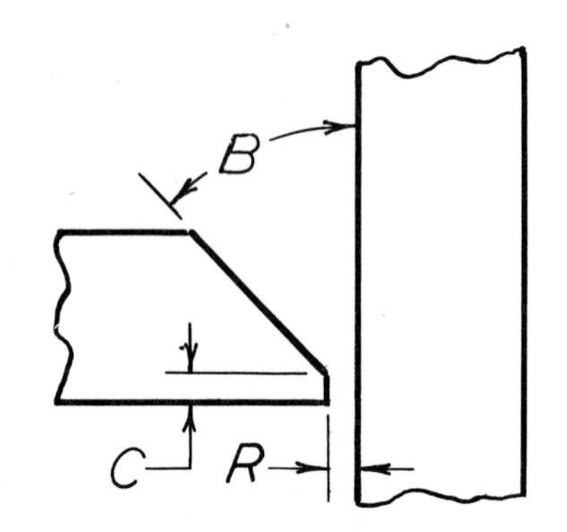

Fig. 17–11 **Dimensions for a bevel-grooved weld.** $B = 45°$ **min.,** $C = 0$ **to** $1/8$ **in.,** $R = 1/8$ **to** $1/4$ **in.**

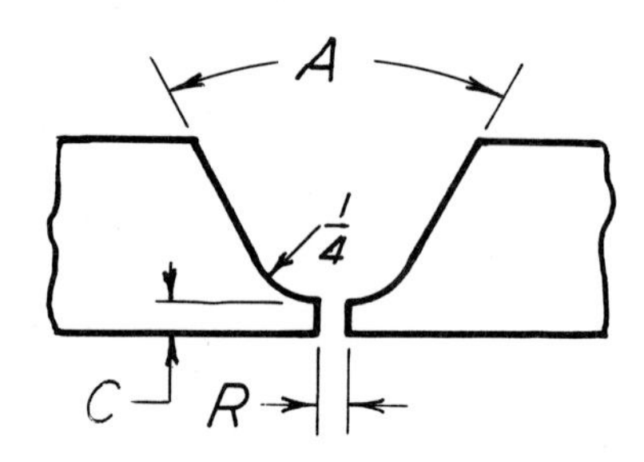

Fig. 17–12 **Dimensions for a U-grooved weld.** $A = 45°$ **min.,** $C = 1/16$ **to** $3/16$ **in.,** $R = 0$ **to** $9/16$ **in.**

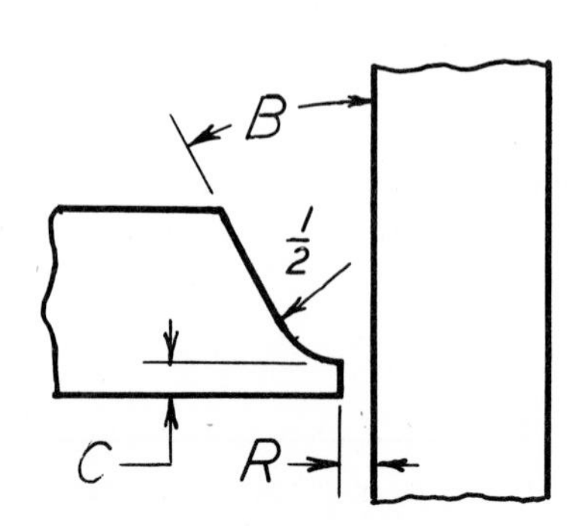

Fig. 17–13 **Dimensions for a J-grooved weld.** $B = 25°$ **min.,** $C = 1/16$ **to** $3/16$ **in.,** $R = 0$ **to** $9/16$ **in.**

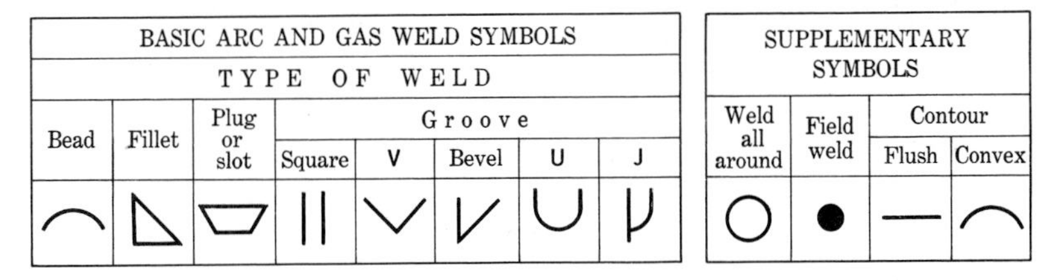

BASIC ARC AND GAS WELD SYMBOLS							
TYPE OF WELD							
Bead	Fillet	Plug or slot	Groove				
			Square	V	Bevel	U	J

SUPPLEMENTARY SYMBOLS			
Weld all around	Field weld	Contour	
		Flush	Convex

Fig. 17–14 **Basic and supplementary arc- and gas-weld symbols.**

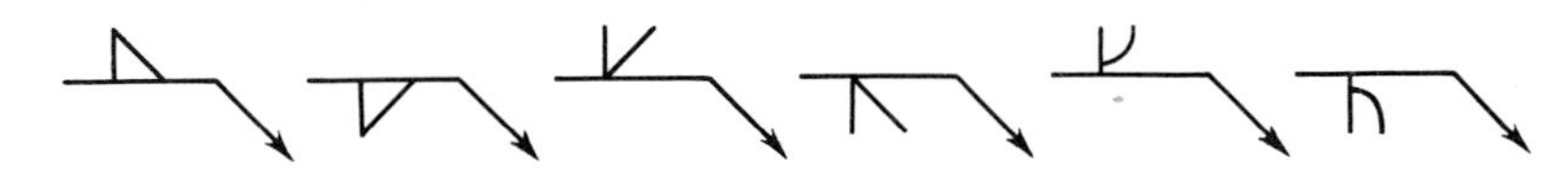

Fig. 17–15 **The perpendicular leg on the weld symbol is always drawn to the left.**

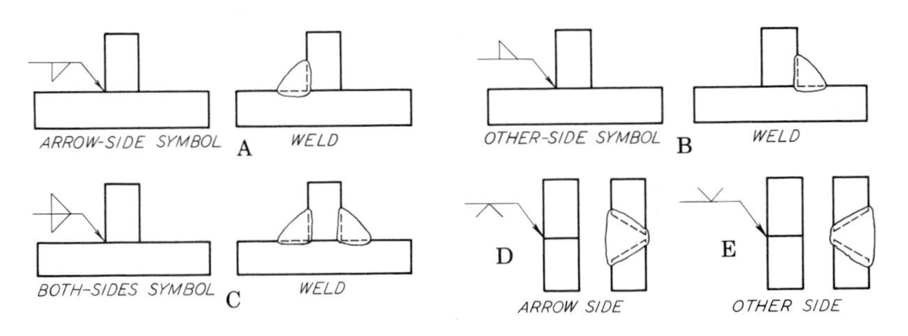

Fig. 17–16 **Arrow side and other side.**

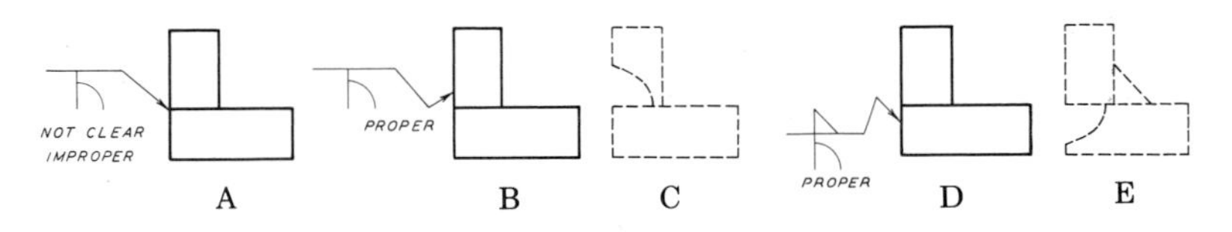

Fig. 17–17 **J-grooved weld indications.**

perpendicular leg of the symbol to the left (Fig. 17–15).

The welding symbol tells what kind of weld to use at a joint and where to place it. You can draw the symbol on either side of the joint as space permits. Always draw an arrow leading from the symbol's reference line to the joint. The side of the joint to which the arrow points is called the *arrow side*. The opposite side is called the *other side*. If the weld is to be on the arrow side of the joint, draw the type-of-weld part of the symbol below the reference line (Fig. 17–16 at A and D). If the weld is to be on the other side, draw the type-of-weld part of the symbol above the reference line (Fig. 17–16 at B and E). If the weld is to be on both sides of the joint, draw the type-of-weld part of the symbol both above and below the reference line (Fig. 17–16 at C).

When the weld is to be a J-grooved weld, the arrow from the welding symbol must be placed correctly or it can be confusing. For example, in Fig. 17–17 at A, it is not clear which piece is to be grooved. At B, the arrow has been redrawn to show clearly that it is the vertical piece that is to be grooved (as at C). At D, two welds are called for. The symbol below the reference line indicates a J-grooved weld on the arrow side. The arrow shows that it is the horizontal piece that is to be grooved. The symbol above the reference line indicates a fillet weld on the other side. The drawing at E shows how the completed welds would look.

In Fig. 17–18 at A, the reference dimensions have been included with the welding symbols. Also, the typical specification A2 has been placed in the tail of the reference line. The drawing at B shows how this data is used.

Figure 17–18 at B shows the joint made according to the reference specifications at A. This joint can be described as follows:

A double filleted-welded, partially grooved, double-J-T-joint with incomplete penetration. The J-groove is of standard proportion. The radius (R) is 13 mm (1/2 in.) and the included angle is 20°. The penetration is 19 mm (3/4 in.) for the other side and 32 mm (1 1/4 in.) deep for the arrow side. There is a continuous 10-mm (3/8-in.) fillet weld on the other side. There is a

13-mm (½-in.) fillet weld on the arrow side. The fillet on the arrow side is 50 mm (2 in.) long. The pitch of 150 mm (6 in.) indicates that it is spaced 150 mm (6 in.) center to center. All fillet welds are standard at 45°.

SUPPLEMENTARY SYMBOLS

In Fig. 17–18 at A, there is a black, solid dot in the elbow of the reference line. This dot is a supplementary symbol for a field weld. This means that the weld is to be made in the field or on the construction site rather than in the shop.

In the tail of the reference line in Fig. 17–18 at A is the typical specification A2. Its meaning is as follows: The work is to be metal-arc process, using a high-grade, covered, mild-steel electrode; the root is to be unchipped and the welds unpeened, but the joint is to be preheated.

In Fig. 17–18 at A, there is a flush contour symbol over the 13-mm (½-in.) fillet weld symbol. This indicates that the contour of this weld is to be flat-faced and unfinished. Over the 10-mm (⅜-in.) fillet weld on the same reference line there is a convex contour symbol. This indicates that this weld is to be finished to a convex contour. Figure 17–14 shows the supplementary welding symbols to be used for finished welding techniques.

TYPICAL GROOVE WELDS

Figure 17–19 shows the five typical groove welds. Each is shown by its symbol and by a picture as it would apply to a butt joint. At A is a square-grooved joint, at B is a squared both-sides joint, at C is a V-grooved joint, at D is a bevel-grooved joint, and at E is a U-grooved joint.

TYPICAL PLUG AND SLOT WELDS

Examples of plug and slot welds are shown in Fig. 17–20. The welding symbol is the same for both types.

BASIC RESISTANCE WELDING

Resistance welding, as described above, is done with heat and pressure. Electric current is passed through the points to be welded. Resistance to the charge generates the welding heat. Then pressure is used to complete the weld. *Flash welding* is a special kind of resistance welding. It is done by placing the parts to be welded in very light contact or by leaving a very small air gap. The electric current then flashes, or arcs. This melts the ends of the parts, and the weld is made.

BASIC RESISTANCE–WELDING SYMBOLS

There are four basic resistance-weld symbols (Fig. 17–21). They signify spot, projection, seam, and flash or upset welds. The basic reference line and arrow are used with resistance-weld symbols as with arc- and gas-weld symbols. However, in general, there is no arrow side or other side. The same supplementary symbols also apply, as Fig. 17–21 shows.

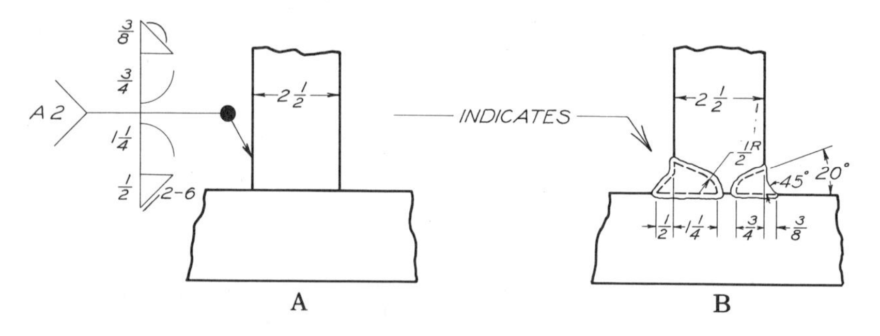

Fig. 17–18 **Reference specifications applied.**

Fig. 17–19 **Five typical groove welds.**

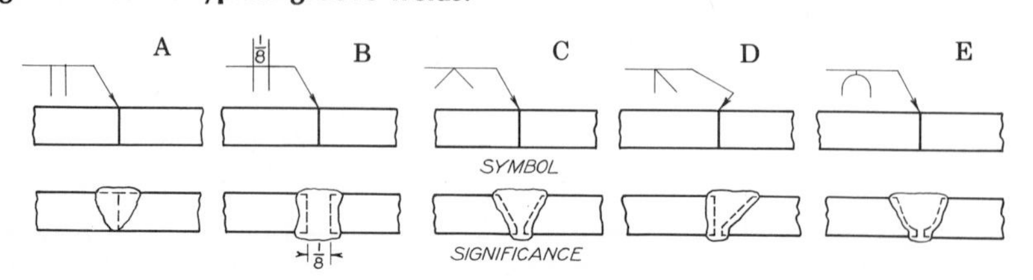

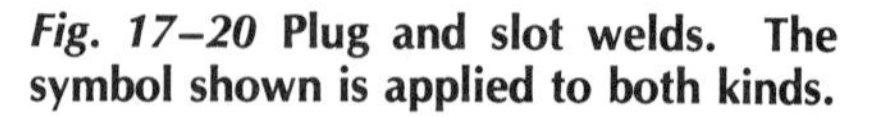

Fig. 17–20 **Plug and slot welds. The symbol shown is applied to both kinds.**

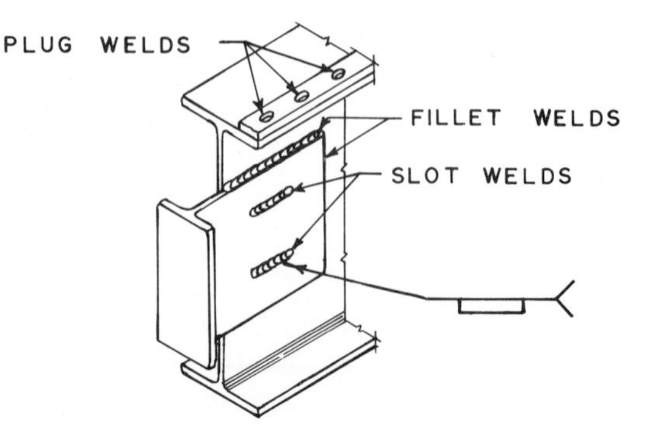

TYPICAL SPOT WELDS

Resistance spot welding is done on lapped parts. The welds are relatively small in area. Figure 17–22 at A shows a reference symbol in the top view. The arrow points to the working centerline of the weld. At A, the minimum diameter of each weld is specified at 7.6 mm (0.30 in.). At B, the minimum shearing strength of each weld is specified at 3.558 kilonewtons (kN) (800 lb.). At C, the reference data indicates that the first weld is centered 25 mm (1 in.) from the left end and the welds are spaced 50 mm (2 in.) from centerline to centerline.

TYPICAL PROJECTION WELDS

A projection weld is identified by strength or size. Figure 17–23 at A and D shows parts set up for such a weld. In each case, one part has a boss projection. At B, the reference 3.10 kN (700 lb.) means that the acceptable shear strength per weld must be at least 3.10 kN (700 lb.). At C, the reference data 2.22 kN (500 lb.) indicates the strength of the weld, the 2 means that the first weld is located 50 mm (2 in.) from the left side, and the 5 indicates a weld every 125 mm (5 in.) center to center. At E, the number 6.0 mm (0.25 in.) indicates the diameter of the weld. At F, the diameter of the weld is 6.0 mm (0.25 in.), there is a weld every 50 mm (2 in.) beginning 25 mm (1 in.) from the left side, and the (5) indicates that there is to be a total of five welds. Notice the arrow-side and other-side indications.

BASIC RESISTANCE WELD SYMBOLS			
TYPE OF WELD			
Spot	Projection	Seam	Flash or upset
○	○	⊖	||

SUPPLEMENTARY SYMBOLS			
Weld all around	Field weld	Contour	
		Flush	Convex
○	●	—	⌒

Fig. 17–21 **Basic resistance-weld symbols.**

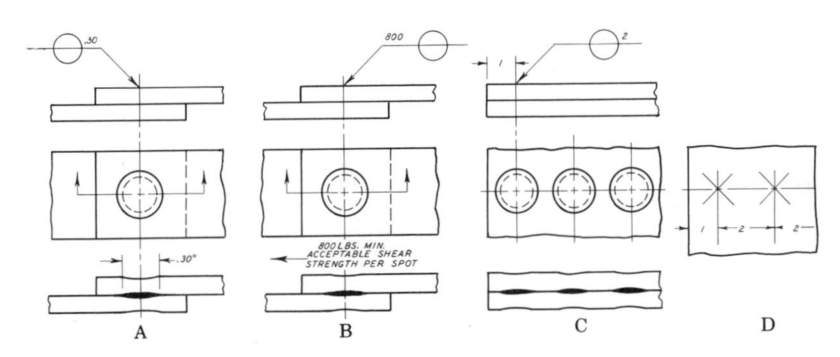

Fig. 17–22 **Examples of spot-welding symbols and their meanings.**

Fig. 17–23 **Examples of projection-welding symbols.**

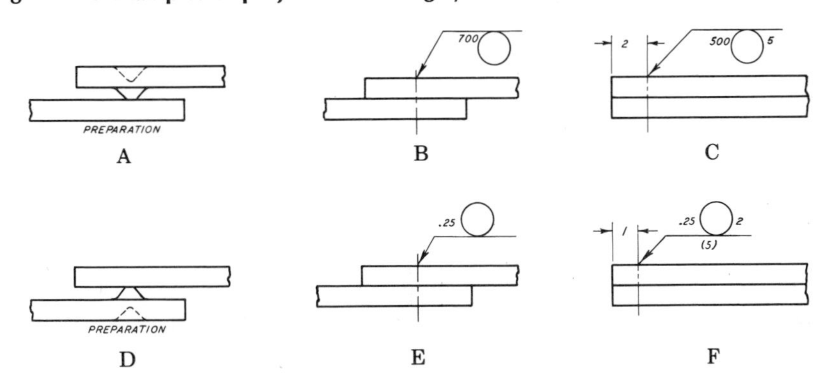

TYPICAL SEAM WELDS

Figure 17–24 at A shows butt-seam and lap-seam welds. The symbol for seam welding is shown at B. The side view shows the two

Fig. 17–24 **Butt-seam and lap-seam welds and symbol.**

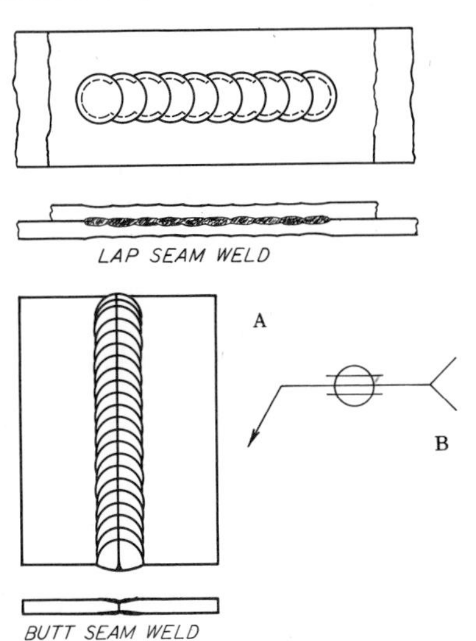

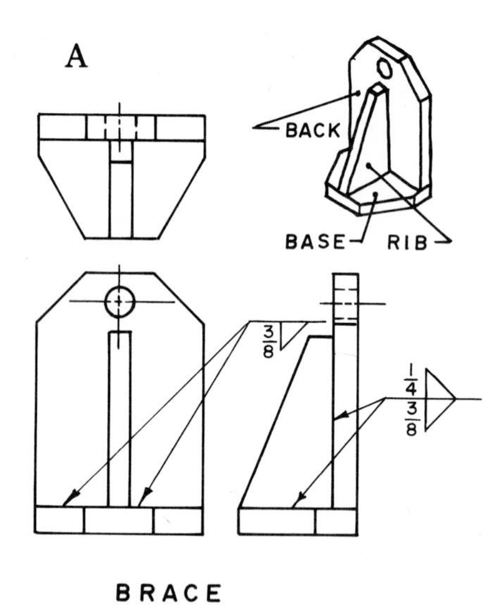

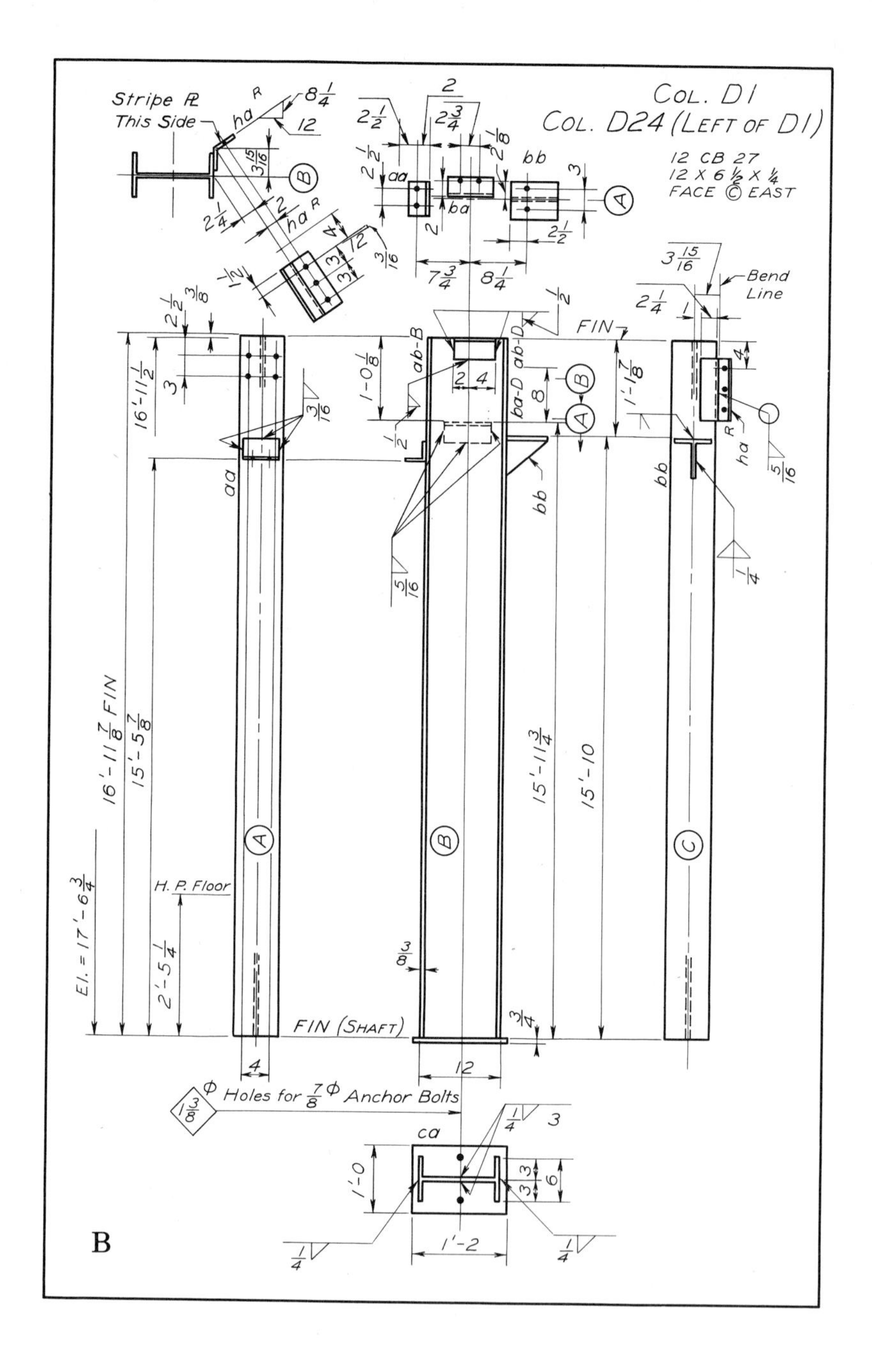

Fig. 17–25 **The application of welding symbols: A, on a machine drawing; B, on a structural drawing.**

pieces positioned edge to edge for butt-seam welding and overlapping for lap-seam welding, which is done with a series of tangent spot welds.

WELDING SYMBOLS

Figure 17–25 shows the use of welding symbols in technical drawings. At A, welding symbols are included on a machine drawing. At B, they are included on a structural drawing.

Review

1. Which kind of welding was developed first, resistance or fusion?

2. The kind of welding that uses only heat is ________.

3. Name the gases that are normally used in gas welding.

4. What is the basic principle of arc welding?

5. What do MIG and TIG mean?

6. Name the basic welding joints.

7. Name the basic types of welds.

8. Identify the several parts of the welding symbol.

Problems

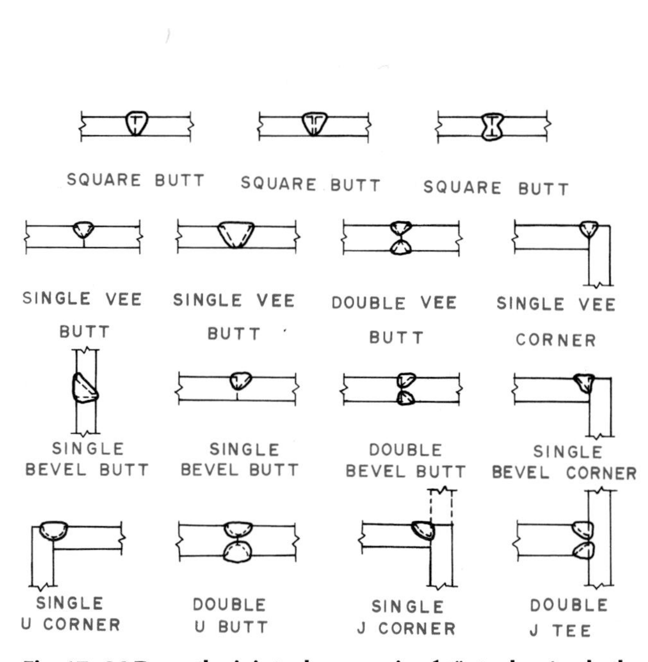

Fig. 17–26 **Draw the joints shown, using ½″ stock. Apply the proper weld symbol.**

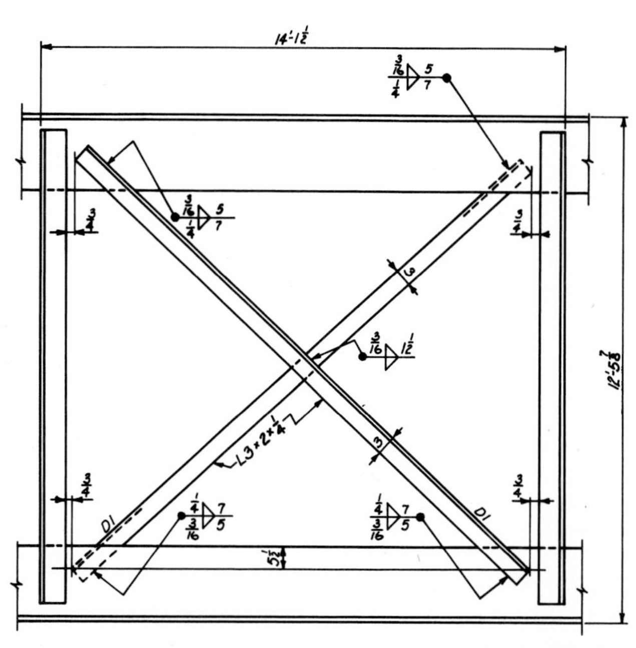

Fig. 17–27 **Make a welding drawing of the double angle cross bracing. Draw the structural members at a suitable scale and note that the horizontal structural members are unidentified. Identify and label structural selections to complement this structure.**

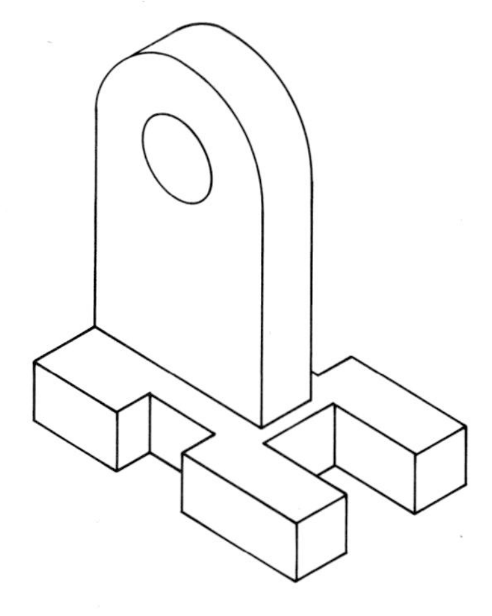

Fig. 17–28 **Make a three-view welding drawing of the lever stand, assigning your own dimensions. Design a weld to support the upright member. Include dimensions and welding symbols on your drawing.**

Fig. 17–29 **Make a three-view welding drawing of the corner lug bracket. Apply weld symbols to indicate permanent assembly. Base = 3″ sq. × ¾″. Uprights = 3″ × 4″ × ½″ and 2″ × 4″ × ½″. Holes = 1″ drill located 1″ from each edge.**

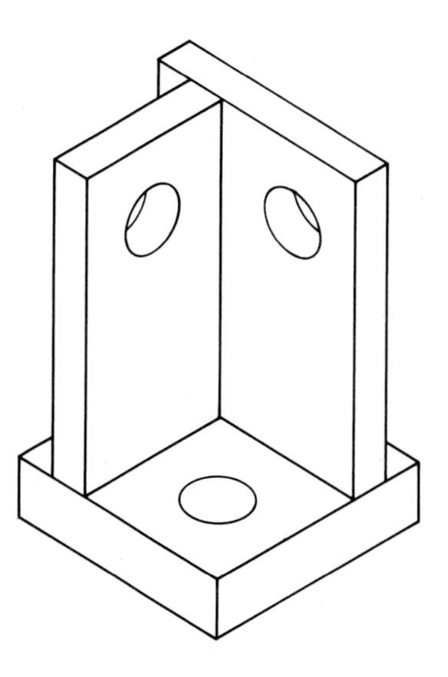

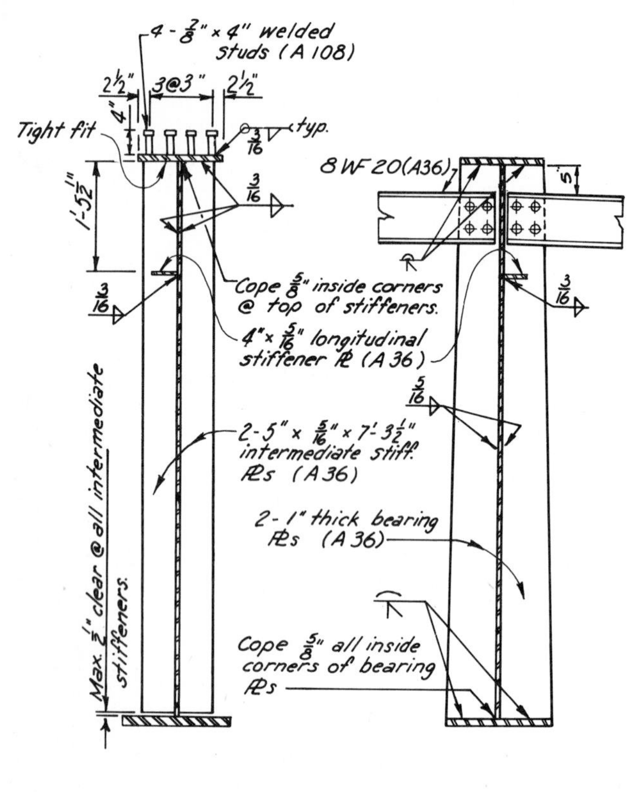

Fig. 17–30 **Make a drawing of each section and place the weld symbols in appropriate locations. Identify the type of weld, and letter the name close to the symbol.**

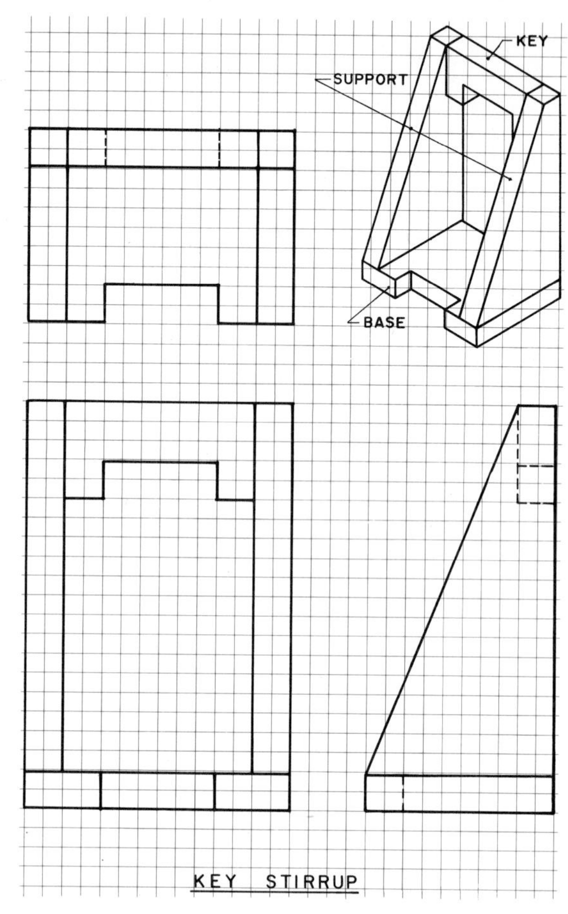

Fig. 17–31 **Make a three-view welding drawing of the key stirrup. Choose appropriate weld symbols to join the four parts. Grid for dimensions = 1/4".**

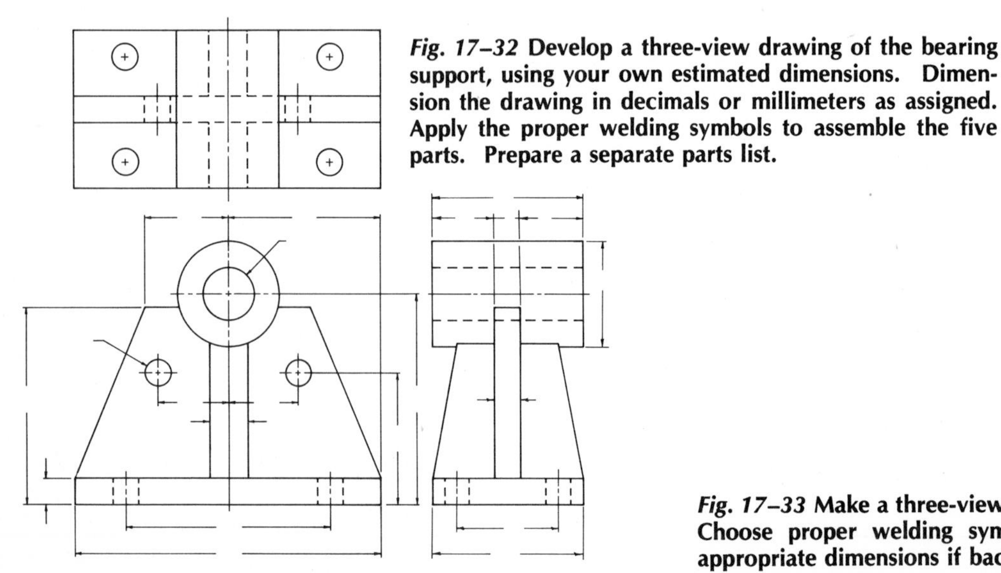

Fig. 17–32 **Develop a three-view drawing of the bearing support, using your own estimated dimensions. Dimension the drawing in decimals or millimeters as assigned. Apply the proper welding symbols to assemble the five parts. Prepare a separate parts list.**

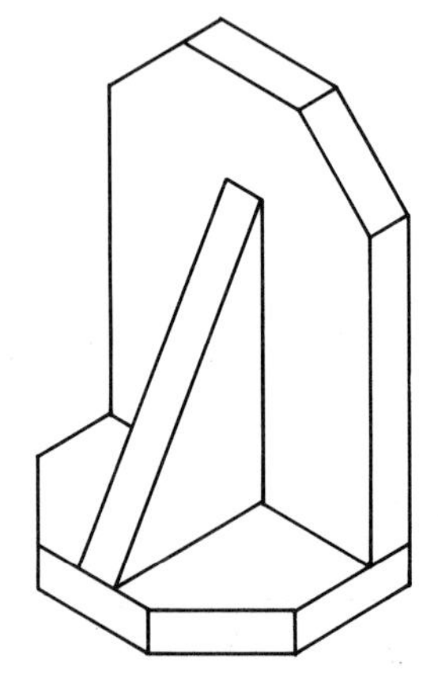

Fig. 17–33 **Make a three-view welding drawing of the brace. Choose proper welding symbols for fabrication. Assign appropriate dimensions if back upright = 3½ × 4¼ × ½.**

Chapter 18

Surface Developments and Intersections

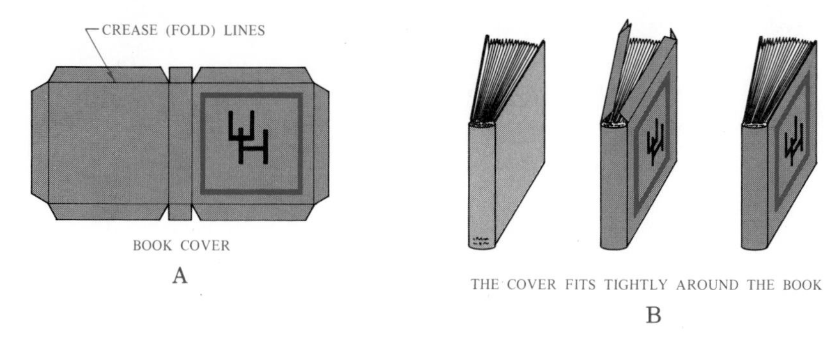

Fig. 18–1 **A book cover is an example of a surface development.**

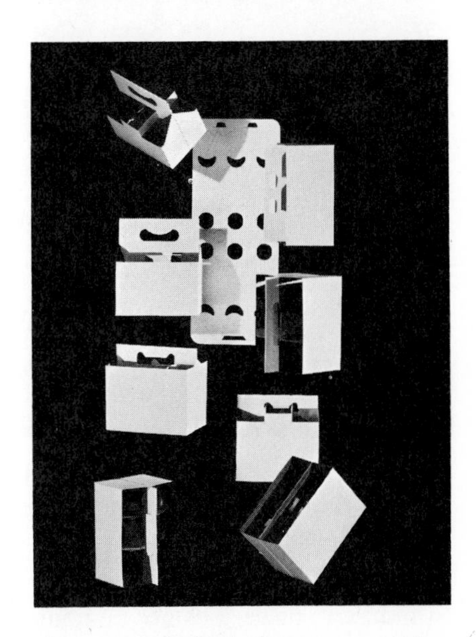

Fig. 18–2 **Soft-drink cartons are cut in a flat pattern and then folded. *(Olin Packaging Division, Olin Mathieson Chemical Corp.)***

SURFACE DEVELOPMENT

The book cover shown in Fig. 18–1 is an example of a surface development. At A, the cover is laid out flat. At B, it is wrapped around a book to make a protective covering. Notice that it fits neatly around all surfaces. It does so because each part has been carefully measured and laid out in relation to other parts. The layout is full size and made on a single flat plane. A surface development is also called a *stretchout,* a *pattern,* or simply a *development.*

Making surface developments is an important part of industrial drafting. Many different industries use surface developments. Many familiar items are made with them, including pipes, ducts for hot- or cold-air systems, parts of buildings, aircraft, automobiles, storage tanks, cabinets, office furniture, boxes and cartons, frozen food packages, and countless other items.

To make any such item, a surface development is drawn as a pattern. Then this pattern is cut on flat sheets of material that can be folded, rolled, or otherwise formed into the required shape. Materials used include paper, various cardboards, plastics and films, metals (such as steel, tin, copper, brass, aluminum), wood, fiberboard, fabrics, and so on.

THE PACKAGING INDUSTRY

Packaging is one very large industry that uses surface developments. Creating packages takes both engineering and artistic skill. The packages must be designed so that they protect their contents during shipment. They must also look attractive for sales appeal. Some packages are meant to be used just briefly and then thrown away. Others are made to last a long time.

Packages and containers for industrial goods are designed to be mass produced at a reasonable cost. A familiar example is the soft-drink carrier (Fig. 18–2). It is made from a flat pattern cut from strong kraftboard. Its surface is white to permit attractive printing and design.

Flat patterns for packages must be printed, cut, creased, folded, glued (if necessary), and completed.

Fig. 18–3 **Twenty-up die on cylinder bed, showing makeready on cylinder. Dies are used to cut the sheet material for making packages.** ***(Olin Packaging Division, Olin Mathieson Chemical Corp.)***

Fig. 18–4 **Phase just prior to glue-lap contact.** ***(Olin Packaging Division, Olin Mathieson Chemical Corp.)***

In industry, these jobs are all done by specially designed machines (Figs. 18–3 and 18–4).

Packages and cartons are made of many materials and in many thicknesses. Some are made of thin or medium-thickness paper stock (Figs. 18–5 and 18–6). This material can be folded easily into the desired form. Some are designed so that no glue is needed. Others may need glue on their tabs.

Packages made of cardboard, corrugated board, and many other materials require allowances for thickness. Examples are boxes made in two parts, a container and a cover (Fig. 18–7), and a slide-in box (Fig. 18–8).

Designing package pattern layouts, lettering, color, and artwork provides jobs for many people. Look at various cartons and packages. You will see many interesting and challenging problems in the development of surfaces.

SHEET–METAL PATTERN DRAFTING

Many different metal objects are made from sheets of metal that are laid out, cut, formed into the re-

Fig. 18–5 **A familiar container made by cutting and folding a flat sheet.**

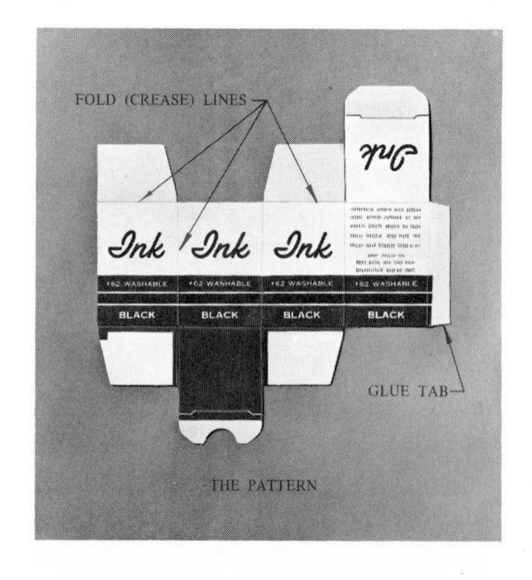

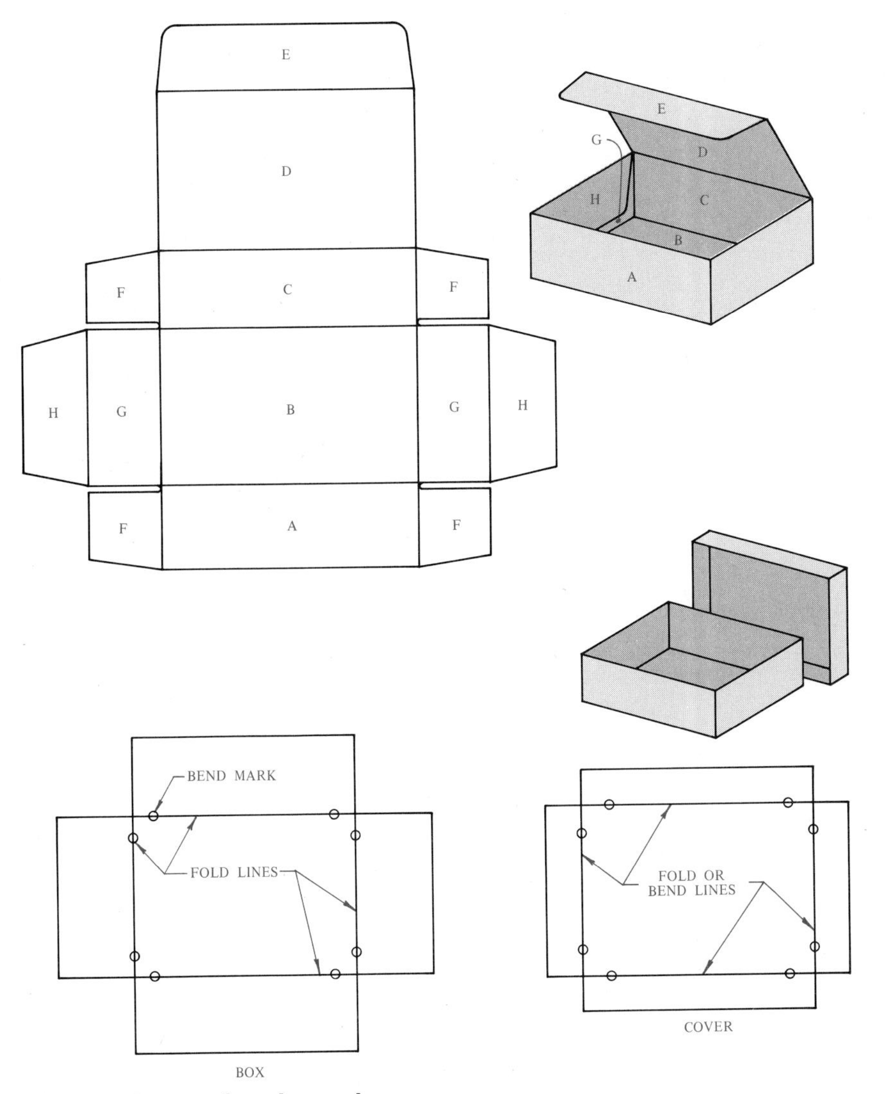

Fig. 18–7 **Patterns for a box and cover.**

quired shape, and fastened together. The shaping is done by bending, folding, or rolling. The fastening is done by riveting, seaming, soldering, or welding.

For each sheet-metal object, two drawings must be made. One is a pictorial drawing of the finished product. The other shows the shape of the flat sheet that, when rolled or folded and fastened, will form the finished object (Fig. 18–9). This second drawing is called the *development,* or *pattern,* of the piece. Drawing it is called *sheet-metal pattern drafting.*

A great many thin-metal objects without seams are formed by *die stamping* (pressing a flat sheet into shape under heavy presses). Examples range from brass cartridge cases and household utensils to steel wheelbarrows and parts of automobiles and aircraft. Other kinds of thin-metal objects are made by spinning. Such objects include some brass- and aluminumware. Stamping and spinning both stretch the metal out of its original shape.

Making sheet-metal objects can involve many operations. These

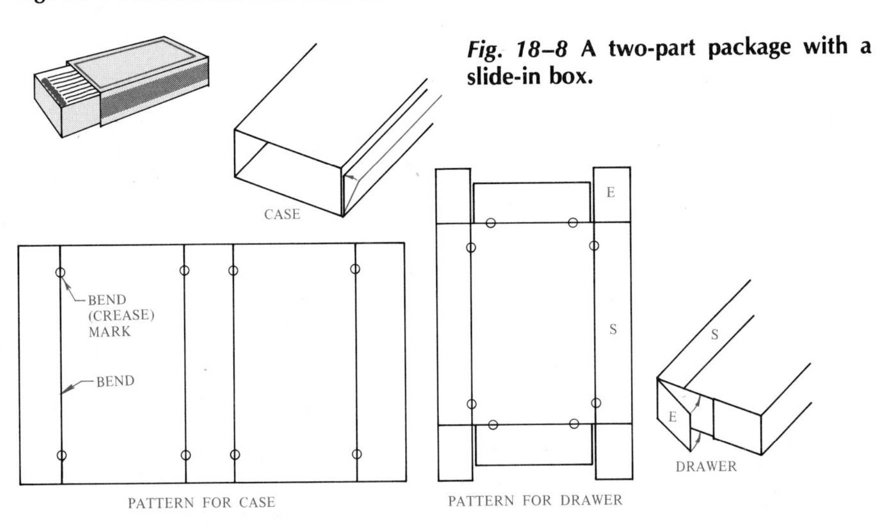

Fig. 18–8 **A two-part package with a slide-in box.**

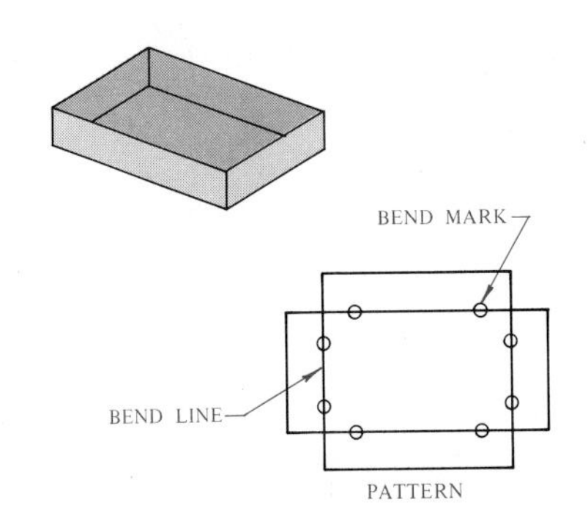

Fig. 18–9 **Pictorial drawing and stretchout of a sheet-metal object.**

Squaring Shears

Used for trimming and squaring sheet metal.

Box and Pan Brake

A small bench-mounted brake for straight bends up to full length of machine or for box and pan work up to 3 inch depth.

A — This shows how bottom edge of body and bottom of can are prepared by Burring Machine for Setting-Down.

B — The Pexto Setting-Down Machine closes the seam as shown here. It works both speedily and accurately.

Setting-Down Machine

The Setting-Down Machine prepares the seams in body of vessels for double seaming.

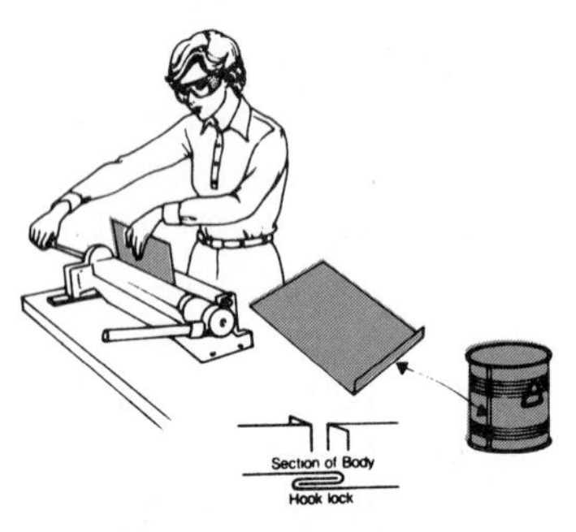

Folding Machine

The Folding Machine is used extensively for edging sheet metal or the forming of locks or angles.

AA — Showing edges turned by Pexto Burring Machine. Note right-angle burr on body of can and a still more pronounced burr on the bottom piece. The edge on bottom is turned smaller than on the body.

Burring Machine

A difficult operation to master, but practice will produce uniform flanges on sheet metal bodies. Prepares the burr for bottoms preparatory to setting down and double seaming.

AA — Seats for Wire — made on the Turning Machine.

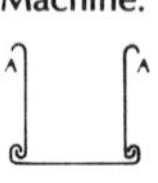

Turning Machine

Used to prepare a seat in bodies to receive a wire. The operation is completed with use of Wiring Machine.

Doubling-Seaming Machine

Offering in various styles and follows the setting-down operation.

Wiring Machine

Works the metal completely and compactly around wire. Depending on shape of work, seats to receive wire are prepared on Folder, Brake or Turning Machine.

Forming Machines

Used for forming flat sheets into cylinders of various diameters, such as stove pipe, the bodies of vessels, cans, etc. Made in a variety of sizes and capacities.

Beading Machine

For ornamenting and stiffening sheet metal bodies.

Fig. 18–10 **Some of the machines used in sheet-metal working.** ***(The Peck, Stow, and Wilcox Co.)***

include cutting, folding, wiring, forming, turning, beading, and so forth. All are done with machines. Some machines and operations are shown in Fig. 18–10. The machines shown are for hand operations. For industrial mass production, large complex automatic equipment is generally used.

DEVELOPMENT

There are two general classes of surfaces: plane and curved. The six faces of a cube are plane surfaces. The top and bottom of a cylinder are also plane surfaces. However, the side surface of the cylinder is curved (Fig. 18–11). There are also different kinds of curved surfaces. Those that can be rolled in contact with a plane surface, such as cylinders and cones, are called *single-curved surfaces.* Exact surface developments can be made for them.

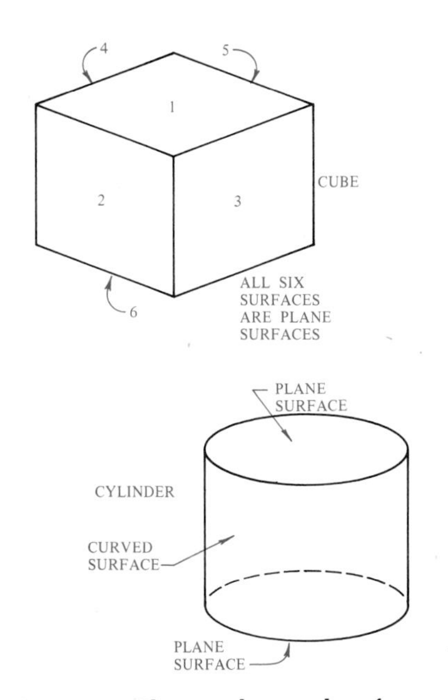

Fig. 18–11 **Plane and curved surfaces.**

Another kind of curved surface is the *double-curved surface.* It is found on spheres and spheroids. Exact surface developments cannot be made for objects with double-curved surfaces. However, drafters can make approximations.

Figure 18–12 shows how to cut a piece of paper so that it can be folded into a cube. The shape cut out is the pattern of the cube. Figure 18–13 shows the patterns for all five of the regular solids. If you wish to understand surface development better, lay these patterns out on rather stiff drawing paper. Then cut them out and fold them to make the solids. Secure the joints with tape.

You can make any solid that has plane surfaces in the same way. Just make sure your pattern shows each plane in its proper relationship to the others.

SEAMS AND LAPS

Drawing developments is only part of sheet-metal pattern drafting. Drafters must also know about the processes of wiring, hemming, and seaming. In addition, they must know how much material should be added for each. *Wiring* involves reinforcing open ends of articles by enclosing a wire in the edge. How this is done is shown in Fig. 18–14 at A. To allow for wiring, a drafter must add a band of material to the pattern equal to 2.5 times the diameter of the wire. *Hemming* is another way of stiffening edges. Single- and double-hemmed edges are shown at B and C. Edges are fastened by soldering on lap seams (D), flat lock seams (E), or grooved seams (F). Other types of seams and laps are shown at G and H. Each has its own general or specific use. How much material is allowed in each case depends on the thickness of material, method of fastening, and application. In most cases, the corners of the lap are notched to make a neater joint.

Fig. 18–12 **Pattern for a cube.**

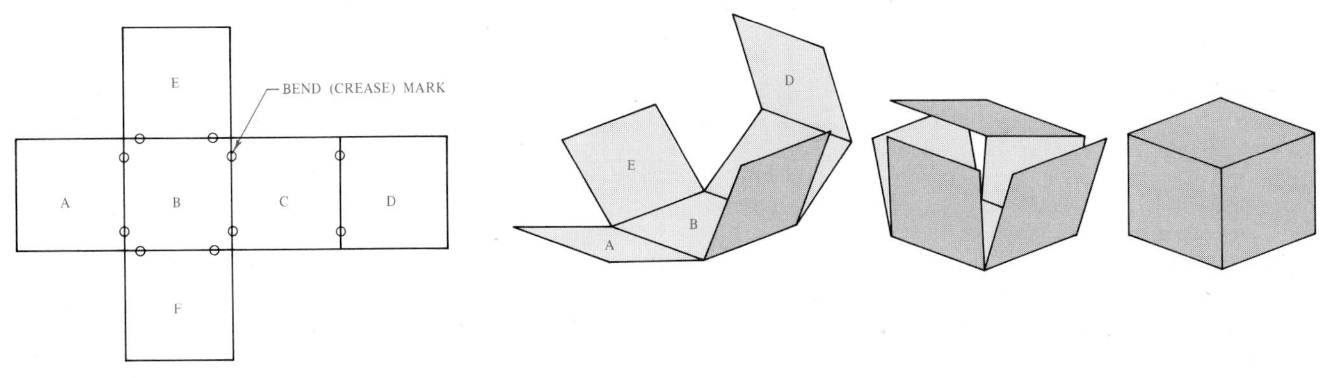

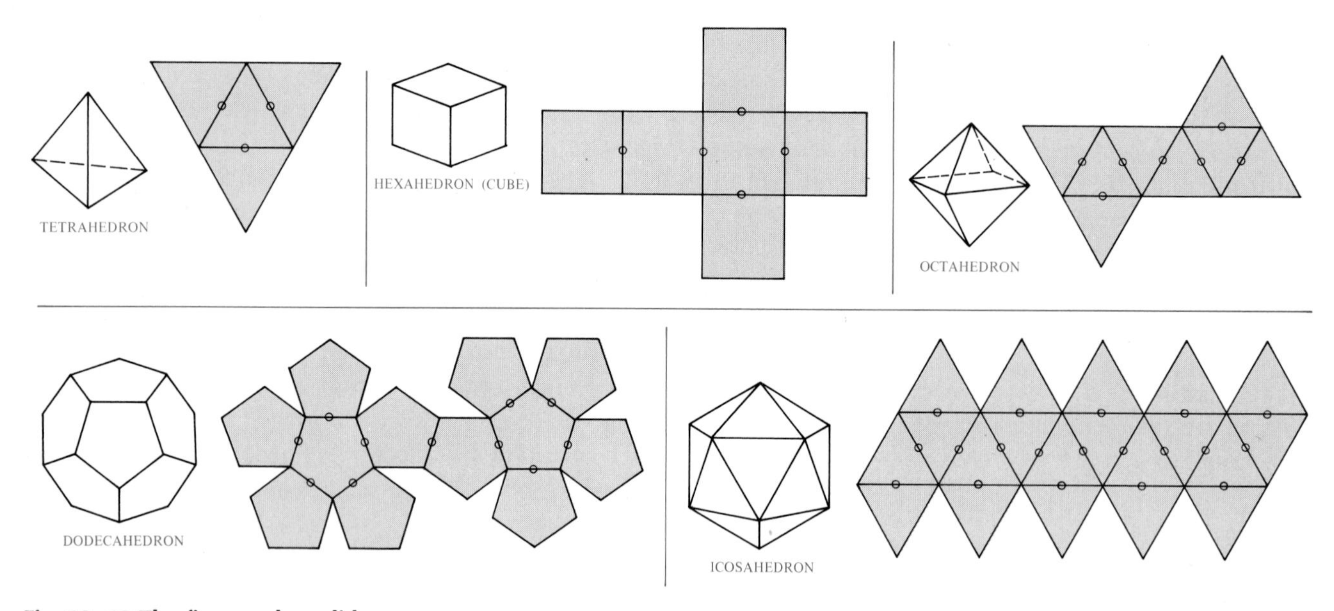

Fig. 18–13 **The five regular solids.**

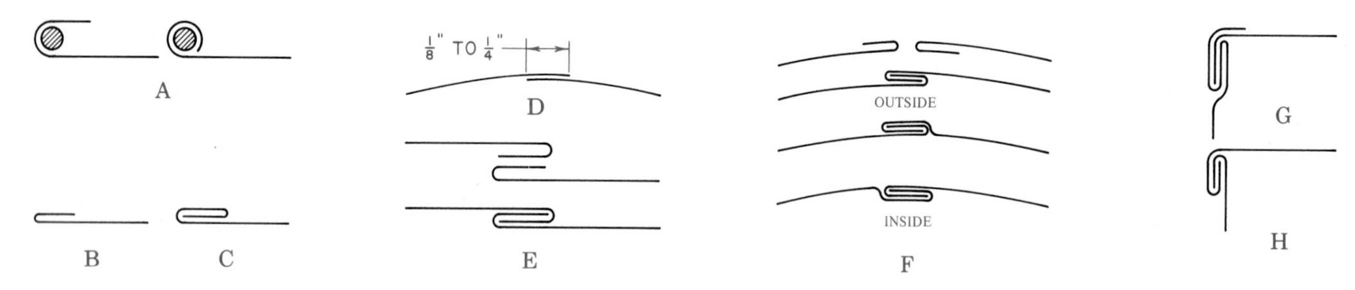

Fig. 18–14 **Wiring, seaming, and hemming.**

PARALLEL–LINE DEVELOPMENT

Parallel-line development is a simple way of making a pattern. It is done by drawing the edges of an object as parallel lines. The patterns in Figs. 18–12 and 18–15 are made in this way. Figure 18–16 is a pictorial view of a rectangular prism. In Fig. 18–17, a pattern for this prism is made by parallel-line development. To draw this pattern, proceed as follows:

A. Draw the front and top views full size. Label the points as shown (18–17A).

B. Draw the *stretchout line,* SL. Find the lengths of sides 1–2, 2–3, 3–4, and 4–1 in the top view. Measure off these lengths on SL (Fig. 18–17B).

C. At points 1, 2, 3, 4, and 1 on the stretchout line, draw vertical *crease* (fold) *lines.* Make them equal in length to the height of the prism (Fig. 18–17C). These are also called *measuring lines.*

D. Project the top line of the pattern from the top of the front view. Make it parallel to SL. Darken all outlines until they are thick and black. You can use a small circle or X to identify a fold line (Fig. 18–17D).

E. Add the top and bottom to the pattern by transferring distances 1–4 and 2–3 from the top view, as shown in Fig. 18–17E. Laps may be added for assembly of the prism.

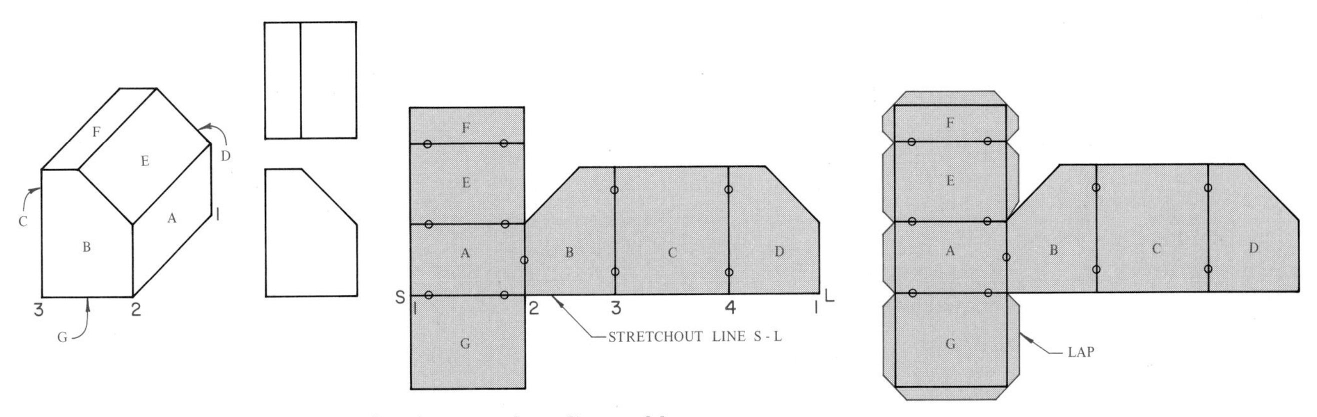

Fig. 18–15 **A pattern for a prism, showing stretchout line and lap.**

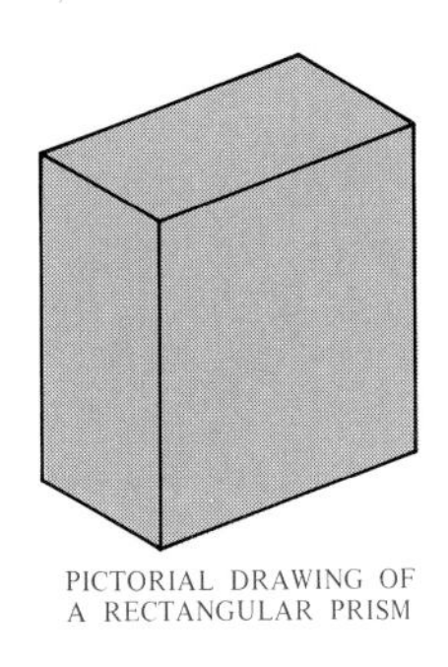

Fig. 18–16 **Pictorial drawing of a rectangular prism.**

Fig. 18–17 **Parallel-line development of a rectangular prism.**

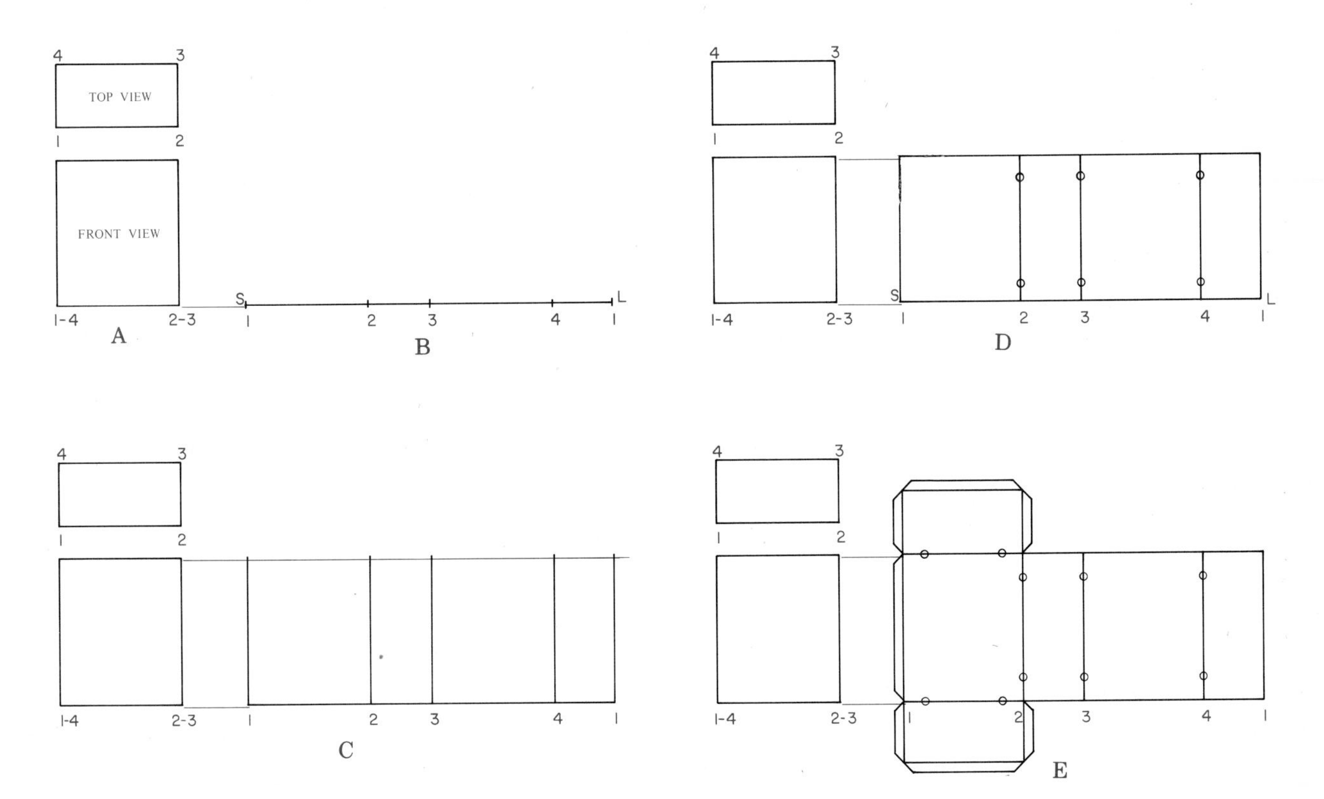

A slight variation is the pattern for a truncated prism shown in Fig. 18–18. To draw it, first make front, top, and auxiliary views full size. Label points as shown. The next two steps are the same as steps B and C in Fig. 18–17. Then, project horizontal lines from points *A–B* and *C–D* on the front view to locate points on the pattern. Connect the points to complete the top line of the pattern. Add the top and bottom as shown.

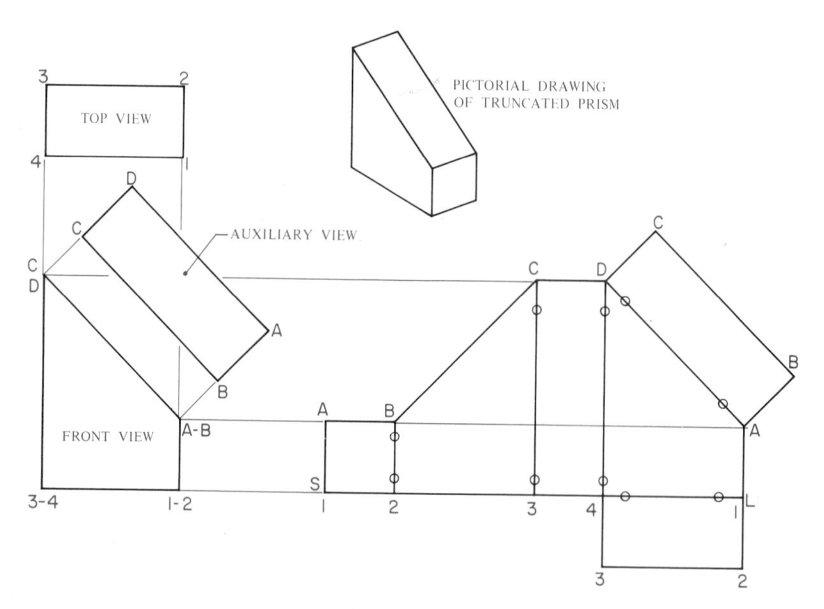

Fig. 18–18 **Development of a pattern for a truncated prism.**

CYLINDERS

Figure 18–19 shows a surface development for a cylinder. It is made by rolling the cylinder out on a plane surface.

In the pattern for cylinders, the stretchout line is straight and equal in length to the circumference of the cylinder. If the base of the cylinder is perpendicular to the axis, its rim will roll out to form the straight line. If the base is not perpendicular to the axis, you will have to make a right section to get the stretchout line.

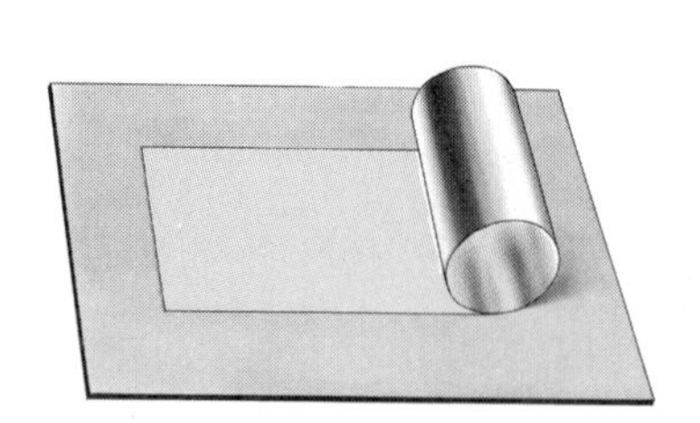

Fig. 18–19 **Developed surface of a right circular cylinder.**

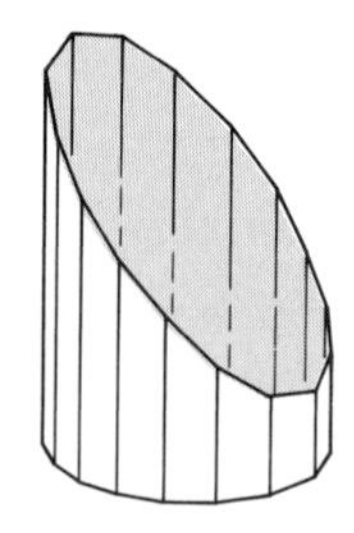

Fig. 18–20 **Pictorial drawing of a truncated right cylinder.**

DEVELOPMENT OF CYLINDERS

Imagine that the cylinder is actually a many-sided prism. Each side forms an edge called an *element.* On the surface of the cylinder, however, these elements seem to form a smooth curve. That is because there are so many of them and they are so close together. Imagining the cylinder this way will help you find the length of the stretchout line. This length will equal the total of the distances between all the elements. Technically, of course, the elements are infinite in number. But, for your purposes, you need only mark off elements at convenient equal spaces around the circumference of the cylinder. Then add up these spaces to make your stretchout line.

Figure 18–20 is a pictorial view of a truncated right cylinder. Figure 18–21 shows how to develop a pattern for this cylinder. To draw this pattern, proceed as follows:

1. Draw the front and top views full size. Divide the top view into a convenient number of equal parts (12 in this case). (See Fig. 18–22 for various methods of dividing a circle.) This will locate a set of equally spaced points (the tips of the elements around the edge in the top view).
2. Draw the stretchout line (SL). Its actual length will be determined later, when you mark off the elements.
3. Using a divider, find the distance between any two elements in the top view. Then mark off this distance along SL as many times as there are parts in the top view. Label the points thus found as

shown. Then, from each point, draw a vertical construction line upward.

4. Project other lines downward from the elements on the top view to the front view. Label the points where they intersect the front view.
5. From these intersection points, project horizontal construction lines toward the development.
6. Locate corresponding points where the horizontal construction lines meet the vertical lines from SL. Connect these points in a smooth curve.
7. Darken outlines and add laps as necessary. The red arrows on the figure show the direction in which the various lines are projected.

Since the surface of a cylinder is a smooth curve, your pattern will not be wholly accurate. This is because it was made by measuring distances on a straight line rather than on a curve. Figure 18–23, which looks like part of the top view of the cylinder discussed above, shows that the distance from point to point is slightly less along a straight line than along the arc. In most cases, however, the difference is so slight that the inaccuracy is not critical. You can find the difference by figuring the actual length of the arc using the formula, circumference = πD. Then, measure this distance out along the stretchout line.

A slightly different method for developing a cylinder is shown in Fig. 18–24. In this case, a front and a half-bottom view are used. Attaching the two views saves time and increases accuracy. Notice that

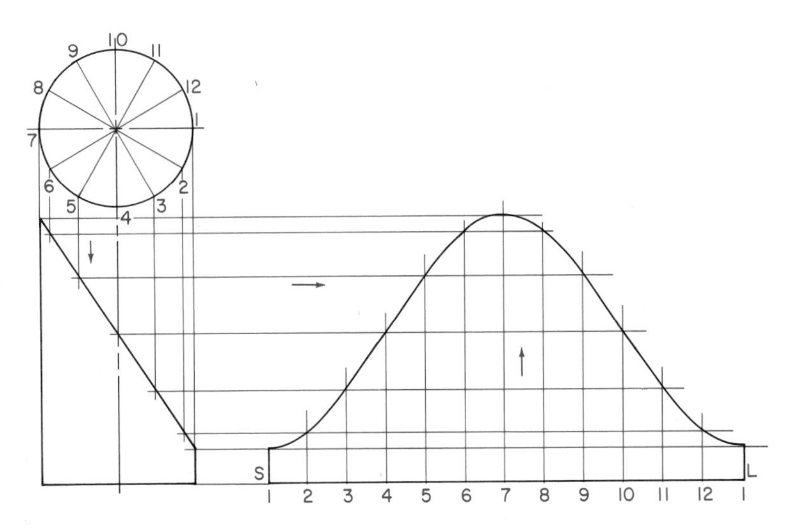

Fig. 18–21 **Development of a pattern for a cylinder.**

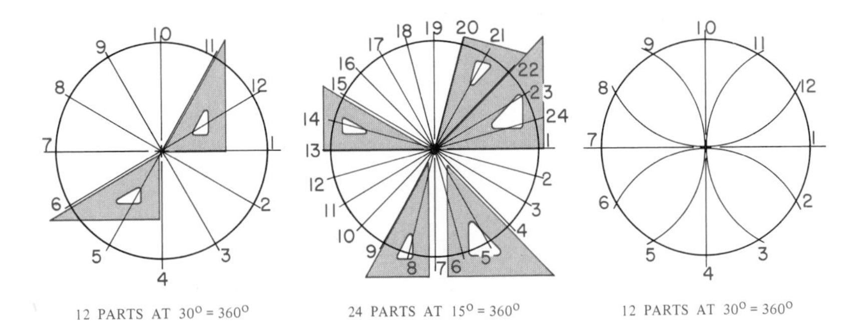

Fig. 18–22 **Dividing a circle.**

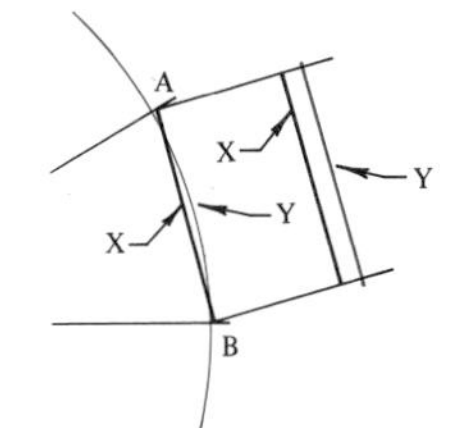

Fig. 18–23 **A straight line is the shortest distance between two points.**

Fig. 18–24 **Development of a pattern for a cylinder, using a front and half-bottom view.**

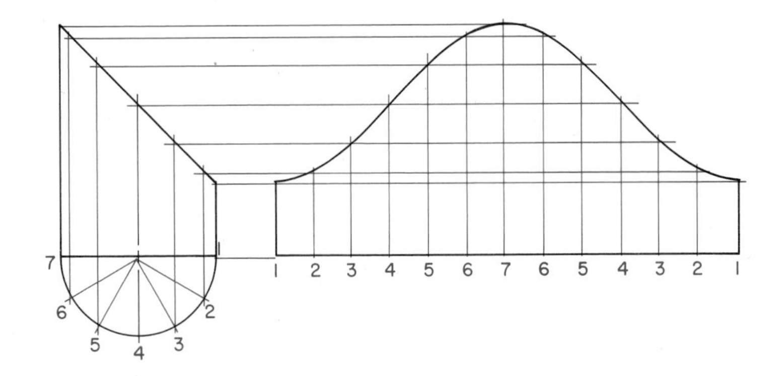

both methods produce the same development.

TO DRAW THE PATTERN FOR A TWO-PIECE, OR SQUARE, ELBOW

This elbow consists of two cylinders cut off at 45°. Therefore, you need only make a pattern for one part, as shown in Fig. 18–25. Allow a lap depending on the type of seam to be made. If you do not have to allow for a lap on the curved edges, you can develop both parts on one stretchout, as shown in Fig. 18–26. Notice that the seam on part A is on the short side, while on part B it is on the long side. In Fig. 18–25, the seam on both pieces would be on the short side. In most cases, this is not critical.

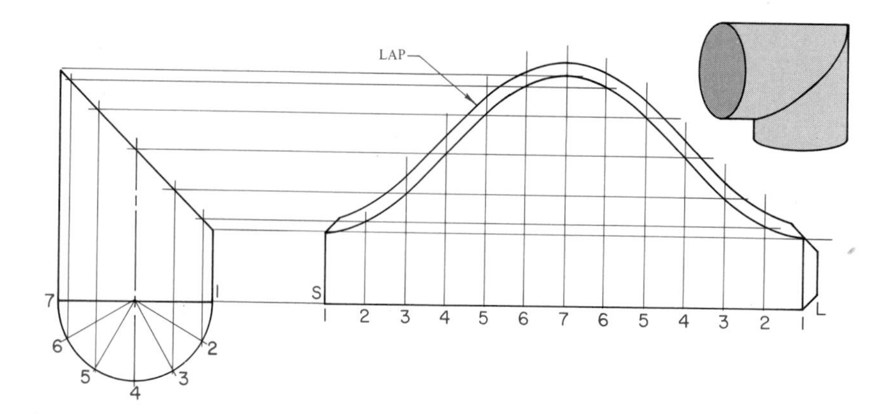

Fig. 18–25 **Pattern for a square elbow.**

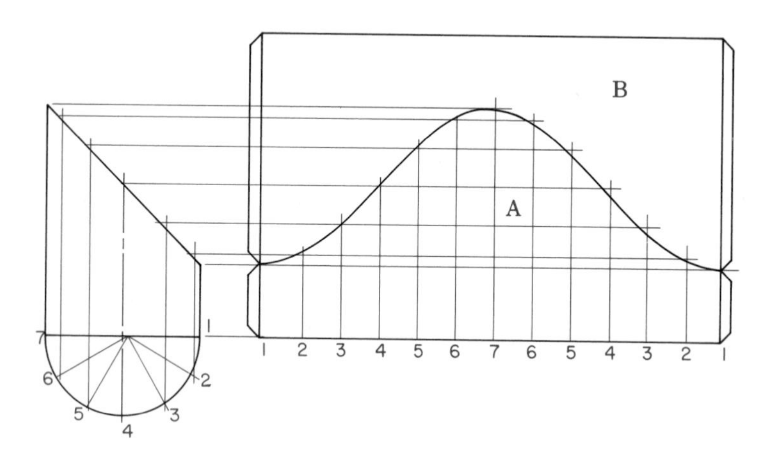

Fig. 18–26 **Both parts of the pattern may be made on one stretchout.**

THE DEVELOPMENT OF A FOUR-PIECE ELBOW

The pattern for a four-piece elbow is shown in Fig. 18–27. To draw it, proceed as follows:

A. Draw arcs with the desired inner and outer radii, as shown at A. Divide the outer circle into six equal parts. Draw radial lines from points *1, 3,* and *5* to locate the joints (seams). Draw tangents to the arcs through points

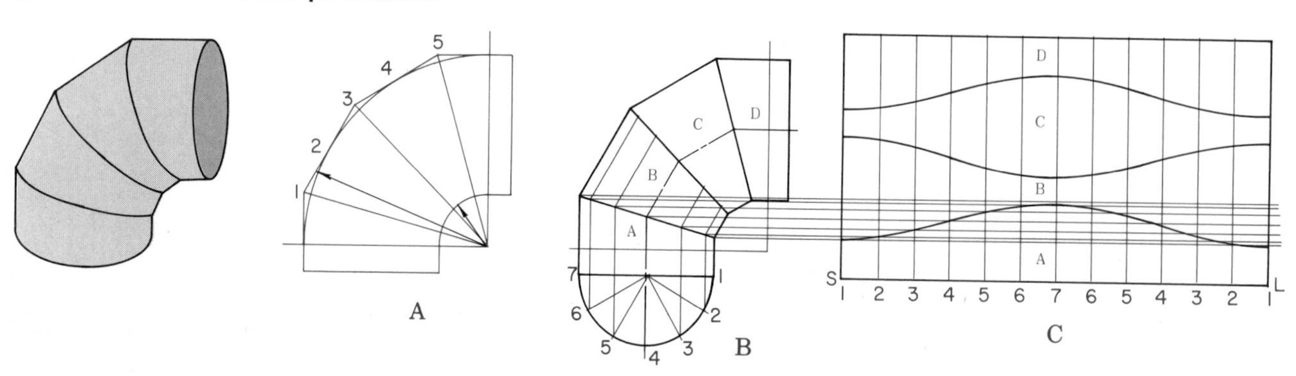

Fig. 18–27 **Pattern for a four-piece elbow.**

2 and *4* on both arcs to complete the front view.

B. Draw a half-bottom view and divide it as shown at B. Project the elements from this view to the front view.

C. Develop the pattern for part *A* just as the pattern for the square elbow was developed in Figs. 18–25 and 18–26. If you do not have to allow for laps on the curved edges, you can draw the patterns for all four parts as at C. To find what lengths to mark off on the vertical lines, work from the front view. In the front view of part *B,* for example, continue the construction lines from the half-bottom view, but make them parallel to the surface lines. Then measure all the lines within part *B*. These lengths, starting with the longest, are the ones to use in the pattern of part *B*. The curves for all the parts are the same. Therefore, you need plot only one of them. You can then use it as a *template* (pattern) for the others.

RADIAL–LINE DEVELOPMENT

In the patterns for prisms and cylinders, the stretchout line is straight, and the measuring lines are perpendicular to it and parallel to each other. Hence the name parallel-line development. On cones and pyramids, however, the edges are not parallel. Therefore the stretchout line will not be a continuous straight line. Also, the measuring lines, instead of being parallel to each other, will *radiate* (project) from a single point. This type of development is called *radial-line development.*

CONES

Imagine the curved surface of a cone as being made up of an infinite number of triangles, each running the height of the cone. To develop the surface, you would roll out each of these triangles, one after another, on a plane. The resulting pattern would look like a sector of a circle. Its radius would be equal to an element of the cone, that is, a line from the cone's tip to the rim of its base. Its arc would be the length of the rim of the cone's base. The developed surface of a cone is shown in Fig. 18–28.

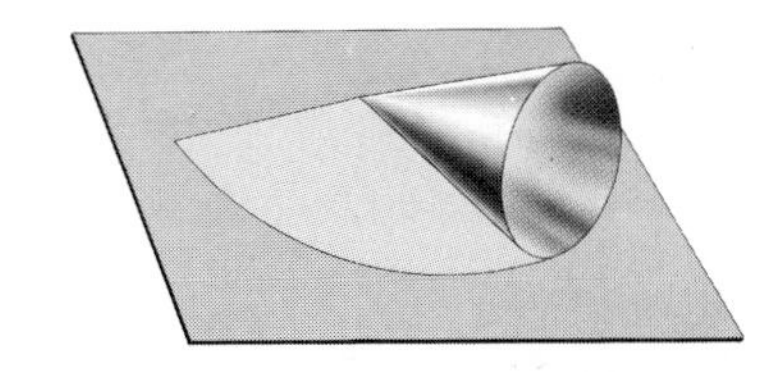

Fig. 18–28 **Developed surface of a cone.**

TO DRAW THE PATTERN FOR A RIGHT CIRCULAR CONE

A right circular cone is one in which the base is a true circle and the tip is directly over the center of the base (Fig. 18–29A). A *frustum* of a cone is made by cutting through the cone on a plane parallel to the

Fig. 18–29 **Development of a pattern for a cone and a frustum of a cone.**

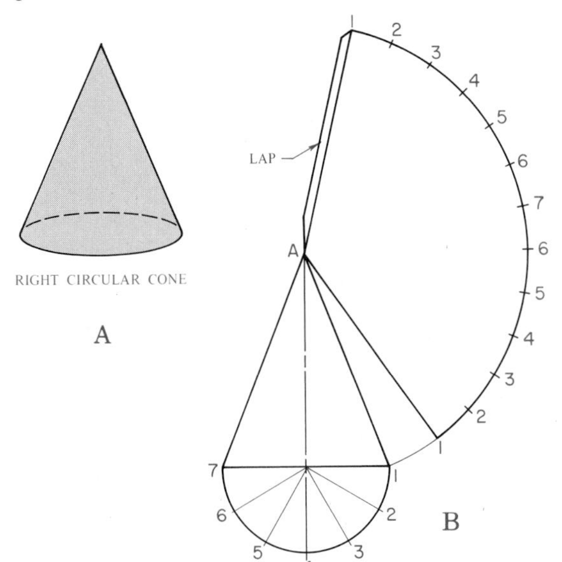

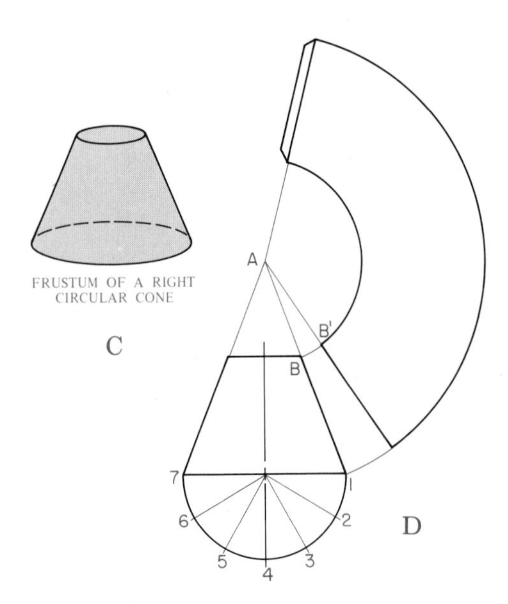

base (Fig. 18–29C). The pattern for a cone is shown in Fig. 18–29B. To draw it, proceed as follows:

1. Draw front and half-bottom views to the desired size.
2. Divide the half-bottom view into several equal parts. Label the division points as shown.
3. On the front view, measure the *slant height* of the cone, that is, the true distance from the *apex* (tip) to the rim of the base (line *A*1). Using this length as a radius, draw an arc of indefinite length as a measuring arc. Draw a line from the apex *(A)* to the arc at any point a short distance from the front view.
4. With a divider, find the straight-line distance between any two division points on the half-bottom view. Then use this length to mark off spaces 1–2, 2–3, 3–4, and so forth, along the arc. Label the points to be sure none have been missed. Complete the development by drawing line *A*1 at the far end.
5. Add laps for the seam as desired. How much to allow for the seam depends on the size of the development and the type of joint to be made.

Figure 18–29D shows the development for a frustum of a cone. To draw it, you use the same method as in Fig. 18–29B, except that you draw a second arc *AB* from point *B* on the front view.

THE PATTERN FOR A TRUNCATED CIRCULAR CONE

A truncated circular cone is a circular cone that has been cut along a plane that is not parallel to the base (Fig. 18–30A). The pattern for such a cone is shown in Fig. 18–30B. To draw it, proceed as follows:

1. Draw the front and top or bottom views. You can use a half-top or bottom view if you wish.
2. Proceed as in Fig. 18–29 to develop the overall layout for the pattern.
3. Project points 1 through 6 from the bottom view to the front view and then to the apex. Label the points where they intersect the *miter* (cut) line to avoid mistakes. These lines, representing elements of the cone, do not show in true length on the front view. Their true length shows only when they are projected to the points on the arc.
4. Project the elements of the cone from the apex to the points on the arc.
5. In the front view, find the points on the miter line that were located in Step C. Project horizontal lines from them to the edge of the front view. Continue these lines as arcs through the development. Mark the points where they intersect the element lines. Join these points in a smooth curve. Complete the pattern by adding a lap.

TO FIND THE TRUE LENGTH OF AN EDGE OF A PYRAMID

In Fig. 18–31, the pyramid at A is shown at B in top and front views. In neither view does the edge *OA* show in true length. However, if the pyramid were in the position shown at C, the front view would show *OA* in true length. At C, the pyramid has been revolved about a vertical axis until *OA* is parallel to the vertical plane. At D, the line *OA* is shown before and after revolving.

Fig. 18–30 **Development of a pattern for a truncated right circular cone.**

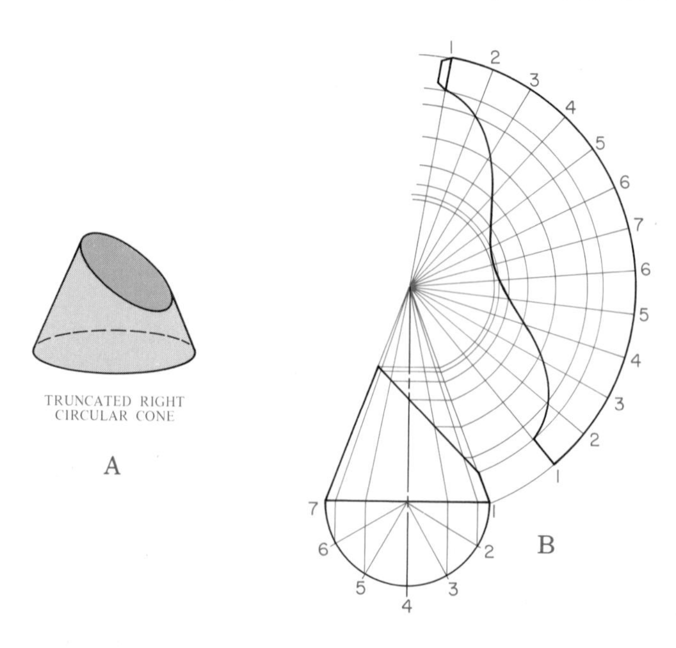

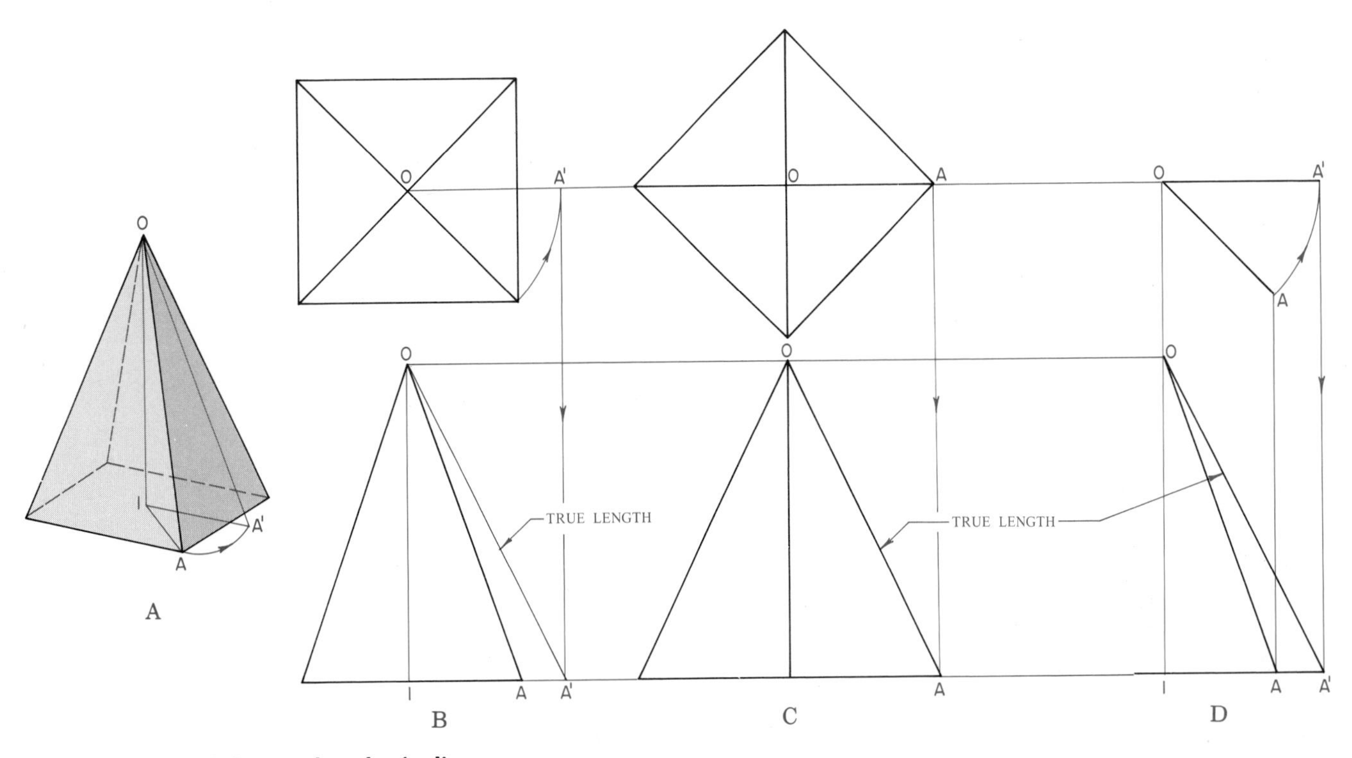

Fig. 18–31 **To find the true length of a line.**

The construction at D is a simple way to find the true length of the edge line *OA*. First, draw the top view of *OA*. Revolve this view to make the horizontal line *OA′*. Project *A′* down to meet a base line projected from the original front view. Draw a line from this intersection point to a new front view of *O*. This line will show the true length of *OA*.

TO DRAW THE PATTERN FOR A RIGHT RECTANGULAR PYRAMID

Figure 18–32 shows the pattern for a right rectangular pyramid. To draw it, proceed as follows:

1. Find the true length of one of the edges (*O*1 in this case) by revolving it until it is parallel to the vertical plane (*O*1′).
2. With the true length as a radius, draw an arc of indefinite length to use as a measuring arc.
3. On the top view, measure the lengths of the four base lines. Mark these lengths off as straight-line distances along the arc.
4. Connect the points and draw crease lines. Mark the crease lines if desired.

TO DEVELOP THE PATTERN FOR AN OBLIQUE PYRAMID OR A TRUNCATED OBLIQUE PYRAMID

Figure 18–33 shows a development of an oblique pyramid. To draw it, proceed as follows:

1. First, find the true lengths (TL) of the lateral edges. Do this by revolving them parallel to the vertical plane, as is shown for edges *O*2 and *O*1. These edges are both revolved in the top view, then projected to locate 2′ and 1′. Lines *O*2′ and *O*1′ in the front view are the true lengths of edges *O*2 and *O*1. Edge *O*2 = edge *O*3. Edge *O*1 = edge *O*4.
2. Start the development by laying off 2–3. Since edge *O*2 = edge *O*3, you can locate point *O* by plotting arcs centered on 2 and 3 and with radii the true length of *O*2 (*O*2′). Point *O* is where the arcs intersect.
3. Construct triangles *O*–3–4, *O*–4–1, and *O*–1–2 with the true lengths of the sides to complete the development of the pyramid as shown.

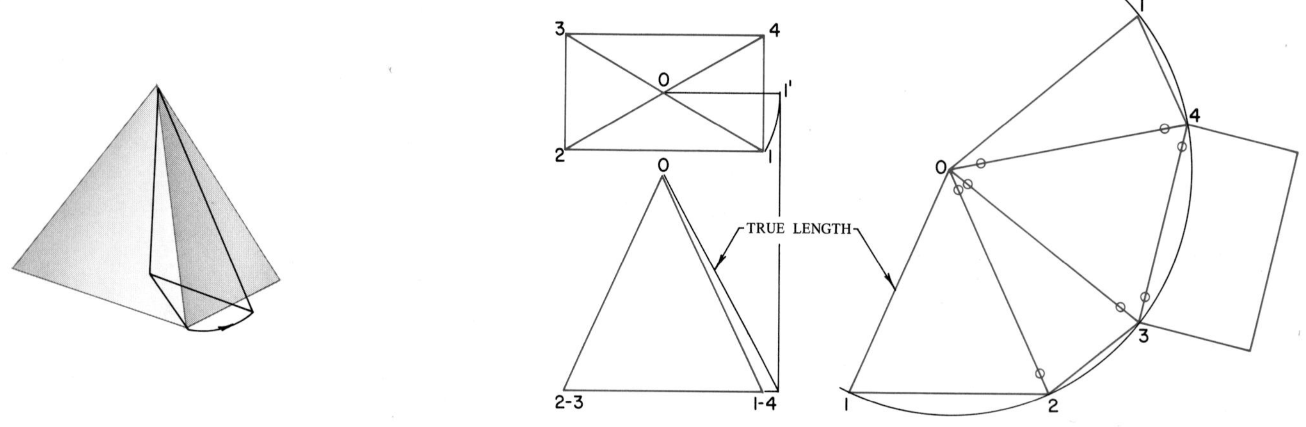

Fig. 18–32 **Development of a pattern for a right rectangular pyramid.**

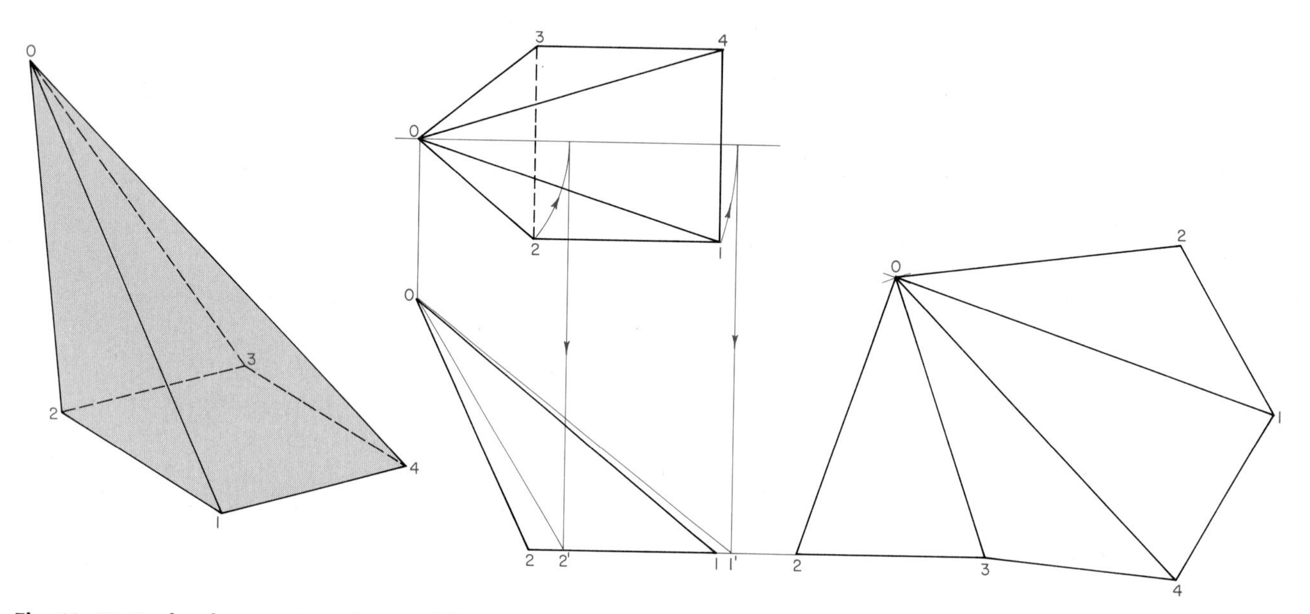

Fig. 18–33 **To develop a pattern for an oblique pyramid.**

Figure 18–34 shows a pattern for a truncated oblique pyramid. The pyramid has an inclined surface *ABCD*. If it were not truncated, it would extend to the apex point *O*. To draw the pattern, first find the true lengths of *OA, OB, OC,* and *OD*. For this pyramid, $OA = OD$ and $OB = OC$. To locate these lengths, locate *B′* and *A′* in the front view. Lines *OB′* and *OA′* will give the true lengths. Then make a development of the pyramid as if it extended to point *O*. Now lay off the true lengths you have found on the corresponding edges of this development. Join them to make the pattern for the edge of the frustrum. The inclined surface is shown in the auxiliary view and could be attached to the rest of the pattern as indicated.

TRIANGULATION

Triangulation is a convenient method for making approximate developments of surfaces that cannot be exactly developed. It involves dividing the surface into triangles, finding the true lengths of the sides, and then constructing the triangles in regular order on a plane. Because the triangles will have one short side, on the plane they will

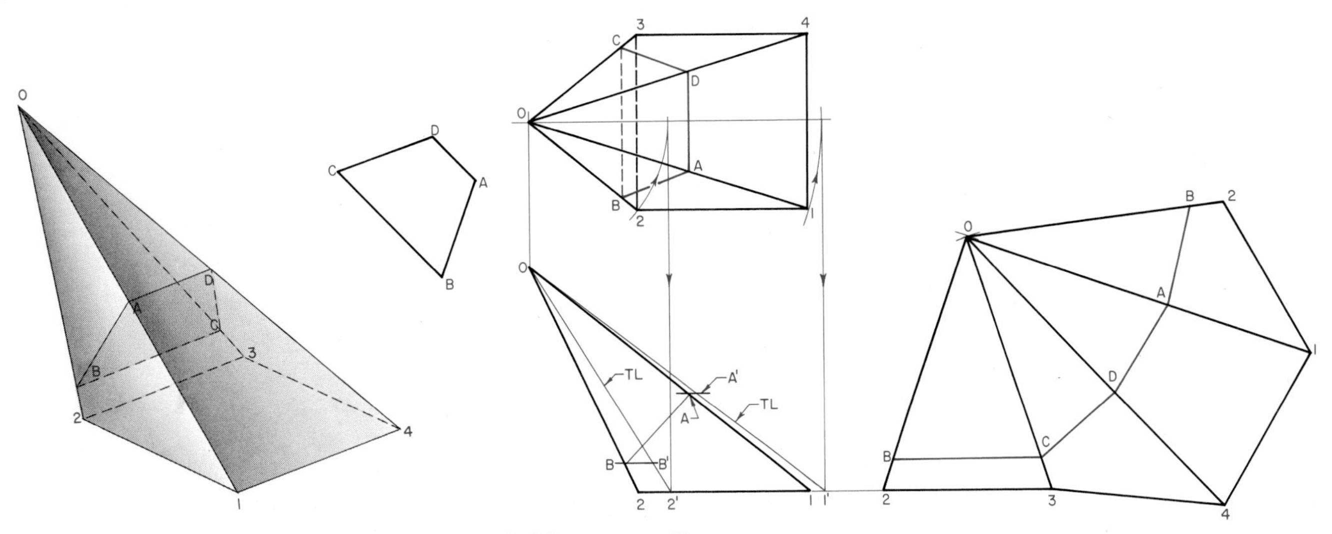

Fig. 18–34 **To develop a pattern for a truncated oblique pyramid.**

approximate the curved surface. Triangulation is also sometimes used for single-curved surfaces.

Figure 18–35 shows how triangulation is used in developing an oblique cone. To draw this development, proceed as follows. Draw elements on the top- and front-view surfaces to create a series of triangles. Number the elements 1, 2, and so forth. For a better approximation of the curve, use more triangles than are shown here. Find the true lengths of the elements by revolving them in the top view until each is horizontal. From the tip of each, project down to the front-view base line to get a new set of points 1, 2, and so forth. Connect these with the front view of point *O* to make a true-length diagram, as at C. To plot the development at D, construct the triangles in the order in which they occur. Take the distances 1–2, 2–3, and so forth, from the top view. Take the distances *O*1, *O*2, and so forth, from the true-length diagram.

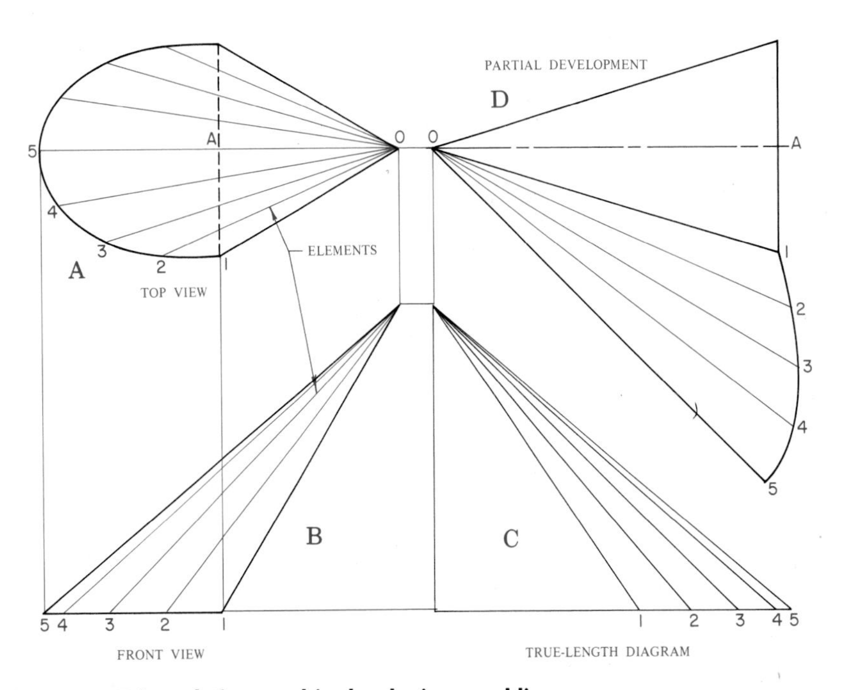

Fig. 18–35 **Triangulation used in developing an oblique cone.**

TRANSITION PIECES

Transition pieces are used to connect pipes or openings of different shapes, sizes, or positions. Transition pieces have a surface that is a combination of different forms, including planes, curves, or both. A few transition pieces are shown in Fig. 18–36.

Figure 18–37 shows the making of a square-to-round transition piece. According to the manufacturer, the piece is easily made in two parts. The illustration shows how one of

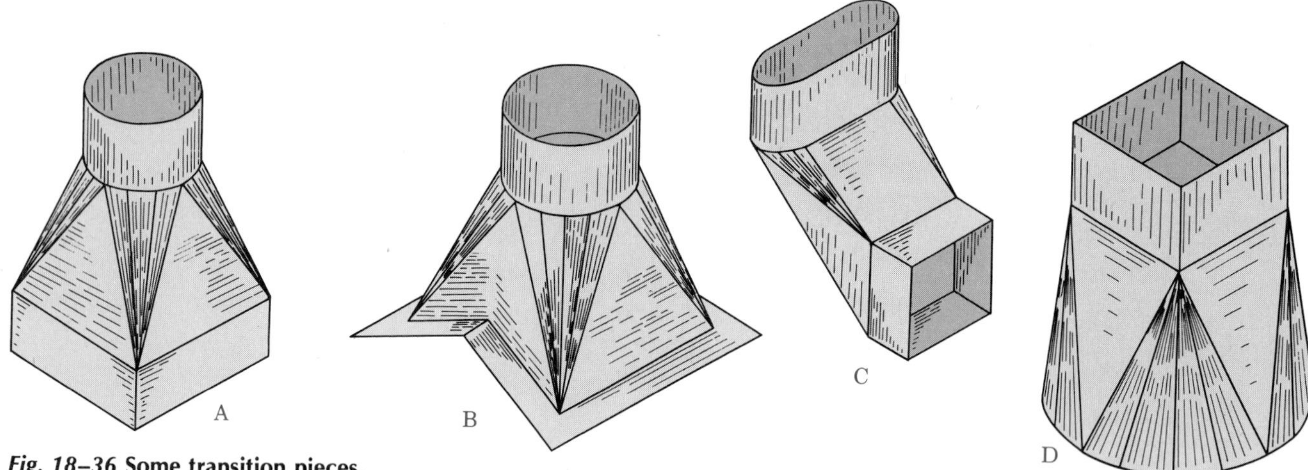

Fig. 18–36 **Some transition pieces.**

the parts is positioned for making the last of eight partial bends on the second conical corner. For this conical bending, one end of the plate is moved out the proper distance for each bend. At the other end, meanwhile, the point for the square corner remains fixed. When completed, the round end is 305 mm (12″) in diameter. The square end is 1220 mm by 1525 mm (4′ by 5′). The height is 1829 mm (6′). The material is 6 mm (1/4″) steel plate. The complete bending process for each section takes about 30 minutes.

DEVELOPMENT OF A TRANSITION PIECE

As you have seen, transition pieces have a surface that is a combination of different forms. These forms are developed by triangulation. This consists of dividing a surface into triangles (exact or approximate), then laying them out on the development in regular order.

Fig. 18–37 **Forming a square-to-round transition piece.** ***(Dreis & Krump Manufacturing Co.)***

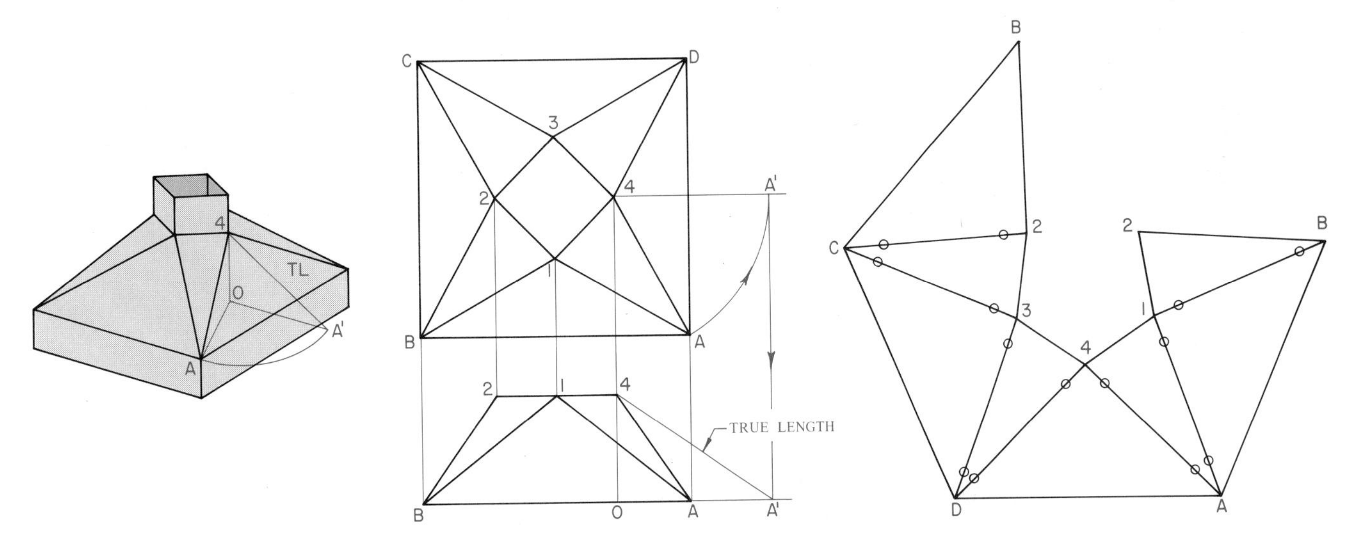

Fig. 18–38 **Development of a square-to-square transition piece.**

Figure 18–38 shows a transition piece connecting two square ducts, one of which is at 45° with the other. This piece is made up of eight triangles, four of one size and four of another.

To draw the development, proceed as follows:

1. On the top and front views, number the various points as shown. Lines 1–2, 2–3, 3–4, and 4–1 show in their true size in the top view. Lines *AB, BC, CD,* and *DA* show in their true size in both views.
2. Find the true length of one of the slanted lines, in this case *A*4. Do this by revolving it in the top view until it is parallel to the vertical plane. Then project down to the front view, where the true length will show in line 4*A'*. This will also be the true length of *D*4, *D*3, and so forth.
3. Start the development by drawing line *DA*.
4. Using *A* and *D* as centers and a radius equal to the true length of *A*4 or any of the other slanted lines, draw intersecting arcs to locate point 4 on the development of the piece.
5. Draw another arc using point 4 as a center and a radius equal to line 4–3. Where this arc intersects the arc drawn in step 4 with *D* as a center, it will locate point 3.
6. Proceed to lay off the remaining triangles until the transition piece is completed.

Figure 18–39 shows a transition piece made to connect a round pipe with a rectangular one. This piece contains four large triangles. Between them are four conical parts with apexes at the corners of the rectangular opening and bases each one-quarter of the round opening. To draw the development, proceed as follows:

1. Start with the cone whose apex is at *A*. Divide its base 1–4 into a number of equal parts, as 1–2, 2–3, 3–4. Then draw the lines *A*2, *A*3 to give triangles approximating the cone.
2. Find the true length of each of these lines. In practice, this is done by constructing a separate diagram, diagram I. The construction is based on the fact that the true length of each line is the hypotenuse of a triangle whose altitude is the altitude of the cone and whose base is the length of the top view of the line.
3. Begin diagram I by drawing the vertical line *AE,* the altitude of the cone, on the front view. On the base *EF,* lay off the distances *A*1, *A*2, and so forth, taken from the top view. In the figure, this is done by swinging each distance about the point *A* in the top view, then dropping perpendiculars to the base *EF.*
4. Connect the points thus found with point *A* in diagram I to find the desired true lengths. Diagram II, constructed in the

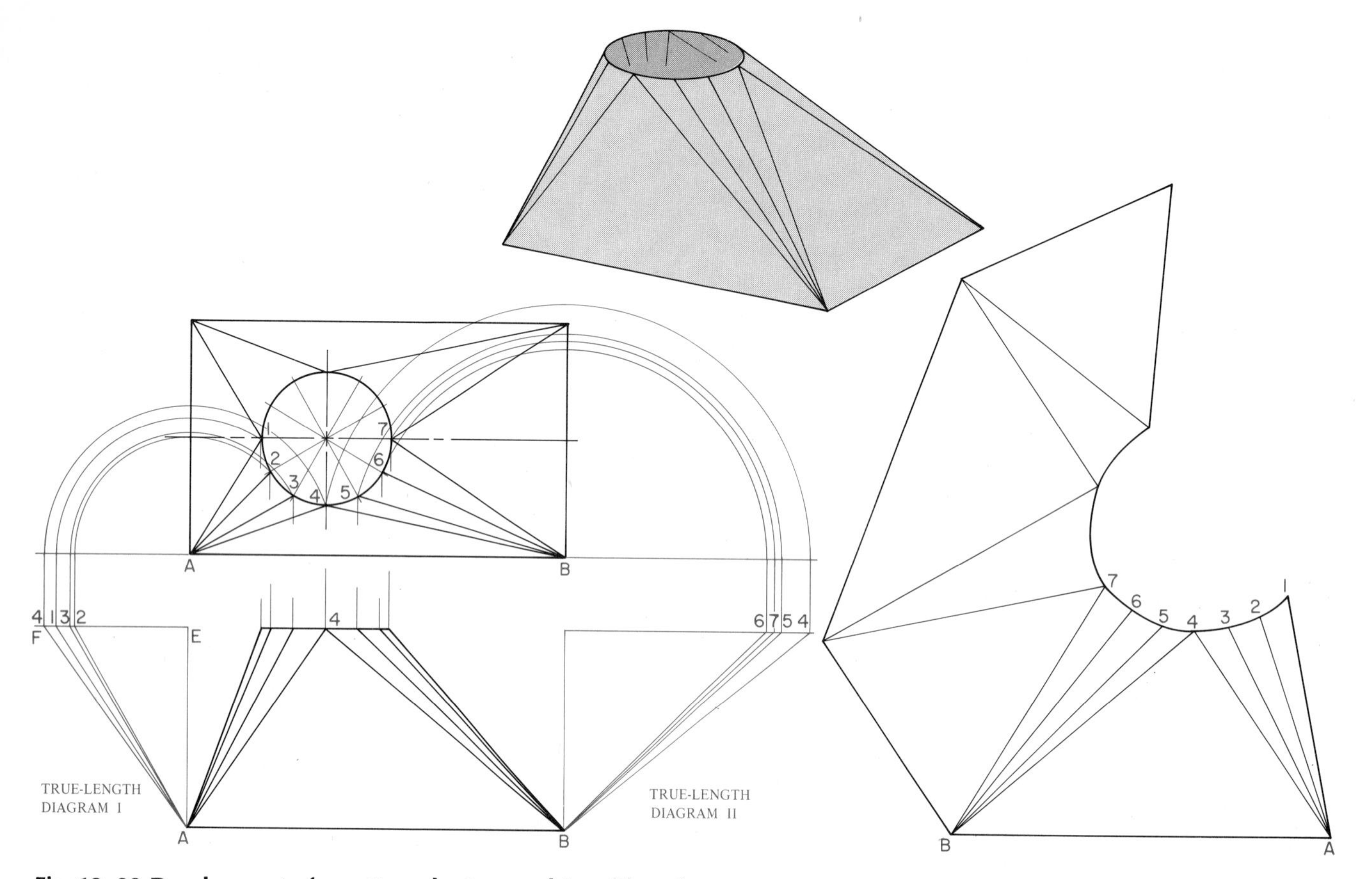

Fig. 18–39 **Development of a rectangular-to-round transition piece.**

same way, gives the true lengths of the triangle lines *B*4, *B*5, and so forth, on the cone whose apex is at *B*.

5. Next, start the development with the seam line *A*1. Draw line *A*1 equal to the true length of *A*1.
6. With 1 as a center and a radius equal to the distance 1–2 in the top view, draw an arc. Then draw another arc using *A* as a center and a radius equal to the true length of *A*2. Where the two arcs intersect will be the point 2 on the development.
7. With 2 as center and radius 2–3, draw an arc. Then draw another with center *A* and radius of the true length of *A*3. Where these two arcs intersect will be point 3.
8. Proceed similarly to find point 4. Then draw a smooth curve through the points 1, 2, 3, and 4 thus found.
9. Next, attach triangle *A*4*B* in its true size. Find point *B* by drawing one arc from *A* with a radius *AB* taken from the top view, and another arc from 4 with a radius the true length of *B*4. Where these arcs intersect will be point *B*.
10. Continue until the development is completed.

INTERSECTIONS

When a line pierces a plane, the point where it does so is called the *point of intersection* (Fig. 18–40). When two plane surfaces meet, the line where they come together, or where one passes through the other, is called the *line of intersection* (Fig. 18–41). When a plane surface meets a curved surface, or where two curved surfaces meet, the line of intersection may be either a straight line or a curved line, depending on the surfaces and/or their relative positions. Package designers, sheet-metal workers, and machine designers all must be able to find the point at which a line pierces a surface and the line where two surfaces intersect.

INTERSECTING PRISMS

Figures 18–42 and 18–43 show some of the ways that different surfaces intersect.

To draw the line of intersection of two prisms, first start the regular top and front views. In Fig. 18–44, a square prism passes through a hexagonal prism. Through the front edge of the square prism, pass a plane parallel to the vertical plane. The top view of this plane appears as a line *AA*. The cut this plane makes in one of the faces of the hexagonal prism shows in the front view as the *cutting line aa.* This line meets the front edge of the square prism at point 1. Point 1 is a point on both prisms and, therefore, a point in the desired line of intersection. Next, through the top edge of the square prism, pass plane *BB,* also parallel to the vertical plane. This plane will make one cutting line through the lateral edges of the hexagonal prism. It will make another through the top edge of the square prism. Where the two lines meet will be point 2 on the front view. Points 1 and 2 are on both prisms. Therefore, a line joining them will be on both prisms and thus a part of the line of intersection. Plane *BB* also determines point 3.

The planes in the above example are called *cutting planes.* You can use them to solve most problems in intersections. For intersecting prisms, pass a plane through each edge on both prisms that touches the line of intersection. Each such plane will make cutting lines on

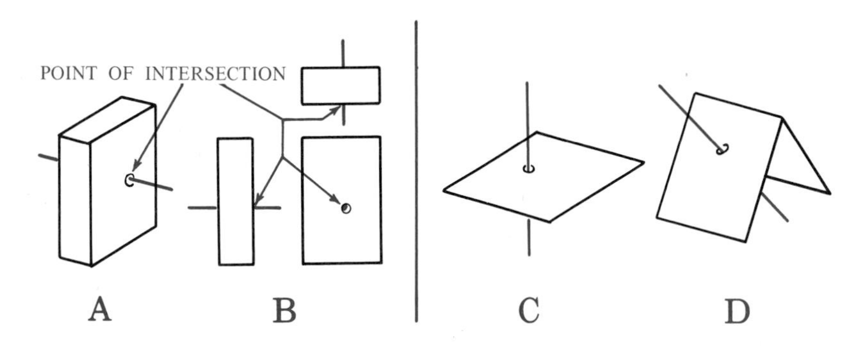

Fig. 18–40 **The intersection of a line and a plane is a point.**

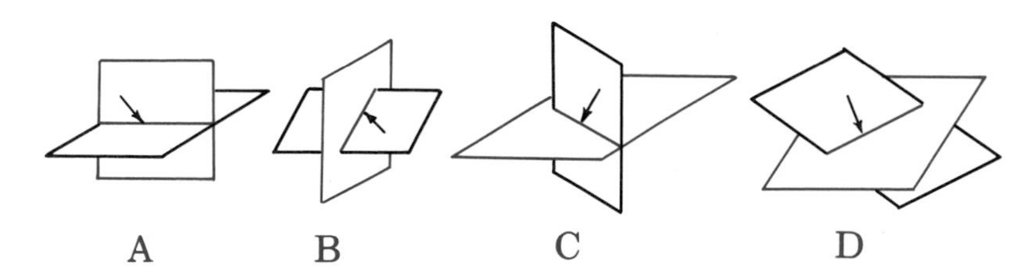

Fig. 18–41 **The intersection of two planes is a line. The arrow points to the line of intersection.**

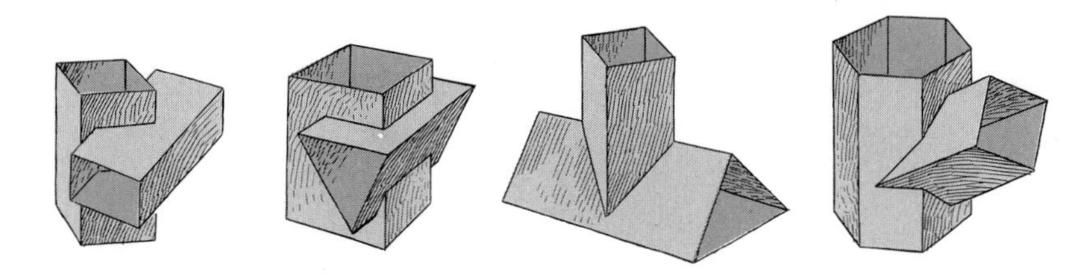

Fig. 18–42 **Some lines of intersection.**

Fig. 18–43 **Some lines of intersection.**

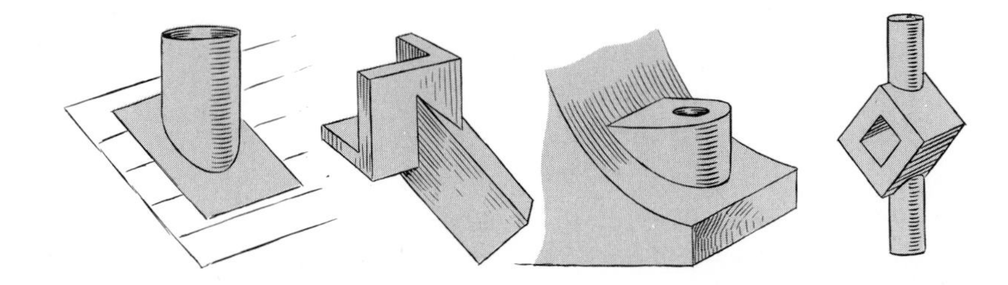

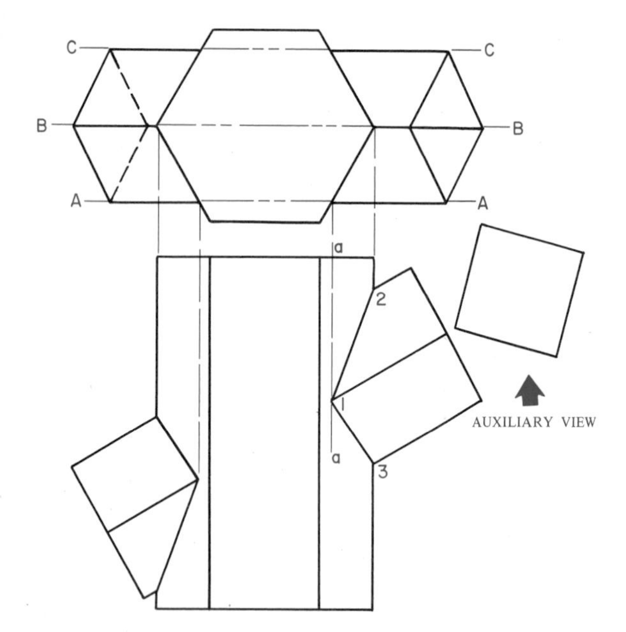

Fig. 18–44 **Intersecting prisms.**

Fig. 18–45 **Intersection of prisms.**

both prisms. Where the cutting line on one prism meets the cutting line on the other, there is a point on the required line of intersection. In Fig. 18–45, four cutting planes are required. No planes are needed in front of *AA* or behind *DD,* because there they would cut only one of the prisms. Planes *AA* and *DD* are thus called *limiting planes.*

INTERSECTING CYLINDERS

Figure 18–46 shows how to draw the line of intersection of two cylinders. Since there are no edges on the cylinders, you have to assume positions for the cutting planes. Draw plane *AA* to contain the front line (element) of the vertical cylinder. This plane will also cut a line (element) on the horizontal cylinder. Where these two lines meet in the front view, there is a point on the required curve. Similarly, the other planes will cut other lines on

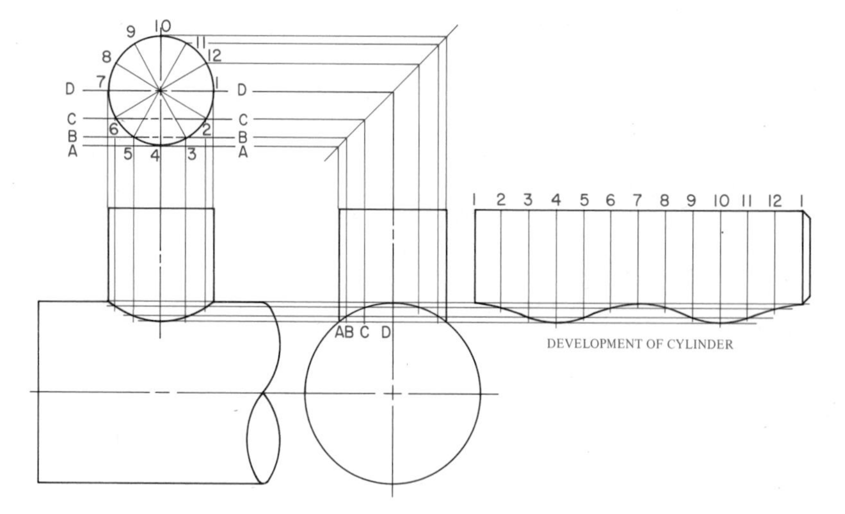

Fig. 18–46 **Intersection of cylinders at a right angle.**

both cylinders that intersect at points common to both cylinders. The figure also shows the development of the vertical cylinder. Figure 18–47 shows how to draw the line of intersection of two cylinders joined at an angle. Here you locate the cutting planes by an auxiliary view. To make the development of the inclined cylinder, take the length of the stretchout line from the circumference of the auxiliary view. If you have chosen the cutting planes so that this circumference is divided into equal parts, the measuring lines will be equally

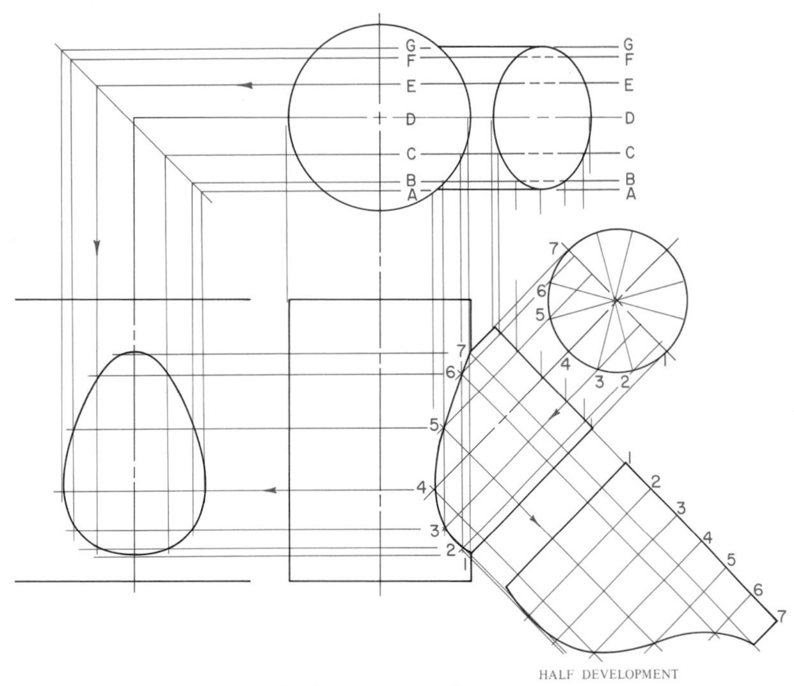

Fig. 18–47 **Intersection of cylinders at an angle.**

spaced along the stretchout line. Project the lengths of the measuring lines from the front view. Join their ends with a smooth curve.

INTERSECTION OF CYLINDERS AND PRISMS

Finding the intersection line of a cylinder and a prism is also done with cutting planes. In Fig. 18–48, a triangular prism intersects a cylinder. The planes *A, B, C,* and *D* cut lines on both the prism and the cylinder. These lines cross in the front view at points that determine the curve of intersection. To make the development of the triangular prism, take the length of the stretchout line from the top view. Take the lengths of the measuring lines from the front view. Note that one plane of the triangular prism (line 1–5 in the top view) is perpendicular to the axis of the cylinder. The curve of the intersection line on that face is made with the radius of the cylinder.

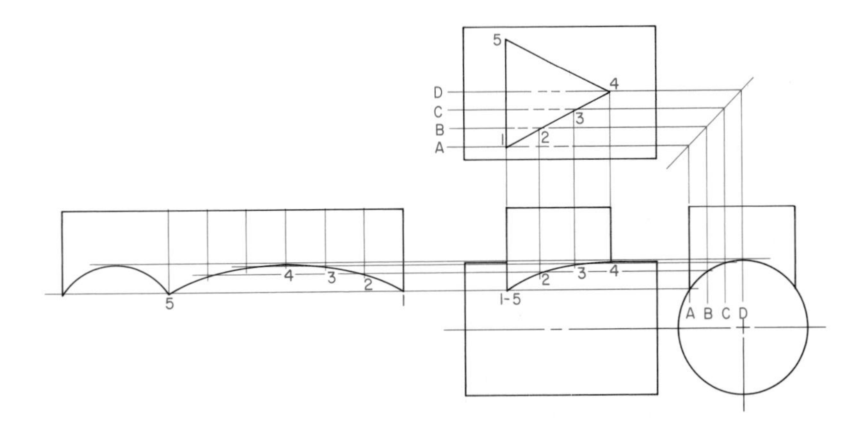

Fig. 18–48 **Intersection of a prism and a cylinder and development of the prism.**

INTERSECTION OF CYLINDERS AND CONES

To find the intersection line of a cylinder and a cone, use horizontal cutting planes, as shown in Fig. 18–49. Each plane will cut a circle on the cone and two straight lines on the cylinder. Where the straight lines of the cylinder cross the circles of the cone in the top view, there are the points of intersection. Project these points onto the front view to get the intersection line. Figure 18–50 shows this construc-

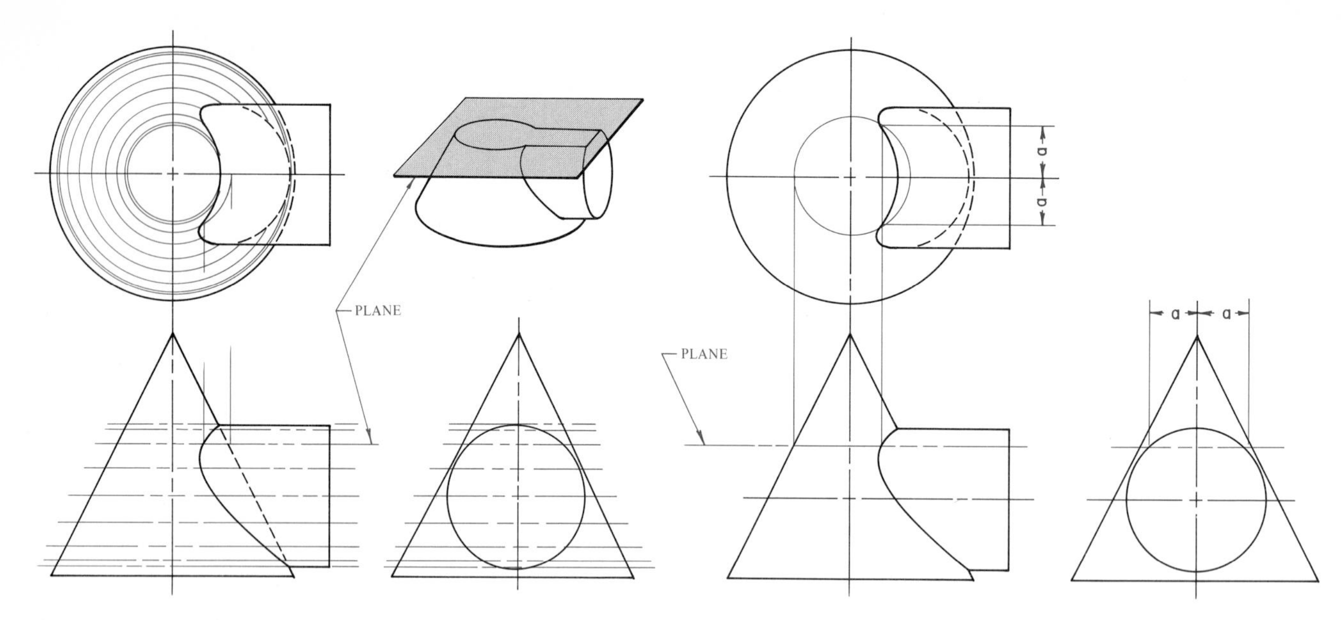

Fig. 18–49 **Intersection of a cylinder and a cone.**

Fig. 18–50 **A cutting plane.**

tion for a single plane. Use as many planes as are needed to make a smooth curve.

INTERSECTION OF PLANES AND CURVED SURFACES

Figure 18–51 shows the intersection of a plane *MM* and the curved surface of a cone. To find the line of intersection, use the horizontal cutting planes *A, B, C,* and *D*. Each plane cuts a circle from the plane *MM*. Thus, you can locate points common to both *MM* and the cone, as in the top view. Project these points onto the front view to get the curve of intersection.

Figure 18–52 shows the intersection at the end of a connecting rod. To find the curve of intersection, use cutting planes perpendicular to the rod's axis. These planes cut circles as shown in the end view. The circles, in turn, cut the "flat" at points that can be projected back as points on the curve.

Fig. 18–51 **Intersection of a plane and a curved surface.**

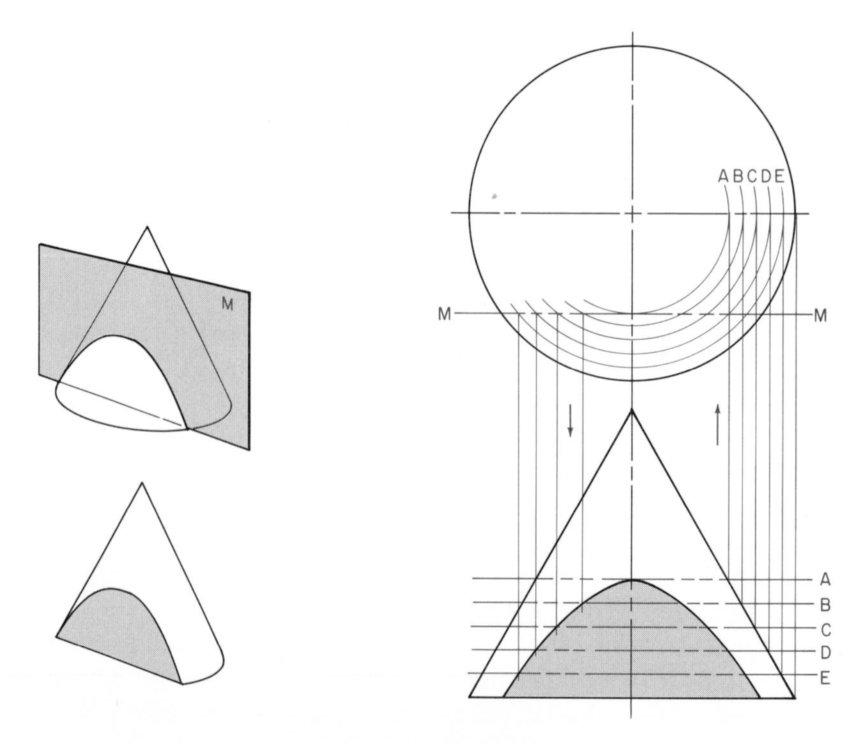

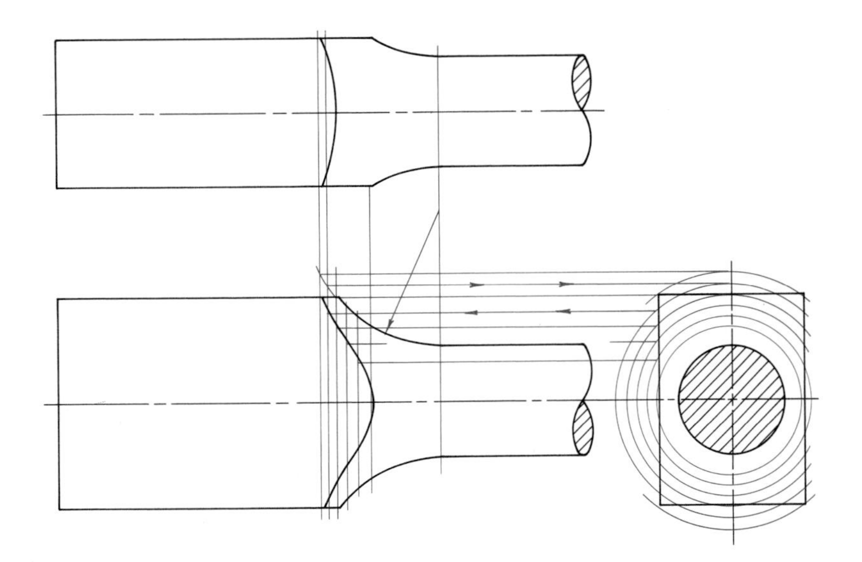

Fig. 18–52 **Intersection of a plane and a turned surface.**

Review

1. Name the three basic types of surface developments.

2. Name the two general classes of surfaces.

3. A sphere has a ________ -curved surface.

4. A drawing that when folded and fastened, shows the shape of an object in three dimensions is called the ________ of the object.

5. Lines on a stretchout that show where to make a fold are called ________ lines.

6. Using a series of triangles to develop a pattern is called ________.

7. Parts used to connect openings of different shapes, sizes, or positions are called ________.

8. A point of intersection is a point where a line passes through a ________.

9. When two plane surfaces meet, the line where they come together is called the ________.

10. Sheet materials are usually fastened together by ________ or ________.

Problems

Fig. 18–53 **Developments. Scale: full size. Problems A through L are planned to fit on an 11″ × 17″ or 12″ × 18″ drawing sheet. Draw the front and top views of the problem assigned. Develop the stretchout (pattern) as shown in the example at the right. For problems A through F, add the top in the position it would be drawn for fabrication. Include dimensions and numbers if instructed to do so. Patterns may be cut out and assembled.**

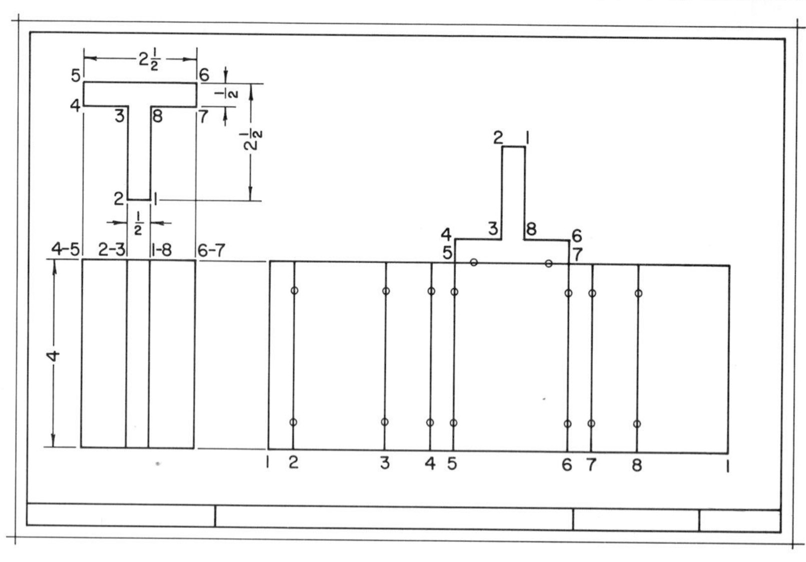

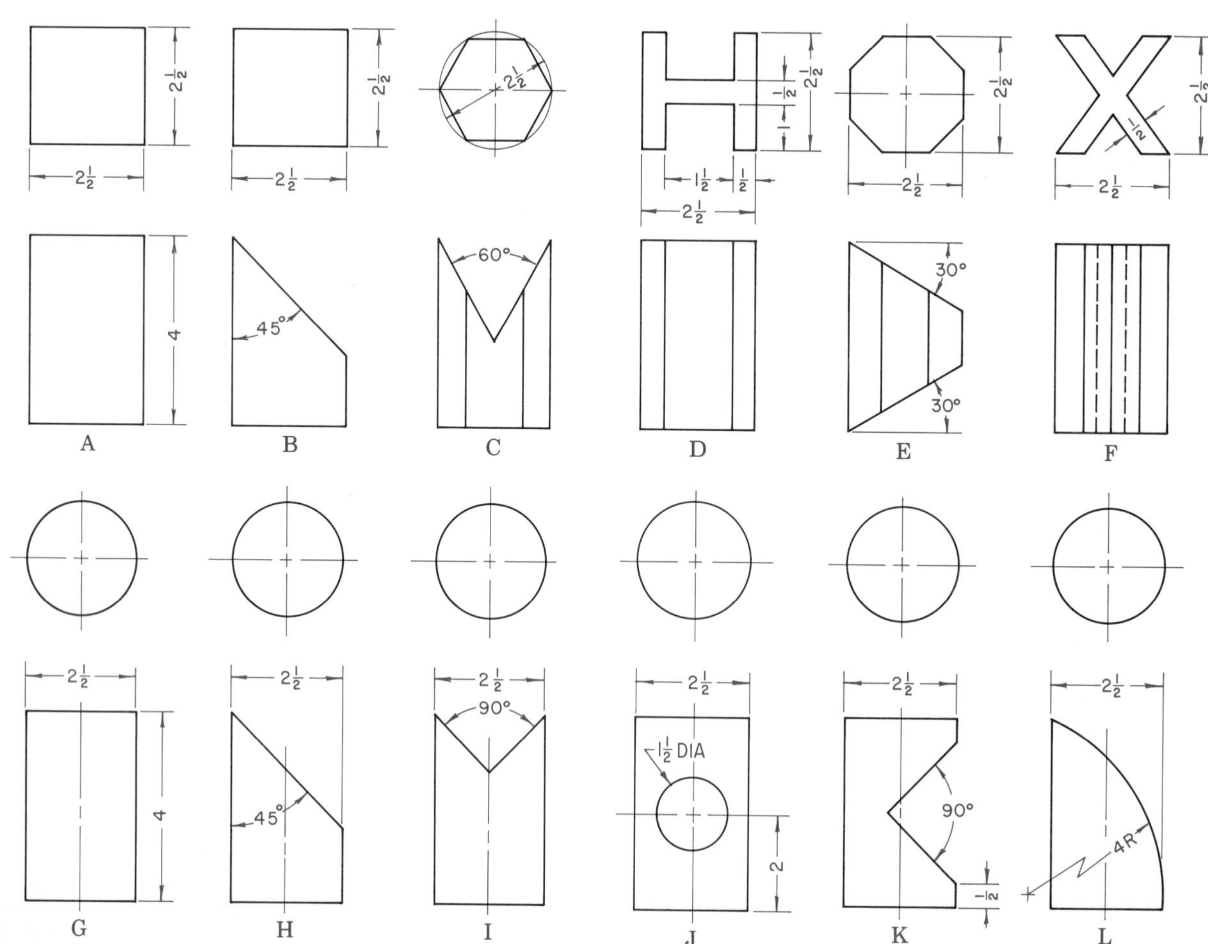

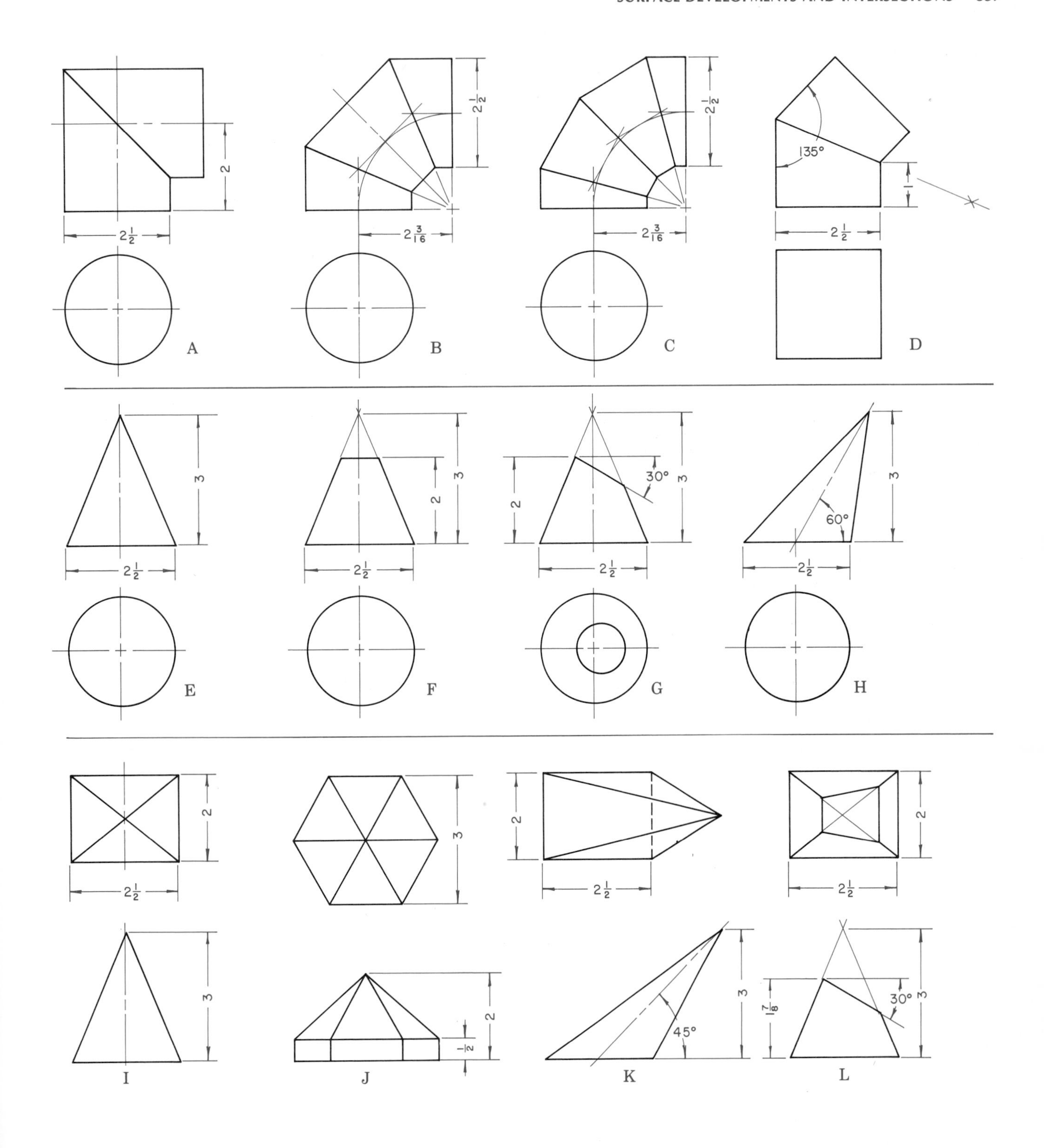

Fig. 18–54 **Make two views of the problem assigned and develop the pattern. Scale: full size.**

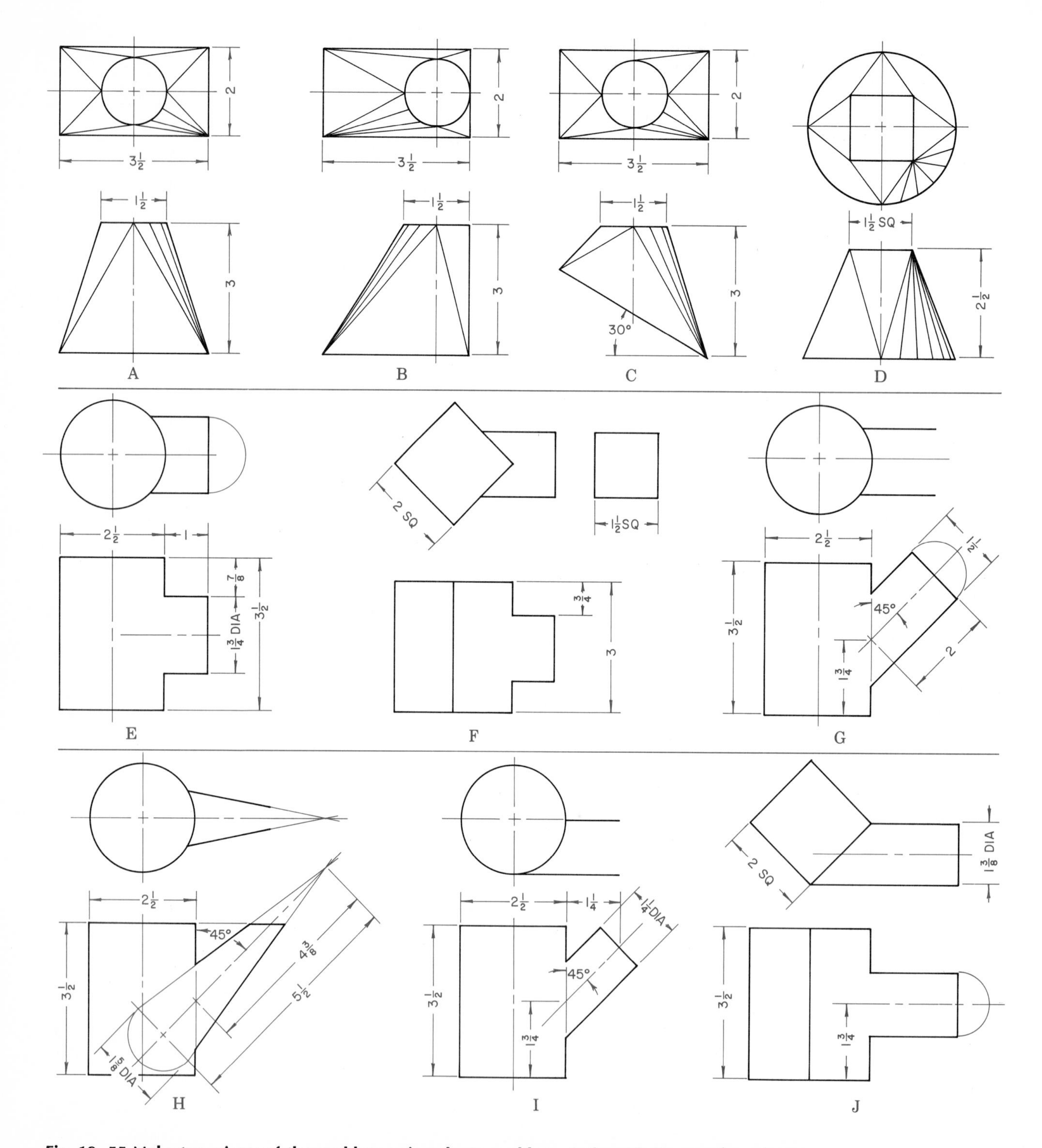

Fig. 18–55 **Make two views of the problem assigned. In problems A through D, complete the top views and develop the pattern. In problems E through J, complete views where necessary by developing the line of intersection and completing the top views in G, H, and I. Develop patterns for both parts in problems E through J.**

Chapter 19

Cams and Gears

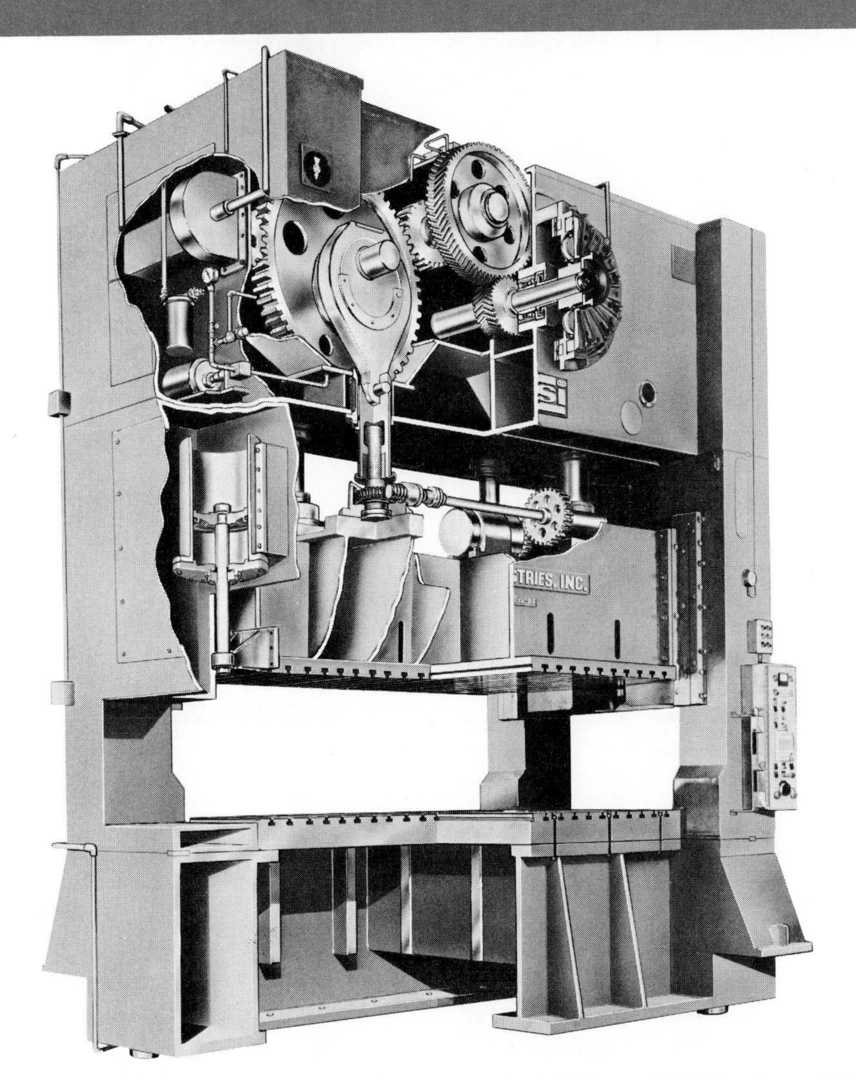

Fig. 19–1 **Cams and gears are necessary parts of machines.** *(USI-Clearing, a U.S. Industries Co.)*

CAMS AND GEARS

Cams and gears are machine parts that do specific jobs. Cams and gears can transmit motion, change the direction of motion, and change the speed of motion in a machine (Fig. 19–1). The ability to draw cams and gears and to understand their function is a first step in becoming a machine designer. This chapter will introduce the skills that are basic to drawing and understanding these important machine parts.

NOTE: metric conversion has not been done in this chapter, because of the complexity of following the fundamentals of cams and gears. However, the principles apply whether customary or metric dimensions are used.

CAMS

A cam is a machine part that usually has an irregular curved outline or a curved groove. When the cam *rotates* (turns), it gives a specific motion to another machine part that is called the follower. The cam and the follower together make up the cam mechanism. The cam drives the follower. The plate cam in Fig. 19–2 would have a follower that moves up and down as the cam turns on the shaft. The *cylindrical* (can-shaped) cam in Fig. 19–3 would have a follower that moves back and forth parallel to the axis of the shaft. The grooved cam in Fig. 19–4 would have a follower that moves in an irregular pattern as the cam turns on a shaft.

It is the design of a cam that provides the predetermined and continuous motion of the follower. This kind of motion is needed in all automatic machinery. Therefore,

Fig. 19–2 **A plate cam.** *(Camco.)*

Fig. 19–3 **A cylindrical cam.** *(Camco.)*

Fig. 19–4 **A grooved cam.** *(Camco.)*

the cam is important to the automatic control and accurate timing that is found in many kinds of machinery. It is the unlimited variety of shapes that makes the cam so useful to the designer.

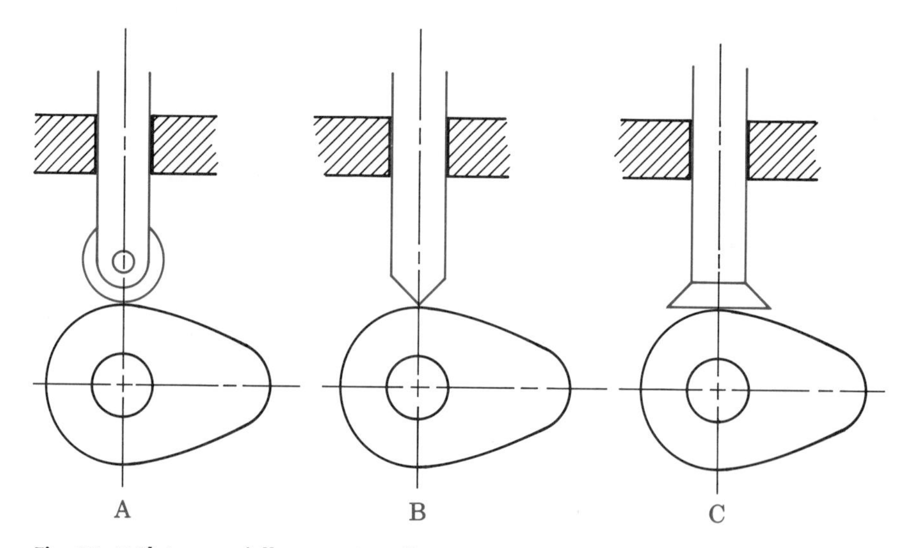

Fig. 19–5 **Plate cam followers: A, roller; B, point; C, flat-surface.** *(Camco.)*

CAM FOLLOWERS

Three types of followers for plate cams are shown in Fig. 19–5. The roller follower is used to reduce friction to a minimum. Therefore, the roller follower transmits force at high speeds. The flat-surface follower and the point follower are made with a hardened surface to reduce wear from friction. Flat-surface and point followers are generally used with cams that rotate slowly.

Plate cams require a spring loaded follower so that contact can be made throughout a full revolution. The *rise* (lifting) of the follower by the cam is made through direct contact of cam and follower. However, contact of cam and follower during a *drop* (fall) or *dwell* (rest) cannot be made unless contact is brought about by an outside pressure. Outside pressure given by a spring pushes the follower against the cam and assures direct contact.

CAM TERMS

The illustrations in Fig. 19–6 are pictorial descriptions of how the cam works. The stroke, or rise, takes place within one-half a revolution or 180°. The movement is repeated every 360°, or one full revolution.

KINDS OF CAMS

A cam for operating the valve of an automobile engine is shown in Fig. 19–7. This cam has a flat follower that rests against the face of the plate cam. In Fig. 19–8, the slider cam at A moves the follower

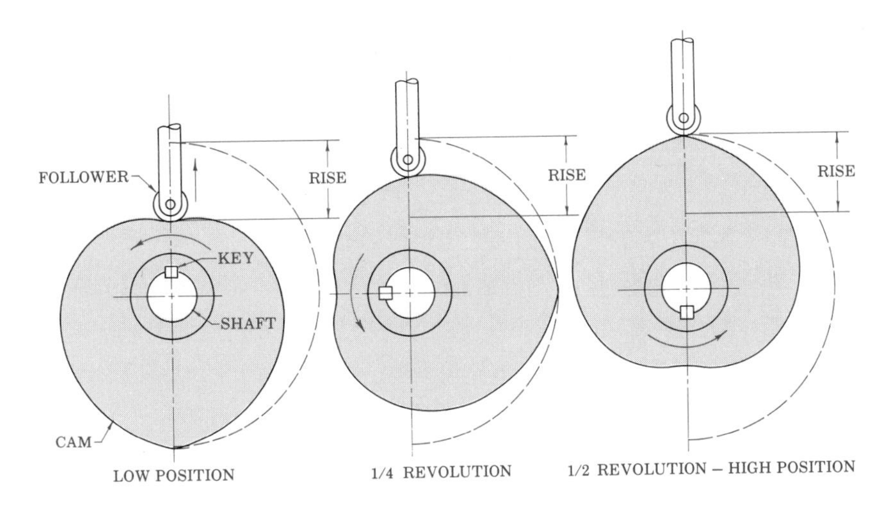

Fig. 19–6 **Cam action and terms.**

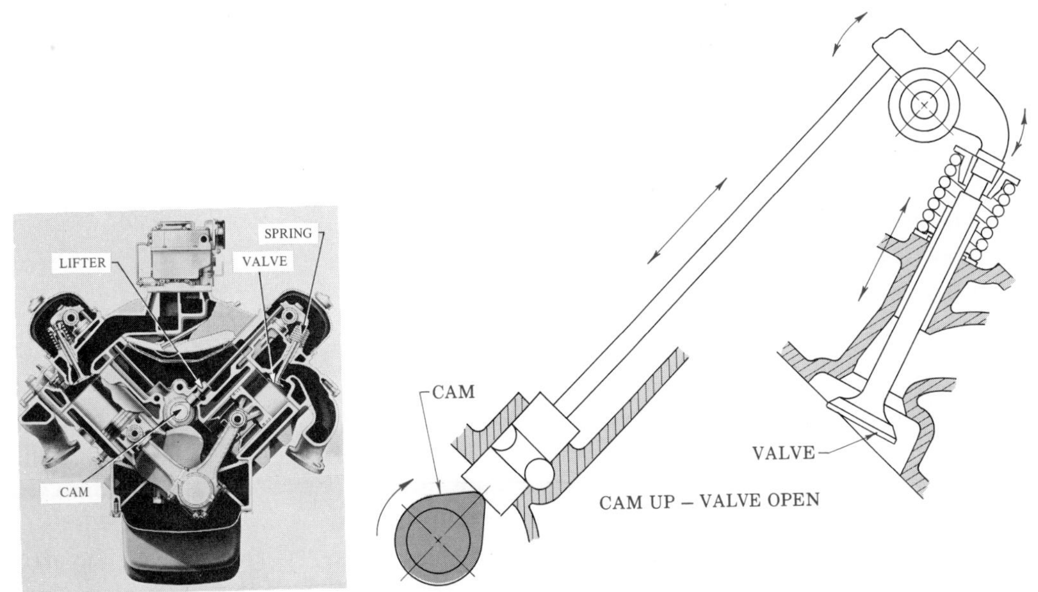

Fig. 19–7 **Automobile valve cam.** ***(Oldsmobile Div., General Motors Corp.)***

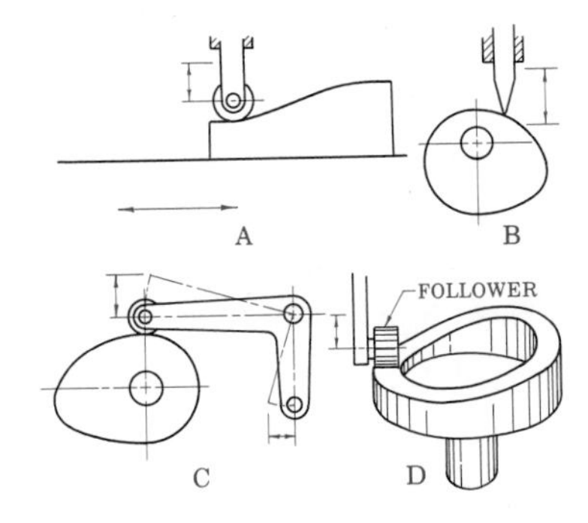

Fig. 19–8 **Kinds of cams: A, slider; B, offset plate; C, pivoted roller; D, cylindrical roller.**

up and down as the cam moves back and forth from right to left. An offset plate cam with a point follower off center is shown at B. A pivoted roller follower cam is shown at C. The cylindrical edge cam with a swinging follower is shown at D. All cams can be thought of as simple inclines that produce predetermined motion.

CAM LAYOUT

The shape of the cam determines the direction of motion in the follower. The displacement diagram in Fig. 19–9 shows the shape of the cam. It also shows the motion the cam will produce through one revolution. The length of the displacement diagram represents one revolution of 360°. The height of the diagram represents the total displacement stroke, 1.875 in., of the follower from its lowest position. The cam is assumed to rotate with a constant speed. The length or time of a revolution is divided into convenient parts. The parts are proportional to the number of degrees for each action. The divisions of the revolution into convenient parts are called *time periods*. These proportional parts are identified A, 1^1, 2^1, 3^1, B^1, C, 4^1, 5^1, D^1, E, 6^1, 7^1, A, as shown. Each proportional part or time period is a 30° angular division of the base circle.

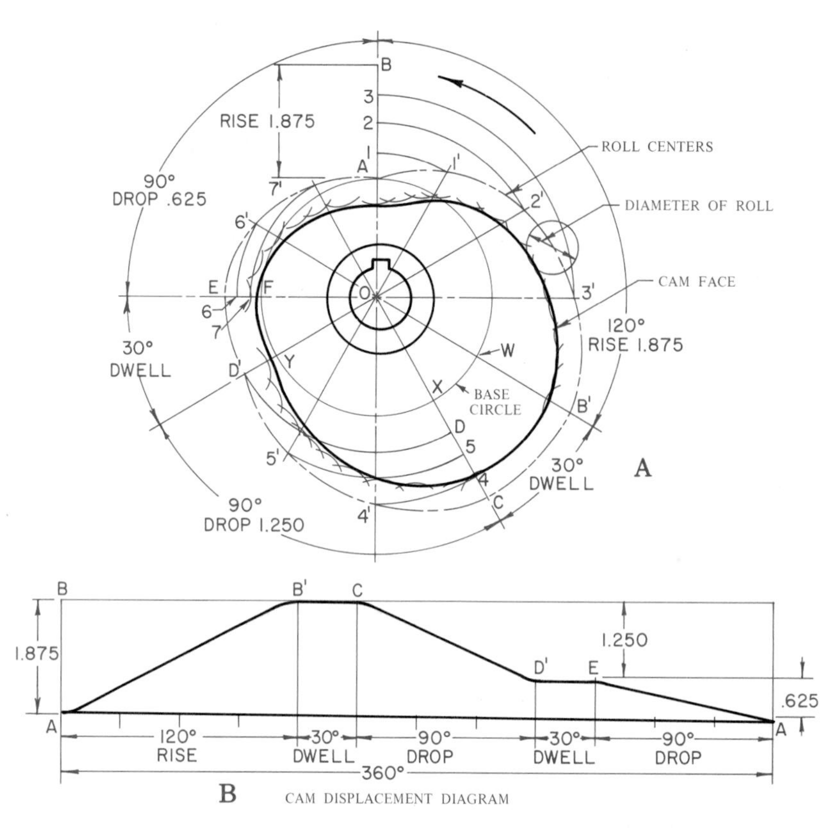

Fig. 19–9 **Cam displacement diagram.**

THE PROBLEM

Given point O, the center of the cam shaft, and point A, the lowest position of the center of the roller follower. The center of the roller follower must be raised 1.875 in. with uniform motion during the first 120° of a revolution of the shaft. It must then dwell for 30°, drop 1.250 in. for 90°, dwell for another 30°, and drop 0.625 in. during the remaining 90°. The shaft is assumed to revolve *uniformly* (with constant speed).

DISPLACEMENT DIAGRAM

The displacement diagram at B in Fig. 19–9 illustrates the solution to the problem. Points A, B^1, C, C^1, and E relate to the travel pattern that will take place in one complete revolution.

TO DRAW THE PROFILE OF THE CAM

The profile of the cam shown at A in Fig. 19–9 has five features that are important.

1. *Rise.* Divide the rise AB, or 1.875 in. into a number of equal parts. Four parts are used, but eight parts could make the layout more accurate. The rise occurs from A to W (120°). You divide it into the same number of equal parts as the rise (four at 30°) with radial lines from O. Using center O, you draw arcs with radii $O1$, $O2$, $O3$, and OB until they locate 1^1, 2^1, 3^1, and B^1 on the four radial lines. Use an irregular curve to draw a smooth line through these four points.
2. *Dwell.* You first draw an arc B^1C (30°) using radius O^1B^1. This will allow the follower to be at rest, because it will stay the same distance from center O.
3. *Drop.* At C, you lay off CD, 1.250 in., on a radial line from O. Divide it into any number of equal parts (three are shown). Next, divide the arc XY (90°) on the base circle into the same number of equal parts (three). Draw three radial lines from center O every 30°. Then you draw arcs with center O and radii $O4$, $O5$, and OD to locate points 4^1, 5^1, and D^1 on the three radial lines. Using an irregular curve, draw a smooth line through points 4^1, 5^1, and D^1.
4. *Dwell.* Draw an arc D^1E^1 (30°) with radius OD. This will provide the constant distance from center O to let the roller follower be at rest.

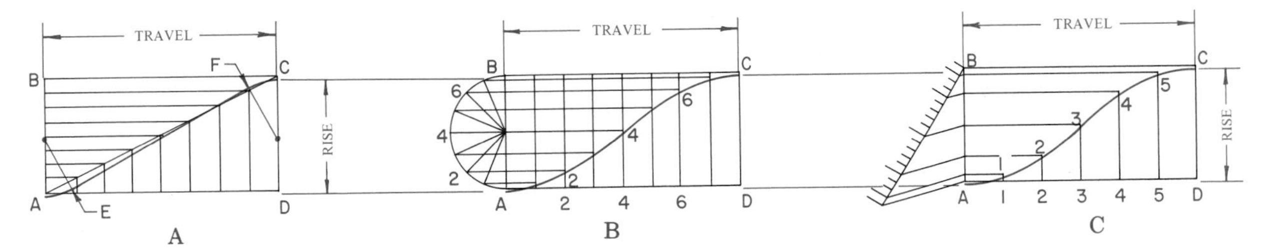

Fig. 19–10 **Kinds of cam motion: A, uniform; B, harmonic; C, uniformly accelerated or decelerated.**

5. *Drop.* In the last 90° of a full revolution, the roller will return to point *A*. It will move through a distance *EF*, or 0.625 in. First divide *EF* into any number of equal parts (three are shown). Then divide arc *FA* into the same number of equal parts (three). Next draw radial lines every 30°. Then you draw arcs with radii *O*6 and *O*7 to locate points 6^1 and 7^1. Using the irregular curve as a guide, draw a smooth curve through points *E*, 6^1, 7^1, and *A*. This finishes the roll centers.

Using the line-of-roll centers as a centerline, draw *successive* arcs (one after the other) with the radius of the roller, as shown. Then use an irregular curve to draw the cam profile. The profile will be a smooth curve tangent to the arcs you drew representing the roller.

CAM MOTION

A cam can be designed so that the followers can have three types of motion. The displacement diagram is used to plot the different kinds of motion.

Uniform Motion

Uniform-motion (steady motion) cams are suitable for high-speed operations. In Fig. 19–10 at A, the thin line represents uniform motion. Equal distances on the rise (eight units) are made for equal distances on the travel (equal intervals of time, eight units). To avoid a sudden jar at the beginning and end of the motion, use arcs to change it slightly. The small change is shown by the heavy line formed by arcs *E* and *F*.

Harmonic Motion

Harmonic-motion (smooth-acting motion) cams are also good to use with high speeds. They are used when you need a smooth start-and-stop motion. The method for plotting a harmonic curve is shown in Fig. 19–10 at B. To draw harmonic motion, first draw a semicircle with the rise as the diameter. Divide the semicircle into eight equal parts using radial lines. Then divide the travel into the same number of equal parts. Project the eight points horizontally from the semicircle until they intersect the corresponding vertical projections as shown. Finally, draw a smooth curve through all eight points with an irregular curve.

Uniformly Accelerated and Decelerated Motion

Cams with *uniformly accelerated and decelerated motion* (steadily increasing and decreasing motion) do not operate at a constant speed. The motion produced is very smooth.

In Fig. 19–10 at C, we are plotting a curve called a parabola, or parabolic curve. A parabola is formed when you slice a cone vertically at any place other than through the center. See Fig. 18–51. If you design a cam in the following manner, it will have a uniform acceleration and deceleration curve. First divide the rise into parts proportional to 1, 3, 5, . . ., 5, 3, 1. Note that the six parts are not equal. Then divide the travel into the same number of parts but divide it equally. Project the six points horizontally from the rise until they intersect the corresponding vertical projections. Finish by drawing a smooth curve through all the points of intersection, as shown.

CAM DRAWINGS

A drawing for a face, or plate, cam is shown in Fig. 19–11. Note that the amount of movement, or rise, is given by showing the radii for the dwells. These are a 4.5-in. radius and a 7.0-in. radius. Harmonic motion is used, and there seem to be two rolls working on this cam. In Fig. 19–12, a drawing for a *barrel* (cylindrical) cam is shown with a displacement diagram. The diagram shows two dwells and two kinds of motion. Note that the distance traveled from center to center is 1.5 in.

GEARS

There are many kinds of gears, a few of which are illustrated in Fig. 19–13. However, one of the most practical and dependable machine parts for transmitting rotary motion from one shaft to another is the spur gear. The operation of simple spur gears can be explained in this way. If the rims of two wheels are in contact, as in Fig. 19–14, both will revolve if only one is turned. If the small friction-wheel is two-thirds the diameter of the larger wheel, it will make one and one-half revolutions for every one revolution of the larger wheel. This assumes no slipping occurs. When the load on the driven wheel gets larger and the wheel is hard to turn, slipping begins to occur. Friction wheels cannot be counted on for a smooth transfer of rotary motion. When teeth are added to the wheels in Fig. 19–14, they become gears. With teeth added to the wheels you get the same kind of motion as rolling friction wheels. Now, however, there is no slipping.

GEAR TEETH

The basic forms used for gear teeth are *involute* and *cycloidal* curves. These curves are explained in the following paragraphs.

The spur gear and pinion gear with parallel shafts shown in Fig. 19–15 are good examples of involute gears. Note that the small spur gear is called the pinion. Gear teeth have a special shape that lets them mesh smoothly. This shape is an involute curve. Figure 19–19 explains an involute curve that is used in drawing gear teeth. An involute of a circle can be thought of as a curve made by a taut string as it unwinds from around the circumference of a cylinder (circle).

A cycloidal curve can be thought of as the path of a curve formed by a point on a rolling circle (Fig. 19–16). The information given in this chapter is for the 14½° involute system. It can be used for a 20° system as well, since the only practical difference is the number of degrees of the pressure angle. The

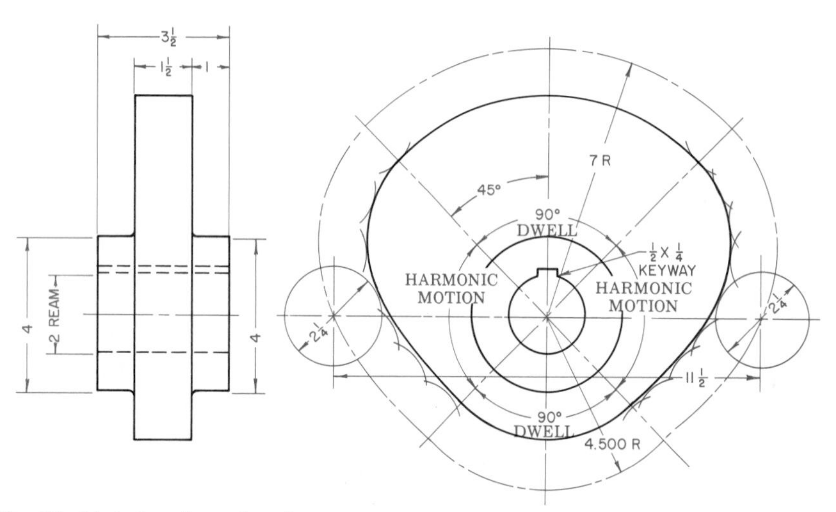

Fig. 19–11 **A drawing of a plate cam.**

Fig. 19–12 **A drawing of a cylindrical cam.**

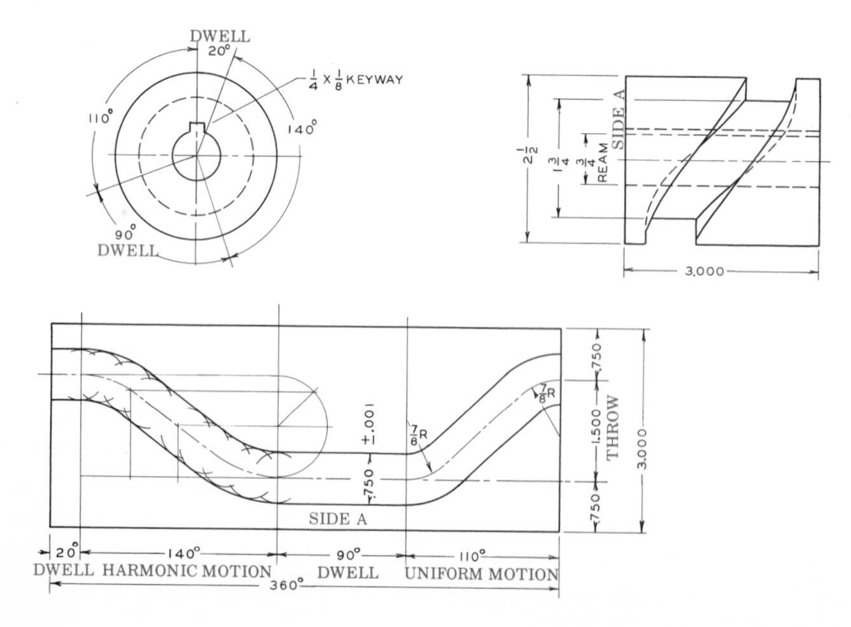

Fig. 19–13 **Several gears that are used as typical machine elements: A, B, and C, spur gears; D, a bevel gear; E, a pinion (spur gear); F, a rack (E and F together are called a rack and pinion); G, an internal gear.** ***(The Fellows Gear Shaper Co.)***

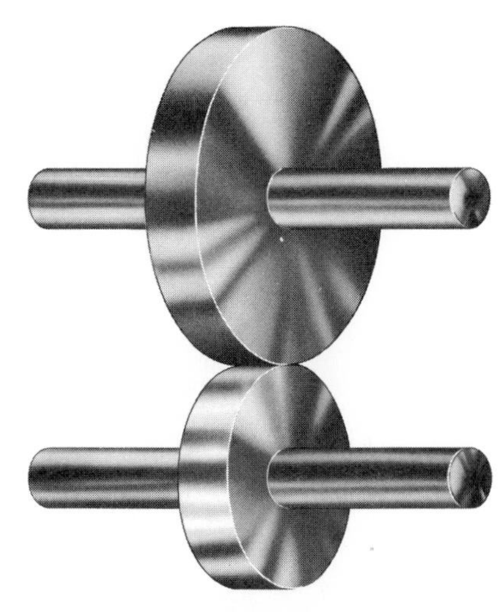

Fig. 19–14 **Friction wheels are a simple means of transmitting rotary motion from one shaft to another.**

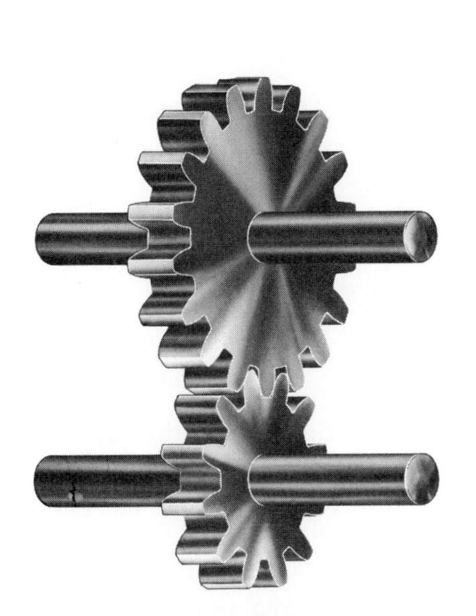

Fig. 19–15 **The spur gear. Teeth added to friction wheels provide a more efficient means of transmitting rotary motion.**

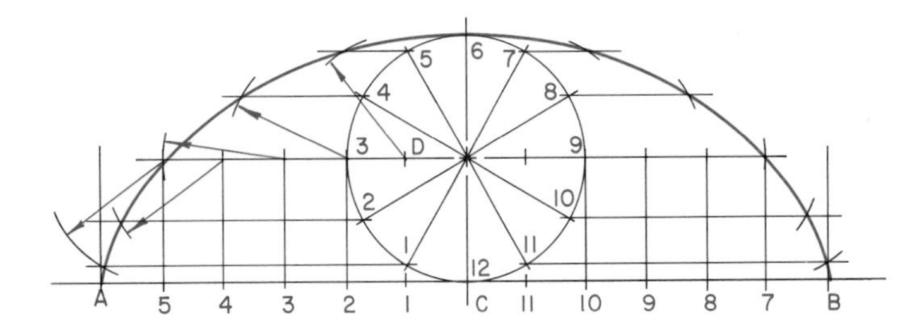

Fig. 19–16 **A cycloidal curve.**

14½° or 20° refers to the pressure angle (Fig. 19–17). The pressure angle and the distance between the centers of mating spur gears determine the diameters of the base circles (Fig. 19–17). Note that the point of tangency of the gears is their pitch diameter. These are equal to the diameters of the rolling friction wheels that are replaced by the gears. The involute is drawn from the base circle, which is smaller than the pitch circle.

In Fig. 19–18, R_A is the radius of the pitch circle of the gear with center at A. R_B is the radius of the pitch circle of the pinion with the center B. The distance between gear centers is $R_A + R_B$. The line of pressure $T_A T_B$ is drawn through O

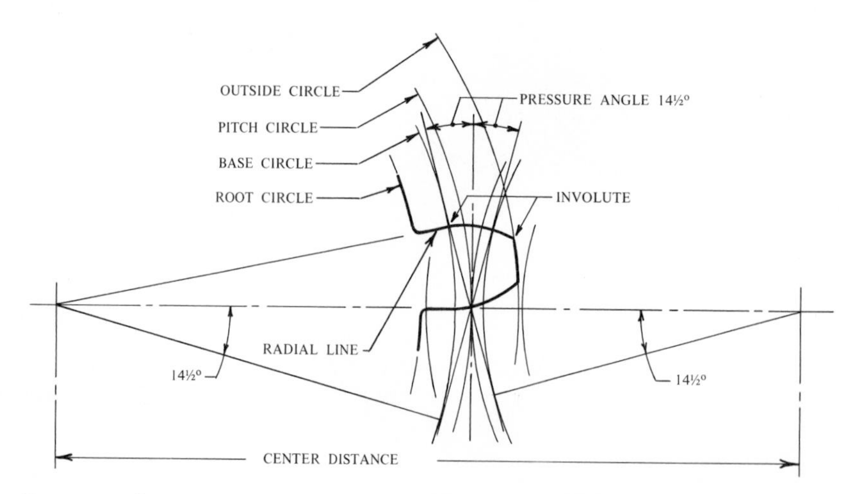

Fig. 19–17 **The pressure angle. Note that the center distance indicates the distance between shafts of mating gears.**

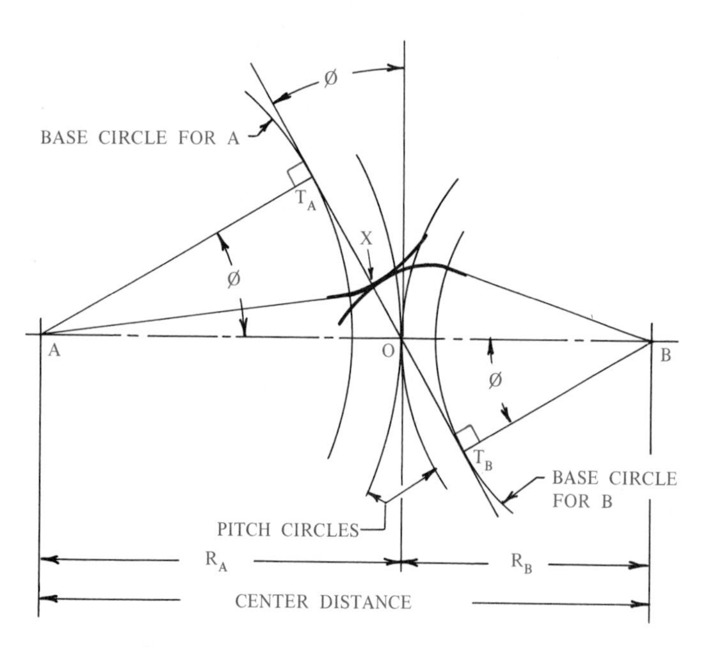

Fig. 19–18 **Gear tooth interaction; the rolling nature of surface contact.**

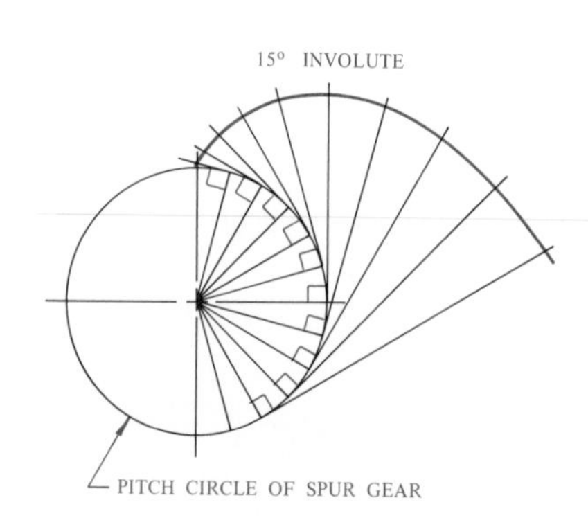

Fig. 19–19 **Involute of a circle.**

(which is the point of tangency of the pitch circles). It makes the pressure angle ϕ (Greek letter phi) with the perpendicular to the line of centers. This angle is 14½°. Note that lines AT_A and BT_B are drawn from centers A and B perpendicular to the pressure-angle line T_AT_B. A point X on a cord (line of pressure T_AT_B) will describe the points that form the involute curve as the cord winds and unwinds. This represents the outlines of gear teeth outside the base circles. The profile of the gear tooth inside the base circle is a radial line. Figure 19–19 illustrates the cord unwinding off the surface of the base circle. To simplify the drawing of the involute curve, note that the radius used in Fig. 19–20 is one-eighth the pitch diameter.

SPUR–GEAR TERMS AND FORMULAS

Some parts of a spur gear are named and illustrated in Fig. 19–21.

You should use the following terms and formulas to find the dimensions you need for standard 14½° involute spur-gear problems.

Spur-Gear Terms

N = number of teeth

N_G = number of teeth of gear

N_P = number of teeth of pinion

D = pitch diameter— diameter of pitch circle

D_G = pitch diameter of gear

D_P = pitch diameter of pinion

P = diametral pitch— number of teeth per inch of pitch diameter

a = addendum— radial distance the tooth extends above the pitch circle

b = dedendum— radial distance the tooth extends below the pitch circle

D_O = outside diameter— pitch diameter plus twice the addendum gives the overall gear size

D_R = root diameter— pitch diameter minus twice the dedendum gives the root diameter

h_t = whole depth— radial distance from the root diameter to the outside diameter, equal to addendum plus dedendum

p = circular pitch— distance from a point on one tooth to the same point on the next tooth measured along the pitch circle; the distance of one tooth and one space

t_c = circular thickness— thickness of a tooth measured along the pitch circle

c = clearance— difference between the addendum and the dedendum. Clearance is the distance between the top of a tooth and the bottom of the mating space when gear teeth are meshing

h_K = working depth— the mating distance a tooth projects into the mating space; twice the radial distance of the addendum

o = pressure angle— the direction of pressure between teeth at the point of contact

D_b = base-circle diameter— the circle from which the involute profile is developed

Fig. 19-20 Simplified method of drawing a gear tooth.

Fig. 19-21 Gear terms illustrated.

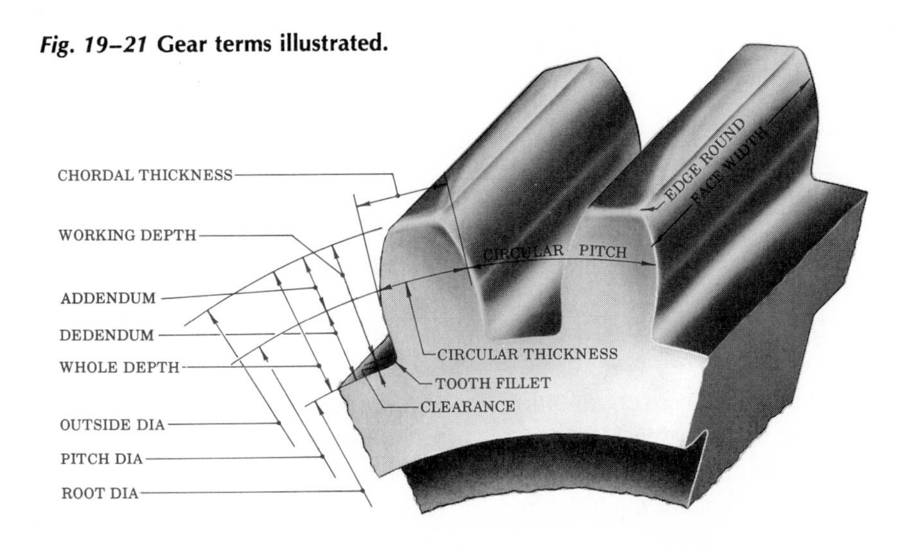

APPLICATION OF GEAR FORMULAS

Before the formulas can be applied, written descriptions must be given. Typical information given to the drafter is as follows: A pair of involute gears is to be drawn according to the specifications that follow.

The pressure angle will be 14½°.

The distance between parallel shaft centers will be 12 in.

The driving shaft will turn at 800 rpm clockwise.

The driven shaft will turn at 400 rpm.

The diametral pitch equals 4 (number of teeth per inch of pitch diameter).

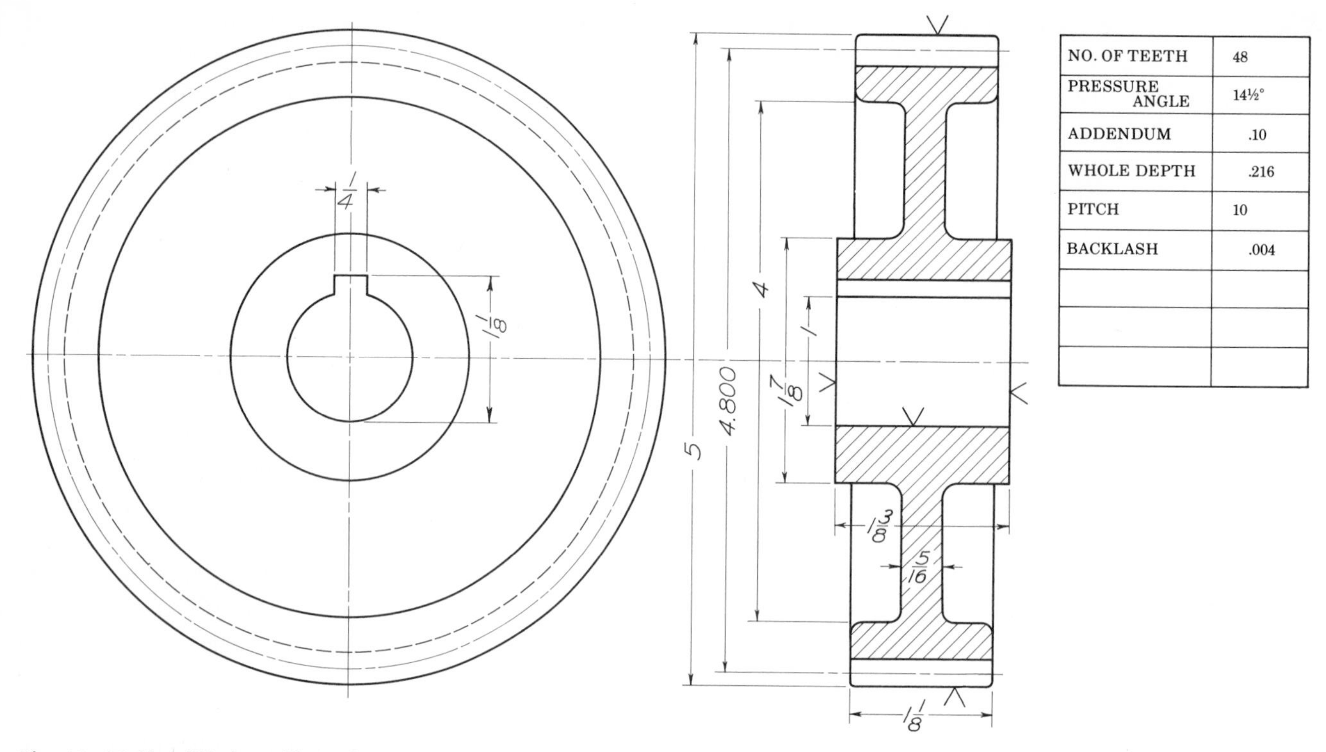

***Fig. 19–22* Simplified profile and cross section of a spur-gear working drawing.**

Typical Computations

You must do the following computations before the formulas can be applied.

1. Pitch radius of the *pinion* (smaller gear)

$$R_p = \frac{400}{400 + 800} \times 12''$$
$$= \text{4-in. radius}$$

2. Pitch radius of the spur gear

$$R_s = \frac{800}{400 + 800} \times 12''$$
$$= \text{8-in. radius}$$

Velocity ratio ½ (4 in.: 8 in)

The first two computations are based on the ratio of the velocity of the two cylinders. One cylinder drives the other. Thus, the ratio is obtained by dividing the velocity (rpm) of the driver by the velocity (rpm) of the driven member.

3. Number of teeth on the pinion $= N_p = DP = 4 \times 8 = 32$

4. Number of teeth on the spurgear $= N_s = DP = 4 \times 16 = 64$

5. Addendum $= a = \dfrac{1}{P} = \dfrac{1}{4}$ in.

6. Dedendum $= b$

$$= \frac{1.157}{p} = \frac{1.157}{4}$$
$$= 0.289 \text{ in.}$$

STANDARD INVOLUTE GEARS

Involute gears are interchangeable when they have set conditions that allow them to mesh properly. The four conditions for interchanging involute gears are the following: the same diametral pitch, the same pressure angle, the same addendum, and the same dedendum.

GEAR DRAWINGS

It is not necessary to show the teeth on typical gear drawings. A drawing for a cut spur gear is shown in Fig. 19–22. The gear blank should be drawn with dimensions for making the pattern and for the machining operations. The spur-gear drawing should include information for cutting the teeth. It should also include information for the tolerances required and a notation of the material to be used. On assembly drawings, you may use a simplified gear, as shown in Fig. 19–23. Even though the drawing is

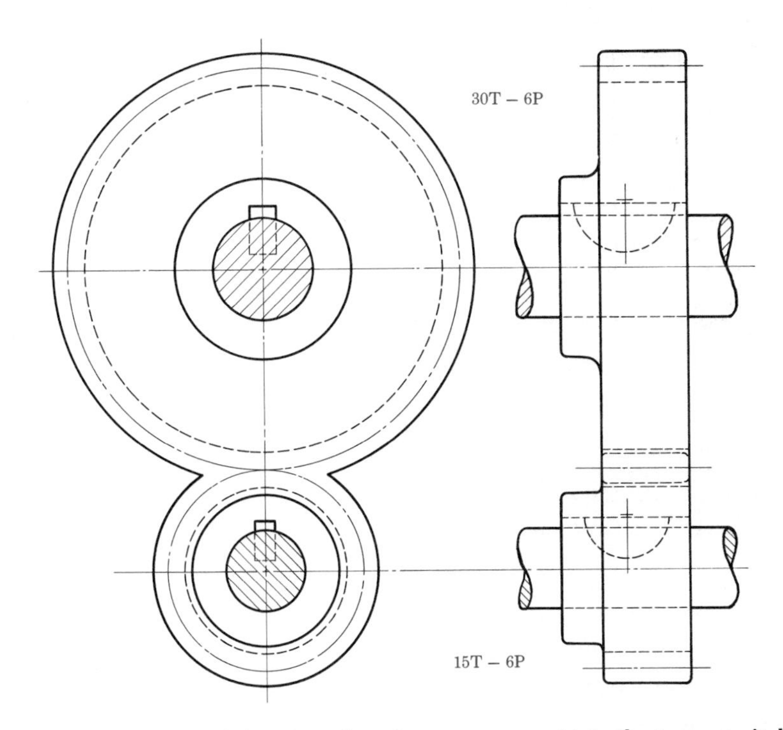

Fig. 19–23 **Simplified drawing of mating spur gears. Note the tangent pitch circles, the number of teeth, and the pitch.**

Fig. 19–24 **A rack and pinion.** ***(Brad Foote Gear Works, Inc.)***

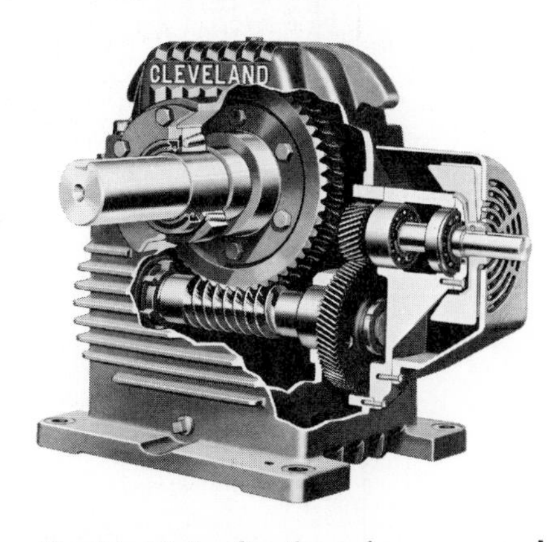

Fig. 19–25 **Application of a worm and wheel.** ***(Industrial Drives Division, Eaton Corp.)***

simplified, however, it should include all necessary notes for making the gear.

Involute Rack and Pinion

A rack and pinion is shown in Fig. 19–24. A rack is simply a gear with a straight pitch line instead of a circular pitch line. The tooth profiles become straight lines. These lines are perpendicular to the line of action.

Worm and Wheel

Figure 19–25 shows how the worm and wheel mesh at right angles. The worm gear is similar to a screw. It may have single or multiple threads. You would use this system to transmit motion between two perpendicular nonintersecting shafts. The wheel is similar to a spur gear in design except that the teeth must be curved to engage the worm gear.

Bevel Gears

When two gear shafts intersect, bevel gears are used to transfer motion. Sometimes bevel gears are referred to as miter gears. This is the case when the gears are the same size, and the shafts are at right angles. Figure 19–26 shows four rolling cones. You can think of bevel gears as replacing the friction cones, just as the spur gear replaced the circular friction wheels. Figure 19–27 illustrates mating beveled gears. The smaller gear is called the pinion.

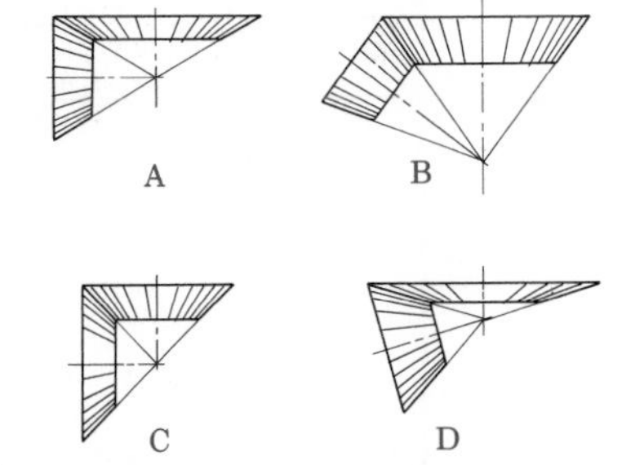

Fig. 19–26 **Rolling cones that represent bevel gearing.**

Fig. 19–27 **Mating bevel gears.** ***(Arrow Gear Co.)***

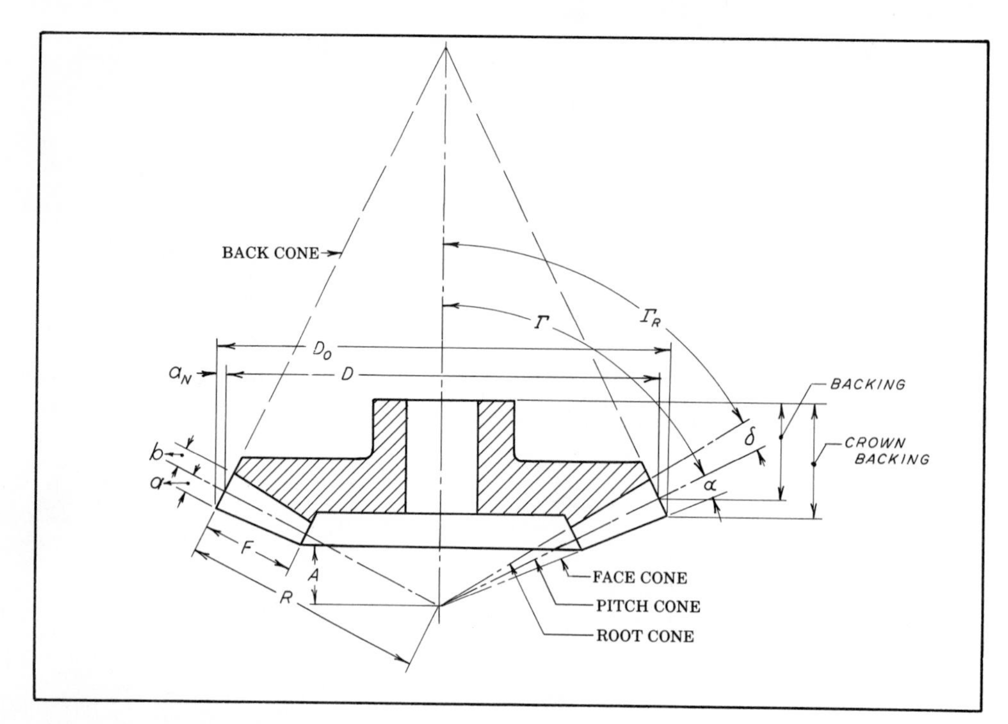

Fig. 19–28 **Bevel-gear terms. Note the cone shape.**

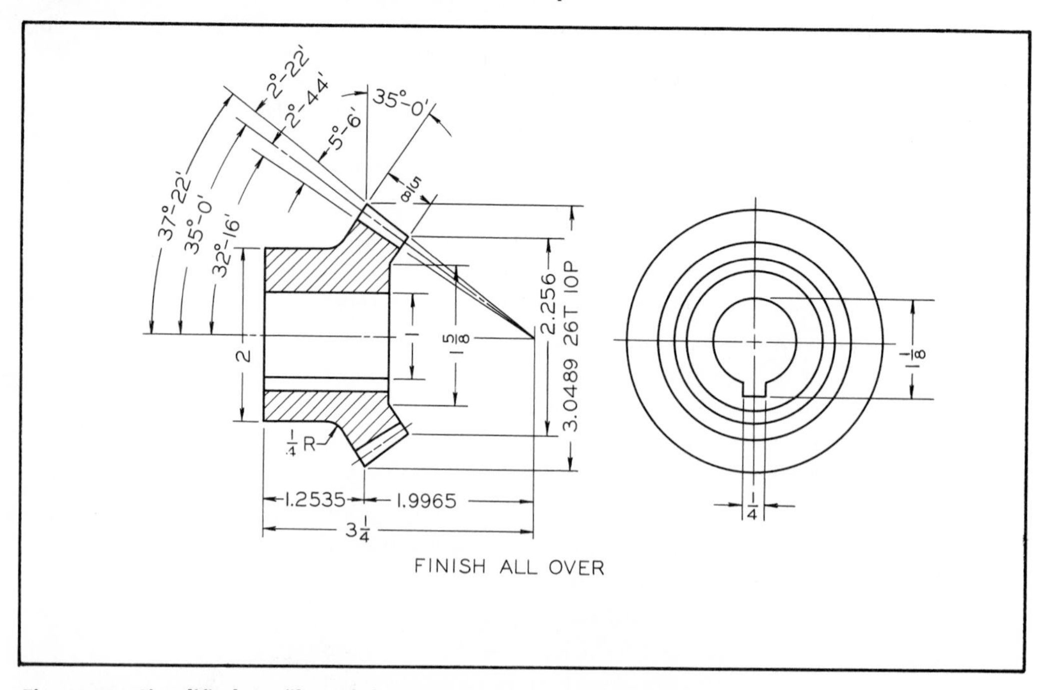

Fig. 19–29 **Simplified profile and detailed section of a bevel gear.**

Bevel-Gear Terms

Some basic information about bevel gears is given in Fig. 19–28. There are three Greek letters used in the following list: α = alpha, δ = delta, and Γ = gamma. Examine Fig. 19–28 for the similarity of design found in spur and bevel gears. Note the differences. The pitch diameter is the diameter of the pitch cone in the bevel-gear design. The circular pitch and the diametral pitch are based on the pitch diameter just as in spur gears. The important features, such as the angles, are listed here.

α = addendum angle
δ = dedendum angle
Γ = pitch angle
Γ_R = root angle
Γ_o = face angle
a = addendum
b = dedendum
a_n = angular dedendum
A = cone distance
F = face
D = pitch diameter
D_O = outside diameter
N = number of teeth
P = diametral pitch
R = pitch radius

Bevel-Gear Drawing

On working drawings of bevel gears, you need to give the needed dimensions for machining the blank as well as all the gear information. An example of a working drawing for a cut bevel gear is shown in Fig. 19–29.

Gear Information

The American National Standards Institute has established standards for satisfactorily detailing gear drawings. ANSI Y14.7, Section Seven, and B6.1 and B6.5 are standard references that can be used for further study of the subject of gear detailing.

Review

1. What other than time is important in the design of a cam?

2. What is the purpose of a displacement diagram?

3. List three types of cams and three types of followers.

4. List three uses of a cam and the kind of cam used.

5. Why is harmonic motion used in a cam?

6. How is the curve on a spur-gear tooth developed?

7. What technical phrase describes the ratio of the number of teeth to the pitch diameter?

8. Rolling cones are used to describe the meshing of what kind of gears?

9. What two bits of information are needed to find circular pitch?

10. List two applications of bevel gears.

11. Name the four circles used by the drafter in drawing gears.

Problems

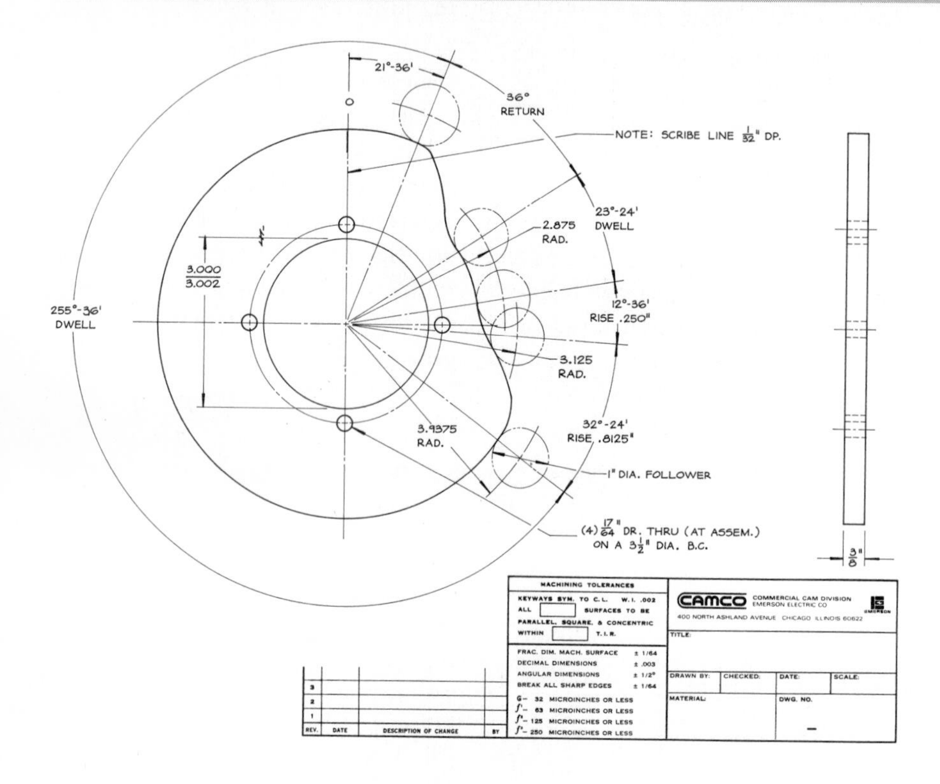

Fig. 19–30 **Prepare a displacement diagram to illustrate the specified motion travel patterns from the original base circle. Make a profile drawing of the radial plate cam.** ***(Camco.)***

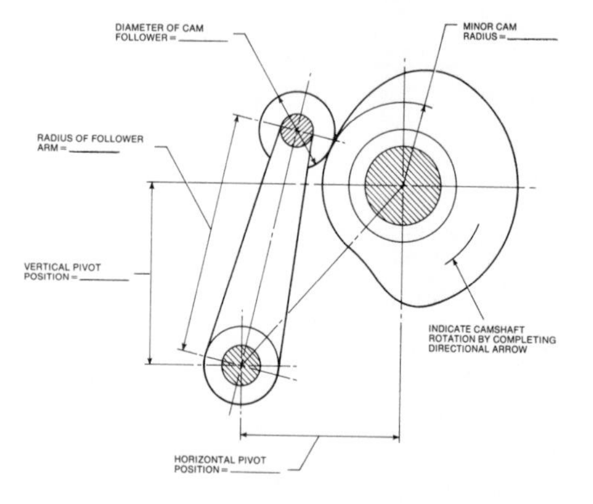

Note:

A. Locate the position of a pertinent timing feature (i.e. keyway, dowel hole, split line, etc.) assuming that the follower as shown is at zero degrees of camshaft rotation.

B. Complete the graph and table below indicating all critical points, transitional motions and timing features.

Fig. 19–31 **Complete the data with assistance from the instructor, and prepare a displacement diagram for the follower arm cam. Develop a profile of the cam and a left-side view showing the cam and follower. Choose appropriate dimensions.** ***(Camco.)***

GRAPH OF FOLLOWER DISPLACEMENT VS. CAMSHAFT ROTATION

CAM FOLLOWER DISPLACEMENT UNITS (   )

0 40 80 120 160 200 240 280 320 360

INSTRUCTIONS

TYPICAL EXAMPLE: CONSTRUCT A 2½" CYCLOIDAL RISE IN 120° CAM ROTATION.

(1) DRAW TWO PARALLEL HORIZONTAL LINES 2½" APART REPRESENTING THE RISE.

(2) DIVIDE THE 120° CAM ANGLE INTO 10 EQUAL PARTS (12°-24°-36° ETC.).

(3) LAY CAM SCALE BETWEEN BASE LINE & 2½" RISE. TRANSFER POINTS FROM CAM SCALE TO PAPER.

(4) PROJECT THESE POINTS HORIZONTALLY TO THEIR RESPECTIVE CAM DIVISIONS AND DRAW THE CURVE.

COPYRIGHT 1962 COMMERCIAL CAM AND MACHINE CO.

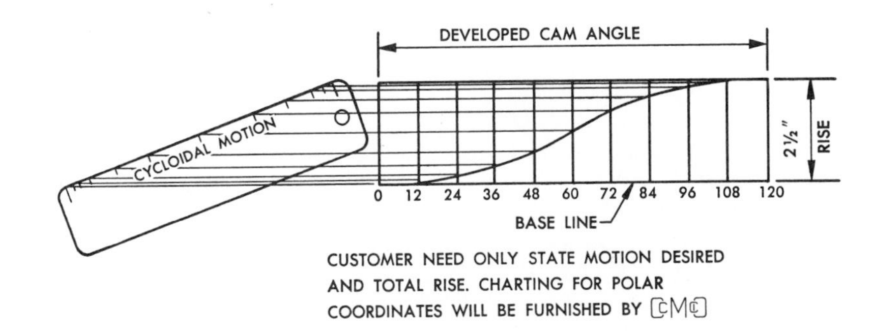

Fig. 19–32 **Examine the displacement diagram carefully and then make a drawing along a base line as shown, with a rise of 2½" in 120°, a 90° dwell, a drop of 2½" in 120°, and a dwell of 30°. Prepare a profile of this plate cam, using a base circle of 3½.** ***(Camco.)***

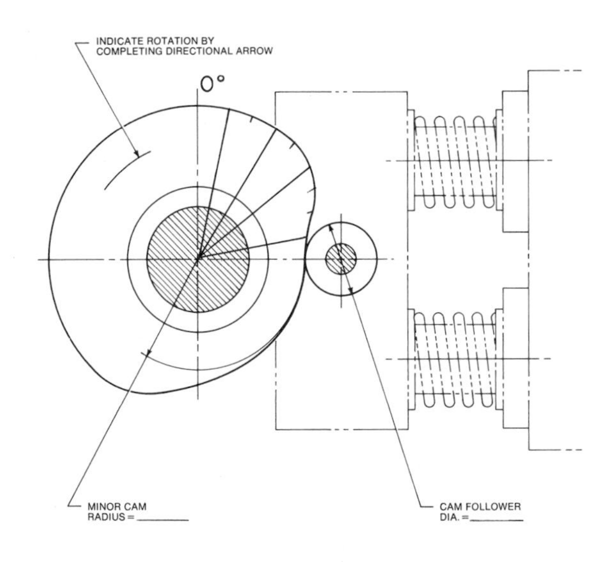

Note:

A. Locate the position of a pertinent timing feature (i.e. keyway, dowel hole, split line, etc.) assuming that the follower as shown is at zero degrees of camshaft rotation.

B. Complete the graph and table below indicating all critical points, transitional motions and timing features.

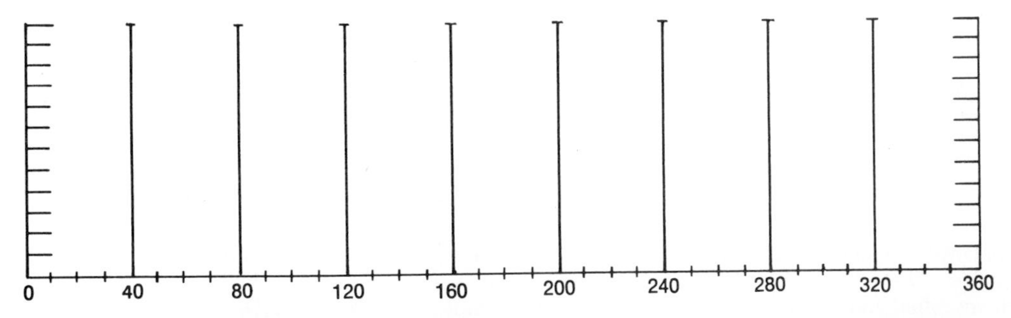

Fig. 19–33 **Develop a displacement diagram for the radial cam. Start at 0° and work clockwise to determine the travel pattern at 10° intervals. Choose the size of the minor cam radius and the cam follower diameter. Complete the drawing with a cam profile, using dimensions of your choice or as supplied by the instructor.** ***(Camco.)***

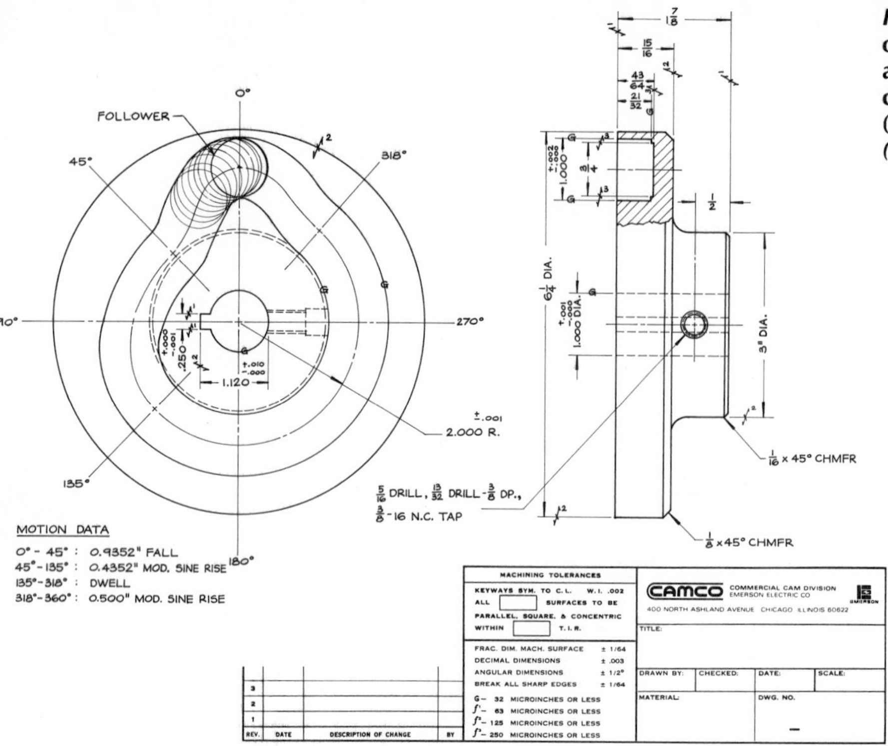

Fig. 19–34 **Develop a working drawing of the grooved cam as shown. Prepare a displacement diagram illustrating the counterclockwise motion of this cam. (Note the motion of the follower.) *(Camco.)***

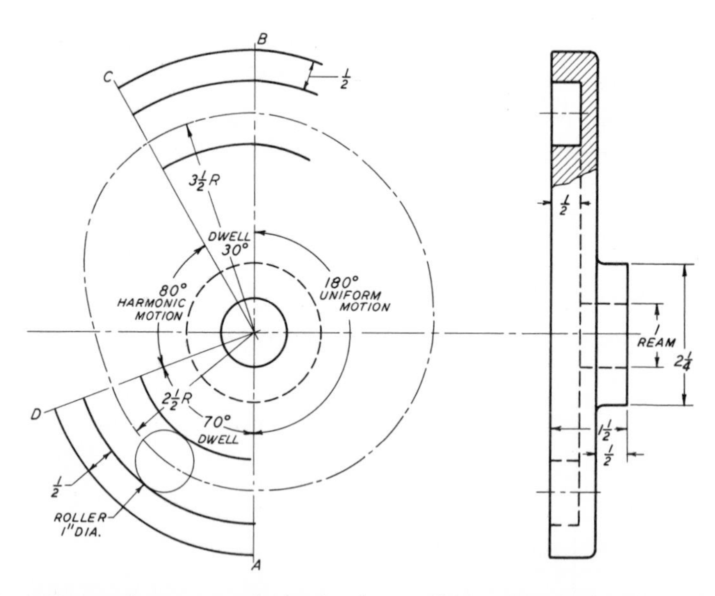

Fig. 19–35 **Draw a box (grooved) cam for the conditions given in the illustration. Lay off the angles. Draw the path of the roll centers. From *A* to *B*, rise from 2½″ to 3½″, with modified uniform motion. From *B* to *C*, dwell, radius 3½″. From *C* to *D*, drop from 3½″ to 2½″, with harmonic motion. From *D* to *A*, dwell, radius 2½″. Draw the groove, using a roller with a 1″ diameter in enough positions to fix outlines for the groove. Complete the cam drawing. Keyway ¼″ × ⅛″. Leave construction lines.**

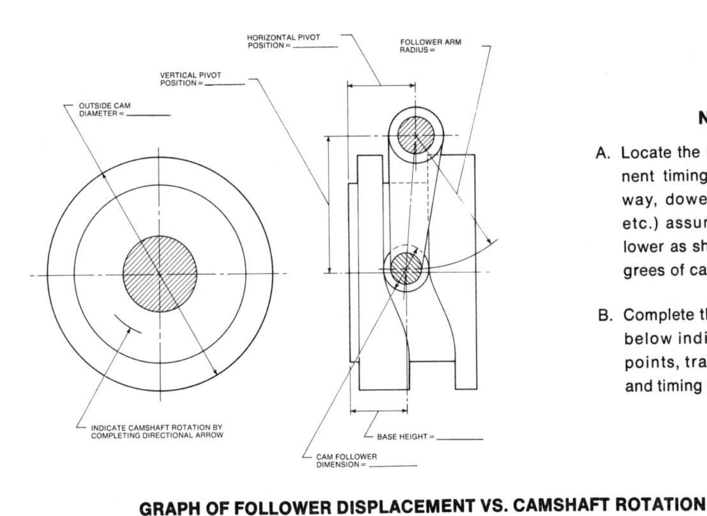

Note:

A. Locate the position of a pertinent timing feature (i.e. keyway, dowel hole, split line, etc.) assuming that the follower as shown is at zero degrees of camshaft rotation.

B. Complete the graph and table below indicating all critical points, transitional motions and timing features.

Fig. 19–36 **As design detailer, determine the data necessary for drawing a displacement diagram of the barrel cam. Prepare the cam profile with adequate detail dimensions.** ***(Camco.)***

GRAPH OF FOLLOWER DISPLACEMENT VS. CAMSHAFT ROTATION

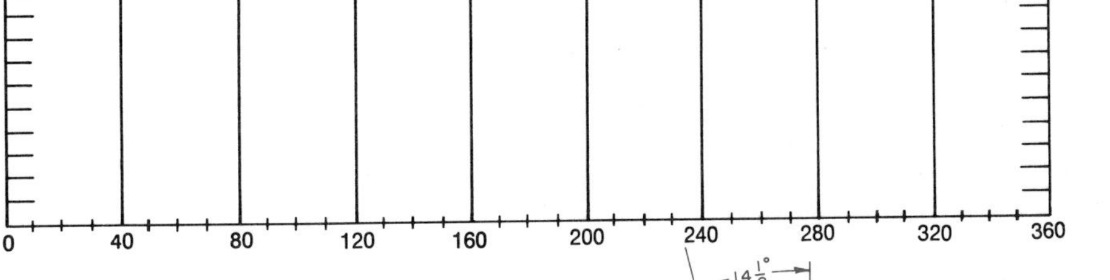

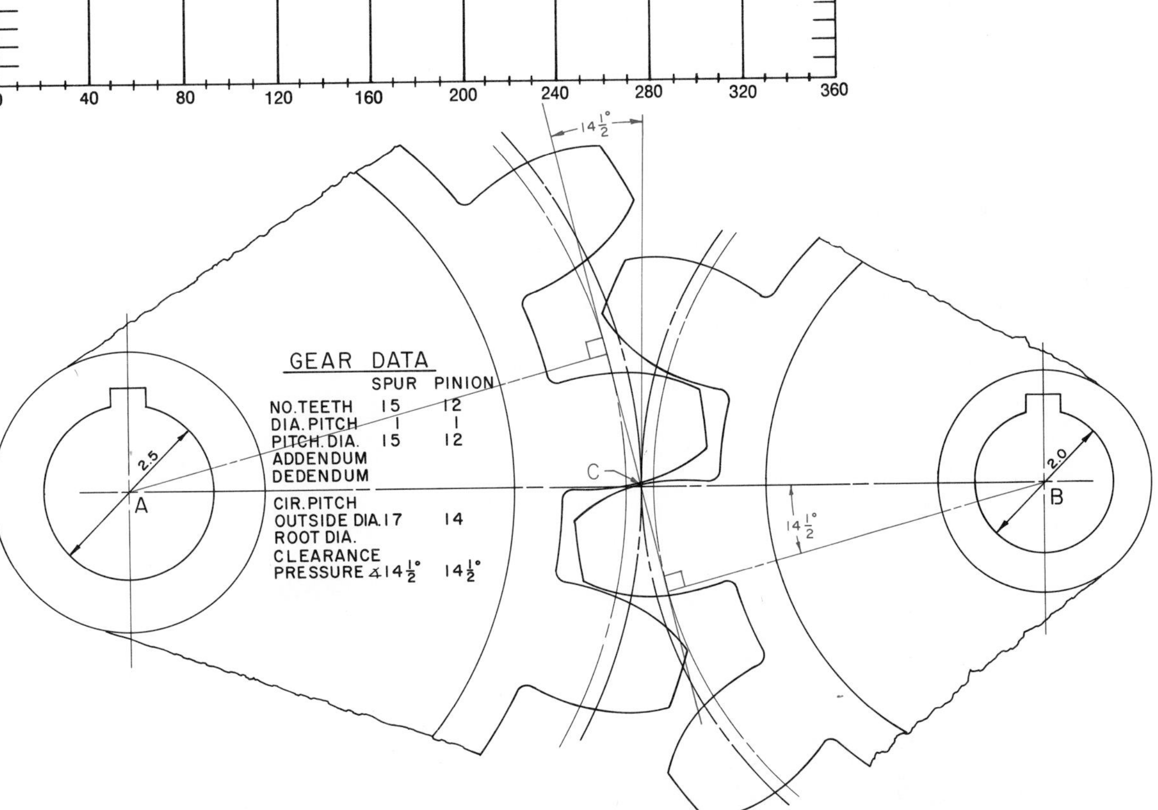

Fig. 19–37 **Complete the gear data table, using formulas from Chapter 19. Make an enlarged drawing of a mating spur and pinion as shown. Select a suitable scale. Use an involute to draw the gear teeth or 1/8 PD, as directed by the instructor.**

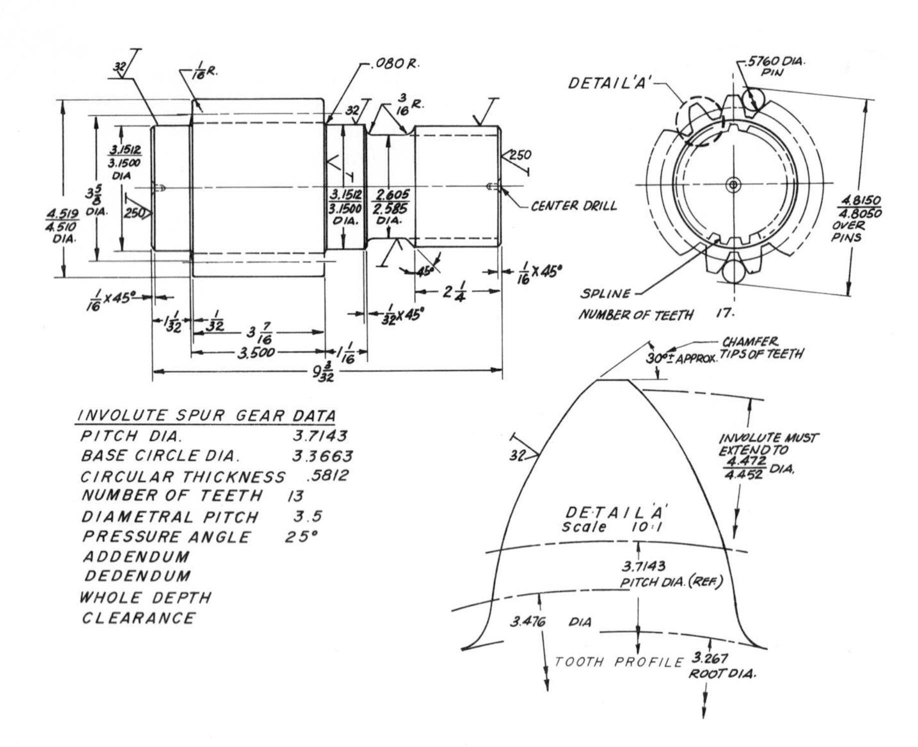

Fig. 19–38 **Examine the spur gear data and drawing carefully before making the following revisions. The 3.500 and the 9 3/32 dimensions are to be reduced by 1.500". Select a suitable scale and make a gear drawing showing two teeth, bottom and top, in profile. Complete the gear data where necessary and list it on the drawing.**

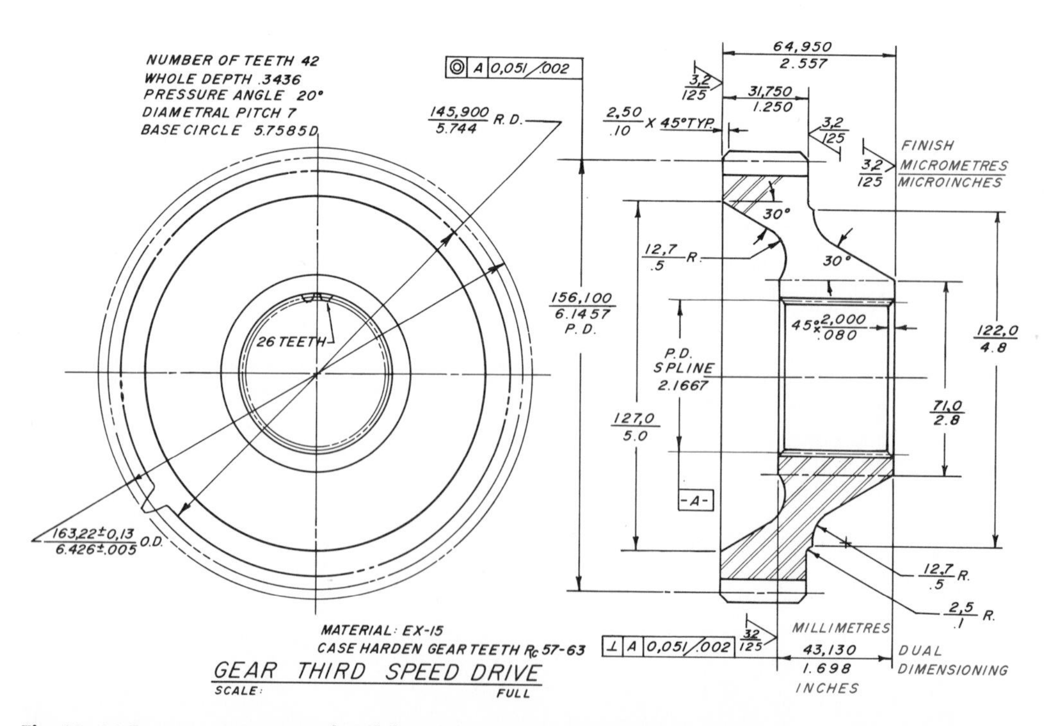

Fig. 19–39 **Prepare a spur gear detail drawing, using the data given to determine necessary dimensions. Note that the gear is designed in metric dimensions (using a comma in place of a decimal point) and is dual dimensioned with decimal inches. *(International Harvester Co.)***

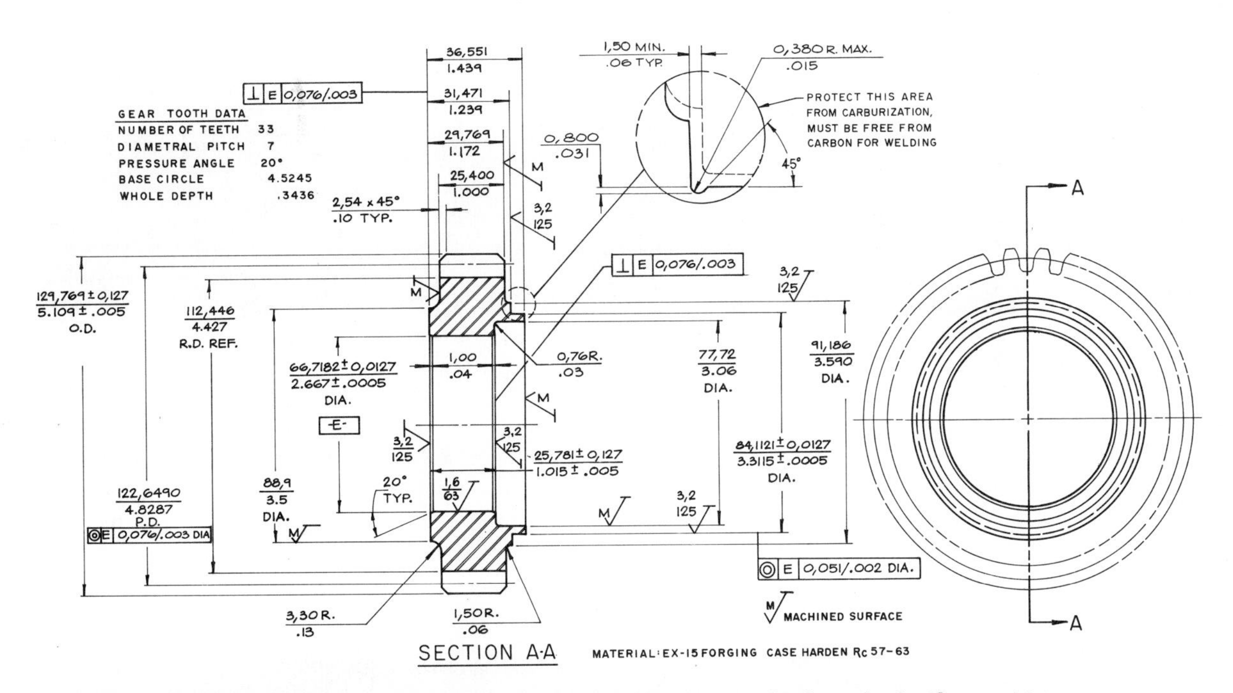

Fig. 19–40 **Develop a detailed working drawing of the spur gear. Use the formulas in Chapter 19 to complete the gear data. Dimension in millimeters, decimal inches, or dually, as directed by the instructor. Note the datum used for accuracy in machine-finish dimensioning.** ***(International Harvester Co.)***

Chapter 20

Architectural Drafting

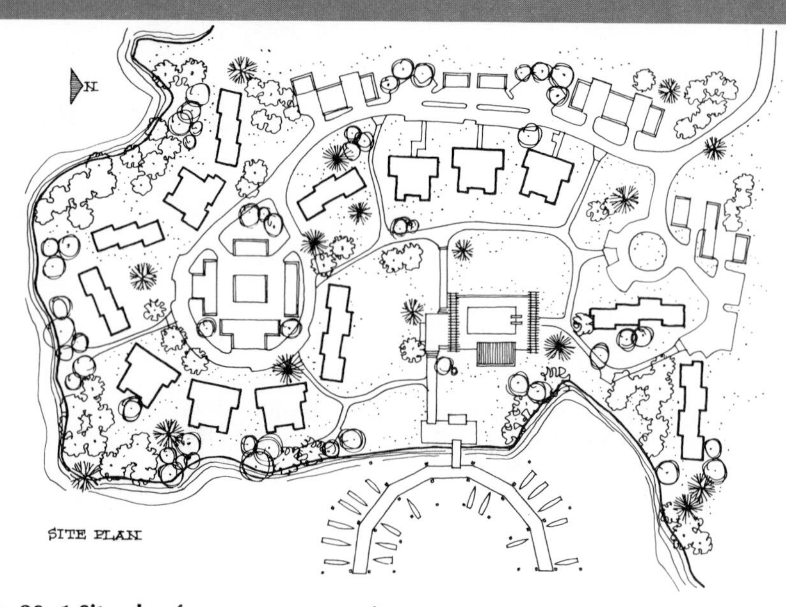

Fig. 20–1 **Site plan for a new community.**

TODAY'S ENVIRONMENT

The work that architects do greatly affects the way we live. Our everyday lives take place within an *environment* (setting) designed by architects. Architectural drawings are the plans for three-dimensional forms called *architecture.* Architects design structures and the spaces around them. These structures and spaces form rural communities, transportation networks, towns, and cities. Architects plan, design, and supervise the building of structures that make progress, society, and even human life itself possible.

NOTE: Because the construction industry is far behind the manufacturing industry in converting to the metric system, metric units are shown after customary units in this chapter.

CAREER OPPORTUNITIES

The growth and changes taking place in human society today demand the creation of new environments for living. New materials and ways of building make this possible. Thus, designers are faced with an exciting and diffucult job. But architecture does not just need design. An architectural firm has many different jobs for many different kinds of workers. Some employees must work with clients. Others must plan design programs. Specifications must be written. Materials have to be chosen. Structural and mechanical needs must be determined. Someone has to figure out costs. Someone else must supervise construction. The many career opportunities in the field of architecture are described in publications that you can get from these groups:

1. *Architectural careers*
   American Institute of Architects (AIA)
   1735 New York Avenue N.W.
   Washington, DC 20006
2. *Urban planning*
   American Institute of Planners
   907 15th Street N.W.
   Washington, DC 20056
3. *Landscape architects*
   American Society of Landscape Architects
   2000 K Street N.W.
   Washington, DC 20006
4. *Construction contracting*
   Associated General Contractors of America, Inc.
   1957 E Street N.W.
   Washington, DC 20006
5. *Careers in building trades*
   Building and Construction Trades Department
   AFL–CIO
   815 16th Street N.W.
   Washington, DC 20005

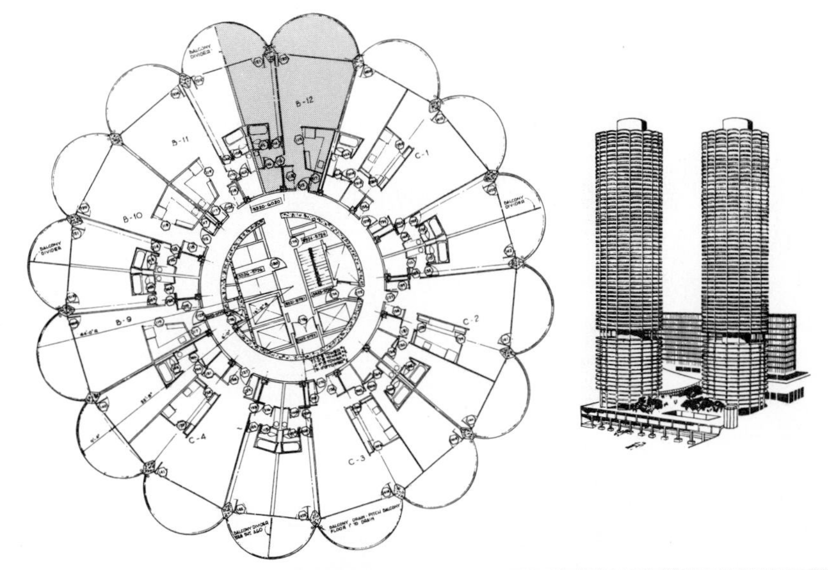

Fig. 20–2 **Typical floor plan for multistory living.** *(Marina City, Bertrand Goldberg Associates.)*

These are only a few of the groups that can provide information about architectural careers. People who enter this challenging field will help plan and build the communities of tomorrow (Fig. 20–1).

COMMUNITY ARCHITECTURE

You can see the effects of the architect's planning by looking carefully at any town or city. Many residential areas have an overall architectural style. Others combine a number of styles. One style may be built in different materials.

There may be wooden, or frame, cottages and brick bungalows. The larger residential buildings usually have rental apartments. Some may be *condominiums.* These are buildings in which the units are individually owned, not rented (Fig. 20–2). Buildings with many stories are usually made of materials such as brick, stone, heavy timber, steel, and concrete. These materials make large buildings strong and let them last for a long time.

You can identify different architectural styles by size, roof shape, and materials. Several kinds of houses are shown in Fig. 20–3. Contemporary Ranch, French Contemporary, Colonial, French Chateau, Georgian, and English Tudor are some of the common styles of residential buildings.

Fig. 20–3 **Several styles of houses.** *(Orrin Dressler, designer-contractor.)*

Fig. 20–3 **(Cont.)**

NEIGHBORHOOD ARRANGEMENTS

Residential buildings are arranged in geometric patterns. The pattern is generally rectangular. The basic unit is called a *block*. Religious buildings may be located on an important street in or near the center of the neighborhood. Many churches and synagogues are on the corner of a block. Parks or recreation centers are conveniently placed throughout each neighborhood. Figure 20–4 shows a contemporary recreation center for a neighborhood that has a lot of space. The more important, busy streets often form the boundaries of a neighborhood. These streets are frequently lined with small stores and with small office and apartment buildings.

URBAN DEVELOPMENT

Neighborhoods combine to make up towns and cities. The main force affecting the human environment today is the movement of people from *rural* (farm) areas to *urban* (city) areas.

In the United States, many people have also moved from "inner" cities and farms to *suburbs*. These are communities on the edges of cities. The growing cities and suburbs need *municipal* (public) buildings to serve their new residents. Municipal buildings often include the city hall and facilities for services such as police, fire protection, water, and sanitation. Health centers such as medical offices and hospitals are also important. Cultural facilities such as libraries, museums, and theaters are significant needs. Architectural services are

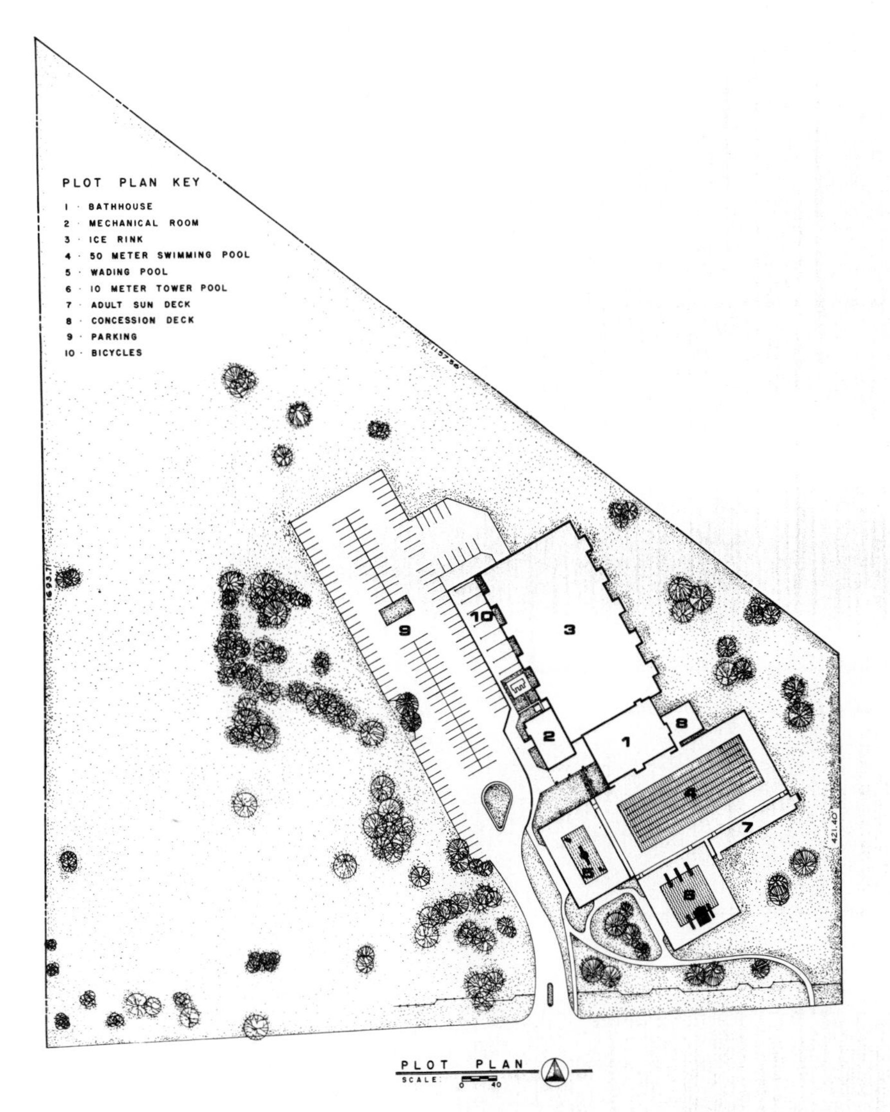

Fig. 20–4 **Site plan for a park and recreation center.** ***(J. E. Barclay, Jr., & Associates.)***

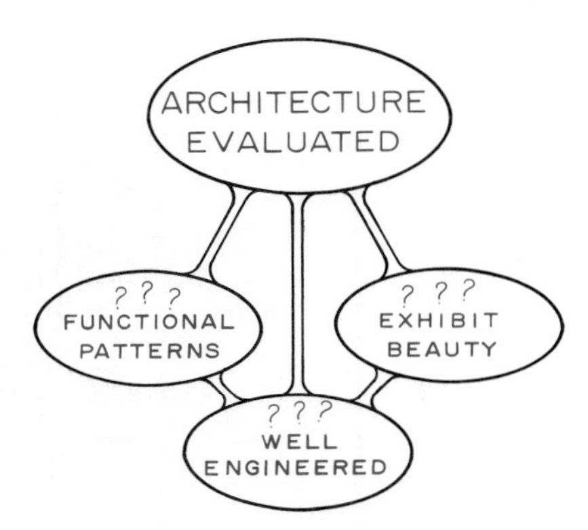

Fig. 20–5 **The basis for evaluating architecture.**

needed to plan and develop these communities.

TRAFFIC PATTERNS

The mobility of a community's residents is reflected in transportation systems. The streets around rectangular blocks or the wandering lanes with "deadends" that form a community are connected to major thoroughfares. These are lined with large stores, shopping centers, and office buildings. Apartment buildings, banks, and hospitals are also generally located on or near key roads that make it easy to get to them. Government bodies known as building and zoning commissions decide what kinds of buildings will be in different areas. Manufacturing plants may be located in the center of town or on the edge. Traffic patterns are designed so as to let industrial employees get to and from work easily. The success of a community may rest on good planning for industry.

ARCHITECTURE DEFINED

Architecture is not just buildings. It is people—how they live, work, play, and worship. Good architectural environments can make any human activity better. Thus, architecture is thought of as any physical environment. It is the sprawling suburb, the crowded slum, the mighty industrial complex, the bright lights of Broadway.

ARCHITECTURE EVALUATED

Architecture affects everyone's life. Those who study an architectural structure look for three things in it (Fig. 20–5):

First, the structure has to satisfy a social purpose. That is, it must show *functional* (useful) patterns for human activity. Begin by studying how people move through the structure to see how its spaces meet their needs (Fig. 20–6).

Second, the structure must be well engineered. Structural members have to be well constructed.

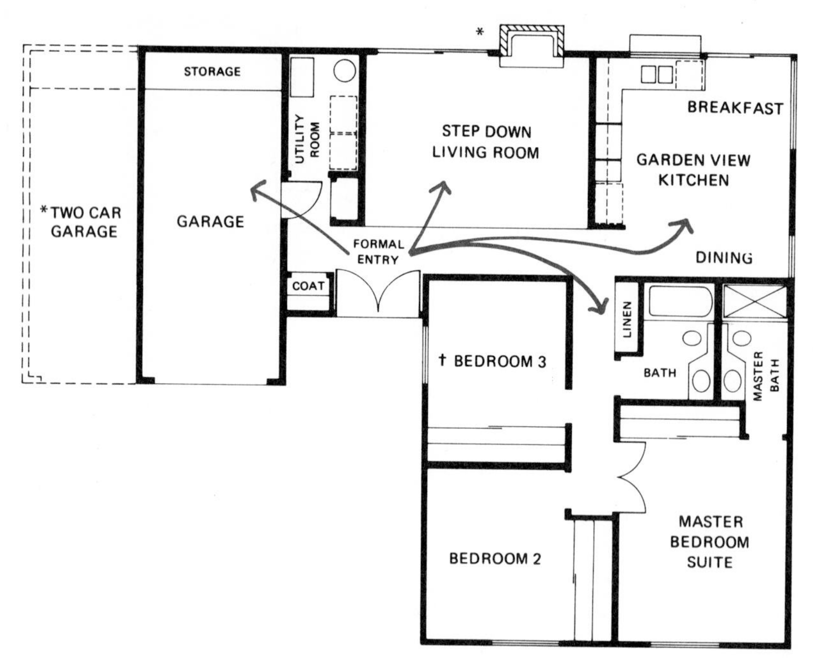

Fig. 20–6 **Functional traffic patterns.** ***(Larwin Co.)***

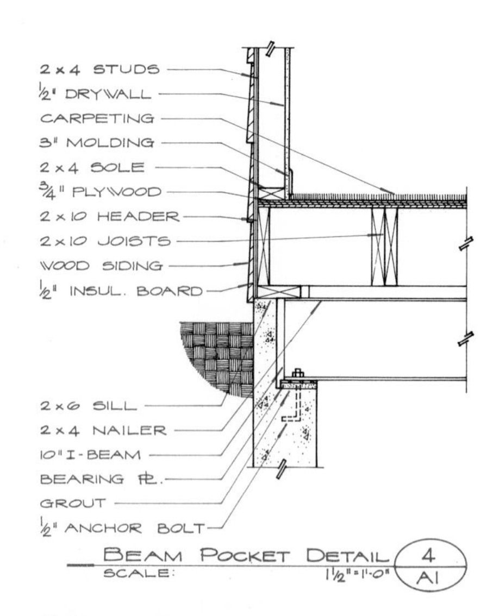

Fig. 20–7 **Structural detail that indicates good engineering.**

Materials must be well selected (Fig. 20–7).

Third, the structure must have *aesthetic value* (beauty). Successful architecture has to have a pleasing form based on appealing design qualities (Fig. 20–8).

Architectural form is easily evaluated by trained architects. Designing a human environment means merging function, structure, and beauty.

THE ARCHITECT'S OFFICE

The typical architectural firm is a *partnership.* It has two to four main partners. The partners employ six to twelve persons. However, some firms may have a hundred or more employees. The large firm may offer all the basic architectural services. These include architectural design, structural design, mechanical engineering, civil engi-

Fig. 20–8 **Residential design in a contemporary form.** ***(Larwin Co.)***

neering, landscape design, interior design, and urban planning. These services may be used in the design and building of one structure or of an entire city. Some firms provide only one service. They may make subcontracts with other specialized firms in working on a project.

Not all architectural firms are partnerships. One kind has just one owner. This is called a *proprietorship.* Another kind is owned by people who buy stock in the business. This kind is a *corporation.* Whatever the form of business organization, however, an architectural office works as a team. Each member of the architectural team has a well-defined position. An office usually has an overall project director. Individual team leaders might help find clients or work on a particular project. The National Council of Architecture Licensing Boards in Washington, DC, promotes professional registration of architects within each state.

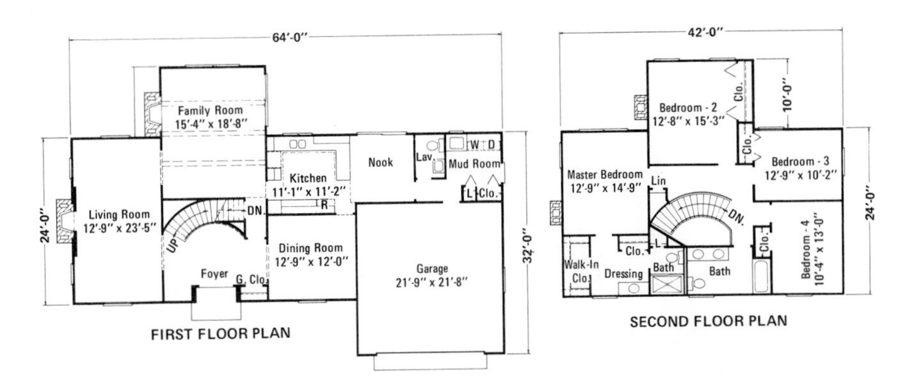

Fig. 20–9 **A plan, the first of the four basic drawings.** ***(Inland Steel Co.)***

Fig. 20–10 **An elevation, the second of the four basic drawings.** ***(Inland Steel Co.)***

THE BASIC DRAWINGS

For centuries, architects have been preparing four basic kinds of drawings. These drawings are considered necessary for communicating the details of a project that is to be built. The four kinds of drawings are plan, elevation, perspective, and section. The projection methods of representation used in mechanical drawing are also used in architectural drawing.

The plan. This is a drawing that shows the horizontal plane on which spaces are arranged for human use (Fig. 20–9). The three main areas of a residential plan are living, sleeping, and service. The plan is a section that cuts through walls and shows room arrangements.

The elevation. The elevation, or facade, defines the structural form and architectural style. The roof shape, sides, and side openings are shown in the architectural style chosen and in relationship to the plan. With such drawings, the need for changes to provide balance or symmetry can be quickly seen. The elevations, however, are only two-dimensional. This limits in-depth study (Fig. 20–10).

Perspective drawing. It is easiest to see architectural form in a pictorial presentation drawing. One-two-, or three-point perspective drawings show a building in a realistic way. Architectural perspective studies do not have to be mechanically accurate. Therefore, freehand techniques, along with good proportions, can be used in making these drawings (Fig. 20–11).

The cross-detail section. In this kind of drawing, a vertical plane might be chosen to cut a section to show construction details. Typically, material details are labeled. A section might be cut across the entire structure to aid in interpreting the relationships of the important spaces. In full sections, it is

Fig. 20–11 **A perspective, the third of the four basic drawings.** ***(Inland Steel Co.)***

easier to see the proportions between spaces and how these relate to construction and use (Fig. 20–12).

Fig. 20–12 **A cross-detail section, the fourth of the four basic drawings.**

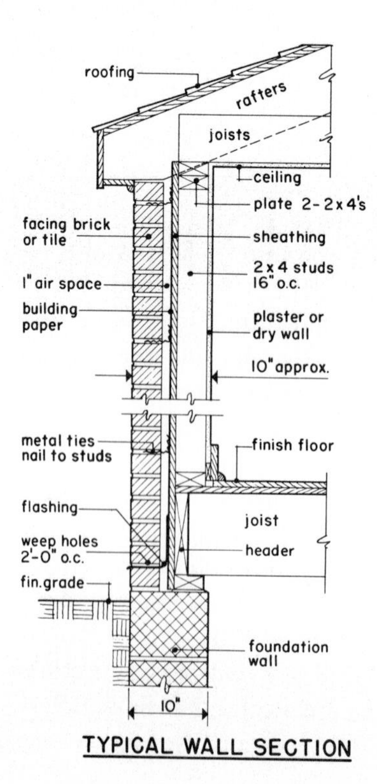

Levels of Communication

The four levels of communication discussed in Chapter 1 apply to architectural drafting (Fig. 20–13).

Level one: creative communication. The preliminary architectural studies are sketched or laid out lightly so they can be studied and changed.

Level two: technical communication. The sketches of refined ideas are approved by professionals.

Level three: market communication. The pictorial drawings and colored perspective-presentation drawings are evaluated and approved by the client.

Level four: construction communication. Detailed plans, elevations, and sections, together with specifications, are studied and approved by builders. Working drawings from the suppliers of materials supplement the architect's detailed drawings.

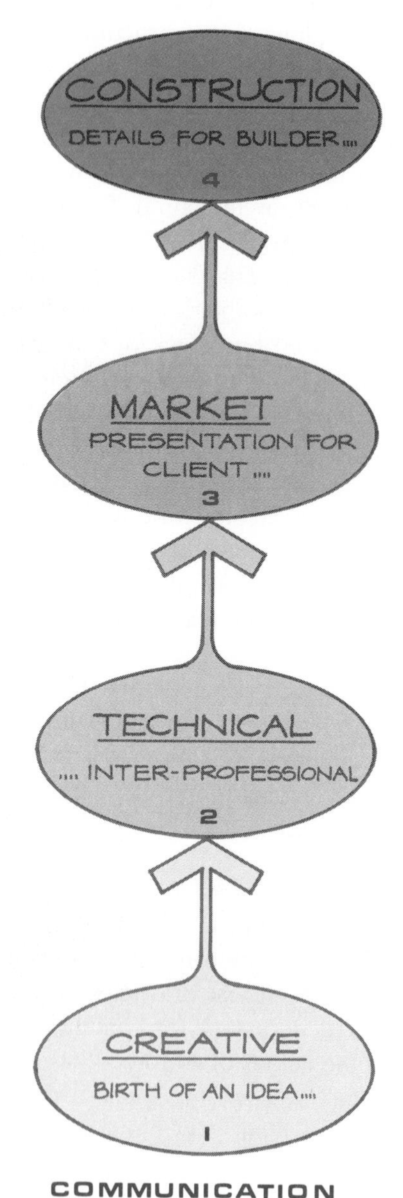

Fig. 20–13 **The four levels of graphic communication.**

ARCHITECTURAL DRAFTING TECHNIQUES

A good line technique is the most important skill for architectural designing and detailing. In Fig. 20–14, a series of lines are shown that intersect and stop after forming a flared corner. The visual weights of lines express spatial relationships. Good lines are needed for reproduction. They are needed to make the design appealing. They are needed so that the third dimension of depth can be imagined on a two-dimensional medium, as shown in Fig. 20–15. In well-prepared architectural details, line tone and texture complement each other.

Lettering

Both traditional and contemporary styles of lettering are commonly and acceptably used by architects. The traditional lettering is based on the Old Roman alphabet, as shown in Fig. 20–16. These letters have *serifs*. Serifs are small lines extending from the main strokes. Elaborate and important titles are often designed with Old Roman letters. However, plain, single-stroke Gothic lettering is better for architectural working drawings (Fig. 20–17). The condensed and extended styles shown are often used by experienced professionals. They use the right style for inserting notes or balancing the drawing.

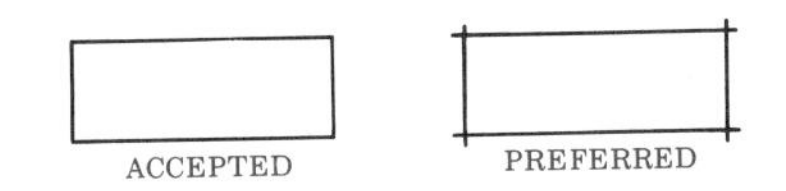

Fig. 20–14 **Architectural line technique.**

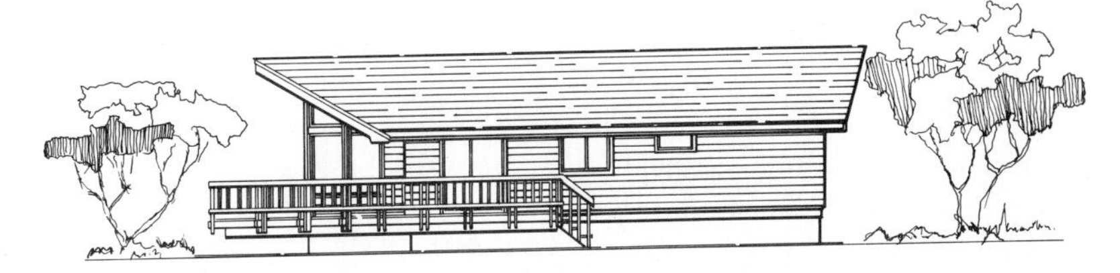

Fig. 20–15 **Techniques that feature the major proportions.** ***(Justus Company, Inc.)***

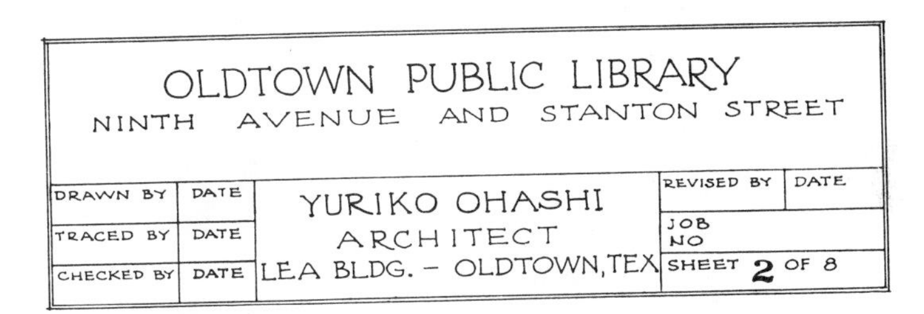
OLDTOWN PUBLIC LIBRARY
NINTH AVENUE AND STANTON STREET

DRAWN BY	DATE	YURIKO OHASHI	REVISED BY	DATE
TRACED BY	DATE	ARCHITECT	JOB NO	
CHECKED BY	DATE	LEA BLDG. – OLDTOWN, TEX	SHEET 2 OF 8	

Fig. 20–16 **Old Roman lettering applied in an architectural style.**

A B C D E F G H I J K L M N O P Q R S T U V W X Y Z

FIRST FLOOR PLAN
SCALE 1/4" = 1'-0"

A B C D E F G H I J K L M N O P Q R S T U V W X Y Z

TYPICAL WALL SECTION
SCALE: 1/2" = 1'-0"

A B C D E F G H I J K L M N O P Q R S T U V W X Y Z

GENERAL NOTES AND MATERIALS

Fig. 20–17 **Single-stroke Gothic lettering in condensed, extended, and the general Gothic stroke.**

Preparing Title Blocks

An architectural firm can create a quickly identified image with a *logo* (symbol), as shown in Fig. 20–18. Some firms use forms with preprinted borders, title, and revision blocks. The blocks are ready to be filled in with black ink. The typical title blocks shown in Fig. 20–19 were created by students preparing to play professional roles. The title block generally gives the following information: the owner's name and address, the type of structure, the architect's name and address, the title of the sheet and the sheet number, the date, the scale, and the initials of the detailing drafter and the supervisor.

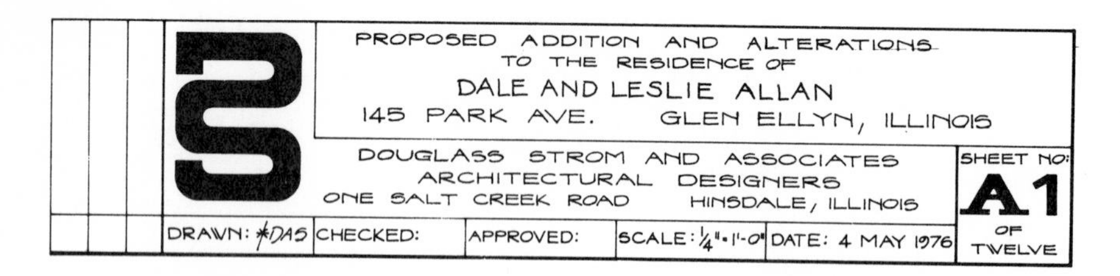

Fig. 20–18 Title block, including the architectural firm's logo.

Material Symbols

Architectural symbols indicate the materials used in building. They are shown in Fig. 20–20. These symbols are used in plan, section, and elevation drawings.

Pressure-sensitive Symbols

There are many graphic aids that help the designer meet professional standards on both plan and elevation drawings. An example of these heat-resistant and pressure-sensitive symbols is shown in Fig. 20–21. Many landscape symbols, such as shrubs, hedges, and trees, are available in both plan and elevation forms. There are also doors, furniture, plumbing and electrical symbols, and many other items for presentation and detail drawings. You transfer these symbols to the vellum, film, or paper by burnishing them with a smooth, blunt instrument.

Architectural Wall Symbols

The details with dimensions in the plan have technical meaning in the building industry. Typical wall symbols for a residential structure are shown in Fig. 20–22. The crosshatched masonry wall at A is dimensioned to the outside face of the wall. This is done with solid brick, concrete, and concrete-block walls. The frame wall at B is dimensioned to the face of the stud. The brick-veneer wall at C with

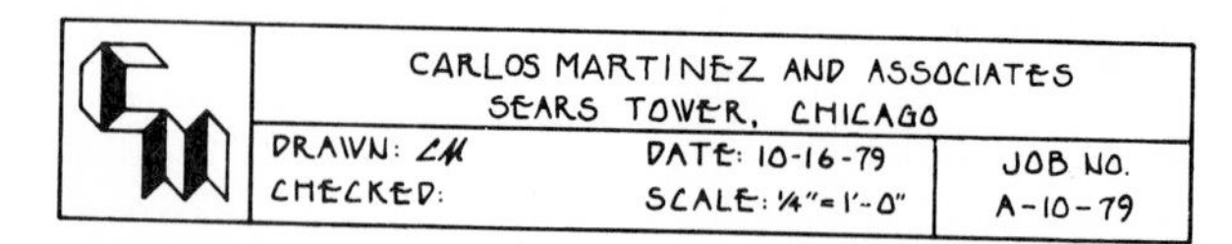

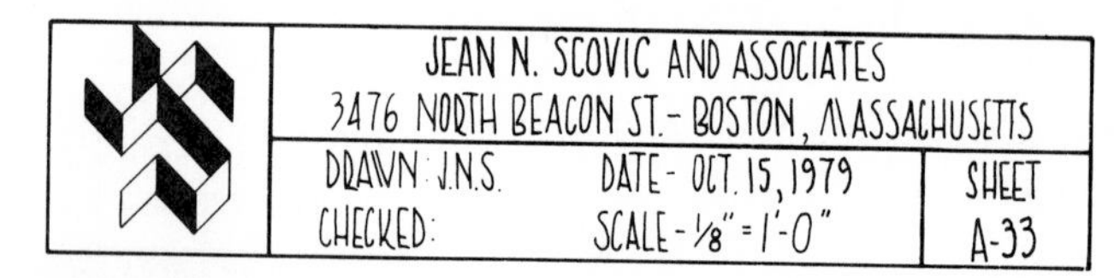

Fig. 20–19 Student-created title blocks.

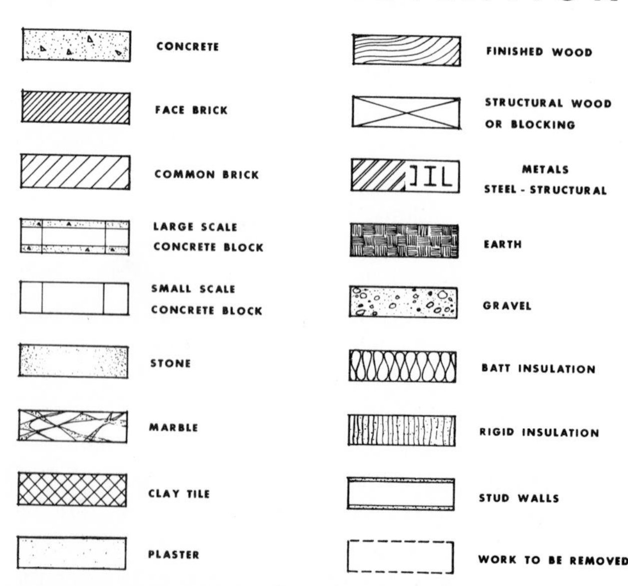

Fig. 20–20 Architectural material symbols.

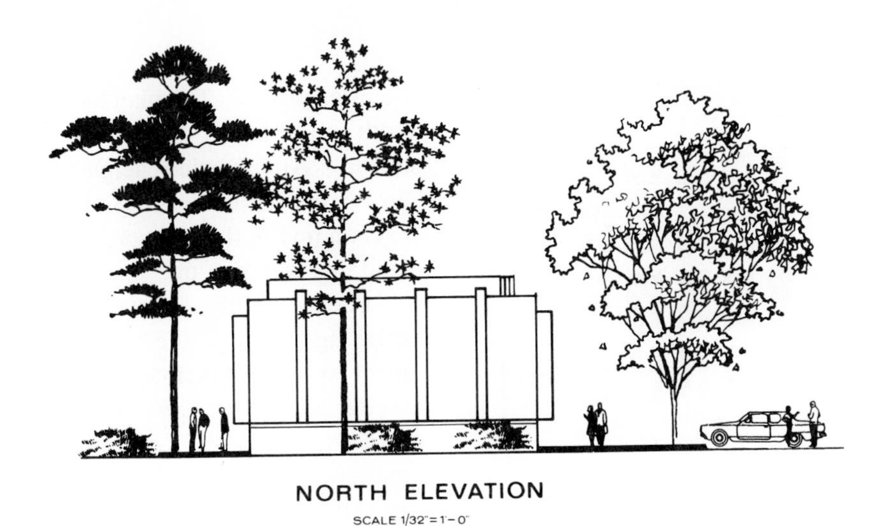

Fig. 20–21 **Examples of architectural pressure-sensitive symbols.** ***(Para-tone, Inc.)***

crosshatched brick and a sole plate is dimensioned to the face of the stud. The remaining material is assumed to be exterior wall facing.

Symbols for Door and Window Openings

Some door symbols are shown in Fig. 20–23. The plan shows the outside door ajar in a single line. An arc indicates the direction in which the door swings. Interior doors are shown in a variety of ways. Typical windows are shown in Fig. 20–24. Symbols help the reader interpret window movement. They also show what type of wall material frames the window.

Templates for Design and Detail

Templates have improved the quality of the drawings developed today in the architect's office. Templates are usually made of lightweight plastic with openings that form structural shapes or fixtures. You follow the outline of the opening with your pencil or pen to make the graphic image, as in Fig. 20–25.

Figure 20–26 shows one of the many special-purpose templates.

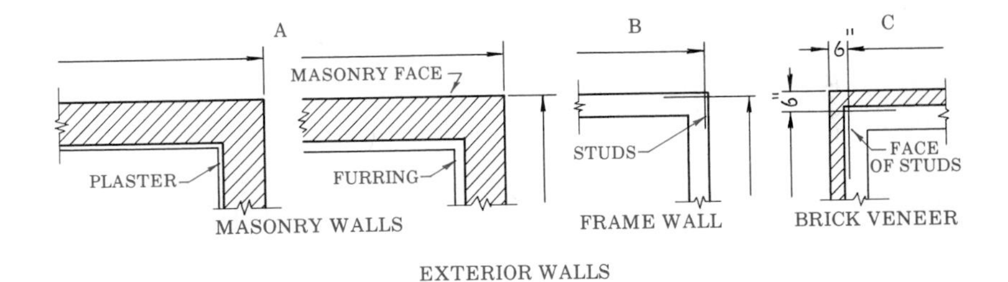

Fig. 20–22 **Wall symbols.**

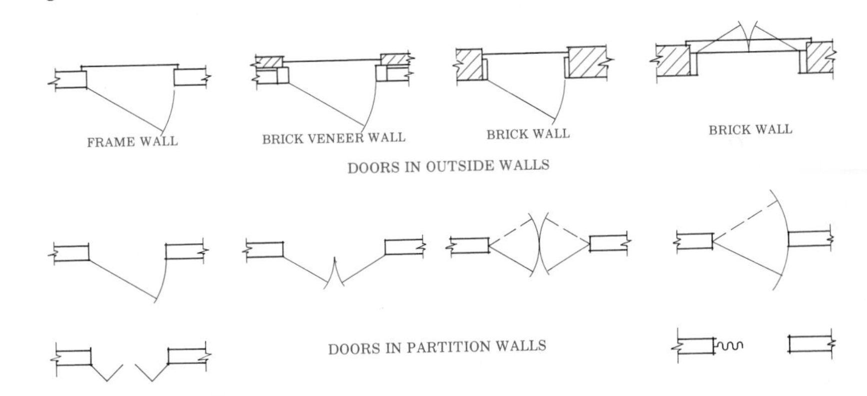

Fig. 20–23 **Door symbols.**

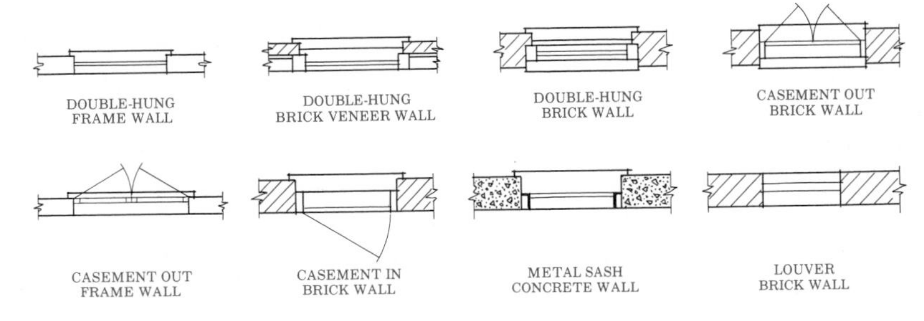

Fig. 20–24 **Window symbols.**

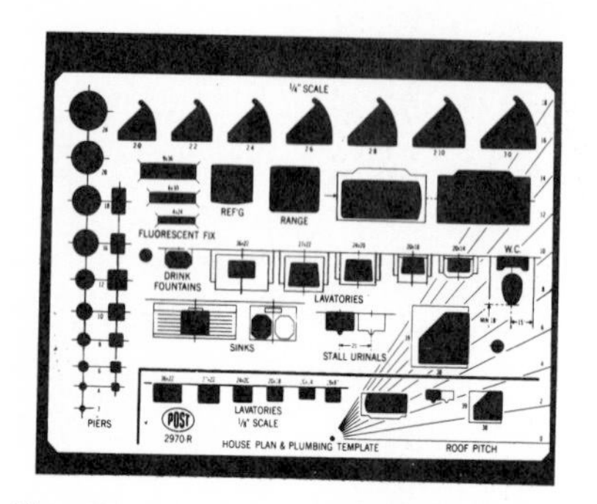

Fig. 20–25 **House plan and plumbing template.** ***(Teledyne Post.)***

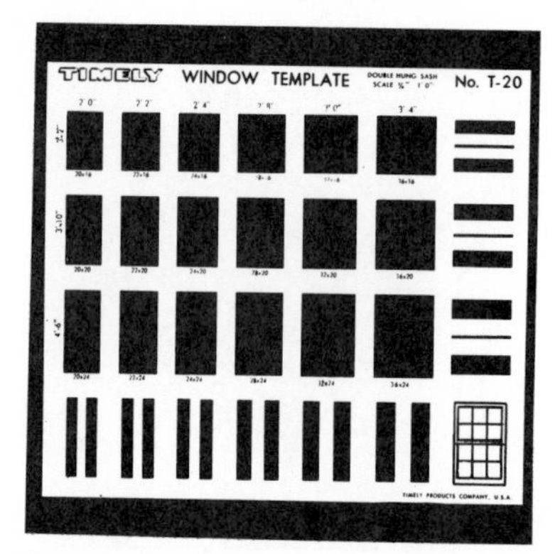

Fig. 20–26 **Window template.** ***(Teledyne Post.)***

These have door, window, landscape, furniture, structural-steel-shape, plumbing, and electrical symbols.

DIMENSIONING TECHNIQUES

Structural wall openings are dimensioned in similar ways on the plan drawings for frame and brick-veneer structures. The general dimensioning techniques discussed in Chapter 6 are used. However, the dimension line is unbroken. In addition, the dimensions appear above the dimension line, as shown in Fig. 20–27. Dimensions end with the usual arrowhead or with several other symbols, as shown in Fig. 20–28. Note that dimension lines may cross when the sizes of interior rooms are given. The plan has only width and depth dimensions, as shown. Any needed heights are shown as floor-to-ceiling dimensions on one of the elevations or on a suitable wall section.

To help simplify and standardize designing buildings, the building industry and the American National Standards Institute established a standard, Modular Coordination, numbered A62.1. This standard was sponsored by the American Institute of Architects. The basic customary-inch module is 4 in. for all U.S. building materials and products. The International Standards Organization has established 100 mm (almost 4 in.) as the module for countries using the metric system. Modular components are typically drawn on a 4-in. or 100-mm centerline. All buildings currently designed for customary modular specifications are planned with multiples of 4 in. In this chapter, the inch-system measurement comes first and the metric follows in parentheses. Figure 20–29 shows Crown Hall School of Architecture at the Illinois Institute of Technology.

Fig. 20–27 **Architectural style of dimensioning.**

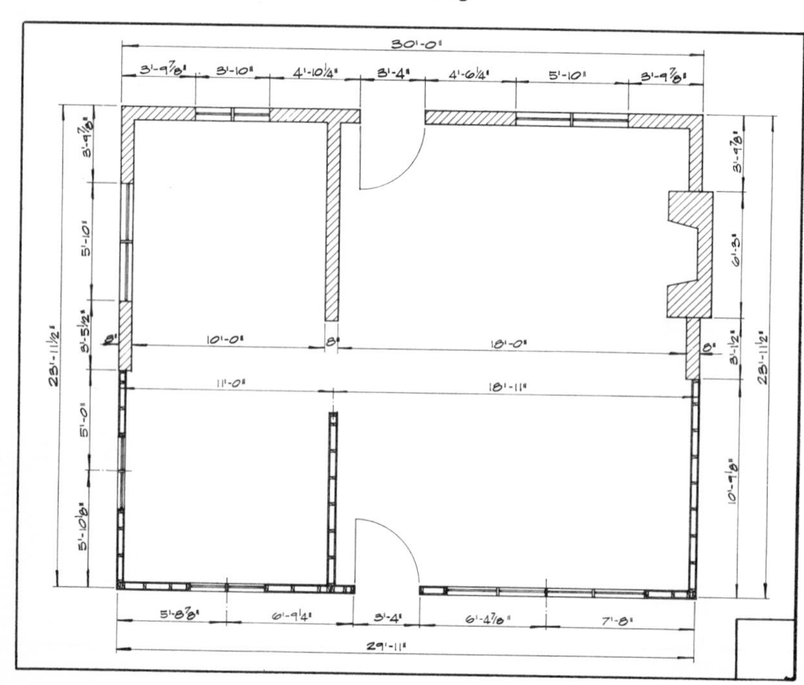

Fig. 20–28 **Dimension-line arrowheads and alternate terminating symbols.**

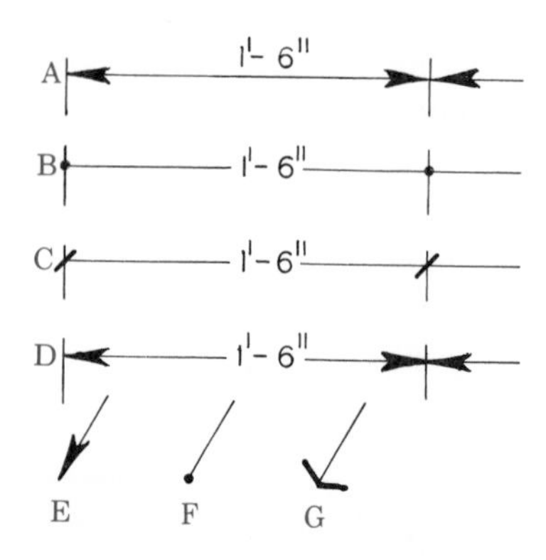

A

Fig. 20–29 **A. Crown Hall School of Architecture; B. Modular floor plan of Crown Hall; C. Schematic line elevations of Crown Hall.** ***(The office of Mies Van der Rohe and Hedrich—Blessing.)***

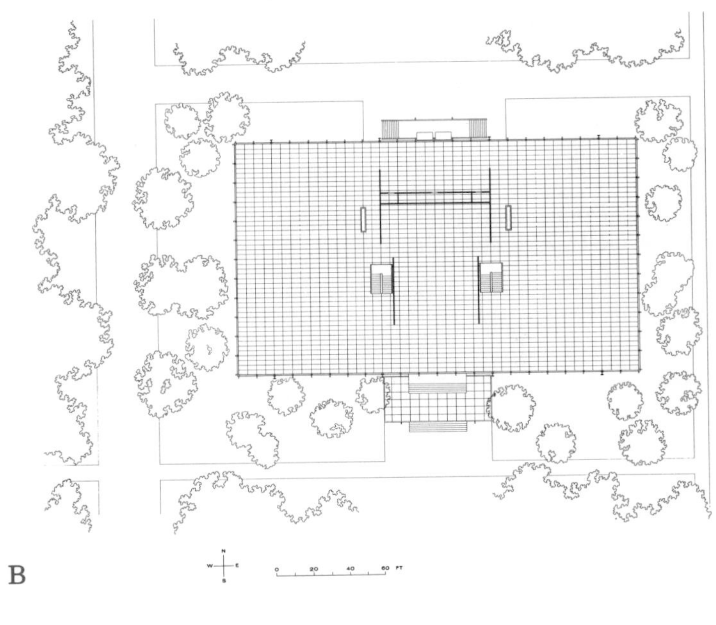

B

C

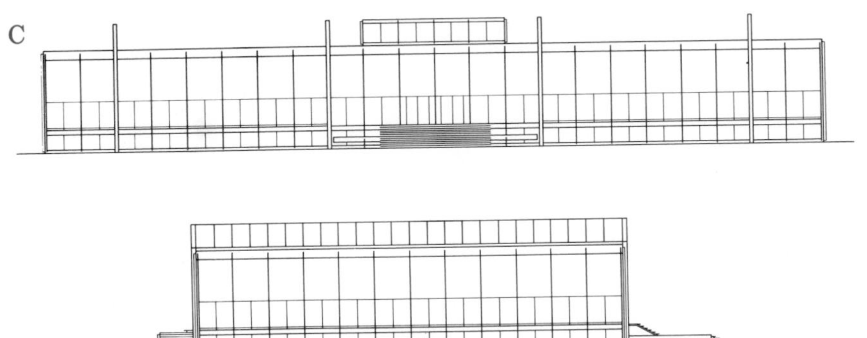

This building was designed by Ludwig Mies Van der Rohe using a modular grid. Note the grid lines on the floor plan at B. Schematic line elevations at C are also drawn on a modular grid. Modular planning reduces building costs and time. This is because standard components can be used. Materials do not have to be cut to many different sizes.

The dimensions used on architectural drawings are either finish or construction details. The elevations generally have the expected finished dimensions. The floor plan and sectional details have either construction or finish dimensions, as shown in Fig. 20–30.

Dimension lines are placed about ⅜ in. (10 mm) apart. They are needed for overall finished building and room dimensions. The structural openings and wall offsets are defined with detailed construction dimensions. Dimension lines showing the distances between centerlines (for modules) or extension lines (for general design) usually end in arrowheads. Dimensions are placed above the line and given in feet and inches, as shown. Note the different ways of showing dimensions that are acceptable for architectural drawings (Fig. 20–31).

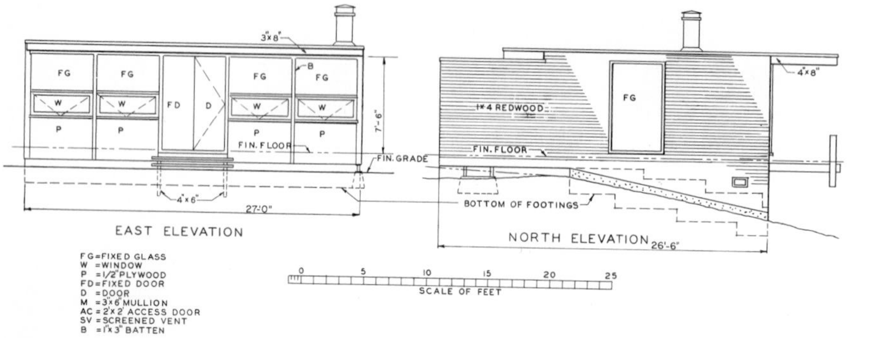

Fig. 20–30 **Dimensions applied to an elevation.** ***(Adapted from drawings of Henry Hill, architect, AIA, and John W. Kruse, AIA, associate, San Francisco, CA.)***

SCALES AND DIMENSIONING RELATIONSHIP

Residential designs and details are usually developed with the aid of reduction scales. The architect's scale is discussed in Chapter 3. The ¼″ = 1′–0 (1:50 in metric) scale is best suited for plans for houses and small buildings. The usual scale for larger buildings is ⅛″ = 1′–0 (1:100 in metric). Plot plans may be

Fig. 20–31 **Dimensions that apply to plan and section drawings.** ***(Adapted from drawings of Henry Hill, architect, AIA, and John W. Kruse, AIA, associate, San Francisco, CA.)***

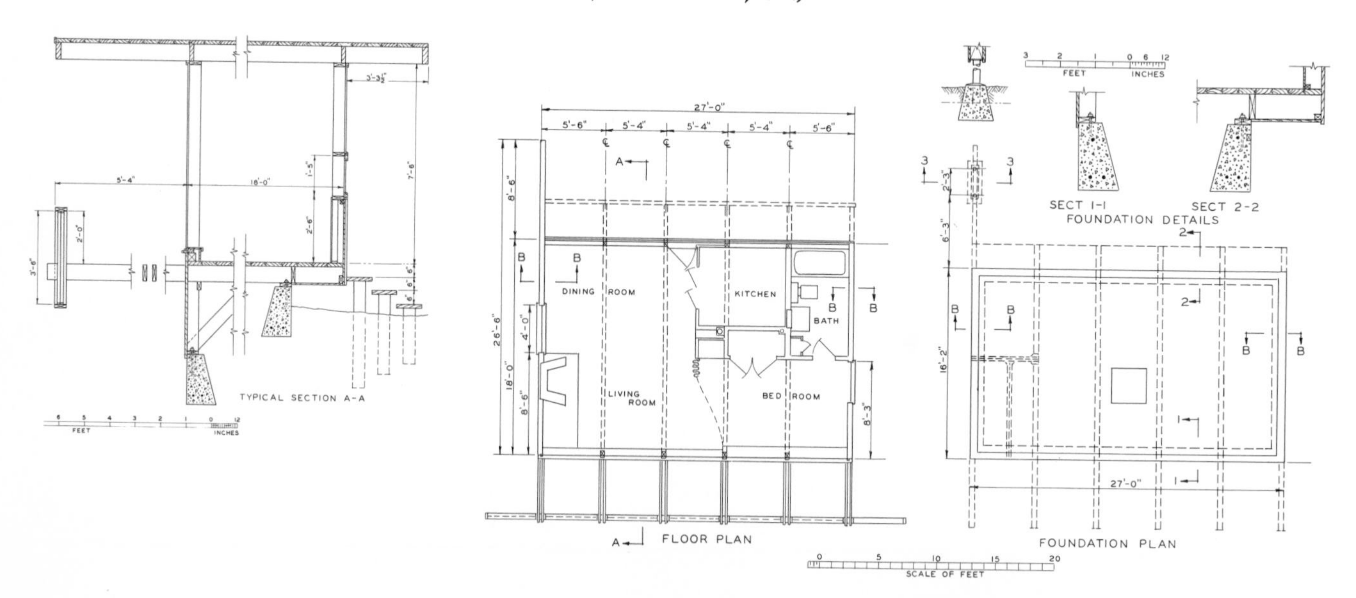

drawn at 1/10″ = 1′–0 or 1/32″ = 1′–0, but it is better to use an engineer's scale at 1″ = 20′, 1″ = 30′, or 1″ = 40′ (1:200, 1:300, or 1:400 in metric). Enlarged details are developed at 1″ = 1′ or 1½″ = 1′. Sectional details are defined at ½″ = 1′ or ¾″ = 1′. Some details may require half-size or full-size drawings. Metric scale for enlarged details may be 1:5, 1:10, or 1:20.

MATERIALS DESIGN AND DETAIL

The new environments designed by architects are created from materials. Detailed drawings and specifications must show how these materials are to be fabricated and constructed. The *Sweets Architectural Catalog* for materials manufacturers is one of the most important tools of the design team. Many materials are available in standard units for the building trades. Others are custom-designed. A few standards are discussed here.

LUMBER					
Nominal Size	2 × 4	2 × 6	2 × 8	2 × 10	2 × 12
Dressed Size	$1\frac{1}{2}'' \times 3\frac{1}{2}''$	$1\frac{1}{2}'' \times 5\frac{1}{2}''$	$1\frac{1}{2}'' \times 7\frac{1}{2}''$	$1\frac{1}{2}'' \times 9\frac{1}{2}''$	$1\frac{1}{2}'' \times 11\frac{1}{2}''$
Nominal Size	4 × 6	4 × 8	4 × 10	6 × 6	6 × 8
Dressed Size	$3\frac{9}{16}'' \times 5\frac{1}{2}''$	$3\frac{9}{16}'' \times 7\frac{1}{2}''$	$3\frac{9}{16}'' \times 9\frac{1}{2}''$	$5\frac{1}{2}'' \times 5\frac{1}{2}''$	$5\frac{1}{2}'' \times 7\frac{1}{2}''$
Nominal Size	6 × 10	8 × 8	8 × 10		
Dressed Size	$5\frac{1}{2}'' \times 9\frac{1}{2}''$	$7\frac{1}{2}'' \times 7\frac{1}{2}''$	$7\frac{1}{2}'' \times 9\frac{1}{2}''$		
BOARDS					
Nominal Size	1 × 4	1 × 6	1 × 8	1 × 10	1 × 12
Actual Size, Common Boards	$\frac{3}{4}'' \times 3\frac{9}{16}''$	$\frac{3}{4}'' \times 5\frac{9}{16}''$	$\frac{3}{4}'' \times 7\frac{1}{2}''$	$\frac{3}{4}'' \times 9\frac{1}{2}''$	$\frac{3}{4}'' \times 11\frac{1}{2}''$
Actual Size, Shiplap	$\frac{3}{4}'' \times 3''$	$\frac{3}{4}'' \times 4\frac{15}{16}''$	$\frac{3}{4}'' \times 6\frac{7}{8}''$	$\frac{3}{4}'' \times 8\frac{7}{8}''$	$\frac{3}{4}'' \times 10\frac{7}{8}''$
Actual Size, Tongue-and-Groove	$\frac{3}{4}'' \times 3\frac{1}{4}''$	$\frac{3}{4}'' \times 5\frac{3}{16}''$	$\frac{3}{4}'' \times 7\frac{1}{8}''$	$\frac{3}{4}'' \times 9\frac{1}{8}''$	$\frac{3}{4}'' \times 11\frac{1}{8}''$

Fig. 20–32 **Standard sizes for lumber and boards.** ***(Customary U.S. measure)***

Lumber

Lumber may be specified by *nominal* dimensions. These differ from the *actual* dimensions of the surfaced wood. For example, most of the lumber and boards for residential construction are surfaced on four sides. The *dressed* (finished) sizes are noted in Fig. 20–32.

Masonry

Figure 20–33 gives the sizes of brick building materials. The common brick has modular dimensions. Brick, block, stone, and stucco may be used for exterior and interior walls and floors. Most of these materials can also serve as structural load-bearing walls for supporting floors and roofs. They generally can be *bonded* (overlapped) so as to increase their structural strength, as shown in Fig. 20–34. Bonding also forms decorative patterns. Well-designed masonry needs little upkeep. Its colors and patterns are important parts of the overall architectural design.

Concrete

Concrete may be thought of as a structural unit. As footing, foundation walls, and poured-in forms, it supports floor loads. Both interior and exterior walls can be formed from concrete. Concrete can be precast to give certain finishes.

Fig. 20–33 **Brick sizes.**

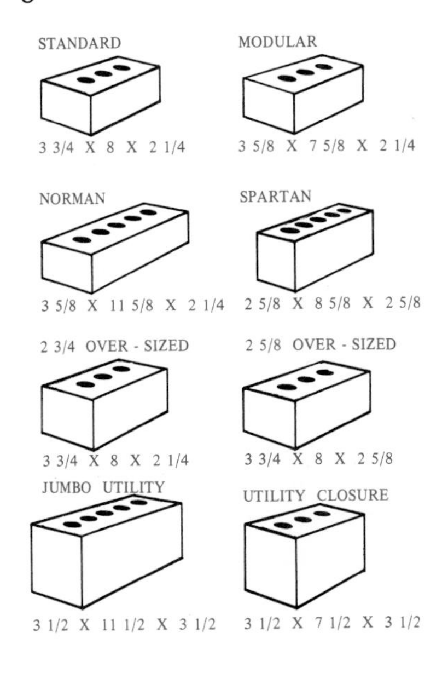

8" ALL ROLOK WALL
COMMON BOND

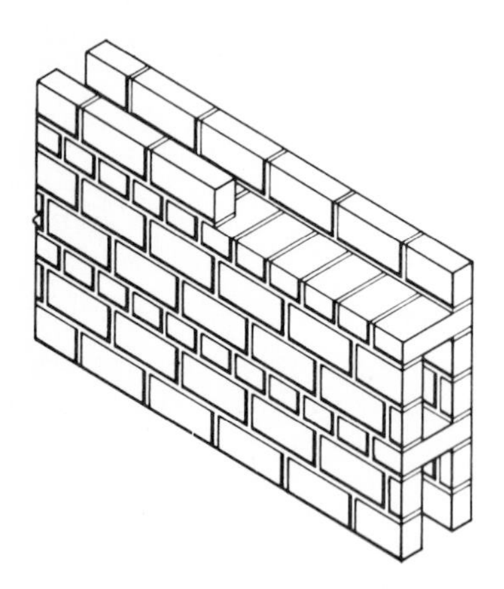

12" ALL ROLOK WALL
FLEMISH BOND

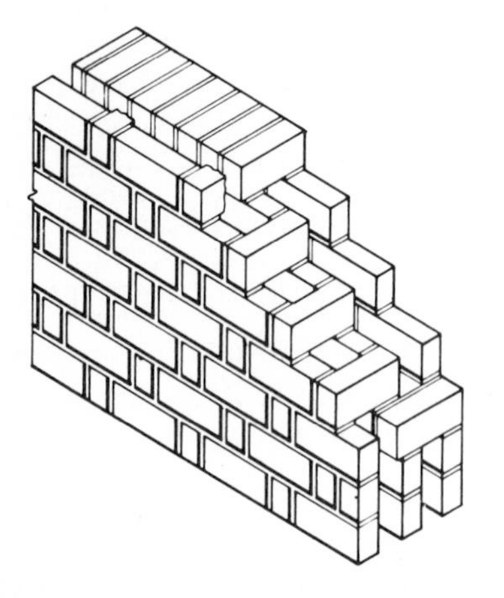

Fig. 20–34 **Common and Flemish bonds.** ***(Structural Clay Products Institute.)***

Fig. 20–35 **Essential parts of a house.** ***(From National Bureau of Standards Circular 489.)***

1. Gable end
2. Louver
3. Interior trim
4. Shingles
5. Chimney cap
6. Flue linings
7. Flashing
8. Roofing felt
9. Roof sheathing
10. Ridge board
11. Rafters
12. Roof valley
13. Dormer window
14. Interior wall finish
15. Studs
16. Insulation
17. Diagonal sheathing
18. Sheathing paper
19. Window frame and sash
20. Corner board
21. Siding
22. Shutters
23. Exterior trim
24. Waterproofing
25. Foundation wall
26. Column
27. Joists
28. Basement floor
29. Gravel fill
30. Heating plant
31. Footing
32. Drain tile
33. Girder
34. Stairway
35. Subfloor
36. Hearth
37. Building paper
38. Finish floor
39. Fireplace
40. Downspout
41. Gutter
42. Bridging

Color, texture, and pattern all work together to create a desired "look" for a wall.

PARTS OF A HOUSE

The main parts of a house are shown in Fig. 20–35. Every house does not have all of these parts. Some parts may be made of different materials. The typical wood framing of an exterior wall is shown in Fig. 20–36. The framing begins on the foundation wall with the sill, header, and floor plate. Then, the stud wall is erected. Next, sheathing, plywood, or insulation board is put on. Many kinds of facing can be used. Horizontal or vertical siding is typical. In addition, shakes or shingles are often used. Also common today is brick veneer, shown in Fig. 20–37.

Housing Frame

The framework of a building must be strong and rigid to ensure low maintenance costs over many years. Even a prefabricated home and a custom-designed home have some features in common.

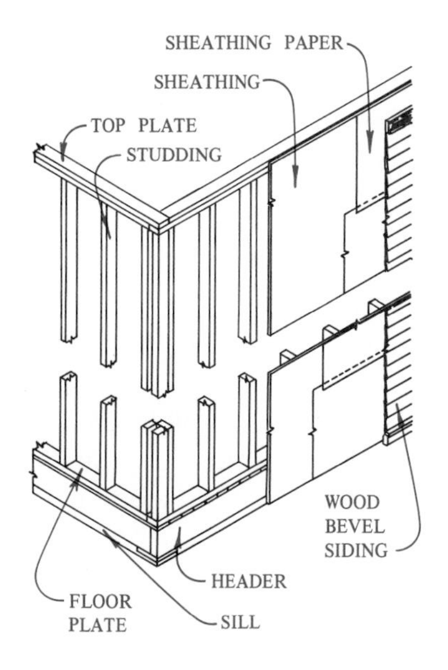

Fig. 20–36 **Frame wall with wood siding.**

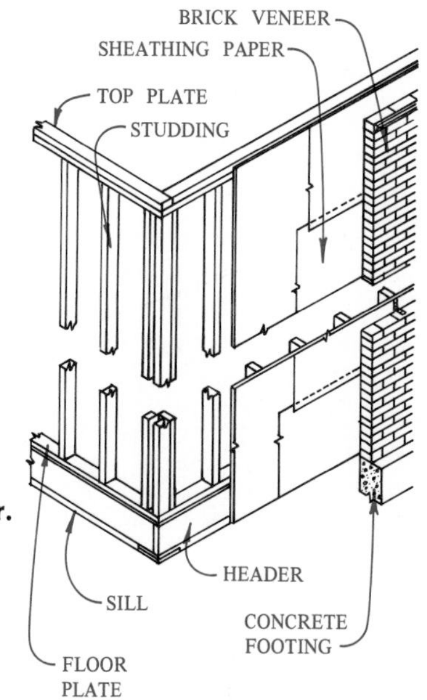

Fig. 20–37 **Frame wall with brick veneer.**

Fig. 20–38 **Western or platform framing.**

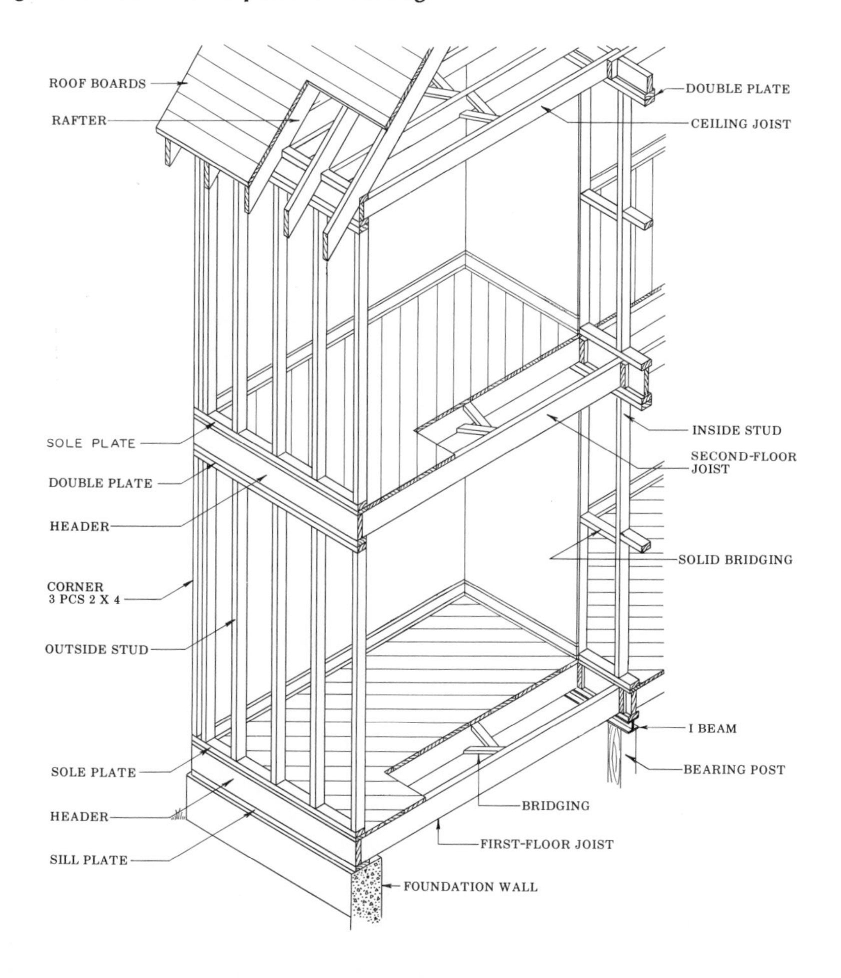

Western Framing

In western, or platform, framing, each floor is framed separately (Fig. 20–38). The first floor is built on top of the foundation wall as a platform. Studs are one story high. They are used to develop and support the framework for the second story and the load-bearing interior walls.

Balloon Framing

In balloon framing, the studs are two stories high, as in Fig. 20–39. A false girt inserted in the stud wall carries the second floor joists. A box sill is used. Diagonal bracing brings rigidity to the corners. This system is not commonly used today. However, architects must know about it for remodeling older homes.

Plank and Beam Framing

Plank and beam framing (Fig. 20–40) uses heavier posts and beams than the other systems.

These members carry a deck of continuous planking. This kind of framing allows ceilings to be higher and more open, with fewer supporting members. It also generally costs less to build.

Sill Construction

Figure 20–41 shows different types of sill and wall construction. At A, the frame wall is set up on a box-sill construction. Note the metal termite shield atop the foundation wall. At B, a brick veneer starts below the floor line on a stepped foundation wall. At C, the slab construction is reinforced with a wire mesh. At D, the sill and foundation wall are laid in one continuous piece, forming a monolithic slab. Study the details at E, F, G, H, and I. The special features of each are identified in the architect's symbols and notes.

Fig. 20–39 **Balloon framing.**

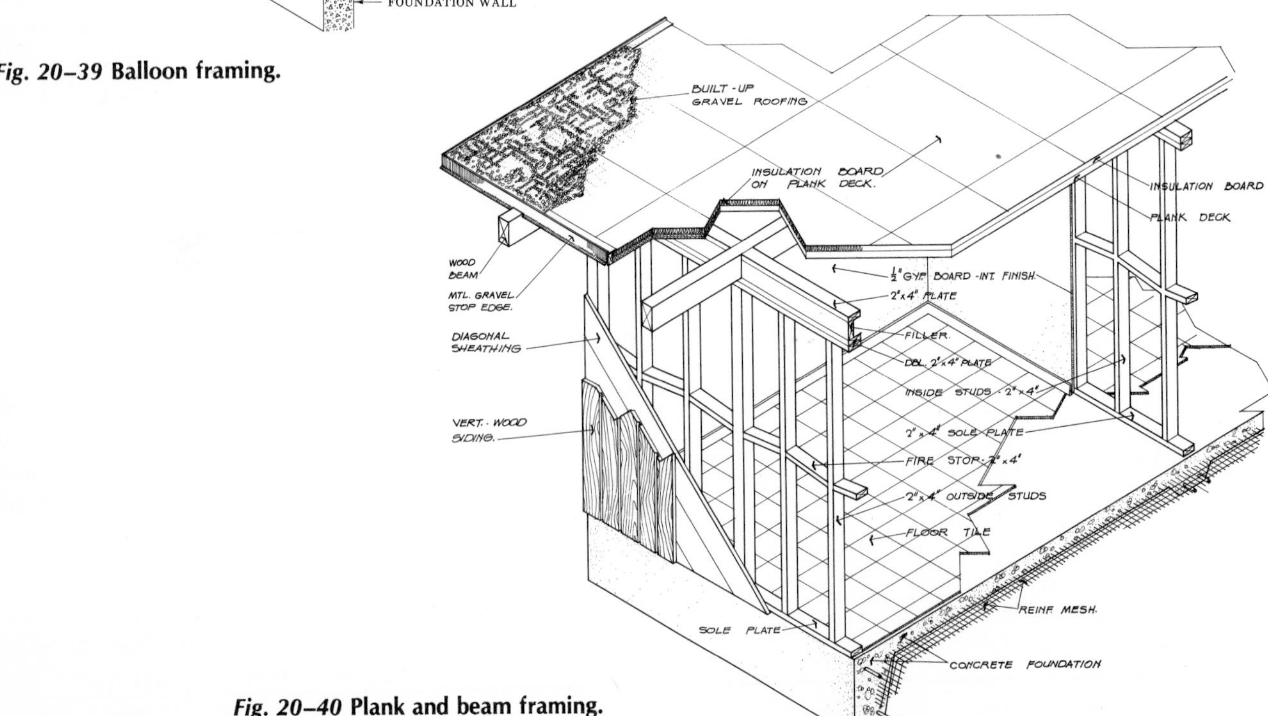

Fig. 20–40 **Plank and beam framing.**

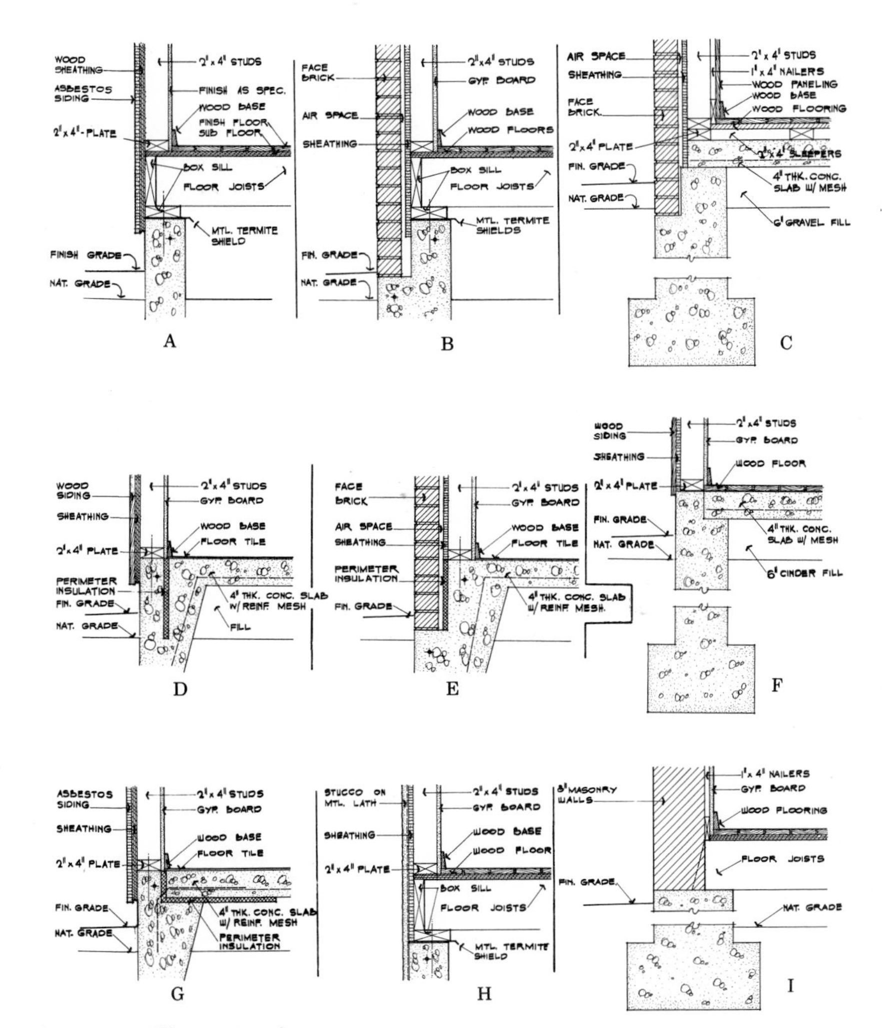

Fig. 20–41 **Sill constructions.**

Fig. 20–42 **Corner studs and sheathing.**

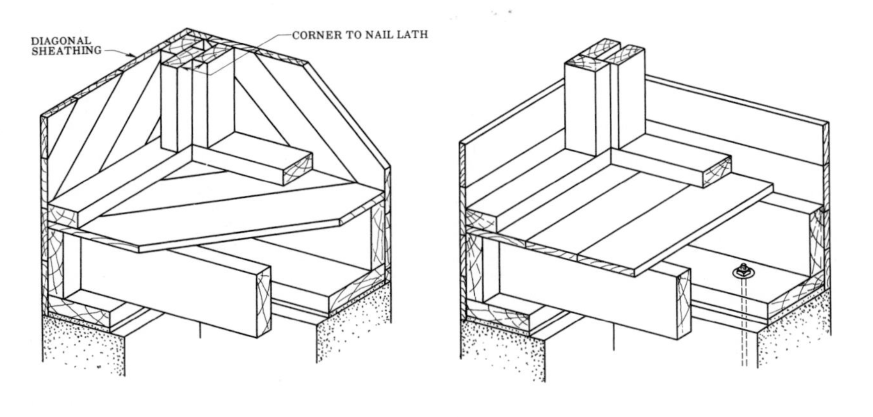

Corner Studs and Sheathing

Some typical corner bracing is shown in Fig. 20–42. Diagonal sheathing was formerly the most common kind of bracing. Today, however, to save labor, builders use horizontal sheathing and plywood along with modular insulation board. Plywood is used not only on exterior walls, but also for *interior decking* (subflooring) and roof sheathing. Anchor bolts secure the superstructure to the substructure. These are normally ½- to ¾-in (12 mm to 19 mm) bolts. They are placed about 8 ft (2440 mm) apart and extend 18 in. (457 mm) into the concrete.

Roof Designs

Some basic roof types are shown and named in Fig. 20–43. The shape of the roof is often the key to a building's architectural style. The common shapes are the gable, hip, flat, and shed. The mansard, gambrel, butterfly combination, clerestory, and A-frame shapes are used more with specific design styles. The vertical measurement of a roof is called the *rise.* The horizontal dimension is called the *run.* Together, they determine the *roof pitch* (slope).

Roof-framing Sections

Figure 20–44 shows some typical cornice details. Note the terms used for the members that finish off the joints between the wall and roof. Some rafters are open at the ends. Others are boxed. There is also built-up flat roofing for plank and beam construction. Aluminum and galvanized gutters remain common. One detail shows a built-in, metal-lined gutter. This is a costly detail often included on formal designs. The note "COND. @ BRICK" (condition at brick) means that a brick-veneer construction may be added.

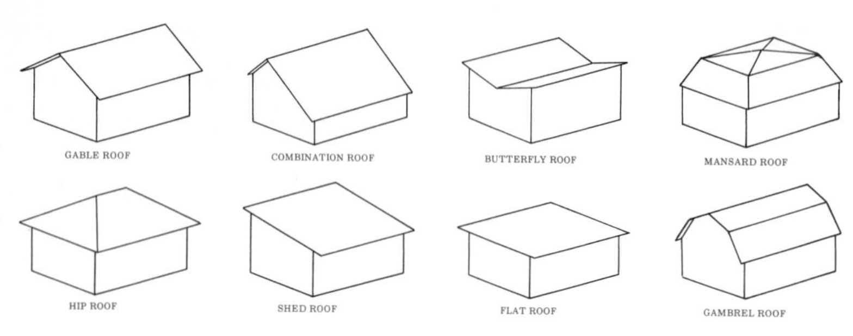

Fig. 20–43 **Some roof types.**

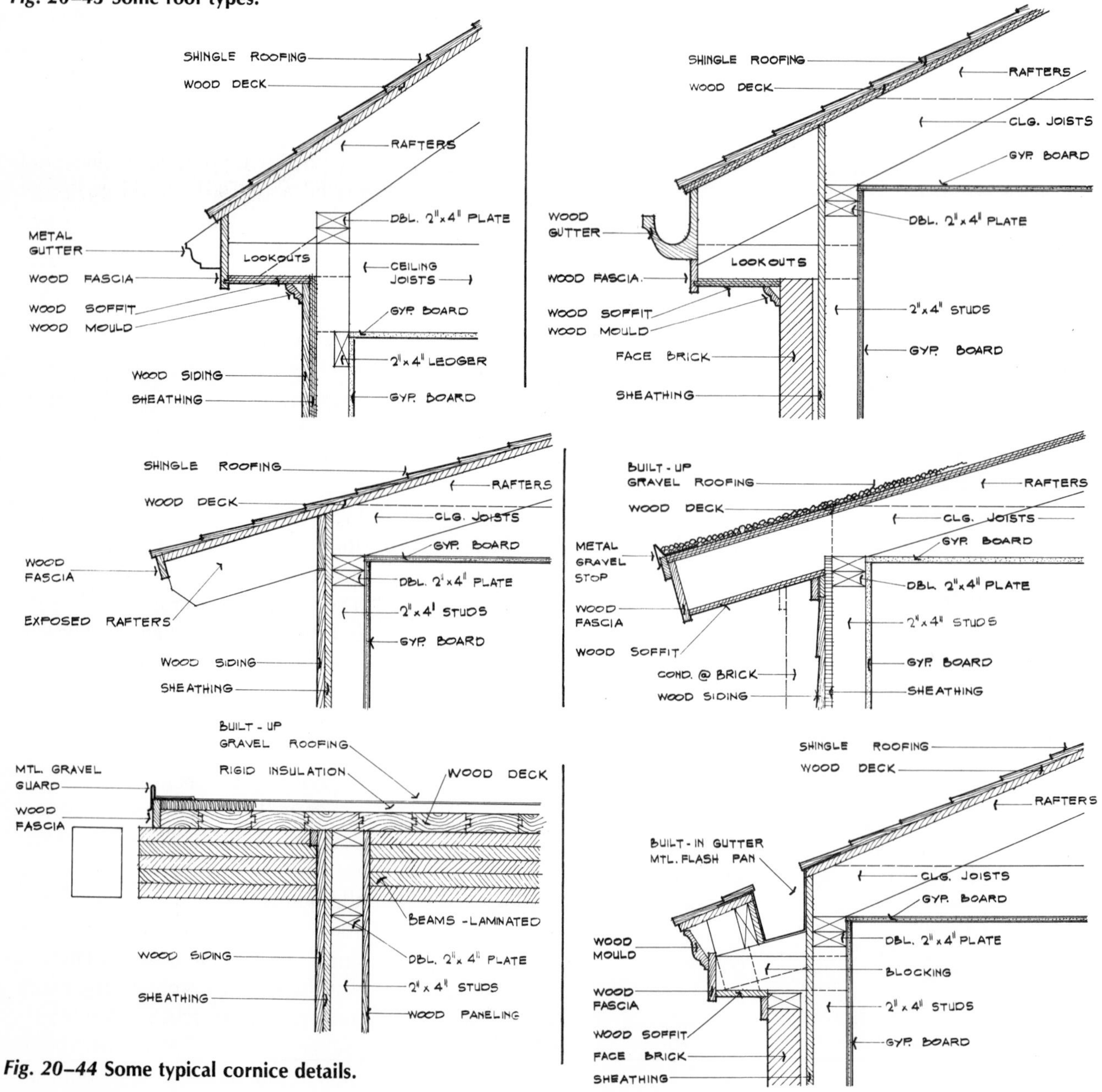

Fig. 20–44 **Some typical cornice details.**

Stairway Framing and Detail

Three types of stairs are (1) the straight run, (2) the platform, and (3) the circular. Stairs are made up of *risers* (the vertical part of each step) and *treads* (the horizontal part of each step). These parts are illustrated in Fig. 20–45. The *rise* of a flight of stairs is the height measured from the top of one floor to the top of the next floor. The *run* is the horizontal distance from the face of the first riser to the face of the last riser in one stairwell. It also equals the sum of the width of the treads.

Risers are generally 6½ to 7½ in. (165 mm to 190 mm) high. The width of treads is such that the sum of one riser and one tread is about 17 (430 mm) to 18 in. (480 mm). A 7-in. (180-mm) riser and an 11-in. (280-mm) tread is considered a general standard). You can easily use a scale to divide the floor-to-floor height into the number of risers. A good stairway feels comfortable and safe to use. A simple rule for building a safe stairway is to keep the angle of incline between 28° and 35°.

On working drawings, stairways are usually not drawn in their entirety. Instead, break lines are used, and the drawing shows what is on the level beneath the stairs.

Doors

Doors are usually 6 ft 8 in. or 7 ft 0 in. (2000 mm or 2100 mm) high. The width may vary from 2 ft 0 in. to 3 ft 0 in., (600 mm to 900 mm). But it is usually 2 ft 8 in. or 3 ft 0 in. (800 mm or 900 mm). The thickness varies from 1⅜ in. to 1¾ in. (35 mm to 45 mm) for interior doors, and from 1¾ in. to 2½ in. (45 mm to 65 mm) for exterior doors. The head, jamb, and sill details may vary depending on whether a swinging, sliding, or folding door is used. The door must fit its frame closely. Yet it must also open and close easily. Figure 20–46 shows various patterns for doors.

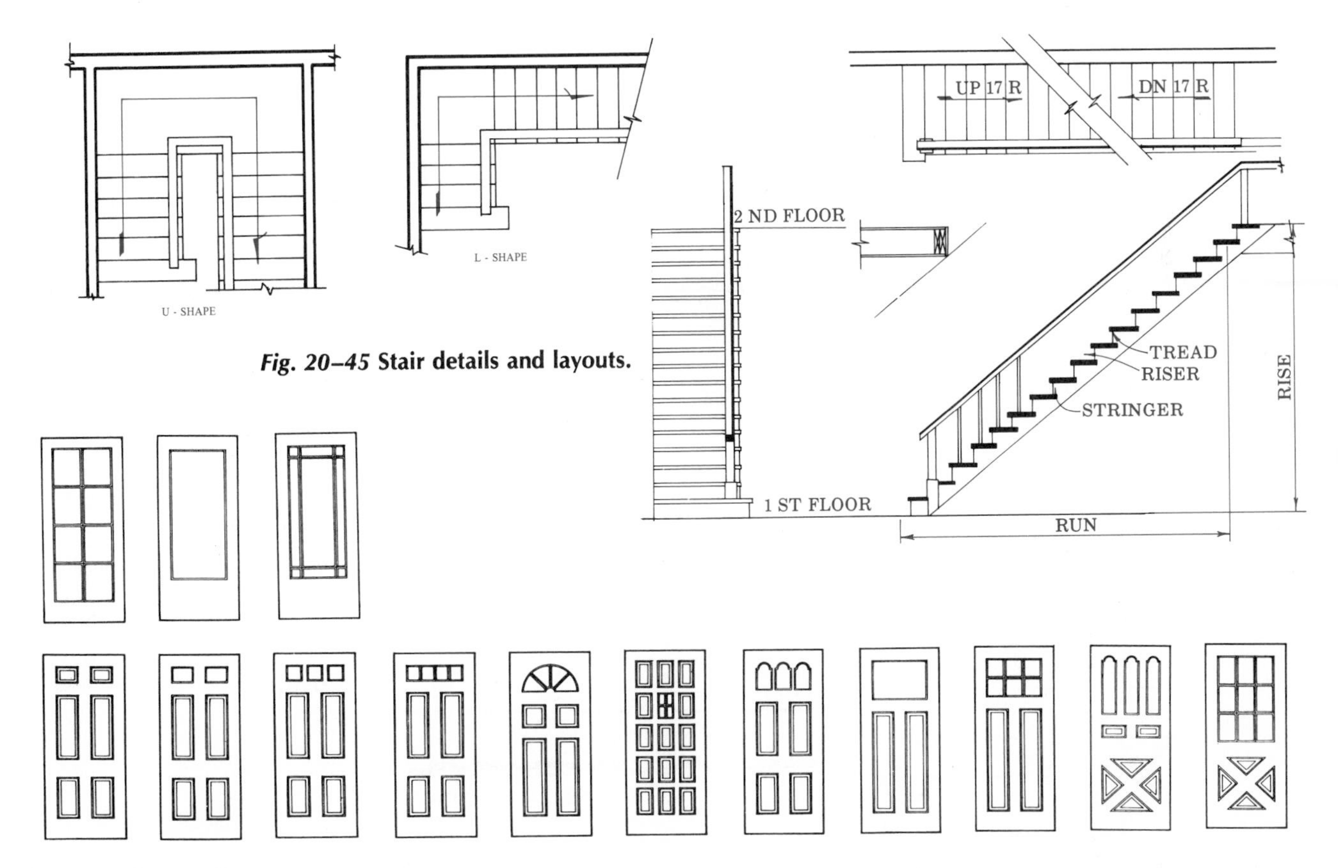

Fig. 20–45 **Stair details and layouts.**

Fig. 20–46 **Door patterns.**

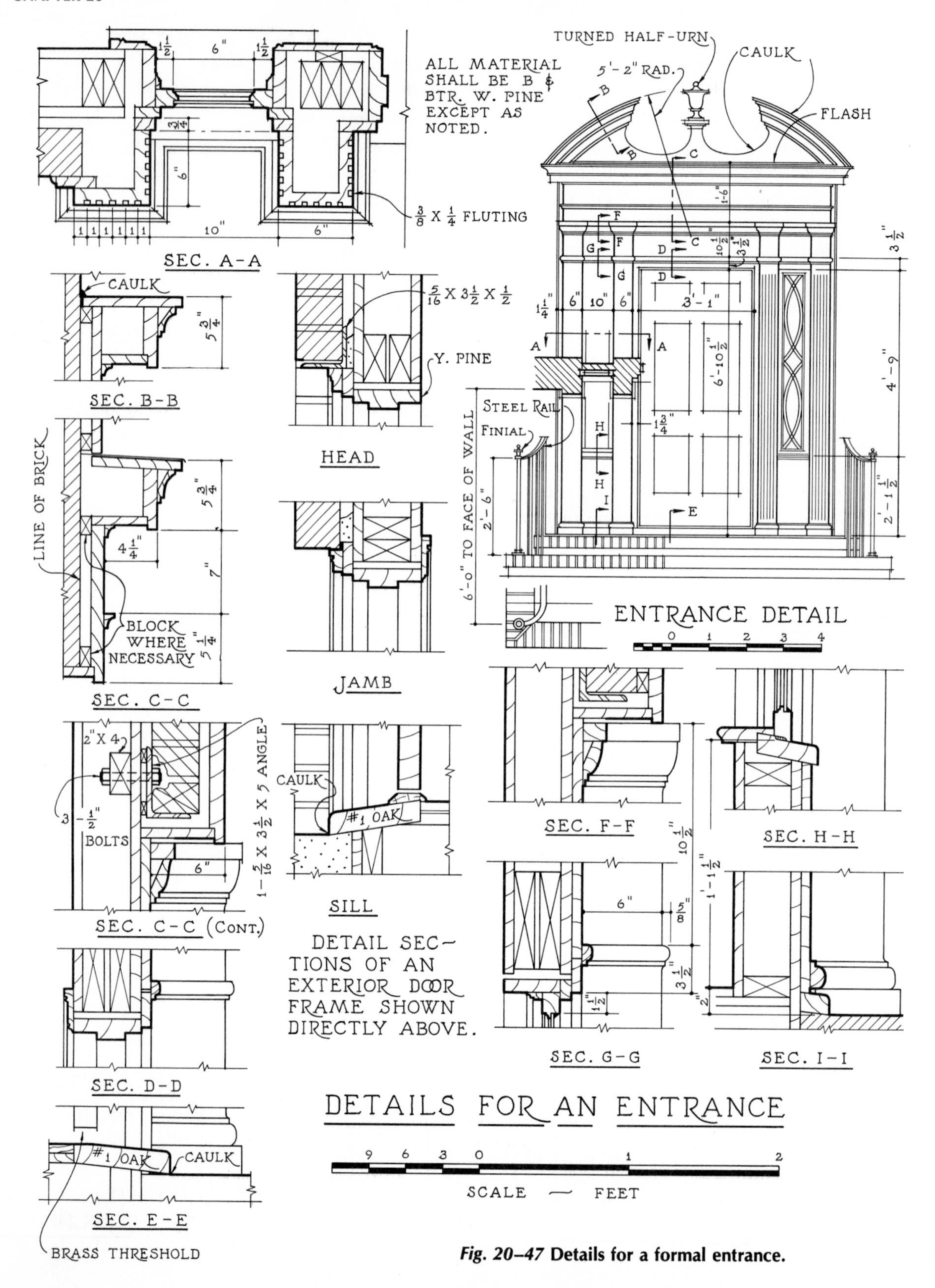

Fig. 20–47 **Details for a formal entrance.**

Doors to the outside are usually larger than others to allow for heavier use and for bringing in furniture. These doors are usually 3 ft 0 in. (900 mm) wide. Bedroom doors are usually 2 ft 6 in. (762 mm). Bathroom doors run 2 ft 4 in. (710 mm). Bifold and folding accordian doors have special features that must be assembled from the manufacturer's individual specifications. Details for a formal entrance are illustrated in Fig. 20–47.

Windows

The style of a house determines what style of window is used and how windows are placed. Double-hung and casement windows are practical in most kinds of houses. However, horizontal sliding windows are very popular. Figure 20–48 shows the various types available to the architect.

Casement windows are popular for French and English designs. They are hinged at the sides to swing open (in or out). Hopper windows are hinged at the bottom. Awning windows are hinged at the top. Both hopper and awning windows are also called projected windows. Sliding windows move sidewise. Double-hung windows are commonly used for Colonial and American-style structures. Each double-hung window contains two independent sashes that can move in a vertical track. A counterbalance holds them at any position desired. Newer windows can have a press-in, spring-loaded track that is very convenient. Fixed windows and *jalousies* (windows made of adjustable glass slats) are special kinds of windows. Figure 20–49 shows sectional details and some technical terms for a window. Figure 20–50 shows sectional details of a window in a brick veneer and a masonry wall. Normally, windows are placed in walls so that their tops line up with the top of the door.

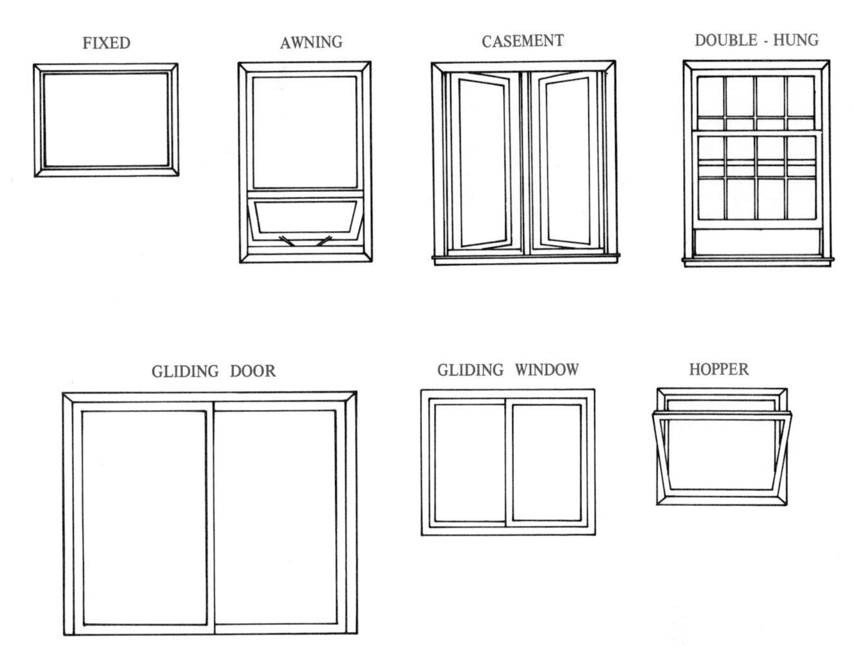

Fig. 20–48 **Types of windows.**

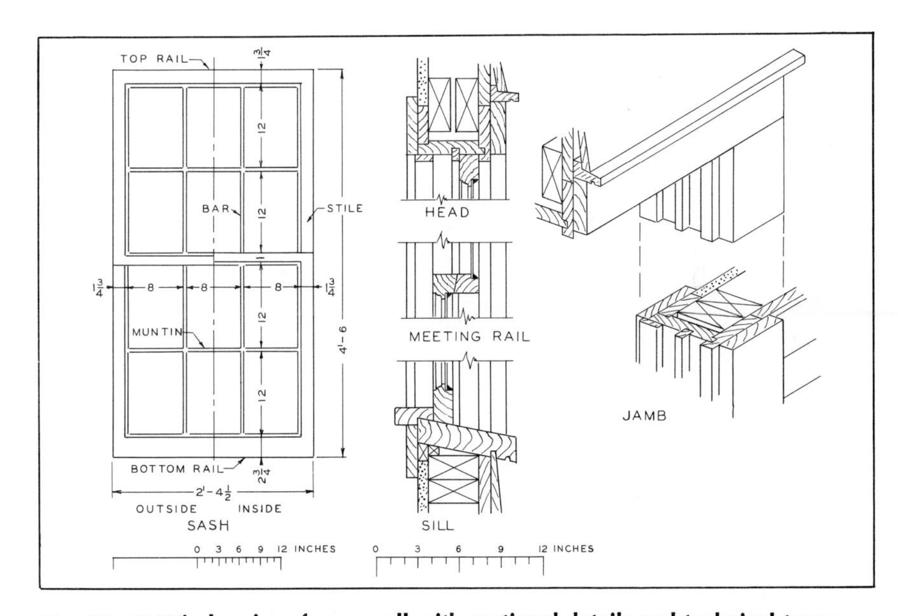

Fig. 20–49 **Window in a frame wall with sectional details and technical terms.**

WORKING DRAWINGS

These are the most important class of drawings. They include plans, elevations, sections, schedules, schematics, and details. Along with the specifications for materials and finish, they are the

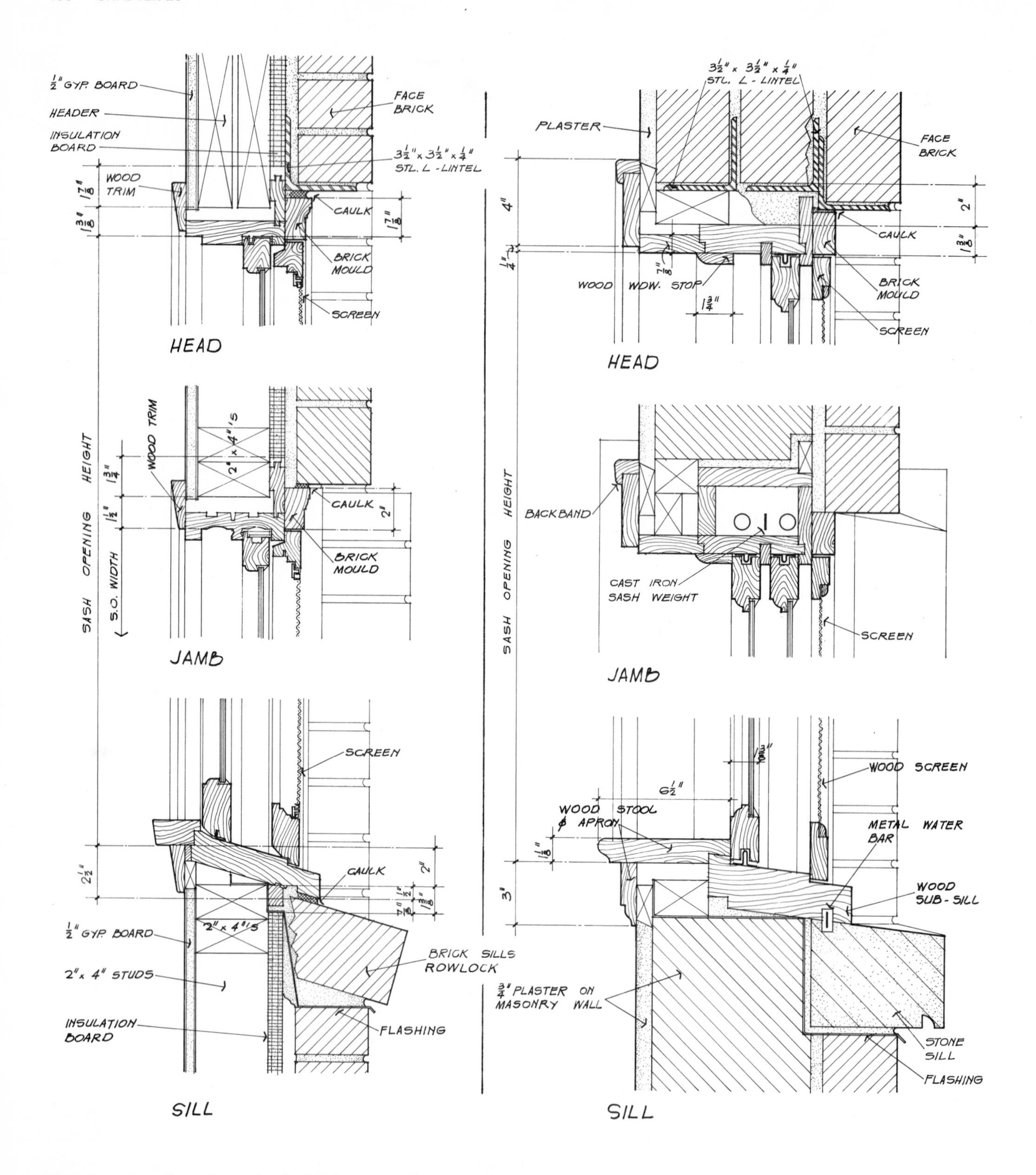

Fig. 20–50 **Detail sections of a double-hung window in a brick veneer and a masonry wall.**

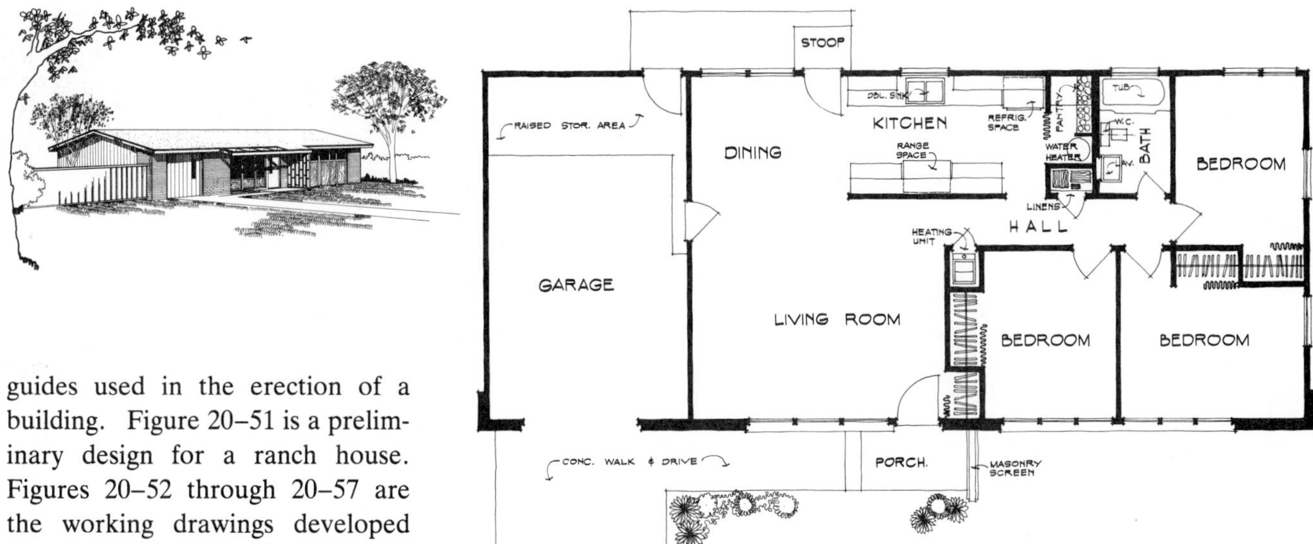

guides used in the erection of a building. Figure 20–51 is a preliminary design for a ranch house. Figures 20–52 through 20–57 are the working drawings developed from this design. They form a complete set of plans for the house.

Fig. 20–51 **A preliminary study developed for client approval.**

Fig. 20–52 **Foundation plan.**

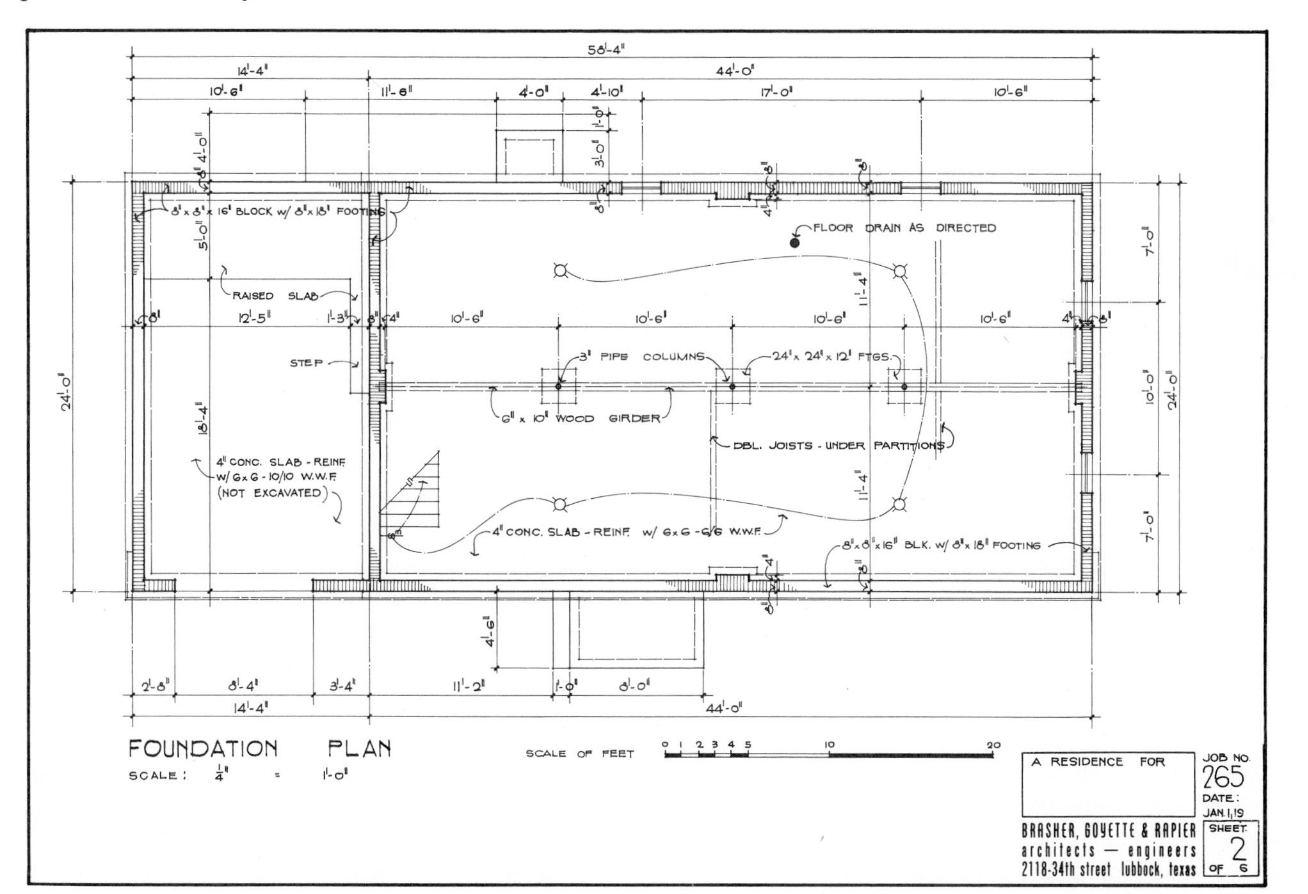

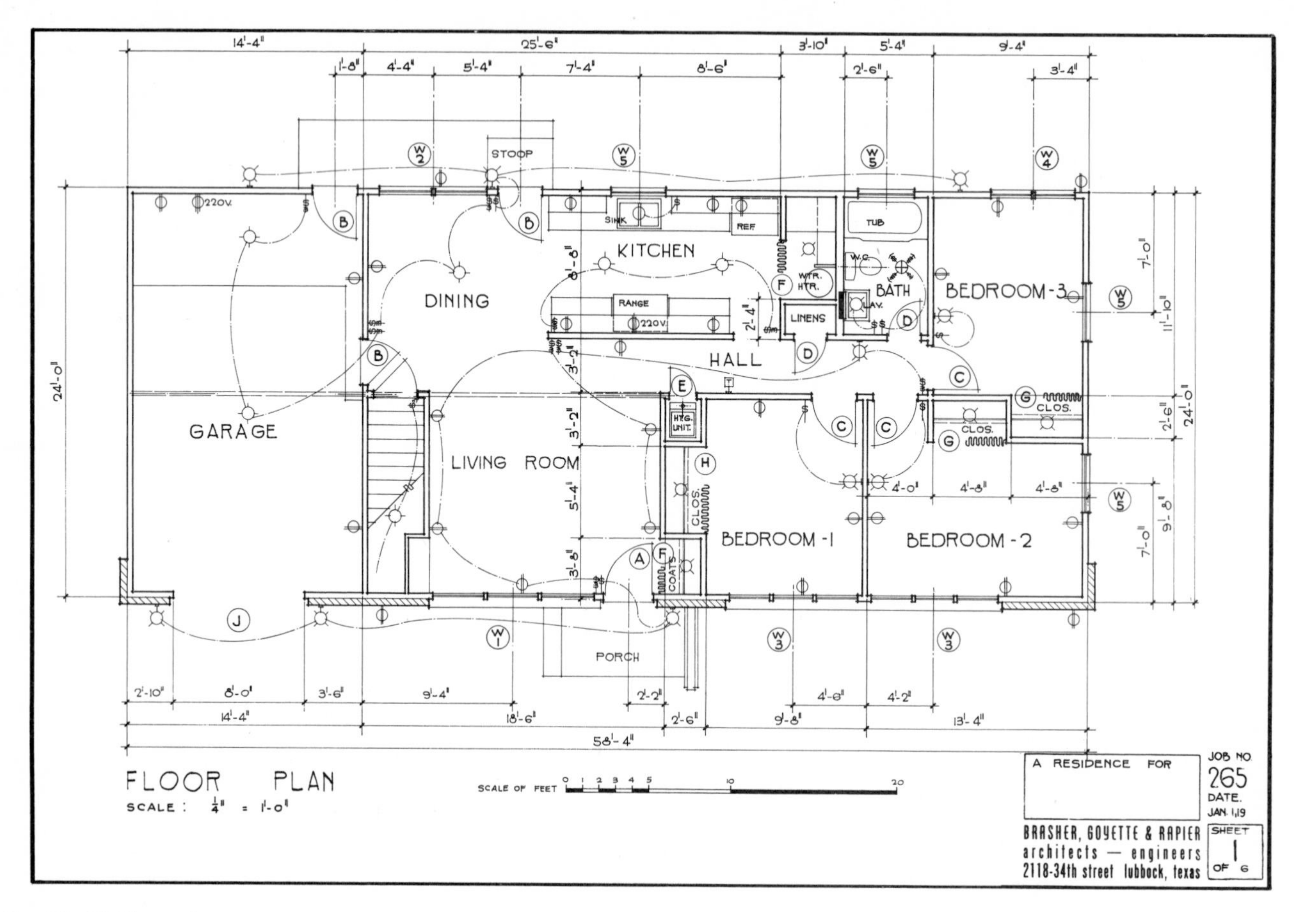

Fig. 20–53 **Floor plan.**

Plans

The basement plan in Fig. 20–52 serves as a guide for constructing the foundation. Therefore, it must be completely dimensioned. It should be checked with the first-floor plan. It can even be traced from it. Note the foundations for the porch and garage. Window placement depends on structural needs.

The first-floor plan in Fig. 20–53 is a horizontal section taken above the floor. It shows all walls, doors, windows, and other structural features. It also shows fixed features such as the cabinets, stairways, heating and plumbing fixtures, lighting outlets in the walls, and ceilings. Frame walls are drawn to what is shown on the scale as 6 in. thick. Windows and doors are located properly, then indicated by conventional symbols. Their sizes are listed on schedules that accompany the plans.

Elevations

Figure 20–54 shows front and side elevations. A complete set of plans would include four elevations. Elevations show the exterior look of a house, floor and ceiling heights, openings for windows, doors, roof pitch, and selected materials. To draw an elevation, start with the grade line. Then, locate the centerlines that indicate the finished working dimensions. You can plot the doors, windows, and other openings from the floor plan.

Sections

Figure 20–55 details a typical wall section from substructure to superstructure. This drawing, to a larger scale than that of the elevation, shows the wall in a much clearer and more detailed view. Door and window details are also included. They are developed as the building progresses. That way the millwork and custom framework will assemble readily.

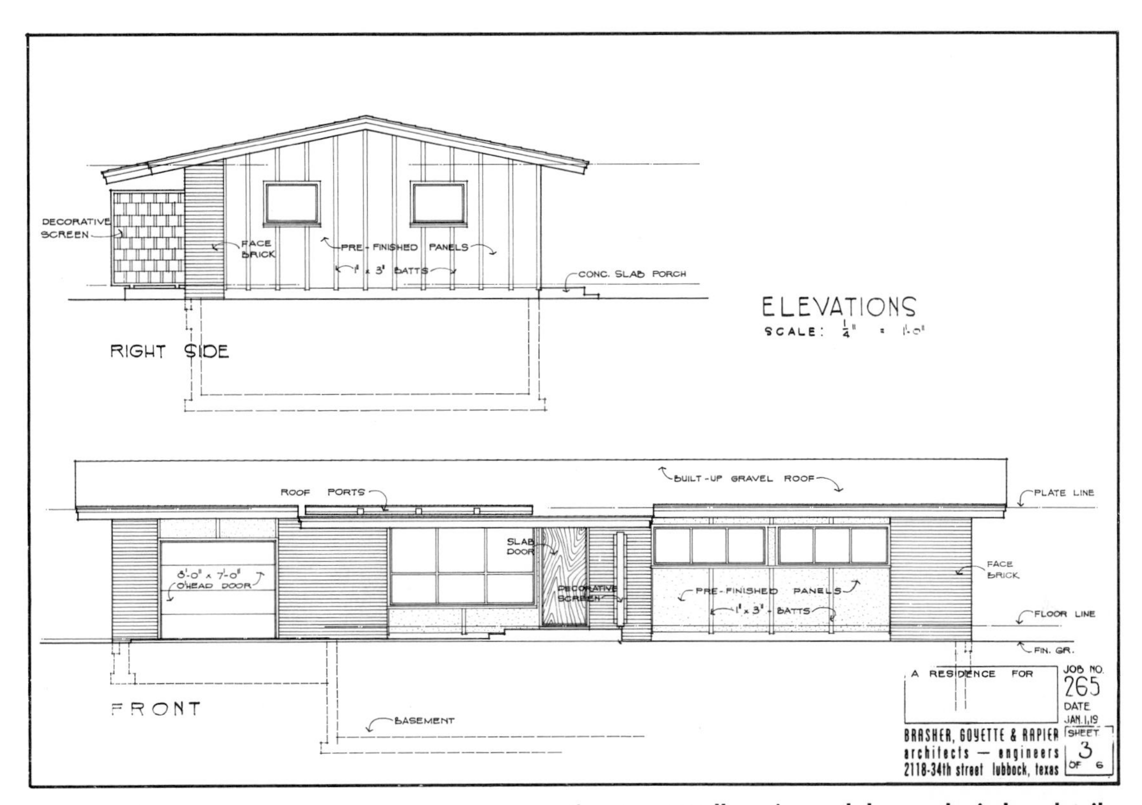

Fig. 20–54 **Elevations.**

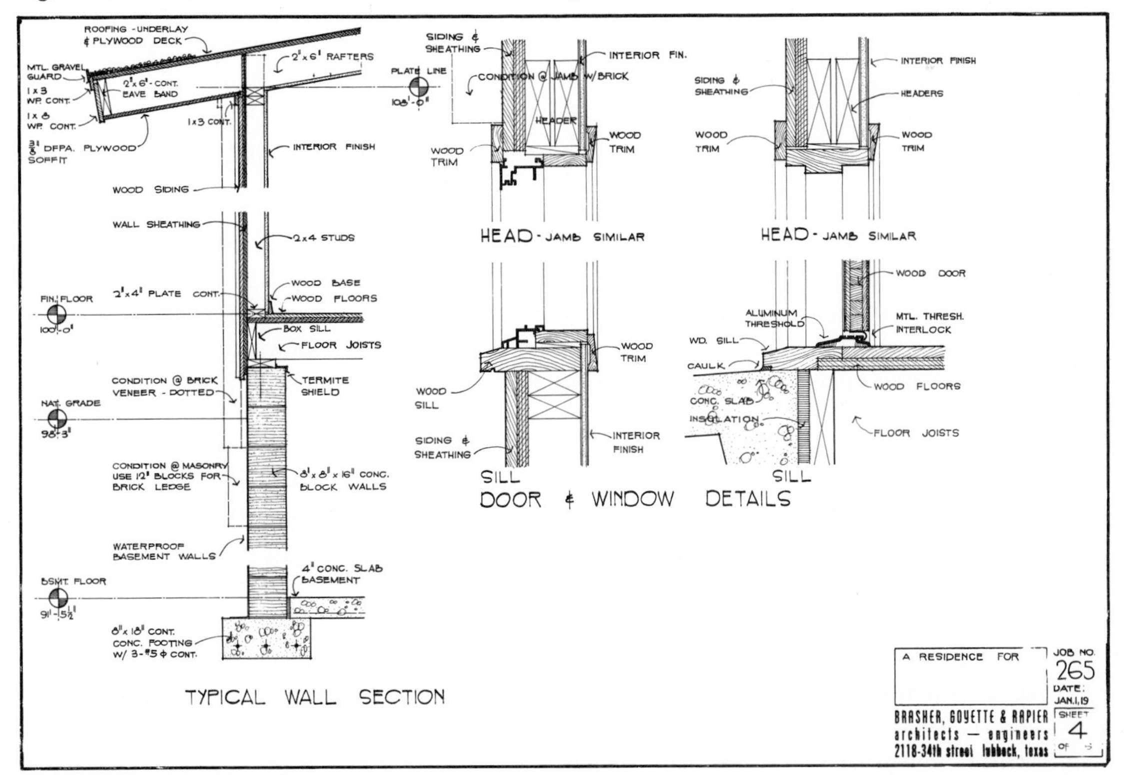

Fig. 20–55 **Wall section and door and window details.**

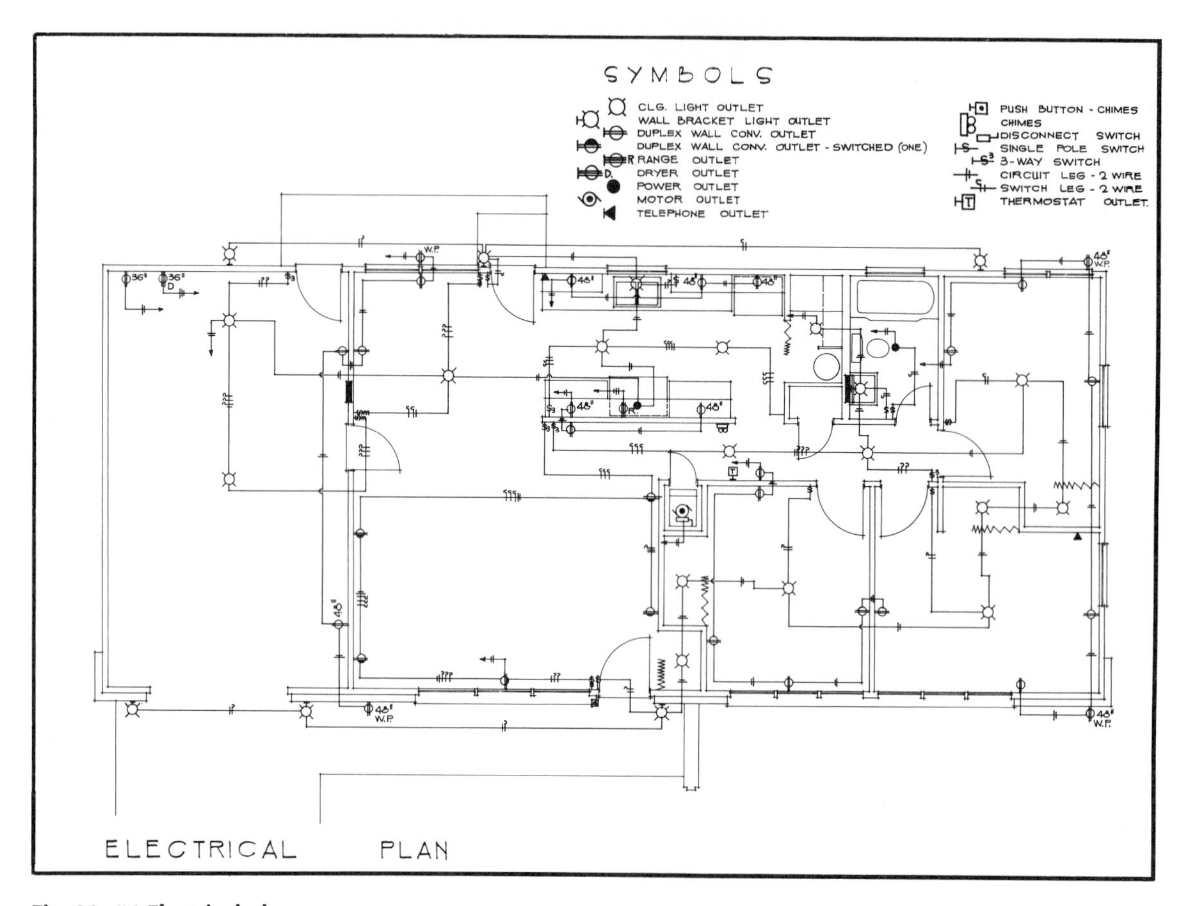

Fig. 20–56 **Electrical plan.**

Electrical Planning

Figure 20–56 is a plan of the electrical wiring. It shows the circuits for 110-V and 220-V service. It also shows the electric outlets and switches located for the major appliances. A key is included for the various symbols. Additional data would then be listed in the formal specifications.

Site Plan and Schedules

Figure 20–57 is the site plan. A site plan shows the lot and locates the house on it. It should give complete and accurate dimensions. It should also show all driveways, sidewalks, and other pertinent information required by the building inspector. The site plan in the figure is for an ordinary urban lot. Note the roof plan. Its center ridge represents a gable roof. What scale would be best for drawing this site plan?

Figure 20–57 also includes schedules describing the five types of windows and nine types of doors used in the house. Another, perhaps more important, schedule lists the finishes required for the floors, walls, and ceilings.

Plumbing Symbols

Some of the standard symbols are shown on the preceding working drawings. There are also standard schematics and diagrams for plumbing. You can find these in the latest edition of *Architectural Graphic Standards.* These standards are periodically revised by the American Institute of Architects.

Electrical Symbols

The many standard symbols are shown in Fig. 20–58. These are typically used on house plans to show the basic circuit layout needed.

CONTEMPORARY HOUSE PLANS

Study and compare the contemporary house plans described below. You should judge each plan by how well it combines function, structure, and beauty.

SPLIT-LEVEL DESIGN

Figure 20–59 shows three views of a split-level town house. The house is of masonry and frame construction. The exterior walls are made of textured cement. The living room looks out onto a garden patio surrounded by walls. The roof is of red clay tile with copper flashing.

The living-room level layout is open, making it easy to pass from one area to another. The main stairway is centrally located for efficient use.

THE CALIFORNIA-STYLED HOUSE

Figure 20–60 shows a California-styled house. The exterior is shown at A. The materials used give the house a distinctive look. The sides are faced with textured vertical boarding. The roof is even more highly textured. The photograph at B shows the dramatic open spaciousness of the interior. The floor plans of the house are shown at C. Note the spacious master bedroom suite.

THE VACATION HOME

Figure 20–61 shows a vacation retreat lodge. This house is ideal for a rolling, wooded lakeside. It is prefabricated from a standard design. A few variations are added to suit the owner. The chief building material is natural Western Red Cedar. This material makes the home warmer in winter and cooler in summer.

THE CONTEMPORARY HOUSE OF WORSHIP

Figure 20–62 shows a community church designed in contemporary style. The site plan at A and the floor plan at B are typical preliminary studies done by the architect. The interior and exterior rendered views at C and D give a feeling of materials; reverence, perhaps; and function.

***Fig. 20–57* Schedules, site, and roof plan.**

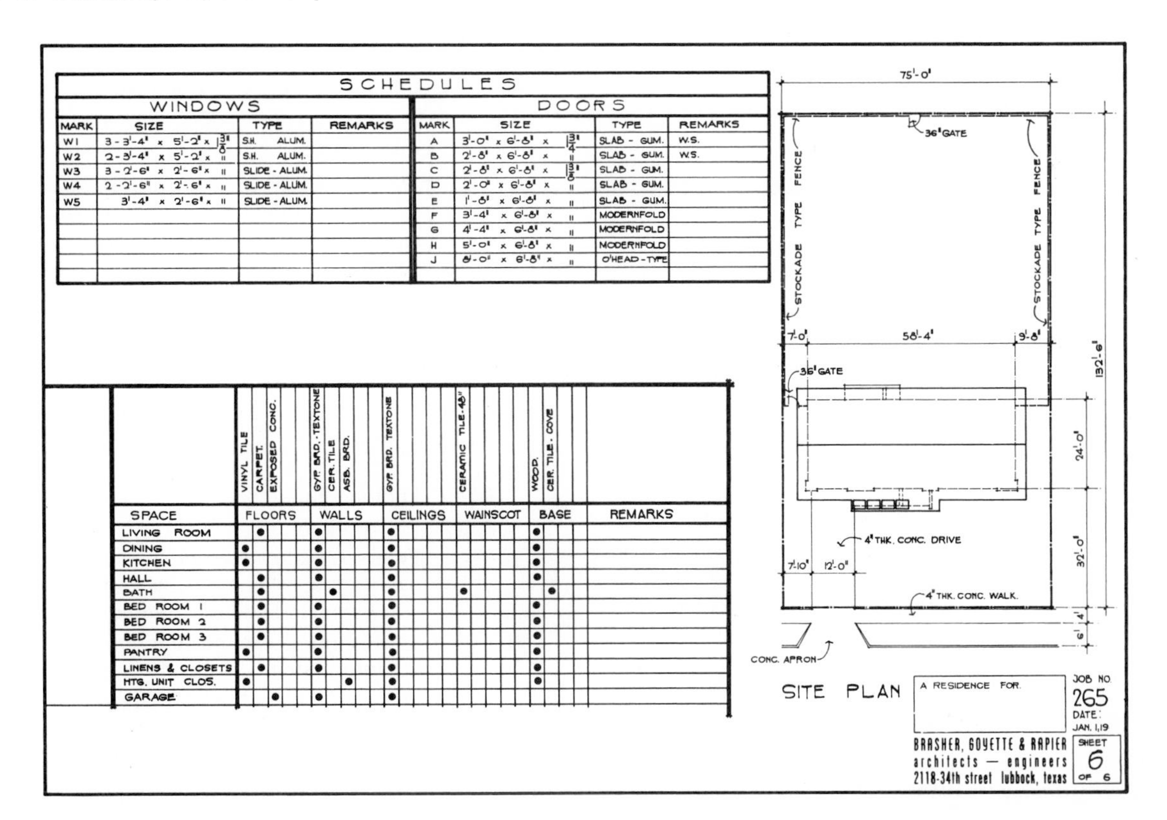

SCHEDULES

WINDOWS

MARK	SIZE	TYPE	REMARKS
W1	3 - 3'-4" x 5'-2" x 1 3/8"	S.H. ALUM.	
W2	2 - 3'-4" x 5'-2" x "	S.H. ALUM.	
W3	3 - 2'-6" x 2'-6" x "	SLIDE - ALUM.	
W4	2 - 2'-6" x 2'-6" x "	SLIDE - ALUM.	
W5	3'-4" x 2'-6" x "	SLIDE - ALUM.	

DOORS

MARK	SIZE	TYPE	REMARKS
A	3'-0" x 6'-8" x 1 3/4"	SLAB - GUM.	W.S.
B	2'-8" x 6'-8" x "	SLAB - GUM.	W.S.
C	2'-8" x 6'-8" x 1 3/8"	SLAB - GUM.	
D	2'-0" x 6'-8" x "	SLAB - GUM.	
E	1'-8" x 6'-8" x "	SLAB - GUM.	
F	3'-4" x 6'-8" x "	MODERNFOLD	
G	4'-4" x 6'-8" x "	MODERNFOLD	
H	5'-0" x 6'-8" x "	MODERNFOLD	
J	8'-0" x 6'-8" x "	O'HEAD - TYPE	

SPACE	FLOORS: VINYL TILE	FLOORS: CARPET.	FLOORS: EXPOSED CONC.	WALLS: GYP. BRD. - TEXTONE	WALLS: CER. TILE	WALLS: ASB. BRD.	CEILINGS: GYP. BRD. TEXTONE	WAINSCOT: CERAMIC TILE - 48"	BASE: WOOD.	BASE: CER. TILE - COVE	REMARKS
LIVING ROOM		●		●			●		●		
DINING	●			●			●		●		
KITCHEN	●			●			●		●		
HALL		●		●			●		●		
BATH		●			●		●	●		●	
BED ROOM 1		●		●			●		●		
BED ROOM 2		●		●			●		●		
BED ROOM 3		●		●			●		●		
PANTRY	●			●			●		●		
LINENS & CLOSETS		●		●			●		●		
HTG. UNIT CLOS.	●					●	●		●		
GARAGE			●	●			●				

Fig. 20–58 **Electrical symbols.**

Ceiling Wall

GENERAL OUTLETS

Outlet
Blanked Outlet
Electric Outlet
For use only when circle used alone might be confused with columns, plumbing symbols, etc.
Fan Outlet
Junction Box
Lamp Holder
Lamp Holder with Pull Switch
Pull Switch
Clock Outlet (Specify Voltage)
Recessed Incandescent Lamp Fixture

CONVENIENCE OUTLETS

Duplex Convenience Outlet
Convenience Outlet other than Duplex
1 = Single, 3 = Triplex, etc.
Weatherproof Convenience Outlet
Range Outlet
Switch and Convenience Outlet
Special Purpose Outlet (Des. in Spec.)
Floor Outlet

SWITCH OUTLETS

S	Single-pole Switch
S_2	Double-pole Switch
S_3	Three-way Switch
S_P	Switch and Pilot Lamp
S_{CB}	Circuit Breaker
S_{WP}	Weatherproof Switch
S_F	Fused Switch

SPECIAL OUTLETS

a, b, c, etc. $S_{a, b, c, etc.}$

Any standard symbol as given above with the addition of a lower-case subscript letter may be used to designate some special variation of standard equipment of particular interest in a specific set of architectural plans.

When used they must be listed in the Key of Symbols on each drawing and if necessary further described in the specifications.

AUXILIARY SYSTEMS

Pushbutton
Buzzer
Bell
Annunciator
Outside Telephone
Interconnecting Telephone
Bell-ringing Transformer
Interconnection Box
Electric -range Outlet
Vent-hood Fan Motor
Television-antenna Outlet
Telephone Outlet
Telephone Jack
Push Button
Door Bell
Ceiling or pendent mounted incandescent-lighting fixture
Wall mounted incandescent-lighting fixture
Surface mounted fluorescent-lighting fixture
Recessed fluorescent-lighting fixture
Conduit run in ceiling and walls
Conduit run under floors and in walls
5,6 — Arrow indicates home run to lighting panel
Hash lines indicate numbers of wires
Numerals indicate circuit numbers
A — Letters indicate type of lighting fixture
(See lighting- fixture schedule)

PANELS, CIRCUITS, AND MISCELLANEOUS

Lighting Panel
Power Panel
Feeders. Note: Use heavy lines and designate by number corresponding to listing in Feeder Schedule.
Underfloor Duct and Junction Box–Triple System
Note: For a double or single systems eliminate one or two lines.
This symbol is equally adaptable to auxiliary system layouts

Fig. 20–59 **A split-level town house. A. Perspective; B. floor plan; and C. interior.** *(Larwin Co.)*

A

B

C

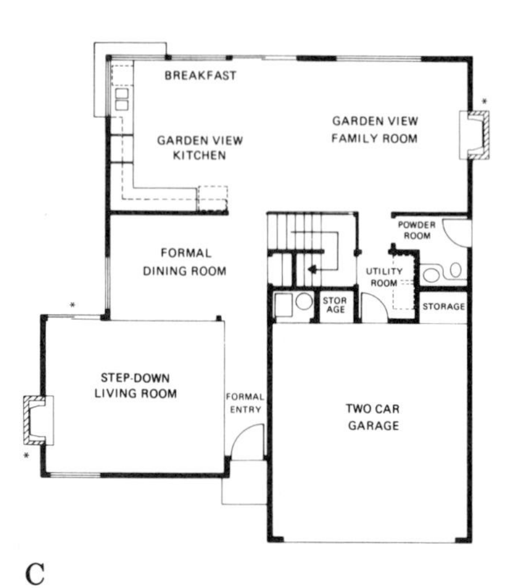

Fig. 20–60 **The California-styled house. A. Exterior; B. interior; C. floor plan. *(Larwin Co.)***

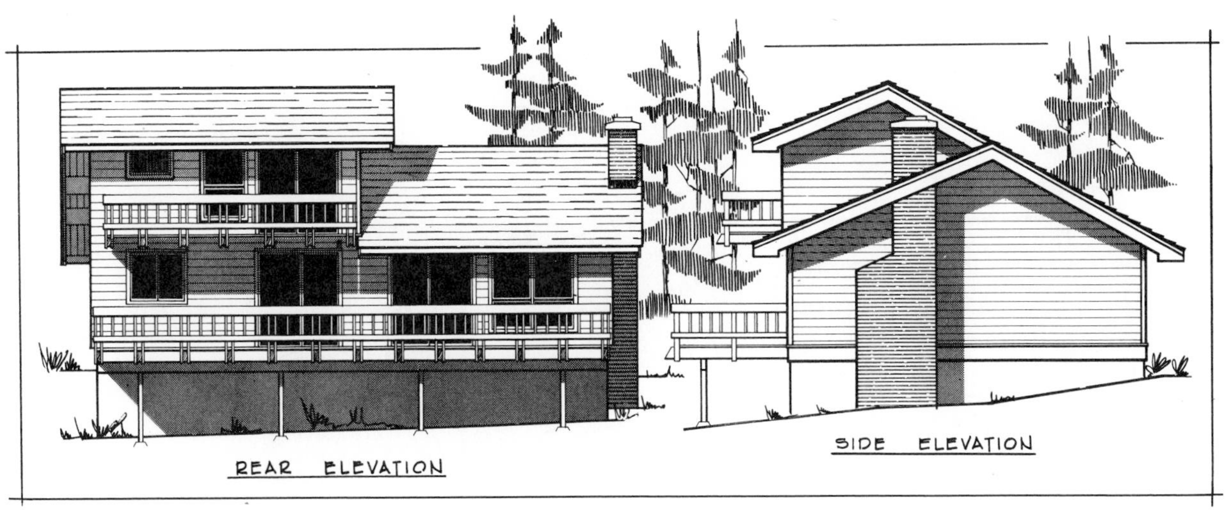

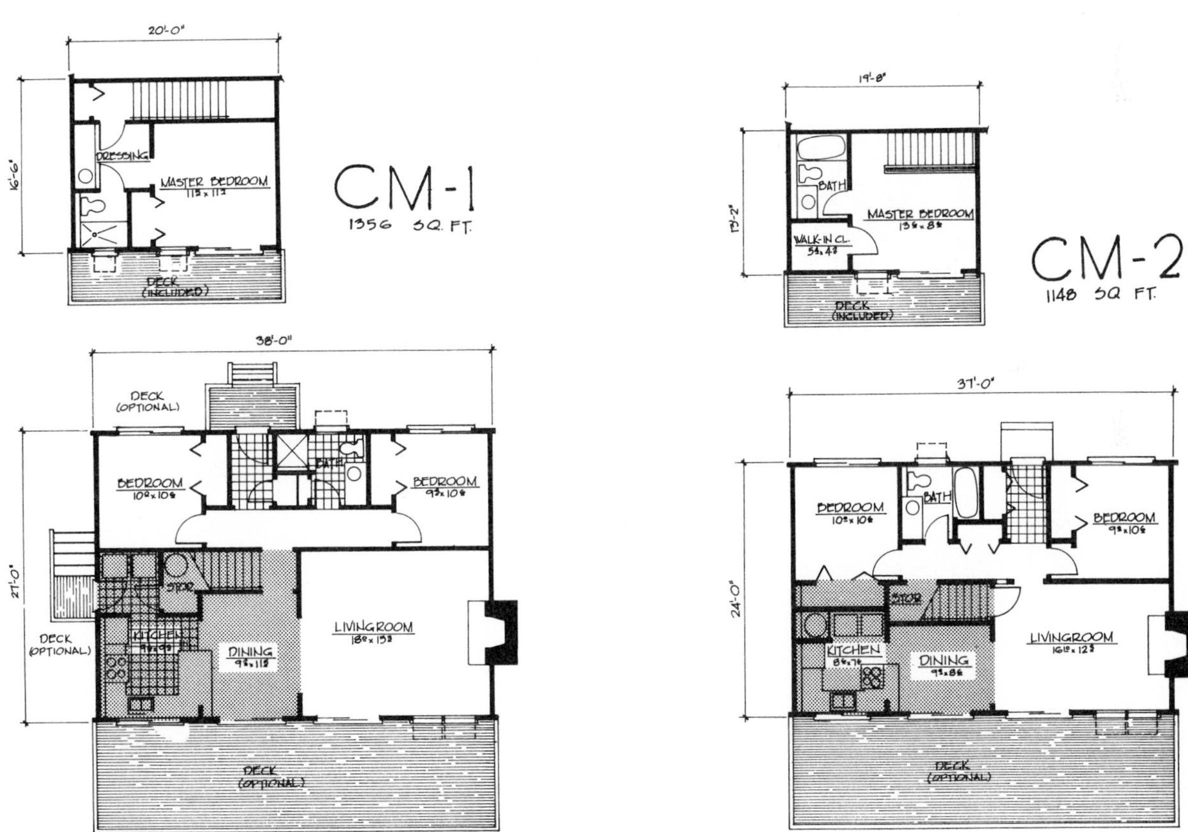

Fig. 20–61 **A vacation home.** ***(Justus Company, Inc.)***

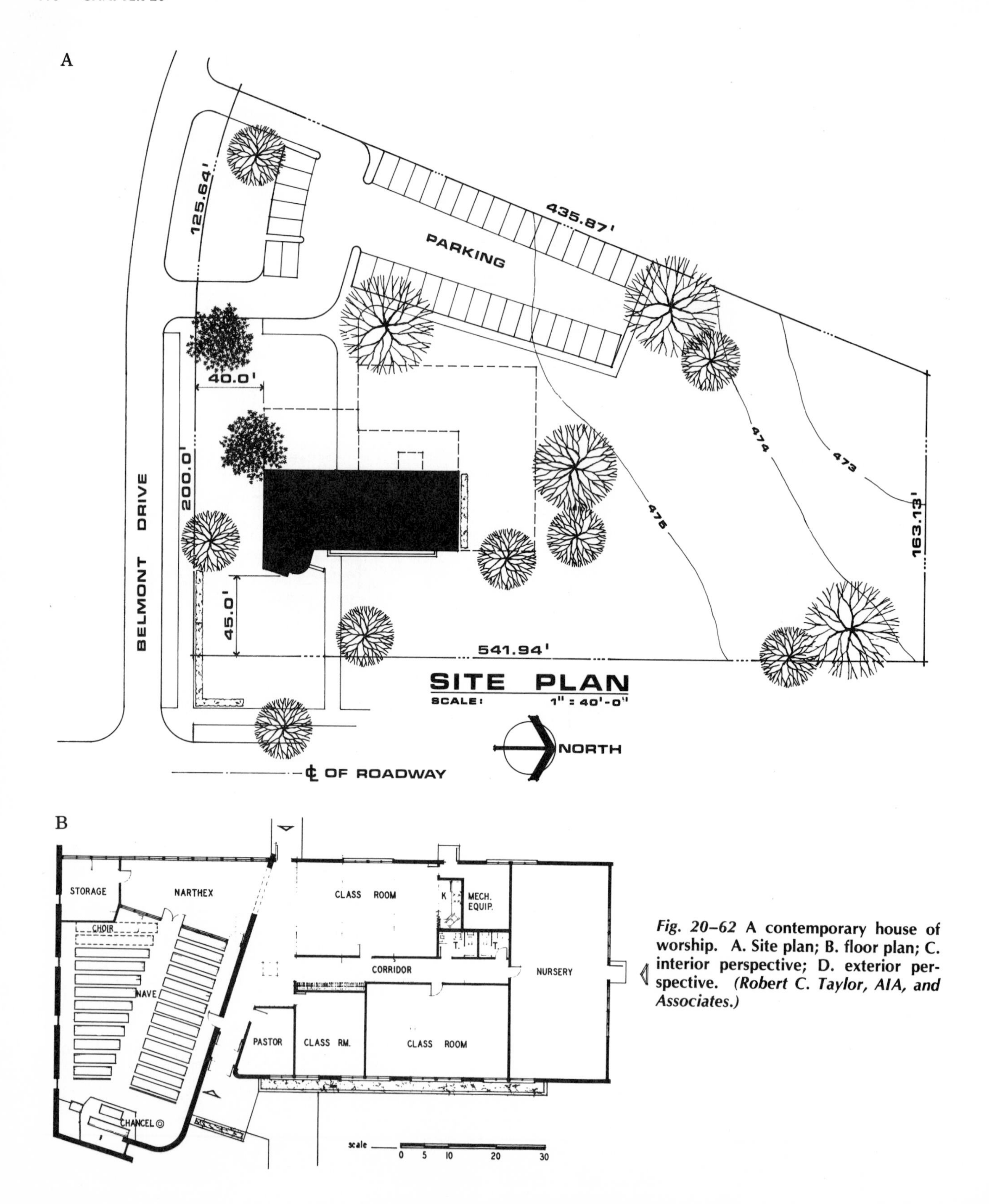

Fig. 20–62 **A contemporary house of worship. A. Site plan; B. floor plan; C. interior perspective; D. exterior perspective.** ***(Robert C. Taylor, AIA, and Associates.)***

Fig. 20–62 **(Cont.)**

WORKING DRAWINGS FOR A CONTEMPORARY MULTILEVEL RESIDENCE

Figure 20–63 shows a home with a French mansard roof. This kind of roof is shaped to fit and hug a building. It is textured in natural wood shades to give a feeling of warmth. The site plan (Fig. 20–64) is contoured to show the gentle roll of the land. The family-room (lower-level), main-level (Fig. 20–65), and upper-level (Fig. 20–66) plans are arranged gracefully around a central core of stairs and hallways. The front elevation (Fig. 20–67) shows the rolling contour of the site. Note the line

Fig. 20–63 **Contemporary French-mansard styled residence.** ***(Orrin Dressler, designer-contractor.)***

Fig. 20–64 **Site plan.** ***(Orrin Dressler, designer-contractor.)***

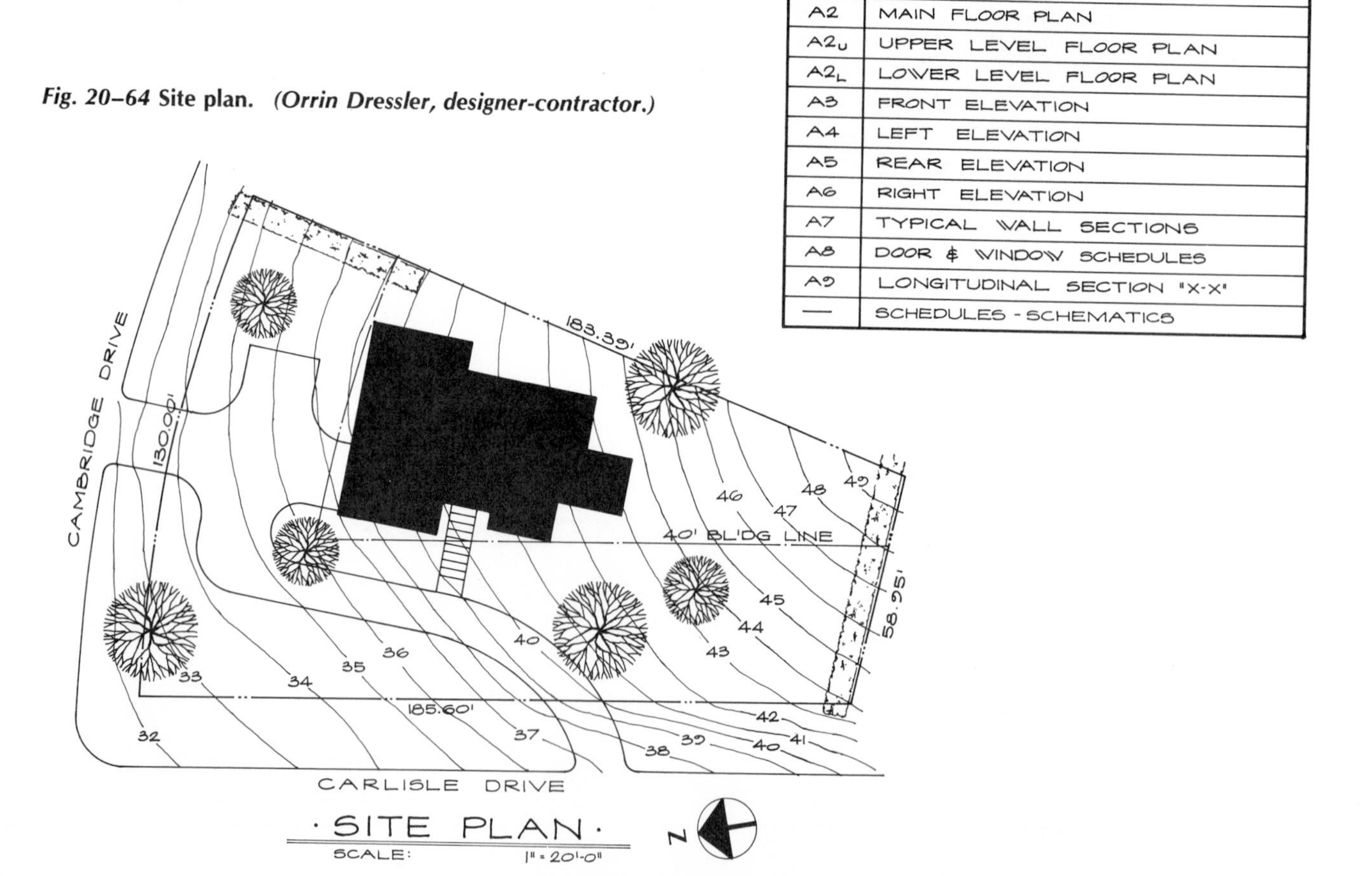

SHEET:	DRAWING DESCRIPTION
A1	SITE PLAN & DRAWING INDEX
A2	MAIN FLOOR PLAN
$A2_U$	UPPER LEVEL FLOOR PLAN
$A2_L$	LOWER LEVEL FLOOR PLAN
A3	FRONT ELEVATION
A4	LEFT ELEVATION
A5	REAR ELEVATION
A6	RIGHT ELEVATION
A7	TYPICAL WALL SECTIONS
A8	DOOR & WINDOW SCHEDULES
A9	LONGITUDINAL SECTION "X-X"
—	SCHEDULES - SCHEMATICS

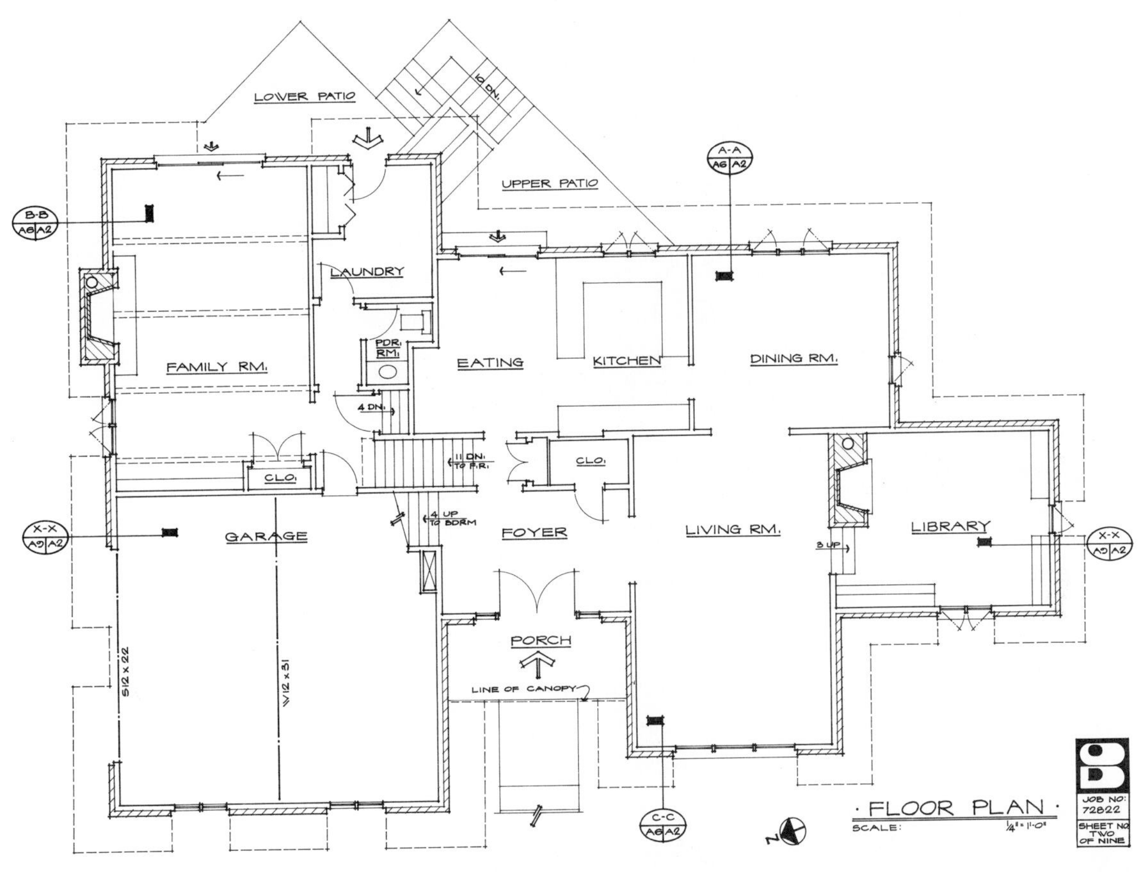

Fig. 20–65 **Lower- and main-level floor plan.** *(Orrin Dressler, designer-contractor.)*

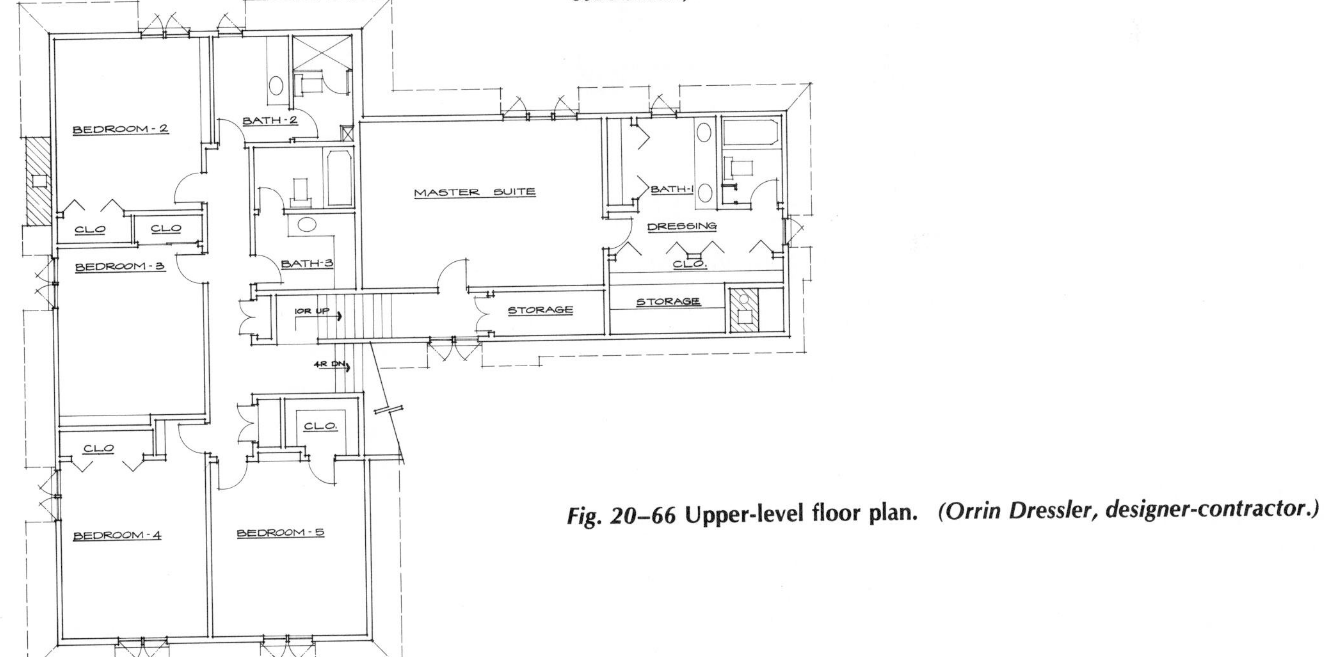

Fig. 20–66 **Upper-level floor plan.** *(Orrin Dressler, designer-contractor.)*

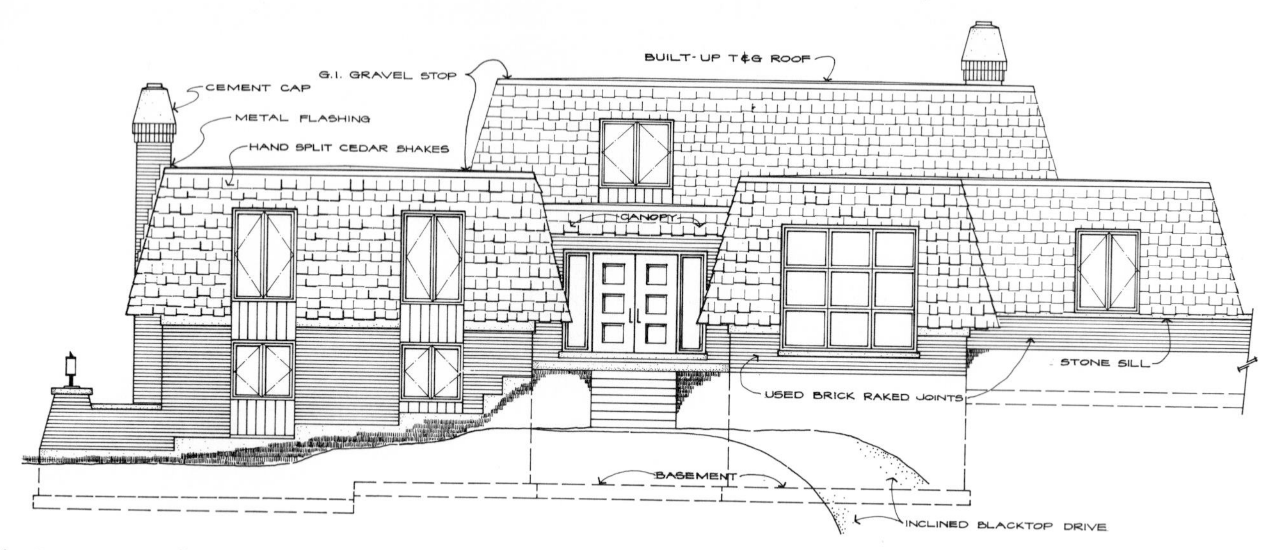

Fig. 20–67 **Front elevation.** ***(Orrin Dressler, designer-contractor.)***

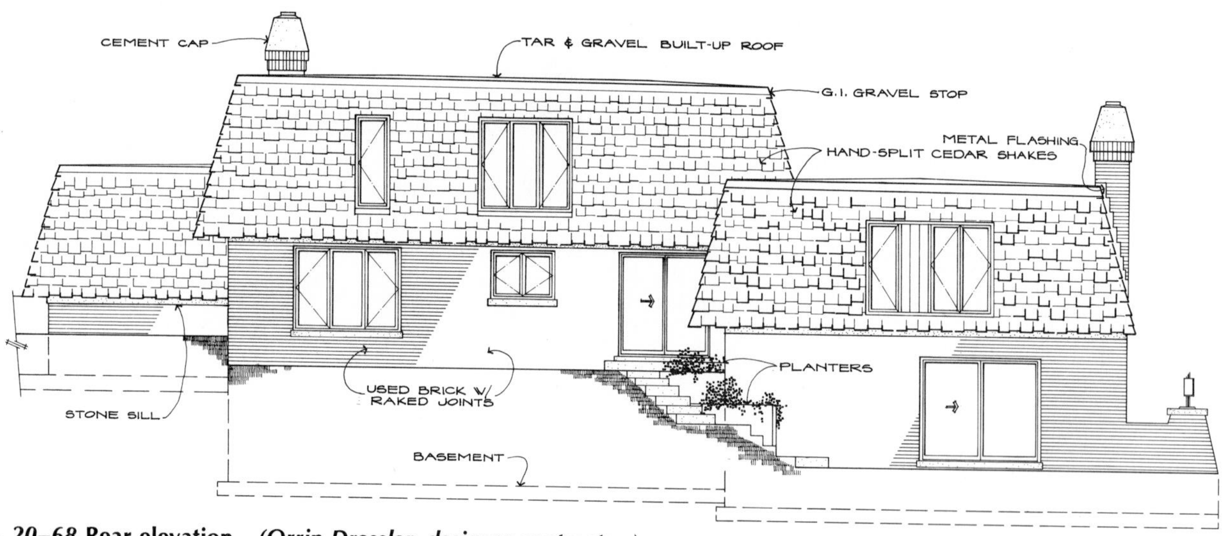

Fig. 20–68 **Rear elevation.** ***(Orrin Dressler, designer-contractor.)***

weights on the cedar shakes that make up the roof. These add depth to the view. The rear elevation (Fig. 20–68) shows contoured stairs between the main and lower levels. The right and left elevations are shown in Figs. 20–69 and 20–70. The longitudinal section (Fig. 20–71) explains the arrangement of levels and rooms. The sections in Fig. 20–72 are identified by symbols on the main floor plan (Fig. 20–65).

Architectural specifications are more detailed written instructions that should accompany any set of plans. They are essential for turning these plans into a building. They note the general conditions of the site. They also specify the materials to the client's and contractor's mutual agreement. Guidelines for specifications have been drawn up by the American Institute of Architects. These guidelines are available through local state chapters of the AIA.

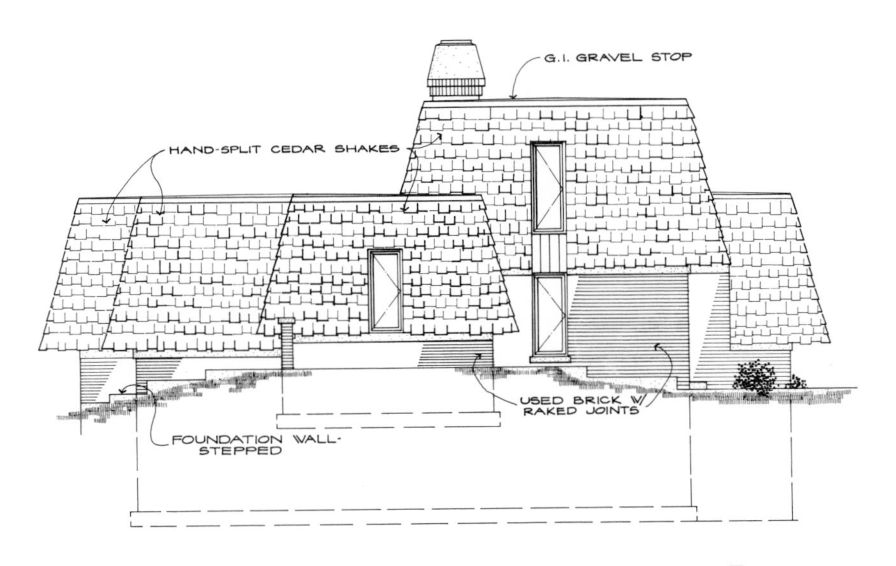

Fig. 20–69 **Right elevation.** *(Orrin Dressler, designer-contractor.)*

Fig. 20–70 **Left elevation.** *(Orrin Dressler, designer-contractor.)*

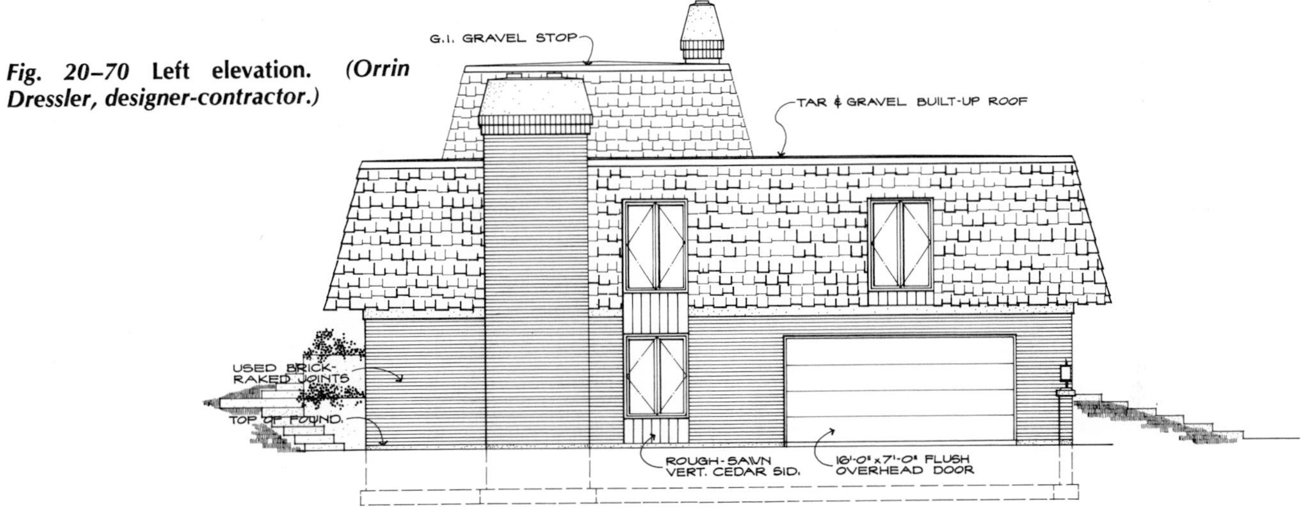

Fig. 20–71 **Longitudinal section.** *(Orrin Dressler, designer-contractor.)*

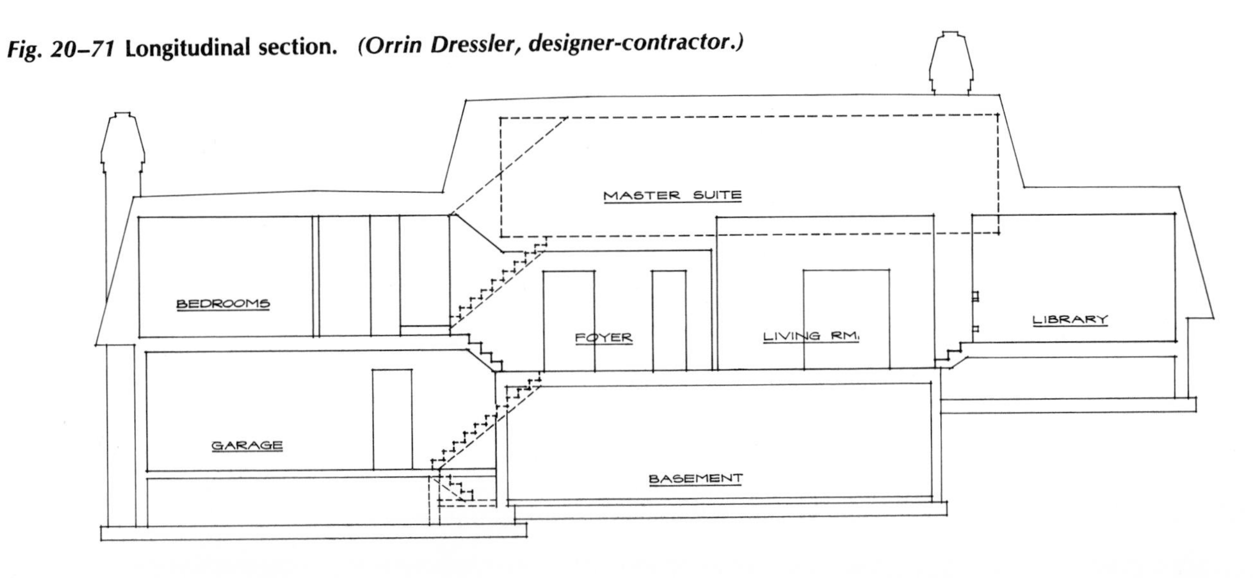

TAR & GRAVEL BUILT-UP ROOF ON ½" EXT. PLY SHEATHING ON 2×6 RIPPED TO 2×4 FOR PITCH @ 24" O.C. ON 2×6 SPACERS @ 6'-0" O.C.
G.I. GRAVEL STOP
6" FIB. GLASS INSUL.
3.3
12
¾" PLASTER WALL
10'-0"
¼" PLY
10'-3"
VENT
2'-3"
⅝" FIN. PLY FL. ON 1×2 STRIPS ON ½" PLY SUB FL.
BRICK VENEER
2×10 FL. JST @ 16" O.C.
½" ANCHOR BOLT
C-C SECTION
A6 A2
SCALE: ½"=1'-0"

Fig. 20–72 **Section views.** ***(Orrin Dressler, designer-contractor.)***

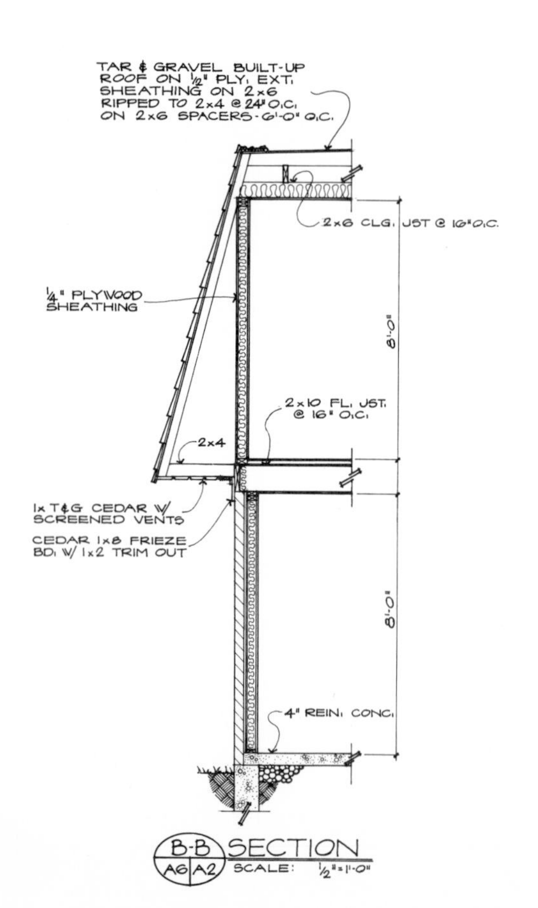

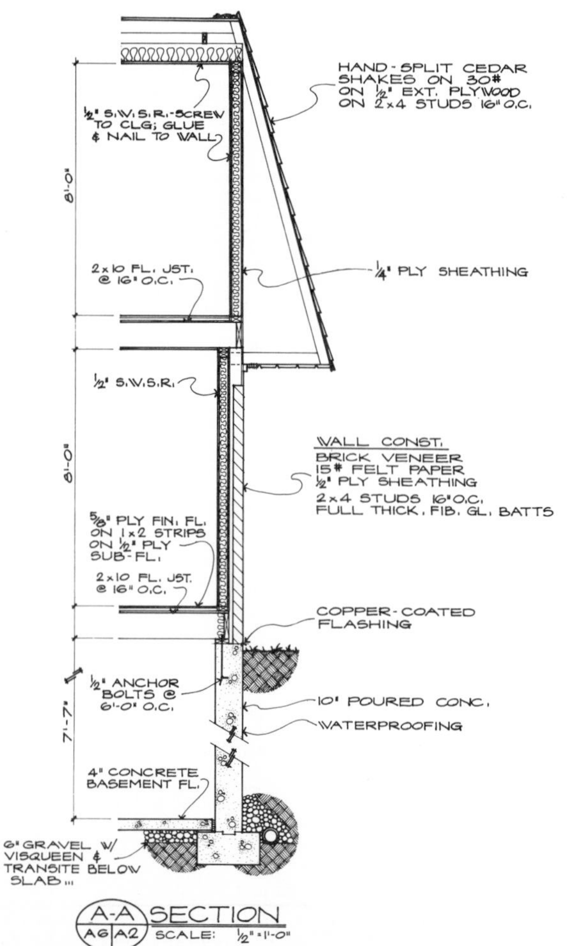

Fig. 20–73 **Farnsworth residence, Plano, IL.** ***(The office of Mies Van der Rohe and Hedrich—Blessing.)***

A famous house design is that for the Farnsworth residence in Plano, Illinois (Fig. 20–73). It was made by the noted architect, Ludwig Mies Van der Rohe. The house is built of structural steel. The floor plan drawn on a modular grid (Fig. 20–74) and the elevation section (Fig. 20–75) show the pure form of the design. The stair detail (Fig. 20–76) is shown as it actually looks in Fig. 20–77. The sectional view of the structural steel wall in Fig. 20–78 illustrates the steel column and channel at work. Figure 20–79 shows the proportions of glass and steel in the exterior curtain wall. Steel and glass curtain walls are used on most of the newer skyscrapers that dominate many large North American cities (Fig. 20–80).

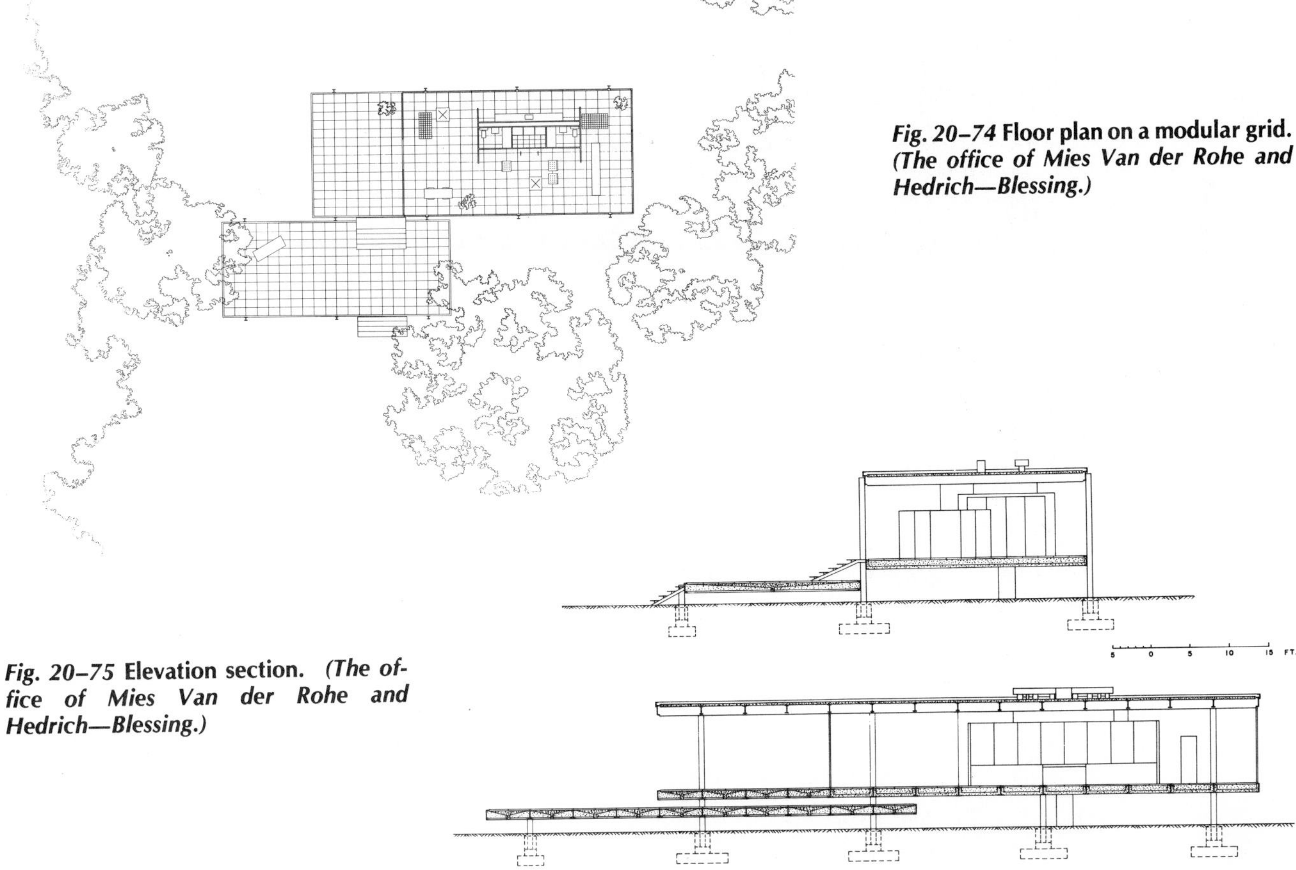

Fig. 20–74 **Floor plan on a modular grid.** ***(The office of Mies Van der Rohe and Hedrich—Blessing.)***

Fig. 20–75 **Elevation section.** ***(The office of Mies Van der Rohe and Hedrich—Blessing.)***

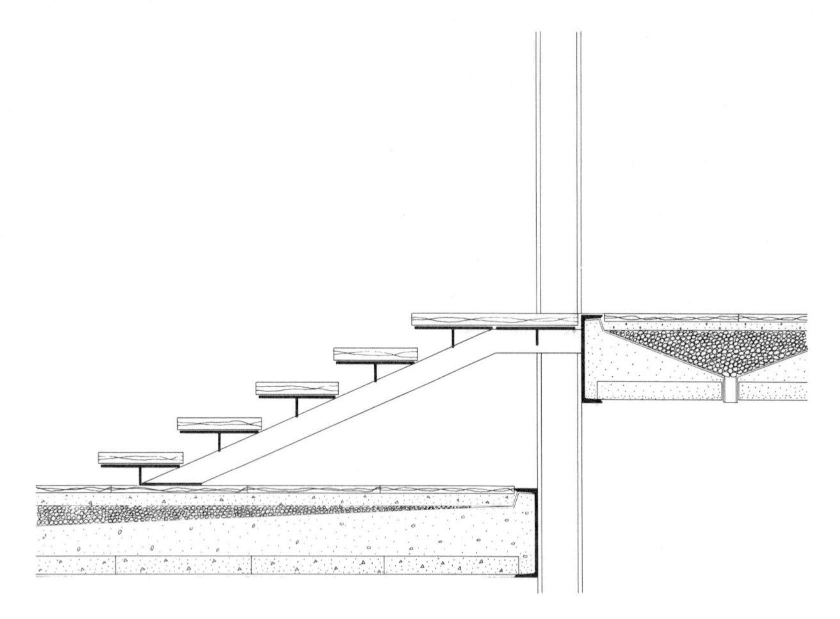

Fig. 20–76 **Stairs in detailed section.** ***(The office of Mies Van der Rohe and Hedrich—Blessing.)***

Fig. 20–77 **The finished stairs complement structural form.** ***(The office of Mies Van der Rohe and Hedrich—Blessing.)***

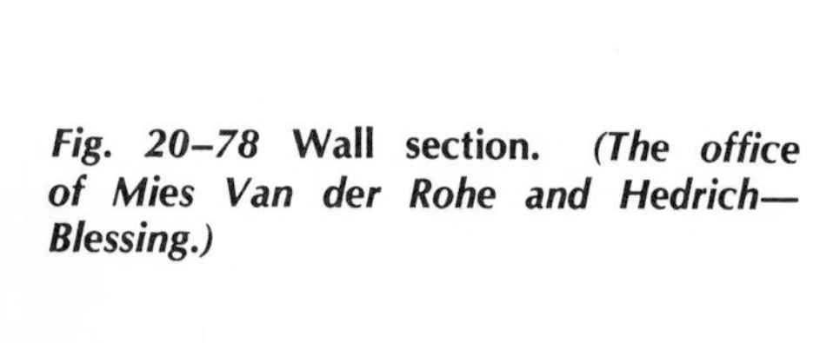

Fig. 20–78 **Wall section.** ***(The office of Mies Van der Rohe and Hedrich—Blessing.)***

Fig. 20–79 **Steel and glass for open design.** *(The office of Mies Van der Rohe and Hedrich—Blessing.)*

Fig. 20–80 **Steel and glass help form a skyline.** *(The office of Mies Van der Rohe and Ron Vickers, Ltd.)*

Review

1. Name the two styles of lettering used in architecture.

2. List four house styles.

3. What are the three main ways to judge a building's architecture?

4. What are the four basic drawings that an architect makes?

5. What are working drawings?

6. What major services does an architect render?

7. Define architecture in your own words.

8. Draw four material symbols used by the architect.

9. Illustrate a line technique important to architectural style.

10. What are pressure-sensitive symbols?

11. List six types of windows used in houses.

12. What are the preferred scales for plan and elevation drawings of small buildings?

13. What scale is preferred for enlarging details?

14. List three types of wall constructions common in residential design.

15. What is modular coordination?

16. What is the basic metric unit?

Problems

INDEX TO DRAWINGS

ARCHITECTURAL

A1 – SITE PLAN, INDEX TO DRAWINGS, ARCHITECTS SYMBOLS
A2 – GROUND FLOOR PLAN
A3 – FIRST FLOOR PLAN
A4 – TYPICAL FLOOR PLAN (2ND THRU 10TH FLRS.)
A5 – PENTHOUSE & ROOF PLANS
A6 – LARGE SCALE CORE PLANS, DETAILS, SECTIONS
A7 – LARGE SCALE APARTMENT PLANS, ELEVATIONS, DETAILS
A8 – EXTERIOR ELEVATIONS
A9 – EXTERIOR ELEVATIONS, STAIR ENTRANCE, DETAILS
A10 – WALL SECTIONS, CROSS SECTION, MISCELLANEOUS DETAILS
A11 – ENTRANCE SECTIONS, MISCELLANEOUS DETAILS

STRUCTURAL

S1 – FOUNDATION & GROUND FLOOR PLAN
S2 – GENERAL NOTES, SECTIONS, DETAILS
S3 – FIRST FLOOR FRAMING PLAN
S4 – TYPICAL FLOOR FRAMING PLAN (2ND THRU 10TH FLRS.)
S5 – ROOF & PENTHOUSE FRAMING PLAN
S6 – COLUMN SCHEDULE & DETAILS
S7 – BEAM SCHEDULE & DETAILS
S8 – STAIR SECTIONS & DETAILS
S9 – PARKING DECK GRADES, BEAM & SLAB DETAILS

Fig. 20–81 **Practice lettering the architectural or structural index to drawings as assigned by the instructor. Use a bold extended style. The letter sizes should be 1/4, 3/16, and 1/8″. Use guide lines.**

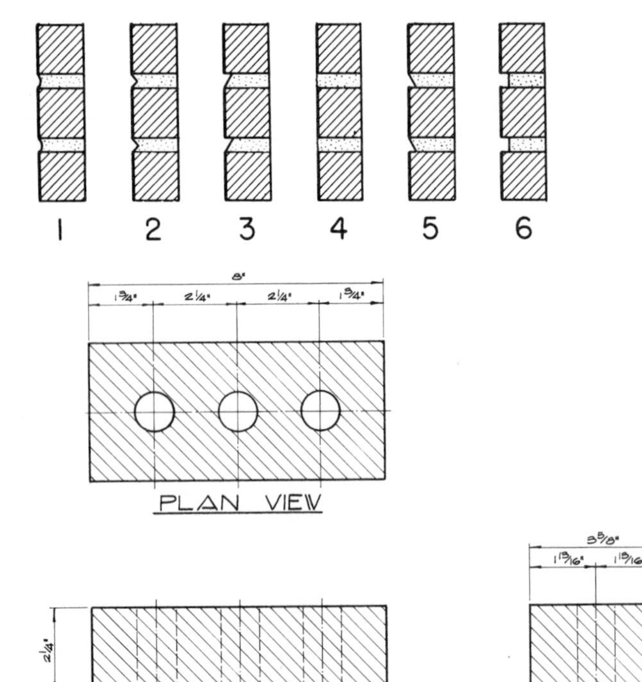

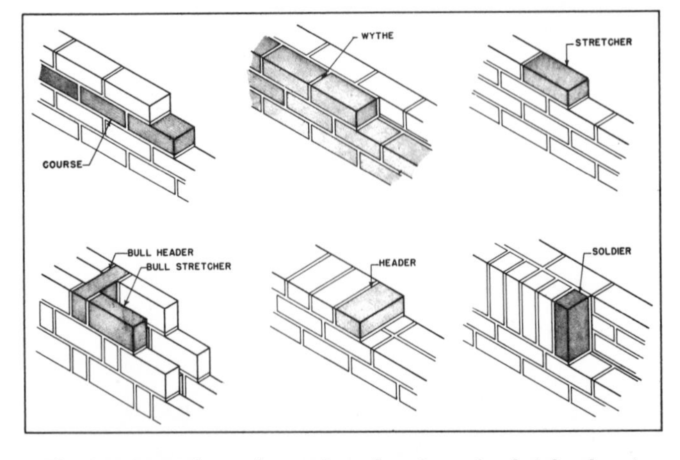

Fig. 20–82 **Make a three-view drawing of a brick, showing dimensions. Label the appropriate surface with "header," "stretcher," and "face." Make a pictorial drawing of the brick bonds as assigned. The nominal dimensions are 2 1/4, 4, and 8″, with 1/2″ for mortar joints. Use a scale of 1 1/2″ = 1′-0″. Draw six sectional views of joint shapes: (1) concave, (2) V-joint, (3) weathered, (4) flush, (5) struck, and (6) raked.**

Fig. 20–83 Prepare an isometric view of a wall 4′ long, showing the section illustrated. The brick bonding may be shown, or it may appear as horizontal lines in pictorial. Scale: 1½″ = 1′-0″. *(Doug Strom.)*

10"
1"
5"
1"
4'-0"
6"
8"
LIMESTONE CAP
16 OZ. COPPER FLASHING
BRICK WALL SECTION
SCALE:
1½" = 1'-0"
FOUNDATION WALL
GRADE

OBLIQUE DETAIL OF CONCRETE BLOCK WALL

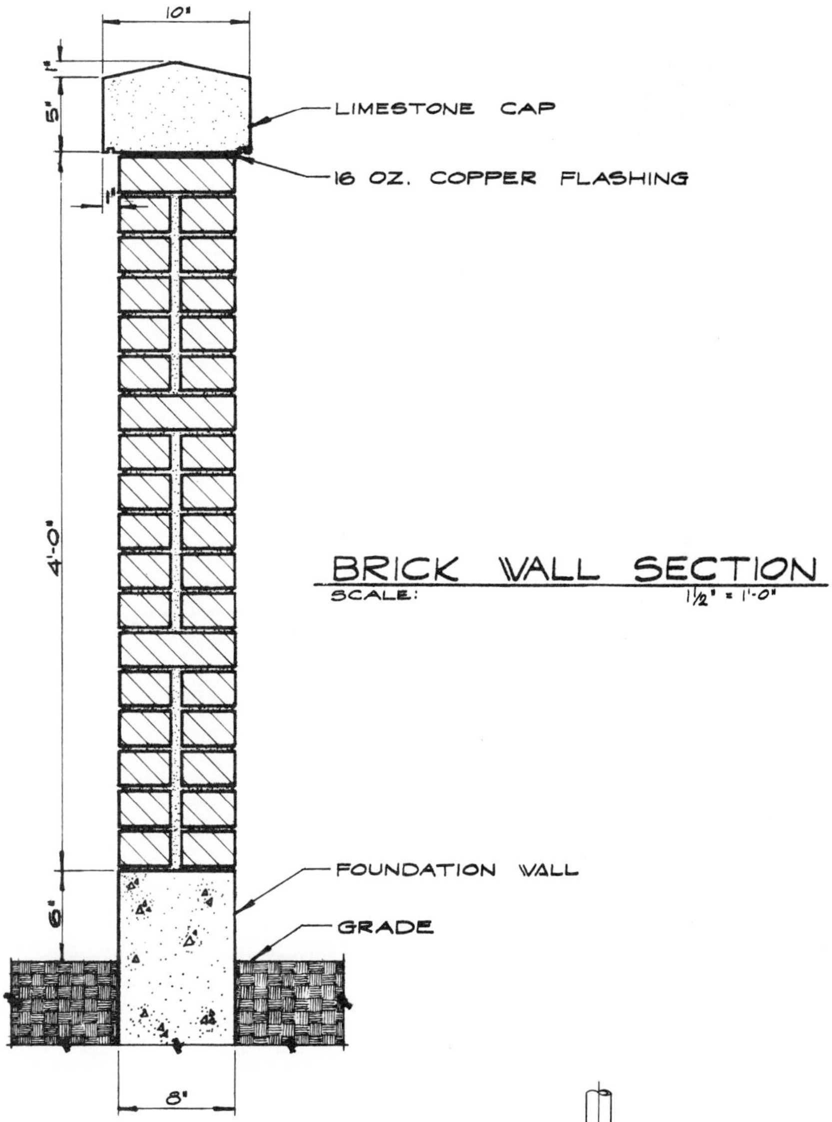

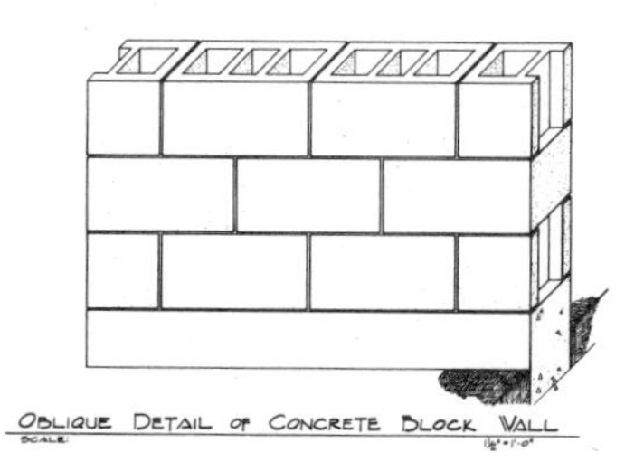

Fig. 20–84 Prepare an isometric detail of a concrete block wall. Investigate and use the nominal sizes of concrete block. Scale: 1½″ = 1′-0″. *(Doug Strom.)*

Fig. 20–85 Draw a footing detail as shown. Use sectional symbols common to architectural detailing. Design your footing to be on load-bearing soil at a depth below a local frost line or as established by building codes. *(Doug Strom.)*

1'-0"
GRADE
FOUNDATION WALL
2-#5 BARS T & B
FOOTING
2'-0"
FOOTING DETAIL
SCALE:
1½"=1'-0"

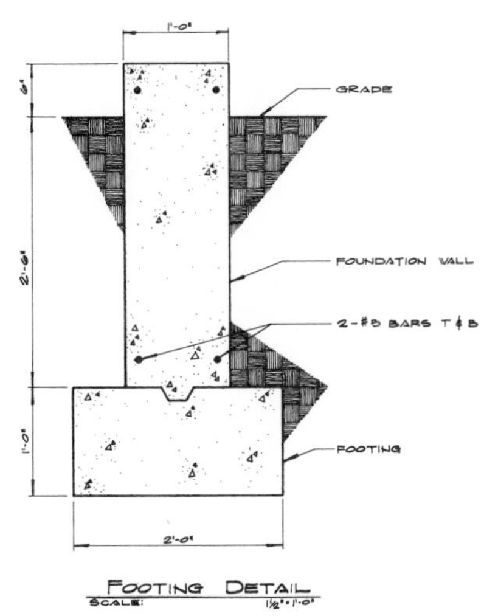

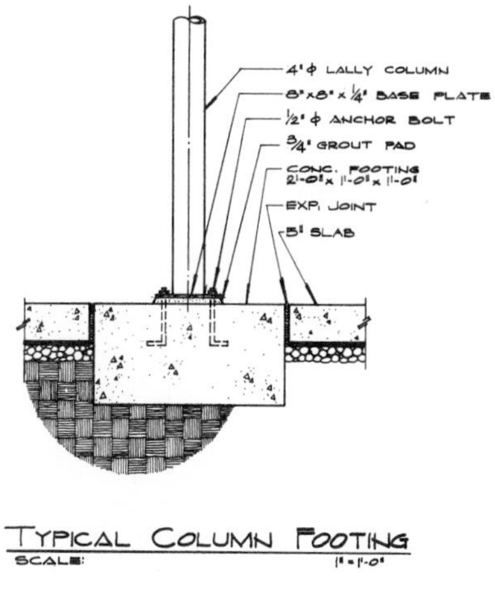

4" ϕ LALLY COLUMN
8"x8"x¼" BASE PLATE
½" ϕ ANCHOR BOLT
¾" GROUT PAD
CONC. FOOTING 2'-0"x 1'-0"x 1'-0"
EXP. JOINT
5" SLAB
TYPICAL COLUMN FOOTING
SCALE:
1"=1'-0"

Fig. 20–86 Prepare an isometric view of a typical column footing using the given sizes of the structural items. *(Doug Strom.)*

Figs. 20–87 through 20–92 **are all part of the same structure. When preparing the details (Figs. 20–89 through 20–92), refer to the framing plan (Fig. 20–88).** ***(Doug Strom.)***

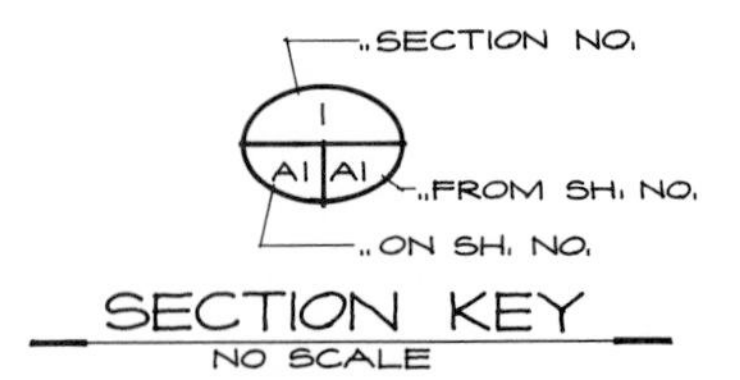

Fig. 20–87 **Prepare a sectional key diagram using an ellipse, circle, or rectangle as suits your style. Note how it applies to the framing plan and details.**

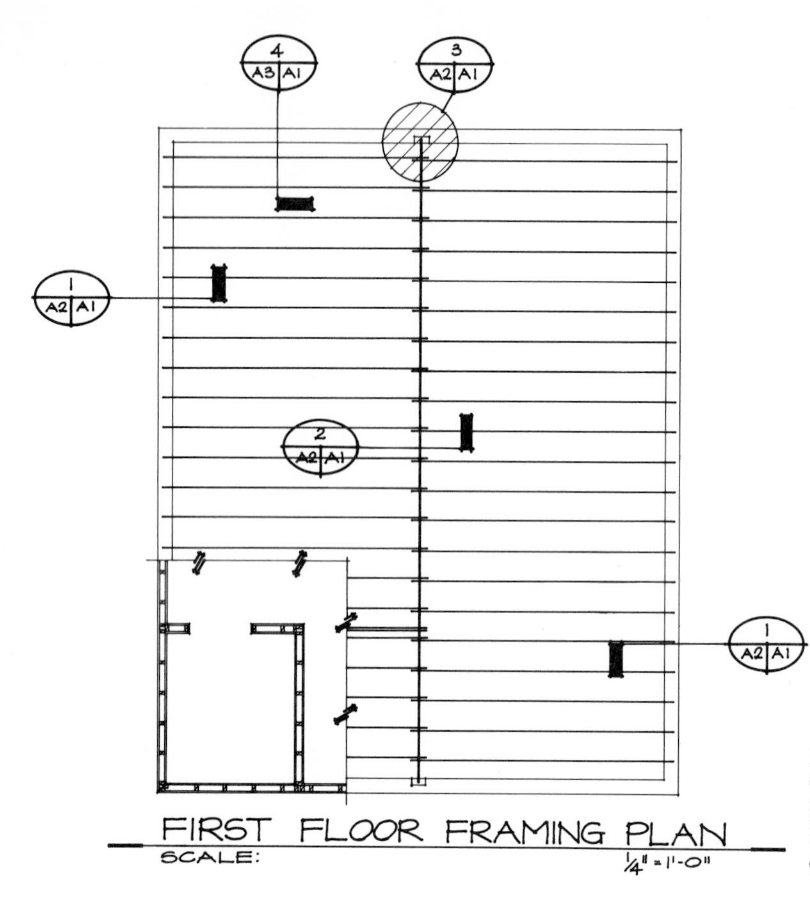

Fig. 20–88 **Prepare a framing plan as shown. Outside building dimensions are 24′-0″ by 30′-0″. Joists are 16″ on center. Include your sectional key diagrams for the enlarged details to be drawn.**

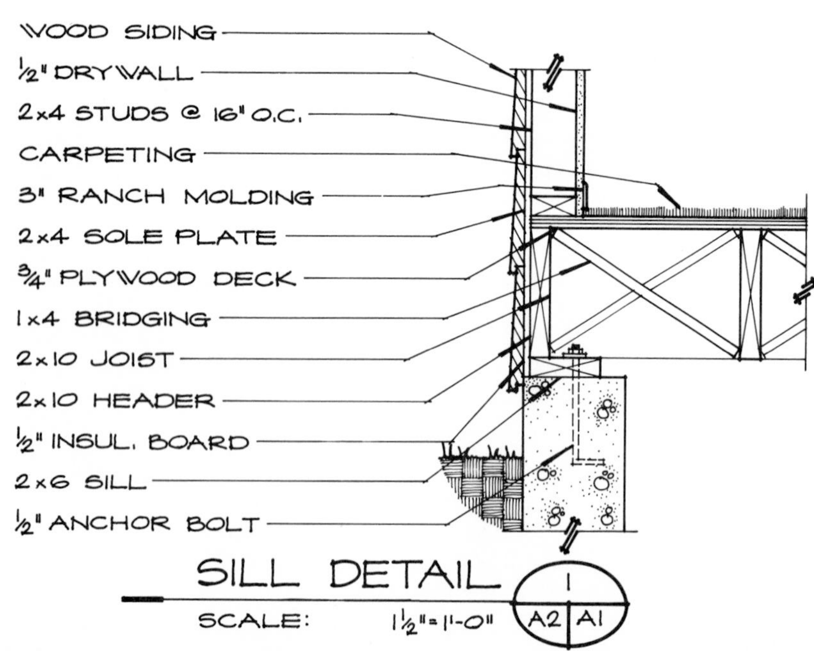

Fig. 20–89 **Using the nominal dimensions of the materials listed, construct a sill detail in section. The lapped, 1″ wood siding has an 8″ exposure. Scale: 1½″ = 1′-0″.**

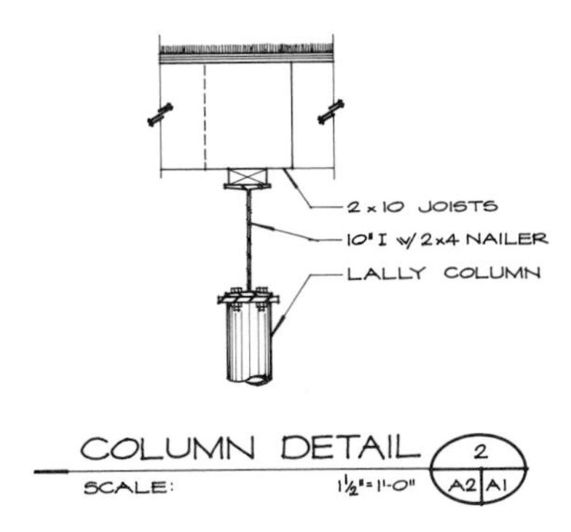

Fig. 20–90 **Prepare a partial column-girder detail. The column is 4" round with a 6" plate welded to the top. The 10" I-beam is bolted to the plate. The 2 × 4 rests on the 4½" flange of the I-beam, supporting the overlapped floor joists.**

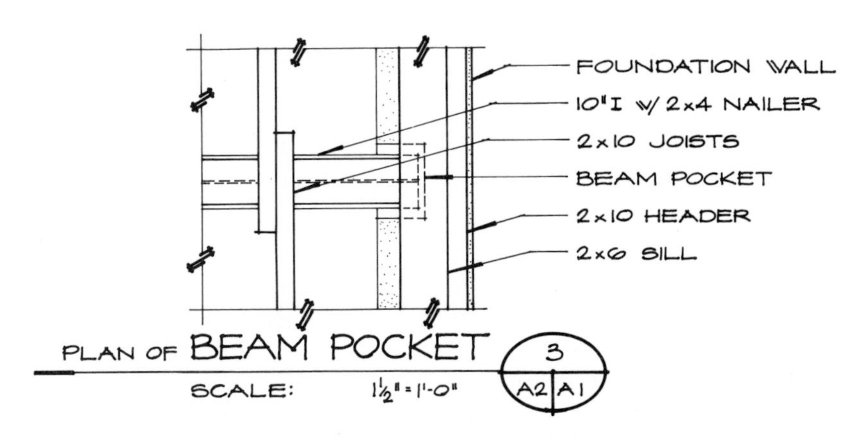

Fig. 20–91 **Prepare a plan of the 6" beam pocket. The beam is to have a 4" bearing on the concrete wall.**

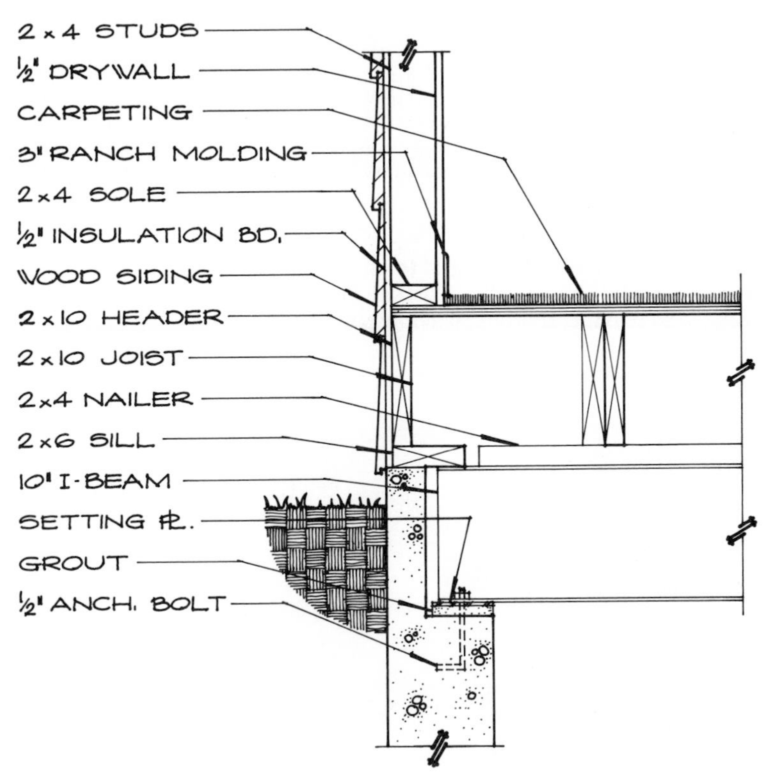

Fig. 20–92 **Prepare a sectional beam pocket detail showing the 10" I-beam on the stepped foundation wall. Use nominal dimensions for all the materials listed. Allow the flange of the I-beam to show as a nominal ½" on the top and bottom of the beam.**

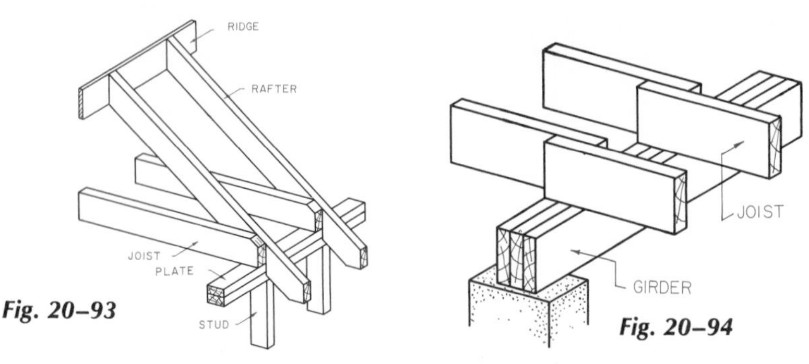

Fig. 20–93

Fig. 20–94

Figs. 20–93 and 20–94 **Prepare a pictorial of typical roof framing. Studs, ceiling joists, and rafters are 16″ on center. Joist and rafter size: 2″ by 6″. Studs and top plate size: 2″ by 4″. Ridge size: 1″ by 8″. Use the nominal sizes at a scale of 1½″ = 1′-0″. Draw a pictorial of the built-up girder to support the floor joist. Girder and joist size: 2″ by 10″.**

Figs. 20–95 through 20–99 **are a partial set of plans for a two-story residence. Make drawings as assigned by the instructor.** ***(A. W. Wendell & Sons, designer-contractor.)***

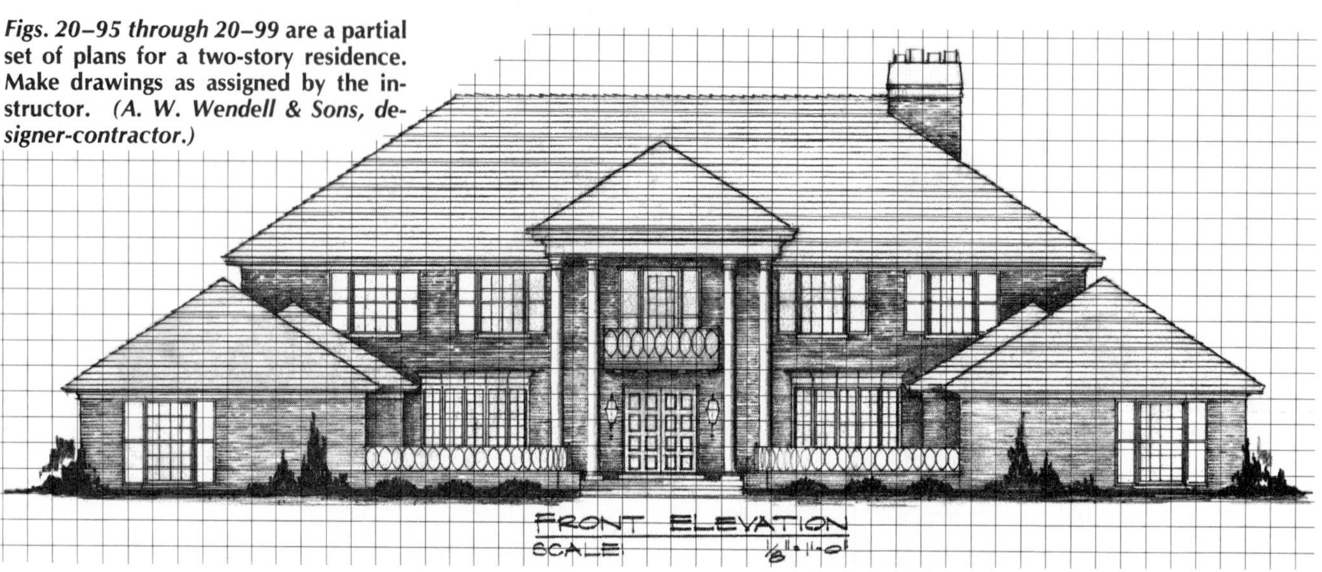

Fig. 20–95 **Draw the front elevation of the two-story residence. Use the modular grid to establish the rectangular proportions, roof slope, and structural openings. Scale: ⅛″ = 1′-0″. Illustrate the brick, masonry walls, and asphalt roofing with horizontal lines. Add window shutters, window panes, and decorative appointments to suit your own design taste.**

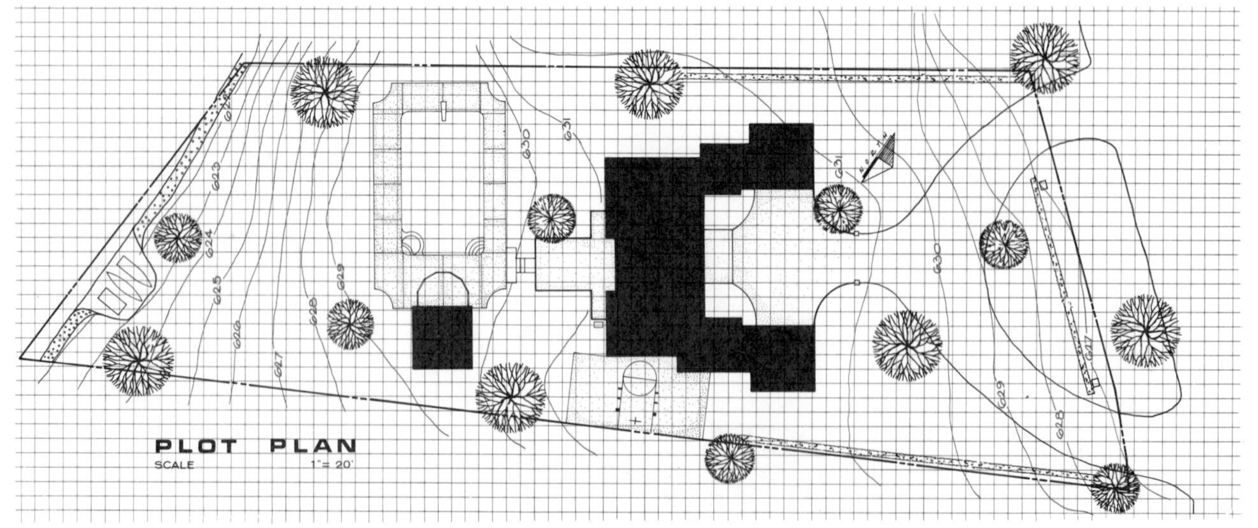

Fig. 20–96

Fig. 20–96 **Examine the gridded plot plan, and develop the boundary lines at a scale of 1″ = 20′-0″, on a C-size sheet. Dimension the length of each boundary line. Complete the plot plan with the plan of the house and your own landscaping design. All radii on the drive must be 18′ minimum. Lay out the approximate contour lines, noting the change of terrain. The basketball court and cabana are optional design appointments. If instructed, prepare a simple plan and elevation of a cabana to fit the space provided.**

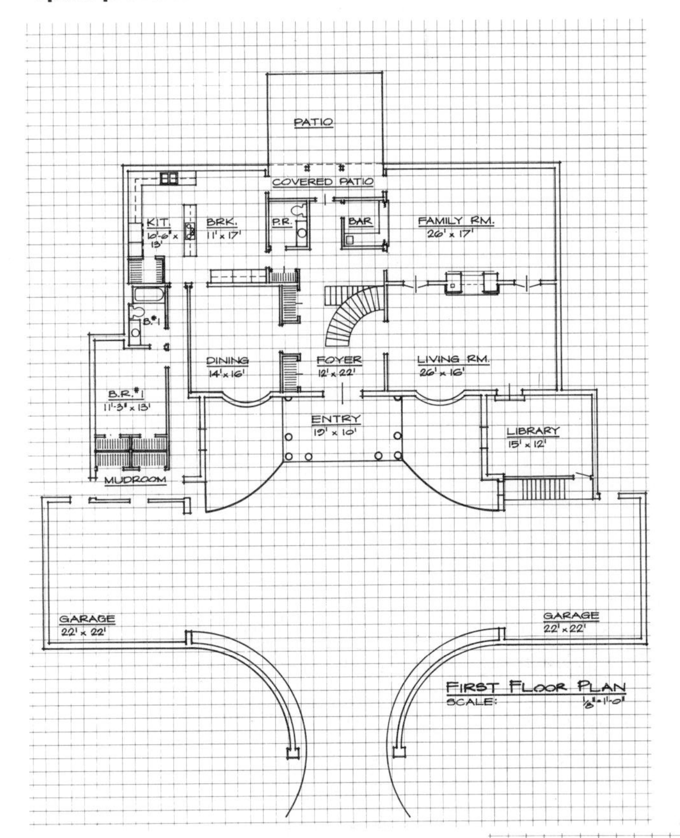

Fig. 20–98 **Prepare a second floor plan to include the roof plan of the single-story area. Scale: 1/8″ = 1′-0″. Locate the missing windows according to the front and right-side elevations. Locate the windows you believe necessary on the left and rear elevation walls. Label the rooms and their approximate sizes. Note the hip roof design and the intersecting inclined planes forming the roof.**

Fig. 20–97 **Draw a first floor plan at a scale of 1/8″ = 1′-0″. By examining the modular grid, locate and include missing windows on your drawing. Label the rooms and their approximate sizes. If instructed, plot the functional traffic patterns on an overlay to the floor plan.**

Fig. 20–99 **Draw a right-side elevation by examining the proportions on the modular grid. Note the centerlines that locate the finished floors and ceilings. Find the common roof pitch and label it on the elevation. Draw in the materials that describe this elevation. Prepare a left-side and rear elevation as assigned by the instructor.**

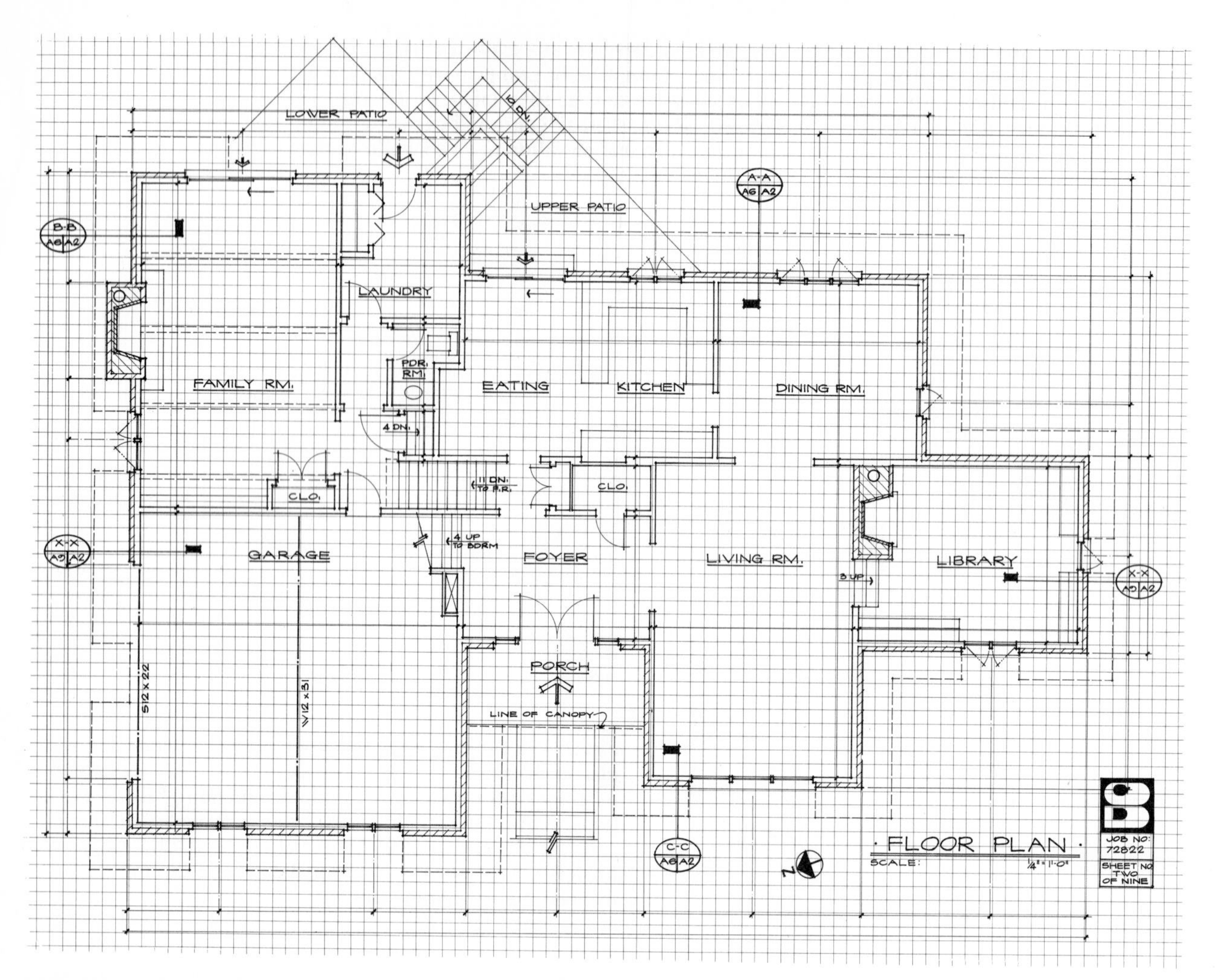

Fig. 20–100 **Note that this floor plan is Fig. 20–65 in the text. Prepare the first floor plan at a scale of 1/4" = 1'-0" on a C-size sheet. Add the dimensions to the plan. Examine the 1/8" modular grid to establish the room sizes, window opening sizes, and the overall building size.** ***(Orrin Dressler.)***

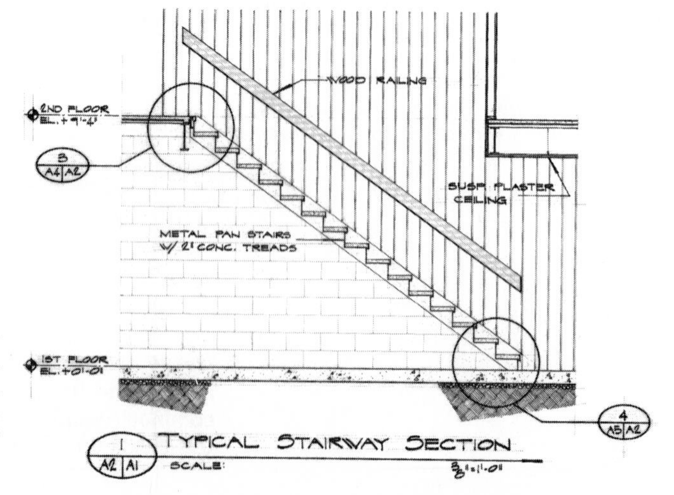

Fig. 20–101 **Prepare a sectional elevation of a typical stairway. Riser 7", tread 11", nosing 1", and stringer 10". Note the elevational readings from finished floor to finished floor. Scale: 3/8" = 1'-0".**

Fig. 20–102

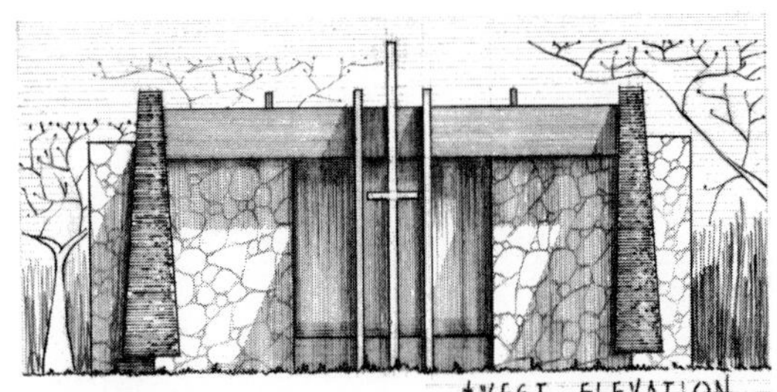

Fig. 20–103

Fig. 20–104

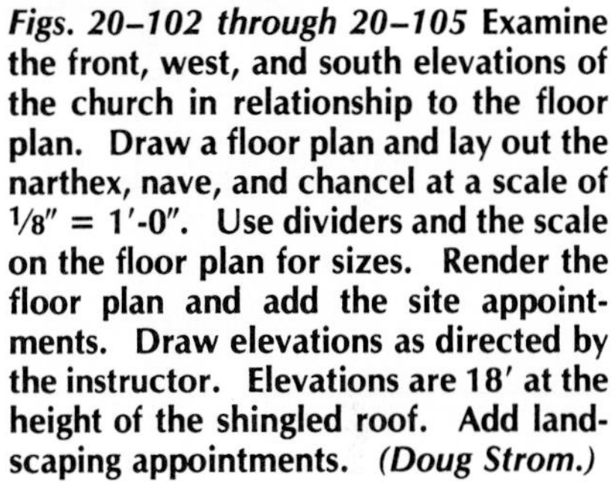

Figs. 20–102 through 20–105 **Examine the front, west, and south elevations of the church in relationship to the floor plan. Draw a floor plan and lay out the narthex, nave, and chancel at a scale of 1/8″ = 1′-0″. Use dividers and the scale on the floor plan for sizes. Render the floor plan and add the site appointments. Draw elevations as directed by the instructor. Elevations are 18′ at the height of the shingled roof. Add landscaping appointments.** ***(Doug Strom.)***

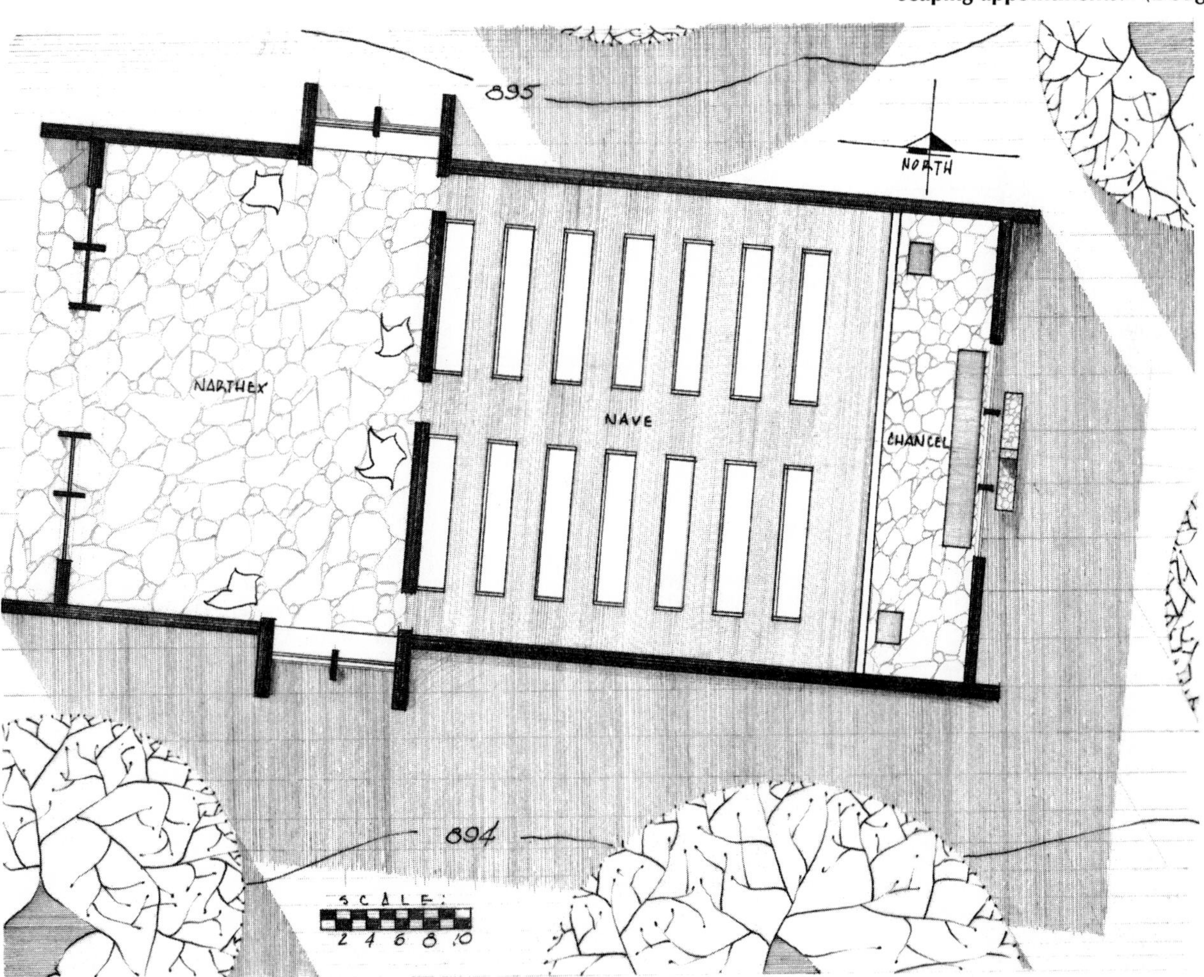

Fig. 20–105

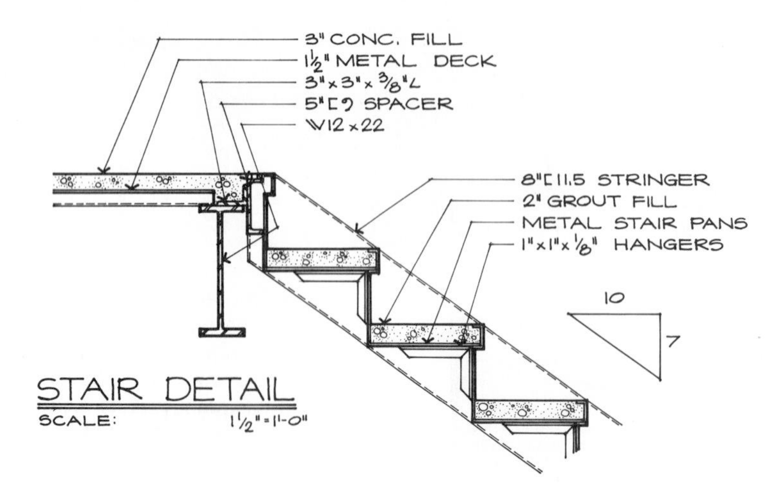

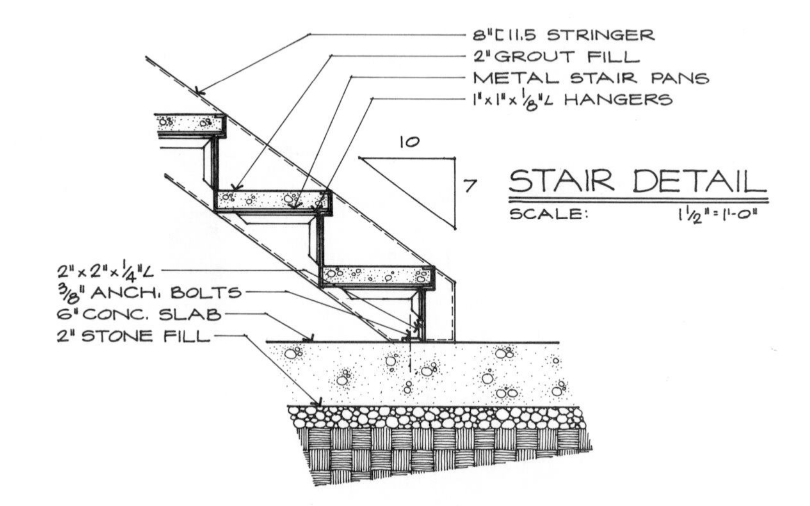

Fig. 20–106 **Problem 1: Draw stair details for a straight run rise of 7′ on the outside of a building. Scale: 1″ = 1′-0″. Include necessary dimensions. Problem 2: Design a run of stairs with a rise of 14′ from a sunken plaza in front of a building to street level, with a U- or L-turn as assigned by the instructor. Include all dimensions and notes.**

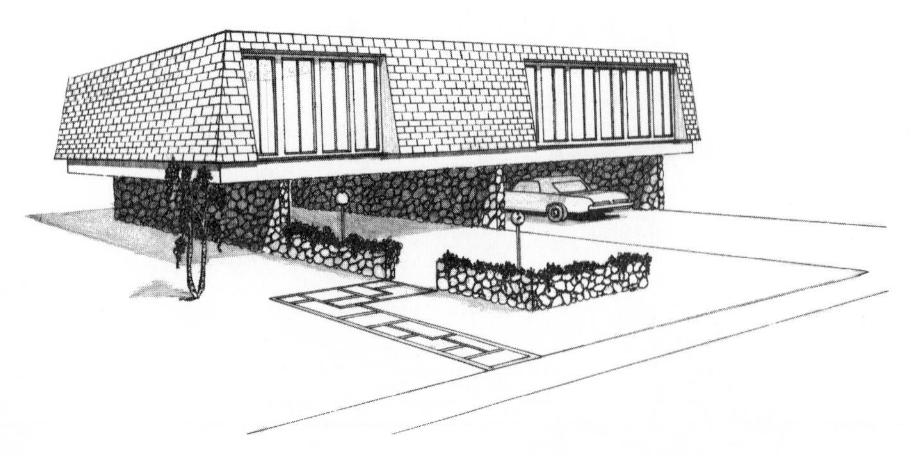

Fig. 20–107 **Prepare a front elevation of the professional office shown in the perspective view. Design a floor plan and office layout for a profession of your choice. Select overall sizes and a suitable scale.**

Fig. 20–108 **The twin towers contain both office space and apartments. The outside dimensions are 60 feet square. Prepare typical floor plans for an office floor and for an apartment floor, designed around a central service core containing elevators, stairwells, and so forth. Mix the kinds of apartments—from one-room efficiencies to two- or three-bedroom apartments.**

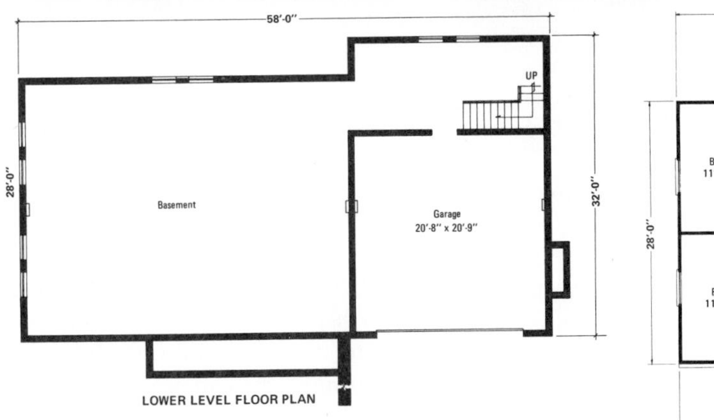

LOWER LEVEL FLOOR PLAN

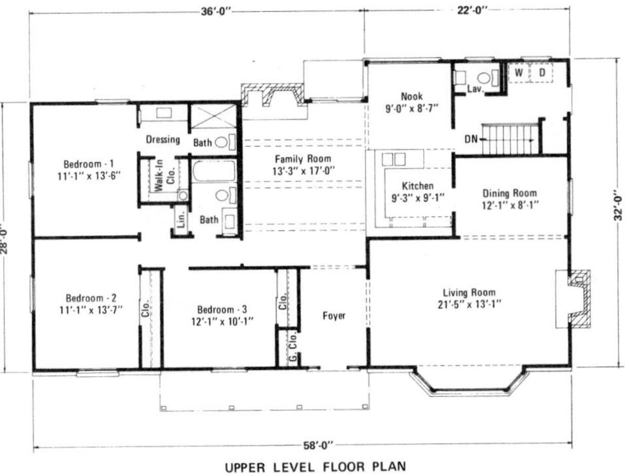

UPPER LEVEL FLOOR PLAN

Fig. 20–109 **Draw a front- and left-side elevation of the split-level house from the floor plans and perspective given. Choose an appropriate scale. Prepare a plot plan with contour lines for this house. Scale: 1/8" = 1'-0".**

Chapter 21

Structural Drafting

Fig. 21–1 **The Deere & Company administrative center, designed by Eero Saarinen and Associates, required many large sheets of structural details.** ***(John Deere & Co.)***

THE STRUCTURAL DRAFTER

Builders must have detailed structural drawings to work from when they are putting up buildings, bridges, dams, storage tanks, communication towers, and many other kinds of structures. Thousands of structural drawings are needed all the time. Generally, these are prepared by structural drafters. The structural drafter is usually a member of an engineering team. This team often works together with other teams under the direction of a project manager or a job superintendent. As an example of this teamwork, structural and architectural designers often combine their efforts. The architectural designer designs the form of a building based on the function it will have (Fig. 21–1). Then, the structural designer designs the frame of the building to fit this form (Fig. 21–2). The work of the structural drafter is very important in engineering. The construction of buildings, bridges, and other structures depends on the detailed instructions in structural drawings.

A structural drafter usually works at one of the following five jobs:

1. Drafting details in an architect's or engineer's office.
2. Preparing construction details for a contractor (making the shop drawings for a construction company.)
3. Drafting structural details for a manufacturer of structural materials.
4. Working for the engineering department of a manufacturing plant that maintains its own engineering operations.
5. Preparing drawings for government or other agencies that regulate the construction and design of public buildings, bridges, dams, and other structures.

Career opportunities for structural drafters generally depend on how much practical experience they gain as junior members of engineering teams. Structural drafters can be promoted to such jobs as structural detail checker, estimator, chief drafter, construction supervisor, or building inspector (Fig. 21–3).

Further career opportunities can be gained through continuing education. Formal training at a junior college or university is very helpful. To succeed as a structural drafter, you must have the ability to communicate and an aptitude for mechanics. High school or college students interested in the construction industry should work on language skills, mathematics, science, and graphic communication.

STRUCTURAL MATERIALS

Designers and detail drafters must be familiar with a great many structural materials. In addition, all members of the engineering team must learn about new construction materials and systems as they are developed. All structural materials have special ways of being assembled and fastened together. These ways must be considered whenever accurate drafting details are prepared.

The basic structural materials used today are steel, wood, concrete, structural clay products, and stone masonry. Different materials have different characteristics. It is the designer's job to choose the right combination of materials to bear the stresses imposed by a building.

STEEL AS A STRUCTURAL MATERIAL

Steel makes a good construction material because of the shapes into which it can be formed at the mill. Steel shapes are produced in rolling mills. They are then shipped to fabrication shops where they are cut to specific lengths and connections are prepared. Some of the basic steel shapes are the wide flanged W-shape used as a beam or column; the S-shape, formerly known as an I-beam; the C-shape, formerly known as the channel; and the L-shape, formerly known as the angle.

These steel shapes have framed the skyscrapers of our cities for nearly three-quarters of a century. The American Institute of Steel Construction maintains regional offices from coast to coast to help provide guidelines for designing and building steel structures.

A Steel-framed Village

The Tempo Bay Resort Hotel and Polynesian Village in Walt Disney World, Florida, is an interesting example of the use of structural steel in contemporary buildings. Figure 21–4 shows the A-shape structure under construction. Structural steel is rising around a concrete elevator core. In the left foreground, complete modular steel-framed rooms are stacked three high. Each room is 2750 mm (9 ft) high and 4570 mm (15 ft) wide.

Fig. 21–2 **A structural detail of the Deere building.** *(John Deere & Co.)*

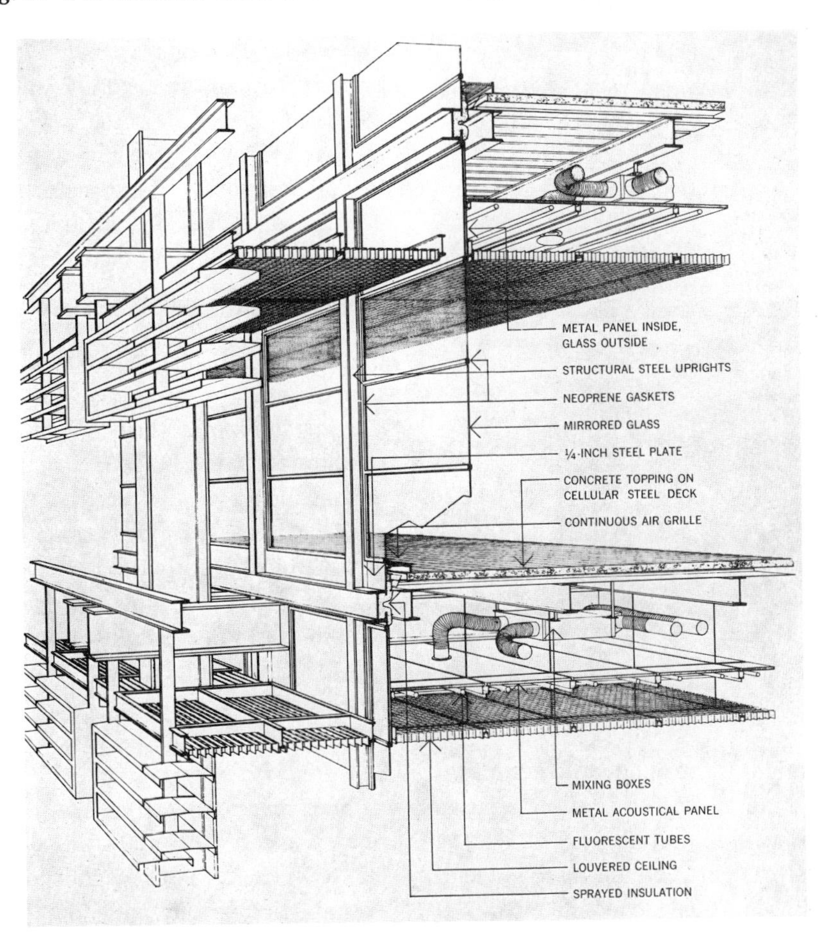

Fig. 21–3 **Career opportunities in structural design.**

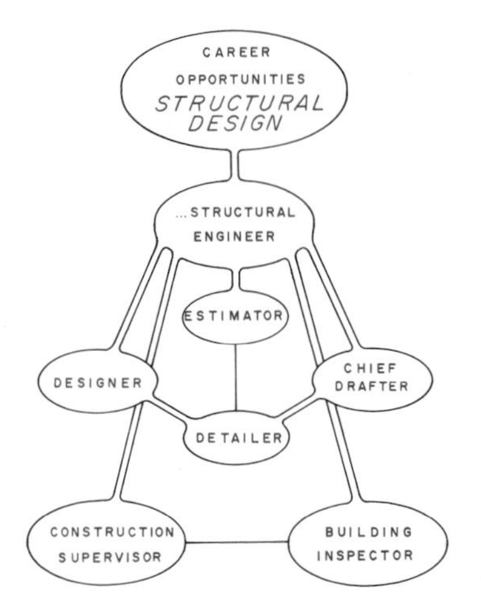

Fig. 21–4* Structural steel begins to rise at Walt Disney World. *(United States Steel Corp.)

Fig. 21–5* Basic building blocks are trucked to the site. *(United States Steel Corp.)

Each unit varies from 9750 to 11 880 mm (32 to 39 ft) long, including a balcony.

In Fig. 21–5, basic building blocks are trucked to the site. At the right is the support structure for a monorail transportation system. The two monorail tracks and the platform between them are made of poured-in-place concrete spanning 20 120 mm (66 ft). Although the monorail passes through the hotel building, the two structures do not touch. This avoids any transfer of vibrations.

The completed A-frame building is shown in Fig. 21–6. Each of the 13 steel A-frames is approximately 67 000 mm (220 ft) across at the base, 41 150 mm (135 ft) across at the top, and 4570 mm (15 ft) across the vertical bents. It is built of tubes, wide-flanged sections, and chords made of 455 × 660 mm (18- by 26-in.) tubes. Each A-frame is assembled on the site and erected in five pieces. All the steel members are connected with high-strength bolts (HSB).

The A-frames were designed with the aid of a computer to hold up under many different combinations of *loads* (weights or pressures borne by a structure). These include wind loads, temperature changes, the *dead load* (weight of the building), and the *live load* (weights added temporarily, such as furniture). If the building is designed correctly, these loads are carried through the structure to the ground. For the building in Florida, the designer had to allow for wind loads from hurricane winds. Allowance must always be made for loads created by geographical location, the prevailing weather, the function of the building, and the size and shape of the form determined by the architectural designer.

A Steel-framed Skyscraper

(See review question 9.) The framework of the 100-story John Hancock Center in Chicago is a fine example of a new design and system

Fig. 21–6 **The completed A-frame form.** ***(United States Steel Corp.)***

Fig. 21–7 **A steel-framed skyscraper, the John Hancock Building, Chicago.** ***(Skidmore, Owings & Merrill; Hedrich—Blessing.)***

Fig. 21–8 **Rio Grande Gorge Bridge.** ***(New Mexico State Highway Department.)***

that was also economical to build. Big John, as the building is called, tapers from a 265 × 165 ft area at the base to a 160 × 100 ft area at the top. The towering giant presses upward to a little over 1100 ft in height. The walls are stiffened by diagonals (Fig. 21–7) that form huge 18-story-high steel "Xs." This unusual design saved about 27 000 tons of structural steel and $15 000 000. The firm of Skidmore, Owings and Merrill and their designer, Bruce Graham, used a computer to help work out the structural design.

The John Hancock project includes 700 apartments ranging from one-room efficiencies to four-bedroom luxury apartments. Of the 2 800 000 sq ft of floor space, there are 812 000 sq ft for commercial entrances, lobbies, and major traffic patterns. Additional facilities include parking within the structure, restaurants, health clubs, a swimming pool, and an ice-skating rink. The three-acre site helped determine the character of this steel structure.

A Steel-framed Bridge

Steel forms the framework of a bridge over a deep gorge of the Rio Grande near Taos, New Mexico (Fig. 21–8). The bridge had a rigid structural framing made of high-strength steel fastened with high-strength bolts and welds. More than 1725 t (metric tons) or 1900 customary tons of structural steel were used in its construction. The center span is 183 m (600 ft) long. The two side spans are each 91.5 m (300 ft) long. The distance from

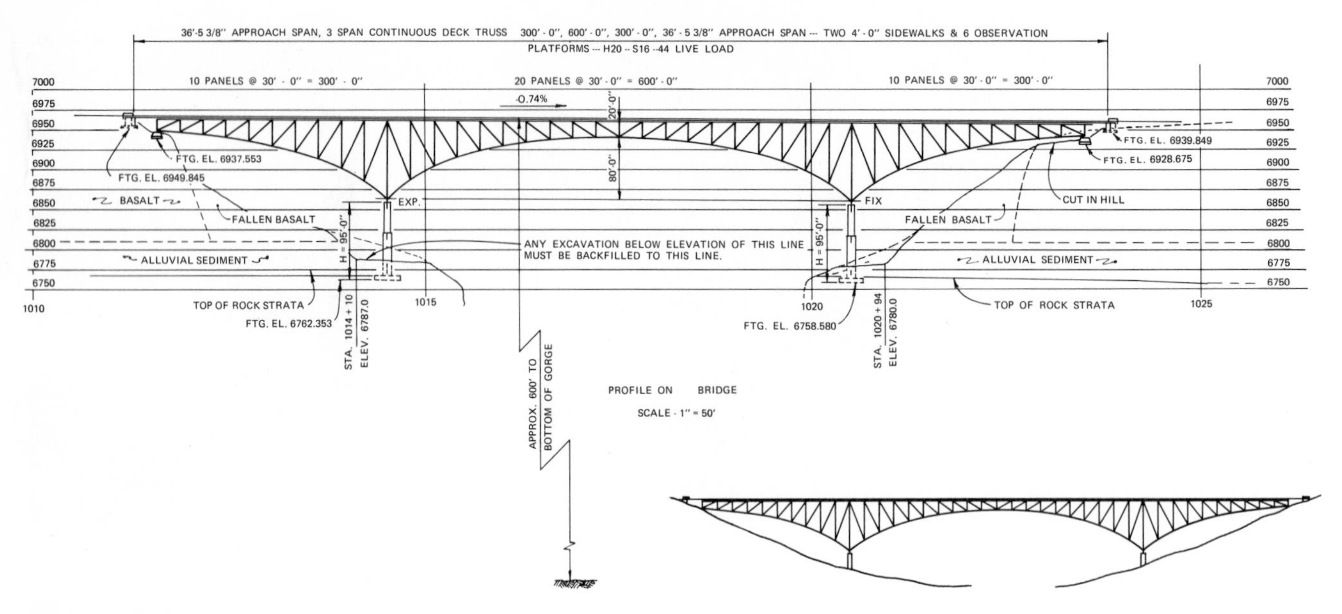

Fig. 21–9 **Schematic of bridge framing.** *(United States Steel Corp.)*

the canyon floor to the bridge floor is 183 m (600 ft) (Fig. 21–9).

A typical welded-steel member of this bridge appears in Fig. 21–10.

Figure 21–11 shows a bolted steel member of the bridge. The diagram calls for the bolting to be done at the construction site.

STEEL SYSTEMS

Using steel, engineers have developed a number of new structural systems. The Unistrut Space

Fig. 21–10 **Welded bridge member.** *(United States Steel Corp.)*

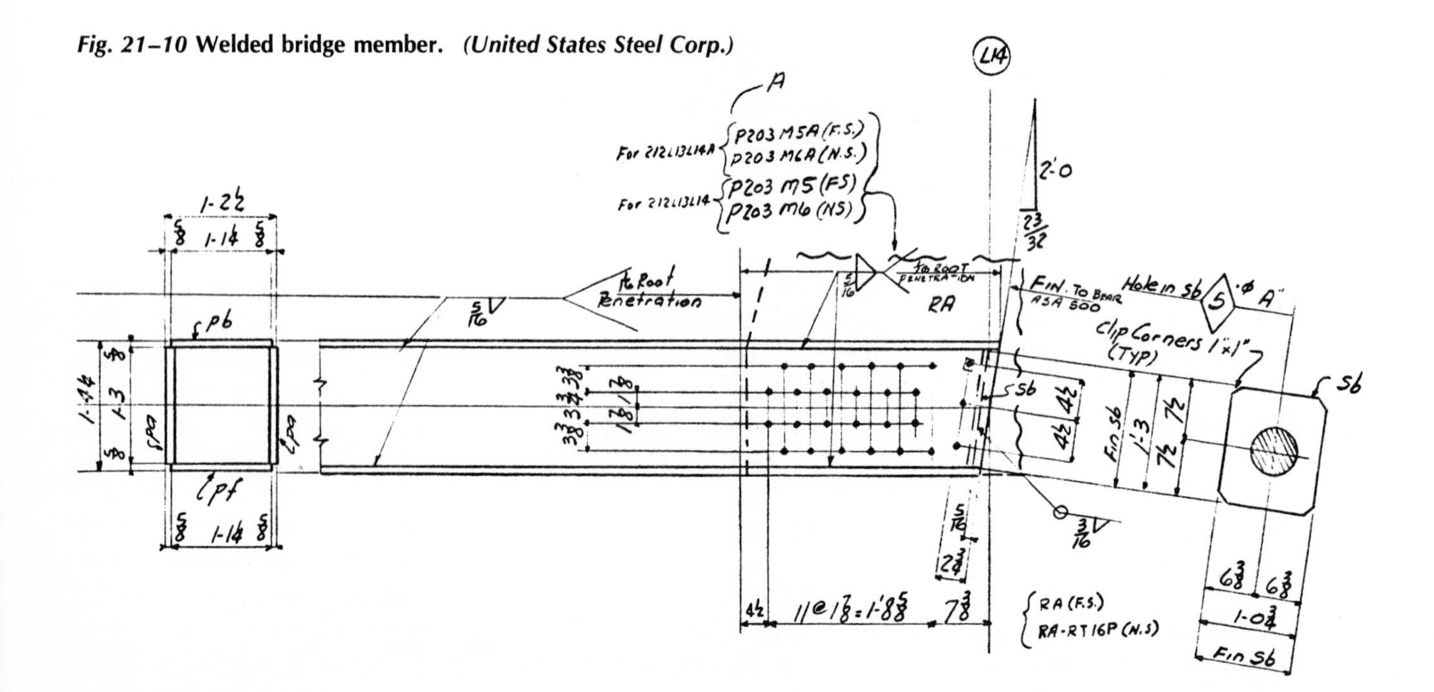

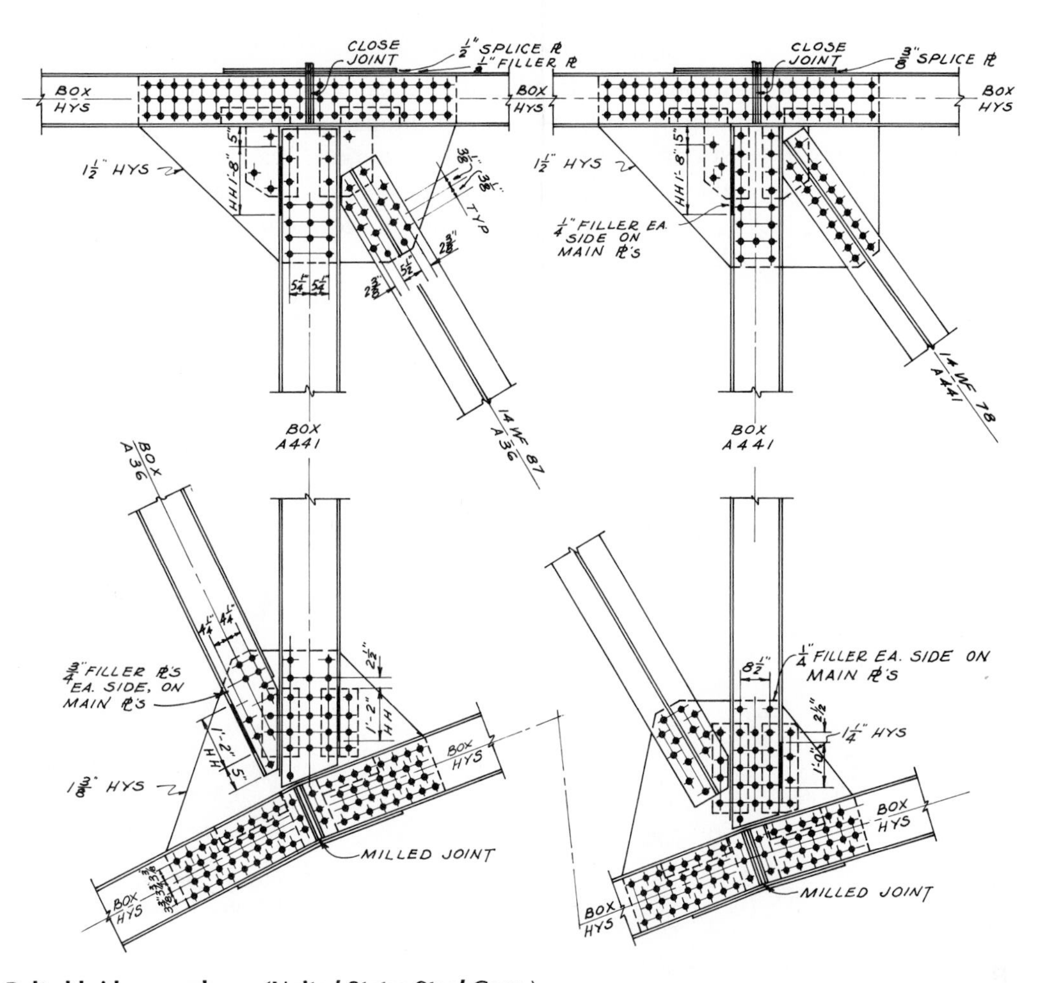

Fig. 21–11 **Bolted bridge member.** ***(United States Steel Corp.)***

Fig. 21–12 **Unistrut geometric form.** ***(Unistrut Corp.)***

Frame shown in Fig. 21–12 consists of four or five modular units in a geometric pattern. The basic unit is made of four or five parts that are bolted together. The system is used mainly in canopies and roofs.

Figure 21–13 shows how a roof system was put up after it was assembled on the ground. Figure 21–14 shows how the finished steel roof looks from the inside. Note how the geometric pattern of the exposed steel becomes a part of the overall design. This roof is in a mall in Columbia, Maryland.

STRUCTURAL DOME SYSTEMS

Figure 21–15 shows a dome over a theater in Reno, Nevada. The geometry used in this structure is called *geodesic.* It is different from other dome geometry in that its strength is in all directions (omnidirectional). The three-dimensional triangulated framing of the dome (Fig. 21–16) makes it exceptionally strong. It also uses less material than other dome frames. Geodesic geometry was invented by Dr. R. Buckminster Fuller.

STRUCTURAL STORAGE SYSTEM

Figure 21–17 shows four stages in the construction of a water tank called a water spheroid because of its shape. This is a typical example of the use of welding in steel construction. The parts are shaped in

Fig. 21–13 **The erection of the unistrut roof.** ***(Unistrut Corp.)***

Fig. 21–14 **Finished interior steel roof frame.** ***(Unistrut Corp.)***

Fig. 21–15 **Dome system for a theater.** ***(Temcor.)***

Fig. 21–16 **Diagram of the dome.** ***(Temcor.)***

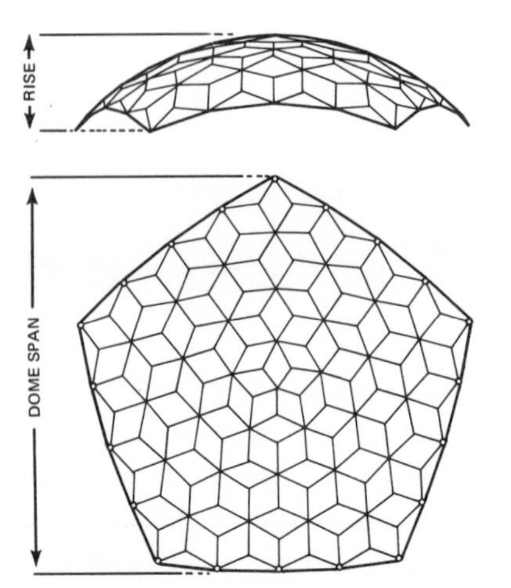

the shop, then welded together in the field. This geometric structure is a clear solution to the storage problem. Building it takes great technical ingenuity.

STRUCTURAL DRAFTING OF STEEL SHAPES

Earlier in this chapter, we learned some of the basic structural steel shapes. The American Institute of Steel Construction (AISC), 101 Park Avenue, New York, NY 10017, publishes the *Manual of Steel Construction (AISC Manual)*. The seventh edition of the *AISC Manual,* or any handbook published by a major steel company, lists all the major shapes available and the great variety of their sizes and weights. The *AISC Manual* contains tables for designing and detailing the various shapes in any combination. Figure 21–18 shows cross sections of various plain steel shapes. These shapes are grouped by the AISC as follows:

1. American Standard beams (S).
2. American Standard channels (C).
3. Miscellaneous channels (MC). These include special-purpose channels that are not standard.
4. Wide-flange shapes (W). These are used as both beams and columns.
5. Miscellaneous shapes (M). These light-weight shapes look in cross section like W shapes.
6. Structural tees (ST, WT, MT). These are made by splitting S, W, and M shapes, usually along the mid-depth of their webs.
7. Angles (L). These consist of two legs of equal or unequal widths. The legs are at right angles to each other.
8. Plates (PL) and flat bars (Bar). These are rectangular in cross section.

These plain shapes are basic to structural detailing. You must be familiar with them in order to make adequate drawings.

Fig. 21–17A* Construction of cylinder. *(Chicago Bridge & Iron Co.)

Fig. 21–17B* Beginning construction of spheroid. *(Chicago Bridge & Iron Co.)

Fig. 21–17C* The partially completed spheroid. *(Chicago Bridge & Iron Co.)

Fig. 21–17D* Spheroid water storage system completed. *(Chicago Bridge & Iron Co.)

SCALES

Structural details are prepared at a scale of 1:10 (1″ = 1′-0″) for beams up to 533 mm (21 in.) in depth. For beams of greater depth, a 1:20 (3/4″ = 1′-0″) scale is preferred. You can shorten the overall length of structural members as long as you show details adequately. You can also exaggerate very small dimensions, such as a clearance, to clarify views.

TYPICAL STEEL DETAILS ON THE DESIGN DRAWINGS

Designers always place on their design drawings all the information needed to prepare shop drawings. Figure 21–19 shows a small part of a designed floor plan. The view is from above. There are enough notes and dimensions on the plan to prepare a shop drawing of the wide-flanged beam (W).

The 20-ft dimension is presumed to be the structural bay, or distance from *A* to *B*. The structural members are at right angles to one another unless noted. The height given on the line diagram of a beam is significant. Height elevations are assumed to be level at the figure given. The figure shows two elevations, 98′-6″ and 98′-9″.

SHOP DRAWINGS

Detail drawings like the one in Fig. 21–20 are essential for making structural pieces. The figure is a

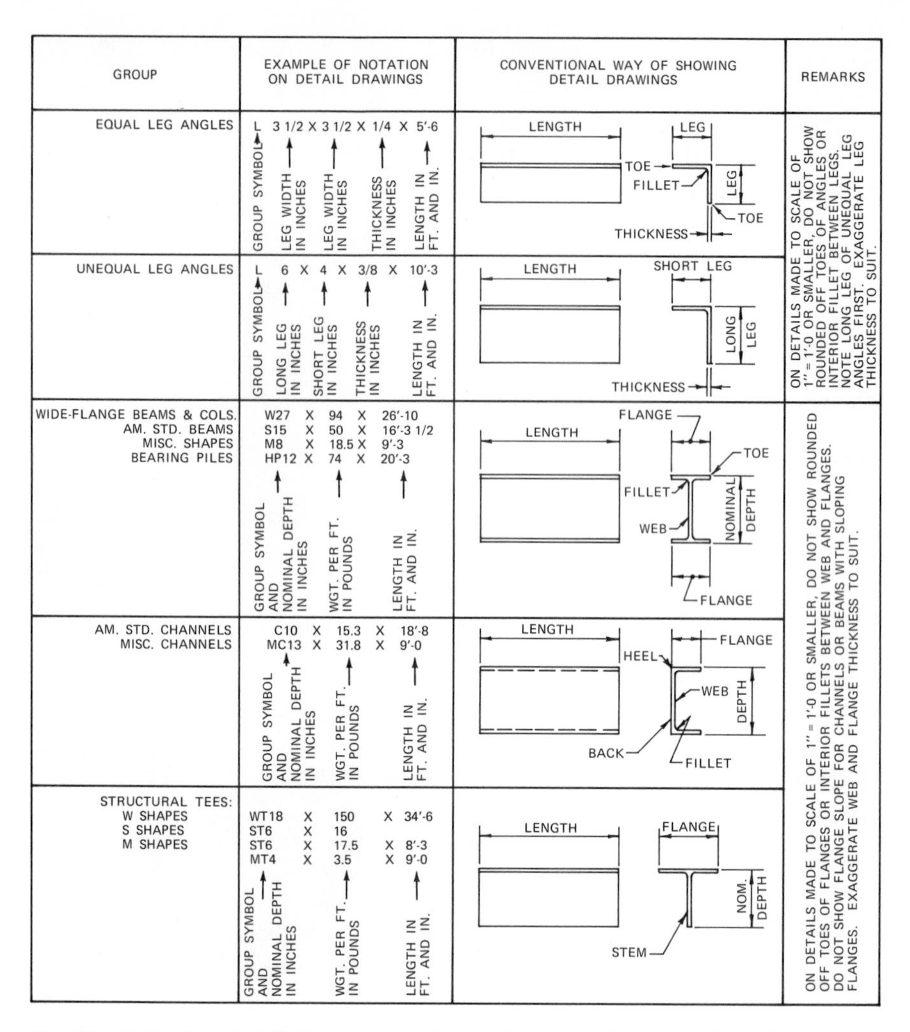

GROUP	EXAMPLE OF NOTATION ON DETAIL DRAWINGS	CONVENTIONAL WAY OF SHOWING DETAIL DRAWINGS	REMARKS
EQUAL LEG ANGLES	L 3 1/2 X 3 1/2 X 1/4 X 5′-6 GROUP SYMBOL — L; LEG WIDTH IN INCHES; LEG WIDTH IN INCHES; THICKNESS IN INCHES; LENGTH IN FT. AND IN.	LENGTH; LEG; TOE; FILLET; LEG; TOE; THICKNESS	ON DETAILS MADE TO SCALE OF 1″ = 1′-0 OR SMALLER, DO NOT SHOW ROUNDED OFF TOES OF ANGLES OR INTERIOR FILLET BETWEEN LEGS. NOTE LONG LEG OF UNEQUAL LEG ANGLES FIRST. EXAGGERATE LEG THICKNESS TO SUIT.
UNEQUAL LEG ANGLES	L 6 X 4 X 3/8 X 10′-3 GROUP SYMBOL — L; LONG LEG IN INCHES; SHORT LEG IN INCHES; THICKNESS IN INCHES; LENGTH IN FT. AND IN.	LENGTH; SHORT LEG; LONG LEG; THICKNESS	
WIDE-FLANGE BEAMS & COLS. AM. STD. BEAMS MISC. SHAPES BEARING PILES	W27 X 94 X 26′-10 S15 X 50 X 16′-3 1/2 M8 X 18.5 X 9′-3 HP12 X 74 X 20′-3 GROUP SYMBOL AND NOMINAL DEPTH IN INCHES; WGT. PER FT. IN POUNDS; LENGTH IN FT. AND IN.	LENGTH; FLANGE; TOE; FILLET; WEB; NOMINAL DEPTH; FLANGE	ON DETAILS MADE TO SCALE OF 1″ = 1′-0 OR SMALLER, DO NOT SHOW ROUNDED OFF TOES OF FLANGES OR INTERIOR FILLETS BETWEEN WEB AND FLANGES. DO NOT SHOW FLANGE SLOPE FOR CHANNELS OR BEAMS WITH SLOPING FLANGES. EXAGGERATE WEB AND FLANGE THICKNESS TO SUIT.
AM. STD. CHANNELS MISC. CHANNELS	C10 X 15.3 X 18′-8 MC13 X 31.8 X 9′-0 GROUP SYMBOL AND NOMINAL DEPTH IN INCHES; WGT. PER FT. IN POUNDS; LENGTH IN FT. AND IN.	LENGTH; HEEL; FLANGE; WEB; DEPTH; BACK; FILLET	
STRUCTURAL TEES: W SHAPES S SHAPES M SHAPES	WT18 X 150 X 34′-6 ST6 X 16 ST6 X 17.5 X 8′-3 MT4 X 3.5 X 9′-0 GROUP SYMBOL AND NOMINAL DEPTH IN INCHES; WGT. PER FT. IN POUNDS; LENGTH IN FT. AND IN.	LENGTH; FLANGE; NOM. DEPTH; STEM	

Fig. 21–18 **Steel material shapes in section.** ***(American Institute of Steel Construction.)***

Fig. 21–19 **Small part of designed floor plan.** ***(Adapted from* AISC Handbook, *with permission of the publisher.)***

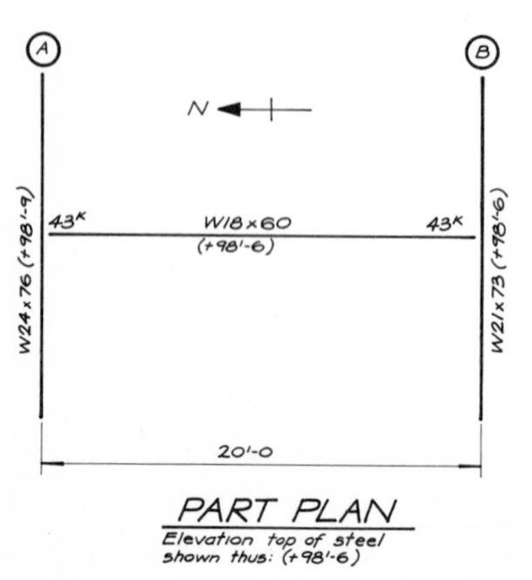

Fig. 21–20 **Typical beam detail.** ***(Adapted from* AISC Handbook, *with permission of the publisher.)***

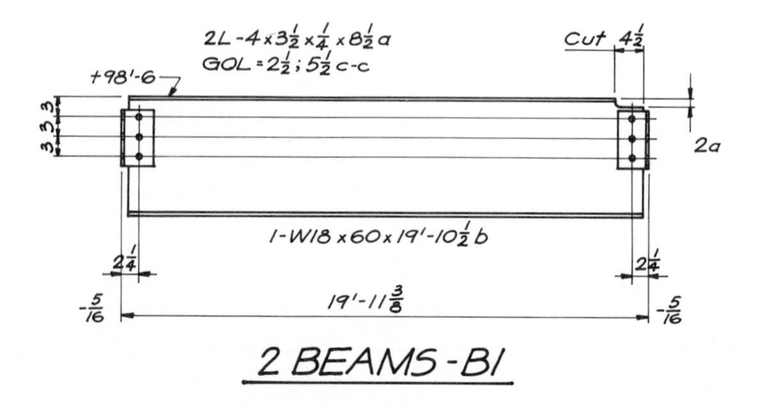

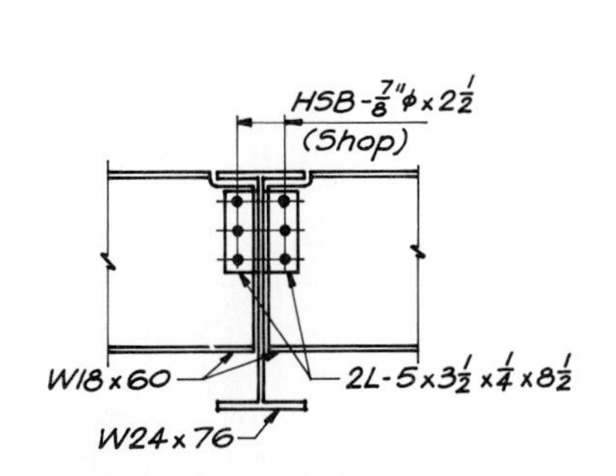

Fig. 21–21 **Typical connection of mating parts.** *(Adapted from* **AISC Handbook,** *with the permission of the publisher.)*

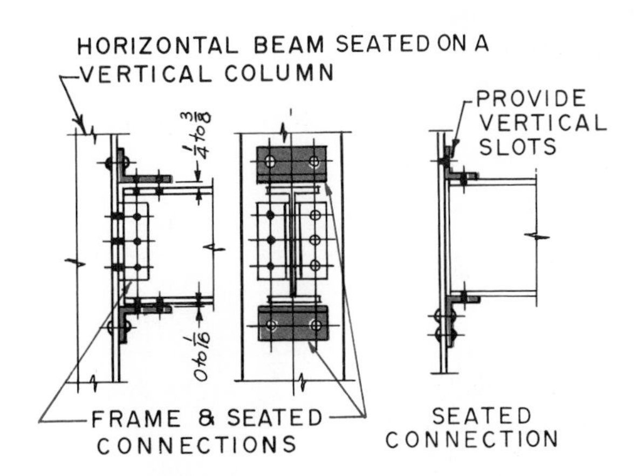

Fig. 21–22 **Frame and seated connections.** *(Adapted from* **AISC Handbook,** *with permission of the publisher.)*

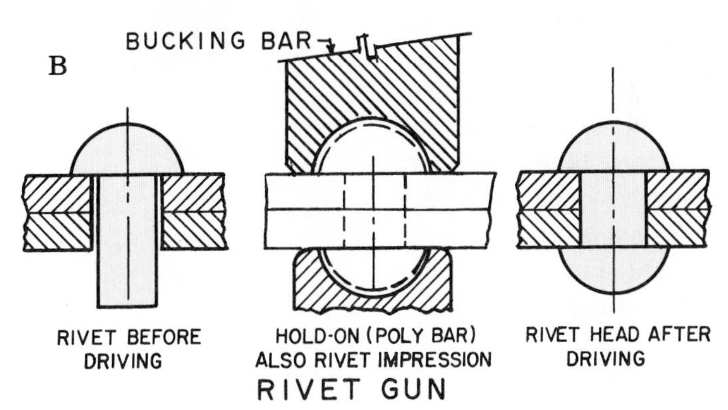

Fig. 21–23 **Rivet symbols and buttonhead rivet.**

A

SHOP RIVETS													FIELD RIVETS			
	COUNTERSUNK AND CHIPPED			COUNTERSUNK NOT OVER 1/8" HIGH			FLATTENED TO 1/4" 1/2" AND 5/8" RIVETS			FLATTENED TO 3/8" 3/4" RIVETS AND OVER				COUNTERSUNK HEADS		
TWO FULL HEADS	NEAR SIDE	FAR SIDE	BOTH SIDES	NEAR SIDE	FAR SIDE	BOTH SIDES	NEAR SIDE	FAR SIDE	BOTH SIDES	NEAR SIDE	FAR SIDE	BOTH SIDES	TWO FULL HEADS	NEAR SIDE	FAR SIDE	BOTH SIDES

detail of a beam. This kind of drawing seldom describes the connections of mating parts (Fig. 21–21). Instead, it just shows those features (for example, connection angles) that will be involved when the piece is used in building.

In preparing structural details, the drafter refers to the handbook and the dimensions for detailing. Figure 21–22 shows both framed and seated connections.

RIVETING

One way to join beams is with rivets. The standard symbols for rivets are seen in Fig. 21–23 at A. A typical buttonhead rivet is shown at B. If the riveting is to be done in the shop, the drafter will use shop rivet symbols. These are open circles the diameter of the rivet head. If the riveting is to be done in the field (on the construction site), the drafter will use field rivet symbols. These are blacked-in circles the diameter of the rivet hole. Lines on which rivets are spaced are called *gage lines.* The distance between rivet centers on the gage lines is called the *pitch.*

STRUCTURAL BOLTING

High-strength-steel bolts are rated by the American Society for Testing and Materials (ASTM). The bolt can be applied in the field or in the shop. The hole into which it fits is normally 2 mm (1/16 in.) larger than the bolt. Figure 21–24 shows two kinds of bolted connections: frame and seated. The bolt transmits the force of the beam load to the column. The stess this creates in the bolt is *shear* (cutting). The stress created in the column is *compressive* (pushing together).

Fig. 21–24 **Structural bolt and bolted connections.**

WELDING STRUCTURAL MEMBERS

Structural steel members are usually welded with the metal-arc process. The fillet weld is the most common on structural connectors. (See Chapter 17 for a review of the standard welding symbols.)

DIMENSIONING

In structural drawings, dimensions are given primarily to working points. For beams, give dimensions to the centerline. For angles and, normally, channels, give dimensions to the backs. Give vertical dimensions on beams and channels to the tops or bottoms. You do not generally dimension the edges of flanges and the toes of angles. Make your dimension lines continuous and unbroken. Place the dimensions above the dimension lines.

When dimensions are in feet and inches, use the foot symbol but not the inch symbol.

Structural Drawings

Figure 21–25 is a structural drawing of a small steel roof truss. This symmetrical piece is detailed about the left of a centerline. Study the drawing closely. You will see that each member is completely dimensioned or described. You will see, too, that the dimensions adequately relate the fixed location of each structural member.

Roof Trusses

Figure 21–26 shows diagrams and names for some roof trusses and bridge trusses. The ones shown are only a few of those available. Each type can also be modified to carry specific loads.

WOOD FOR CONSTRUCTION

Wood is commonly used for the frames of homes and other small structures. Details for wood construction have been drawn up by the National Forest Products Association. They are now used as a standard method of construction.

Structural timber can be manufactured in many forms by *laminating* (cutting wood into thin slabs and gluing them together). Builders can buy "factory grown" timbers in any size or shape (Fig. 21–27). Some of the forms available include tudor arches, radial arches, parabolic arches, A-frames, and tapered beams (Fig. 21–28). The American Institute of Timber Construction (AITC) has set up guidelines for makers of structural glued laminated timber.

Figure 21–29 shows some of the construction details that must be used with structural timber. These

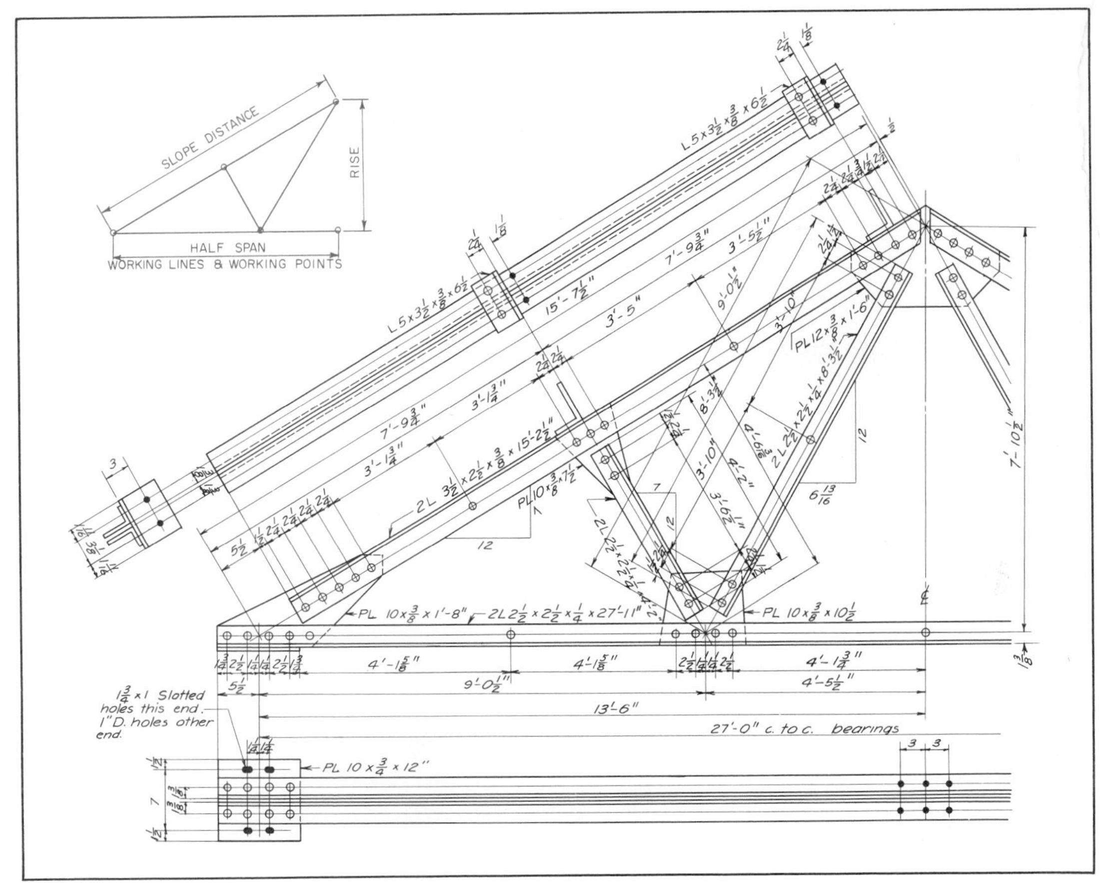

Fig. 21–25 **Roof truss detail.**

Fig. 21–26 **Roof and bridge truss diagrams.**

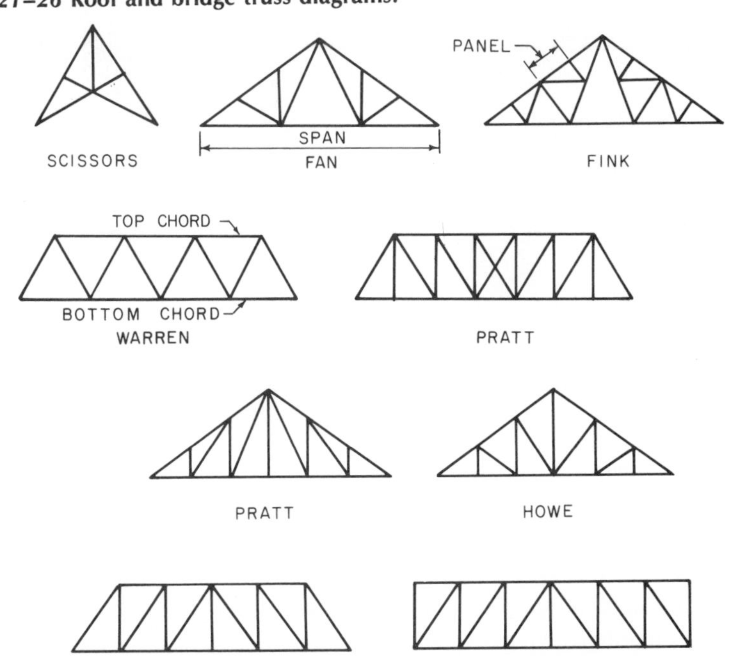

Fig. 21–27 **Laminated wood forms take any size or shape.** ***(American Institute of Timber Construction.)***

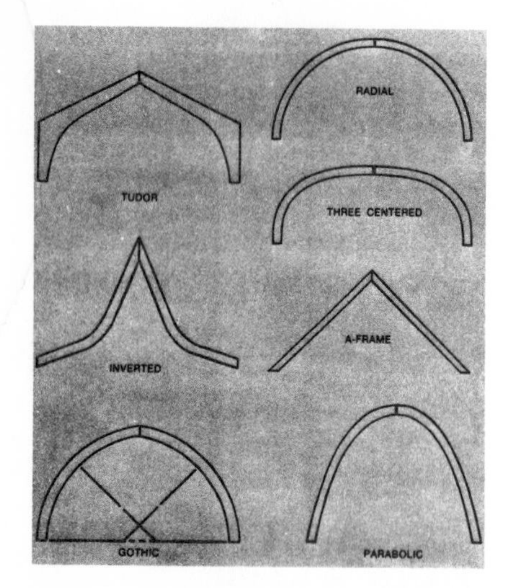

Fig. 21–28 **Typical shapes of laminated wood forms.** ***(American Institute of Timber Construction.)***

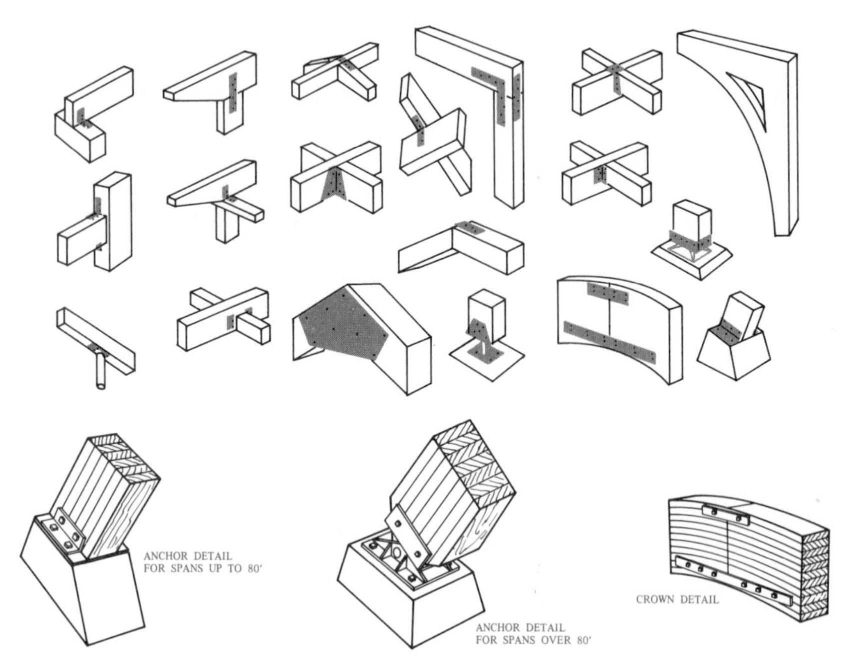

Fig. 21–29 **Construction details for timber construction.** ***(American Institute of Timber Construction.)***

detail drawings show how timbers are joined together and how structural members are anchored to foundations.

Systems for Timber Construction

Structural timber is graded by how much stress it can bear and by the connections through which stress can be transferred to other members. Wood commonly used for building includes pine, ash, birch, cedar, cypress, elm, oak, and redwood. Design handbooks list the specific grades and related structural properties.

Figure 21–30 shows the use of laminated wood forms in construction. Notice the wrappings used for protection. In Fig. 21–31, a worker lays out templates for glued laminated wood arches.

Fig. 21–30 **Erection of laminated wood forms.** ***(American Institute of Timber Construction.)***

Fig. 21–31 **Preparing templates for glued laminated wood arches.** ***(American Institute of Timber Construction.)***

CONCRETE SYSTEMS

Many of our buildings, bridges, and dams have only been made possible by concrete. The American Concrete Institute has prepared a manual of standard practice for concrete structures.

Concrete has only limited strength unless it is specially prepared. It is made from a mixture of gravel, sand, water, and portland cement. Various grades are produced, depending on the proportions of these ingredients. The concrete can also be reinforced or prestressed.

Reinforced concrete has steel bars embedded in it. These bars are arranged to bear the structural loads that the concrete could not support by itself. When concrete and steel are combined, they can be used in the monolithic form shown in Fig. 21–32. Figure 21–33 shows a typical reinforced concrete detail.

In *prestressed* concrete, the reinforcing bars are stretched before the concrete is poured over them. The prestressed form will then accept a predesigned load. This combination of concrete and stretched steel is stronger than either plain concrete or reinforced concrete.

Detailed Drawings

Concrete forms designed by a structural engineer are drawn for the manufacturer's use only. Construction drawings of these forms are prepared by the manufacturer. These drawings are made to show the contractor the location, placement, and connections. See Fig. 21–34 for typical details.

Fig. 21–32 **Monolithic concrete form.** ***(Ceco Steel Products Corp.)***

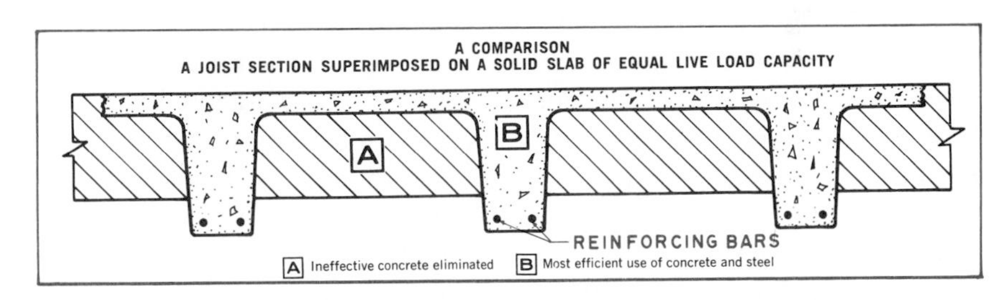

Fig. 21–33 **Reinforced concrete detail.** ***(Ceco Steel Products Corp.)***

Fig. 21–34 **Concrete placement drawing.** ***(Ceco Steel Products Corp.)***

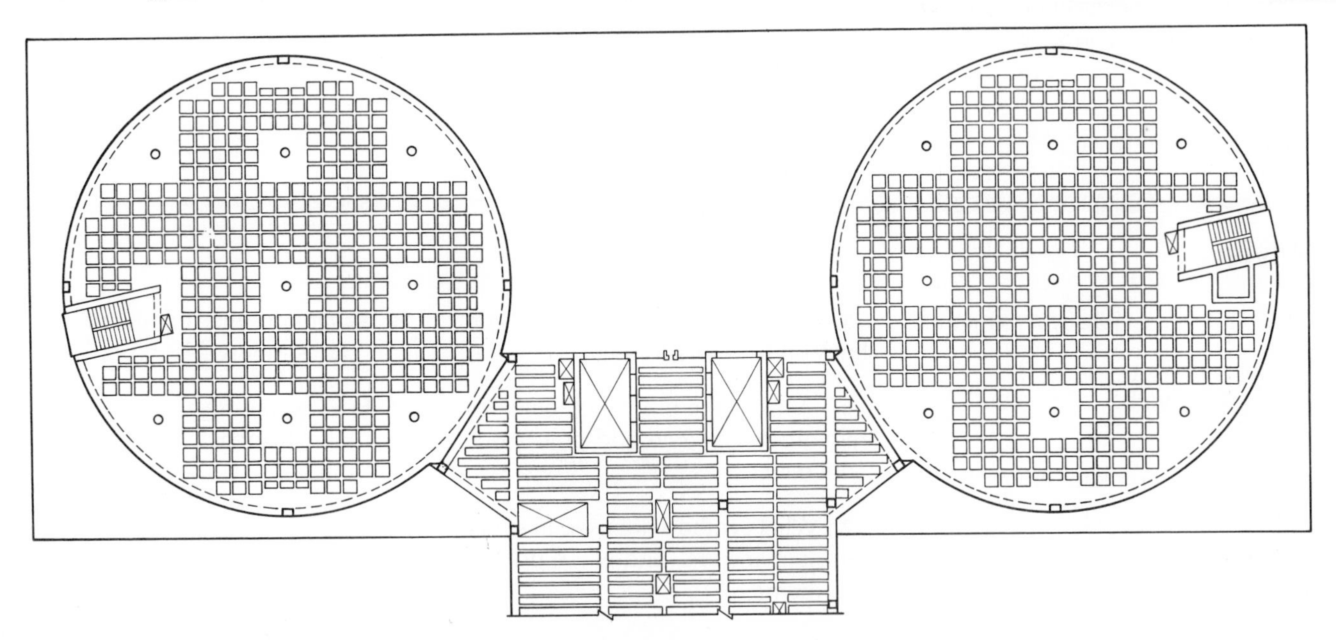

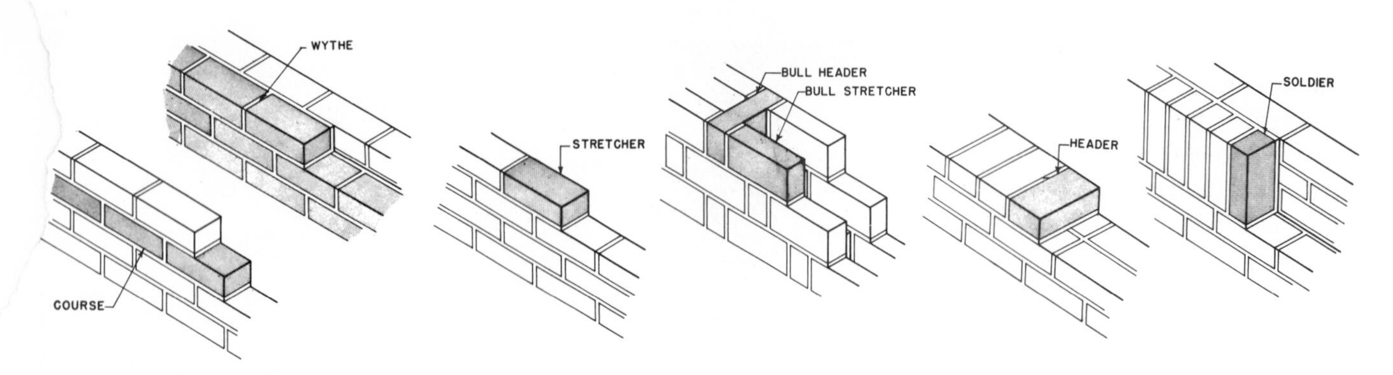

Fig. 21–35 **Bonds and structural patterns of common brick.** ***(Structural Clay Products Institute.)***

STRUCTURAL CLAY SYSTEMS

The solid brick wall is the oldest type of masonry construction known. Bricks are made from different types of clay in many shapes, forms, and colors. The common brick size is 57 mm × 95 mm × 203 mm (2¼″ × 3¾″ × 8″). Brick walls are made to support floors and roofs. For the vertical walls to be able to carry the horizontal floors and roofs, the bricks must have high compressive strength.

Structural Bonding

Bricks in construction are arranged in overlapping and interlocking patterns and fastened together with connecting mortar joints. This produces a structural assembly that acts as a single structural unit. Figure 21–35 shows some of the common bonds and structural patterns used with bricks. Note the terms applied to the brick.

How strong a structural clay system is generally depends on the strength of the mortar joints. When the design limits this strength, the designer can call for *reinforced masonry.* This is masonry with steel rods or wire embedded in the mortar.

Brick or concrete masonry can also be used to enclose a structural steel framework. This provides enough fireproofing to meet standard building codes.

Natural stone and manufactured stone are also used for masonry. However, these materials are rarely used today for structural purposes. Manufactured stone, along with marble, granite, limestone, and sandstone, is used for ornamental facing on buildings. This use of stone is known as curtain-wall construction. Curtain-wall construction is used on many skyscrapers built today.

Review

1. Name five jobs for structural drafters.

2. Name five basic structural steel shapes.

3. What is the major advantage of a geodesic dome?

4. How do you dimension beams, angles, and channels?

5. How do you draw dimension lines on structural drawings?

6. What is the difference between reinforced concrete and prestressed concrete?

7. What symbols are used in structural dimensions, and when do you use them?

8. What does the strength of structural clay systems depend on?

9. Review the details of the 100-story John Hancock Center in Chicago and translate the data to metric dimensions.

10. What are the metric dimensions of a common brick?

Problems

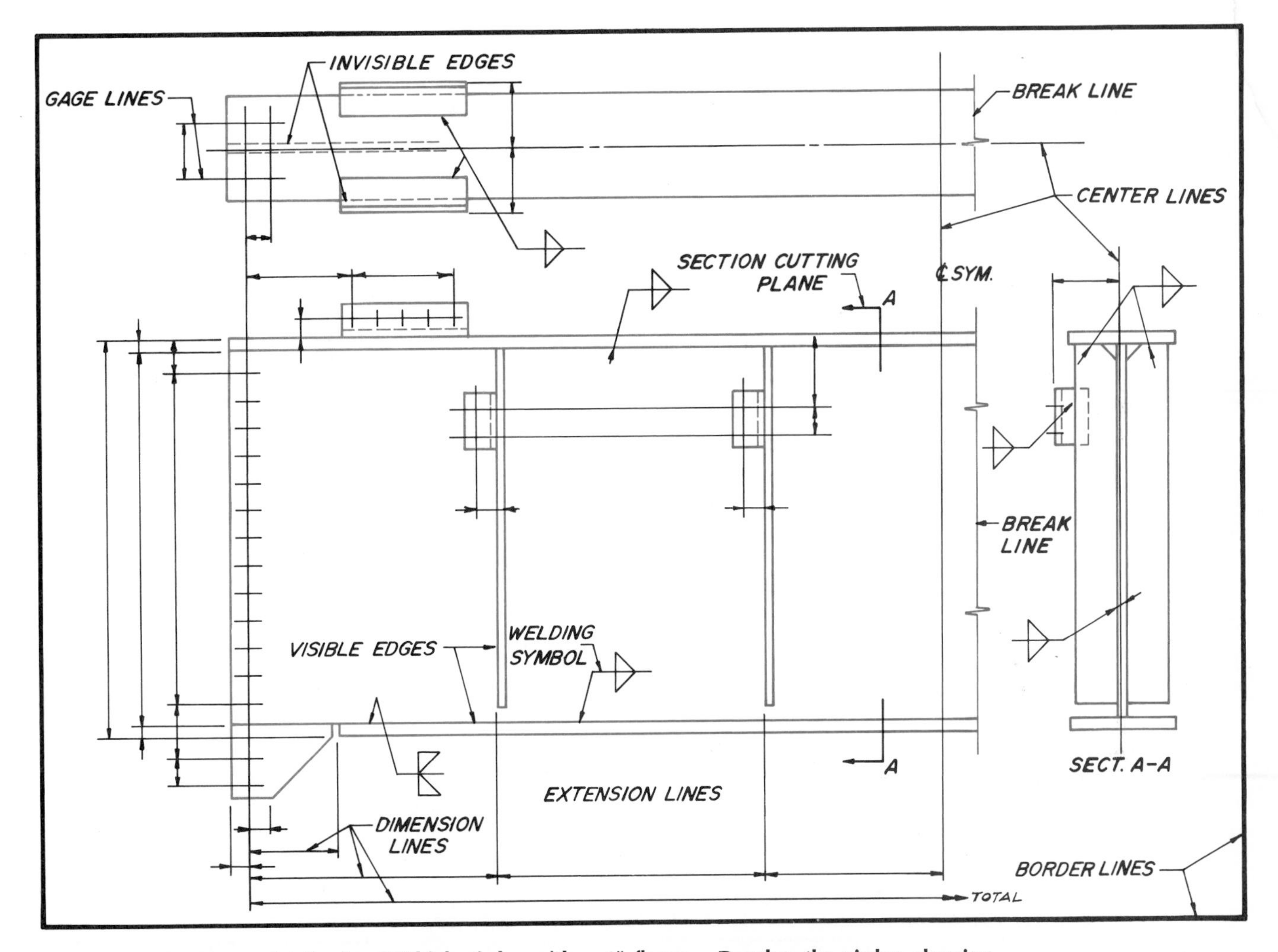

Fig. 21-36 Prepare a detail of a 22″-high girder with a 6″ flange. Develop the girder, showing conventional lines and their relative weights. Identify angles, welds, stiffeners, and field bolts. Fill in missing dimensions with appropriate selection. Scale: 1½″ = 1′-0″.

Fig. 21-37 Draw the detail of a framed connection on an S-beam that is 8″ by 18.4 lbs. The flange width is 4″ and the web is .270″. The 3″ by 3″ connection angle is 6″ long. Pitch = 1½″. Gage = 1¾″. Rivet = ½″. Prepare details at half scale.

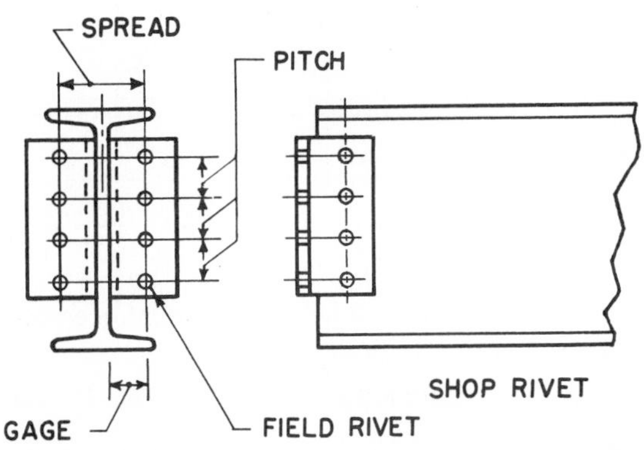

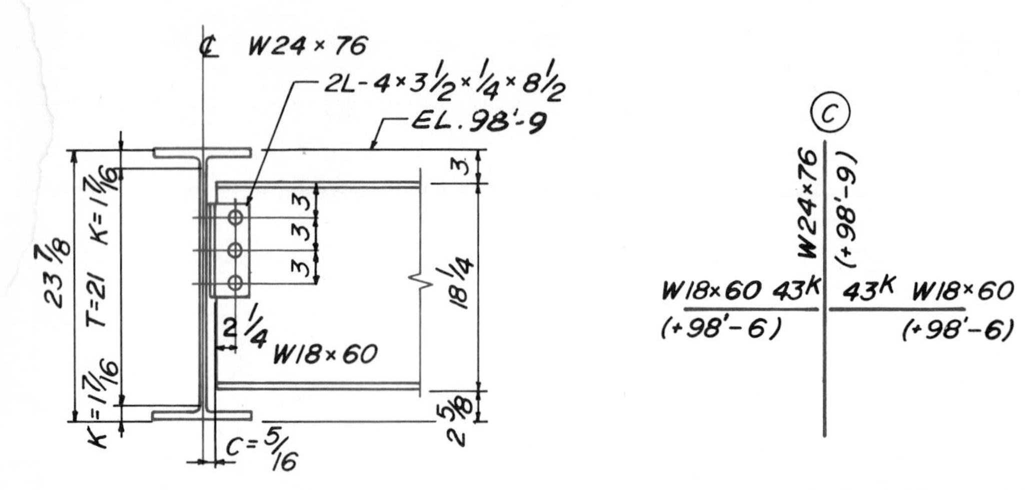

Fig. 21–38 **Prepare a partial detail of the framed connection of the 18″-wide flanged beam with a 24″-wide flanged beam. Scale: 3″ = 1′-0″. Prepare a simple erection diagram, no scale assigned.**

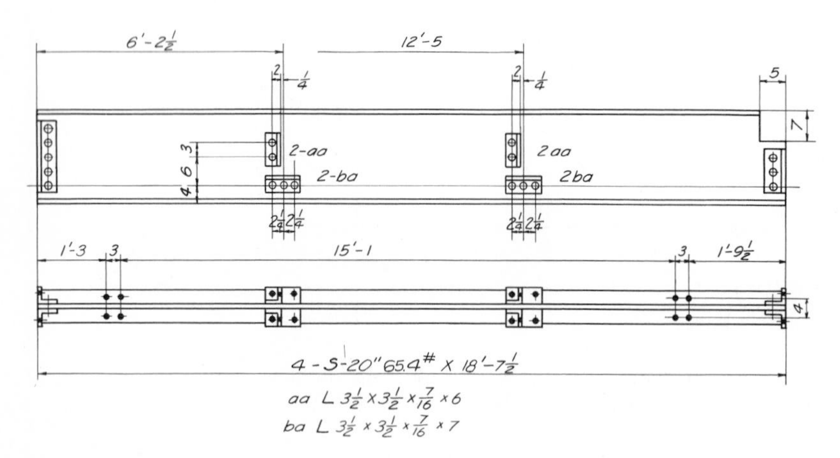

Fig. 21–39 **Prepare a detail drawing of a standard S-beam, 20″ by 65.4 lbs. Flange = 6¼″ wide with web thickness of ½″. Scale: ¾″ = 1′-0″.**

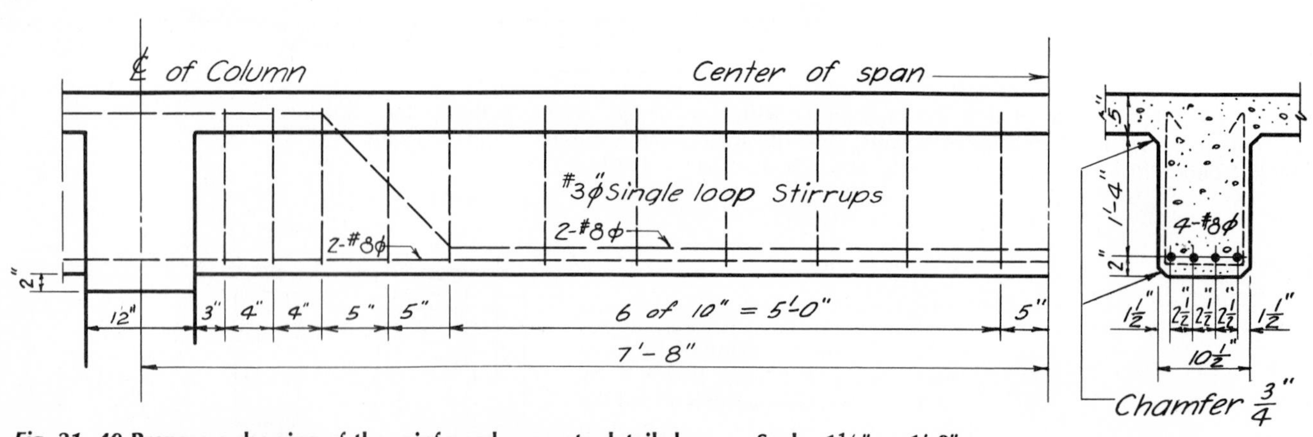

Fig. 21–40 **Prepare a drawing of the reinforced concrete detail shown. Scale: 1½″ = 1′-0″.**

Fig. 21–41 **Prepare a truss detail of Fig. 21–25 at a scale of 1:12 (1″ = 1′).**

Chapter 22

Map Drafting

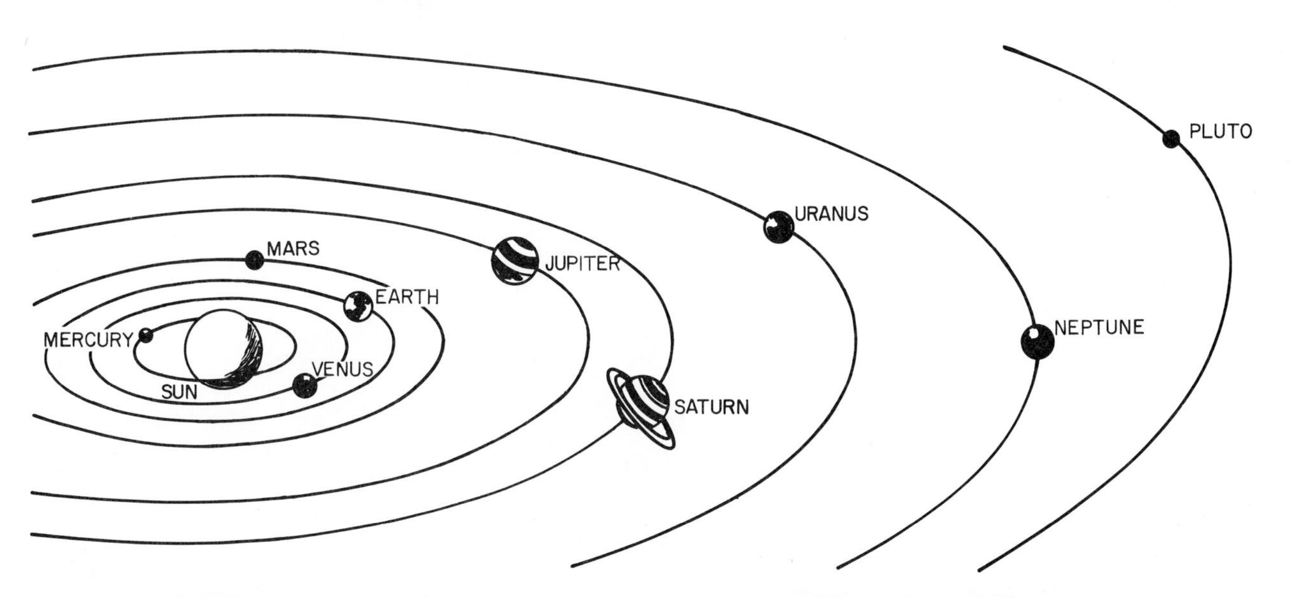

Fig. 22–1 **The solar system. The Sun is a star located in the center of the system, with nine planets revolving around it in noncircular orbits. All the planets move in the same direction around the Sun and are located in nearly the same plane. The Earth is 145 million kilometers (90 million miles) from the Sun. The planet Pluto is 5.6 billion kilometers (3.5 billion miles) from the Sun.**

MAPPING—A CHANGING INDUSTRY

Mapmakers are people who have been trained to gather information and prepare maps. Mapmakers are also called *cartographers.* The skilled map drafter prepares maps in detail for the civil engineer, scientist, geographer, and geologist.

Map making is a pictorial method of representing facts about the surface of the earth or other bodies in the solar system (Fig. 22–1). Rockets fired from Cape Canaveral (Fig. 22–2) to the moon or to Mars, as in the Mariner program, carry the necessary equipment for mapping the moon and the planets. A method called *photogrammetry* (aerial photography) is used in today's modern mapping industry. It is this method or technique that has been used in mapping the moon and the planets of our solar system.

Maps were used when streams and rivers were the earliest sources of transportation in the field of

Fig. 22–2 **Missiles can make surveys.** ***(Official U.S. Air Force photo.)***

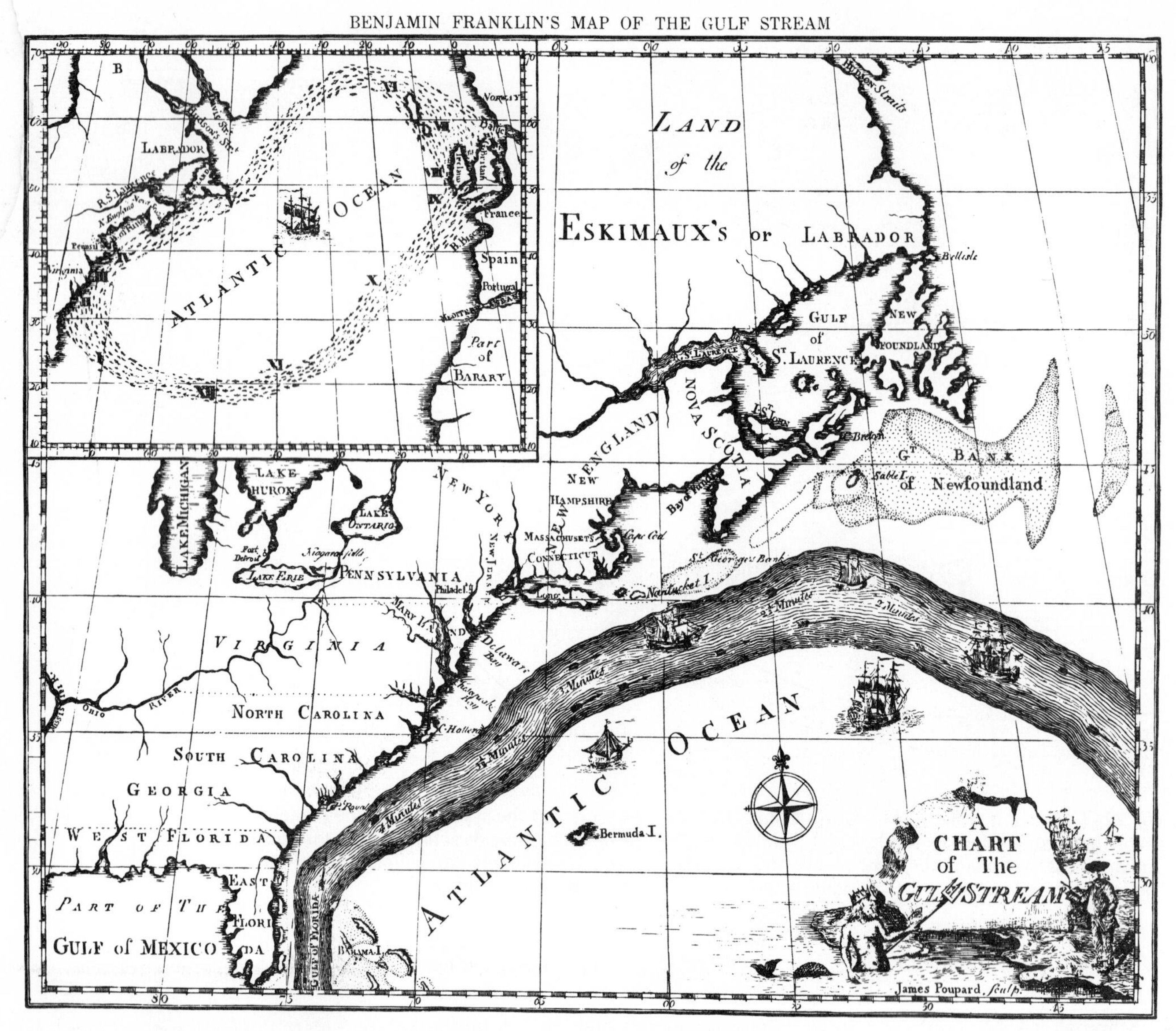

Fig. 22–3 **Benjamin Franklin's map of the Gulf Stream.** ***(U.S. Coast and Geodetic Survey.)***

commerce. However, many early maps were artistic representations (Fig. 22–3). They lacked the accuracy found in present-day maps.

CAREER OPPORTUNITIES

Many job areas are available for those who are able to prepare maps. The field of civil engineering is always expanding. The planning of railroads, highways, harbor facilities, airports, and space stations are just a few areas in this broad industry where map planning is being used.

The drafter may prepare maps and charts under the direction of the design engineer or cartographer. There may be opportunities for advancement in job areas such as photogrammetry, surveying with laser beams, or research and development projects with the geographer. Additional information about careers in civil engineering and map making is available from:

Association of American Geographers
1146 16th Street N.W.
Washington, DC 20036

American Society of Certified Engineering Technicians
2029 K Street N.W.
Washington, DC 20036

American Geographical Union
1145 19th Street N.W.
Washington, DC 20036

MAP SIZES

As the world has entered the space age, there are maps that show enormous distances. The scale (size) is quite small on maps that cover large distances. Some maps that indicate ownership of property, such as city plats, must be very accurate. They may be drawn to a large scale in order to note all the information of physical property. Maps showing the geography of states or countries, which show boundary lines, streams, lakes, or coastlines, may use a scale of several miles to the inch.

SCALES USED ON MAPS

The civil engineer's scale is used for map drawings. Distances are given in decimals of a foot or meter, such as tenths, hundredths, and so forth. (See the maps in your geography and history books.) Practically all the countries of the world use the metric system. In this system, distances are commonly measured in kilometers instead of miles.

The earth's spherical measurements provided the basis for the metric scale. The *meter,* the basic unit of measure, is one ten-millionth of the distance between the equator and the poles. The meter was adopted by the French in 1791 as their official measure. The rest of the metric system is based on multiples of ten. A *kilometer,* for example, is 1000 meters.

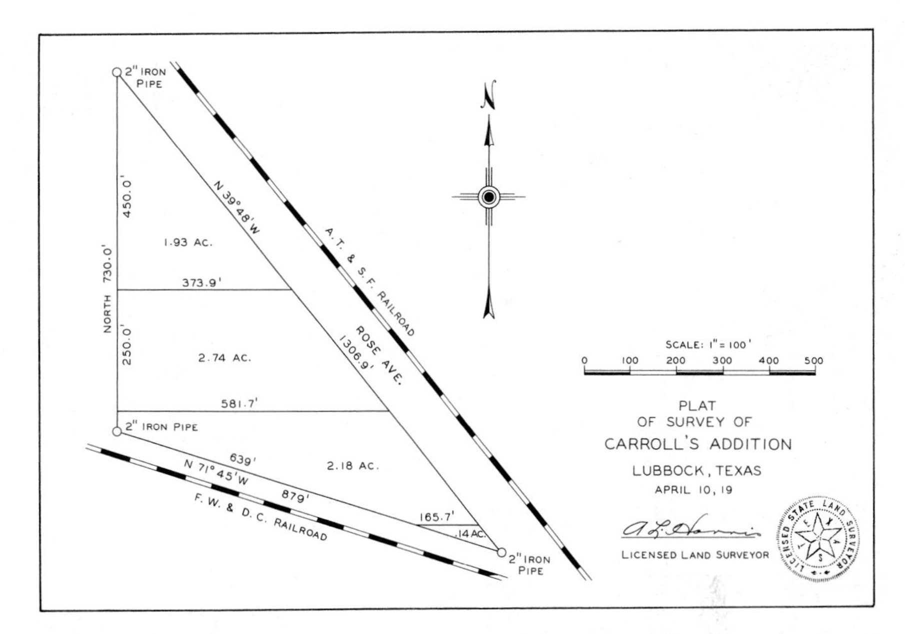

Fig. 22–4 **Plat of a survey. Note parts that make up the plat: the acreage of each part, the iron pipe locating the corners of the graphics scale, the signature of the surveyor, and the official seal.**

The scale of a map is generally noted as 1 in. equals 500 ft, or 1 part equals 6000 parts, noted as 1:6000. The scale of 1 in. equals 1 mile can also be shown as 1:63 360. Graphic scales should be shown on maps as part of their basic information.

PLATS OF A SURVEY

A map used to show the boundaries of a piece of land and to identify it is called a *plat.* The amount and kind of information presented depend upon the purpose of the map. The plat of a plane survey that was made to accompany the legal description of a property is shown in Fig. 22–4. Accuracy of information on a plat is all-important; it must agree with the legal description.

CITY PLAT

Maps of cities are made for the following reasons: to keep a record of street improvements, to show the location of utilities, and to record sizes and location of property for tax purposes. A part of such a city plat is shown in Fig. 22–5. Notice the numbering of the lots and the location of streets, sidewalks, and other details found in a city.

OPERATIONS MAPS

Operations maps are maps that show the relationship between the land's physical features and the operation that is to be performed. Such a map is found in Fig. 22–6. Operations maps can greatly help engineering, management, or government groups in the presentation of a project. A presentation well done will aid in the selling of a program.

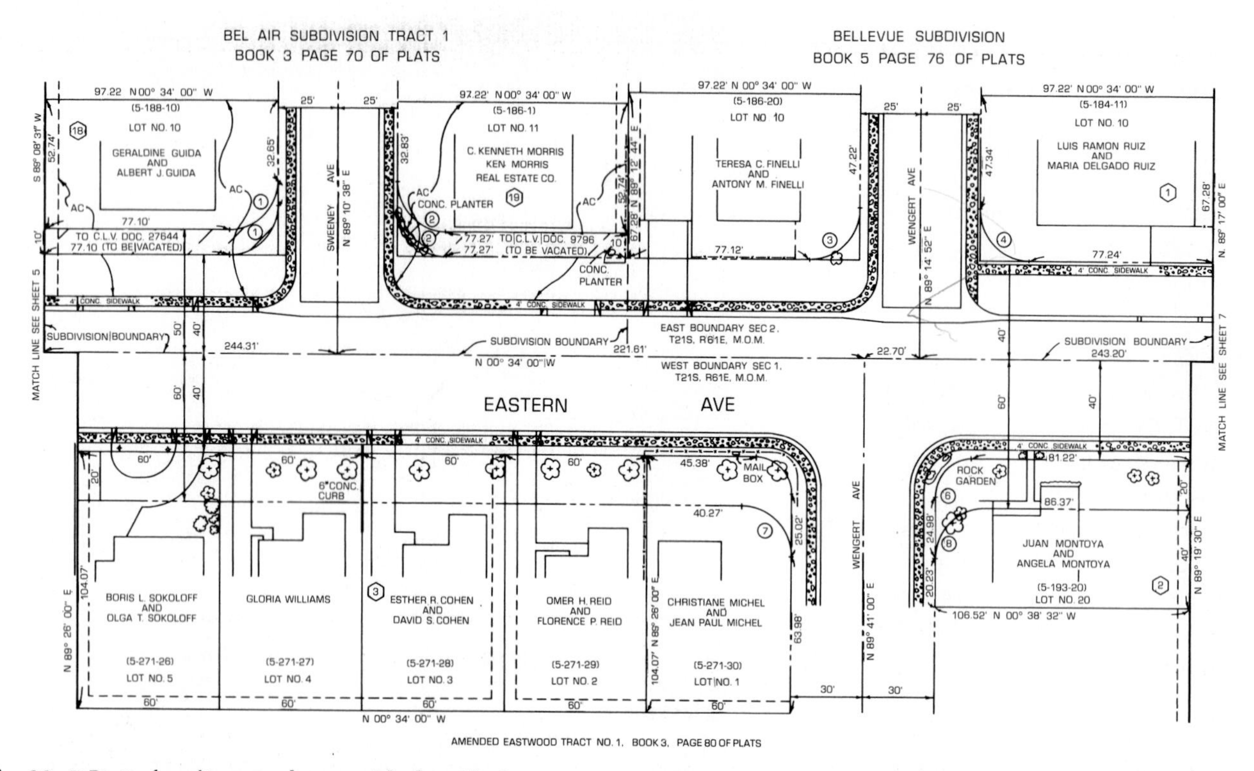

Fig. 22–5 **Part of a city map drawn with the aid of a computer.** ***(City of Las Vegas Department of Public Works, and Calcomp.)***

Fig. 22–6 **An oil-field operations map. Note scale of kilometers. One kilometer = 0.621 statute mile.**

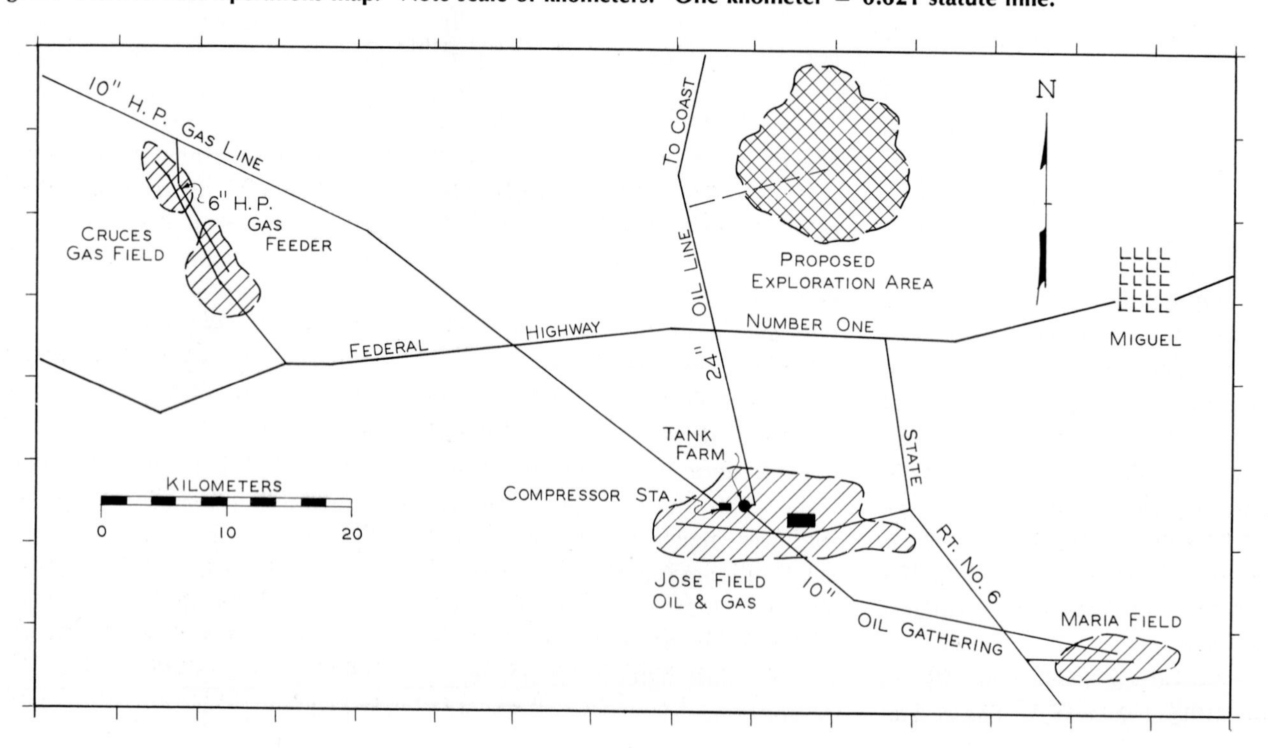

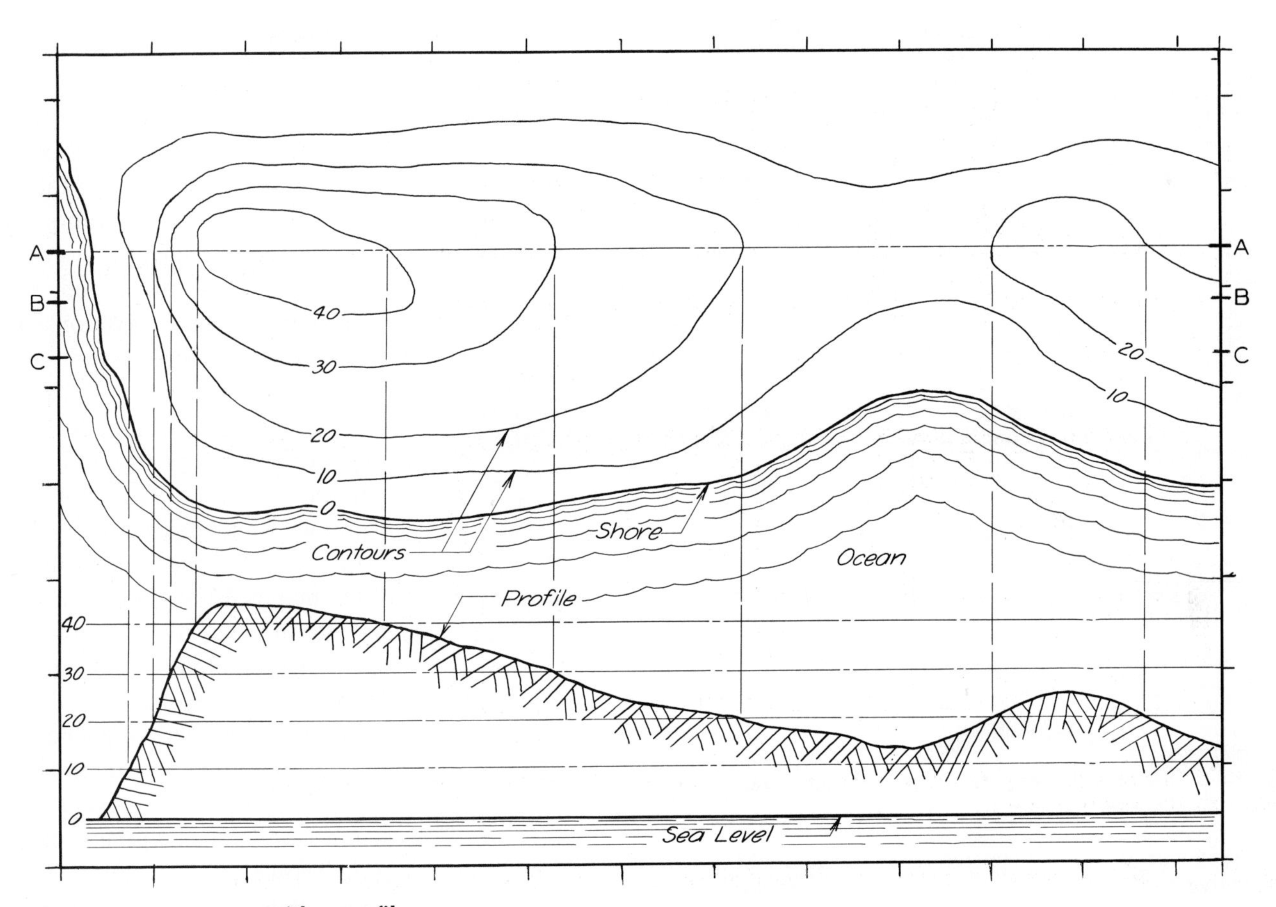

Fig. 22–7 **A contour map with a profile.**

CONTOURS

Since maps are one-view drawings, vertical distances and differences in ground level do not show. They can, however, be shown by lines of constant level called *contours*. This is illustrated in Fig. 22–7. The contour lines indicate the height of the ground above sea level. Contour lines that are close together indicate a steeper slope than lines that are far apart. This can be seen by projecting the intersections of the horizontal level lines with the profile section as shown in Fig. 22–7.

Note that the contour map and the profile correspond to the plan and section of an ordinary drawing. Note the horizontal line, *A-A*, or cutting plane, on the contour map. This line shows the position, or line, on which the profile is taken. You can see how the profile would change if the cutting plane were moved toward the ocean or to some other new position. The distance between contour lines will change according to needs and the scale of maps. A 10-ft distance may be quite satisfactory, while a 5-ft distance may be used on maps that require a larger scale. For close detail work, such as an irrigation project, the contour distances may be reduced to 0.5 ft, 1 ft, or 2 ft. On small-scale maps with a high degree of relief, distances may be increased from 20 to 200 ft or more. As an aid to reading the map, every fifth contour is usually emphasized by drawing a much heavier line (refer to Fig. 22–11).

A technical pen is satisfactory for inking contour lines. It is replacing the contour pen (Fig. 22–8). The contour pen has blades that swivel so that contour lines can be easily followed.

There are many forms of flexible curves and splines that can be bent to match curves to be drawn. The rubber-covered curve in Fig. 22–9 has a core made of a strip of lead and two strips of steel. The spline in Fig. 22–10 is a metal strip that can be bent to fit a desired curve. It is held in place by "ducks" (metal weights), as shown.

A contour map is shown in Fig. 22–11. It uses a *contour distance* (vertical distance between contour lines) of 20 ft. The elevation in feet is marked in a break in each contour line. Notice that the drainage is shown as intermittent streams.

Before a contour map can be drawn, elevations must be obtained in the field for several key points that control the drawing of the contours. The following methods are used: a grid system where all intersection elevations are obtained, along with important elevations on grid lines; points located by transit (Fig. 22–12) and stadia rod, with the corresponding elevations figured by plane table; and aerial photographic surveys (Fig. 22–13). The finished map in Fig. 22–13 was produced by photogrammetric methods. The equipment for making photogrammetric maps is shown in Fig. 22–14. The actual drawing of the map is usually done by scribing lines in the coating on a piece of film. An example of scribing on film is shown in Fig. 22–15.

Experience in surveying is needed in all of these mapping methods.

Topographic maps present complete pictorial descriptions of the areas shown (Fig. 22–16). These maps show such information as boundaries, natural features, struc-

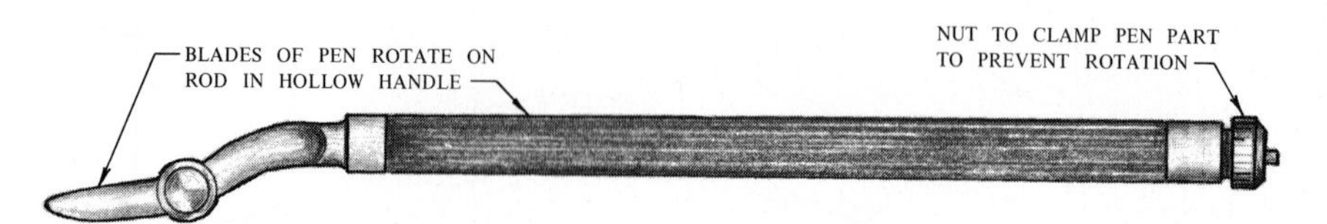

Fig. 22–8 **A contour pen.**

Fig. 22–9 **Part of a rubber-covered flexible curve.**

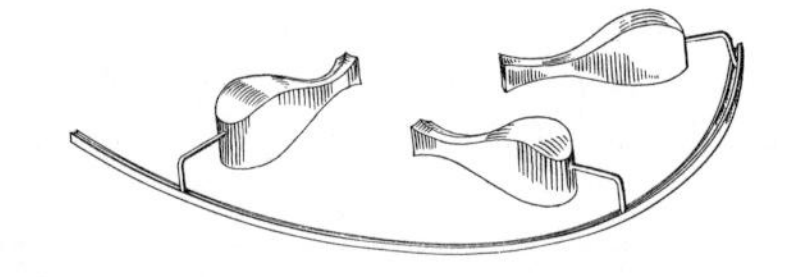

Fig. 22–10 **A metal spline held in place by "ducks."**

Fig. 22–11 **A contour map.**

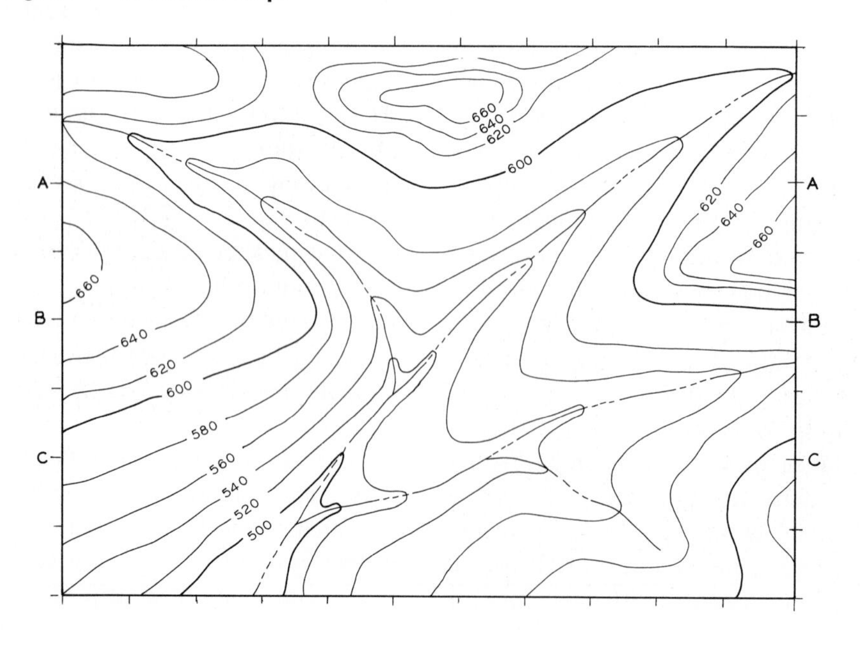

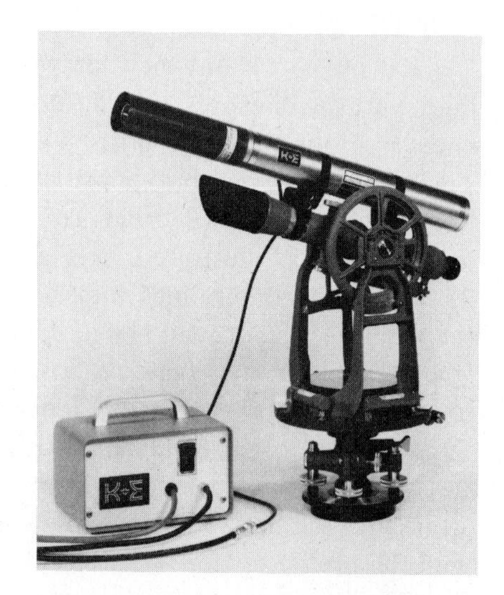

Fig. 22–12 **A laser-beam transit. It assists the field engineer with its extreme accuracy.** ***(Keuffel & Esser Co.)***

Fig. 22–13 Portions of an aerial photo and a topographic map of Concepción, Chile, compiled by photogrammetric methods. The photo and map are part of a Chilean-OAS program to speed reconstruction of earthquake-damaged areas and to advance a broad plan for national growth. Actual map sheets were done at a scale of 1:2000 with 1-meter contours. Photos from which maps were prepared are at a scale of 1:10,000. This project is being carried out by a group of companies under the direction of Aero Service Corporation, Philadelphia, a division of Litton Industries.

Fig. 22–14 Photogrammetric equipment with computerized storage. *(Keuffel & Esser Co.)*

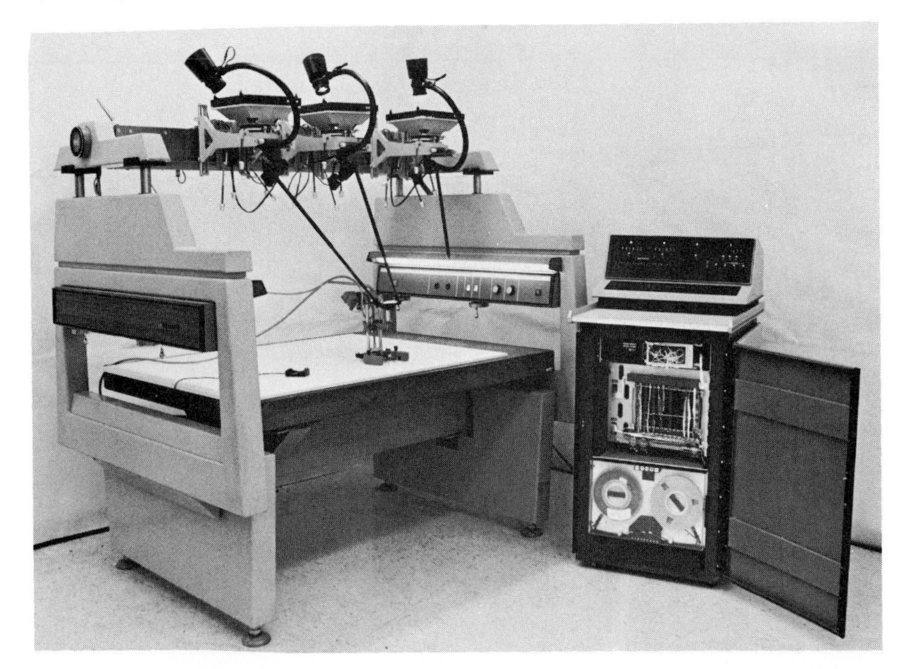

Fig. 22–15 Scribing with a jewel point on coated film provides a very accurate and durable map. *(Keuffel & Esser Co.)*

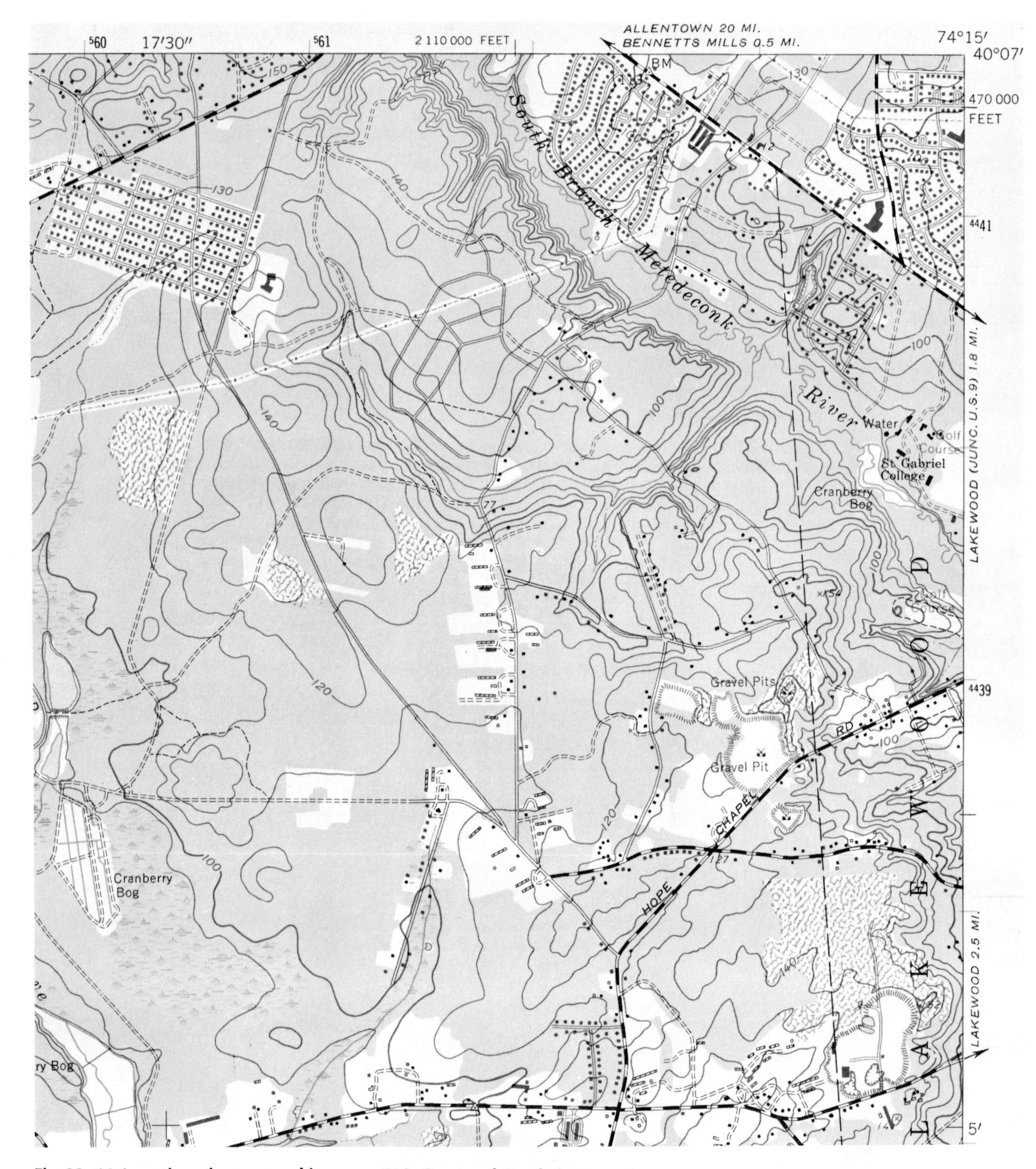

Fig. 22–16 **A portion of a topographic map.** *(U.S. Coast and Geodetic Survey.)*

TOPOGRAPHIC MAP SYMBOLS

VARIATIONS WILL BE FOUND ON OLDER MAPS

Hard surface, heavy-duty road
Hard surface, medium-duty road
Improved light-duty road
Unimproved dirt road
Trail
Railroad: single track
Railroad: multiple track
Bridge
Drawbridge
Tunnel
Footbridge
Overpass—Underpass
Power transmission line with located tower
Landmark line (labeled as to type) TELEPHONE

Dam with lock
Canal with lock
Large dam
Small dam: masonry — earth
Buildings (dwelling, place of employment, etc.)
School—Church—Cemeteries Cem
Buildings (barn, warehouse, etc.)
Tanks; oil, water, etc. (labeled only if water) Water Tank
Wells other than water (labeled as to type) Oil Gas
U.S. mineral or location monument — Prospect
Quarry — Gravel pit
Mine shaft—Tunnel or cave entrance
Campsite — Picnic area
Located or landmark object—Windmill
Exposed wreck
Rock or coral reef
Foreshore flat
Rock: bare or awash

Horizontal control station
Vertical control station BM 671 × 672
Road fork — Section corner with elevation 429 + 58
Checked spot elevation × 5970
Unchecked spot elevation × 5970

Boundary: national
State
county, parish, municipio
civil township, precinct, town, barrio
incorporated city, village, town, hamlet
reservation, national or state
small park, cemetery, airport, etc.
land grant
Township or range line, U.S. land survey
Section line, U.S. land survey
Township line, not U.S. land survey
Section line, not U.S. land survey
Fence line or field line
Section corner: found—indicated
Boundary monument: land grant—other

Index contour — Intermediate contour
Supplementary cont. — Depression contours
Cut — Fill — Levee
Mine dump — Large wash
Dune area — Tailings pond
Sand area — Distorted surface
Tailings — Gravel beach

Glacier — Intermittent streams
Perennial streams — Aqueduct tunnel
Water well—Spring — Falls
Rapids — Intermittent lake
Channel — Small wash
Sounding—Depth curve 10 — Marsh (swamp)
Dry lake bed — Inundated area

Woodland — Mangrove
Submerged marsh — Scrub
Orchard — Wooded marsh
Vineyard — Bldg. omission area

tures, vegetation, and relief (elevations and depressions).

Symbols are used for many of the features shown on topographic maps. Some of these are given in Fig. 22–17. Maps using topographic symbols can be obtained at a low cost from the Director, U.S. Geological Survey, Department of the Interior, or from the U.S. Coast and Geodetic Survey, Department of Commerce, Washington, DC. Naval charts (maps) come from the Hydrographic Office, Bureau of Navigation, Department of the Navy.

Aeronautical maps (Fig. 22–18) use special symbols that need to be understood in order to be read. Some symbols from the U.S. Coast and Geodetic Survey are shown in Fig. 22–19.

Fig. 22–17 **Some conventional symbols used on maps.** ***(U.S. Coast and Geodetic Survey.)***

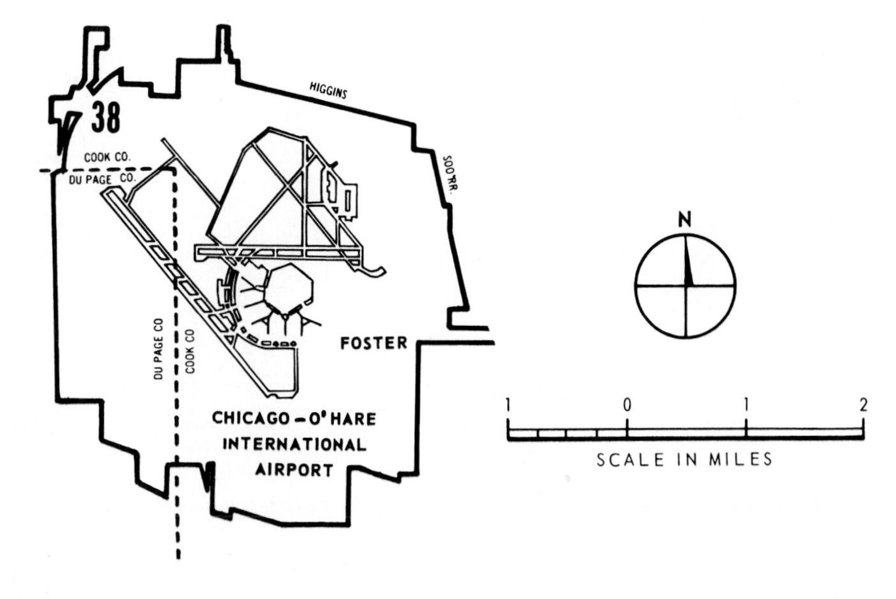

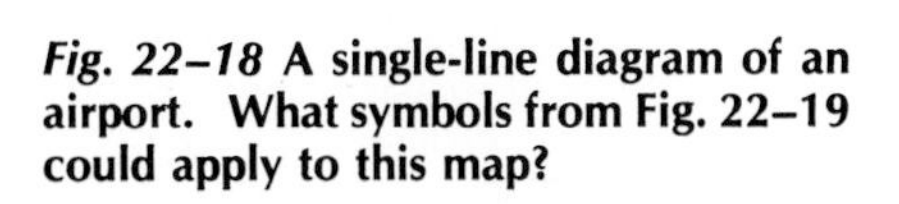

Fig. 22–18 **A single-line diagram of an airport. What symbols from Fig. 22–19 could apply to this map?**

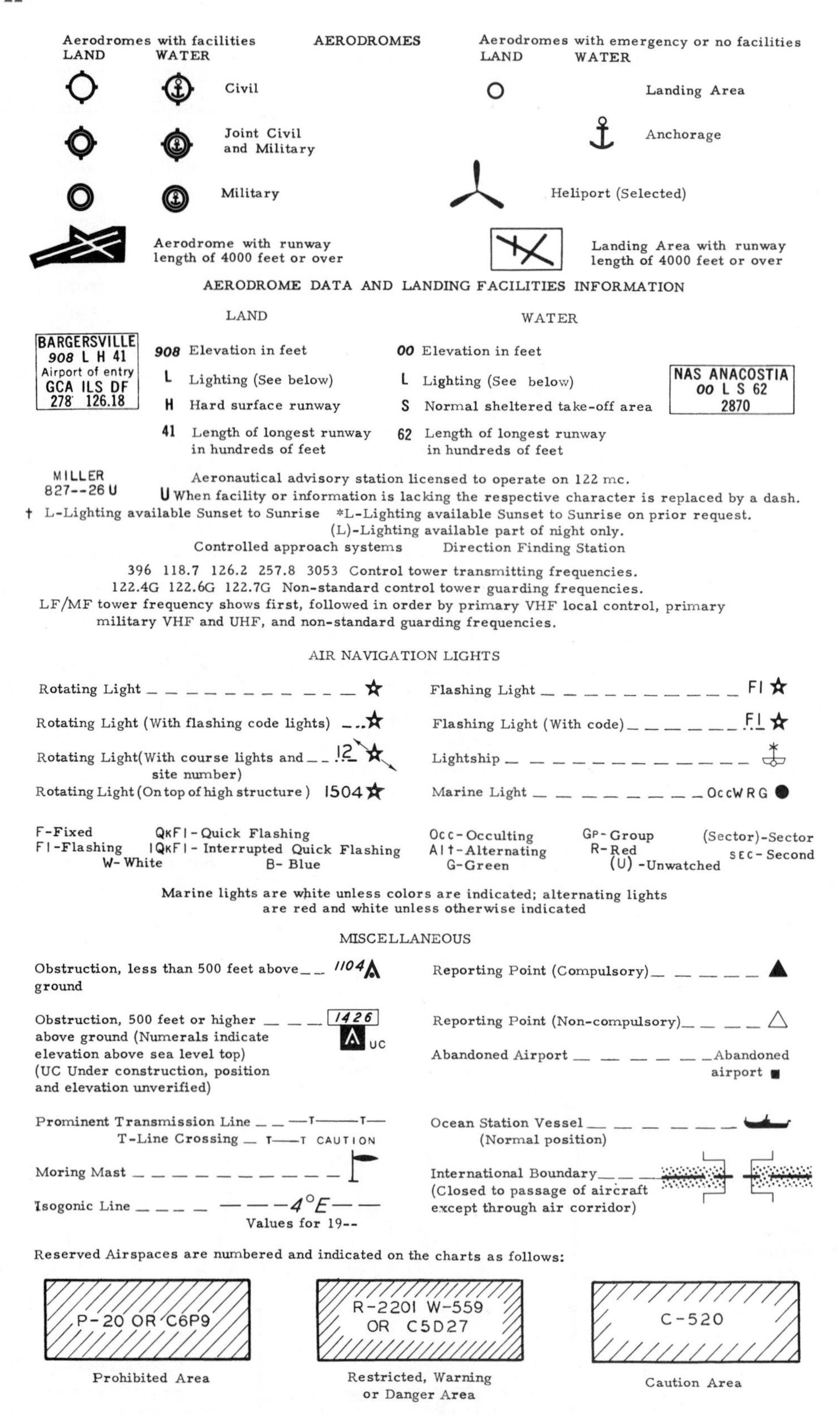

Fig. 22–19 Some aeronautical symbols. *(U.S. Coast and Geodetic Survey.)*

BLOCK DIAGRAMS

Discussion thus far has been given to mapping in the horizontal plane and the vertical plane by profiles or sections. To help you see the three-dimensional problem, a *block diagram* is also used. It is a three-dimensional projection using the isometric view (Fig. 22–20). This block diagram has been developed from Fig. 22–11. Keep in mind that each contour represents a level plane, similar to a card in a deck of cards. True lengths are measured on the isometric axes.

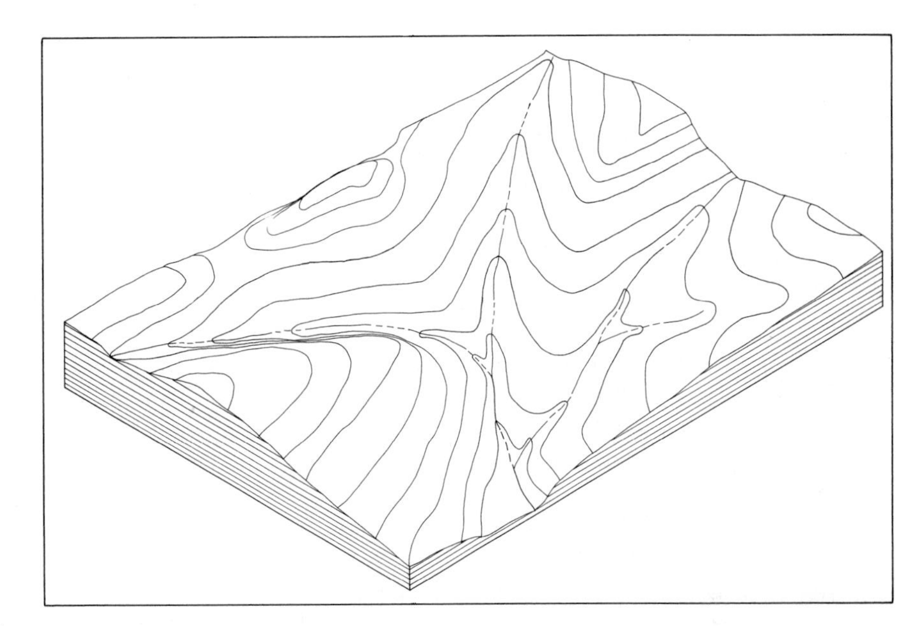

Fig. 22–20 **Block diagram shows a block of earth in an isometric view. This picture was made from Fig. 22–11.**

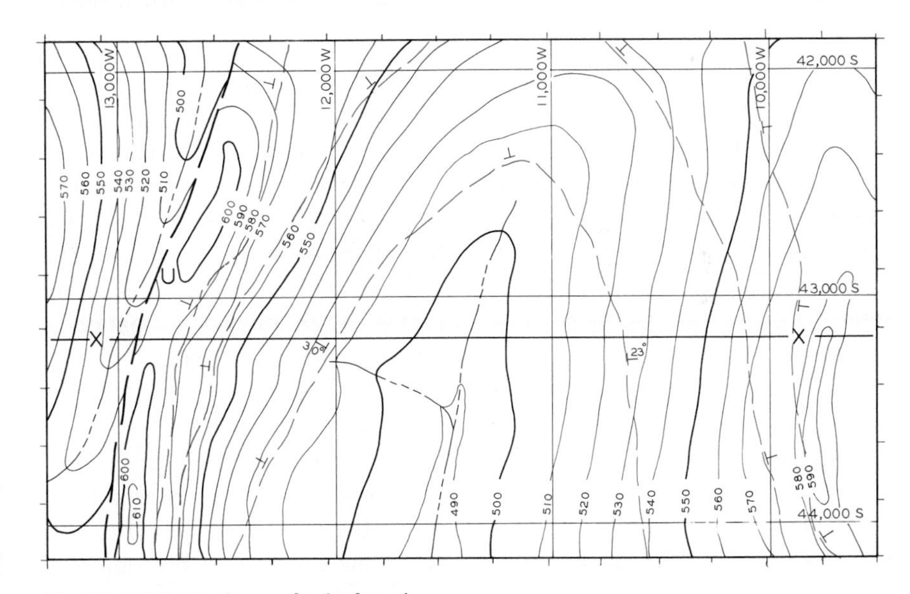

Fig. 22–21 **Part of a geological surface map.**

GEOLOGICAL MAPPING

Geology is the science dealing with the makeup and structure of the earth's surface and interior depths.

The crust of the earth is made up of three groups of rock: igneous or crystalline, sedimentary, and metamorphic. *Igneous* rocks, for purposes of this general discussion, are the basic materials that make up the earth's crustal ring. This rock was once *molten* (melted). It has cooled, but it has not been eroded nor has its makeup changed. *Sedimentary* rocks, as a rule, are deposited in water in layers of different thicknesses similar to the layers of an onion. If the onion is cut perpendicular to its axis, a series of concentric rings will be noted. In a slice of the earth's crust made in a sedimentary area, a similar pattern can be seen. A series of layers that can be identified by texture, color, and material is visible (refer to pictures of the Grand Canyon). *Metamorphic* rocks are generally considered to be sedimentary rocks. These rocks have been deeply buried, heated to high temperatures, and recomposed so that they can no longer be identified as sedimentary rocks.

Nature, being ever-changing, folds, tips, and slices these sedimentary layers in a number of ways. Sometimes sedimentary layers include large areas of crystalline rock. At times, sedimentary layers are formed on top of crystalline rocks. Often, the whole mass may be tipped, perhaps for miles. This tipping raises the mass of rock high above sea level. Sometimes the mass is dropped thousands of feet. The geologist making investigations has the problem of representing what has happened or what a particular area looks like.

Figure 22–21 is part of a geological surface map. The red lines represent the line of surface exposure of the contact between two *formations* (like the line of two contacting layers of a cut onion). The geologist locates this in the field and notes the observation point with the "tee" symbol. The *strike* (direction of this contact) is shown by the top of the tee. The *dip* (slope of the contact) is indicated by the figures at the tee, such as 23° on the tee and the right-hand side and below the section line *X–X*. Since the stem of the tee, in this case, is pointing to the east, or to the right, the dip is 23° to the east. In other words, this formation slopes 23° below the horizontal and to the east. Another tee symbol on the left side shows a 30° dip to the west. The heavy, broken line near the left edge represents a *fault trace* (the line along which the layer broke).

GEOLOGICAL SECTIONS

These sections help in the interpretation of the surface map. An example (Fig. 22–22) shows what the geologist believes the area below the surface is like. This is a section along line *X–X* of Fig. 22–21. The dips that the geologist noted are used in developing the curvature of the folds. By means of a type section of the region, the geologist can determine the various normal thicknesses of each formation or stratum. These values are used in making this section. The fault, as indicated, shows the area at the right to be upthrown. The displacement is easily seen by comparing the position of formation *A* on either side of the fault.

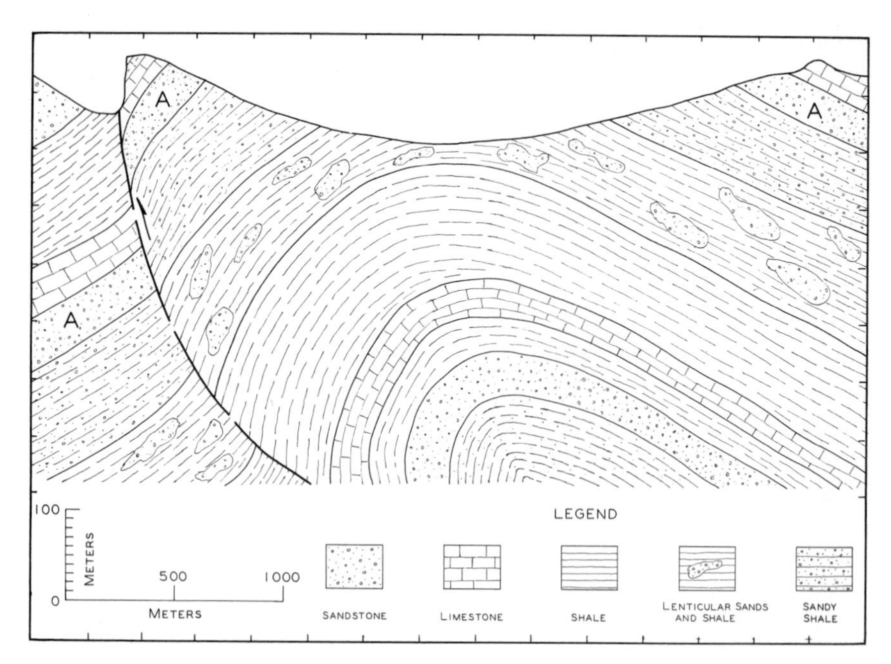

***Fig. 22–22* A geological section along line *X-X* of Fig. 22–21.**

SUBSURFACE MAPPING

This kind of mapping is a means of showing details of strata lying below the surface of the earth. It can show the top or bottom of a given formation or possibly an assumed horizon. Information for constructing such a map is obtained from many sources. These sources may include core holes, electrically recorded logs, seismograph surveys, and so forth. An example of information that was obtained from electrically recorded logs taken in a series of oil wells is shown in Fig. 22–23. The wells are located on a grid pattern. Producing wells are indicated by a solid black circle. Dry holes are indicated by an open circle with outward-extending rays. The top of a producing sand that is cut by a fault on the west is shown. Notice that the contours are numbered with negative values, or depths that are below sea level. The greater the value, the deeper the point below sea level. Section *X–X* shows the thickness of the sand and the level of the *oil-water contact.*

Geological maps and sections are greatly improved by the use of colors. In Fig. 22–21 colors may be applied to each of the formations that show between the red formation-contact lines. This can also be done in Fig. 22–22 by applying the same colors to the corresponding formations. The use of colors helps

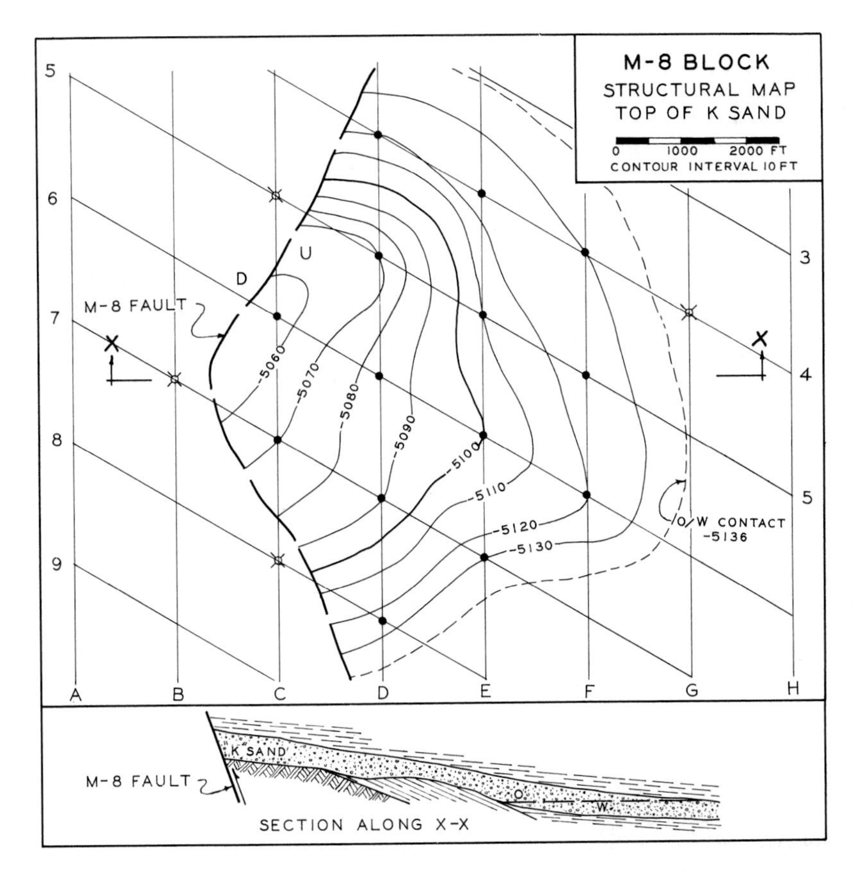

Fig. 22–23 **A structural map showing strata details below the surface.**

in bringing out the three-dimensional relationship of the surface and the shape of the structure. Color also helps in understanding the geology of the area. Paper prints of the tracing are colored and then rubbed carefully to give smooth, even color texture. Examples of color use may be seen on a U.S. Geological Survey map.

The making of geological maps and drawings is an important part of the extractive minerals industry, particularly for petroleum. With the aid of maps, it is possible to keep proper records and information so that activity in this economic field can be continued. Standards for records are different from company to company. However, general standards are well covered in technical literature such as publications of the AIME, petroleum branch, the AAPG, U.S. Geological Survey, U.S. Bureau of Mines, and others.

NOTES AND DEFINITIONS

Maps are important in understanding the news, in studying geography and history, in making car trips, and in surveying, geology, civil engineering, petroleum engineering, and space exploration.

Brief definitions for some of the terms used in map drafting are listed here. For more complete descriptions refer to books on surveying and geology.

1. *Bearing:* The direction of a line as shown by its angle with a north-south line (meridian).
2. *Meridian:* North-south line.

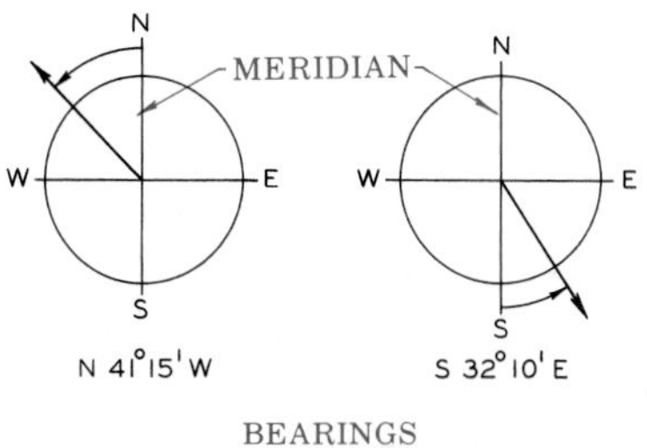

3. *Grid:* A series of uniformly spaced horizontal and perpendicular lines used to locate points by coordinates, or for enlarging or reducing a figure.

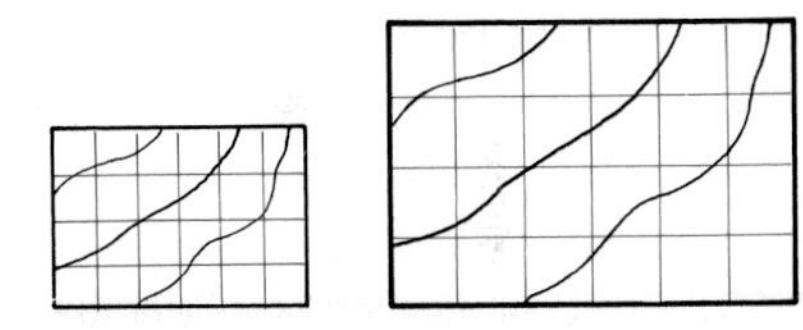

4. *Profile:* A section of the earth on a vertical plane showing the intersection of the surface of the earth and the plane (see Fig. 22–7).

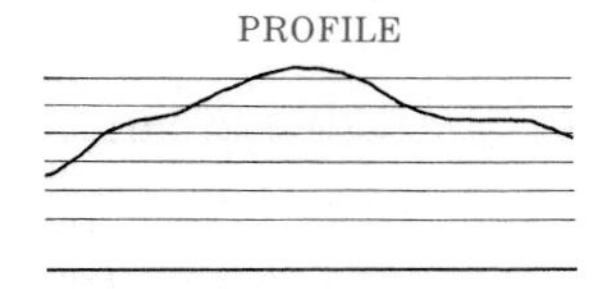

5. *Cut:* Earth to be removed to prepare for construction, such as a desired level or slope for a road.

6. *Fill:* Earth to be supplied and put in place to prepare for a construction in order to obtain a desired level or slope.
7. *Grade:* A particular level or slope, as a downgrade.

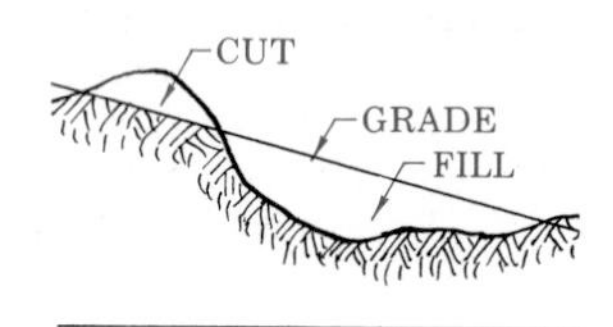

8. *Coordinate system:* A system for locating points by reference to lines that are generally at right angles.

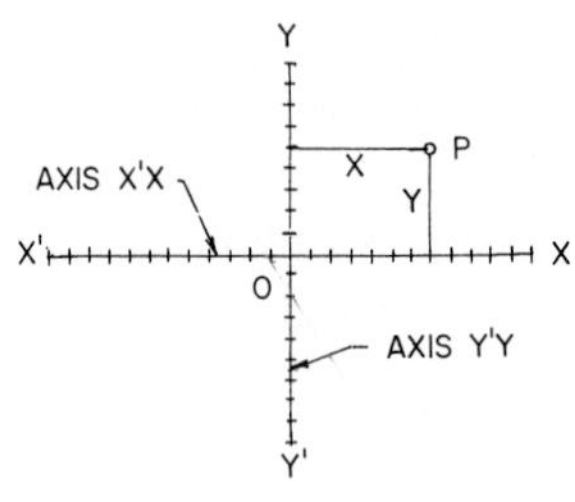

COORDINATE SYSTEM

9. *Contour:* A line of constant level showing where a level plane cuts through the surface of the earth (see Fig. 22–7).
10. *Block diagram:* A pictorial drawing (generally isometric) of a block of earth showing profiles and contours (see Fig. 22–20).

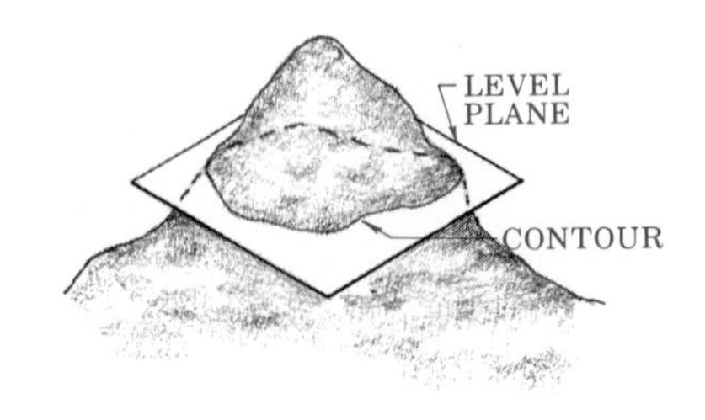

11. *Fault:* A break in the earth's crust with a movement of one side of the break parallel to the line of the break.

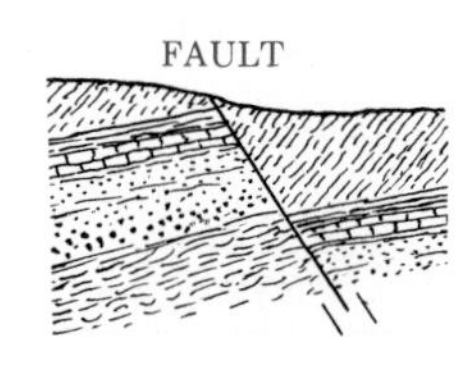

12. *Stratum:* A layer of rock, earth, sand, and the like, horizontal or inclined, arranged in flat form clearly different from the matter next to it. (Plural is strata.)

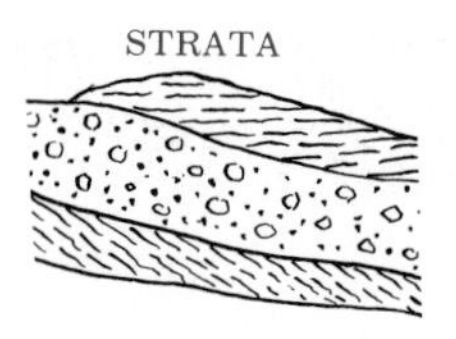

13. *Fold:* A bend in a layer or layers of rock brought about by forces acting upon the rock after it has been formed.

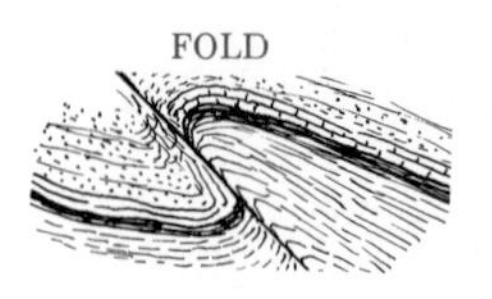

14. *Dip:* The angle that an inclined stratum, or like geological feature, makes with a horizontal plane. The direction of dip is perpendicular to the strike.

15. *Strike:* The direction at the surface of the intersection of a stratum with a horizontal plane.

Review

1. If a map covers a large area the scale is ________.
2. If a map covers a small area the scale is ________.
3. Define a plat.
4. Define a contour line.
5. Name some reasons for making and using a contour map.
6. What is a topographic map?
7. What is a geological map?
8. What is a cartographer?

Problems

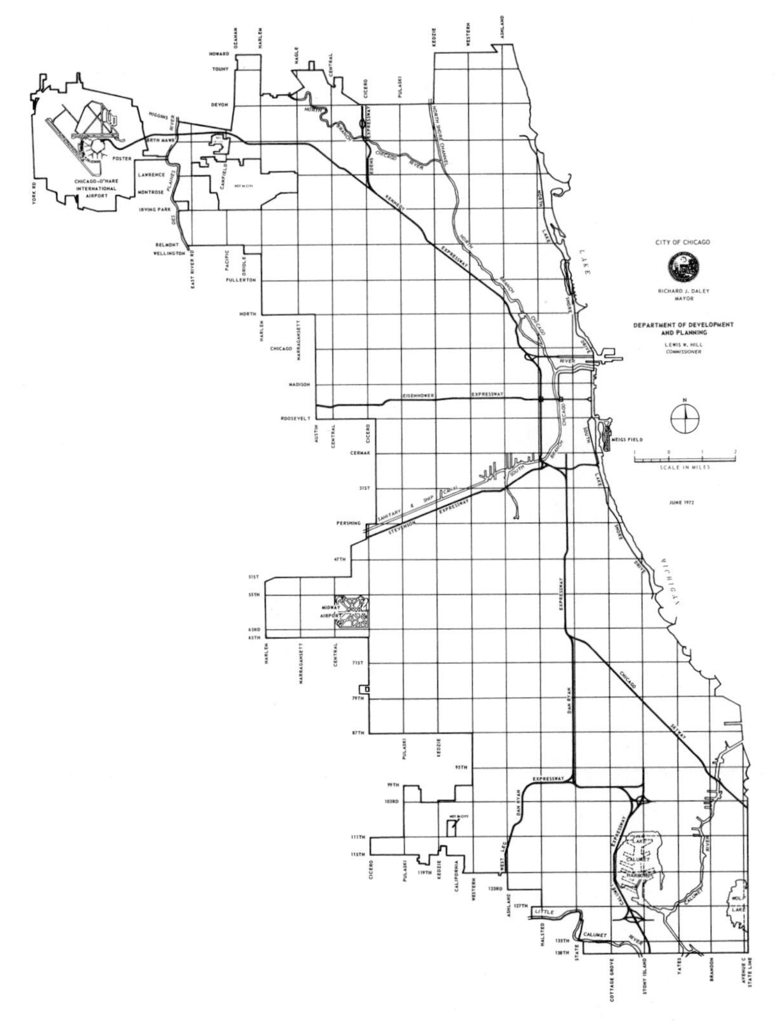

Fig. 22–24 **Plot the map of Chicago on a C-size sheet. Scale: 3/4″ = 1 mile. Calculate the approximate number of square miles, the number of acres, or the number of square kilometers that make up this metropolitan center. (1 square mile = 2 589 988 square meters.)**

Fig. 22–25 **Make a drawing of a plat of a survey as shown in Fig. 22–4. Scale: 1:1200 or 1″ = 100′.**

Fig. 22–26 **Make a contour drawing with the profile shown in Fig. 22–7. Use a grid or dividers to enlarge the figure. Use a vertical scale of 1:240 or 1′ = 20′. Use a horizontal scale of 1:6000 or 1″ = 500′.**

Fig. 22–27 **Problem 1: Make a city map, using the data provided in Fig. 22–5. Problem 2: Enlarge the city map, Fig. 22–5, to include five complete blocks. Begin with the data provided and complete the map with data of your choice. Name all lots and streets added and provide dimensions on the drawing.**

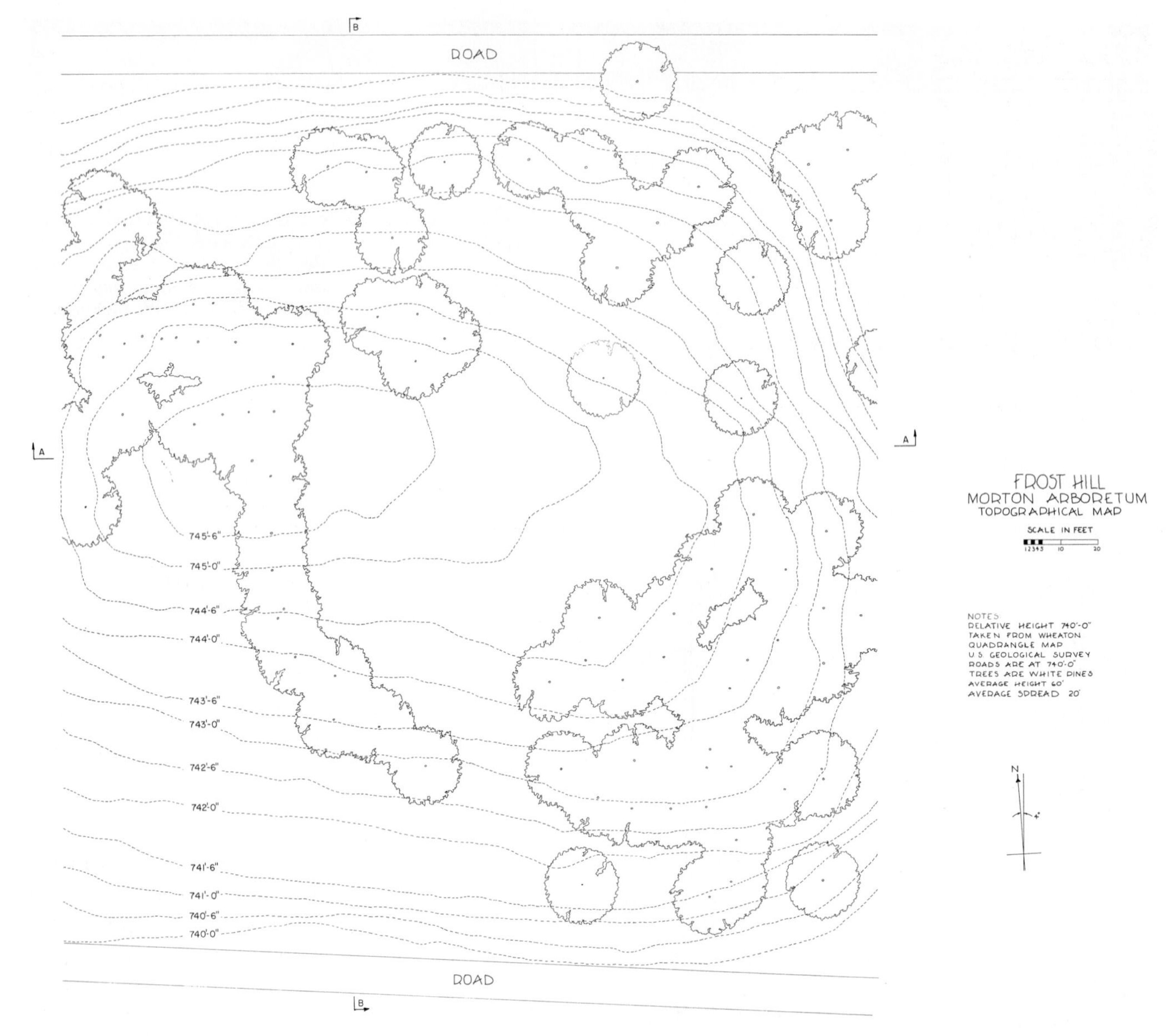

Fig. 22–28 **Draw the contour map of Frost Hill, Morton Arboretum. Prepare a contour profile at *AA* or *BB* as instructed. Choose a suitable scale. Locate all trees that appear within 20 feet of the profile. *(Joseph DeSalvo.)***

Chapter 23

Graphic Charts and Diagrams

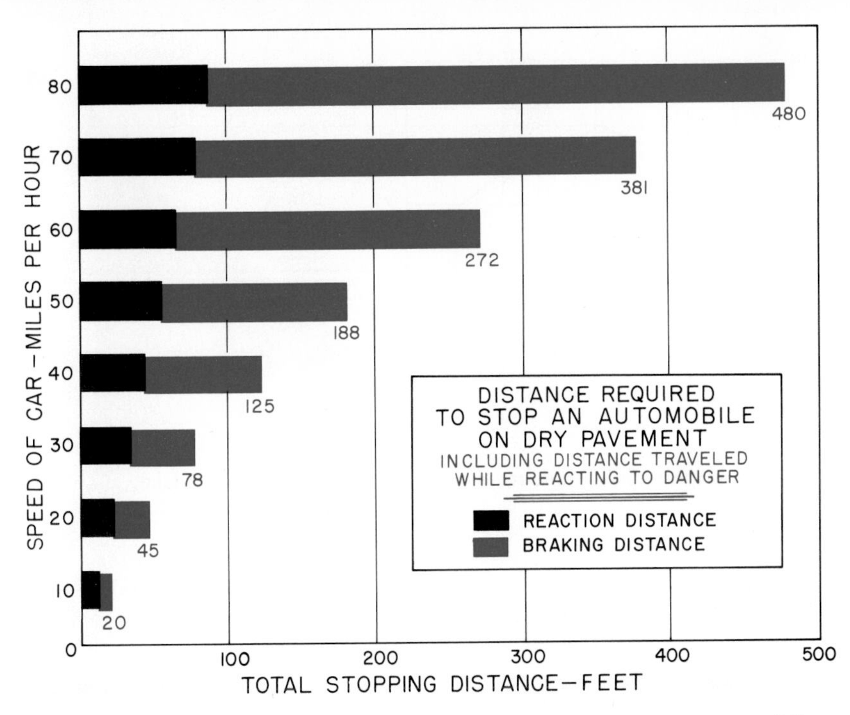

Fig. 23–1 **Chart showing stopping distances at different speeds for automobiles.**

DISTANCE REQUIRED TO STOP AN AUTOMOBILE ON DRY PAVEMENT			
MILES PER HOUR	REACTION DISTANCE FT.	BRAKING DISTANCE FT.	TOTAL STOPPING DISTANCE FT.
10	11	9	20
20	22	23	45
30	30	45	78
40	44	81	125
50	55	133	188
60	66	206	272
70	77	304	381

Fig. 23–2 **The same information takes longer to read and understand in this form.**

IMPORTANCE OF GRAPHIC CHARTS AND DIAGRAMS

Graphic charts and diagrams are an important part of technical drawing. They are important to scientists, engineers, mathematicians, and nearly everyone else in everyday life.

Scientists use charts and diagrams to record and study the results of research. Engineers use them to record information about materials and conditions. Mathematicians use them to record facts and numerical information. Doctors use charts to record body temperature, heart action, and other body functions. Most people use and read charts and diagrams to learn about the weather, the stock market, finances, and for many other purposes. Because Fig. 23–1 is a chart, it can readily be seen that driver reaction distance and automobile braking distance increase as speed is increased. Figure 23–2 contains the same information, but it takes more time to read, study, and understand it. You can see and understand the relationship of speed and distance more easily in Fig. 23–1 because of its *graphic* (pictorial) presentation.

DEFINITIONS

Graphic charts and diagrams are pictures of numerical information that show the relationship of one thing to another. Sometimes they show trends in such areas as economics. You can tell at a glance whether the cost of living and wages are rising or falling over a period of time. Information like this can be determined quite easily because of the pictorial nature of a graph or chart.

Charts can also show ratios. An example of a chart that shows the ratio of speed versus distance is Fig. 23–1. Charts can show percentages of a whole. This type of chart is called a *bar chart* or a *pie chart.*

Charts can also be used to explain information that is not numerical. For instance, a *flow chart* shows sequential information. That means that the chart shows which operation comes first, second, third, and so on.

Graphic charts may also be used to solve various kinds of mathematical problems. You should be able to answer the following questions easily by studying the chart in Fig. 23–3.

1. What is the normal water level in the lake?
2. When does the highest water level occur?
3. At what times is the water level lowest?

You probably had no trouble answering the questions because charts of this type are easily understood by almost everyone.

The curve on a chart is not necessarily a curved line. As shown in Fig. 23–4, a curve on a chart may be a straight line, a curved line, a broken line, or a stepped line. It may also be a straight line or curved line adjusted to plotted points.

The selection of the proper scales (vertical and horizontal squares) is important. The vertical and horizontal scales must give a true pictorial impression by the angle of slope

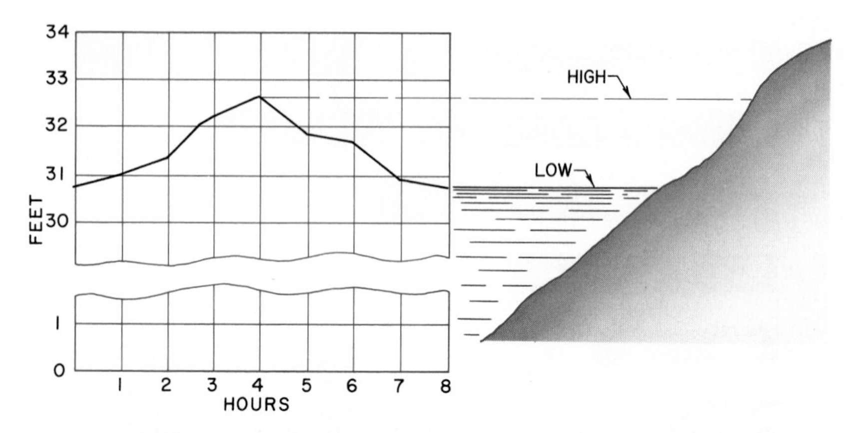

Fig. 23–3 **A graphic chart showing the rise and fall of water in a lake.**

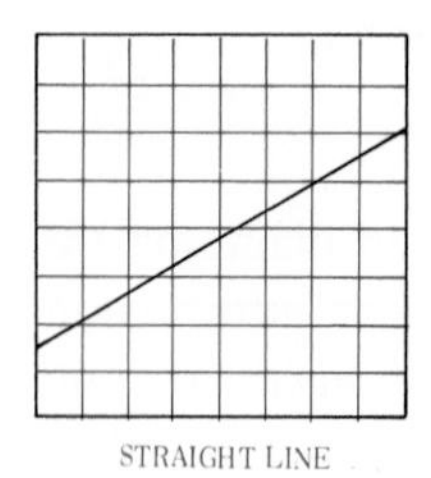
STRAIGHT LINE

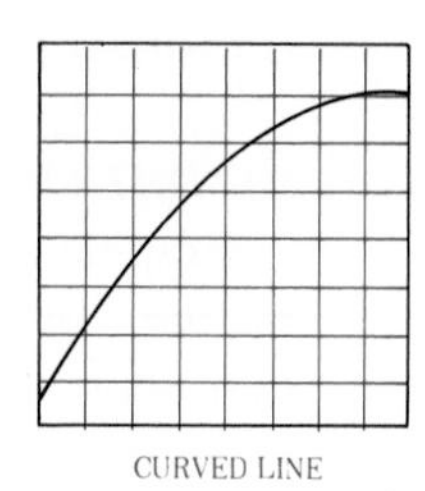
CURVED LINE

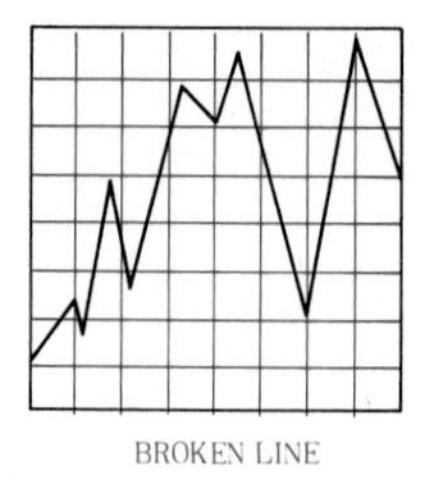
BROKEN LINE

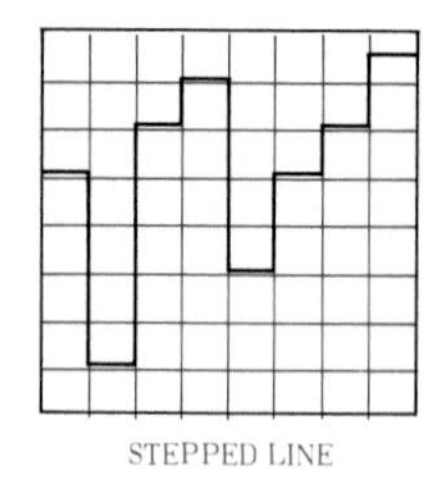
STEPPED LINE

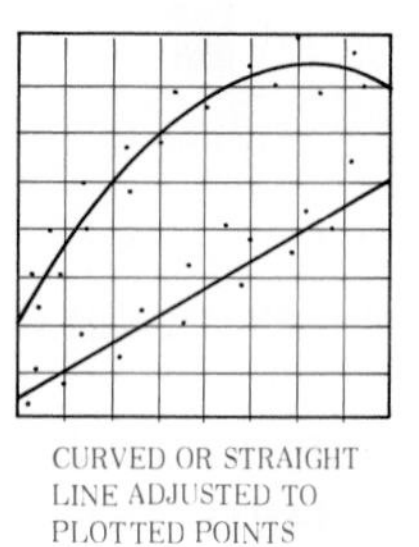
CURVED OR STRAIGHT LINE ADJUSTED TO PLOTTED POINTS

Fig. 23–4 **Curves on graphs may have different forms.**

Fig. 23–5 **A false impression may result if vertical and horizontal scales are not properly selected.**

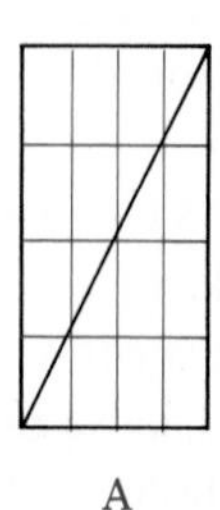
A

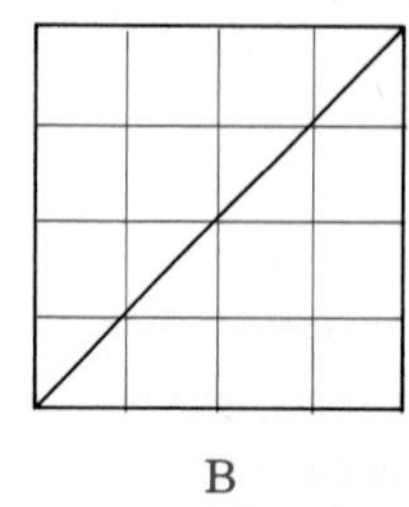
B

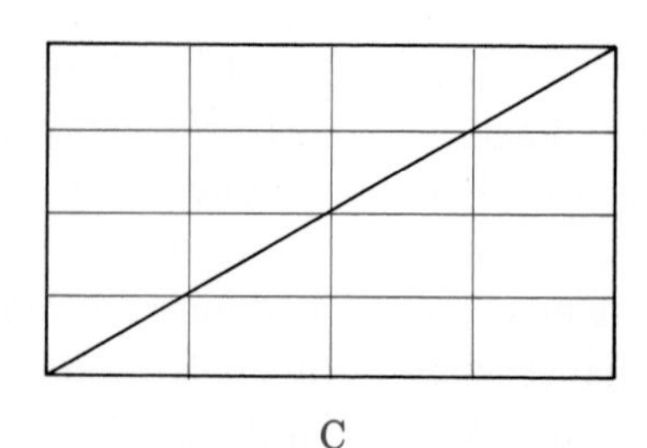
C

of the curve. In Fig. 23–5, you will notice that different impressions are given by the three charts. A presents a very abrupt change. B presents a normal change. At C, a very slow or gradual change is indicated. You should choose the scale that gives the most accurate pictorial impression.

Printed grid or graph paper is available in many forms. It may be purchased with lines ruled for drafting use at 2 mm or 4, 5, 8, 10, 16, and 20 to the inch, and in many other forms (Fig. 23–6). Graph paper that has certain lines printed more heavily than others is also available. This type of graph paper makes it easy to plot points and to read the finished chart. Figure 23–6 shows every tenth line in heavy print. In this case, the heavy lines are 25 mm (1 in.) apart. The lines on grid paper may form squares or rectangles, as shown in Fig. 23–5.

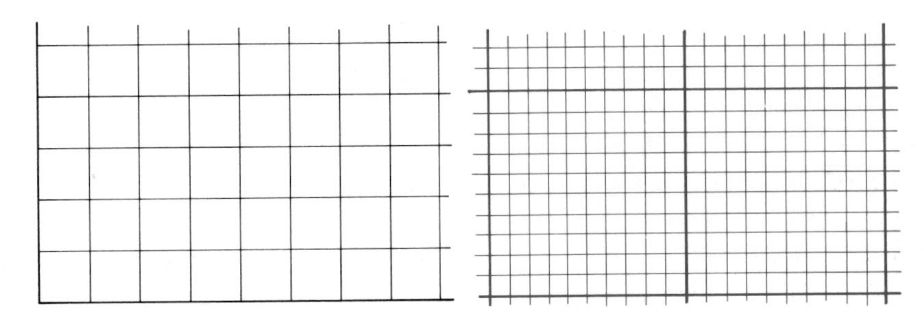

Fig. 23–6 **Graph paper is available in many forms and is printed in different colors.**

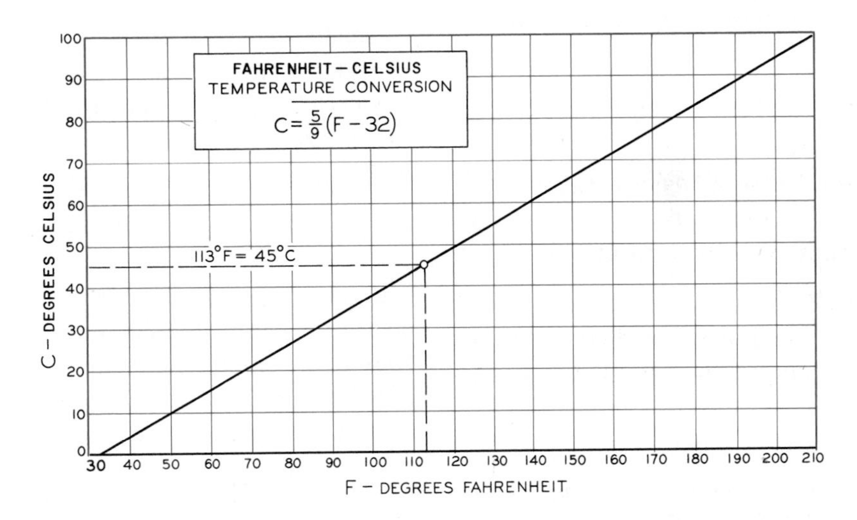

Fig. 23–7 **A conversion chart.**

LINE CHARTS

Line charts are most often used to show *trends* or changes. For example, changes in the weather, ups and downs in sales, or trends in population growth can be plotted and shown graphically on a line chart. A line chart can have one or several curves. A conversion chart with one curve (Fig. 23–7) is often convenient for changing from one value to another. The chart in Fig. 23–7 indicates the conversion from Fahrenheit to Celsius degrees. Figure 23–8 contains four curves that compare sales in various parts of the country.

Fig. 23–8 **A multiline chart.**

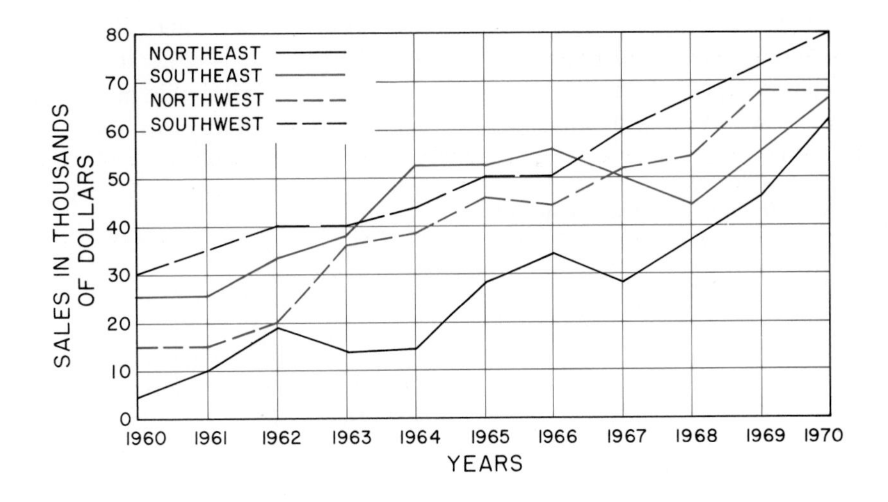

SCORING INFORMATION EASTERN HIGH SCHOOL	
GAME NUMBER	POINTS SCORED
1	38
2	20
3	50
4	40
5	40
6	30
7	10
8	40
9	55
10	45

***Fig. 23–9* Information to be presented in a line chart.**

TO DRAW A LINE CHART

You can make a line chart by following steps 1 through 9.

1. Prepare and list the information to be presented (Fig. 23–9).
2. Select plain or ready-ruled graph paper.
3. Select a suitable size and proportion for your chart so that the overall design will be effective.
4. Select the proper scale.
5. If graph paper is not used, lay off and draw thin horizontal lines (called *X* axis or abscissa lines). Then draw thin vertical lines (called *Y* axis or ordinate lines). These steps are shown in Fig. 23–10A. You will notice that the intersection of *X* and *Y* is zero.
6. Lay off the scale divisions on the *X* axis and the *Y* axis (Fig. 23–10B).
7. Mark the scale values on the *X* and *Y* axes (Fig. 23–10B)
8. Plot the points accurately by using the information you have listed (Fig. 23–9). You will find that it is usually better to use small circles, triangles, or squares rather than crosses or dots for plotting points (Fig. 23–10C).
9. Connect the points to complete the line chart (Fig. 23–10D).

If you draw more than one curve on a chart, use different types of lines or different colors for each curve (Fig. 23–11). Use a full, continuous line of the brightest color for the most important curve. In general, the scales and other identifying notes, or captions, are placed below the *X* axis and to the left of the *Y* axis. For large charts, you may sometimes want to show the *Y*-axis scale at both the right and the left sides. You may also want to show the *X*-axis scale at the top and bottom. This will make the chart more convenient to read.

ENGINEERING CHARTS

Experimental information may be plotted from tests and used to obtain an unknown value. In Fig. 23–12, the results of tests have been plotted. These results show as a straight-line curve when drawn in an adjusted position. (Ω is the Greek letter omega.) Values can be taken from two points and inserted in the formula. In this way, the value of the unknown resistance can be obtained and checked. Notice that tests were made on two occasions, and the results were plotted on the chart. A straight-line curve has been drawn along the center of the path made by the dots.

Nomograms are charts that show the solutions to problems containing three or more variables (kinds of information). Figure 23–13 is an example of this type of chart. A straight line from values on the outside scales will cross the inside scale. The solution to the equation can be read at this point of intersection. *Nomography* is a special kind of chart construction that requires more than simple mathematics.

BAR CHARTS

A bar chart is probably the most familiar kind of graphic chart. Bar charts are easily read and under-

***Fig. 23–10* Steps in drawing a line chart.**

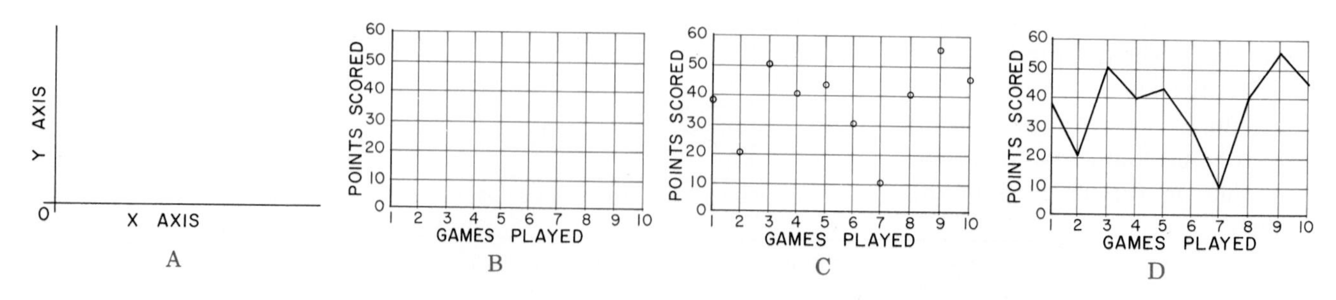

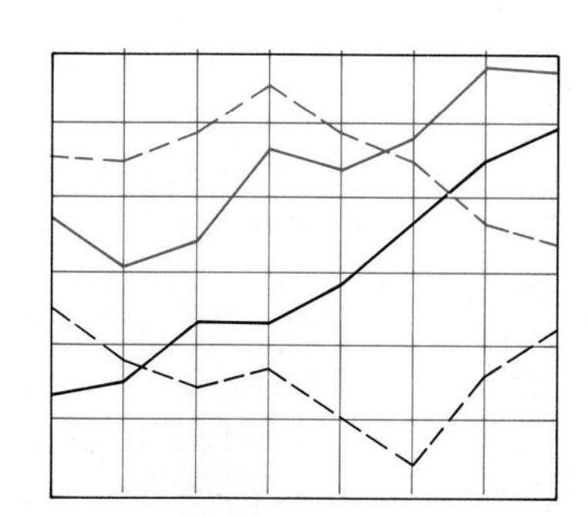

Fig. 23–11 **Use different types of lines and different colors to distinguish curves on a multiline chart.**

Fig. 23–12 **An engineering test chart.**

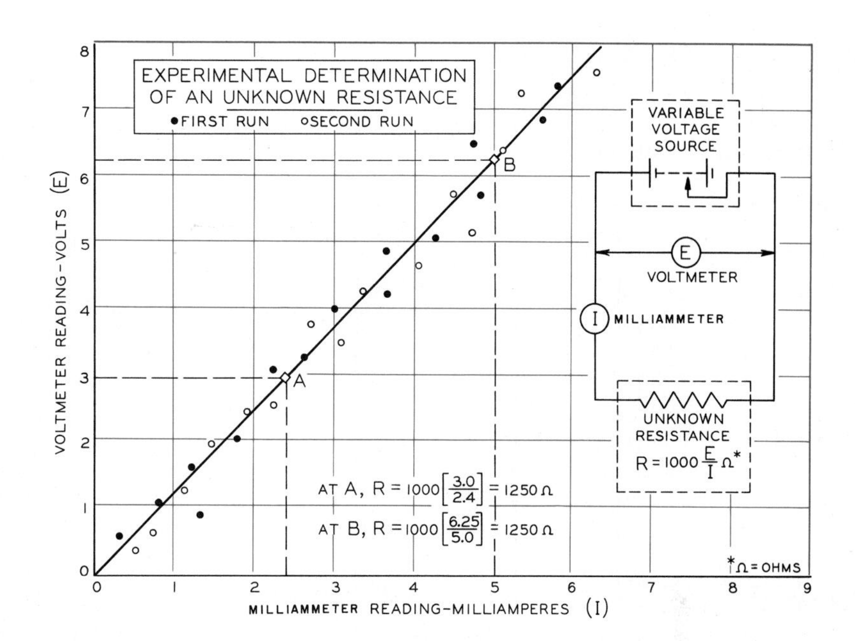

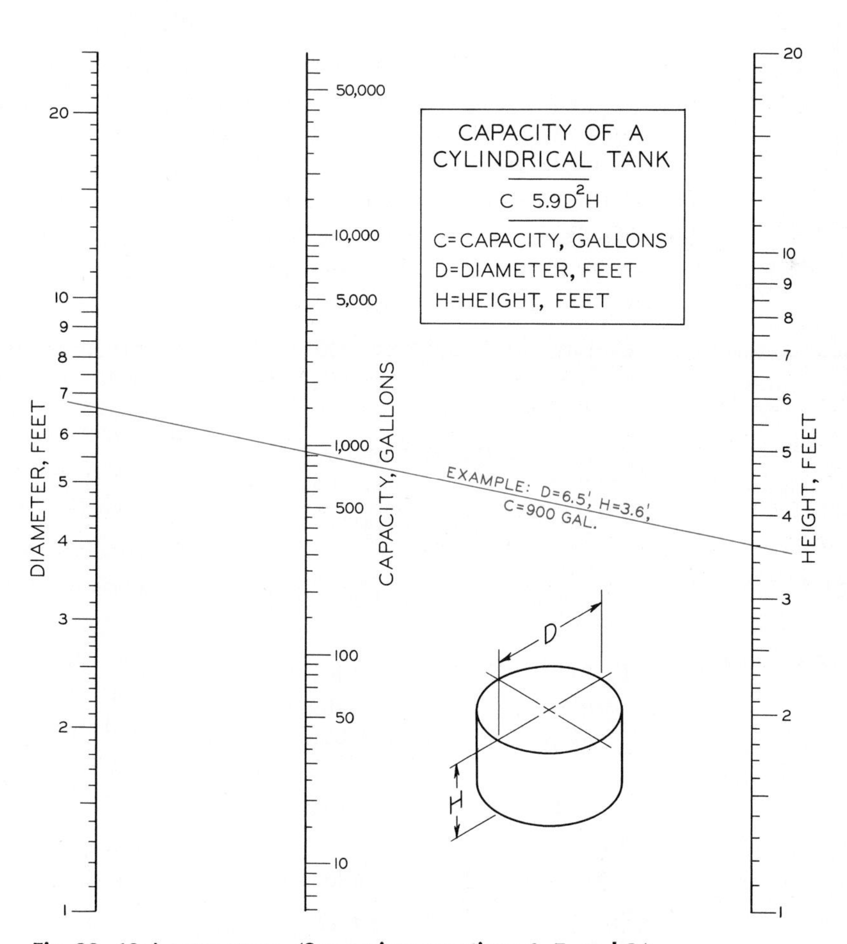

Fig. 23–13 **A nomogram. (See review questions 6, 7, and 8.)**

	QUANTITY	PERCENTAGE
GAMES WON	18	60
GAMES LOST	9	30
GAMES TIED	3	10
GAMES PLAYED	30	100

A

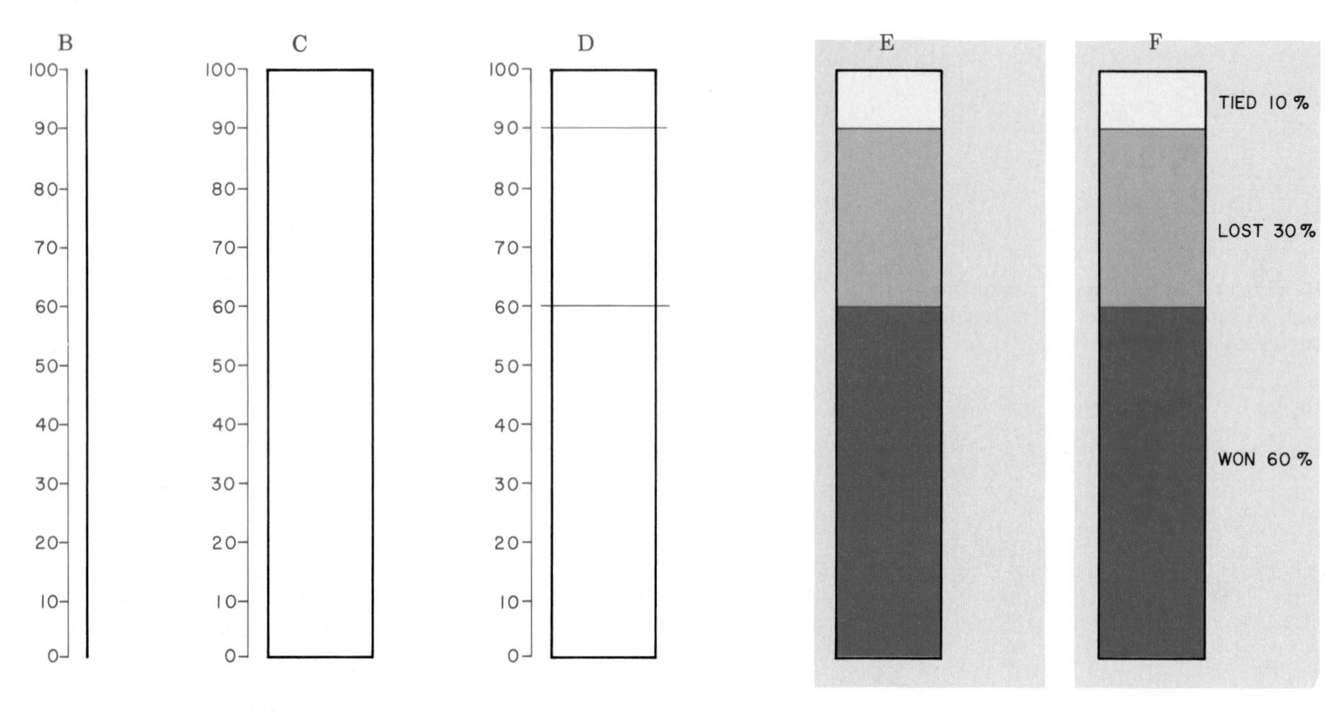

Fig. 23–14 **Steps in drawing a one-column bar chart.**

stood. A bar chart may consist of a single rectangle representing 100 percent (Fig. 23–14F). The chart pictured in Fig. 23–14F represents the total number of games won, lost, and tied.

TO DRAW A ONE-COLUMN BAR CHART

You can make a one-column bar chart by following steps A through F listed below:

A. Prepare and list the information to be presented (Fig. 23–14A).

B. Lay off the long side equal to 100 units. You can find an example of how this is done in Fig. 23–14B.

C. Lay off a suitable width and complete the rectangle (Fig. 23–14C).

D. Lay off the percentage of the parts and draw lines parallel to the base (Fig. 23–14D).

E. Crosshatch, shade, or color the various parts, as shown in Fig. 23–14E.

F. Letter all necessary information in or near the parts so that it can be read easily (Fig. 23–14F).

TO DRAW A MULTIPLE-COLUMN BAR CHART

You can make a multiple-column bar chart by following steps A through D.

A. Prepare and list the information to be presented (Fig. 23–15A).

B. Select a suitable scale and lay off the *X* and *Y* axes. Lay off the scale divisions (Fig. 23–15B).

C. Block in the bars using the information gathered in Fig. 23–15A. Allow enough space between the bars for all necessary lettering. Make the bars any convenient width so that the overall appearance is pleasing (Fig. 23–15C).

D. Complete the bar chart by adding shading or color to the bars, lettering, and any other lines and information. This will add

PERSONAL SAVINGS IN DOLLARS		
1968	—	$ 65.00
1969	—	78.00
1970	—	52.00
1971	—	85.00
1972	—	60.00

A

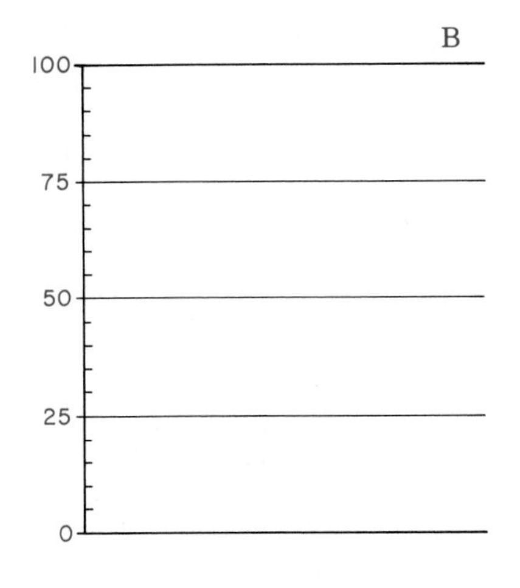

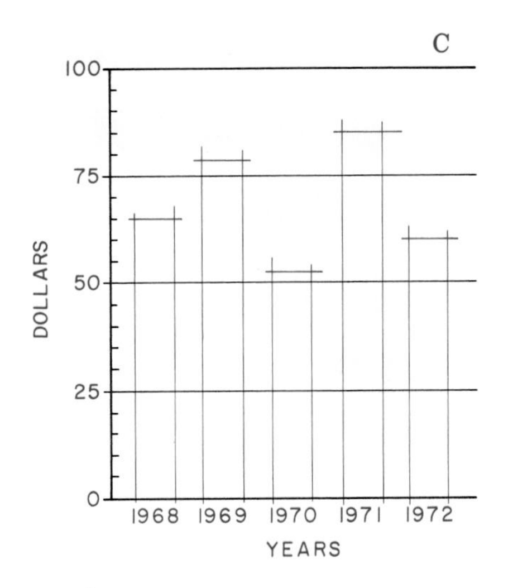

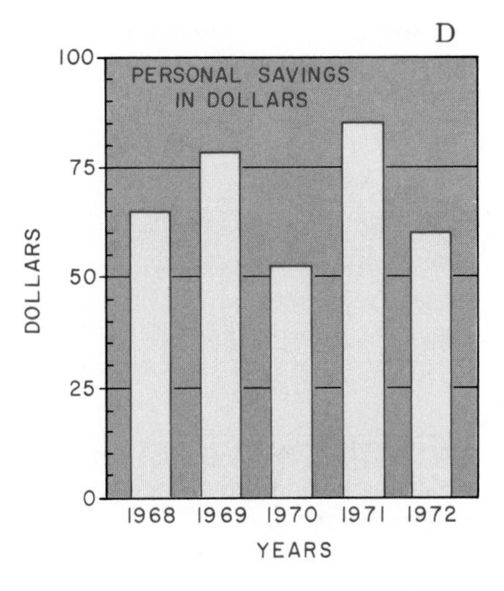

***Fig. 23–15* Steps in drawing a multiple-bar graph.**

to the appearance of the chart and the ease with which it can be read and understood.

***Fig. 23–16* This horizontal-bar chart is a form of progressive chart.** *(Caterpillar Tractor Co.)*

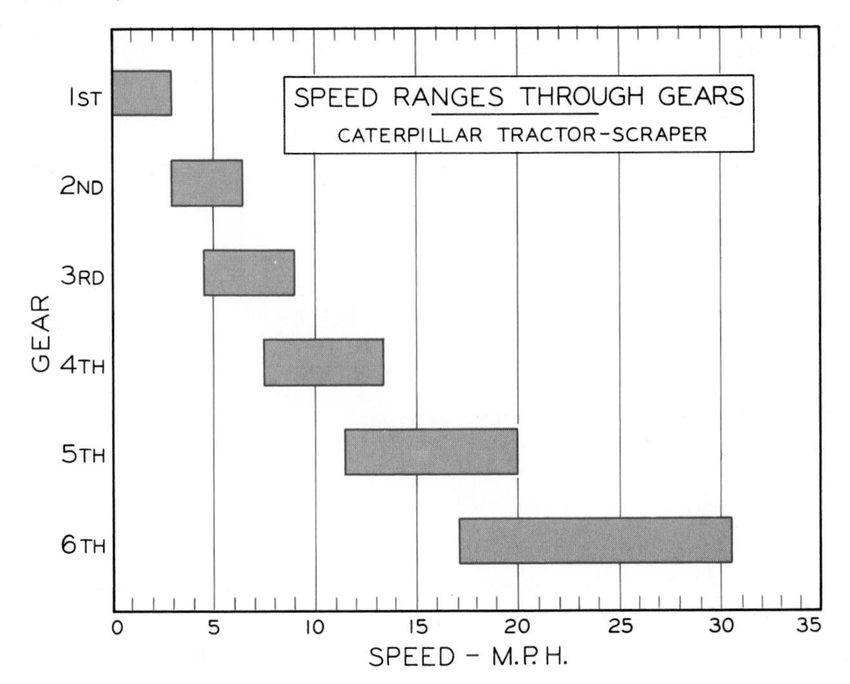

A bar chart with horizontal bars is shown in Fig. 23–16. It gives speed ranges for Caterpillar tractors. Note that the bars do not start at the same line because they show different speed ranges. This kind of chart is called a *progressive chart.*

A *compound-bar chart* is shown in Fig. 23–17, where the total length of the bars is made up of two parts. The black portion is the distance traveled at a given speed before application of the brakes. The blue portion is the distance traveled after applying the brakes. The black and blue portions are added together graphically to give the total distance involved.

The multiple-bar chart shown in Fig. 23–18 has minus values. These values are represented by red bars and set below the X axis. The Y

axis must have values less than 0 and be drawn past the X axis. The same information is given in Fig. 23–19. The vertical scale is the same as in Fig. 23–18. The dashed line shows the net gain. It can be found by laying off the minus values down from the total-gain values.

PIE CHARTS

In this type of chart, a circle represents 100 percent. Various sectors represent parts of the whole (Fig. 23–20).

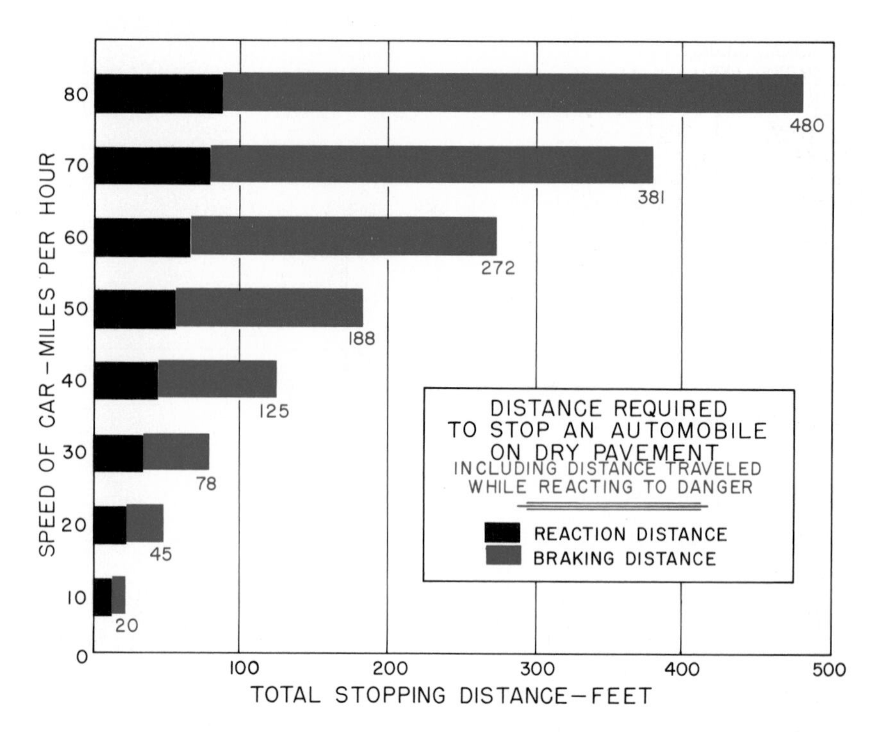

Fig. 23–17 **A compound-bar chart in which the total length of each bar is the sum of two parts.**

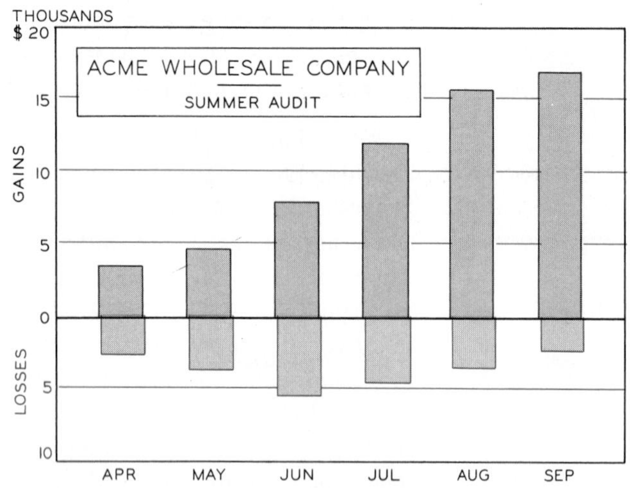

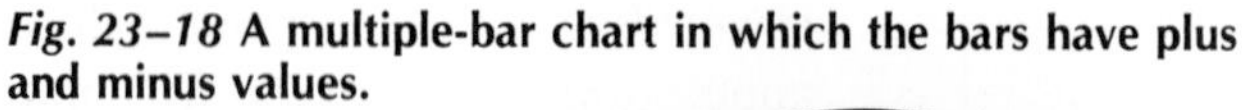

Fig. 23–18 **A multiple-bar chart in which the bars have plus and minus values.**

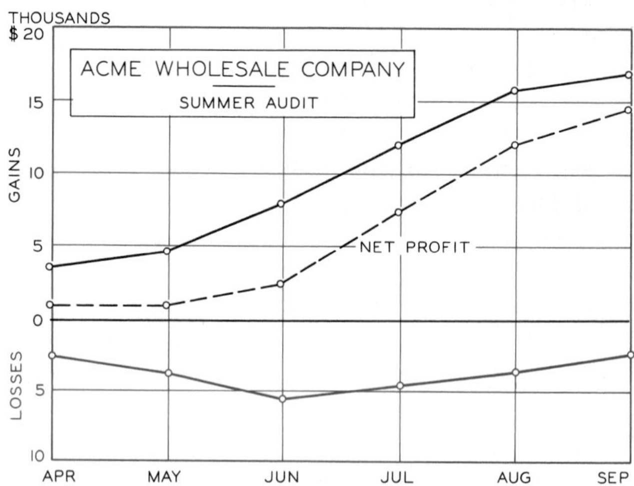

Fig. 23–19 **A line chart for the same information as that in Fig. 23–18.**

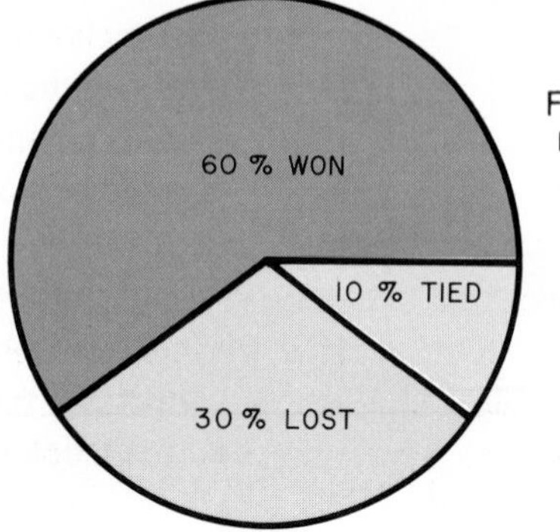

FOOTBALL SEASON RECORD
RELATIONSHIP OF GAMES WON, LOST, AND TIED

Fig. 23–20 **A 100-percent circular chart, or pie chart.**

DISTRIBUTION OF CLASS TREASURY		
ITEM	COST	%
DANCE	$ 72.00	40
PARTY	36.00	20
PICNIC	32.40	18
CLASS PLAY	21.60	12
PHOTOGRAPHS	18.00	10
TOTAL	$180.00	100 %

A

B

C

DANCE 40 %
PHOTOS 10 %
PARTY 20 %
CLASS PLAY 12 %
PICNIC 18 %

D

DISTRIBUTION OF CLASS TREASURY

Fig. 23–21 **Steps in drawing a pie chart.**

TO DRAW A PIE CHART

You can make a pie chart by following steps A through D.

A. Prepare and list the information to be presented (Fig. 23–21A).

B. Draw a circle of the desired size. Lay off and draw the radial lines representing the amount or percentage for each part on the circumference of the circle (Fig. 23–21B). If a protractor is used, 3.6° = 1 percent. Thirty percent is 30 × 3.6°, or 108°, and so on. If a circle is to be divided into a 24-hour day, each hour represents 15° on the circle. This can be drawn by using a T-square and triangles.

C. Crosshatch, shade, or color the various parts, as shown in Fig. 23–21C.

D. Complete the pie chart by adding all necessary information (Fig. 23–21D).

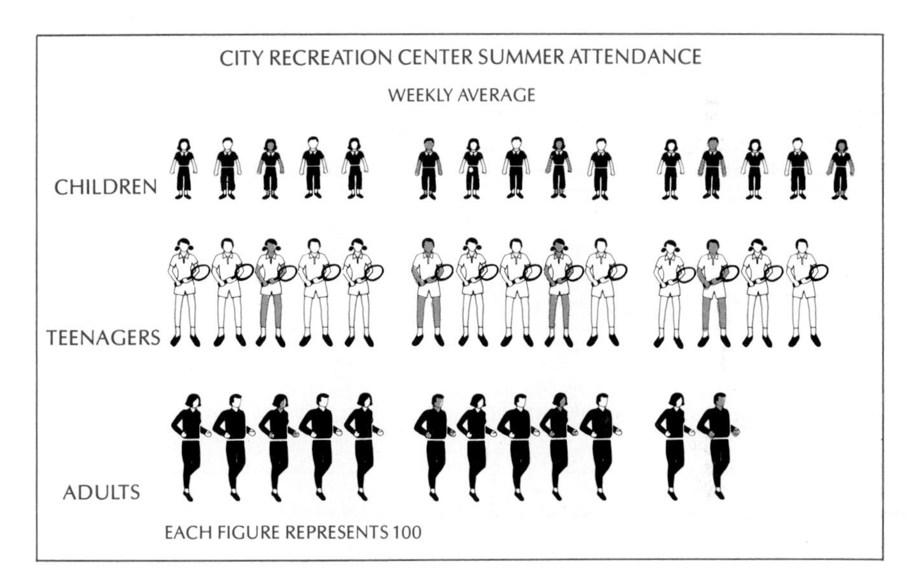

Fig. 23–22 **A pictorial graphic chart (also called a pictograph).**

PICTORIAL CHARTS, OR PICTOGRAPHS

Pictorial charts, or *pictographs,* are similar to bar charts. Pictures or symbols are used instead of bars. Figure 23–22 illustrates a pictorial graphic chart, or pictograph. The chart pictured is a multiple-bar chart. In this chart, each figure represents 100 people. However, it is understood that a symbol or picture may represent any number that the maker of the chart assigns to it. There are three kinds of individuals represented. Adhesive symbols may be used on graphic charts. They are available in many forms (Fig. 23–23). These symbols make pictorial charts easy to construct.

ORGANIZATION AND FLOW CHARTS

There are many kinds of organization charts. Most, however, have the features of a flow chart. Figure 23–24 is an example. It shows the path, or flow, of drawings from the top engineer to the shop. It also shows the organization of the drafting department.

A flow chart may show the path or series of operations that it takes

Fig. 23–23 **Many styles of adhesive symbols are available for use on graphic charts. *(Chart-Pak, Inc.)***

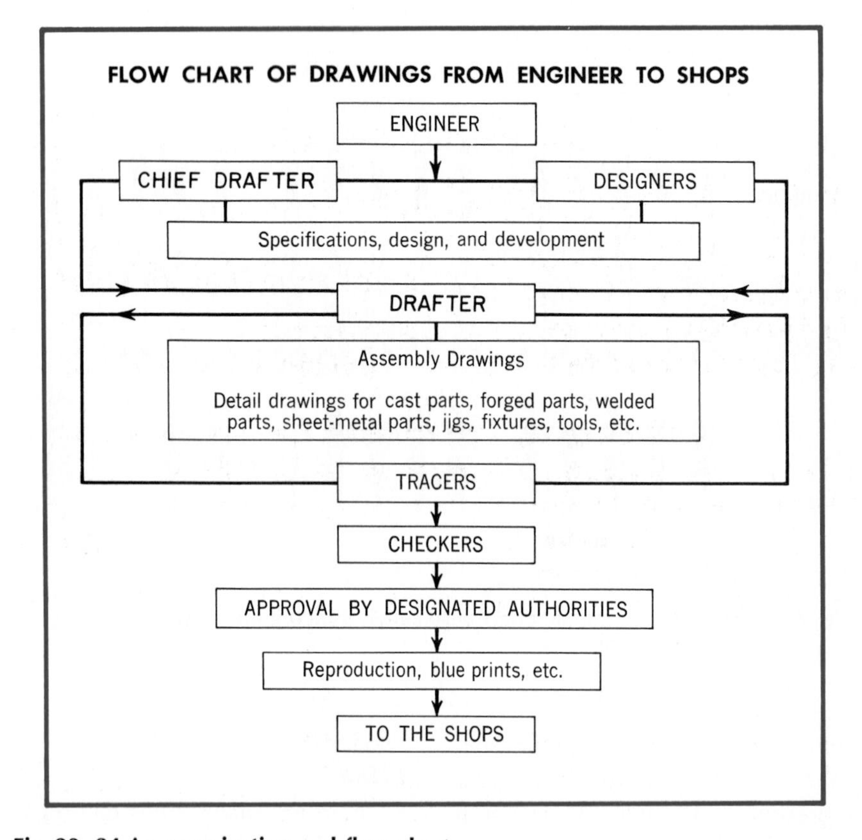

Fig. 23–24 **An organization and flow chart.**

to manufacture a product or a material. An example of this is the flow chart of steelmaking (Fig. 23–25).

TAPE DRAFTING

Tape drafting is a convenient method of preparing graphic charts. Adhesive tape comes in many colors, designs, and widths. It is applied from a roll dispenser (Fig. 23–26) and pressed onto the chart in the desired position. Tapes of different widths provide a quick and simple way of making bar charts.

THE USE OF COLOR

Black-and-white charts are often used by scientists and mathematicians for recording information. Black-and-white charts can also be found in newspapers. The use of color, however, has become quite common in the preparation of charts and diagrams for magazines, books, pamphlets, and other publications. Color is also used a great deal in making charts for display purposes.

The use of color adds a great deal to the appearance and emphasis of the chart and makes it easier to understand. (Fig. 23–27). Color may be added in a variety of ways. Colored pencils, felt-tipped pens, watercolors, or other similar materials are easy to use. Most of these can be found in the drafting room, art room, or at home. Commercially prepared pressure-sensitive materials are available at art and engineering supply stores.

CHARTS AND ELEMENTS OF CHARTS

A great variety of graphic charts can be made for visual communication. The general characteristics and uses of a few types have been discussed in this chapter.

A variety of charts is shown in Figs. 23–28 to 23–36. Some of the elements of chart making and the effects these elements have are presented in these illustrations.

Remember these important points when you make a chart: Every chart should have a suitable title, well lettered and placed within the area of the chart. Every chart should also have a key to tell what the elements represent.

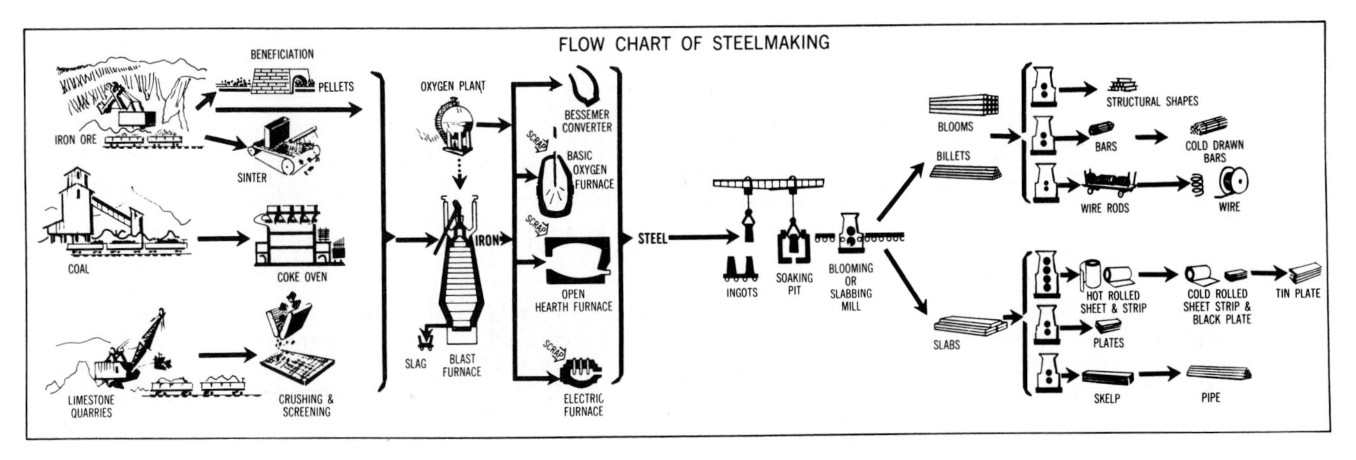

Fig. 23–25 A flow chart of steelmaking. *(American Iron and Steel Institute.)*

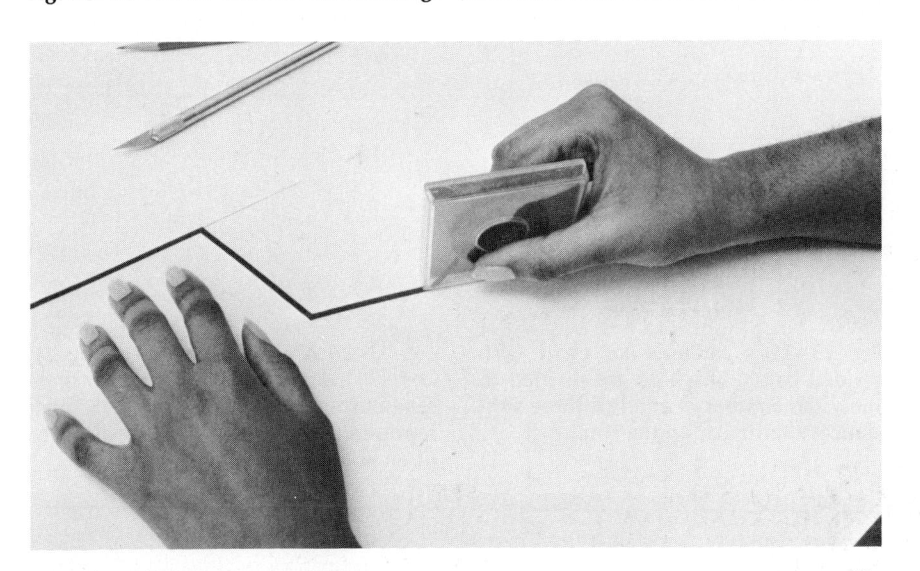

Fig. 23–26 Applying adhesive tape to a graphic chart. *(R. J. Capece/McGraw-Hill.)*

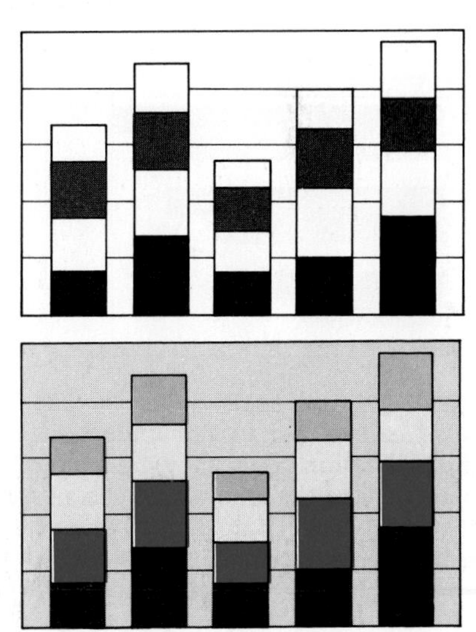

Fig. 23–27 A multiple-bar chart in black and white and the same one in color.

Fig. 23–28 Scales should be selected with care so that the appearance of the curve plotted from the data will aid the understanding of the information.

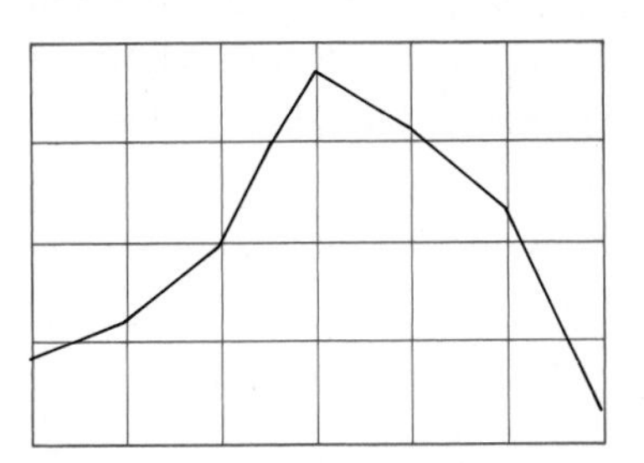

Fig. 23–29 The vertical scale used here is too small. It gives the effect of very little change in values. The movement is slow.

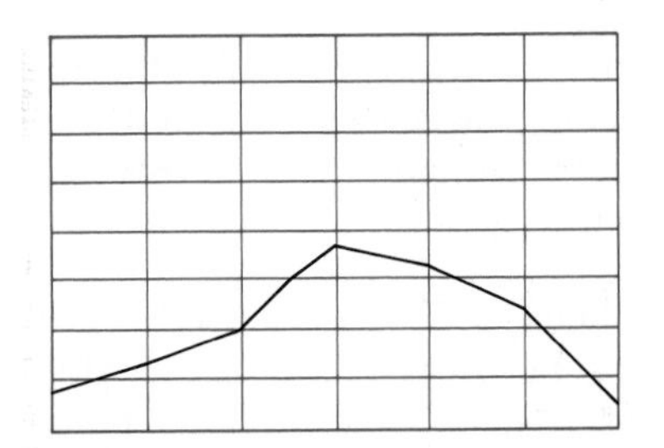

Fig. 23–30 This curve is plotted from the same data used for Figs. 23–28 and 23–29. The horizontal scale is too small (if Fig. 23–28 is a correct picture). The movement is fast.

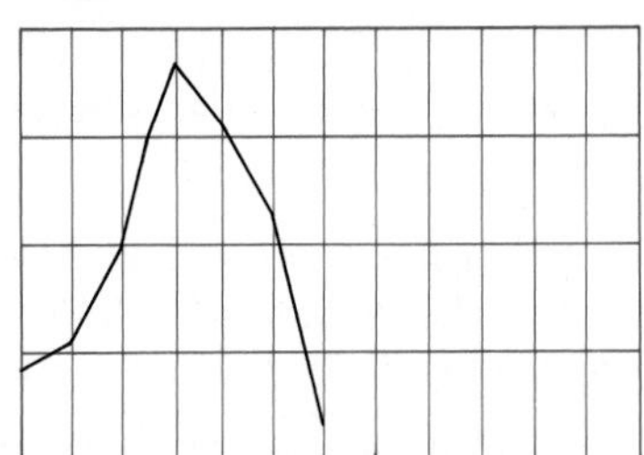

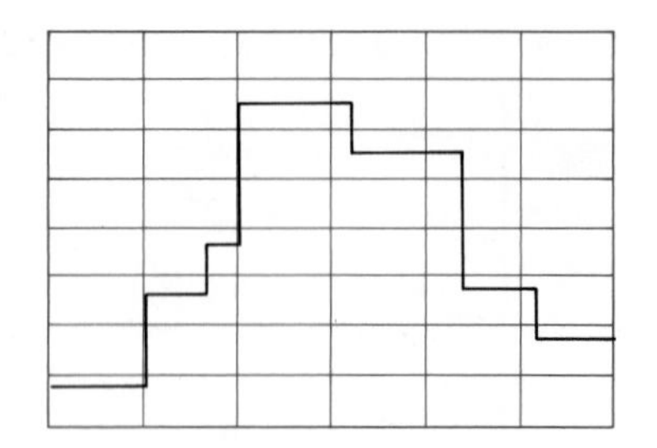

Fig. 23–31 **A step chart shows data that remains constant during regular or irregular intervals. This figure might show time periods during which a price remained constant or was raised or was lowered.**

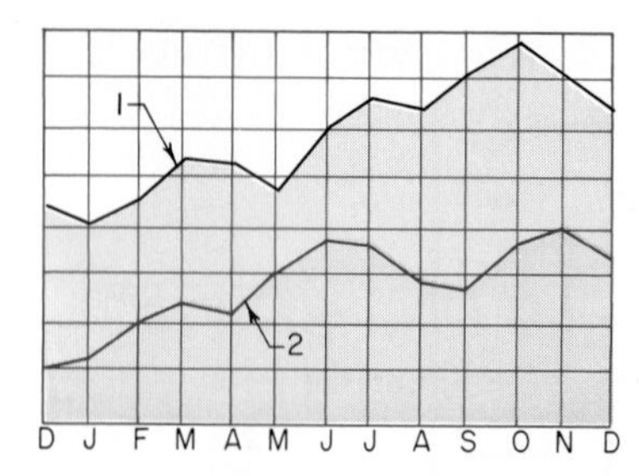

Fig. 23–32 **A shaded-surface, or strata, chart uses shaded areas for contrast. This illustration might show the total amount of each of two materials used each month.**

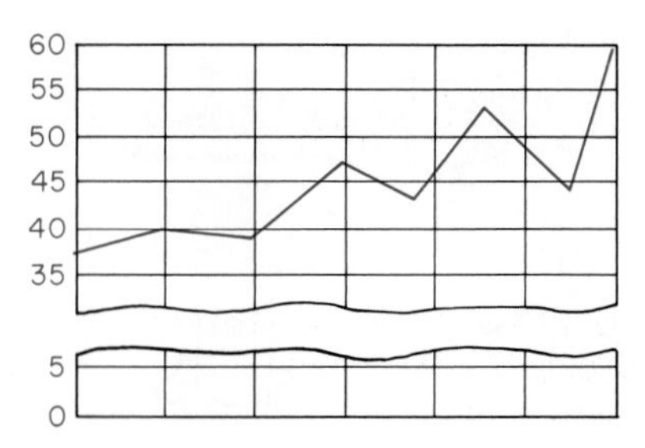

Fig. 23–33 **An omission chart may be used for some purposes, as shown, in order to use a larger vertical scale. There are no values below 35, and so a portion of the chart is broken out.**

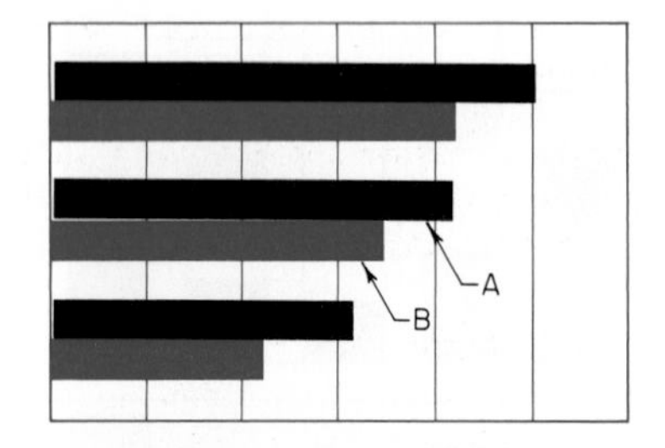

Fig. 23–34 **A comparison-bar chart that might be used for two or three values. The illustration might show the amount made (A) and the amount sold (B) of an item by various companies in one year or by one company for several years.**

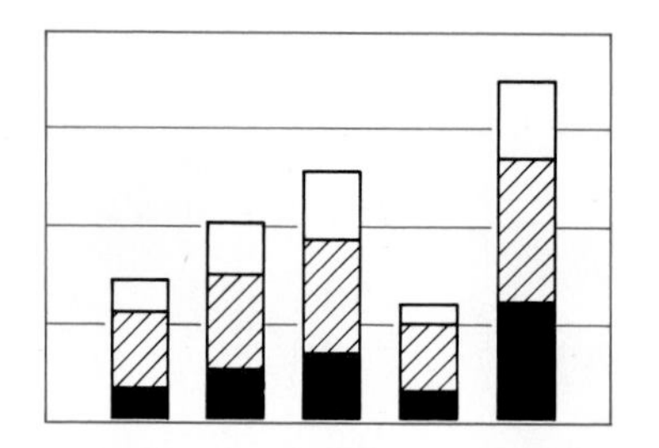

Fig. 23–35 **A multiple-bar chart with divided bars. The bars are divided to show the amount of each of three substances that make up the total.**

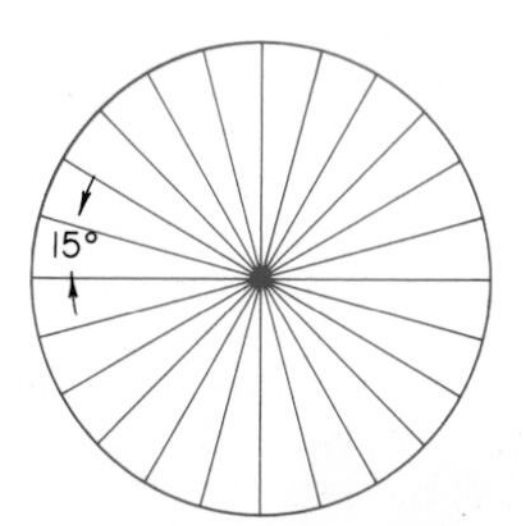

Fig. 23–36 **A pie chart representing 24 hours in a day may be used to show time relationships. Each division is 15° and represents one hour. Fractional parts of an hour may be estimated.**

Review

1. What kind of chart is used to show the solutions to problems that contain three or more kinds of information?

2. Is the curve on a chart always a curved line?

3. Refer to the chart in Fig. 23–8. What was the dollar volume of sales in the Northeast Region in 1968?

4. Refer to the chart in Fig. 23–17. How far does a car travel during reaction time at a speed of 60 mph?

5. In Fig. 23–17, what is the total stopping distance at a speed of 60 mph? Convert your answer to km/h.

6. Refer to Fig. 23–13. What is the capacity in gallons of a cylindrical tank 10 ft in diameter and 17 ft high? Translate your answer to metric units.

7. In Fig. 23–13, what is the height of a cylindrical tank that holds 100 gallons and is 2 ft in diameter?

8. In Fig. 23–13, a tank holds 3000 gallons and is 10 ft high. What is its diameter?

9. What is another name for a pictorial graphic chart?

Problems

Fig. 23–37 Draw a pie chart to show the population distribution in the following four regions of the United States:

Northeast	24.2%
North Central	27.8%
South	31.2%
West	16.8%

Use color or various types of crosshatching for contrast.

Fig. 23–38 Draw a pie chart showing that 67.4% of the population of the United States lives within metropolitan areas, while only 32.6% lives outside metropolitan areas. Use different colors or crosshatch areas for contrast.

Fig. 23–39 Make a one-column bar chart to show how your allowance was spent last month.

Fig. 23–40 Make a multiple bar chart showing a comparison of how you spent your allowance over a period of four months.

Fig. 23–41 Assignment 1: The average cost per pound for beef varied over a period of 10 years as follows:

Year	*Cost Per Pound*
1964	$0.65
1965	0.68
1966	0.70
1967	0.76
1968	0.72
1969	0.78
1970	0.80
1971	0.78
1972	0.95
1973	1.35

Make a line chart showing this relationship. Assignment 2: The average cost of pork varied somewhat differently. Add a second line to your chart showing a comparison of the cost of beef to the cost of pork over the same period of time. Use colored pencils or different types of lines for contrast.

1964	$0.60
1965	0.65
1966	0.75
1967	0.72
1968	0.69
1969	0.70
1970	0.80
1971	0.85
1972	0.90
1973	1.10

Assignment 3: Add a line to the chart showing the average cost of poultry for the same ten years.

1964	$0.22
1965	0.26
1966	0.30
1967	0.20
1968	0.23
1969	0.35
1970	0.40
1971	0.33
1972	0.38
1973	0.45

Fig. 23–42 Draw a flow chart showing how to mass-produce a project of your choice in the school shop.

Fig. 23–43 Make an organizational chart showing the administrative structure of your school.

Fig. 23–44 Assignment 1: Draw a pie chart or a one-column bar chart showing a breakdown of the average person's income if 24.5% goes to federal taxes, 5.8% goes to state taxes, and 1.3% goes to local taxes. Be sure the figures total 100%. Assignment 2: Compute the dollar value of each category for a gross income of $12,500. Mark each on your chart. Be sure the figures total $12,500.

Fig. 23–45 Assignment 1: Make a pictorial chart (pictograph) showing male and female population in your grade in school. Assignment 2: Make a bar chart showing male and female population in your drawing class.

Fig. 23–46 Draw a pictorial chart showing the enrollment of technical drawing classes in your school. Your instructor can supply the information.

Fig. 23–47 Assignment 1: Make a line graph showing the hourly change in outside temperature for a 12-hour period during any day. Assignment 2: Record similar information for several days and make a multiline chart to show a comparison.

Fig. 23–48 Draw a bar chart to show home consumption of electricity for 1 year as follows:

Month	*Kilowatt Hours Used*
January	900
February	885
March	800
April	783
May	722
June	600
July	494
August	478
September	525
October	650
November	735
December	820

Fig. 23–49 Plot the batting averages of the players on your favorite baseball team.

Fig. 23–50 The number of cars per mile of road in the United States is growing. Draw a pictorial chart from the data below to show the growth and anticipated increase.

Year	*Cars Per Mile of Road*	*(or)*	*Cars Per Kilometer of Road*
1930	9		2
1950	15		5
1960	20		8
1970	26		12
1980 (est.)	38		17

Fig. 23–51 **Draw a vertical bar chart showing the following student attendance for a given week of school. The total school enrollment is 925.**

Day	*Attendance*
Monday	625
Tuesday	715
Wednesday	800
Thursday	775
Friday	695

Fig. 23–52 **Assignment 1: Compute your daily calorie intake for 1 week. Make a line chart representing this information. Assignment 2: Use the same information and prepare a horizontal bar chart.**

Fig. 23–53 **Make a multiline chart representing individual game scores of the top five players on the school basketball team for any given season.**

Fig. 23–54 **Assignment 1: From the stock-market listings in the newspaper, select any stock and record its daily status for 10 days. Plot the information on a line chart. Assignment 2: Select several stocks and make a multiline chart showing a comparison of growth and decline.**

Fig. 23–55 **Make a pictorial chart or a pie chart showing a breakdown of the source of each dollar received by the federal government. Use the information given below.**

Individual income tax	$0.38
Employment tax	0.26
Corporate income tax	0.14
Borrowing	0.10
Excise tax	0.07
Other (miscellaneous)	0.05

Fig. 23–56 **Make a pictorial chart or a pie chart showing a breakdown of the expenditure of each dollar by the federal government. Use the information given below.**

National defense	$0.31
Income security	0.27
Interest	0.08
Health	0.07
Commerce, transportation, housing	0.06
Veterans	0.05
Education	0.04
Agriculture	0.03
Other (miscellaneous)	0.09

Fig. 23–57 **The information listed in the next column includes five common foods and the number of calories and grams of carbohydrates in a 4-ounce serving of each. Make a bar chart illustrating these facts.**

Food	*Calories*	*Carbohydrates*
Chocolate ice cream	150	14
Peas	75	14
Pizza	260	29
Milk	85	6
Strawberries	30	6

Fig. 23–58 **Assignment 1: The data in the list below represent a percentage breakdown for a family budget. Make a pie chart illustrating this information.**

Food	23.1%
Housing	24.0%
Transportation	8.8%
Clothing	10.9%
Medical Care	5.6%
Income Tax	12.5%
Social Security	3.8%
Miscellaneous	11.3%

Assignment 2: Compute the dollar value of each category for a gross income of $15,000. Mark each on your chart. Be sure the figures total $15,000.

Fig. 23–59 **Accidents involving children occur in various places. Draw a horizontal bar chart or a pie chart, using the places and percentages given below.**

At home	25%
Between home and school	8%
On school grounds	15%
In school buildings	21%
In other places	31%

Chapter 24

Electrical and Electronics Drafting

THE ELECTRICAL INDUSTRY

Progress in making electric power started with Thomas Alva Edison's electricity generating station in New York City in 1882. Since then electricity has become one of humanity's practical servants. It has completely changed the manufacturing, communication, and utility industries. You can see how quickly the electrical industry has moved ahead by how communication and appliances have changed. A good example is the tiny electronic parts used in controlled systems. Tiny circuits placed on a single chip of silicon are taking the place of transistors. Of course, a few years ago the transistor took the place of the vacuum tube.

CAREER OPPORTUNITIES

The electronics industry is one of the biggest manufacturing industries in the country. This industry offers special opportunities to young women and men who can make the freehand and formal drawings it needs. Electronic drafting is based upon the same basic rules of orthographic projection and dimensioning as all other drafting. However, in electronic drafting, you must also be able to prepare schematic diagrams, block diagrams, and technical illustrations.

PREPARING FOR OPPORTUNITIES

Industry provides training programs in electrical and electronics drafting. However, it helps if you have learned drafting and electronics in high school before entering such programs. A few courses at technical college or junior college in electronic drafting and in *electromechanical systems* (moving things with electric power) will help you if you are looking for a technician rating on the design team. To learn more about careers, write to the following groups:

IEEE — Institute of Electrical and Electronic Engineers
345 East 47 Street
New York, NY 10017

EIA — Electronic Industries Association
2001 I Street N.W.
Washington, D.C. 2006

ASCET — American Society of Certified Engineering Technicians
2029 K Street N.W.
Washington, DC 20006

AFTE — American Federation of Technical Engineers
900 F Street N.W.
Washington, DC 20004

THE ELECTRICAL OR ELECTRONIC DRAFTER

An electrical or electronic drafter on an engineering design team must always develop new skills. The drafter must always be learning new things about the field he or she works in. The drafter must show good judgment, skill, and originality when working from design sketches or written instructions. The electrical or electronic drafter may be able, after training and a few years on the job, to be a technician on a design team.

Design teams may develop electrical equipment. This equipment can be used to make, send, or distribute electrical power. Design teams may also make electronic instruments or appliances. There are career opportunities in making equipment such as computers or other complex information systems. Future electronic developments will be more spectacular than those of today. Scientists, along with electrical engineers, designers, technicians, and drafters, seem to be building an electrical and electronic world.

ELECTRONIC ENVIRONMENT

People have always tried to control their *environment* (surroundings). Today, electronic devices

Fig. 24–1 **A room specially designed for an electronic environment.** ***(Motorola, Inc.)***

control the air we breathe at home and at school. Electronic devices have been made that control the cooling, heating, lighting, and sound systems for different ways of living.

The electronic room shown in Fig. 24–1 is inside a 7620 mm (25 ft) fiberglass dome. The room has a recessed living area in the center. This arrangement makes it easier to relax. The room also has an electronic sight and sound system. *Mood lighting* (lighting that changes brightness or color) responds to the remote-controlled color television, the television recorder, and the sound system.

The floor plan (Fig. 24–2) shows the major *components* (parts) of the electronic environment. The center table at 1 is the control center for the components of the room. The rectangle at 2 is a color television set. The rectangle at 3 is an EVR (electronic video recorder) that tapes television shows. The rectangles numbered 4 are speaker towers. They also contain the mood lighting that changes with the sound. The sketch (Fig. 24–3) helps to identify the parts. Figure 24–4 is an overall view of this electronic environment. The center table rises mechanically to show the controls for the sound system, color television, and EVR. The overhead lighting is *kinetic* (moving). It changes with the beat of the music.

Fig. 24–2 **The floor plan of the electronic room.** ***(Motorola, Inc.)***

Fig. 24–3 **A sketch of the electronic room.** ***(Motorola, Inc.)***

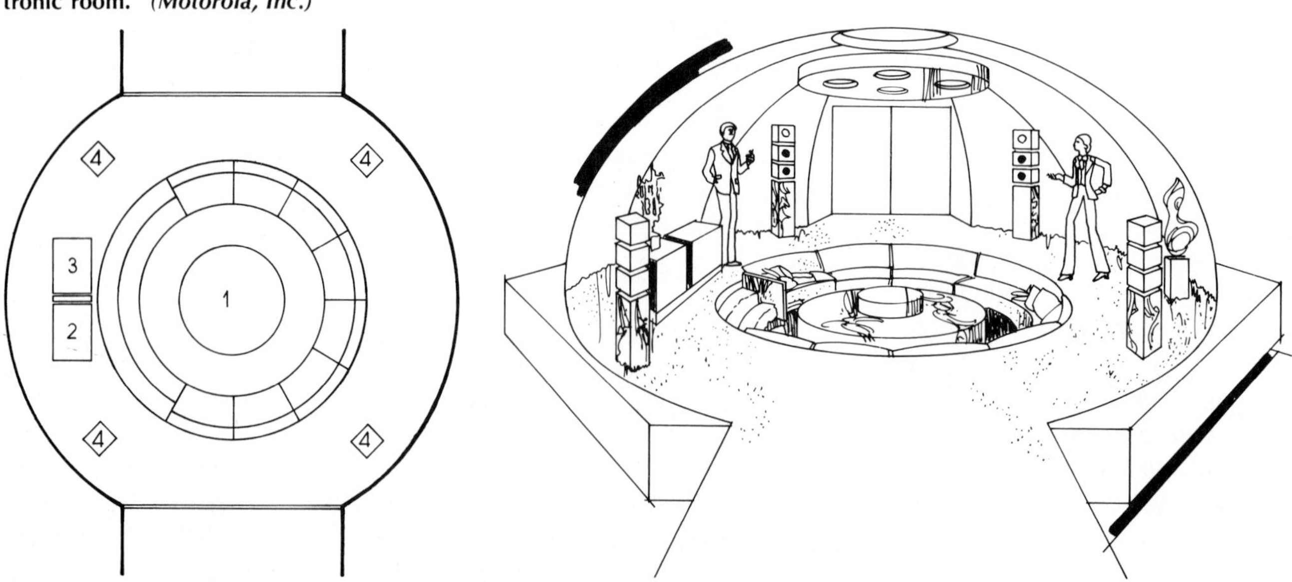

You can see three of the four sound towers that hold speakers and special lighting. Each tower has three speakers, one each for high-, low-, and middle-range sound.

The electric circuits for this room must be planned as a wiring diagram. The electrical design engineer makes this diagram. The electronic components for the room must be designed by the electronic design team.

Fig. 24–4 **An overall view taken through a fish-eye lens.** ***(Motorola, Inc.)***

ELECTRICAL AND ELECTRONIC DRAFTING

If you have had a basic course in electricity, you will find it easier to understand and make electrical or electronic drawings. This chapter will introduce you to electrical and electronic symbols, wiring diagrams, and circuit diagrams. The chapter will also relate these to electrical and electronic drafting. If you have not had a basic course in electricity, the paragraphs that follow will give you a brief explanation of the subject.

ELECTRICITY

The source of electrical energy is the tiny *atom*. Normally, all atoms are made up of many kinds of *particles*. One of these particles is the *electron*. The electron is the most important particle in the study of electricity and electronics.

The electrons in an atom rotate around the *nucleus,* or center, of the atom in definite paths. These paths are called *orbits* (Fig. 24–5). All electrons in all atoms are the same. Each electron has what is called a *negative charge* of electricity.

Different kinds of atoms have different numbers of electrons and other particles. When atoms have the same number of electrons and the same number of *protons* (a particle in the nucleus), they are the same *element*. Copper, gold, and lead are examples of elements.

When atoms join together, they form *molecules*. When different kinds of atoms join, they form *compounds*. Water, acids, and salt are common compounds.

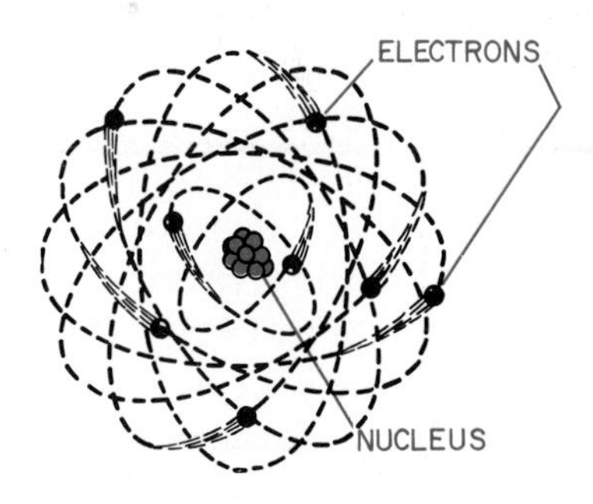

Fig. 24–5 **The structure of an atom.**

VOLTAGE AND CURRENT

Sometimes, electrons can be made to leave their "parent" atoms, the atoms of which they were originally part. This happens, for example, when you connect a piece of wire across the *terminals* (electrical connections) of a battery. The battery produces an electrical pressure called *voltage*. The symbol for voltage is V. The voltage causes a steady stream of electrons to flow through the wire. Connect a light bulb (load) to the wire. Electrons will move through the *lamp filament* (thin wire in the bulb) from the battery (power source) (Fig. 24–6).

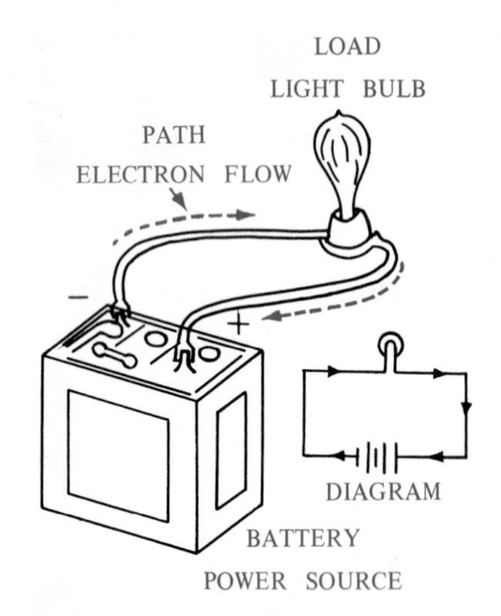

Fig. 24–6 **A simple electric circuit.**

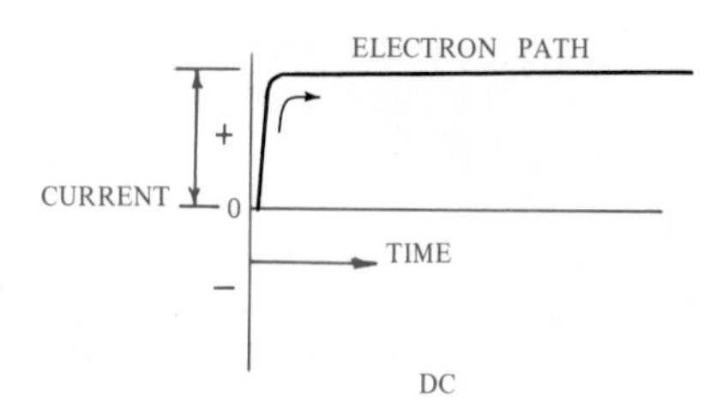

Fig. 24–7 **Direct current attains magnitude and keeps it as long as the circuit is complete.**

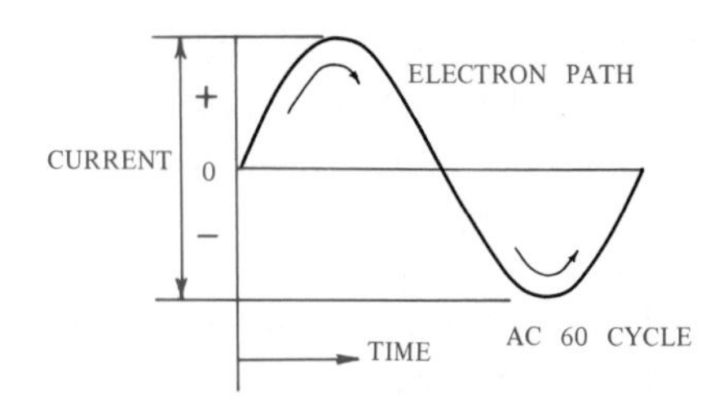

Fig. 24–8 **Alternating current builds up from zero to a maximum in a positive direction, falls to zero, and then builds to a maximum in a negative direction and falls back to zero.**

The energy of the moving electrons is changed into heat energy as the filament becomes white hot. The glow of the filament produces the light.

The electron pathway is formed by the battery, the wire, and the lamp filament. This is a simple form of *electrical circuit.* In other circuits, electrical energy is changed into other kinds of energy. Some of these kinds of energy are magnetism, sound, and light.

A *direct current* (dc) is a flow of electrons through a circuit in one direction only (Fig. 24–7). An *alternating current* (ac) is a flow of electrons in one direction during a fixed time period, and then in the opposite direction during a similar time period (Fig. 24–8). One complete alternation is called a *cycle.* The number of times this cycle is repeated in one second is called the *frequency* of the alternating current, such as 60 cycle. We measure frequency in *Hertz,* for which the symbol is Hz. We measure the amount of current in *amperes,* for which the symbol is I.

RESISTANCE

Electrons can move through some materials more easily than through other materials. Electrical current will flow more easily through a copper wire than through a steel wire of the same size. We say that the steel offers more *resistance* than copper. Materials with small resistance to the flow of electrons are called *conductors.* Silver is the best conductor known. However, it costs too much for general use. Copper and aluminum are good conductors. They are the most widely used. Materials through which electrons will not flow easily are called *insulators.* The insulators used most often are glass, porcelain, plastics, and rubber compounds. We measure resistance in *ohms,* and the symbol is R.

ELECTRICITY AND ELECTRONICS

Electricity has to do with the flow of electrons moving through wires or other metal conductors. Common examples are house wiring systems, generators, and transformers. Electricity refers to an energy source.

Electronics has to do with the flow of electrons moving through metal conductors and conductors other than metals. Some of these other conductors are gases, vacuums, and materials called *semiconductors.* The most common semiconductors are transistors and diodes made of germanium or silicon. Electronics refers to devices that make use of electricity.

Both electricity and electronics deal with electrons flowing through circuits. These circuits carry energy for a definite purpose.

BASIC ELECTRICAL UNITS

As we have seen, volts are used to measure pressure, ohms to measure resistance, and amperes to measure current. In addition, we use *watts* (symbol: w) to measure power. You can often tell the value of a unit used to show an electrical *quantity* (amount) by looking at the *prefix* (beginning) of the word. Take, for example, the unit kilovolt (kv). *Kilo* means a thousand. Thus, one kilovolt equals 1000 volts. Similarly, 1 kilowatt (kw) equals 1000 watts, and 1 kilohm (kohm) equals 1000 ohms. Another prefix is *milli,* which means a thousandth. Thus, 1 milliampere (ma) equals one one-thousandth (0.001) ampere. For other unit prefixes, see Appendix C on the metric system.

BASIC FORMULAS

The amounts of voltage, current, and resistance in a circuit are related. This relationship is called *Ohm's law.* Ohm's law may be expressed as shown in the 12 formulas of Fig. 24–9. V = volts (pressure), I = amperes (current), and R = ohms (resistance). These letters are used in mathematical formulas to mean unknown amounts. They are not SI metric symbols. The SI unit symbols are V for electrical pressure, A for current, and R for resistance. These SI unit symbols should only be used with known amounts.

BASIC ELECTRIC CIRCUITS

These include series circuits, parallel circuits, and series and parallel circuits together. These terms are explained in the following paragraphs.

Series Circuits

Series circuits are those where the current flows from the source (battery, generator, and so forth) through one resistance (lamp, motor, and so forth) after another. This is shown in Figs. 24–10, 24–11, and 24–12.

In Fig. 24–10, a bell *(A)* gets its power from a battery *(C)* when the circuit is closed (when the electron pathway is complete). The circuit is normally open. The pushbutton *(B)* closes the circuit.

In Fig. 24–11, a buzzer *(A)* gets its power by the current from the transformer *(C)*. What is item *B?* What does it do in this circuit?

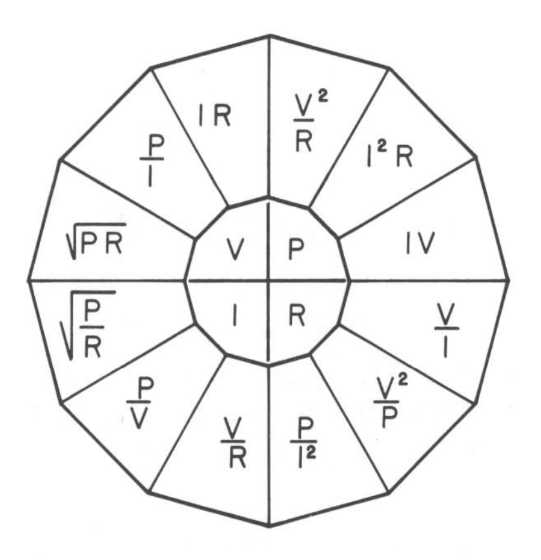

Fig. 24–9 **Power expressed as formulas in Ohm's law.**

In Fig. 24–12, four lamps *(C, D, E,* and *F)* get power from a generator *(A)* when the fused switch *(B)* is closed. All the lights must be on. If any one is not, the circuit will be open. Some strings of Christmas tree lights go out completely when just one lamp burns out. They are connected in series.

Parallel Circuits

Parallel circuits let the current flow through more than one path. This is shown in Figs. 24–13 and 24–14.

You can see three separate branches, or paths *(C, D,* and *E),* with lamps in Fig. 24–13. Each lamp is separate from the others. If one lamp is burned out, the others will still work. With a parallel string of lights on your Christmas tree, the remaining lights will still burn if some are missing, loose, or burned out.

You can see a siren in Fig. 24–14. It can be turned on by any of the pushbuttons *A, B, C,* or *D.* These buttons are all connected in parallel. A good use of this would be in an alarm system to warn of an

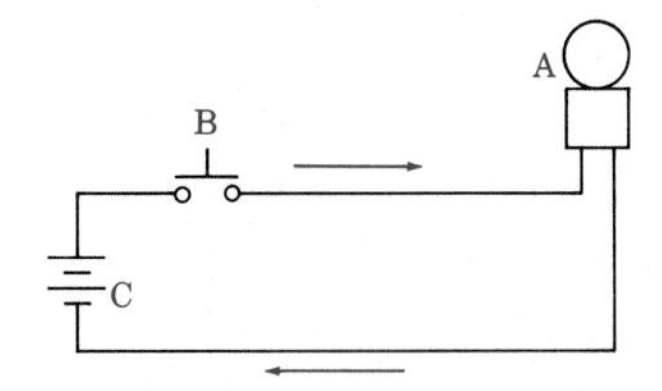

Fig. 24–10 **A series-circuit diagram.**

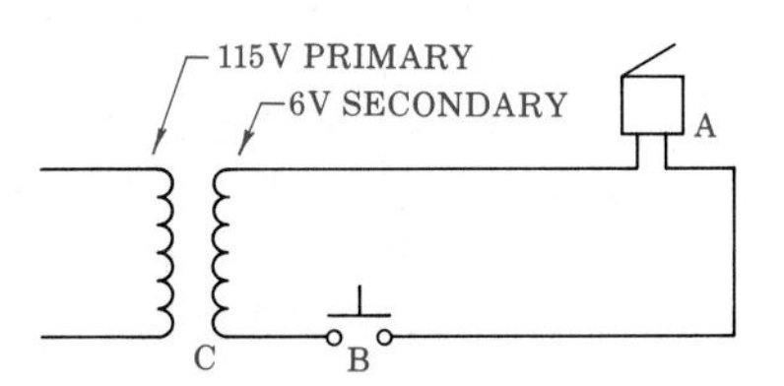

Fig. 24–11 **A series-circuit diagram.**

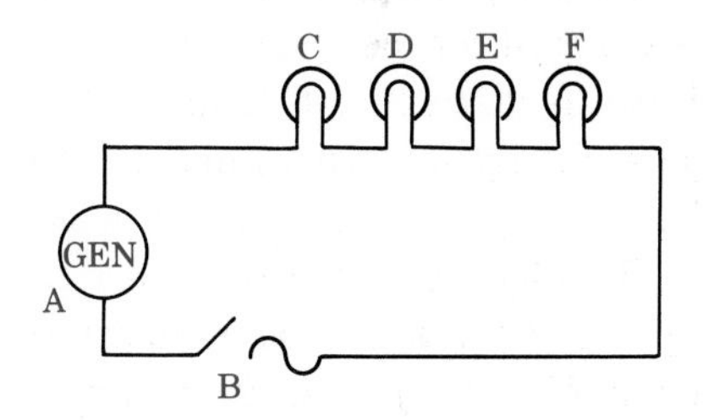

Fig. 24–12 **A series-circuit diagram.**

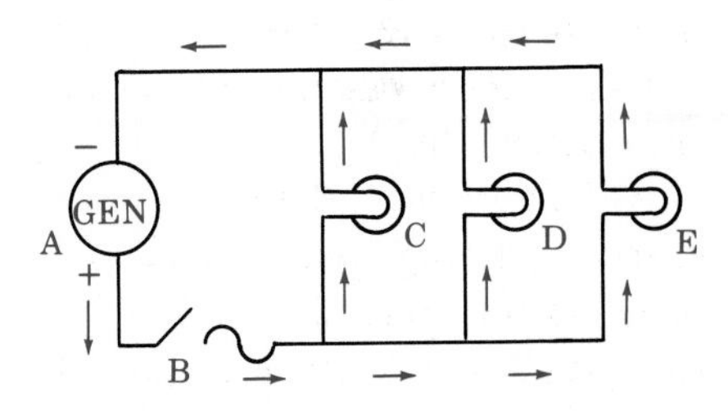

Fig. 24–13 **A parallel-circuit diagram.**

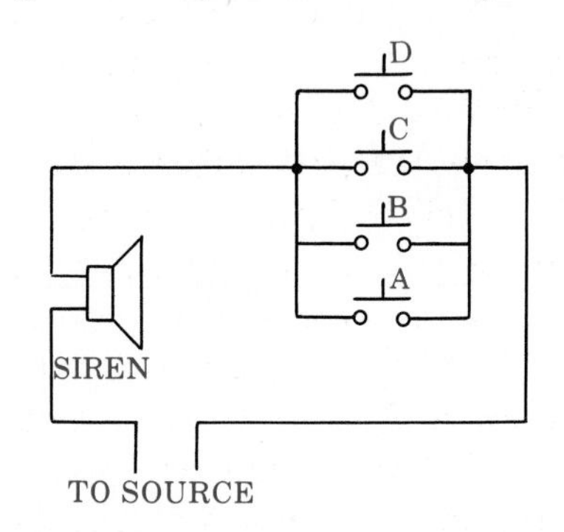

Fig. 24–14 **A parallel-circuit diagram.**

attempted holdup in a store. The pushbuttons, connected in parallel, would be under counters and in the cashier's office.

Notice that the symbol for the siren is the same as for a loudspeaker. So they will not be confused, a note "SIREN" has been added.

Combination Circuits

Combining series and parallel circuits lets you have many different arrangements. In Fig. 24–15, lamps *C* and *D* are in series. Lamps *E* and *F* are in parallel. Both lamps *C* and *D* must be on if switch *A* is closed, since they are in series. When switches *A* and *B* are closed, all the lamps (*C, D, E,* and *F*) are lighted. Lamps *E* and *F* will work separately. If one fails, the other will stay lighted because they are in parallel. However, because lamps *C* and *D* are in series, when one fails, the other will not light, as we have learned from the Christmas tree lights.

ELECTRICAL INSTRUMENTS

You can use many kinds of electrical instruments to measure electricity. Two main ones are the ammeter and the voltmeter. The *ammeter* measures electric current in amperes. To measure the amount of current flowing through a resistance (light, motor, and so on), connect the ammeter directly in series with the resistance you want to measure. This is shown in Fig. 24–16.

The *voltmeter* measures the electromotive force (pressure) in volts. Connect the voltmeter in parallel with the part of a circuit where you want to measure the voltage. This is shown in Fig. 24–17.

Figure 24–18 shows both an ammeter and voltmeter connected in a circuit. The ammeter measures the current flowing through the resistance R. The voltmeter measures voltage flowing across the resistance. The amperes and the volts are then measured.

GRAPHIC SYMBOLS

We use graphic symbols on electrical and electronic diagrams to show the components and workings in a circuit. You can draw symbols quickly and easily by using templates (Fig. 24–19). You can also use grooved templates when you draw electrical symbols. These are like the templates used with scriber guides for lettering. (See Fig. 16–15.)

The graphic symbols in Fig. 24–20 and the following sections about using graphic symbols are adapted from the *American National Standard Graphic Symbols for Electrical and Electronics Diagrams* (ANSI Y32.2) by permission of The Institute of Electrical and Electronic Engineers, Inc. (IEEE).

Graphic symbols for electrical engineering are a shorthand way to show through drawings how a circuit works or how the parts of the circuit are connected. A graphic symbol shows what a part in the circuit does. Drafters use graphic symbols on single-line (one-line) diagrams, on schematic diagrams, or on connection or wiring diagrams. You can relate graphic symbols with parts lists, descriptions, or instructions by marking the symbol.

Drafting and Graphic Symbols

1. A symbol is made up of all its various parts.

Fig. 24–15 A combination series and parallel circuit.

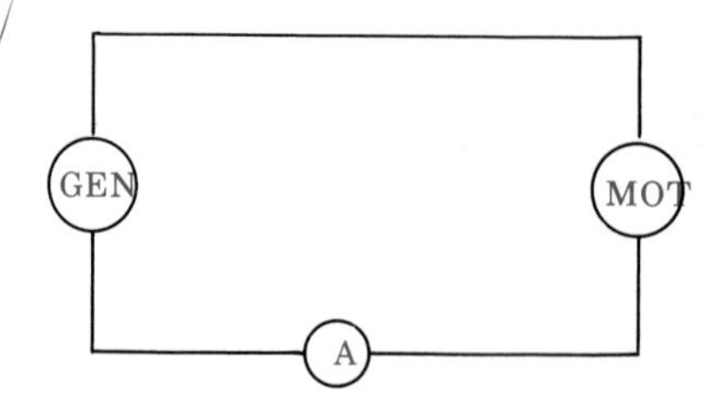

Fig. 24–16 Ammeter connection in series, in a circuit.

Fig. 24–17 Voltmeter connection in parallel, in a circuit.

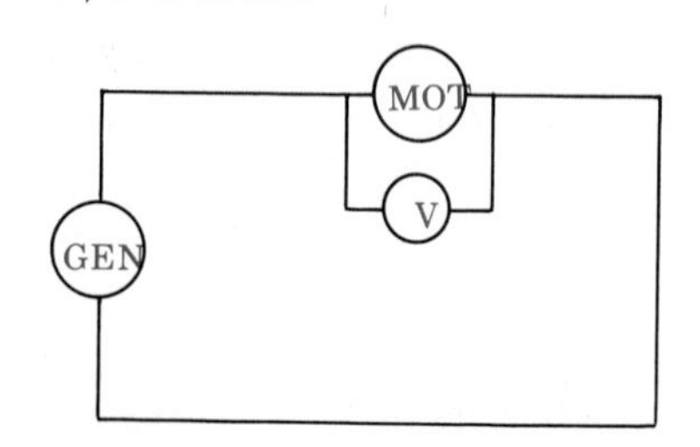

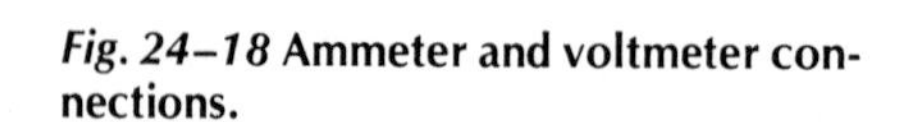

Fig. 24–18 Ammeter and voltmeter connections.

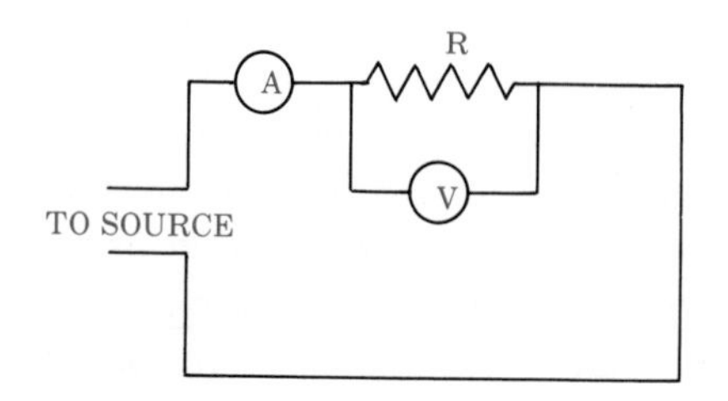

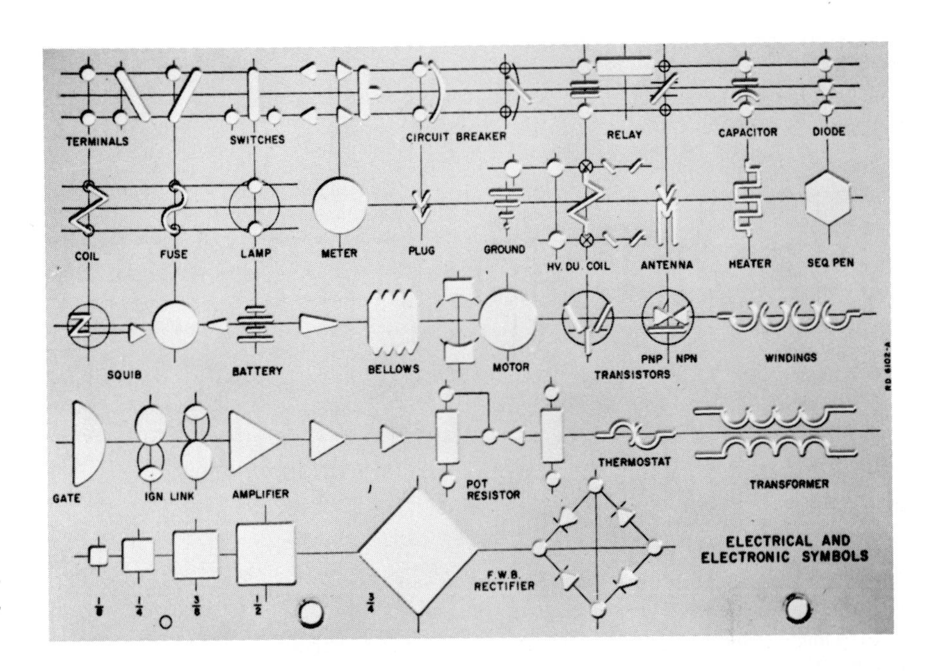

Fig. 24–19 **Template for drawing electrical and electronic symbols.** ***(RapiDesign, Inc.)***

2. The direction a symbol is facing on a drawing does not change its meaning. This is true even if the symbol is drawn backwards.

3. The width of a line does not affect the meaning of the symbol. Sometimes, however, you can use a wider line to show that something is important.

4. Each symbol shown in this standard is the right size related to all the other symbols. That is, if you draw one of these symbols twice as big as shown in the standard, any other symbol you draw should also be twice as big.

5. You can draw a symbol any size you need. Symbols are not drawn to scale. However, their size must fit in with the rest of the drawing. Sometimes, to set something off, you can draw a symbol smaller than the other symbols on a diagram.

6. You can draw the arrowhead of a symbol closed ➞ or open →.

7. You can add the standard symbol for a *terminal* (○) to any one of the graphic symbols used where connecting lines are attached. These are not part of the graphic symbol unless the terminal symbol is part of the symbol shown in this standard.

8. To make a diagram simpler, you may draw a symbol for a device in parts. If you do this, you must show how the parts are related.

9. Most of the time, it does not matter at what angle you draw a line connected to a graphic symbol.

10. Sometimes, you may want to draw paths and equipment that will be added to the circuit later, or that are connected to the circuit but are not part of it. You do this by drawing lines made up of short dashes: - - - -.

11. If you need details of type, impedance, rating, and so on, you may draw them next to a symbol. The abbreviations you use should be from the *American National Standard Abbreviations for Use on Drawings* (Y1.1–1972). Letters that are joined together and used as parts of graphic symbols are not abbreviations.

CIRCUIT COMPONENTS

Figure 24–21 shows and names some of the electrical and electronic components most often used. A symbol for a component should look like that component. You should know what a component does and how it works.

ELECTRICAL DIAGRAMS

There are many kinds of electrical diagrams. Each kind of diagram suits its purpose. The definitions that follow are from the *American National Standard Drafting Manual, Electrical Diagrams* (ANSI Y14–15–1966) with the permission of the publisher, The American Society of Mechanical Engineers, 345 East 47 Street, New York, NY.

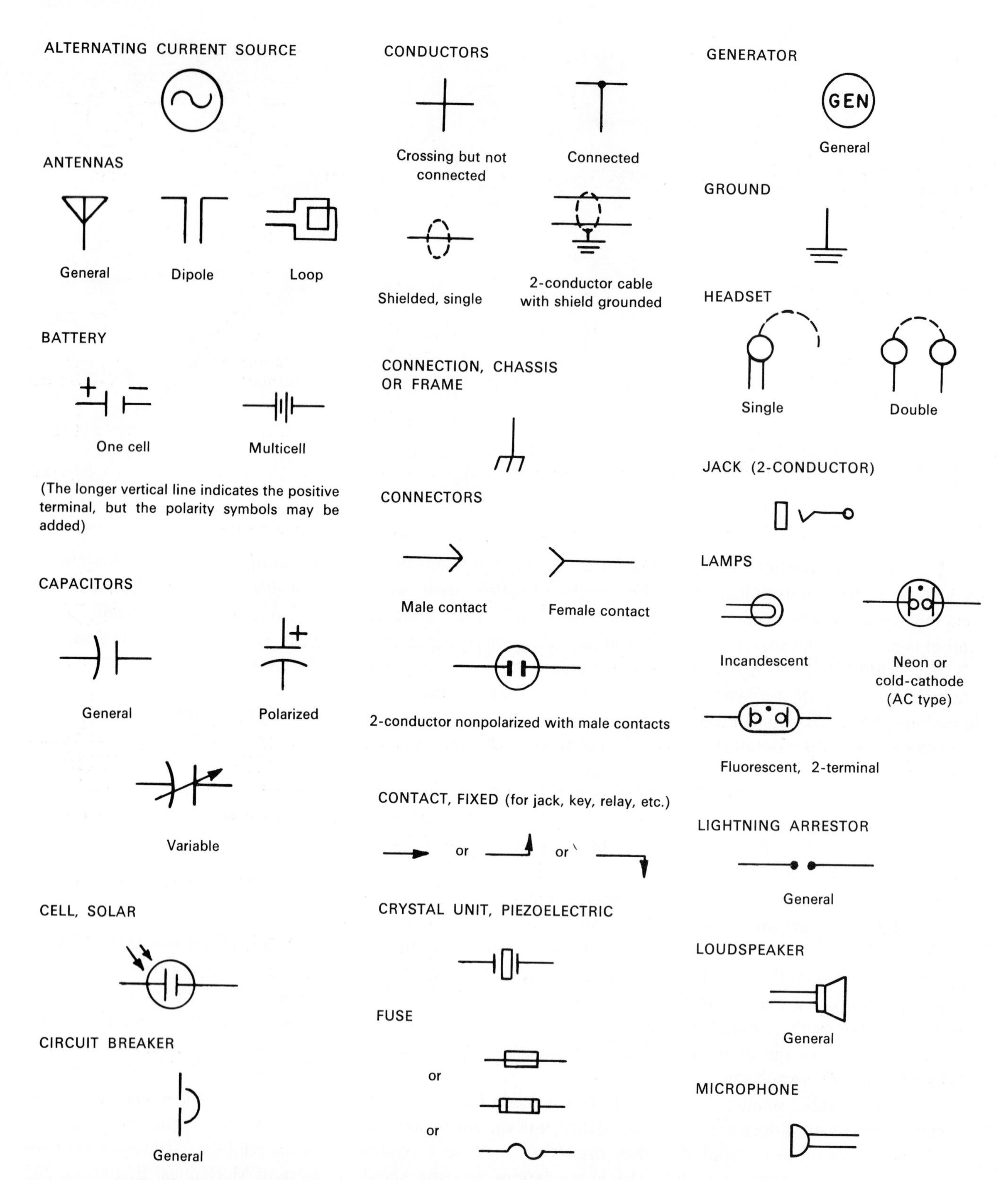

Fig. 24–20 **"American National Standard Graphic Symbols for Electrical and Electronics Diagrams," ANSI Y32.2–1962.** ***(With permission of the Institute of Electrical and Electronic Engineers, Inc.)***

METER

To indicate a specific type of meter, replace the asterisk by one of the following letters or letter combinations

A	Ammeter
F	Frequency meter
G	Galvanometer
UA	Microammeter
MA	Milliammeter
OHM	Ohmmeter
V	Voltmeter
W	Wattmeter
WH	Watthour meter

MOTOR

General

PLUG (2-CONDUCTOR)

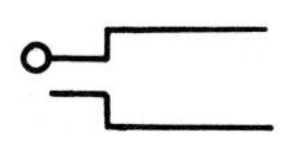

RECTIFIER (SEMICONDUCTOR DIODE OR METALLIC

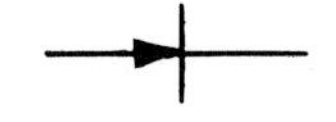

RESISTOR (general)

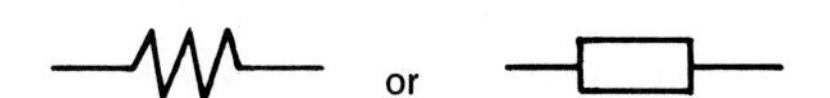

(When the rectangular symbol is used, always add identification within or adjacent to the rectangle)

RESISTOR (with adjustable contact)

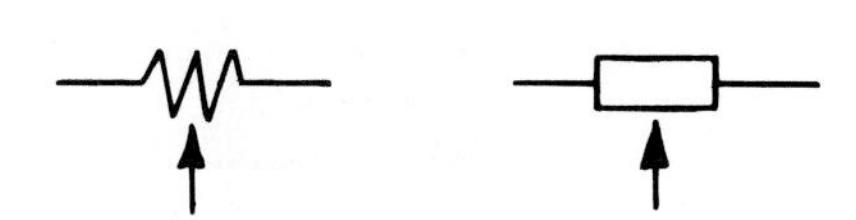

SWITCH, PUSHBUTTON

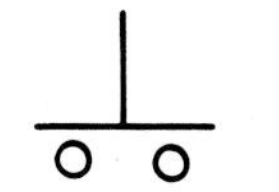

Circuit closing (make)

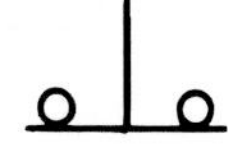

Circuit closing (break)

SWITCHES

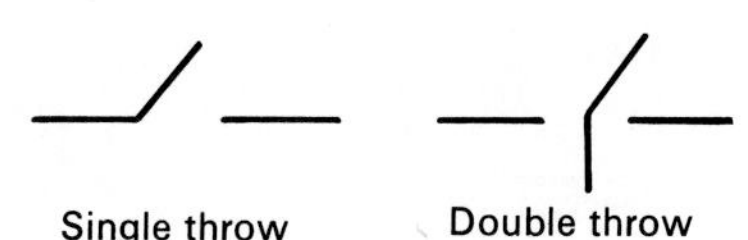

Single throw (general)

Double throw (general)

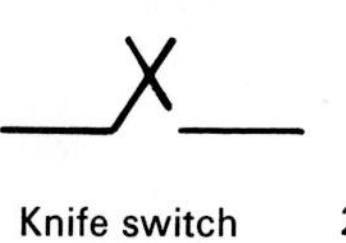

Knife switch (General)

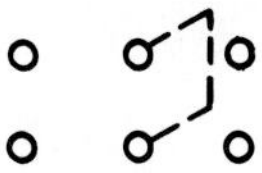

2-pole double throw with terminals shown

THERMISTOR RESISTOR, THERMAL

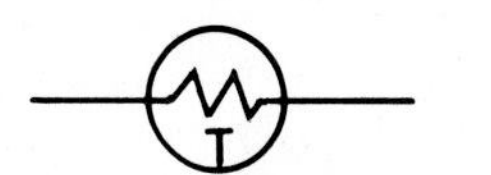

THERMOCOUPLE, TEMPERATURE MEASURING

THERMOSTAT (with break contact)

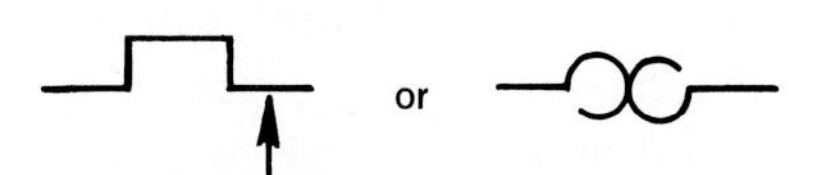

TRANSFORMER

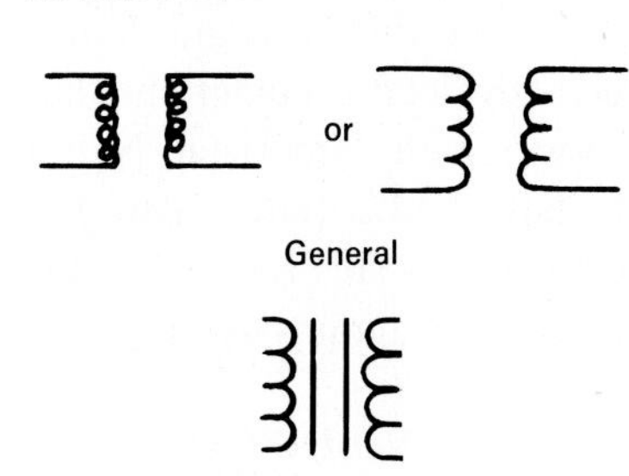

General

Magnetic (iron) Core

TRANSISTORS

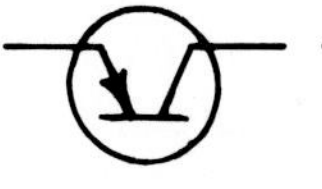

PNP 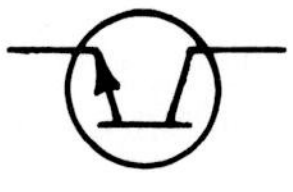 NPN

TUBES, ELECTRON

(a) Components or parts of

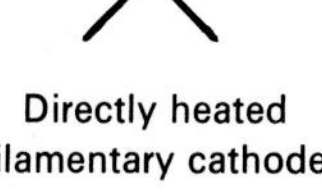

Directly heated filamentary cathode

Indirectly heated cathode

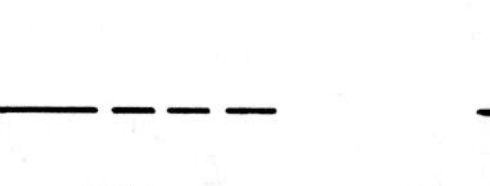

Grid

Plate or Anode

(b) Examples

Diode

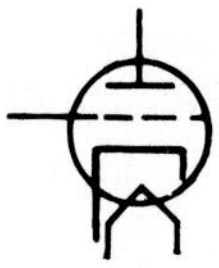

Triode

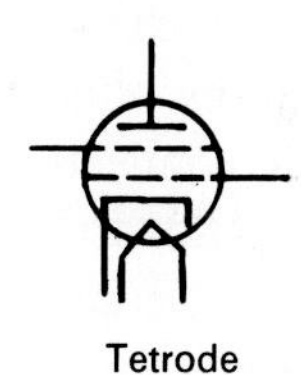

Tetrode

Pentode

WINDING, INDUCTOR, or REACTOR (Coil)

General

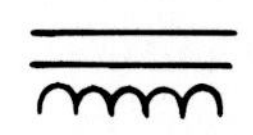

Magnetic (iron) Core

Fig. 24–20 **(Cont.)**

Single-line (one-line) diagram. A diagram that shows, using single lines and graphic symbols, the course of an electric circuit and the parts of the circuit.

Schematic, or elementary, diagram. A diagram that shows, using graphic symbols, the ways a circuit is connected and what the circuit does. The schematic diagram does not have to show the size or shape of the parts of the circuit. It does not have to show where the parts of the circuit actually are.

Connection or wiring diagram. A diagram that shows how the components of a circuit are connected. It may cover connections inside or outside the components. It has as much detail as is needed to make or trace connections. The connection diagram usually shows how a component looks and where it is placed.

Interconnection diagram. A kind of connection or wiring diagram that shows only connections outside a component. An interconnection diagram shows connections between components. The connections inside the component are usually left out.

Fig. 24–21 Some electrical and electronic components, with their names and appearance. ***(Heath Company, a subsidiary of Daystrom, Inc.)***

LINE CONVENTIONS AND LETTERING

Sometimes the size of a drawing is changed. When this is done, remember to choose a line thickness and letter size that will still let people understand the drawing after it is made larger or smaller. A guide for drawing lines on electrical diagrams is shown in Fig. 24–22.

You should draw a line of medium thickness for general use on electrical diagrams. You may use a thin line for brackets, leader lines, and so on. When you need to set off something special, such as main or transmission paths, use a line thick enough to show the difference. Line thickness and lettering used with electrical diagrams should conform with American National Standard Y14.2 (latest issue) and local needs. This is so that microfilm of the diagrams can be made.

SYMBOLS AND LAYOUTS

You can draw a symbol any size you need. However, its size must fit in with the rest of the drawing. Keep in mind whether the drawing

Fig. 24–22 **Line conventions for electrical diagrams.**

LINE APPLICATION	LINE THICKNESS
FOR GENERAL USE	MEDIUM
MECHANICAL CONNECTION, SHIELDING, & FUTURE CIRCUITS LINE	MEDIUM
BRACKET-CONNECTING DASH LINE	MEDIUM
USE OF THESE LINE THICKNESSES OPTIONAL	
BRACKETS, LEADER LINES, ETC.	THIN
BOUNDARY OF MECHANICAL GROUPING	THIN
FOR EMPHASIS	THICK

will be made larger or smaller. For most electrical diagrams meant to be used for manufacturing, or for use in a smaller form, you should draw symbols about 1.5 times the size of those shown in American National Standard Y32.2.

Layout of electrical diagrams. You should lay out electrical diagrams so that the main parts are easily seen. The parts of the diagram should have space between them. This is so there will be an even balance between blank spaces and lines. You should allow enough blank area around symbols so that notes or reference information will not be crowded. Avoid large spaces, however. Only allow large spaces if circuits will be added there later.

SINGLE-LINE DIAGRAMS

The single-line diagram (Fig. 24–23) tells in a basic way how a circuit works. It leaves out much of the detailed information usually shown on schematic or connection diagrams. Single-line diagrams let you draw complex circuits in a simple way. You can draw diagrams of communication or power systems where a single line means a multi-conductor communication or power circuit.

Most of the time, you can draw a single-line diagram with the same methods used to draw schematic diagrams.

SCHEMATIC DIAGRAMS

Following are some guidelines for making schematic diagrams.

Layout. Use a layout that follows the circuit, signal, or transmission path either from input to output, power source to load, or in the order that the equipment works. Do not use long interconnecting lines between parts of the circuit.

Connecting lines. It is better to draw connecting lines horizontally or vertically. Use as few bends and crossovers as possible. Do not connect four or more lines at one point if you can just as easily draw it another way. When you draw connecting lines parallel (side by side), the spacing between lines after making the drawing smaller should be no less than 2 mm (1/16 in.). You should draw parallel lines in groups. It is best to draw them in groups of three. You should allow double

Fig. 24–23 **Single-line diagram of an audio system.**

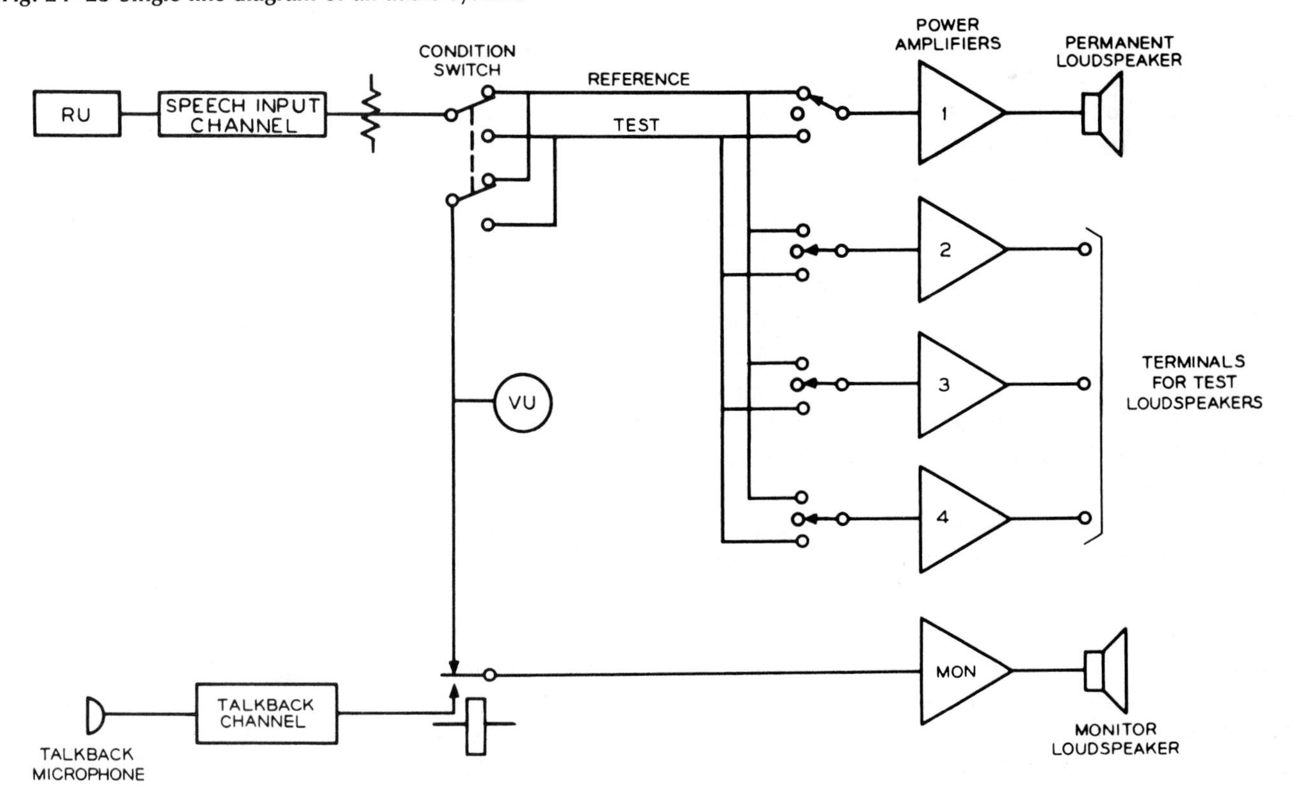

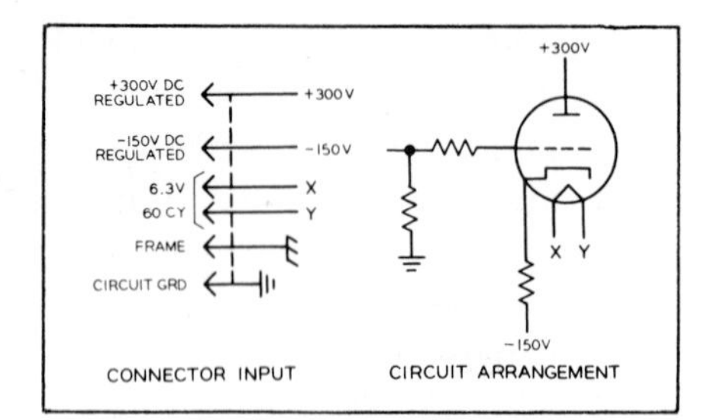

Fig. 24–24 **Identification of interrupted lines. At left, a group of lines interrupted on the diagram. At right, single lines interrupted on the diagram.**

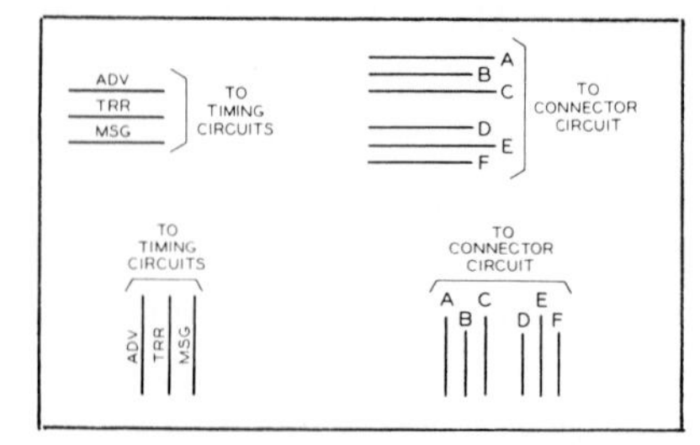

Fig. 24–25 **Typical arrangement of line identifications and circuit destinations.**

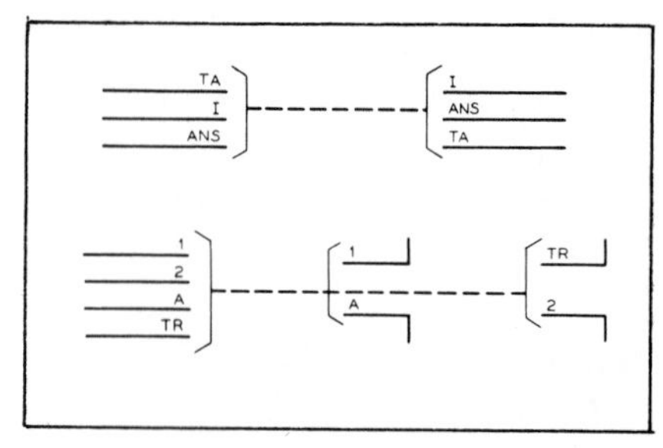

Fig. 24–26 **Typical interrupted lines interconnected by dashed lines. The dashed line shows the interrupted paths that are to be connected. Individual line indentifications indicate matching connections.**

spacing between groups of lines. When you group parallel lines, group them together according to what they do.

Interrupted single lines. When a single line is interrupted, you can show where the line is going in the same place you *identify* (name) it. This is shown in Fig. 24–24 for the power and filament circuit paths. The following section on interrupted group lines tells how to identify grouped and bracketed lines. You should do the same for single interrupted lines.

Interrupted grouped lines. When interrupted lines are grouped and bracketed, you should identify the lines as shown in Fig. 24–25. You can show at the brackets where lines are meant to go or where they are meant to be connected. Do this by using notes outside the brackets, as shown in Fig. 24–25, or as shown in Fig. 24–26 by using a dashed line between brackets. When you use the dashed line to connect brackets, draw it so that it will not be mistaken for part of one of the bracketed lines. You should begin the dashed line in one bracket and end it in no more than two brackets. When drawing schematics, carefully follow the rules given above.

SCHEMATIC DIAGRAMS FOR ELECTRONICS AND COMMUNICATIONS

The following tells in detail about schematic diagrams used with electronic and communication equipment. You should use this material with the general standards of schematic diagrams already learned.

Layout. In general, you should lay out schematic diagrams so that they can be read from left to right. Complex diagrams should generally be laid out to read from upper left to lower right. You may lay them out in two or more layers. Each layer should be read from left to right. The circuit layout should follow the signal, or transmission path, from input to output, or in the order that the circuit works. Where possible, you should draw endpoints for outside connections at the outer edges of the circuit layout.

Ground symbols. Use the ground symbol ⊥ only when the circuit ground is at a potential level equivalent to that of earth potential. Use the symbol ⏊ when you do not get an earth potential from connecting the ground wire to the structure that houses or supports the circuit parts.

SOME ELECTRICAL CIRCUITS

Look at Fig. 24–27. The bell *(C)* and the buzzer (*E*) get power separately from the same battery *(A)* by the pushbuttons (*B* and *D*).

In Fig. 24–28 the current is from an outside source. Three-way switches (*X* and *Y*) are used. This is done so that the light can be turned off or on by either of the switches. Switch *X* might be at the garage. Switch *Y* might be at the house. Each switch has three terminals. If either switch is opened, the light will be turned off. However, the light can be turned on by the switch at the other end.

You can see a circuit diagram in Fig. 24–29 for an *annunciator* (a device for signaling from different places to a station or post). When any of the buttons is pushed, a buzzer in the annunciator sounds. Each button releases or allows a tab to drop down. The tab shows the place where the button is pushed. Trace the circuits that begin with

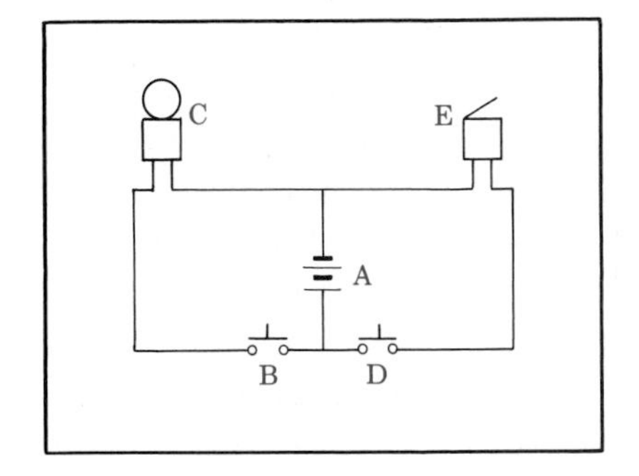

Fig. 24–27 **Bell and buzzer circuit.**

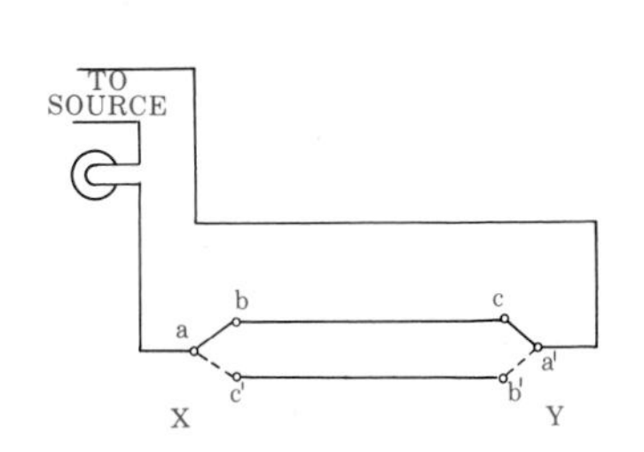

Fig. 24–28 **Three-way switch diagram.**

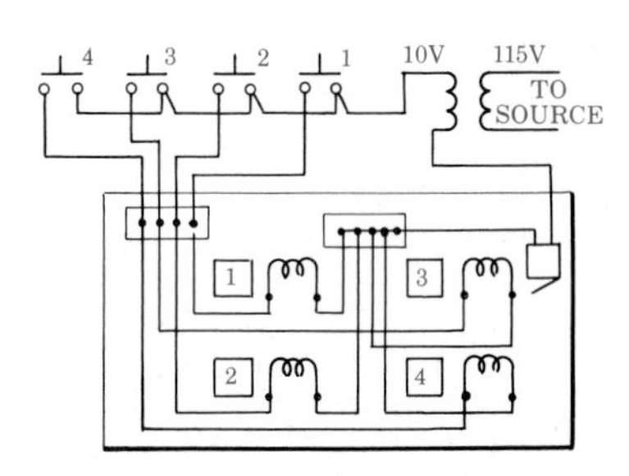

Fig. 24–29 **Annunciator diagram.**

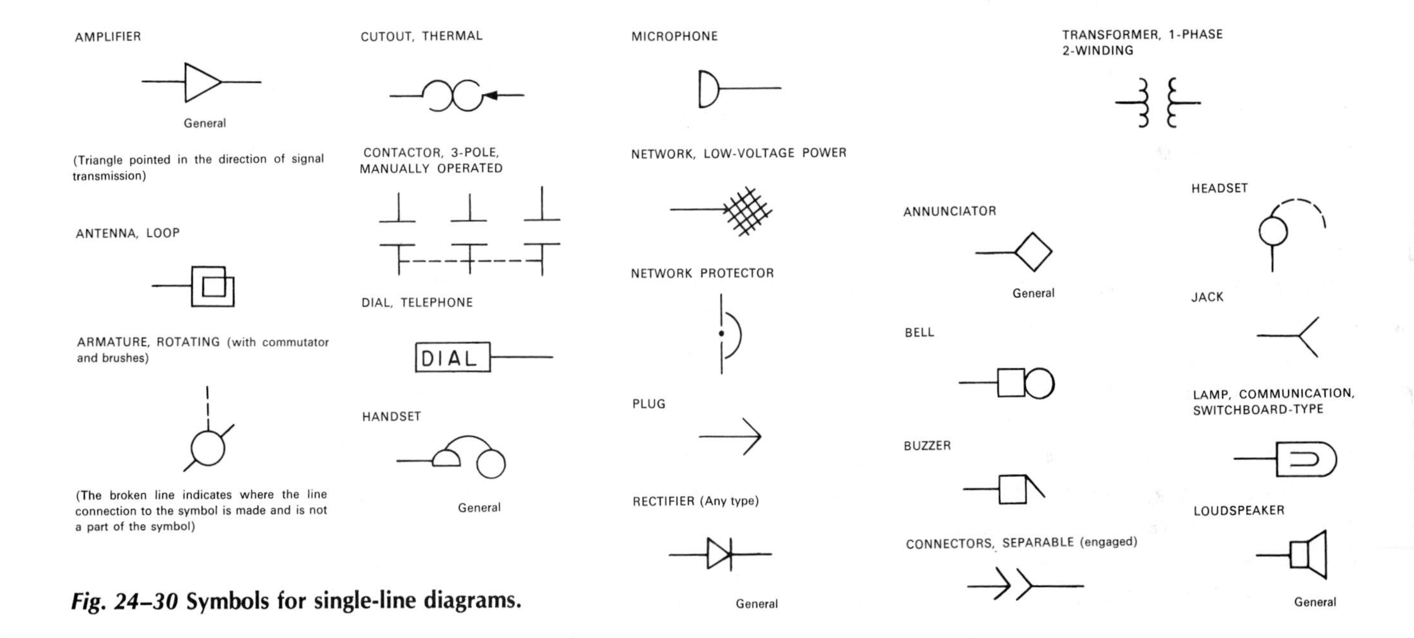

Fig. 24–30 **Symbols for single-line diagrams.**

each of the buttons. The source of the current is a transformer.

Some single-line graphic symbols are shown in Fig. 24–30. You can see a single-line, or one-line, diagram in Fig. 24–31. This diagram shows the component parts of a circuit. It uses single lines and graphic symbols. The single lines mean two or more conductors. The diagram shows all you usually need to know about how the circuit works. However, it is not as detailed as a schematic diagram. A single-line diagram is a simpler way to show circuits. It is often used in the fields of communications and electrical power *transmission* (sending) (Fig. 24–31). Drafters usually draw the highest voltage at the top or left side of the diagram. They draw the lower-voltage lines in the order of their value below or on the right side of the diagram. Drafters give other information at the proper places on the diagram. Such information includes where lines are, what equipment is used, and so on.

You can see a wiring diagram for a motor-starter at A in Fig. 24–32. You can also see a schematic, or one-line, diagram at B for the same circuit. The uses of a motor starter are given below:

1. To protect the circuit against burnouts caused by long *overloads* (too much power on the circuit). This is called thermal

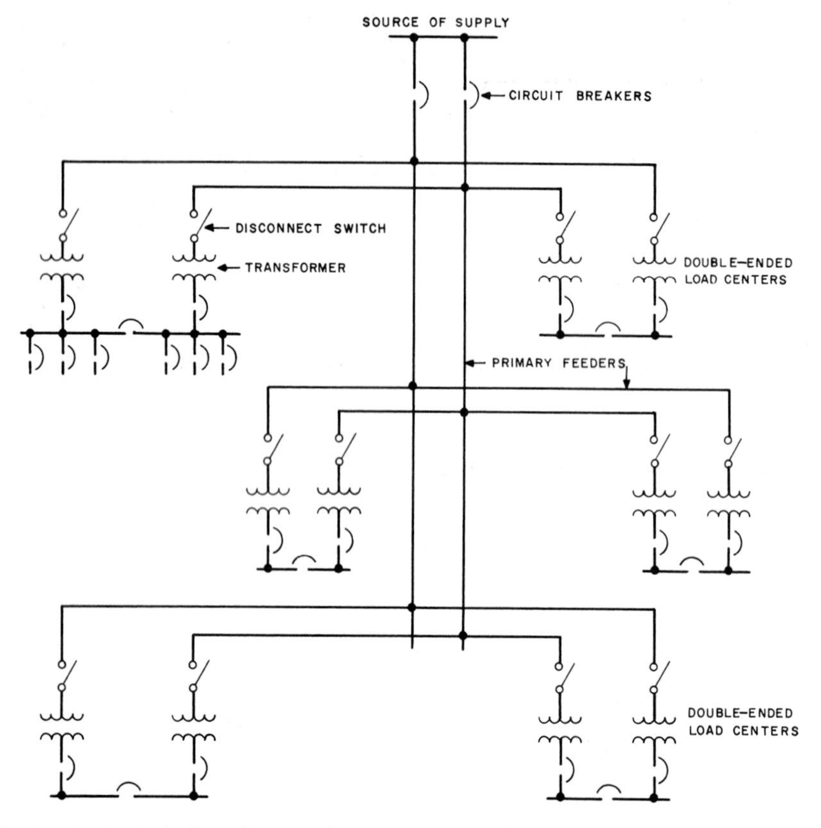

Fig. 24–31 **A single-line diagram for a power distribution system.**

overload protection. Ordinary fuses cannot protect the circuit in this way.

2. To allow remote control by start-stop buttons or automatic devices, such as thermostats.
3. To allow sequence control so that the order in which equipment is started can be controlled. This is illustrated in Fig. 24–33.

Figure 24–33 shows a wiring diagram at A. You can also see a schematic, or one-line, diagram at B (page 491) for the same circuit. This figure shows the sequence control for a conveyor system. The system has three separate motors. Note, at A, that a thin line is used to show the pilot circuit (low voltage). A thick line shows the line-voltage (high voltage) part of the circuit. The low voltage comes from the step-down transformer (SDT). A transformer changes the voltage of alternating current. Alternating current is the kind of current used in

Fig. 24–32 **A single-phase starter.**

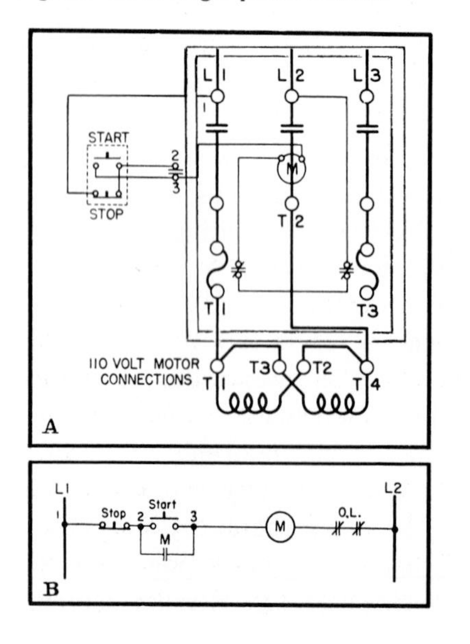

Fig. 24–33 **A three-way conveyor system.**

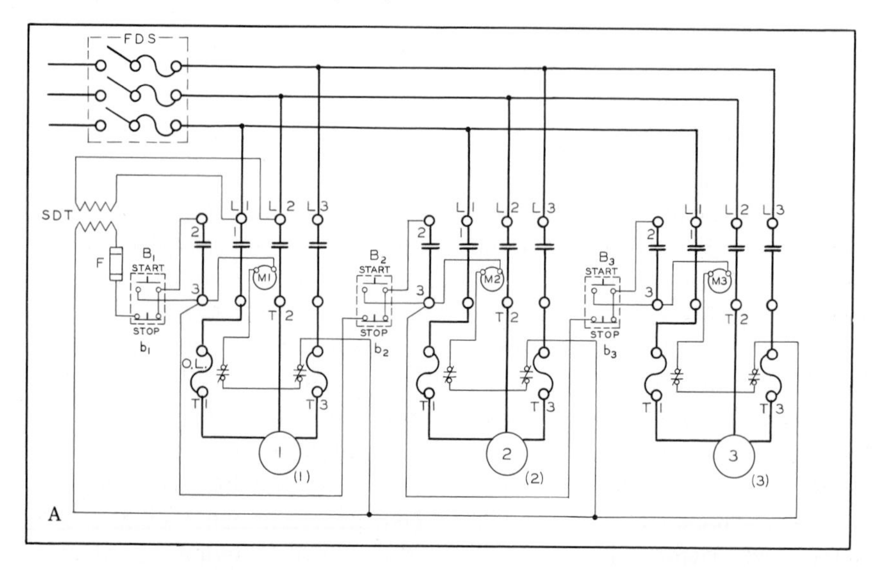

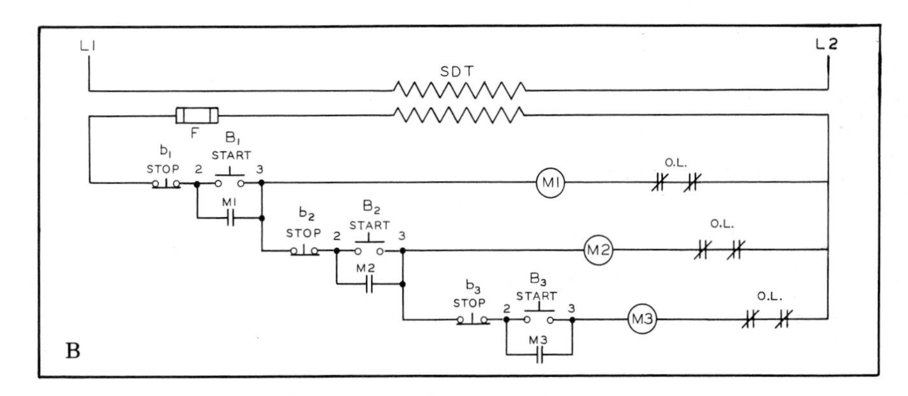

Fig. 24–33 **(Cont.)**

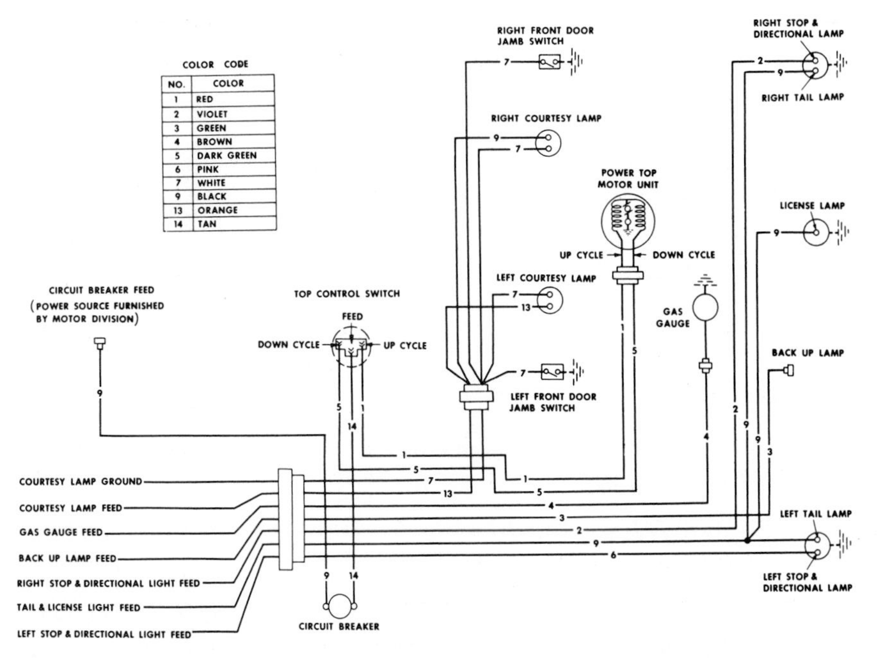

NO.	COLOR
1	RED
2	VIOLET
3	GREEN
4	BROWN
5	DARK GREEN
6	PINK
7	WHITE
9	BLACK
13	ORANGE
14	TAN

Fig. 24–34 **Wire color scheme used on an automobile body wiring circuit.** ***(Chevrolet Motor Division, General Motors Corp.)***

most electrical systems. In this case, the stepdown transformer changes the current from a higher voltage to a lower voltage.

In Fig. 24–33 at A, start button B_1 will start the No. 1 conveyor motor at (1). However, start button B_2 will not start the No. 2 conveyor motor at (2) unless motor No. 1 has started. Start button B_3 starts conveyor motor No. 3 at (3). It will start the motor, however, only after motor No. 2 has started. The conveyor system may be completely stopped by the fused disconnect switch (FDS), by stop button b_1 or by its overload (OL). The stop button b_2 or the No. 2 motor overload will stop both conveyors Nos. 2 and 3. The stop button b_3, or overload, will only stop the No. 3 conveyor. The fuse, *F,* protects the low-voltage control circuit from burning out.

In Fig. 24–33 at B, the same circuit is shown by a one-line diagram. The fuse, *F,* protects the low-voltage control circuit.

COLOR CODES

Color codes are an easy way to show information when drawing circuit diagrams. Color codes are also used on the actual wiring of the circuit. In electrical and electronic work, drafters use a color code to show certain characteristics of components, to identify wire leads, and to show where wires are connected. A color code may be included on the diagram (Fig. 24–34).

When using a color code, you should look up the Electronic Industries Association (EIA) standards. Also look up any other codes you might need. Note the different code used in Fig. 24–34. It is not the same code as the EIA standard code used in Fig. 24–35.

Color codes are used to give exact information about electrical equipment. You can find out about such uses in published standards and in textbooks.

EIA Standard Color Code

Color	Abbre-viation	Number
Black	BLK	0
Brown	BRN	1
Red	RED	2
Orange	ORN	3
Yellow	YEL	4
Green	GRN	5
Blue	BLU	6
Violet	VIO	7
Gray	GRA	8
White	WHT	9

BLOCK DIAGRAMS

A block diagram (Fig. 24–36) is usually made up of squares or rectangles, or "blocks." They are joined by single lines. The blocks show how the components or stages are related when the circuit is working. You can see arrowheads at the terminal ends of the lines. These arrowheads show which way the signal path travels from input to output, reading the diagram from left to right.

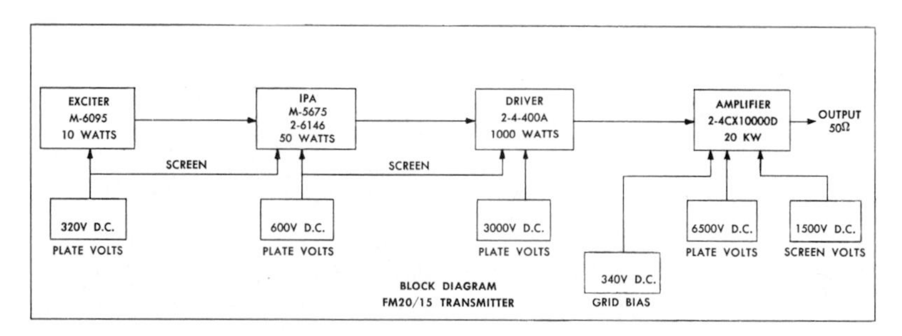

Fig. 24–36 **Block diagram of 20,000-watt broadcast transmitter.** ***(Gates Radio Co., a subsidiary of Harris-Intertype Corp.)***

The identification of the stage is lettered within the block or just outside it. The blocks are often drawn along with symbols and a schematic diagram (see Figs. 24–44 through 24–47).

Engineers often draw or sketch block diagrams as a first step in designing new equipment. Because blocks are easy to sketch, the engineer can try many different layouts before deciding which to use. The overlay method of sketching discussed in Chapter 2 can be used with great success in drawing or sketching block diagrams.

Block diagrams are used in catalogs, descriptive folders, and advertisements for electrical equipment. They are also used in technical-service literature to aid in the repair of equipment.

ELECTRICAL LAYOUTS FOR BUILDINGS

You can see in Fig. 24–37 the usual way an architect shows where electrical outlets and switches are to be placed. This plan only shows where the lights, base plugs, and switches are to be placed. You cannot build a good electrical system using this diagram. For this, you need a complete and detailed set of electrical drawings. These must be made by someone who knows the engineering needs of the system. A list of the symbols used is shown in Fig. 24–38.

THE INTERCONNECTION DIAGRAM

Figure 24–39 is a connection, or wiring diagram. It shows the electrical connections between the different components of an electrical or electronics system. Generally, the connections inside each component are not shown. The name of each component is given. Each component is shown on the diagram by a rectangle.

CONNECTION OR WIRING DIAGRAMS

Figure 24–40 shows wiring connections in a simple way. This is done so that you can easily follow the connections of the circuit system. Connections both inside and outside the components may be shown. The components are named. They are drawn as pictures instead of as blocks. Auxiliary devices are also shown. Color coding is important for repairing electrical equipment and is shown on Fig. 24–40.

Connection or wiring diagrams give information needed for making, installing, and fixing electrical equipment. They are also used with schematic diagrams.

Fig. 24–37 **Electrical plan for a ranch house. (See Chapter 20.)**

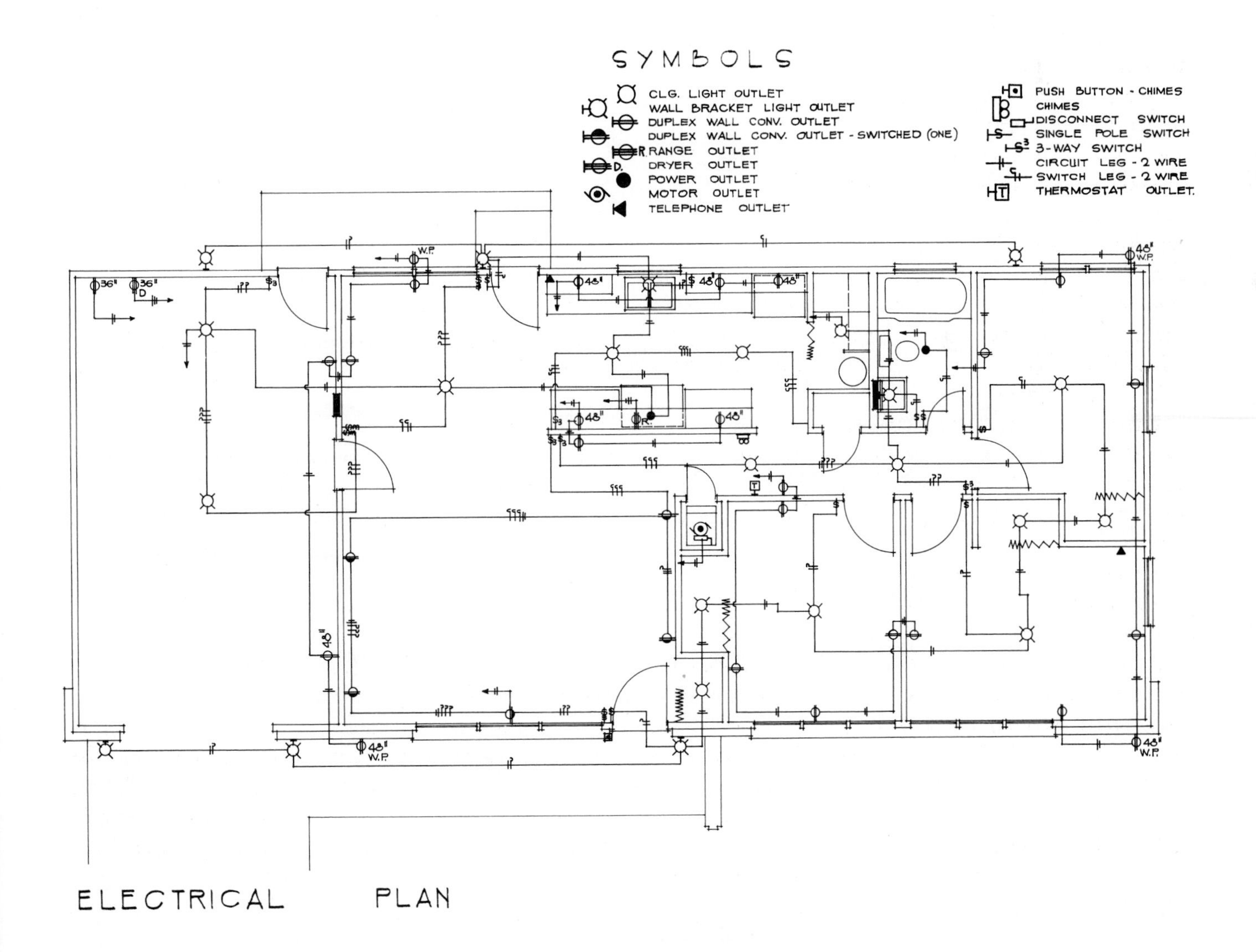

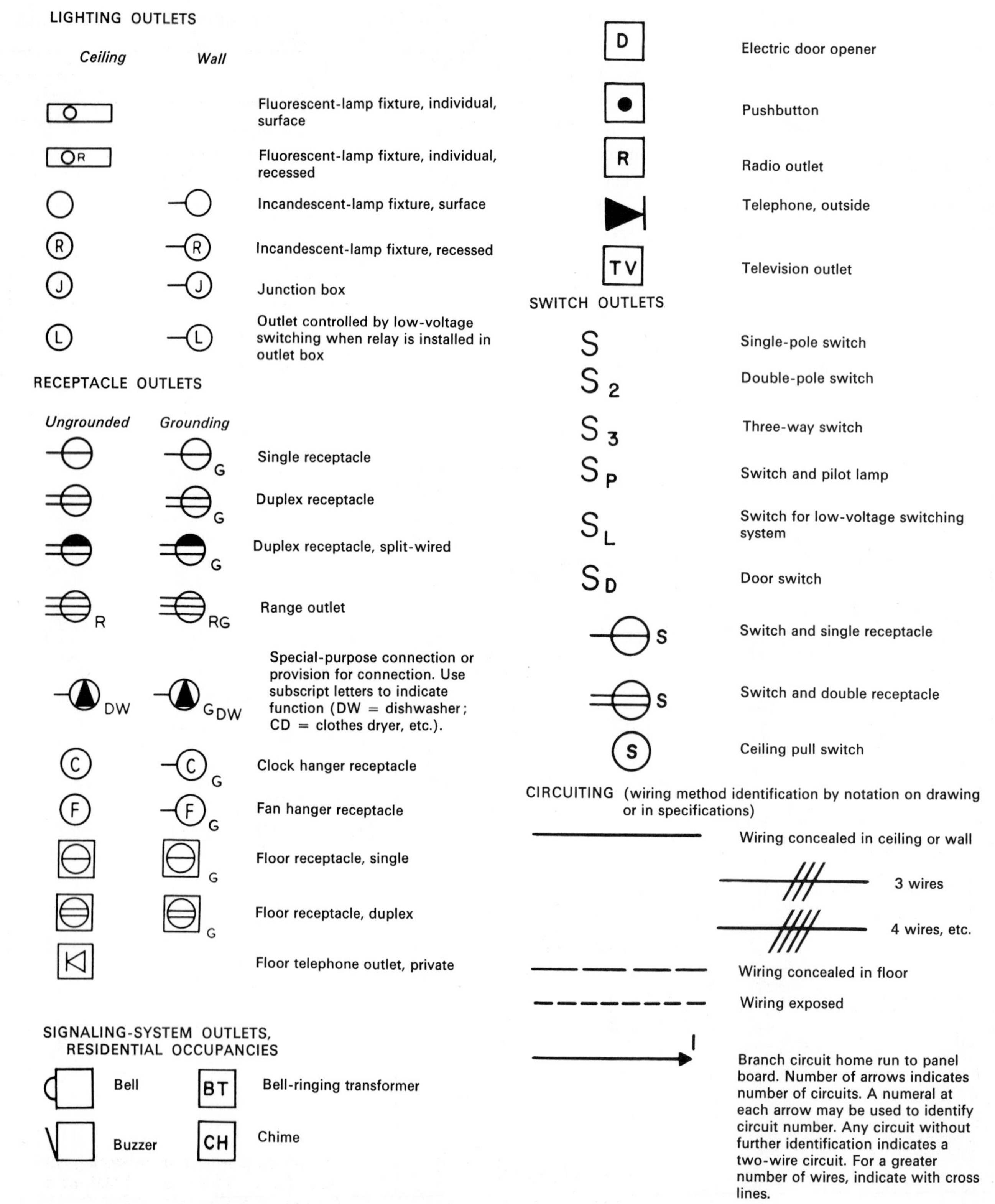

Fig. 24–38 **Electrical wiring symbols for architectural layout.**

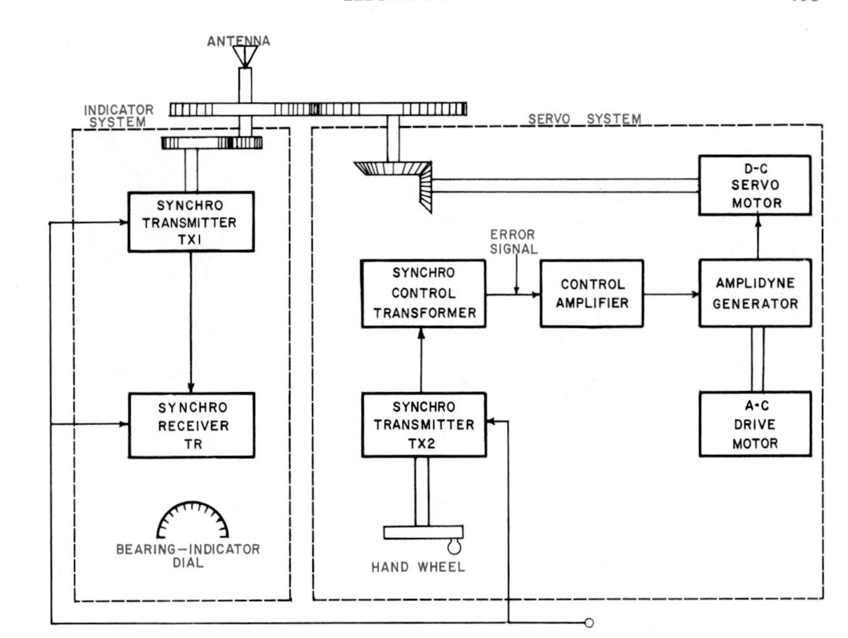

Fig. 24–39 **Interconnection diagram showing the different units of a typical dc servo system used for rotating a search radar system.**

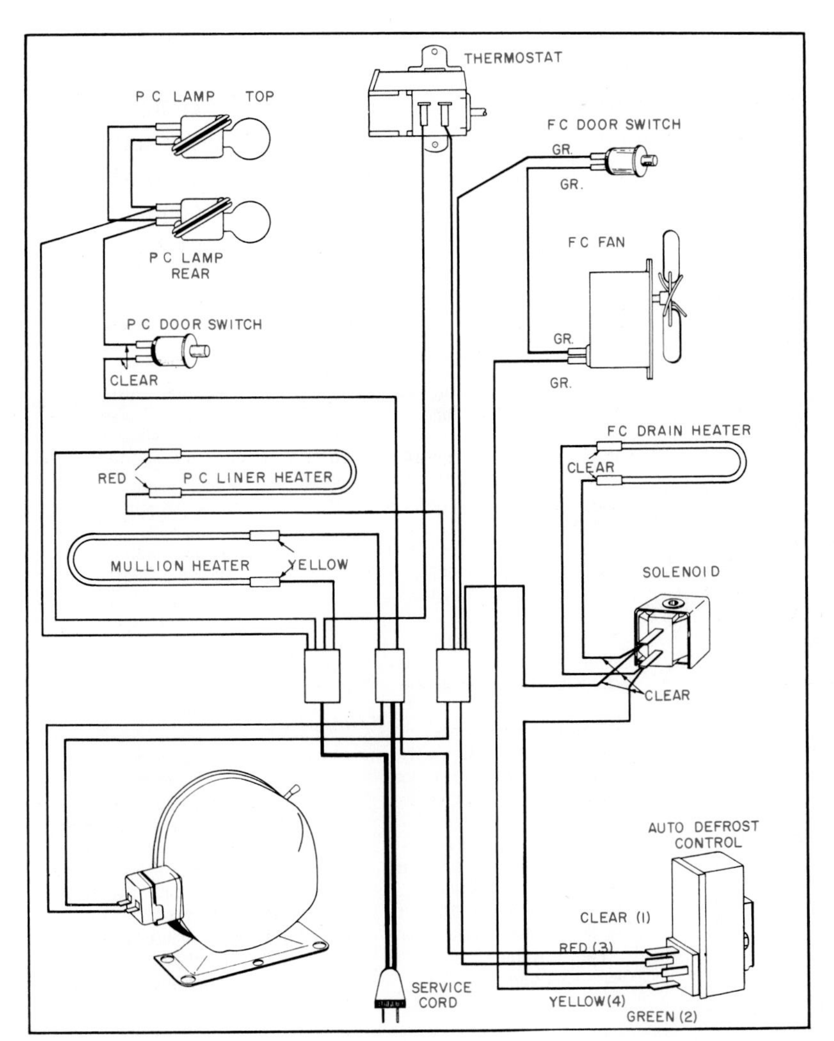

Fig. 24–40 **Connection or wiring diagram for a refrigerator.** ***(Kelvinator Division, American Motors Corp.)***

PRINTED CIRCUIT DRAWINGS

These are used in making actual printed circuit boards. These boards are used in electronic equipment. The drawings are exact layouts of the pattern of the circuit needed. The drawing is made actual size or larger. If drawn larger, it can be made smaller by photography. The lines (conductors) on the pattern should be at least 1 mm (1/32 in.) wide. They should be spaced at least 1 mm (1/32 in.) apart. The circuit layout pattern is transferred to a copper-clad insulating base. This is done by photography or in some other way. Etching is one way to remove the copper from all areas of the insulating base except for the circuits needed. There are many other methods.

The components may be shown on the printed circuit board by using symbols or other markings. This information is transferred to the printed circuit diagram from a component identification overlay (Fig. 24–41).

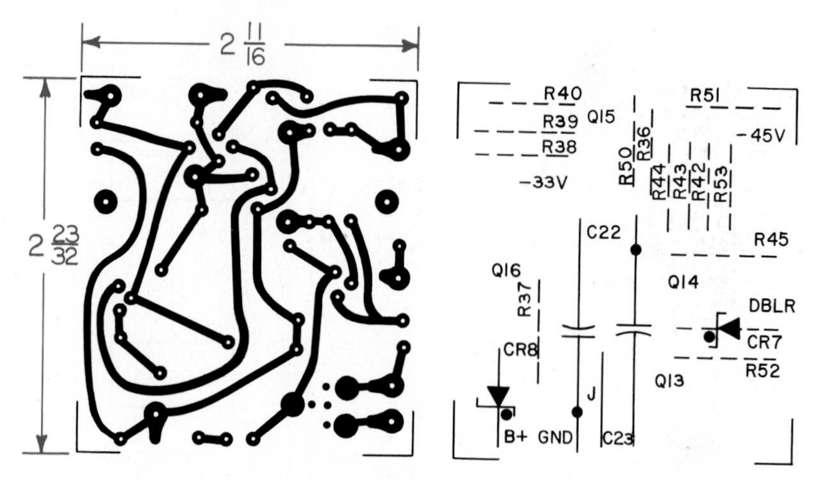

Fig. 24–41 **A printed circuit and a component identification overlay.** ***(Gates Radio Co., a subsidiary of Harris-Intertype Corp.)***

CIRCUIT DIAGRAMS

You can see the circuit diagram for a high-fidelity audio amplifier in Fig. 24–42. It has conventional tubes and electrical components.

Fig. 24–42 **A circuit diagram of a high-fidelity audio amplifier.** ***(RCA.)***

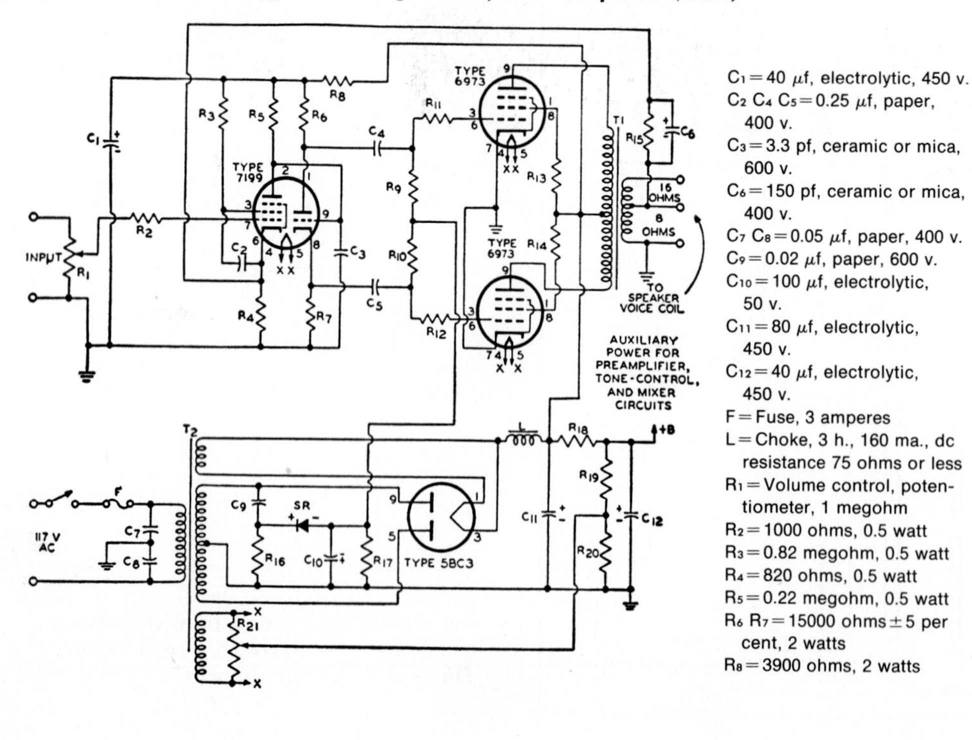

C_1 = 40 μf, electrolytic, 450 v.
C_2 C_4 C_5 = 0.25 μf, paper, 400 v.
C_3 = 3.3 pf, ceramic or mica, 600 v.
C_6 = 150 pf, ceramic or mica, 400 v.
C_7 C_8 = 0.05 μf, paper, 400 v.
C_9 = 0.02 μf, paper, 600 v.
C_{10} = 100 μf, electrolytic, 50 v.
C_{11} = 80 μf, electrolytic, 450 v.
C_{12} = 40 μf, electrolytic, 450 v.
F = Fuse, 3 amperes
L = Choke, 3 h., 160 ma., dc resistance 75 ohms or less
R_1 = Volume control, potentiometer, 1 megohm
R_2 = 1000 ohms, 0.5 watt
R_3 = 0.82 megohm, 0.5 watt
R_4 = 820 ohms, 0.5 watt
R_5 = 0.22 megohm, 0.5 watt
R_6 R_7 = 15000 ohms $\pm$ 5 per cent, 2 watts
R_8 = 3900 ohms, 2 watts
R_9 R_{10} = 0.1 megohm, 0.5 watt
R_{11} R_{12} = 1000 ohms, 0.5 watt
R_{13} R_{14} = 100 ohms, 0.5 watt
R_{15} = 8200 ohms, 0.5 watt
R_{16} = 15000 ohms, 1 watt
R_{17} = 68000 ohms, 0.5 watt
R_{18} = 4700 ohms, 2 watts
R_{19} = 0.27 megohm, 1 watt
R_{20} = 47000 ohms, 0.5 watt
R_{21} = Hum balance adjustment, potentiometer, 100 ohms, 0.5 watt
SR = Selenium rectifier, 20 ma., 135 volts rms
T_1 = Output transformer, (having 8-ohm tap for feedback connection) for matching impedance of voice coil to 6600-ohm plate-to-plate tube load; 50 watts; frequency response, 10 to 50000 cps; Stancor A-8056 or equiv.
T_2 = Power transformer, 360-0-360 volts rms, 120 ma.; 6.3 v., 3.5 a; 5v., 3a; Stancor 8410 or equiv.

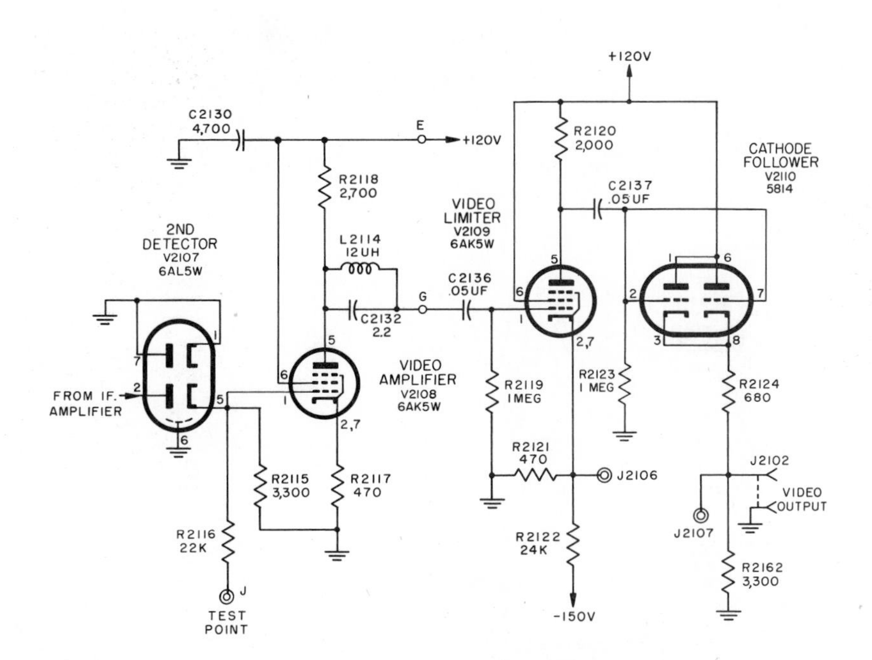

The schematic diagram in Fig. 24–43 shows part of a storm-detector radar. The circuit is a second detector and a video amplifier and limiter.

Fig. 24–43 **A circuit diagram of a second detector and a video amplifier and limiter. *(Raytheon Co.)***

Review

1. Describe a series circuit in technical terms.

2. Draw a simple parallel circuit.

3. Explain how a block diagram and a schematic diagram are different.

4. How are voltage and amperage related in Ohm's law?

5. What is the difference between a conductor and a nonconductor? List two of each.

6. What is the important difference between mechanical drawing and electrical drawing?

Problems

Figs. 24–44 through 24–47 are component diagrams of an AM-FM stereo unit containing integrated circuits. All diagrams should be drawn at least twice the size shown. *(Motorola, Inc.)*

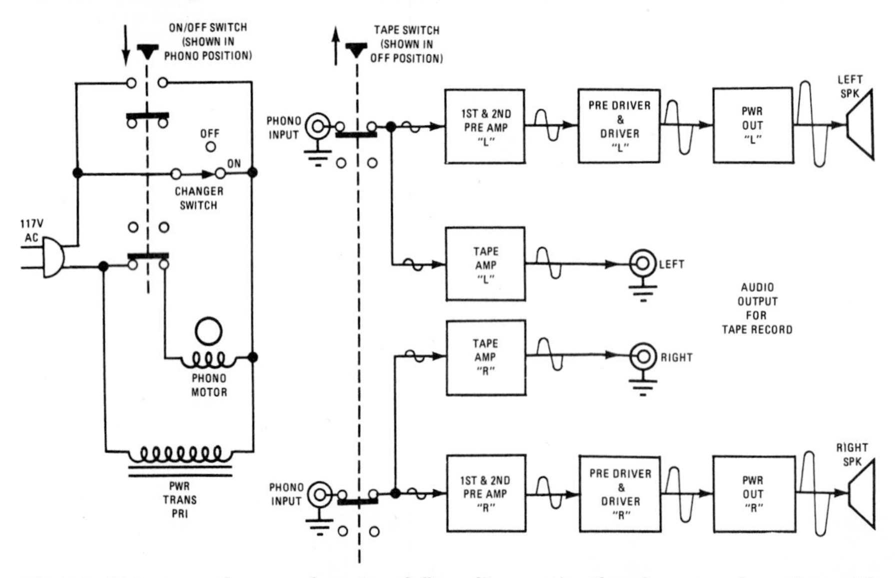

Fig. 24–44 Prepare the complete signal-flow diagram for the phonograph section of the stereo system. Note that only three blocks per channel are shown. Estimate the sizes required, and use a template if one is available.

Fig. 24–45 Draw the tape player signal-flow diagram. This diagram may be drawn as an overlay to Fig. 24–44 as directed by the instructor. Note that when the tape button is depressed, all other functions, AM, FM, and phono, are disabled. Also, the changer switch cannot turn on the phono motor. Since there is no output from the left and right audio-output jacks, no direct tape recording can take place. The tape preamp is grounded to prevent pickup by the tape player.

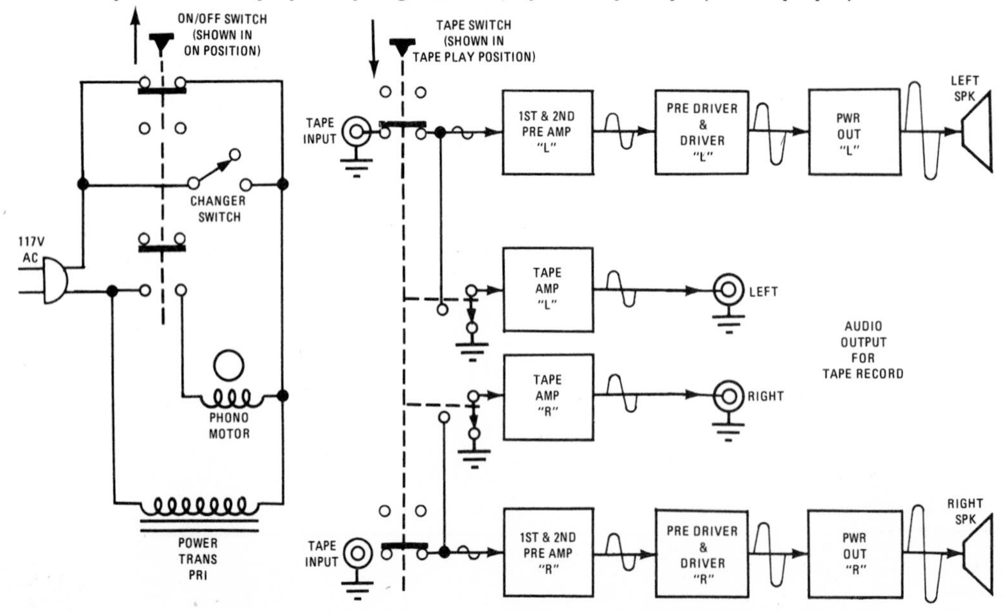

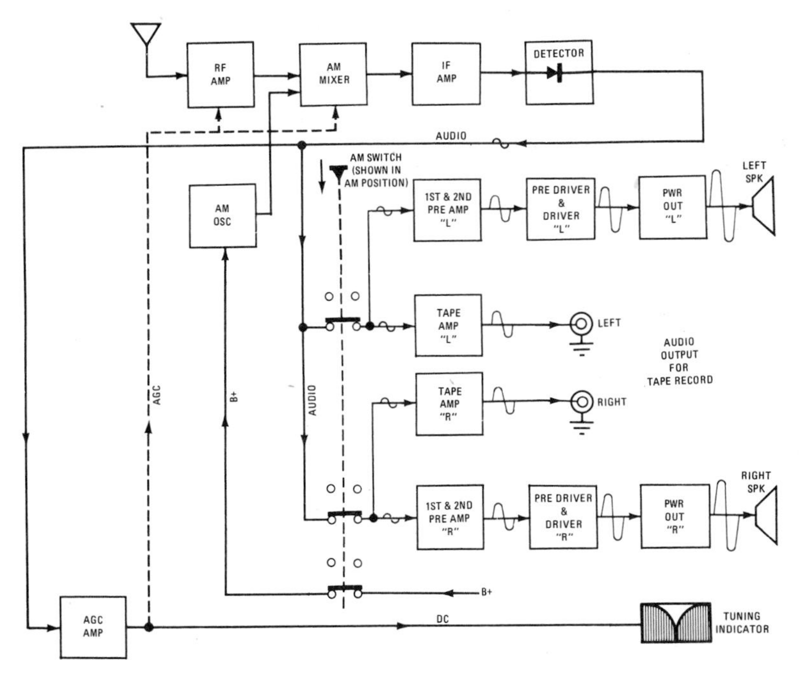

Fig. 24–46 **Prepare a block diagram for the complete signal flow of the AM radio. A loop antenna picks up the signal and presents it to the RF amplifier. The mixer receives both the selected AM and oscillator signals. At the AM detector, the audio is recovered and applied to the stereo channels. The stereo channels are connected in parallel to the AM switch, providing monaural operation.**

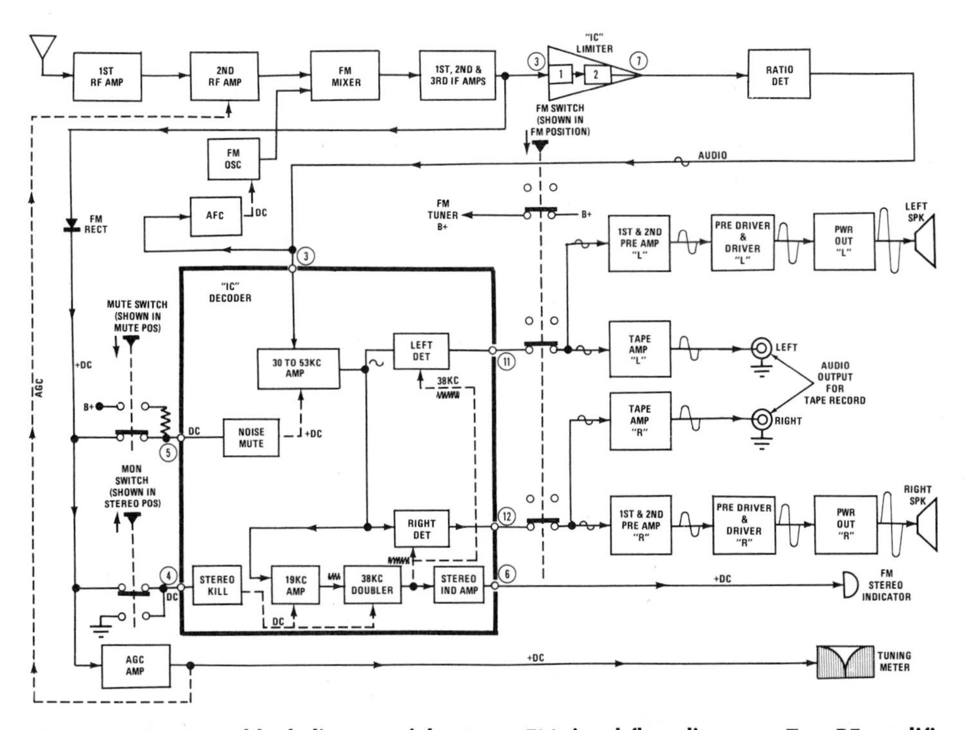

Fig. 24–47 **Prepare a block diagram of the stereo FM signal-flow diagram. Two RF amplifiers are used to provide optimum FM selectivity. Selectivity and gain are further improved by three IF amplifier stages. A ratio detector receives the signal and reduces any AM noise that may be present. The audio signal then enters the "IC" FM decoder through pin #3. The primary functions of the decoder are to pass a monaural FM signal into both right and left channels and to separate a stereo FM signal into a right and left channel.**

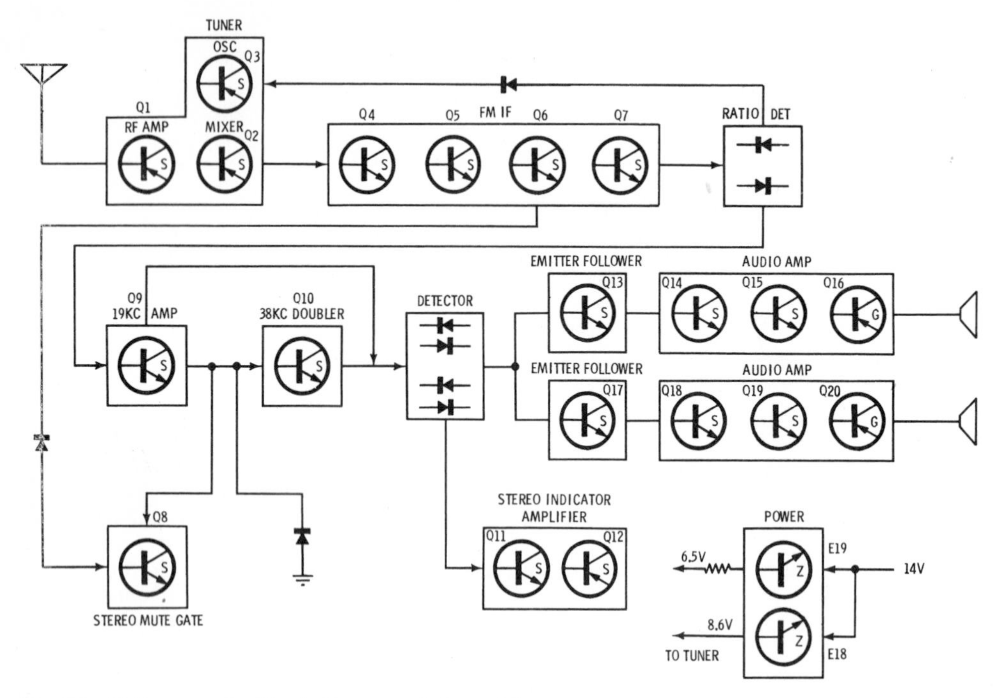

Fig. 24–48 **Prepare the block diagram of the solid-state FM stereo auto radio. The main components employed are 14 n-p-n silicon transistors, 4 p-n-p silicon transistors, 2 p-n-p germanium transistors, 9 diodes, and 2 Zener diodes.** *(Motorola, Inc.)*

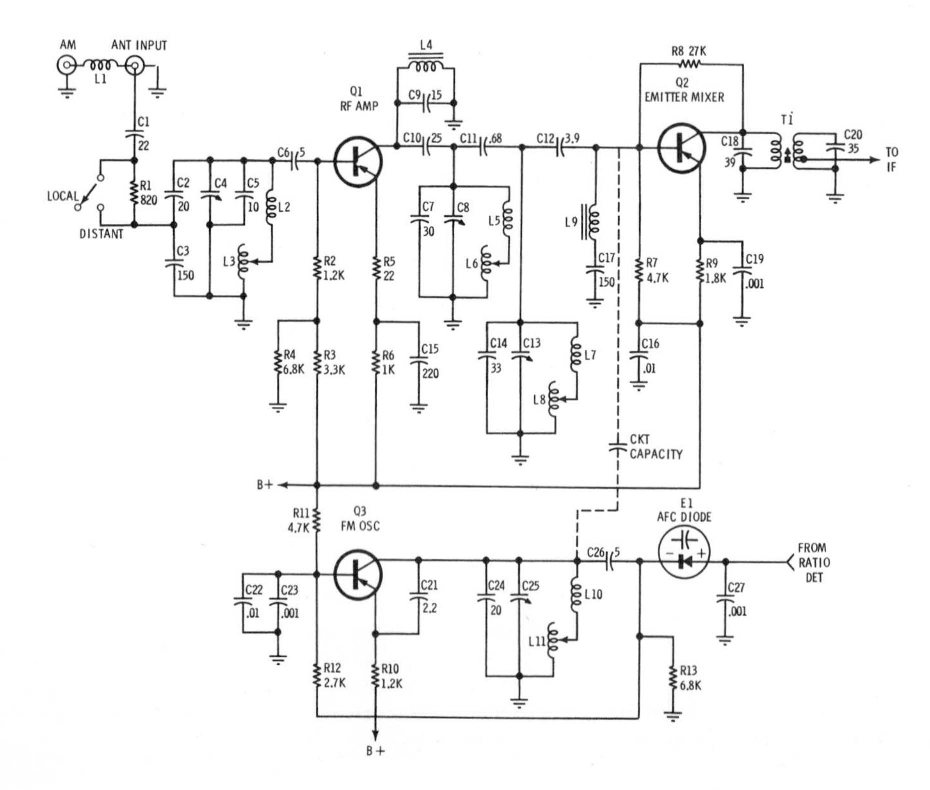

Fig. 24–49 **Prepare a schematic diagram of the tuner used in the FM stereo auto radio. Prepare a parts list alphabetically.** *(Motorola, Inc.)*

Chapter 25
Aerospace Drafting

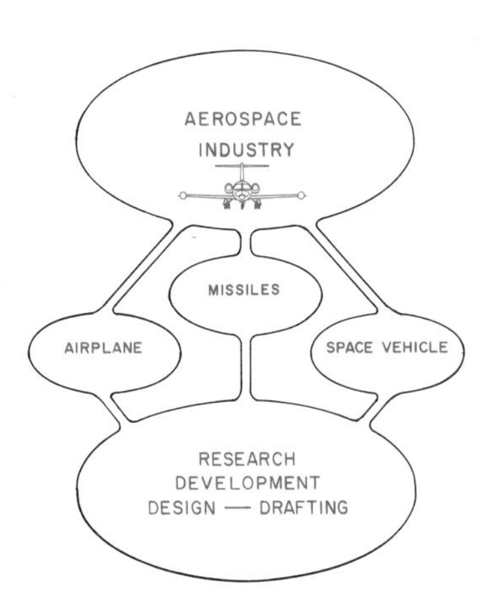

Fig. 25–1 **The aerospace industry.**

THE AEROSPACE INDUSTRY

New ways of making airplanes, missiles, and spaceships are always being found. This means challenging opportunities exist for young men and women interested in aerospace design and drafting (Fig. 25–1). Aerospace design and drafting deals with all kinds of flying vehicles. These vehicles fly at all speeds and at all *altitudes* (heights). The basic aerospace team is made up of hundreds of scientists, engineers, designers, drafters, and technicians and thousands of skilled workers. The government works with industry to develop aircraft. The aircraft industry has many interests. Some of these are hovering helicopters, stunt biplanes, and vehicles that can fly faster than sound for thousands of miles. The aerospace industry even makes spacecraft that orbit the earth or fly millions of miles to the planets Jupiter and Saturn. (Fig. 25–2).

CAREER OPPORTUNITIES

The aviation industry has grown with the speed of the jet. New careers and jobs have developed that did not exist thirty years ago. The knowledge needed for aircraft design and drafting is learned in school and on the job. Those who work in this field usually begin as technicians or engineering aides. If interested, you should study science and mathematics in high school and junior college. You should also take a course in technical drawing or engineering graphics. As a member of a design team, the drafting technician may make drawings of mechanical or electrical systems. These are drawn from the designer's sketch.

CAREER ADVANCEMENT

Drafting technicians may move up as they gain experience in a certain area. With some schooling

Fig. 25–2 **The aircraft industry developed the space lab concept.** ***(NASA.)***

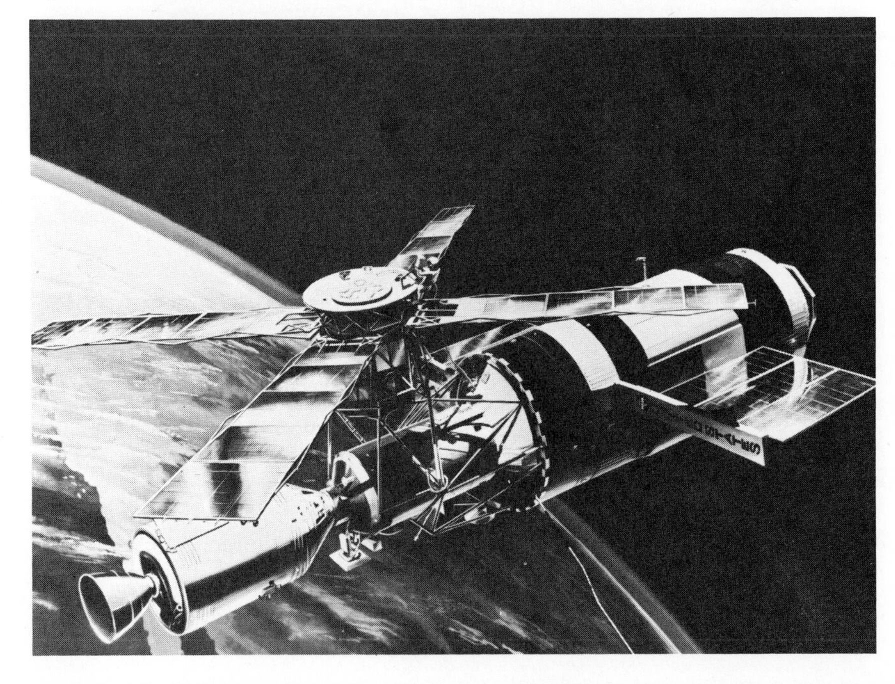

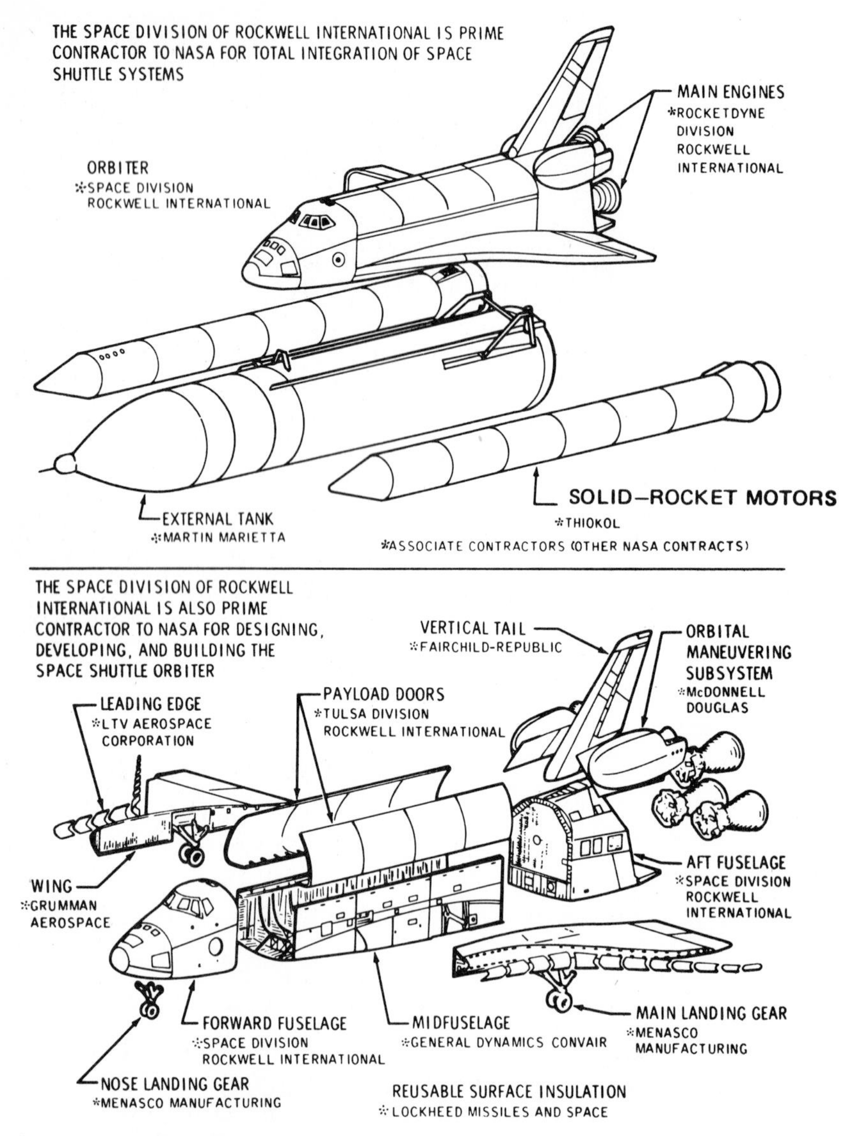

Fig. 25–3* An aircraft company must rely on many other manufacturing plants to make all the parts for an air- or spacecraft. *(NASA.)

and on-the-job training, they may become design technicians. Design technicians have more responsibilities than drafting technicians. The young designer who has just been promoted will probably change and improve designs that have already been made.

The design team may study any part of aircraft design. They do this to make a better aircraft. To learn more about careers, write to the groups listed below:

American Institute of Aeronautics and Astronautics
1290 Avenue of the Americas
New York, NY 10019

Engineers Council for Professional Development
345 East 47th Street
New York, NY 10017

Many *specialists* (people with special skills) are needed in aerospace industries, from the person on the drawing board to the person testing models in the wind tunnel. The big aircraft companies usually rely on many other manufacturing plants across the country to make parts for different aircraft (Fig. 25–3). The largest companies are those that make the power plants and *framework* (skeleton) of the aircraft. Thousands of parts are made by smaller manufacturers.

TESTING AND RESEARCH

The federal government, through many different agencies and the National Aeronautics and Space Administration (NASA), helps industry with vehicle testing and research. *Aeronautics* is the science of designing, building, and operating aircraft. Aircraft experiments explore how far we can travel into space. Today's leisure aircraft normally fly at *subsonic* (below the speed of sound) *Mach numbers* (Fig. 25–4). Mach 1 is the speed of sound, about 1 195 km/h or 742 mph at sea level. Tomorrow's leisure aircraft may be able to fly at *supersonic speeds* (faster than the speed of sound) and *hypersonic speeds* (five times the speed of sound).

AIRCRAFT MATERIALS

Today's high-speed and supersonic aircraft have parts made of many different materials. Aluminum, magnesium, and lightweight steel are usually strong enough for the inside, or substructure, parts. The superstructure (skeleton) and skins become very hot during supersonic flight. For this reason, they must be made of materials that are strong and can stand the heat. Designers today are interested in making lightweight structures. They are also interested in electronic equipment that works faster and better than the human mind.

Fig. 25–4 **Leisure aircraft cruise at subsonic speeds.** ***(Lear Jet Corp.)***

MAJOR AIRCRAFT COMPONENTS

Figure 25–5 shows the McDonnell Douglas F-4C Phantom jet. It has 62 basic parts. The *fuselage* (or central body), part 14, holds the pilot and passenger compartments. The fuselage has three parts. The fuselage structure is made up of a set of shaped *bulkheads* (walls) and rings. The fuselage also has *longitudinal* (lengthwise) members. Together these form a strong framework. The outer skin is made of sheet metal. It is fastened to the framework with rivets or machine screw fasteners.

The Wings and Airfoil

The wings are made up of ribs. These ribs form the shape of the wing. The ribs are connected to *spars* (beams). The spars run toward the fuselage and away from it. The wing skins are attached to the ribs and spars with metal fasteners. Besides the wings, other parts control the aircraft. These are called *airfoils*. They include ailerons, rudders, stabilizers, flaps, and tabs. A *chord line* is a straight line between the leading edge and the trailing edge of an airfoil. The general three-view drawing shows the sleek form of an aircraft that is made to be flown at high speed and high altitude (Fig. 25–6).

The Landing Gear

The landing gear is raised and lowered by *hydraulics* (fluid under pressure) or electricity. Hydraulic shock absorbers ease the shock of landing.

The Power Plant

Piston or jet engines are often placed in the lower part of the fuselage. They can also be placed under the wings.

Many other systems are also found on aircraft. These include air conditioning, compartment pressurization (to keep normal air pressure), radar, radio (communications), hydraulic, electronic, and plumbing systems. Many trades and industries help make the complex aerospace designs of today.

The parts and systems ready to be put together are generally shown on a plan. This plan has many details. Many kinds of drawings are needed before the aircraft is built. The design-team director keeps track of all the drawings. To understand how to build an aircraft from these drawings, you must understand multiview and pictorial drawings.

AIRCRAFT DRAFTING PRACTICES

The bigger aircraft companies have engineering manuals for their workers. Using the manuals, the workers may follow the methods that are best for their company's product. The manuals are carefully made to save time and money in manufacturing. A designer and drafter must know the company's manual. They must also know the standards used by all companies. The Society of Automotive Engineers (SAE) publishes a book of Aerospace-Automotive Drawing Standards.

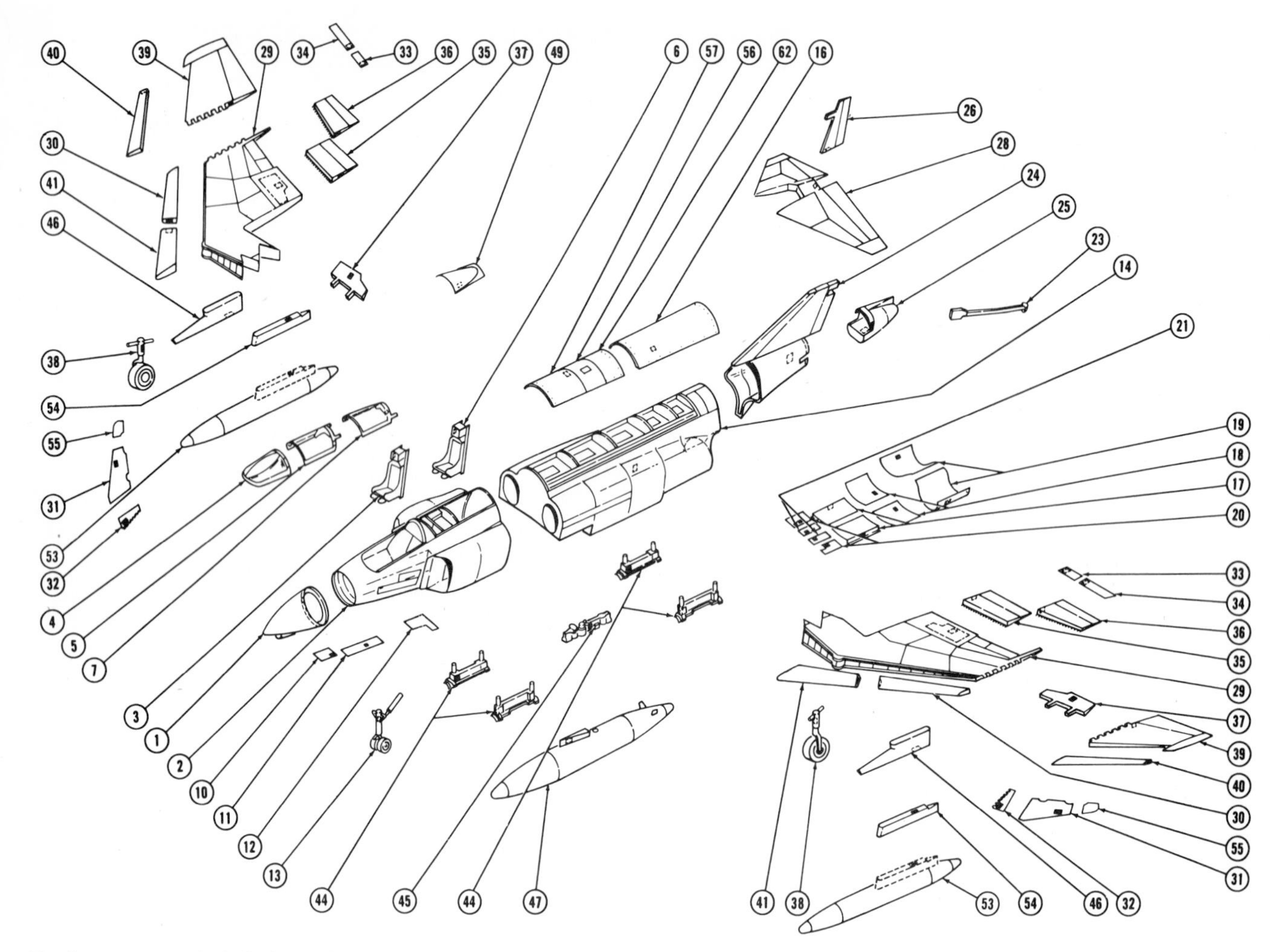

***Fig. 25–5* An exploded view of parts of McDonnell Douglas F-4C Phantom.** *(McDonnell Douglas Corp.)*

1. Radome
2. Forward fuselage
3. Pilot seat
4. Windshield
5. Forward canopy
6. Radar operation seat
7. Aft canopy
10. Nose landing gear door, forward
11. Nose landing gear door, aft
12. Hydraulic compartment access door
13. Nose landing gear shock strut
14. Center fuselage
16. Fuel tank door
17. Engine access door
18. Engine access door
19. Engine access door
20. Engine access door
21. Auxiliary engine air door
23. Arresting hook
24. Aft fuselage
25. Tail cone
26. Rudder
28. Stabilator
29. Center section wing
30. Leading edge flap
31. Main landing gear strut door
32. Main landing gear inboard door
33. Inboard spoiler
34. Outboard spoiler
35. Flap
36. Aileron
37. Speed brake
38. Main landing gear shock strut
39. Outer wing
40. Leading edge flap, outboard
41. Leading edge flap, inboard
44. Missile rack
45. Bomb rack
46. Missile pylon
47. External centerline fuel tank
49. Data link access door
53. External wing fuel tank
54. External wing fuel tank pylon
55. Landing gear door, outboard
56. Boom IFR receptacle access door
57. Fuel cell access door
62. Fuel cell access door

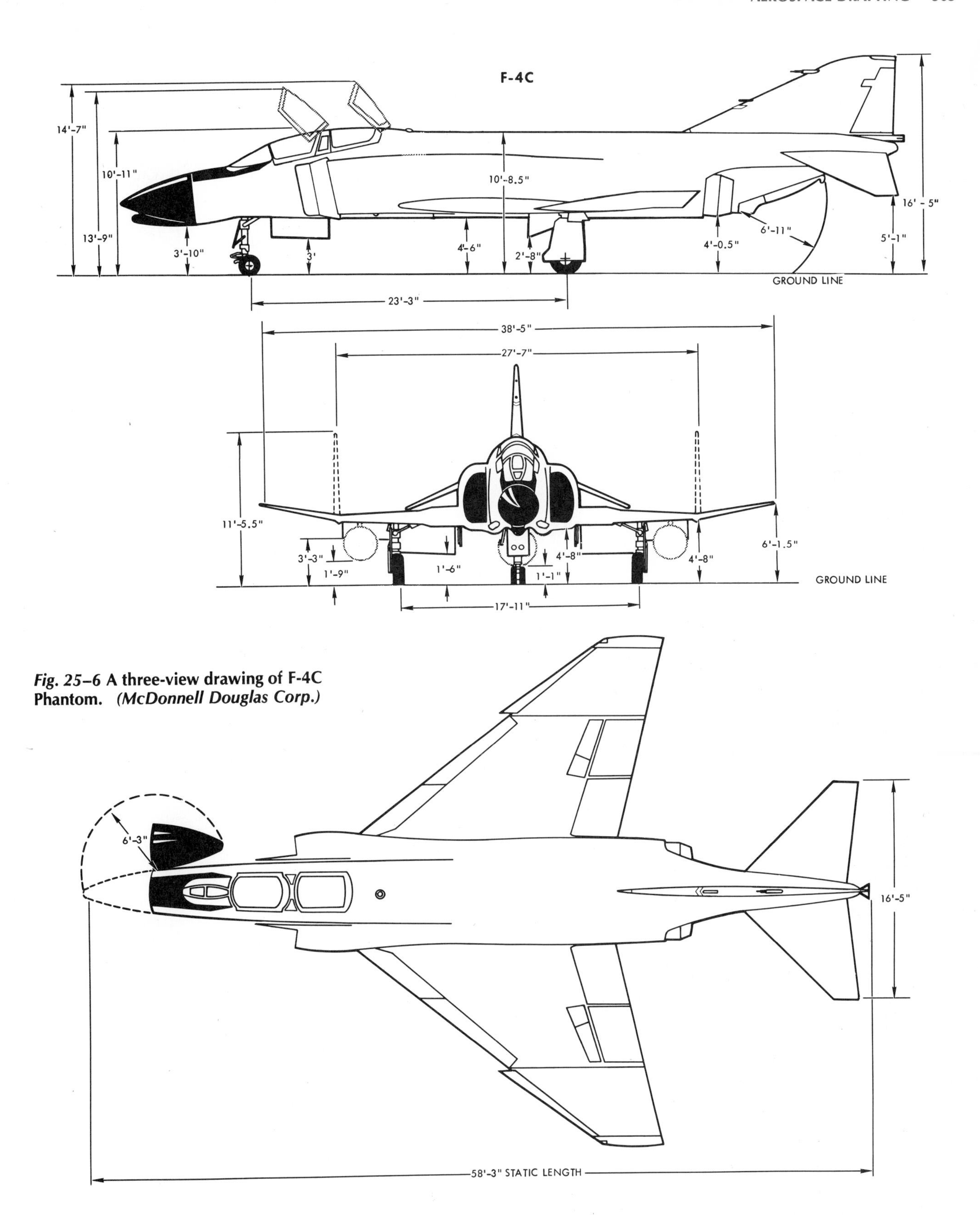

Fig. 25–6 **A three-view drawing of F-4C Phantom.** ***(McDonnell Douglas Corp.)***

METHODS OF UNDIMENSIONED DRAWINGS

The undimensioned drawing is a way of drafting in the aerospace industry. It frees designers and drafters from the detail dimensioning needed for most finished layouts. This method takes less time to turn finished drawings into actual parts ready to be put together. Automatic drafting equipment makes undimensioned drafting possible.

The drawings are made with devices that control drawing accuracy. Some methods are listed for full-size drawings.

1. Lines are drawn on matte-surface (dull and rough) plastic film.
2. Drawings are made with a technical pen that controls line widths.
3. Drawing is accurate. The accuracy needed depends on the part being drawn.

Undimensioned drawings are useful for several reasons.

1. Less time is needed to make finished drawings.
2. Drawings are used to check fit. Full-size drawings are a natural place to test whether parts fit. If parts do not fit on full-size drawings, they will probably not fit on the final product.
3. Flat-pattern drawings can be easily checked. A flat pattern gives a design team time to check out problems.
4. A contact photo makes the master layout. The drawing can be transferred with a photograph.
5. Changes caused by mistakes are less common. By using undimensioned drawings with full-size drawings, and by not includ-

Fig. 25–7 **Undimensioned flat-pattern development.** ***(McDonnell Douglas Corp.)***

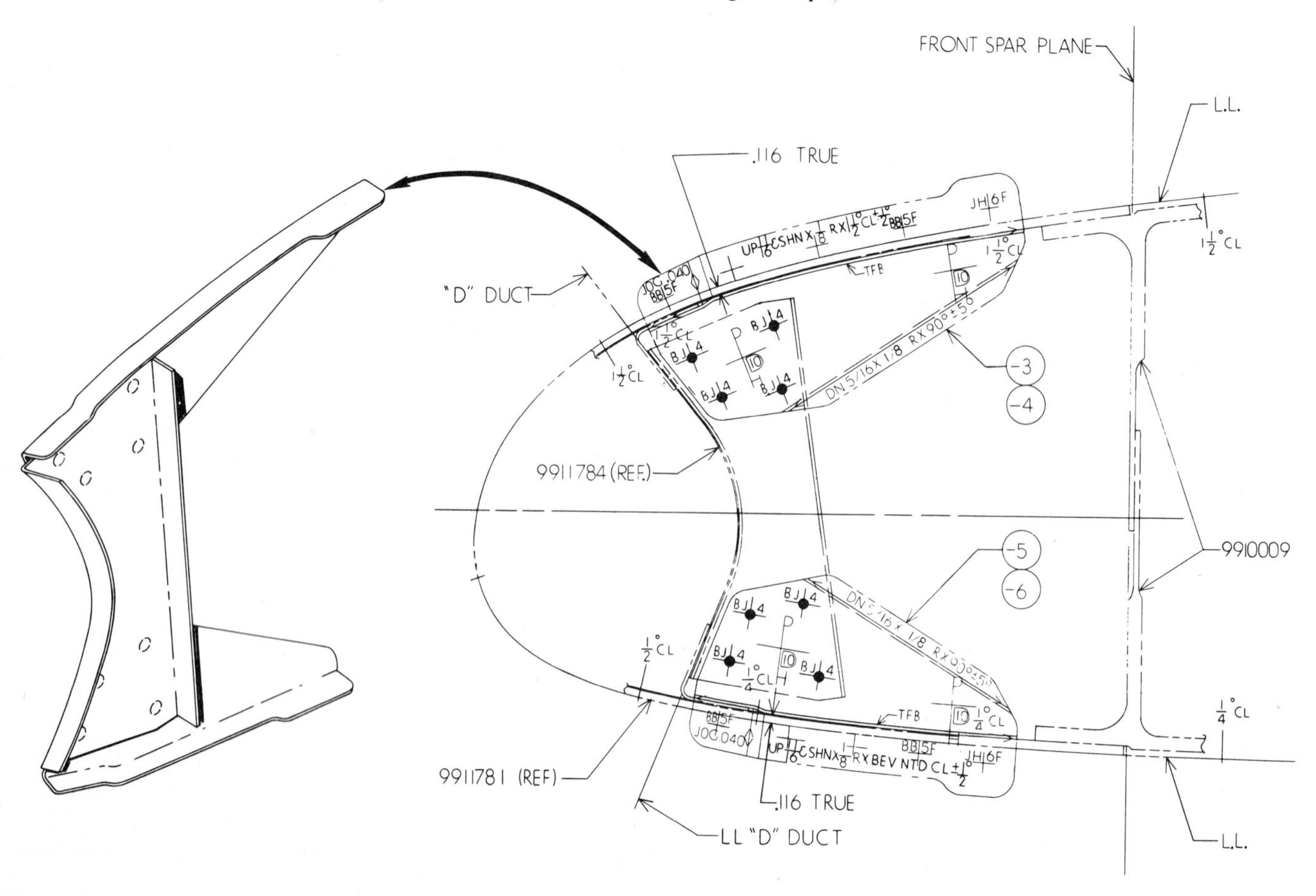

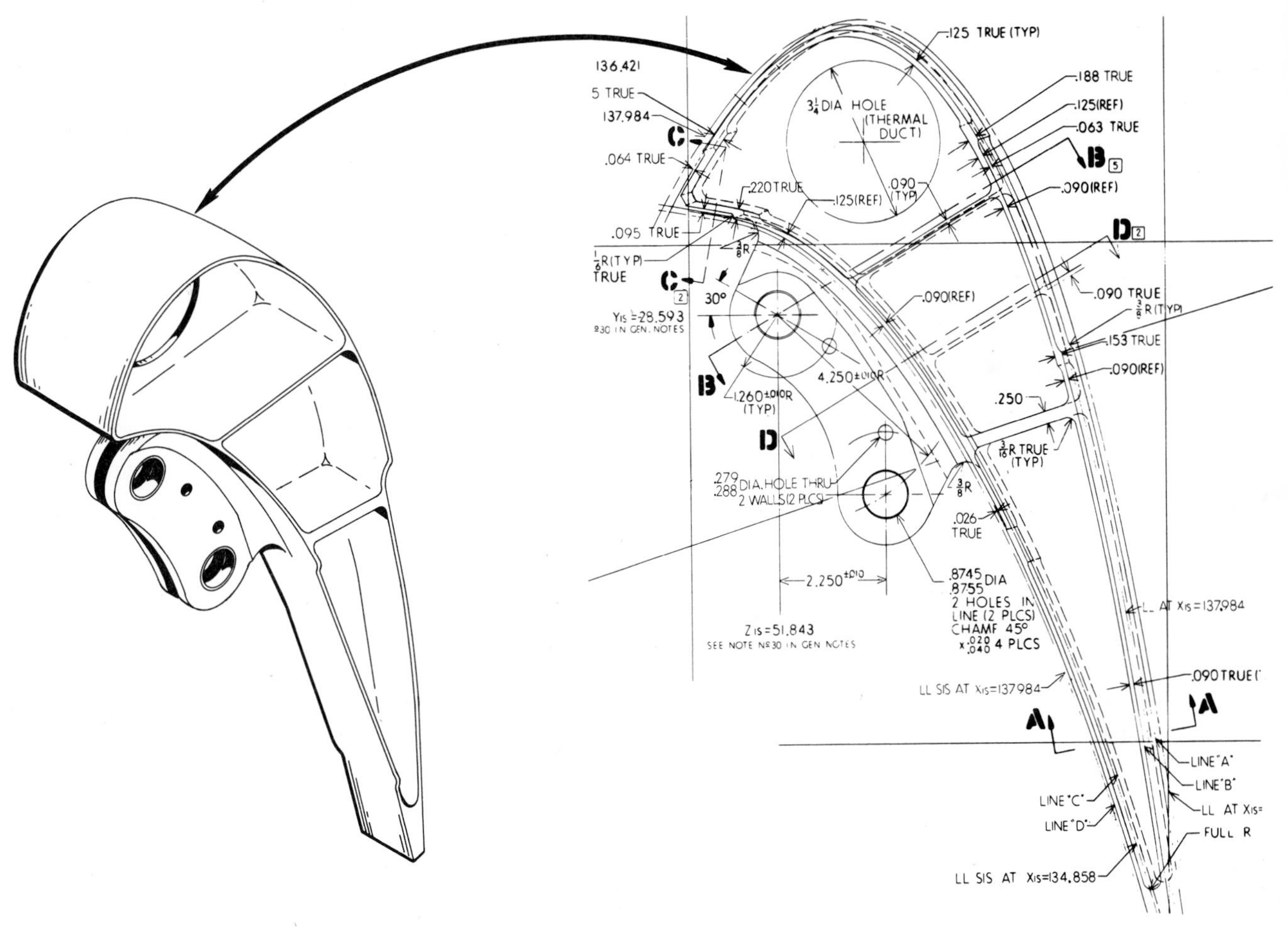

Fig. 25–8* Undimensioned machined parts. *(McDonnell Douglas Corp.)

ing measurements that may be wrong or unclear, drawing mistakes have been lowered by 40 percent.

6. They may be drawn by automation. *Automation* means electronic equipment doing jobs that people once did. In the future age of automation, the full-size undimensioned drawing will be ideal. The instructions for the drawing may be put onto a computer tape. Then the tape can be played back and changed to make new, modified drawings. The program will be used for the numerically controlled machine in the areas where parts are made automatically.

The kinds of drawings that can be made in the undimensioned way are the following:

1. Parts that have flat patterns that can be made in a hydropress, power press brake, or stretch brake. You can see an example of flat-pattern development in Fig. 25–7.
2. Machined parts that are drawn as undimensioned drawings on a piece must be made by profile machine methods (Fig. 25–8).
3. Parts needing a plaster pattern with a three-dimensional form. (Fig. 25–9).
4. Drawings needing artwork layout. Circuits and lighting panels are natural for undimensioned drawing. The artwork is drawn twice the size. Then it is photoreduced to the right size for making the needed part. Photoreduction makes lines clean and sharp (Fig. 25–10).

Fig. 25–9 **Undimensioned drawing for a plaster pattern.** ***(McDonnell Douglas Corp.)***

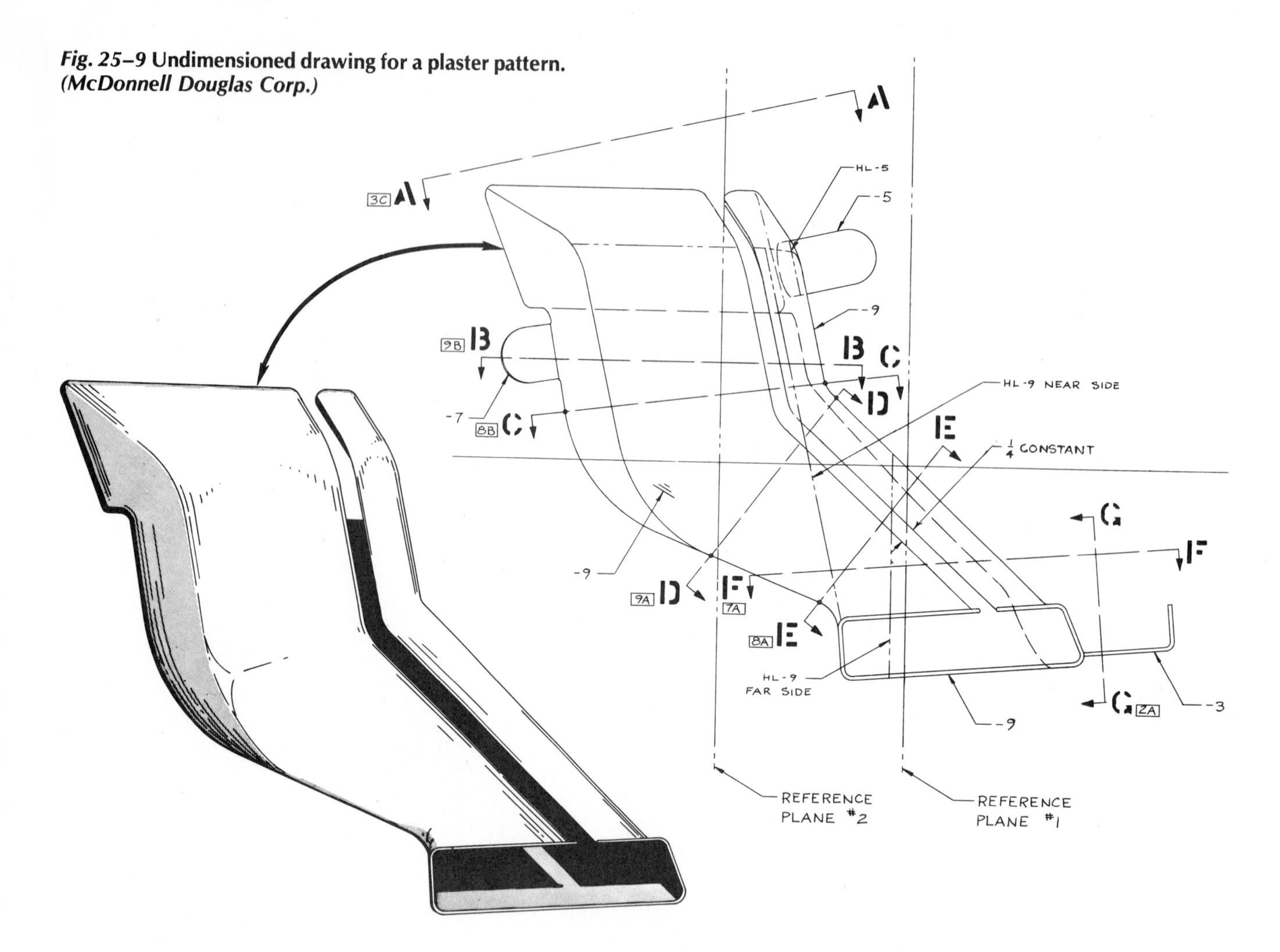

Fig. 25–10 **Undimensioned artwork layout.** ***(McDonnell Douglas Corp.)***

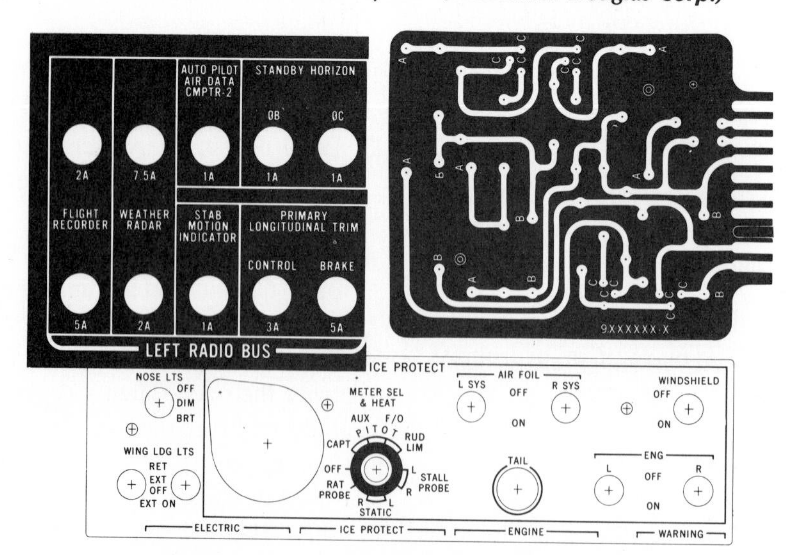

DRAWINGS FOR LARGE AIRCRAFT

You can see many kinds of aircraft drawings in the DC-10 study in Figs. 25–11 to 25–19. The assembly drawing is used to show the inboard profile of the DC-10. This is shown in Fig. 25–11. The cutaway pictorial of a wing engine in Fig. 25–12 shows the huge size of a power plant compared to a human. Figure 25–13 shows the giant transport in a three-view orthographic projection. You can see the wing in a cutaway pictorial in Fig. 25–14. In Fig. 25–15, the exploded pictorial of the outboard aileron damper

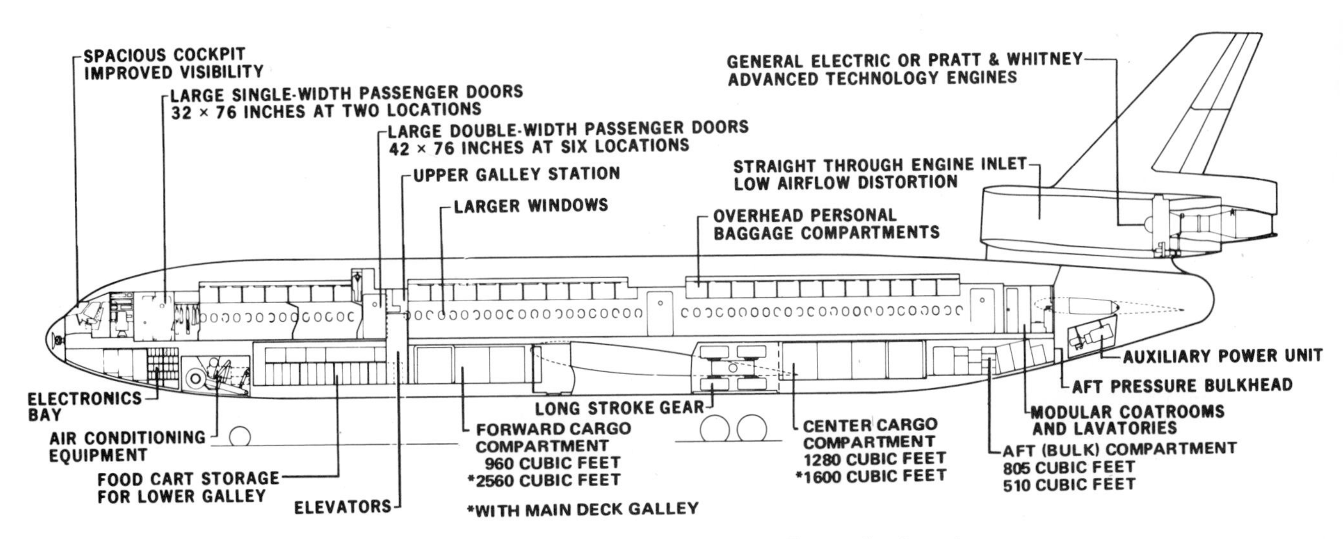

Fig. 25–11 **The inboard profile of a commercial jet transport.** *(McDonnell Douglas Corp.)*

Fig. 25–12 Cutaway of wing-mounted jet engine. *(McDonnell Douglas Corp.)*

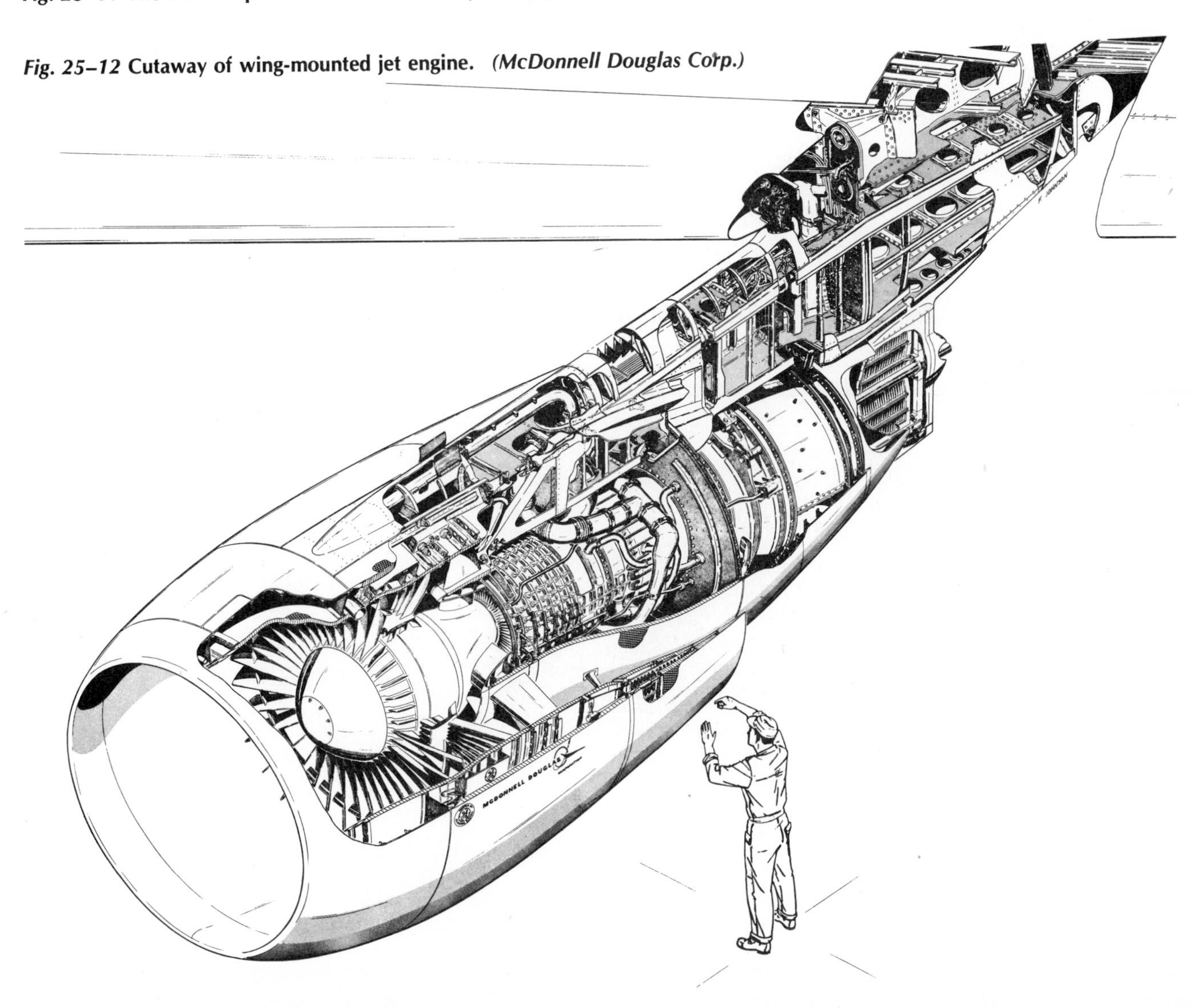

shows its parts. The parts are then listed by name in Fig. 25–16. You can see the hydraulic slat control in a pictorial view (Fig. 25–17) and in schematic form in Fig. 25–18. Figure 25–19 shows the pictorial of a DC-10 *galley* (kitchen). This was drawn by the design illustrator. These kinds of drawings are needed to completely understand aircraft design, building, and flight check-out.

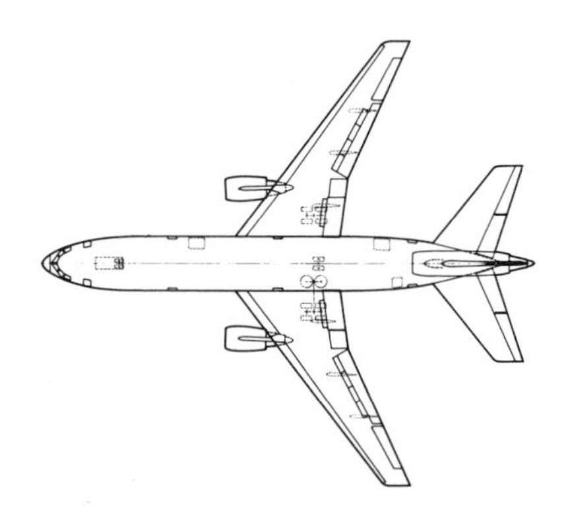

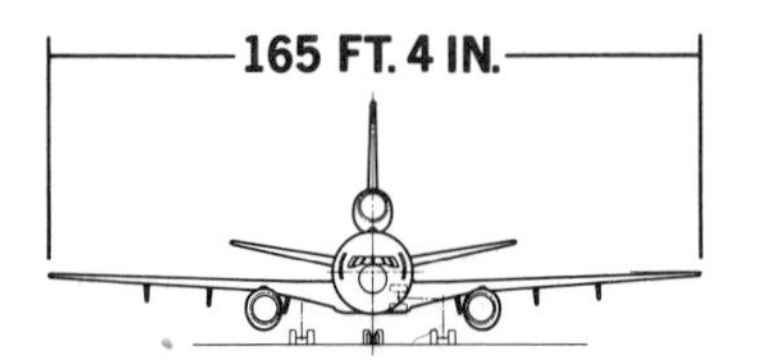

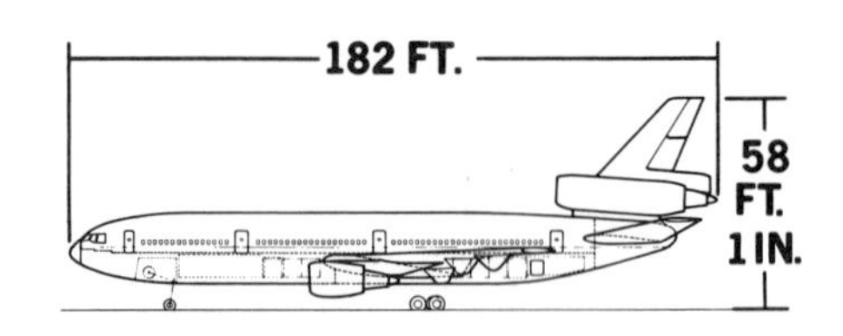

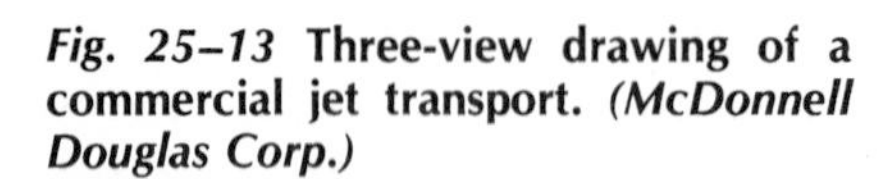

Fig. 25–13 **Three-view drawing of a commercial jet transport.** ***(McDonnell Douglas Corp.)***

Fig. 25–14 **Pictorial drawing of wing section including interior detail.** ***(McDonnell Douglas Corp.)***

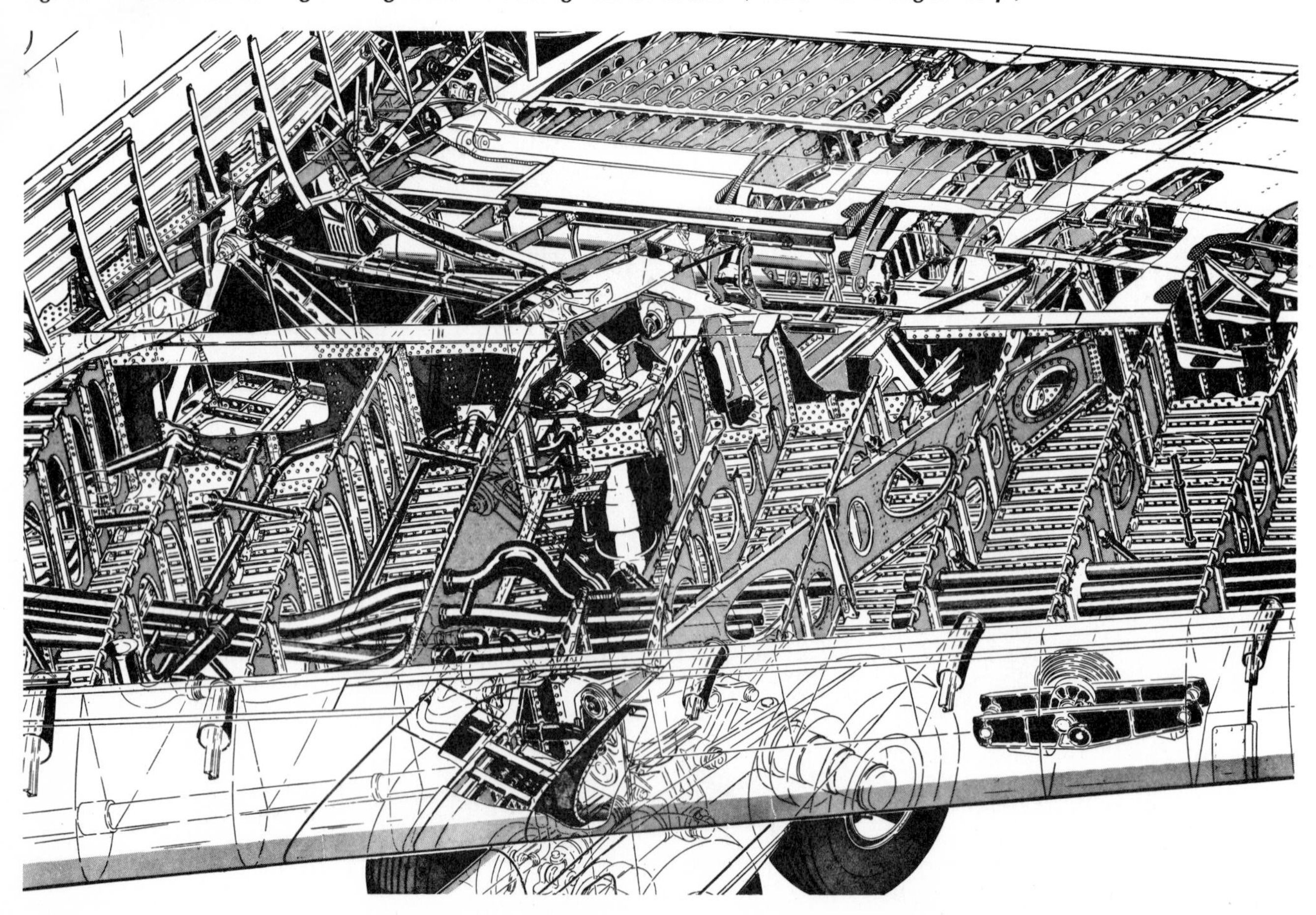

DC-10
ILLUSTRATED PARTS CATALOG

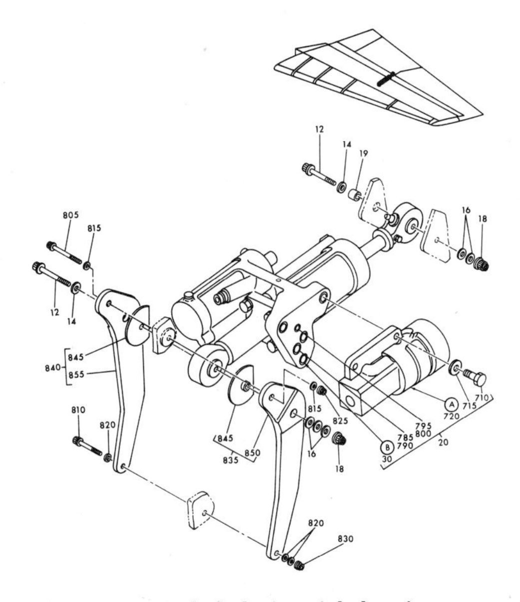

Fig. 25–15 **Exploded pictorial drawing of aileron control.** *(McDonnell Douglas Corp.)*

FIG. ITEM	PART NUMBER	1 2 3 4 5 6 7 NOMENCLATURE
1 - 1	NRG6025	DAMPER INSTL-OUTBD AILERON SEE 57-00-05-01 FOR NHA CONFIG AND DET LOCATIONS
12	72214-6D-31	.BOLT-(V56878)
14	MS20002C6	.WASHER
16	MS20002-6	.WASHER
18	RME9868-6	.NUT-(V72962)
19	4931500-6-030	.BUSHING
20	ARG7231-5001	.DAMPER ASSY- SEE 11-65-01-01 FOR DET
30	ARG7231-5003	..CYLINDER ASSY
40	AN814-2DL	...PLUG
50	NAS1612-2	...PACKING
60	ALG7015-1	...NUT-RSVR RETAINING
70	ALG7008-1	...GLAND-RSVR HOUSING
80	ALG7026-1	...WASHER-RSVR
90	ALG7014-1	...SPRING-RSVR HELICAL CPRS
100	NAS1611-112	...PACKING
110	3891431-112	...RING
120	MS21250-04010	...BOLT
130	MS20002C4	...WASHER
140	ALG7010-1	...BUSHING-RSVR CLAMP UP
150	ALG7009-1	...SPRING-SEAL RSVR HELICAL CPRSN
160	ALG7011-1	...RETAINER-RSVR ADAPTOR
170	ALG7012-1	...ADAPTER-RSVR SEAL
180	2922858-19	...PACKING
190	ALG7013-1	...PISTON ASSY-RSVR
200	MS21209F4-20	....INSERT
210	ALG7013-3	....PISTON
220	WC1203-001	...SETSCREW-(V70318)
230	ALG7041-1	...SCREEN-THERMAL RELIEF AN CHECK VALVE
240	ALG7029-1	...RETAINER-THERMAL RELIEF AND CHECK VALVE
250	ALG7037-1	...SPRING-THERMAL RELIEF AN CHECK VALVE HELICAL CPRSN
260	ALG7028-1	...STEM-THERMAL RELIEF AND CHECK VALVE
270	NAS1611-012	...PACKING
280	3891430-012	...RING
285	NAS620A416L	...WASHER
290	ALG7031-501	...VALVE ASSY-THERMAL RELIEF AND CHECK
300	ALG7030-1	....PIN
310	ALG7042-501	....CAP
320	ALG7027-501	....POPPET
330	NAS6204H8	...BOLT
340	AN960-416	...WASHER
		(CONTINUED)

- ITEM NOT ILLUSTRATED

Fig. 25–16 **Parts list and identification for the drawing in Fig. 25–15.** *(McDonnell Douglas Corp.)*

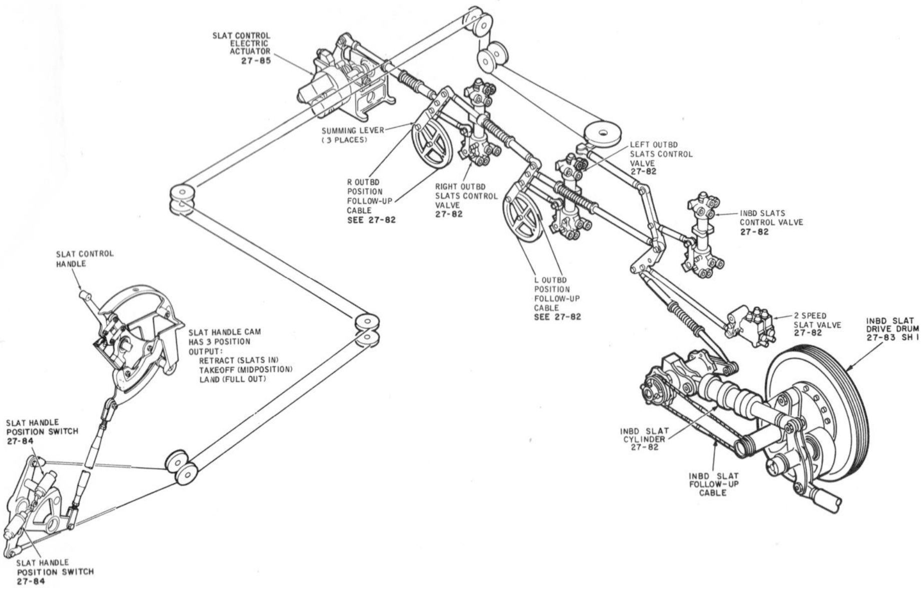

SLAT CONTROL ELECTRIC ACTUATOR 27-85
SUMMING LEVER (3 PLACES)
LEFT OUTBD SLATS CONTROL VALVE 27-82
R OUTBD POSITION FOLLOW-UP CABLE SEE 27-82
RIGHT OUTBD SLATS CONTROL VALVE 27-82
INBD SLATS CONTROL VALVE 27-82
SLAT CONTROL HANDLE
L OUTBD POSITION FOLLOW-UP CABLE SEE 27-82
2 SPEED SLAT VALVE 27-82
INBD SLAT DRIVE DRUM 27-83 SH I
SLAT HANDLE CAM HAS 3 POSITION OUTPUT: RETRACT (SLATS IN) TAKEOFF (MIDPOSITION) LAND (FULL OUT)
SLAT HANDLE POSITION SWITCH 27-84
INBD SLAT CYLINDER 27-82
INBD SLAT FOLLOW-UP CABLE
SLAT HANDLE POSITION SWITCH 27-84

Fig. 25–17 **Pictorial of a hydraulic system.** ***(McDonnell Douglas Corp.)***

What Makes an Airplane Fly?

Figure 25–20 shows why an airplane can fly. Any wing passing through the air at an angle pushes the air downward. This causes an equal and opposite upward push. Lift (flying), then, is caused by the angle of the wing as it moves through air. Look at the shape and position of the wing as you try to figure out how well an aircraft can fly.

SMALLER AIRCRAFT

The new *navigational* (direction-finding) equipment used today has meant challenges for the pilots of small aircraft. The Beechcraft *Bo-*

Fig. 25–18 **Schematic of the hydraulic system in Fig. 25–17.** ***(McDonnell Douglas Corp.)***

SYSTEM SCHEMATIC
HYDRAULIC POWER 29-10
PRESSURE
RETURN
SLATS EXTEND
SLATS RETRACT
FIXED PIVOT
SLAT HANDLE INPUT 27-81
SLATS EXT
OUTBD SLAT VALVES TWO-WAY RESTRICTORS 27-82-07
OUTBD SLAT VALVES ONE-WAY RESTRICTORS 27-82-08
25 lb BUNGEE
8 lb BUNGEE
EXT
HYD SYS 1
HYD SYS 3
AUTO SLAT ACTUATOR INPUT SEE 27-81-0
PRESSURE CHECK VALVE 27-82-05
INBD SLAT VALVE TWO-WAY RESTRICTORS 27-82-06
RIGHT OUTBD SLAT CONTROL VALVE 27-82-03
LEFT OUTBD SLAT CONTROL VALVE 27-82-03
INBOARD SLAT CONTROL VALVE 27-82-03
SLAT TWO SPEED VALVES 27-82-04
RIGHT OUTBD POSITION FOLLOW-UP CABLE
LEFT OUTBD POSITION FOLLOW-UP CABLE
OUTBD SLAT CYLINDER 27-82-11
INBD SLAT CYLINDER 27-82-11
15 lb BUNGEE
CABLE DRUM
RIGHT OUTBOARD SLAT DRIVE MECHANICAL ASSY 27-83 SH 2
LEFT OUTBOARD SLAT DRIVE MECHANICAL ASSY 27-83 SH 2
INBOARD SLAT DRIVE MECHANICAL ASSY 27-83 SH I
INBD SLAT CONTROL VALVE RECEIVES INPUT FROM PILOT'S CONTROL HANDLE.
OUTBD SLAT CONTROL VALVES RECEIVE INPUT FROM MECHANICAL FOLLOW-UP OF INBD SLATS.
EACH SLAT CONTROL VALVE IS NULLED BY ITS OWN FOLLOW-UP MECHANISM.

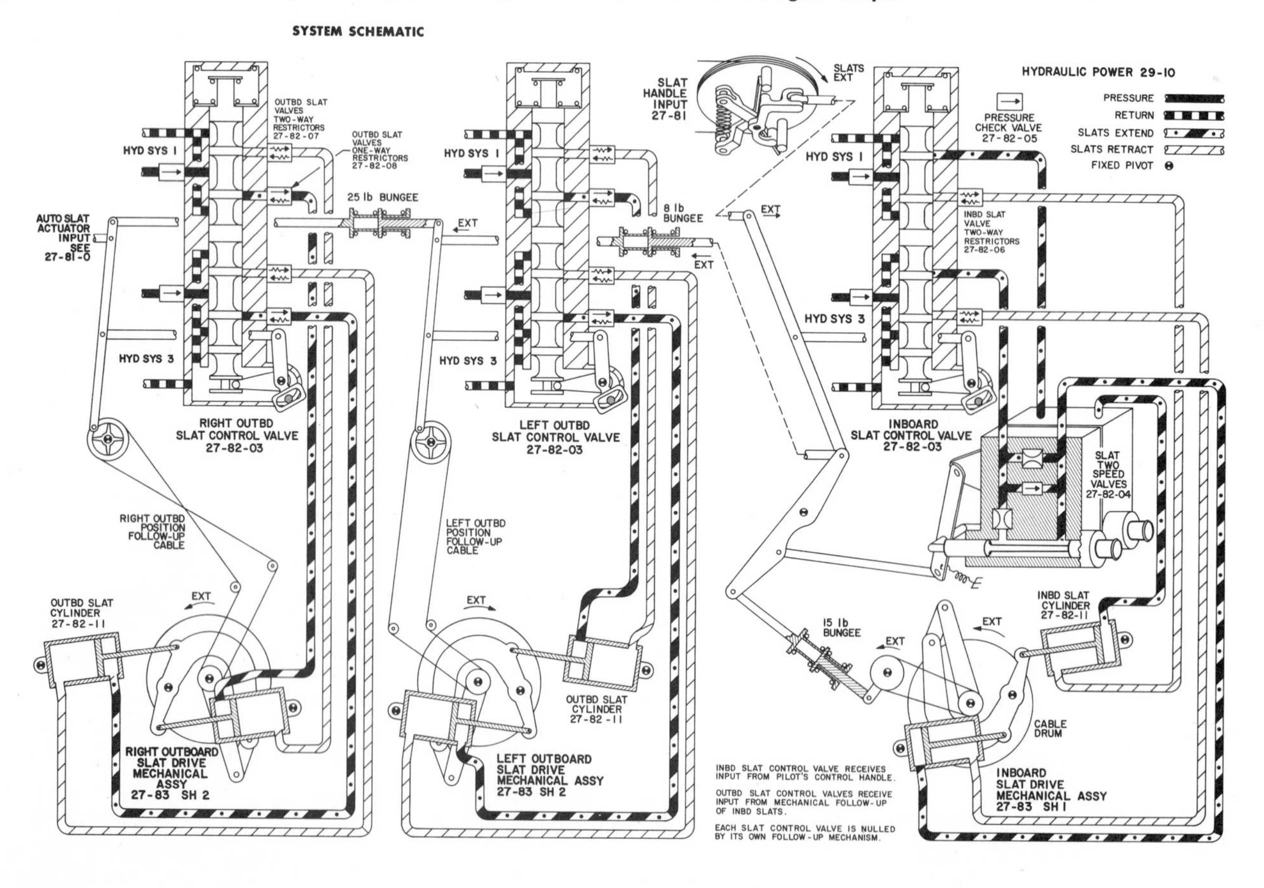

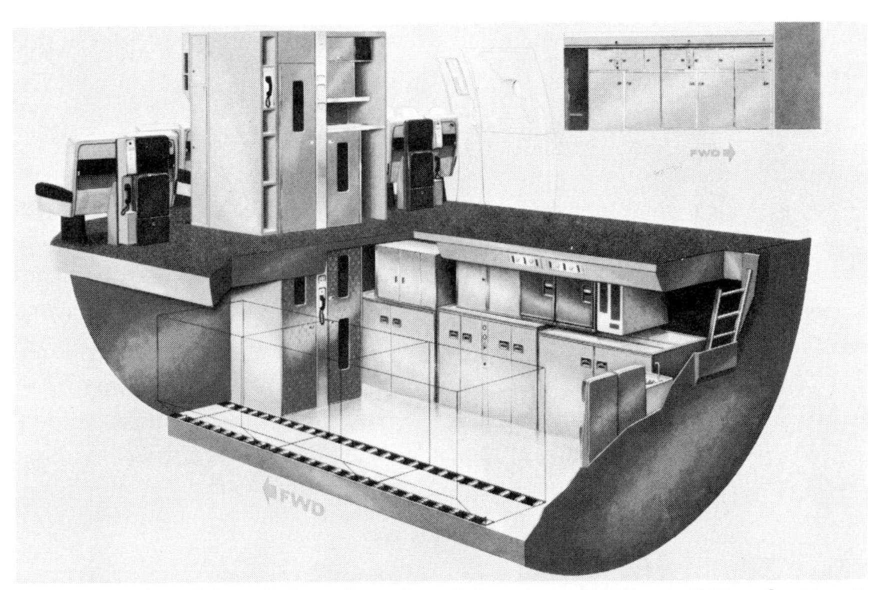

Fig. 25–19 **Pictorial rendering of an aircraft interior. (*McDonnell Douglas Corp.*)**

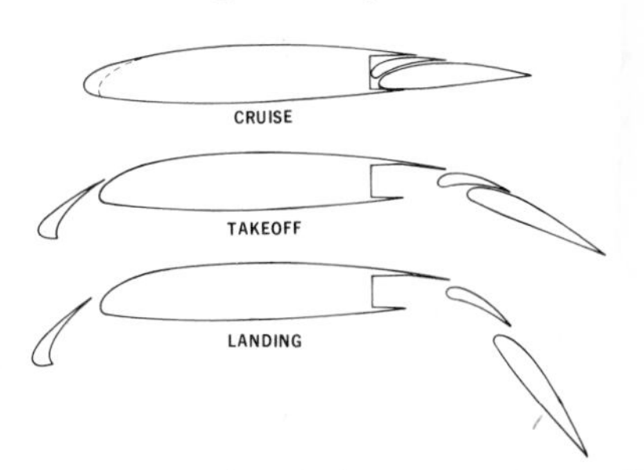

Fig. 25–20 **Wing designs. (*McDonnell Douglas Corp.*)**

Fig. 25–21 **Three views of a small aircraft. (*Beech Aircraft Corp.*)**

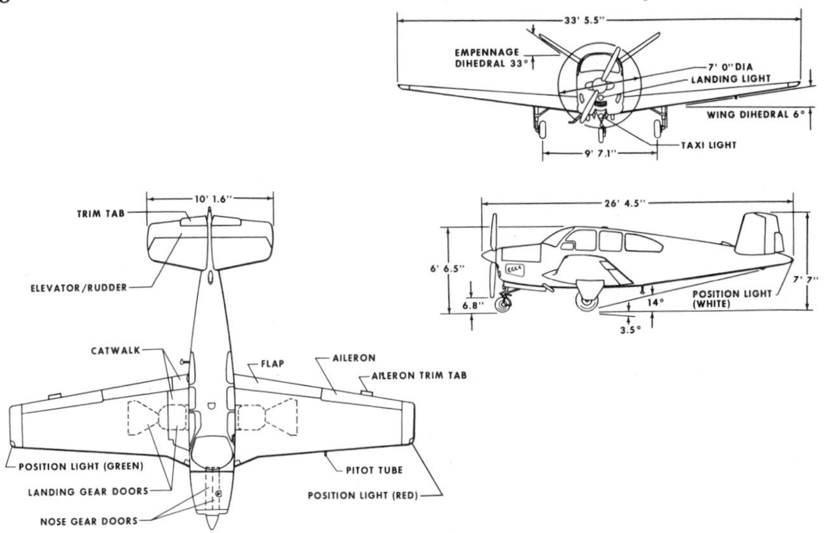

Fig. 25–22 **This small aircraft has an all-metal superstructure with fiberglass components. (*Piper Aircraft Corp.*)**

nanza in Fig. 25–21 has been around for a while. The design today includes a Bendix automatic pilot and flight director system. This system, with communications and navigational equipment, lets the aircraft cruise safely from 280 to 320 km/h (175 to 200 mph). This plane has a *dihedral stabilizer* (V-tail). This design has been used only on this small aircraft.

The Piper Aircraft Corporation designed and made the *Navajo* (Fig. 25–22) as an all-metal superstructure. However, fiberglass parts are used where the toughness of such a material is needed. Such places are the nose cone, wing, rudder and vertical fin tips, door frame, and windshield channel.

The wing of this plane has a stepped-down main spar, a front and rear spar, *lateral* (crosswise) stringers, *longitudinal* (lengthwise) ribs, and stressed skin sheets. The wings are joined together with heavy steel plates. This makes a main spar that runs unbroken from wing tip to wing tip. Besides the sturdy splice joint, the wing is attached to the fuselage at the front, center, and trailing edge. Flat rivets are used forward of the main spar. This allows the air to flow more smoothly over the wing. The wing root fillet and the swept leading edge between the fuselage and the *nacelle* (which holds the engine) also smooth the airflow.

"New" adventures in the world of flying have resulted from the rebirth of the biplane and stunt aircraft. *Citabria* is a popular sport model. It is shown in Fig. 25–23. The tube-shaped steel body has a light fabric cover. This makes it more durable. The Bellanca Aircraft Company is only one of the many companies that have designed old styles with new methods.

New Material for Small Aircraft

A new material has been used in small leisure aircraft in recent years. Windecker Industries, Inc., has made a plastic plane of a material called Fibaloy. The *Eagle* (Fig. 25–24) is built of plastic made stronger with fiberglass. The aircraft has no riveted sections or seams formed by lapping skins. The aircraft becomes stronger after it is made because of a curing process inside the plastic. A chemical reaction causes the plastic to become hard like steel.

Typical Drawings

Some of the common engineering drawings for a smaller aircraft are shown. A torque link from a small airplane is a common dimensioned machine drawing (Fig. 25–25). The forging blank drawing is shown in Fig. 25–26. This drawing tells all that the forger and inspector of the blank need to know. A forging machine drawing gives information for the machinist, set-up person, inspector, and others who may put together bushings, bearings, etc.

Lofting Layouts

Full-size drawings for large jobs are made by *lofting*. This word comes from the ship loft, where the exact lines of the shapes of ships are worked out. Lofting is important to aircraft-design layout. Contours of wing sections are drawn without mistakes by lofting. The curves are

Fig. 25–23 **Citabria, an aerobatic aircraft. (*Bellanca Aircraft Corp.*)**

Fig. 25–24 **An aircraft made from plastic. (*Windecker Research, Inc.*)**

Fig. 25–25 **A typical dimensioned drawing for machining. (*North American Rockwell Corp.*)**

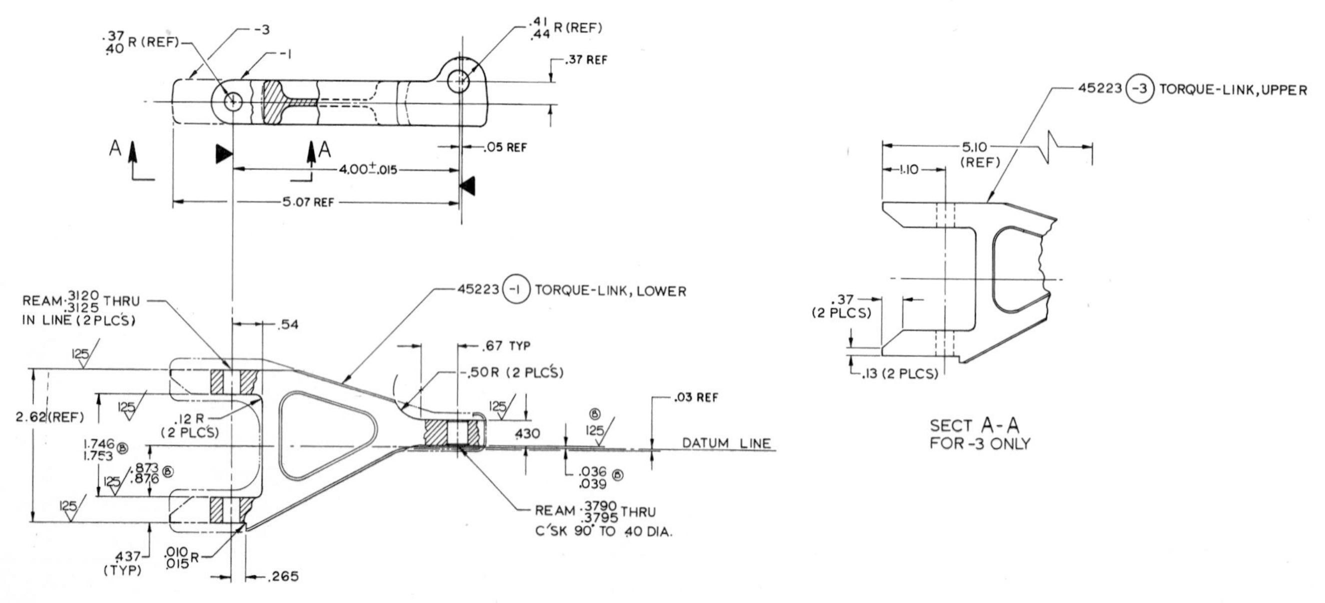

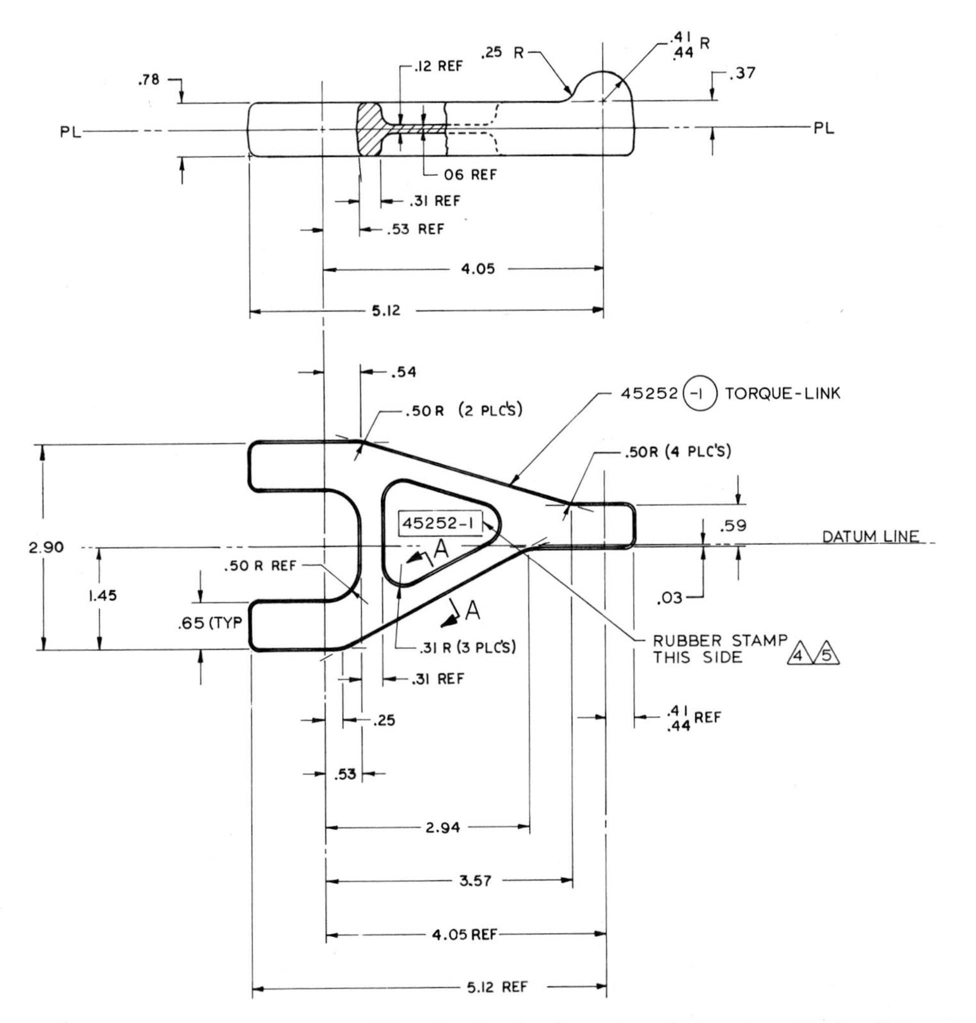

Fig. 25–26* A typical dimensioned drawing for forging.** (North American Rockwell Corp.***)

faired (adjusted or smoothed out) to get smooth surfaces. Templates may be made when needed. The drawing board is generally too small for such work. Layouts are made on special loft floors that have all the space needed. Ribs are drawn with trace chords and a loft line. The drawings of ribs are undimensional. This method is used for sheet aluminum and steel parts. The drawing is transferred to the material by a photo process.

Light Jet Helicopter

Another small aircraft is the light jet helicopter. This craft set records for speed, distance, climbing ability, and altitude. The three-view drawing (Fig. 25–27) shows the structural form. The overall measurements are in United States customary and metric measure. The electronic components for navigation and communications (called *avionics*) let the helicopter make many different movements.

BUSINESS JETS

The aircraft manufacturers have designed a few business jets that are each one-of-a-kind. The jet shown in Fig. 25–28 can travel at speeds from 820 to 880 km/h (508 to 548 mph). The design has a wing with eight spars. The wing has twice the needed strength and is designed not to fail. A spar is the principal part of the framework in an airplane wing. It runs from tip to tip or from root to tip. The highly polished

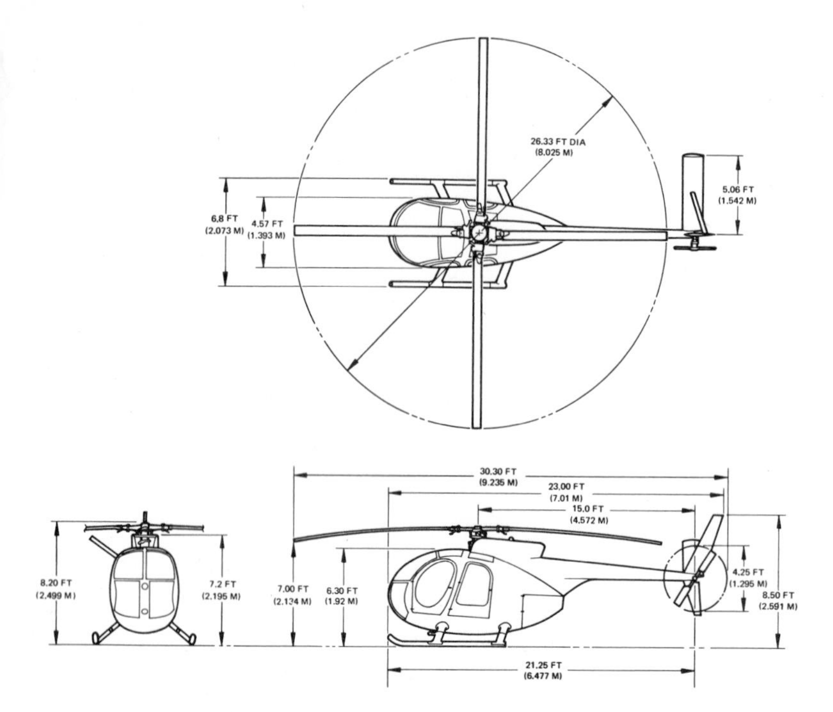

Fig. 25–27 **A small jet helicopter. (*Hughes Tool Co.*)**

aluminum skin brings out the sleekly designed lines of the jet.

The three-view drawing of the business jet in Fig. 25–29 shows another style. The assembly-line photo (Fig. 25–30) shows the final stages of assembly. The complex control panel in Fig. 25–31 is the final thing to be put together and checked in the cockpit.

One of the designs for the jet in Fig. 25–28 was based on a Federal Aviation Regulation, Part 25. This aviation handbook tells what is needed for airframes to pass federal tests. It also tells how an aircraft should perform. The design team prepares drawings that can make a lightweight, strong airframe that follows federal regulations.

THE SPACE SHUTTLE

The newest kind of spacecraft in the NASA space program is the

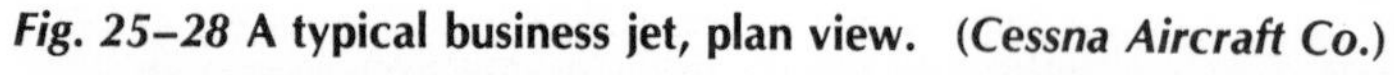

Fig. 25–28 **A typical business jet, plan view. (*Cessna Aircraft Co.*)**

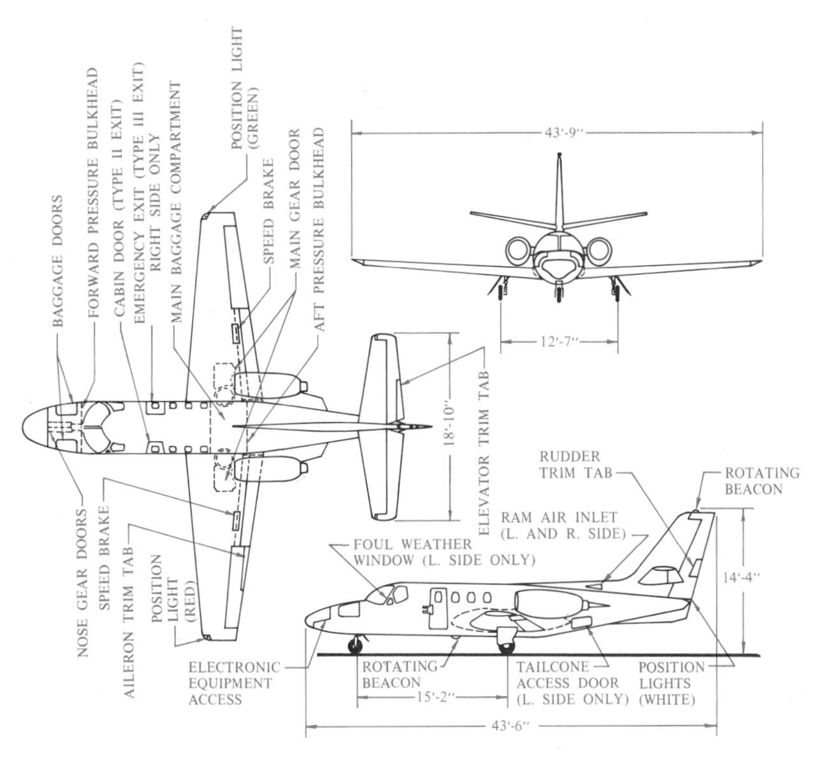

***Fig. 25–29* A three-view drawing of a business jet. (*Cessna Aircraft Co.*)**

Space Shuttle (Fig. 25–32). It consists of a vehicle called the orbiter and two solid rocket boosters to lift it into orbit. It also includes a large external liquid *propellant* (fuel) tank. The Space Shuttle is designed for scientific, civilian, or military use.

The shuttle orbiter is different from older spacecraft in that when its mission in space is over, it can land back on Earth like an airplane. It can then be made ready for other missions in the earth's orbit. The orbiter is about the size of a small jet passenger airplane.

KINDS OF AIRCRAFT DRAWINGS

Many different kinds of drawings are made for many different kinds of aircraft. From helicopters, leisure aircraft, military jets, jet trans-

***Fig. 25–30* Final assembly line. (*Cessna Aircraft Co.*)**

***Fig. 25–31* Final assembly checkpoint is the cockpit. (*Cessna Aircraft Co.*)**

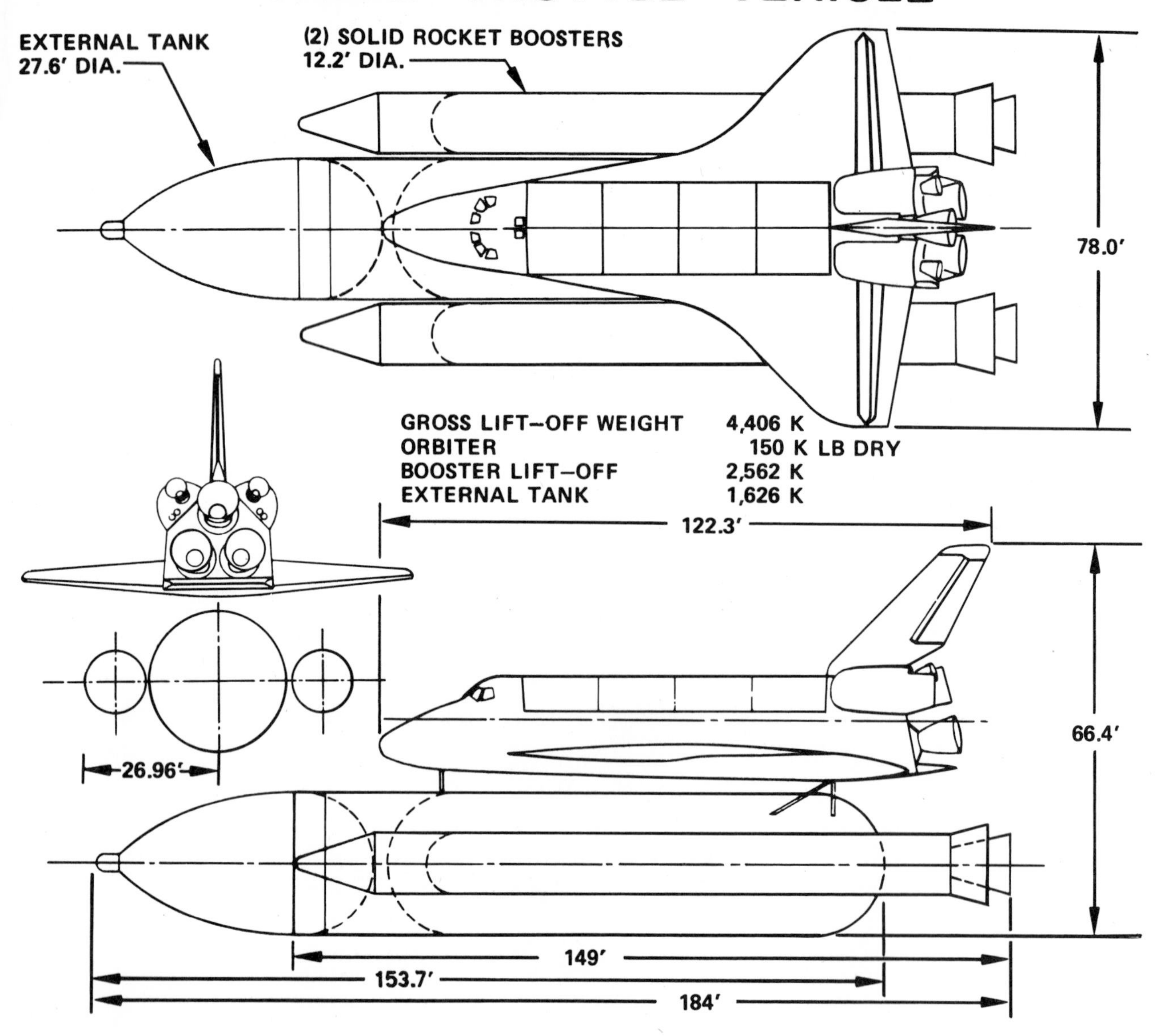

Fig. 25–32 **Three-view drawing of the Space Shuttle, and data on the Shuttle system.** (*NASA.*)

ports, biplanes, to Space Shuttles, drawings are needed. Drawings are used for castings, forgings, sheet-metal layout, schematics, and lofting. Sketching is used for changes and new designs. Of these, one of the important drafting areas is in forming sheet metal.

Sheet-metal drawings are based upon the rules of intersections and development. You learned about these in Chapter 18. Sheet metal is used for forming parts of aircraft. It is also used for the curved skin covering. The designer and drafter consider many things when choosing the right material. Sketching is used in many ways in aircraft design. Views of small parts, or the overall outline of the aircraft foils, may be sketched. Then the sketches are looked at to see how those small parts can be designed in better ways.

SPACE SHUTTLE SYSTEM

PARAMETER	METRIC VALUE	ENGLISH VALUE
OVERALL SPACE SHUTTLE SYSTEM		
LENGTH	56.08 m	184 ft.
HEIGHT	23.2 m	76 ft
WEIGHT AT LAUNCH	1,860,000 kg	4,100,000 lbs.
PAYLOAD WEIGHT INTO ORBIT		
INCLINATION (LOWEST) 28.5 deg	29,500 kg	65,000 lbs.
INCLINATION (HIGHEST) 104 deg	14,500 kg	32,000 lbs.
SOLID ROCKET BOOSTER		
DIAMETER	3.6 m	11.8 ft.
LENGTH	44.2 m	145.1 ft.
WEIGHT		
LAUNCH	527,760 kg	1,163,500 lbs.
INERT	70,000 kg	154,300 lbs.
THRUST AT LAUNCH, EACH	11,120,500 N	2,500,000 lbs.
EXTERNAL TANK		
DIAMETER	8.4 m	27.58 ft.
LENGTH	47 m	153.68 ft.
WEIGHT		
LAUNCH	738,000 kg	1,628,000 lbs.
DRY	35,000 kg	78,000 lbs.
ORBITER		
LENGTH	37.5 m	123 ft.
WING SPAN	23.8 m	78 ft.
HEIGHT TO EXTENDED LANDING GEAR	17.4 m	57 ft.
PAYLOAD BAY		
DIAMETER	4.6 m	15 ft.
LENGTH	18.3 m	60 ft.
CROSS RANGE	2,038 km	1,100 nm
MAIN ENGINES (3)		
VACUUM THRUST, EACH	2,090,700 N	470,000 lbs.
ORBITAL MANEUVERING SUBSYSTEM		
ENGINES (2)		
VACUUM THRUST, EACH	26,700 N	6,000 lbs.
REACTION CONTROL SYSTEM		
ENGINES (40)		
THRUST, EACH	4,003.4	900 lbs.
VERNIER ENGINES (6)		
VACUUM THRUST, EACH	111.2 N.	25 lbs.
WEIGHT		
DRY	68,040 kg	150,000 lbs.
LANDING	81,650 kg	180,000 lbs.

Fig. 25–32 **(Cont.)**

Review

1. Describe the basic structure of an airplane fuselage and wing.

2. Name some good points of the undimensioned drawing.

3. What kinds of drawings can be made in the undimensioned way?

4. Tell about some career opportunities in the aerospace industry.

5. List and tell about opportunities in jobs that are close to the aerospace industry, such as hydraulics engineer, electronics engineer, or interior decorator.

Problems

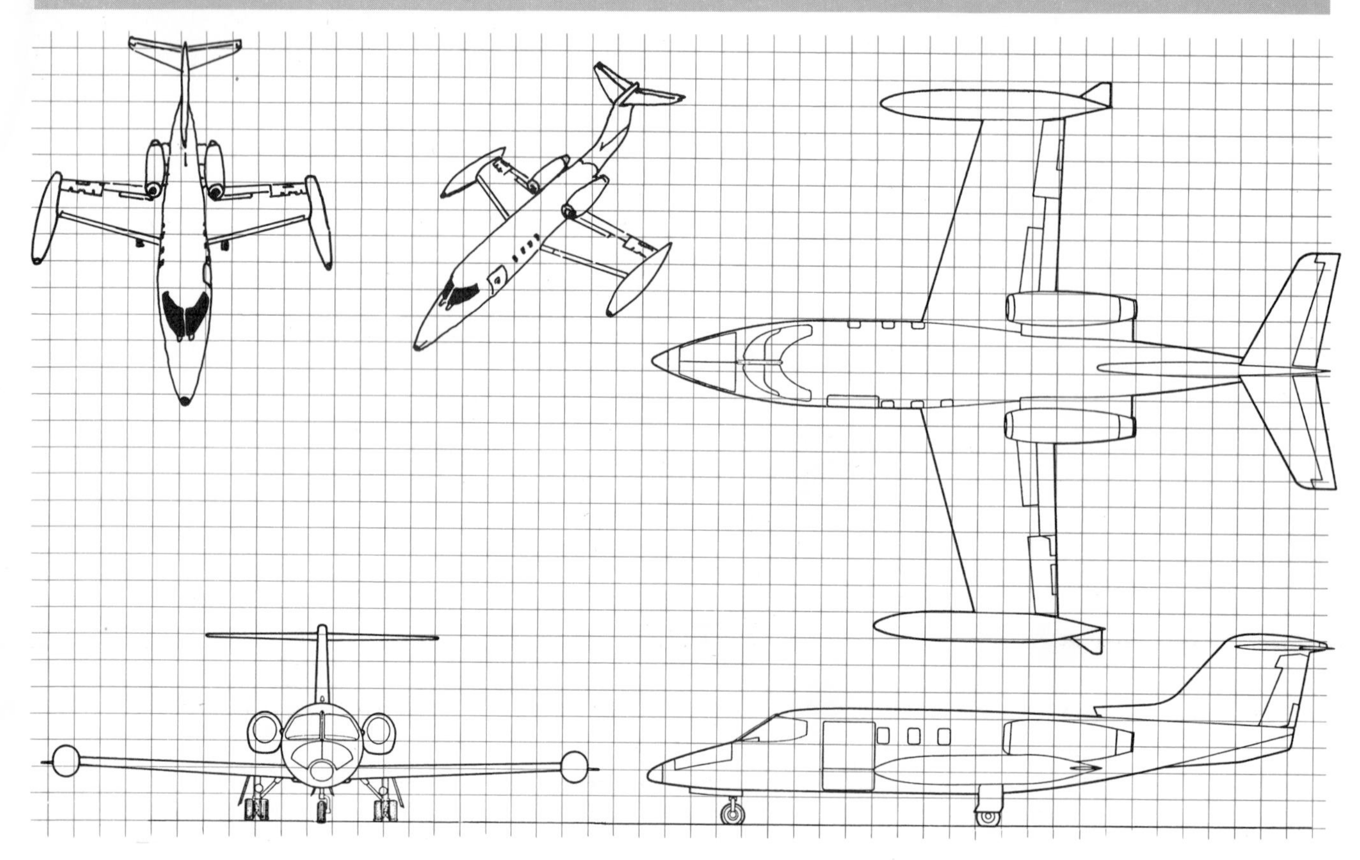

Fig. 25–33 **Using grid paper, sketch any two views of the executive jet. Prepare one pictorial view of the jet.**

Fig. 25–34 **Make a working drawing of the landing gear and lift strut fitting with the following changes. The V-tongue is to be 1¼ instead of 1 9/16. The 1 3/16 location dimension is to be 1¾. Note that the ⅛ label applies to the V-tongue and other radii. List the sequence of the operations necessary to make this fitting.**

.250 PILOT HOLE
2 REQD.
1.000
1 3/8
11/16 R
20°
1/4
1R
22°15'
9/16
.500
.644 HOLE
.500R
1 3/16
9/16

NOTE:
ALL RADII 1/8 R. UNLESS NOTED

.125 X 3 5/8 X 4 5/8 #1025 STEEL
FULL SIZE
FINISH—PIPER SPECIFICATION #10
MANUFACTURING PRACTICES—PIPER SPECIFICATION #9
FITTING – LANDING GEAR & LIFT STRUT
L-14 2 90346
PIPER AIRCRAFT CORP.
LOCK HAVEN, PENNA.
D. M. B. 10-24-
SAFF 10-31-
90344

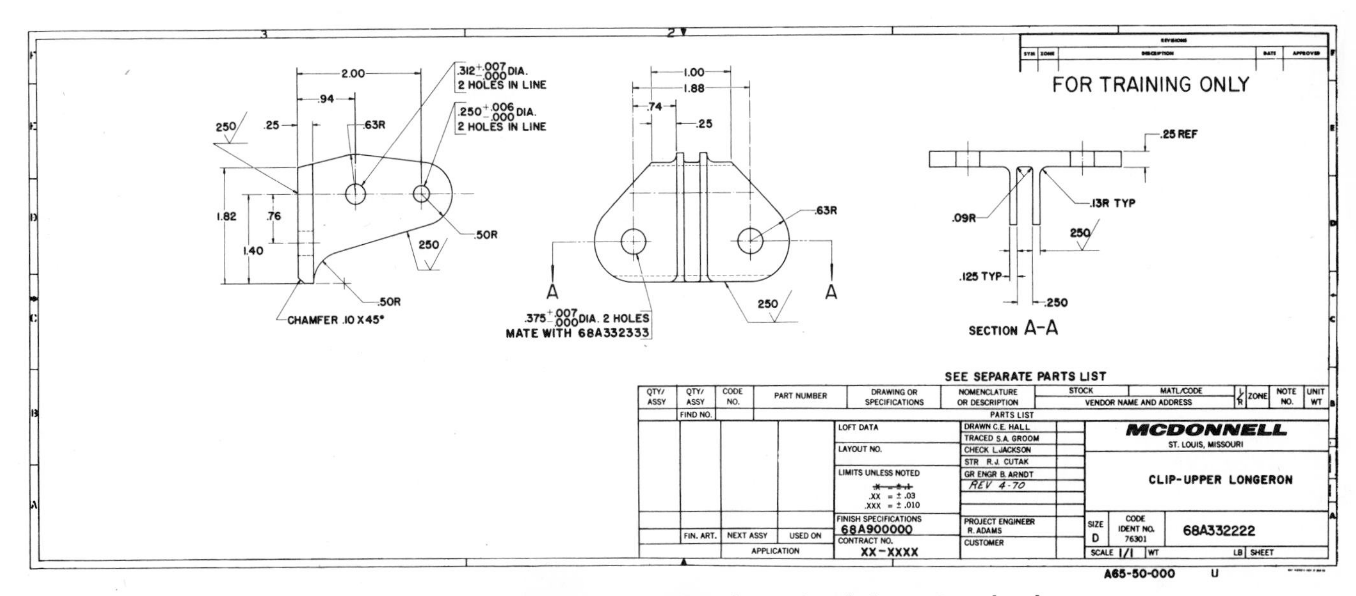

Fig. 25–35 **Problem 1. Make a detail drawing of the clip, including the sectional view. Cross hatching may be omitted from the section. Title block and border are optional. Problem 2. Note the zone markings on the border of the illustration. Change order: change 2.00 dimension in zone F-3 to 2.50.**

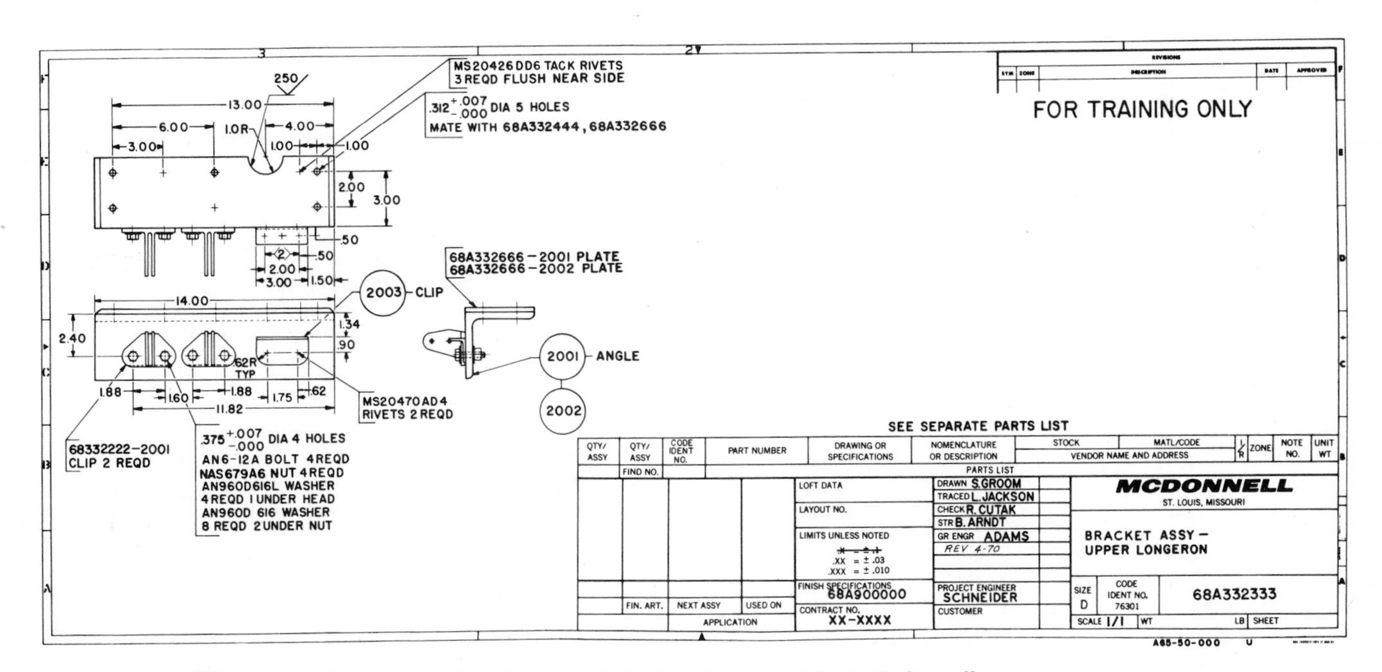

Fig. 25–36 **Problem 1. Make an assembly drawing of the bracket assembly, including all notes as shown. Title block and border may be omitted. Examine the details of the clip in Fig. 25–35 before completing this drawing. Problem 2. Make up a parts list for the bracket assembly.**

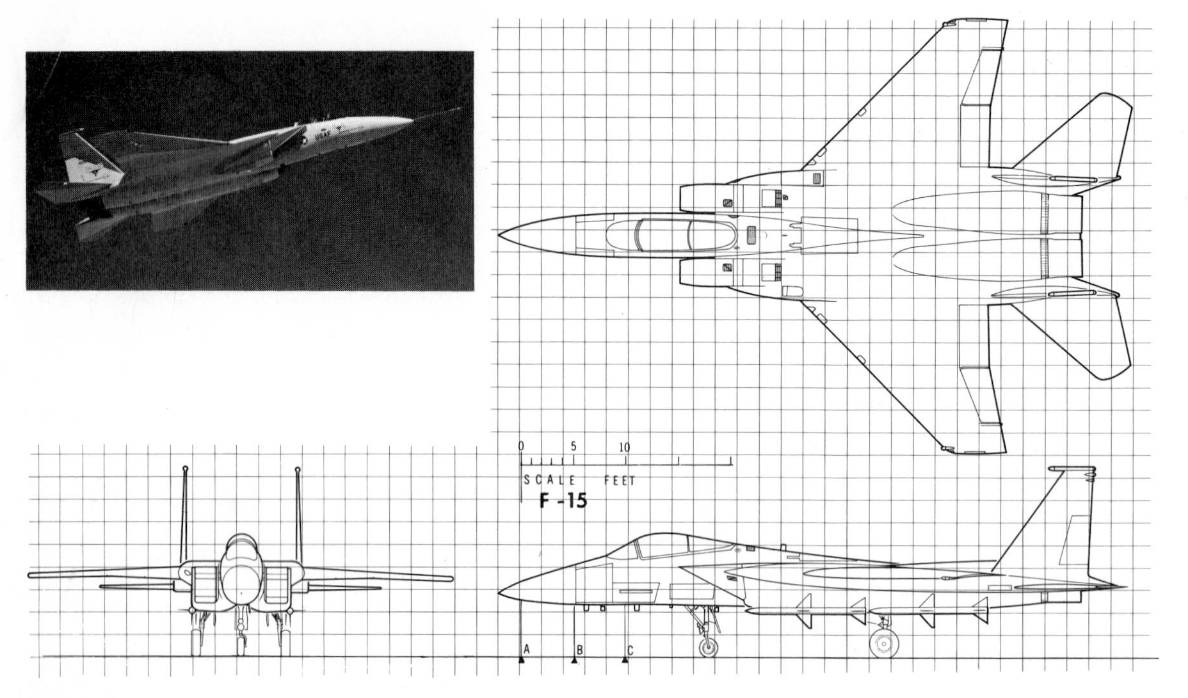

Fig. 25–37 **Problem 1. Draw one of the views of the F-15 aircraft on a grid. Make the grid 60′ across the longest elevation and 20′ high from the grade line to the top of the tail. Take dimensions from the scale. Problem 2. Using dividers, locate the contours above the grade line at *A, B,* and *C*.**

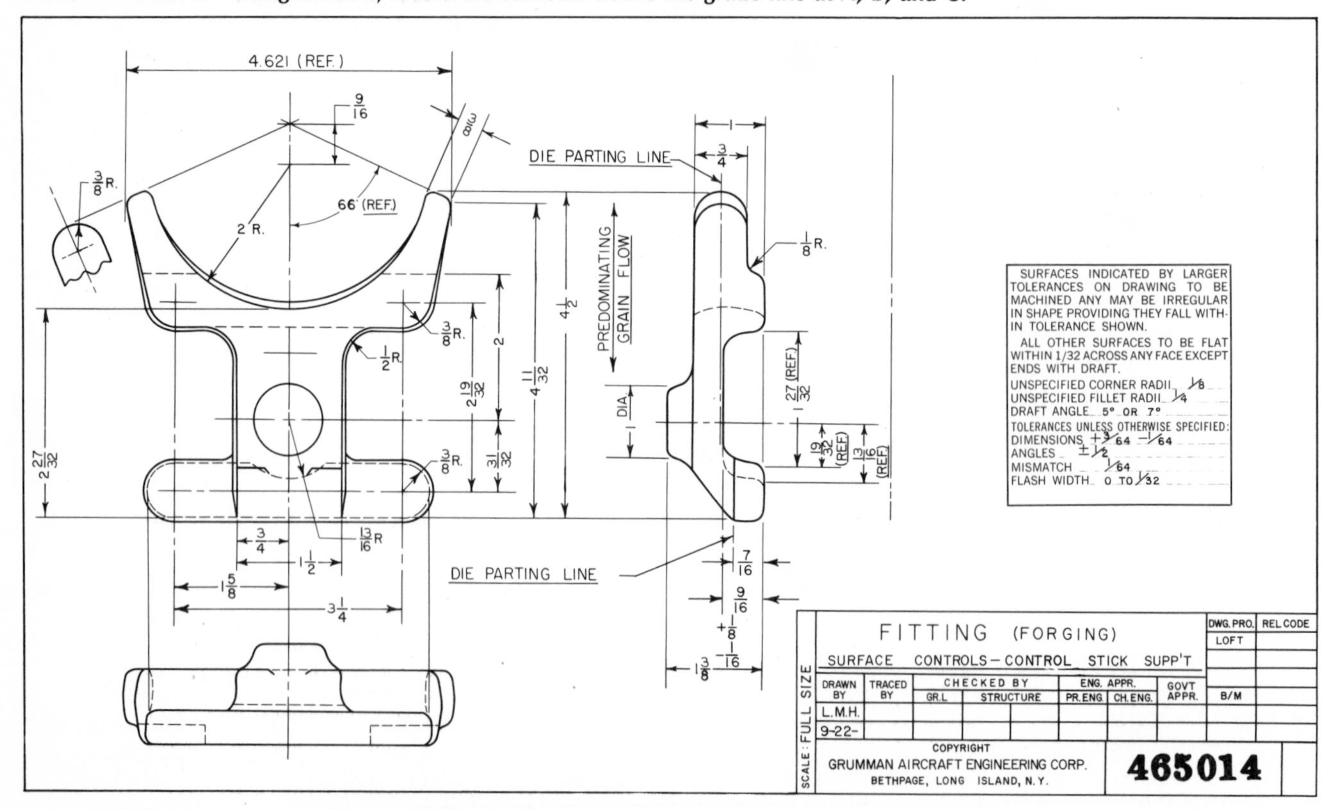

Fig. 25–38 **Make a working drawing of the control-stick support fitting.**

Chapter 26

Technical Illustration

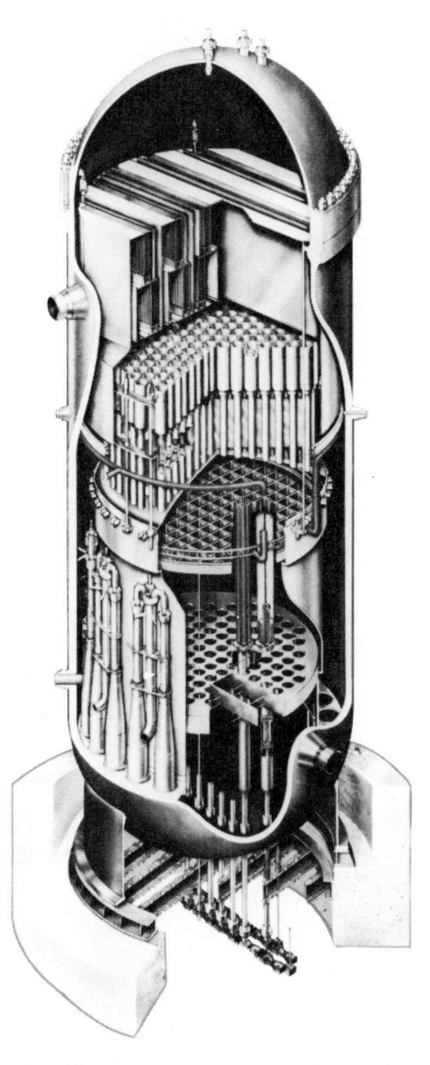

Fig. 26–1 **This illustration of a nuclear reactor is a cutaway view showing the relationship of parts and interior details.** ***(General Electric Co.)***

TECHNICAL ILLUSTRATION

Technical illustration has an important place in all areas of engineering and science. Technical illustrations form a necessary part of the technical and service manuals for machine tools, automobiles, machines, and appliances In technical illustration, pictorial drawings are used to describe parts and the methods for making them. Pictorial drawings show how the parts fit together. They also show the steps that need to be followed to complete the product on the assembly line. Technical illustrations may be used to set up the assembly line. They are useful for industrial, engineering, and scientific purposes.

Technical illustration drawings can range from simple sketches to rather detailed shaded drawings. They may be based on any of the pictorial methods: isometric, perspective, oblique, and so forth. The complete project or parts of groups of parts may be shown. The views may be exterior, interior, sectional, cutaway, or phantom (Fig. 26–1). The purpose in all cases is to provide a clear and easily understood description. Chapter 2, "Sketching and Lettering," and Chapter 12, "Pictorial Drawing," provide the basis for making technical illustrations.

PART NO.	PART NAME	NO. REQD
1	BASE	1
2	MOVABLE JAW	1
3	MOVABLE JAW PLATE	1
4	MACHINE SCREW	1
5	LOCKING PIN	1
6	HANDLE STOP	2
7	HANDLE	1
8	CLAMP SCREW	1
9	JAW FACE	2
10	CAP SCREW	2

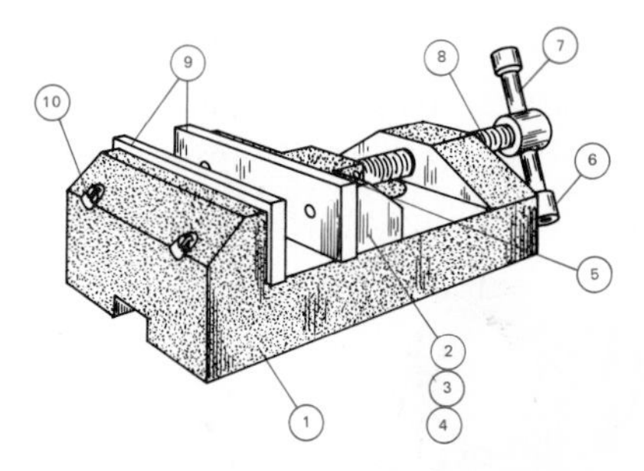

Fig. 26–2 **An illustrated parts list.**

Drawings for use within a company's plant can sometimes be made by a drafter with artistic talent. However, the special skills needed for such drawings call for the work of a professional technical illustrator. Technical illustrations have been used for many years in illustrated parts lists, operation and service manuals, and process manuals (Fig. 26–2). The aircraft industry in particular has found production illustration especially valuable. In aircraft construction, pictorial drawings are used when the plane is first designed. These drawings are also used throughout the many states of a plane's production and completion on the assembly line. When the plane is delivered to the customer, the service, repair, and operation manuals are also illustrated with pictorial drawings.

DEFINITION

Generally, a technical illustration is a pictorial drawing that provides technical information by visual methods. This is usually done by turning a multiview drawing into a three-dimensional pictorial drawing. It must show shapes and relative positions in a clear and accurate way. The proper amount of shading may be used to bring out the shape. A technical illustration, however, is not a work of art. Therefore, the shading must serve a practical end, not an artistic one.

In addition to pictorials, technical illustrations include graphic charts, schematics, flow charts, diagrams, and sometimes circuit layouts. Dimensions are not a part of technical illustrations because they are not working drawings.

TOOLS AND TIPS

As a technical illustrator, you would use most of the tools and regular drafting equipment described in Chapters 3 and 13. An H or 2H pencil, with the point kept well sharpened, is the most useful tool. A few other useful items include a crow quill pen, a felt-tip pen, technical pens, masking tape, Scotch tape, X-Acto knife, paper stomps, two or three brushes, airbrush, and a reducing glass. It is important to keep all your tools clean.

You should also become familiar with the use of Craftint, Zip-a-tone, Chart-Pak, and similar press-on section linings, screen tints, letters, and the like. You should know about the various methods of reproduction. In addition, you should be aware of the effect of reduction when your drawing is to be used in a smaller size. Lines must be firm and black. Erasures must be clean. The part of a drawing not being worked on should be kept covered with tracing paper or sheet plastic.

LETTERING

Lettering is an important element of technical illustration. In some cases, templates or scriber guides can be used. Good freehand lettering is a required skill for a technical illustrator.

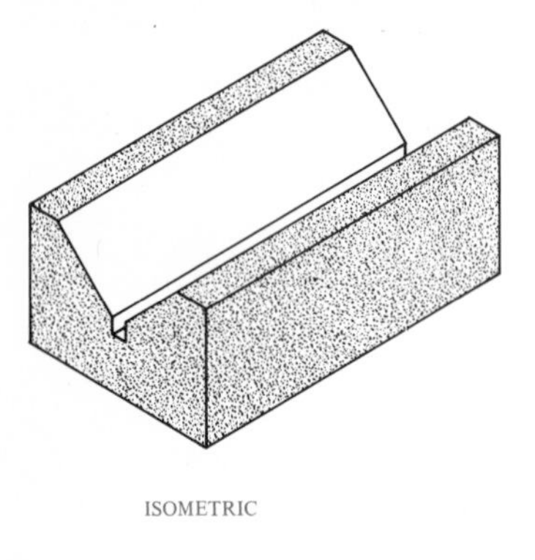

ISOMETRIC

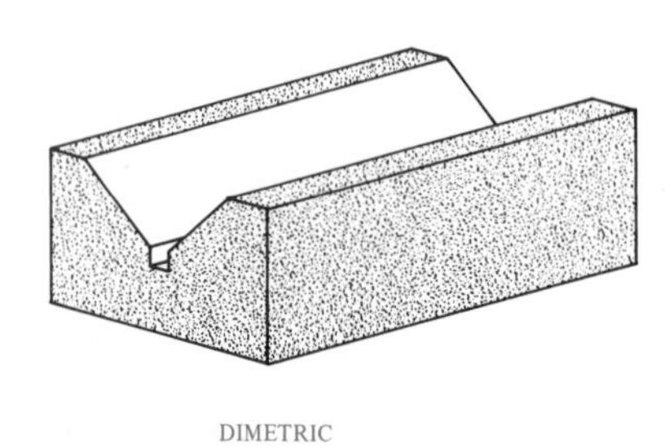

DIMETRIC

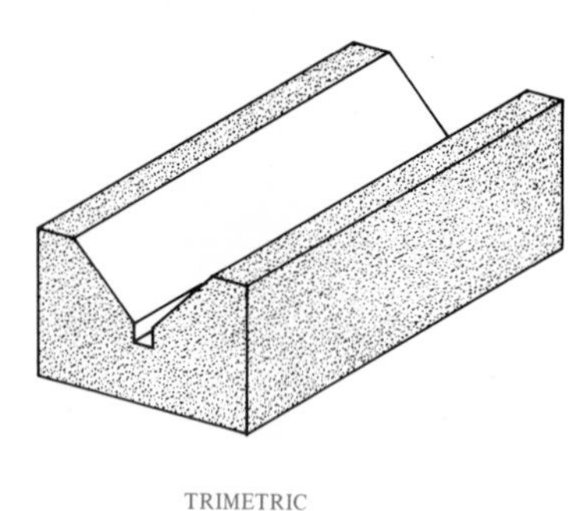

TRIMETRIC

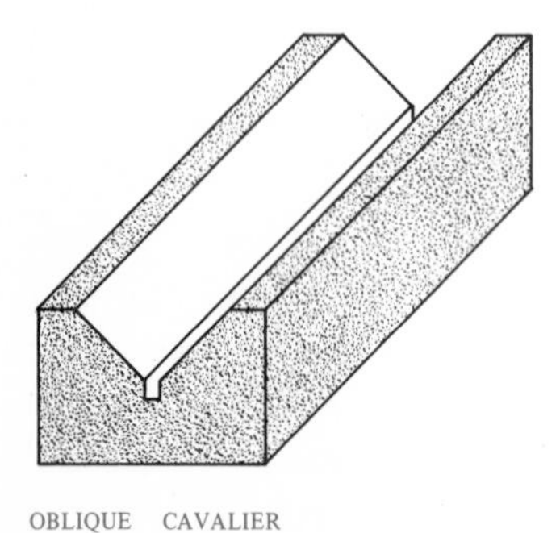

OBLIQUE CAVALIER

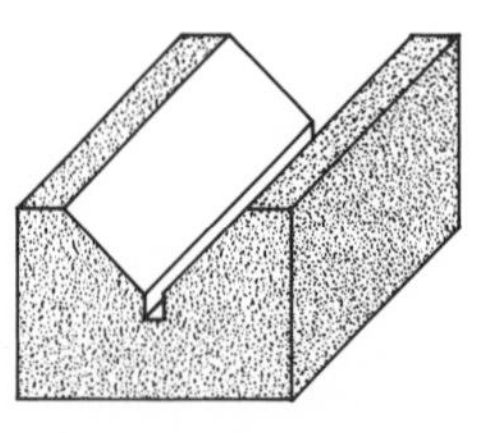

OBLIQUE CABINET

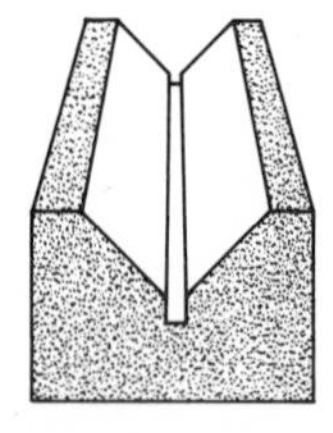

SINGLE - POINT (PARALLEL) PERSPECTIVE

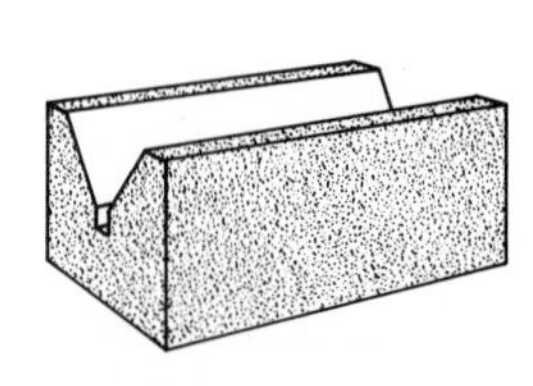

TWO - POINT (ANGULAR) PERSPECTIVE

***Fig. 26–3* A V-block in various types of pictorial drawing.**

PICTORIAL LINE DRAWINGS

Basically, all technical illustrations are pictorial line drawings. Therefore, you should have a complete understanding of the various types of pictorial line drawings and their uses. Chapter 12 describes the various types of pictorial drawings. It also describes the procedure for drawing each.

Usually, any type of pictorial drawing can be used as the basis for a technical illustration. However, some types are more suitable than others. This is especially true if the illustration is to be *rendered* (shaded). Figure 26–3 shows a V-block drawn in several types of pictorial drawing. Notice the difference in the appearance of each. Isometric is the least natural in appearance. Perspective is the most natural. This might suggest, then, that all technical illustrations should be drawn in perspective. This is not necessarily true. While perspective is more natural than isometric in appearance, it takes more time to do. It is also more difficult to draw. Thus, it is a costly method to use.

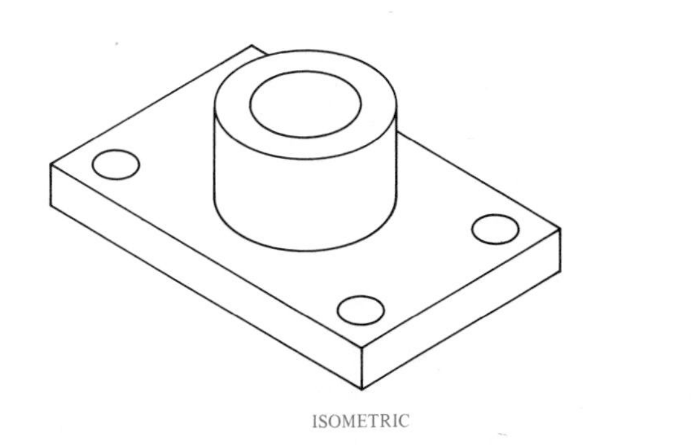

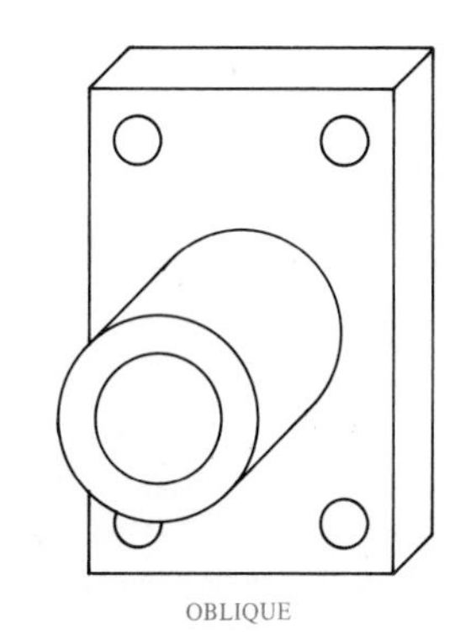

Fig. 26–4 **The shape of an object helps to determine the most suitable types of pictorial drawing to use.**

The shape of the object also helps to determine the type of pictorial drawing you should use. Figure 26–4 shows a pipe bracket drawn in isometric and oblique. The shape of this object is most easily and quickly drawn in oblique. In many cases, it will look more natural than in isometric.

If an illustration is to be used only in-plant, the illustrator will usually make the pictorial drawing in isometric or oblique. These are quickest and the least costly to make. If the illustration is to be used in a publication such as a journal, operator's manual, technical publication, and so on, dimetric, trimetric, or perspective may be used.

EXPLODED VIEWS

Perhaps the easiest way to understand an exploded view is this: Take an object and separate it into its individual parts, as in Fig. 26–5. Three views are shown at A, and a pictorial view is shown at B. At C, an "explosion" has projected the

Fig. 26–5 **How a view is exploded.**

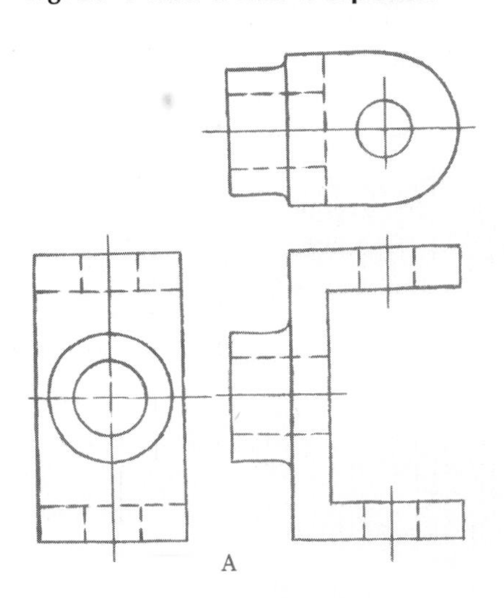

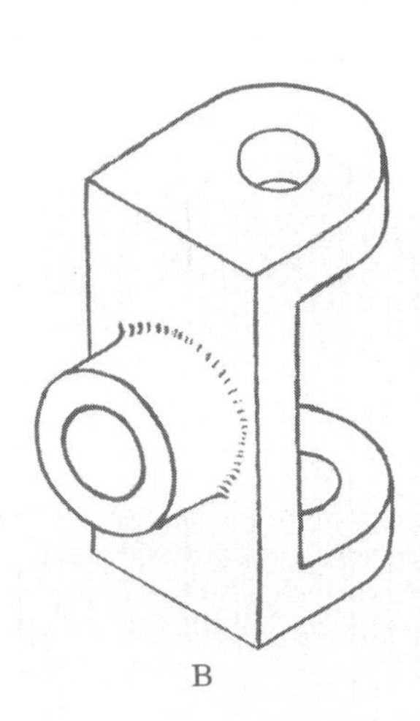

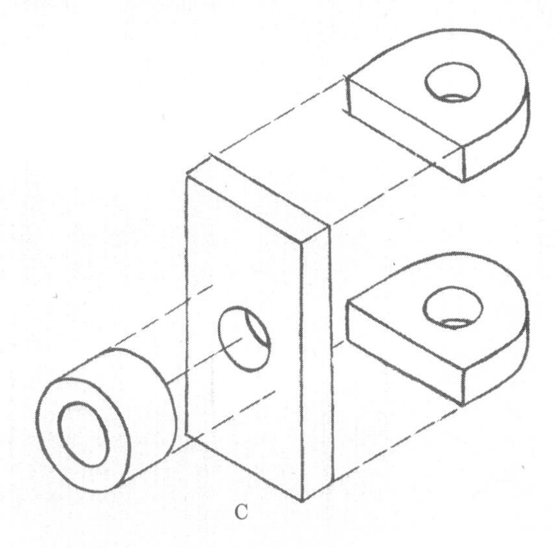

Fig. 26–6 **The exterior of a high-pressure pump.** ***(Industrial Division, Standard Precision, Inc.)***

elementary parts away from each other. This illustrates the principle of exploded views.

All such views are based upon the same principle: projecting the parts from the positions they occupy when put together. Simply, they are just pulled apart. The exterior of a high-pressure piston pump is shown in Fig. 26–6. An exploded illustration of the pump is shown in Fig. 26–7. Note that all parts are easily identifiable.

IDENTIFICATION ILLUSTRATIONS

Pictorial drawings are very useful for identifying parts. They help save time when the parts are manufactured or assembled in place. They are useful for illustrating operating instruction manuals and spare parts catalogs and are also used for many other purposes.

Identification illustrations are usually presented in exploded views. If there are only a few parts, they can be identified by names and pointing arrows. The identification illustration in Fig. 26–8 is an example showing numbers for the parts. A tabulation shows names and quantities.

RENDERING

Surface shading or rendering of some kind may be used when shapes are difficult to read or for other purposes. For most industrial illustrations, accurate descriptions of shapes and positions are more important than fine artistic effects.

1. PUMP BODY (M - 10091)
2. CYLINDER HEADS (M - 10095)
3. PISTON (M - 10097)
4. VALVE ASSEMBLY (M - 10147)
5. PUMP SHAFT (10 - 10050)
6. OUTER BALL BEARING (10 - 10050)
7. WASHER (10 - 10050)
8. INNER ROLLER BEARING (10 - 10050)
9. GREASE ZERK (¼ - 28)
10. "O" RINGS (125)
11. "O" RINGS (132)
12. "O" RINGS (220)
13. BACK-UP RING (9)
14. HEAD BOLTS (¼ - 20 X 1 ¼)

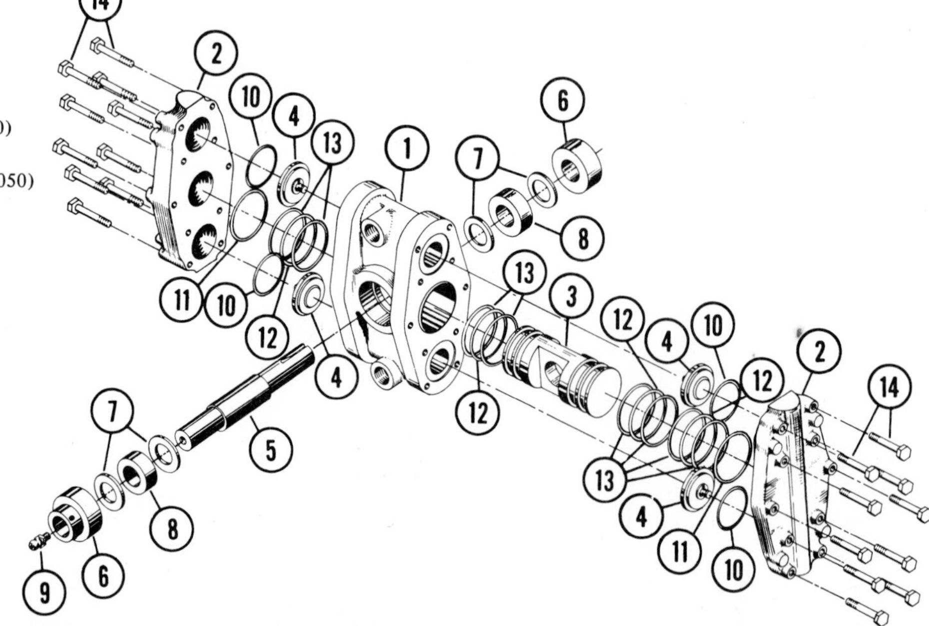

Fig. 26–7 **An exploded view that shows and identifies the parts of a high-pressure piston pump. It is a single-piston, double-ended displacement pump with pressure capabilities up to 1000 psi and a capacity of 2 gpm at 650 psi. (Psi = pounds per square inch; gpm = gallons per minute.)** ***(Industrial Division, Standard Precision, Inc.)***

Fig. 26–8 An identification illustration. ***(The R. K. LeBlond Machine Tool Co.)***

PART NO.	PART NAME	QTY.
403	QUICK CHANGE BOX	1
404	COVER, TOP	1
405	GASKET, COVER	1
406	SCREW, SOCKET HEAD CAP	8
407	SCREW	2
408	SHAFT, SHIFTER	1
409	LINK, SHIFTER	1
410	PIN	1
411	SHOE, SHIFTER	1
412	GASKET (MAKE IN PATTERN SHOP-BOX TO BED)	2
413	"O" RING	2
414	SHAFT, SHIFTER	1
415	PIN, TAPER	2
416	LINK, SHIFTER	1
417	SHOE, SHIFTER	1
418	COVER, SLIP GEAR	1
419	SCREW	4
420	PLUG	2
421	SCREW	3
422	SCREW	1
423	PLUG (NOT USED WITH SCREW REVERSE)	1
424	SCREW	3
425	PIN	2
426	SCREW	6
427	COLLAR	2
428	PLUNGER	2
429	SPRING	2
430	KNOB	2
431	LEVER	2
432	PLATE, FEED-THD.	1
433	PLATE, COMPOUND	1
434	PLATE, ENGLISH INDEX	1
435	COVER	1
436	SCREW	7

Satisfactory results can often be obtained without any shading. In general, you should limit surface shading. Shade the least amount necessary to define the shapes that are being illustrated.

Line drawings are used most for both pictorials and schematics. In addition, there are halftone renderings, photographs, and isometric, oblique, and perspective pictorials (Chapter 12). Different ways of rendering technical illustrations include the use of screen tints, pen and ink, wash, stipple, felt-tip pen and ink, smudge, edge emphasis, and other means. These various means of rendering can be seen in the technical illustrations of aircraft companies, automobile manufacturers, and machine-tool makers. They can even be found in the directions that come with your TV set.

OUTLINE SHADING

Outline shading may be done mechanically or freehand. Sometimes a combination of both methods is used. The light is generally considered to come from in back of and above the left shoulder of the observer. It also comes across the diagonal of the object, as at A in Fig. 26–9. This is a convention, or a standard method, used by drafters and renderers. At B, the upper left and top edges are in the light. They are drawn with thin lines. The lower right and bottom edges should be shaded. They should be drawn with thick lines. At C, the edges meeting in the center are made with thick lines to accent the shape. At D, the edges meeting at the center are made with thin lines. Thick lines are used on the other edges to bring out the shape.

An example of the use of a small amount of line shading is shown and described in Figs. 26–10 and 26–11.

SURFACE SHADING

With the light rays coming in the usual conventional direction, as at A in Fig. 26–12, the top and front surfaces should be lighted. The right-hand surface should then be shaded, as at B. The front surface can have light shading with heavy shading on the right-hand side, as at C. Solid black may be used on the right-hand side, as at D.

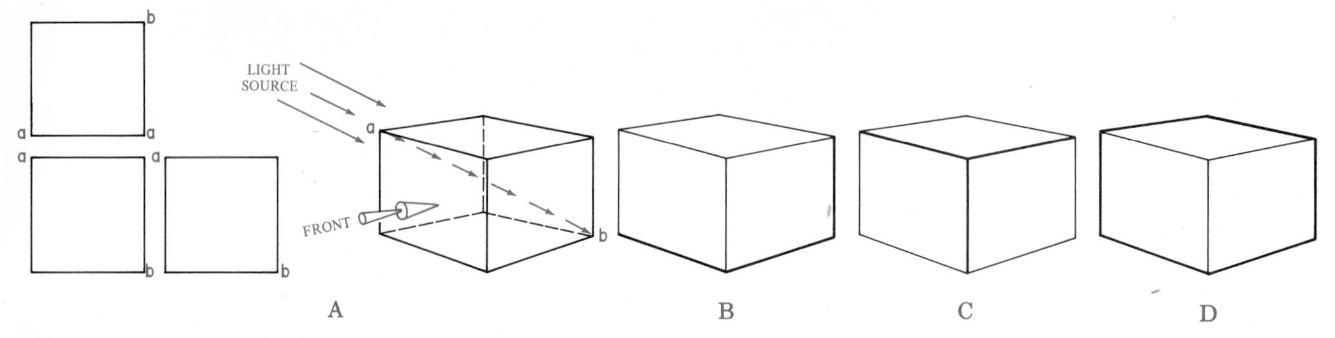

Fig. 26–9 **Light source and line-shaded cubes.**

SOME SHADED SURFACES

Some shaded surfaces are shown in Fig. 26–13. An unshaded view is shown at A for comparison. Ruled-surface shading is shown at B, freehand shading at C, stippled shading at D, and pressure-sensitive overlay shading at E and F.

Stippling, at D, consists of dots. Short, crooked lines can also be used to produce a shaded effect. It is a good method when it is well done, but it takes quite a bit of time. Pressure-sensitive overlays (used at E and F) are available in a great variety of patterns. These overlays can be applied quite easily (See Chapter 16).

AIRBRUSH RENDERING

Airbrush rendering produces illustrations that resemble photographs (Fig. 26–14). The airbrush (Fig. 26–15) is a miniature spray gun. It is used primarily to render illustrations and to retouch photographs. Compressed air is used to spray a solution (usually watercolor) that gives various shading effects.

Fig. 26–10 **A maintenance manual illustration. Notice that only the necessary detail is shown and that just enough shading is used to emphasize and give form to the parts. This apparently simple form of shading is effective for many purposes but must be handled carefully.** ***(courtesy of the Technical Illustrators Association.)***

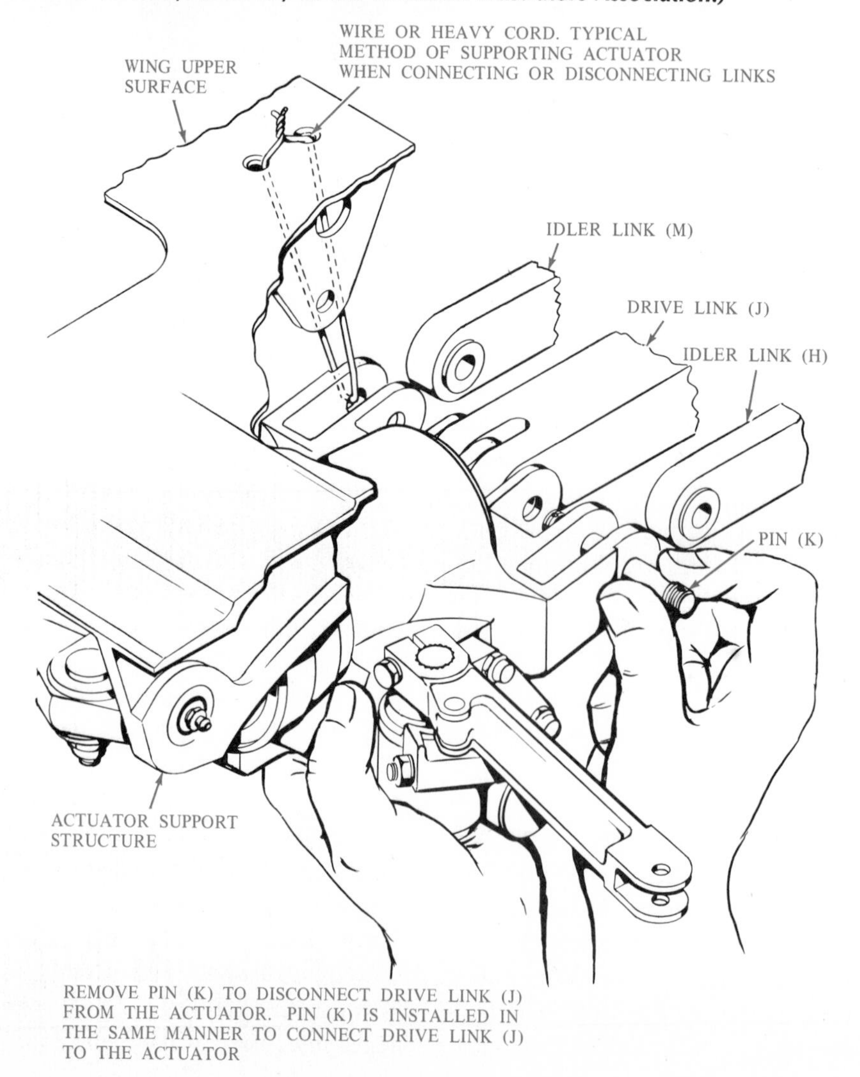

Fig. 26–11 **Outline emphasis by a thick black or white line is an effective method of making a shape stand out.** *(Rockford Clutch Division, Borg-Warner.)*

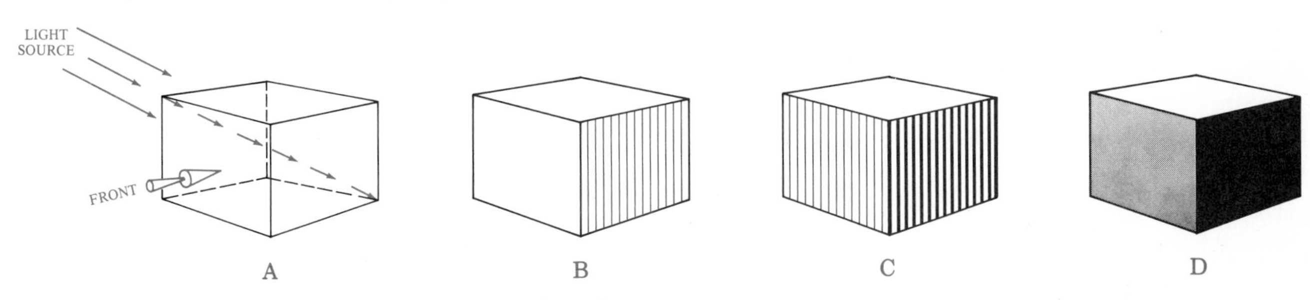

Fig. 26–12 **Some methods of rendering the faces of a cube.**

Fig. 26–13 **Examples of various kinds of rendering.**

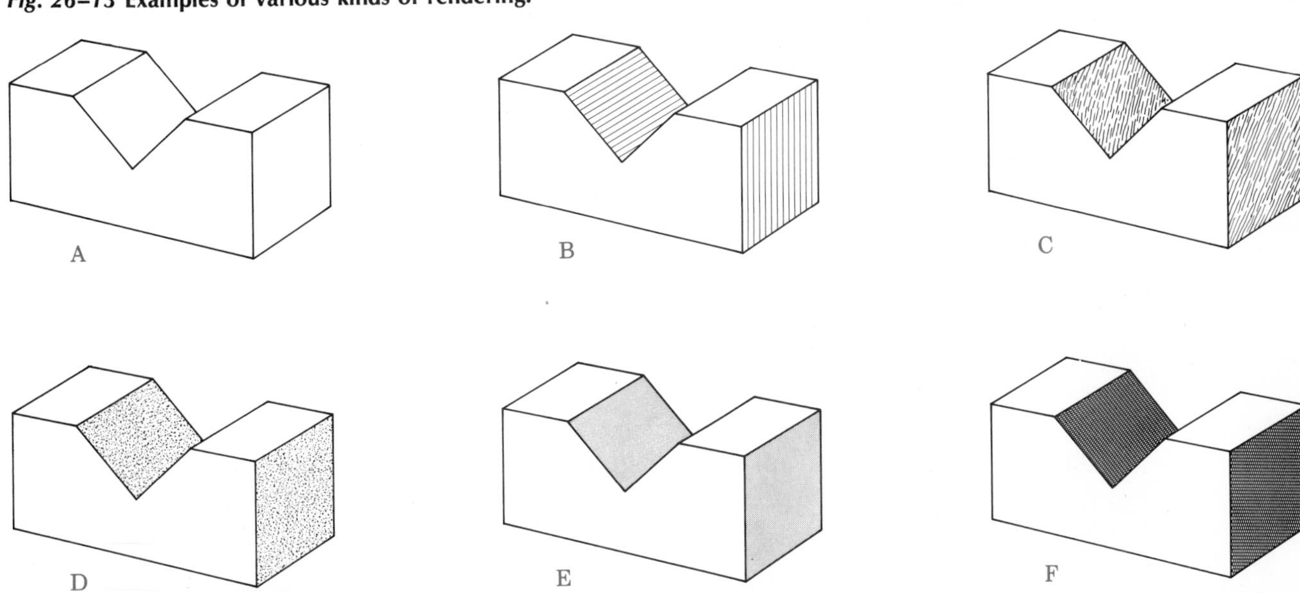

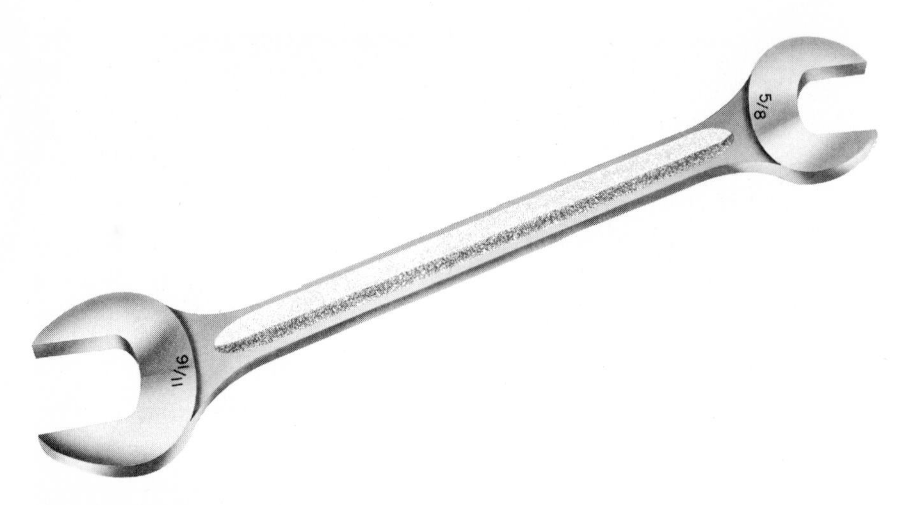

Fig. 26–14 Airbrush rendering.

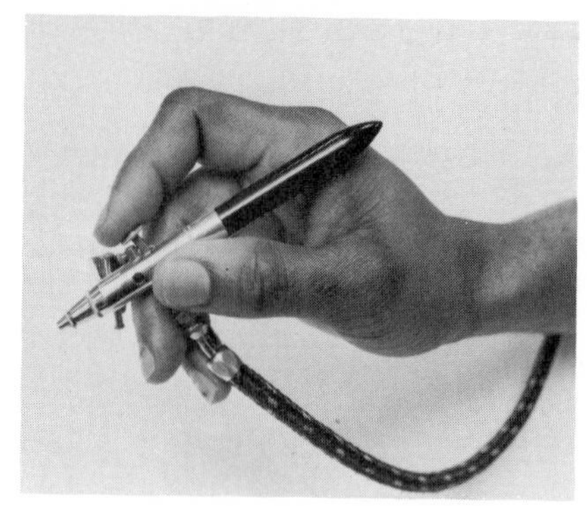

Fig. 26–15 An airbrush. (R. J. Capece.)

Types of Airbrushes

Airbrushes may be classified according to the size of their spray pattern and by the type of spray-control mechanism. Care should be taken in the selection of an airbrush. The size and style should match the work that needs to be done.

The smallest spray pattern can be obtained from an *oscillating* (vibrating) needle airbrush. This type is capable of spraying very thin lines (hairlines) and small dots. It is the most expensive and is used only by professionals who do highly detailed rendering and retouch work. A slightly larger airbrush, often called the pencil-type, is a general-purpose illustrator's airbrush. It can be adjusted to spray thin lines. In addition, it can be opened up to spray large surfaces and backgrounds. This airbrush is most popular for student use. The largest airbrush suitable for use in technical illustrating will spray a pattern large enough to do posters, displays, models, etc. This large brush is often called a poster-type airbrush.

Airbrushes are also classified by the type of spray-control mechanism. That is, an airbrush spray control is either single action or double action. Oscillating-needle and most pencil-type brushes have double-action mechanisms. Most poster-type brushes have single-action mechanisms. Single action simply means that when the finger lever is pressed, both color and air are expelled at the same time. Double action means that two motions are necessary. When the lever is pressed, air is released. When the lever is then pulled back, color is released. Much greater control is possible with a double-action airbrush. It is the kind usually used for rendering technical illustrations.

Air Supply

A constant supply of clean air is needed to produce a high-quality spray pattern. An air supply can be obtained from a carbonic gas unit (CO_2) or an air compressor. Up to 220 kPa (32 pounds) of pressure is required.

An air transformer (regulator and filter) must be installed between the air supply and the airbrush. The regulated pressure for most airbrush rendering is 220 kPa (32 pounds). Less pressure may be used for special effects.

Supplies and Materials

A variety of supplies and materials is needed for airbrush work. Some are common art supplies. Others are special and you can get them through an art or engineering supply house. The following list is in addition to standard drafting supplies and equipment:

- Airbrush and hose
- Rubber cement
- Rubber cement pickup
- Razor knife (X-Acto or similar)
- White illustration board (hot pressed)
- Frisket paper
- Watercolor brushes
- Medicine dropper
- Designer's watercolors (black and white)
- Palette
- Photo retouch set

Procedure for Airbrushing

The following procedure is generally used for airbrushing. Special effects are done well by experimenting with equipment and materials.

1. Prepare a line drawing of the desired object. Transfer it to the surface of the illustration board. Do not use standard typing carbon paper for transferring the image. Either buy a special transfer sheet or make one. This can be done by blackening one side of a sheet of tracing vellum with a soft lead pencil.
2. Cover the image area with frisket paper. This material is available in two forms: prepared and unprepared. Prepared frisket paper has one adhesive side protected with wax paper. Unprepared frisket paper must be coated with thinned rubber cement. Prepared frisket paper is more convenient to use. It is recommended for the beginner. Cut and remove frisket paper from the area to be airbrushed first (Fig. 26–16). Use the rubber-cement pickup to remove any particles of rubber cement left on the surface. Cover all other areas of the illustration board not covered by the frisket paper.
3. Mix the black watercolor in the palette. Water is placed in the palette cup with the medicine dropper. Squeeze a small amount of black watercolor onto the edge of the palette. Use a watercolor brush to mix the color into the water.
4. Transfer the mixed color from the palette to the color cup on the airbrush. You can use the watercolor brush to do this. Fill the cup about half full.
5. Render the exposed surface as needed.
6. Open a second portion of the frisket and cover the rendered surface. Continue in this way until all surfaces are rendered. Remove the frisket. Figure 26–17 shows the finished rendering. Figure 26–18 shows examples of other objects rendered. Highlights may be added using white watercolor.

Fig. 26–16 **Opening the area to be airbrushed. (R. J. Capece.)**

Fig. 26–17 **Finished rendering of a cube.**

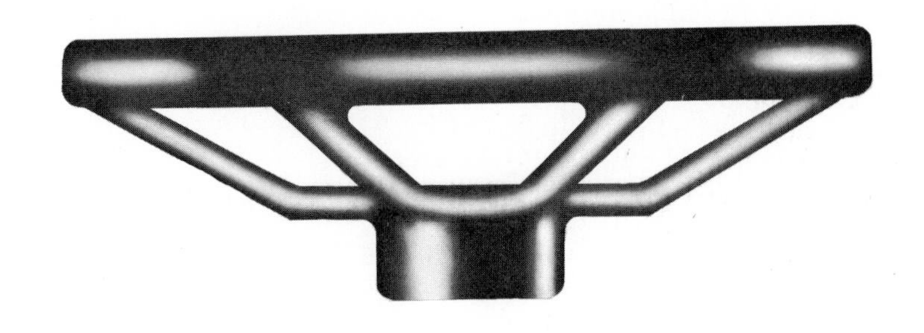

Fig. 26–18 **Examples of airbrushed objects.**

Photo Retouching

Photo retouching is a process used to change details on a photograph. Details may be added, removed, or simply repaired. This process is often needed in preparing photographs for use in publications. It can also be used for changing the appearance of some detail.

Photo retouching is usually done on a glossy photograph. The same basic procedure outlined above for standard airbrush work is used. Care must be taken not to damage the finish on the photo when cutting frisket paper. The gray tones are obtained by using the tones of gray from a photo retouching kit. A watercolor brush may also be used for touching up fine details. Figure 26–19 shows a before and after example of a retouched photograph.

WASH RENDERING

Wash rendering (also called wash drawing) is a form of watercolor rendering. It is done with watercolor and watercolor brushes. It is commonly used for rendering architectural drawings (Fig. 26–20). It is also used for advertising furniture and similar products in newspapers (Fig. 26–21). This technique is highly specialized and is usually done by a commercial artist. However, some technical illustrators and drafters are, at times, required to do this kind of illustrating.

SCRATCHBOARD

Scratchboard drawing is a form of line rendering. Scratchboard is coated with India ink, and a sharp instrument is used to make the lines. This is done by drawing the image on the inked surface, then scratching through the ink to expose the lines or surfaces (Fig. 26–22).

PHOTOGRAPH BEFORE RETOUCHING

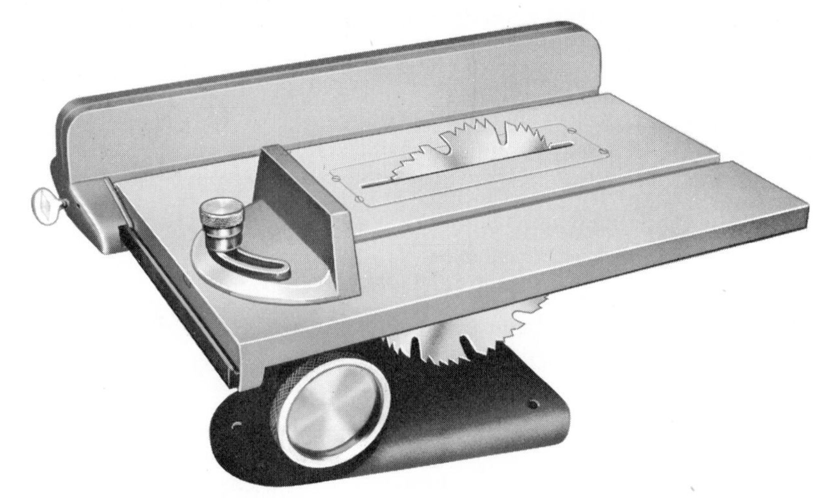

PHOTOGRAPH AFTER RETOUCHING

Fig. 26–19 **Before and after retouching.**

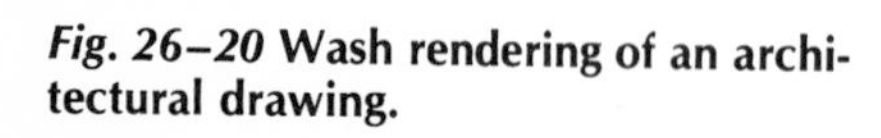

Fig. 26–20 **Wash rendering of an architectural drawing.**

Fig. 26–21 **Wash rendering of a stereo cabinet.**

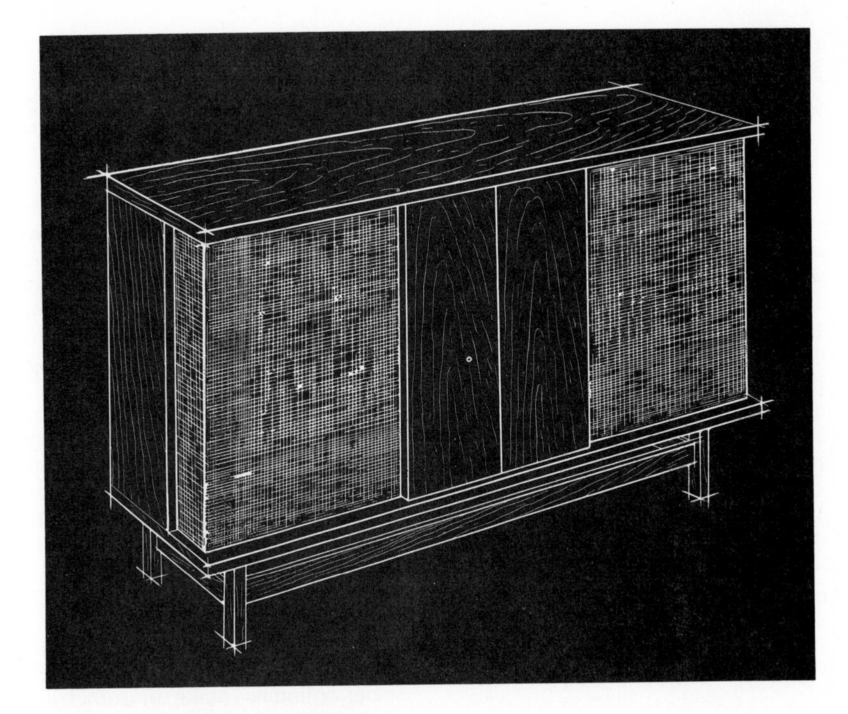

Fig. 26–22 **Scratchboard rendering.**

Review

1. Another name for surface shading is ________.

2. Name four techniques for shading drawings. ________

3. Airbrush rendering requires air pressure up to kPa or ________ pounds.

4. Changing details on a photograph is called ________.

5. Name two major uses for wash rendering.

6. Define technical illustration.

7. Which type of pictorial drawing is most natural in appearance?

8. Which type is the least natural in appearance?

9. When an object is separated into its various parts and the parts drawn in the correct relative positions, what kind of drawing results?

10. Exploded-view drawings with parts numbered or labeled are called ________.

Problems

Fig. 26–23 Pictorial sketching and drawing. Assignment 1. Scale: optional. Most technical illustrations are basically pictorial line drawings. In order to develop a good understanding of the relationship of the various types, make an isometric, oblique-cavalier, oblique-cabinet, single-point perspective, and two-point perspective sketch of the tool support in Fig. 12–71. Assignment 2. Scale: optional. Make instrument drawings of the same tool support (Fig. 12–71) in isometric, oblique-cavalier, oblique cabinet, single-point perspective, and two-point perspective. Compare the sketches with the instrument drawings. Are they similar? Which type of pictorial drawing gives the most natural appearance?

Fig. 26–24 Exploded-view drawings. Make an isometric exploded-view drawing of the letter holder in Fig. 5–79. Scale: optional. Draw your own initials as an overlay 1.0 mm (1/32″) thick. Estimate the height and width.

Fig. 26–25 Surface shading. Scale: optional. Make a pictorial line drawing of the toy boat in Fig. 12–57. Redesign as desired. Refer to Fig. 26–12 and render the boat using the technique shown at *C* or *D*. Maintain sharp clean lines for contrast.

Fig. 26–26 Pictorial assembly drawing. Scale: optional. Make a pictorial assembly drawing of the trammel in Fig. 11–19. Number the parts and add a parts list for identification. Do not show sectional views. Redesign as desired. Add outline shading for accent if instructed to do so. Ink tracing is optional.

Fig. 26–27 Identification illustration. Scale: optional. Make an oblique-cavalier drawing of the mini-sawhorse in Fig. 5–82. Add part numbers and a parts list to make an identification illustration. Render if required. Trace in ink if required.

Fig. 26–28 Airbrush rendering. Make two-point-perspective line drawings of several basic geometric shapes (solids). Examples: cube, cylinder, sphere, cone, etc. Scale: optional. Transfer each to a piece of hot-pressed illustration board and render them in watercolor, using an airbrush.

Fig. 26–29 Scratchboard rendering. Scale: optional. Make a two-point-perspective drawing of the knife rack in Fig. 5–80. Transfer the line drawing to a piece of scratchboard coated with India ink. Use any sharp instrument to scratch through the ink to expose the desired lines.

Fig. 26–30 Exploded-view drawing. Scale: optional. Make an isometric exploded-view drawing of the tic-tac-toe board in Fig. 5–85. Use outline shading to add contrast.

Fig. 26–31 Identification illustration. Make an isometric exploded-view drawing of the note box in Fig. 5–83. Scale: optional. Add part numbers and a parts list to make an identification illustration. Your initial should be designed as an inlay attached to a circular disk. Redesign the note box as desired. Render if required.

Fig. 26–32 Wash rendering. Make a one-point-perspective or a two-point-perspective drawing of any piece of wood furniture. Scale: optional. Transfer the line drawing to a piece of cold-pressed illustration board and render it, using watercolor and a watercolor brush. Use a touch of white or black to sharpen the edges. Keep wood grain and other fine detail lines sharp and clean.

Fig. 26–33 Assignment 1: Identification illustration. Scale: optional. Make an isometric assembly drawing of the trammel in Fig. 11–19. Show the full length of the beam and show two complete assemblies of the point, body, and knurled screw on the beam. Estimate any sizes not given. Redesign as desired. Add part numbers and a parts list to make an identification illustration. Assignment 2: Render the trammel in pencil or transfer the line drawing to hot-pressed illustration board and render in watercolor with an airbrush. Add the part numbers and parts list on an overlay sheet.

Fig. 26–34 Photo retouching. Obtain a glossy photograph of a simple machine or machine part. Study the individual features of the object and list all imperfections. Retouch the photograph using a hand watercolor brush or an airbrush, or both. If possible, have two copies of the original photograph so that a *before* and *after* comparison can be made.

Fig. 26–35 Airbrush rendering. Scale: optional. Make a two-point-perspective drawing of the hammer head in Fig. 5–86. Transfer the line drawing to a piece of hot-pressed illustration board and render it in watercolor, using an airbrush. Add a touch of white or black to edges for contrast.

Fig. 26–36 Surface shading. Scale: optional. Make a single-point-perspective drawing of one of your own initials. Make it 7 units high and 2 units thick. Refer to Fig. 2–15 for the width. Render it using the technique shown at *C* or *D* in Fig. 26–12. This problem may also be used for practice in wash rendering or airbrush rendering. For wash rendering, the line drawing must be transferred to cold-pressed illustration board. For airbrush rendering, transfer the line drawing to hot-pressed illustration board.

Fig. 26–37 Exploded-view drawing. Scale: optional. Make an isometric exploded-view drawing of the garden bench in Fig. 5–84. Use 1/2″ DIA threaded rods, flat washers, and hex nuts to fasten the parts together. If assigned, add part numbers and a parts list to make an identification illustration. This problem may also be used for practice in wash rendering.

Fig. 26–38 Surface shading. Scale: optional. Make a two-point-perspective drawing of the V-Block in Fig. 12–72. Refer to Fig. 26–13 and render the line drawing, using the technique assigned. Keep all edges crisp and sharp.

Fig. 26–39 Wash rendering. Make a two-point-perspective drawing of a house. Scale: 1:50 or 1:100 (1/4″ = 1′-0″ or 1/8″ = 1′-0″). Transfer the line drawing to a piece of cold-pressed illustration board and render it, using watercolor and a watercolor brush.

Fig. 26–40 Scratchboard rendering. Scale: optional. Make a two-point-perspective drawing of any piece of wood furniture. Examples: coffee table, end table, desk, bench, etc. Transfer the line drawing to a piece of scratchboard coated with India ink. Use any sharp instrument to scratch through the ink to expose the desired lines. See Fig. 26–22.

Fig. 26–41 Airbrush rendering. Scale: optional. Make a single-point-perspective drawing of the object of your choice in Figs. 5–79 through 5–86. Transfer it to a piece of hot-pressed illustration board and render it in watercolor, using an airbrush. This problem may also be used for practice in outline or surface shading.

Fig. 26–42 Wash rendering Scale: 1:50 or 1:100 (1/4″ = 1′-0″ or 1/8″ = 1′-0″). Make a line drawing of the front elevation of a house similar to the one shown in Fig. 20–10. Transfer the line drawing to a piece of cold-pressed illustration board and render it, using watercolor and a watercolor brush. This problem may also be used for practice in surface shading in pencil or charcoal.

Fig. 26–43 Surface shading. Scale: optional. Make line drawings of geometric solids assigned from Fig. 8–2. Render each of the solids, using your choice of surface rendering techniques.

Fig. 26–44 Airbrush rendering. Scale: optional. Make a pictorial sketch or instrument drawing of the jet fighter in Fig. 25–6. Redesign as desired. Transfer the line drawing to a piece of hot-pressed illustration board and render it in watercolor, using an airbrush. Add detail and decorate the body and wings as desired. This problem may also be used for practice in pencil rendering.

Fig. 26–45 Surface shading. Scale: optional. Make a pictorial line sketch or instrument drawing of the edge protector in Fig. 5–65. Add surface shading as assigned by your instructor. Keep all edges sharp and crisp. This problem may also be used for practice in airbrush rendering.

Fig. 26–46 Design and surface shading. Scale: optional. Design and make a pictorial sketch of a "car of the future." Render in pencil.

Fig. 26–47 Airbrush rendering. Scale: optional. Make a pictorial assembly drawing of the coupler in Fig. 11–18. Transfer the drawing to hot-pressed illustration board. Render in watercolor, using an airbrush. Add a touch of white or black to keep the edges crisp and sharp.

Appendix A

Reference Tables

AMERICAN NATIONAL STANDARDS

A few standards that are useful for reference are listed below. Standards are subject to revisions and latest issues should be consulted. A catalog with prices is published by the American National Standards Institute, Inc., 1430 Broadway, New York, NY 10018.

Abbreviations	Y1.1
Acme Screw Threads	B1.5
Graphical Electrical Symbols for Architectural Plans	Y32.9
Graphical Symbols for Electrical and Electronics Diagrams	Y32.2
Graphical Symbols for Heating, Ventilating and Air Conditioning	Z32.2.4
Graphical Symbols for Pipe Fittings, Valves and Piping	Z32.2.3
Graphical Symbols for Plumbing	Y32.4
Graphical Symbols for Welding	Y32.3
Hexagon Head Cap Screws, Slotted Head Cap Screws, Square Head Set Screws, Slotted Headless Set Screws	B18.6.2
Keys and Keyseats	B17.1
Large Rivets	B18.4
Machine Tapers	B5.10
Pipe Threads	B2.1
Preferred Limits and Fits for Cylindrical Parts	B4.1
Round Head Bolts	B18.5
Slotted and Recessed Head Tapping Screws and Metallic Drive Screws	B18.6.4
Slotted and Recessed Head Wood Screws	B18.6.1
Small Solid Rivets	B18.1
Socket Cap, Shoulder, and Set Screws	B18.3
Square and Hex Bolts and Screws, Including Hex Cap Screws and Lag Screws	B18.2.1
Square and Hex Nuts	B18.2.2
Unified Screw Threads	B1.1
Woodruff Keys and Keyseats	B17.2

American National Drafting Standards Manual

Section 1.	Size and Format	Y14.1
Section 2.	Line Conventions, Sectioning and Lettering *(ISO R 128)*	Y14.2
Section 3.	Multi and Sectional View Drawings	Y14.3
Section 4.	Pictorial Drawing	Y14.4
Section 5.	Dimensioning and Tolerancing for Engineering Drawings *(ISO R 129; R 406)*	Y14.5
Section 6.	Screw Threads	Y14.6
Section 7.	Gears, Splines, and Serrations	Y14.7
Section 9.	Forging	Y14.9
Section 10.	Metal Stamping	Y14.10
Section 11.	Plastics	Y14.11
Section 14.	Mechanical Assemblies	Y14.14
Section 15.	Electrical and Electronics Diagrams	Y14.15
Section 17.	Fluid Power Diagrams	Y14.17

TABLE A–1 DECIMAL EQUIVALENTS OF COMMON FRACTIONS

Fraction	Decimal	Fraction	Decimal	Fraction	Decimal	Fraction	Decimal
1/64	0.015625	17/64	0.265625	33/64	0.515625	49/64	0.765625
1/32	0.03125	9/32	0.28125	17/32	0.53125	25/32	0.78125
3/64	0.046875	19/64	0.296875	35/64	0.546875	51/64	0.796875
1/16	0.0625	5/16	0.3125	9/16	0.5625	13/16	0.8125
5/64	0.078125	21/64	0.328125	37/64	0.578125	53/64	0.828125
3/32	0.09375	11/32	0.34375	19/32	0.59375	27/32	0.84375
7/64	0.109375	23/64	0.359375	39/64	0.609375	55/64	0.859375
1/8	0.1250	3/8	0.3750	5/8	0.6250	7/8	0.8750
9/64	0.140625	25/64	0.390625	41/64	0.640625	57/64	0.890625
5/32	0.15625	13/32	0.40625	21/32	0.65625	29/32	0.90625
11/64	0.171875	27/64	0.421875	43/64	0.671875	59/64	0.921875
3/16	0.1875	7/16	0.4375	11/16	0.6875	15/16	0.9375
13/64	0.203125	29/64	0.453125	45/64	0.703125	61/64	0.953125
7/32	0.21875	15/32	0.46875	23/32	0.71875	31/32	0.96875
15/64	0.234375	31/64	0.484375	47/64	0.734375	63/64	0.984375
1/4	0.2500	1/2	0.5000	3/4	0.7500	1	1.0000

TABLE A–2 AMERICAN NATIONAL STANDARD UNIFIED AND AMERICAN THREAD SERIES

Threads per inch for coarse, fine, extra-fine, 8-thread, 12-thread, and 16-thread, and 16-thread series[b] [tap-drill sizes for approximately 75 per cent depth of thread (not American Standard)]

Nominal size (basic major dia.)	Coarse-thd. series UNC and NC[c] in classes 1A, 1B, 2A, 2B, 3A, 3B, 2, 3		Fine-thd. series UNF and NF[c] in classes 1A 1B, 1A, 2B, 3A, 3B, 2, 3		Extra-fine thd. series UNEF and NEF[d] in classes 2A, 2B, 2, 3		8-thd. series 8N[c] in classes 2A, 2B, 2, 3		12-thd. series 12UN and 12N[d] in classes 2A, 2B, 2, 3		16-thd. series 16UN and 16N[d] in classes 2A, 2B, 2, 3	
	Thd. /in.	Tap drill	Thd. /in.	Tap drill	Thd. /in.	Tap drill	Thd. /in.	Tap drill	Thd. /in.	Tap drill	Thd. /in.	Tap drill
0(0.060)	....	.....	80	3/64								
1(0.073)	64	No. 53	72	No. 53								
2(0.086)	56	No. 50	64	No. 50								
3(0.099)	48	No. 47	56	No. 45								
4(0.112)	**40**	No. 43	48	No. 42								
5(0.125)	40	No. 38	44	No. 37								

TABLE A–2 (CONT.)

Nominal size (basic major dia.)	Coarse-thd. series UNC and NC[c] in classes 1A, 1B, 2A, 2B, 3A, 3B, 2, 3		Fine-thd. series UNF and NF[c] in classes 1A 1B, 1A, 2B, 3A, 3B, 2, 3		Extra-fine thd. series UNEF and NEF[d] in classes 2A, 2B, 2, 3		8-thd. series 8N[c] in classes 2A, 2B, 2, 3		12-thd. series 12UN and 12N[d] in classes 2A, 2B, 2, 3		16-thd. series 16UN and 16N[d] in classes 2A, 2B, 2, 3	
	Thd. /in.	Tap drill	Thd. /in.	Tap drill	Thd. /in.	Tap drill	Thd. /in.	Tap drill	Thd. /in.	Tap drill	Thd. /in.	Tap drill
6(0.138)	**32**	No. 36	40	No. 33								
8(0.164)	**32**	No. 29	36	No. 29								
10(0.190)	**24**	No. 25	**32**	No. 21								
12(0.216)	24	No. 16	28	No. 14	32	No. 13						
1/4	**20**	N0. 7	**28**	No. 3	32	7/32						
5/16	**18**	Let. F	**24**	Let. I	32	9/32						
3/8	**16**	5/16	**24**	Let. Q	32	11/32						
7/16	**14**	Let. U	**20**	25/64	**28**	13/32						
1/2	**13**	27/64	**20**	29/64	**28**	15/32	..	...	**12**	27/64		
9/16	**12**	31/64	**18**	33/64	24	33/64	..	...	**12**	31/64		
5/8	**11**	17/32	**18**	37/64	24	37/64	..	...	12	35/64		
11/16	....	.....	..	.....	24	41/64	..	...	12	39/64		
3/4	**10**	21/32	**16**	11/16	**20**	45/64	..	...	12	43/64	**16**	11/16
13/16	....	.....	..	.....	**20**	49/64	..	...	12	47/64	**16**	3/4
7/8	**9**	49/64	**14**	13/16	**20**	53/64	..	...	12	51/64	**16**	13/16
15/16	....	.....	..	.....	**20**	57/64	..	...	**12**	55/64	**16**	7/8
1	....	.....	14	15/16	..	.....	8	7/8				
1	**8**	7/8	**12**	59/64	**20**	61/64	..	...	**12**	59/64	**16**	15/16
1 1/16	....	.....	..	.....	18	1	..	...	**12**	63/64	**16**	1
1 1/8	7	63/64	**12**	1 3/64	18	1 5/64	8	1	**12**	1 3/64	**16**	1 1/16
1 3/16	....	.....	..	.....	18	1 9/64	..	...	**12**	1 7/64	**16**	1 1/8
1 1/4	7	1 7/64	**12**	1 11/64	18	1 3/16	8	1 1/8	**12**	1 11/64	**16**	1 3/16
1 5/16	....	.....	..	.....	18	1 17/64	..	...	**12**	1 15/64	**16**	1 1/4
1 3/8	6	1 7/32	**12**	1 19/64	18	1 5/16	8	1 1/4	**12**	1 19/64	**16**	1 5/16
1 7/16	....	.....	..	.....	18	1 3/8	..	...	**12**	1 23/64	**16**	1 3/8
1 1/2	6	1 11/32	**12**	1 27/64	18	1 7/16	8	1 3/8	**12**	1 27/64	**16**	1 7/16
1 9/16	....	.....	..	.....	18	1 1/2	..	...	..	.....	16	1 1/2
1 5/8	....	.....	..	.....	18	1 9/16	8	1 1/2	12	1 35/64	16	1 9/16
1 11/16	....	.....	..	.....	18	1 5/8	..	...	..	.....	16	1 5/8
1 3/4	5	1 9/16	..	.....	**16**	1 11/66	8[e]	1 5/8	**12**	1 43/64	**16**	1 11/16
1 13/16	....	.....	..	.....	..	.....	..	...	..	.....	16	1 3/4
1 7/8	....	.....	..	.....	..	.....	8	1 3/4	12	1 51/64	16	1 13/16
1 15/16	....	.....	..	.....	..	.....	..	...	..	.....	16	1 7/8
2	**4 1/2**	1 25/32	..	.....	**16**	1 15/16	8[e]	1 7/8	**12**	1 59/64	**16**	1 15/16

TABLE A–2 (CONT.)

Nominal size (basic major dia.)	Coarse-thd. series UNC and NC[c] in classes 1A, 1B, 2A, 2B, 3A, 3B, 2, 3		Fine-thd. series UNF and NF[c] in classes 1A, 1B, 2A, 2B, 3A, 3B, 2, 3		Extra-fine thd. series UNEF and NEF[d] in classes 2A, 2B, 2, 3		8-thd. series 8N[c] in classes 2A, 2B, 2, 3		12-thd. series 12UN and 12N[d] in classes 2A, 2B, 2, 3		16-thd. series 16UN and 16N[d] in classes 2A, 2B, 2, 3	
	Thd. /in.	Tap drill	Thd. /in.	Tap drill	Thd. /in.	Tap drill	Thd. /in.	Tap drill	Thd. /in.	Tap drill	Thd. /in.	Tap drill
2 1/16	....	.....	..	.....	..	.....	..	...	..	.....	16	2
2 1/8	....	.....	..	.....	..	.....	8	2	12	2 3/64	16	2 1/16
2 3/16	....	.....	..	.....	..	.....	..	...	..	.....	16	2 1/8
2 1/4	**4 1/2**	2 1/32	..	.....	..	.....	8[e]	2 1/8	**12**	2 11/64	**16**	2 3/16
2 5/16	....	.....	..	.....	..	.....	..	...	..	.....	16	2 1/4
2 3/8	....	.....	..	.....	..	.....	..	...	12	2 19/64	16	2 5/16
2 7/16	....	.....	..	.....	..	.....	..	...	..	.....	16	2 3/8
2 1/2	**4**	2 1/4	..	.....	..	.....	8[e]	2 3/8	**12**	2 27/64	**16**	2 7/16
2 5/8	....	.....	..	.....	..	.....	..	...	12	2 35/64	16	2 9/16
2 3/4	**4**	2 1/2	..	.....	..	.....	8[e]	2 5/8	**12**	2 43/64	**16**	2 11/16
2 7/8	....	.....	..	.....	..	.....	..	...	12	2 51/64	16	2 13/16
3	**4**	2 3/4	..	.....	..	.....	8[e]	2 7/8	**12**	2 59/64	**16**	2 15/16
3 1/8	....	.....	..	.....	..	.....	..	...	12	3 3/64	16	3 1/16
3 1/4	**4**	3	..	.....	..	.....	8[e]	3 1/8	**12**	3 11/64	**16**	3 3/16
3 3/8	....	.....	..	.....	..	.....	..	...	12	3 19/64	16	3 5/16
3 1/2	**4**	3 1/4	..	.....	..	.....	8[e]	3 3/8	**12**	3 27/64	**16**	3 7/16
3 5/8	....	.....	..	.....	..	.....	..	...	12	3 35/64	16	3 9/16
3 3/4	**4**	3 1/2	..	.....	..	.....	8[e]	3 5/8	**12**	3 43/64	**16**	3 11/16
3 7/8	....	.....	..	.....	..	.....	..	...	12	3 51/64	16	3 13/16
4	**4**	3 3/4	..	.....	..	.....	8[e]	3 7/8	**12**	3 59/64	**16**	3 15/16
4 1/4	....	.....	..	.....	..	.....	8[e]	4 1/8	**12**	4 11/64	**16**	4 3/16
4 1/2	....	.....	..	.....	..	.....	8[e]	4 3/8	**12**	4 27/64	**16**	4 7/16
4 3/4	....	.....	..	.....	..	.....	8[e]	4 5/8	**12**	4 43/64	**16**	4 11/16
5	....	.....	..	.....	..	.....	8[e]	4 7/8	**12**	4 59/64	**16**	4 15/16
5 1/4	....	.....	..	.....	..	.....	8[e]	5 1/8	**12**	5 11/64	**16**	5 3/16
5 1/2	....	.....	..	.....	..	.....	8[e]	5 3/8	**12**	5 27/64	**16**	5 7/16
5 3/4	....	.....	..	.....	..	.....	8[e]	5 5/8	**12**	5 43/64	**16**	5 11/16
6	....	.....	..	.....	..	.....	8[e]	5 7/8	**12**	5 59/64	**16**	5 15/16

[a] ANSI B1.1—1960. Dimensions are in inches. [b] Bold type indicates unified combinations.
[c] Limits of size for classes are based on a length of engagement equal to the nominal diameter.
[d] Limits of size for classes are based on length of engagement equal to nine times the pitch.
[e] These sizes, with specified limits of size, based on a length of engagement of 9 threads in classes 2A and 2B, are designated UN.
Note. If a thread is in both the 8-, 12-, or 16-thread series and the coarse, fine, or extra-fine-thread series, the symbols and tolerances of the latter series apply.

TABLE A–3 ISO METRIC SCREW THREADS

Nominal size dia (mm)		Series with graded pitches				Series with constant pitches															
		Coarse		Fine		4		3		2		1.5		1.25		1		0.75		0.5		0.35	
Preferred		Thread pitch	Tap drill size	Thread pitch	Tap drill size	Thread pitch	Tap drill size	Thread pitch	Tap drill size	Thread pitch	Tap drill size	Thread pitch	Tap Drill size	Thread pitch	Tap drill size	Thread pitch	Tap drill size	Thread pitch	Tap drill size	Thread pitch	Tap drill size	Thread pitch	Tap drill size
1.6		0.35	1.25																				
	1.8	0.35	1.45																				
2		0.4	1.6																				
	2.2	0.45	1.75																				
2.5		0.45	2.05																			0.35	2.15
3		0.5	2.5																			0.35	2.65
	3.5	0.6	2.9																			0.35	3.15
4		0.7	3.3																	0.5	3.5		
	4.5	0.75	3.7																	0.5	4.0		
5		0.8	4.2																	0.5	4.5		
6		1	5.0															0.75	5.2				
8		1.25	6.7	1	7.0											1	7.0	0.75	7.2				
10		1.5	8.5	1.25	8.7									1.25	8.7	1	9.0	0.75	9.2				
12		1.75	10.2	1.25	10.8							1.5	10.5	1.25	10.7	1	11						
	14	2	12	1.5	12.5							1.5	12.5	1.25	12.7	1	13						
16		2	14	1.5	14.5							1.5	14.5			1	15						
	18	2.5	15.5	1.5	16.5					2	16	1.5	16.5			1	17						
20		2.5	17.5	1.5	18.5					2	18	1.5	18.5			1	19						
	22	2.5	19.5	1.5	20.5					2	20	1.5	20.5			1	21						
24		3	21	2	22					2	22	1.5	22.5			1	23						
	27	3	24	2	25					2	25	1.5	25.5			1	26						
30		3.5	26.5	2	28					2	28	1.5	28.5			1	29						
	33	3.5	29.5	2	31					2	31	1.5	31.5										
36		4	32	3	33					2	34	1.5	34.5										
	39	4	35	3	36					2	37	1.5	37.5										
42		4.5	37.5	3	39	4	38	3	39	2	40	1.5	40.5										
	45	4.5	39	3	42	4	41	3	42	2	43	1.5	43.5										
46		5	43	3	45	4	44	3	45	2	46	1.5	46.5										

TABLE A–4 SIZES OF NUMBERED AND LETTERED DRILLS

No.	Size	No.	Size	No.	Size	Letter	Size
80	0.0135	53	0.0595	26	0.1470	*A*	0.2340
79	0.0145	52	0.0635	25	0.1495	*B*	0.2380
78	0.0160	51	0.0670	24	0.1520	*C*	0.2420
77	0.0180	50	0.0700	23	0.1540	*D*	0.2460
76	0.0200	49	0.0730	22	0.1570	*E*	0.2500
75	0.0210	48	0.0760	21	0.1590	*F*	0.2570
74	0.0225	47	0.0785	20	0.1610	*G*	0.2610
73	0.0240	46	0.0810	19	0.1660	*H*	0.2660
72	0.0250	45	0.0820	18	0.1695	*I*	0.2720
71	0.0260	44	0.0860	17	0.1730	*J*	0.2770
70	0.0280	43	0.0890	16	0.1770	*K*	0.2810
69	0.0292	42	0.0935	15	0.1800	*L*	0.2900
68	0.0310	41	0.0960	14	0.1820	*M*	0.2950
67	0.0320	40	0.0980	13	0.1850	*N*	0.3020
66	0.0330	39	0.0995	12	0.1890	*O*	0.3160
65	0.0350	38	0.1015	11	0.1910	*P*	0.3230
64	0.0360	37	0.1040	10	0.1935	*Q*	0.3320
63	0.0370	36	0.1065	9	0.1960	*R*	0.3390
62	0.0380	35	0.1100	8	0.1990	*S*	0.3480
61	0.0390	34	0.1110	7	0.2010	*T*	0.3580
60	0.0400	33	0.1130	6	0.2040	*U*	0.3680
59	0.0410	32	0.1160	5	0.2055	*V*	0.3770
58	0.0420	31	0.1200	4	0.2090	*W*	0.3860
57	0.0430	30	0.1285	3	0.2130	*X*	0.3970
56	0.0465	29	0.1360	2	0.2210	*Y*	0.4040
55	0.0520	28	0.1405	1	0.2280	*Z*	0.4130
54	0.0550	27	0.1440				

TABLE A–5 ACME AND STUB ACME THREADS*

ANSI preferred diameter-pitch combinations.							
Nominal (major) dia.	Threads /in.	Nominal (major) dia.	Threads /in.	Nominal (major) dia.	Threads /in.	Nominal (major) dia.	Threads /in.
1/4	16	3/4	6	1 1/2	4	3	2
5/16	14	7/8	6	1 3/4	4	3 1/2	2
3/8	12	1	5	2	4	4	2
7/16	12	1 1/8	5	2 1/4	3	4 1/2	2
1/2	10	1 1/4	5	2 1/2	3	5	2
5/8	8	1 3/8	4	2 3/4	3		

*ANSI B1.5 and B1.8-1952. Diameters in inches.

TABLE A–6 METRIC TWIST DRILL SIZES

Metric drill sizes (mm)[a]		Decimal equivalent in inches	Metric drill sizes (mm)[a]		Decimal equivalent in inches
Preferred	Available	(ref)	Preferred	Available	(ref)
	0.40	.0157	1.70		.0669
	0.42	.0165		1.75	.0689
	0.45	.0177	1.80		.0709
	0.48	.0189		1.85	.0728
0.50		.0197	1.90		.0748
	0.52	.0205		1.95	.0768
0.55		.0217	2.00		.0787
	0.58	.0228		2.05	.0807
0.60		.0236	2.10		.0827
	0.62	.0244		2.15	.0846
0.65		.0256	2.20		.0866
	0.68	.0268		2.30	.0906
0.70		.0276	2.40		.0945
	0.72	.0283	2.50		.0984
0.75		.0295	2.60		.1024
	0.78	.0307		2.70	.1063
0.80		.0315	2.80		.1102
	0.82	.0323		2.90	.1142
0.85		.0335	3.00		.1181
	0.88	.0346		3.10	.1220
0.90		.0354	3.20		.1260
	0.92	.0362		3.30	.1299
0.95		.0374	3.40		.1339
	0.98	.0386		3.50	.1378
1.00		.0394	3.60		.1417
	1.03	.0406		3.70	.1457
1.05		.0413	3.80		.1496
	1.08	.0425		3.90	.1535
1.10		.0433	4.00		.1575
	1.15	.0453		4.10	.1614
1.20		.0472	4.20		.1654
1.25		.0492		4.40	.1732
1.30		.0512	4.50		.1772
	1.35	.0531		4.60	.1811
1.40		.0551	4.80		.1890
	1.45	.0571	5.00		.1969
1.50		.0591		5.20	.2047
	1.55	.0610	5.30		.2087
1.60		.0630		5.40	.2126
	1.65	.0650	5.60		.2205
				5.80	.2283

[a] Metric drill sizes listed in the "Preferred" column are based on the R'40 series of preferred numbers shown in the ISO standard R497. Those listed in the "Available" column are based on the R80 series from the same document.

TABLE A–7 AMERICAN NATIONAL STANDARD REGULAR HEXAGON BOLTS

Diameter	Flats	Height		
		Unfinished	Semifinished	Finished
1/4	7/16	11/64	5/32	5/32
5/16	1/2	7/32	13/64	13/64
3/8	9/16	1/4	15/64	15/64
7/16	5/8	19/64	9/32	9/32
1/2	3/4	11/32	5/16	5/16
9/16	13/16	....	....	23/64
5/8	15/16	27/64	25/64	25/64
3/4	1 1/8	1/2	15/32	15/32
7/8	1 5/16	37/64	35/64	35/64
1	1 1/2	43/64	39/64	39/64
1 1/8	2 11/16	3/4	11/16	11/16
1 1/4	1 7/8	27/32	25/32	25/32
1 3/8	1 1/16	29/32	27/32	27/32
1 1/2	2 1/4	1	15/16	15/16
1 3/4	2 5/8	1 5/32	1 3/32	1 3/32
2	3	1 11/32	1 7/32	1 7/32
2 1/4	3 3/8	1 1/2	1 3/8	1 3/8
2 1/2	3 3/4	1 21/32	1 17/32	1 17/32
2 3/4	4 1/8	1 13/16	1 11/16	1 11/16
3	4 1/2	2	1 7/8	1 7/8
3 1/4	4 7/8	2 3/16	2	
3 1/2	5 1/4	2 5/16	2 1/8	
3 3/4	5 5/8	2 1/2	2 5/16	
4	6	2 11/16	2 1/2	

TABLE A–8 REGULAR METRIC HEXAGON BOLTS

Nominal size (mm)	Width across flats	Thickness
1.6	3.2	1.1
2	4	1.4
2.5	5	1.7
3	5.5	2
4	7	2.8
5	8	3.5
6	10	4
8	13	5.5
10	17	7
12	19	8
14	22	9
16	24	10
18	27	12
20	30	13
22	32	14
24	36	15
27	41	17
30	46	19
33	50	21
36	55	23
39	60	25

TABLE A—9 AMERICAN NATIONAL STANDARD REGULAR HEXAGON NUTS

Diameter	Unfinished		Semifinished		Finished	
	Flats	Thickness	Flats	Thickness	Flats	Thickness
1/4	7/16	7/32	7/16	13/64	7/16	7/32
5/16	9/16	17/64	9/16	1/4	1/2	17/64
3/8	5/8	21/64	5/8	5/16	9/16	21/64
7/16	3/4	3/8	3/4	23/64	11/16	3/8
1/2	13/16	7/16	13/16	27/64	3/4	7/16
9/16	7/8	1/2	7/8	31/64	7/8	31/64
5/8	1	35/64	1	17/32	15/16	35/64
3/4	1 1/8	21/32	1 1/8	41/64	1 1/8	41/64
7/8	1 5/16	49/64	1 5/16	3/4	1 5/16	3/4
1	1 1/2	7/8	1 1/2	55/64	1 1/2	55/64
1 1/8	1 11/16	1	1 11/16	31/32	1 11/16	31/32
1 1/4	1 7/8	1 3/32	1 7/8	1 1/16	1 7/8	1 1/16
1 3/8	2 1/16	1 13/64	2 1/16	11 11/64	2 1/16	1 11/64
1 1/2	2 1/4	1 5/16	2 1/4	1 9/32	2 1/4	1 9/32
1 5/8	. . .	. . .	2 7/16	1 25/64		
1 3/4	. . .	. . .	2 5/8	1 1/2	2 5/8	1 1/2
1 7/8	. . .	. . .	2 13/16	1 39/64		
2	. . .	. . .	3	1 23/32	3	1 23/32
2 1/4	. . .	. . .	3 3/8	1 59/64	3 3/8	1 59/64
2 1/2	. . .	. . .	3 3/4	2 9/64	3 3/2	2 9/64
2 3/4	. . .	. . .	4 1/8	2 23/64	4 1/8	2 23/64
3	. . .	. . .	4 1/2	2 37/64	4 1/2	2 37/64

TABLE A-10 REGULAR METRIC HEXAGON NUTS

Nominal size (mm)	Distance across flats	Thickness		
		Regular	Jamb	Thick
1.6	3.2	1.3		
2	4.0	1.6	1.2	
2.5	5.0	2.0		
3	5.5	2.4	1.6	4.0
4	7.0	3.2	2.0	5.0
5	8.0	4.0	2.5	5.0
6	10	5.0	3.0	6.0
8	13	6.5	5.0	8.0
10	17	8.0	6.0	10
12	19	10	7.0	12
14	22	11	8.0	14
16	24	13	8.0	16
18	27	15	9.0	18.5
20	30	16	9.0	20
22	32	18	10	22
24	36	19	10	24
27	41	22	12	27
30	46	24	12	30
33	50	26		
36	55	29		
39	60	31		

TABLE A-11 AMERICAN NATIONAL STANDARD REGULAR SQUARE BOLTS AND NUTS

Diameter	Bolthead		Nut	
	Flats	Height of head	Flats	Thickness of nut
1/4	3/8	11/64	7/16	7/32
5/16	1/2	13/64	9/16	17/64
3/8	9/16	1/4	5/8	21/64
7/16	5/8	19/64	3/4	3/8
1/2	3/4	21/64	13/16	7/16
5/8	15/16	27/64	1	35/64
3/4	1 1/8	1/2	1 1/8	21/32
7/8	1 5/16	19/32	1 5/16	49/64
1	1 1/2	21/32	1 1/2	7/8
1 1/8	1 11/16	3/4	1 11/16	1
1 1/4	1 7/8	27/32	1 7/8	1 3/32
1 3/8	2 1/16	29/32	2 1/16	1 13/64
1 1/2	2 1/4	1	2 1/4	1 5/16
1 5/8	2 7/16	1 3/32		

TABLE A-12 AMERICAN NATIONAL STANDARD HEAVY HEXAGON BOLTS

Diameter	Flats	Height of head		Diameter	Flats	Height of head	
	Unfinished semi-finished finished	Unfinished	Semi-finished finished		Unfinished semi-finished finished	Unfinished	Semi-finished
1/2	7/8	7/16	13/32	1 5/8	2 9/16	1 9/32	1 7/32
5/8	1 1/16	17/32	1/2	1 3/4	1 3/8	1 3/8	1 5/16
3/4	1 1/4	5/8	19/32	1 7/8	2 15/16	1 15/32	1 13/32
7/8	1 7/16	23/32	11/16	2	3 1/8	1 9/16	1 7/16
1	1 5/8	13/16	3/4	2 1/4	3 1/2	1 3/4	1 5/8
1 1/8	1 13/16	29/32	27/32	2 1/2	3 7/8	1 15/16	1 13/16
1 1/4	2	1	15/16	2 3/4	4 1/4	2 1/8	2
1 3/8	2 3/16	1 3/32	1 1/32	3	4 5/8	2 5/16	2 3/16
1 1/2	2 3/8	1 3/16	1 1/8				

Note: 1 5/8, 1 7/8 not in unfinished or semifinished bolts.

TABLE A–13 AMERICAN NATIONAL STANDARD HEAVY NUTS, SQUARE AND HEXAGON

Diameter	Flats: Unfinished semifinished square and hexagon	Thickness of nut: Unfinished square and hexagon	Thickness of nut: Semifinished hexagon
1/4	1/2	1/4	15/64
5/16	9/16	5/16	19/64
3/8	11/16	3/8	23/64
7/16	3/4	7/16	27/64
1/2	7/8	1/2	31/64
9/16	15/16	....	35/64
5/8	1 1/16	5/8	39/64
3/4	1 1/4	3/4	47/64
7/8	1 7/16	7/8	55/64
1	1 5/8	1	63/64
1 1/8	1 13/16	1 1/8	1 7/64
1 1/4	2	1 1/4	1 7/32
1 3/8	2 3/16	1 3/8	1 11/32
1 1/2	2 3/8	1 1/2	1 15/32
1 5/8	2 9/16	....	1 19/32
1 3/4	2 3/4	1 3/4	1 23/32
1 7/8	2 15/16	....	1 27/32
2	3 1/8	2	1 31/32
2 1/4	3 1/2	2 1/4	2 13/64
2 1/2	3 7/8	2 1/2	2 29/64
2 3/4	4 1/4	2 3/4	2 45/64
3	4 5/8	3	2 61/64
3 1/4	5	3 1/4	3 3/16
3 1/2	5 3/8	3 1/2	3 7/16
3 3/4	5 3/4	3 3/4	3 11/16
4	6 1/8	4	3 15/16

Note: 9/16, 1 5/8, 1 7/8 not in unfinished nuts.

TABLE A–14 AMERICAN NATIONAL STANDARD SQUARE-HEAD SETSCREWS AND POINTS

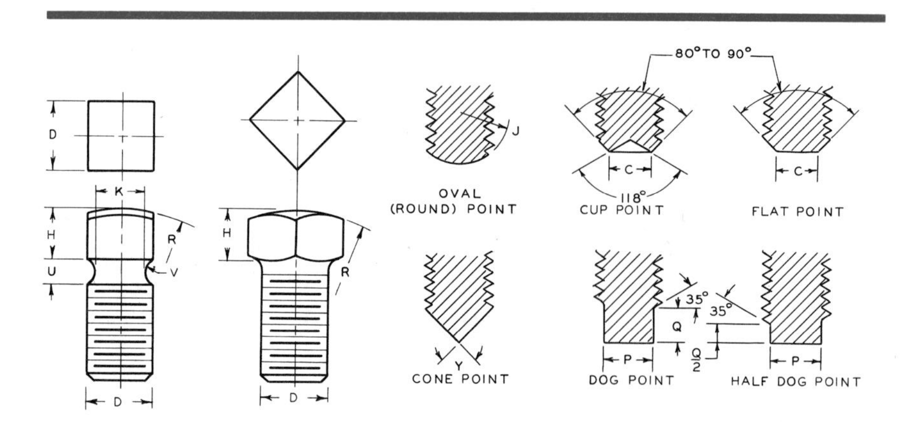

D	H nom.	R nom.	K max.	U min.	V max.	C nom.	J nom.	P max.	Q
10 (0.190)	9/64	15/32	0.145	0.083	0.027	3/32	0.141	0.127	0.090
12 (0.216)	5/32	35/64	0.162	0.091	0.029	7/64	0.156	0.144	0.110
1/4	3/16	5/8	0.185	0.100	0.032	1/8	0.188	0.156	0.125
5/16	15/64	25/32	0.240	0.111	0.036	11/64	0.234	0.203	0.156
3/8	9/32	15/16	0.294	0.125	0.041	13/64	0.281	0.250	0.188
7/16	21/64	1 3/32	0.345	0.143	0.046	15/64	0.328	0.297	0.219
1/2	3/8	1 1/4	0.400	0.154	0.050	9/32	0.375	0.344	0.250
9/16	27/64	1 13/32	0.454	0.167	0.054	5/16	0.422	0.391	0.281
5/8	15/32	1 9/16	0.507	0.182	0.059	23/64	0.469	0.469	0.313
3/4	9/16	1 7/8	0.620	0.200	0.065	7/16	0.563	0.563	0.375
7/8	21/32	2 3/16	0.731	0.222	0.072	33/64	0.656	0.656	0.438
1	3/4	2 1/2	0.838	0.250	0.081	19/32	0.750	0.750	0.500
1 1/8	27/32	2 13/16	0.939	0.283	0.092	43/64	0.844	0.844	0.562
1 1/4	15/16	3 1/8	1.064	0.283	0.092	3/4	0.938	0.938	0.625
1 3/8	1 1/32	3 7/16	1.159	0.333	0.109	53/64	1.031	1.031	0.688
1 1/2	1 1/8	3 3/4	1.284	0.333	0.109	29/32	1.125	1.125	0.750

Note: Threads may be coarse-, fine-, or 8-threaded series, class 2A. Coarse thread normally used on 1/4 in. and larger. When length equals nominal diameter or less $Y = 118°$. When length exceeds nominal diameter $Y = 90°$.

TABLE A-15 AMERICAN NATIONAL STANDARD SLOTTED-HEAD CAP SCREWS

Nominal diameter D	A max.	B max.	C max.	E max.	F max.	G max.	H average	I max.	J max.	K max.	M max.
1/4	0.375	0.216	0.172	0.075	0.097	0.500	0.140	0.068	0.437	0.191	0.117
5/16	0.437	0.253	0.203	0.084	0.115	0.625	0.177	0.086	0.562	0.245	0.151
3/8	0.562	0.314	0.250	0.094	0.142	0.750	0.210	0.103	0.625	0.273	0.168
7/16	0.625	0.368	0.297	0.094	0.168	0.812	0.210	0.103	0.750	0.328	0.202
1/2	0.750	0.413	0.328	0.106	0.193	0.875	0.210	0.103	0.812	0.354	0.218
9/16	0.812	0.467	0.375	0.118	0.213	1.000	0.244	0.120	0.937	0.409	0.252
5/8	0.875	0.521	0.422	0.133	0.239	1.125	0.281	0.137	1.000	0.437	0.270
3/4	1.000	0.612	0.500	0.149	0.283	1.375	0.352	0.171	1.250	0.546	0.338
7/8	1.125	0.720	0.594	0.167	0.334	1.625	0.423	0.206			
1	1.312	0.803	0.656	0.188	0.371	1.875	0.494	0.240			
1 1/8	....	....	....	0.196	....	2.062	0.529	0.257			
1 1/4	....	....	....	0.211	....	2.312	0.600	0.291			
1 3/8	....	....	....	0.226	....	2.562	0.665	0.326			
1 1/2	....	....	....	0.258	....	2.812	0.742	0.360			

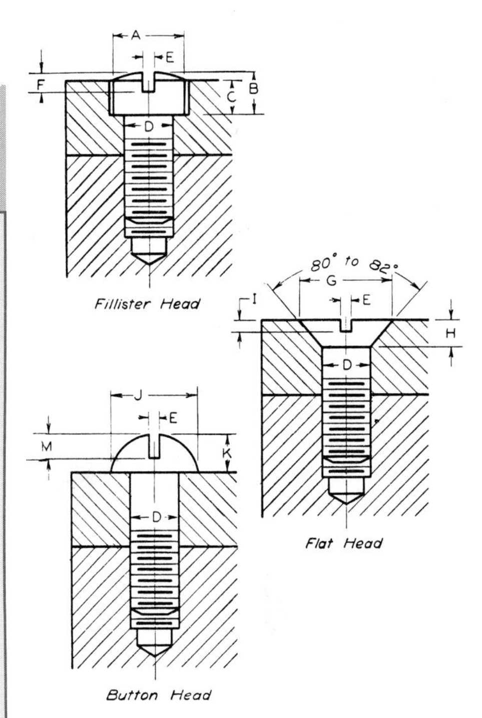

TABLE A-16 AMERICAN NATIONAL STANDARD MACHINE SCREWS

Diameter: Nominal	Diameter: Max.	A	B	C	E	F	G	H	I	J	K	M
		Maximum Dimensions										
0	0.060	0.119	0.035	0.056	0.113	0.053	0.096	0.045	0.059	....	....	0.023
1	0.073	0.146	0.043	0.068	0.138	0.061	0.118	0.053	0.071	....	....	0.026
2	0.086	0.172	0.051	0.080	0.162	0.069	0.140	0.062	0.083	0.167	0.053	0.031
3	0.099	0.199	0.059	0.092	0.187	0.078	0.161	0.070	0.095	0.193	0.060	0.035
4	0.112	0.225	0.067	0.104	0.211	0.086	0.183	0.079	0.107	0.219	0.068	0.039
5	0.125	0.252	0.075	0.116	0.236	0.095	0.205	0.088	0.120	0.245	0.075	0.043
6	0.138	0.279	0.083	0.128	0.260	0.103	0.226	0.096	0.132	0.270	0.082	0.048
8	0.164	0.332	0.100	0.152	0.309	0.120	0.270	0.113	0.156	0.322	0.096	0.054
10	0.190	0.385	0.116	0.176	0.359	0.137	0.313	0.130	0.180	0.373	0.110	0.060
12	0.216	0.438	0.132	0.200	0.408	0.153	0.357	0.148	0.205	0.425	0.125	0.067
1/4	0.250	0.507	0.153	0.232	0.472	0.175	0.414	0.170	0.237	0.492	0.144	0.075
5/16	0.3125	0.635	0.191	0.290	0.590	0.216	0.518	0.211	0.295	0.615	0.178	0.084
3/8	0.375	0.762	0.230	0.347	0.708	0.256	0.622	0.253	0.355	0.740	0.212	0.094
7/16	0.4375	0.812	0.223	0.345	0.750	0.328	0.625	0.265	0.368	....	....	0.094
1/2	0.500	0.875	0.223	0.354	0.813	0.355	0.750	0.297	0.412	....	....	0.106
9/16	0.5625	1.000	0.260	0.410	0.938	0.410	0.812	0.336	0.466	....	....	0.118
5/8	0.625	1.125	0.298	0.467	1.000	0.438	0.875	0.375	0.521	....	....	0.133
3/4	0.750	1.375	0.372	0.578	1.250	0.547	1.000	0.441	0.612	....	....	0.149

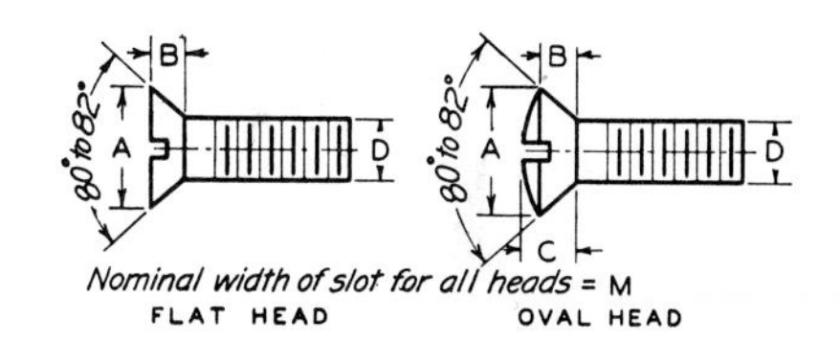

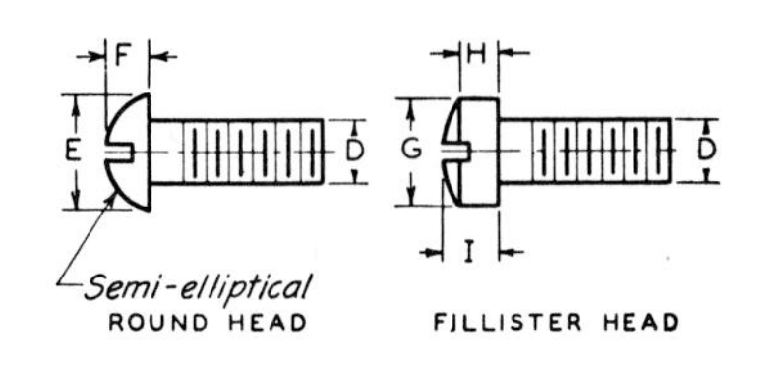

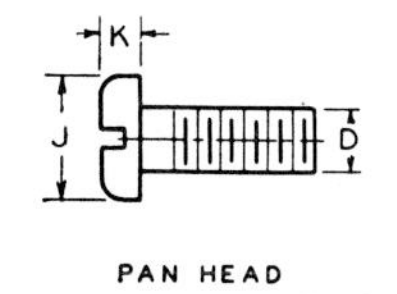

TABLE A-17 AMERICAN NATIONAL STANDARD HEXAGON-HEAD CAP SCREWS

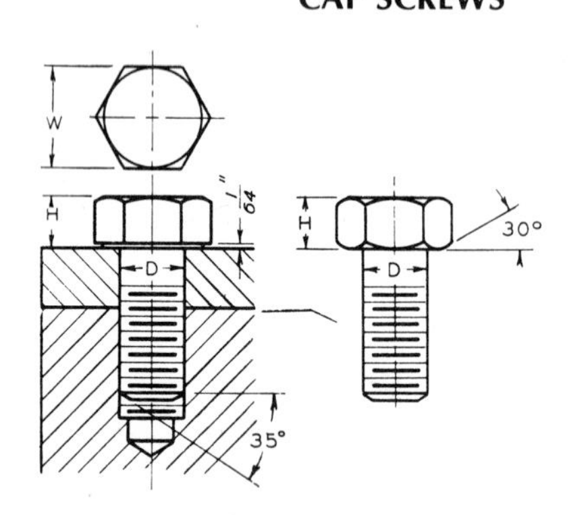

D	W	H	D	W	H
1/4	7/16	5/32	3/4	1 1/8	15/32
5/16	1/2	13/64	7/8	1 5/16	35/64
3/8	9/16	15/64	1	1 1/2	39/64
7/16	5/8	9/32	1 1/8	1 11/16	11/16
1/2	3/4	5/16	1 1/4	1 7/8	25/32
9/16	13/16	23/64	1 3/8	2 1/16	27/32
5/8	15/16	25/64	1 1/8	2 1/4	15/16

Note: Bearing surfaces shall be flat and either washer faced or with chamfered corners. Minimum thread length shall be twice the diameter plus 1/4 in. for lengths up to and including 6 in.; twice the diameter plus 1/2 in. for lengths over 6 in.

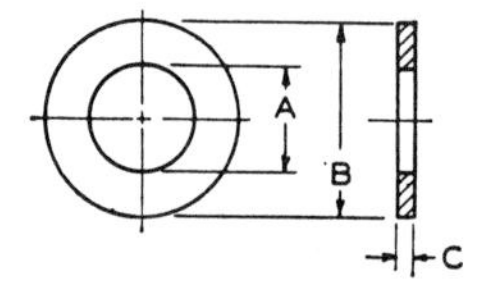

TABLE A-18 AMERICAN NATIONAL STANDARD PLAIN WASHERS

Inside diameter A	Outside diameter B	Thickness, C Gage	Thickness, C Nom.
5/64	3/16	25	0.020
3/32	7/32	25	0.020
3/32	1/4	25	0.020
1/8	1/4	24	0.022
1/8	5/16	21	0.032
5/32	5/16	20	0.035
5/32	3/8	18	0.049
11/64	13/32	18	0.049
3/16	3/8	18	0.049
3/16	7/16	18	0.049
13/64	15/32	18	0.049
7/32	7/16	18	0.049
7/32	1/2	18	0.049
15/64	17/32	18	0.049
1/4	1/2	18	0.049
1/4	9/16	18	0.049
1/4	9/16	16	0.065
17/64	5/8	18	0.049
9/32	5/8	16	0.065
5/16	3/4	16	0.065

Inside diameter A	Outside diameter B	Thickness, C Gage	Thickness, C Nom.
5/16	7/8	16	0.065
11/32	11/16	16	0.065
3/8	3/4	16	0.065
3/8	7/8	14	0.083
3/8	1 1/8	16	0.065
13/32	13/16	16	0.065
7/16	7/8	14	0.083
7/16	1	14	0.083
7/16	1 3/8	14	0.083
15/32	59/64	16	0.065
1/2	1 1/8	14	0.083
1/2	1 1/4	14	0.083
1/2	1 5/8	14	0.083
17/32	1 1/16	13	0.095
9/16	1 1/4	12	0.109
9/16	1 3/8	12	0.109
9/16	1 7/8	12	0.109
19/32	1 3/16	13	0.095
5/8	1 3/8	12	0.109
5/8	1 1/2	12	0.109

Inside diameter A	Outside diameter B	Thickness, C Gage	Thickness, C Nom.
5/8	2 1/8	10	0.134
21/32	1 5/16	13	0.095
11/16	1 1/2	10	0.134
11/16	1 3/4	10	0.134
11/16	2 3/8	8	0.165
13/16	1 1/2	10	0.134
13/16	1 3/4	9	0.148
13/16	2	9	0.148
13/16	2 7/8	8	0.165
15/16	1 3/4	10	0.134
15/16	2	8	0.165
15/16	2 1/4	8	0.165
15/16	3 3/8	7	0.180
1 1/16	2	10	0.134
1 1/16	2 1/4	8	0.165
1 1/16	2 1/2	8	0.165
1 1/16	3 7/8	4	0.238
1 3/16	2 1/2	8	0.165
1 1/4	2 3/4	8	0.165
1 5/15	2 3/4	8	0.165

Inside diameter A	Outside diameter B	Thickness, C Gage	Thickness, C Nom.
1 3/8	3	8	0.165
1 7/16	3	7	0.180
1 1/2	3 1/4	7	0.180
1 9/16	3 1/4	7	0.180
1 5/8	3 1/2	7	0.180
1 11/16	3 1/2	7	0.180
1 3/4	3 3/4	7	0.180
1 13/16	3 3/4	7	0.180
1 7/8	4	7	0.180
1 15/16	4	7	0.180
2	4 1/4	7	0.180
2 1/16	4 1/4	7	0.180
2 1/8	4 1/2	7	0.180
2 3/8	4 3/4	5	0.220
2 5/8	5	4	0.238
2 7/8	5 1/4	3	0.259
3 1/8	5 1/2	2	0.284

TABLE A-19 AMERICAN NATIONAL STANDARD COTTER PINS

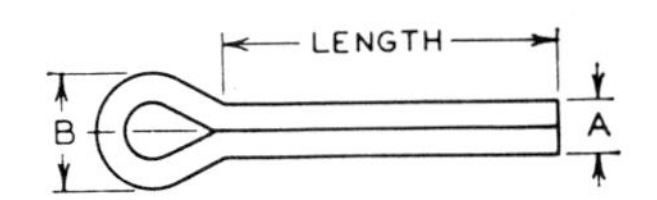

Design of head may vary but outside diameters should be adhered to.

A nominal	B min.	Hole sizes recommended
0.031	1/16	3/64
0.047	3/32	1/16
0.062	1/8	5/64
0.078	5/32	3/32
0.094	3/16	7/64
0.109	7/32	1/8
0.125	1/4	9/64
0.141	9/32	5/32
0.156	5/16	11/64
0.188	3/8	13/64
0.219	7/16	15/64
0.250	1/2	17/64
0.312	5/8	5/16
0.375	3/4	3/8
0.438	7/8	7/16
0.500	1	1/2
0.625	1 1/4	5/8
0.750	1 1/2	3/4

TABLE A-20 AMERICAN NATIONAL STANDARD SLOTTED-HEAD WOOD SCREWS

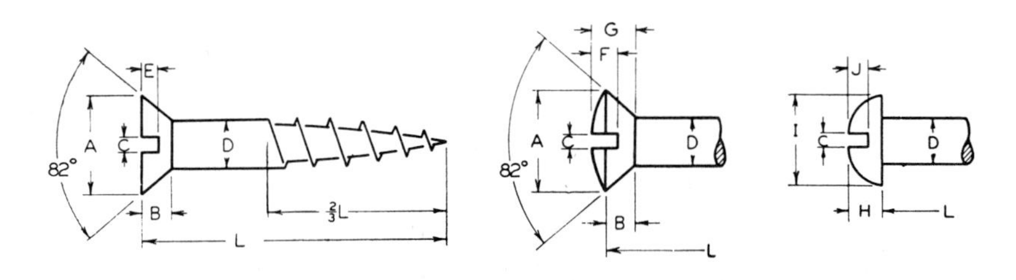

Nominal size	Maximum Dimensions D	A	B	C	E	F	G	H	I	J	Number threads per inch
0	0.060	0.119	0.035	0.023	0.015	0.030	0.056	0.053	0.113	0.039	32
1	0.073	0.146	0.043	0.026	0.019	0.038	0.068	0.061	0.138	0.044	28
2	0.086	0.172	0.051	0.031	0.023	0.045	0.080	0.069	0.162	0.048	26
3	0.099	0.199	0.059	0.035	0.027	0.052	0.092	0.078	0.187	0.053	24
4	0.112	0.225	0.067	0.039	0.030	0.059	0.104	0.086	0.211	0.058	22
5	0.125	0.252	0.075	0.043	0.034	0.067	0.116	0.095	0.236	0.063	20
6	0.138	0.279	0.083	0.048	0.038	0.074	0.128	0.103	0.260	0.068	18
7	0.151	0.305	0.091	0.048	0.041	0.081	0.140	0.111	0.285	0.072	16
8	0.164	0.332	0.100	0.054	0.045	0.088	0.152	0.120	0.309	0.077	15
9	0.177	0.358	0.108	0.054	0.049	0.095	0.164	0.128	0.334	0.082	14
10	0.190	0.385	0.116	0.060	0.053	0.103	0.176	0.137	0.359	0.087	13
12	0.216	0.438	0.132	0.067	0.060	0.117	0.200	0.153	0.408	0.096	11
14	0.242	0.491	0.148	0.075	0.068	0.132	0.224	0.170	0.457	0.106	10
16	0.268	0.544	0.164	0.075	0.075	0.146	0.248	0.187	0.506	0.115	9
18	0.294	0.597	0.180	0.084	0.083	0.160	0.272	0.204	0.555	0.125	8
20	0.320	0.650	0.196	0.084	0.090	0.175	0.296	0.220	0.604	0.134	8
24	0.372	0.756	0.228	0.094	0.105	0.204	0.344	0.254	0.702	0.154	7

TABLE A-21 AMERICAN NATIONAL STANDARD TAPER PINS

TAPER PINS *Maximum length for which standard reamers are available. Taper 1/4 in. per ft.

LENGTH

Size No.	0000000	000000	00000	0000	000	00	0
Size (large end)	0.0625	0.0780	0.0940	0.1090	0.1250	0.1410	0.1560
Maximum length*	0.625	0.750	1.000	1.000	1.000	1.250	1.250
Size No.	1	2	3	4	5	6	7
Size (large end)	0.1720	0.1930	0.2190	0.2500	0.2890	0.3410	0.4090
Maximum length*	1.250	1.500	1.750	2.000	2.250	3.000	3.750
Size No.	8	9	10	11	12	13	14
Size (large end)	0.4920	0.5910	0.7060	0.8600	1.032	1.241	1.523
Maximum length*	4.500	5.250	6.000	(Special sizes. Special lengths.)			

Table A-22 AMERICAN NATIONAL STANDARD SQUARE- AND FLAT-STOCK KEYS AND SHAFT DIAMETERS

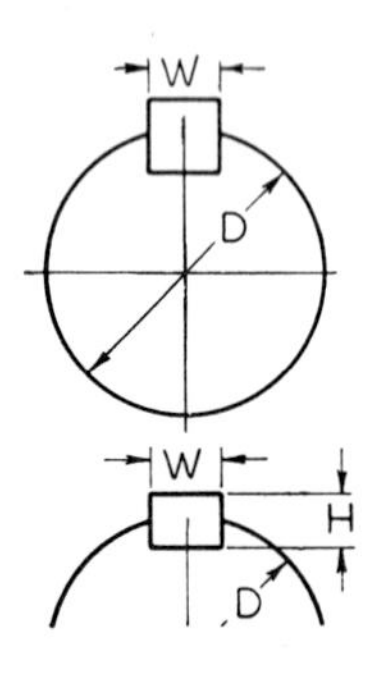

Diameter of shaft D inclusive	Square keys W	Flat keys W × H
1/2 – 9/16	1/8	1/8 × 3/32
5/8 – 7/8	3/16	3/16 × 1/8
15/16 – 1 1/4	1/4	1/4 × 3/16
1 5/16 – 1 3/8	5/16	5/16 × 1/4
1 7/16 – 1 3/4	3/8	3/8 × 1/4
1 13/16 – 2 1/4	1/2	1/2 × 3/8
2 5/16 – 2 3/4	5/8	5/8 × 7/16
2 7/8 – 3 1/4	3/4	3/4 × 1/2
3 3/8 – 3 3/4	7/8	7/8 × 5/8
3 7/8 – 4 1/2	1	1 × 3/4
4 3/4 – 5 1/2	1 1/4	1 1/4 × 7/8
5 3/4 – 6	1 1/2	1 1/2 × 1

TABLE A-23 WOODRUFF KEYS

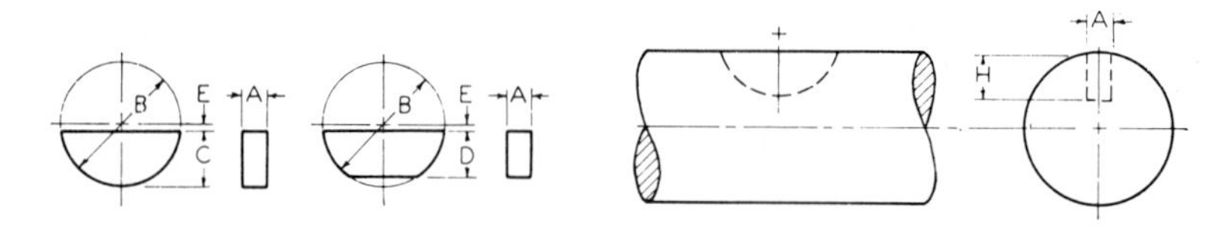

Key No.	Dimensions in Inches					
	Nominal			Maximum		
	A	B	E	C	D	H
204	1/16	1/2	3/64	0.203	0.194	0.1718
304	3/32	1/2	3/64	0.203	0.194	0.1561
305	3/32	5/8	1/16	0.250	0.240	0.2031
404	1/8	1/2	3/64	0.203	0.194	0.1405
405	1/8	5/8	1/16	0.250	0.240	0.1875
406	1/8	3/4	1/16	0.313	0.303	0.2505
505	5/32	5/8	1/16	0.250	0.240	0.1719
506	5/32	3/4	1/16	0.313	0.303	0.2349
507	5/32	7/8	1/16	0.375	0.365	0.2969
606	3/16	3/4	1/16	0.313	0.303	0.2193
607	3/16	7/8	1/16	0.375	0.365	0.2813
608	3/16	1	1/16	0.438	0.428	0.3443
609	3/16	1 1/8	5/64	0.484	0.475	0.3903
807	1/4	7/8	1/16	0.375	0.365	0.2500
808	1/4	1	1/16	0.438	0.428	0.3130
809	1/4	1 1/8	5/64	0.484	0.475	0.3590
810	1/4	1 1/4	5/64	0.547	0.537	0.4220
811	1/4	1 3/8	3/32	0.594	0.584	0.4690
812	1/4	1 1/2	7/64	0.641	0.631	0.5160
1008	5/16	1	1/16	0.438	0.428	0.2818
1009	5/16	1 1/8	5/64	0.484	0.475	0.3278
1010	5/16	1 1/4	5/16	0.547	0.537	0.3908
1011	5/16	1 3/8	3/32	0.594	0.584	0.4378
1012	5/16	1 1/2	7/64	0.641	0.631	0.4848
1210	3/8	1 1/4	5/64	0.547	0.537	0.3595
1211	3/8	1 3/8	3/32	0.594	0.584	0.4060
1212	3/8	1 1/2	7/64	0.641	0.631	0.4535

Note: Nominal dimensions are indicated by the key number in which the last two digits give the diameter (B) in eighths and the ones in front of them give the width (A) in thirty-seconds. For example, No. 809 means B = 9/8 or 1 1/8 and A = 8/32 or 1/4.

TABLE A-24 WIRE AND SHEET-METAL GAGES

Dimensions in Decimal Parts of an Inch							
No. of wire gage	American, or Brown & Sharpe	Birmingham, or Stubs wire	Washburn & Moen or American Steel & Wire Co.	W. & M. steel music wire	New American S. & W. Co. music wire gage	Imperial wire gage	U.S. Standard gage for sheet and plate iron and steel
00000000	....	.....	....	0.0083			
0000000	....	.....	....	0.0087			
000000	....	.....	....	0.0095	0.004	0.464	0.46875
00000	....	.....	....	0.010	0.005	0.432	0.4375
0000	0.460	0.454	0.3938	0.011	0.006	0.400	0.40625
000	0.40964	0.425	0.3625	0.012	0.007	0.372	0.375
00	0.3648	0.380	0.3310	0.0133	0.008	0.348	0.34375
0	0.32486	0.340	0.3065	0.0144	0.009	0.324	0.3125
1	0.2893	0.300	0.2830	0.0156	0.010	0.300	0.28125
2	0.25763	0.284	0.2625	0.0166	0.011	0.276	0.265625
3	0.22942	0.259	0.2437	0.0178	0.012	0.252	0.250
4	0.20431	0.238	0.2253	0.0188	0.013	0.232	0.234375
5	0.18194	0.220	0.2070	0.0202	0.014	0.212	0.21875
6	0.16202	0.203	0.1920	0.0215	0.016	0.192	0.203125
7	0.14428	0.180	0.1770	0.023	0.018	0.176	0.1875
8	0.12849	0.165	0.1620	0.0243	0.020	0.160	0.171875
9	0.11443	0.148	0.1483	0.0256	0.022	0.144	0.15625
10	0.10189	0.134	0.1350	0.027	0.024	0.128	0.140625
11	0.090742	0.120	0.1205	0.0284	0.026	0.116	0.125
12	0.080808	0.109	0.1055	0.0296	0.029	0.104	0.109375
13	0.071961	0.095	0.0915	0.0314	0.031	0.092	0.09375
14	0.064084	0.083	0.0800	0.0326	0.033	0.080	0.078125
15	0.057068	0.072	0.0720	0.0345	0.035	0.072	0.0703125
16	0.05082	0.065	0.0625	0.036	0.037	0.064	0.0625
17	0.045257	0.058	0.0540	0.0377	0.039	0.056	0.05625
18	0.040303	0.049	0.0475	0.0395	0.041	0.048	0.050
19	0.03589	0.042	0.0410	0.0414	0.043	0.040	0.04375
20	0.031961	0.035	0.0348	0.0434	0.045	0.036	0.0375
21	0.028462	0.032	0.03175	0.046	0.047	0.032	0.034375
22	0.025347	0.028	0.0286	0.0483	0.049	0.028	0.03125
23	0.022571	0.025	0.0258	0.051	0.051	0.024	0.028125
24	0.0201	0.022	0.0230	0.055	0.055	0.022	0.025
25	0.0179	0.020	0.0204	0.0586	0.059	0.020	0.021875
26	0.01594	0.018	0.0181	0.0626	0.063	0.018	0.01875
27	0.014195	0.016	0.0173	0.0658	0.067	0.0164	0.0171875
28	0.012641	0.014	0.0162	0.072	0.071	0.0149	0.015625
29	0.011257	0.013	0.0150	0.076	0.075	0.0136	0.0140625
30	0.010025	0.012	0.0140	0.080	0.080	0.0124	0.0125
31	0.008928	0.010	0.0132	......	0.085	0.0116	0.0109375
32	0.00795	0.009	0.0128	......	0.090	0.0108	0.01015625
33	0.00708	0.008	0.0118	......	0.095	0.0100	0.009375
34	0.006304	0.007	0.0104	......	.....	0.0092	0.00859375
35	0.005614	0.005	0.0095	......	.....	0.0084	0.0078125
36	0.005	0.004	0.0090	......	.....	0.0076	0.00703125
37	0.004453	.....	....	......	.....	0.0068	0.006640625
38	0.003965	.....	....	......	.....	0.0060	0.00625
39	0.003531	.....	....	......	.....	0.0052	
40	0.003144	.....	....	......	.....	0.0048	

TABLE A-25 AMERICAN NATIONAL STANDARD WELDED AND SEAMLESS STEEL PIPE

Nominal pipe size	Nominal wall thickness				Threads per inch
	Out side dia.	Standard wall	Extra strong wall	Double extra strong wall	
1/8	0.405	0.068	0.095	.....	27
1/4	0.540	0.088	0.119	.....	18
3/8	0.675	0.091	0.126	.....	18
1/2	0.840	0.109	0.147	0.294	14
3/4	1.050	0.113	0.154	0.308	14
1	1.315	0.133	0.179	0.358	11 1/2
1 1/4	1.660	0.140	0.191	0.382	11 1/2
1 1/2	1.900	0.145	0.200	0.400	11 1/2
2	2.375	0.154	0.218	0.436	11 1/2
2 1/2	2.875	0.203	0.276	0.552	8
3	3.500	0.216	0.300	0.600	8
3 1/2	4.000	0.226	0.318	.....	8
4	4.500	0.237	0.337	0.674	8
5	5.563	0.258	0.375	0.750	8
6	6.625	0.280	0.432	0.864	8
8	8.625	0.322	0.500	0.875	8
10	10.750	0.365	0.500	.....	8
12	12.750	0.375	0.500	.....	8
14	14.000	0.375	0.500	.....	
16	16.000	0.375	0.500	.....	8
18	18.000	0.375	0.500	.....	8
20	20.000	0.375	0.500	.....	8
24	24.000	0.375	0.500	.....	8

Note: To find the inside diameter, subtract twice the wall thickness from the outside diameter. Schedule numbers have been set up for wall thicknesses for pipe and the American Standard should be consulted for complete information. Standard wall thicknesses are for Schedule 40 up to and including nominal size 10. Extra strong walls are Schedule 80 up to ae including size 8, and Schedule 60 for size 10.

TABLE A-26 STEEL-WIRE NAILS

		American Steel & Wire Company Gage					
Size	Length	Common wire nails and brads		Casing nails		Finishing nails	
		Gage, dia.	No. to pound	Gage, dia.	No. to pound	Gage, dia.	No. to pound
2*d*	1	15	876	15½	1010	16½	1351
3*d*	1¼	14	568	14½	635	15½	807
4*d*	1½	12½	316	14	473	15	584
5*d*	1¾	12½	271	14	406	15	500
6*d*	2	11½	181	12½	236	13	309
7*d*	2¼	11½	161	12½	210	13	238
8*d*	2½	10¼	106	11½	145	12½	189
9*d*	2¾	10¼	96	11½	132	12½	172
10*d*	3	9	69	10½	94	11½	121
12*d*	3¼	9	64	10½	87	11½	113
16*d*	3½	8	49	10	71	11	90
20*d*	4	6	31	9	52	10	62
30*d*	4½	5	24	9	46		
40*d*	5	4	18	8	35		
50*d*	5½	3	14				
60*d*	6	2	11				

Table A-27 AMERICAN NATIONAL STANDARD LARGE RIVETS

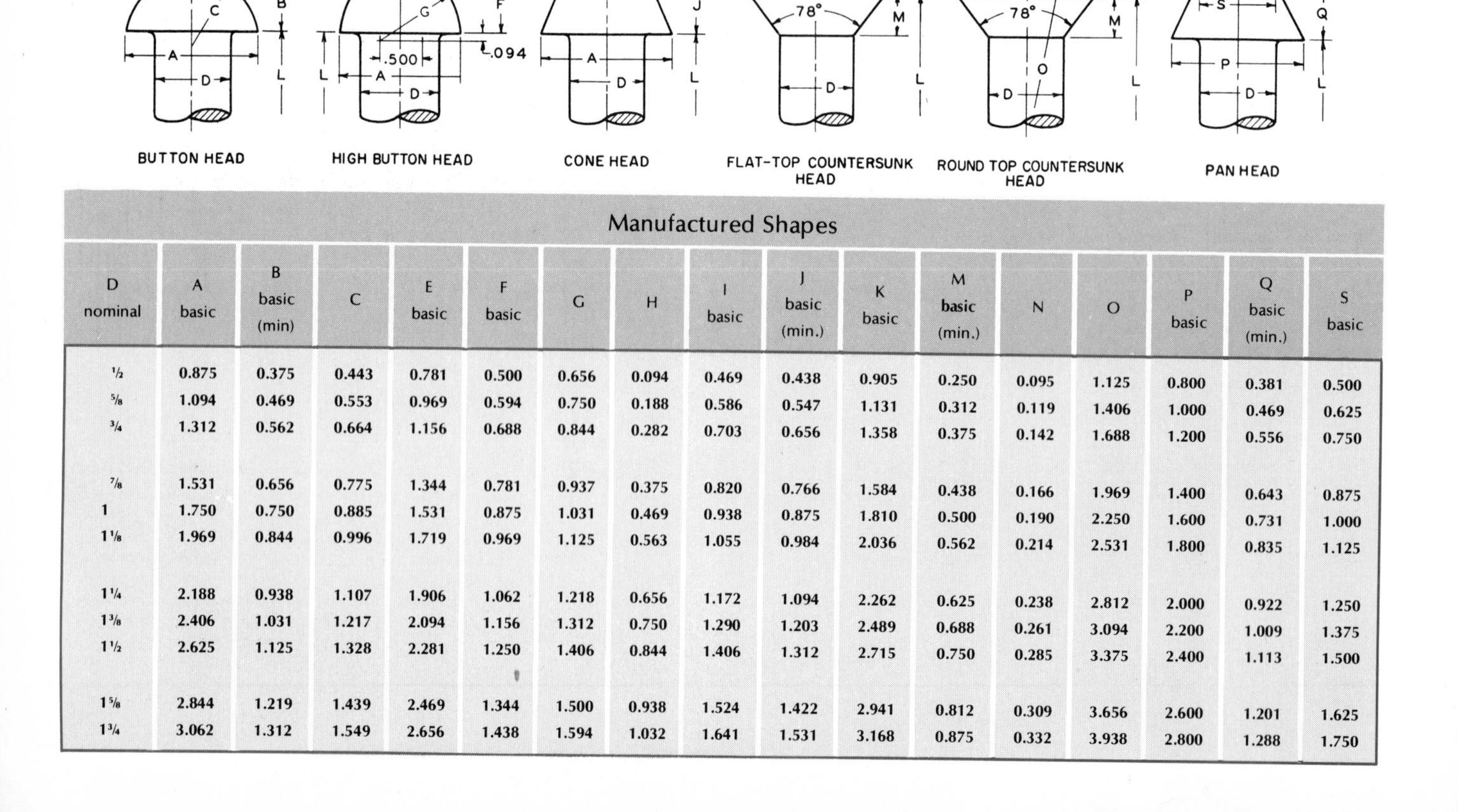

Manufactured Shapes																	
D nominal	A basic	B basic (min)	C	E basic	F basic	G	H	I basic	J basic (min.)	K basic	M basic (min.)	N	O	P basic	Q basic (min.)	S basic
½	0.875	0.375	0.443	0.781	0.500	0.656	0.094	0.469	0.438	0.905	0.250	0.095	1.125	0.800	0.381	0.500
⅝	1.094	0.469	0.553	0.969	0.594	0.750	0.188	0.586	0.547	1.131	0.312	0.119	1.406	1.000	0.469	0.625
¾	1.312	0.562	0.664	1.156	0.688	0.844	0.282	0.703	0.656	1.358	0.375	0.142	1.688	1.200	0.556	0.750
⅞	1.531	0.656	0.775	1.344	0.781	0.937	0.375	0.820	0.766	1.584	0.438	0.166	1.969	1.400	0.643	0.875
1	1.750	0.750	0.885	1.531	0.875	1.031	0.469	0.938	0.875	1.810	0.500	0.190	2.250	1.600	0.731	1.000
1⅛	1.969	0.844	0.996	1.719	0.969	1.125	0.563	1.055	0.984	2.036	0.562	0.214	2.531	1.800	0.835	1.125
1¼	2.188	0.938	1.107	1.906	1.062	1.218	0.656	1.172	1.094	2.262	0.625	0.238	2.812	2.000	0.922	1.250
1⅜	2.406	1.031	1.217	2.094	1.156	1.312	0.750	1.290	1.203	2.489	0.688	0.261	3.094	2.200	1.009	1.375
1½	2.625	1.125	1.328	2.281	1.250	1.406	0.844	1.406	1.312	2.715	0.750	0.285	3.375	2.400	1.113	1.500
1⅝	2.844	1.219	1.439	2.469	1.344	1.500	0.938	1.524	1.422	2.941	0.812	0.309	3.656	2.600	1.201	1.625
1¾	3.062	1.312	1.549	2.656	1.438	1.594	1.032	1.641	1.531	3.168	0.875	0.332	3.938	2.800	1.288	1.750

Table A-28 AMERICAN WELDING SOCIETY STANDARD WELDING SYMBOLS

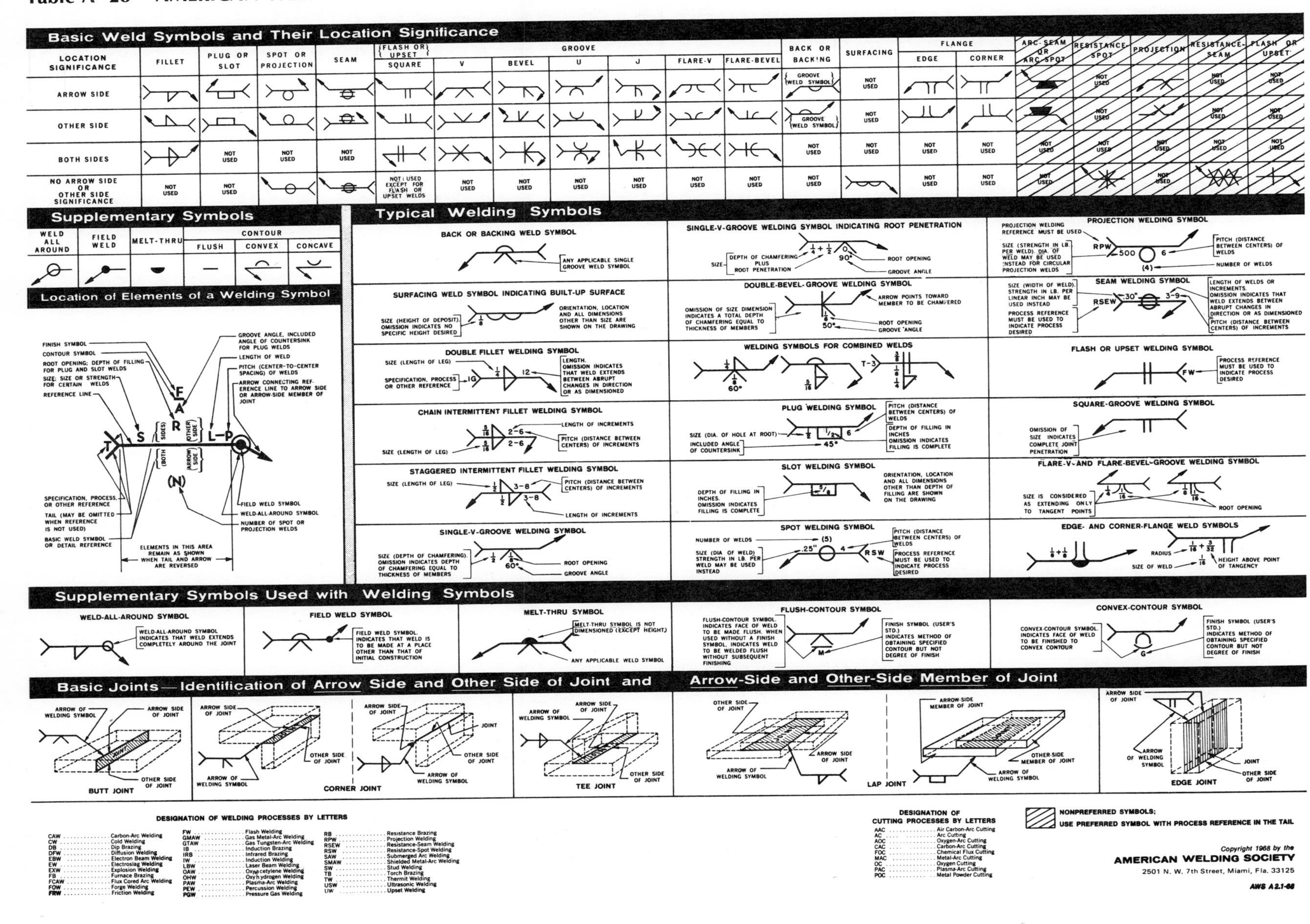

Appendix B

Glossary

(*v*) = Verb (*n*) = Noun

acme ***(n)*** A screw-thread form.

addendum ***(n)*** The distance a gear tooth extends above the pitch circle.

airbrush ***(n)*** A miniature spray gun that sprays a solution (usually watercolor) to give shading effects.

allowance ***(n)*** The smallest space permitted between mating parts.

alloy ***(n)*** Two or more metals combined to form a new metal.

anneal ***(v)*** To heat slowly to a critical temperature, then gradually cool. Used to soften and to remove internal stresses.

assembly drawing ***(n)*** A drawing of several parts to show how they are put together.

auxiliary view ***(n)*** An additional view, usually of a slanted surface, showing that surface in true shape.

azimuth ***(n)*** In descriptive geometry, the measure of how far a line is off due north.

babbitt or **babbitt metal** ***(n)*** A friction bearing metal, invented by Isaac Babbitt. Composed of antimony, tin, and copper.

bearing ***(n)*** Any part that bears up, or supports, another part. In particular, the support for a turning shaft.

bearing ***(n)*** In descriptive geometry, the angle a line in a top view makes with a north-south line.

bevel ***(n)*** A surface slanted to another surface. Called a *miter* when the angle is 45°.

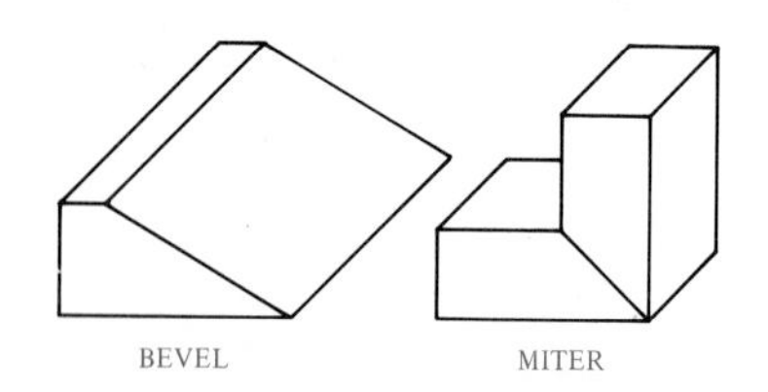

blueprint ***(n)*** A copy of a drawing made by a machine. It has white lines on a blue background.

bolt circle ***(n)*** A circular centerline locating the centers of holes arranged in a circle.

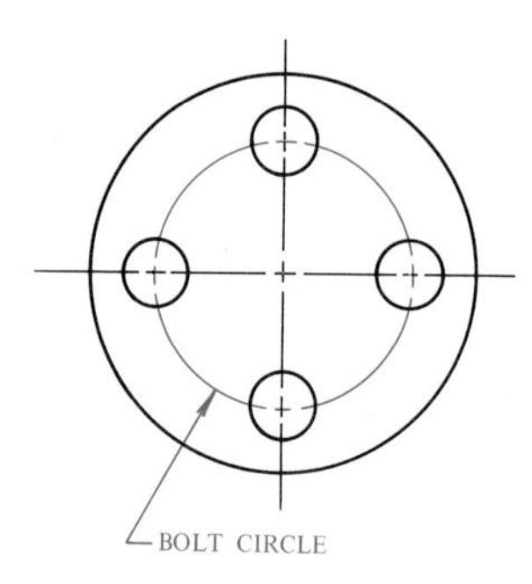

bore ***(v)*** To enlarge or to finish a hole with a cutting tool called a *boring bar,* used in a boring mill or lathe.

boss ***(n)*** A raised circular surface as used on a casting or forging.

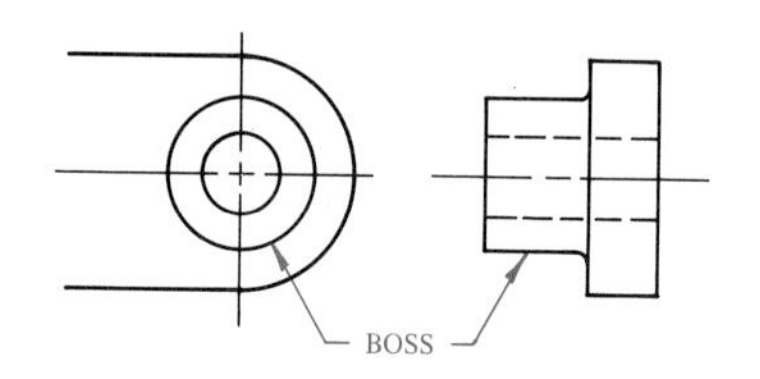

brass ***(n)*** An alloy of copper and zinc or of copper, zinc, and lead.

braze ***(v)*** To join two pieces of metal by using hard solder, such as brass or zinc.

broach ***(v)*** To machine (finish) and change the forms of holes or outside surfaces to a desired shape, generally other than round. ***(n)*** A tool with a series of chisel edges used to broach.

bronzes ***(n)*** Alloys of copper and tin in varying proportions, mostly copper. Sometimes other metals, such as zinc, are added.

buff ***(v)*** To polish a surface on a fabric wheel by rubbing it with some rough material.

burnish ***(v)*** To smooth or polish by a sliding or rolling motion.

burr ***(n)*** A rough or jagged edge caused by cutting or punching.

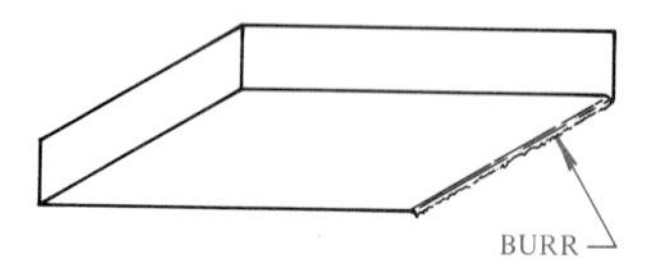

bushing ***(n)*** A hollow cylindrical sleeve used as a bearing or as a guide for drills or other tools.

cabinet drawing ***(n)*** A kind of oblique pictorial drawing in which receding edges are drawn to one-half their proportional length.

caliper ***(n)*** A measuring device with two adjustable legs. It is used for measuring thicknesses or diameters.

cam ***(n)*** A machine part mounted on a turning shaft, used to change rotary (turning) motion into back-and-forth motion. See Chapter 19.

caseharden (carbonize or carburize) ***(v)*** To heat-treat steel. The outside surface is hardened by making it absorb carbon. It is then cooled rapidly by quenching in oil.

casting ***(n)*** A part formed by pouring molten metal into a hollow form (mold) of the desired shape, then allowing it to harden.

cavalier drawing ***(n)*** A kind of oblique pictorial drawing in which receding edges are drawn to their full proportional length.

center drill ***(n)*** A special combination drill and countersink. It is used to drill the ends of stock to be turned between centers on a lathe.

chamfer ***(v)*** To bevel an edge. ***(n)*** An edge that has been beveled.

circular pitch *(n)* The distance from a point on one gear tooth to the same point on the next tooth measured along the pitch circle.

clearance *(n)* The space that a moving part needs in order to function. On gears, the distance between the top of a tooth and the bottom of the mating space.

core *(v)* To form the hollow part of a casting. A part made of sand and shaped like the hollow part (called a *core*) is placed in the mold (see **casting**). The core is broken up and removed after the casting is cool.

counterbore *(v)* To make a hole deeper and wider. *(n)* A counterbored hole.

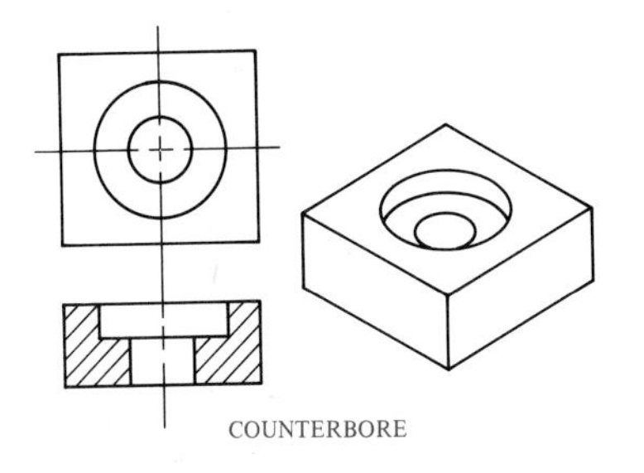

countersink *(v)* To drill a cone shape at the end of a hole.

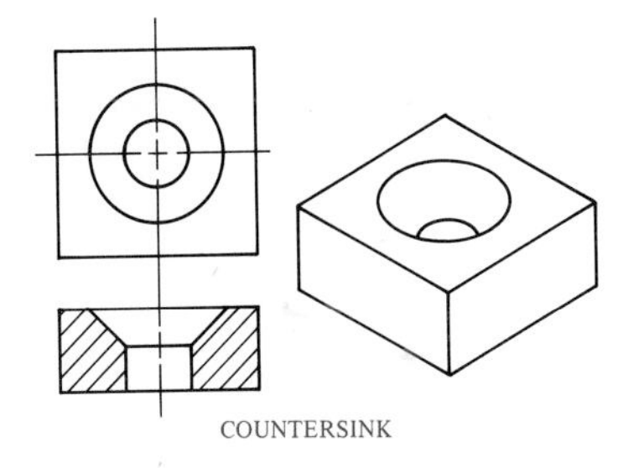

crown *(n)* The contour of the face of a belt pulley, rounded or angular, used to keep the belt in place. The belt tends to climb to the highest place.

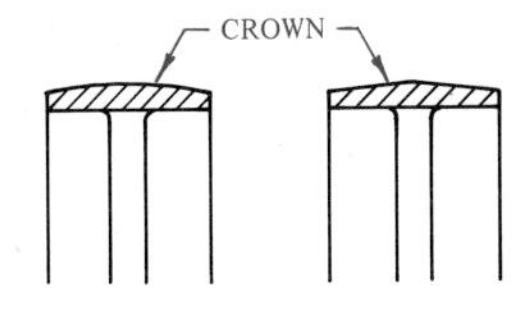

cutting plane *(n)* An imaginary plane passed through an object to cut away part of it.

dedendum *(n)* The distance a gear tooth extends below the pitch circle.

detail drawing *(n)* A drawing of one part of a machine or structure.

development *(n)* A drawing of all the surfaces of an object in true shape.

diametral pitch *(n)* The number of gear teeth per inch of pitch diameter.

diazo *(n)* A print with dark lines on a white background. The opposite of a blueprint.

die *(n)* A hardened metal block used to make a desired shape by cutting or pressing. Also, a tool used to cut external screw threads.

die casting *(n)* A casting made by pouring molten alloy (or plastic composition) into a metal mold or die, generally forced under pressure. Die castings are smooth and accurate.

die stamping *(n)* A piece that has been formed or cut from sheet material, generally metal.

dimension line *(n)* A line with arrowheads at either end to show the distance between two points.

dividers *(n)* A tool with two legs joined at the top. Each leg ends in a needle point. Dividers are used to find and lay off measurements.

dowel *(n)* A cylindrical-shaped pin used to fasten parts together.

draft *(n)* The tapered sides on a foundry pattern that allow it to be easily pulled from the sand.

drill *(v)* To make a cylindrical hole using a revolving tool with cutting edges, called a *drill,* generally a twist drill.

drop forging *(n)* A piece formed between dies while hot, using pressure or a drop hammer.

elevation *(n)* A drawing of a facade of a structure.

exploded view *(n)* A drawing showing how an object would look if an "explosion" had propelled its component parts away from each other. Used to show how the parts relate to each other.

face *(v)* To machine (finish) a flat surface on a lathe with the surface perpendicular to the axis of rotation.

FAO Finished all over.

file *(v)* To smooth, finish, or shape with a file.

fillet *(n)* The rounded-in corner between two surfaces.

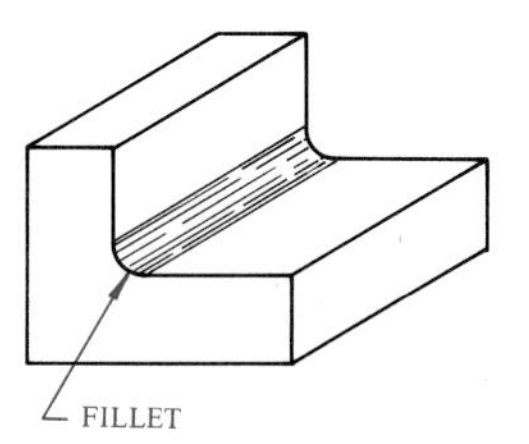

fit *(n)* The tightness or looseness between two mating parts.

fixture *(n)* A device used to hold a workpiece during machining (finishing).

flange *(n)* A rim extension, as at the end of a pipe.

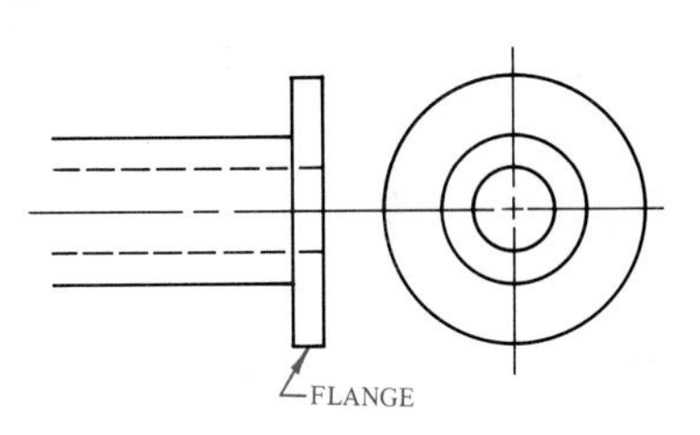

forge *(v)* To form hot metal into a desired shape by hammering or by pressure.

French curve *(n)* A tool used to draw curved lines that are not arcs. Also called an irregular curve.

functional drawing *(n).* A drawing that uses the fewest number of views and the fewest number of lines to give the exact information required.

gage *(n)* A device to find whether a specified dimension on an object is within specified limits.

gage line *(n)* A line along which rivets are spaced.

galvanize *(v)* To give a coating of zinc or zinc and lead.

gasket *(n)* A thin piece of material placed between two surfaces to produce a tight joint.

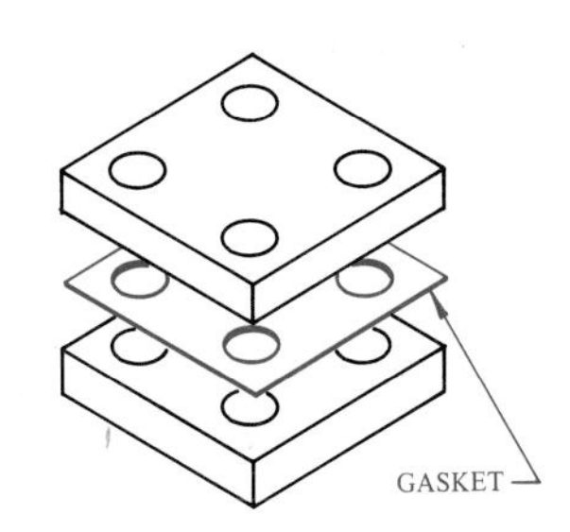

gear *(n)* A toothed wheel used to transmit power or motion from one shaft to another. A machine part used to transmit motion or force.

grind *(v)* To used an *abrasive* (rough-surfaced) wheel to polish or to finish a surface.

heat-treat *(v)* To change the properties of metal by carefully controlled heating and cooling.

isometric drawing *(n)* A kind of pictorial drawing based on height, width, and depth axes in equal 120° angles with each other.

jig *(n)* A special device used to guide a cutting tool; it may also hold the workpiece.

kerf *(n)* A slot or groove made by a cutting tool.

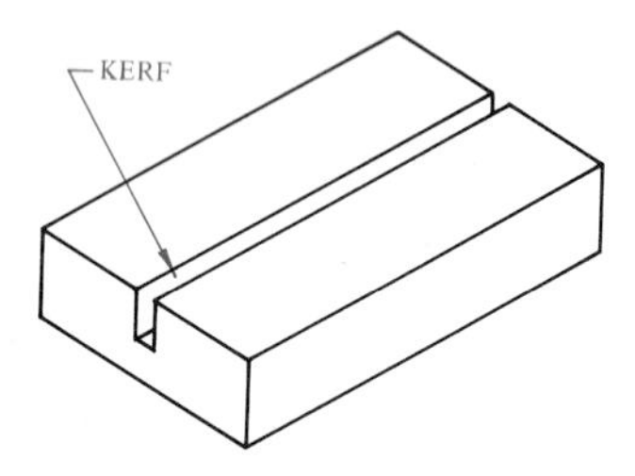

key *(n)* A piece used to fasten a hub to a shaft, or for a similar purpose.

keyway or **keyseat** *(n)* A groove or slot in a shaft or hub into which a key fits.

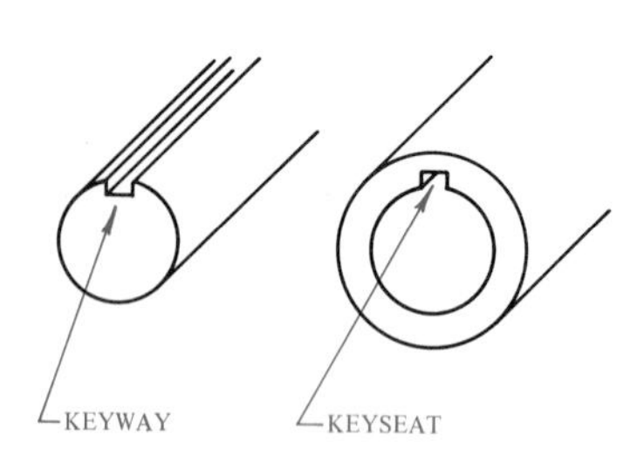

knurl *(v)* To roughen or indent a rounded surface so that it can be held or turned by hand.

lead *(n)* The distance a screw moves along its axis and against a fixed mating part when given one full turn.

limit *(n)* A boundary. Indication of only the largest and smallest permissible dimensions. See Chapter 6.

load *(n)* The weight or pressure borne by a structure.

lug *(n)* An "ear" forming a part of, and extending from, a part.

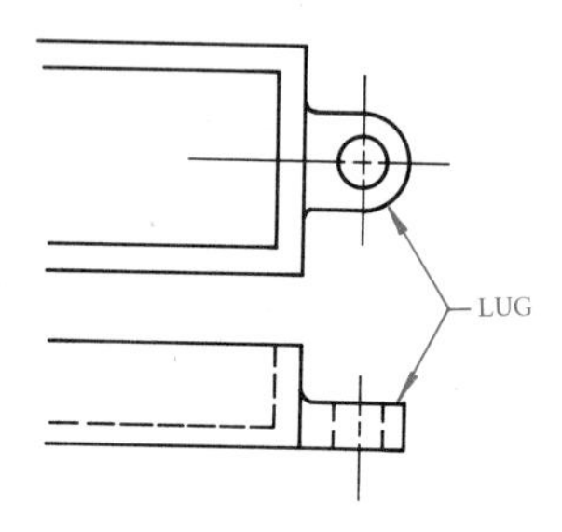

mallable casting *(n)* A casting toughened by annealing.

matte *(adj)* Dull and rough, as a surface.

micrometer caliper *(n)* A device used to find exact measurements of diameter thicknesses.

mill *(v)* To machine (finish) a part on a milling machine, using a rotating toothed cutter.

neck *(v)* To cut a groove around a cylindrical part, generally at a change in diameter.

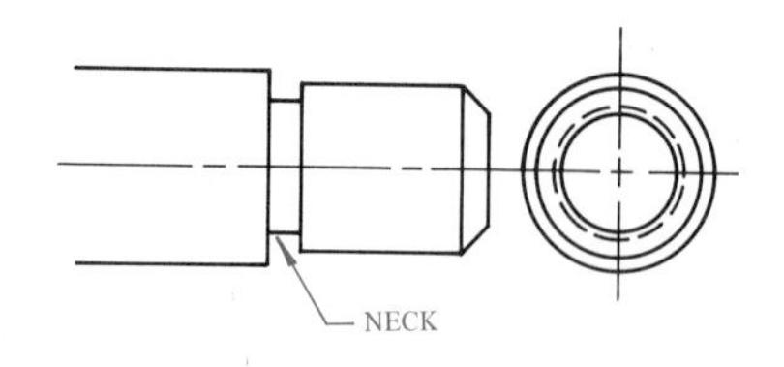

oblique drawing *(n)* A kind of pictorial drawing that shows one face of the object in true shape, but other faces on a distorted angle.

orthographic projection *(n)* A system of showing an object in several views.

overlay *(n)* A translucent sheet placed over a drawing. It can be used to trace the drawing. It can also contain additional material to be superimposed on the drawing.

pattern *(n)* See **development.**

peen *(v)* To stretch or bend material with the peen, or ball, end of a machinist's hammer.

perspective drawing *(n)* A kind of pictorial drawing that shows objects as they look to the eye.

photodrawing *(n)* A drawing prepared from a photograph, or a photograph on which dimensions, changes, or additions have been drawn.

pickle *(v)* To clean a metal object by using a weak sulfuric acid bath.

pictorial drawing *(n)* A drawing that looks like a picture.

pitch *(n)* On a screw, the distance from one point on the thread form to the corresponding point on the next form, measured in line with the axis of the screw. Also, on any slanting structure, the degree of slant.

plate *(v)* To electrochemically coat a metal object with another metal.

polish *(v)* To smooth with a very fine *abrasive* (rough surface).

projector *(n)* In pictorial drawing, a line along a receding axis.

punch *(v)* To make a hole in thin metal by pressing a tool of the desired shape through it.

ream *(v)* To make a hole the exact size by finishing with a turning cutting tool that has may cutting edges.

rendering *(n)* Surface shading used in a drawing.

rib *(n)* A thin, flat part of an object used to brace or strengthen another part.

rise *(n)* In a slanting structure, such as a roof or staircase, the vertical distance from the bottom to the top.

rivet *(n)* A metal rod with a preformed head on one end. A rivet is used to join metal parts together permanently.

round *(n)* The rounded-over corner of two surfaces.

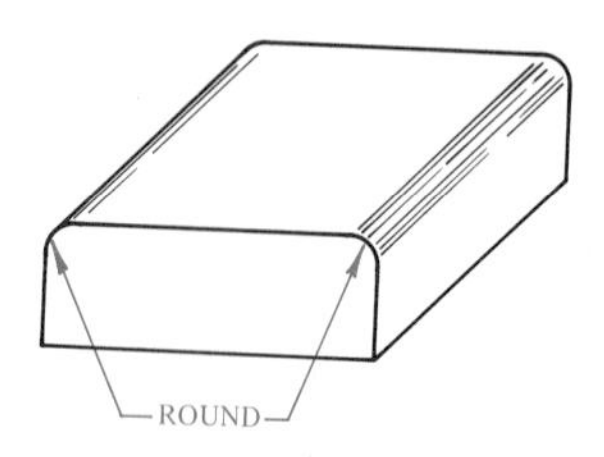

run *(n)* In a slanting structure, such as a roof or staircase, the horizontal distance from the bottom to the top.

schematic *(n)* Usually, an electric wiring diagram using symbols.

sectional view *(n)* A drawing of an object as if part of it were cut away to show the inside.

section lines *(n)* In a sectional view, thin, evenly spaced lines that mark the cut surface.

shaft *(n)* Round stock to which gears, pulleys, or turning pieces are attached for support or to transmit power.
shear *(v)* To cut material between two blades.
shim *(n)* A thin plate of metal used between two surfaces to adjust the distance between them.
skew lines *(n)* Lines that are not parallel and do not intersect.
specifications *(n)* In construction and manufacturing, a detailed list of facts concerning materials and measurements.
spline *(n)* A metal strip that drafters use to trace a curve.
spline *(n)* A long keyway.
spot-face *(v)* To finish a circular spot slightly below a rough surface on a casting. This provides a smooth, flat seat for a bolthead or other fastener.
steel casting *(v)* A part made of cast iron to which scrap steel has been added when melted.
tap *(v)* To cut threads in a hole with a threading tool called a tap. *(n)* A hardened screw, fluted to provide cutting edges.
taper pin *(v)* A piece of round stock that narrows gradually and uniformly in diameter.
technical illustration *(v)* A pictorial drawing made to simplify and interpret technical information.
technical pen *(n)* A pen with a point of a specific size, to draw lines of a specific width.
temper *(v)* To make hardened steel less brittle by heating it in various ways, as in a bath of oil, salt, sand, or lead, to a specified temperature, and then cooling.
template or **templet** *(n)* A flat form or pattern of full size, used to lay out a shape and to locate holes or other features.
tolerance *(n)* The distance allowed away from the exact size before the size is wrong. See Chapter 6.
turn *(v)* To machine (finish) a piece on a lathe by turning the piece against a cutting tool (as when forming a cylindrical surface).
upset *(v)* To make an enlarged section or shoulder on a rod, bar, or similar piece while forging.
vanishing point *(n)* In perspective drawing, the point at which receding axes converge.
vernier *(n)* A small auxiliary scale to find fractional parts of a major scale.
washer *(n)* A ring of metal used as a seat for a bolt or nut.
web *(n)* A thin, flat part of an object used to brace or strengthen another part.
weld *(v)* To join pieces of metal by heating them to the melting point, then pressing or hammering them together. See Chapter 17.
working drawing *(n)* A drawing that contains all the information needed to make an object.

Appendix C

The Metric System

U.S. Customary Measurements

The measuring standards used in the United States today derive from long-established custom, not systematic planning. Their origins are diverse, and they do not form one unified measuring system. Some, such as the foot, inch, and mile, were created by the ancient Romans. Others originated in England. The United States has also set up its own standards, some of which are interchangeable with the English standards and some of which are not.

Development of the Metric System

The metric system of measurement was developed in France in the late 1700s. Within a century, most other countries had adopted it. This helped their trade, because they now manufactured to the same standards and could used each other's goods more easily. The main holdout countries were Great Britain and the United States. Great Britain converted to metrics during the 1960s and 1970s. The United States is now doing the same.

The metric system is a very practical system built around units of ten. It was to be a *natural measurement* based on the Earth's own dimensions. The meter was originally defined as one ten-millionth of the distance from the North Pole to the equator. Measures of volume and weight were related to linear distance. No such unified system had ever been made before. Today the meter is more preceisely defined as a wavelength of the red-orange light of krypton 86.

In 1960, the International Organization of Weights and Measures, representing countries from all over the world, gave the metric system its formal title of Système International d-Unités, which is abbreviated SI. The basic SI units are shown in Fig. C–1.

Unit	Name	Symbol
Length	Meter	m
Mass	Kilogram	kg
Time	Second	s
Electric current	Ampere	A
Temperature	Kelvin	K
Luminous intensity	Candela	cd

***Fig. C–1* The basic SI units.**

Using the Metric System

The most common measurements of the metric system are the *meter,* for length; the *gram,* for mass or weight; and the *liter,* for volume.

The parts and multiples of these units are based on the number 10. This is similar to our money system and our number system, which are also based on 10. It is easier to figure with units of 10 than to divide by 12 (inches) or by 16 (ounces).

Special names are given to certain multiples and divisions of the basic unit. These names become a prefix to the name of the unit. The prefixes apply to any of the unit names—meter, gram, or liter. The prefixes and their corresponding amounts are shown in Fig. C–2.

Prefix	Amount	Fraction	Decimal
Milli	One-thousandth	$\frac{1}{1000}$	0.001
Centi	One-hundredth	$\frac{1}{100}$	0.01
Deci	One-tenth	$\frac{1}{10}$	0.1
Deka	Ten	10	10.0
Hecto	Hundred	100	100.0
Kilo	Thousand	1000	1000.0

***Fig. C–2* Metric prefixes and their corresponding amounts.**

Adding the prefixes to the word *meter* gives the following:

millimeter (mm) = one-thousandth of a meter
centimeter (cm) = one-hundredth of a meter
decimeter (dm) = one-tenth of a meter
dekameter (dam) = ten meters
hectometer (hm) = one hundred meters
kilometer (km) = one thousand meters

Adding the prefixes to the word *gram* gives the following:

milligram (mg) = one-thousandth of a gram
centigram (cg) = one-hundredth of a gram
decigram (dg) = one-tenth of a gram
dekagram (dag) = ten grams
hectogram (hg) = one hundred grams
kilogram (kg) = one thousand grams

Adding the prefixes to the word *liter* gives the following:

milliliter (ml) = one-thousandth of a liter
centiliter (cl) = one-hundredth of a liter
deciliter (dl) = one-tenth of a liter
dekaliter (dal) = ten liters
hectoliter (hl) = one hundred liters
kiloliter (kl) = one thousand liters

As a basis for comparison, approximate United States Customary and metric equivalents to remember are: one liter is a little larger than a quart; one kilogram is a little greater than two pounds; one meter is a little longer than a yard; and one kilometer is about five-eighths of a miles. Figure C–3 is a list of metric system equivalents.

Figure C–4 shows the actual size as well as the relationship between metric measures of length, volume, and mass.

The Metric System in Drafting

In drafting, the primary dimension is length. On drawings, the millimeter is used for dimensions. This is true whether the drawing is prepared with all metric dimensions or whether it is dual dimensioned, with sizes expressed in both metric and United States Customary units. In cases where drawings are not dual dimensioned, a metric equivalent chart (Fig. C–5) may be used to make the conversions. However, not all dimensions are found on the chart. Therefore, it is useful to know how to convert mathematically from one system to the other.

Inches to millimeters

One inch is equal to approximately 25.400 millimeters. Multiplying an inch dimension by 25.4 will give an answer in millimeters. To simplify the calculation, the inch dimension should be in decimal form.

Length

Centimeter	= 0.3937 inch
Meter	= 3.28 feet
Meter	= 1.094 yards
Kilometer	= 0.621 statute mile
Kilometer	= 0.5400 nautical mile
Inch	= 2.54 centimeters
Foot	= 0.3048 meter
Yard	= 0.9144 meter
Statute mile	= 1.61 kilometers
Nautical mile	= 1.852 kilometers

Area

Square centimeter	= 0.155 square inch
Square meter	= 10.76 square feet
Square meter	= 1.196 square yards
Hectare	= 2.47 acres
Square kilometer	= 0.386 square miles
Square inch	= 6.45 square centimeters
Square foot	= 0.0929 square meter
Square yard	= 0.836 square meter
Acre	= 0.405 hectare
Square mile	= 2.59 square kilometers

Volume

Cubic centimeter	= 0.0610 cubic inch
Cubic meter	= 35.3 cubic feet
Cubic meter	= 1.308 cubic yards
Cubic inch	= 16.39 cubic centimeters
Cubic foot	= 0.0283 cubic meter
Cubic yard	= 0.765 cubic meter

Capacity

Milliliter	= 0.0338 U.S. fluid ounce
Liter	= 1.057 U.S. liquid quarts
Liter	= 0.908 U.S. dry quart
U.S. fluid ounce	= 29.57 milliliters
U.S. liquid quart	= 0.946 liter
U.S. dry quart	= 1.101 liters

Mass or Weight

Gram	= 15.43 grains
Gram	= 0.0353 avoirdupois ounce
Kilogram	= 2.205 avoirdupois pounds
Metric ton	= 1.102 short, or net, tons
Grain	= 0.0648 gram
Avoirdupois ounce	= 28.35 grams
Avoirdupois pound	= 0.4536 kilogram
Short, or net, ton	= 0.907 metric ton

Fig. C–3 **Metric system equivalents.**

Example: How many millimeters equal 2⅝ inches? Convert 2⅝ to a decimal, 2.625, then multiply by 25.4:

```
  2.625
  .25.4
 ------
  10500
 13125
 5250
 ------
 666750
```

Step off four decimal places to get the result of 66.6750 mm.

Millimeters to inches

One millimeter is equal to approximately 0.0394 inch. Multiplying a millimeter dimension by 0.0394 will give the answer in inches.

Example: How many inches equal 66.675 mm? Multiply 66.675 by 0.0394:

```
   66.675
    .0394
 --------
   266700
  599075
 200025
 --------
 26259950
```

Step off seven decimal places to get the result of 2.6259950, or 2.625 inches (2⅝″).

Rounding Off

It is sometimes necessary to round numbers to fewer digits. For example, 25.63220 mm rounded to three decimal places would be 25.632 mm. The same number rounded to two places would be 25.63 mm. The procedure is as follows:

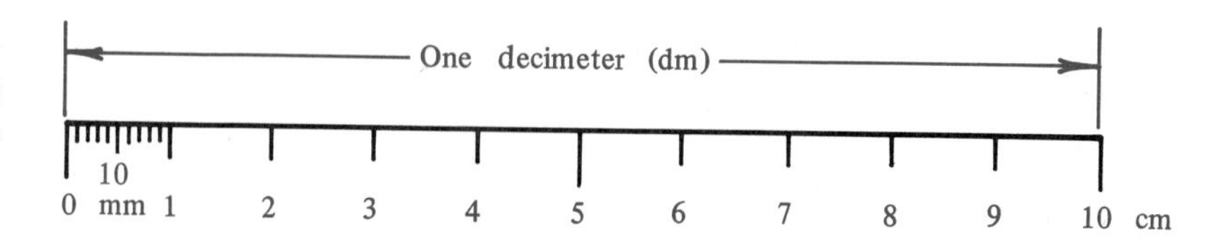

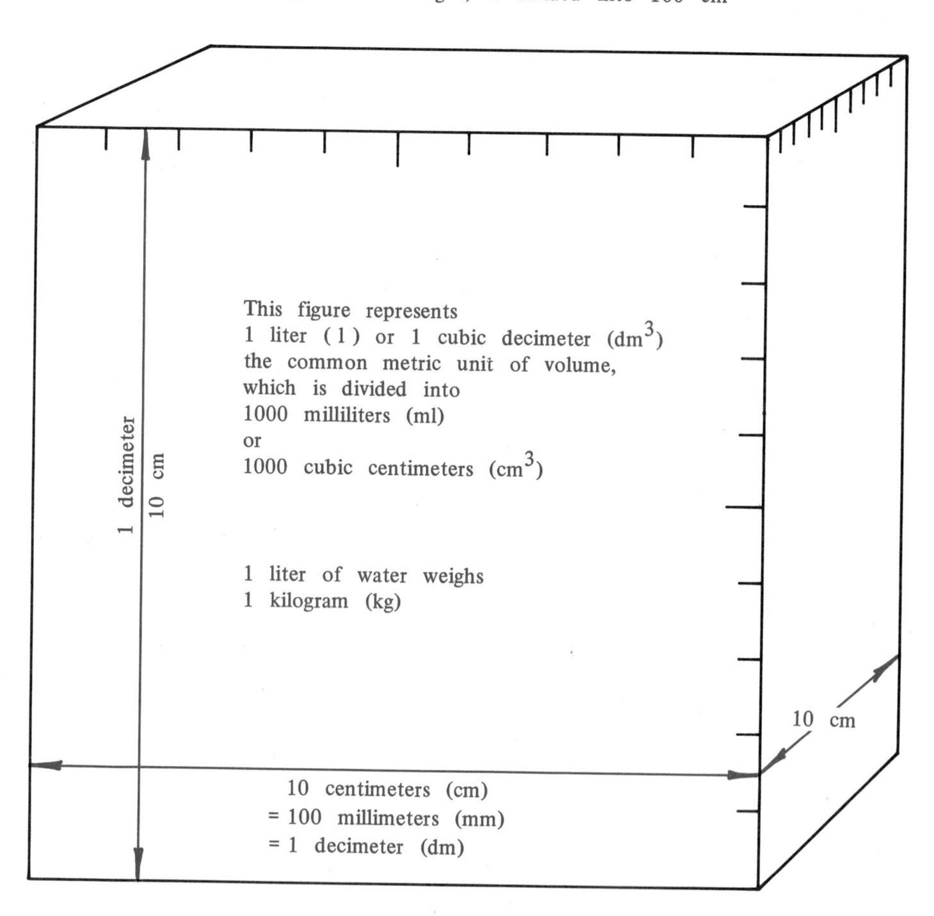

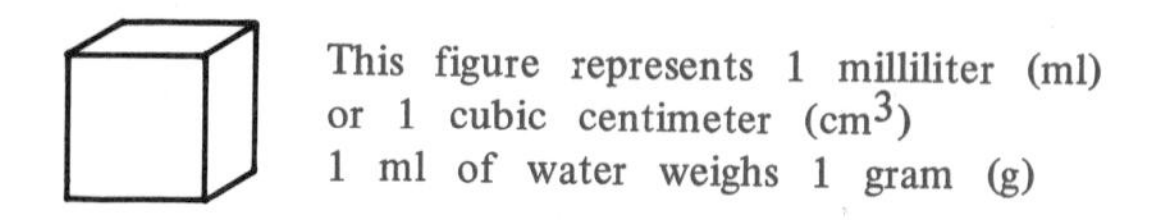

Fig. C–4 **Relationship of metric measures of length, volume, and mass.** *(Metric Association.)*

A. When the first number (digit) dropped is less than 5, the last number kept does not change. For example, 15.232 rounded to two decimal places would be 15.23.

B. When the first number dropped is greater than 5, the last number kept should be increased by one. For example, 6.436 rounded to two decimal places would be 6.44.

C. When the first number dropped is 5 followed by at least one number greater than 0, the last number kept should be increased by one. For example, 8.4253 rounded to two decimal places would be 8.43.

D. When the first number dropped is 5 followed by zeroes, the last number kept is increased by one if it is an odd number. If it is an even number, no change is made. For example, 3.23500 rounded to two places would be 3.24. However, 3.22500 would be 3.22.

Mm	In.*	Mm	In.
1 =	0.0394	17 =	0.6693
2 =	0.0787	18 =	0.7087
3 =	0.1181	19 =	0.7480
4 =	0.1575	20 =	0.7874
5 =	0.1969	21 =	0.8268
6 =	0.2362	22 =	0.8662
7 =	0.2756	23 =	0.9055
8 =	0.3150	24 =	0.9449
9 =	0.3543	25 =	0.9843
10 =	0.3937	26 =	1.0236
11 =	0.4331	27 =	1.0630
12 =	0.4724	28 =	1.1024
13 =	0.5118	29 =	1.1418
14 =	0.5512	30 =	1.1811
15 =	0.5906	31 =	1.2205
16 =	0.6299	32 =	1.2599

In.	Mm†	In.	Mm
1/32 (0.03125) =	0.794	17/32 (0.53125) =	13.493
1/16 (0.0625) =	1.587	9/16 (0.5625) =	14.287
3/32 (0.09375) =	2.381	19/32 (0.59375) =	15.081
1/8 (0.1250) =	3.175	5/8 (0.6250) =	15.875
5/32 (0.15625) =	3.968	21/32 (0.65625) =	16.668
3/16 (0.1875) =	4.762	11/16 (0.6875) =	17.462
7/32 (0.21875) =	5.556	23/32 (0.71875) =	18.256
1/4 (0.2500) =	6.349	3/4 (0.7500) =	19.050
9/32 (0.28125) =	7.144	25/32 (0.78125) =	19.843
5/16 (0.3125) =	7.937	13/16 (0.8125) =	20.637
11/32 (0.34375) =	8.731	27/32 (0.84375) =	21.431
3/8 (0.3750) =	9.525	7/8 (0.8750) =	22.225
13/32 (0.40625) =	10.319	29/32 (0.90625) =	23.018
7/16 (0.4375) =	11.112	15/16 (0.9375) =	23.812
15/32 (0.46875) =	11.906	31/32 (0.96875) =	24.606
1/2 (0.5000) =	12.699	1 (1.0000) =	25.400

* Calculated to *nearest* fourth decimal place. † Calculated to *nearest* third decimal place.

Fig. C–5 **Inch-millimeter equivalent chart.**

Dual Dimensions

The dimensions in the units to which the product was designed are given first. They are placed above the line. The converted dimensions either follow or are put below the line, and are enclosed in parentheses. (Fig. C–6.) More information may be found in Chapter 6, Dimensioning.

Angular Dimensions

Angular dimensions do not need to be converted. Angles dimensioned in degrees, minutes, and seconds are common to both systems.

Fig. C–6 **Dual dimensioning.**

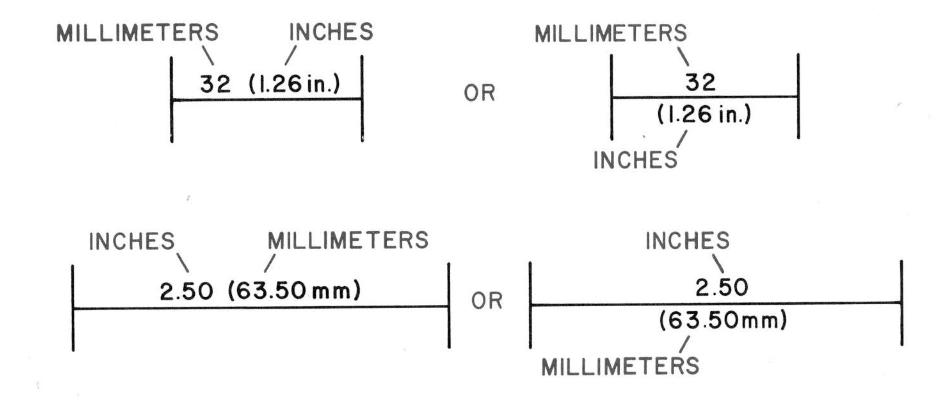

Index